1000	1200	1600	1800

1000–1200

Iraq: Al-Karajī, induction, and Pascal triangle

Egypt: Ibn al-Haytham, sums of powers, and volumes of paraboloids

Iran: Omar Khayyam and the geometric solution of cubic equations

India: Al-Birūni and spherical trigonometry; Bhaskara and the Pell equation

China: Pascal triangle used to solve equations

Spain: Arabic works translated in Latin; Abraham ibn Ezra and combinatorics

Italy: Leonardo of Pisa and introduction of Islamic mathematics

United States: Astronomical alignments in Anasazi buildings in the Southwest

Zimbabwe: Construction of Great Zimbabwe structures

1200–1400

Iran: Nasīr al-Dīn al-Ṭūsī and trigonometry

France: Jordanus and advanced algebra; Levi ben Gerson and induction; Oresme and kinematics

England: Velocity, acceleration, and the mean speed theorem

China: Chinese remainder theorem; Solution of polynomial equations

Peru: Quipus used for record keeping

1400–1600

India: Discovery of power series for sine, cosine, and arctangent

Italy: Algebraic solution of the cubic equation

Germany: Perspective and geometry

England: New algebra and trigonometry texts

Poland: Copernicus and the heliocentric system

France: Viète and algebraic symbolism

1600–1700

Kepler, Newton, and celestial physics

Descartes, Fermat, and analytic geometry

Napier, Briggs, and logarithms

Girard, Descartes, and the theory of equations

Pascal, Fermat, and elementary probability

Pascal, Desargues, and projective geometry

Newton, Leibniz, and the invention of calculus

1700–1800

Development of techniques for solving ordinary and partial differential equations

Development of the calculus of functions of several variables

Attempts to give logically correct foundations to the calculus

Lagrange and the analysis of the solution of polynomial equations

1800–1900

Algebraic number theory

Galois theory

Groups and fields

Quaternions and the discovery of noncommutative algebra

Theory of matrices

The arithmetization of analysis

Development of complex analysis

Vector analysis

Differential geometry

Non-Euclidean geometry

Projective geometry

Foundations of geometry

1900–2000

Set theory

Growth of topology

Algebraization of mathematics

Influence of computers

1000	1200	1600	1800

A HISTORY
OF MATHEMATICS

AN INTRODUCTION

A HISTORY
OF MATHEMATICS
AN INTRODUCTION

Victor J. Katz

University of the District of Columbia

HarperCollins*College*Publishers

To Phyllis, for long talks, long walks, and afternoon naps

Sponsoring Editor: George Duda
Project Coordination, Text and Cover Design: Elm Street Publishing Services, Inc.
Compositor: Weimer Incorporated
Printer and Binder: R. R. Donnelley & Sons Company
Cover Printer: Lehigh Press Lithographers

A History of Mathematics
Copyright © 1993 by HarperCollins College Publishers

Library of Congress Cataloging-in-Publication Data

Katz, Victor J.
    A history of mathematics / Victor J. Katz.
        p.   cm.
    Includes bibliographical references and index.
    ISBN 0-673-38039-4
    1. Mathematics—History.   I. Title
QA21.K33   1992
510'.9—dc20   92-20989

92   1

The cover: Astronomical instruments and timepieces (geometrical shapes in perspective), from *Jean de Dinteville and Georges de Selve ("The Ambassadors")* by Hans Holbein the Younger. Reproduced by courtesy of the Trustees, The National Gallery, London. The entire picture is shown on the frontispiece.

Brief Contents

Detailed Contents **vi**

Preface **x**

Part One
Mathematics
Before the Sixth
Century **1**

Chapter 1   Ancient Mathematics **1**

Chapter 2   The Beginnings of Mathematics in Greece **42**

Chapter 3   Archimedes and Apollonius **95**

Chapter 4   Mathematical Methods in Hellenistic Times **127**

Chapter 5   The Final Chapters of Greek Mathematics **157**

Part Two
Medieval
Mathematics:
500–1400 **181**

Chapter 6   Medieval China and India **181**

Chapter 7   The Mathematics of Islam **223**

Chapter 8   Mathematics in Medieval Europe **266**

Interchapter   Mathematics Around the World **302**

Part Three
Early Modern
Mathematics:
1400–1700 **313**

Chapter 9   Algebra in the Renaissance **313**

Chapter 10   Mathematical Methods in the Renaissance **354**

Chapter 11   Geometry, Algebra, and Probability in the Seventeenth Century **394**

Chapter 12   The Beginnings of Calculus **428**

Part Four
Modern
Mathematics:
1700–2000 **494**

Chapter 13   Analysis in the Eighteenth Century **494**

Chapter 14   Probability, Algebra, and Geometry in the Eighteenth Century **539**

Chapter 15   Algebra in the Nineteenth Century **585**

Chapter 16   Analysis in the Nineteenth Century **635**

Chapter 17   Geometry in the Nineteenth Century **691**

Chapter 18   Aspects of the Twentieth Century **726**

Answers to Selected Problems **773**

General References in the History of Mathematics **778**

Index and Pronunciation Guide **780**

Detailed Contents

Preface   **x**

***Part One*
Mathematics
Before the Sixth
Century   1**

Chapter 1   Ancient Mathematics   1
1.1 Ancient Civilizations   **2**
1.2 Counting   **4**
1.3 Arithmetic Computations   **7**
1.4 Linear Equations   **13**
1.5 Elementary Geometry   **18**
1.6 Astronomical Calculations   **21**
1.7 Square Roots   **23**
1.8 The Pythagorean Theorem   **26**
1.9 Quadratic Equations   **31**

Chapter 2   The Beginnings of Mathematics in Greece   42
2.1 The Earliest Greek Mathematics   **43**
2.2 The Time of Plato   **48**
2.3 Aristotle   **50**
2.4 Euclid and the *Elements*   **54**
2.5 Euclid's Other Works   **88**

Chapter 3   Archimedes and Apollonius   95
3.1 Archimedes and Physics   **96**
3.2 Archimedes and Numerical Calculations   **101**
3.3 Archimedes and Geometry   **104**
3.4 Conic Sections Before Apollonius   **108**
3.5 *The Conics* of Apollonius   **112**

Chapter 4   Mathematical Methods in Hellenistic Times   127
4.1 Astronomy Before Ptolemy   **128**
4.2 Ptolemy and the *Almagest*   **136**
4.3 Practical Mathematics   **147**

Chapter 5   The Final Chapters of Greek Mathematics   157
5.1 Nicomachus and Elementary Number Theory   **158**
5.2 Diophantus and Greek Algebra   **162**
5.3 Pappus and Analysis   **173**

Part Two
**Medieval
Mathematics:
500–1400   181**

Chapter 6   **Medieval China and India   181**
6.1 Introduction to Medieval Chinese Mathematics   **181**
6.2 The Mathematics of Surveying and Astronomy   **182**
6.3 Indeterminate Analysis   **187**
6.4 Solving Equations   **191**
6.5 Introduction to the Mathematics of Medieval India   **199**
6.6 Indian Trigonometry and Surveying   **200**
6.7 Indian Indeterminate Analysis   **204**
6.8 Algebra and Combinatorics   **211**
6.9 The Hindu-Arabic Place-Value System   **215**

Chapter 7   **The Mathematics of Islam   223**
7.1 Decimal Arithmetic   **225**
7.2 Algebra   **228**
7.3 Geometry   **247**
7.4 Trigonometry   **253**

Chapter 8   **Mathematics in Medieval Europe   266**
8.1 Geometry and Trigonometry   **269**
8.2 Combinatorics   **276**
8.3 Medieval Algebra   **282**
8.4 The Mathematics of Kinematics   **289**

Interchapter   **Mathematics Around the World   302**
I.1 Mathematics at the Turn of the Fourteenth Century   **302**
I.2 Mathematics in America, Africa, and the Pacific   **304**

Part Three
**Early Modern
Mathematics:
1400–1700   313**

Chapter 9   **Algebra in the Renaissance   313**
9.1 The Italian Abacists   **314**
9.2 Algebra in France, Germany, England, and Portugal   **319**
9.3 The Solution of the Cubic Equation   **328**
9.4 The Work of Viète and Stevin   **337**

Chapter 10   **Mathematical Methods in the Renaissance   354**
10.1 Perspective   **357**
10.2 Geography and Navigation   **360**
10.3 Astronomy and Trigonometry   **364**
10.4 Logarithms   **379**
10.5 Kinematics   **383**

Chapter 11   Geometry, Algebra, and Probability in the Seventeenth Century   394
11.1 Analytic Geometry   **395**
11.2 The Theory of Equations   **406**
11.3 Elementary Probability   **409**
11.4 Number Theory   **418**
11.5 Projective Geometry   **420**

Chapter 12   The Beginnings of Calculus   428
12.1 Tangents and Extrema   **429**
12.2 Areas and Volumes   **434**
12.3 Power Series   **450**
12.4 Rectification of Curves and the Fundamental Theorem   **453**
12.5 Isaac Newton   **460**
12.6 Gottfried Wilhelm Leibniz   **472**
12.7 First Calculus Texts   **482**

Part Four
**Modern
Mathematics:
1700–2000   494**

Chapter 13   Analysis in the Eighteenth Century   494
13.1 Differential Equations   **495**
13.2 Calculus Texts   **506**
13.3 Multiple Integration   **519**
13.4 Partial Differential Equations: The Wave Equation   **522**
13.5 The Foundations of Calculus   **525**

Chapter 14   Probability, Algebra, and Geometry in the Eighteenth Century   539
14.1 Probability   **540**
14.2 Algebra and Number Theory   **551**
14.3 Geometry   **560**
14.4 The French Revolution and Mathematics Education   **572**
14.5 Mathematics in the Americas   **576**

Chapter 15   Algebra in the Nineteenth Century   585
15.1 Number Theory   **587**
15.2 Solving Algebraic Equations   **596**
15.3 Groups and Fields—The Beginning of Structure   **604**
15.4 Symbolic Algebra   **611**
15.5 The Theory of Matrices   **621**

Chapter 16   Analysis in the Nineteenth Century   635
16.1 Rigor in Analysis   **637**
16.2 The Arithmetization of Analysis   **658**
16.3 Complex Analysis   **665**
16.4 Vector Analysis   **675**
16.5 Probability and Statistics   **681**

Chapter 17   **Geometry in the Nineteenth Century   691**
17.1 Differential Geometry   **693**
17.2 Non-Euclidean Geometry   **696**
17.3 Projective Geometry   **709**
17.4 Geometry in *N* Dimensions   **715**
17.5 The Foundations of Geometry   **718**

Chapter 18   **Aspects of the Twentieth Century   726**
18.1 Set Theory: Problems and Paradoxes   **728**
18.2 Topology   **735**
18.3 New Ideas in Algebra   **742**
18.4 Computers and Applications   **751**

Answers to Selected Problems   773
General References in the History of Mathematics   778
Index and Pronunciation Guide   780

Preface

APPROACH AND GUIDING PHILOSOPHY

In *A Call for Change: Recommendations for the Mathematical Preparation of Teachers of Mathematics*, the Mathematical Association of America's (M.A.A.) Committee on the Mathematical Education of Teachers recommends that all prospective teachers of mathematics in schools

> develop an appreciation of the contributions made by various cultures to the growth and development of mathematical ideas; investigate the contributions made by individuals, both female and male, and from a variety of cultures, in the development of ancient, modern, and current mathematical topics; [and] gain an understanding of the historical development of major school mathematics concepts.

According to the M.A.A., knowledge of the history of mathematics shows students that mathematics is an important human endeavor. Mathematics was not discovered in the polished form of our textbooks but often developed in intuitive and experimental fashion out of a need to solve problems. The actual development of mathematical ideas can be effectively used to excite and motivate students today.

This new textbook on the history of mathematics grew out of the conviction that not only prospective school teachers of mathematics but also prospective college teachers of mathematics need a background in history to more effectively teach the subject to their students. It is therefore designed for mathematics majors who intend to teach in college or high school and thus concentrates on the history of those topics typically covered in an undergraduate curriculum or in elementary or high school. Because the history of any given mathematical topic often provides excellent ideas for teaching the topic, there is sufficient detail in each explanation of a new concept for the future (or present) teacher of mathematics to develop a classroom lesson or series of lessons based on history. In fact, many of the problems ask the student to develop a particular lesson. My hope is that the student and prospective teacher will gain from this book a knowledge of how we got here from there, a knowledge that will provide a deeper understanding of many of the important concepts of mathematics.

DISTINGUISHING FEATURES
Flexible Organization

Although the chief organization of the book is by chronological period, within each period the material is organized topically. By consulting the detailed subsection headings, the reader can choose to follow a particular theme throughout history. For example, to study equation solving one could consider ancient Egyptian and Babylonian methods, the geometrical solution methods of the Greeks, the numerical methods of the Chinese, the Islamic solution methods for cubic equations by use of conic sections, the Italian discovery of an algorithmic solution of cubic and quartic equations, the work of Lagrange in developing criteria for methods of solution of higher degree polynomial equations, the work of Gauss in solving cyclotomic equations, and the work of Galois in using permutations to formulate what is today called Galois theory.

Focus on Original Textbooks/Sources

There is an emphasis throughout the book on the important textbooks of various periods. It is one thing to do mathematical research and discover new theorems and techniques. It is quite another to elucidate these in a way that others can learn them. In nearly every chapter, therefore, there is a discussion of one or more important texts of the time. These will be the works from which students learned the important ideas of the great mathematicians. Today's students will see how certain topics were treated and will be able to compare these treatments to those in current texts and see the kinds of problems students of years ago were expected to solve.

Astronomy and Mathematics

Two chapters are devoted entirely to mathematical methods; that is, to the ways in which mathematics was used to solve problems in other areas of endeavor. A substantial part of both of these chapters, one for the Greek period and one for the Renaissance, deals with astronomy. In fact, in ancient times astronomers and mathematicians were usually the same people. It is crucial to the understanding of a substantial part of Greek mathematics to understand the Greek model of the heavens and how mathematics was used in applying this model to give predictions. Similarly, we will discuss the Copernicus-Kepler model of the heavens and see how mathematicians of the Renaissance applied mathematics to its study.

Nonwestern Mathematics

A special effort has been made to consider mathematics developed in parts of the world other than Europe. Thus, there is substantial material on mathematics in China, India, and the Islamic world. There is also an "interchapter" in which a comparison is made of the mathematics in the major civilizations at about the turn of the fourteenth century. That comparison is followed by a discussion of the mathematics of various other societies around the world. The reader will see how certain mathematical ideas have occurred in many places, although not perhaps in the context of what we in the West call "mathematics."

Topical Exercises

Each chapter contains many exercises, collected by topic for easy access. Some of the exercises are simple computational ones while others help to fill the gaps in the mathematical arguments presented in the text. **For Discussion** exercises are open-ended questions for discussion, which may involve some research to find answers. Many of these ask students to think about how they would use historical material in the classroom. (Answers to most of the computational exercises are provided at the end of the book.) Even if readers do not attempt many of the exercises, they should at least read them to gain a fuller understanding of the material of the chapter.

Focus Essays

Biographies

For easy reference, many biographies of the mathematicians whose work is discussed appear in separate boxes. In particular, although for various reasons women have not participated in large numbers in mathematical research, biographies of several important female mathematicians are included. These are women who succeeded, usually against heavy odds, in contributing to the mathematical enterprise.

Side Bars

Scattered throughout the book are Side Bars, which focus on special topics. These include such items as a discussion of the idea of a function in the work of Ptolemy, a comparison of various notions of continuity, and a special essay on the work of Nicolas Bourbaki. There are also boxes containing important definitions collected together for easy reference.

Additional Pedagogy

Each chapter begins with a relevant quotation and a description of an important mathematical "event." At the end of each chapter, a brief chronology of the mathematicians discussed will help students organize their knowledge. Each chapter also contains an annotated list of references to both primary and secondary sources from which the students can obtain more information. There is a **time line** of the history of mathematics on the inside front cover and a **map** on the inside back cover indicating the location of some of the important places mentioned in the text. Finally, given that students may have difficulty pronouncing the names of some mathematicians, the **index** has a special feature—a **phonetic pronunciation guide.**

PREREQUISITES

A working knowledge of one year of calculus is sufficient to understand the first twelve chapters of the text. The mathematical prerequisites for the later chapters are somewhat more demanding, but the titles of the various sections indicate clearly what kind of mathematical knowledge is required. For example, a full understanding of Chapters 14 and 15 will require that the student has studied abstract algebra.

COURSE FLEXIBILITY

There is far more material in this text than can be included in a typical one-semester course in the history of mathematics. In fact, there is adequate material for a full-year course, the first half being devoted to the period through the invention of calculus in the late seventeenth century and the second half covering the mathematics of the eighteenth, nineteenth, and twentieth centuries. For those instructors who have only one semester, however, there are several ways to use this book. First, one could cover most of the first twelve chapters and simply conclude with calculus. Second, one could choose to follow one or two particular themes through history. Some possible themes with the appropriate section numbers are as follows.

> **Equation Solving:** 1.4, 1.9, 2.4.3, 5.2, 6.3, 6.4, 6.7, 6.8, 7.2, 8.3, 9.3, 9.4, 11.2, 14.2.4, 15.2.
>
> **Ideas of Calculus:** 2.3.3, 2.4.9, 3.3, 7.2.4, 7.3.4, 8.4, 10.5, 12, 13.3, 13.5, 16.1, 16.2, 18.1.
>
> **Concepts of Geometry:** 1.5, 1.8, 2.2, 2.4, 3.3, 3.4, 3.5, 4.3, 7.3, 8.1, 11.1, 11.5, 14.3, 17, 18.2.
>
> **Trigonometry, Astronomy, and Surveying:** 1.6, 4.1, 4.2, 6.2, 6.6, 7.4, 8.1, 10.3, 13.1.3.
>
> **Probability and Statistics:** 6.8, 8.2, 11.3, 14.1, 16.5.
>
> **Linear Algebra:** 1.4, 14.2.1, 15.5, 17.4.
>
> **Number Theory:** 2.4.7, 5.1, 11.4, 14.2.3, 15.1.
>
> **"Modern" Algebra:** 6.8, 7.2, 8.3, 9.1, 9.2, 14.2, 15.3, 15.4, 18.3.

Third, one could cover in detail most of the first ten chapters and then pick selected ideas from the later chapters, again following a particular theme. One could also assign various sections for individual or small group reading assignments and reports.

ACKNOWLEDGMENTS

Like any book, this one could not have been written without the help of many people. The following people have read large sections of the book at my request and have offered many valuable suggestions: Marcia Ascher (Ithaca College), J. Lennart Berggren (Simon Fraser University), Robert Kreiser (A.A.U.P.), Robert Rosenfeld (Nassau Community College), and John Milcetich (University of the District of Columbia).

The many reviewers of sections of the manuscript have also provided great help with their detailed critiques and have made this a much better book than it otherwise could have been. They include Duane Blumberg, University of Southwestern Louisiana; Walter Czarnec, Framington State University; Joseph Dauben, Herbert H. Lehman College–CUNY; Harvey Davis, Michigan State University; Joy Easton, West Virginia University; Carl FitzGerald, University of California–San Diego; Basil Gordon, University of California–Los Angeles; Mary Gray, American University; Branko Grunbaum, University of Washington; William Hintzman, San Diego State University; Barnabus Hughes, California State University–Northridge; Israel Kleiner, York University; David E. Kullman, Miami University; Robert L. Hall, University of Wisconsin, Milwaukee; Richard Marshall, Eastern Michigan University; Jerold Mathews, Iowa State University; Willard Parker, Kansas State University; Clinton M. Petty, University of Missouri–Columbia; Howard Prouse, Mankato State University; Helmut Rohrl, University of California–San Diego; David Wilson, University of Florida; and Frederick Wright, University of North Carolina–Chapel Hill. I greatly appreciate the time and effort they put into their work.

I have also benefited greatly from conversations with many historians of mathematics at various forums. In particular, those who have regularly attended the annual History of Mathematics seminars organized by Uta Merzbach, former curator of mathematics at the National Museum of American History, may well recognize some of the ideas discussed there.

The book has also profited from discussions over the years with, among others, Charles Jones (Ball State University), V. Frederick Rickey (Bowling Green State University), Florence Fasanelli (M.A.A.), Israel Kleiner (York University), Abe Shenitzer (York University), Ubiratan D'Ambrosio (Univ. Estadual de Campinas), and Frank Swetz (Pennsylvania State University). My students in History of Mathematics (and other) classes at the University of the District of Columbia have also helped me clarify many of my ideas. Naturally, I welcome any additional comments and correspondence from students and colleagues elsewhere in an effort to continue to improve this book.

Special thanks are due to the librarians at the University of the District of Columbia and especially to Clement Goddard, who never failed to secure any of the obscure books I requested on interlibrary loan. Leslie Overstreet of the Smithsonian Institution Libraries' Special Collections Department was extremely helpful in finding sources for pictures. Thanks are due also to editors Steve Quigley and Don Gecewicz, who cajoled and threatened, and to George Duda, my editor at HarperCollins, who finally saw the book to completion. And Cathy Crow of Elm Street Publishing Services has very cheerfully and efficiently handled the production aspects of this project.

My family has been very supportive during the many years of writing the book. I thank my parents for their patience and their faith in me. I thank my children, Sharon, Ari, and Naomi, for help at various times and especially for allowing me to use our computer. And last, I thank my wife, Phyllis, for long discussions at any hour of the day or night and for being there when I needed her. I owe her much more than I can ever repay.

Victor J. Katz

Chapter 1

Ancient Mathematics

*"Accurate reckoning. The entrance into the knowledge of all existing things
and all obscure secrets."*
(Introduction to Rhind Mathematical Papyrus)[1]

$\mathcal{I}$n a scribal school in the city of Babylon some 3800 years ago, a teacher is trying to develop mathematics problems to assign to his students so they can practice the ideas just introduced on the relationship among the sides of a right triangle. The teacher not only wants the computations to be difficult enough to show him who really understands the material but also wants the answers to come out as whole numbers so the students will not be frustrated. After playing for several hours with the few triples (a,b,c) of numbers he knows which satisfy $a^2 + b^2 = c^2$, a new idea occurs to him. With a few deft strokes of his stylus, he quickly scratches on a wet clay tablet some calculations and convinces himself that he has discovered how to generate as many of these triples as necessary. After organizing his thoughts a bit longer, he takes a fresh tablet and carefully records a table listing not only fifteen such triples but also a brief indication of some of the preliminary calculations. He does not, however, record the details of his new method. Those will be saved for his lecture to his colleagues and the principal. Perhaps now his long sought promotion will be forthcoming.

The opening quote from one of the few documentary sources on Egyptian mathematics and the fictional story of the Babylonian scribe illustrate some of the difficulties in giving an accurate picture of ancient mathematics. Mathematics certainly existed in virtually every ancient civilization of which there are records. But in every one of these civilizations, mathematics was in the domain of specially trained priests and scribes, government officials whose job it was to develop and use mathematics for the benefit of that government in such areas as tax collection, measurement, building, trade, calendar making, and ritual practices. But although the origins of many mathematical concepts stem from their usefulness in these areas, mathematicians always exercised their curiosity by extending these ideas far beyond the limits of practical necessity. Nevertheless, since mathematics was a tool of power, its methods were passed on only to the privileged few, often through an oral tradition. Hence the written records are generally sparse and usually do not provide great detail.

In recent years, however, a great deal of scholarly effort has gone into reconstructing the mathematics of ancient civilizations from whatever clues can be found. Naturally all scholars do not agree on every point, but there is enough agreement so that a reasonable picture can be presented of the mathematical knowledge of the ancient civilizations in Egypt, Mesopotamia, China, and India. In order to see the similarities and differences in the mathematics of these civilizations, we will not treat each civilization separately but will organize the discussion of ancient mathematics around the following topics: counting, arithmetic computations, linear equations, elementary geometry, astronomical and calendrical computations, square roots, the "Pythagorean" theorem, and quadratic equations. To place the story in context, we begin with a brief description of the civilizations themselves and the sources from which our knowledge of their mathematics is derived.

1.1 ANCIENT CIVILIZATIONS

FIGURE 1.1
Hammurapi on a stamp of Iraq.

FIGURE 1.2
Babylonian clay tablet on a stamp of Austria.

Probably the oldest of the world's civilizations is that of Mesopotamia, which began in the Tigris and Euphrates river valleys sometime around 3500 B.C.E. There were many kingdoms which came to the fore in this area over the next 3000 years, but for the purposes of mathematical history, it is customary to denote the Mesopotamian civilization as Babylonian, after the city whose ruler Hammurapi conquered the entire area around 1700 B.C.E. (Figure 1.1). By this time the area had developed a national political loyalty and a supreme god called Marduk; a bureaucracy and a professional army had come into existence; written communication had become well developed, in part to aid in central control over a large area; and a middle class of merchants and artisans had grown up between the masses of peasants and the royal officials. The writing which developed was done by means of a stylus on clay tablets, thousands of which have been excavated during the past 150 years (Figure 1.2). It was Henry Rawlinson (1810–1895) who, by the mid-1850s, was first able to translate this cuneiform writing by comparing the Persian and Babylonian inscriptions of King Darius I of Persia on a rockface at Behistun (in modern Iran) describing a military victory.

A large number of these tablets are mathematical in nature, containing mathematical problems and solutions or mathematical tables. Several hundreds of these have been copied, translated, and explained. These are our source materials on Babylonian mathematics. The great majority of these tablets date from the time of Hammurapi (the Old Babylonian period), while a smaller group dates from some 1400 years later, the Seleucid era. The Seleucid tablets, however, are of the same character and often contain problems virtually identical to the earlier ones. Mathematics, like the rest of Babylonian culture, displayed a remarkable stability over this long period of time.

Agriculture emerged in the Nile valley in Egypt close to 7000 years ago, but the first dynasty to rule both Upper Egypt (the river valley) and Lower Egypt (the delta) dates from about 3100 B.C.E. The legacy of the first pharaohs included an elite of officials and priests, a rich court, and for the kings themselves a role as intermediary between mortals and gods. This role fostered the development of Egypt's monumental architecture, includ-

FIGURE 1.3
The pyramids at Gizeh.

FIGURE 1.4
Jean Champollion and
the Rosetta Stone.

FIGURE 1.5
Amenhotep, an Egyptian
high official and scribe
(15th century, B.C.E.).

ing the pyramids, built as royal tombs, and the great temples at Luxor and Karnak (Figure 1.3). The scribes gradually developed the hieroglyphic writing which dots the tombs and temples. Jean Champollion (1790–1832) is chiefly responsible for the first translation of this writing early in the 19th century through the help again of a multi-lingual inscription—the Rosetta stone—in hieroglyphics and Greek as well as the later demotic writing, a form of the hieratic writing of the papyri (Figure 1.4).

Much of our knowledge of the mathematics of ancient Egypt, however, comes not from the hieroglyphs in the temples but from two papyri containing collections of mathematical problems with their solutions: the *Rhind Mathematical Papyrus*, named for the Scotsman A. H. Rhind (1833–1863) who purchased it at Luxor in 1858, and the *Moscow Mathematical Papyrus*, purchased in 1893 by V. S. Golenishchev (d. 1947) who later sold it to the Moscow Museum of Fine Arts. The former papyrus was copied about 1650 B.C.E. by the scribe A'h-mose from an original about 200 years earlier and is approximately 18 feet long and 13 inches high. The latter papyrus dates from roughly the same period and is over 15 feet long, but only some 3 inches high. The many other fragments of papyri, generally written in the hieratic script as are the Rhind and Moscow papyri, confirm the basic outline of Egyptian mathematics to be presented (Figure 1.5).

Although there are legends which date Chinese civilization back 5000 years or more, the earliest solid evidence of such a civilization is provided by the excavations at Anyang, near the Huang River, which are dated to about 1600 B.C.E. It is to the society centered there, the Shang dynasty, that the "oracle bones" belong, curious pieces of bone inscribed with very ancient writing which were used for divination by the priests of the period. The bones are the source of our knowledge of early Chinese number systems. Around the beginning of the first millenium B.C.E. the Shang were replaced by the Zhou dynasty, which in turn dissolved into numerous warring feudal states. In the sixth century B.C.E. there was a great period of intellectual flowering, in which the most famous philosopher was Confucius. Academies of scholars were founded in several of the states. Individual scholars were hired by other feudal lords to advise them in a time of technological growth caused by the coming of iron.

The feudal period ended as the weaker states were gradually absorbed by the stronger until ultimately China was unified under the Emperor Qin Shi Huangdi in 221 B.C.E. Under his leadership, China was transformed into a highly centralized bureaucratic state. He enforced a severe legal code, levied taxes evenly, and demanded the standardization of weights, measures, money, and especially the written script. Legend holds that this emperor ordered the burning of all books from earlier periods to suppress dissent, but there is some reason to doubt that this was actually carried out. The emperor died in 210 B.C.E. and his dynasty was soon overthrown and replaced by that of the Han, which was to last about 400 years. The Han completed the establishment of a trained civil service, for which a system of education was necessary. Among the texts used for this purpose were two mathematical works, probably compiled early in the Han dynasty, the *Zhoubi suanjing (Arithmetical Classic of the Gnomon and the Circular Paths of Heaven)* and the *Jiuzhang suanshu (Nine Chapters on the Mathematical Art)*. It is unfortunately impossible to date exactly the first discoveries of the mathematics contained there. Nevertheless, since there are fragmentary records of older sources similar to the *Nine Chapters*, it is generally believed that at least some of the material was extant in China near the beginning of the Zhou period. Of course it should be kept in mind that, even with this dating, the Chinese mathematical developments to be discussed are at least several hundred years later than those in Babylonia and Egypt. Whether there was any transmission from these civilizations to China is not known.

A civilization called the Harappan arose in India on the banks of the Indus River in the third millenium B.C.E., but there is no direct evidence of its mathematics. The earliest Indian civilization for which there is such evidence was formed along the Ganges River by Aryan tribes migrating from the Asian Steppes late in the second millenium B.C.E. From about the eighth century B.C.E. there were monarchical states in the area which had to deal with such activities as fortifications, administrative centralization, and large scale irrigation works. These states had a highly stratified social system headed by the king and the priests (brahmins). The literature of the brahmins was oral for many generations, expressed in lengthy verses called Vedas; some of the Vedic verses describe the intricate sacrificial system of the priests (Figure 1.6). Although these verses probably achieved their current form by 600 B.C.E., there are no written records dating back beyond the current era. Curiously, however, although the mathematics of these Vedic works, the *Sulvasutras*, deals with the theoretical requirements for building altars out of bricks, as far as is known the early Vedic civilization did not have a tradition of brick technology while the Harappan culture did. Thus there is a strong possibility that the mathematics in the *Sulvasutras* was created in the Harappan period, although the mechanism of its transmission to the later period is currently unknown. In any case, it is the *Sulvasutras* which are the sources for our knowledge of ancient Indian mathematics.

Although there were civilizations in other parts of the world before the first millenium B.C.E., the data uncovered so far gives us few clues to their mathematical knowledge. Any discussion, therefore, must await new archaeological evidence.

FIGURE 1.6
A Vedic manuscript.

1.2 COUNTING

The simplest mathematical idea—and one which probably existed even before civilization—is that of counting, in words and in more permanent form as written symbols. Although an interesting study can be made of number words in various languages (Side Bar 1.1), we restrict ourselves here to a discussion of number symbols. Several methods of organization can be distinguished in the writing of these symbols. One method is called the grouping method. In its simplest form, a stroke / is used to represent the number 1; appropriate repetitions are made to represent larger numbers. One of the earliest dated occurrences of such a representation of numbers is on a fossilized bone discovered at Ishango in Zaire and carbon-dated to some time around 8000 B.C.E. It is not clear what the strokes or notches represent, but one scholar who has studied the bone in detail believes they represent a count of certain periods of the moon.[2] That early peoples used this elementary form of numerical representation to deal with astronomical phenomena would confirm what will be seen in much later time periods, that the development of mathematics goes hand in hand with the development of astronomy. Artifacts similar to the Ishango bone and dating from roughly the same time, with regular groupings of notches perhaps representing astronomical observations, have been found in central Europe as well.

Side Bar 1.1   Number Words in Various Languages

	18			**40**	
English	eighteen	8, 10 (ten becomes teen)	English	forty	4 × 10 (ten becomes ty)
Welsh	deu naw	2 × 9 (deu from dau = 2, naw = 9)	Welsh	de-ugeint	2 × 20 (de from dau = 2, ugeint = 20)
Hebrew	shmona-eser	8, 10 (shmona = 8, eser = 10)	Hebrew	arba-im	4's (arba = 4, im is the plural ending)
Yoruba	eeji din logun	20 less 2 (ogun = 20, eeji = 2)	Yoruba	ogoji	20 × 2 (from ogun = 20, eeji = 2)
Chinese	shih-pa	10, 8 (shih = 10, pa = 8)	Chinese	szu-shih	4 × 10 (szu = 4, shih = 10)
Sanskrit	asta-dasa	8, 10 (asta = 8, dasa = 10)	Sanskrit	catvarim-sat	4 × 10 (catvarah = 4, sat from dasa = 10)
Mayan	uaxac-lahun	8, 10 (uaxac = 8, lahun = 10)	Mayan	ca-ikal	2 × 20 (ca = 2, kal is suffix for 20)
Latin	duodeviginti	2 from 20 (duo = 2, viginti = 20)	Latin	quadraginta	4 × 10 (quad = 4, ginta from decem = 10)
Greek	okto kai deka	8 and 10 (okto = 8, deka = 10)	Greek	tettarrakonto	4 × 10 (tettara = 4, kunta from deka = 10)

This table displays the words for 18 and 40 in nine ancient and modern languages along with a linguistic analysis of word derivations.[3]

A somewhat more sophisticated example of the grouping method of number representation was developed by the Egyptians some 5000 years ago. In this hieroglyphic system, each of the first several powers of ten was represented by a different symbol, beginning with the familiar vertical stroke for 1. Thus 10 was represented by ∩, 100 by ?, 1000 by ℒ, and 10,000 by ℓ. Arbitrary whole numbers were then represented by appropriate repetitions of the symbols. For example, to represent 12,643 the Egyptians would write ‖‖∩∩???? ℒℒℓ. (Note that the usual practice was to put the smaller digits on the left.)

The hieroglyphic number system was used for writing on temple walls or carving on columns. But when the scribes wrote on papyrus, they needed a form of handwriting. For this purpose they developed the hieratic system, an example of a ciphered system. Here each number from 1 to 9 had a specific symbol, as did each multiple of 10 from 10 to 90 and each multiple of 100 from 100 to 900, and so on. A given number, for example 37, was written by putting the symbol for 7 next to that for 30. Since the symbol for 7 was ℓ and that for 30 was ⅄, 37 was written ℓ⅄. Again, since 3 was written as ⅢⅠ, 40 as ∸, and 200 as ⌣, the symbol for 243 was Ⅲ∸⌣. Similar ciphered systems were used in written Hebrew and Greek. In those two languages, the specific symbols were simply the letters of the alphabet.

Side Bar 1.2   **The Chinese Counting Board**

The Chinese counting board was a table on which counting rods (small bamboo rods about 10 cm long) were manipulated to perform various calculations. There were two possible arrangements of the rods to represent integers less than 10:

1	2	3	4	5	6	7	8	9
I	II	III	IIII	IIIII	⊤	⊤⊤	⊤⊤⊤	⊤⊤⊤⊤
—	=	≡	≣	≣	⊥	⊥	⊥	⊥

To represent numbers greater than 10, the Chinese used a decimal place value system on the counting board. The rods were set up in columns with the rightmost column holding the units, the next the tens, the next the hundreds, and so on. A blank column in a given arrangement represented a zero. To help one read the numbers easily, the two arrangements of rods were alternated. The vertical arrangement was used in the units column, the hundreds column, and the ten thousands column, while the horizontal arrangement was used in the other columns. Thus 1156 was represented by — I ≡ ⊤ and 6083 by ⊥ ≣ III.

Arithmetic operations were performed by setting up the numbers involved on different rows of the counting board and making appropriate manipulations. For example, to add 6 and 9, that is ⊤ and ⊤⊤⊤, one noted that the two horizontal rods go together to make ten while the vertical ones make five. Addition and subtraction of multi-digit numbers were generally carried out from left to right. For multiplication, practitioners needed to memorize the basic multiplication facts, but then the process was carried out from left to right with additions being performed after each multiplication step. Divisions were performed analogously.

Negative numbers were represented on the counting board by using some feature to distinguish "negative" rods from "positive." One way was to use red rods for positive numbers and black ones for negative numbers. Calculations with these were performed as described later in the chapter. Manipulations on the counting board were eventually extended to such procedures as solving systems of linear equations and finding numerical solutions to polynomial equations.

The Chinese from earliest recorded times used a multiplicative system of writing numbers, again based on powers of 10. That is, they developed symbols for the numbers 1 through 9 as well as for each of the powers of 10. Then, for example, the number 659 would be written using the symbol for 6 (⇑) attached to that for 100 (◎), then the 5 (⊠) attached to the symbol for 10 (/), and finally the symbol ⧢ for 9: ⇑⊠⧢. The development of this system was probably related to the early use of a counting board, in which rods were arranged in vertical columns standing for the various powers of 10 (Side Bar 1.2). Therefore it was natural to think of 659 as 6 in the 100's column, 5 in the 10's column, and 9 in the 1's column and to represent each column in written form by a special symbol.

Each of the above systems except the first is organized around the base 10. Namely, there are always distinguished symbols for the integral powers of 10 around which symbols for other numbers are organized. The Babylonians organized their number symbols around two separate bases. First, they used a grouping system based on 10 to represent numbers up to 59. Thus a vertical stylus stroke ▼ on a clay tablet represented 1 and a tilted stroke ◀ represented 10. By grouping they would, for example, represent 37 by ◀◀◀▼▼▼▼▼▼▼. But second, for numbers greater than 59, there developed in Mesopotamia sometime in the third millenium B.C.E. the first positional or place-value number system. In such a system, the powers of the base—in this case 60—are represented by "places" rather than symbols, while the digit in each place represents the number of each power to be counted. Hence $3 \times 60^2 + 42 \times 60 + 37$ (or 13,357) was represented by the Babylonians as ▼▼▼ ⟨⟨▼▼ ⟨▼▼▼▼. (This will be written from now on as 3,42,37 rather than with the Babylonian strokes.) The Babylonians did not originally use a symbol for 0, but often left a space if a given number was missing a particular power. It is, however, not always easy to distinguish the numbers $3 \times 60^2 + 42 \times 60 + 37$ from $3 \times 60^3 + 42 \times 60^2 + 37 \times 60$ or from $3 \times 60^3 + 42 \times 60^2 + 0 \times 60 + 37$ on the tablets themselves. A symbol for 0 was finally developed around 300 B.C.E.

An obvious question here is why the Babylonians took 60 as their base rather than the 10 of today. There have been several theories given as answers, but most probably the choice was related to the Babylonian system of measurement in which often one "large" unit was equal to 60 "small" units. It was natural then to have the same digit 1 represent 60 as well. Of course, this answer only brings up a new question. Why did the Babylonians use 60 in their measurement system? Again, there is no definitive answer, but one possibility is that 60 is evenly divisible by many small integers. Therefore, fractional values of the "large" unit could easily be expressed as integral values of the "small." The Babylonian base-60 place-value system is still in use in our units for angle and time measurement, units preserved over the centuries in astronomical contexts and today an irreplaceable part of world culture.

There is no record of the written number system of ancient India, but there is literary evidence that numerical symbols did exist. It is only from about the third century B.C.E. that examples of written numbers are available. Originally the system was mixed. There was a ciphered system similar to the hieratic with separate symbols for the numbers 1 through 9 and 10 through 90. For larger numbers, the system was a multiplicative one similar to the Chinese. For example, the symbol for 200 was a combination of the symbol for 2 and that for 100, while the symbol for 70,000 combined the symbols for 70 and 1000. As will be discussed in Chapter 6, it was in or near India that the modern base-10 place-value system developed, but not until about the seventh century C.E.

1.3 ARITHMETIC COMPUTATIONS

Once there is a system of writing numbers, it is only natural that a civilization devise rules for computation with these numbers. The particular rules used are closely related to the systems of writing numbers. All of the civilizations under discussion devised rules for the basic arithmetic operations—addition, subtraction, multiplication and division—and as a consequence of the last operation, rules for writing and operating with fractions. These rules may be considered as some of the earliest algorithms developed.

An **algorithm** is a definite procedure designed to produce an answer to a given type of problem. Ancient peoples produced algorithms of all sorts to handle many different problems. In fact, ancient mathematics can be characterized as algorithmic, in contrast to the Greek emphasis on theory. In most of the available documents of ancient mathematics, the author describes a problem to be solved and then proceeds to use a rule, either explicit or implicit, which provides the solution. There is little concern in the documents as to how the rule was discovered, why the rule works, or what the limitations of the rule are. In most cases there are simply many examples of the use of the rule, often in increasingly complex situations. Nevertheless, in the discussion of these rules, possible origins and justifications of each will be indicated, the possible answers the Babylonian or Chinese or Egyptian scribes gave to their students who asked the eternal question "why?"

In the Egyptian hieroglyphic grouping system, addition is quite simple: combine the units, then the tens, then the hundreds, and so on. Whenever a group of ten of one type of symbol appears, replace it by one of the next. Hence to add 275 and 783, put $\mathrm{IIInnn_{99}^{II\,nnn}}$ and $\mathrm{III^{nnnn????}_{nnnn\,???}}$ together to get $\mathrm{IIIII^{nnnnnnnn?????}_{III\ nnnnnn\ ????}}$. Since there are 15 ∩'s, replace ten of them by one ?. This then gives ten of the latter. Replace these by one $\mathscr{L}$. The final answer is $\mathrm{IIIII^{nnn}_{IIII\,nn}}\mathscr{L}$ or 1058. Subtraction is done similarly. In this case, of course, whenever "borrowing" is needed, one of the symbols would have to be converted to 10 of the next lower symbol.

Such a simple algorithm for addition and subtraction is not possible in the hieratic system. For these operations, the mathematical papyri give us no good evidence. The answers to addition and subtraction problems are merely written down. Most probably, then, the scribes had addition tables. At some point these were in written form, but a competent scribe, of course, memorized them. The scribes presumably used the addition tables in reverse for subtraction problems.

The Egyptian algorithm for multiplication was based on a continual doubling process. To multiply two numbers, the scribe would first decide which number would be the multiplicand. He would then double this repeatedly, all the while recording the partial multipliers, until the next doubling would exceed the original multiplier. For example, to multiply 12 by 13 the scribe would set down the following lines:

$$
\begin{array}{ll}
1 & 12 \\
2 & 24 \\
4 & 48 \\
8 & 96
\end{array}
$$

At this point, he noticed that the next doubling would give him 16, which is larger than 13. He would then check off those multipliers which added to 13, namely 1, 4, and 8, and add the corresponding numbers in the other column. The result would be written as: Totals   13   156.

As before, there is no record of how the scribe did the doubling. The answers are simply written down. Perhaps the scribe had memorized an extensive 2 times table. On the other hand, the scribes were aware that every positive integer could be uniquely expressed as the sum of powers of 2. That fact provides the justification for the procedure. How was it discovered? The best guess is that it was discovered by experimentation and then passed down as tradition.

Since division is the inverse of multiplication, a problem such as 156 ÷ 12 would be stated as, "multiply 12 so as to get 156." The scribe would then write down the same lines as above. This time, however, he would check off the lines having the numbers in the right hand column which added to 156; here that would be 12, 48, and 96. Then the sum of the corresponding numbers on the left, namely 1, 4, and 8, would give the answer 13. Of course, division does not always "come out even." When it did not, the Egyptians resorted to fractions.

The Egyptians only dealt with unit fractions (fractions with numerator 1), with the single exception of 2/3, perhaps because these fractions are the most "natural." The fraction 1/*n* is represented in hieroglyphics by the symbol for the integer *n* with the symbol ⌒ above. In the hieratic a dot is used instead. So 1/8 is denoted in the former system by ⅏ and in the latter by ≛. The single exception, 2/3, had a special symbol: ☥ in hieroglyphic and Υ in hieratic. (The former symbol is indicative of the reciprocal of 1 1/2.) In what follows, however, the notation $\bar{n}$ will be used to represent 1/*n* and $\bar{\bar{3}}$ to represent 2/3.

Because fractions show up as the result of divisions which do not come out evenly, surely there is a need to deal with fractions other than unit fractions. It was in this connection that the most intricate of the Egyptian mathematical techniques developed, the representation of any fraction in terms of unit fractions. The Egyptians did not put the question this way, however. Whenever we would use a non-unit fraction, they simply wrote a sum of unit fractions. For example, problem 3 of the *Rhind Mathematical Papyrus* asks how to divide 6 loaves among 10 men. The answer is given that each man gets $\bar{2}\ \overline{10}$ loaves (that is, 1/2 + 1/10). The scribe checks this by multiplying this value by 10. Although today the Egyptian technique is regarded as cumbersome, it was in fact used in the Mediterranean basin for over 2000 years.

In multiplying whole numbers, the important step is the doubling step. So too in multiplying fractions, the scribe had to be able to express the double of any unit fraction. For example, in the problem above, the check of the solution is written as follows:

1	$\bar{2}\ \overline{10}$
'2	$1\ \bar{5}$
4	$2\ \bar{3}\ \overline{15}$
'8	$4\ \bar{\bar{3}}\ \overline{10}\ \overline{30}$
10	6

How are these doubles formed? To double $\bar{2}\ \overline{10}$ is easy; since each denominator is even, each is merely halved. In the next line, however, $\bar{5}$ must be doubled. It was here that the scribe had to use a table to get the answer $\bar{3}\ \overline{15}$ (that is, 2 · 1/5 = 1/3 + 1/15). In fact, the first section of the *Rhind Papyrus* is a table of the division of 2 by every odd integer from 3 to 101 (Figure 1.7), and the Egyptian scribes realized that the result of multiplying $\bar{n}$ by 2 is the same as that of dividing 2 by n. It is not known how the division table was constructed, but there are several scholarly accounts giving hypotheses for the scribes' methods.[4] In any case, the solution of problem 3 depends on using that table twice, first as already indicated and second, in the next step, where the double of $\overline{15}$ is given as $\overline{10}\ \overline{30}$ (or 2 · 1/15 = 1/10 + 1/30). The final step in this problem involves the addition of $1\ \bar{5}$ to $4\ \bar{3}\ \overline{10}\ \overline{30}$ and here the scribe just gave the answer. Again, the conjecture is that for such addition problems an extensive table existed. The *Egyptian Mathematical Leather Roll*, which dates from about 1600 B.C.E., contains a short version of such an addition table. There are also extant several other tables for dealing with unit fractions and a multiplication table for the special fraction 2/3. It thus appears that the arithmetic algorithms used

2 DIVIDED BY 3, 5, AND 7

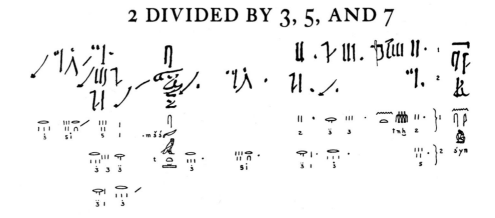

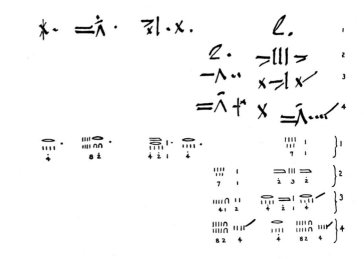

FIGURE 1.7
Transcription and
hieroglyphic translation
of $2 \div 3$, $2 \div 5$, and $2 \div 7$
from the Rhind
Mathematical Papyrus.
(Source: *The Rhind
Mathematical Papyrus*,
N.C.T.M.)

by the Egyptian scribes involved extensive knowledge of basic tables for addition, subtraction, and doubling and then a definite procedure for reducing multiplication and division problems into steps, each of which could be done using the tables.

Often in dealing with division, the scribes replaced the doubling procedure by halving. For example, in determining $2 \div 7$, the first steps are

$$
\begin{array}{cc}
1 & 7 \\
\overline{2} & 3\,\overline{2} \\
\overline{4} & 1\,\overline{2}\,\overline{4}
\end{array}
$$

To get 2 as a total in the right hand column requires the addition of $\overline{4}$ to $1\,\overline{2}\,\overline{4}$ in the third line. The scribe knew that $4 \times 7 = 28$ and therefore realized that 1/28 of 7 is 1/4. It followed that $2 \div 7 = \overline{4}\,\overline{28}$.

As an example of many other modifications that the scribes made to their basic procedure, consider problem 69 of the *Rhind Papyrus*, which includes the division of 80 by $3\,\bar{2}$ and its subsequent check:

1	$3\,\bar{2}$	'1	$22\,\bar{\bar{3}}\,\bar{7}\,\overline{21}$
10	35	'2	$45\,\bar{3}\,\bar{4}\,\overline{14}\,\overline{28}\,\overline{42}$
'20	70	'$\bar{2}$	$11\,\bar{3}\,\overline{14}\,\overline{42}$
'2	7	$3\,\bar{2}$	80
'$\bar{3}$	$2\,\bar{3}$		
'$\overline{21}$	$\bar{6}$		
'$\bar{7}$	$\bar{2}$		
$22\,\bar{\bar{3}}\,\bar{7}\,\overline{21}$	80		

In the second line the scribe has taken advantage of the decimal nature of his notation to give immediately the product of $3\,\bar{2}$ by 10. In the fifth line he has used the 2/3 multiplication table mentioned earlier. The scribe's ingenuity is evident in his choices of multipliers in the last two lines and in his probably prior choices of the numbers needed in the second column to give the desired sum of 80. The check shows several uses of the table of division by 2 as well as great facility in addition.

That the Babylonians used tables in the process of performing arithmetic computations is proved by extensive direct evidence. Many of the preserved tablets are in fact multiplication tables. No addition tables have turned up, however. Since over 200 Babylonian table texts have been analyzed, it may be assumed that these did not exist. The Babylonian scribes evidently learned their addition facts by heart and merely wrote down the answers when needed. Of course, they may well have used "scratch tablets" of some sort in complicated addition or subtraction problems. And because the Babylonian number system was a place-value system, the actual algorithms for addition and subtraction, including carrying and borrowing, may well have been similar to modern ones. For example, to add 23,37 ($= 1417$) to 41,32 ($= 2492$), one first adds 37 and 32 to get 1,9 ($= 69$). One writes down 9 and carries 1 to the next column. Then $23 + 41 + 1 = 1,5$ ($= 65$), and the final result is 1,5,9 ($= 3909$).

Because the place-value system was based on 60, the multiplication tables were extensive. Any given one listed the multiples of a particular number, say 9, from 1×9 to 20×9 and then gave 30×9, 40×9, and 50×9. If one needed the product 34×9, one simply added the two results 30×9 and 4×9. One might think that for a complete system of tables, the Babylonians would have one for each integer from 2 to 59. Such was not the case, however. In fact, although there are no tables for 11, 13, or 17, for example, there are tables for 1,30, 3,45, and 44,26,40. It appears that, in general, the Babylonians created multiplication tables only for "reciprocal numbers," numbers found in the reciprocal tables. A **reciprocal table** lists integers along with their multiplicative inverses. Multiplicative inverses of integers, of course, are fractions. The Babylonians, however, treated these not as common fractions but as sexagesimal fractions, analogous to our use of decimal fractions. Namely, the first place after the "sexagesimal point," denoted by ";", represents 60ths, the next place 3600ths, and so on. For example, 1;15 represents 1 15/60 or 1 1/4, and 0;0,44,26,40 represents $44/60^2 + 26/60^3 + 40/60^4$ or 1/81. Because the Babylonians did not have a symbol for 0, or for the sexagesimal point, this last number would just be written as 44,26,40. For actually performing the multiplication algorithm, naturally, this representation was adequate. The only question, as in today's decimal calculations, was the placement of the sexagesimal point. The Babylonians evidently understood this placement by context. In any case, a table of reciprocals, part of one of which is reproduced here, is a list of pairs of numbers whose product is 1,0 or 1,0,0 or 1,0,0,0.

2	30	16	3,45	50	1,12
3	20	25	2,24	1,4	56,15
10	6	40	1,30	1,21	44,26,40

The reciprocal tables were used in conjunction with the multiplication tables to do division. Thus the multiplication table for 1,30 (= 90) served not only to give multiples of that number, but also, since 40 is the reciprocal of 1,30, to do divisions by 40. In other words, the Babylonians considered the problem 50 ÷ 40 to be equivalent to 50 × 1/40, or, in sexagesimal notation, to 50 × 0;1,30. The multiplication table for 1,30, part of which appears here, then gives 1,15 (or 1,15,0) as the product (Figure 1.8). The appropriate placement of the sexagesimal point gives 1;15 (= 1 1/4) as the correct answer to the division problem.

1	1,30	10	15	30	45
2	3	11	16,30	40	1
3	4,30	12	18	50	1,15

The reciprocal tables only list integers whose multiplicative inverses can be exactly expressed in a finite number of sexagesimal places. The number 7, for example, is not such a number. If one divides 1 by 7, one gets a periodic sexagesimal 0;8,34,17,8,34,17, ...

FIGURE 1.8
YBC 9883: Multiplication table for 1,30.
(Source: Yale Babylonian Collection)

The Babylonians did, however, have multiplication tables for 7, which thus completed these tables for integers up to 10. If one needed a multiplication not provided in a table, for example 33 × 17, one found the two products 30 × 17 = 8,30 and 3 × 17 = 51 in the "30" table and the "3" table, respectively, and added: 8,30 + 51 = 9,21. And if one wanted to divide by 7 (or certain other non-reciprocal numbers), there did exist tables giving three or four place approximations to their reciprocals. There are also examples where the answer is merely guessed and then checked.

In ancient China, arithmetic calculations were made on the counting board. In general, whenever fractions were needed, they were expressed as common fractions. In fact, the Chinese used our modern rules of calculation with such fractions, including our device of common denominators. There is some early evidence, however, of the use of decimal fractions simply as additional columns on the counting board, in particular, in dealing with measures of length and weight. A fully developed decimal fraction system was not in place until much later.

1.4 LINEAR EQUATIONS

Most of the mathematical sources from ancient times are concerned with the solution of problems, to which various mathematical techniques are applied. Our study of these problems begins with several methods for solving what are today known as linear equations. Many sources, in fact, take the solution of single linear equations for granted. Whenever such an equation appears as part of a more complex problem, the answer is merely stated without any method of solution being mentioned. But the Egyptian papyri deal explicitly with such equations.

For example, the *Moscow Papyrus* uses the usual current technique to find the number such that if it is taken 1 1/2 times and then 4 is added, the sum is 10. In modern notation, the equation is simply $(1\ 1/2)x + 4 = 10$. The scribe proceeds as we are taught today. Namely, he first subtracts 4 from 10 to get 6, then multiplies 6 by 2/3 (the reciprocal of 1 1/2) to get 4 as the solution. Similarly, problem 31 of the *Rhind Papyrus* asks to find a quantity such that the sum of itself, its 2/3, its 1/2, and its 1/7 become 33, that is, to find x such that $x + (2/3)x + (1/2)x + (1/7)x = 33$. The problem is not conceptually difficult, but it is arithmetically challenging. It and the three following problems were probably put in to demonstrate methods of division, for the scribe solved the problem by dividing 33 by $1 + 2/3 + 1/2 + 1/7$. His answer, and this should be checked, is written as 14 $\overline{4}$ $\overline{56}$ $\overline{97}$ $\overline{194}$ $\overline{388}$ $\overline{679}$ $\overline{776}$ (or, in modern notation, 14 28/97). These two problems are presented as purely abstract with no reference to real quantities such as areas or loaves of bread. In fact, it would be difficult to find a real-life problem related to the second of these examples. The scribe is simply showing that his technique works for any division problem, no matter how difficult. Problem 35, on the other hand, does have a practical orientation. It asks to find the size of a scoop which requires 3 1/3 trips to fill a 1 hekat measure. The scribe solves the equation which would today be written as $(3\ 1/3)x = 1$ by dividing 1 by 3 1/3. He writes the answer as $\overline{5}$ $\overline{10}$ and proceeds to prove that the result is correct.

A second technique of solving a linear equation is demonstrated in problem 26 of the *Rhind Papyrus*, which seeks to find a quantity such that when it is added to 1/4 of itself the result is 15. The problem is solved by the method of false position, that is, by assuming a convenient but incorrect answer and then adjusting it appropriately. The scribe's solution is as follows: "Assume [the answer is] 4. Then 1 $\overline{4}$ of 4 is 5 . . . Multiply 5 so as to get 15. The answer is 3. Multiply 3 by 4. The answer is 12."[5] In modern notation, the problem is to solve $x + (1/4)x = 15$. The first guess is 4, since 1/4 of that is an integer. But then the scribe notes that $4 + 1/4 \cdot 4 = 5$. To find the correct answer, he must multiply 4 by the quotient of 15 by 5, namely 3. The *Rhind Papyrus* has several similar problems, all solved using false position. The step-by-step procedure of the scribe can therefore be considered as an algorithm for the solution of a linear equation of this type. There is, however, no discussion of how the algorithm was discovered or why it works. But it is evident that the Egyptian scribes understood the basic idea of a linear relationship between two quantities, that a multiplicative change in the first quantity implies the same multiplicative change in the second.

This understanding is further exemplified in the solution of proportion problems. For example, problem 75 asks for the number of loaves of *pesu* 30 which can be made from the same amount of flour as 155 loaves of *pesu* 20. (*Pesu* is the Egyptian measure for the inverse "strength" of bread and can be expressed as *pesu* = [number of loaves]/[number of hekats of grain], where a hekat was a dry measure approximately equal to 1/8 bushel.) The problem is thus to solve the proportion $x/30 = 155/20$. The scribe accomplished this by dividing 155 by 20 and multiplying the result by 30 to get 232 1/2. Similar problems occur elsewhere in the *Rhind Papyrus* and in the *Moscow Papyrus*.

Chapter 2 of the *Jiuzhang* has many problems on simple proportions as well. Thus, given that 50 measures of millet can be exchanged for 24 measures of polished rice, problem 3 asks how much rice does one get for 4 5/10 measures of millet. In modern notation, the proportion to be solved is 4 5/10 : x = 50 : 24. The Chinese author reduces the right side to 25 : 12 and solves the problem, as we would, by multiplying 12 by 4 5/10 and then dividing by 25.

There are few extant single linear equations in Babylonian texts. Even where a relatively complex one appears, such as $(x + x/7) + \frac{1}{11}(x + x/7) = 60$, the scribe just presents the answer, here x = 48 1/8. On the other hand, there is more detail given in the solution of simultaneous linear equations in two unknowns. And one of the methods used, making a convenient guess and then adjusting it, shows that the Babylonians too understood linearity. Here is an example from an Old Babylonian text: One of two fields yields 2/3 sila per sar, the second yields 1/2 sila per sar (sila and sar are measures for volume and area respectively). The yield of the first field was 500 sila more than that of the second; the areas of the two fields were together 1800 sar. How large is each field? It is easy enough to translate the problem into a system of two equations with x and y representing the unknown areas:

$$\frac{2}{3}x - \frac{1}{2}y = 500$$
$$x + y = 1800.$$

A modern solution might be to solve the second equation for x and substitute the result in the first. But the Babylonian scribe here made the initial assumption that x and y were both equal to 900. He then calculated that $2/3 \cdot 900 - 1/2 \cdot 900 = 150$. The difference between the desired 500 and the calculated 150 is 350. To adjust the answers the scribe presumably realized that every unit increase in the value of x and consequent unit decrease in the value of y gave an increase in the "function" $(2/3)x - (1/2)y$ of $2/3 + 1/2 = 7/6$. He therefore needed only to solve the equation $(7/6)s = 350$ to get the necessary increase $s = 300$. Adding 300 to 900 gave him 1200 for x while subtracting gave him 600 for y, the correct answers.

The Chinese were also interested in systems of linear equations and used two basic algorithms to handle them. The first method, used chiefly for solving problems we would translate into systems of two equations in two unknowns, is called the method of surplus and deficiency and is found in Chapter 7 of the *Jiuzhang*. The methodology, which like that of the Babylonians begins with the "guessing" of possible solutions and concludes by adjusting the guess to get the correct solution, shows that the Chinese also understood the concept of a linear relationship.

Consider problem 17: "The price of 1 acre of good land is 300 pieces of gold; the price of 7 acres of bad land is 500. One has purchased altogether 100 acres; the price was 10,000. How much good land was bought and how much bad?"[6] A modern translation of this problem would be as a system of two equations in two unknowns:

$$x + y = 100$$
$$300x + \frac{500}{7}y = 10,000.$$

The Chinese rule for the solution states: "Suppose there are 20 acres of good land and 80 of bad. Then the surplus is 1714 2/7. If there are 10 acres of good land and 90 of bad, the deficiency is 571 3/7." The solution procedure, as explained by the Chinese author, is to multiply 20 by 571 3/7, 10 by 1714 2/7, add the products, and finally divide this sum by the sum of 1714 2/7 and 571 3/7. The result, 12 1/2 acres, is the amount of good land. The amount of bad land, 87 1/2 acres, is then easily found. We can express this algorithm by the formula

$$x = \frac{b_1 x_2 + b_2 x_1}{b_1 + b_2},$$

where b_1 is the surplus determined by the guess x_1 and b_2 is the deficiency determined by the guess x_2.

The Chinese author does not explain how he arrived at his algorithm, an algorithm which was to turn up in the Islamic world and then in western Europe over a thousand years later. One conjecture as to how this algorithm was found begins by noting that the change from the correct but unknown x to the guessed value 20 involves a change in the value of the "function" $300x + (500/7)y$ of 1714 2/7 while a change from 10 to x involves a change in the function value of 571 3/7. Since linearity implies that the ratios of each pair of changes are equal, we derive the proportion

$$\frac{20 - x}{1714\frac{2}{7}} = \frac{x - 10}{571\frac{3}{7}},$$

or, in the general case,

$$\frac{x_1 - x}{b_1} = \frac{x - x_2}{b_2}.$$

The desired solution for x then follows.

Each of the 20 problems in Chapter 7 is solved by one or another modification of this algorithm of "surplus and deficiency." For example, two different guesses may both give a surplus. In every case, the author gave an explanation of the appropriate calculation. Since it is possible using modern symbolism to write each of these problems in the same form and give a single (algebraic) solution, the reader must always keep in mind that neither the Chinese nor any of the other ancient peoples used the symbolism which today enables such problems to be solved with little effort. All of the problems and their solutions were written out in words. Even so, the scribes did not hesitate to present problems with unwieldy solutions, perhaps because they wanted to convince their students that a thorough mastery of the methods would enable even difficult problems to be solved.

Chapter 8 of the *Jiuzhang* describes a second method of solving systems of linear equations, again by presenting various examples with slightly different twists. In this case, however, the modern methods are no simpler. In fact, the Chinese solution procedure is virtually identical to the method of Gaussian elimination and is presented in matrix form. As an example, here is problem 1 of that chapter. "There are three classes of grain, of which three bundles of the first class, two of the second, and one of the third make 39 measures. Two of the first, three of the second, and one of the third make 34 measures. And one of the first, two of the second and three of the third make 26 measures. How many measures of grain are contained in one bundle of each class?" The problem can be translated into modern terms as the system:

$$3x + 2y + z = 39$$
$$2x + 3y + z = 34$$
$$x + 2y + 3z = 26.$$

The rule, or algorithm, for the solution is then stated: "Arrange the 3, 2, and 1 bundles of the three classes and the 39 measures of their grains at the right. Arrange other conditions at the middle and at the left." This arrangement is presented in the diagram:

$$
\begin{array}{ccc}
1 & 2 & 3 \\
2 & 3 & 2 \\
3 & 1 & 1 \\
26 & 34 & 39
\end{array}
$$

The text continues: "With the first class on the right column multiply currently the middle column and directly leave out." This means to multiply the middle column by 3 (the first class on the right) and then subtract off a multiple (in this case, 2) of the right hand column so that the first number in the middle column becomes 0. The same operation is then performed with respect to the left column. The results are presented as:

$$
\begin{array}{ccc}
1 & 0 & 3 \\
2 & 5 & 2 \\
3 & 1 & 1 \\
26 & 24 & 39
\end{array}
\qquad
\begin{array}{ccc}
0 & 0 & 3 \\
4 & 5 & 2 \\
8 & 1 & 1 \\
39 & 24 & 39
\end{array}
$$

"Then with what remains of the second class in the middle column, directly leave out." That is, perform the same operations using the middle column and the left column. The result is:

$$\begin{array}{ccc} 0 & 0 & 3 \\ 0 & 5 & 2 \\ 36 & 1 & 1 \\ 99 & 24 & 39 \end{array}$$

Since this diagram is equivalent to the triangular system

$$\begin{aligned} 3x + 2y + z &= 39 \\ 5y + z &= 24 \\ 36z &= 99 \end{aligned}$$

the author explains how to solve that system by what is today called "back substitution" beginning with $z = 99/36 = 2\ 3/4$.[7]

As is normal in all the sources, there is no explanation of why this algorithm works or how it was derived. It can only be surmised that the Chinese discovered that subtracting multiples of equations from other equations produces a new system with the same solutions as the original one. What is clear is that the written method corresponds to the actual working out of the solution on a counting board with counting rods in the various boxes. One might wonder what happened when such a matrix manipulation led to a negative quantity in one of the boxes. A glance at problem 3 of the same chapter shows that this was not a limitation. The method was carried through perfectly correctly for the system

$$\begin{aligned} 2x + y \quad\quad &= 1 \\ 3y + z &= 1 \\ x \quad\quad + 4z &= 1. \end{aligned}$$

In fact, the author gave the rules for adding and subtracting with positive and negative quantities: "For subtraction—with the same signs, take away one from the other; with different signs add one to the other; positive taken from nothing makes negative, negative from nothing makes positive. For addition—with different signs subtract one from the other; with the same signs add one to the other; positive and nothing makes positive; negative and nothing makes negative."[8]

As an example with a different difficulty, consider finally problem 13, a system of 5 equations in 6 unknowns:

$$\begin{aligned} 2x + y \quad\quad\quad\quad\quad\quad &= s \\ 3y + z \quad\quad\quad\quad &= s \\ 4z + u \quad\quad &= s \\ 5u + v &= s \\ x \quad\quad\quad\quad + 6v &= s. \end{aligned}$$

There is only one answer presented. The matrix method, however, leads ultimately to the equation $v = 76s/721$. If $s = 721$, then $v = 76$. This is the answer given. Unfortunately, it is not known if the Chinese considered other possibilities for s or considered the implications of an infinite number of solutions. For the most part, however, both the Babylonians and Chinese only dealt with systems with the same number of equations and unknowns. And there are no records of any discussion of why that situation produces a unique solution or what happens in other situations.

1.5 ELEMENTARY GEOMETRY

With regard to elementary geometry, the records show that all of the ancient peoples under discussion knew how to calculate areas of simple rectilinear figures. Examples of uses of the standard formulas $A = bh$ and $A = (1/2)bh$ for the areas of rectangles and triangles respectively abound in the texts, although it is not always clear whether the lines whose measures are referred to here as b and h are always perpendicular. It does appear that in many cases the authors of the documents were content with reasonable approximations to the actual areas, particularly when they were computing areas of fields by use of the measured sides. There is good evidence, however, in Egypt, Babylonia, and China, of the use of the correct rule for finding the area of a trapezoid.

The measure of the area and circumference of a circle is a harder problem. Circles certainly occur in all of these civilizations, but there is no easy way to measure the circumference of a circle of given diameter or to calculate the area exactly. The modern formulas $C = \pi d$ and $A = \pi r^2$ both involve the same constant π. It may well be "obvious" that the circumference is proportional to the diameter and the area to the square of the radius, but it is by no means obvious that the proportionality constants (today written as π) are the same. That ancient peoples were aware of the proportionality of the circumference to the diameter is evident from various texts. As far as the area is concerned, the Egyptians calculated it independently of the circumference. The Babylonians and Chinese, conversely, were aware of the relationships among the area, circumference, and diameter of a circle.

In many Babylonian tablets, the circumference of the circle is given as 3 times the diameter. Similarly, problem 31 of Chapter 1 of the *Jiuzhang* states: "One has a round field; the circumference is 30 steps, the diameter 10 steps." And in the Hebrew Bible, I Kings 7:23, dealing with the reign of Solomon in about 950 B.C.E., is written, "And he made a molten sea of ten cubits from brim to brim, round in compass . . . and a line of thirty cubits did compass it round about." Again, the circumference is taken as three times the diameter. In some sense this is a curious result, because even crude measurements show that the circumference is larger than three times the diameter. Perhaps this value was used for so long because it was easy to calculate with, and because, for practical purposes, it gave accurate enough results. Or perhaps it had been handed down by tradition and was difficult to change. Nevertheless, there is an Old Babylonian tablet which gives a different value, based on the comparison of the circle to the inscribed hexagon, whose perimeter was itself calculated to be three times the diameter. The text states that the circumference of the hexagon is 0;57,36 (= 24/25) times the circumference of the circle. Therefore, $3d = (24/25)C$ or $C = (25/8)d$; in other words, the proportionality constant is 3 1/8, a much better approximation to the true value.

In most problems dealing with the circle, however, it is not the circumference which is sought, but the area. And there is a wide variety of results. Problem 50 of the *Rhind Papyrus* reads, "Example of a round field of diameter 9. What is the area? Take away 1/9 of the diameter; the remainder is 8. Multiply 8 times 8; it makes 64. Therefore, the area is 64."[9] In other words, the Egyptian scribe is using the formula $A = (d - d/9)^2 = [(8/9)d]^2$. A comparison with the formula $A = (\pi/4)d^2$ shows that the Egyptian value for the constant π in the case of area was $256/81 = 3.16049\ldots$. Where did the Egyptians

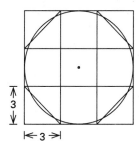

FIGURE 1.9
Octagon inscribed in a
square, from Rhind
Papyrus, problem 48.

get this value, and why was the answer expressed as the square of $(8/9)d$ rather than in modern terms as a multiple (here 64/81) of the square of the diameter? A hint is given by problem 48 of the same papyrus, in which is shown the figure of an octagon inscribed in a square of side 9 (Figure 1.9). There is no statement of the problem, however, only a bare computation of $8 \times 8 = 64$ and $9 \times 9 = 81$.

If the scribe had inscribed a circle in the same square, he would have seen that its area was approximately that of the octagon. Since the octagon has area 7/9 that of the square, the scribe might have simply put $A = (7/9)d^2 [= (63/81)d^2]$. But apparently the scribe really wanted the answer as a square. He was interested in the problem of squaring the circle, of finding a square whose area was equal to a given circle. He therefore needed the side of a square whose area was $(7/9) \cdot 81 = 63$. One way to find this is to start with the octagon in the square of side 9, and draw in the 81 small squares (Figure 1.10). The top two shaded corners are equal to the top row of small squares, while the bottom two are equal to the left hand column. Removing that row and column (removing one square twice) leaves a square of side 8/9 of the original which closely approximates the area of the octagon and hence the circle. This reconstruction perhaps clarifies why the scribe insists on writing "take away 1/9 of the diameter" and then squares the remainder. It should be noted that problem 50 is not an isolated problem. In two other problems, where the circle is part of a more complex situation, the same procedure is used.

In the Indian *Sulvasutras* the question of squaring the circle is also solved, in connection with the construction of a square altar of the same area as a circular one: "If you wish to turn a circle into a square, divide the diameter into 8 parts, and again one of these eight parts into 29 parts; of these 29 parts remove 28, and moreover the sixth part (of the one part left) less the eighth part (of the sixth part)."[10] The Indian priest meant by this that the side of the required square is equal to

$$\frac{7}{8} + \frac{1}{8 \times 29} - \frac{1}{8 \times 29 \times 6} + \frac{1}{8 \times 29 \times 6 \times 8}$$

of the diameter of the circle. This is equivalent to taking 3.0883 for π.

The Babylonians and Chinese treated the area problem differently. In both civilizations, the area of the circle is calculated by the formula $A = Cd/4 = (C/2)(d/2)$ where C is the circumference. Both also use the result $A = C^2/12$, easily derivable from the former by replacing d by $C/3$. The Chinese sometimes used the formula $A = 3d^2/4$ which could be derived from $A = Cd/4$ by replacing C by $3d$, but could also be found by averaging the areas of the inscribed and circumscribed squares.

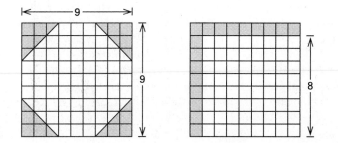

FIGURE 1.10
Dissecting an octagon.

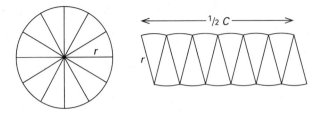

FIGURE 1.11
Dissecting a circle.

How did the Babylonians and Chinese discover the result $A = (C/2)(d/2)$ and therefore connect the calculation of the area to that of the circumference? As usual, there is no answer to this question given in the texts. One possible explanation is that they considered the slicing of a circle into sectors and their rearrangement into an approximate rectangle (Figure 1.11).

A final result dealing with plane figures is the formula for the area of a segment of a circle (often called a bow). In this case, the *Jiuzhang* gives the rule $A = (sp + p^2)/2$ where s is the length of the chord and p the length of the "arrow." This formula is certainly not correct in general. It is, however, the correct formula for a trapezoid of height p with bases p and s. The formula may therefore stem from approximating the desired area by that trapezoid (Figure 1.12). It is interesting that the same formula appears in a third century B.C.E. Egyptian papyrus and is referred to by a later writer as the method of "the ancients." There are indications of problems on segments in Babylonian texts as well, but these have not yet been interpreted.

In the case of formulas from solid geometry, there is also widespread knowledge. The rule giving the volume of a rectangular block as $V = lwh$ was well known as was the rule for a cylinder. In fact, there are several problems in the *Rhind Papyrus* where the scribe uses the rule $V = Bh$ for a cylinder where B, the area of the base, is calculated by the circle rule already discussed.

In both Babylonian texts and the *Jiuzhang* there are problems concerned with the volumes of walls and dams and the number of workers needed to build them. In many cases the cross section of the wall is given as a trapezoid; the volume is then calculated by multiplying the length of the wall by the area of the trapezoid.

Since one of the prominent forms of building in Egypt was the pyramid, one might expect to find a formula for its volume. Unfortunately such a formula does not appear in any extant document. The *Rhind Papyrus* does have several problems dealing with the *seked* (slope) of a pyramid; this is measured as so many horizontal units to one vertical unit rise. The workers building the pyramids, or at least their foremen, had to be aware of this value as they built. Since the *seked* is in effect the cotangent of the angle of slope of the pyramid's faces, one can easily calculate the angles given the values appearing in the problems. It is not surprising that these calculated angles closely approximate the actual angles used in the construction of the three major pyramids at Gizeh.

The *Moscow Papyrus*, however, does have a fascinating formula related to pyramids, namely the formula for the volume of a truncated pyramid: "If it is said to thee, a truncated pyramid of 6 cubits in height, of 4 cubits of the base by 2 of the top; reckon thou with this 4, squaring. Result 16. Double thou this 4. Result 8. Reckon thou with this 2, squaring. Result 4. Add together this 16 with this 8 and with this 4. Result 28. Calculate thou

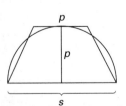

FIGURE 1.12
Approximation to the area of a segment of a circle.

$\frac{1}{3}$ of 6. Result 2. Calculate thou with 28 twice. Result 56. Lo! It is 56. Thou has found rightly."[11] If this algorithm is translated into a formula, with the length of the lower base denoted by a, that of the upper base by b, and height by h, it gives the correct result $V = \frac{h}{3}(a^2 + ab + b^2)$. Although no papyrus gives the formula for a completed pyramid of square base a and height h, it is a simple matter to derive it from the given formula by simply putting $b = 0$. On the other hand, it takes a higher level of algebraic skill to go the other way around. Still, although many ingenious suggestions involving dissection have been given, no one knows for sure how the Egyptians found their formula.

The *Jiuzhang* presents the same formula as the *Moscow Papyrus*, along with the formula for complete pyramids. A third century commentator gave a proof of the former result, by a clever decomposition of the solid, but had to use the pyramid formula $V = \frac{1}{3}a^2h$ in his argument, which formula he also demonstrated. Of course, it is still not known how the Chinese mathematician who developed the formula demonstrated the result.

The Babylonians also considered the volumes of truncated pyramids, but the textual evidence in this case is mixed. Namely, tablets exist in which the volume of a truncated pyramid with square base a^2, square top b^2 and height h is calculated by the rule $V = \frac{1}{2}(a^2 + b^2)h$, while at least one tablet appears to have the formula $V = \left[\left(\frac{a+b}{2}\right)^2 + \frac{1}{3}\left(\frac{a-b}{2}\right)^2\right]h$. One can see that the first rule is a simple but incorrect generalization of the rule for the area of the trapezoid, while the second is equivalent to the correct Egyptian formula. There is also a Babylonian formula for the volume of a truncated cone analogous to the incorrect truncated pyramid formula. It is well to remember, however, that even if the formulas are incorrect, the calculated answers would not be very different from the correct ones. It is difficult to see how anyone would realize that the answers were wrong in any case, because there was no accurate method for measuring the volume empirically. Since the problems in which these formulas occurred were practical ones and often related to the number of workmen needed to build a particular structure, the slight inaccuracy from the formula would have little effect on the final answer. A similar argument applies to the Chinese use of the formula $V = \frac{9}{16}d^3$ to calculate the volume of a sphere.

1.6 ASTRONOMICAL CALCULATIONS

Volume problems were important to both the Egyptians and Babylonians because these problems had real application in the building of pyramids, temples, and palaces. It was important for the architects and engineers to determine the amount of material necessary for the construction, so they could calculate the number of workmen needed and the amount of bread necessary to feed them. Many of these great civil engineering works were built for the housing of the rulers, but a significant number were built for ritual purposes. Such religious monuments were also built in many other places around the globe. To construct these monuments required a considerable amount of technological and organizational skill. But even before the general engineering problems were solved, the builders first attacked the problem of siting. In particular, many of these monuments are aligned with significant astronomical events. It may be concluded that the architects were familiar

FIGURE 1.13
Stonehenge. (Source:
British Tourist Authority)

with the basics of astronomy. This knowledge was used not only in the construction of monuments, but also in the creation of calendars. The needs of astronomy were crucial in the development of certain mathematical tools.

What did these ancient peoples know about the heavens? The most important heavenly bodies were the sun and the moon. It was obvious that both rose in the east and set in the west, but the actual movements of each were considerably more subtle. For example, the sun rises at exactly the east point on the spring equinox, well north of east through the summer, due east again at the autumn equinox, and south of east during the winter. It was observed everywhere that this sun cycle repeated itself at intervals. Wherever there are records of the calculation, the length of this interval, the year, is specified to be about 365 days.

If one wants to identify the important days in this yearly calendar, one needs to be able to observe the sun's position. It was in part for such observation that the great stone temple at Stonehenge in England was constructed beginning in the third millenium B.C.E. (Figure 1.13). Many similar but smaller such structures were built elsewhere in England and other parts of northern Europe (Figure 1.14). Although the reasons for the construction of these structures are not entirely clear, most scholars believe that among these reasons was the determination of the farthest north and farthest south sunrise and sunset positions.[12] Many of the alignments between stones in the construction or between a stone and a prominent natural landmark on the horizon mark precisely the directions of these four important events. Namely, when the sun rose (or set) at the horizon point indicated by the alignment, it was certain that the date was the summer (or winter) solstice.

In theory, one can construct a calendar based on the sunrise positions of the sun. But in most civilizations of which records exist, it was the motions of the moon which determined the important intervals within the year, the months. The moon, like the sun, rises at varying positions on the eastern horizon. Patient observations over a period of many years evidently enabled the builders of Stonehenge to mark the most northerly and southerly positions of moonrise. They also may have noted the existence of an 18.6 year cycle of the moonrise positions which could have been used to help predict lunar eclipses. Eclipses, both lunar and solar, were of great significance to ancient peoples. The ability to

FIGURE 1.14
Alignment of stones at
Carnac, France.

predict such striking phenomena and by appropriate ritual to cause the heavenly body to reappear after being "consumed" was an important function of the priestly classes.[13]

The most prominent feature of the moon's appearance in the sky is not its position of rising, however, but its phases. All early civilizations noted the times it took for the moon to change from tiny crescent to full moon to invisibility and back to tiny crescent again. Such observations may well have been the basis of the earliest numerical markings yet found. The Egyptians and Babylonians both used the phases of the moon to establish the months of their years, but in different ways. It was easy enough to determine that the time from the appearance of the moon's crescent in the western sky through all the phases to the next appearance of the crescent was about 29 1/2 days. Unfortunately, there is no integral multiple of 29 1/2 which equals 365, the number of days in the solar year, so there was no simple way of constructing a calendar incorporating both the moon's phases and the sun's control of the seasons. The Egyptians from a fairly early period simplified matters entirely. They employed a twelve month calendar of 30 days each with an additional 5 days tacked on at the end to give the 365 day year. By necessity, this calendar ignored the moon's cycles. In addition, since the year is in fact 365 1/4 days long, eventually even the yearly calendar was out of step with the seasons. In other words, as the Egyptian priests were well aware, the beginning of the year would slip 1/4 day per year, and thus would in 1460 years (4 × 365) make a complete cycle through the seasons. Thus for various religious purposes the priests did keep track of the actual lunar months. They also discovered that the annual Nile flood, that most important agricultural event which brought rich silt to the fields, always began just after the bright star Sirius first appeared in the eastern sky shortly before dawn after a period of invisibility. They were thus able to make the accurate predictions which helped to justify their power.

The calendrical situation in Babylon was different. The priests there wanted to accommodate the calendar to both the sun and the moon so that given agricultural events would always occur in the same month. Hence the months generally alternated in length between 29 and 30 days, a new month always starting with the first appearance of the crescent moon in the evening. Since twelve of these months equal 354 days, they decided to add an extra month every several years. In earliest times, this was done by decree whenever it was believed necessary, but eventually the Babylonians codified the calendar into a system of 7 leap years each 19 years, each leap year consisting of 13 months. The lengths of the months were occasionally adjusted too so that in each 19 year cycle of 235 months there were 6940 days. In fact, the Babylonians were aware that the mean value for the length of the moon's cycle was equal to about 29.53 days, which is in turn equal to 6940/235. The current Jewish calendar preserves the essence of the Babylonian calendar, with some minor modifications to keep it in agreement with Jewish law.

1.7 SQUARE ROOTS

Recall the Egyptian example of a square root calculation in the problem of squaring a circle. The idea there was, in effect, to find a square of area 63/81. One way to look at this is to rewrite that square as a rectangle 7/9 × 9/9. To turn the rectangle into a square, cut off a square of side 7/9, divide the remaining rectangle (7/9 × 2/9) into two parts and then bring one of them around to a side of the square (Figure 1.15). This results in a **gnomon**

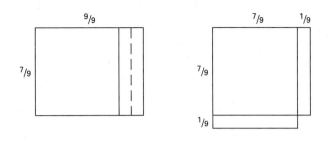

FIGURE 1.15
Turning a rectangle into a square.

figure, a square minus a corner. The Egyptian scribe was evidently content to assume that the large square itself, of side 8/9, was a sufficiently good approximation to this gnomon. After all, he was not interested in $\sqrt{63/81}$ as such, but in its use in approximating the area of a circle. Since he did not know by how much even the area approximation was in error, it did not matter if he ignored the little square of size 1/81.

Unfortunately, there are no other examples of the scribes actually calculating a square root. Whenever square roots are needed, the problems are fixed so that they come out even. That is not to say that all such square roots are integral. In one papyrus, the scribe writes that the square root of 6 1/4 is 2 1/2. What is probably true is that the scribes used square root tables, which are merely tables of squares read in reverse and are easy enough to construct. If the table is large enough, one can even interpolate to find square roots not given explicitly, but no such examples are known.

The Babylonians also had extensive square and square root tables, as well as similar ones for cubes and cube roots, examples of which still exist. Usually, when square roots are needed in solving problems, the problems are arranged so that the square root is one which is listed in a table and is a rational number. But there are cases where an irrational square root is needed, in particular, $\sqrt{2}$. When this particular value occurs, the result is generally written as 1;25 (= 1 5/12). There is an interesting tablet, however, on which is drawn a square with side indicated as 30 and two numbers, 1;24,51,10 and 42;25,35 written on the diagonal (Figure 1.16). The product of 30 by 1;24,51,10 is precisely 42;25,35. It is then a reasonable assumption that the last number represents the length of the diagonal and that the other number represents $\sqrt{2}$.

Whether $\sqrt{2}$ is given as 1;25 or as 1;24,51,10, there is no record as to how the value was calculated. But since the scribes were surely aware that the square of either of these was not exactly 2, or that these values were not exactly the length of the side of a square of area 2, they must have known that these values were approximations. How were they determined? One possible method, a method for which there is some textual evidence, begins with the algebraic identity $(x + y)^2 = x^2 + 2xy + y^2$, whose validity was probably discovered by the Babylonians from its geometric equivalent. Now given a square of area N for which one wants the side $\sqrt{N}$, the first step would be to choose a value a close to, but less than, the desired result. Setting $b = N - a^2$, the next step is to find c so that $2ac + c^2$ is as close as possible to b (Figure 1.17). If a^2 is "close enough" to N, then c^2 will be small in relation to $2ac$, so c can be chosen to equal $b/2a$, that is, $\sqrt{N} = \sqrt{a^2 + b} \approx a + b/2a$. A similar argument shows that $\sqrt{a^2 - b} \approx a - b/2a$. In the particular case of $\sqrt{2}$, one begins with $a = 1;20$ (= 4/3). Then $a^2 = 1;46,40$ and $b = 0;13,20$,

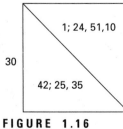

FIGURE 1.16
$\sqrt{2}$ on a Babylonian tablet.

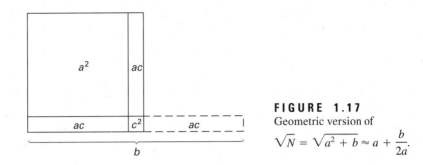

FIGURE 1.17
Geometric version of
$$\sqrt{N} = \sqrt{a^2 + b} \approx a + \frac{b}{2a}.$$

so $\sqrt{2} = \sqrt{1;46,40 + 0;13,20} \approx 1;20 + \frac{0;13,20}{2;40} = 1;25$ (or 17/12). Of course, one doesn't really need a procedure to get $\sqrt{2} \approx 1;25$. One simply needs a good guess or a table in which 1;25 appears as one of the numbers to be squared. After all, $(1;25)^2$ differs from 2 by only $0;0,25 = 1/144$.

The Indian *Sulvasutras*, interestingly enough, approximate $\sqrt{2}$ as

$$1 + \frac{1}{3} + \frac{1}{3 \times 4} - \frac{1}{3 \times 4 \times 34} = \frac{17}{12} - \frac{1}{12 \times 34}.$$

This is easily derived from the formula for $\sqrt{a^2 - b}$ with $a = 17/12$. For then

$$\sqrt{2} = \sqrt{\left(\frac{17}{12}\right)^2 - \frac{1}{144}} \approx \frac{17}{12} - \frac{\frac{1}{144}}{\frac{34}{12}}$$

$$= \frac{17}{12} - \frac{1}{12 \times 34}.$$

Whether the Indians used this method is not known. In any case, this value occurs in connection with the construction of a square altar double the size of a given one.

The only one of these civilizations for which there is an explicit square root algorithm written down is the Chinese. It too is based on the algebraic formula $(x + y)^2 = x^2 + 2xy + y^2$. The description of the procedure is given in words in Chapter 4 of the *Jiuzhang*. But specialists in Chinese mathematics have concluded that the original author probably had a diagram like Figure 1.18 in mind.[14] To explain the Chinese algorithm, we use the

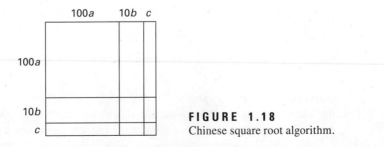

FIGURE 1.18
Chinese square root algorithm.

example of problem 12, where $\sqrt{55{,}225}$ is calculated. The idea is to find digits a, b, c so that the answer can be written as $100a + 10b + c$. (Recall that the Chinese used a decimal system.) First find the largest digit a so that $(100a)^2 < 55{,}225$. In this case, $a = 2$. The difference between the large square (55,225) and the square on $100a$ (40,000) is the large gnomon in Figure 1.18. If the outer thin gnomon is neglected, it is clear that b must satisfy $55{,}225 - 40{,}000 > 2(100a)(10b)$ or $15{,}225 > 4000b$. So certainly $b < 4$. To check that $b = 3$ is correct, that is, that with the square on $10b$ included, the area of the large gnomon is still less than 15,225, it is necessary to check that $2(100a)(10b) + (10b)^2 < 15{,}225$. Since this is in fact true, the same procedure can be repeated to find c: $55{,}225 - 40{,}000 - 30(2 \times 200 + 30) > 2 \times 230c$ or $2325 > 460c$. Evidently, $c < 6$. An easy check shows that $c = 5$ gives the correct square root: $\sqrt{55{,}225} = 235$.

The Chinese algorithm for calculating square roots is similar to one which was taught in schools in recent years. This method gives a series of answers, in this case 200, 230, 235, each a better approximation to the true result than the one before. So it is an example of the determination of a convergent sequence of numbers, each generated from the previous one by an explicit algorithm, that is, what is today called a **recursive algorithm**. Although it appears clear to a modern reader that, if the answer is not a whole number, the procedure could continue indefinitely using decimal fractions, the Chinese author used common fractions as a remainder in the cases where there is no integral square root.

Two further notes are in order here. First, a close examination of the algorithm shows that the solution of a quadratic equation (or, at least, a quadratic inequality) is part of the process. Second, the *Jiuzhang* also contains a recursive algorithm for finding cube roots, perhaps derived from a consideration of actual cubes as this one was from squares. The Chinese ultimately developed these ideas into a detailed procedure for solving polynomial equations of any degree, a procedure to be discussed in Chapter 6.

A reasonable assumption is that the Babylonian scribes also had some algorithm to determine their value $\sqrt{2} \approx 1;24,51,10$. The Chinese procedure, if transformed to base 60, involves some difficult arithmetic, but the scribes were excellent calculators. Alternatively, they could simply have continued the application of the approximations $\sqrt{a^2 \pm b} \approx a \pm \frac{b}{2a}$ through two more steps, starting with the value 1;25. A third possibility is that they used the method of the mean. Namely, since $(1;25)^2 > 2$, it follows that $\left(\frac{2}{1;25}\right)^2 < 2$. Then the mean μ of 1;25 and $\frac{2}{1;25}$ provides a closer approximation. This process can be continued by next taking the mean of μ and $\frac{2}{\mu}$. Each of these methods naturally leads to 1;24,51,10, but the important point is that this value for the square root, together with the approximation for 1/7 mentioned earlier, gives us grounds for assuming that the Babylonians had at least an elementary notion of convergent sequences and recursive algorithms as well.

1.8 THE PYTHAGOREAN THEOREM

One of the Babylonian square root problems was connected to the relation between the side of a square and its diagonal. That relation is a special case of the result known as the Pythagorean theorem: In any right triangle, the sum of the squares on the legs equals the square on the hypotenuse. This theorem, named after the sixth century B.C.E.

Greek philosopher and mathematician, is arguably the most important elementary theorem in mathematics, since its consequences and generalizations have wide-ranging application. Nevertheless, it is one of the earliest theorems known to ancient civilizations. In fact there is evidence that it was known at least 1000 years before Pythagoras.

Some scholars have argued that the astronomically related stone temples in England built in the third millenium B.C.E. were constructed using a knowledge of the Pythagorean theorem and, in particular, Pythagorean triples, triples of integers (a,b,c) such that $a^2 + b^2 = c^2$. But the evidence for this is rather tenuous.[15] There is much more substantial evidence of interest in Pythagorean triples, however, in the Babylonian tablet labeled Plimpton 322 (Figure 1.19), which dates from approximately 1700 B.C.E.[16] The extant piece of the tablet consists of four columns of numbers. Other columns were possibly broken off on the left. The tablet follows, reproduced in modern decimal notation with the few corrections that recent editors have made and with a conjectured fifth column on the left.

y	$\left(\dfrac{d}{y}\right)^2$	x	d	#
120	1.9834028	119	169	1
3456	1.9491586	3367	4825	2
4800	1.9188021	4601	6649	3
13,500	1.8862479	12,709	18,541	4
72	1.8150077	65	97	5
360	1.7851929	319	481	6
2700	1.7199837	2291	3541	7
960	1.6845877	799	1249	8
600	1.6426694	481	769	9
6480	1.5861226	4961	8161	10
60	1.5625	45	75	11
2400	1.4894168	1679	2929	12
240	1.4500174	161	289	13
2700	1.4302388	1771	3229	14
90	1.3871605	56	106	15

It was a major piece of mathematical detective work for modern scholars first to decide that this was a mathematical work rather than a list of orders from a pottery business and second to find a reasonable mathematical explanation. But find one they did. The columns headed x and d (whose headings in the original have been translated as "width" and "diagonal") contain in each row two of the three numbers of a Pythagorean triple. It is easy enough to subtract the square of column x from the square of column d. In each case a perfect square results, whose square root is indicated in the reconstructed column y. Finally, the other column represents the quotient $\left(\frac{d}{y}\right)^2$.

How and why were these triples derived? One cannot find Pythagorean triples of this size by trial and error. A hint as to the why of the table is found in the decreasing sequence in the $\left(\frac{d}{y}\right)^2$ column, beginning with a number nearly 2. If these numbers represent the sides

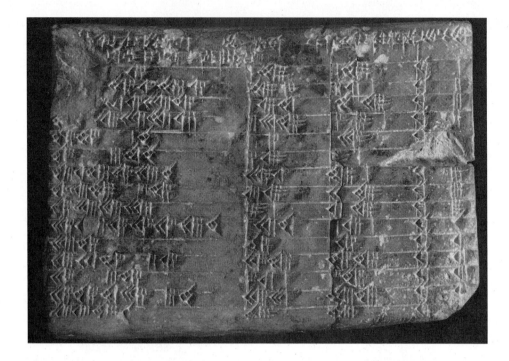

of right triangles, then the triangle of row 1 is very nearly a 45°–45° right triangle. The shapes of the triangles change regularly until the final one is nearly a 30°–60° triangle. What the Babylonians may have been interested in is finding right triangles of varying shape, all of whose sides were integral. Such triangles could then be used in constructing problems for students for which the instructor would know that the solution would be possible in integers or finite sexagesimal fractions. And in fact virtually all such problems in Babylonian problem texts have answers of this form.

The $\left(\frac{d}{y}\right)^2$ column may also give a hint as to how the table was constructed. To find integer solutions to the equation $x^2 + y^2 = d^2$, one can divide by y and first find solutions to $\left(\frac{x}{y}\right)^2 + 1 = \left(\frac{d}{y}\right)^2$, or, setting $u = \frac{x}{y}$ and $v = \frac{d}{y}$, to $u^2 + 1 = v^2$. This latter equation is equivalent to $(v + u)(v - u) = 1$. So given a choice for $v + u$, one can find $v - u$ in the table of reciprocals and then solve for u and v. Multiplication by a suitable number y would then give the integral Pythagorean triple. For example, if $v + u = 2;15 (= 2 \ 1/4)$, the reciprocal $v - u$ is $0;26,40 (= 4/9)$. Solving for v and u gives $v = 1;20,50 = 1 \ 25/72$ and $u = 0;54,10 = 65/72$. Multiplying each value by $1;12 = 72$ gives the values 65 and 97 for x and d respectively given in line 5 of the table.

But whether or not this was the Babylonian method, the fact remains that the scribes were well aware of the Pythagorean relationship. And although this particular table had no indication of a geometrical relationship except for the headings of two of the columns, there are problems in Old Babylonian tablets making explicit geometrical use of the Pythagorean theorem. One example occurs in a tablet found at Susa in modern Iran. The problem is to calculate the radius of a circle circumscribed about an isosceles triangle with altitude 40 and base 60. By considering the right triangle ABC (Figure 1.20), whose hypotenuse is the desired radius, the scribe derived the equation $r^2 = 30^2 + (40 - r)^2$

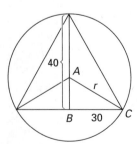

FIGURE 1.20
Circumscribing a circle about an isosceles triangle.

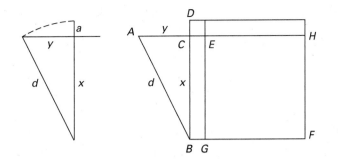

FIGURE 1.21
The length of a reed in a pond, from the *Jiuzhang,* Chapter 9, problem 6.

from the Pythagorean theorem. He then calculated that $1,20r = 30^2 + 40^2 = 41,40$ and, using reciprocals, that $r = (0;0,45)(41,40) = 31;15$.

The Pythagorean theorem is mentioned in one of the Indian *Sulvasutras* and examples of Pythagorean triples, including (5,12,13), (8,15,17), (7,24,25), and (12,35,37) are given. There is even a construction shown of a square equal to the sum of two given squares, which explicitly uses the theorem. Even more detail on the theorem is found in Chinese sources, especially in Chapter 9 of the *Jiuzhang*, a chapter devoted to problems on right triangles, in all of which the Pythagorean theorem is presupposed. Thus in problem 6 we are given a square pond of side 10 feet, with a reed growing in the center whose top is 1 foot out of the water. If the reed is pulled to the shore, the top just reaches the shore. The problem is to find the depth of the water and the length of the reed. In Figure 1.21, $y = 5$ and $x + a = d$, where, in this case, $a = 1$. A modern solution might begin by setting $d^2 = x^2 + y^2$ and substituting for d. A brief algebraic calculation gives $x = \frac{y^2 - a^2}{2a}$. With the given numerical values, $x = 12$ and therefore $d = 13$. The Chinese rule states: "Multiply half of the side of the pond by itself; decrease this by the product of the length of the reed above the water with itself; divide the difference by twice the length of the reed above the water. This gives the depth. Add this to the length of the reed above the water. This gives the length of the reed."[17] A translation of this rule into a formula gives the same $x = \frac{y^2 - a^2}{2a}$ already derived. It is not clear, however, whether the Chinese author found the solution algebraically as above or by the equivalent geometric method illustrated, where $y^2 = AC^2 = AB^2 - BC^2 = BD^2 - EG^2 = DE^2 + 2 \times CE \times BC = a^2 + 2ax$. But what is certain is that the author was fluent in the use of the Pythagorean theorem.

As a further note, in problem 6 as in all problems of Chapter 9 of the *Jiuzhang*, the answers are rational numbers. Because in every problem a right triangle is involved, it follows that, as in the Babylonian texts, the problems were made up so that all these right triangles would have rational sides. Not only do such familiar triples as (3,4,5) and (5,12,13) occur, but also less obvious ones such as (55,48,73) and (91,60,109). How did the author calculate these triples so his problems come out even?

Problem 14 of the chapter gives a clue. There are two people A, B who start to walk from the same place. The speed of A is 7 and that of B is 3. B goes east; A first goes 10 *pu* south and then towards the northeast until he meets B. How far did each of A and B travel? Setting up a right triangle with legs x and y and hypotenuse z, we have $y = 10$ and $z + y = (7/3)x$. The author then calculates

$$z = \frac{7^2 + 3^2}{2}v, \qquad y = \frac{7^2 - 3^2}{2}v, \qquad x = 7 \cdot 3v,$$

where v is an arbitrary constant. Since $y = 10$, v must be 1/2, and therefore $z = 14\ 1/2\ pu$ and $x = 10\ 1/2\ pu$ give the solution. The important point here is that the author has shown how to calculate Pythagorean triples in general. For if $z + y = \frac{a}{b}x$ and $z^2 - y^2 = x^2$, then $z - y = \frac{b}{a}x$ and

$$z = \frac{a^2 + b^2}{2ab}x, \qquad y = \frac{a^2 - b^2}{2ab}x.$$

It follows that the formulas

$$x = ab, \qquad y = \frac{a^2 - b^2}{2}, \qquad z = \frac{a^2 + b^2}{2}$$

with a, b odd and $a > b$ always determine a Pythagorean triple. For example, the triple (55,48,73) can be obtained by taking $a = 11$ and $b = 5$, while the triple (91,60,109) comes from $a = 13$, $b = 7$.

Pythagorean triples are useful for finding right triangles with integral sides. If in fact this was the reason for their development, it follows that the geometric version of the Pythagorean theorem was already known. The natural question then is how the theorem was discovered and "proved." As in many other areas, there is no record of the discovery. In all the texts mentioned, the theorem is taken as known. There are, however, hints of "proofs."

The Indian *Sulvasutras* state, among their rules for altar construction, that "the cord which is stretched across a square produces an area of double the size."[18] This statement suggests Figure 1.22a, from which the proof of the quoted statement is obvious. Since this is a special case of the Pythagorean theorem, it may be conjectured that the general case was discovered by modifying this diagram. In fact, in the Chinese *Zhoubi suanjing*, which dates back to at least several hundred years B.C.E., such a modification does appear (Figure 1.22b). The accompanying commentary is as follows:

> Thus let us cut a rectangle [diagonally] and make the width 3 [units] and length 4 [units]. The diagonal between the [opposite] corners will then be 5 [units]. Now after drawing a square on this diagonal, circumscribe it by half-rectangles like that which has been left outside so as to form a [square] plate. Thus the [four] outer half-rectangles of width 3, length 4, and diagonal 5, together make two rectangles [of area 24]. Then [when this is subtracted from the square plate of area 49], the remainder is of area 25. This [process] is called "piling up the rectangles."[19]

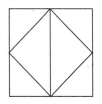

FIGURE 1.22a
Indian proof of a special case of the Pythagorean theorem.

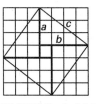

FIGURE 1.22b
Chinese proof of the Pythagorean theorem.

Although the commentary and picture are given for the specific case of a (3,4,5) triangle, the proof (given in the last two lines) is quite general. Denote the width by a, the length by b, and the diagonal by c. The argument is then as follows: $(a + b)^2 - 2ab = c^2$; since $(a + b)^2 = a^2 + b^2 + 2ab$, the Pythagorean result $a^2 + b^2 = c^2$ is immediate. Looking at it geometrically, the argument depends simply on dissecting the large square in two ways, first as the square on a plus the square on b plus twice the rectangle ab, and second as the square on c plus twice the rectangle ab. Again the result follows immediately.

Is the argument just given a proof? To meet modern standards, it would be necessary to show either that the inscribed figure (the square on c) or the circumscribed figure (the square on $a + b$) is in fact a square. To the ancients, however, and probably to most students today, this was obvious. The Chinese had no notion of an axiomatic system from which theorems could be derived. "Proof" means here simply a convincing argument. In fact, the Greek word theorem is derived from *theorein*, to look at. If one looks at the diagram, one sees the theorem at once. The earliest Babylonian records of the theorem have no argument for it at all. Nevertheless, the above argument would certainly have been accessible to the scribes.

1.9   QUADRATIC EQUATIONS

Problems involving the product of two unknowns or the square of one lead to equations known today as quadratic equations. For example, problems involving the theorem of Pythagoras often lead to such equations. These equations were a major area of study of the ancient Babylonians and also occur in the Chinese sources. In both civilizations, the methodology of the solutions was based on geometric ideas, that is, geometric squares and rectangles rather than arithmetic squares and products.

Chapter 9 of the Chinese *Jiuzhang*, the chapter on right triangles, contains several problems translatable as quadratic equations. For example, problem 20 requires the solution of $x^2 + 34x = 71,000$. Unfortunately, the Chinese author merely states the solution $x = 250$ without presenting any method. As will be seen in Chapter 6, however, the Chinese method of solution assumed here by the author is closely related to the Chinese square root algorithm discussed earlier, an algorithm which had a geometric origin. Namely, it is a recursive procedure which at each step gives a closer approximation to the correct answer.

Most of the quadratic problems of the *Jiuzhang* are ones we would translate into systems of two equations. Thus problem 11 presents a door whose height is 6.8 more than its width. The distance between the corners is given as 10, and the question is to find the height and width. The problem, given the Pythagorean theorem, translates into the system

$$x - y = 6.8 \qquad x^2 + y^2 = 100.$$

The Chinese solution appears to be based on the Chinese "proof" of the Pythagorean theorem presented earlier. If we rewrite this problem in the generic form $x - y = d$, $x^2 + y^2 = c^2$, the figure used in that proof shows that $(x + y)^2 = 4xy + (x - y)^2$ and also that $c^2 = 2xy + (x - y)^2$, or, that $4xy = 2c^2 - 2(x - y)^2$. It follows that $(x + y)^2 = 2c^2 - (x - y)^2$ or that $x + y = \sqrt{2c^2 - (x - y)^2}$ or, finally, that

$$\frac{x + y}{2} = \sqrt{\frac{c^2}{2} - \left(\frac{d}{2}\right)^2}.$$

It is the steps determined by this formula which enable the author first to determine $x + y = 12.4$ and then, by combining this with $x - y = 6.8$, to get the solution $x = 9.6$, $y = 2.8$.

Although the *Jiuzhang* contains a few other problems which can be translated into systems of linear and quadratic equations, it is the Babylonian sources which contain most of the examples of quadratic equations from ancient times.[20] In fact, many Old Babylonian tablets contain extensive lists of quadratic problems. The standard form

$$x + y = a \qquad xy = b$$

of some of these Babylonian problems suggests that originally the Babylonians wanted to deal with the relation between the area and perimeter of a rectangle. It appears that in ancient times many believed that the area of a field, say, only depended on its perimeter. There are various stories indicating that those who knew better used this knowledge to take advantage of those having that belief. It is thus conceivable that the Babylonian scribes, in order to demonstrate that rectangles of the same perimeter could have very different areas, constructed tables of areas b for a given perimeter $2a$ using different values for the length x and the width y. A study of such tables relating the varying lengths $x = \frac{a}{2} + z$ and widths $y = \frac{a}{2} - z$ to the areas $b = \left(\frac{a}{2} + z\right)\left(\frac{a}{2} - z\right) = \left(\frac{a}{2}\right)^2 - z$ could well have led to their noticing that $z = \sqrt{(a/2)^2 - b}$ and therefore that

$$x = \frac{a}{2} + \sqrt{\left(\frac{a}{2}\right)^2 - b} \qquad y = \frac{a}{2} - \sqrt{\left(\frac{a}{2}\right)^2 - b}$$

is the solution to the given system. In any case, it is the algorithm described by these modern formulas which the Babylonian scribes used to solve this type of problem.

Naturally, the Babylonians do not give a formula. Each problem is presented with numbers assigned to the length, width, and area and specific numerical calculations are indicated which we can interpret in terms of the above formula. For example, consider the problem $x + y = 6\ 1/2$, $xy = 7\ 1/2$ from tablet YBC 4663. The scribe first halves 6 1/2 to get 3 1/4. Next he squares 3 1/4, getting 10 9/16. From this is subtracted 7 1/2, leaving 3 1/16, and then the square root is extracted to get 1 3/4. The length is thus 3 1/4 + 1 3/4 = 5, while the width is given as 3 1/4 − 1 3/4 = 1 1/2. Whatever the ultimate origin of this method, a close reading of the wording of the tablets seems to indicate that the scribe had in mind a geometric procedure (Figure 1.23), where for the sake of generality the sides have been labeled in accordance with the generic system $x + y = a$, $xy = b$.[21] The scribe begins by halving the sum a and then constructing the square on it. Since $\frac{a}{2} = x - \frac{x - y}{2} = y + \frac{x - y}{2}$, the square on $\frac{a}{2}$ exceeds the original rectangle of area b by the square on $\frac{x - y}{2}$. Figure 1.23 then shows that if one adds the side of this square, namely $\sqrt{(a/2)^2 - b}$, to $a/2$ one finds the length x, while if one subtracts it from $a/2$,

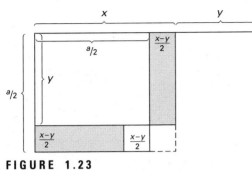

FIGURE 1.23
Geometric procedure for solving the system
$x + y = a, xy = b$.

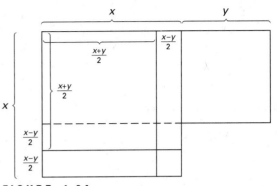

FIGURE 1.24
Geometric procedure for solving the system
$x - y = a, x^2 + y^2 = b$.

one gets the width y. The algorithm is therefore exactly that indicated in the previous paragraph.

Similar geometric interpretations can be given to the algorithms which the Babylonians developed for other classes of quadratic problems. For example, the solution of the system

$$x - y = a \qquad x^2 + y^2 = b$$

was found by a procedure describable by the modern formula

$$x = \sqrt{\frac{b}{2} - \left(\frac{a}{2}\right)^2} + \frac{a}{2} \qquad y = \sqrt{\frac{b}{2} - \left(\frac{a}{2}\right)^2} - \frac{a}{2}.$$

Although this problem is of the same form as the Chinese one about the door, it appears that the Babylonians developed the solution by using a different geometric idea. Figure 1.24 shows that $\frac{1}{2}(x^2 + y^2) = \left(\frac{x+y}{2}\right)^2 + \left(\frac{x-y}{2}\right)^2$. It follows that $\frac{b}{2} = \left(\frac{x+y}{2}\right)^2 + \left(\frac{a}{2}\right)^2$ and therefore that $\frac{x+y}{2} = \sqrt{\frac{b}{2} - \left(\frac{a}{2}\right)^2}$. Since $x = \frac{x+y}{2} + \frac{x-y}{2}$ and $y = \frac{x+y}{2} - \frac{x-y}{2}$, the result follows.

The Babylonians solved single quadratic equations as well as systems. Several such problems are given on tablet BM 13901, including the following: The sum of the area of a square and 4/3 of the side is 11/12. Find the side. In modern terms, the equation to be solved is $x^2 + (4/3)x = 11/12$. For the solution, the scribe tells us to take half of 4/3, giving 2/3, square the 2/3, giving 4/9, then add this result to 11/12, giving 1 13/36. This value is the square of 7/6. Subtracting 2/3 from 7/6 gives 1/2 as the desired side. The Babylonian rule is easily translated into a modern formula for solving $x^2 + ax = b$, namely $x = \sqrt{(a/2)^2 + b} - a/2$, recognizable as a version of the quadratic formula. The question, however, is how the Babylonians interpreted their procedure. At first glance, it would appear that the statement of the problem is not a geometric one, since we are asked to add a multiple of a side to an area. But the geometric language of the solution seems to indicate that this multiple is to be considered as a rectangle with length x and width 4/3, a rectangle which is added to the square of side x (Figure 1.25). Under this interpretation, the proce-

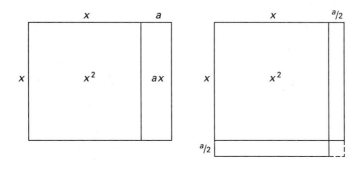

FIGURE 1.25
Geometric version of the quadratic formula for solving $x^2 + ax = b$.

dure amounts to cutting half of the rectangle off from one side of the square and moving it to the bottom. Adding a square of side $a/2$ "completes the square." It is then evident that the unknown length x is equal to the difference of the side of the new square and $a/2$, exactly as the formula implies.

One can similarly discover a geometric argument which leads to a Babylonian procedure for the solution $x = \sqrt{(a/2)^2 + b} + a/2$ to an equation of the type $x^2 - ax = b$. One should, however, keep in mind that the "quadratic formula" did not mean the same thing to the Babylonian scribes as it means to us. First, the scribes gave different procedures for solving the two types $x^2 + ax = b$ and $x^2 - ax = b$ because the two problems were different; they had different geometric meanings. To a modern mathematician, on the other hand, these problems are the same because the coefficient of x can be taken as positive or negative. Second, the modern quadratic formula in these two cases gives a positive and a negative solution to each equation. The negative solution, however, makes no geometrical sense and was completely ignored by the Babylonians. Interestingly, the cases of quadratic equations of the form $x^2 + b = ax$ having two positive solutions were also ignored by the Babylonians. Such an equation does not appear in the tablets, even though problems which we would put in that form do occur. The scribes evidently could not believe that a single equation could have two different values for the same unknown. Therefore, they used all sorts of ingenious devices to prevent this possibility. The simplest way of doing this was to rewrite the problem into the form $x + y = a$, $xy = b$ so that the two solutions would be for two different unknowns. The only case where the equation $x^2 + b = ax$ is treated is the case where $\left(\frac{a}{2}\right)^2 = b$, for in that case there is only one solution.

There are many Babylonian tablets consisting of long sets of quadratic problems. The problems of such a set are often quite involved algebraically. Some problem sets are written in "abstract" form, but others appear as "real-world" problems in the context, for example, of engineering situations. These latter quadratic problems are, in reality, just as artificial as the ones found in most current algebra texts. That the authors knew they were artificial is shown by the fact that, typically, all problems of a given set have the same answer. It thus appears that the tablets were used to develop techniques of solution. In other words, the purpose of solving the various problems was not to determine the answer, but to learn various methods of reducing complicated problems to simpler ones. One can speculate, therefore, that the mathematical tablets in general, and the ones on quadratic equations in particular, were used to train the minds of future leaders of the country. In other words, it was not really that important to solve quadratic equations—there were few real situations which required them. What was important was for the students to develop skills in solving problems in general, skills which could be used in dealing with the everyday problems which a nation's leaders need to solve. These skills included not only the following of well-established procedures—algorithms—but also knowing how and when one could modify the methods and how one could reduce more complicated problems to ones already solved. Today's students are often told that mathematics is studied to "train the mind." It seems that teachers have been telling their students the same thing for the past 4000 years.

Exercises

Counting

1. Determine the words for the numbers 18 and 40 in all languages you and your classmates know. Compare the construction of these words. Are there any forms essentially different from the ones in the box in the text?

2. Represent 125 in Egyptian hieroglyphics and Babylonian cuneiform.

3. The Greeks used a ciphered system based on their alphabet to represent numbers, at least from about 450 B.C.E. The representation was as follows:

α	1	ι	10	ρ	100
β	2	κ	20	σ	200
γ	3	λ	30	τ	300
δ	4	μ	40	υ	400
ε	5	ν	50	φ	500
ς	6	ξ	60	χ	600
ζ	7	ο	70	ψ	700
η	8	π	80	ω	800
θ	9	ϙ	90	ϡ	900

where the letters ς (digamma) for 6, ϙ (koppa) for 90, and ϡ (sampi) for 900 are letters which by this time were no longer in use. Hence 754 was written ψνδ and 293 was written σϙγ. To represent thousands a mark was made to the left of the letters α through θ; e.g., ‚θ represented 9000. Larger numbers still were written using the letter M to represent myriads (10,000) with the number of myriads written above: $M^\delta = 40,000$, $M^{\varsigma\rho o\epsilon} = 71,750,000$. Represent 125, 62, 4821, and 23,855 in Greek alphabetic notation.

4. The basic Chinese symbols for numbers from the Shang period are

1	2	3	4	5	6	7	8	9	10	100	1000

There were compound symbols for 20, 30, 40 (namely, ∪ ∪∪ ∪∪∪), but in general notation followed the plan indicated in the text. Hence 88 is)( )(and 162 is ⓐ⇑=. Write the Chinese form of 56, 554, 63, and 3282.

Arithmetic Computations

5. Justify each line of the division of 80 by $3\,\overline{2}$ given in the text.

6. Multiply $7\,\overline{2}\,\overline{4}\,\overline{8}$ by $12\,\overline{\overline{3}}$ using the Egyptian multiplication technique. Note that it is necessary to multiply each term of the multiplicand by $\overline{\overline{3}}$ separately.

7. A part of the *Rhind Mathematical Papyrus* table of division by 2 follows: $2 \div 11 = \overline{6}\,\overline{66}$, $2 \div 13 = \overline{8}\,\overline{52}\,\overline{104}$, $2 \div 23 = \overline{12}\,\overline{276}$, $2 \div 29 = \overline{24}\,\overline{58}\,\overline{174}\,\overline{232}$, $2 \div 99 = \overline{66}\,\overline{198}$. The calculation of $2 \div 13$ is given as follows:

1	13
$\overline{2}$	$6\,\overline{2}$
$\overline{4}$	$3\,\overline{4}$
$\overline{8}$	$1\,\overline{2}\,\overline{8}$ '
$\overline{52}$	4 '
$\overline{104}$	8 '
$\overline{8}\,\overline{52}\,\overline{104}$	$1\,\overline{2}\,\overline{4}\,\overline{8}\,\overline{8}$
	2

Perform similar calculations for the other divisions indicated to check the results.

8. Given the value $2 \div 13 = \overline{8}\,\overline{52}\,\overline{104}$ above, the unit fraction values of 3, 4, 5, . . . , 12 divided by 13 can be easily found. For example, $3 \div 13 = \overline{8}\,\overline{13}\,\overline{52}\,\overline{104}$ (since $3 = 1 + 2$) and $4 \div 13 = \overline{4}\,\overline{26}\,\overline{52}$ (since $4 = 2 \times 2$). Similarly, calculate the unit fraction values of $5 \div 13$, $6 \div 13$, and $8 \div 13$.

9. The second part of problem 79 of the *Rhind Papyrus* reads

Houses	7
Cats	49
Mice	343
Spelt	2401
Hekat	16,807
Total	19,607

It has been surmised that this was a problem similar to the Old English children's rhyme, "As I was going to St. Ives." Thus the complete problem may have read: "An estate has 7 houses, each house has 7 cats, each cat catches 7 mice, each mouse eats 7 spelt, each spelt was capable of producing 7 hekats of grain. How many things were there in the estate?" The first part of the problem shows that the product of 2801 by 7 is 19,607. Show that this is the correct answer for the sum of the geometric series $7 + 49 + 343 + 2401 + 16,807$.

10. Show that $1 \div 7$ gives the periodic sexagesimal fraction $0;8,34,17,8,34,17\ldots$ by dividing in base 60.

11. Find the reciprocals in base 60 of 18, 27, 32, 54, 64 ($=1,4$), and 108 ($=1,48$). What is the condition on the integer n which insures that its reciprocal is a finite sexagesimal fraction?

12. In the Babylonian system, multiply 25 by 1,4 and 18 by 1,21. Divide 50 by 16 and 1,21 by 25. Use our standard multiplication algorithm modified for base 60.

Linear Equations

13. Solve by the method of false position: A quantity and its 1/7 added together become 19. What is the quantity?

14. Verify the solution to problem 31 of the *Rhind Mathematical Papyrus*, that is, check that $33 \div 1\,\overline{3}\,\overline{2}\,\overline{7} = 14\,\overline{4}\,\overline{56}\,\overline{97}\,\overline{194}\,\overline{388}\,\overline{679}\,\overline{776}$.

15. Problem 72 of the *Rhind Mathematical Papyrus* reads "100 loaves of *pesu* 10 are exchanged for loaves of *pesu* 45. How many of these loaves are there?" The solution is given as, "Find the excess of 45 over 10. It is 35. Divide this 35 by 10. You get $3\,\overline{2}$. Multiply $3\,\overline{2}$ by 100. Result 350. Add 100 to this 350. You get 450. Say then that the exchange is 100 loaves of *pesu* 10 for 450 loaves of *pesu* 45."[22] Translate this solution into modern terminology. Compare the method here to the solution in the text of problem 75. How does this solution demonstrate "linearity"?

16. Solve problem 3 of Chapter 3 of the *Jiuzhang*: Three people, who have 560, 350, and 180 coins respectively, are required to pay a total tax of 100 coins in proportion to their wealth. How much does each pay?

17. Find the solution to problem 3 of Chapter 8 of the *Jiuzhang* using the Chinese method: The yields of 2 bundles of the best grain, 3 bundles of ordinary grain, and 4 bundles of the worst grain are neither sufficient to make a whole measure. If we add to the 2 bundles of good grain 1 bundle of the ordinary, to the 3 bundles of ordinary 1 bundle of the worst, and to the 4 bundles of the worst 1 bundle of the best, then each yield is exactly one measure. How many measures does 1 bundle of each of the three types of grain contain? Show that the solution according to the Chinese method involves the use of negative numbers.

18. Solve problem 1 of Chapter 7 of the *Jiuzhang* using the method of surplus and deficiency: Several people purchased in common one item. If each person paid 8 coins, the surplus is 3; if each paid 7, the deficiency is 4. How many people were there and what is the price of the item?

19. Solve problem 26 of Chapter 6 of the *Jiuzhang*: There is a reservoir with five channels bringing in water. If only the first channel is open, the reservoir can be filled in 1/3 of a day. The second channel by itself will fill the reservoir in 1 day, the third channel in 2 1/2 days, the fourth one in 3 days, and the fifth one in 5 days. If all the channels are open together, how long will it take to fill the reservoir? (This problem is the earliest known one of this type. Similar problems appear in later Greek, Indian, and Western mathematics texts.)

20. Solve problem 27 of Chapter 6 of the *Jiuzhang*: A man is carrying rice on a journey. He passes through five customs stations. At the first, he gives up 1/3 of his rice, at the second 1/5 of what was left, and at the third, 1/7 of what remains. After passing through all three customs stations, he has left 5 pounds of rice. How much did he have when he started? (Versions of this problem occur in later sources in various civilizations.)

Elementary Geometry

21. Given a circle of radius 1 and a chord cutting off a central angle of 90°, show that s, the length of the chord, is $\sqrt{2}$ and that p, the length of the "arrow," is $\frac{2-\sqrt{2}}{2}$. Calculate the area of the segment using the Chinese formula given in the text and by using modern methods. Compare the answers. Do the same for a segment in an angle of 60° and one in an angle of 45°.

22. Various conjectures have been made for the derivation of the Egyptian formula $A = \left(\frac{8}{9}d\right)^2$ for the area A of a circle of diameter d. One of these uses circular counters, known to have been used in ancient Egypt. Show by experiment using pennies, for example, whose diameter can be taken as 1, that a circle of diameter 9 can essentially be filled by 64 circles of diameter 1. (Begin with one penny in the center; surround it with a circle of 6 pennies, and so on.) Use the obvious fact that 64 circles of diameter 1 also fill a square of side 8 to show how the Egyptians may have derived their formula.[23]

23. For the truncated pyramid from the *Moscow Papyrus*, compare the correct volume given in the text with the volume calculated by means of the incorrect Babylonian formula $V = \frac{1}{2}(a^2 + b^2)h$. Find the percentage error. Do the same for a truncated pyramid of lower base 10, upper base 8, and height 2.

24. In the Indian *Sulvasutras*, the priests gave the following procedure for finding a circle whose area was equal to a given square. In square $ABCD$, let M be the intersection of the diagonals (Figure 1.26). Draw the circle with M as center and MA as radius; let ME be the radius of the circle perpendicular to the side AD and cutting AD in G. Let $GN = \frac{1}{3}GE$. Then MN is the radius of the desired circle. Show that if $AB = s$ and $2MN = d$, then $\frac{d}{s} = \frac{2 + \sqrt{2}}{3}$. Replace $\sqrt{2}$ by the approximation $1 + \frac{1}{3} + \frac{1}{3 \cdot 4} - \frac{1}{3 \cdot 4 \cdot 34}$ given in the text and show that the reciprocal is $\frac{s}{d} = \frac{7}{8} + \frac{1}{8 \cdot 29} - \frac{1}{8 \cdot 29 \cdot 6} + \frac{1}{8 \cdot 29 \cdot 6 \cdot 8} - \frac{41}{8 \cdot 29 \cdot 8 \cdot 1393}$, the value (except for the last term) given in the text for the ratio of the side of a square equal in area to a circle of diameter d.

Astronomical Calculations

25. Look up the 18.6 year cycle of moonrise positions in an astronomy text. Discuss its astronomical basis and how it can be used in predicting eclipses. Do you think it reasonable that the priests at Stonehenge could have made such predictions?

26. Look up details on the ancient Chinese calendar. How did it reconcile the lunar and solar cycles?

27. Consult a reference on the details of the Jewish calendar. Make a short report on how the length of each year is determined.

Square Roots

28. Square 1;24,51,10 using the sexagesimal system. Convert your answer to decimals and determine the accuracy of the approximation.

29. Calculate $\sqrt{2}$ to 3 sexagesimal places by using each of the three approximation techniques mentioned in the text and show that each answer is equal to 1;24,51,10: (a) Find the mean μ of 1;25 and $\frac{2}{1;25}$, the mean of μ and $\frac{2}{\mu}$, and so on; (b) Use the approximation formula $\sqrt{a^2 \pm b} \approx a \pm \frac{b}{2a}$; (c) Use the Chinese square root algorithm converted to base 60. Using any of these methods, find a fourth sexagesimal place in the approximation.

30. Show that $12\ \overline{\overline{3}}\ \overline{15}\ \overline{24}\ \overline{32}$ is a good approximation to $\sqrt{164}$. (This value appears in a late Greek-Egyptian papyrus.)

31. Use the Chinese square root algorithm (which is the same as the one taught currently) to derive a sequence of decimal approximations to $\sqrt{2}$ to 5-place accuracy. Use $\sqrt{a^2 \pm b} \approx a \pm \frac{b}{2a}$ to do the same. Use the principle of the mean to do the same. Which algorithm works best, that is, which produces 5-place accuracy with the least amount of calculation?

Pythagorean Theorem

32. Assuming that d and y represent the lengths of the hypotenuse and side respectively of a right triangle, calculate for each line of Plimpton 322 the angle opposite the side of length y.

33. Show that taking $v + u = 2;24$ ($= 2\ 2/5$) leads to line 1 of Plimpton 322. Show that taking $v + u = 1;48$ ($= 1\ 4/5$) leads to line 15. Find the values for $v + u$ which lead to lines 6 and 13 of that tablet.

34. Solve problem 8 of Chapter 9 of the *Jiuzhang*: The height of a wall is 10 *ch'ih*. A pole of unknown length leans against the wall so that its top is even with the top of the wall. If the bottom of the pole is moved 1 *ch'ih* further

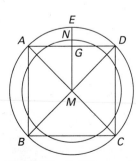

FIGURE 1.26
Indian procedure for "circling" the square.

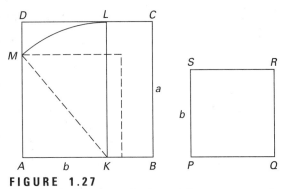

FIGURE 1.27
Procedure in *Sulvasutras* for determining a square equal
to the difference of two squares.

from the wall, the pole will fall to the ground. What is
the length of the pole?

35. Prove that the construction given in the *Sulvasutras* for
constructing a square equal to the difference of two
squares is correct (Figure 1.27): Let *ABCD* be the larger
square with side equal to *a*, and *PQRS* the smaller square
with side equal to *b*. Cut off $AK = b$ from *AB* and draw
KL perpendicular to *AK* intersecting *DC* in *L*. With *K* as
center and radius *KL*, draw an arc meeting *AD* at *M*. Then
the square on *AM* is the required square.

Quadratic Equations

36. Solve the problem from the Old Babylonian tablet BM
13901: The sum of the areas of two squares is 1525. The
side of the second square is 2/3 of that of the first plus 5.
Find the sides of each square.

37. Solve the problem from the *Berlin Papyrus*: If the area of
a square of 100 square cubits is equal to the sum of the
areas of two smaller squares, and if the side of one is $\overline{2}\,\overline{4}$
($= 3/4$) times the side of the other, then find the sides of
the two unknown squares.

38. Solve problem 20 of Chapter 9 of the *Jiuzhang*: A square
walled city of unknown dimensions has four gates, one at
the center of each side. A tree stands 20 *pu* from the north
gate. One must walk 14 *pu* southward from the south gate
and then turn west and walk 1775 *pu* before one can see
the tree. What are the dimensions of the city?

39. Give a geometric argument to justify the Babylonian "quadratic formula" which solves the equation
$x^2 - ax = b$.

40. Consider the system of equations taken from an Old
Babylonian text:

$$x = 30 \qquad xy - (x - y)^2 = 500.$$

Show that the substitution of the first equation into the
second leads to a quadratic equation in *y* which has two
positive roots, a type the Babylonians did not deal with.
Show that subtraction of the second equation from the
square of the first gives the equation $(x - y)^2 +
30(x - y) = 400$, a quadratic in $x - y$ which has only
one positive root.

41. Solve the following Babylonian problem:

$$x + y = 5\frac{5}{6} \qquad \frac{x}{7} + \frac{y}{7} + \frac{xy}{7} = 2$$

by first multiplying the second equation by 7 and then
subtracting off the first equation, thus reducing the system to a standard form.

42. Solve the following problem from a Seleucid text:

$$xy = 1 \qquad x + y = 2;0,0,33,20$$

using standard Babylonian techniques. (The hard part of
this problem is calculating the square root.)

FOR DISCUSSION . . .*

43. Discuss the pros and cons of using a grouping system, a
ciphered system, and a place-value system (all base 10)
in terms of brevity of expression, amount of memorization required, and ease of arithmetic calculation.

44. Compare a base 10 and a base 60 place-value system in
terms of the criteria of problem 43. Is there a base which
is better than either of these two? Explain.

45. Devise a lesson on place value using the Babylonian
system.

46. Devise a lesson teaching the quadratic formula using geometric arguments similar to the (assumed) Babylonian
ones.

47. Compare the standard division algorithm with the Babylonian method of using reciprocals in terms of ease of
use. Determine the algorithm which your calculator uses
to perform division.

48. Devise a lesson which explains the basic idea of linearity
using examples from Egyptian and Chinese sources.

49. Devise a lesson to convince students of the correctness of the formula $A = \pi r^2$ for the area of a circle, where π is defined as $\frac{c}{d}$. How would a value for π be determined? How would you convince students that there is a constant of proportionality between the circumference and diameter of every circle?

50. Why is it useful to have a calendar whose months are determined by the cycles of the moon? Do we lose anything today by not having such a calendar?

51. Devise a lesson teaching the Pythagorean theorem using material from Chinese sources.

―――――――

*Discussion problems asking to "devise a lesson" are intended to help future teachers (at the appropriate level) to use history in teaching mathematics. These problems are included in most of the exercise sets in this book. It may be useful for a future teacher to present such a lesson in class.

References and Notes

Basic information on the ancient civilizations discussed can be found, for example, in William McNeill, *The Rise of the West* (Chicago: University of Chicago Press, 1970) and *Peoples and Places of the Past* (Washington: National Geographic Soc., 1983). The standard accounts of Babylonian mathematics are Otto Neugebauer, *The Exact Sciences in Antiquity* (Princeton: Princeton University Press, 1951; New York: Dover, 1969) and B. L. Van der Waerden, *Science Awakening I* (New York: Oxford University Press, 1961). A somewhat different point of view is found in Evert Bruins, "Interpretation of Cuneiform Mathematics," *Physis* 4 (1962), 277–371. Translations and analyses of the Babylonian tablets themselves are found principally in Otto Neugebauer, *Mathematische Keilschrift-Texte* (New York: Springer, 1973, reprint of 1935 original), Otto Neugebauer and Abraham Sachs, *Mathematical Cuneiform Texts* (New Haven: American Oriental Society, 1945), and Evert Bruins and M. Rutten, *Textes mathématiques de Suse* (Paris: Paul Geuthner, 1961). The best source on Egyptian mathematics is Richard J. Gillings, *Mathematics in the Time of the Pharaohs* (Cambridge: MIT Press, 1972). See also Gillings, "The Mathematics of Ancient Egypt," *Dictionary of Scientific Biography* (New York: Scribners, 1978), vol. 15, 681–705. The *Rhind Mathematical Papyrus* is available in an edition by Arnold B. Chace (Reston, Va: National Council of Teachers of Mathematics, 1967). (This work is an abridgement of the original publication by the Mathematical Association of America in 1927 and 1929.) For a general discussion of Chinese mathematics, see J. Needham, *Science and Civilization in China* (Cambridge: Cambridge University Press, 1959), vol. 3 and Li Yan and Du Shiran, *Chinese Mathematics—A Concise History*, translated by John N. Crossley and Anthony W. C. Lun (Oxford: Clarendon Press, 1987). A brief survey of Chinese mathematics is found in an article by Frank Swetz, "The Evolution of Mathematics in Ancient China," *Mathematics Magazine* 52 (1979), 10–19. An English translation of Chapter 9 of the *Jiuzhang suanshu* with commentary has been published by Frank Swetz and T. I. Kao as *Was Pythagoras Chinese?* (Reston, Va: N.C.T.M., 1977). A German translation of the entire work is *Neun Bücher arithmetischer Technik*, translated by Kurt Vogel (Braunschweig: F. Vieweg und Sohn, 1968). There are no comprehensive histories of ancient Indian mathematics, but two approximations toward these are B. Datta and A. N. Singh, *History of Hindu Mathematics* (Bombay: Asia Publishing House, 1961) (reprint of 1935–1938 original) and C. N. Srinivasiengar, *The History of Ancient Indian Mathematics* (Calcutta: The World Press Private Ltd., 1967). A more general work on ancient Indian science, which contains much interesting material on Indian mathematics is Debiprasad Chattopadhyaya, *History of Science and Technology in Ancient India—The Beginnings* (Calcutta: Firma KLM Pvt. Ltd., 1986). Finally, a book which discusses the mathematics of these ancient societies comparatively and also deals with the questions of transmission and a possible single origin of mathematics is B. L. Van der Waerden, *Geometry and Algebra in Ancient Civilizations* (New York: Springer, 1983).

1. Chace, *Rhind Mathematical Papyrus*, p. 27.

2. A discussion is found in Alexander Marshack, *The Roots of Civilization* (New York: McGraw-Hill, 1972).

3. For more information on number words, the standard reference is Karl Menninger, *Number Words and Number Symbols* (Cambridge: MIT Press, 1977).

4. See Gillings, *Mathematics in the Time of the Pharaohs*, pp. 45–80. Also see Paul J. Campbell, *Bibliography of Algorithms for Egyptian Fractions* (1981, available from the author) and Wilbur Knorr, "Techniques of Fractions in Ancient Egypt and Greece," *Historia Mathematica* 9 (1982), 133–171.

5. Chace, *Rhind Mathematical Papyrus*, p. 69.

6. Vogel, *Neun Bücher*, p. 78.

7. Yoshio Mikami, *Mathematics in China and Japan* (New York: Chelsea, 1974) (reprint of 1913 original), p. 18. This work also provides a good, if dated, survey of Chinese mathematics and virtually the only survey of Japanese mathematics available in English.

8. Vogel, *Neun Bücher*, p. 82.

9. Gillings, *Mathematics in the Time of the Pharaohs*, p. 139. For further analysis, see Hermann Engels, "Quadrature of the Circle in Ancient Egypt," *Historia Mathematica* 4 (1977), 137–140.

10. Abraham Seidenberg, "The Ritual Origin of Geometry," *Archive for History of Exact Sciences* 1 (1962), 488–527, p. 515. This article presents an interesting account of ancient Indian geometry and its relationship to geometry elsewhere. Also, see Seidenberg, "On the Area of a Semicircle," *Archive for History of Exact Sciences* 9 (1973): 171–211 and "The Origin of Mathematics," *Archive for History of Exact Sciences* 18 (1978), 301–342.

11. Gillings, *Mathematics in the Time of the Pharaohs*, p. 188.

12. For Stonehenge astronomy, see Euan W. MacKie, *Science and Society in Prehistoric Britain* (London: Paul Elek, 1977). Also see Gerald Hawkins, *Stonehenge Decoded* (New York: Doubleday, 1965) and Fred Hoyle, *On Stonehenge* (San Francisco: Freeman, 1977).

13. For an amusing fictional example of this, see Mark Twain, *A Connecticut Yankee in King Arthur's Court*. The books by Hawkins and Hoyle from the previous note have an extensive discussion of the 18.6 year cycle and its use in eclipse predictions.

14. See Wang Ling and J. Needham, "Horner's Method in Chinese Mathematics: Its Origins in the Root Extraction Procedures of the Han Dynasty," *T'oung Pao* 43 (1955), 345–401.

15. For a discussion of the Pythagorean theorem in these stone circles, see Van der Waerden, *Geometry and Algebra*. Also see the review by Frank J. Swetz in *Historia Mathematica* 13 (1986), 83–85 and W. R. Knorr, "The Geometer and the Archaeoastronomer," *British Journal of the History of Science* 18 (1985), 202–211. The latter article argues strongly against the view that the English stone circle builders knew the Pythagorean theorem.

16. See Jöran Friberg, "Methods and Traditions of Babylonian Mathematics I: Plimpton 322, Pythagorean Triples, and the Babylonian Triangle Parameter Equations," *Historia Mathematica* 8 (1981), 277–318, for a detailed discussion of the possible methods of construction of the tables on this tablet as well as a discussion of the possible reasons for the errors. See also R. C. Buck, "Sherlock Holmes in Babylon," *American Mathematical Monthly* 87 (1980), 335–345. The designations on the tablets refer to the collection in which they are housed. For example, Plimpton refers to the Plimpton Collection of Columbia University, YBC refers to the Yale Babylonian Collection, and BM to the British Museum.

17. Swetz, *Was Pythagoras Chinese?*, p. 30.

18. Jerold Mathews, "A Neolithic Oral Tradition for the van der Waerden/Seidenberg Origin of Mathematics," *Archive for History of Exact Sciences* 34 (1985), 193–220, p. 203. This paper discusses the Chinese and Babylonian use of the Pythagorean theorem as well as other geometric concepts.

19. Needham, *Science and Civilization*, p. 22.

20. A detailed discussion of the quadratic equation in Babylonian tablets is found in Solomon Gandz, "The Origin and Development of the Quadratic Equations in Babylonian, Greek, and Early Arabic Algebra," *Osiris* 3 (1937), 405–557 and in Solomon Gandz, "Studies in Babylonian Mathematics III: Isoperimetric Problems and the Origin of the Quadratic Equation," *Isis* 32 (1947), 103–115. See also Philip Jones, "Recent Discoveries in Babylonian Mathematics," *Mathematics Teacher* 50 (1957), 162–165, 442–444, 570–571.

21. For a fresh look at the geometric bases of Babylonian methods, see Jens Høyrup, "Algebra and Naive Geometry: An Investigation of Some Basic Aspects of Old Babylonian Mathematical Thought," *Altorientalische Forschungen* 17 (1990). Some of the geometric ideas mentioned in this section are adapted from this work.

22. Gillings, *Mathematics in the Time of the Pharaohs*, p. 134.

23. This suggestion comes from Paulus Gerdes, "Three Alternate Methods of Obtaining the Ancient Egyptian Formula for the Area of a Circle," *Historia Mathematica* 12 (1985), 261–267. Two other possibilities are also presented in that article.

Summary of Mathematical Achievements of Ancient Societies

Egypt—c. 1800 B.C.E.

Unit fractions

Measurement of circle

Frustum of a pyramid

Linear equations

Lunar-Solar calendar

Babylonia—c. 1700 B.C.E.

Base 60 place-value system

Measurement of circle

Lunar-Solar calendar

Square root calculations

Pythagorean triples

Systems of two linear equations

Quadratic equations and systems

Square and cube root tables

Pythagorean theorem

Volume of frustum of a pyramid (incorrect)

India—c. 500 B.C.E.

Pythagorean theorem

Square root calculations

Measurement of circle

China—c. 200 B.C.E.

Counting board in base 10

Measurement of circle

Square and cube root algorithm

Pythagorean triples

Systems of up to five linear equations

Volume of a pyramid

Pythagorean theorem

Quadratic equations and systems

(Note: The dates given above are those by which all of the mathematical ideas listed were known. The dates when the discoveries were made in each civilization are not known.)

Chapter 2

The Beginnings of Mathematics in Greece

"Thales was the first to go to Egypt and bring back to Greece this study [geometry]; he himself discovered many propositions, and disclosed the underlying principles of many others to his successors, in some case his method being more general, in others more empirical."
(Proclus' Summary *(c. 450 C.E.) of Eudemus'* History *(c. 320 B.C.E.))*[1]

A report from a visit to Egypt with Plato by Simmias of Thebes in 379 B.C.E. (from a dramatization by Plutarch of Chaeronea (first/second century C.E.)): "On our return from Egypt a party of Delians met us . . . and requested Plato, as a geometer, to solve a problem set them by the god in a strange oracle. The oracle was to this effect: the present troubles of the Delians and the rest of the Greeks would be at an end when they had doubled the altar at Delos. As they not only were unable to penetrate its meaning, but failed absurdly in constructing the altar . . . , they called on Plato for help in their difficulty. Plato . . . replied that the god was ridiculing the Greeks for their neglect of education, deriding, as it were, our ignorance and bidding us engage in no perfunctory study of geometry; for no ordinary or near-sighted intelligence, but one well versed in the subject, was required to find two mean proportionals, that being the only way in which a body cubical in shape can be doubled with a similar increment in all dimensions. This would be done for them by Eudoxus of Cnidus . . . ; they were not, however to suppose that it was this the god desired, but rather that he was ordering the entire Greek nation to give up war and its miseries and cultivate the Muses, and by calming their passions through the practice of discussion and study of mathematics, so to live with one another that their relationships should be not injurious, but profitable."[2]

As the quotation and the (probably) fictional account indicate, a new attitude toward mathematics appeared in Greece before the fourth century B.C.E. It was no longer sufficient merely to calculate numerical answers to problems. One now had to prove that the results were correct. To double a cube, that is, to find a new cube whose volume was twice that of the original one, is equivalent to determining the cube root of 2, and that was not a difficult problem numerically. The oracle, however, was not concerned with numerical calculation, but with geometric construction. That in turn depended on geometric proof by some logical argument, the earliest manifestation of such in Greece being attributed to Thales.

This change in the nature of mathematics, beginning around 600 B.C.E., was related to the great differences between the emerging Greek civilization and those of Egypt and Babylonia, from whom the Greeks learned. The physical nature of Greece with its many mountains and islands is such that large scale agriculture was not possible. Perhaps because of this, Greece did not develop a central government. The basic political organization was the *polis* or city-state. The governments of the city-state were of every possible variety, but in general controlled populations of only a few thousands. Whether the governments were democratic or monarchical, they were not arbitrary. Each government was ruled by law and therefore encouraged its citizens to be able to argue and debate. It was perhaps out of this characteristic that there developed the necessity for proof in mathematics, that is, for argument aimed at convincing others of a particular truth.

Because virtually every city-state had access to the sea, there was constant trade, both in Greece itself and with other civilizations. As a result, the Greeks were exposed to many different peoples and, in fact, themselves settled in areas all around the eastern Mediterranean. In addition, a rising standard of living helped to attract able people from other parts of the world. Hence the Greeks were able to study differing answers to fundamental questions about the world. They began to create their own answers. In many areas of thought they learned not to accept what had been handed down from ancient times. Instead, they began to ask, and to try to answer, "why?" Greek thinkers gradually came to the realization that the world around them was knowable, that they could discover its characteristics by rational inquiry. Hence they were anxious to discover and expound theories in such fields as physics, biology, medicine, and politics. The Greeks felt, however, that mathematics was one of the primary sciences, the basis for all study of the physical world. And although Western civilization owes a great debt to Greek society in literature, art, and architecture, it is to Greek mathematics that we owe the idea of mathematical proof, an idea at the basis of modern mathematics and, by extension, at the foundation of our modern technological civilization.

This chapter will begin with the contributions of the earliest Greek mathematicians in the sixth century B.C.E. It will then deal with the beginnings of the Greek approach to problem solving, before proceeding to the work of Plato and Aristotle in the fourth century on the nature of mathematics and the idea of logical reasoning. Finally, there will be a detailed analysis of one of the earliest mathematical treatises to survive the ages, the *Elements* of Euclid, written around 300 B.C.E. That analysis will also deal with the work of various earlier Greek mathematicians whose contributions are included in Euclid's text.

2.1 THE EARLIEST GREEK MATHEMATICS

There are no complete texts of Greek mathematics earlier than about 300 B.C.E. There do, however, exist numerous fragments of mathematical manuscripts as well as many references in later works to early Greek mathematics. The most complete reference to this early work is in the commentary to Book I of Euclid's *Elements* written in the fifth century C.E. by Proclus, some 800–1000 years after the fact. This account of the early history of Greek mathematics is generally thought to be a summary of a formal history written by Eudemus of Rhodes in about 320 B.C.E., the original of which is lost. In any case, the earliest Greek mathematician mentioned is Thales (c. 624–547 B.C.E.), from Miletus in Asia Minor. There are many stories recorded about him, all written down several hundred years after his death. These include his prediction of a solar eclipse in 585 B.C.E. and his application of the angle-side-angle criterion of triangle congruence to the problem of measuring the distance to a ship at sea. Thales is also credited with discovering the theorems that the base angles of an isosceles triangle are equal, that vertical angles are equal, and that the diameter of a circle divides the circle into two equal parts. Although exactly how Thales "proved" any of these results is not known, it does seem clear that he advanced some logical arguments.

Aristotle relates the story that Thales was once reproved for wasting his time on idle pursuits. Therefore, noticing from certain signs that a bumper crop of olives was likely in a particular year, he quietly cornered the market on oil presses. When the large crop in fact was harvested, the olive growers all had to come to him for presses. He thus demonstrated that a philosopher or a mathematician could in fact make money if he thought it worthwhile. Whether this or any of the other stories is literally true is not known. In any case, the Greeks of the fourth century B.C.E. and later credited Thales with beginning the Greek mathematical tradition. In fact, he is generally credited with beginning the entire Greek scientific enterprise, including recognizing that material phenomena are governed by discoverable laws.

FIGURE 2.1
Pythagoras on a Greek coin.

2.1.1 Pythagoras and his School

Pythagoras (c. 572–497 B.C.E.), the next mathematician about whom there are extensive stories, spent much time not only in Egypt, where Thales was said to have visited, but also in Babylonia (Figure 2.1). Around 530 B.C.E., after having been forced to leave his native Samos, an island off the coast of Asia Minor (Figure 2.2), he settled in Crotona, a Greek town in southern Italy. There he gathered around him a group of disciples, later known as the Pythagoreans, which was both a religious order and a philosophical school. From the surviving biographies, all written centuries after his death, we can infer that Pythagoras was probably more of a mystic than a rational thinker, but one who commanded great respect from his followers. Since there are no extant works ascribed to Pythagoras or the Pythagoreans, the mathematical doctrines of his school can only be surmised from the works of later writers, including the "neo-Pythagoreans."

One important such mathematical doctrine was that "number was the substance of all things," that numbers, that is, positive integers, formed the basic organizing principle of the universe. What the Pythagoreans meant by this was not only that all known objects have a number, or can be ordered and counted, but also that numbers are at the basis of all physical phenomena. For example, a constellation in the heavens could be characterized by both the number of stars which compose it and its geometrical form, which itself could be thought of as represented by a number. The motions of the planets could be expressed in terms of ratios of numbers. Musical harmonies depend on numerical ratios: two plucked strings with ratio of length 2:1 give an octave, with ratio 3:2 give a fifth, and with ratio 4:3 give a fourth. Out of these intervals an entire musical scale can be created. Finally, the fact that triangles whose sides are in the ratio of 3:4:5 are right-angled established a

FIGURE 2.2
The island of Samos, Pythagoras' birthplace.

connection of number with angle. Given the Pythagoreans' interest in number as a funda-mental principle of the cosmos, it is only natural that they studied the properties of positive integers, what we would call the elements of the theory of numbers.

The starting point of this theory was the dichotomy between the odd and the even. The Pythagoreans probably represented numbers by dots or, more concretely, by pebbles. Hence an even number would be represented by a row of pebbles which could be divided into two equal parts. An odd number could not be so divided because there would always be a single pebble left over. It was easy enough using pebbles to verify some simple theorems. For example, the sum of any collection of even numbers is even, while the sum of an even collection of odd numbers is even and that of an odd collection is odd (Figure 2.3).

Among other simple corollaries of the basic results above were the theorems that the square of an even number is even, while the square of an odd number is odd. Squares themselves could also be represented using dots, providing simple examples of "figurate" numbers. If one represents a given square in this way, for example the square of 4, it is easy to see that the next higher square can be formed by adding a row of dots around two sides of the original figure. There are $2 \cdot 4 + 1 = 9$ of these additional dots. The Pythagoreans generalized this observation to show that one can form squares by adding the successive odd numbers to 1. For example, $1 + 3 = 2^2$, $1 + 3 + 5 = 3^2$, and $1 + 3 + 5 + 7 = 4^2$. The added odd numbers were in the L-shape generally called a gnomon (Figure 2.4).

Other examples of figurate numbers include the triangular numbers, produced by successive additions of the natural numbers themselves. Similarly, oblong numbers, num-bers of the form $n(n + 1)$, are produced by beginning with 2 and adding the succes-sive even numbers (Figure 2.5). The first four of these are 2, 6, 12, and 20, that is, 1×2,

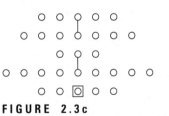

FIGURE 2.3a
The sum of even numbers is even.

FIGURE 2.3b
An even sum of odd numbers is even.

FIGURE 2.3c
An odd sum of odd numbers is odd.

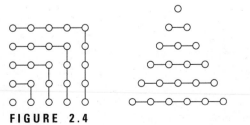

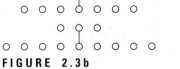

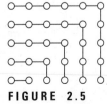

FIGURE 2.4
Square and triangular numbers.

FIGURE 2.5
Oblong numbers.

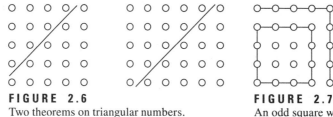

FIGURE 2.6
Two theorems on triangular numbers.

FIGURE 2.7
An odd square which is the difference
of two squares.

2×3, 3×4, and 4×5. Figure 2.6 provides easy demonstrations of the results that any oblong number is the double of a triangular number and that any square number is the sum of two consecutive triangular numbers.

Another number theoretical problem of particular interest to the Pythagoreans was the construction of Pythagorean triples. There is documentation that they showed that for n an odd number, the triple $(n, \frac{n^2-1}{2}, \frac{n^2+1}{2})$ is a Pythagorean triple, while if m is even, $(m, (m/2)^2 - 1, (m/2)^2 + 1)$ is such a triple. An explanation of how the Pythagoreans may have proved the first of these results from their dot configurations begins with the remark that any odd number is the difference of two consecutive squares. Hence if the odd number is itself a square, then three square numbers have been found such that the sum of two equals the third (Figure 2.7). To find the sides of these squares, the Pythagorean triple itself, note that the side of the gnomon is given since it is the square of an odd number. The side of the smaller square is found by subtracting 1 from the gnomon and halving the remainder. The side of the larger square is 1 more than that of the smaller. A similar proof can be given for the second result. Although there is no explicit testimony to additional results involving Pythagorean triples, it seems probable that the Pythagoreans considered the odd and even properties of the terms of these triples. For example, it is not difficult to prove that in a Pythagorean triple, if one of the terms is odd, then two of them must be odd and one even.

The geometric theorem out of which the study of Pythagorean triples grew, namely that in any right triangle the square on the hypotenuse is equal to the sum of the squares on the legs, has long been attributed to Pythagoras himself, but there is no direct evidence of this. The theorem was known in other cultures long before Pythagoras lived. Nevertheless, it was the knowledge of this theorem by the fifth century B.C.E. which led to the first discovery of what is today called the irrational.

For the Pythagoreans, number always was connected with things counted. Because counting requires that the individual units must remain the same, the units themselves can never be divided or joined to other units. In particular, for the Pythagoreans, and throughout Greek mathematics, a number meant a "multitude composed of units," that is, a counting number. Furthermore, since the unit 1 was not a multitude composed of units, it was not considered a number in the same sense as the other positive integers. Even Aristotle noted that two was the smallest "number."

Because the Pythagoreans considered number as the basis of the universe, everything could be counted, including lengths. In order to count a length, of course, one needed a measure. The Pythagoreans thus assumed that one could always find such. Once such a measure was found in a particular problem, it became the unit and could not be divided.

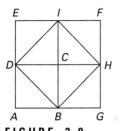

FIGURE 2.8
The incommensurability
of the side and diagonal
of a square (first
possibility).

The Pythagorean failure to recognize the fundamental distinction between number and magnitude, or between the indivisibility of the unit for number and the infinite divisibility of measures of magnitudes such as length, was to lead to trouble.

Since all lengths could be counted, the Pythagoreans assumed that one could find a measure by which both the side and diagonal of a square could be counted. In other words, there should exist a length such that the side and diagonal were integral multiples of it. Unfortunately, this turned out not to be true. The side and diagonal of a square are incommensurable; there is no common measure. Whatever unit of measure is chosen such that an exact number will fit the length of one of these lines, the other line will require some number plus a portion of the unit, and one cannot divide the unit. The discovery of this result, in approximately 430 B.C.E., forced a change in the basic Greek philosophy that all things were made up of numbers.

The question naturally occurs as to how this fundamental incommensurability was discovered. The only hint is in the work of Aristotle, who notes that if the side and diagonal are assumed commensurable, then one may deduce that odd numbers equal even numbers. One possibility as to the form of the discovery is the following: Assume that the side *BD* and diagonal *DH* in Figure 2.8 are commensurable, that is, that each is represented by the number of times it is measured by their common measure. It may be assumed that at least one of these numbers is odd, for otherwise there would be a larger common measure. Then the squares *DBHI* and *AGFE* on the side and diagonal respectively, represent square numbers. The latter square is clearly double the former, so it represents an even square number. Therefore its side *AG = DH* also represents an even number and the square *AGFE* is a multiple of four. Since *DBHI* is half of *AGFE*, it must be a multiple of two, that is, it represents an even square. Hence its side *BD* must also be even. But this contradicts the original assumption that one of *DH*, *BD* must be odd. Therefore, the two lines are incommensurable.

It must be realized that a proof like this presupposes that by this time the notion of proof was ingrained into the Greek conception of mathematics. Although there is no evidence that the Greeks of the fifth century B.C.E. possessed the entire mechanism of an axiomatic system and had explicitly recognized that certain statements need to be accepted without proof, they certainly had decided that some form of logical argument was necessary for determining the truth of a particular result. Furthermore, this entire notion of incommensurability represents a break from the Babylonian and Egyptian concepts of calculation with numbers. There is naturally no question that one can assign a numerical value to the length of the diagonal of a square of side one unit, as the Babylonians did, but the notion that no "exact" value can be found is first formally recognized in Greek mathematics.

2.1.2 Squaring the Circle and Doubling the Cube

The idea of proof and the change from numerical calculation are further exemplified in the mid-fifth century attempts to solve two geometric problems, problems which were to occupy Greek mathematicians for centuries: the squaring of the circle (already attempted in Egypt) and the duplication of the cube (as noted in the oracle). The multitude of attacks on these particular problems and the slightly later one of trisecting an arbitrary angle serve to remind us that a central goal of Greek mathematics was geometrical problem solving,

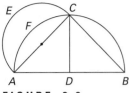

FIGURE 2.9
Hippocrates' lune on a
side of a square inscribed
in a circle.

and that, to a large extent, the great body of theorems found in the major extant works of Greek mathematics served as logical underpinnings for these solutions.

Hippocrates of Chios (mid-fifth century B.C.E.) (no connection to the famous physician) was among the first to attack the cube and circle problems. As to the first of these, Hippocrates perhaps realized that the problem was analogous to the simpler problem of doubling a square of side a. That problem could be solved by constructing a mean proportional b between a and $2a$, a length b such that $a : b = b : 2a$, for then $b^2 = 2a^2$. From the fragmentary records of Hippocrates' work, it is evident that he was familiar with performing such constructions. In any case, ancient accounts record that Hippocrates was the first to come up with the idea of reducing the problem of doubling the cube of side a to the problem of finding two mean proportionals b, c between a and $2a$. For if $a : b = b : c = c : 2a$, then

$$a^3 : b^3 = (a : b)^3 = (a : b)(b : c)(c : 2a) = a : 2a$$

and $b^3 = 2a^3$. Hippocrates was not, however, able to construct the two mean proportionals using the geometric tools at his disposal. As will be discussed in Chapter 3, it was left to some of his successors to find this construction.

Hippocrates similarly made progress in the squaring of the circle, essentially by showing that certain lunes, figures bounded by arcs of two circles, could be "squared," that is, that their areas could be shown equal to certain regions bounded by straight lines. To do this, he first had to prove that the areas of circles are to one another as the squares on their diameters. How he accomplished this is not known. In any case, he could now square the lune on a quadrant of a circle (Figure 2.9):

Suppose that AB is the diameter of a circle with center D and AC, CB are two sides of an inscribed square, and let AEC be the semicircle with diameter AC. Since $AB^2 = 2AC^2$, semicircle $ACB = 2$(semicircle AEC). Since, in addition, semicircle $ACB = 2$(quadrant ADC), it follows that semicircle $AEC =$ quadrant ADC. Subtracting off the common part, segment AFC, shows that the lune $AECF$ is equal to the triangle ADC.

Although Hippocrates gave constructions for squaring other lunes or combinations of lunes, he was unable actually to square a circle. Nevertheless, it is apparent that his attempts on the squaring problem and the doubling problem were based on a large collection of geometric theorems, theorems which he organized into the first recorded book on the elements of geometry.

2.2 THE TIME OF PLATO

The time of Plato (429–347 B.C.E.) (Figure 2.10) saw significant efforts made toward solving the problems of doubling the cube and squaring the circle and toward dealing with incommensurability and its impact on the theory of proportion. These advances were achieved partly because Plato's Academy, founded in Athens around 385 B.C.E., drew together scholars from all over the Greek world. These scholars conducted seminars in mathematics and philosophy with small groups of advanced students and also conducted research in mathematics, among other fields. There is an unverifiable story, dating from some 700 years after the school's founding, that over the entrance to the Academy was

FIGURE 2.10
Plato and Aristotle
(a detail of Raphael's
painting, The School
of Athens).

inscribed the Greek phrase ΑΓΕΩΜΕΤΡΗΤΟΣ ΜΗΔΕΙΣ ΕΙΣΙΤΩ, meaning approximately "Let no one ignorant of geometry enter here." A student "ignorant of geometry" would also be ignorant of logic and hence unable to understand philosophy.

The mathematical syllabus inaugurated by Plato for students at the Academy is described by him in his most famous work, *The Republic*, where he discussed the education which should be received by the philosopher-kings, the ideal rulers of a state. The mathematical part of this education was to consist of five subjects, arithmetic (that is, the theory of numbers), plane geometry, solid geometry, astronomy, and harmonics (music). The leaders of the state are "to practice calculation, not like merchants or shopkeepers for purposes of buying and selling, but with a view to war and to help in the conversion of the soul itself from the world of becoming to truth and reality. . . . It will further our intentions if it is pursued for the sake of knowledge and not for commercial ends. . . . It has a great power of leading the mind upwards and forcing it to reason about pure numbers, refusing to discuss collections of material things which can be seen and touched."[3] In other words, arithmetic is to be studied for the training of the mind (and incidentally for its military usefulness). The arithmetic of which Plato writes includes not only the Pythagorean number theory already discussed but also additional material which is included in Books VII–IX of Euclid's *Elements* and will be considered later.

Again, a limited amount of geometry is necessary for practical purposes, particularly in war, when a general must be able to lay out a camp, or extend army lines. But even though mathematicians talk of operations in geometry such as squaring or adding, the object of geometry, according to Plato, is not to *do* something but to gain knowledge, "knowledge, moreover, of what eternally exists, not of anything that comes to be this or that at some time and ceases to be."[4] So, as in arithmetic, the study of geometry—and for Plato, this means theoretical, not practical, geometry—is for "drawing the soul towards truth." It is worth mentioning that Plato distinguished carefully between, for example, the real geometric circles drawn by people and the essential or ideal circle, held in the mind, which is the true object of geometric study. In practice, one cannot draw a circle and its tangent with only one point in common, although this is the nature of the mathematical circle and the mathematical tangent.

The next subject of mathematical study should be solid geometry. Plato complains in the *Republic* that this subject has not been sufficiently investigated. This is because "no state thinks [it] worth encouraging" and because "students are not likely to make discoveries without a director, who is hard to find."[5] Nevertheless, Plato felt that new discoveries would be made in this field, and, in fact, much was done between the dramatic date of the dialogue (about 400 B.C.E.) and the time of Euclid, some of which is included in Books XI–XIII of the *Elements*. In any case, a decent knowledge of solid geometry was necessary for the next study, that of astronomy, or, as Plato puts it, "solid bodies in circular motion." Again, in this field Plato distinguishes between the stars as material objects with motions showing accidental irregularities and variations and the ideal abstract relations of their paths and velocities expressed in numbers and perfect figures such as the circle. It is this mathematical study of ideal bodies which is the true aim of astronomical study. Thus this study should take place by means of problems and without attempting actually to follow every movement in the heavens. Similarly, a distinction is made in the final subject, of harmonics, between material sounds and their abstraction. The Pythagoreans had discovered the harmonies which occur when strings are plucked together with lengths in the

FIGURE 2.11
Bust of Aristotle.

ratios of certain small positive integers. But in encouraging his philosopher-kings in the study of harmonics, Plato meant them to go beyond the actual musical study, using real strings and real sounds, to the abstract level of "inquiring which numbers are inherently consonant and which are not, and for what reasons."[6] That is, they should study the mathematics of harmony, just as they should study the mathematics of astronomy, and should not be overly concerned with real stringed instruments or real stars. It turns out that a principal part of the mathematics necessary in both studies is the theory of ratio and proportion, the subject matter of Euclid's *Elements*, Book V.

Although it is not known whether the entire syllabus discussed by Plato was in fact taught at the Academy, it is certain that Plato brought in the best mathematicians of his day to teach and do research, including Theaetetus (c. 417–369 B.C.E.) and Eudoxus (c. 408–355 B.C.E.). The most famous person associated with the Academy, however, was Aristotle.

2.3 ARISTOTLE

FIGURE 2.12
Painting of Alexander on horseback.

Aristotle (384–322 B.C.E.) (Figure 2.11) studied at Plato's Academy in Athens from the time he was 18 until Plato's death in 347. Shortly thereafter, he was invited to the court of Philip II of Macedon to undertake the education of Philip's son Alexander, who soon after his own accession to the throne in 335 began his successful conquest of the Mediterranean world (Figure 2.12). Meanwhile, Aristotle returned to Athens where he founded his own school, the Lyceum, and spent the rest of his days writing, lecturing, and holding discussions with his advanced students. Although Aristotle wrote on many subjects, including politics, ethics, epistemology, physics, and biology, as far as mathematics is concerned, his strongest influence was in the area of logic.

2.3.1 Logic

Although there is only fragmentary evidence of logical argument in mathematical works before the time of Euclid, some appearing in the work of Hippocrates already mentioned, it is apparent that from at least the sixth century, the Greeks were developing the notions of logical reasoning. The active political life of the city-states encouraged the development of argumentation and techniques of persuasion. And there are many examples from philosophical works, including especially those of Parmenides (late sixth century) and his disciple Zeno of Elea (fifth century), which demonstrate various detailed techniques of argument. In particular, there are examples of such techniques as *reductio ad absurdum*, in which one assumes that a proposition to be proved is false and then derives a contradiction, and *modus tollens*, in which one shows first that if *A* is true, then *B* follows, shows next that *B* is not true, and concludes finally that *A* is not true. It was Aristotle, however, who took the ideas developed over the centuries and first codified the principles of logical argument.

Aristotle believed that logical arguments should be built out of **syllogisms**, where "a syllogism is discourse in which, certain things being stated, something other than what is stated follows of necessity from their being so."[7] In other words, a syllogism consists of certain statements which are taken as true and certain other statements which are then necessarily true. For example, the argument "if all monkeys are primates, and all primates are mammals, then it follows that all monkeys are mammals," exemplifies one type of syllogism, while the argument "if all Catholics are Christians and no Christians are Moslem, then it follows that no Catholic is Moslem," exemplifies a second type.

After clarifying the principles of dealing with syllogisms, Aristotle notes that syllogistic reasoning enables one to use "old knowledge" to impart new. If one accepts the premises of a syllogism as true, then one must also accept the conclusion. One cannot, however, obtain every piece of knowledge as the conclusion of a syllogism. One has to begin somewhere with truths that are accepted without argument. Aristotle distinguishes between the basic truths which are peculiar to each particular science and the ones which are common to all. The former are often called **postulates**, while the latter are known as **axioms**. As an example of a common truth he gives the axiom "take equals from equals and equals remain." On the other hand, for geometry his examples of peculiar truths are "the definitions of line and straight." By these he presumably means that one postulates the existence of straight lines. Only for the most basic ideas does Aristotle permit the postulation of the object defined. In general, however, whenever one defines an object, one must in fact prove its existence. "For example, arithmetic assumes the meaning of odd and even, square and cube, geometry that of incommensurable, . . . , whereas the existence of these attributes is demonstrated by means of the axioms and from previous conclusions as premises."[8] Aristotle also lists certain basic principles of argument, principles which earlier thinkers had used intuitively. One such principle is that a given assertion cannot be both true and false. A second principle is that an assertion must be either true or false; there is no other possibility.

For Aristotle, logical argument according to his methods is the only certain way of attaining scientific knowledge. There may be other ways of gaining knowledge, but demonstration through a series of syllogisms is the one way by which one can be sure of the results. Because one cannot prove everything, however, one must always be careful that the premises, or axioms, are true and well-known. As Aristotle says, "syllogism there may indeed be without these conditions, but such syllogism, not being productive of scientific knowledge, will not be demonstration."[9] In other words, one can choose any axioms one wants and draw conclusions from them, but if one wants to attain knowledge, one must start with "true" axioms. The question then becomes, how can one be sure that one's axioms are true? Aristotle answers that these primary premises are learned by induction, by drawing conclusions from our own sense perception of numerous examples. This question of the "truth" of the basic axioms has been discussed by mathematicians and philosophers ever since Aristotle's time. On the other hand, Aristotle's rules of attaining knowledge by beginning with axioms and using demonstrations to gain new results has become the model for mathematicians to the present day.

Although Aristotle emphasized the use of syllogisms as the building blocks of logical arguments, Greek mathematicians apparently never used them. They used other forms, as have most mathematicians down to the present. Why Aristotle therefore insisted on syllogisms is not clear. The basic forms of argument actually used in mathematical proof were analyzed in some detail in the third century B.C.E. by the Stoics, of whom the most prominent was Chrysippus (280–206 B.C.E.). This form of logic is based on **propositions**, statements which can be either true or false, rather than on the Aristotelian syllogisms.

The basic rules of inference dealt with by Chrysippus, with their traditional names, are the following, where p, q, and r stand for propositions:

(1) *Modus ponens*	(2) *Modus tollens*
If p, then q.	If p, then q.
p.	Not q.
Therefore q.	Therefore, not p.

(3) *Hypothetical syllogism*	(4) *Alternative syllogism*
If p, then q.	p or q.
If q, then r.	Not p.
Therefore, if p, then r.	Therefore q.

For example, from the statements "if it is daytime, then it is light" and "it is daytime," one can conclude by *modus ponens* that "it is light." From "if it is daytime, then it is light" and "it is not light," one concludes by *modus tollens* that "it is not daytime." Adding to the first hypothesis the statement "if it is light, then I can see well," one concludes by the hypothetical syllogism that "if it is daytime, then I can see well." Finally, from "either it is daytime or it is nighttime" and "it is not daytime," the rule of the alternative syllogism allows us to conclude that "it is nighttime."

2.3.2 Number vs. Magnitude

Another of Aristotle's contributions was the introduction into mathematics of the distinction between number and magnitude. The Pythagoreans had insisted that all was number, but Aristotle rejected that idea. Although he placed number and magnitude in a single category, "quantity," he divided this category into two classes, the discrete (number) and the continuous (magnitude). As examples of the latter, he cited lines, surfaces, volumes, and time. The primary distinction between these two classes is that a magnitude is "that which is divisible into divisibles that are infinitely divisible,"[10] while the basis of number is the indivisible unit. Thus magnitudes cannot be composed of indivisible elements, while numbers inevitably are.

Aristotle further clarified this idea in his definition of "in succession" and "continuous." Things are **in succession** if there is nothing of their own kind intermediate between them. For example, the numbers 3 and 4 are in succession. Things are **continuous** when they touch and when "the touching limits of each become one and the same."[11] Line segments are therefore continuous if they share an endpoint. Points cannot make up a line, because they would have to be in contact and share a limit. Since points have no parts, this is impossible. It is also impossible for points on a line to be in succession, that is, for there to be a "next point." For between two points on a line is a line segment, and one can always find a point on that segment.

Today, a line segment is considered to be composed of an infinite collection of points, but to Aristotle this would make no sense. He did not conceive of a completed or actual infinity. Although he used the term "infinity," he only considered it as potential. For example, one can bisect a continuous magnitude as often as one wishes, and one can count these bisections. But in neither case does one ever come to an end. One can simply continue as long as one wishes. So to Aristotle the infinite occurs in two senses, with respect to division (in magnitudes) and with respect to addition (in numbers). It is interesting that Aristotle notes that "the infinite turns out to be the contrary of what it is said to be. It is not what has nothing outside it that is infinite, but what always has something outside it."[12] Furthermore, although it appears that Aristotle is rejecting the actual infinite in the physical sense, he also rejects it mathematically. Mathematicians, he says, really do not need infinite quantities such as infinite straight lines. They only need to postulate the existence of, for example, arbitrarily long straight lines.

2.3.3 Zeno's Paradoxes

One of the reasons Aristotle has such an extended discussion of the notions of infinity, indivisibles, continuity, and discreteness was that he wanted to refute the famous paradoxes of Zeno. Zeno stated these paradoxes, perhaps in an attempt to show that the then current notions of motion were not sufficiently clear, but also to show that any way of dividing up space or time must lead to problems. The first paradox, the *Dichotomy*, "asserts the non-existence of motion on the ground that that which is in locomotion must arrive at the half-way stage before it arrives at the goal."[13] (Of course, it must also then cover the half of the half before it reaches the middle, and so on.) The basic contention here is that an object cannot cover a finite distance by moving during an infinite sequence of time intervals. The second paradox, the *Achilles*, asserts the same point: "In a race, the quickest runner can never overtake the slowest, since the pursuer must first reach the point whence the pursued started, so that the slower must always hold a lead."[14] Aristotle, in refuting the paradoxes, concedes that time, like distance, is infinitely divisible. But he is not bothered by an object covering an infinity of intervals in a finite amount of time. For "while a thing in a finite time cannot come in contact with things quantitatively infinite, it can come in contact with things infinite in respect to divisibility, for in this sense time itself is also infinite."[15] In fact, given the motion in either of these paradoxes, one can calculate when one will reach the goal or when the fastest runner will overtake the slowest.

Zeno's third and fourth paradoxes show what happens when one asserts that a continuous magnitude is composed of indivisible elements. The *Arrow* states that "if everything when it occupies an equal space is at rest, and if that which is in locomotion is always occupying such a space at any moment, the flying arrow is therefore motionless."[16] In other words, if there are such things as indivisible instants, the arrow cannot move during that instant. Since if, in addition, time is composed of nothing but instants, then the moving arrow is always at rest. Aristotle refutes this paradox by noting that not only are there no such things as indivisible instants, but motion itself can only be defined in a period of time. A modern refutation, on the other hand, would deny the first premise because motion is now defined by a limit argument.

The paradox of the *Stadium* supposes that there are three sets of identical objects, the A's at rest, the B's moving to the right past the A's, and the C's moving to the left with equal velocity. Suppose the B's have moved one place to the right and the C's one place to the left, so that B_1, which was originally under A_4, is now under A_5, while C_1, originally under A_5, is now under A_4 (Figure 2.13). Zeno supposes that the objects are indivisible elements of space and that they move to their new positions in an indivisible unit of time.

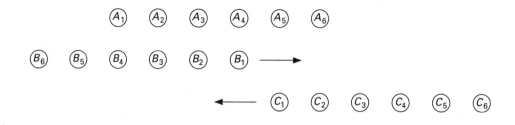

FIGURE 2.13
Zeno's paradox of the
Stadium.

But since there must have been a moment at which B_1 was directly over C_1, there are two possibilities. Either the two objects did not cross, and so there was no motion at all, or in the indivisible instant, each object had occupied two separate positions, so that the instant was in fact not indivisible. Aristotle believed that he had refuted this paradox because he had already denied the original assumption, that time is composed of indivisible instants.

Controversy regarding these paradoxes has lasted throughout history. The ideas contained in Zeno's statements and Aristotle's attempts at refutation have been extremely fruitful in forcing mathematicians to the present day to think carefully about their assumptions in dealing with the concepts of the infinite or the infinitely small. And in Greek times they were probably a significant factor in the development of the distinction between continuous magnitude and discrete number so important to Aristotle and ultimately to Euclid.

2.4 EUCLID AND THE *ELEMENTS*

The most important mathematical text of Greek times, and probably of all time, the *Elements* of Euclid, written about 2300 years ago, has appeared in more editions than any work other than the *Bible*. It has been translated into countless languages and has been continuously in print in one country or another nearly since the beginning of printing. Yet to the modern reader the work is incredibly dull. There are no examples; there is no motivation; there are no witty remarks; there is no calculation. There are simply definitions, axioms, theorems, and proofs. Nevertheless, the book has been intensively studied. Biographies of many famous mathematicians indicate that Euclid's work provided their initial introduction into mathematics, that it in fact excited them and motivated them to become mathematicians. It provided them with a model of how "pure mathematics" should be written, with well thought out axioms, precise definitions, carefully stated theorems, and logically coherent proofs. Although there were earlier versions of *Elements* before that of Euclid, his is the only one to survive, perhaps because it was the first one written after both the foundations of proportion theory and the theory of irrationals had been developed in Plato's school and the careful distinctions always to be made between number and magnitude had been propounded by Aristotle. It was therefore both "complete" and well-organized. Since the mathematical community as a whole was of limited size, once Euclid's work was recognized for its general excellence, there was no reason to keep another inferior work in circulation. The detailed analysis of Euclid's work which follows will not only give the reader insight into the excellence of the *Elements* but also explain the origins of the various parts of Greek mathematics included. Virtually every aspect of the *Elements* to be discussed here turned out to be of great importance for future developments.

Very little is known about the life of the author of the *Elements*. What is written down about Euclid dates from some 750 years after he lived, in the *Commentary* of Proclus (410–485 C.E.):

> Not long after these men [Hermotimus of Colophon and Philippus of Mende, pupils of Plato] came Euclid, who brought together the *Elements*, systematizing many of the theorems of Eudoxus, perfecting many of those of Theaetetus, and putting in irrefutable demonstrable form propositions that had been rather loosely established by his predecessors. He lived in the time of Ptolemy the First, for Archimedes, who lived after the time of the first Ptolemy, mentions Euclid. It is also reported that Ptolemy once asked Euclid if there was not a shorter road to geometry than through the *Elements*, and Euclid replied that there was no royal road to geometry. He was therefore later than Plato's group but earlier than Eratosthenes and Archimedes.[17]

In any case, evidently Euclid taught and wrote at the Museum and Library at Alexandria (Figure 2.14). This complex was founded around 300 B.C.E. by Ptolemy I Soter, the Macedonian general of Alexander the Great who became ruler of Egypt after the death of Alexander in 323 B.C.E. "Museum" here means a "Temple of the Muses," that is, a location where scholars meet and discuss philosophical and literary ideas. The Museum was to be, in effect, a university. The Fellows of the Museum received stipends and free board and were exempt from taxation. In this way Ptolemy I and his successors hoped that men of eminence would be attracted there from the entire Greek world. In fact, the Museum and Library soon became a focal point of the highest developments in Greek scholarship, both in the humanities and the sciences. The Fellows were initally appointed to carry on research, but since younger students gathered there as well, the Fellows soon turned to teaching. The aim of the Library was to collect the entire body of Greek literature in the best available copies and to organize it systematically. Ship captains who sailed from Alexandria were instructed to bring back scrolls from every port they touched until their return. The story is told that Ptolemy III, who reigned from 247–221 B.C.E., borrowed the authorized texts of the playwrights Aeschylus, Sophocles, and Euripides from Athens, against a large deposit. But rather than return the originals, he returned only copies. He was quite willing to forfeit the deposit. The Library ultimately contained over 500,000 volumes in every field of knowledge. Although parts of the library were destroyed in various wars, some of it remained intact until the fourth century C.E.

Euclid wrote his text about 2300 years ago. There are, however, no copies of the work dating from that time. The earliest extant fragments include some potsherds discovered in Egypt dating from about 225 B.C.E., on which are written what appear to be notes on two propositions from Book XIII, and pieces of papyrus containing parts of Book II dating from about 100 B.C.E. Copies of the work were, however, made regularly from the time of Euclid. Various editors made emendations, added comments, or put in new lemmas. In particular, Theon of Alexandria (fourth century C.E.) was responsible for one important new edition. Most of the extant manuscripts of Euclid's *Elements* are copies of this edition. The earliest such copy now in existence is in the Bodleian Library at Oxford University and dates from 888. There is, however, one manuscript in the Vatican Library, dating from

FIGURE 2.14
Euclid (detail from Raphael's painting, The School of Athens). (Note that there is no evidence of Euclid's actual appearance.)

the tenth century, which is not a copy of Theon's edition but of an earlier version. It was from a detailed comparison of this manuscript with several old manuscript copies of Theon's version that the German scholar J. L. Heiberg compiled a definitive Greek version in the 1880s, as close to the Greek original as possible. (Heiberg did the same for several other important Greek mathematical texts.) The extracts to be discussed here are all adapted from Thomas Heath's 1908 English translation of Heiberg's Greek.

Euclid's *Elements* is a work in thirteen books, but it is certainly not a unified work. It is clear both from the internal structure of the work and from the external sources on the history of Greek mathematics that the *Elements* is a compendium, that it was organized by Euclid from many existing works on the various parts of mathematics included in the text. But Euclid did give the work an overarching structure, one in keeping with Aristotle's fundamental distinction between number and magnitude. Namely, the first six books form a relatively complete treatment of two-dimensional geometric magnitudes, while Books VII–IX deal with the theory of numbers. In fact, Euclid includes two entirely separate treatments of proportion theory—in Book V for magnitudes and in Book VII for numbers. Book X then provides the link between the two concepts, because it is here that Euclid introduces the notions of commensurability and incommensurability and shows that, in regard to proportions, commensurable magnitudes may be treated as if they were numbers. The book continues by presenting a detailed study of irrational magnitudes. Euclid deals in Books XI and XII with three-dimensional geometric objects, while in Book XIII he uses the results of Book X in his construction of the five regular polyhedra.

It is also useful to note that nearly all of the ancient mathematics discussed in Chapter 1 is included in one form or another in Euclid's masterwork, with the exception of actual methods of arithmetic computation. The methodology, however, is entirely different. Namely, mathematics in earlier cultures always involved numbers and measurement. Numerical algorithms for solving various problems are prominent. The mathematics of Euclid, however, is completely nonarithmetical. There are no numbers used in the entire work aside from a few small positive integers. There is also no measurement. Various geometrical objects are compared, but not by use of numerical measures. There are no cubits or acres or degrees. The only measurement standard—for angles—is the right angle. Nevertheless, the question must be asked whether the material in Euclid which is related to ideas in earlier cultures was developed in Greece or was adapted from those cultures. Although the Greeks did record that some of their earliest sages studied in Egypt or Babylonia, the answer to this question is not clearcut. Certain pieces of evidence in this regard will be discussed later.

2.4.1 Definitions and Postulates

As Aristotle suggested, a scientific work needs to begin with definitions and axioms. Euclid therefore prefaced several of the thirteen books with definitions of the mathematical objects discussed. The study of the *Elements* begins with the definitions for Book I (Side Bar 2.1).

According to modern standards of definition, Euclid's first several definitions really are not definitions. It is impossible to define everything in terms of concepts already understood. In particular, one would not attempt today to define such terms as point or line. But Euclid, following Aristotle, is using these definitions not just to explain certain

---

Side Bar 2.1   **Definitions from Euclid's *Elements*, Book I**

---

1. A *point* is that which has no part.

2. A *line* is breadthless length.

3. The extremities of a line are points.

4. A *straight line* is a line which lies evenly with the points on itself.

5. A *surface* is that which has length and breadth only.

6. The extremities of a surface are lines.

7. A *plane surface* is a surface which lies evenly with the straight lines on itself.

8. A *plane angle* is the inclination to one another of two lines in a plane which meet one another and do not lie in a straight line.

9. And when the lines containing the angle are straight, the angle is called *rectilinear*.

10. When a straight line meeting another straight line makes the adjacent angles equal to one another, each of the equal angles is *right*, and the first straight line is called a *perpendicular* to the second line.

15. A *circle* is a plane figure contained by one line such that all the straight lines meeting it from one point among those lying within the figure are equal to one another.

16. And the point is called the *center* of the circle.

17. A *diameter* of the circle is any straight line drawn through the center and terminated in both directions by the circumference of the circle, and such a straight line also bisects the circle.

18. A *semicircle* is the figure contained by the diameter and the circumference cut off by it. And the center of the semicircle is the same as that of the circle.

23. *Parallel* straight lines are straight lines which, being in the same plane and being produced indefinitely in both directions, do not meet one another in either direction.

---

terms, but also to assert the existence of the objects so defined. Thus his definitions of point and line, as well as straight line, surface, plane surface, and plane angle not only help us to conceptualize the ideas involved but also allow us to assert the existence of these objects. On the other hand, definitions 3 and 6 are put in to relate the notions of point and line, and line and surface, respectively.

Definition 9 tells us that when Euclid writes of a line, he is not necessarily thinking of a straight line. Thus, angles can even be formed by arcs of circles. The angles in Book I, however, are the rectilinear angles of that definition. Similarly, Euclid uses "line" in definition 15 to designate the circumference of a circle. In definition 17, Euclid states the theorem due to Thales, that the diameter of the circle bisects the circle. Euclid needs this result to make the following definition (*semi*circle) make sense. The other definitions are relatively straightforward. Most of them probably date from well before Euclid's time, because the objects defined are discussed in many earlier works. The definition of parallel lines, however, is somewhat curious. It is not clear how one would use it in practice. How would one ever know whether two lines would in fact meet? In any case, Euclid clearly has in mind a Platonic conception of ideal lines in an ideal plane, for one cannot produce an earthly line indefinitely.

Again, as Aristotle stated, one needs to accept certain statements as true. Euclid gives these in two groups. The first group, the postulates, are the truths peculiar to the science of geometry:

1. To draw a straight line from any point to any point.
2. To produce a finite straight line continuously in a straight line.
3. To describe a circle with any center and distance.
4. That all right angles are equal to one another.
5. That, if a straight line intersecting two straight lines make the interior angles on the same side less than two right angles, the two straight lines, if produced indefinitely, meet on that side on which the angles are less than two right angles.

The first three of these postulates are the basis of the constructions Euclid makes in the *Elements*. A great many of the propositions of this work ask for the construction of figures satisfying certain properties. These are, in fact, examples of the types of problems the Greek mathematicians wanted to solve, and it is quite possible that it was the very search for these solutions which led to the discovery of the theorems which make up the rest of the propositions. In any case, whenever Euclid claims that a particular line may be drawn satisfying certain conditions, or a particular circle can be drawn, or a special type of figure actually exists, he demonstrates the claim by showing how the line or circle or figure is constructed, using as a basis the first three postulates.

It is well known that Euclidean constructions are based on the straightedge and compass. Postulates 1 and 2 assert that a straightedge may be used to draw a line between two points or extend a given line, while postulate 3 says that one can use a compass to draw a circle centered at any given point with any given radius. The question then is why Euclid restricted constructions to these two devices, especially because other mathematicians, both before and after Euclid, used different types of constructions to solve problems. No answer for Euclid can be given definitively, however, other than to note that these constructions are all he needed to develop what he considered the basic results, the "elements." Other constructions belonged to "advanced" mathematics.

Postulates 4 and 5 are somewhat more difficult to understand than the first three. Of course, there is no quarrel with the truth of postulate 4. The only question is why it is necessary. The answer is that Euclid is here setting up the right angle as his standard for measurement of angles. The postulate then asserts that this is a valid standard. Such a standard is necessary to make sense of the next postulate.

Postulate 5, the so-called "parallel postulate," has always been the most intriguing of Euclid's assumptions. It is certainly not as self-evident as the first four. The postulate asserts that if the line l crosses the lines m and n making the sum of angles 1 and 2 less than two right angles, then the lines m and n will eventually meet in the direction of A (Figure 2.15).

Whatever the origin of the first four postulates, the fact that Aristotle notes that in his time the theory of parallels was not on a firm footing lends credence to the belief that Euclid is responsible for the fifth postulate, a starting point for this theory. Nevertheless, ever since Euclid's time various mathematicians have attempted to prove this postulate as

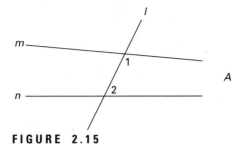

FIGURE 2.15
Euclid's *Elements*, Postulate 5, the parallel postulate.

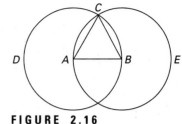

FIGURE 2.16
Elements, Proposition I-1.

a theorem. All such attempts ultimately failed, so Euclid's decision to include this postulate gives us a strong indication of his mathematical genius. On the other hand, modern mathematicians have discovered that Euclid assumed other postulates without stating them, some of which will be mentioned in the appropriate place. In the overall scheme of things, these missing assumptions are of minor importance. In general, the logical structure of the *Elements* remains as a model for mathematics.

After the postulates, Euclid included what he called "common notions," truths common to all of the sciences:

1. Things which are equal to the same thing are also equal to one another.
2. If equals are added to equals, the wholes are equal.
3. If equals are subtracted from equals, the remainders are equal.
4. Things which coincide with one another are equal to one another.
5. The whole is greater than the part.

These axioms certainly appear to be self-evident. The first three are used often today in elementary algebra. Euclid uses these and the other two in various geometrical arguments.

2.4.2 Basic Propositions

Many of the theorems of Book I will be familiar to the reader. In fact, most of the results themselves, if not the rigorous proofs, probably date back to the earliest period of Greek geometry. It is only possibly the organization of the book and, in particular, the inclusion of the Pythagorean theorem, which is due to Euclid.

The first three propositions of Book I are constructions. In particular, proposition 1 provides the construction of an equilateral triangle on a given straight line (Figure 2.16): Let *AB* be the given finite straight line. With center *A* and radius *AB* draw the circle *BCD*. Again, with center *B* and radius *BA* draw the circle *ACE*. From the point *C*, in which the circles intersect, draw the straight lines *CA* and *CB* to form the triangle *ABC*. To prove that this constructed triangle is equilateral, Euclid notes that since *A* is the center of circle

CDB, *AC* is equal to *AB*. Again, since *B* is the center of the circle *CAE*, *BC* is equal to *AB*. Since both *AC* and *BC* are equal to *AB* and since things which are equal to the same thing are also equal to one another, Euclid concludes that *AC*, *AB*, and *BC* are all equal to one another and therefore that triangle *ABC* is equilateral.

Because this is the first proposition, Euclid only has the definitions, postulates, and axioms to appeal to in the proof. Therefore, he uses postulate 3 to allow him to construct the two circles and postulate 1 to draw the two straight line segments. In the main body of the proof, he appeals to the definition of a circle (definition 15) to conclude that *AC* equals *AB* and *BC* equals *BA*. Finally, common notion 1 allows him to conclude that all three sides of the constructed triangle are equal. If one attempts to analyze the logic of Euclid's proof, one concludes that Euclid does not use Aristotelian syllogisms to get from one statement to the next. Instead he uses the logic of propositions with a form of *modus ponens*.

Syllogisms or propositions aside, many commentators have noted that there is a logical gap in this proof. How does Euclid know that the two circles *BCD* and *ACE* intersect? Some postulate of continuity is necessary. This was supplied in the nineteenth century and will be discussed in Chapter 17. Presumably, such a postulate was so obvious to Euclid that he did not think to state it.

Propositions 2 and 3 are constructions which show how to draw line segments equal to given ones. Proposition 4 is the first theorem in the book, that is, an assertion of a truth about a particular geometric configuration. It is the first of the three triangle congruence theorems, the one usually abbreviated as side-angle-side or *SAS*:

Proposition I-4.   *If two triangles have two sides equal to two sides respectively, and have the angles contained by the equal sides also equal, then the two triangles are congruent.*

The word "congruent" is used here as a modern shorthand for Euclid's conclusion that each part of one triangle is equal to the corresponding part of the other. Euclid proves this theorem, as also the second triangle congruence theorem, side-side-side or *SSS* (proposition 8), by superposition. Namely, in the present case he imagines the first triangle being moved from its original position and placed on the second triangle with one side placed on the corresponding equal side and the angles also matching. Euclid here tacitly assumes that such a motion is always possible without deformation. Rather than supply such a postulate, nineteenth-century mathematicians tended to assume this theorem itself as a postulate. For future reference, note that the third triangle congruence theorem, angle-side-angle or *ASA*, in which the hypothesis is that two angles and one side of one triangle are equal respectively to two angles and one side of a second, is proved as proposition 26.

Proposition I-5.   *In isosceles triangles the angles at the base are equal to one another, and, if the equal straight lines are produced further, the angles under the base will be equal to one another.*

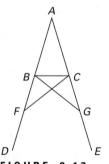

FIGURE 2.17
Elements, Proposition I-5.

To prove this result, Euclid assumes that *ABC* is an isosceles triangle with *AB* = *AC* (Figure 2.17). He extends the equal sides to lines *AD*, *AE* and, taking an arbitrary point *F* on *BD*, constructs *AE* equal to *AF* (proposition 3) and draws the lines *FC*, *GB*. The proof proceeds to show that triangle *AFC* is congruent to triangle *AGB* by using proposition 4: *AF* = *AG*, *AB* = *AC*, and the included angles are identical. It follows again by proposition 4 that triangles *BFC* and *CGB* are congruent and therefore that the corresponding angles, *FBC* and *GCB*, are equal. Since also angle *BCF* equals angle *CBG* and angle *ACF* equals angle *ABG*, Euclid concludes from common notion 3 that the base angles, *ACB* and *ABC*, are equal.

Euclid's proof structure can be analyzed in terms of the hypothetical syllogism. If *p* stands for the statement "*ABC* is an isosceles triangle" and *q* for "the base angles are equal," then the theorem to be proved is "If *p*, then *q*." But if p_1 stands for "triangle *AFC* is congruent to triangle *AGB*" and p_2 stands for "triangle *BFC* is congruent to triangle *CGB*," Euclid has shown that "If *p*, then p_1," "If p_1, then p_2," and "If p_2, then *q*." Two applications of the rule of hypothetical syllogism let him then conclude "If *p*, then *q*." Chains of implications like this are typical in Euclidean proofs, but will not be detailed in the other theorems to be studied.

However one analyzes the logic of this or other Euclidean arguments, it is well to remember that the validity of the argument does not show us how Euclid, or one of his predecessors, thought up the argument to begin with. Euclid himself does not generally present his analysis of a given theorem. He does not say why he proceeded in a certain way. Since the "why" of a result is never made clear, it is somewhat difficult to use his work as a text. Presumably, however, Euclid and his successors at Alexandria and other schools taught their students how to analyze problems and discover proofs. The only detailed discussion in the ancient literature of the method of analysis, the method of assuming that a given theorem is true or that a desired construction has been made and then deriving consequences until one reaches a statement already known to be true, is given by Pappus in the third century C.E., a discussion which will be dealt with in Chapter 5.

Euclid nevertheless does make explicit use of one form of analysis, in his use of the proof method of *reductio ad absurdum*. In this method one assumes that the proposed theorem *p* is false, or that not *p* is true, and then draws various conclusions until one reaches a contradiction. This contradiction generally takes one of three forms. Either one derives *p* (which contradicts not *p*), or one derives *q* and not *q* for some other proposition *q*, or one derives some false statement *r*. By *modus tollens*, one then concludes *p* as desired.

Euclid's first example of a *reductio* proof is in

Proposition I-6.   *If in a triangle two angles are equal to one another, then the opposite sides are also equal.*

Given triangle *ABC* with angle *B* equal to angle *C*, Euclid assumes that *AB* is not equal to *AC* (Figure 2.18). There are two remaining possibilities, either *AB* > *AC* or *AB* < *AC*. Euclid begins by supposing the first alternative. He then finds *D* on *AB* so that *DB* = *AC* and connects *DC*. Since ∠*DBC* = ∠*ACB*, it follows from *SAS* that triangles *BDC* and *CAB* are congruent. But then triangle *BDC*, the part, is equal to triangle *CAB*, the whole, in contradiction to common notion 5. A similar argument shows that "*AB* < *AC*" is also false, and the equality of the two line segments follows. Not only is

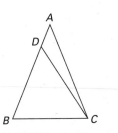

FIGURE 2.18
Elements, Proposition I-6.

this a *reductio* argument, but also it is typical of Euclid's proofs of the equality of two lines or two angles. In such proofs he generally assumes, without explicit mention, the trichotomy principle, that of two quantities *a* and *b*, one of the conditions $a = b$, $a < b$, or $a > b$ must hold. In other words, he assumes that the statement "not ($a = b$)" is equivalent to "($a < b$) or ($a > b$)," and is then able to conclude his desired result as shown.

Propositions 9 through 12 give more necessary constructions. Euclid shows how to bisect a given rectilinear angle, to bisect a straight line segment, to draw a perpendicular to a line from a point on it, and to draw a perpendicular to a line from a point not on it. Proposition 13, on which much of the remainder of the book depends, is the familiar result that if a given straight line meets another straight line, then the two angles on one side either are both right angles or have their sum equal to two right angles. Proposition 14 is the converse of proposition 13, while proposition 15 states that if two straight lines intersect one another, the vertical angles are equal.

Proposition 16, that in any triangle an exterior angle is greater than either of the interior and opposite angles, is of interest because in its proof Euclid again makes an unstated assumption, namely that a line can be extended to any arbitrary length. An immediate corollary to proposition 16 is proposition 17, that two angles of any triangle are always less than two right angles. As will be discussed in Chapter 14, this proposition was important in the developments leading to the discovery of non-Euclidean geometry.

After dealing with triangles in several further propositions, Euclid begins at proposition 27 to deal with the important concept of parallel lines. It is in this group of propositions that Euclid first uses his controversial postulate 5, and virtually every proposition in Book I after proposition 28 depends on this postulate.

Proposition I-29.   *A straight line intersecting two parallel straight lines makes the alternate angles equal to one another, the exterior angle equal to the interior and opposite angle, and the interior angles on the same side equal to two right angles.*

The second and third statements are easy consequences of the first. To prove that part, Euclid uses a *reductio* argument beginning with the assumption that $\angle AGH > \angle GHD$ (Figure 2.19). It follows that the sum of angles *AGH* and *BGH* is greater than the sum of angles *GHD* and *BGH*. The first sum is equal to two right angles by proposition 13. So the second sum must be less than two right angles. Postulate 5 then asserts that the lines *AB*

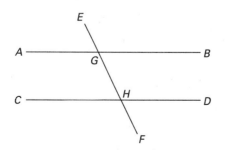

FIGURE 2.19
Elements, Proposition I-29.

and *CD* must ultimately meet. But by hypothesis, they are parallel and hence cannot meet. This contradiction and a similar contradiction to the assumption that $\angle AGH < \angle GHD$ prove the desired result.

Among other results depending on proposition 29 is the theorem that the three interior angles of any triangle are together equal to two right angles, like proposition 17 a result important to the development of non-Euclidean geometry.

Proposition 33 begins the study of parallelograms. Although Euclid does not deal with measurement or formulas as such, he does prove results allowing one to compare the areas of parallelograms and triangles. For example, he shows that parallelograms with equal bases and in the same parallels (that is, with the same altitude) are equal (proposition 36) and that a parallelogram on the same base and in the same parallels as a triangle is double the triangle (proposition 41).

Book I concludes with probably the most famous theorems in the text, the Pythagorean theorem and its converse (Figure 2.20):

Proposition I-47. *In right-angled triangles the square on the hypotenuse is equal to the sum of the squares on the legs.*

Euclid proves the result by constructing a perpendicular from the right angle *A* to the base *DE* of the square on the hypotenuse and then showing that rectangle *BL* is equal to the square on *AB* and rectangle *CL* is equal to the square on *AC* (Figure 2.21). These equalities can be easily proved using the concept of similarity. In fact, the similarity of triangles *ABN*, *CAN*, and *CBA* implies that the square on *AB* equals the rectangle with sides *BC*, *BN* (that is, the rectangle *BL*) and the square on *AC* equals the rectangle with sides *BC*, *NC* (that is, the rectangle *CL*). Euclid, however, wanted to place this result as early as possible in his work. Because the material on similarity was not to come until Books V and VI, he had to devise another method to show the equality of the squares on the legs with the two rectangles making up the square on the hypotenuse. He therefore used the result that parallelograms are double the triangles with the same base and between

FIGURE 2.20
The Pythagorean theorem on a Greek stamp.

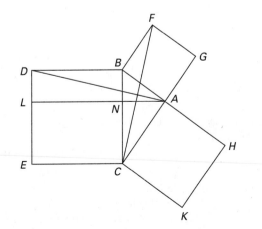

FIGURE 2.21
The Pythagorean theorem in Euclid's *Elements.*

the same parallels. In this case, rectangle *BL* is double triangle *ABD*, and square *AF* is double triangle *FBC*. But the two triangles are congruent by *SAS*. Since an analogous result is true for the other rectangle and square, the theorem is proved.

2.4.3 Geometric Algebra

Book II of the *Elements* is quite different in tone from Book I. It deals with the relationships between various rectangles and squares, most of which can be interpreted in modern terms by the use of algebraic notions. In fact, it is the propositions in Book II, together with propositions 43–45 of Book I and propositions 27–30 of Book VI, which form the content of what is called "geometric algebra," the representation of algebraic concepts through geometric figures. The entire subject of geometric algebra has stirred a major debate in recent years among those interested in Greek mathematics, a debate centering around the questions of whether the Greeks were interested in algebra at all and of whether this aspect of Greek mathematics was directly or indirectly borrowed from the Babylonians. On the one hand, this collection of results does form a relatively coherent structure as a body of geometric knowledge. On the other hand, however, it is quite easy to translate these various results into simple algebraic rules as well as into the standard Babylonian procedures for solving quadratic equations.[18]

Euclid begins Book II with a definition:

*Any rectangle is said to be **contained** by the two straight lines forming the right angle.*

This definition shows Euclid's geometric usage. The statement does not mean that the area of a rectangle is the product of the length by the width. Euclid never multiplies two lengths together, because he has no way of defining such a process for arbitrary lengths. At various places, he multiplies lengths by numbers (that is, positive integers), but otherwise he only writes of rectangles contained by two lines. One question then is whether one should interpret Euclid's "rectangle" as meaning simply a "product."

As an example of Euclid's use of this definition consider

Proposition II-1.   *If there are two straight lines, and one of them is cut into any number of segments whatever, the rectangle contained by the two straight lines is equal to the sum of the rectangles contained by the uncut straight line and each of the segments.*

As an algebraic result, this proposition simply says that given a length l and a width w cut into several segments, say $w = a + b + c$, the area of the rectangle determined by those lines, namely lw, equals the sum of the areas of the rectangles determined by the length and the segments of the width, namely $la + lb + lc$ (Figure 2.22). In other words, this theorem states the familiar distributive law: $l(a + b + c) = la + lb + lc$. Curiously, however, this result is already used in the proof of the Pythagorean theorem in showing that the square on *BC* is equal to the area of the rectangles *BL* and *CL*. Why did Euclid feel compelled to prove it in Book II?

Proposition II-4.   *If a straight line is cut at random, the square on the whole is equal to the squares on the segments and twice the rectangle contained by the segments.*

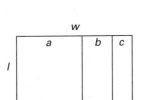

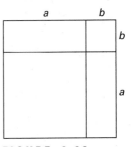

FIGURE 2.22
Elements, Proposition II-1: $l(a + b + c) = la + lb + lc$.

FIGURE 2.23
Elements, Proposition II-4: $(a + b)^2 = a^2 + b^2 + 2ab$.

Algebraically this proposition is simply the rule for squaring a binomial, $(a + b)^2 = a^2 + b^2 + 2ab$, the basis for the square root algorithms discussed in Chapter 1 (Figure 2.23). Euclid's proof is quite complex, since he needs to prove that the various figures in the diagram are in fact squares and rectangles. A modern proof would reduce the result to proposition 1. Should one consider this result as algebra or as geometry?

The next two propositions can be interpreted as the geometric justifications of the standard algebraic solutions of the quadratic equation.

Proposition II-5. *If a straight line is cut into equal and unequal segments, the rectangle contained by the unequal segments of the whole together with the square on the straight line between the points of section is equal to the square on the half.*

Proposition II-6. *If a straight line is bisected and a straight line is added to it, the rectangle contained by the whole with the added straight line and the added straight line together with the square on the half is equal to the square on the straight line made up of the half and the added straight line.*

Figure 2.24 should help clarify these propositions. If AB is labeled in each diagram as b, AC and BC as $b/2$, and DB as x, proposition 5 translates into $(b - x)x + (b/2 - x)^2 = (b/2)^2$, while proposition 6 gives $(b + x)x + (b/2)^2 = (b/2 + x)^2$. The quadratic

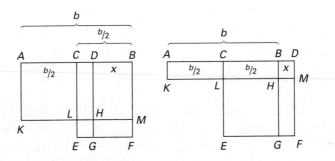

FIGURE 2.24
Elements, Propositions II-5 and II-6.

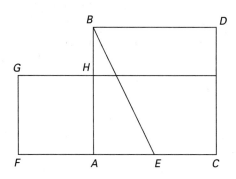

FIGURE 2.25
Elements, Proposition II-11.

equation $bx - x^2 = c$ [or $(b - x)x = c$] can be solved using the first equality by writing $(b/2 - x)^2 = (b/2)^2 - c$ and then getting

$$ x = \frac{b}{2} - \sqrt{\left(\frac{b}{2}\right)^2 - c}. $$

Similarly, the equation $bx + x^2 = c$ (or $(b + x)x = c$) can be solved from the second equality by using an analogous formula. Alternatively, one can label AD as y and DB as x in each diagram and translate the first result into the standard Babylonian system $x + y = b, xy = c$ and the second into the system $y - x = b, yx = c$.

Euclid, of course, does not do either of these translations. He just uses the construc-tions indicated to prove the equalities of the appropriate squares and rectangles. But as early as the ninth century, the Islamic mathematician Thābit ibn Qurra quoted these theorems in his proof of what amounts to the Babylonian formulas for the solutions of the given quadratic equations.

What did these theorems then mean for Euclid? Consider how they are used. Propo-sition 6 is used in the proof of proposition 11, a problem which can easily be translated into an equation.

Proposition II-11.   *To cut a given straight line so that the rectangle contained by the whole and one of the segments is equal to the square on the remaining segment.*

The goal of the proposition is to find a point H on the line so that $AB \times HB$ equals the square on AH (Figure 2.25). To translate this problem into algebra, let the line AB be a and let AH be x. Then $HB = a - x$. The equation to be solved is

$$ a(a - x) = x^2 \qquad \text{or} \qquad x^2 + ax = a^2. $$

The Babylonian solution is

$$ x = \sqrt{\left(\frac{a}{2}\right)^2 + a^2} - \frac{a}{2}. $$

Euclid's proof seemingly amounts to precisely this formula. To get the square root of the sum of two squares, the obvious method is to use the hypotenuse of a right triangle whose sides are the given roots, in this case *a* and *a*/2. So Euclid draws the square on *AB* and then bisects *AC* at *E*. It follows that *EB* is the desired hypotenuse. To subtract *a*/2 from this length, he draws *EF* equal to *EB* and subtracts off *AE* to get *AF*; this is the needed value *x*. Since he wants the length marked off on *AB*, he simply chooses *H* so that *AH* = *AF*. To prove that this choice of *H* is correct, Euclid then appeals to proposition 6, which provides the needed justification.

Euclid has apparently solved a quadratic equation, albeit in geometric dress, in the same manner as the Babylonians. Interestingly enough, he solves the same problem again in the *Elements* as proposition VI-30. There he wants to cut a given straight line in "extreme and mean ratio," that is, given a line *AB* to find a point *H* such that *AB* : *AH* = *AH* : *HB*. Naturally, this translates algebraically into the same equation as given above. The ratio *a* : *x* from that equation, namely $\sqrt{5} + 1 : 2$, is generally known as the **golden ratio**. Much has been written about its importance from Greek times to today.[19]

Before considering an example of the use of proposition II-5, a slight digression back to Book I is necessary.

Proposition I-44.   *To a given straight line to apply, in a given rectilinear angle, a parallelogram equal to a given triangle.*

The aim of the construction is to find a parallelogram of given area with one angle given and one side equal to a given line segment. That is, the parallelogram is to be "applied" to the given line segment. This notion of the "application" of areas is, according to some sources, due to the Pythagoreans. That this too can be interpreted algebraically is easily seen if the given angle is a right angle. If the area of the triangle is taken to be *c* and the given line segment to have length *a*, the goal of the problem is to find a line segment *b* such that the rectangle with length *a* and width *b* has area *c*, that is, to solve the equation *ax* = *c*. Given that Euclid does not deal with "division" of magnitudes, a solution for him amounts to finding the fourth proportional in the proportion *a* : 1 = *c* : *x*, where 1 is some given unit length. But since he could not use the theory of proportions in Book I, he was forced to use a more complicated method involving areas.

In proposition I-45, Euclid demonstrates how to construct a rectangle equal to any given rectilinear figure, by simply dividing the figure into triangles and using the result of I-44, among others. This proposition is then used in the first step of the solution of

Proposition II-14.   *To construct a square equal to a given rectilinear figure.*

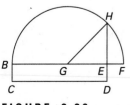

FIGURE 2.26
Elements, Proposition
II-14.

In algebraic terminology, Euclid's goal is to solve the equation $x^2 = c$. He begins by using I-45 to find a rectangle with area *c* (Figure 2.26). Placing the sides of the rectangle *BE*, *EF* in a straight line and bisecting *BF* at *G*, he constructs the semicircle *BHF* of radius *GF*, where *H* is the intersection of that semicircle with the perpendicular to *BF* at *E*. Proposition II-5 now shows that the rectangle contained by *BE* and *EF* together with the square on *EG* is equal to the square on *GF*. But since *GF* = *GH* and the square on *GH* equals the sum of the squares on *GE* and *EH*, it follows that the square on *EH* satisfies the

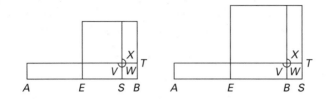

FIGURE 2.27
Elements, Propositions
VI-28 and VI-29.

condition of the problem. Although one can think of II-14 as the solving of a quadratic equation, it seems to strain the terminology to consider the proof and the use of II-5 as algebraic.

Our final bit of evidence in this controversy comes from Book VI.

Proposition VI-28.   *To a given straight line to apply a parallelogram equal to a given rectilinear figure and deficient by a parallelogram similar to a given one; thus the given rectilinear figure must not be greater than the parallelogram described on the half of the straight line and similar to the defect.*

Proposition VI-29.   *To a given straight line to apply a parallelogram equal to a given rectilinear figure and exceeding by a parallelogram similar to a given one.*

The notion of applying a parallelogram to a given line already occurred in I-44. In these propositions from Book VI, Euclid deals with applications "deficient" and "exceeding." In the first case, he proposes to construct a parallelogram of given area whose base is less than the given line segment AB. The parallelogram on the deficiency, the line segment SB, is to be similar to a given one (Figure 2.27). In the second case, the constructed parallelogram of given area has base greater than the given line segment AB, while the parallelogram on the excess, the line segment BS, is to be similar to a given one. The importance of these ideas of deficient and exceeding will be apparent in the discussion of conic sections in Chapter 3. For the discussion of geometric algebra, it is simplest to assume that the given parallelogram in each case is a square. The propositions and their proofs will be translated into algebra under that assumption. Thus the constructed parallelogram must in each case be a rectangle.

Designate AB in both cases by b, and the area of the given rectilinear figure by c. The problems reduce to finding a point S on AB (proposition 28) or on AB extended (proposition 29) so that $x = BS$ satisfies $x(b - x) = c$ in the first case and $x(b + x) = c$ in the second. That is, it is necessary to solve the quadratic equations $bx - x^2 = c$ and $bx + x^2 = c$ respectively. In each case, Euclid finds the midpoint E of AB and constructs the square on BE, whose area is $(b/2)^2$. In the first case, S is chosen so that ES is the side of a square whose area is $(b/2)^2 - c$. That is why the condition is stated in the proposition that in effect c cannot be greater than $(b/2)^2$. This choice for ES implies that

$$x = BS = BE - ES = \frac{b}{2} - \sqrt{\left(\frac{b}{2}\right)^2 - c}.$$

In the second case, S is chosen so that ES is the side of a square whose area is $(b/2)^2 + c$. Then

$$x = BS = ES - BE = \sqrt{\left(\frac{b}{2}\right)^2 + c} - \frac{b}{2}.$$

In both cases, Euclid proves that his choice is correct by showing that the desired rectangle equals the gnomon XWV and that the gnomon is in turn equal to the given area c. Algebraically, that amounts in the first case to showing that

$$x(b - x) = \left(\frac{b}{2}\right)^2 - \left[\left(\frac{b}{2}\right)^2 - c\right] = c$$

and in the second that

$$x(b + x) = \left[\left(\frac{b}{2}\right)^2 + c\right] - \left(\frac{b}{2}\right)^2 = c.$$

Of course, in these cases propositions II-5 and II-6 respectively could be quoted, since they provide precisely the justifications needed. It is also possible to translate proposition 28 into the standard Babylonian system $x + y = b$, $xy = c$ and proposition 29 into $y - x = b$, $xy = c$. Here too Euclid's method is essentially the same as that of the Babylonians.

The argument that geometric algebra actually stems from a translation of Babylonian results into geometry is very persuasive, particularly when one considers the similarity of the geometric procedures to the algebraic ones, at least in the special cases discussed. One can then argue that the Greek adaptation into their geometric viewpoint, given the necessity of proof, was related to the discovery that not every line segment could be represented by a "number." One can further argue that, once one has translated the material into geometry, one might just as well state and prove the results for parallelograms as for rectangles, since little extra effort is required. A further argument supporting the transmission and translation is that the original Babylonian methodology itself may well have been couched in a "naive" geometric form, a form well suited to a translation into the more sophisticated Greek geometry.

On the other hand, there is no direct evidence of any transmission of Babylonian mathematics to Greece during or before the fourth century B.C.E., when the major mathematical developments included in Euclid's text were worked out, nor is it known whether algebra in any form was taught in Greek schools. In fact, since the discovery of incommensurability only dates to about 430 B.C.E., the argument presented would presume that the translation took place during the fourth century. But there is no record of any numerical algebra during the preceding two centuries either. Even the Pythagorean number theory was represented in a somewhat geometrical form. One could then argue that although the Greeks did employ what we think of as algebraic procedures, their mathematical thought was so geometrical that all such procedures were automatically expressed that way. In any case, it is certain that the Greeks of the period up to 300 B.C.E. had no algebraic notation and therefore no way of manipulating expressions which stood for magnitudes, except by thinking of them in geometric terms. It is also certain that Greek mathematicians became very proficient in manipulating geometric entities. Finally, there was no way the Greeks could express, other than geometrically, irrational solutions of quadratic equations.

A clear answer to the related questions of whether Babylonian algebra was transmitted in some form to Greece by the fourth century B.C.E. and whether the theorems discussed in this section should be considered as "algebra" cannot be given here. The interested reader should consult some of the references listed and also carefully read the original sources.

2.4.4 Circles and the Pentagon Construction

Books I and II dealt with properties of rectilinear figures, that is, figures bounded by straight line segments. In Book III, Euclid turns to the properties of the most fundamental curved figure, the circle. The Greeks were greatly impressed with the symmetry of the circle, the fact that no matter how you turned it, it always appeared the same. They thought of it as the most perfect of plane figures. Similarly, they felt the three-dimensional analogue of the circle, the sphere, was the most perfect of solid figures. These philosophical ideas provided the basis for the Greek ideas on astronomy which will be discussed in Chapter 4. Many of the theorems in this book and the next date from the earliest period of Greek mathematics. For example, Hippocrates' work on lunes shows that he was well aware of the important properties of the circle. Although many of the propositions of Book III are independent of one another and seem to appear only for their intrinsic interest, the organizing principle of Book III seems to be its use in the construction of regular polygons, both inscribed in and circumscribed about circles, in Book IV. In particular, most of the propositions from the last half of Book III are used in the most difficult construction of Book IV, the construction of the regular pentagon. This construction in turn is used in Euclid's construction of some of the regular solids in Book XIII.

After presenting a few relevant definitions (Side Bar 2.2), Euclid begins Book III with some elementary constructions and propositions. He then shows how to construct a tangent to a circle:

Proposition III-16.   *The straight line drawn at right angles to the diameter of a circle from its extremity will fall outside the circle, and into the space between the straight line and the circumference another straight line cannot be interposed.*

This proposition asserts that the line perpendicular to the diameter at its extremity is what is today called a **tangent**. Euclid only remarks in a corollary that it "touches" the circle, as in definition 2. But the statement that no straight line can be interposed between the curve and the line ultimately became part of the definition of a tangent before the introduction of calculus. Euclid's proof of this result, as to be expected, is by a *reductio* argument.

Propositions 18 and 19 give partial converses to proposition 16. The former shows that the line from the center of a circle which meets a tangent is perpendicular to the tangent; the latter demonstrates that a perpendicular from the point of contact of a tangent goes through the center of the circle. Propositions 20 and 21 also give familiar results, that the angle at the center is double the angle at the circumference, if both angles cut off the

---

Side Bar 2.2   **Definitions from Book III**

---

2. A straight line is said to *touch a circle* which, meeting the circle and being produced, does not cut the circle.

3. A *segment of a circle* is the figure contained by a straight line and a circumference of a circle.

8. An *angle in a segment* is the angle which, when a point is taken on the circumference of the segment and straight lines are joined from it to the extremities of the straight line which is the *base of the segment*, is contained by the straight lines so joined.

---

same arc, and that angles in the same segment are equal. The proofs of both are clear from Figure 2.28 as is the proof of proposition 22, that the opposite angles of quadrilaterals inscribed in a circle are equal to two right angles.

Proposition 31 asserts that the angle in a semicircle is a right angle. One could conclude this immediately from proposition 20, if one is prepared to consider a straight angle as an angle. For then the angle in a semicircle is half of the straight angle of the diameter, which is in turn equal to two right angles. Euclid, however, does not consider a straight angle as an angle, so he gives a different proof.

In Book IV the treatment of the pentagon construction is preceded by simpler constructions. Euclid shows how to inscribe triangles and squares in circles and circles in triangles and squares, and how to circumscribe triangles and squares about circles, and circles about triangles and squares.

Euclid then divides his construction of a regular pentagon into two steps, the first being the construction of an isosceles triangle with each of the base angles double the vertex (IV-10), and the second being the actual inscribing of the pentagon in the circle (IV-11). As usual, Euclid does not show how he arrived at the construction, but a close reading of it gives a clue to his analysis of the problem. We will therefore assume the construction made and try to see where that assumption leads.

So suppose *ABCDE* is a regular pentagon inscribed in a circle (Figure 2.29) and draw the diagonals *AC* and *CE*. Since angles *CEA* and *CAE* each subtend an arc double that

FIGURE 2.28
Elements, Propositions III-20, III-21, and III-22.

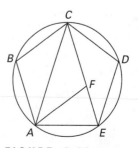

FIGURE 2.29
Construction of a regular pentagon.

subtended by angle ACE, it follows that triangle ACE is an isosceles triangle with base angles double those of the vertex. We have therefore reduced the pentagon construction to the construction of that triangle. Assume then that ACE is such an isosceles triangle and let AF bisect angle A. It follows that triangles AFE and CEA are similar, so $EF : AF = EA : CE$. But triangles AFE and AFC are both isosceles, so $EA = AF = FC$. Therefore $EF : FC = FC : CE$, or, in modern terminology, $FC^2 = EF \cdot CE$. The construction is therefore reduced to finding a point F on a given line segment CE such that the square on CF is equal to the rectangle contained by EF and CE. But this is precisely the construction of proposition II-11. Once F is found, the isosceles triangle with base angles double the vertex angle can be constructed by drawing a circle centered on C with radius CE and another circle centered on E with radius CF. The intersection A of the two circles is the third vertex of the desired triangle.

Euclid constructs this isosceles triangle in exactly this way in proposition IV-10, but, as in the case of the Pythagorean theorem, he cannot use similarity arguments in his proof of the validity of the construction. He therefore uses alternatives based on some of the more technical results of Book III. With the isosceles triangle constructed, the inscribing of the regular pentagon in a circle is now straightforward. Euclid first inscribes the isosceles triangle ACE in the circle. Next, he bisects the angles at A and E. The intersection of these bisectors with the circle are points D and B respectively. Then A, B, C, D, E are the vertices of a regular pentagon.

Euclid completes Book IV with the construction of a regular hexagon and a regular fifteen-gon in a circle, but does not mention the construction of other regular polygons. Presumably he was aware that the construction of a polygon of $2^n k$ sides ($k = 3, 4, 5$) was easy, beginning with the constructions already made, and even that, in analogy with his fifteen-gon construction, that it was straightforward to construct a polygon of kl sides (k, l relatively prime) if one can construct one of k sides as well as one of l sides. Whether he was aware of a construction for the heptagon, however, is not known. In any case, that construction, the first record of which is in the work of Archimedes, would for Euclid be part of advanced mathematics, rather than part of the "elements," because it requires tools other than a straightedge and compass.

2.4.5 Ratio and Proportion

Several important results in the first four books of the *Elements* could have been proved more simply by using elementary ideas of similarity. But Euclid evidently organized the work so that as much as possible could be done without introducing that notion. It is only in Book VI that he began the study of similar figures. As a preliminary to that study, however, Euclid needed a treatment of the basic notions of ratio and proportion. Book V is devoted to those topics.

In fact, Book V greatly influenced the direction of mathematical research well into modern times. Throughout the centuries Euclid's theory of proportion determined the nature of such diverse studies as the theory of equations, the properties of fractions, and the nature of the real number system. An understanding of Euclid's material is thus important for an understanding of much of this later history as well.

The central concept of Book V is the notion of **equal ratio**. Today, the equality $a : b = c : d$ is generally considered equivalent to the equality of the fractions a/b and c/d. But the Greeks of Euclid's time and earlier did not use fractions at all in their formal

> ## *Biography*   Theaetetus (417–369 B.C.E.)
>
> Because Plato dedicated a dialogue to him, something is known about Theaetetus' life. He was born near Athens into a wealthy family and was educated there. A meeting with Theodorus of Cyrene before he was 20 excited him about studying mathematics. Theodorus showed him the demonstration that not only was the square root of 2 incommensurable with
>
> 1, but also so were the square roots of the other non-square integers up to 17. Theaetetus then began research on this issue of incommensurability, both in Heraclea (on the Black Sea) and after 375 B.C.E. in Athens at the Academy. In 369 he was drafted into the army during a war, was wounded in battle at Corinth, and soon after died of dysentery.

work. Recall that the unit could not be divided. In computational work, naturally, fractions existed (as will be seen in the work of Archimedes in Chapter 3), but often the Greeks used the traditional Egyptian unit fractions, always expressed as a "part." In general, the replacement for common fractions in formal Greek mathematics was the ratio of two quantities.

To define equal ratio of two pairs of numbers is easy enough, and Euclid did this in definition 20 of Book VII (see Side Bar 2.5). Recall, however, that late in the fifth century B.C.E. it was discovered that there was no way to interpret both the side and diagonal of a square as numbers. Hence the Greeks sought a definition which would apply to all quantities, even incommensurable ones. As far as can be determined, the process of finding such a definition began with the Euclidean algorithm, the familiar process for finding the greatest common divisor of two numbers. This algorithm, known long before Euclid, is presented in propositions 1 and 2 of Book VII. Given two numbers, a, b with $a > b$, one divides a by b; if b does not divide a evenly, one considers the remainder c, which is less than b, and divides b by c. Continuing in this manner, one eventually comes either to a number m which "measures" the one before (proposition VII-2) or to the unit (1) (proposition VII-1). In the first case, Euclid proves that m is the greatest common measure (divisor) of a and b. In the second case, he shows that a and b are prime to one another. For example, given the two numbers 18 and 80, first divide 80 by 18 and get the remainder 8. Next divide 18 by 8 and get remainder 2. Finally, note that 8 is evenly divisible by 2. It then follows that 2 is the greatest common divisor of 18 and 80. Although this is the only numerical algorithm Euclid presented, he gave no actual examples of the algorithm in use.[20]

It was probably Theaetetus who provided the next stage in the development of a definition of proportionality for magnitudes by investigating the possibilities of applying the Euclidean algorithm to such quantities. The results appear as propositions 2 and 3 of Book X. Theaetetus showed how to determine whether the two magnitudes A and B have a common measure (are commensurable) or do not (are incommensurable). This use of the Euclidean algorithm for magnitudes is generally called *antenaresis* (reciprocal subtraction). Thus supposing that $A > B$, one first subtracts from A the greatest multiple of B which is less than A, getting a remainder b which is less than B. One next subtracts from B the greatest multiple of b which is less than B, getting a remainder b_1 less than b. Euclid shows in proposition X-2 that if this process never ends, then the original two magnitudes are incommensurable. If, on the other hand, one of the magnitudes of this sequence

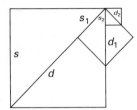

FIGURE 2.30
The incommensurability
of the side and diagonal
of a square (second
possibility).

measures the previous one, then that magnitude is the greatest common measure of the original two (proposition X-3). A natural question here is how one can tell whether or not the process ends. In general, that is difficult. But in certain cases, one observes a repeated pattern in the remainders which shows that the process cannot end. For example, to show by this method that the side s and the diagonal d of a square are incommensurable, note first that s may be subtracted twice from $d + s$, leaving remainder $s_1 = d - s$ (Figure 2.30). Since $s = s_1 + d_1$, the next step is to subtract the side s_1 of the small square twice from $s_1 + d_1$, leaving remainder $s_2 = d_1 - s_1$. That is, at every stage one is subtracting a side of a square twice from the sum of the side and the diagonal. Since this process never ends, $d + s$ and s are incommensurable and therefore also d and s. It is thought by some historians that the original discovery of the incommensurability of d and s is due to this procedure rather than the one outlined earlier.

Given the *antenaresis* procedure, Theaetetus was able to give a definition of equal ratio which applied to all magnitudes. Suppose there are two pairs of magnitudes A, B and C, D. Applying *antenaresis* to each pair gives two sequences of equalities:

$$A = n_0 B + b \quad (b < B) \qquad C = m_0 D + d \quad (d < D)$$
$$B = n_1 b + b_1 \quad (b_1 < b) \qquad D = m_1 d + d_1 \quad (d_1 < d)$$
$$b = n_2 b_1 + b_2 \quad (b_2 < b_1) \qquad d = m_2 d_1 + d_2 \quad (d_2 < d_1)$$
$$\cdots \qquad\qquad\qquad \cdots$$
$$\cdots \qquad\qquad\qquad \cdots$$

If the two sequences of numbers $(n_0, n_1, n_2, \ldots)$, $(m_0, m_1, m_2, \ldots)$ are equal term by term and both end at, say, $n_k = m_k$, then one can check that the ratios $A : B$ and $C : D$ are both equal to the same ratio of integers. Hence Theaetetus could give the general definition that $A : B = C : D$ if the (possibly never ending) sequences $(n_0, n_1, n_2, \ldots)$, $(m_0, m_1, m_2, \ldots)$ are equal term by term. Of course, for specific magnitudes this definition may be awkward to use, but there are interesting cases in which the sequence $n_0, n_1, n_2, \ldots$ is relatively simple to determine. It turned out, however, that the proofs of certain desirable properties of proportion were very difficult using this definition. Therefore, a new definition was sought and eventually found by Eudoxus.[21]

It is not known what inspired Eudoxus to his new definition of equal ratio, but a reasonable guess can be made. Theaetetus' definition shows, for example, that if $A : B = C : D$, then $A > n_0 B$ while $C > n_0 D$ (since $m_0 = n_0$). Since $n_1 A = n_1 n_0 B + n_1 b = (n_1 n_0 + 1)B - b_1$, also $n_1 A < (n_1 n_0 + 1)B$ and similarly $n_1 C < (n_1 n_0 + 1)D$. A comparison of further multiples of A and B and corresponding multiples of C and D shows that for various pairs r, s of numbers, $rA > sB$ whenever $rC > sD$ and $rA < sB$ whenever $rC < sD$. Thus Eudoxus took for his definitions of **same ratio** and **proportional** the ones now reproduced as definitions 5 and 6 of Book V (Side Bar 2.3) and used them to prove rigorously the various theorems on proportion which make up Book V, thereby giving a firm foundation to the geometric theory of similarity given in Book VI.

The first two definitions of Book V are easily understood. The important idea to notice is that Euclid is dealing with *magnitudes*. Book V is to be a treatment of proportion theory for magnitudes, that is, continuous quantities. Book VII is a treatment of proportion theory for *numbers*, that is, discrete quantities. Even though in modern mathematics the latter could easily be subsumed under the former, this is not possible for Euclid. The two

---

Biography   **Eudoxus (408–355 B.C.E.)**

Eudoxus studied medicine in his youth in Cnidus, an island off the coast of Asia Minor. On a visit to Athens, he was attracted to the lectures at the Academy in philosophy and mathematics and began the study of these subjects. Later he visited Egypt and was able to make numerous astronomical observations and study the Egyptian calendar. Returning to his home, he opened a school and conducted his own research. Although he returned at least one other time to Athens, this time with his own students, he spent most of the remainder of his life in Cnidus. He is famous not only for his work in geometry, but also for his application of spherical geometry to astronomy.

Side Bar 2.3   **Definitions from Book V**

---

1. A magnitude is a *part* of a magnitude, the less of the greater, when it measures (divides) the greater.

2. The greater is a *multiple* of the less when it is measured by the less.

3. A *ratio* is a sort of relation in respect to quantity between two magnitudes of the same kind.

4. Magnitudes are said to *have a ratio* to one another which are capable, when multiplied, of exceeding one another.

5. Magnitudes are said to *be in the same ratio*, the first to the second and the third to the fourth, when, if any equal multiples whatever are taken of the first and third, and any equal multiples whatever of the second and fourth, the former multiples alike exceed, are alike equal to, or alike fall short of, the latter multiples respectively taken in corresponding order.

6. Let magnitudes which have the same ratio be called *proportional.*

7. When, of the equimultiples, the multiple of the first magnitude exceeds the multiple of the second, but the multiple of the third does not exceed the multiple of the fourth, then the first is said to *have a greater ratio* to the second than the third has to the fourth.

9. When three magnitudes are proportional, the first is said to have to the third the *duplicate ratio* of that which it has to the second.

10. When four magnitudes are continuously proportional, the first is said to have to the fourth the *triplicate ratio* of that which it has to the second.

---

concepts are separate and distinct and must therefore be dealt with separately. In any case, in these first two definitions, the meaning of one magnitude measuring another is that an integral multiple of the first is equal to the second.

The third and fourth definitions, although a bit vague, help us to understand what Euclid means by the term **ratio**. Two magnitudes can only have a ratio if they are of the same *kind*, that is both lines or both surfaces or both solids, and if there is a multiple of each which is greater than the other. For example, since no multiple of the angle between the circumference of a circle and the tangent line can exceed a given rectilinear angle, there can be no ratio between these two angles.

Definition 5 encompasses Eudoxus' significant breakthrough. Translated into algebraic symbolism, it says that $a : b = c : d$ if, given any positive integers m, n, whenever $ma > nb$, also $mc > nd$, whenever $ma = nb$, also $mc = nd$, and whenever $ma < nb$, also $mc < nd$. Another way of stating this in modern terms is that for every fraction n/m, the quotients a/b and c/d are alike greater than, equal to, or less than that fraction.

Definition 9 is Euclid's version of what is today called the square of a ratio, or, equivalently, the ratio of the squares. Euclid states that if $a : b = b : c$, then $a : c$ is the duplicate of the ratio $a : b$. A modern form would be $a : c = (a : b)(b : c) = (a : b)(a : b) = (a : b)^2 = a^2 : b^2$, or, in fractions, $a/c = (a/b)^2 = a^2/b^2$. Euclid, however, does not multiply ratios, much less fractions, just as he does not multiply magnitudes. He only multiplies magnitudes by numbers. Similarly, he never divides magnitudes. One cannot interpret Euclid's ratio $a : b$ as a fraction corresponding to a particular point on a number line to which can be applied the standard arithmetical operations. On the other hand, Euclid does use the equivalence between the duplicate ratio of two quantities and the ratio of their squares in the cases where it makes sense to speak of the "square" of a quantity, or between the triplicate ratio of two quantities and the ratio of their cubes.

Proposition V-1.   *If there are any number of magnitudes whatever which are, respectively, equal multiples of the same number of other magnitudes, then, whatever multiple one of the magnitudes is of one, that multiple will all be of all.*

In modern symbols, the proposition simply asserts that if $ma_1, ma_2, \ldots, ma_n$ are equal multiples of $a_1, a_2, \ldots, a_n$, then $ma_1 + ma_2 + \cdots + ma_n = m(a_1 + a_2 + \cdots + a_n)$. Similarly, proposition V-2 asserts in effect that $ma + na = (m + n)a$, while the next result can be translated as $m(na) = (mn)a$. In other words, these first propositions of Book V give versions of the modern distributive and associative laws.

Proposition V-4 is the first in which the definition of same ratio is invoked. The result states that if $a : b = c : d$, then $ma : nb = mc : nd$, where m, n are arbitrary numbers. To show that equality, Euclid needs to show that if $p(ma)$, $p(mc)$ are equal multiples of ma, mc, and $q(nb)$, $q(nd)$ are equal multiples of nb, nd, then according as $p(ma) > = < q(nb)$, so is $p(mc) > = < q(nd)$. But since $a : b = c : d$, the associative law and the definition of same ratio for the original magnitudes allow Euclid to conclude the equality of the ratios for the multiples.

Proposition 7 shows that if $a = b$, then $a : c = b : c$ and $c : a = c : b$, while proposition 8 asserts that if $a > b$, then $a : c > b : c$ and $c : b > c : a$. The proof of the first part of the latter shows Euclid's use of definitions 4 and 7. Since $a > b$, there is an integral multiple, say m, of $a - b$ which exceeds c (by definition 4). Let q be the first multiple of c which equals or exceeds mb. Then $qc \geq mb > (q - 1)c$. Since $m(a - b) = ma - mb > c$, it follows that $ma > mb + c > qc$. Since also $mb \leq qc$, definition 7 implies that $a : c > b : c$. A similar argument gives the second conclusion.

Among other results of Book V are proposition 11, which asserts the transitive law, if $a : b = c : d$ and $c : d = e : f$, then $a : b = e : f$, and proposition 16, which states that if

---

Side Bar 2.4   Definitions from Book VI

1. *Similar rectilinear figures* are such as have their angles respectively equal and the sides about the equal angles proportional.

3. A straight line is said to have been *cut in extreme and mean ratio* when, as the whole line is to the greater segment, so is the greater to the less.

4. The *height* of any figure is the perpendicular drawn from the vertex to the base.

$a : b = c : d$, then $a : c = b : d$. The remaining results give other properties of magnitudes in proportion, in particular results dealing with adding or subtracting quantities to the antecedents or consequents in various proportions.

2.4.6 Similarity

Book VI uses the results of the general theory of magnitudes in proportion to prove results in the theory of similar rectilinear figures, defined in the first definition of that book (Side Bar 2.4). It is generally thought that the results of Book VI were known to the Pythagoreans, since there are references to various aspects of similarity dating back to the fifth century B.C.E. The foundation of the idea of similarity, however, the notion of equal ratio, was originally based on the idea that all quantities could be thought of as numbers. Once this notion was destroyed, the foundation for these results no longer existed. That is not to say that mathematicians ceased to use them. Intuitively, they knew that the concept of equal ratio made perfectly good sense, even if they could not provide a formal definition. In Greek times as also in modern times, mathematicians often ignored foundational questions and proceeded to discover new results. The working mathematician knew that eventually the foundation would be strengthened. Once this occurred, around 360 B.C.E., the actual similarity results could be organized into a logically acceptable treatise. It is not known who provided this final organization. What is probably true is that there was actually very little to redo except for the proof of the first proposition of the book. That is the only one which depends directly on Eudoxus' definition.

Proposition VI-1.   *Triangles and parallelograms which have the same height are to one another as their bases.*

Given triangles *ABC*, *ACD* with the same height, Euclid needs to show that as *BC* is to *CD*, so is the triangle *ABC* to the triangle *ACD*. Proceeding as required by Eudoxus' definition, he extends the base *BD* to both right and left so that he can take arbitrary multiples of both *BC* and *CD* along that line (Figure 2.31). Because triangles with equal heights and equal bases are equal, it follows that whatever multiple the base *HC* is of the base *BC*, the triangle *AHC* is the same multiple of triangle *ABC*. The same holds for triangle *ALC* with respect to triangle *ACD*. Since again triangles *AHC* and *ALC* have the same heights, the former is greater than, equal to, or less than the latter precisely

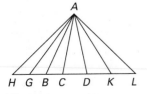

FIGURE 2.31
Elements, Proposition VI-1.

when *HC* is greater than, equal to, or less than *CL*. Equal multiples having been taken of base *BC* and triangle *ABC*, and other equal multiples of base *CD* and triangle *ACD*, and the results compared as required by Eudoxus' definition, it follows that $BC : CD = ABC : ACD$ as desired. The result for parallelograms is immediate, since each parallelogram is double the corresponding triangle.

After showing in proposition 2 that a line parallel to one of the sides of a triangle cuts the other two sides proportionally, and conversely, and in the following proposition that the bisector of an angle of a triangle cuts the opposite side into segments in the same ratio as that of the remaining sides, and conversely, Euclid next gives various conditions under which two triangles are similar. Since the definition of similarity requires both that corresponding angles are equal and that corresponding sides are proportional, Euclid shows that one or the other of these two conditions is sufficient. He also states the conditions under which the equality of only one pair of angles and the proportionality of two pairs of sides guarantees similarity. Proposition 8 then shows that the perpendicular to the hypotenuse from the right angle of a right triangle divides the triangle into two triangles each similar to the original one.

Among the useful constructions of Book VI are the finding of proportionals. Given line segments *a*, *b*, *c*, Euclid shows how to determine *x* satisfying $a : b = b : x$ (proposition 11), $a : b = c : x$ (proposition 12), and $a : x = x : b$ (proposition 13). This last result is equivalent to finding a square root, that is, to solving $x^2 = ab$, and is therefore nearly identical to the result of proposition II-14. In fact, the proofs are even related.

Proposition 16 is in essence the familiar one that in a proportion the product of the means is equal to the product of the extremes. But since Euclid never multiplies magnitudes, he could not have stated this result in terms of Book V. In the geometry of Book VI, however, he has the equivalent of multiplication, for line segments only.

Proposition VI-16.   *If four straight lines are proportional, the rectangle contained by the extremes is equal to the rectangle contained by the means; and if the rectangle contained by the extremes is equal to the rectangle contained by the means, the four straight lines will be proportional.*

Proposition 19 is of fundamental importance later. It also illustrates Euclid's notion of duplicate ratio.

Proposition VI-19.   *Similar triangles are to one another in the duplicate ratio of the corresponding sides.*

A modern statement of this result would replace "in the duplicate ratio" by "as the square of the ratio." But Euclid does not multiply either magnitudes or ratios. Ratios are not quantities; they are not to be considered as numbers in any sense of the word. Hence for this particular proposition, Euclid needs to construct a point *G* on *BC* so that $BC : EF = EF : BG$ (Figure 2.32). The ratio $BC : BG$ is then the duplicate of the ratio $BC : EF$ of the corresponding sides. To prove the result, he shows that the triangles *ABG*, *DEF* are equal. Since triangle *ABC* is to triangle *ABG* as *BC* is to *BG*, the conclusion follows immediately. Proposition 20 extends this result to similar polygons. In particular, the duplicate ratio of two line segments is equal to the ratio of the squares on the segments.

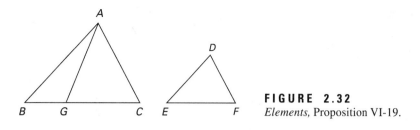

FIGURE 2.32
Elements, Proposition VI-19.

Two parallelograms, of course, can be equiangular without being similar. Euclid is also able to deal with the ratio of such figures, but only by using a concept not formally defined.

Proposition VI-23. *Equiangular parallelograms have to one another the ratio compounded of the ratios of the sides.*

The proof shows what Euclid means by the term "compounded," at least in the context of ratios of line segments. If the two ratios are $a : b$ and $c : d$, one first constructs a segment e such that $c : d = b : e$. The ratio **compounded** of $a : b$ and $c : d$ is then the ratio $a : e$. In modern terms, the fraction a/e is simply the product of the fractions a/b and $c/d = b/e$. Interestingly enough, although Euclid never considers compounding again, this notion became quite important in later Greek times as well as in the medieval period.

2.4.7 Number Theory

Book VII of the *Elements* is the first of three dealing with the elementary theory of numbers. There is no mention of the first six books in Books VII, VIII, and IX; these three books form an entirely independent unit. Only in later books is there some connection made between the three arithmetic books and the earlier geometric ones. The new start that Euclid makes in Book VII is evidence of his desire to stick with Aristotle's clear separation between magnitude and number. The first six books dealt with magnitudes, in particular lengths and areas. The fifth book dealt with the general theory of magnitudes in proportion. But in Books VII–IX Euclid deals only with numbers. He does not consider these as types of magnitudes, but as entirely separate entities. Therefore, although there are many results in Book VII which appear to be merely special cases of results in Book V, for Euclid they are quite different. One should not be misled by the line segments Euclid uses in these books to represent numbers. He does not use the fact of the representation in his proofs. Perhaps this representation was the only one which occurred to him.

It is reasonably certain that many of the propositions in the arithmetic books date back to the Pythagoreans. But from the use of Book VII in Book X, it appears that the details of the compilation of that book are due to the same mathematician who is responsible for Book X, Theaetetus. Namely, Theaetetus took the loosely structured number theory of the Pythagoreans and made it rigorous by introducing precise definitions and detailed proofs. It is these which Euclid included in his version of the material.

Book VII, like most of Euclid's books, begins with definitions (Side Bar 2.5). The first definition is, like the beginning definitions of Book I, mathematically useless in

---

Side Bar 2.5   Definitions from Book VII

1. A *unit* is that by virtue of which each of the things that exist is called one.

2. A *number* is a multitude composed of units.

3. A number is a *part* of a number, the less of the greater, when it measures the greater;

4. but *parts* when it does not measure it.

5. The greater number is a *multiple* of the less when it is measured by the less.

11. A *prime number* is that which is measured by the unit alone.

12. Numbers *prime to one another* are those which are measured by the unit alone as a common measure.

15. A number is said to *multiply* a number when that which is multiplied is added to itself as many times as there are units in the other, and thus some number is produced.

20. Numbers are *proportional* when the first is the same multiple, or the same part, or the same parts, of the second that the third is of the fourth.

---

modern terms. For Euclid, however, the definition appears as the mathematical abstraction of the concept of "thing." What is more interesting is the second definition, that a number is a **multitude** of units. Since "multitude" means plurality, and the unit is not a plurality, it appears that for Euclid, as for the Pythagoreans earlier, 1 is not a number.

Definitions 3 and 5 are virtual word for word repetitions of definitions 1 and 2 in Book V, while definition 4 would make no sense in the context of arbitrary magnitudes. Definitions 11 and 12 are essentially modern definitions of prime and relatively prime, with the note that for Euclid a number does not measure itself. Definition 15 is somewhat curious in that this is the only arithmetic operation defined by Euclid. He assumes that addition and subtraction are known. Note that there is no analogue of this definition in Book V.

Recall that the first two propositions of Book VII deal with the Euclidean algorithm. Several of the next propositions are direct analogues of propositions in Book V. For example, Euclid proves in propositions 5 and 6 what amounts to the distributive law $\frac{m}{n}(b + d) = \frac{m}{n}b + \frac{m}{n}d$. He had proved this for magnitudes as proposition V-1, except that there the result dealt with (integral) multiples rather than the parts—here represented as fractions—of Book VII. Even the proofs of these results are virtually identical. That Euclid does not simply quote results from Book V is evidence that for Euclid number is not a type of magnitude.

Propositions 11–22 include various standard results on numbers in proportion, several of which Euclid proved for magnitudes in Book V. Most are used again in the following two books. In particular, proposition 16 proves the commutativity of multiplication, a nontrivial result given Euclid's definition of multiplication.

Proposition 19 gives the usual test for proportionality, that $a : b = c : d$ if and only if $ad = bc$. Recall that Euclid had already proved an analogue for line segments (proposition VI-16). The proof here, however, is quite different. Given that $a : b = c : d$, it follows that $ac : ad = c : d = a : b$. Also, $a : b = ac : bc$. Therefore, $ac : ad = ac : bc$. Hence, $ad = bc$. The converse is proved similarly.

Propositions 23–32 deal with primes and numbers relatively prime to one another. In particular, propositions 31 and 32 give a version of the fundamental theorem of arithmetic, that every number can be expressed as a product of prime numbers.

Proposition VII-31. *Any composite number is measured by some prime number.*

Proposition VII-32. *Any number either is prime or is measured by some prime number.*

The latter proposition is an easy consequence of the former. That one in turn is proved by a technique Euclid uses often in the arithmetic books, the least number principle. He begins with a composite number a, which is therefore measured (divided) by another number b. If b were prime, the result would follow. If not, then b is in turn measured by c, which will then measure a, and c is in turn either prime or composite. As Euclid then says, "if the investigation is continued in this way, some prime number will be found which will measure the number before it, which will also measure a. For, if it is not found, an infinite series of numbers will measure the number a, each of which is less than the other; which is impossible in numbers." One can again note the distinction between number and magnitudes. Any decreasing sequence of numbers has a least element, but the same is not true for magnitudes.

Book VIII primarily deals with numbers in continued proportion, that is with sequences $a_1, a_2, \ldots, a_n$ such that $a_1 : a_2 = a_2 : a_3 = \cdots$. In modern terms, such a sequence is called a **geometric progression**. It is generally thought today that much of the material in this book is due to Archytas (fifth century B.C.E.), the person from whom Plato received his mathematical training. In particular, proposition 8 is a generalization of a result due to Archytas and coming out of his interest in music. The original result is that there is no mean proportional between two numbers whose ratio in lowest terms is equal to $n + 1 : n$. Recall that the ratio of two strings whose sound is an octave apart is $2 : 1$. This ratio is the compound of $4 : 3$ and $3 : 2$. Thus, the octave is composed of a fifth and a fourth. Archytas' result then states that the octave cannot be divided into two equal musical intervals. Of course, in this case, the result is equivalent to the incommensurability of $\sqrt{2}$ with 1. But the result also shows that one cannot divide a whole tone, whose ratio of lengths is $9 : 8$, into two equal intervals.

Proposition VIII-8. *If between two numbers there are numbers in continued proportion with them, then, however many numbers are between them in continued proportion, so many will also be in continued proportion between numbers which are in the same ratio as the original numbers.*

Euclid concerns himself in several other propositions of Book VIII with determining the conditions for inserting mean proportional numbers between given numbers of various types. Proposition 11 in particular is the analogue for numbers of the special case of VI-20. Namely, Euclid shows that between two square numbers there is one mean proportional and that the square has to the square the ratio duplicate of that which the side has to the side. Similarly, in proposition 12 Euclid shows that between two cube numbers there are two mean proportionals and the cube has to the cube the triplicate ratio of that which the side has to the side. This is, of course, the analogue in numbers of Hippocrates' reduction of the problem of doubling the cube to that of finding two mean proportionals.

The final book on number theory is Book IX. Proposition 20 of that book shows that there are infinitely many prime numbers:

Proposition IX-20.   *Prime numbers are more than any assigned multitude of prime numbers.*

Given the "assigned multitude" of primes $p_1, p_2, \ldots, p_n$, Euclid tells us to consider the number $N = p_1 p_2 \cdots p_n + 1$. If N is prime, a prime other than those given has been found. If N is composite, then it is divisible by a prime p. Euclid shows that p is distinct from the given primes since none of those can divide N evenly. It follows again that a new prime p has been found.

Propositions 21–34 form a nearly independent unit of very elementary results about even and odd numbers. They probably represent a remnant of the earliest Pythagorean mathematical work. This section includes such results as the sum of even numbers is even, an even sum of odd numbers is even, and an odd sum of odd numbers is odd. These elementary results are followed by two of the most significant results of the entire number theory section of the *Elements*.

Proposition IX-35.   *If as many numbers as we please are in continued proportion, and there is subtracted from the second and the last numbers equal to the first, then, as the excess of the second is to the first, so will the excess of the last be to all those before it.*

In effect, this result determines the sum of a geometric progression. Represent the sequence of numbers in "continued proportion" by $a, ar, ar^2, ar^3, \ldots, ar^n$ and the sum of "all those before [the last]" by S_n (since there are n terms before ar^n). Euclid's result states

$$(ar^n - a) : S_n = (ar - a) : a.$$

The modern form for this sum is:

$$S_n = \frac{a(r^n - 1)}{r - 1}.$$

The final proposition of Book IX, proposition 36, shows how to find **perfect numbers**, those which are equal to the sum of all their factors. The result states that if the sum of the sequence $1, 2, 2^2, \ldots, 2^n$ is prime, then the product of that sum and 2^n is perfect. For example, $1 + 2 + 2^2 = 7$ is prime; therefore, $7 \times 4 = 28$ is perfect. And, in fact, $28 = 1 + 2 + 4 + 7 + 14$. Other perfect numbers known to the Greeks were 6, corresponding to $1 + 2$; 496, corresponding to $1 + 2 + 4 + 8 + 16$; and 8128, corresponding to $1 + 2 + 4 + 8 + 16 + 32 + 64$. Although several other perfect numbers have been found by using Euclid's criterion, it is still not known whether there are any perfect numbers which do not meet it. In particular, it is not known whether there are any odd perfect numbers. It is curious, perhaps, that Euclid devoted the culminating theorem of the number theory books to the study of a class of numbers only four of which were known. Nevertheless, the theory of perfect numbers has always proved a fascinating one for mathematicians.

2.4.8 Irrational Magnitudes

Many historians consider Book X the most important of the *Elements*. It is the longest of the thirteen books and probably the best organized. The purpose of Book X is evidently the classification of irrational magnitudes. One of the motivations for the book was the desire to characterize the edge lengths of the regular polyhedra, whose construction in Book XIII forms a fitting climax to the *Elements*. Euclid needed a non-numerical way of comparing the edges of the icosahedron and the dodecahedron to the diameter of the sphere in which they were inscribed. In a manner familiar in modern mathematics, this simple question was to lead to the elaborate classification scheme of Book X, far past its direct answer. It is generally believed that much of this book is due to Theaetetus, since he is credited with some of the polyhedral constructions of Book XIII and since it was in Plato's dialogue bearing his name that the question of determining which numbers have square roots incommensurable with the unit was brought up. It is the answer to that question, given early in Book X, which then leads up to the general classification.

The introductory definitions give Euclid's understanding of the basic terms "incommensurable" and "irrational" (Side Bar 2.6). The first two definitions are relatively straightforward. The third one, on the other hand, needs some comment. First of all, it includes a theorem, which is proved subsequently in Book X. But secondly, note that Euclid's use of the term "rational" is different from the modern usage. For example, if the assigned straight line has length 1, then not only are lines of length a/b called rational, but also lines of length $\sqrt{a/b}$ (where a and b are positive integers).

The first proposition of Book X is fundamental, not only in that Book but also in Book XII.

Proposition X-1. *Two unequal magnitudes being given, if from the greater there is subtracted a magnitude greater than its half, and from that which is left a magnitude greater than its half, and if this process is repeated continually, there will be left some magnitude less than the lesser of the given magnitudes.*

The result depends on definition 4 of Book V, the criterion that two given magnitudes have a ratio. That definition requires that some multiple n of the lesser magnitude exceeds the greater. Then n subtractions of magnitudes greater than half of what is left at any stage gives the desired result.

Propositions 2 and 3 are the results on *antenaresis* discussed in section 2.4.5. But since Euclid has used the same procedure for magnitudes as he did for numbers in Book VII, he can now connect these two distinct concepts. Namely, Euclid shows in propositions 5 and 6 that magnitudes are commensurable precisely when their ratio is that of a number to a number. So even though number and magnitude are distinct notions, one can now apply the machinery of numerical proportion theory to commensurable magnitudes. The more complicated Eudoxian definition is then only necessary for incommensurable magnitudes.

Side Bar 2.6   **Definitions from Book X**

---

1. Those magnitudes are said to be *commensurable* which are measured by the same measure, and those *incommensurable* which cannot have any common measure.

2. Straight lines are *commensurable in square* when the squares on them are measured by the same area, and *incommensurable in square* when the squares on them cannot possibly have any area as a common measure.

3. With these hypotheses, it is proved that there exist straight lines infinite in multitude which are commensurable and incommensurable respectively, some in length only, and others in square also, with an assigned straight line. Let then the assigned straight line be called *rational*, and those straight lines which are commensurable with it, whether in length and in square or in square only, *rational*, but those which are incommensurable with it *irrational*.

Proposition 9 is the result due to Theaetetus which provides the generalization of the Pythagorean discovery of the incommensurability of the diagonal of a square with its side, or, in modern terms, of the irrationality of $\sqrt{2}$. Namely, Euclid shows here in effect that the square root of every nonsquare integer is incommensurable with the unit. In Euclid's terminology, the theorem states that two sides of squares are commensurable in length if and only if the squares have the ratio of a square number to a square number. The more interesting part is the "only if" part. Suppose the two sides *a*, *b* are commensurable in length. Then $a : b = c : d$ where *c*, *d* are numbers. Hence the duplicates of each ratio are equal. But Euclid has already shown (VI-20) that the square on *a* is to the square on *b* in the duplicate ratio of *a* to *b* as well as (VIII-11) that c^2 is to d^2 in the duplicate ratio of *c* to *d*. The result then follows.

After some further preliminaries on criteria for incommensurability, Euclid proceeds to the major task of Book X, the classification of irrational lengths, lengths which are neither commensurable with a fixed unit length nor commensurable in square with it. The entire classification is too long to discuss here, so only a few of the definitions, those which are of use in Book XIII, will be mentioned to provide some of the flavor of Theaetetus' accomplishments. It is well to note that although each of these irrational lengths can be expressed today as a solution of a polynomial equation, Euclid did not use any algebraic machinery. Everything is done geometrically. Nevertheless, for ease of understanding, numerical examples of each definition are presented.

A **medial** straight line is one which is the side of a square equal to the rectangle contained by two rational straight lines commensurable in square only. For example, since the lengths 1, $\sqrt{5}$ are commensurable in square only, and since the rectangle contained by these two lengths has an area equal to $\sqrt{5}$, the length equal to $\sqrt[4]{5}$ would be medial. A **binomial** straight line is the sum of two rational straight lines commensurable in square only. So the length $1 + \sqrt{5}$ is a binomial. Similarly, the difference of two rational straight lines commensurable in square only is called an **apotome**. The length $\sqrt{5} - 1$ provides a

---

Side Bar 2.7   **Definitions from Book XI**

---

12. A *pyramid* is a solid figure, contained by planes, which is constructed from one plane to one point.

13. A *prism* is a solid figure contained by planes two of which, namely those which are opposite, are equal, similar and parallel, while the rest are parallelograms.

14. When, the diameter of a semicircle remaining fixed, the semicircle is carried round and restored again to the same position from which it began to be moved, the figure so comprehended is a *sphere*.

18. When, one leg of a right triangle remaining fixed, the triangle is carried around and restored again to the same position from which it began to be moved, the figure so comprehended is a *cone*. And if the fixed leg is equal to the other leg, the cone will be *right-angled*; if less, *obtuse-angled*; and if greater, *acute-angled*.

---

simple example. A final, more complicated example is given by Euclid's definition of a **minor** straight line. Such a line is the difference $x - y$ between two straight lines such that x, y are incommensurable in square, such that $x^2 + y^2$ is rational, and such that xy is a medial area, that is equal to the square on a medial straight line. For example, if $x = \sqrt{5 + 2\sqrt{5}}$ and $y = \sqrt{5 - 2\sqrt{5}}$, then $x - y$ is a minor.

2.4.9 Solid Geometry

Book XI of the *Elements* is the first of three books dealing with solid geometry. This book contains the three-dimensional analogues of many of the two-dimensional results of Books I and VI. The introductory definitions include such notions as pyramids, prisms, and cones (Side Bar 2.7). The only definition which is somewhat unusual is that of a sphere, which is defined not by analogy to the definition of a circle, but in terms of the rotation of a semicircle about its diameter. Presumably Euclid used this definition because he did not intend to discuss the properties of a sphere as he had discussed the properties of a circle in Book III. The elementary properties of the sphere were in fact known in Euclid's time and dealt with in other texts, including one due to Euclid himself. In the *Elements*, however, Euclid only deals with spheres in Book XII, where he considers the volume, and in Book XIII, where he constructs the regular polyhedra and shows how they fit into the sphere. His constructions in Book XIII, in fact, show how these polyhedra are inscribed in a sphere by rotating a semicircle around them, as in his definition.

The propositions of Book XI include some constructions analogous to those of Book I. For example, proposition 11 shows how to draw a straight line perpendicular to a given plane from a point outside it, while proposition 12 shows how to draw such a line from a point in the plane. There is also a series of theorems on parallelepipeds. In particular, by analogy with proposition I-36, Euclid shows that parallelepipeds on equal bases and with the same height are equal, and then, in analogy with VI-1, that parallelepipeds of the same height are to one another as their bases. Also, in analogy with VI-19, 20, he shows in proposition 33 that similar parallelepipeds are to one another in the triplicate ratio of their sides. Hence the volumes of two similar parallelepipeds are in the ratio of the cubes of any pair of corresponding sides. As before, Euclid computes no volumes. Nevertheless, one can easily derive from these theorems the basic results on volumes of parallelepipeds. The "formulas" for volumes of other solids are included in Book XII.

The central feature of Book XII, which distinguishes it from the other books of the *Elements*, is the use of a limiting process, generally known as the method of exhaustion. This process, developed by Eudoxus, is used to deal with the area of a circle, as well as the volumes of pyramids, cones, and spheres. "Formulas" giving some of these areas and volumes were known much earlier, but for the Greeks a proof was necessary, and Eudoxus' method provided a proof. What it did not provide was a way of discovering the formulas to begin with.

The main results of Book XII are the following:

Proposition XII-2.   *Circles are to one another as the squares on the diameters.*

Proposition XII-7 (Corollary).   *Any pyramid is a third part of the prism which has the same base with it and equal height.*

Proposition XII-10.   *Any cone is a third part of the cylinder which has the same base with it and equal height.*

Proposition XII-18.   *Spheres are to one another in the triplicate ratio of their respective diameters.*

The first of these results is Euclid's version of the ancient result on the area of a circle, a version already known to Hippocrates 150 years earlier. In modern terms it states that the area of a circle is proportional to the square on the diameter. It does not state what the constant of proportionality is, but the proof does provide a method for approximating this. Proposition XII-1, that similar polygons inscribed in circles are to one another as the squares on the diameters, serves as a lemma to this proof. This result in turn is a generalization of the result of VI-20 that similar polygons are to one another in the duplicate ratio of the corresponding sides. It is not difficult to show first of all, that one can take any corresponding lines in place of the "corresponding sides," even the diameter of the circle, and secondly that one can replace "duplicate ratio" by "squares."

The main idea of the proof of XII-2 is to "exhaust" the area of a particular circle by inscribing in it polygons of increasingly many sides. In particular, Euclid shows that one can inscribe in the given circle a polygon whose area differs from that of the circle by less than any given value. His proof of the theorem begins by assuming that the result is not true. Namely, if the two circles C_1, C_2 have areas A_1, A_2 respectively and diameters d_1, d_2, he assumes that $A_1 : A_2 \neq d_1^2 : d_2^2$. Therefore, there is some area S, either greater or less than A_2, such that $d_1^2 : d_2^2 = A_1 : S$. Suppose first that $S < A_2$ (Figure 2.33). Then beginning with an inscribed square and continually bisecting the subtended arcs, inscribe in C_2 a polygon P_2 such that $A_2 > P_2 > S$. In other words, P_2 is to differ from A_2 by less than the difference between A_2 and S. This construction is possible by proposition X-1, since at

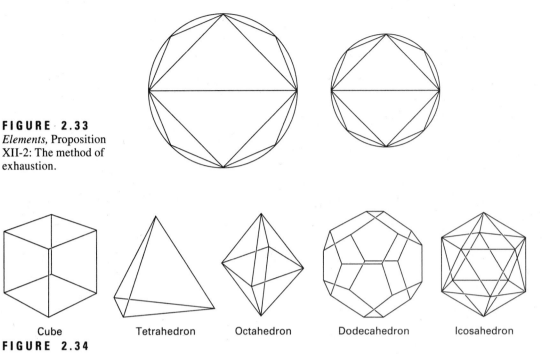

FIGURE 2.33
Elements, Proposition XII-2: The method of exhaustion.

| Cube | Tetrahedron | Octahedron | Dodecahedron | Icosahedron |

FIGURE 2.34
The five regular polyhedra: tetrahedron, octahedron, cube, icosahedron, dodecahedron.

each bisection one is increasing the area of the polygon by more than half of the difference between the circle and the polygon. Next inscribe a polygon P_1 in C_1 similar to P_2. By proposition XII-1, $d_1^2 : d_2^2 = P_1 : P_2$. By assumption, this ratio is also equal to $A_1 : S$. Therefore, $P_1 : A_1 = P_2 : S$. But clearly, $A_1 > P_1$. It follows that $S > P_2$, contradicting the assumption that $S < P_2$. Therefore S cannot be less than A_2. Euclid proves that S also is not greater than A_2 by reducing it to the case already dealt with. It then follows that the ratio of the circles must be equal to the ratio of the squares on the diameters, as asserted.

The other important results of Book XII are proved similarly. In each case, Euclid uses a *reductio* argument. Assuming the falsity of the given assertion, he proceeds to construct inside the given solid other solids, whose properties are already known, such that the difference between the given solid and the constructed one is less than a given "small" value, the "error" defined by the false assumption. That is, he exhausts the solid. The known properties of the constructed figure then lead him to a contradiction as in the proof of XII-2.

The final book of the *Elements,* Book XIII, is devoted to the construction of the five regular polyhedra and their "comprehension" in a sphere (Figure 2.34). This book is the three-dimensional analogue to Book IV. The study of the five regular polyhedra—the cube, tetrahedron, octahedron, dodecahedron, and icosahedron—and the proof that these are the only regular polyhedra, are due to Theaetetus. The first three solids were known in pre-Greek times, and there is archaeological evidence of bronze dodecahedra dating back perhaps to the seventh century B.C.E. The icosahedron, however, was evidently first studied by Theaetetus. It was also he who recognized that these five were the only regular polyhedra, and that in fact the properties of the regular polyhedra were something to study.

Euclid proceeds systematically in Book XIII to construct each of the polyhedra, to demonstrate that each may be comprehended (inscribed) in a sphere, and to compare the edge length of the polyhedron with the diameter of the sphere. For the tetrahedron, Euclid shows that the square on the diameter is 1 1/2 times the square on the edge. In the cube, the square on the diameter is triple the square on the edge while in the octahedron the square on the diameter is double that on the edge. The other two cases are somewhat trickier. Euclid proves that the edge of the dodecahedron is an apotome equal in length to the greater segment of the edge of the inscribed cube when that edge is cut in extreme and mean ratio. Thus, if the diameter of the sphere is 1, then the edge of the cube is $c = \sqrt{3}/3$. Therefore the edge length of the dodecahedron is the positive root of $x^2 + cx - c^2 = 0$ or $(c/2)(\sqrt{5} - 1) = (1/6)(\sqrt{15} - \sqrt{3})$. Since both $\sqrt{15}$ and $\sqrt{3}$ are rational by Euclid's definition, and since they are commensurable in square only, the edge length is in fact an apotome.

For the icosahedron, Euclid proves that the side is a minor straight line. In this case the square on the diameter of the sphere is five times the square on the radius r of the circle circumscribing the five upper triangles of the icosahedron. The bases of these five triangles form a regular pentagon, each edge of which is an edge of the icosahedron. The side of a pentagon inscribed in a circle of radius r is equal to

$$\frac{r}{2}\sqrt{5 + 2\sqrt{5}} - \frac{r}{2}\sqrt{5 - 2\sqrt{5}} = \frac{r}{2}\sqrt{10 - 2\sqrt{5}}.$$

Given that r is rational if the diameter of the sphere is rational (say, equal to 1), this value is indeed a minor straight line. Since r in this case equals $\sqrt{5}/5$, the edge length of the icosahedron can be written as

$$\frac{\sqrt{5}}{10}\sqrt{10 - 2\sqrt{5}} = \frac{1}{10}\sqrt{50 - 10\sqrt{5}}.$$

In a fitting conclusion to Book XIII and the *Elements*, Euclid constructs the edges of the five regular solids in one plane figure, thereby comparing them to each other and the diameter of the given sphere. He then demonstrates that there are no regular polyhedra other than these five.

2.5 EUCLID'S OTHER WORKS

Euclid wrote several mathematics books more advanced than the *Elements*. The most important of the ones which have survived is the *Data*.[22] This was in effect a supplement to Books I–VI of the *Elements*. Each proposition of the *Data* takes certain parts of a geometric configuration as known and shows that therefore certain other parts are determined. Thus the *Data* in essence transformed the synthetic purity of the *Elements* into a manual appropriate to one of the goals of Greek mathematics, the solution of new problems.

Only two propositions from the *Data* will be considered here as examples of the general method. These are closely related to proposition VI-29 of the *Elements* and deal again with geometric algebra.

Proposition 84.   *If two straight lines contain a given space in a given angle, and if the one is greater than the other by a given line, then each of the straight lines is given (i.e., determined).*

If, as in the discussion of VI-29, it is assumed that the given angle is a right angle—and the diagram in the medieval manuscripts which survive shows such an angle—the problem is related to one of the standard Babylonian problems: Find x, y, if the product and difference are given. That is, solve the system

$$xy = c$$
$$x - y = b.$$

Euclid begins by setting up the rectangle, one of whose sides is x, the other y, and laying off y on the line of length x. But now he has a given area c applied to a given line b, exceeding by a square figure. He can then apply proposition 59.

Proposition 59.   *If there is applied to a given straight line a given space which exceeds it by a figure given in species (i.e., one whose angles and the ratio of whose sides are given), the sides of the excess space are given.*

It is here that Euclid really solves the problem of proposition 84, using the same diagram as in the discussion of VI-29. As there, he halves the length b, constructs the square on $b/2$, notes that the sum of that square and the area c is equal to the square on $y + b/2$ (or $x - b/2$), and thereby shows how either of those quantities can be determined as the side of that square. Algebraically, this amounts to the standard Babylonian formula

$$y = \sqrt{\left(\frac{b}{2}\right)^2 + c} - \frac{b}{2}$$

$$x = \sqrt{\left(\frac{b}{2}\right)^2 + c} + \frac{b}{2}.$$

As before, Euclid deals only with geometric figures and never actually writes out a rule like the above. Nevertheless, one can easily translate the geometry into algebra. It also seems that the problem is in fact to find two lengths satisfying certain conditions. In other words, even its formulation is nearly identical to the Babylonian formulation. On the other hand, as in VI-29, the statement of the result enables one to deal with parallelograms as well as the rectangles discussed by the Babylonians. Euclid treats other similar geometric-algebra problems in the *Data*, including, in propositions 85 and 58 the geometric equivalent of the system

$$xy = c$$
$$x + y = b.$$

Whoever Euclid was, it appears from the texts attributed to him, including works in such fields as optics, music, and the conic sections, that he saw himself as a compiler of the Greek mathematical tradition to his time. Certainly this would be appropriate if he was the first mathematician called to the Museum at Alexandria. It would therefore have been his aim to demonstrate to his students not only the basic results known to that time but also some of the methods as in the *Data* by which new problems could be approached. The two mathematicians in the third century B.C.E. who most advanced the field of mathematics, Archimedes and Apollonius, probably received their earliest mathematical training from the students of Euclid, training which in fact enabled them to solve many problems left unsolved by Euclid and his predecessors.

Exercises

Problems from Thales

1. Thales is said to have invented a method of finding distances of ships from shore by use of the angle-side-angle theorem. One possible method is as follows: Suppose A is a point on shore and S is a ship (Figure 2.35). Measure the distance AC along a line perpendicular to AC and bisect it at B. Draw CE at right angles to AC and pick point E on it which is in a straight line with B and S. Show that $\triangle EBC \cong \triangle SBA$ and therefore that $SA = EC$.

2. A second possibility for Thales' method is the following: Suppose Thales was atop a tower on the shore with an instrument made of a straight stick and a crosspiece AC which could be rotated to any desired angle and then would remain where it was put. One rotates AC until one sights the ship S, then turns and sights an object T on shore without moving the crosspiece. Show that $\triangle AET \cong \triangle AES$ and therefore that $SE = ET$.

Pythagorean Number Theory Problems

3. Show that the nth triangular number is represented algebraically as $T_n = \frac{n(n+1)}{2}$ and therefore that an oblong number is double a triangular number.

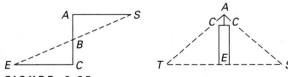

FIGURE 2.35
Possible methods Thales could have used to determine the distance to a ship at sea.

4. Show algebraically that any square number is the sum of two consecutive triangular numbers.

5. Show using dots that eight times any triangular number plus 1 makes a square. Conversely, show that any odd square diminished by 1 becomes eight times a triangular number. Show these results algebraically as well.

6. Show that in a Pythagorean triple, if one of the terms is odd, then two of them must be odd and one even.

7. Show that in a Pythagorean triple, if the largest term is divisible by 4, then so are the other two terms.

8. State and prove the analogue of problem 7, replacing "4" by "3."

9. Construct 5 Pythagorean triples using the formula $\left(n, \frac{n^2-1}{2}, \frac{n^2+1}{2}\right)$ where n is odd. Construct 5 different ones using the formula $\left(m, \left(\frac{m}{2}\right)^2 - 1, \left(\frac{m}{2}\right)^2 + 1\right)$, where m is even.

10. Show that $\sqrt{3}$ is incommensurable with 1 by an argument similar to the proposed Pythagorean argument that $\sqrt{2}$ is incommensurable with 1.

Problems from Euclid's **Elements**

11. Look up Euclid's proposition I-16, that in any triangle the exterior angle is greater than either of the opposite interior angles. Show that the proof requires the construction of a line of arbitrary length.

12. Prove proposition I-32, that the three interior angles of any triangle are equal to two right angles. Show that the proof depends on I-29 and therefore on postulate 5.

13. Prove proposition I-42, that a parallelogram can be constructed equal in area to a given triangle with one angle equal to a given angle.

14. Solve the (modified) problem of proposition I-44, to apply to a given straight line AB a rectangle equal to a given area c. Assume the result of I-42 and use Figure 2.36, where $BEFG$ is a rectangle of area c, D is the intersection of the extension of the diagonal HB and the extension of the line FE, and $ABML$ is the rectangle to be constructed.

15. Translate proposition II-8 into an algebraic result and show that it is valid: If a straight line is cut at random, four times the rectangle contained by the whole and one of the segments together with the square on the remaining segment is equal to the square on the whole and the former segment taken together.

16. Show that proposition II-13 is equivalent to the law of cosines for an acute-angled triangle: In acute-angled triangles, the square on the side opposite the acute angle is less than the sum of the squares on the other two sides by twice the rectangle contained by one of the sides about the acute angle, namely that on which the perpendicular falls, and the line segment between the angle and the perpendicular. (Proposition II-12 gives the law of cosines for an obtuse-angled triangle.)

17. Provide the details of the proof of proposition III-20: In a circle, the angle at the center is double the angle at the circumference, when the angles cut off the same arc.

18. Provide the details of the proof of proposition III-21: In a circle, the angles in the same segment are equal to one another.

19. Prove proposition III-31, that the angle in a semicircle is a right angle.

20. Find a construction for inscribing a regular hexagon in a circle.

21. Given that a pentagon and an equilateral triangle can be inscribed in a circle, show how to inscribe a regular 15-gon in a circle.

22. Prove that the last nonzero remainder in the Euclidean algorithm applied to the numbers a, b is in fact the greatest common divisor of a and b.

23. Use the Euclidean algorithm to find the greatest common divisor of 963 and 657; of 2689 and 4001.

24. Suppose that a line of length 1 is divided in extreme and mean ratio, that is, that the line is divided at x so that $1/x = x/(1 - x)$. Show by the method of the Euclidean algorithm that 1 and x are incommensurable. (Use proposition X-2.)

25. Use Theaetetus' definition of equal ratio to show that $46 : 6 = 23 : 3$ and that $\sqrt{2} : 1 = \sqrt{6} : \sqrt{3}$.

26. Compare Euclid's treatment of the distributive law in V-1 with the treatment in II-1 in terms of the types of magnitudes involved and the proofs given.

27. Prove proposition V-12 both by using Eudoxus' definition and by modern methods: If any number of magnitudes are proportional, as one of the antecedents is to one of the consequents, so will all of the antecedents be to all of the consequents. (In algebraic notation, this says that if $a_1 : b_1 = a_2 : b_2 = \cdots = a_n : b_n$, then $(a_1 + a_2 + \cdots + a_n) : (b_1 + b_2 + \cdots + b_n) = a_1 : b_1$.)

28. Use Eudoxus' definition to prove proposition V-16: If $a : b = c : d$, then $a : c = b : d$.

29. Construct geometrically the solution of $8 : 4 = 6 : x$.

30. Solve geometrically the equation $9/x = x/5$ by beginning with a semicircle of diameter $9 + 5 = 14$.

31. Solve geometrically (using VI-28) the quadratic equation $x^2 + 10 = 7x$. There are two positive solutions to this equation. Modify the diagram so that both solutions are evident.

32. Solve geometrically (using VI-29) the quadratic equation $x^2 + 10x = 39$.

33. Prove proposition VIII-8 and Archytas' special case that there is no mean proportional between $n + 1$ and n.

34. Find the one mean proportional between two squares guaranteed by proposition VIII-11.

35. Find the two mean proportionals between two cubes guaranteed by proposition VIII-12.

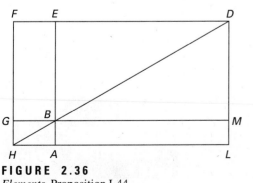

FIGURE 2.36
Elements, Proposition I-44.

36. Prove proposition VIII-14: If a^2 measures b^2, then a measures b and conversely.

37. Turn Euclid's proof of XII-2 into a recursive algorithm for calculating the area of a circle. Use the algorithm several times to approximate the area of a circle of radius 1.

38. Prove XIII-9: If the side of the hexagon and the side of the decagon inscribed in the same circle are placed together in a single straight line, then the meeting point divides the entire line segment in extreme and mean ratio, with the greater segment being the side of the hexagon.

39. Prove XIII-10: If an equilateral pentagon, hexagon, and decagon are each inscribed in a given circle, then the square on the side of the pentagon equals the sum of the squares on the sides of the hexagon and the decagon.

40. Solve geometrically, using the propositions from the *Data*, the system $x - y = 7$; $xy = 18$.

FOR DISCUSSION . . .

41. Eratosthenes of Cyrene (276–194 B.C.E.) is credited with measuring the earth by an argument from parallel lines. Namely, he found that at noon on the summer solstice the sun was directly overhead at Syene, a place on the tropic of Cancer, while at the same time at Alexandria, approximately 5000 stades due north, the sun was at 7 1/5° from the zenith. Given that the rays from the sun to the earth are all parallel, he concluded that $\angle SOA = 7\ 1/5°$ (Figure 2.37). Calculate, therefore, Eratosthenes' value for the circumference of the earth in stades. If the length of a stade is taken to be 516.7 feet (= 300 Royal Egyptian cubits), calculate Eratosthenes' values for the circumference and diameter of the earth. How accurate is the method described? How would Eratosthenes know the distance from Alexandria to Syene?

42. Outline the evidence which supports and that which contradicts the statement: The Greeks developed geometric algebra by translating the algebraic techniques of the Babylonians into geometric form.

43. Discuss the advantages and disadvantages of a geometric approach relative to a purely algebraic approach in the teaching of the quadratic equation in school.

44. Prepare a lesson proving a number of simple algebraic identities geometrically. (For example, prove $(a + b)^2 = a^2 + 2ab + b^2$ and $(a + b)(a - b) = a^2 - b^2$.)

45. Compare Euclid's treatment of triangle congruence with that of a modern high school geometry text. Which method is easier to teach?

46. Discuss whether Euclid's *Elements* fits Plato's dictums that the study of geometry is for "drawing the soul toward truth" and that it is to gain knowledge "of what eternally exists."

47. Discuss the military usefulness of such studies as arithmetic and geometry, as suggested by Plato in *The Republic*. Why would a general need to be expert in mathematics? And why is military usefulness the only "practical" application of mathematical studies mentioned by Plato?

48. Should one base the study of geometry in high school on Euclid's *Elements* as was done for many years? Discuss the pros and cons of Euclid versus a "modern" approach.[23]

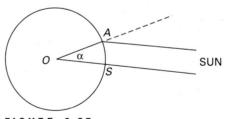

FIGURE 2.37
Eratosthenes' determination of the size of the earth.

References and Notes

A good source of basic information on Greek civilization is H. D. F. Kitto, *The Greeks* (London: Penguin, 1951). Two excellent general works on early Greek science are by G. E. R. Lloyd, *Early Greek Science: Thales to Aristotle* (New York: Norton, 1970) and *Magic, Reason and Experience* (Cambridge: Cambridge University Press, 1979). The latter work, in particular, deals with the beginnings of logical reasoning in Greece and the emergence of the idea of mathematical proof. The standard reference on Greek mathematics is Thomas Heath, *A History of Greek Mathematics* (New York: Dover,

1981, reprinted from the 1921 original). However, many of Heath's conclusions have been challenged in the more recent works listed below as well as in the notes to the next three chapters. Heath also prepared the standard modern English version of Euclid's *Elements* (New York: Dover, 1956). Many of the available fragments from earlier Greek mathematics are collected in Ivor Bulmer-Thomas, *Selections Illustrating the History of Greek Mathematics* (Cambridge: Harvard University Press, 1941). B. L. Van der Waerden's *Science Awakening I* (Groningen: Noordhoff, 1954) also contains an extensive treatment of Greek mathematics. A detailed, if somewhat controversial, discussion of the background of Euclid's *Elements* is Wilbur Knorr, *The Evolution of the Euclidean Elements* (Dordrecht: Reidel, 1975). Knorr's more recent book, *The Ancient Tradition of Geometric Problems* (Boston: Birkhäuser, 1986) contains a lengthy argument supporting the idea that geometric problem solving was the motivating factor for much of Greek mathematics. I have also used material from the doctoral dissertation of Charles Jones, *On the Concept of One as a Number* (University of Toronto, 1979) and I. Mueller, *Philosophy and Deductive Structure in Euclid's Elements* (Cambridge: MIT Press, 1981) in organizing the treatment of Euclid's ideas in this chapter. Other useful works on Greek mathematics include F. Lasserre, *The Birth of Mathematics in the Age of Plato* (Larchmont, N.Y.: American Research Council, 1964), J. Klein, *Greek Mathematical Thought and the Origin of Algebra* (Cambridge: MIT Press, 1968), and Asger Aaboe, *Episodes from the Early History of Mathematics* (Washington: M.A.A., 1964). Finally, a survey of the recent literature on Greek mathematics is available by J. L. Berggren, "History of Greek Mathematics: A Survey of Recent Research," *Historia Mathematica* 11 (1984), 394–410.

1. From Proclus' *Summary*, translated in Bulmer-Thomas, *Selections Illustrating,* I, p. 147.

2. *Plutarch's Moralia*, translated by Phillip H. De Lang and Benedict Einarson (Cambridge: Harvard University Press, 1959), VII, pp. 397–399.

3. Plato, *Republic* VII, 525. The translation used is that of Frances Cornford (London: Oxford University Press, 1941), but the references are to lines of the standard Greek text and can be checked in any modern translation.

4. Ibid. VII, 527.

5. Ibid. VII, 528.

6. Ibid. VII, 531.

7. Aristotle, *Prior Analytics* I, 1, 24^b, 19. The translations of Aristotle's works used are those in the *Great Books* edition (Chicago: Encyclopedia Britannica, 1952), but as in the case of Plato, the references are to line numbers in the standard Greek version.

8. Aristotle, *Posterior Analytics* I, 10, 76^a 40–76^b 10.

9. Ibid. I, 2, 71^b 23.

10. Aristotle, *Physics* VI, 1, 231^b 15.

11. Ibid. V, 3, 227^a 12.

12. Ibid. III, 6, 206^b 32.

13. Ibid. VI, 9, 239^b 11. For more on Zeno's paradoxes, see F. Cajori, "History of Zeno's Arguments on Motion," *American Mathematical Monthly* 22 (1915), 1–6, 39–47, 77–82, 109–115, 145–149, 179–186, 215–220, 253–258, 292–297, and also H. D. P. Lee, *Zeno of Elea* (Cambridge: Cambridge University Press, 1936).

14. Ibid. VI, 9, 239^b 15.

15. Ibid. VI, 2, 233^a 26–29.

16. Ibid. VI, 9, 239^b 6.

17. Proclus, *Commentary on the First Book of Euclid's Elements*, translated by G. P. Morrow, (Princeton: Princeton Univ. Press, 1970), p. 56. This edition also contains a valuable introduction.

18. The debate over geometric algebra was renewed with a vengeance in an article by Sabetai Unguru entitled "On the need to rewrite the history of Greek mathematics," *Archive for History of Exact Sciences* 15 (1975), 67–114. He was answered by several other historians over the next two years. The most important responses were by B. L. Van der Waerden, "Defence of a shocking point of view," *Archive for History of Exact Sciences* 15 (1976), 199–210, and by Hans Freudenthal, "What is algebra and what has it been in history?" *Archive for History of Exact Sciences* 16 (1977), 189–200. A reply to these was offered by Unguru and David Rowe, "Does the Quadratic Equation Have Greek Roots? A Study of Geometric Algebra, Application of Areas, and Related Problems," *Libertas Mathematica* 1 (1981) and 2 (1982). These articles are recommended as examples of the strong feelings historical controversy can bring out.

19. For more details on division in extreme and mean ratio, see Roger Herz-Fischler, *A Mathematical History of Division in Extreme and Mean Ratio* (Waterloo, Ont.: Wilfrid Laurier Univ. Press, 1987).

20. See D. H. Fowler, *The Mathematics of Plato's Academy* (Oxford: Clarendon Press, 1987) for a detailed discussion

of a nontraditional view of the Euclidean algorithm as a basis for much of early Greek mathematics. See also D. H. Fowler, "Ratio in Early Greek Mathematics," *Bulletin of the American Mathematical Society* 1 (1979), 807–847.

21. The discussion of the origins of Eudoxus' theory of proportion is adapted from the treatment in Knorr, *The Evolution of the Euclidean Elements.*

22. The most accessible version of Euclid's *Data* is by Shuntaro Ito, *The Medieval Latin Translation of the Data of Euclid* (Boston: Birkhäuser, 1980).

23. See Paul Daus, "Why and How We Should Correct the Mistakes in Euclid," *Mathematics Teacher* 53 (1960), 576–581.

Summary of Greek Mathematics to c. 300 B.C.E.

624–547 B.C.E.	Thales	"Proofs" of certain theorems
572–497	Pythagoras	All is Number
Fifth century	Hippocrates of Chios	Lunes, duplication of cube
Fifth century	Zeno	Paradoxes of motion
Fifth century	Archytas	Music and number theory
430	Discovery of incommensurability of side and diagonal of square	
c. 400	Theodorus	Incommensurables
429–347	Plato	Academy founded in 385
417–369	Theaetetus	Incommensurables, proportion theory
408–355	Eudoxus	Theory of proportions
384–322	Aristotle	Logic of syllogisms
c. 300	Founding of Museum and Library in Alexandria	
c. 300	Euclid	The *Elements*
280–206	Chrysippus	Logic of propositions

(Note: Most dates listed are only approximate.)

Chapter 3

Archimedes and Apollonius

"The third book [of Conics] *contains many remarkable theorems useful for the syntheses of solid loci . . . ; the most and prettiest of these theorems are new, and it was their discovery which made me aware that Euclid did not work out the synthesis of the locus with respect to three and four lines . . . ; for it was not possible for the said synthesis to be completed without the aid of the additional theorems discovered by me."*
(Preface to Book I of Apollonius' Conics)[1]

The following story is told by Vitruvius: "After his victories [Hiero, the king of Syracuse in the third century B.C.E.] determined to set up in a certain temple a gold wreath vowed to the immortal gods. He let out the execution as far as the craftsman's wages were concerned, and weighed the gold out to the contractor to an exact amount. At the appointed time the man presented the work finely wrought for the king's acceptance, and appeared to have furnished the weight of the crown to scale. However, information was given that gold had been withdrawn and that the same amount of silver had been added in the making of the crown. Hiero was indignant that he had been made light of, and failing to find a method by which he might detect the theft, asked Archimedes to undertake the investigation. While Archimedes was considering the matter, he happened to go to the baths. When he went down into the bathing pool he observed that the amount of water which flowed outside the pool was equal to the amount of his body that was immersed. Since this fact indicated the method of explaining the case, he did not linger, but moved with delight he leapt out of the pool, and going home naked, cried aloud that he had found exactly what he was seeking. For as he ran he shouted in Greek: 'Eureka, eureka (I have found it).' "[2]

Greek mathematics in the third and early second centuries B.C.E. was dominated by two major figures, Archimedes of Syracuse (c. 287–212 B.C.E.) and Apollonius of Perge (c. 250–175 B.C.E.), each heir to a different aspect of fourth century Greek mathematics. The former took over the "limit" methods of Eudoxus and succeeded not only in applying them to determine areas and volumes of new figures, but also in developing new techniques which enabled the results to be discovered in the first place. Archimedes, unlike his predecessors, was neither reluctant to share his methods of discovery nor afraid of performing numerical calculations and exhibiting numerical results. Furthermore,

Archimedes wrote several treatises presenting mathematical models of certain aspects of what we would call theoretical physics and applied his physical principles to the invention of various mechanical devices.

Apollonius, on the other hand, was instrumental in extending the domain of analysis to new and more difficult geometric construction problems. As a foundation for these new approaches, he created his magnum opus, the *Conics*, a work in eight books developing synthetically the important properties of this class of curves, properties which were central in developing new solutions to such problems as the duplication of the cube and the trisection of the angle.

This chapter will survey the extant works of both of these mathematicians, as well as the work of certain others who considered similar problems.

3.1 ARCHIMEDES AND PHYSICS

Archimedes was the first mathematician to derive quantitative results from the creation of mathematical models of physical problems on earth. In particular, Archimedes is responsible for the first proof of the law of the lever (Figure 3.1) and its application to finding centers of gravity as well as the first proof of the basic principle of hydrostatics and some of its important applications.

3.1.1 The Law of the Lever

Everyone is familiar with the principle of the lever from having played on seesaws as a child. Equal weights at equal distances from the fulcrum of the lever balance, and a lighter child can balance a heavier one by being farther away. The ancients were well aware of this principle as well. The law even appears in writing in a work on mechanics attributed to Aristotle: "Since the greater radius is moved more quickly than the less by an equal weight, and there are three elements in the lever, the fulcrum . . . and two weights, that which moves and that which is moved, therefore the ratio of the weight moved to the moving weight is the inverse ratio of their distances from the fulcrum."[3]

As far as is known, no one before Archimedes had created a mathematical model of the lever, by which one could derive a mathematical proof of the law of the lever. In general, a difficulty in attempting to apply mathematics to physical problems is that the physical situation is often quite complicated. Therefore the situation needs to be idealized. One ignores those aspects which appear less important and concentrates on only the essential variables of the physical problem. This idealization is referred to today as the creation of a mathematical model. The lever is a case in point. To deal with it as it actually occurs, one would need to consider not only the weights applied to the two ends and their distances from the fulcrum, but also the weight and composition of the lever itself. It may be heavier at one end than the other. Its thickness may vary. It may bend slightly—or even break—when certain weights are applied at certain points. In addition, the fulcrum is also a physical object of a certain size. The lever may slip somewhat along the fulcrum, so it may not be clear from what point the distance of the weights should be measured. To include all of these factors in a mathematical analysis of the lever would make the mathe-

FIGURE 3.1
Archimedes and the law of the lever.

Biography   Archimedes (287–212 B.C.E.)

More biographical informaton about Archimedes survives than about any other Greek mathematician. Much is found in Plutarch's biography of the Roman general Marcellus, who captured Syracuse, the major city of Sicily, after a siege in 212 B.C.E. during the Second Punic War. Other Greek and Roman historians also discuss aspects of Archimedes' life.

Archimedes was the son of the astronomer Phidias and perhaps a relative of King Hiero II of Syracuse, under whose rule from 270 to 216 B.C.E. the city greatly flourished. It is also probable that he spent time in his youth in Alexandria, for he is credited with the invention there of the Archimedean screw, a machine for raising water used for irrigation (Figure 3.2). Moreover, the prefaces of many of his works are addressed to scholars at Alexandria, including one of the chief librarians, Eratosthenes. Most of his life, however, was spent in his native Syracuse, where he was repeatedly called upon to use his mathematical talents to solve various practical problems for Hiero and his successor. Many stories are recorded about his intense dedication to his work. Plutarch wrote that on many occasions his concentration on mathematics "made him forget his food and neglect his person, to that degree that when he was carried by absolute violence to bathe or have his body anointed, he used to trace geometrical figures in the ashes of the fire, and diagrams in the oil on his body, being in a state of entire preoccupation, and in the truest sense, divine possession with his love and delight in science."[4] And it was this dedication which ultimately cost him his life.

It was his genius as a military engineer that kept the Roman army under Marcellus at bay for months during the siege of Syracuse. Finally, however, probably through treachery, the Romans were able to enter the city. Marcellus gave explicit orders that Archimedes be spared, but Plutarch relates that "as fate would have it, he was intent on working out some problem with a diagram and, having fixed his mind and his eyes alike on his investigation, he never noticed the incursion of the Romans nor the capture of the city. And when a soldier came up to him suddenly and bade him follow to Marcellus, he refused to do so until he had worked out his problem to a demonstration; whereat the soldier was so enraged that he drew his sword and slew him."[5]

FIGURE 3.2
Archimedes and the Archimedean screw.

matics extremely difficult. Archimedes therefore simplified the physical situation. He assumed that the lever itself was rigid, but weightless, and that the fulcrum and the weights were mathematical points. He was then able to develop the mathematical principles of the lever.

Archimedes dealt with these principles at the beginning of his treatise *On the Equilibrium of Planes*, or *The Centers of Gravity of Planes*. Being well trained in Greek geometry, he began by stating seven postulates he would assume, four of which are reproduced here.

1. Equal weights at equal distances are in equilibrium, and equal weights at unequal distances are not in equilibrium but incline toward the weight which is at the greater distance.

2. If, when weights at certain distances are in equilibrium, something is added to one of the weights, they are not in equilibrium but incline toward the weight to which the addition was made.

3. Similarly, if anything is taken away from one of the weights, they are not in equilibrium but incline toward the weight from which nothing was taken.

6. If magnitudes at certain distances are in equilibrium, other magnitudes equal to them will also be in equilibrium at the same distances.

These postulates come from basic experience with levers. The first postulate, in fact, is an example of what is usually called the **Principle of Insufficient Reason.** That is, one assumes that equal weights at equal distances balance because there is no reason to make any other assumption. The lever cannot incline to the right, for example, since what is the right side from one viewpoint is the left side from another. The second and third postulates are equally obvious. The sixth appears to be virtually meaningless. In Archimedes' use of it, however, it appears that the second clause means "other equal magnitudes, the centers of gravity of which lie at the same distances from the fulcrum, will also be in equilibrium." Namely, the influence of a magnitude on the lever depends solely on its weight and the position of its center of gravity.

Although Archimedes uses the term "center of gravity" in the title of the work, as well as in many of its propositions, he never gives a definition. Presumably he felt that the concept was so well known to his readers that a definition was unnecessary. There are, however, later Greek texts which do give a definition, perhaps the one that was even used in Archimedes' time: "We say that the center of gravity of any body is a point within that body which is such that, if the body be conceived to be suspended from that point, the weight carried thereby remains at rest and preserves the original position."[6] But it was also clear to Archimedes, and this is what he expressed in postulate 6, that the downward tendency of gravitation may be thought of as being concentrated in that one point. Note that in neither the postulates nor the theorems is there any mention of the lever itself. It is just there. Its weight does not enter into the calculations. Archimedes has in effect assumed that the lever is weightless and rigid. Its only motion is inclination to one side or the other.

The first two in Archimedes' sequence of propositions leading to the law of the lever are very easy:

Proposition 1.   *Weights which balance at equal distances are equal.*

Proposition 2.   *Unequal weights at equal distances will not balance but will incline toward the greater weight.*

The proof of the first result is by *reductio ad absurdum.* For if the weights are not equal, take away from the greater the difference between the two. By postulate 3, the remainders will not balance. This contradicts postulate 1, since now we have equal weights at equal distances. Our original assumption must then be false. To prove proposition 2, again take away from the greater weight the difference between the two. By postulate 1, the remainders will balance. So if this difference is added back, the lever will incline toward the greater by postulate 2.

Proposition 3.   *Suppose A and B are unequal weights with A > B which balance at point C (Figure 3.3). Let AC = a, BC = b. Then a < b. Conversely, if the weights balance and a < b, then A > B.*

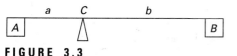
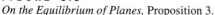

FIGURE 3.3
On the Equilibrium of Planes, Proposition 3.

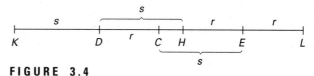

FIGURE 3.4
On the Equilibrium of Planes, Proposition 6.

The proof is again by contradiction. Suppose $a \not< b$. Subtract from A the difference $A - B$. By postulate 3, the lever will incline toward B. But if $a = b$, the equal remainders will balance, and if $a > b$, the lever will incline toward A by postulate 1. These two contradictions imply that $a < b$. The proof of the converse is equally simple.

In propositions 4 and 5 Archimedes shows that the center of gravity of a system of two (and three) equally spaced equal weights is at the geometric center of the system. These results are extended in the corollaries to any system of equally spaced weights provided that those at equal distance from the center are equal. The law of the lever itself is stated in propositions 6 and 7.

Propositions 6, 7.   *Two magnitudes, whether commensurable (Proposition 6) or incommensurable (Proposition 7), balance at distances inversely proportional to the magnitudes.*

First assume that the magnitudes A, B are commensurable; that is $A : B = r : s$ where r, s are numbers. Archimedes' claim is that if A is placed at E and B at D, and if C is taken on DE with $DC : CE = r : s$, then C is the center of gravity of the two magnitudes A, B (Figure 3.4). To prove the result, assume that units have been chosen so that $DC = r$ and $CE = s$. Choose H on DE so that $HE = r$ and extend the line past E to L so that EL also equals r. Also extend the line in the opposite direction to K, making $DK = HD = s$. Then C is the midpoint of LK. Now break A into $2r$ equal parts and B into $2s$ equal parts. Space the first set equally along LH and the second along HK. Since $A : B = r : s = 2r : 2s$, it follows that each part of A is equal to each part of B. From the corollary mentioned above, the center of gravity of the parts of A will be at the midpoint E of HL, while the center of gravity of the parts of B will be at the midpoint D of KH. By postulate 6, nothing is changed if A itself is considered situated at E and B at D. On the other hand, the total system consists of $2r + 2s$ equal parts equally spaced along the line KL. Hence the center of gravity of the system is at the midpoint C of that line. Therefore, weight A placed at E and weight B placed at D balance about the point C.

Archimedes concludes the proof in the incommensurable case by a *reductio* argument using the fact that if two magnitudes are incommensurable, one can subtract from the first an amount smaller than any given quantity such that the remainder is commensurable with the second. Interestingly enough, Archimedes makes no use here of the Eudoxian proportion theory for incommensurables of *Elements*, Book V nor even of Theaetetus' earlier version based on the Euclidean algorithm. He instead makes use essentially of a continuity argument. But even so, his proof is somewhat flawed.

Nevertheless, Archimedes uses the law of the lever in the remainder of the treatise to find the centers of gravity of various geometrical figures. He proves that the center of gravity of a parallelogram is at the intersection of its diagonals, of a triangle at the intersection of two medians, and of a parabolic segment at a point on the diameter three-fifths of the distance from the vertex to the base.

3.1.2 Applications to Engineering

Not only are there geometric consequences of the law of the lever, but also there are physical consequences. In particular, given any two weights A and B and any lever, there is always a point C at which the weights balance. If A is much heavier than B, they will balance when A is sufficiently close to C and B is sufficiently far away. But then any additional weight added to B will incline the lever in that direction. That is, any additional weight added to B will cause weight A to be lifted. Archimedes therefore was able to boast that "any weight might be moved and . . . if there were another earth, by going into it he could move this one."[7] When King Hiero heard of this boast, he asked Archimedes to demonstrate his principles in actual experiment. Archimedes complied, but instead of using a lever, he probably made use of some kind of pulley or tackle system, which also provided a great mechanical advantage. Plutarch writes that "he fixed accordingly upon a ship of burden out of the king's arsenal, which could not be drawn out of the dock without great labor and many men; and loading her with many passengers and a full freight, sitting himself the while far off, with no great endeavor, but only holding the head of the pulley in his hand and drawing the cords by degrees, he drew the ship in a straight line, as smoothly and evenly as if she had been in the sea."[8] Other sources give a variant of Plutarch's story, to the effect that Archimedes was responsible for the construction of a magnificent ship, named the "Syracusa," and singlehandedly launched this 4200-ton luxury vessel.

Archimedes enjoyed the greatest fame in antiquity, however, for his design of various engines of war. These engines enabled Syracuse to hold off the Roman siege for many months. Archimedes devised various missile launchers as well as huge cranes by which he was able to lift Roman ships out of the water and dash them against the rocks or simply dump out the crew. In fact, he was so successful that any time the Romans saw a little rope or piece of wood come out from the walls of the city, they fled in panic.

Plutarch relates that Archimedes was not particularly happy as an engineer: "He would not deign to leave behind him any commentary or writing on such subjects; but, repudiating as sordid and ignoble the whole trade of engineering, and every sort of art that lends itself to mere use and profit, he placed his whole affection and ambition in those purer speculations where there can be no reference to the vulgar needs of life."[9] In fact, however, there is evidence that Archimedes did write on certain mechanical subjects, including a book *On Sphere Making* in which he described his planetarium, a mechanical model of the motions of the heavenly bodies, and another one on water clocks.

The incident of the gold wreath and the bath led Archimedes to the study of an entirely new subject, that of hydrostatics, in which he discovered its basic law, that a solid heavier than a fluid will, when weighed in the fluid, be lighter than its true weight by the weight of the fluid displaced. It is, however, not entirely clear how Archimedes' noticing the water being displaced in his bath led him to the concept of weight being lessened. Perhaps he also noticed that his body felt lighter in the water.

As in his study of levers, Archimedes began the mathematical development of hydrostatics, in his treatise *On Floating Bodies*, by giving a simplifying postulate. He was then able to show, among other results, that the surface of any fluid at rest is the surface of a sphere whose center is the same as that of the earth. He could then deal with solids floating or sinking in fluids by assuming that the fluid was part of a sphere. Archimedes was able to solve the wreath problem by using the basic law, proved as proposition 7. One way by which he could have applied the law is suggested by Heath, based on a description in a Latin poem of the fifth century C.E.[10] Suppose the wreath is of weight W, composed of unknown weights w_1 and w_2 of gold and silver respectively. To determine the ratio of gold to silver in the wreath, first, weigh it in water and let F be the loss of weight. This amount can be determined by weighing the water displaced. Next take a weight W of pure gold and let F_1 be its weight loss in water. It follows that the weight of water displaced by a weight w_1 of gold is $\frac{w_1}{W}F_1$. Similarly, if the weight of water displaced by a weight W of pure silver is F_2, the weight of water displaced by a weight w_2 of silver is $\frac{w_2}{W}F_2$. Therefore, $\frac{w_1}{W}F_1 + \frac{w_2}{W}F_2 = F$. Thus the ratio of gold to silver is given by

$$\frac{w_1}{w_2} = \frac{F - F_2}{F_1 - F}.$$

Vitruvius himself provides a somewhat different suggestion for solving the wreath problem, more clearly based on the story of the bath, but not on the basic law of hydrostatics. He also records that Archimedes found that the goldsmith had cheated the king. What happened to the smith, however, is not mentioned.

3.2 ARCHIMEDES AND NUMERICAL CALCULATIONS

The brief treatise, *On the Measurement of the Circle*, contains numerical results, unlike anything found in Euclid's work. Its first proposition, in addition, gives Archimedes' answer to the question of squaring the circle, by showing that the area of a circle of given radius can be found once the circumference is known.

Proposition 1. *The area A of any circle is equal to the area of a right triangle in which one of the legs is equal to the radius and the other to the circumference.*

Archimedes' result is equivalent to the Babylonian result that $A = (C/2)(d/2)$, where d is the length of the diameter and C the length of the circumference. Archimedes, however, gives a rigorous proof, using a Eudoxian exhaustion argument. Namely, if K is the area of the given triangle, Archimedes first supposes that $A > K$. By inscribing in the circle regular polygons of successively more sides, he eventually determines a polygon of area P such that $A - P < A - K$. Thus $P > K$. Now the perpendicular from the center of the circle to the midpoint of a side of the polygon is less than the radius, while the perimeter of the polygon is less than the circumference. It follows that $P < K$, a contradiction.

Similarly, the assumption that $A < K$ leads to another contradiction and the result is proved.

The third proposition of this treatise complements the first by giving a numerical approximation to the length of the circumference.

Proposition 3.   *The ratio of the circumference of any circle to its diameter is less than 3 1/7 but greater than 3 10/71.*

Archimedes' proof of this statement provides algorithms for determining the perimeter of certain regular polygons circumscribed about and inscribed in a circle. Namely, Archimedes begins with regular hexagons, the ratios of whose perimeters to the diameter of the circle are known from elementary geometry. He then in effect uses the following lemmas to calculate, in turn, the ratios to the diameter of the perimeters of regular polygons with 12, 24, 48, and 96 sides respectively.

Lemma 1.   *Suppose OA is the radius of a circle and CA is tangent to the circle at A. Let DO bisect $\angle COA$ and intersect the tangent at D. Then $DA/OA = CA/(CO + OA)$ and $DO^2 = OA^2 + DA^2$ (Figure 3.5).*

Lemma 2.   *Let AB be the diameter of a circle and ACB a right triangle inscribed in the semicircle. Let AD bisect $\angle CAB$ and meet the circle at D. Connect DB. Then $AB^2/BD^2 = 1 + (AB + AC)^2/BC^2$ and $AD^2 = AB^2 - BD^2$.*

Archimedes uses the first lemma repeatedly to develop a recursive algorithm for determining the desired ratio using circumscribed polygons. He begins by assuming that $\angle COA$ is one-third of a right angle (30°), so CA is half of one side of a circumscribed regular hexagon. Therefore CA and CO are known. Since $\angle DOA = 15°$, it follows that DA is half of one side of a regular 12-gon. DA and DO are then calculated by use of the lemma. Next, $\angle DOA$ is bisected to get an angle of $7\frac{1}{2}°$. The piece of the tangent subtending that angle is then half of one side of a regular 24-gon. Its length can be calculated as well. If r is the radius of the circle, t_i half of one side of a regular 3×2^i-gon ($i \geq 1$), and u_i the length of the line from the center of the circle to a vertex of that polygon, the lemma can be translated into the recursive formulas:

$$t_{i+1} = \frac{rt_i}{u_i + r} \qquad u_{i+1} = \sqrt{r^2 + t_{i+1}^2}.$$

The ratio of the perimeter of the ith inscribed polygon to the diameter of the circle is then $6(2^i t_i) : 2r = 3(2^i t_i) : r$.

Archimedes developed a similar algorithm for inscribed polygons by use of the second lemma and in both cases provided explicit numerical results at each stage. What is not known is exactly how Archimedes found these results, many of which require square roots. For example, Archimedes noted that the ratio $OA : AC$ of lemma 1 when $\angle COA$ is 30°, namely $\sqrt{3} : 1$, is greater than $265 : 163$ and less than $1351 : 780$. What is sure, however, is that Archimedes, like many great mathematicians of later times, was a superb calculator. After four steps of both algorithms, he concluded that the ratio of the perimeter of the circumscribed 96-sided polygon to the diameter is less than $14688 : 4673\frac{1}{2} =$

$3 + \frac{667\ 1/2}{4673\ 1/2} < 3\frac{1}{7}$ while the ratio of the inscribed 96-sided polygon to the diameter is greater than $6336 : 2077\frac{1}{4} > 3\frac{10}{71}$, thus proving the theorem.

This proof provides the first recorded method for actually computing π. Once Archimedes' method was known, it was merely a matter of patience to calculate π to as great an accuracy as desired. Archimedes does not tell us why he stopped at 96-sided polygons. But his value of 3 1/7 has become a standard approximation for π to the present day.

It was Nicomedes (late third century B.C.E.), a successor of Archimedes, who used an entirely new method to determine the length of the circumference of a circle and, therefore, by proposition 1 above, to square the circle. Namely, he used the **quadratrix**, a curve probably introduced a century earlier, defined through a combination of two motions: In the square $ABCD$, imagine that the ray AB rotates uniformly around A from its beginning position to the ending position on AD, while at the same time, the line BC moves parallel to itself from BC to AD (Figure 3.6). The quadratrix BZK is then the curve traced out by the moving intersection point. It follows from this definition that a point Z on the quadratrix satisfies the proportion $ZL : BA = $ arc $DG :$ arc BD, or $ZL :$ arc $DG = AB :$ arc BD. In modern notation, if the polar equation of the curve is given by $\rho = \rho(\theta)$, ρ satisfies the equation

$$\frac{\rho(\theta)\sin\theta}{a\theta} = \frac{a}{\frac{1}{2}\pi a},$$

where a is the length of a side of the square.

If we take the limit of the left side of the equation as θ approaches 0, we get the result

$$\frac{\rho(0)}{a} = \frac{a}{\frac{1}{2}\pi a}.$$

Naturally, the Greeks did not present such a limit argument, but the result, in the form $AK : AB = AB :$ arc BD, was proved, probably by Nicomedes, through a double *reductio*

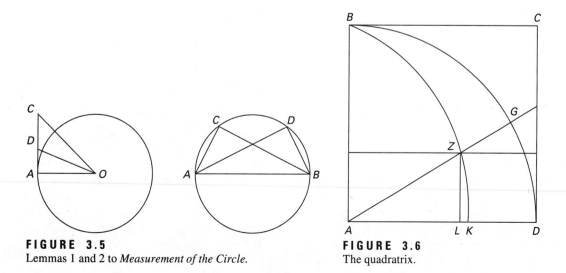

FIGURE 3.5
Lemmas 1 and 2 to *Measurement of the Circle.*

FIGURE 3.6
The quadratrix.

argument. It then follows that arc *BD*, a quarter of the circumference of the circle, is a third proportional to the known lines *AK* and *AB* and thus can be constructed by Euclidean means. (It should be noted that even in ancient times this construction was criticized, because the actual position of the terminal point *K* is not determined by the definition of the curve. It can only be approximated.)

3.3 ARCHIMEDES AND GEOMETRY

What distinguishes Archimedes' work in geometry from that of Euclid is that Archimedes often presents his method of discovery of the theorem and/or his analysis of the situation before presenting a rigorous synthetic proof. The methods of discovery of several of his results are collected in a treatise called *The Method*, which was unexpectedly discovered in 1899 in Jerusalem. The manuscript dates from the tenth century, but the writing had been partially washed out in the thirteenth century and the parchment reused for a religious work. (Parchment was a very valuable commodity in the middle ages.) Fortunately, the old writing is in large part still readable. Heiberg inspected it in Constantinople in 1906 and soon after published the Greek text.

3.3.1 Archimedes' Method of Discovery

The Method contains Archimedes' method of discovery by mechanics of many important results on areas and volumes, most of which are rigorously proved elsewhere. The essential feature of the method is the balancing of cross sections of a given figure against corresponding cross sections of a known figure, using the law of the lever. Archimedes knew that he could not give a rigorous proof by this means, since there was no way he could show that a given figure was composed of its various cross sections, that is, that a plane figure was composed of lines or a solid of plane slices. Therefore, he noted in his preface, the theorems "had to be demonstrated by geometry afterwards because their investigation by the said method did not furnish an actual demonstration." He also noted that "it is of course easier, when we have previously acquired, by the method, some knowledge of the questions, to supply the proof than it is to find it without any previous knowledge."[11]

The first proposition of *The Method*, that a segment *ABC* of a parabola is 4/3 of the triangle *ABC*, is presented here as a typical example of that work. Given the parabolic segment *ABC* with vertex *B*, draw a tangent at *C* meeting the axis produced at *E* and a line through *A* parallel to the axis meeting the tangent line at *F* (Figure 3.7). Produce *CB* to meet *AF* in *K* and extend it to *H* so that *CK* = *KH*. Archimedes now considers *CH* as a lever with midpoint *K*. The idea of his demonstration is to show that triangle *CFA* placed where it is in the figure balances the segment *ABC* placed at *H*. He does this, line by line, by beginning with an arbitrary line segment *MO* of triangle *CFA* parallel to *ED* and showing that it balances the line *PO* of segment *ABC* placed at *H*. To show the balancing, two properties of the parabola are needed, first that *EB* = *BD*, and second that *MO* : *PO* = *CA* : *AO*. (It is evident that Archimedes was quite familiar with the elementary properties of parabolas.) From *EB* = *BD* it follows that *FK* = *KA* and *MN* = *NO* and from the proportion and the fact that *CK* bisects *AF*, it follows from *Elements* VI-2

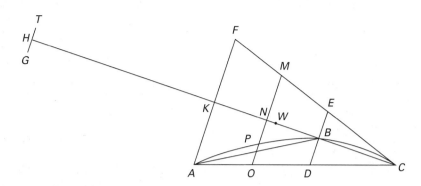

FIGURE 3.7
Balancing a parabolic segment, from Archimedes' *Method.*

that $MO : PO = CK : KN = HK : KN$. If a line TG equal to PO is placed with its center at H, this latter proportion becomes $MO : TG = HK : KN$. Therefore, by the law of the lever, MO and TG will be in equilibrium about K.

Archimedes continues, "since the triangle CFA is made up of all the parallel lines like MO, and the segment CBA is made up of all the straight lines like PO within the curve, it follows that the triangle, placed where it is in the figure, is in equilibrium about K with the segment CBA placed with its center of gravity at H."[12] Because nothing is changed by considering the triangle as located at its center of gravity, the point W on CK two-thirds of the way from C to K, Archimedes derives the proportion $\triangle ACF$: segment $ABC = HK : KW = 3 : 1$. Therefore, segment $ABC = (1/3)\triangle ACF$. But $\triangle ACF = 4\triangle ABC$. Hence segment $ABC = (4/3)\triangle ABC$ as asserted. Archimedes concludes this demonstration with a warning: "Now the fact here stated is not actually demonstrated by the argument used; but that argument has given a sort of indication that the conclusion is true. Seeing then that the theorem is not demonstrated, but at the same time suspecting that the conclusion is true, we shall have recourse to the geometrical demonstration which I myself discovered and have already published."[13]

3.3.2 Sums of Series

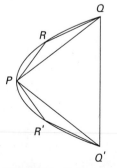

FIGURE 3.8
Area of a parabolic segment by summation of a geometric series.

The geometrical proof which Archimedes did consider valid occurs in his treatise *Quadrature of the Parabola* and is based on Eudoxus' method of exhaustion. The idea as before is to construct rectilinear figures inside the parabolic segment whose total area differs from that of the segment by less than any given value. The figures Archimedes used for this purpose are triangles. Thus in each of the two parabolic segments PRQ, $PR'Q'$ left by the original triangle he constructed a triangle PRQ, $PR'Q'$; in each of the four segments left by these triangles, he constructed new triangles, and so on (Figure 3.8). (The vertex of a segment, and of the corresponding triangle, is that point whose perpendicular distance to the chord of the segment is greatest. Archimedes showed that it is also the intersection point with the curve of the line through the midpoint of the chord parallel to the axis of the parabola.)

Archimedes next calculated that the total area of the triangles constructed at each stage is 1/4 of the area of the triangles constructed in the previous stage. The more steps taken, the more closely the sum of the areas approaches the area of the parabolic segment.

Therefore, to complete the proof, Archimedes in effect needed to find the sum of the geometric series $a + \frac{1}{4}a + \left(\frac{1}{4}\right)^2 a + \cdots + \left(\frac{1}{4}\right)^n a + \cdots$, where a is the area of the original triangle. Archimedes gave that sum in the form

$$a + \frac{1}{4}a + \left(\frac{1}{4}\right)^2 a + \cdots + \left(\frac{1}{4}\right)^n a + \frac{1}{3}\left(\frac{1}{4}\right)^n a = \frac{4}{3}a,$$

and completed the argument through a double *reductio ad absurdum*, just as Euclid did in similar situations. He assumed that $K = (4/3)\triangle PQR$ is not equal to the area B of the segment. If K is less than this area, then triangles can be inscribed as above so that $B - T < B - K$, where T is the total area of the inscribed triangles. But then $T > K$. This is impossible since the summation formula shows that $T < (4/3)\triangle PQR = K$. On the other hand, if $K > B$, n is determined so that $\left(\frac{1}{4}\right)^n a < K - B$. Since also $K - T = \frac{1}{3}\left(\frac{1}{4}\right)^n a < \left(\frac{1}{4}\right)^n a$, it follows that $B < T$, which is again impossible. Hence, $K = B$.

The important lemma to this proof shows how to find the sum of a geometric series. Archimedes' demonstration of this result was given for a series of five numbers, because he had no notation to express a series with arbitarily many numbers. But since his method generalizes easily, we will use modern notation with n denoting an arbitrary positive integer. Archimedes began by noting that $\left(\frac{1}{4}\right)^n a + \frac{1}{3}\left(\frac{1}{4}\right)^n a = \frac{1}{3}\left(\frac{1}{4}\right)^{n-1} a$. Then he calculated:

$$a + \frac{1}{4}a + \left(\frac{1}{4}\right)^2 a + \cdots + \left(\frac{1}{4}\right)^n a + \frac{1}{3}\left[\frac{1}{4}a + \left(\frac{1}{4}\right)^2 a + \cdots + \left(\frac{1}{4}\right)^n a\right]$$

$$= a + \left(\frac{1}{4}a + \frac{1}{3}\cdot\frac{1}{4}a\right) + \left[\left(\frac{1}{4}\right)^2 a + \frac{1}{3}\left(\frac{1}{4}\right)^2 a\right] + \cdots + \left[\left(\frac{1}{4}\right)^n a + \frac{1}{3}\left(\frac{1}{4}\right)^n a\right]$$

$$= a + \frac{1}{3}a + \frac{1}{3}\cdot\frac{1}{4}a + \cdots + \frac{1}{3}\left(\frac{1}{4}\right)^{n-1} a$$

$$= a + \frac{1}{3}a + \frac{1}{3}\left[\frac{1}{4}a + \cdots + \left(\frac{1}{4}\right)^{n-1} a\right].$$

Subtracting equals and rearranging gives the desired result:

$$a + \frac{1}{4}a + \left(\frac{1}{4}\right)^2 a + \cdots + \left(\frac{1}{4}\right)^n a + \frac{1}{3}\left(\frac{1}{4}\right)^n a = \frac{4}{3}a.$$

Another formula for a sum led to another area result in *On Spirals*, a result again proved by Eudoxian methods. In proposition 10 of that book Archimedes demonstrated a formula for determining the sum of the first n integral squares:

$$(n + 1)n^2 + (1 + 2 + \cdots + n) = 3(1^2 + 2^2 + \cdots + n^2),$$

as a corollary to which he showed that

$$3(1^2 + 2^2 + \cdots + (n - 1)^2) < n^3 < 3(1^2 + 2^2 + \cdots + n^2).$$

Archimedes needed the last inequality to determine the area bounded by one turn of the "Archimedean spiral," the curve given in modern polar coordinates by the equation $r = a\theta$. In proposition 24 of *On Spirals* he demonstrated that the area R bounded by one complete circuit of that curve and the radius line AL to its endpoint equals one third of the

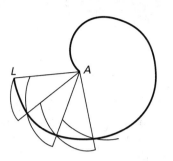

FIGURE 3.9
Area of the Archimedean spiral.

area C of the circle with that line as radius. Archimedes first noted that one can inscribe and circumscribe figures about the region R whose areas differ by less than any assigned area ϵ (Figure 3.9). By continued bisection (according to *Elements* X-1), one can determine an integer n such that the circular sector with radius AL and angle $(360/n)°$ has area less than ϵ. Then, inscribing a circular arc in and circumscribing a circular arc about the part of the spiral included in each of the n sectors with this angle, one notes that the difference between the complete circumscribed figure and the complete inscribed figure is equal to the area of the sector chosen initially and thus is less than ϵ.

The proof of the area result by a double *reductio* argument is now straightforward. For suppose that $R \neq \frac{1}{3}C$. Then either $R < \frac{1}{3}C$ or $R > \frac{1}{3}C$. In the first case, circumscribe a figure F about R as described above so that $F - R < \frac{1}{3}C - R$. Therefore $F < \frac{1}{3}C$. From the defining equation of the curve, it follows that the radii of the sectors making up F are in arithmetic progression, which can be considered as $1, 2, \ldots, n$. Since $n \cdot n^2 < 3(1^2 + 2^2 + \cdots + n^2)$ and since the areas of the sectors (and the circle itself) are proportional to the squares of their radii, it follows that $C < 3F$ or $\frac{1}{3}C < F$, a contradiction. A similar argument using an inscribed figure shows that $R > \frac{1}{3}C$ also leads to a contradiction, and the proposition is proved.

3.3.3 Analysis

Our final examples of Archimedes' work show again his concern that his readers learn not only the solution to a geometric problem but also how the solution was found. In this case, proposition 3 of *On the Sphere and Cylinder II*, he provides his procedure in the context of a formal proof.

Problem. *To cut a given sphere by a plane so that the surfaces of the segments may have to one another a given ratio.*

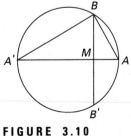

FIGURE 3.10
On the Sphere and Cylinder II,
Proposition 3.

Archimedes' procedure, the method of analysis, is to assume the problem solved and then deduce consequences until he reaches a result already known. Thus, he assumes that the plane BB' cuts the sphere so that the surface of BAB' is to the surface of $BA'B'$ as H is to K (Figure 3.10). He had already shown in *On the Sphere and Cylinder I* that the areas of such segments equal the area of the circles on the radii AB, $A'B$. Hence, he concludes that $AB^2 : A'B^2 = H : K$ and therefore that $AM : A'M = H : K$ (since the areas of the triangles are as the bases). But the dividing of a line segment in a given ratio is a known

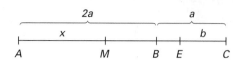

FIGURE 3.11

On the Sphere and Cylinder II, Proposition 4.

procedure. Archimedes can therefore solve the original problem by beginning with that step and proceeding in reverse. Namely, he takes M so that $AM : MA' = H : K$. The same theorem already quoted then shows that $AM : MA' = AB^2 : A'B^2 =$ (circle with radius AB) : (circle with radius $A'B$) = (surface of segment BAB') : (surface of segment $BA'B'$). The problem is solved.

Archimedes presented the analysis of a more complex problem in proposition 4 of the same book, where he proposed to cut a given sphere by a plane so that the volumes of the segments are in a given ratio. In this case, his analysis reduced the problem to the following: Given a straight line ABC with $AB = 2BC$ and given a point E on BC, to cut AB in a point M such that $AB^2 : AM^2 = MC : EC$ (Figure 3.11). If one sets $AB = 2a$, $BC = a$, $EC = b$, and $AM = x$, the problem can be translated algebraically into $(2a)^2 : x^2 = (3a - x) : b$, or $3ax^2 - x^3 = 4a^2b$. Hence, Archimedes needed to solve a cubic equation. He proceeded to do so by finding the desired point M as the intersection of a parabola and a hyperbola.

Archimedes' mathematical genius was far-reaching. Only a few items from some of the fourteen extant treatises have been discussed here. Among other results, Archimedes proved that the volume of a sphere is four times that of the cone with base equal to a great circle of the sphere and height equal to its radius, that the volume of a segment of a paraboloid of revolution is 3/2 that of the cone with the same base and axis, and that the surface of a sphere is four times the greatest circle in it.

Archimedes was buried near one of the gates of Syracuse. He had requested that his tomb include a cylinder circumscribing a sphere, together with an inscription of what he evidently thought one of his most important theorems, that a cylinder whose base is a great circle in the sphere with height equal to the diameter is 3/2 of the sphere in volume and also has surface area 3/2 of the surface area of the sphere (Figure 3.12). The tomb was found neglected by Cicero when he served as an official in Sicily about 75 B.C.E. and was restored. Unfortunately, however, it no longer exists.

FIGURE 3.12
Cylinder with base equal to a great circle on a sphere on stamp of Archimedes from San Marino.

3.4 CONIC SECTIONS BEFORE APOLLONIUS

The exact origins of the theory of conic sections are somewhat hazy, but they may well be connected to the problem of doubling the cube. Recall that Hippocrates in the fifth century B.C.E. reduced the problem of constructing a cube double the volume of a given cube of side a to the finding of two mean proportionals x, y between the lengths a and $2a$, that is, of determining x, y such that $a : x = x : y = y : 2a$. In modern terms, this is equivalent to solving simultaneously any two of the three equations $x^2 = ay$, $y^2 = 2ax$, and $xy = 2a^2$, equations which represent parabolas in the first two instances and a hyperbola in the third.

It was Menaechmus (fourth century B.C.E.) who first constructed curves which satisfy these algebraic properties and thus showed that the point of intersection of these curves

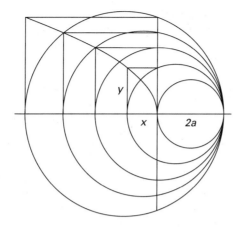

FIGURE 3.13
Euclidean pointwise construction
of a parabola.

would give the desired two means and solve the problem of doubling the cube. It is not
known how he constructed these curves, but a pointwise construction was certainly possible using Euclidean methods. To construct the points of a curve satisfying $y^2 = 2ax$, one
would merely have to apply repeatedly the method of *Elements* VI-13 (Figure 3.13). First,
put segments of length $2a$ and x together into a single line. Then, draw a semicircle having
that line as diameter, and erect a perpendicular at the join of the two segments. This
perpendicular has length y satisfying the equation. If this is done for various lengths x and
the endpoints of the perpendiculars are connected, the desired curve is drawn.[14] We note
that although each point of this curve has been constructed using Euclidean tools, the
completed curve is not a proper construction in Euclid's sense. In any case, it does appear
that the conic sections were introduced as tools for the solution of certain geometric
problems. We have already seen that Archimedes used them in solving a problem on
spheres.

There can only be speculation as to how the Greeks realized that curves useful in
solving the cube doubling problem could be generated as sections of a cone. Someone,
perhaps Menaechmus himself, may have noticed that the circle diagram above could be
thought of as a diagram of level curves of a certain cone. Hence, the curve could be
generated by a section of such a cone. Another possibility is that these curves appeared as
the path of the moving shadow of the gnomon on a sundial as the sun traveled through its
circular daily path, which in turn was one base of a double cone whose vertex was the tip
of the gnomon. In this suggestion, the plane in which the shadow falls would be the cutting
plane. It might further have been noted that the apparent shape of a circle viewed from a
point outside its plane was an ellipse, and this shape comes from a plane cutting the cone
of vision. In any case, by the end of the fourth century there were in existence two
extensive treatises on the properties of the curves obtained as sections of cones, one by
Aristaeus (fourth century B.C.E.) and one by Euclid. Although neither is still available, a
good deal about their contents can be inferred from Archimedes' extensive references to
basic theorems on conic sections.

Recall that Euclid (in Book XI of the *Elements*) defined a cone as a solid generated by
rotating a right triangle about one of its legs. He then classified the cones in terms of their

vertex angles as right-angled, acute-angled, or obtuse-angled. A section of such a cone can be formed by cutting the cone by a plane at right angles to the generating line, the hypotenuse of the given right triangle. The "section of a right-angled cone" is today called a parabola, the "section of an acute-angled cone" an ellipse, and the "section of an obtuse-angled cone" a hyperbola. The names in quotation marks are those generally used by Archimedes and his predecessors. From these definitions, the Greek mathematicians derived the "symptom" of the curve, the characteristic relation between the ordinate and abscissa of an arbitrary point on the curve that can easily be translated into an algebraic equation.

The three types of cones will be considered together. In each case let the plane of the paper represent a plane through the generator. The angle at the vertex V is therefore a right angle (Figure 3.14), an obtuse angle (Figure 3.15), or an acute angle (Figure 3.16). The cutting plane, perpendicular to the generator, is indicated by the line AB. The desired curve lies in this plane. For simplicity, AB will also represent the axis of the curve. Pick an arbitrary point K on the curve and drop a perpendicular to the axis, meeting it at D, and then cut the cone by another plane at right angles to the axis VW of the cone which also passes through K. This plane will cut a circular section of the cone, the ends of whose diameter are designated by M and N. Therefore in each case $KD^2 = MD \cdot ND$. The aim is to find a relation between the ordinate KD of the point on the curve and the abscissa AD.

In the case of the right-angled cone, draw AP parallel to NM and let Z be its intersection with the axis of the cone. The similarity of $\triangle NDA$ to $\triangle WAZ$ implies that $ND : AD = AW : AZ$. Also, since $MDAP$ is a parallelogram, $MD = 2AZ$. It follows that

$$KD^2 = 2AZ\frac{AD \cdot AW}{AZ} = 2AW \cdot AD.$$

If $KD = y$, $AD = x$, and $2AW = p$, the equation becomes $y^2 = px$, the standard equation of the parabola. In fact, Archimedes calls the parameter p "the double of the distance to the axis." In modern terminology, it is the length of the **latus rectum** of the parabola, the chord through the focus perpendicular to the axis.

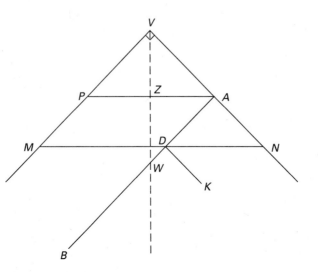

FIGURE 3.14
Section of a right-angled cone.

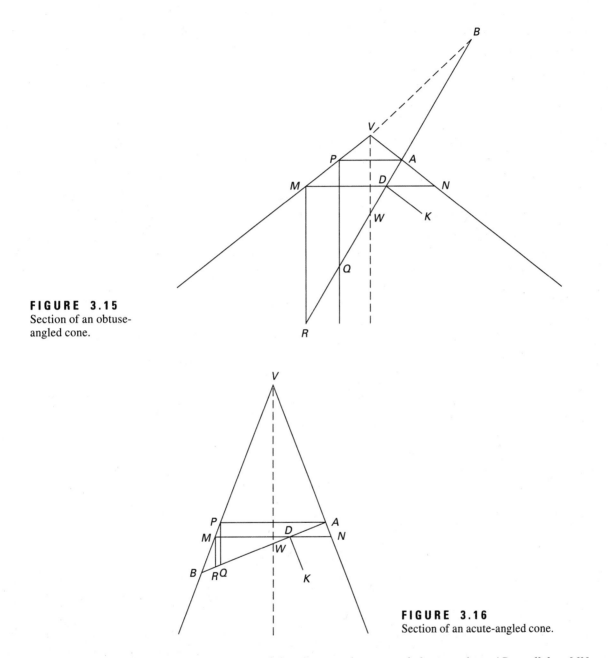

FIGURE 3.15
Section of an obtuse-
angled cone.

FIGURE 3.16
Section of an acute-angled cone.

In the case of the obtuse and acute angled cones, draw AP parallel to MN, and PQ, MR parallel to the axis of the cone meeting AB (or AB extended) in Q, R respectively. The similarity of $\triangle AND$ to $\triangle MRD$ implies that $ND : RD = AD : MD$. Also, since $\triangle RMD$ is similar to $\triangle QPA$ and $\triangle MDB$ is similar to $\triangle PAB$, the proportions $RD : QA = MD : PA = BD : BA$ hold. Therefore, $RD : BD = QA : BA$. It follows that

$$KD^2 = MD \cdot ND = AD \cdot RD = AD\frac{BD \cdot QA}{AB} = AD \cdot BD\frac{2AW}{AB}.$$

If $KD = y$, $AD = x_1$, $BD = x_2$, $2AW = p$, and $AB = 2a$, the equations of both the hyperbola and ellipse turn out to be

$$y^2 = \frac{p}{2a}x_1 x_2.$$

This latter equation may not be familiar, but it is essentially the form used by Archimedes. Note also that at this time the hyperbola was only considered to have one branch. The second abscissa x_2 was thus measured from a point outside the curve.[15]

At some time before 200 B.C.E., it was probably noticed that it would be more useful to state the symptoms of the hyperbola and ellipse somewhat differently. Namely, since $BD = AB \pm AD$ (in the hyperbola and ellipse respectively), the equation can be rewritten as

$$KD^2 = AD(2AW \pm \frac{2AW}{AB}AD).$$

So if AD is now called x, the equation becomes

$$y^2 = x(p + \frac{p}{2a}x)$$

for the hyperbola and

$$y^2 = x(p - \frac{p}{2a}x)$$

for the ellipse. These equations can easily be transformed into the standard equations used today.

3.5 THE *CONICS* OF APOLLONIUS

Apollonius, in his *Conics*, took this latest form of the definitions of the curves. He noted further that it was not necessary to restrict oneself to a cutting plane perpendicular to a generator, nor even to a right circular cone, to determine the curves. In fact, he generalized the notion of a cone as follows:

> If from a point a straight line is joined to the circumference of a circle which is not in the same plane as the point, and the line extended in both directions, and if, with the point remaining fixed, the straight line is rotated about the circumference of the circle . . . , then the generated surface composed of the two surfaces lying vertically opposite one another . . . [is] a **conic surface**. The fixed point [is] the **vertex** and the straight line drawn from the vertex to the center of the circle [is] the **axis**. . . . The circle [is] the **base** of the cone.[16]

For Apollonius, a conic surface is what is today called a double oblique cone. In general, its axis is not perpendicular to the base circle.

To define the three curves, Apollonius first cuts the cone by a plane through the axis. The intersection of this plane with the base circle is a diameter CD of that circle. The resulting triangle VCD is called the axial triangle. The parabola, ellipse, and hyperbola are

---

***Biography*   Apollonius (250–175 B.C.E.)**

Apollonius was born in Perge, a town in southern Asia Minor, but few details are known about his life. Most of the reliable information comes from the prefaces to the various books of his magnum opus, the *Conics* (Figure 3.17). These indicate that he went to Alexandria as a youth to study with successors of Euclid and probably remained there for most of his life, studying, teaching, and writing. He became famous in ancient times first for his work on astronomy, but later for his mathematical work, most of which is known today only by titles and summaries in works of later authors. Fortunately, seven of the eight books of the *Conics* do survive, and these represent in some sense the culmination of Greek mathematics. It is difficult for us today to comprehend how Apollonius could discover and prove the hundreds of beautiful and difficult theorems without modern algebraic symbolism. Nevertheless, he did so, and there is no record of any later Greek mathematical work which approaches the complexity or intricacy of the *Conics*.

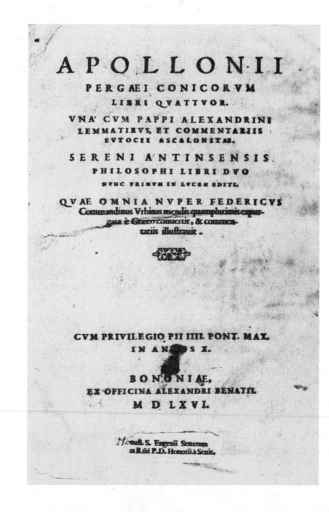

FIGURE 3.17
Title page of the first Latin printed edition of Apollonius' *Conics*, 1566. (Source: Smithsonian Institution, Photo No. 86-4346)

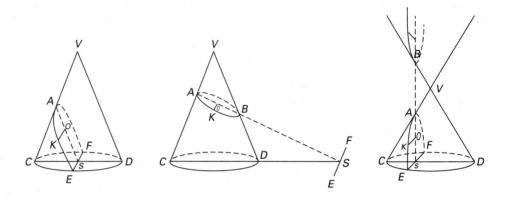

FIGURE 3.18
Conic sections of
Apollonius.

then defined as sections of this cone by certain planes which cut the plane of the base circle
in the straight line *EF* perpendicular to *CD* or *CD* produced (Figure 3.18). The straight
line *AS* is the intersection of the cutting plane with the axial triangle. If *AS* is parallel to
one side of the axial triangle, the section is a parabola. If *AS* intersects both sides of the
axial triangle, the section is an ellipse. Finally, if *AS* intersects one side of the axial triangle
and the other side produced beyond *V*, the section is a hyperbola. In this situation, there
are two branches of the curve, unlike in the earlier obtuse-angled cone.

For each case, Apollonius derives the symptoms of the curve in ways similar to those
already described. To express the symptoms as Apollonius did, begin as before by picking
an arbitrary point *K* on the section and passing a plane through *K* parallel to the base
circle. Let *O* be the intersection of the plane with diameter *AS*. For the case where *AS* is
parallel to a side of the axial triangle, Apollonius shows (proposition I-11) that the square
on *KO* is equal to the rectangle on *AO* and a line *N* where *N* depends only on the curve and
the cutting plane. Since the length of *N* may be considered as a constant *p*, if $KO = y$ and
$AO = x$, the same equation $y^2 = px$ results. The name "parabola" comes from the Greek
word *paraboli* (applied), since the square on the ordinate *y* is equal to the rectangle *applied*
to the abscissa *x*.

In the other two cases, let *B* be the intersection of *AS* with the second side of the axial
triangle (ellipse) or with the second side produced (hyperbola). Apollonius proves in these
cases (propositions I-12, I-13) that the square on *KO* is equal to a rectangle applied to a
line *N* with width equal to *AO* and exceeding (*yperboli*) or deficient (*ellipsis*) by a rectangle
similar to the one contained by *AB* and *N*, thus indicating the reason for the curves' names.
As before, *N* does not depend on the point *K*, so it will be denoted by *p*. If *KO* is denoted
by *y*, *AO* by *x*, and *AB* by *2a*, Apollonius' symptoms in these cases can be translated into
the equations $y^2 = px \pm \frac{p}{2a}x^2$, exactly the same as earlier.

The symptom of a hyperbola can be expressed in terms of its asymptotes as well as in
terms of its parameter and axis. Thus in Book II Apollonius deals with the idea of asymptotes. These are constructed in proposition II-1 (Figure 3.19). Drawing a tangent to the
vertex *A* of the hyperbola and laying off on this tangent two segments *AL*, *AL'* (in opposite
directions from the vertex) such that $AL^2 = AL'^2 = \frac{pa}{2}(= b^2)$, Apollonius shows that the

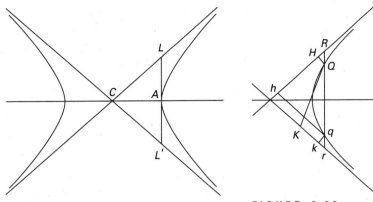

FIGURE 3.19
Constructing asymptotes to a hyperbola.

FIGURE 3.20
The symptom of a hyperbola using asymptotes.

lines CL, CL' drawn to L, L' from the center of the hyperbola do not meet either branch of the curve. (The word *asymptotos* in Greek means "not capable of meeting.") Furthermore, in proposition II-14 Apollonius shows that the distance between the curve and these asymptotes, if both are extended indefinitely, becomes less than any given distance.

In proposition II-4, Apollonius shows how to construct a hyperbola given a point on the hyperbola and its asymptotes. In II-8, he then establishes the fact that segments cut off by a secant of a hyperbola between the hyperbola and the two asymptotes are equal. And in II-12 he expresses the new symptom. First, take Q, q on the hyperbola such that Qq is perpendicular to the diameter (Figure 3.20). If R, r are the intersection points of Qq with the two asymptotes, and if we write $Q = (x, y)$, then, since $b^2 = \frac{pa}{2}$,

$$QR \cdot Qr = \left(\frac{b}{a}(x + a) - y\right)\left(\frac{b}{a}(x + a) + y\right) = \frac{b^2}{a^2}(x + a)^2 - y^2$$

$$= \frac{b^2 x^2}{a^2} + \frac{2b^2 ax}{a^2} + b^2 - px - \frac{p}{2a}x^2$$

$$= \left(\frac{b^2}{a^2} - \frac{p}{2a}\right)x^2 + \left(\frac{2b^2}{a} - p\right)x + b^2 = b^2.$$

Apollonius then proves that $QR \cdot Qr = b^2$ for any points Q, q on the hyperbola and also that $Rq \cdot qr = b^2$. It follows that $RQ : Rq = qr : Qr$. If one draws from Q, q a pair of parallel lines to each of the asymptotes, intersecting one at H, h respectively and the second at K, k, one sees that $RQ : Rq = HQ : hq$ and $qr : Qr = qk : QK$. It is then immediate that $HQ : hq = qk : QK$ or that $HQ \cdot QK = hq \cdot qk$. In other words, the product of the lengths of the two lines drawn from any point of the hyperbola in given directions to the asymptotes is a constant. In modern notation, this result shows that a hyperbola can be defined by the equation $xy = d$.

3.5.1 Tangents and Normals

In deriving the properties of the conics, Apollonius generally uses the symptoms of the curves, rather than the original definition, just as in modern practice these properties are derived from the equation. Although Apollonius always uses geometric language, much of his work can be characterized as geometric algebra. Namely, the symptom of a curve can be thought of as an algebraic characterization of Apollonius' geometric derivation. Therefore, in our brief survey of some of the theorems of the *Conics* familiar to modern readers, algebra will be used to simplify some of the statements and proofs.

The problem of drawing tangents to the conic sections is discussed in Book I.

Proposition I-33.   *Let C be a point on the parabola CET with CD perpendicular to the diameter EB. If the diameter is extended to A with AE = ED, then line AC will be tangent to the parabola at C (Figure 3.21).*

Set $DC = y$, $DE = x$, and $AE = t$. The theorem says that if $t = x$, then line AC is tangent to the curve at C. In other words, the tangent can be found by simply extending the diameter past E a distance equal to x and connecting the point so determined with C. For Apollonius, as for Euclid, a tangent line is a line which touches but does not cut the curve. So to prove the result, Apollonius uses a *reductio* argument and assumes that the line through A and C does cut the curve again, say at K. Then the line segment from C to K lies within the parabola. Pick F on that segment and drop a perpendicular from F to the axis, meeting the axis at B and the curve at G. Then $BG^2 : CD^2 > BF^2 : CD^2 = AB^2 : AD^2$. Also, since G and C lie on the curve, the symptom shows that $BG^2 = p \cdot EB$ and $CD^2 = p \cdot ED$, so $BG^2 : CD^2 = BE : DE$. Therefore, $BE : DE > AB^2 : AD^2$. So also $4BE \cdot EA : 4DE \cdot EA > AB^2 : AD^2$, and therefore $4BE \cdot EA : AB^2 > 4DE \cdot EA : AD^2$. But since $AE = DE$, we have $AD^2 = 4DE \cdot EA$, and since $BE > AE$, also $4BE \cdot EA < AB^2$, a contradiction.

In the next proposition, Apollonius shows how to construct a tangent to an ellipse or hyperbola. The proof is similar to the previous one.

Proposition I-34.   *Let C be a point on an ellipse or hyperbola, CB the perpendicular from that point to the diameter. Let G and H be the intersections of the diameter with the curve and choose A on the diameter or the diameter extended so that AH : AG = BH : BG. Then AC will be tangent to the curve at C.*

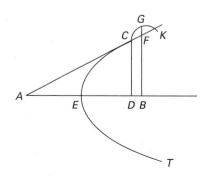

FIGURE 3.21
Conics, Proposition I-33.

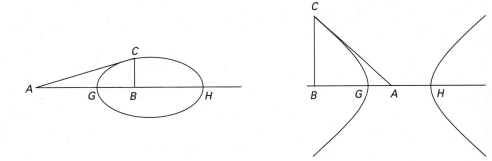

FIGURE 3.22
Conics, Proposition I-34.

This result can be stated algebraically by letting $AG = t$ and $BG = x$. In the case of the ellipse, $BH = 2a - x$ and $AH = 2a + t$, while in the case of the hyperbola, $BH = 2a + x$ and $AH = 2a - t$ (Figure 3.22). Therefore, for the ellipse $(2a + t)/t = (2a - x)/x$ and for the hyperbola $(2a - t)/t = (2a + x)/x$. Solving these for t gives $t = ax/(a - x)$ for the ellipse and $t = ax/(a + x)$ for the hyperbola. The tangent line can now be constructed. Apollonius completes his treatment of the tangents by proving the converses of these results as propositions I-35 and I-36. Since the symptoms of the ellipse and hyperbola are virtually identical, Apollonius often proves analogous properties about the two curves, even stating them in the same proposition. Parabolas, on the other hand, are generally treated separately.

Apollonius considers normals to the conics in Book V. Only the parabola is considered here.

Propositions V-8, V-13, V-27.   *In a parabola with vertex A and symptom $y^2 = px$, let G be a point on the axis such that $AG > p/2$. Let N be taken between A and G so that $NG = p/2$. Then, if NP is drawn perpendicular to the axis meeting the curve in P, PG is the minimum straight line from G to the curve. Conversely, if PG is the minimum straight line from G to the curve, and GN is drawn perpendicular to the axis, $NG = p/2$. Finally, PG is perpendicular to the tangent at P (Figure 3.23).*[17]

For the proof, suppose P' is another point on the parabola with abscissa AN'. By the defining property of the parabola, we have $P'N'^2 = p \cdot AN' = 2NG \cdot AN'$. Also $N'G^2 = NN'^2 + NG^2 \pm 2NG \cdot NN'$ (with the sign depending on the position of N'). Adding these two equations together and using the Pythagorean theorem gives $P'G^2 = 2NG \cdot AN +$

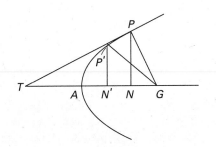

FIGURE 3.23
Conics, Propositions V-8, V-13, and V-27.

$NN'^2 + NG^2 = PN^2 + NG^2 + NN'^2 = PG^2 + NN'^2$. Thus PG is the minimum straight line from G to the curve. The converse is proved by a *reductio* argument. Finally, to show that PG is perpendicular to the tangent TP, note that $AT = AN$. Therefore $NG : p = 1/2 = AN : NT$, so $TN \cdot NG = p \cdot AN = PN^2$. Since the angle at N is a right angle, so is the angle TPG as desired.

3.5.2 Foci

In Book III Apollonius deals with the focal properties of the ellipse and hyperbola. In proposition III-45, for example, he defines the foci of an ellipse as the points F, G on the axis AB such that the rectangle on AF, FB equals one-fourth of the rectangle on the parameter N and the axis AB, and similarly for the rectangle AG, GB. (Apollonius only calls these the "points arising from the application" of the rectangle to the axis. The term "foci" was first used by Johannes Kepler in 1604.) In algebraic terms, if the distances from F and G to the center O are equal to c, Apollonius' condition may be translated as the equation

$$(a - c)(a + c) = \frac{1}{4} \cdot 2ap \qquad \text{or} \qquad a^2 - c^2 = \frac{pa}{2}.$$

Given this definition, Apollonius then presents a series of propositions which culminate in the well-known result that the lines from the two foci to any point on the ellipse make equal angles with the tangent to the ellipse at that point.

Although Apollonius presents a similar result for the hyperbola, he does not deal with the focal properties of the parabola, perhaps because he had discussed these in an earlier work now lost. In any case, the analogous property for a parabola, that any line from the focus to a point on the parabola makes an angle with the tangent at that point equal to the one made by a line parallel to the axis, was probably first proved by Diocles (early second century B.C.E.), a contemporary of Apollonius, in a treatise *On Burning Mirrors*, perhaps written a few years before the *Conics*. It is in fact that property of the parabola which gives this treatise its name. The problem is to find a mirror surface such that when it is placed facing the sun, the rays reflected from it meet at a point and thus cause burning. Diocles shows that this would be true for a paraboloid of revolution. There are stories told about Archimedes and others that such a mirror was used to set enemy ships on fire. However, there is no reliable evidence for the veracity of these stories.

To complete this topic of foci, then, we consider Diocles' proof of the focal property of the parabola from proposition 1 of his treatise. Given a parabola LBM with axis BW, lay off BE along the axis equal to half the parameter and bisect BE at D (Figure 3.24). It is this point D, whose distance from the vertex is $p/4$, which is today called the focus. Pick an arbitrary point K on the parabola, draw a tangent line AKC through K meeting the axis extended at A, draw KS parallel to the axis, and connect DK. The proposition then asserts that $\angle AKD = \angle SKC$.

To prove this, first drop a perpendicular from K to G. By *Conics* I-33, $AB = BG$. Next draw a line perpendicular to AK from K which meets the axis at Z. So $GZ = p/2$. It follows that $GZ = BE$, and therefore $GB = EZ$, $AB = EZ$, and finally $AD = DZ$. Because triangle AKZ is a right triangle whose hypotenuse is bisected at D, we have $AD = DK = DZ$. Therefore, $\angle DZK = \angle DKZ$. Since KS is parallel to AZ, it also follows that $\angle ZKS =$

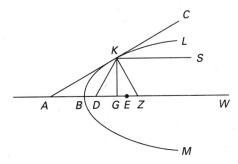

FIGURE 3.24
The focus of a parabola, from Diocles'
On Burning Mirrors.

∠*DKZ*. Subtracting these equal angles from the right angles *ZKC* and *ZKA*, we obtain the desired result. Diocles concludes the proposition by showing how to construct the burning mirror by rotating *LBM* about the axis *AZ* and covering the resulting surface with brass.

In propositions 4 and 5 of his brief treatise, Diocles shows how to construct a parabola with given focal length. His construction in effect uses the focus-directrix property of a parabola, that the points of the parabola are equally distant from the focus and a given straight line called the **directrix**. There is no earlier reference in antiquity to this particular property of the parabola, although it is discussed by the fourth century commentator Pappus. In fact, Pappus also notes that an ellipse is determined as the locus of a point moving so that the ratio of its distances from a fixed point (the focus) and a fixed line (the directrix) is a constant less than 1, while a hyperbola is described if this constant ratio is greater than 1. These latter properties were probably also discovered around the time of Diocles and Apollonius.

In the *Conics*, however, there is only the two-focus property of the ellipse and hyperbola. Proposition III-51 states that in a hyperbola, if one connects an arbitrary point to each focus, "the greater of the two straight lines exceeds the less by exactly as much as the axis." Proposition III-52 shows that in an ellipse, the sum of these two straight lines equals the axis. In other words, if *P* is a point on the curve and *D, E* are the two foci, then $PD - PE = 2a$ for the hyperbola and $PD + PE = 2a$ for the ellipse. This property is, in fact, the standard defining property for these two curves in current textbooks.

3.5.3 Problem Solving Using Conic Sections

Apollonius' aim in the *Conics* was not so much to develop the properties of the conic sections for their own intrinsic beauty, but to develop the theorems necessary for the application of these curves to the solution of geometric problems. We will conclude this chapter with three examples of how the conics were used in Greek times.

We first consider the angle trisection problem. Let angle *ABC* be the angle to be trisected (Figure 3.25). Draw *AC* perpendicular to *BC* and complete the rectangle *ADBC*. Extend *DA* to the point *E* which has the property that if *BE* meets *AC* in *F*, then the segment *FE* is equal to twice *AB*. It then follows that ∠*FBC* = (1/3)∠*ABC*. For if *FE* is bisected at *G*, then *FG* = *GE* = *AG* = *AB*. Therefore ∠*ABG* = ∠*AGB* = 2∠*AEG* = 2∠*FBC* and the trisection is demonstrated. To complete the proof, however, it is necessary to show how to construct *BE* satisfying the given condition. Again an analysis will help.

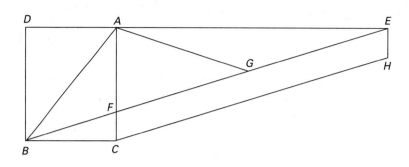

Assuming $FE = 2AB$, draw CH and EH parallel to FE and AC respectively. It follows that H lies on the circle of center C and radius FE ($= 2AB$). Moreover, since $DE : DB = BC : CF$, or $DE : AC = DA : EH$, we have $DA \cdot AC = DE \cdot EH$, so H also lies on the hyperbola with asymptotes DB, DE and passing through C. Therefore if one constructs the hyperbola and the circle and drops a perpendicular from the intersection point H to DA extended, the foot E of the perpendicular is the point needed to complete the solution.

Cube duplication constructions virtually all begin with Hippocrates' reduction of the problem to the construction of two mean proportionals between given lines AB and AC. One of them from the time of Apollonius begins with the completion of the rectangle on these two lines, the drawing of the diagonal AD, and the construction of the circle with diameter AD passing through B (Figure 3.26). Now, let F be the intersection of the circle with a hyperbola through D with asymptotes AB and AC. Extend line DF to meet AB produced in E and AC produced in G. By *Conics* II-8, $FE = DG$, and therefore $DE = FG$. Furthermore, since F, D, C, A, and B all lie on the same circle, *Elements* III-36 (which states that the product of a secant to a circle from a point outside, with that segment exterior to the circle, equals the square on the tangent from the given point)

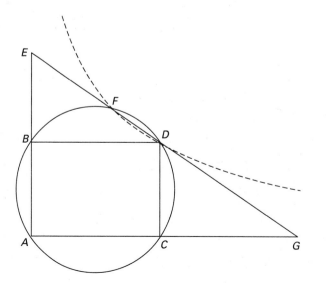

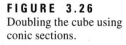

implies that $GA \cdot GC = GF \cdot GD$ and $EA \cdot EB = ED \cdot EF$. Therefore $GA \cdot GC = EA \cdot EB$ or $GA:EA = EB:GC$. By similarity, $GA:EA = DB:BE = AC:BE$, and also $GA:EA = GC:DC = GC:AB$. It follows that $AC:BE = BE:GC = GC:AB$, so that BE, GC are the two desired mean proportionals.

A final problem to be dealt with here, a problem which had reverberations down to the seventeenth century, is the three- and four-line locus problem. The problem in its most elementary form can be stated as follows: Given three fixed straight lines, to find the locus of a point moving so that the square of its distance to one line is in a constant ratio to the product of its distance to the other two lines. (Here distance is to be measured at a fixed angle to each line.) If one takes the special case where two of the lines are parallel and the third perpendicular to the first two, it is easy to see analytically that the given locus is a conic section. Recall that the original equation of the ellipse and hyperbola was $y^2 = \frac{p}{2a}x_1x_2$, where y is the distance of a given point from the diameter of the conic, and x_1, x_2 are the distances of that point from the endpoints of the diameter. If tangents are drawn to the conic at those two endpoints, these conics then provide a solution to the three-line locus problem with respect to the diameter and the two tangents.

The problem for the Greek mathematicians was to generalize this solution, that is, to show that the locus was a conic whatever the position of the three lines. Apollonius wrote (see the opening quotation) that the three-line locus problem had been only partially worked out by Euclid, but that his new results in Book III would enable the problem to be completely solved. The text of Book III doesn't mention the problem as such, but in fact one can derive from theorem III-16 the result that a conic has the property of the three-line locus relative to two tangents to the curve from a given point and the secant joining the two points of tangency. Other theorems in Book III enable one to show that a conic also solves the four-line locus problem, to find the locus of a point such that the product of its distances to one pair of lines is in a constant ratio to the product of its distances to the other pair. In later Greek times, an attempt was made, without great success, to find the locus with regard to greater numbers of lines. It was this problem which Descartes and Fermat both demonstrated they could solve through their new method of analytic geometry in the seventeenth century, a method whose germ came from a careful reading of Apollonius' work. Descartes in fact was able to derive the equations of curves which satisfied analogous conditions for various numbers of lines and to classify the solutions. As should be evident from our description of many of the Greek problems in modern notation, the Greek tradition of geometric problem solving, which was carried on in the Islamic world long after its demise in the Hellenic world, ultimately led to new advances in mathematical technique, advances which finally reduced much of this kind of Greek mathematics to mere textbook exercises.

Exercises

Archimedes' On the Equilibrium of Planes

1. Find where to place the fulcrum in a lever of length 10 m so that a weight of 14 kg at one end will balance a weight of 10 kg at the other.

2. If a weight of 8 kg is placed 10 m from the fulcrum of a lever and a weight of 12 kg is placed 8 m from the fulcrum in the opposite direction, toward which weight will the lever incline?

3. Prove proposition 4 by using proposition 3 and *reductio ad absurdum*.

Proposition 4.   *If two equal weights do not have the same center of gravity, the center of gravity of both taken together is at the midpoint of the line joining their centers of gravity.*

4. Prove proposition 5.

Proposition 5.   *If three equal magnitudes have their centers of gravity on a straight line at equal distances, the center of gravity of the system will coincide with that of the middle magnitude.*

Archimedes' On Floating Bodies

5. An alternative method by which Archimedes could have solved the wreath problem is given by Vitruvius in *On Architecture*.[2] Assume as in the text that the wreath is of weight W, composed of weights w_1 and w_2 of gold and silver respectively. Assume that the crown displaces a certain quantity of fluid, V. Furthermore, suppose that a weight W of gold displaces a volume V_1 of fluid while a weight W of silver displaces a volume V_2 of fluid. Show that $V = \frac{w_1}{W}V_1 + \frac{w_2}{W}V_2$ and therefore that $\frac{w_1}{w_2} = \frac{V_2 - V}{V - V_1}$.

Archimedes' On the Measurement of the Circle

6. Prove the two lemmas (see page 102) which Archimedes used to derive his algorithms for calculating π.

7. Use a calculator (or program a computer) to calculate π by iterating the algorithm of Archimedes given by lemma 1.

8. Translate lemma 2 into a recursive algorithm for calculating π. Iterate this algorithm to get an approximate value for π.

9. Show that if a is the nearest positive integer to the square root of $a^2 \pm b$, then

$$a \pm \frac{b}{2a} > \sqrt{a^2 \pm b} > a \pm \frac{b}{2a \pm 1}.$$

Beginning with $2^2 - 1 = 3$, and therefore, as a first approximation, $2 - 1/4 > \sqrt{3}$, show first that $\sqrt{3} > 5/3$, second that $\sqrt{3} < 26/15$, third that $\sqrt{3} < 1351/780$, and fourth that $\sqrt{3} > 265/153$. Note that these last two approximations are the values Archimedes uses in his *Measurement of the Circle*.

Archimedes' The Method

10. Given the parabolic segment with MO parallel to the axis of the segment and MC tangent to the parabola, show analytically that $MO : OP = CA : AO$ (see Figure 3.7).

11. The proof of proposition 2 of Archimedes' *The Method* is outlined here.

Proposition 2.   *Any sphere is (in respect of solid content) four times the cone with base equal to a great circle of the sphere and height equal to its radius.*

Let $ABCD$ be a great circle of a sphere with perpendicular diameters AC, BD. Describe a cone with vertex A and axis AC and produce its surface to the circle with diameter EF. On the latter circle erect a cylinder with height and axis AC. Finally, produce AC to H such that $HA = AC$. Certain pieces of the figures described are to be balanced using CH as the lever (Figure 3.27).

Let MN be an arbitrary line in the plane of the circle $ABCD$ and parallel to BD with its various intersections marked as in the diagram. Through MN, draw a plane at right angles to AC. This plane will cut the cylinder in a circle with diameter MN, the sphere in a circle with diameter OP, and the cone in a circle with diameter QR.

a. Show that $MS \cdot SQ = OS^2 + SQ^2$.

b. Show that $HA : AS = MS : SQ$. Then, multiplying both parts of the last ratio by MS, show that $HA : AS = MS^2 : (OS^2 + SQ^2) = MN^2 : (OP^2 + QR^2)$.

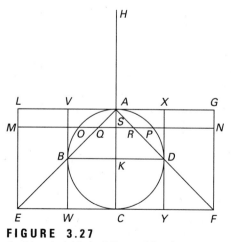

FIGURE 3.27
Archimedes' *Method*, Proposition 2.

Show that this last ratio equals that of the circle with diameter MN to the sum of the circle with diameter OP and that with diameter QR.

c. Conclude that the circle in the cylinder, placed where it is, is in equilibrium about A with the circle in the sphere together with the circle in the cone, if both the latter circles are placed with their centers of gravity at H.

d. Archimedes concludes from the above that the cylinder, placed where it is, is in equilibrium about A with the sphere and the cone together, when both are placed with their center of gravity at H. Show therefore that HA : AK = (cylinder) : (sphere + cone AEF).

e. From the fact that the cylinder is three times the cone AEF and the cone AEF is eight times the cone ABD, conclude that the sphere is equal to four times the cone ABD.

Archimedes' Other Works

12. Using calculus, prove Archimedes' result that the area of a parabolic segment is four-thirds of the area of the inscribed triangle.

13. Using calculus, prove Archimedes' result that the volume of a segment of a paraboloid of revolution is 3/2 that of the cone with the same base and axis.

14. Using calculus, prove Archimedes' result that the area bounded by one complete turn of the spiral given in polar coordinates by $r = a\theta$ is one-third of the area of the circle with radius $2\pi a$.

15. Generalize Archimedes' calculation of the sum of a geometric series with ratio 1/4 to give the sum of a geometric series with ratio $1/n$ and with ratio $m/n < 1$.

16. Consider proposition 1 of *On the Sphere and Cylinder II*: Given a cylinder, to find a sphere equal to the cylinder. Provide the analysis of this problem. That is, assume that V is the given cylinder and that a new cylinder P has been constructed of volume $(3/2)V$. Assume further that another cylinder Q has been constructed equal to P but with height equal to its diameter. The sphere whose diameter equals the height of Q would then solve the problem, because the volume of the sphere is 2/3 that of the cylinder. So given the cylinder P of diameter AB and height OD, determine how to construct a cylinder Q of the same volume but whose height and diameter are equal.

17. Determine the equations of the two conic sections whose intersection provides the solution x to the cubic equation

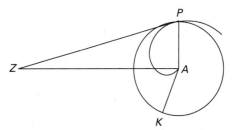

FIGURE 3.28
Archimedes' *On Spirals,* Proposition 20.

$3ax^2 - x^3 = 4a^2b$ needed by Archimedes to solve proposition 4 of *On the Sphere and Cylinder II*. Sketch the two curves on the same pair of axes.

18. Prove proposition 20 of *On Spirals,* here modified somewhat for clarity: Suppose PZ is a tangent to the spiral $r = a\theta$ at P, and construct a circle centered at the origin A with radius AP (Figure 3.28). Draw AZ perpendicular to AP, meeting the tangent at Z. Further, let AK be the tangent to the spiral at the origin A, meeting the circle at K. Prove that AZ is equal to the arc KP of the circle. (This proposition implies that an arc of the circle can be expressed as the length of a particular straight line. This in turn, as noted in the text, implies that the circle can be squared.)

19. Show that Archimedes' result on the sum of the first n squares may be rewritten in the form

$$\sum_{i=1}^{n} i^2 = \frac{n(n+1)(2n+1)}{6}.$$

From Apollonius' Conics

20. Show that in the curve $y^2 = px$, the value p represents the length of the latus rectum, the straight line through the focus perpendicular to the axis.

21. Rewrite the equations $y^2 = x\left(p + \frac{p}{2a}x\right)$ and $y^2 = x\left(p - \frac{p}{2a}x\right)$ for the hyperbola and ellipse respectively in the current standard forms for those equations. What point is the center of the curve? What is the relationship of b, the length of the semiminor axis of the ellipse, to p and a?

22. Show that $(a + b)^2 \geq 4ab$ with equality if and only if $a = b$. (Note that this result is used in the proof of *Conics* I-33.)

23. Use calculus to prove *Conics* I-33.

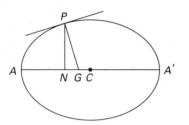

FIGURE 3.29
Finding the perpendicular to an ellipse.

24. Use calculus to prove *Conics* I-34.

25. Given an ellipse with diameter $AA' = 2a$, center C, and symptom $y^2 = x\left(p - \frac{p}{2a}x\right)$, let G be any point on AA' such that $AG > p/2$ (Figure 3.29). Choose N on AG so that $NG : CN = p : 2a$. Prove analytically that if NP is drawn perpendicular to the axis and meets the curve at P, then PG is the minimum straight line from G to the curve. Also show that PG is perpendicular to the tangent at P.

26. Convert Apollonius' treatment of the normal line to a point on the ellipse to the analogous case of the normal line to a point on the hyperbola.

27. Show how to construct (using straightedge and compass) as many points as desired on the ellipse $y^2 = x\left(p - \frac{p}{2a}x\right)$ and on the hyperbola $xy = k$.

28. Show analytically that the solution to the three-line locus problem is a conic section in the case where two of the lines are parallel and the third is perpendicular to the other two. Characterize the curve in reference to the distance between the two parallel lines and the given ratio.

29. Show analytically that the solution to the general three-line locus problem is always a conic section.

30. Use the quadratrix to trisect an arbitrary angle. That is, given the quadratrix BZK and $\angle BAZ$, show how to construct $\angle BAX$ such that $3\angle BAX = \angle BAZ$ (see Figure 3.6). More generally, show how to construct an angle bearing any given rational ratio to $\angle BAZ$.

31. Fill in the details of the following solution to the angle trisection problem (given in Pappus but probably dating from much earlier):[18] Let the given angle AOG be placed at the center of the circle, cutting off the arc AG on the circumference (Figure 3.30). To trisect this angle, it is sufficient to trisect arc AG, that is, to find a point B on the

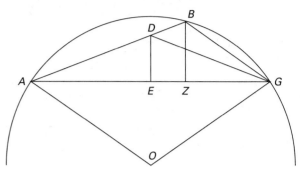

FIGURE 3.30
Angle trisection by way of conic sections, second method.

circle such that arc BG is one-half of arc AB. Using the method of analysis, suppose that this has been done. Then $\angle BGA = 2\angle BAG$. Draw GD to bisect $\angle BGA$ and draw DE, BZ perpendicular to AG. Use *Elements* VI-3 and similarity to show that $BG : EZ = AG : AE = 2 : 1$. Use the focus-directrix property to conclude that B lies on a particular hyperbola, and then complete the synthesis.

FOR DISCUSSION . . .

32. Design a lesson for a precalculus course which will demonstrate the formulas for the sum of a geometric series as in Archimedes' work. Discuss whether one can adapt Archimedes' procedure for determining the area of a parabolic segment and/or the area bounded by one turn of the spiral to introduce a precalculus class (or even a calculus class) to the calculation of areas bounded by curves.

33. Design a series of lessons for a precalculus course which will demonstrate the basic tangent and focal properties of the conic sections.

34. How is Apollonius' treatment of the conic sections similar to a modern analytic geometry treatment of the same subject? Can one consider Apollonius as an inventor of analytic geometry?

35. In what sense can one consider Archimedes as an inventor of the integral calculus?

36. Why did the Greeks continue to search for solutions to the three problems of squaring the circle, trisecting an angle, and doubling the cube, when solutions had already been found?

References and Notes

Many of the books on Greek mathematics referred to in Chapter 2 have sections on Archimedes and Apollonius. In particular, Thomas Heath's *A History of Greek Mathematics*, B. L. Van der Waerden's *Science Awakening*, and Wilbur Knorr's *The Ancient Tradition of Geometric Problems* are good sources of further reading on the material of this chapter. Selections from the works of these two mathematicians as well as others discussed in this chapter can be found in Ivor Bulmer-Thomas, *Selections Illustrating the History of Greek Mathematics*. A complete translation of the extant works of Archimedes, slightly edited for modern readers, is found in Thomas Heath, *The Works of Archimedes* (New York: Dover, 1953). The most detailed discussion of Archimedes' works, however, is in E. J. Dijksterhuis, *Archimedes* (Princeton: Princeton University Press, 1987). This new edition of Dijksterhuis' work has a bibliographic essay by Wilbur Knorr which gives the latest information on research on the work of Archimedes. In particular, several of Knorr's articles are of interest. These include "Archimedes and the Measurement of the Circle: A New Interpretation," *Archive for History of Exact Sciences* 15 (1976), 115–140, "Archimedes and Spirals: The Heuristic Background," *Historia Mathematica* 5 (1978), 43–75, and "Archimedes and the *Elements*: Proposal for a revised chronological ordering of the Archimedean corpus," *Archive for History of Exact Sciences* 19 (1978), 211–290. The only available English translation of part of Apollonius' *Conics* is by R. Catesby Taliaferro (the first three books) and appears in the *Great Books* (Chicago: Encyclopedia Britannica, 1952). Thomas Heath's *Apollonius of Perga* (Cambridge: W. Heffer and Sons, 1961) contains all seven extant books of the *Conics*. But since Heath modifies the order and often combines several theorems, this cannot be considered a literal translation. Nevertheless, it is still the only complete version of Apollonius' major work, with commentary, available in English.

1. From the Preface to Book I of Apollonius' *Conics*, translated in Heath, *Apollonius of Perga*, pp. lxx–lxxi.

2. Vitruvius, *On Architecture* (Cambridge: Harvard University Press, 1934), IX, 9–10.

3. A discussion of the pseudo-Aristotelian *Mechanica* can be found in Thomas Heath, *A History of Greek Mathematics*, I, pp. 344–346, while a translation of sections of this work is found in Bulmer-Thomas, *Selections Illustrating*, I, p. 431.

4. Plutarch, *The Lives of the Noble Grecians and Romans* (Dryden translation), in the *Great Books*, p. 254. This reference and the succeeding ones are taken from the section on Marcellus.

5. Plutarch, *Great Books*, p. 252.

6. Dijksterhuis, *Archimedes*, p. 299. The reference is to a translation of Pappus, *Collectio* VIII, 5, 11.

7. Plutarch, *Great Books*, p. 252.

8. Ibid., p. 253.

9. Ibid., p. 253.

10. The discussion of the wreath problem is from Heath, *The Works of Archimedes*, pp. 259–260. Heath's introduction provides insight into the various mathematical techniques of Archimedes.

11. Heath includes as an appendix to the previous work a translation of *The Method of Archimedes*. The present quotation is found on p. 13. A valuable discussion of this work is also found in Asger Aaboe, *Episodes from the Early History of Mathematics* (Washington, M.A.A., 1964). A brief account is in S. H. Gould, "The Method of Archimedes," *American Mathematical Monthly* 62 (1955), 473–476.

12. Ibid., p. 17.

13. Ibid.

14. The picture and discussion are adapted from Wilbur Knorr, *The Ancient Tradition of Geometric Problems*. Knorr has an extensive discussion of the contributions of Apollonius to Greek geometric problem solving.

15. The previous discussion is adapted from Gerald Toomer, *Diocles on Burning Mirrors* (New York: Springer, 1976) and from Dijksterhuis, *Archimedes*. The first named work provides a complete translation as well as a discussion of the importance of Diocles' work.

16. The quotes from the first three books of Apollonius' *Conics* are taken from the R. Catesby Taliaferro translation, *Great Books*.

17. This theorem appears in Thomas Heath's *Apollonius of Perga*, pp. 143, 152.

18. Knorr, *The Ancient Tradition*, p. 128.

Summary of Work of Archimedes, Apollonius and Their Predecessors

Fifth century B.C.E.	Hippocrates	Problem of doubling the cube
Mid-fourth century B.C.E.	Menaechmus	First construction of conic sections
Late fourth century B.C.E.	Aristaeus	Early text on conics
287–212 B.C.E.	Archimedes	Mathematical models, areas, and volume
Late third century B.C.E.	Nicomedes	Quadratrix and squaring the circle
250–175 B.C.E.	Apollonius	Conic sections
Early second century B.C.E.	Diocles	Burning mirrors

Chapter 4

Mathematical Methods in Hellenistic Times

"Plato . . . set the mathematicians the following problem: What circular motions, uniform and perfectly regular, are to be admitted as hypotheses so that it might be possible to save the appearances presented by the planets?"
(Simplicius' Commentary on Aristotle's On the Heavens*)*[1]

$\mathcal{A}$ classified advertisement in an Alexandria newspaper (c. 150 C.E.): Calculators wanted to perform extensive but routine calculations to create tables necessary for major work on astronomy. Must be able to follow detailed instructions with great accuracy. Compensation: Room and board plus the gratitude of the thousands of people who will use these tables for the next 1200 years. Contact: Claudius Ptolemy at the Observatory.

Although such an advertisement did not actually appear, Claudius Ptolemy did write a major work answering Plato's challenge, a work studied and extensively imitated for over a thousand years, a work in which Ptolemy not only used ideas from plane and spherical geometry but also devised ways to perform the extensive numerical calculations necessary to make his book a useful one. Ptolemy's text, and other ancient astronomical works from Babylonia and Egypt, were heavily used in astrology. Nevertheless, the evidence from all of these civilizations indicates that the primary reason for the study of astronomy was the solving of problems connected with the calendar, problems such as the determination of the seasons, the prediction of eclipses, and the establishment of the beginning of the lunar month.

In the process of using mathematics to study astronomy, the Greeks created plane and spherical trigonometry and also developed a mathematical model of the universe, a model which they modified many times during the five centuries between the times of Plato and Ptolemy. Among the major contributors to the development of mathematical astronomy whose ideas will be discussed in this chapter are Eudoxus in the fourth century B.C.E., Apollonius late in the third century, Hipparchus in the second century, Menelaus around 100 C.E., and finally Ptolemy. The chapter then concludes with a survey of other work in "practical mathematics" developed in the Greek world, mathematics applicable to problems on earth rather than the heavens. In particular, the work of Heron in the first century C.E. and Rabbi Nehemiah in the second century will be considered.

4.1 ASTRONOMY BEFORE PTOLEMY

Centuries of observation of the heavens had enabled the Babylonians to make relatively accurate predictions of the recurrence of various celestial phenomena, from such simple ones as the time of sunrise and sunset to such complicated ones as the times of lunar eclipses. But they had never apparently applied more than arithmetic and simple algebra to this study. The Pythagoreans too explained celestial phenomena by way of number. Neither the Babylonians nor the Pythagoreans, however, developed a model to connect the various celestial phenomena. The initial creation of such a model was a product of fourth century Greece, the time of Plato's Academy.

The basic model developed at that time is one of two concentric spheres, the sphere of the earth and the sphere of the stars. The immediate evidence of our senses indicates that the earth is flat, but more sophisticated observations, including the facts that the hull of a ship sailing away disappears before the top of the mast and that the shadow of the earth on the moon during a lunar eclipse has a circular edge, convinced the Greeks of the earth's sphericity. Their sense of esthetics—that a sphere was the most perfect solid shape—added to this conviction. That the shape of the heavens should mirror the shape of the earth was also only natural.

The evidence of the senses, and some logical argument as well, further convinced the Greeks that the earth was stationary in the middle of the celestial sphere. The second part of this conclusion came from the general symmetry of the major celestial phenomena, while the first part came from the lack of any sensation of motion of the earth. The Greeks also noted that if the earth rotated on its axis once a day, its motion would of necessity be so swift that "objects not actually standing on the earth would appear to have the same motion, opposite to that of the earth; neither clouds nor other flying or thrown objects would ever be seen moving toward the east, since the earth's motion toward the east would always outrun and overtake them, so that all other objects would seem to move in the direction of the west and the rear."[2] With the earth considered immovable, the observed daily motion in the sky must be due to the rotation of the celestial sphere, to which were firmly attached the so-called fixed stars, grouped into patterns called constellations. These never change their positions with respect to each other and form the fixed background for the "wandering stars" or planets (Side Bar 4.1).

The seven wanderers—the sun, the moon, Mercury, Venus, Mars, Jupiter, and Saturn—were more loosely attached to the celestial sphere. That they were attached was obvious; in general they participated in the daily east to west rotation of the celestial sphere. But they also had their own motion, usually in the opposite direction (west to east) at much slower speeds. It is these motions that the Greek astronomers (and indeed all earlier astronomers) attempted to make sense of. They were limited in their attempts at solution, however, by an overriding philosophical consideration. Namely, since the universe outside the earth was thought to be unchanging and perfect, according to Aristotle,

---

Side Bar 4.1 **Precursors of Copernicus**

Some ancient astronomers asserted a theory contrary to the immovable, central earth theory discussed in the text. Heraclides of Pontus (c. 388–310 B.C.E.) is credited with having the earth's rotation account for the daily motion of the heavens, while Aristarchus of Samos (c. 310–230 B.C.E.), as reported by Archimedes, hypothesized "that the fixed stars and the sun remain unmoved [and] that the earth revolves about the sun in the circumference of a circle, the sun lying in the middle of the orbit."[3] The chief objection to Aristarchus' theory was that it implied that the appearance of the fixed stars would change as one viewed them from different parts of the earth's orbit. Aristarchus met this objection by further assuming that the distance to the fixed stars was so enormous that this effect would be unnoticeable. Other astronomers at the time could not bring themselves to believe that these huge distances were possible. In addition, certain thinkers charged Aristarchus with impiety for having "set in motion the hearth of the universe"[4] in order to save the appearances. Conflicts between science and religion evidently date back to ancient times.

---

the only movements in the heavens were the "natural" movements of these perfect bodies. Because the bodies were spherical, the natural movements were circular. Thus the astronomers and mathematicians (usually the same people) attempted to solve Plato's problem quoted at the opening of the chapter, that is, to develop a model which would explain the phenomena in the heavens ("save the appearances") through a combination of geometrical constructs using circular and uniform motion. It was not the business of the astronomer-mathematicians to decide if or how such motions were physically possible, for celestial physics as we know it was never dealt with in ancient Greece. But they did in fact succeed in finding several different systems which met Plato's challenge.

Because the basic Greek model of the heavens consisted of spheres, the first element of the study of celestial motion was the study of the properties of the sphere. Recall that Euclid's *Elements* contained virtually nothing about these properties. There were, however, other texts written in the fourth century on the general subject of Spherics, including ones by Autolycus of Pitane (c. 300 B.C.E.) and by Euclid himself, which did cover the basics, mostly in the context of results immediately useful in astronomy. These books contained such definitions as that of a **great circle** (a section of a sphere by a plane through its center) and its **poles** (the extremities of the diameter of the sphere perpendicular to this plane). The texts also included three important theorems very useful in what follows. One, any two points on the sphere which are not diametrically opposite determine a unique great circle. Two, any great circle through the poles of a second great circle is perpendicular to the original one, and, in this case, the second circle also contains the poles of the first. Finally, any two great circles bisect one another.

There are several great circles on the celestial sphere which are important for astronomy. For example, the sun's path in its west to east movement through the stars is a great circle. This great circle, called the ecliptic, passes through the twelve constellations of the Zodiac (Figure 4.1). (These constellations were first mentioned in Babylonian astronomy and appear in Greek sources as early as 300 B.C.E.) The diameter of the earth through the north and south pole, extended to the heavens, is the axis around which the daily rotation of the celestial sphere takes place. The great circle corresponding to the poles of that axis

FIGURE 4.1
Mosaic of the Zodiac on an Israeli souvenir sheet.

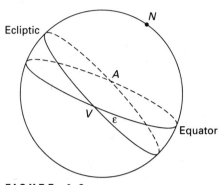

FIGURE 4.2
Ecliptic and equator on the celestial sphere.

is called the **celestial equator.** The equator and ecliptic intersect at two diametrically opposite points, the vernal and autumnal equinoxes, for on those dates the sun is located on those intersections (Figure 4.2). Since the Greeks knew that the earth was so small that it could in effect be considered as a point with respect to the sphere of the stars, they assumed that the horizon plane passed through the center of the celestial sphere and hence that the horizon itself was also a great circle. The horizon intersects the equator at the east and west points. Finally, the local meridian is the great circle which passes through the north and south points of the horizon and the point directly overhead, the local zenith. Since the meridian circle is perpendicular to both the horizon and the celestial equator, it also passes through the north and south poles of the latter. The angle ϵ between the equator and the ecliptic can be determined by taking half the distance (in degrees) between the noon altitudes of the sun at the summer and winter solstices. This value was measured to be $24°$ by the time of Euclid and was taken to be $23°51'20''$ by Ptolemy. (In fact, this value is slowly decreasing and is now about $23\ 1/2°$.) The angle between the horizon and the

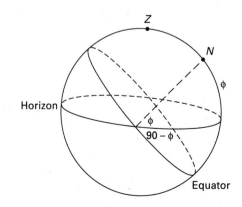

FIGURE 4.3
Horizon, equator, and latitude.

equator is $90° − \phi$, where ϕ is the geographical latitude of the observer (Figure 4.3). The latitude may also be thought of as the measure of the arc between the north celestial pole and the horizon.

4.1.1 Eudoxus and Spheres

FIGURE 4.4
Spheres of Eudoxus (with earth at center) on lower left side of Liberian stamp.

Eudoxus, famous for his work on ratios and the method of exhaustion, was the person largely responsible for turning astronomy into a mathematical science. He was probably the inventor of the two-sphere model as well as of the modifications necessary to account for the various motions of the sun, moon, and planets, nevertheless keeping to Plato's dictum to use only circular motion. In his scheme each of the heavenly bodies was placed on the inner sphere of a set of two or more interconnected spheres, all centered on the earth, whose simultaneous rotation about different axes produced the observed motion (Figure 4.4). For example, the sun requires two spheres to account for its two basic motions. The outer sphere represents the sphere of the stars; it rotates westward about its axis once in a day. The inner sphere, which contains the sun, is attached to the outer sphere so that its axis is inclined at angle ϵ to the axis of the outer sphere. If this sphere now rotates slowly eastward so that it makes a complete revolution in one year, the combination of the two motions will produce the apparent motion of the sun (Figure 4.5). In the case of the moon, the second sphere makes a complete eastward revolution in 27 1/2 days, the time it takes the moon to make one complete journey through the ecliptic. Unlike the sun, however, the moon deviates somewhat from the ecliptic in this journey. Thus Eudoxus postulated a third inner sphere, to which the moon is attached, to produce, at least qualitatively, these north and south deviations. For the even more complicated motion of the planets, including not only their general eastward movement but also their occasional retrograde (westward) motion, Eudoxus required four spheres.[5]

Although Eudoxus' plan was ingenious, in all probability he did not even try to incorporate detailed numerical data into his construction, but merely attempted to find a general geometric model. In fact, it was noted early on that the theory did not account for several of the observed phenomena. For example, the moon's motion to the north and south of the ecliptic is considerably more complicated than could be explained by only three spheres.

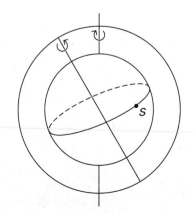

FIGURE 4.5
Eudoxus' spheres for the sun.

And the four-sphere theory of the planets did not account for their brightening during retrogression. Nevertheless, the notion of the sun, moon, and planets being carried around on spheres became part of Aristotle's detailed cosmology and as such was part of the general conception of the heavens in western civilization through the sixteenth century.

4.1.2 Apollonius: Eccenters and Epicycles

About 150 years after Eudoxus, Apollonius attempted a new answer to Plato's challenge. It had by then been known for 200 years that the seasons of the year were not equal in length; for example, the time from the vernal equinox to the summer solstice is two days longer than the time from the summer solstice to the autumnal equinox. Therefore, a simple model of the sun revolving in a circle centered on the earth at constant speed, even if the sun were attached to one of Eudoxus' spheres, could not account for this phenomenon. Because nonuniform motion would not satisfy Plato's rules, Apollonius or one of his predecessors proposed the following solution. Place the center of the sun's orbit at a point (called the **eccenter**) displaced away from the earth. Then if the sun moves uniformly around the new circle (called the **deferent circle**), an observer on earth will see more than a quarter of the circle against the spring quadrant (the upper right) than against the summer quadrant (the upper left) (Figure 4.6a). The distance ED, or better, the ratio of ED to DS, is known as the **eccentricity** of the deferent. If line ED is extended to the deferent circle, the intersection point closest to the earth is called the **apogee** of the deferent, while the one farthest from the earth is called the **perigee**. Assuming that one can determine the correct parameters in this model (the length and direction of ED) so that the seasonal lengths come out right, the question in using the model is where the sun will be seen on a particular day. To answer this question, one needs to find angle DES. This requires solving triangle DES, which in turn requires trigonometry. In fact, it was the necessity for introducing numerical parameters into these geometric models which led to the invention of trigonometry.

Apollonius also noticed that one can replace this eccentric model by another geometric model, the epicyclic one. That is, instead of considering the sun as traveling on the eccentric circle, it may be imagined as traveling on a small circle, the **epicycle**, whose center travels on the original earth-centered circle (Figure 4.6b). If the epicycle rotates

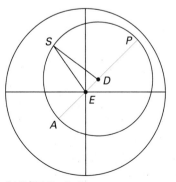

FIGURE 4.6a
Apollonius' eccenter model for sun.

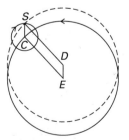

FIGURE 4.6b
Apollonius' epicycle model for sun.

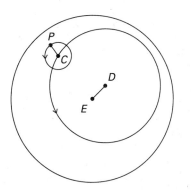

FIGURE 4.7
Apollonius' model for planetary motion.

once clockwise in the same time as its center rotates around the earth, that is, if the two motions always keep *DECS* a parallelogram, the actual path of the sun will be the same as it was using the deferent circle above.

It then turns out that if one combines epicycles and eccentric circles, one can produce the more complicated motions of the planets. In fact, Apollonius initiated the study of this model. The planet *P* travels uniformly counterclockwise on an epicycle with center *C*. This latter point travels in the same direction on a deferent circle with center *D* at a distance *DE* from the earth (Figure 4.7). If the speeds along these circles are set appropriately, the planet as seen from the earth will in general travel eastward along the ecliptic, but during certain periods will travel in the opposite direction (when the planet is on the inner part of the epicycle). To use this model it is again necessary to find the various parameters involved, such as the lengths *PC* and *ED* and their relative directions. Once these are established for a given planet, however, the position of the planet at any time can be found by solving certain triangles.

4.1.3 Hipparchus and the Beginning of Trigonometry

Apollonius himself did not possess the trigonometric machinery necessary to complete the solution of these problems. It was Hipparchus of Bithynia (190–120 B.C.E.) who systematically carried out numerous observations of planetary positions, introduced a coordinate system for the stellar sphere, and began the tabulation of trigonometric ratios necessary to enable one easily to solve right triangles and successfully deal with Apollonius' questions (Figure 4.8).

To deal quantitatively with the positions of the stars and planets, one needs both a unit of measure for arcs and angles as well as a method of specifying where a particular body is located on the celestial sphere, that is, a system of coordinates. Euclid's unit of angle measure was simply a right angle. Other angles were referred to as parts or multiples of this angle. The Babylonians, however, sometime before 300 B.C.E., initiated the division of the circumference of the circle into 360 parts, called degrees, and within the next two centuries this measure, along with the sexagesimal division of degrees into minutes and seconds, was adopted in the Greek world. Hipparchus was one of the first to make use of this measure, although he also used arcs of 1/24 of a circle and 1/48 of a circle, so-called "steps" and "half-steps," in some of his work. Why the Babylonians divided the circle into 360 parts is not known. Perhaps it was because 360 is easily divisible by many small

FIGURE 4.8
Hipparchus on a Greek stamp.

integers or because it is the closest "round" number to the number of days in the year. The latter reason gives us the convenient approximation that the sun travels 1° along the ecliptic each day.

It was also the Babylonians who first introduced coordinates into the sky. The system they used, later taken over by Ptolemy, is known as the ecliptic system. Positions of stars are measured both along and perpendicular to the ecliptic. The coordinate along the ecliptic (measured in degrees counterclockwise from the vernal point as seen from the north pole) is called the **longitude** λ; the perpendicular coordinate, measured in degrees north or south of the ecliptic, is called the **latitude** β (Figure 4.9a). This coordinate system is particularly useful when dealing with the sun, moon, and planets. The sun, since it travels along the ecliptic, always has latitude 0°. Its longitude increases daily by approximately 1° from 0° at the vernal equinox to 90° at the summer solstice, 180° at the autumn equinox, and 270° at the winter solstice. Often, however, in both the Babylonian sources and the later Greek ones, longitudes were counted using the zodiacal signs. Namely, the ecliptic was divided into twelve intervals of 30° each, named by the zodiacal constellations. For example, Aries included longitudes from 0° to 30° and Taurus from 30° to 60°. Thus, if one noted that the sun had longitude Taurus 5°, one meant it had ecliptic longitude 35°.

In place of this ecliptic coordinate system, Hipparchus used a system based on the celestial equator. The coordinate along the equator, also measured counterclockwise from the vernal point, is called the **right ascension** α. The perpendicular coordinate, measured north and south from the equator, is called the **declination** δ (Figure 4.9b). Hipparchus drew up a catalogue of fixed stars in which he described some of their positions in terms of this coordinate system.

To be able to relate the coordinates of a point in one coordinate system to its coordinates in another—and this is necessary to solve astronomical problems—one needs spherical trigonometry. But before this could be developed, it was necessary to understand plane trigonometry. Hipparchus was evidently the first to attempt the detailed tabulation of lengths which would enable plane triangles to be solved. Although there are no explicit documents giving Hipparchus' table or his method, enough has been pieced together from various sources to give us a reasonable picture of his work.

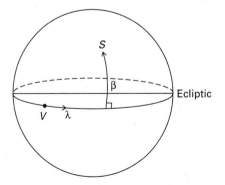

FIGURE 4.9a
Ecliptic coordinate system on the celestial sphere.

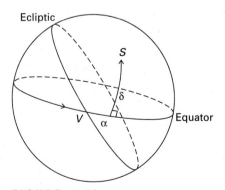

FIGURE 4.9b
Equatorial coordinate system on the celestial sphere.

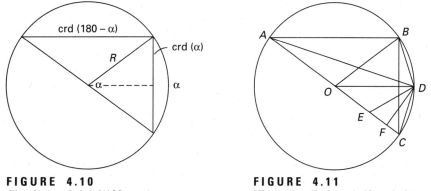

FIGURE 4.10
Chord(α) and chord($180 - \alpha$).

FIGURE 4.11
Hipparchus-Ptolemy half-angle formula.

The basic element in Hipparchus' (and also, later, in Ptolemy's) trigonometry was the chord subtending a given arc (or central angle) in a circle of fixed radius. Namely, both men gave a table listing α and chord(α) for various values of the arc α. Note that chord(α), henceforth abbreviated crd(α), is simply a length (Figure 4.10). If the radius of the circle is denoted by R, then the chord is related to the sine by the equations

$$\frac{1}{2}\text{crd}(\alpha)/R = \sin\frac{\alpha}{2} \quad \text{or} \quad \text{crd}(\alpha) = 2R\sin\frac{\alpha}{2}.$$

Because the angle or arc was to be measured in degrees and minutes, Hipparchus decided to use the same measure for the radius of the circle. Knowing that the circumference equalled $2\pi R$, and taking for π the sexagesimal approximation 3;8,30 (which is close to the mean between the two Archimedean values of 3 10/71 and 3 1/7), he calculated the radius R as $\frac{60 \cdot 360}{2\pi} = \frac{6,0,0}{6;17} = 57,18 = 3438'$ to the nearest integer. In a circle of this radius, the measure of an angle (defined as length cut off on the circumference divided by the radius) equals its radian measure.

To calculate a table of chords, Hipparchus began with a 60° angle. In this case the chord equals the radius, or crd(60°) $= 3438' = 57,18$. For a 90° angle, the chord is equal to $R\sqrt{2} = 4862' = 81,2$. (Note that the mixed decimal and sexagesimal notation used here is common in both Greek and modern angle measure.) To calculate chords of other angles Hipparchus used two geometric results. First, it is clear from Figure 4.10 that crd($180 - \alpha$) $= \sqrt{(2R)^2 - \text{crd}^2(\alpha)}$. Because crd($180 - \alpha$) $= 2R\cos\frac{\alpha}{2}$, this result is equivalent to $\sin^2\alpha + \cos^2\alpha = 1$. Second, Hipparchus calculated crd$\left(\frac{\alpha}{2}\right)$ from a version of the half-angle formula. (It is conjectured that he used the method given later by Ptolemy.) Suppose $\alpha = \angle BOC$ is bisected by OD (Figure 4.11). To express crd$\left(\frac{\alpha}{2}\right) = DC$ in terms of crd(α) $= BC$, choose E on AC so that $AE = AB$. Then $\triangle ABD$ is congruent to $\triangle AED$ and $BD = DE$. Since $BD = DC$, also $DC = DE$. If DF is drawn perpendicular to EC, then $CF = \frac{1}{2}CE = \frac{1}{2}(AC - AE) = \frac{1}{2}(AC - AB) = \frac{1}{2}(2R - \text{crd}(180 - \alpha))$. But also, triangles ACD and DCF are similar, so $AC : CD = CD : CF$. Therefore,

$$\text{crd}^2\left(\frac{\alpha}{2}\right) = CD^2 = AC \cdot CF = R(2R - \text{crd}(180 - \alpha)).$$

Putting this into modern notation gives

$$\left(2R \sin\frac{\alpha}{4}\right)^2 = R\left(2R - 2R \cos\frac{\alpha}{2}\right),$$

or, replacing α by 2α,

$$\sin^2\frac{\alpha}{2} = \frac{1 - \cos\alpha}{2},$$

the standard half-angle formula.

Hipparchus could now easily calculate the chord for every angle from 7 1/2° to 180° in "half-steps" of 7 1/2°. For example, by applying the formula three times to crd(60°), one finds crd(7 1/2°). By complements, one then finds crd(172 1/2°). This limited table enabled Hipparchus to make some progress in solving triangles and applying the results toward completing the models of the heavens. Since, however, the actual works of Hipparchus are lost, it is necessary to turn to the most influential astronomical work of antiquity, the *Almagest* of Claudius Ptolemy.

4.2 PTOLEMY AND THE *ALMAGEST*

FIGURE 4.12
Ptolemy (with crown and globe) (detail from Raphael's painting, The School of Athens). The crown represents Raphael's mistaken assumption that Ptolemy was related to the rulers of Egypt.

Very little is known of the personal life of Claudius Ptolemy (c. 100–178 C.E.) other than that he made numerous observations of the heavens from locations near Alexandria and wrote several important books (Figure 4.12). The *Geography*, for example, was a compilation of places in the known world along with their latitudes and longitudes, which included a discussion of the projections needed for map making. He also wrote works on astrology, music, and optics and attempted a proof of Euclid's parallel postulate. He is most famous today, however, for the *Mathematical Collection*, a work in thirteen books which contained a complete mathematical description of the Greek model of the universe with parameters for the various motions of the sun, moon, and planets. The book was the culmination of Greek astronomy. Like Euclid's *Elements*, it replaced all earlier works on its subject. It was the most influential astronomical work from the time it was written until at least the sixteenth century, being copied and commented on countless times. More than any other book it gave impetus to the notion of a mathematical model, that one could in fact devise a quantitative, mathematical description of natural phenomena which would yield reliable predictions. Virtually all subsequent astronomical works, both in the Islamic world and in the West, up to and including the work of Copernicus, were based on Ptolemy's masterpiece. Many centuries after it was written, Islamic scientists began calling it *al-magisti,* "the greatest," to distinguish it from lesser astronomical works. Ever since, the book has been known as the *Almagest* (Figure 4.13).

4.2.1 Chord Tables

Ptolemy began the *Almagest* with a basic introduction to the Greek concept of the cosmos followed by strictly mathematical material detailing the plane and spherical trigonometry necessary for the computation of the planetary positions. The first order of business for

FIGURE 4.13
Woodcut from early printing of a summary of the *Almagest* (1496). (Source: Smithsonian Institution Photo No. 76-14409)

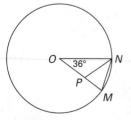

FIGURE 4.14
Ptolemy's calculation of crd(36°).

Ptolemy was the construction of a table of chords more complete than that of Hipparchus. To construct this table of chords of all arcs from 1/2° to 180° in intervals of 1/2°, as well as a scheme for interpolating between the computed values, he needed somewhat more geometry than Hipparchus. Also, instead of taking $R = 57,18$, a rather difficult value to compute with, he took $R = 60$, a unit in the sexagesimal system in which all of Ptolemy's computations are made.

Ptolemy's first calculation established the chord of 36°. Recall that the Euclidean construction of the pentagon involved the construction of an isosceles triangle whose base angles are double that at the vertex. If this triangle is placed with its vertex at the center of the circle, the base MN is crd(36°) (Figure 4.14). If NP bisects the angle at N, then triangles NPM and ONM are similar. Letting x denote crd(36°) and $R = 60$ the radius of the circle, it follows that $R : x = x : (R - x)$ or $x^2 = R^2 - Rx$. Ptolemy in effect solved this quadratic equation to get

$$x = \text{crd}(36°) = \sqrt{R^2 + \left(\frac{1}{2}R\right)^2} - \frac{1}{2}R = \sqrt{3600 + 900} - 30 = 37;4,55.$$

Ptolemy next noted that since the square on the side of a regular pentagon ($= \text{crd}(72°)$) equals the sum of the squares on the side of a regular decagon ($= \text{crd}(36°)$) and the side of a regular hexagon ($= R$) (*Elements* XIII-10), it followed that $\text{crd}(72°) = \sqrt{R^2 + \text{crd}^2(36°)} = 70;32,3$. Furthermore, because $\text{crd}(90°) = \sqrt{2R^2} = \sqrt{7200} = 84;51,10$ and $\text{crd}(60°) = R = 60$, he also easily calculated $\text{crd}(120°) = \sqrt{3R^2} = 103;55,23$. Finally, since $\text{crd}^2(180 - \alpha) = (2R)^2 - \text{crd}^2\,\alpha$, he could also calculate the chord of the supplement to any arc whose chord was known. For example, $\text{crd}(144°) = 114;7,37$. He was therefore well started on a chord table simply from propositions of Euclidean geometry and the ability to calculate square roots.

Ptolemy, like Archimedes four centuries earlier, never mentioned how he calculated these square roots, but merely presented the results. A commentary on Ptolemy's work by Theon in the late fourth century gave a method Ptolemy could well have used: "If we seek the square root of any number, we take first the side of the nearest square number, double it, divide the product into the remainder reduced to minutes, and subtract the square of the quotient; proceeding in this way, we reduce the remainder to seconds, divide it by twice the quotient in degrees and minutes, and we shall have the required approximation to the side of the square area."[6]

The method proposed by Theon to give two sexagesimal place approximations is quite similar to the Chinese square root algorithm discussed in Chapter 1. For example, to calculate $\sqrt{7200}$, note first that $84^2 = 7056$ and $85^2 = 7225$, so the answer must be of the form $84;x,y$. Since $7200 - 84^2 = 144$, we divide $144 \cdot 60$ by $2 \cdot 84$ and get 51 as the nearest integer. Therefore the answer is now known to be of the form $84;51,y$. Finally, $7200 - (84;51)^2 = 0;28,39$, which, converted to seconds, is 1719. Dividing this by $2 \cdot 84;51 (= 169;42)$ gives 10 to the nearest integer. The desired square root approximation is thus $84;51,10$ as noted. The relative complexity of this operation, and the fact that Ptolemy simply stated the results of large numbers of such calculations, leads us to believe that Ptolemy must have had the assistance of numerous "calculators" who performed these tedious but necessary calculations. In particular, these calculators were necessary to help Ptolemy complete his chord table, using the basic values above, the half-angle formula due to Hipparchus, and a new theorem from which certain sum and difference formulas could be derived.

Ptolemy's Theorem.   *Given any quadrilateral inscribed in a circle, the product of the diagonals equals the sum of the products of the opposite sides.*

To prove that $AC \cdot BD = AB \cdot CD + AD \cdot BC$ in quadrilateral $ABCD$ choose E on AC so that $\angle ABE = \angle DBC$ (Figure 4.15). Then $\angle ABD = \angle EBC$. Also $\angle BDA = \angle BCA$ since they both subtend the same arc. Therefore $\triangle ABD$ is similar to $\triangle EBC$. Hence $BD : AD = BC : EC$ or $AD \cdot BC = BD \cdot EC$. Similarly, since $\angle BAC = \angle BDC$, $\triangle ABE$ is similar to $\triangle DBC$. Hence $AB : AE = BD : CD$ or $AB \cdot CD = BD \cdot AE$. Adding equals to equals gives $AB \cdot CD + AD \cdot BC = BD \cdot AE + BD \cdot EC = BD(AE + EC) = BD \cdot AC$ and the theorem is proved.

To derive a formula for the chord of a difference of two arcs α, β, Ptolemy used the theorem with $AC = \text{crd}\,\alpha$ and $AB = \text{crd}\,\beta$ given. Applying the result to quadrilateral $ABCD$ gives $AB \cdot CD + AD \cdot BC = AC \cdot BD$ (Figure 4.16). Since $BC = \text{crd}(\alpha - \beta)$,

$$120\,\text{crd}(\alpha - \beta) = \text{crd}\,\alpha \cdot \text{crd}(180 - \beta) - \text{crd}\,\beta \cdot \text{crd}(180 - \alpha).$$

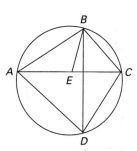

FIGURE 4.15
Ptolemy's Theorem.

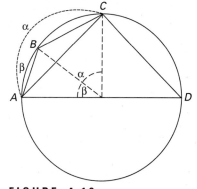

FIGURE 4.16
The difference formula for chords.

This is easily translated into the modern difference formula for the sine:

$$\sin(\alpha - \beta) = \sin \alpha \cos \beta - \cos \alpha \sin \beta.$$

A similar argument shows that

$$120 \operatorname{crd}(180 - (\alpha + \beta)) = \operatorname{crd}(180 - \alpha) \operatorname{crd}(180 - \beta) - \operatorname{crd} \beta \cdot \operatorname{crd} \alpha,$$

a formula equivalent to the sum formula for the cosine:

$$\cos(\alpha + \beta) = \cos \alpha \cos \beta - \sin \alpha \sin \beta.$$

Using the difference formula and the half-angle formula, Ptolemy then calculated $\operatorname{crd}(12°) = \operatorname{crd}(72° - 60°)$, $\operatorname{crd}(6°) = \operatorname{crd}\left(\frac{1}{2} \cdot 12°\right)$, $\operatorname{crd}(3°)$, $\operatorname{crd}\left(1\frac{1}{2}°\right)$, and $\operatorname{crd}\left(\frac{3°}{4}\right)$. His last two results are $\operatorname{crd}\left(1\frac{1}{2}°\right) = 1;34,15$ and $\operatorname{crd}\left(\frac{3°}{4}\right) = 0;47,8$. Using the addition formula he could have built up the table in intervals of $1\frac{1}{2}°$ or even $\frac{3°}{4}$. Since, however, he wanted his table to be in intervals of $\frac{1}{2}°$, and since, "if a chord such as the chord of $1\frac{1}{2}°$ is given, the chord corresponding to an arc which is one-third of the previous one cannot be found by geometrical methods (if this were possible we should immediately have the chord of $\frac{1}{2}°$)," Ptolemy could only find $\operatorname{crd}(1°)$ and $\operatorname{crd}\left(\frac{1°}{2}\right)$ by a procedure which, although "it cannot in general exactly determine the size [of chords], in the case of such very small quantities can determine them with a negligibly small error."[7] In other words, Ptolemy was convinced, although he offered no proof, that Euclidean tools (geometrical methods) are not sufficient rigorously to determine $\operatorname{crd}\left(\frac{1°}{2}\right)$. An alternative method is therefore necessary.

This alternative, an approximation procedure, is based on the lemma that if $\alpha < \beta$, then $\operatorname{crd} \beta : \operatorname{crd} \alpha < \beta : \alpha$, or, in modern notation, that $\frac{\sin x}{x}$ increases as x approaches 0. Applying this lemma first to $\alpha = \frac{3°}{4}$ and $\beta = 1°$, Ptolemy found $\operatorname{crd}(1°) < \frac{4}{3}\operatorname{crd}\left(\frac{3°}{4}\right) = \frac{4}{3}(0;47,8) = 1;2,50,40$. Applying it next to $\alpha = 1°$ and $\beta = 1\frac{1}{2}°$, he found $\operatorname{crd}(1°) > \frac{2}{3}\operatorname{crd}\left(1\frac{1}{2}°\right) = \frac{2}{3}(1;34,15) = 1;2,50$. Since all calculated values were rounded off to two sexagesimal places, it appears that to that number of places, $\operatorname{crd}(1°) = 1;2,50$ and therefore $\operatorname{crd}\left(\frac{1°}{2}\right) = 0;31,25$. The addition formula now enabled Ptolemy to build up his table in steps of $\frac{1}{2}^\circ$ from $\mathrm{crd}\left(\frac{1}{2}^\circ\right)$ to $\mathrm{crd}(180^\circ)$. To aid in interpolation for calculating chords of any integral number of minutes, he appended a third column to his table containing one-thirtieth of the increase from $\mathrm{crd}\ \alpha$ to $\mathrm{crd}\left(\alpha + \frac{1}{2}^\circ\right)$. A small portion of the table, whose accuracy is roughly equivalent to that of a modern five-decimal-place table, follows:

Arcs	Chords	Sixtieths	Arcs	Chords	Sixtieths
$\frac{1}{2}$	0;31,25	0;1,2,50	6	6;16,49	0;1,2,44
1	1;2,50	0;1,2,50	47	47;51,0	0;0,57,34
$1\frac{1}{2}$	1;34,15	0;1,2,50	49	49;45,48	0;0,57,7
2	2;5,40	0;1,2,50	72	70;32,3	0;0,50,45
$2\frac{1}{2}$	2;37,4	0;1,2,48	80	77;8,5	0;0,48,3
3	3;8,28	0;1,2,48	108	97;4,56	0;0,36,50
4	4;11,16	0;1,2,47	120	103;55,23	0;0,31,18
$4\frac{1}{2}$	4;42,40	0;1,2,47	133	110;2,50	0;0,24,56

4.2.2 Solving Plane Triangles

Given his chord table, Ptolemy could now solve plane triangles. Although he never stated a systematic procedure for doing so, he does seem to apply fixed rules. One difference to keep in mind when comparing Ptolemy's method to a modern one is that Ptolemy's table contains lengths of chords when the radius is 60 rather than ratios. Therefore, he always has to adjust his tabular values in a given problem to the actual length of the radius. We consider three examples of his procedures.

First, to calculate the length CF of the noon shadow of a pole CE of length 60 at Rhodes (latitude 36°) at the vernal equinox, Ptolemy began by noting that at that time the sun is 36° below the zenith (that is, $\angle AEB = 36^\circ$) (Figure 4.17). Ptolemy considered CF as the chord of the circle circumscribing triangle ECF. Because the angle at the center is double the angle at the circumference, $CF = \mathrm{crd}(72^\circ) = 70;32,3$. Then $CE = \mathrm{crd}(180^\circ - 72^\circ) = \mathrm{crd}(108^\circ) = 97;4,56$. Since Ptolemy wanted the shadow when $CE = 60$, he reduced this calculated value by the ratio $\frac{60}{97;4,56}$. Thus the desired shadow is $\frac{60}{97;4,56} \cdot (70;32,3) = 43;36$. This calculation for the finding of the leg a of a right triangle, given α and b, can be rewritten as

$$a = b \cdot \frac{\mathrm{crd}(2\alpha)}{\mathrm{crd}(180 - 2\alpha)} = b \cdot \frac{2R \sin \alpha}{2R \cos \alpha} = b \tan \alpha$$

in agreement with modern procedure. It is Ptolemy's lack of a tangent function and his need to use actual chords in circles which forced him to calculate the chords of double both the given angle and its complement as well as their quotient.

A second example shows how Ptolemy calculated the parameters for the eccentric model of the sun.[8] The calculation amounts to solving the right triangle LDE where D represents the center of the sun's orbit and E represents the earth (Figure 4.18). Divide the ecliptic into four quadrants by perpendicular lines through E and similarly divide the eccentric circle. To find LD and LE, one must first calculate the arcs $\theta = \frac{1}{2}UU'$ and

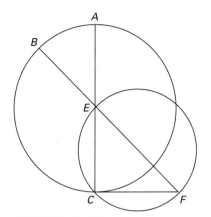

FIGURE 4.17
Calculating the length of a shadow.

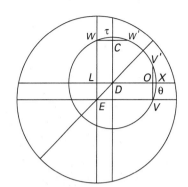

FIGURE 4.18
Calculating the parameters in the eccentric model of the sun.

$\tau = \frac{1}{2}WW'$ using the known inequalities of the seasons. Given that the spring path of the sun is 94.5 days while that of the summer is 92.5 days, and supposing that v is the mean daily angular velocity of the sun, the diagram shows that $90 + \theta + \tau = 94.5v$ for the spring while $90 + \theta - \tau = 92.5v$ for the summer. Because v equals the length of the year (observed to be 365;14,48 days) divided by 360°, or 0°59'8" per day, it follows that $90° + \theta + \tau = 93°9'$ while $90° + \theta - \tau = 91°11'$. A simple calculation then shows that $\theta = 2°10'$ and $\tau = 0°59'$.

The sides of the triangle DLE can now be determined under the assumption that the radius DX of the deferent is 60. Since DX bisects arc UU', it is evident that $LE = OU = \frac{1}{2}UU' = \frac{1}{2}\text{crd } 2\theta = \frac{1}{2}\text{crd}(4°20') = 2;16$. Similarly, $DL = \frac{1}{2}\text{crd } 2\tau = \frac{1}{2}\text{crd}(1°58') = 1;2$. By the Pythagorean theorem, $DE^2 = LE^2 + DL^2 = 6;12,20$ and $DE = 2;29,30$ or, approximately, $2;30 = 2\ 1/2$. In modern terminology, Ptolemy has simply calculated $LE = OU = R\sin\theta$ and $DL = CW = R\sin\tau$. The necessity of calculating half the chord of double the angle so often led later astronomers to tabulate this quantity, the modern sine function.

To complete the solution of the triangle, Ptolemy calculated $\angle LED$ by circumscribing a circle around $\triangle LDE$. Since $LD = 1;2$ when $DE = 2;29,30$, it would be 49;46 if DE were 120. Using the table of chords in reverse, Ptolemy read off that the corresponding arc is about 49°, hence $\angle LED$ is half of that, or 24°30'. Then $\angle LDE = 65°30'$ and the triangle is solved. Again, in modern terminology, Ptolemy first calculated $120a/c = 2R\sin\alpha$ or $\sin\alpha = a/c$ and then used the inverse sine relation to determine α.

A final example is provided by Ptolemy's solution of an oblique triangle. The problem here is to find the direction $\angle DES$ of the sun, from the eccentric model, given that $DE = 2;30$ if DS is arbitrarily picked to be 60 (Figure 4.19). For a given day, the angle PDS is known from the speed of the sun in its orbit and hence the angle EDS is known. Ptolemy made the calculation where $\angle PDS = 30°$ and $\angle EDS = 150°$.

Ptolemy first constructed the perpendicular EK to SD extended. Considering as before the circle about triangle DKE, he concluded that arc $DK = 120°$. From the table he noted

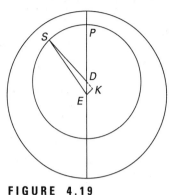

FIGURE 4.19
Finding the position of the sun.

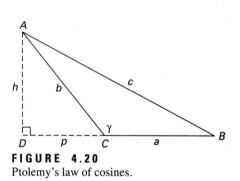

FIGURE 4.20
Ptolemy's law of cosines.

that if the radius were 60 (or $DE = 120$), then DK would be crd($120°$) = 103;55. Since, however, $DE = 2;30$, by proportionality $DK = 2;10$. Then $SK = SD + DK = 62;10$. Since $\angle KDE = 30°$, also $EK = \frac{1}{2}DE = 1;15$. Applying the Pythagorean theorem to $\triangle SKE$ gives $SE = 62;11$. Next, consider the circle circumscribing $\triangle SKE$. Since $KE = 1;15$ when $SE = 62;11$, it would be 2;25 if SE were 120. The chord table is now used in reverse to find that 2;25 corresponds to an arc of $2°18'$. It follows that $\angle KSE = 1°9'$ and therefore that $\angle DES$ is $180° - 150° - 1°9' = 28°51'$.

Ptolemy's procedure can be translated as follows. Given $\triangle ABC$ with a, b, and $\gamma > 90°$ known, drop AD perpendicular to BC extended (Figure 4.20). If $AD = h$ and $CD = p$, then $p = \frac{\text{crd}(2\gamma - 180) \cdot b}{2R}$ and $h = \frac{\text{crd}(360 - 2\gamma) \cdot b}{2R}$. It follows that

$$c^2 = h^2 + (a + p)^2$$

$$= a^2 + \left(\frac{\text{crd}^2(360 - 2\gamma)}{4R^2} + \frac{\text{crd}^2(2\gamma - 180)}{4R^2}\right)b^2 + \frac{2ab\,\text{crd}(2\gamma - 180)}{2R}$$

$$= a^2 + b^2 + 2ab\frac{\text{crd}(2\gamma - 180)}{2R}$$

or

$$c^2 = a^2 + b^2 - 2ab\cos\gamma,$$

precisely the law of cosines for the case where two sides and the included angle are known. To find the angles, Ptolemy then noted that crd(2β) = $\frac{h \cdot 2R}{c}$ and found β from the table. This translates as $\sin\beta = \frac{h}{c} = \frac{b\sin\gamma}{c}$. Hence Ptolemy has also used the equivalent of the law of sines.

It should be noted that in giving the above example Ptolemy explicitly provides an algorithm for calculating c and β given values of a, b, and γ. In fact, such algorithms are common in the *Almagest*. These algorithms of plane trigonometry can therefore be translated into modern formulas without doing injustice to Ptolemy's own procedure.

4.2.3 Solving Spherical Triangles

Ptolemy dealt even more extensively with algorithms for solving spherical triangles. Although spherical geometry had been studied as early as 300 B.C.E., the earliest work on spherical trigonometry appears to be the *Spherica* of Menelaus (c. 100 C.E.). A major result of that work, today known as Menelaus' theorem, gives the relationships among the arcs of great circles in the configuration on a spherical surface illustrated in Figure 4.21. Two arcs *AB*, *AC* are cut by two other arcs *BE*, *CD* which intersect at *F*. With the arcs labeled as in the figure, and further with $AB = m$, $AC = n$, $CD = s$, and $BE = r$, Menelaus' theorem, written using sines rather than chords, states that

$$\frac{\sin(n_2)}{\sin(n_1)} = \frac{\sin(s_2)}{\sin(s_1)} \cdot \frac{\sin(m_2)}{\sin(m)} \tag{4.1}$$

and

$$\frac{\sin(n)}{\sin(n_1)} = \frac{\sin(s)}{\sin(s_1)} \cdot \frac{\sin(r_2)}{\sin(r)}. \tag{4.2}$$

Menelaus proved these results (and the same proof also appears in the *Almagest*) by first proving them for a similar plane configuration and then projecting the spherical diagram onto a plane.[9] Ptolemy then used Menelaus' theorem to solve spherical right triangles, triangles composed of arcs of great circles where two of the arcs meet in a right angle. Given such a triangle with the right angle at *C*, and the sides opposite angles *C*, *B*, *A* labeled *c*, *b*, and *a* respectively (Figure 4.22), Ptolemy constructed a Menelaus configuration containing it. For example, if *ABC* is the right triangle, let *A* be a pole for the great circle *PM*, *B* a pole of *QN*, and extend each leg of the triangle to meet both of those great circles. There are then two Menelaus configurations, one with vertex at *M*, the other with vertex at *N*. Since the length of an arc on a great circle subtended by an angle at a pole of that circle is equal to the degree measure of the angle, and since *P* and *Q* are poles

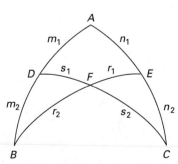

FIGURE 4.21
Menelaus configuration.

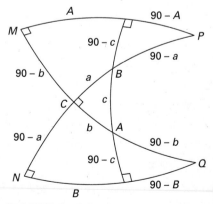

FIGURE 4.22
Ptolemy's double Menelaus configuration.

of QM, PN respectively, the two theorems can be simplified considerably to get results relating the angles and sides of the given triangle.

First, if one uses the configuration with vertex M, equation 4.1 becomes

$$\frac{\sin(90 - A)}{\sin A} = \frac{\sin(90 - a)}{\sin a} \cdot \frac{\sin b}{\sin 90} \qquad \text{or} \qquad \tan A = \frac{\tan a}{\sin b}. \qquad (4.3)$$

Equation 4.2 becomes

$$\frac{\sin 90}{\sin A} = \frac{\sin 90}{\sin a} \cdot \frac{\sin c}{\sin 90} \qquad \text{or} \qquad \sin A = \frac{\sin a}{\sin c}. \qquad (4.4)$$

Second, if one uses the configuration with vertex N, equation 4.1 becomes

$$\frac{\sin a}{\sin(90 - a)} = \frac{\sin c}{\sin(90 - c)} \cdot \frac{\sin(90 - B)}{\sin 90} \qquad \text{or} \qquad \cos B = \frac{\tan a}{\tan c} \qquad (4.5)$$

while equation 4.2 becomes

$$\frac{\sin 90}{\sin(90 - a)} = \frac{\sin 90}{\sin (90 - c)} \cdot \frac{\sin(90 - b)}{\sin 90} \qquad \text{or} \qquad \cos c = \cos a \cdot \cos b. \qquad (4.6)$$

Ptolemy's first application of these results is to find the declination δ and right ascension α of the sun, given its longitude λ (Figure 4.23). Here, VA is the equator, VB the ecliptic and V the vernal point. The angle ϵ between the equator and ecliptic, according to Ptolemy, is $23°51'20''$. Suppose the sun is at H, a point with longitude λ. To determine $HC = \delta$ and $VC = \alpha$, the right triangle VHC must be solved. From equation 4.4, $\sin \epsilon = \frac{\sin \delta}{\sin \lambda}$ or $\sin \delta = \sin \epsilon \sin \lambda$. Ptolemy performed this calculation with both $\lambda = 30°$ and $\lambda = 60°$ to get in the first case, $\delta = 11°40'$ and in the second, $\delta = 20°30'9''$. Having thus demonstrated the algorithm, he presumably set his calculators to work to produce a table for δ given each integral value of λ from $1°$ to $90°$. Similarly, from equation 4.5, $\cos \epsilon = \frac{\tan \alpha}{\tan \lambda}$ or $\tan \alpha = \cos \epsilon \tan \lambda$. Again, Ptolemy calculated the value of α corresponding to $\lambda = 30°$ to be $27°50'$ while that corresponding to $\lambda = 60°$ to be $57°44'$. He then listed the values of α corresponding to other values of λ.

Many of the other problems solved by Ptolemy are closely related to the determination of the "rising time" of an arc of the ecliptic. Namely, at a given geographical latitude Ptolemy wanted to determine the arc of the celestial equator which crosses the horizon at the same time as a given arc of the ecliptic. Since it is sufficient to determine this for arcs one endpoint of which is the vernal point, it is only necessary to determine the length VE of the equator which crosses the horizon simultaneously with the given arc VH of the ecliptic (Figure 4.24). This arc length is called the "rising time" because time is measured by the uniform motion of the equator around its axis. One complete revolution takes 24 hours, so $15°$ along the equator corresponds to 1 hour, and $1°$ corresponds to 4 minutes. In any case, to solve Ptolemy's problem it suffices to solve the triangle HCE for $EC = \sigma(\lambda)$ and then subtract that value from $VC = \alpha(\lambda)$ already determined. For example, suppose that the latitude $\phi = 36°$ and that $\lambda = 30°$. By the calculation above, $\delta = 11°40'$. Equation 4.3 then gives $\sin \sigma = \frac{\tan \delta}{\tan(90 - \phi)} = \tan \delta \tan \phi$ and, therefore, $\sigma = 8°38'$. Since $\alpha = 27°50'$, the rising time $VE = 27°50' - 8°38' = 19°12'$. Ptolemy (or his staff) calculated the rising time $\rho(\lambda, \phi)$ for values of λ in $10°$ intervals from $10°$ to $360°$ at eleven

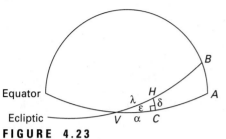

FIGURE 4.23
Method for determining the declination and
right ascension of the sun, given its longitude.

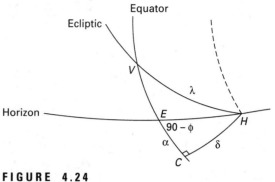

FIGURE 4.24
Calculating the rising time.

different latitudes ϕ and presented the results in an extensive table, the part of which for
latitude 36° is displayed:

Longitude	Time	Longitude	Time	Longitude	Time
10	6;14	130	118;50	250	265;42
20	12;35	140	131;13	260	277;29
30	19;12	150	143;32	270	288;45
40	26;13	160	155;45	280	299;19
50	33;46	170	167;54	290	309;6
60	41;58	180	180;0	300	318;2
70	50;54	190	192;6	310	326;14
80	60;41	200	204;15	320	333;47
90	71;15	210	216;28	330	340;48
100	82;31	220	228;47	340	347;25
110	94;18	230	241;10	350	353;46
120	106;30	240	253;30	360	360;0

As an easy application of the table, Ptolemy showed how to calculate the length of
daylight at any date at any given latitude. If the sun is at longitude λ, the point at longitude
$\lambda + 180$ is rising when the sun is setting. Hence one simply needs to subtract the rising
time of λ from that of $\lambda + 180$. For example, when $\phi = 36°$ and $\lambda = 90°$ (the summer
solstice), $\rho(270,36) - \rho(90,36) = 288°45' - 71°15' = 217°30'$, which corresponds to
14 1/2 hours.

Using the same diagram, Ptolemy could also calculate the position of the sun when it
rises, that is, the length of arc $EH = \beta$. To determine this at latitude 36° when $\lambda = 30°$,
one uses equation 4.4 to get

$$\sin \beta = \frac{\sin \delta}{\sin(90 - \phi)} = \frac{\sin 11°40'}{\sin 54°} = 0.25$$

and $\beta = 14°28'30''$. Therefore the sun will rise on that day $14°28'30''$ north of the east point on the horizon.

As a final application of spherical trigonometry, the distance of the sun from the zenith at noon is calculated. The sun on any given day is always at a distance δ from the equator. Hence at noon, when it crosses the meridian, it is (assuming $\delta > 0$) between the north pole N and the intersection T of the meridian with the equator at a distance δ from that intersection (Figure 4.25). Because arc $NT = 90°$ and arc $NY = \phi$, it follows that arc $SZ = 90° - (90° - \phi) - \delta = \phi - \delta$. Note that if $\phi - \delta > 0$, or $\phi > \delta$, the sun will be in the south at noon and hence shadows will point north. Because the maximum value of δ is $23°51'20''$, this will always be the case for latitudes greater than that value. On the other hand, when $\phi = \delta$, the sun is directly overhead at noon. The dates on which that occurs and also the dates when the sun is in the north at noon can easily be calculated for a given latitude. In any case, given the angular distance of the sun from the zenith, Ptolemy was able to calculate shadow lengths as previously described. He presented his results in a long table. For 39 different parallels of latitude he gave the length of the longest day as well as the shadow lengths of a pole of length 60 at noon on the summer solstice, the equinoxes, and the winter solstice.

The examples above deal only with the sun and are taken from the first three books of the *Almagest*. In the remainder of the work Ptolemy discussed the moon and the planets. For each heavenly body, he gave first a brief qualitative sketch of the phenomena to be explained, then an account of the postulated geometrical model, which combined epicycles and eccenters, and finally a detailed deduction of the parameters of the model from certain observations which he had personally made or of which he had records. He generally concluded by showing that his model with the calculated parameters in fact predicted a new planetary position, which was verified by observation. Ptolemy is thus the first mathematical scientist of whom there is documented evidence of the use of mathematical models in actually "doing" science. He began with a model and then used observations to improve it to the point that it predicted observed phenomena to within the limits of his observational accuracy.

Ptolemy was proud of his accomplishments in "saving the appearances," that is, in showing that for all seven of the wandering heavenly bodies "their apparent anomalies can be represented by uniform circular motions, since these are proper to the nature of divine beings. . . . Then it is right that we should think success in such a purpose a great thing and truly the proper end of the mathematical part of theoretical philosophy. But, on many grounds, we must think that it is difficult and that there is good reason why no one before us has yet succeeded in it."[10] Ptolemy, however, overcame the difficulties and gave to posterity a masterful mathematical work which did predict the celestial phenomena, a work not superseded for 1400 years (Side Bar 4.2).

FIGURE 4.25
Calculation of the distance of the sun from the zenith.

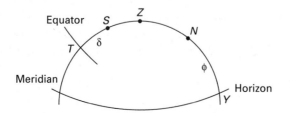

Side Bar 4.2   Ptolemy and the Idea of a Function

As a mathematical work, Ptolemy's *Almagest* raises the interesting question of whether one can see in it the germ of the modern idea of a function. First, there are many examples of tables displaying a functional relationship between sets of quantities. The Babylonians much earlier had compiled tables for square roots and reciprocals, for example, as well as astronomical ones giving the predicted time of various celestial phenomena. In general, however, they were only interested in discrete values. Ptolemy took the enormous step of providing a basis for the computational treatment of continuous phenomena not only by presenting tables, but also by showing how to interpolate to provide functional values for any given value of the "independent variable." Thus the chord is expressed as a function crd(α) of the arc, the declination of the sun as a function $\delta(\lambda)$ of the longitude, and the rising time $\rho(\lambda,\phi)$ as a function of the two variables representing the length of arc λ along the ecliptic and the geographical latitude ϕ. Ptolemy often used his tables in reverse as well, finding, for example, the arc from the chord, and thereby using what we would call the inverse function.

Second, however, given that Ptolemy's general aim was to predict planetary positions, in many places he wrote down an explicit algorithm describing how to do this for a given time. For example, to calculate the sun's position at any given time, Ptolemy described the various steps required: First calculate the time t from epoch (the starting point for all calculations—February 26, 747 B.C.E.) to the desired time;

next obtain the mean motion $\mu(t)$ from the "mean motion" table; add $\mu(t)$ to 265°15′ and subtract multiples of 360° to get a value $\overline{\lambda}$ less than 360°; enter $\overline{\lambda}$ in the table of the sun's anomaly (an entry of which was calculated in the example of Ptolemy's solution of an oblique triangle) to get $\theta(\overline{\lambda})$; and then add $\theta(\overline{\lambda})$ to $\overline{\lambda}$ and 65°30′ to get the final result. In modern symbols, we can write this result as $p(t) = \theta(\overline{\lambda}(t)) + \overline{\lambda}(t) + 65°30′$ (mod 360°), where $\overline{\lambda}(t) \equiv \mu(t) + 265°15′$ (mod 360°) and where θ, μ are themselves defined by tables derived from functional procedures. Although Ptolemy did not use modern symbolism, it is clear that he was well aware of the modern idea of a functional relationship. In many of his procedures, he even used appropriate symmetries to simplify his calculations.

Ptolemy did not, however, discuss the general notion of function. In fact, he apparently took the procedures for dealing with functions for granted. One concludes that such methods may well have been familiar to his readers and must have been used, at least by astronomers, before his time. Nevertheless, there is no evidence that any Greek mathematician wrote on the subject of functions, perhaps because there were no good theoretical methods of dealing with functions or their properties. There were no relevant postulates. It is, however, important to realize that behind the "geometrical facade of official Greek mathematics"[11] there existed areas of practical mathematics, the mathematics necessary to solve problems, both in the heavens and on earth.

4.3 PRACTICAL MATHEMATICS

Hipparchus' and Ptolemy's trigonometry enabled the Greeks to "measure" triangles in the heavens as well as those on the earth related to occurrences in the heavens. But surely the Greeks needed to be able to solve ordinary triangles on earth in order to make indirect measurements of distance and height. It would seem natural that, at least after the time of Hipparchus, they would use trigonometrical methods, that is, methods involving the table of chords. But the available historical evidence shows that they did not do so.

Naturally, before the time of Hipparchus one would only expect methods of indirect measurement coming directly from the notion of similarity. And this is exactly what is found in Euclid's *Optics*. This treatise is basically a work on the geometrical principles of vision, based on the assumption that light rays travel in straight lines. But Euclid does include several results on indirect measurement. For example, proposition 18 asks "to find the magnitude of a given height, the sun being visible."[12] In other words, with the sun at Γ, Euclid wanted to determine the height of a tower AB whose shadow has length $B\Delta$ (Figure 4.26). Placing an object of known height EZ in such a way that its shadow also has tip Δ and therefore length $E\Delta$, Euclid concluded from the similarity of triangles $AB\Delta$ and $ZE\Delta$ that the height AB was determined.

4.3.1 The Work of Heron

Some 350 years after Euclid, there is a more detailed work on indirect measurement in the *Dioptra* of Heron of Alexandria (first century C.E.). (The dioptra is a sighting instrument.) Heron too used similar triangles even though it appears from another of his books that he was familiar with a table of chords. Thus Heron showed how to determine the distance from the observer (at A) to an inaccessible point B by first choosing Γ so that ΓAB is a straight line, then constructing the perpendicular ΓE to ΓAB, and finally sighting B from E, thereby establishing a point Δ on BE such that $A\Delta$ is also perpendicular to $BA\Gamma$ (Figure 4.27). Since triangles $AB\Delta$ and ΓBE are similar, $\Gamma E : A\Delta = \Gamma B : BA$. The first ratio is known, since each length can be measured. Therefore the second ratio is known. But $\Gamma B : BA = (\Gamma A + AB) : BA = \Gamma A : BA + 1$, and since ΓA is known, BA can be determined.

Heron used analogous methods to determine such quantities as the distance between two inaccessible points, the height of a tower (without using shadows), and the depth of a valley. He also showed how to determine the direction to dig from each end in order to construct a straight tunnel through a mountain.

Heron's many works include other significant ideas in applied mathematics. Thus his *Catoptrica* contains an interesting proof that for light rays impinging on a mirror, the angle of incidence equals the angle of reflection. Although the result was known earlier, Heron

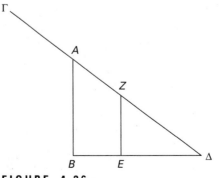

FIGURE 4.26
Calculating heights using the sun, from Euclid's *Optics*.

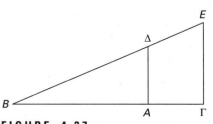

FIGURE 4.27
Calculating distances, from Heron's *Dioptra*.

based his proof on the hypothesis that "Nature does nothing in vain,"[13] that is, that the path of the light ray from object C by way of the mirror to the eye D must be the shortest possible. For suppose A is the point on the mirror GE which makes $\angle CAE = \angle DAG$ (Figure 4.28). Then extend DA to meet CE extended at F. It follows easily that $\triangle AEF$ is congruent to $\triangle AEC$ and therefore that the light path $DA + AC$ is equal to the straight line DAF. Now suppose B is any other point on the mirror. Connect BF, BD, and BC. Since $BF = BC$, we have $DB + BC = DB + BF > DAF$. Therefore any other proposed light ray path is longer than the one making the angle of incidence equal the angle of reflection.

In Heron's *Mechanics,* there appears what is today called the parallelogram of veloci-ties, although this idea too had appeared earlier in the work on mechanics attributed to Aristotle. Namely, suppose a point moves with uniform velocity along a straight line AB from A to B while at the same time the line AB moves with uniform velocity parallel to itself, ending on the line $\Gamma\Delta$ (Figure 4.29). Suppose EZ is any intermediate position of line AB and that G is the position of the moving point on it. Then $AE : A\Gamma = EG : EZ$ (by definition of the motion), so $AE : EG = A\Gamma : EZ = A\Gamma : \Gamma\Delta$ and G therefore lies on the diagonal $A\Delta$. In other words, the diagonal is the actual path of the moving point. In modern terms, the "velocity vector" $\overrightarrow{A\Delta}$ is the vector sum of the "velocity vectors" $\overrightarrow{AB}$ and $\overrightarrow{A\Gamma}$.

Naturally, the Greeks did not themselves consider "velocity vectors." Velocity to the Greeks was not considered as an independent quantity capable of being measured. There was no such concept as "miles per hour." Recall that according to *Elements* V, definition 3, ratios can only be taken between magnitudes of the same kind. One could not, therefore, consider the ratio of a distance to a time. One could only compare distances or compare times. Thus an early definition of velocity by Autolycus states, "a point is said to be moved with equal movement when it traverses equal and similar quantities in equal time. When any point on an arc of a circle or on a straight line traverses two lines with equal motion, the proportion of the time in which it traverses one of the two lines to the time in which it traverses the other is as the proportion of one of the two lines to the other."[14] In modern terms, Autolycus states that the velocity of a point is uniform when it covers equal dis-tances in equal times, and further that if the point covers distance s_1 in time t_1 and distance s_2 in time t_2, then $s_1 : s_2 = t_1 : t_2$. It is from this definition that the initial proportions in the previous paragraph as well as those in the discussion of the quadratrix in Chapter 3 stem.

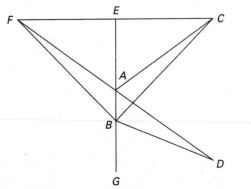

FIGURE 4.28
The angle of incidence equals the angle of reflection.

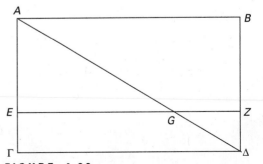
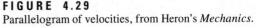

FIGURE 4.29
Parallelogram of velocities, from Heron's *Mechanics.*

The Greeks certainly observed that falling bodies did not move with uniform velocity. Thus they were aware of the notion of acceleration. One of the few extant explicit comments on accelerated motion, however, is from a sixth century C.E. commentary on the lost treatise *On Motion* by the physicist Strato (third century B.C.E.). Strato asserted first of all that a falling body "completes the last stage of its motion in the shortest time" and further that it traverses "each successive space more swiftly."[15] In other words, accelerated motion implies that successive equal distances are covered in shorter and shorter times and therefore with increasing velocities. It is not clear from the brief fragment, however, whether Strato meant to imply that the velocity of a falling body was proportional to distance fallen. A third century C.E. commentator on Aristotle did claim, however, that "bodies move downward more swiftly in proportion to their distance from above."[16]

Although the Greeks were familiar with the basic notions of kinematics, there is no evidence that they performed numerical calculations using them, as was done in the field of astronomy. On the other hand, the *Metrica* of Heron is an example of a handbook of practical mensuration, a book which enabled its readers to learn how to measure areas and volumes of various types of figures. Here, Heron showed how to arrive at numerical answers, even where "irrational" quantities were involved. Heron sometimes gave proofs, but always his aim was to calculate. In some sense this work is reminiscent of Chinese and Babylonian texts, but Heron often quoted the work of men such as Archimedes and Eudoxus in justifying his rules.

Book I of the *Metrica* gives procedures for calculating areas of plane figures and surface areas of solids. After the easy cases of the rectangle and the right and isosceles triangles, Heron dealt with finding the area of a scalene triangle whose sides are given. He presented two methods. The first method is based on *Elements* II-12, 13: Given a triangle *ABC*, drop the perpendicular *AD* to *BC* (or *BC* extended) and use the quoted theorems to show that $c^2 = a^2 + b^2 \mp 2a \cdot CD$ (Figure 4.30). It follows that *CD* is known, hence that $AD = h$ is known. The area is then $\frac{1}{2}ah$.

The second method is known today as Heron's formula. Namely, if $s = \frac{1}{2}(a + b + c)$, then the area equals $\sqrt{s(s - a)(s - b)(s - c)}$. As Heron stated it, "let the sides of the triangle be 7, 8, and 9. Add together 7, 8, and 9; the result is 24. Take half of this, which gives 12. Take away 7; the remainder is 5. Again, from 12 take away 8; the remainder is 4. And again 9; the remainder is 3. Multiply 12 by 5; the result is 60. Multiply this by 4; the result is 240. Multiply this by 3; the result is 720. Take the square root of this and it will be the area of the triangle."[17]

FIGURE 4.30
Area of a triangle from
Heron.

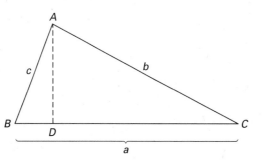

Heron gave here a correct geometrical proof of this area result. The formula and proof, probably due originally to Archimedes, are unusual in Greek times in that they involve the product of four lengths, a completely "ungeometrical" concept. Heron made no special note of this seeming aberration, so presumably it was already present in his source. Although in the *Elements* only two or three lengths could be multiplied to give a rectangle or a rectangular parallelepiped, the practical requirements of such aspects of Greek mathematics as are discussed in this chapter led certain mathematicians to consider lengths as "numbers" and, as such, to multiply them. Naturally, this new concept violated Aristotle's basic philosophical tenets as to how mathematics should be understood. It does show again, however, that there was much going on in Greek mathematics behind its "geometrical facade."

Heron continued in this passage to show how to calculate the necessary square roots:

> Since 720 has not a rational square root, we shall make a close approximation to the root in this manner. Since the square nearest to 720 is 729, having a root 27, divide 27 into 720; the result is 26 2/3; add 27; the result is 53 2/3. Take half of this; the result is 26 5/6. Therefore the square root of 720 will be very nearly 26 5/6. For 26 5/6 multiplied by itself gives 720 1/36; so that the difference is 1/36. If we wish to make the difference less than 1/36, instead of 729 we shall take the number now found, 720 1/36, and by the same method we shall find an approximation differing by much less than 1/36.[18]

This square root algorithm is another piece of practical mathematics which is, interestingly enough, quite different from Theon's description of Ptolemy's algorithm. Perhaps Heron's method was the procedure when calculating in base ten, while Ptolemy's was the method in astronomical sexagesimal calculation. It is also quite possible that one or both of these algorithms were used by the Babylonians.

The *Metrica* continued with formulas for the area A_n of a regular polygon of n sides of length a where n ranges from 3 to 12. For example, Heron showed that $A_3 \approx \frac{13}{30}a^2$, $A_5 \approx \frac{5}{3}a^2$, and $A_7 \approx \frac{43}{12}a^2$. In each case he used approximations to the various square roots which appeared in the geometrical derivations. It was in his derivation of the formula for the regular 9-gon that Heron appealed to a "table of chords" in which he found that the chord of a central angle of 40° is equal to one-third of the diameter of the circle. Therefore, $AC^2 = 9AB^2$, $BC^2 = 8AB^2$ (Figure 4.31), and $A_9 = 9\triangle ABO = \frac{9}{2}\triangle ABC = \frac{9}{4}BC \cdot AB = \frac{9}{4}\sqrt{8}a^2 \approx \frac{9}{4} \cdot \frac{17}{6}a^2 = \frac{51}{8}a^2$.

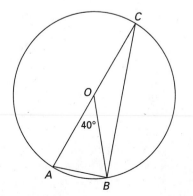

FIGURE 4.31
Calculating the area of a regular 9-gon.

For the circle, Heron used Archimedes' value 22/7 for π, thus giving the area of the circle as $\frac{11}{14}d^2$ where d is the diameter. He then quoted "the ancients" on the formula for the area of a segment of a circle, $A = \frac{1}{2}(b + h)h$, where b is the base and h the height of the segment. A more accurate value, he said, is given by adding the extra term $\frac{1}{14}\left(\frac{b}{2}\right)^2$. This new formula is certainly accurate for the semicircle, given that $\pi = 22/7$, but is only approximate for other segments. Heron even noted that it is only "reasonably" accurate when $b \leq 3h$.

For surface areas of spheres and segments of a sphere, as well as volumes of certain solids, Heron merely quoted Archimedes' results. Among other results, he gave formulas for the volume of a torus $(2\pi^2 ca^2)$, where a is the radius of a circular section and c the distance of the center of the section from the center of the torus, and of a regular octahedron $\left(\frac{1}{3}\sqrt{2}a^3\right)$, where a is the edge length.

4.3.2 The *Mishnat ha-Middot*

As noted, Heron's texts were circulated throughout the Greco-Roman world. In addition other similar works, even further removed from the Euclidean model, were produced. One particularly interesting example is the Hebrew work, the *Mishnat ha-Middot,* whose exact date and author are unknown. It is thought, however, that it was written by Rabbi Nehemiah in the middle of the second century, probably for use by Jewish farmers and artisans in Palestine. Like Heron's *Metrica,* it is a compilation of rules for areas and volumes. Unlike the *Metrica,* however, it contains no proofs.

A few paragraphs will give the flavor of this text:[19]

> II-3 How is it with the circle? . . . Multiply the diameter into itself and throw away from it one-seventh and the half of one-seventh; the rest is the area. For instance: The diameter is 7 long, its multiplication [into itself] is 49; a seventh and half a seventh is 10 1/2; the area is thus 38 1/2.

> II-4 How is it with the bow? . . . Add the arrow to the chord, [take] both together and multiply them into the half of the arrow and put them aside; and again . . . take the half of the chord, multiply it into itself and divide it upon 14 and the result let him add to the one standing [aside]; the resulting [sum] is the area.

In other words, the directions for the area of a circle give the formula $A = d^2 - \frac{1}{7}d^2 - \frac{1}{14}d^2 = \frac{11}{14}d^2$, and for the bow, that is, the segment of a circle, the result $A = (b + h)\frac{h}{2} + \frac{1}{14}\left(\frac{b}{2}\right)^2$. Both of these formulas are identical with those of Heron.

> II-10, 11 How do you compute [the volume of a frustum of a pyramid]? Take for example a square pillar, its bottom is four cubits in four cubits, [and its height is ten cubits], decreasing gradually as it rises, and its top is square, two cubits upon two cubits. . . . You can figure it out in numbers: [The ratio of] the length of the pillar which is the half of the ascension to the whole pillar is equal to the ratio of two to four. You find thus that the whole pillar till the top is finished is twenty cubits, and till [the beginning of] the frustum is ten cubits.

Rabbi Nehemiah completed the computation by using the standard formula for a pyramid to calculate that the volume of the entire pyramid of height 20 is $1/3 \cdot 4^2 \cdot 20 = 106 \ 2/3$ and that the volume of the (missing) top pyramid is 13 1/3. The frustum thus has volume 106 2/3 − 13 1/3 = 93 1/3.

III-4 What is that [quadrilateral] with equal sides and unequal angles? For example: five to each side, two narrow angles and two broad angles, and two threads [diagonals] cutting each other in the midst, the one eight and other one six. If one wants to measure, let him multiply one thread into the half of the other one, and the result is the area, 24 cubits.

The area of the rhombus with diagonals d_1, d_2 is thus calculated as $A = \frac{1}{2}d_1d_2$. This formula is also found in the work of Heron. Similarly, Nehemiah uses Heron's formula to determine the area of a triangle.

The final example demonstrates an attempt by Rabbi Nehemiah to overcome a potential conflict between science and religion:

V-3 . . . If you want to know the circumference all around [of a circle], multiply the diameter into 3 and one-seventh . . .

V-4 Now it is written: "And he made a molten sea of ten cubits from brim to brim, round in compass," and [nonetheless] its circumference is thirty cubits, for it is written: "And a line of thirty cubits compassed it round about." What is the meaning of the verse "and a line of thirty cubits" etc.? Nehemiah says, Since the people of the world say that the circumference of a circle contains three times and a seventh of the diameter, take off from that one-seventh for the thickness of the sea on the two brims, then there remain "thirty cubits [that compass it round about]."

The Biblical quotation, as noted in Chapter 1, is from I Kings 7:23. Nehemiah's solution to the conflict between the Bible and the "people of the world" was to note that the diameter of 10 cubits must have included the walls of the sea, while the circumference of 30 must be the inner circumference, excluding the walls. Unfortunately, it was not always to prove so simple to resolve apparent conflicts between religious beliefs and mathematical or scientific research.

Exercises

From Ptolemy's **Almagest**

1. Calculate crd(30°), crd(15°), and crd$\left(7\frac{1}{2}°\right)$ using the half-angle formula of Hipparchus. Also calculate crd(120°), crd(150°), crd(165°), and crd$\left(172\frac{1}{2}°\right)$ using his formula for crd(180° − α).

2. Use Theon's method to calculate $\sqrt{4500}$ to two sexagesimal places. The answer is 67;4,55.

3. Prove the sum formula,

$$120 \, \text{crd}(180 - (\alpha + \beta)) = \text{crd}(180 - \alpha)\,\text{crd}(180 - \beta) \\ - \text{crd}\,\alpha\,\text{crd}\,\beta,$$

using Ptolemy's theorem on quadrilaterals inscribed in a circle.

4. Use Ptolemy's difference formula to calculate crd(12°) and then apply the half-angle formulas to calculate crd(6°), crd(3°), crd$\left(1\frac{1}{2}°\right)$, and crd$\left(\frac{3°}{4}\right)$. Compare your results to Ptolemy's.

5. Write a computer program to do the calculations (in sexagesimal notation) of problem 4.

6. Compare the derivation of Hipparchus' half-angle formula to the method used by Archimedes in his lemmas in *Measurement of a Circle*.

7. Prove that crd β : crd α < β : α or, equivalently, that sin β/sin α < β/α for 0 < α < β.

8. Derive the law of cosines for the case where γ is acute by a method analogous to that of Ptolemy in his algorithm for finding the direction of the sun.

9. Calculate, using Ptolemy's methods, the length of a noon shadow of a pole of length 60 at the vernal equinox at a place of latitude 40° and one of latitude $23\frac{1}{2}°$. If G represents the position of the sun at noon on the summer solstice and L the position at noon on the winter solstice, it is known that arc GB = arc BL = $23\frac{1}{2}°$. Therefore, calculate the shadow lengths at those dates for latitudes

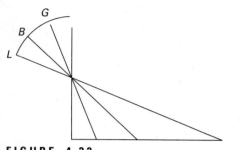

FIGURE 4.32
Shadow lengths.

$23\frac{1}{2}°$, 36°, and 40° (Figure 4.32). (Use modern methods for these calculations.)

10. Prove the Menelaus theorem in the plane (Figure 4.33). Show that

$$\frac{n}{n_1} = \frac{s}{s_1} \cdot \frac{r_2}{r}$$

and

$$\frac{n_2}{n_1} = \frac{s_2}{s_1} \cdot \frac{m_2}{m}.$$

Hint: Draw a line through E parallel to DC meeting AB at F to prove the first result, and a line through A parallel to BE meeting CD extended at G to prove the second.

11. Calculate the declination and right ascension of the sun when it is at longitude 90° (summer solstice), 120°, and 45°. By symmetry, find the declination at longitudes 270°, 240°, and 315°.

12. Write a computer program to calculate rising times $\rho(\lambda,\phi)$ for any values of the longitude λ and the geographic latitude ϕ.

13. Check the tabulated rising times $\rho(\lambda, \phi)$ for $\phi = 36°$ and $\lambda = 60°, 90°$, and 120°. Use symmetry to check $\rho(\lambda, 36°)$ for $\lambda = 300°, 270°$, and 240°.

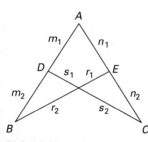

FIGURE 4.33
Menelaus' theorem in the plane.

14. Calculate the rising times $\rho(\lambda, \phi)$ for $\phi = 45°$ and $\lambda = 60°, 90°, 120°, 240°, 270°$, and 300°.

15. Calculate the length of daylight on a day when $\lambda = 60°$ at latitude 36°. Calculate the local time of sunrise and sunset. Calculate the length of daylight at latitude 45° on a day when $\lambda = 60°$. Calculate the local time of sunrise and sunset in that case.

16. Suppose that the maximum length of day at a particular location is known to be 15 hours. Calculate the latitude of that location and the position of the sun at sunrise on the summer and winter solstices.

17. The formula $\sin \sigma = \tan \delta \tan \phi$ only makes sense if the right hand side is less than or equal to 1. Since the maximum value of δ is 23°51′20″, show that the right hand side will be greater than 1 whenever $\phi > 66°8′40″$. Interpret the formula in this case in terms of the length of daylight.

18. Calculate the angular distance of the sun from the zenith at latitude 45° when $\lambda = 45°, 90°$, and 120°.

19. Calculate the position of the sun at sunrise at latitude 45° on the date when $\lambda = 30°$. Calculate the same position at latitudes 75°, 36°, and 20°.

20. Calculate the sun's maximal northerly sunrise point for latitudes 75°, 45°, 36°, and 20°. At approximately what date does the "midnight sun" begin at latitude 75°?

From Heron's Works

21. Show how to calculate the distance between two inaccessible points A, B by the use of similar triangles. (Assume, for example, that the two points are on the bank of a river opposite your position.)

22. Calculate the area of a triangle with sides of lengths 4, 7, and 10 using both of Heron's methods.

23. In Heron's formula $A_3 = \frac{13}{30}a^2$ for the area of an equilateral triangle with side a, what approximation has he used for $\sqrt{3}$? Derive this value by his square-root algorithm.

24. Derive a formula for the area A_5 of a regular pentagon with side a (using plane geometry). Discuss the differences between Heron's formula $A_5 = \frac{5}{3}a^2$ and your formula.

25. Heron derived his formula for the area A_7 of a regular heptagon of side a, $A_7 = \frac{43}{12}a^2$, by assuming that $a = \frac{7}{8}r$ where r is the radius of the circumscribed circle. Use this approximation to derive Heron's result. What square root approximation is necessary here?

26. Derive 17/6 as an approximation to $\sqrt{8}$ to complete the proof of Heron's formula for A_9.

27. Using trigonometry, derive a general formula for the area A_n of a regular n-gon of side a.

28. Derive Heron's formula for the volume $\frac{1}{3}\sqrt{2}a^3$ of a regular octahedron of edge length a.

FOR DISCUSSION...

29. Outline a trigonometry course following Ptolemy's order of presentation. Namely, derive the major formulas as tools for producing a sine table. Discuss the advantages and disadvantages of this approach compared to the standard textbook approach today.

30. Ptolemy must have been aware of a method of trisecting angles by the use of conic sections. Such a method would have enabled him to construct the chord of 1/2° given that he knew the chord of 1 1/2°. Why would Ptolemy not have considered this to be a construction by "geometrical methods"? Perform this construction using conic sections as in Chapter 3 and use it to calculate numerically the chord of 1/2°.

31. Discuss the potential for including some spherical trigonometry in courses on trigonometry, following the general lines of Ptolemy's approach.

32. Outline a lesson using the basic formulas of spherical trigonometry to calculate some simple astronomical phenomena.

33. List evidence which convinces you that the earth (a) rotates on its axis once a day and (b) revolves around the sun once a year. Would this evidence have convinced the Greeks? How would you refute the reasons Ptolemy gives for the earth's immovability?

34. Look up in an astronomy work the "equation of time," and discuss why the times of sunrise and sunset calculated by the methods in the text are likely to be incorrect by several minutes.

35. Read the recent article in *Mathematics Magazine* to learn why the extrema of sunrise and sunset do not occur on the solstices, even though those days are in fact the longest and shortest of the year.[20] Prepare a brief report to explain this surprising phenomenon.

References and Notes

Thomas Heath's *A History of Greek Mathematics* and B. L. Van der Waerden's *Science Awakening*, referred to in Chapter 2, have sections on the material of this chapter. Selections from the works of Ptolemy and others can be found in Ivor Bulmer-Thomas, *Selections Illustrating the History of Greek Mathematics*. The standard reference on the subject of the applications of mathematics to astronomy from Babylonian times to the sixth century C.E., however, is Otto Neugebauer, *A History of Ancient Mathematical Astronomy* (New York: Springer, 1975). This work provides a detailed study of the mathematical techniques used by Ptolemy and other astronomers as they worked out their versions of the system of the heavens. The best available English translation of Ptolemy's *Almagest* is by Gerald T. Toomer: *Ptolemy's Almagest* (New York: Springer, 1984). An earlier translation by R. Catesby Taliaferro is available in the *Great Books* (see Chapter 3 references).

1. Simplicius' commentary on Aristotle's *On the Heavens*, quoted in Pierre Duhem, *To Save the Phenomena* (Chicago: University of Chicago Press, 1969), p. 5. Duhem's work provides a detailed look at how the Greeks attempted to "save the phenomena."

2. Ptolemy, *Almagest*, I, 7.

3. Thomas Heath, *The Works of Archimedes* (New York: Dover, 1953), p. 222.

4. Plutarch's *On the Face of the Moon*, in Bulmer-Thomas, *Selections Illustrating*, II, p. 5.

5. The discussion is adapted from Thomas Kuhn, *The Copernican Revolution* (Cambridge: Harvard University Press, 1957), p. 58. This work provides excellent background reading for the nature of astronomy in Greek times. A detailed mathematical description of the motion of the planetary spheres, with several diagrams, is found in Neugebauer, *History of Ancient*, pp. 677–685.

6. Theon's *Commentary on Ptolemy's Syntaxis*, quoted in Bulmer-Thomas, *Selections Illustrating*, I, p. 61.

7. Ptolemy, *Almagest* I, 10.

8. The following discussion is adapted from that of J. L. Berggren in "Mathematical Methods in Ancient Science: Astronomy," an article in *History in Mathematics Education*, I. Grattan-Guinness, ed. (Paris: Belin, 1987).

9. For a proof of Menelaus' theorem, see Ptolemy, *Almagest*, I, 13 or Neugebauer, *History of Ancient*, pp. 27–28.

10. Ptolemy, *Almagest*, IX, 2.

11. Olaf Pedersen, *A Survey of the Almagest* (Odense: University Press, 1974), p. 93. This work is an excellent companion to a translation of Ptolemy's work. It provides background and commentary on all of Ptolemy's mathematical and astronomical material.

12. Euclid, *L'Optique et la Catoptrique*, translated into French by Paul Ver Eecke (Paris: Descleé de Brouwer, 1938), p. 13.

13. Bulmer-Thomas, *Selections Illustrating*, II, p. 497.

14. Quoted in Marshall Clagett, *The Science of Mechanics in the Middle Ages* (Madison: University of Wisconsin Press, 1961), p. 165. Although this work is chiefly concerned with medieval mechanics, there are summaries of Greek work at the beginning of most chapters.

15. Simplicius' *Commentary on Aristotle's Physics*, in Morris Cohen and I. E. Drabkin, *A Source Book in Greek Science* (Cambridge: Harvard University Press, 1948), p. 211. This book is an excellent source of original materials in Greek mathematics, astronomy, physics, and the other sciences.

16. Quoted in Marshall Clagett, *Greek Science in Antiquity* (New York: Collier, 1963), p. 92. Clagett provides here a succinct treatment of various aspects of Greek science from its beginnings through its effect on Latin science up through the early middle ages.

17. Bulmer-Thomas, *Selections Illustrating*, II, p. 471.

18. Ibid., II, p. 471.

19. From the *Mishnat Ha-Middot* from *Studies in Hebrew Astronomy and Mathematics* by Solomon Gandz. Copyright © 1970 KTAV Publishing House, Inc. Reprinted by permission. Gandz provides an introduction giving his reasons for placing this work in the second century. A more recent study by Gad B. Sarfatti, "The Mathematical Terminology of the *Mishnat ha-Middot*," *Leshonenu* 23, 156–171, particularly of the language used, casts some doubts on Gandz's conclusion and suggests that the work was not written until the ninth century. However, the close resemblance between it and Heron's *Metrica* does lead one to believe that the author was familiar with that work and might have adapted it to the conditions of second century Palestine.

20. Stan Wagon, "Why December 21 is the Longest Day of the Year," *Mathematics Magazine* 63 (1990), 307–311.

Summary of Astronomy and Practical Mathematics

408–355 B.C.E.	Eudoxus	Two-sphere model of universe
c. 300 B.C.E.	Autolycus	Text on spherics
c. 300 B.C.E.	Euclid	Texts on spherics and optics
310–230 B.C.E.	Aristarchus	Heliocentric theory
250–175 B.C.E.	Apollonius	Eccenters and epicycles
190–120 B.C.E.	Hipparchus	Beginnings of trigonometry
First century C.E.	Heron	Measurement, mechanics, optics
c. 100 C.E.	Menelaus	Spherical trigonometry
100–178 C.E.	Ptolemy	The *Almagest*
Second century C.E.	Rabbi Nehemiah	The *Mishnat ha-Middot*
330–405 C.E.	Theon	Commentary on the *Almagest*

Chapter 5
The Final Chapters of Greek Mathematics

"This tomb holds Diophantus . . . [and] tells scientifically the measure of his life. God granted him to be a boy for the sixth part of his life, and adding a twelfth part to this, he clothed his cheeks with down. He lit him the light of wedlock after a seventh part, and five years after his marriage He granted him a son. Alas! late-born wretched child; after attaining the measure of half his father's life, chill Fate took him. After consoling his grief by this science of numbers for four years, he ended his life."
(Epigram 126 of Book XIV of the Greek Anthology *(c. 500 C.E.))*[1]

*A*lexandria, March, 415 C.E.: "A rumor was spread among the Christians that [Hypatia], the daughter of Theon, was the only obstacle to the reconciliation of the prefect [Orestes] and the archbishop [Cyril]. On a fatal day in the holy season of Lent, Hypatia was torn from her chariot, stripped naked, dragged to the church, and inhumanly butchered by the hands of Peter the reader and a troop of savage and merciless fanatics. . . . The murder of Hypatia has imprinted an indelible stain on the character and religion of Cyril of Alexandria."[2]

Although the Egyptian-Greek empire of the Ptolemies, under whose rule the Museum and Library flourished, collapsed under the onslaught of Roman might in 31 B.C.E., the intellectual-scientific tradition of Alexandria continued to exist for many centuries. In Chapter 4, we discussed the work of three prominent mathematicians who flourished under Roman rule in Egypt—Heron, Menelaus, and Ptolemy—as well as that of a rabbi from Roman-ruled Judaea. There were other mathematicians in the first centuries of the common era whose works also had influence stretching into the Renaissance. This chapter will deal with four of them.

We will first discuss the works of Nicomachus of Gerasa, a Greek town in Judaea. He wrote in the late first century an *Introduction to Arithmetic,* based on his understanding of Pythagorean number philosophy. Besides Books VII–IX of Euclid's *Elements,* this is the only extant number theory work from Greek antiquity. The second mathematician to be studied is Diophantus of Alexandria. In the mid-third century, he wrote another important work entitled *Arithmetica,* which was destined to be of far more importance than

Nicomachus' book. Despite its title, this was a work in algebra, consisting mostly of an organized collection of problems translatable into what are today called indeterminate equations, all to be solved in rational numbers. Like Heron's *Metrica,* the style of the *Arithmetica* is that of a Chinese or Babylonian problem text rather than a classic Greek geometrical work. The third mathematician to be considered is also from Alexandria, the geometer Pappus of the early fourth century. He is best known not for his original work, but for his commentaries on various aspects of Greek mathematics and in particular for his discussion of the Greek method of geometric analysis. The chapter will be concluded with a brief discussion of the work of Hypatia, the first woman mathematician of whom any details are known. It was her death at the hands of an enraged mob which marked the effective end of the Greek mathematical tradition in Alexandria (Side Bar 5.1).

5.1 NICOMACHUS AND ELEMENTARY NUMBER THEORY

Almost nothing is known about the life of Nicomachus, but since his work is suffused with Pythagorean ideas, it is likely that he studied in Alexandria, the center of mathematical activity and of Neo-Pythagorean philosophy. Two of his works survive, the *Introduction to Arithmetic* and the *Introduction to Harmonics.* From other sources it appears that he also wrote introductions to geometry and astronomy, thereby completing a series on Plato's basic curriculum, the so-called quadrivium.

Nicomachus' *Introduction to Arithmetic* was probably one of several works written over the years to explain Pythagorean number philosophy, but it is the only one still extant. Since no text exists from the time of Pythagoras, it is the source of some of the ideas about Pythagorean number theory already discussed in Chapter 2. Because the work was written some 600 years after Pythagoras, however, it must be considered in the context of its time and compared with the only other treatise on number theory available, Books VII–IX of Euclid's *Elements.*

Nicomachus began this brief work, written in two books, with a philosophical introduction. Like Euclid, he follows the Aristotelian separation of the continuous "magnitude" from the discontinuous "multitude." Like Aristotle, he notes that the latter is infinite by increasing indefinitely, while the former is infinite by division. Continuing the distinction in terms of the four elements of the quadrivium, he distinguishes arithmetic and music, which deal with the discrete, the former absolutely, the latter relatively, from geometry and astronomy, which deal with the continuous, the former at rest and the latter in motion. Of these four subjects, the one which must be learned first is arithmetic, "not solely because . . . it existed before all the others in the mind of the creating God like some universal and exemplary plan, relying upon which as a design and archetypal example the creator of the universe sets in order his material creations and makes them attain to their proper ends, but also because it is naturally prior in birth inasmuch as it abolishes other sciences with itself, but is not abolished together with them."[3] In other words, arithmetic is necessary for each of the other three subjects.

Most of Book I of Nicomachus' *Introduction* is devoted to classification of integers and their relations. For example, the author divides the even integers into three classes, the even-times-even (those which are powers of two), the even-times-odd (those which are doubles of odd numbers), and the odd-times-even (all the others). The odd numbers are

---

Side Bar 5.1 **Mathematics in Rome**

It has long been recognized that there was no "Roman mathematics." Of course there were original mathematicians who worked in the Roman empire, primarily in Alexandria, but all of them were part of the continuing Greek tradition. There is no record, however, of any mathematicians who lived and worked at the center of the empire or who wrote in Latin. The great orator Cicero even admitted that the Romans were not interested in mathematics: "The Greeks held the geometer in the highest honor; accordingly, nothing made more brilliant progress among them than mathematics. But we have established as the limit of this art its usefulness in measuring and counting."[4]

In actual fact, the Romans did have somewhat more to do with mathematics than "measuring and counting." As Vitruvius wrote in *On Architecture,* "Mathematics furnishes many resources to architecture. It teaches the use of rule and compass, and this facilitates the laying out of buildings in their sites by the use of set squares, levels, and alignments. By optics in buildings, lighting is daily drawn from certain aspects of the sky. By arithmetic, the cost of building is summed up; the methods of mensuration are indicated, while the difficult problems of symmetry are solved, by geometrical rules and methods."[5] Still, it only appears that the architect needed the rudiments of arithmetic, geometry, and optics. All the necessary knowledge could be gleaned from works such as Heron's. The few mathematical ideas actually mentioned in Vitruvius' work include the use of certain proportions in building, the problem of doubling the square and the cube, and the use of a set square made of three rods of length 3, 4, 5 to get exact right angles.

What mathematics was necessary for surveying?[6] Roman surveyors laid out roads and aqueducts throughout a huge empire, many of which still survive. But an inspection of the extant surveying manuals shows that the Roman surveyors used only very elementary mathematics. For example, a surveyor's manual by Marcus Junius Nipsius uses the Pythagorean theorem to solve the problem of finding the two legs of a right triangle whose area is 60, hypotenuse 17, and sum of the two legs 23. Another manual, by Hygmus Giomaticus, gives a method for finding true north. One draws a circle on a flat space on the ground and places a sundial gnomon in the center, long enough so that its shadow sometimes falls outside the circle. One then marks where the moving shadow crosses the circle both in the morning and in the afternoon. If one draws a straight line connecting the two points and then constructs the perpendicular bisector of the line, that bisector will point due north and south (Figure 5.1).

The manual by Nipsius displays a method for measuring the width of a river by using congruent triangles (Figure 5.2). The distance BC is to be found. The point A is sighted in line with BC, and line AD is drawn at right angles to AC and bisected at G. Line DH is drawn perpendicular to AD to the point H from which G and C are sighted in a straight line. Then BC is equal to $DH - AB$. Thus, the Roman surveyors apparently used methods even more elementary than those of Heron.

It appears that the mathematics used in surveying, in architecture, or in the other activities necessary to administer the empire, all taken from earlier discoveries, was sufficient to solve whatever problems arose. With no need for more and no official encouragement of those whose intellectual curiosity ran toward that particular domain, the Roman Empire of the West survived for 500 years without making any contributions to the world's store of mathematical knowledge.

---

divided into the primes and the composites. Nicomachus takes what appears to us an inordinate amount of space discussing these classes and showing how the various members are formed. But it must be remembered that he was writing an introduction for beginners, not a text for mathematicians.

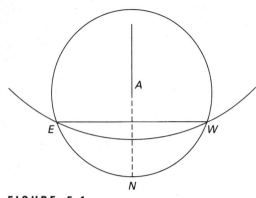

FIGURE 5.1
Determining true north.

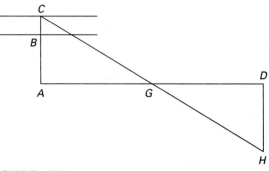

FIGURE 5.2
Finding the distance across a river.

Nicomachus discusses the Euclidean algorithm of repeated subtraction to find the greatest common measure of two numbers and to determine if two numbers are relatively prime. He also deals with the perfect numbers, giving the Euclidean construction (*Elements* IX-36) and, unlike Euclid, actually calculating the first four, 6, 28, 496, and 8128. However, also unlike Euclid, Nicomachus presents no proofs. He just gives examples.

The final six chapters of the first book are devoted to an elaborate scheme for naming ratios of unequal numbers. The basic classification is tenfold, five of which classes are obtained from the other five by adjoining the prefix "sub." One begins by reducing the ratio $A : B$ to its lowest terms, $a : b$. Then, if $a = nb$, the ratio of A to B is called **multiple**; if $b = na$, the ratio is called **submultiple**. If $a = b + 1$ (or $a/b = 1 + 1/b$), the ratio $A : B$ is called **superparticular**, while if $a + 1 = b$, it is called **subsuperparticular**. If $a = b + k$ $(1 < k < b)$, the ratio $A : B$ is called **superpartient**, while if $a + k = b$, it is called **subsuperpartient**. Combining the first of these classes with the next two gives the last two. Namely, if $a = nb + 1$, the ratio is called a **multiple superparticular** and if $a = nb + k$ $(1 < k < b)$, it is called a **multiple superpartient**. The addition of "sub" to these last two names, when the roles of a and b are reversed, completes the classification. Nicomachus further refines the scheme by giving special names to the ratios for various values of n, b, and k. For example, the ratio of $9 : 3$ is called the **triple**, the ratio of $3 : 2$ the **sesquialter**, that of $4 : 3$ the **sesquitertian**, that of $5 : 3$ the **superbipartient** (since $5 = 3 + 2$), and that of $7 : 4$ the **supertripartient**. Finally, the ratio $7 : 3$ is called the **double sesquitertian**, while that of $8 : 3$ is called the **double superbipartient**.

There are remnants of this naming scheme in *Elements* Book VII, but its origins are probably to be found in the early history of ratio in Pythagorean times. Some of the terms in fact stem from early music theory. This scheme was in common use in medieval and Renaissance arithmetics, and there is even a discussion of it in editions of Euclid's *Elements* into the eighteenth century.

In Book II, Nicomachus discusses plane and solid numbers, again in great detail but without proofs. This material is not mentioned at all by Euclid. Nicomachus not only deals with triangular and square numbers (see Chapter 2), but also considers pentagonal, hexagonal, and heptagonal numbers and shows how to extend this series indefinitely. For

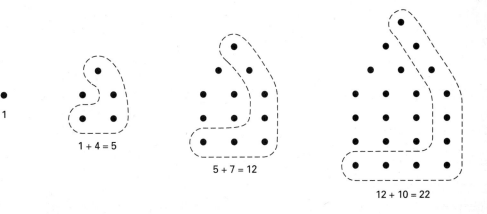

FIGURE 5.3
Pentagonal numbers.

example, the pentagonal numbers are the numbers 1, 5, 12, 22, 35, 51, . . . (although Nicomachus notes here that 1 is only the side of a "potential" pentagon). Each of these numbers can be exhibited, using the dot notation of Chapter 2, as a pentagon with equal sides (Figure 5.3). Beginning with 5, each is formed from the previous one in the sequence by adding the next number in the related sequence 4, 7, 10, ... So $5 = 1 + 4$, $12 = 5 + 7$, $22 = 12 + 10$, and so on. This is in perfect analogy to the series of triangular numbers 1, 3, 6, 10, . . . , each of which comes from the previous one by adding numbers of the sequence 2, 3, 4, . . . , and the series of squares, 1, 4, 9, 16, . . . , each of which results from the previous one by adding numbers of the sequence 3, 5, 7, ... Nicomachus continues this analogy and displays the first 10 numbers of each of the polygonal classes mentioned.

Nicomachus further explores the solid numbers. A pyramidal number, on a given polygonal base of side n, is formed by adding together the first n polygonal numbers of that shape. For example, the pyramidal numbers with triangular base are 1, $1 + 3 = 4$, $1 + 3 + 6 = 10$, $1 + 3 + 6 + 10 = 20$, . . . , while those with square base are 1, $1 + 4 = 5$, $1 + 4 + 9 = 14$, $1 + 4 + 9 + 16 = 30$, ... One can similarly construct pyramidal numbers on any polygonal base.

Another form of solid number is the cubic number. Nicomachus notes, again without proof, that the cubes are formed from odd numbers, not even. Thus, the first (potential) cube 1 equals the first odd number, the second cube, 8, equals the sum of the next two odd numbers, the third cube, 27, equals the sum of the next three odd numbers, and so on. Thus the cubes are closely related to the squares, which are also formed by adding odd numbers. And, Nicomachus concludes, these two facts show that the odd numbers, not the even, are the cause of "sameness."

The final topic of the treatise is proportion. Nicomachus, referring to pre-Euclidean terminology, uses the word proportion in a different sense from Euclid's definition 2 of *Elements,* Book VII. For Euclid, three numbers are in proportion if the first is the same multiple (or part or parts) of the second that the second is of the third. Nicomachus notes that "the ancients" considered not only this type (the type he calls geometric), but also two others, the arithmetic and the harmonic. For Nicomachus, an **arithmetic** proportion of three terms is a series in which each consecutive pair of terms differs by the same quantity. For example, 3, 7, 11 are in arithmetic proportion. Among the properties of such a proportion are that the product of the extremes is smaller than the square of the mean by the square of the difference. In a **geometric** proportion, "the only one in the strict sense of the word to be called a proportion,"[7] the greatest term is to the next greatest as that one is to the next. For example, 3, 9, 27 are in geometric proportion. Among the properties of such a proportion is that the product of the extremes equals the square of the mean. Nicomachus quotes two results of Euclid in this regard, namely that only one mean term lies between two squares while two lie between two cubes.

The third type of proportion among three terms, the **harmonic**, is that in which the greatest term is to the smallest as the difference between the greatest and mean terms is to the difference between the mean and the smallest terms. For example, 3, 4, 6 are in harmonic proportion because $6 : 3 = (6 - 4) : (4 - 3)$. Among the properties of this proportion is that when the extremes are added together and multiplied by the mean, the result is twice the product of the extremes. Nicomachus gives as a possible reason for the term "harmonic" that 6, 4, 3 come from the most elementary harmonies. The ratio $6 : 4 = 3 : 2$ gives the musical fifth; the ratio $4 : 3$ gives the fourth, and the ratio $6 : 3 = (4 : 3)(3 : 2) = 2 : 1$ gives the octave. Today, it is more common to use the names "arithmetic," "geometric," and "harmonic" for means rather than for proportions. Thus 7 is the arithmetic mean of 3 and 11, 9 the geometric mean of 3 and 27, and 4 the harmonic mean of 3 and 6.

Nicomachus concludes the work by dealing with seven other forms of proportion which "do not occur frequently in the ancient writings, but are included merely for the sake of our own acquaintance with them, and, so to speak, for the completeness of our reckoning."[8] As an example, there is the **subcontrary**, which occurs when in three terms the greatest is to the smallest as the difference of the smaller terms is to the difference of the greater. The numbers 3, 5, 6 are in the subcontrary proportion.

The *Introduction to Arithmetic* was obviously just that, a basic introduction to elementary ideas about the positive integers. Although it has some points in common with Euclid's *Elements,* it is written at a far lower level. There are no proofs at all, just a large number of examples. The book was therefore suitable for use by beginners in schools. It was in fact used extensively during ancient times, was translated into Arabic in the ninth century, and was used, in a Latin paraphrase by Boethius (c. 480–524) throughout the early Middle Ages in Europe. For these reasons, copies still exist. That it was so popular and that no more advanced work on the subject, including Euclid's *Elements,* was studied during much of the period in Europe, shows the level to which mathematical study there fell from its Greek heights. These elementary number properties were for many centuries the summit of the arithmetic curriculum.

5.2 DIOPHANTUS AND GREEK ALGEBRA

Little is known about Diophantus' life, other than what is found in the epigram at the beginning of the chapter, except that he lived in Alexandria. It is through his major work, the *Arithmetica,* that his influence has reached modern times. Diophantus tells us in his introduction that the *Arithmetica* is divided into thirteen books. Only six have survived in

---

Side Bar 5.2   Diophantus' Terms and Symbolism

"All numbers are made up of some multitude of units. . . . Among them are—

   squares, which are formed when any number is multiplied by itself; the number itself is called the *side of the square*; *cubes,* which are formed when squares are multiplied by their sides;
   square-squares, which are formed when squares are multiplied by themselves;
   square-cubes, which are formed when squares are multiplied by the cubes formed from the same side;
   cube-cubes, which are formed when cubes are multiplied by themselves;

and it is from the addition, subtraction, or multiplication of these numbers, or from the ratio which they bear one to another or to their own sides, that most arithmetical problems are formed; you will be able to solve them if you follow the method shown below.

   Now each of these numbers, which have been given abbreviated names, is recognized as an element in arithmetical science; the *square* [of the unknown quantity] is called *dynamis* and its sign is Δ with the index Y, that is Δ^Y; the cube is called *kubos* and has for its sign K with the index Y, that is K^Y; the square multiplied by itself is called *dynamo-dynamis* and its sign is two *deltas* with the index Y, that is $\Delta^Y\Delta$; the square multiplied by the cube formed from the same root is called *dynamo-kubos* and its sign is ΔK with the index Y, that is ΔK^Y; the cube multiplied by itself is called *kubo-kubos* and its sign is two *kappas* with the index Y, $K^Y K$.

   The number which has none of these characteristics, but merely has in it an undetermined multitude of units, is called *arithmos*, and its sign is ς. There is also another sign denoting the invariable element in determinate numbers, the unit, and its sign is M with the index O, that is $\overset{o}{M}$."[9]

Greek. Four others were recently discovered in an Arabic version. From internal references it appears that these form books four through seven of the complete work, while the final three Greek books come later.[10] We will refer to the Greek books as I–VI and the Arabic ones as A, B, C, D. The style of the Arabic books is somewhat different from that of the Greek in that each step in the solution of a problem is explained more fully. It is quite possible, therefore, that the Arabic work is a translation not of Diophantus' original, but of a commentary on the *Arithmetica,* written by Hypatia around 400.

   Before dealing with the problems of the *Arithmetica,* it is worthwhile to discuss Diophantus' major advance in the solution of equations, his introduction of symbolism. The Egyptians and Babylonians wrote out equations and solutions in words. Diophantus, on the other hand, introduced symbolic abbreviations for the various terms involved in equations (Side Bar 5.2). And in a clear break with traditional Greek usage, he dealt with powers higher than the third.

   Note that all of Diophantus' symbols are abbreviations, including the final two: ς is a contraction of the first two letters of αριθμος (*arithmos* or number) while $\overset{o}{M}$ stands for μονας (*monas* or unit). Thus the manuscripts contain expressions such as $\triangle^Y$γςιβ$\overset{o}{M}$θ, which stands for 3 squares, 12 numbers, and 9 units, or, as we will write it, $3x^2 + 12x + 9$. (Recall that the Greeks used an alphabetic cipher for representing numbers in which, for example, γ = 3, ιβ = 12, and θ = 9.) Diophantus further used the above symbols with the mark χ to designate reciprocals. For example, $\triangle^{Yx}$ represented $1/x^2$. In addition, the symbol Λ, perhaps coming from an abbreviation for λεῦψις (*lepsis* or wanting or negation), is used for "minus" as in KᵞαγϚ ΛΔᵞγM̊α for $x^3 - 3x^2 + 3x - 1$. (Negative terms are always collected, so a single Λ suffices for all terms following it.) In the discussion of Diophantus' problems, however, we will use modern notation.

Diophantus was also aware of the rules for multiplying with the minus: "A minus multiplied by a minus makes a plus, a minus multiplied by a plus makes a minus."[11] Of course, Diophantus is not here dealing with negative numbers, which do not exist for him. He is simply stating the rules necessary for multiplying algebraic expressions involving subtractions. Diophantus assumes that the rules for adding and subtracting with positive and negative terms are known. Near the conclusion of his introduction he states,

> It is well that one who is beginning this study should have acquired practice in the addition, subtraction, and multiplication of the various species [types of terms]. He should know how to add positive and negative terms with different coefficients to other terms, themselves either positive or likewise partly positive and partly negative, and how to subtract from a combination of positive and negative terms other terms either positive or likewise partly positive and partly negative.

The basic rules for solving equations are then succinctly stated:

> If a problem leads to an equation in which certain terms are equal to terms of the same species but with different coefficients, it will be necessary to subtract like from like on both sides, until one term is found equal to one term. If by chance there are on either side or on both sides any negative terms, it will be necessary to add the negative terms on both sides, until the terms on both sides are positive, and then again to subtract like from like until one term only is left on each side. This should be the object aimed at in framing the hypotheses of propositions, that is to say, to reduce the equations, if possible, until one term is left equal to one term; but I will show you later how, in the case also where two terms are left equal to one term, such a problem is solved.[12]

In other words, Diophantus' general method of solving equations is designed to lead to an equation of the form $ax^n = bx^m$, where, in the first three books at least, m and n are no greater than 2. On the other hand, he does know how to solve quadratic equations of, for example, the form $ax^2 + c = bx$.

5.2.1 Linear and Quadratic Equations

Most of Diophantus' problems are indeterminate, that is, can be written as a set of k equations in more than k unknowns. Often there are infinitely many solutions. For these problems, Diophantus generally gives only one solution explicitly, but one can easily extend the method to give other solutions. For determinate problems, once certain quantities are made explicit, there is only one solution. Examples of both of these types will be described in what follows.[13]

Problem I-1.   *To divide a given number into two having a given difference.*

Diophantus presents the solution for the case where the given number is 100 and the given difference is 40. If x is the smaller of the two numbers of the solution, then $2x + 40 = 100$, so $x = 30$, and the required numbers are 30 and 70. This problem is determinate, once the "given" numbers are specified, but Diophantus' method works for any pair. If a is the given number and $b < a$ the given difference, then the equation would be $2x + b = a$, and the required numbers would be $\frac{1}{2}(a - b)$ and $\frac{1}{2}(a + b)$.

Problem I-5.   *To divide a given number into two numbers such that given fractions (not the same) of each number when added together produce a given number.*

In modern notation, we are given a, b, r, s ($r < s$) and asked to find u, v such that $u + v = a$, $\frac{1}{r}u + \frac{1}{s}v = b$. Diophantus notes that for this problem to be solvable, it is necessary that $\frac{1}{s}a < b < \frac{1}{r}a$. He then presents the solution in the case where $a = 100$, $b = 30$, $r = 3$, and $s = 5$: Let the second part (of 100) be $5x$. Therefore, the first part is $3(30 - x)$. Hence $90 + 2x = 100$ and $x = 5$. The required parts are then 75 and 25.

Like problem I-1, once the "given" numbers are specified, this problem is determinate, but the method works for any choice of the "givens" meeting the required condition. In the present case, Diophantus takes for his unknown 1/5 of the second part. This allows him to avoid fractions in the rest of his calculation because 1/3 of the first part must then equal $30 - x$ and the first part must be $3(30 - x)$. The remainder of the solution is clear. To check the generality, let sx represent the second part of a and $r(b - x)$ the first. The equation becomes

$$sx + r(b - x) = a \qquad \text{or} \qquad br + (s - r)x = a.$$

Thus,

$$x = \frac{a - br}{s - r},$$

a perfectly general solution. Since x must be positive, $a - br > 0$ or $b < \frac{1}{r}a$, the first half of Diophantus' necessary condition. The second half, that $\frac{1}{s}a < $ b, or $a < sb$, comes from the necessity that $sx < a$ or $s\left(\frac{a - br}{s - r}\right) < a$. In this particular problem, as in most of the problems in Book I, the given values are picked to ensure that the answers are integers. But in the other books, the only general condition on solutions is that they be positive rational numbers. Evidently Diophantus began with integers merely to make these introductory problems easier. In what follows, then, the word "number" should always be interpreted as "rational number."

Problem I-28.   *To find two numbers such that their sum and the sum of their squares are given numbers.*

It is a necessary condition that double the sum of the squares exceeds the square of the sum by a square number. In the problem presented, the given sum is 20 and the sum of the squares is 208.

This problem is of the general form $x + y = a$, $x^2 + y^2 = b$, a type solved by the Babylonians. Three other Babylonian types are exemplified in I-27, I-29, and I-30, namely $x + y = a$, $xy = b$; $x + y = a$, $x^2 - y^2 = b$; and $x - y = a$, $xy = b$ respectively. Diophantus' solution to the present problem, as to the others, is similar to the Babylonian methods. Namely, he takes the two unknowns as $10 + z$ and $10 - z$. The equation on the squares becomes $200 + 2z^2 = 208$, so $z = 2$ and the required two numbers are 12 and 8.

Diophantus' method, applicable to any system of the given form, can be translated into the modern formula

$$x = \frac{a}{2} + \frac{\sqrt{2b - a^2}}{2} \qquad y = \frac{a}{2} - \frac{\sqrt{2b - a^2}}{2}.$$

His condition is then necessary to ensure that the solution is rational. As in the first example, however, the problem is determinate.

   Did Diophantus take his solutions from Babylonian material? Or perhaps did he "translate" Euclid's geometric solutions to these problems from the *Elements* or the *Data*? These questions cannot be answered. It is, however, apparent that there is no geometric methodology in Diophantus' procedures. Perhaps by this time the Babylonian algebraic methods, stripped of their geometric origins, were known in the Greek world.

Problem II-8.   *To divide a given square number into two squares.*

   Let it be required to divide 16 into two squares. And let the first square $= x^2$; then the other will be $16 - x^2$; it shall be required therefore to make $16 - x^2 =$ a square. I take a square of the form $(ax - 4)^2$, a being any integer and 4 the root of 16; for example, let the side be $2x - 4$, and the square itself $4x^2 + 16 - 16x$. Then $4x^2 + 16 - 16x = 16 - x^2$. Add to both sides the negative terms and take like from like. Then $5x^2 = 16x$, and $x = 16/5$. One number will therefore be 256/25, the other 144/25, and their sum is 400/25 or 16, and each is a square.[14] (Figure 5.4)

   This is an example of an indeterminate problem. It translates into one equation in two unknowns, $x^2 + y^2 = 16$. This problem also demonstrates one of Diophantus' most common methods. In many problems from Book II onward, Diophantus requires a solution, expressed in the form of a quadratic polynomial, which must be a square. To ensure a rational solution, he chooses his square in the form $(ax \pm b)^2$ with a and b selected so that either the quadratic term or the constant term is eliminated from the equation. In this case, where the quadratic polynomial is $16 - x^2$, he uses $b = 4$ and the negative sign, so the constant term will be eliminated and the resulting solution will be positive. The rest of the solution is then obvious. The method can be used to generate as many solutions as desired to $x^2 + y^2 = 16$, or, in general, to $x^2 + y^2 = b^2$. Take any value for a and set $y = ax - b$. Then $b^2 - x^2 = a^2x^2 - 2abx + b^2$ or $2abx = (a^2 + 1)x^2$, so $x = \frac{2ab}{a^2 + 1}$.

   As another example where Diophantus needs a square consider

Problem II-19.   *To find three squares such that the difference between the greatest and the middle has a given ratio to the difference between the middle and the least.*

   Diophantus assumes that the given ratio is $3 : 1$. If the least square is x^2, then he takes $(x + 1)^2 = x^2 + 2x + 1$ as the middle square. Since the difference between these two squares is $2x + 1$, the largest square must be $x^2 + 2x + 1 + 3(2x + 1) = x^2 + 8x + 4$. To make that quantity a square, Diophantus sets it equal to $(x + 3)^2$, in this case choosing the coefficient of x so that the x^2 terms cancel. Then $8x + 4 = 6x + 9$, so $x = 2 \ 1/2$ and the desired squares are 6 1/4, 12 1/4, 30 1/4. One notices, however, that given his initial choice of $(x + 1)^2$ as the middle square, 3 is the only integer b Diophantus could use in $(x + b)^2$ which would give him a solution. Of course, with other values of the initial ratio, there would be more possibilities as there would with a different choice for

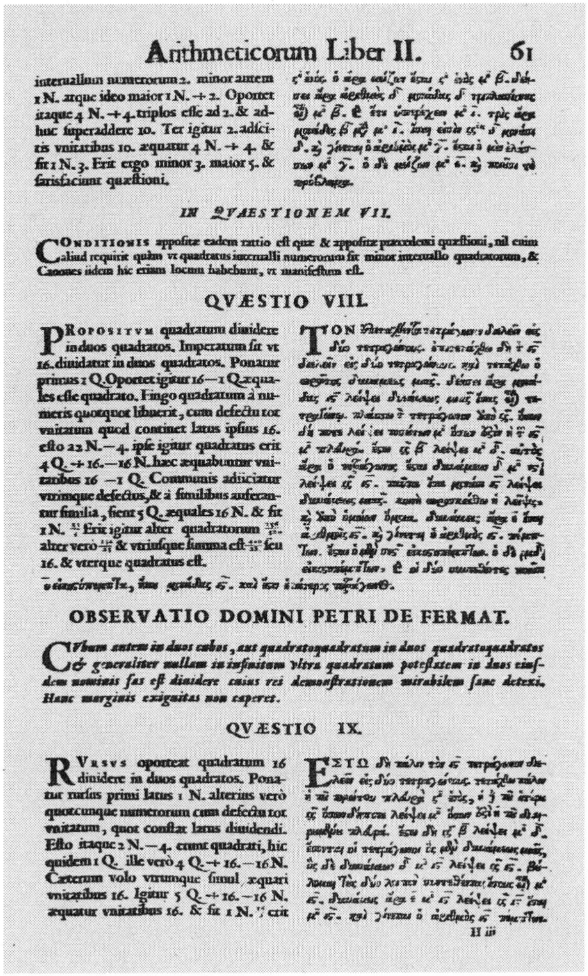

Arithmeticorum Liber II. 61

interuallum numerorum 2. minor autem 1 N. atque ideo maior 1 N. + 2. Oportet itaque 4 N. + 4. triplos esse ad 2. & adhuc superaddere 10. Ter igitur 2. adscitis vnitatibus 10. æquatur 4 N. + 4. & fit 1 N. 3. Erit ergo minor 3. maior 5. & satisfaciunt quæstioni.

IN QVÆSTIONEM VII.

CONDITIONIS appositæ eadem ratio est quæ & appositæ præcedenti quæstioni, nil enim aliud requirit quàm vt quadratus interualli numerorum sit minor interuallo quadratorum, & Canones iidem hic etiam locum habebunt, vt manifestum est.

QVÆSTIO VIII.

PROPOSITVM quadratum diuidere in duos quadratos. Imperatum sit vt 16. diuidatur in duos quadratos. Ponatur primus 1 Q. Oportet igitur 16 − 1 Q. æquales esse quadrato. Fingo quadratum à numeris quotquot libuerit, cum defectu tot vnitatum quod continet latus ipsius 16. esto 2 N. − 4. ipse igitur quadratus erit 4 Q. + 16. − 16 N. hæc æquabuntur vnitatibus 16 − 1 Q. Communis adiiciatur vtrimque defectus, & à similibus auferantur similia, fient 5 Q. æquales 16 N. & fit 1 N. 16/5. Erit igitur alter quadratorum 256/25. alter verò 144/25. & vtriusque summa est 400/25. seu 16. & vterque quadratus est.

OBSERVATIO DOMINI PETRI DE FERMAT.

Cubum autem in duos cubos, aut quadratoquadratum in duos quadratoquadratos & generaliter nullam in infinitum vltra quadratum potestatem in duos eiusdem nominis fas est diuidere cuius rei demonstrationem mirabilem sane detexi. Hanc marginis exiguitas non caperet.

QVÆSTIO IX.

RVRSVS oporteat quadratum 16 diuidere in duos quadratos. Ponatur rursus primi latus 1 N. alterius verò quotcunque numerorum cum defectu tot vnitatum, quot constat latus diuidendi. Esto itaque 2 N. − 4. erunt quadrati, hic quidem 1 Q. ille verò 4 Q. + 16. − 16 N. Cæterum volo vtrumque simul æquari vnitatibus 16. Igitur 5 Q. + 16. − 16 N. æquatur vnitatibus 16. & fit 1 N. 16/5 erit

H iii

FIGURE 5.4 Page 61 from 1670 edition of the *Arithmetica* of Diophantus. This page contains problem II-8 and the note of Fermat in which he states the impossibility of dividing a cube into a sum of two cubes or, in general, any nth power ($n > 2$) into a sum of two nth powers. (Source: Smithsonian Institution Photo No. 92-337)

the second square. In any case, in this problem as in all Diophantus' problems, only one solution is requested.

Problem II-11 introduces another general method, that of the double equation.

Problem II-11. *To add the same (required) number to two given numbers so as to make each of them a square.*

Diophantus takes the given numbers as 2 and 3. If his required number is x, he needs both $x + 2$ and $x + 3$ to be squares. He therefore must solve $x + 3 = u^2, x + 2 = v^2$ for x, u, v. Again, this is an indeterminate problem. Diophantus describes his method as follows: "Take the difference between the two expressions and resolve it into factors. Then take either (a) the square of half the difference between these factors and equate it to the lesser expression or (b) the square of half the sum and equate it to the greater."[15]

Since the difference between the expressions is $u^2 - v^2$ and this factors as $(u + v)$ $(u - v)$, the difference of these two factors is $2v$ while the sum is $2u$. What Diophantus does not mention explicitly is that the initial factoring must be carefully chosen so that the solution for x is a positive rational number. In the present case, the difference between the two expressions is 1. Diophantus factors that as $4 \times 1/4$. Thus $u + v = 4$ and $u - v = 1/4$, so $2v = 15/4$, $x + 2 = v^2 = 225/64$, and $x = 97/64$. Note, for example, that the factorization $2 \times 1/2$ would not give a positive solution, nor would the factorization $3 \times 1/3$. The factorization $1 = a \cdot 1/a$ needs to be chosen so that $\left[\frac{1}{2}\left(a - \frac{1}{a}\right)\right]^2 > 2$.

5.2.2 Higher Degree Equations

Book A begins with a new introduction. Since the problems now begin to involve cubes and even higher powers, Diophantus describes the rules for multiplying such powers:

> When I then multiply x^3 by x, the result is the same as when x^2 is multiplied by itself, and it is called x^4. If x^4 is divided by x^3, the result is x, namely the root of x^2; if it is divided by x^2, the result is x^2; if it is divided by x, namely the root of x^2, the result is x^3 ... When x^5 is then multiplied by x, the result is the same as when x^3 is multiplied by itself and when x^2 is multiplied by x^4, and it is called x^6. If x^6 is divided by x, namely the root of x^2, the result is x^5; if it is divided by x^2, the result is x^4; if it is divided by x^3, the result is x^3; if it is divided by x^4, the result is x^2; if it is divided by x^5, the result is x, namely the root of x^2.[16]

It is perhaps a bit misleading to translate Diophantus' rules using the symbolism of x^n, since that symbolism makes the rules almost self-evident. Diophantus' own symbols, recall, are Δ^Y, K^Y, $\Delta^Y\Delta$, ΔK^Y, and K^YK for x^2 through x^6. So his readers would read statements like ΔK^Y multiplied by ς equals K^Y multiplied by itself, equals Δ^Y multiplied by $\Delta^Y\Delta$, and all equal K^YK.

Diophantus, continuing his introduction, explains that as before, his equations will end up with a term in one power equaling a term in another, that is $ax^n = bx^m$ $(n < m)$. Now, however, m may be any number up to 6, and to solve, one must use the rules to divide both sides by the lesser power and end up with one "species" equal to a number, that is, $a = bx^{m-n}$. The latter equation is easily solved. Speaking to the reader, he writes further, "when you are acquainted with what I have presented, you will be able to find the answer to many problems which I have not presented, since I shall have shown to you the procedure for solving a great many problems and shall have explained to you an example of each of their types."[17]

As an example of Diophantus' use of higher powers of x, consider

Problem A-25. *To find two numbers, one a square and the other a cube, such that the sum of their squares is a square.*

The goal is to find x, y, and z such that $(x^2)^2 + (y^3)^2 = z^2$. Diophantus sets x equal to $2y$ (the 2 is arbitrary) and performs the exponentiation to conclude that $16y^4 + y^6$ must be a square which he takes to be the square of ky^2. So $16y^4 + y^6 = k^2y^4$, $y^6 = (k^2 - 16)y^4$, and $y^2 = k^2 - 16$. It follows that $k^2 - 16$ must be a square. Diophantus chooses the easiest value, namely $k^2 = 25$, so $y = 3$. It then follows that the desired numbers are $y^3 = 27$ and $(2y)^2 = 36$. This solution is easily generalized. Take $x = ay$ for any positive a. Then k and y must be found so that $k^2 - a^4 = y^2$ or so that $k^2 - y^2 = a^4$. Diophantus had, however, already demonstrated in problem II–10 that one can always find two squares whose difference is given.

Problem B-7 shows that Diophantus knew the expansion of $(x \pm y)^3$. As he puts it, "whenever we wish to form a cube from some side made up of the sum of, say, two different terms—so that a multitude of terms does not make us commit a mistake—we have to take the cubes of the two different terms, and add to them three times the results of the multiplication of the square of each term by the other."[18]

Problem B-7. *To find two numbers such that their sum and the sum of their cubes are equal to two given numbers.*

The problem asks to solve $x + y = a$, $x^3 + y^3 = b$. This system of two equations in two unknowns is determinate. It is a generalization of the "Babylonian" problem I-28, $x + y = a$, $x^2 + y^2 = b$. Diophantus' method of solution generalizes his method there. Letting $a = 20$ and $b = 2240$, he begins as before by letting the two numbers be $10 + z$ and $10 - z$. The second equation then becomes $(10 + z)^3 + (10 - z)^3 = 2240$ or, using the expansions mentioned,

$$2000 + 60z^2 = 2240 \qquad \text{or} \qquad 60z^2 = 240, z^2 = 4, \text{ and } z = 2.$$

Diophantus gives, of course, a condition for a rational solution, namely, that $\frac{4b - a^3}{3a}$ is a square (equivalent to the more natural condition that $\frac{b - 2(a/2)^3}{3a}$ is a square). It is interesting that the answers here are the same as in I-28, namely, 12 and 8.

When reading through the *Arithmetica,* one never quite knows what to expect next. There are a great variety of problems. Often there are several similar problems grouped together, one involving a subtraction where the previous one involved an addition, for example. But then one wonders why other similar problems were not included. The first four problems of Book A ask for (1) two cubes whose sum is a square, (2) two cubes whose difference is a square, (3) two squares whose sum is a cube, and (4) two squares whose difference is a cube. What is missing from this list is, first, to find two squares whose sum is a square—but that had been solved in II-8—and second, to find two cubes whose sum is a cube. This latter problem is impossible to solve, and there are records stating this impossibility dating back to the tenth century. Probably Diophantus was also aware of the impossibility. At the very least, he must have tried the problem and failed to solve it. But he did not mention anything about it in his work. A similar problem with fourth powers occurs as V-29: To find three fourth powers whose sum is a square. Although Diophantus solves that problem, he does not mention the impossibility of finding two fourth powers whose sum is a square. Again, one assumes that he tried the latter problem and failed to solve it.

In his discussion of problem D-11, he does write of an impossibility. After solving that problem, to divide a given square into two parts such that the addition of one part to the square gives a square and the subtraction of the other part from the square also gives a square, he continues, "since it is not possible to find a square number such that, dividing it into two parts and increasing it by each of the parts, we obtain in both cases a square, we shall now present something which is possible."[19]

Problem D-12. *To divide a given square into two parts such that when we subtract each from the given square, the remainder is (in both cases) a square.*

Why is the quoted case impossible? To solve $x^2 = a + b, x^2 + a = c^2, x^2 + b = d^2$ would imply that

$$3x^2 = c^2 + d^2 \quad \text{or} \quad 3 = \left(\frac{c}{x}\right)^2 + \left(\frac{d}{x}\right)^2.$$

It is, in fact, impossible to decompose 3 into two rational squares. One can show this easily by congruence arguments modulo 4. Diophantus himself does not give a proof, nor later, when he states in VI-14 that 15 is not the sum of two squares, does he tell why. The solution of D-12, however, is very easy.

5.2.3 The Method of False Position

In Book IV, Diophantus begins use of a new technique, a technique reminiscent of the Egyptian "false position."

Problem IV-8. *To add the same number to a cube and its side and make the first sum the cube of the second.*

To solve $x^3 + y = (x + y)^3$, Diophantus begins by assuming $x = 2y$. Thus,

$$8y^3 + y = (3y)^3 = 27y^3, \quad \text{or} \quad y = 19y^3 \quad \text{or} \quad 19y^2 = 1.$$

But, writes Diophantus, "19 is not a square." So he needs to find a square to replace 19. Retracing his steps, he notes that 19 is $3^3 - 2^3$ and 3 comes from the assumed value $x = 2y$ by increasing the coefficient by 1. Hence he needs to find two consecutive numbers such that their cubes differ by a square. To solve this subsidiary problem, he denotes these by $z, z + 1$. Then $(z + 1)^3 - z^3 = 3z^2 + 3z + 1$ must be a square. By his usual square-finding technique, he sets this equal to $(1 - 2z)^2$. So $z = 7$, and $z + 1 = 8$. Returning to the beginning of the problem, he starts again by setting $x = 7y$. Then $343y^3 + y = (8y)^3 = 512y^3$ or $1 = 169y^2$. It follows that y, the added number, is 1/13. The desired cube has side 7/13. One might note that in his finding a square, his choice of $m = 2$ in $1 - mz$ is the only choice which gives a positive integral solution for z.

In problem IV-31, Diophantus finds again that his original assumption does not work. But here the problem is that a mixed quadratic equation, the first one to appear in the *Arithmetica,* fails to have a rational solution.

Problem IV-31. *To divide unity into two parts so that, if given numbers are added to them respectively, the product of the two sums is a square.*

Diophantus sets the given numbers at 3, 5 and the parts of unity as $x, 1 - x$. Therefore $(x + 3)(6 - x) = 18 + 3x - x^2$ must be a square. Since neither of his usual techniques for determining a square will work here (neither 18 nor -1 are squares), he tries $(2x)^2 = 4x^2$ as the desired square. But the resulting quadratic equation, $18 + 3x = 5x^2$ "does not give a rational result." He needs to replace $4x^2$ by a square of the form $(mx)^2$ which does give a rational solution. Thus, since $5 = 2^2 + 1$, he notes that the quadratic equation will be solvable if $(m^2 + 1) \cdot 18 + (3/2)^2$ is a square. This implies that $72m^2 + 81$ is a square, say $(8m + 9)^2$. (Here, his usual technique succeeds.) Then $m = 18$ and, returning to the beginning, he sets $18 + 3x - x^2 = 324x^2$. He then simply presents the solution: $x = 78/325 = 6/25$, and the desired numbers are 6/25, 19/25. Although he does not give details here on the solution of the quadratic, he does give them in problem IV-39. His words in that problem are easily translated into the formula

$$x = \frac{\frac{b}{2} + \sqrt{ac + \left(\frac{b}{2}\right)^2}}{a}$$

for solving the equation $c + bx = ax^2$. This formula is the same as the Babylonian formula, assuming that one first multiplies the equation through by a and solves for ax. That technique is in fact used in Babylonian problems. Diophantus is sufficiently familiar with this formula and its variants that he used it in various later problems not only to solve quadratic equations but also to solve quadratic inequalities.

Problem V-10.   *To divide unity into two parts such that, if we add different given numbers to each, the results will be squares.*

In this problem the manuscripts have, for one of only two times in the entire work, a diagram (Figure 5.5). Diophantus assumes that the two given numbers are 2 and 6. He represents them, as well as 1, by setting $DA = 2$, $AB = 1$, and $BE = 6$. The point G is chosen so that $DG (= AG + DA)$ and $GE (= BG + BE)$ are both squares. Since $DE = 9$, the problem is reduced to dividing 9 into two squares such that one of them lies between 2 and 3. If that square is x^2, the other is $9 - x^2$. Unlike the situation in previous problems, Diophantus cannot simply put $9 - x^2$ equal to $(3 - mx)^2$ with an arbitrary m, for he needs x^2 to satisfy the inequality condition. So he sets it equal to $(3 - mx)^2$ without specifying m. Then,

$$x = \frac{6m}{m^2 + 1}.$$

Rather than substitute the expression for x into $2 < x^2 < 3$ and attempt to solve a fourth degree inequality, he picks two squares close to 2 and 3 respectively, namely $289/144 = (17/12)^2$ and $361/144 = (19/12)^2$ and substitutes the expression into the inequality $17/12 < x < 19/12$. Therefore,

$$\frac{17}{12} < \frac{6m}{m^2 + 1} < \frac{19}{12}.$$

The left inequality becomes $72m > 17m^2 + 17$. Although the corresponding quadratic equation has no rational solution, nevertheless Diophantus uses the quadratic formula and

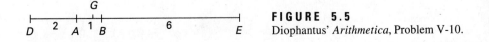

FIGURE 5.5
Diophantus' *Arithmetica*, Problem V-10.

shows that since $\sqrt{(72/2)^2 - 17^2} = \sqrt{1007}$ is between 31 and 32, the number m must be chosen so that $m \leq 67/17$. The right inequality similarly shows that $m \geq 66/19$. Diophantus therefore picks the simplest m between these two limits, namely 3 1/2. So,

$$9 - x^2 = \left(3 - 3\frac{1}{2}x\right)^2 \qquad \text{and} \qquad x = \frac{84}{53}.$$

Then $x^2 = 7056/2809$ and the desired segments of 1 are 1438/2809 and 1371/2809.

The final extant book of the *Arithmetica* deals with Pythagorean triples. Each of the problems asks to find such a triple which satisfies some other condition. For example, consider

Problem VI-16.   *To find a right triangle such that the length of the bisector of an acute angle is rational.*

For the solution it is necessary that not only a, b, c be rational, but also d, where d represents the length of the bisector AD of angle A (Figure 5.6). Diophantus sets $d = 5x$, $DC = 3x$, and $a = 3$. It follows that $b = 4x$ and $BD = 3 - 3x$. By *Elements* VI-3, $AB : BD = AC : CD$ or $c : (3 - 3x) = 4x : 3x$; therefore, $c = 4 - 4x$. By the Pythagorean theorem, $16 - 32x + 16x^2 = 16x^2 + 9$ and $x = 7/32$. Diophantus multiplies through by 32. The resulting triangle is represented by the triple (28, 96, 100) while the bisector is 35. Note that Diophantus' method in this problem can easily be generalized to give other solutions by taking $\triangle ACD$ to have sides (rx, sx, tx), where (r, s, t) is any Pythagorean triple, and a to be any positive integer. On the other hand, it is interesting to note that Diophantus does not mention the analogous problem where the right angle is bisected. In that case, it is easy to see that the problem is impossible.

Diophantus' work, the only example of a genuinely algebraic work surviving from ancient Greece, was highly influential. Not only was it commented on in late antiquity, but it was also studied by Islamic authors. Many of its problems were taken over by Rafael Bombelli and published in his *Algebra* of 1572 while the initial printed Greek edition of Bachet, published in 1621, was carefully studied by Pierre Fermat and led him to numerous general results in number theory of which Diophantus himself only hinted. Perhaps more important, however, is the fact that this work, as a work of algebra, was in effect a treatise on the analysis of problems. Namely, the solution of each problem began with the assumption that the answer x, for example, had been found. The consequences of this fact

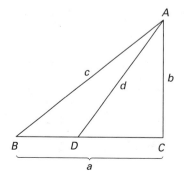

FIGURE 5.6
Diophantus' *Arithmetica*, Problem VI-16.

were then followed to the point where a numerical value of x could be determined by solving a simple equation. The synthesis, which in this case is the proof that the answer satisfies the desired conditions, was never given by Diophantus since it only amounted to an arithmetic computation. Thus Diophantus' work is at the opposite end of the spectrum from the purely synthetic work of Euclid.

5.3 PAPPUS AND ANALYSIS

Although analysis and synthesis had been used by all of the major Greek mathematicians, there was no systematic study of the methodology published, as far as is known, until the work of Pappus, who lived in Alexandria early in the fourth century. Pappus was one of the last mathematicians in the Greek tradition. He was familiar with the major and minor works of the men already discussed and even extended some of their work in certain ways. He is best known for his *Collection,* a group of eight separate works on various topics in mathematics, probably put together shortly after Pappus' death by an editor attempting to preserve Pappus' papers. The books of the collection vary greatly in quality, but most of the material consists of surveys of certain mathematical topics collected from the works of Pappus' predecessors.

The preface to Book 3 provides an interesting sidelight to the work. Pappus addresses the preface to Pandrosian, a woman teacher of geometry. He complains that "some persons professing to have learned mathematics from you lately gave me a wrong enunciation of problems."[20] By that Pappus meant that these people attempted to solve problems by methods that could not work, for example, to solve the problem of finding two mean proportionals between two given lines using only circles and straight lines. There is no indication of how Pappus knew that such a construction was impossible. From his remark, however, we learn that women were involved in mathematics in Alexandria.[21]

Book 5, the most polished book of the *Collection,* deals with isoperimetric figures, figures of different shape but with the same perimeter. Pappus' introduction provides a counterpoint to the pure mathematics of the text as he writes of the intelligence of bees:

> [The bees], believing themselves, no doubt, to be entrusted with the task of bringing from the gods to the more cultured part of mankind a share of ambrosia in this form, . . . do not think it proper to pour it carelessly into earth or wood or any other unseemly and irregular material, but, collecting the fairest parts of the sweetest flowers growing on the earth, from them they prepare for the reception of the honey the vessels called honeycombs, [with cells] all equal, similar and adjacent, and hexagonal in form.
>
> That they have contrived this in accordance with a certain geometrical forethought we may thus infer. They would necessarily think that the figures must all be adjacent one to another and have their sides common, in order that nothing else might fall into the interstices and so defile their work. Now there are only three rectilineal figures which would satisfy the condition, I mean regular figures which are equilateral and equiangular, inasmuch as irregular figures would be displeasing to the bees. . . . [These being] the triangle, the square and the hexagon, the bees in their wisdom chose for their work that which has the most angles, preceiving that it would hold more honey than either of the two others.
>
> Bees, then, know just this fact which is useful to them, that the hexagon is greater than the square and the triangle and will hold more honey for the same expenditure of material in constructing each.[22]

5.3.1 The Domain of Analysis

The most influential book of Pappus' *Collection,* however, is Book 7, *On the Domain of Analysis.* It is here that there occurs the most explicit discussion from Greek times of the method of analysis, of the methodology which Greek mathematicians used to solve problems. The central ideas are spelled out in Pappus' introduction to Book 7:

> That which is called the Domain of Analysis . . . is, taken as a whole, a special resource . . . for those who want to acquire a power in geometry that is capable of solving problems set to them; and it is useful for this alone. It was written by three men, Euclid the writer of the *Elements,* Apollonius of Perga, and Aristaeus the elder, and proceeds by analysis and synthesis.
>
> Now analysis is the path from what one is seeking, as if it were established, by way of its consequences, to something that is established by synthesis . . . There are two kinds of analysis; one of them seeks after truth and is called "theorematic," while the other tries to find what was demanded, and is called "problematic." In the case of the theorematic kind, we assume what is sought as a fact and true, then advance through its consequences, as if they are true facts according to the hypothesis, to something established; if this thing that has been established is a truth, then that which was sought will also be true, and its proof the reverse of the analysis; but if we should meet with something established to be false, then the thing that was sought too will be false. In the case of the problematic kind, we assume the proposition as something we know, then proceed through its consequences, as if true, to something established; if the established thing is possible and obtainable, which is what mathematicians call "given," the required thing will also be possible, and again the proof will be the reverse of the analysis; but should we meet with something established to be impossible, then the problem too will be impossible.[23]

According to Pappus, then, to solve a problem or prove a theorem by analysis, begin by assuming what is required, p, for example, and then prove that p implies q_1, q_1 implies $q_2, \ldots, q_n$ implies q, where q is something known to be true. To give the formal synthetic proof of the theorem, or solve the problem, reverse the process beginning with q implies q_n. This method of reversal has always been a controversial point; after all, not all theorems have valid converses. In fact, however, most important theorems from Euclid and Apollonius do have at least partial converses. Thus the method does often provide the desired proof or solution, or at least demonstrates, when there are only partial converses, the conditions under which a problem can be solved.

There are few examples in the extant literature of theorematic analysis, since Euclid, for example, never shares his method of discovery of his proofs. But some of the manuscripts of *Elements,* Book XIII contain, evidently as an interpolation made in the early years of the common era, an analysis of each of the first five propositions. Consider

Proposition XIII-1.   *If a straight line is cut in extreme and mean ratio, the square on the sum of the greater segment and half of the whole is five times the square on the half.*

Let AB be divided in extreme and mean ratio at C, AC being the greater segment, and let $AD = \frac{1}{2}AB$ (Figure 5.7). To perform the analysis, assume the truth of the conclusion, namely $CD^2 = 5AD^2$, and determine its consequences. Since also $CD^2 = CA^2 + AD^2 + 2CA \cdot AD$, therefore $CA^2 + 2CA \cdot AD = 4AD^2$. But $AB \cdot AC = 2CA \cdot AD$ and, since $AB:AC = AC:BC$, also $CA^2 = AB \cdot BC$. Therefore $AB \cdot BC + AB \cdot AC = 4AD^2$ or

FIGURE 5.7

Analysis of *Elements* XIII-1.

$AB^2 = 4AD^2$, or, finally, $AB = 2AD$, a result known to be true. The synthesis can then proceed by reversing each step: Since $AB = 2AD$, we have $AB^2 = 4AD^2$. Since also $AB^2 = AB \cdot AC + AB \cdot BC$, it follows that $4AD^2 = 2DA \cdot AC + AC^2$. Adding to each side the square on AD gives the result $CD^2 = 5AD^2$.

More important for Greek mathematics than theorematic analysis is the problematic analysis. We have already discussed several examples of this type of analysis, including the problems of angle trisection and cube duplication and Archimedes' problems on the division of a sphere by a plane. And although Euclid does not present the analysis as such, one can carry out the procedure in solving Euclid VI-28, the geometric algebra problem leading to the solution of the quadratic equation $x^2 + c = bx$. The analysis there shows that an additional condition is required for the solution, namely that $c \leq (b/2)^2$.

Pappus' Book 7, then, is a companion to the Domain of Analysis, which consists of several geometric treatises, all written many centuries before Pappus. These works, Apollonius' *Conics* as well as six other books (all but one lost), Euclid's *Data* and two other lost works, and single works (both lost) by Aristaeus and Eratosthenes, even though the last named author is not mentioned in Pappus' introduction, provided the Greek mathematician with the tools necessary to solve problems by analysis. For example, to deal with problems which result in conic sections, one needs to be familiar with Apollonius' work. To deal with problems solvable by "Euclidean" methods, the material in the *Data* is essential.

Pappus' work does not include the Domain of Analysis itself. It is designed only to be read along with these treatises. Therefore, it includes a general introduction to most of the individual books along with a large collection of lemmas which are intended to help the reader work through the actual texts. Pappus evidently decided that the texts themselves were too difficult for most readers of his day to understand as they stood. The teaching tradition had been weakened through the centuries, and there were few, like Pappus, who could appreciate these several-hundred-year-old works. Pappus' goal was to increase the numbers who could understand the mathematics in these classical works by helping his readers through the steps where the authors wrote "clearly . . .!" He also included various supplementary results as well as additional cases and alternative proofs.

Among these additional remarks is the generalization of the three- and four-line locus problems discussed by Apollonius. Pappus notes that in that problem itself the locus is a conic section. But, he says, if there are more than four lines, the loci are as yet unknown, that is, "their origins and properties are not yet known." He is disappointed that no one has given the construction of these curves that satisfy the five- and six-line locus. The problem in these cases is, given five (six) straight lines, to find the locus of a point such that the rectangular parallelepiped contained by the lines drawn at given angles to three of these lines has a given ratio to the rectangular parallelepiped contained by the remaining two lines and some given line (remaining three lines). Pappus notes that one can even generalize the problem further to more than six lines. But, he complains, "[geometers] have by no means solved [the multi-line locus problem] to the extent that the curve can be recognized. . . . The men who study these matters are not of the same quality as the ancients and the best writers. Seeing that all geometers are occupied with the first principles of mathematics . . . and being ashamed to pursue such topics myself, I have proved propositions of much greater importance and utility."[24]

Pappus concludes Book 7 by stating one of the "important" results he had proved, that "the ratio of solids of complete revolution is compounded of that of the revolved figures and that of the straight lines similarly drawn to the axes from the center of gravity in them."[25] The modern version of this theorem is that the volume of a solid formed by revolving a region Ω around an axis not intersecting Ω is the product of the area of Ω and the circumference of the circle traversed by the center of gravity of Ω. Unfortunately, there is no record of Pappus' proof. There is some indication that it is in one of the books of the *Collection* now lost.

Much of the explicit analysis in Greek mathematics has to do with material we generally think of as algebraic. The examples from *Elements* XIII-1 and VI-28 are clearly such. The examples using the conic sections are ones which today would be solved using analytic geometry, a familiar application of algebra. It is somewhat surprising, then, that Pappus does not mention the strictly algebraic *Arithmetica* of Diophantus as a prime example of analysis, because, in effect, every problem in Diophantus' work is solved according to Pappus' model. Possibly Pappus did not include this work because it was not on the level of the classic geometric works. In any case, it was the algebraic analysis of Diophantus and the "quasi-algebraic" analysis of many of the other mentioned works, rather than the pure geometric analysis, which provided the major impetus for sixteenth and seventeenth century European mathematicians to expand on the notion of algebra and develop it into a major tool to solve even purely geometric problems.[26]

5.3.2 Hypatia and the End of Greek Mathematics

Pappus' aim of reviving Greek mathematics was unsuccessful, probably in part because the increasingly confused political and religious situation affected the stability of the Alexandrian Museum and Library. In his time, Christianity was changing from a persecuted sect into the official religion of the Roman Empire. In 313 the Emperor Galerius issued an edict of toleration in the eastern empire, and two years later the same was done in the West by Constantine. The latter was in fact converted to Christianity before his death in 337. Within 60 years, Christianity became the state religion of the empire and the ancient worship of the Roman gods was banned. Of course, the banning of paganism did not cause everyone to adopt Christianity. In fact, around 400, Hypatia (c. 370–415), the daughter of Theon of Alexandria, became head of the Neoplatonic school in Alexandria and continued the teaching of the philosophic doctrines dating back to Plato's Academy.

Although there is some evidence of earlier women being involved in Greek mathematics, it is only about Hypatia that the evidence is substantial enough to give some indication of her mathematical accomplishments. Hypatia was given a very thorough education in mathematics and philosophy by her father. She became a popular and respected teacher in Alexandria, having for students many distinguished men, including Synesius of Cyrene (in Libya), who later became a bishop. Although the only surviving documents with a clear reference to Hypatia are Synesius' letters to her requesting scientific advice, recent detailed textual studies of Greek, Arabic, and medieval Latin manuscripts lead to the conclusion that she was responsible for many mathematical works. These include several parts of her father's commentary on Ptolemy's *Almagest,* the edition of Archimedes' *Measurement of the Circle* from which most later Arabic and Latin translations

---

Side Bar 5.3   **The Decline of Greek Mathematics**

Why did Greek mathematics decline so dramatically from its height in the fourth and third centuries B.C.E.? Among the several answers to the question, the most important was the change in the socio-political scene in the region surrounding the Mediterranean.

A consideration of the mathematical development in the various ancient societies already studied shows that mathematical creativity requires some sparks of intellectual curiosity, whether or not these are stimulated by practical concerns. But this spark of curiosity needs a climate of government encouragement for its flames to spread. The Babylonians used their most advanced techniques, not for everyday purposes, but for solving intellectually challenging problems. The government encouraged the use of these mathematical problems to help train the minds of its future leaders. In Greek civilization, the intellectual curiosity ran even deeper. In the Greek homeland, the socio-political system provided philosophy and mathematics with encouragement. The Ptolemies continued this encouragement in Egypt after 300 B.C.E.

But even in Greek society, the actual number of those who understood theoretical mathematics was small. There were never many who could afford to spend their lives as mathematicians or astronomers and persuade the rulers to provide them with stipends. The best of the mathematicians wrote works that were discussed and commented on in the various mathematical schools, but not everything could be learned from the texts. An oral teaching tradition was necessary to keep mathematics progressing because, in general, one could not master Euclid's *Elements* or Apollonius' *Conics* on one's own. A break of a generation in this tradition thus meant that the entire process of mathematical research would be severely damaged.

One factor certainly weakening the teaching tradition, if not breaking it entirely, was the political strife around the eastern Mediterranean in the years surrounding the beginning of the common era. More importantly, because the Roman imperial government evidently decided that mathematical research was not an important national interest, it did not support it. There was little encouragement of mathematical studies in Rome. Few Greek scholars were imported to teach mathematics to the children of the elite. Soon, no one in Rome could even understand, let alone extend, the works of Euclid or Apollonius. The Greek tradition did continue for several centuries, however, under the Roman governors of Egypt, particularly because the Alexandrian Museum and Library remained in existence. Anyone interested could continue to study and interpret the ancient texts. With fewer and fewer teachers, however, less and less new work was accomplished. The virtual destruction of the great library by the late fourth century finally severed the tenuous links with the past. Although there continued to be some limited mathematical activity for a while in Athens and elsewhere, wherever copies of the classic works could be found, by the end of the fifth century there were too few people devoting their energies to mathematics to continue the tradition and Greek mathematics ceased to be.

---

stem, a work on areas and volumes reworking Archimedean material, and a text on isoperimetric figures related to Pappus' Book V.[27] She is also responsible for commentaries on Apollonius' *Conics* and, as noted earlier, on Diophantus' *Arithmetica*.

Unfortunately, the Christian community of Alexandria, under the leadership of the patriarch Cyril, felt that her beliefs provided a pagan influence running counter to Christian doctrine. In addition, she was a friend of Orestes, the Roman prefect and Cyril's political rival for control of Alexandria. Caught in the middle of this pagan–Christian turmoil, Hypatia was murdered as already described. Her death effectively ended the Greek mathematical tradition of Alexandria (Side Bar 5.3).

Exercises

From Nicomachus' Introduction to Arithmetic

1. Devise a formula for the nth pentagonal number and for the nth hexagonal number.

2. Derive an algebraic formula for the pyramidal numbers with triangular base and one for the pyramidal numbers with square base.

3. Show that in a subcontrary proportion the product of the greater and mean terms is twice the product of the mean and smaller terms.

4. Show that in a harmonic proportion the sum of the extremes multiplied by the means is twice the product of the extremes.

5. Nicomachus' "fifth proportion" exists whenever among three terms the middle term is to the lesser as their difference is to the difference between the greater and the mean. Show that 2, 4, 5 are in fifth proportion. Find another triple in this proportion. Show that if three numbers are in this proportion, then the product of the greatest by the middle term is double that of the greatest by the smallest.

6. Nicomachus' "seventh proportion" exists among three terms whose greatest term is to the least as their difference is to the difference of the lesser terms. Show that 6, 8, 9 are in seventh proportion. Find two other triples of numbers in seventh proportion.

From Diophantus' Arithmetica

7. Determine Diophantus' age at his death from his epigram at the opening of the chapter.

8. Solve Diophantus' problem I-27 by the method of I-28: To find two numbers such that their sum and product are given. Diophantus gives the sum as 20 and the product as 96.

9. Solve Diophantus' problem II-10: To find two square numbers having a given difference. Diophantus puts the given difference as 60. Also, give a general rule for solving this problem given any difference.

10. Generalize Diophantus' solution to II-19 by choosing an arbitrary ratio $n : 1$ and the value $(x + m)^2$ for the second square.

11. Solve Diophantus' problem II-13 by the method of the double equation: From the same (required) number to subtract two given numbers so as to make both remain-

ders square. (Take 6, 7 for the given numbers. Then solve $x - 6 = u^2, x - 7 = v^2$.)

12. Solve Diophantus' problem B-8: To find two numbers such that their difference and the difference of their cubes are equal to two given numbers. (Write the equations as $x - y = a, x^3 - y^3 = b$. Diophantus takes $a = 10$, $b = 2170$.) Derive necessary conditions on a and b which ensure a rational solution.

13. Solve Diophantus' problem D-12: To divide a given square into two parts such that when we subtract each from the given square, the remainder (in both cases) is a square. Note that the solution follows immediately from II-8.

14. Solve Diophantus' problem IV-9: To add the same number to a cube and its side and make the second sum the cube of the first. (The equation is $x + y = (x^3 + y)^3$. Diophantus begins by assuming that $x = 2z$ and $y = 27z^3 - 2z$.)

15. Show that there is no integral-sided right triangle such that the length of the bisector of the right angle is rational.

16. Solve Diophantus' problem V-10 for the two given numbers 3, 9. Show that a necessary condition for a solution is that the sum of the two given numbers plus 1 must be equal to the sum of two squares.

Problems Related to Pappus' Collection

17. Carry out the analysis of *Elements* VI-28: To a given straight line to apply a parallelogram equal to a given rectilinear figure and deficient by a parallelogram similar to a given one. Just consider the case where the parallelograms are all rectangles. Begin with the assumption that such a rectangle has been constructed and derive the condition that "the given rectilinear figure must not be greater than the rectangle described on the half of the straight line and similar to the defect."

18. Provide the analysis for *Elements* XIII-3: If a straight line is cut in extreme and mean ratio, the square on the sum of the lesser segment and half of the greater segment is five times the square on half of the greater segment.

19. Write an equation for the locus described by the problem of five lines. Assume for simplicity that all the lines are either parallel or perpendicular to one of them and that all the given angles are right.

20. Show that a regular hexagon of given perimeter has a greater area than a square of the same perimeter.

From the Greek Anthology (c. 500 C.E.)

21. Solve Epigram 116: Mother, why do you pursue me with blows on account of the walnuts? Pretty girls divided them all among themselves. For Melission took two-sevenths of them from me, and Titane took the twelfth. Playful Astyoche and Philinna have the sixth and third. Thetis seized and carried off twenty, and Thisbe twelve, and look there at Glauce smiling sweetly with eleven in her hand. This one nut is all that is left to me. How many nuts were there originally?[28]

22. Solve Epigram 130: Of the four spouts, one filled the whole tank in a day, the second in two days, the third in three days, and the fourth in four days. What time will all four take to fill it? (Note the similarity of this problem to the problem from the sixth chapter of the *Jiuzhang* in the exercises to Chapter 1.)

23. Solve Epigram 145: *A.*: Give me ten coins and I have three times as many as you. *B.*: And if I get the same from you, I have five times as much as you. How many coins does each have?

FOR DISCUSSION . . .

24. Why would surveyors in Roman times have been ignorant of trigonometry and not have made use of it in their work?

25. Read Nicomachus' complete discussion of his naming scheme and devise a verbal "formula" for finding the name of any given ratio of integers.

26. "Quadratic equations were totally useless in solving problems necessary to the running of the Roman Empire." Give arguments for and against.

27. What factors influence mathematical development in a particular civilization? Give examples from the civilizations already studied.

28. Why were there so few women involved in mathematics in Greek times?

References and Notes

Thomas Heath's *A History of Greek Mathematics* and B. L. Van der Waerden's *Science Awakening,* referred to in Chapter 2, have sections on the material discussed in this chapter. A translation of Nicomachus' major work is found in M. L. D'Ooge, F. E. Robbins, and L. C. Karpinski, *Nicomachus of Gerasa: Introduction to Arithmetic* (New York: Macmillan, 1926). This translation can also be found in the *Great Books.* The six books of Diophantus still extant in Greek are found in Thomas L. Heath, *Diophantus of Alexandria: A Study in the History of Greek Algebra* (New York: Dover, 1964). Heath does not, however, translate Diophantus literally. He generally just outlines Diophantus' arguments. More literal translations of certain of the problems are found in Bulmer-Thomas, *Selections Illustrating the History of Greek Mathematics.* A translation and commentary on the four newly discovered books of Diophantus' *Arithmetica* is J. Sesiano, *Books IV to VII of Diophantos' Arithmetica in the Arabic Translation of Qusṭā ibn Lūqā* (New York: Springer, 1982). A brief survey of Diophantus' work is in J. D. Swift, "Diophantus of Alexandria," *American Mathematical Monthly* 63 (1956), 163–170. The entire extant text of Pappus' *Collection* is translated into French in Paul Ver Eecke, *Pappus d'Alexandrie, La Collection Mathematique* (Paris: Desclée, De Brouwer et Cie., 1933). A recent English translation of Book 7, with commentary, is provided by Alexander Jones, *Pappus of Alexandria: Book 7 of the Collection* (New York: Springer, 1986). There are several recent brief studies of the life of Hypatia. Probably the most easily accessible is the chapter in Lynn M. Osen, *Women in Mathematics* (Cambridge: MIT Press, 1974).

1. W. R. Paton, trans., *The Greek Anthology* (Cambridge: Harvard University Press, 1979), Volume V, pp. 93–94 (Book XIV, Epigram 126).

2. Edward Gibbon, *The Decline and Fall of the Roman Empire* (Chicago: Encyclopedia Britannica, 1952) (*Great Books* edition), Chapter 47.

3. Nicomachus, *Introduction to Arithmetic*, I, IV, 2.

4. Cicero, *Tusculan Disputations* (Cambridge: Harvard University Press, 1927), I, 2.

5. Vitruvius, *On Architecture* (Cambridge: Harvard University Press, 1934), I.

6. For a good treatment of Roman surveying see O. A. W. Dilke, *The Roman Land Surveyors: An Introduction to the Agrimensores* (Newton Abbot, U.K.: David-Charles, 1971).

7. Nicomochus, *Introduction,* II, XXIV, 1.

8. Ibid., II, XXVII, 1.

9. Bulmer-Thomas, *Selections Illustrating*, II, pp. 519–523.

10. Details of this argument are presented in J. Sesiano, *Books IV to VII of Diophantos'*, pp. 71–75.

11. Bulmer-Thomas, *Selections Illustrating*, p. 525.

12. Thomas L. Heath, *Diophantus of Alexandria*, pp. 130–131.

13. The problems from Books I–VI are adapted from Heath, *Diophantus*, while those from Books A–D are taken from Sesiano, *Books IV to VII of Diophantos'*.

14. Bulmer-Thomas, *Selections Illustrating*, II, p. 553.

15. Heath, *Diophantus*, p. 146.

16. Sesiano, *Books IV to VII of Diophantos'*, p. 88.

17. Ibid., p. 87.

18. Ibid., p. 130.

19. Ibid., p. 165.

20. Bulmer-Thomas, *Selections Illustrating*, II, p. 567.

21. The arguments for concluding that Pandrosian is a woman are given in Jones, *Pappus of Alexandria*.

22. Bulmer-Thomas, *Selections Illustrating*, II, pp. 589–593.

23. This translation is adapted from one in Michael Mahoney, "Another Look at Greek Geometrical Analysis," *Archive for History of Exact Sciences* 5 (1968), 318–348 and from Jones, *Pappus of Alexandria*. Mahoney's conclusions about Greek analysis are disputed in some respects in J. Hintikka and U. Remes, *The Method of Analysis: Its Geometrical Origin and Its General Significance* (Boston: Reidel, 1974).

24. Jones, *Pappus of Alexandria*, pp. 120–122 and Bulmer-Thomas, *Selections Illustrating*, II, p. 601.

25. Jones, *Pappus of Alexandria*, p. 122.

26. An extensive discussion of the algebraic analysis of Diophantus and its effects on the development of algebra is found in J. Klein, *Greek Mathematical Thought and the Origin of Algebra* (Cambridge: MIT Press, 1968).

27. Details on the attribution of various mathematical works to Hypatia are found in Wilbur Knorr, *Textual Studies in Ancient and Medieval Geometry* (Boston: Birkhäuser, 1989).

28. This problem and the next two are taken from Paton, *The Greek Anthology*.

Summary of Greek Algebra and Analysis

Late first century C.E.	Nicomachus	Elementary number theory
Mid-third century C.E.	Diophantus	Indeterminate equations
Early fourth century C.E.	Pappus	Summary of Greek mathematics
c. 370–415 C.E.	Hypatia	Commentary on Apollonius, Diophantus

Chapter 6

Medieval China and India

"Now the science of mathematics is considered very important. This book [Precious Mirror of the Four Elements, by Zhu Shijie] . . . therefore will be of great benefit to the people of the world. The knowledge for investigation, the development of intellectual power, the way of controlling the kingdom and of ruling even the whole world, can be obtained by those who are able to make good use of the book. Ought not those who have great desire to be learned take this with them and study it with great care?"
(Introduction to Precious Mirror of the Four Elements, *1303)*[1]

*A*strologers had predicted that the daughter Lilāvati of the Indian mathematician Bhāskara (1114–1185) would not wed. But her father, being an expert astronomer and astrologer himself, divined the one lucky moment for her marriage. The time was kept by a water clock, but shortly before the exact hour, while Lilāvati was looking into the clock, a pearl from her headdress accidentally dropped into the clock unnoticed and stopped the flow of water. By the time it was discovered, the designated moment had passed. To console his daughter, Bhāskara named the chapter on arithmetic of his major work, the *Siddhāntasiromani,* after her.

6.1 INTRODUCTION TO MEDIEVAL CHINESE MATHEMATICS

The Han dynasty in China disintegrated early in the third century C.E., and China broke up into several warring kingdoms. The period of disunity lasted until 581, when the Sui dynasty was established, followed 37 years later by the Tang dynasty which was to last nearly 300 years. Although another brief period of disunity followed, much of China was again united under the Song dynasty (960–1279), a dynasty itself overthrown by the Mongols under Ghengis Khan. Throughout this thousand year period, however, it is possible to speak of "Chinese" mathematics. Despite the numerous wars and dynastic conflicts, a true Chinese culture was developing throughout most of east Asia, with a common language and common values. The Han dynasty had already instituted a civil service based

181

on examination results rather than family ties. This system of imperial examinations lasted—with various short periods of disruption—into the twentieth century. Although the examination was chiefly based on Chinese literary classics, the demands of the empire for administrative services, including surveying, taxation, and calendar making, required that many civil servants be competent in certain areas of mathematics. The Chinese imperial government therefore, unlike its Roman counterpart, at least encouraged the study of applicable mathematics, as indicated in the opening quotation. In fact, there was an imperial Office of Mathematics where officials were trained in "practical mathematics." Hence it is not surprising that mathematical texts were written as collections of problems from which candidates for office, as well as the officeholders themselves, could learn methods of solution. It is also not surprising that new methods were rarely introduced. As in literature, the "classics" were studied and commented on. The examination system often required recitation of relevant passages from the mathematics texts, as well as the solving of problems in the same manner as described in these texts. There was no particular incentive for mathematical creativity.

Nevertheless, creative mathematicians did appear in China, mathematicians who applied their talents not only to improving old methods of solution to practical problems, but also to extending these methods far beyond the requirements of practical necessity. The first half of this chapter is therefore devoted to aspects of the work of some of these Chinese mathematicians. In particular, we will begin by considering the work of Liu Hui in the third century on surveying problems and then deal with the importation of trigonometric methods from India in the eighth century to help with calendrical problems. Next, we will consider the extensive work of Chinese authors on solving what are today called linear congruences. These problems first appear in the work of Sun Zi in the fourth century and Zhang Qiujian in the fifth century, but are not completely dealt with until the extensive work of Qin Jiushao in the thirteenth century. The final section on Chinese mathematics will treat some of the additional work of the Chinese mathematicians of the thirteenth century in the area of equation solving. Qin Jiushao was himself influential in this regard, while Li Ye, Yang Hui, and Zhu Shijie also made significant contributions.

6.2 THE MATHEMATICS OF SURVEYING AND ASTRONOMY

Liu Hui (third century) came from the northern kingdom of Wei, during the period shortly after the breakup of the Han Empire, but virtually nothing is known of his life. His major mathematical work was a commentary on the *Jiuzhang suanshu*. Since the last several problems of that book were elementary surveying problems, Liu decided to add an addendum on more complicated problems of that type. This addendum ultimately became a separate mathematical work, the *Haidao suanjing* (*Sea Island Mathematical Manual*).

This brief work, in the continuing tradition of problem texts, was simply a collection of nine problems with solutions, derivations, illustrations, and commentary. Unfortunately, all that remains today are the problems themselves with the computational directions for finding the solutions. No reasons are given why these particular computations are to be performed, so in the discussion to follow we will present some possible methods by which Liu Hui worked out his rules.

The first of the nine problems, for which the text is named, shows how to find the distance and height of a sea island. The others demonstrate how to determine such items as the height of a tree, the depth of a valley, and the width of a river. The sea island problem reads, "to measure a sea island, erect two poles, each 5 feet high, the distance between the two poles being 1000 feet. Assume that the two poles are aligned with the island. When a man walks 123 feet back from the nearest pole, the peak of the island is just visible through the top of the pole, if he tries to see with his eye at ground level. When he walks 127 feet back from the further pole, he observes the peak of the island in line with the top of that one, again from ground level. What is the height of the island and how far is it from the first pole?"[2]

Liu Hui's answer is that the height of the island is 1255 feet while its distance from the pole is 30,750 feet. He also presents the rule for the solution (Figure 6.1):

Multiply the distance between poles by the height of the pole, giving the *shi*. Take the difference in distances from the points of observations as the *fa* to divide the *shi*. Add what is thus obtained to the height of the pole; the result is the height of the island. [Thus the height h is given by the formula $h = a + \frac{ab}{c - d}$, where a is the height of the pole, b the distance between the poles, and c and d the respective distances from the poles to the observation points.] To find the distance of the island, multiply the distance backward from the front pole by the distance between the poles, giving the *shi*. Take the difference in distances from the points of observation as the *fa* to divide the *shi*. The result is the distance of the island. [The distance s is given by $s = \frac{bd}{c - d}$.][3]

Liu Hui called his method the method of "double differences," because two differences are used in the solution procedure. A modern derivation of the method would use similar triangles: Construct MT parallel to EK. Then $\triangle AEM$ is similar to $\triangle MTR$ and $\triangle ABM$ is similar to $\triangle MNR$. Therefore $ME : TR = AM : MR = AB : MN$, so

$$AB = \frac{ME \cdot MN}{TR} = \frac{FN \cdot EF}{TR}$$

and the height $h \,(= AB + BC)$ of the island is

$$h = \frac{FN \cdot EF}{TR} + EF = \frac{ab}{c - d} + a,$$

as noted above. A similar argument gives Liu Hui's result for the distance s of the island.

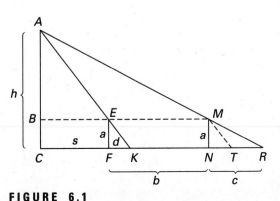

FIGURE 6.1
Problem 1 of the *Sea Island Mathematical Manual.*

However, there are other ways of deriving Liu Hui's formula. In the mid-thirteenth century, Yang Hui commented on this particular problem and gave a justification using only congruent triangles and area relationships. Since triangles *APR* and *ACR* are congruent as are triangles *ALM* and *ABM*, trapezoid *LPRM* has the same area as trapezoid *BMRC*. Subtracting off the congruent triangles *MQR* and *MNR* shows that rectangles *LPQM* and *BMNC* are also equal in area. By a similar argument, rectangles *DGHE* and *BECF* are also equal in area. It follows that rectangle *EMNF* (= rectangle *BMNC* − rectangle *BECF*) = rectangle *LPQM* − rectangle *DGHE*. Writing each of the areas of the rectangles as products gives

$$FN \cdot EF = PQ \cdot QM - GH \cdot HE = PQ \cdot RN - PQ \cdot FK$$
$$= PQ(RN - FK) = AB(RN - FK).$$

Therefore, $AB = \frac{FN \cdot EF}{RN - FK}$ and the height $h = AC$ is given by

$$h = AC = AB + BC = \frac{FN \cdot EF}{RN - FK} + EF$$

as desired. One can solve for the distance $s = CF$ by beginning with the equality of the areas of rectangles *DGHE* and *BCFE*, that is, with $CF \cdot BC = DE \cdot EH$, and replacing $DE = AB$ by the value already found. Of course, in some sense the solution by area manipulation is equivalent to that by similar triangles, but the geometric knowledge assumed is somewhat different.

In problem 4, Liu Hui calculates the depth of a valley from two observations made along the valley wall. Figure 6.2 illustrates the situation, where x is the desired depth and the measurements are in feet. A modern solution would use the theory of similar triangles to solve the problem. Namely,

$$\frac{6}{8.5} = \frac{y + 30}{z} \quad \text{and} \quad \frac{6}{9.1} = \frac{y}{z}.$$

It follows that $6z = 8.5(y + 30) = 9.1y$. So $0.6y = 8.5(30)$ and $y = \frac{8.5(30)}{0.6} = 425$. Liu Hui gives precisely this calculation and then notes that the valley depth is 6 feet less than

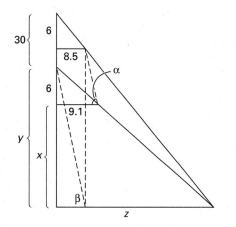

FIGURE 6.2
Problem 4 of the *Sea Island Mathematical Manual.*

this value, or 419 feet. An area manipulation similar to that of problem 1 can also be carried out in this case to give the same result.

Calculations using similar triangles may often be thought of as "trigonometry" calculations. It is therefore instructive to see whether there is indication of any trigonometry in these problems. One can consider the instructions in problem 4 as instructions for finding y by multiplying 8.5 by the tangent $\left(\frac{30}{0.6}\right)$ of angle α or angle β (Figure 6.2). But Liu Hui mentions nothing about angles. Since several problems in the *Haidao suanjing* similarly involve multiplying lengths by tangents of angles, it has been argued that the *Sea Island Mathematical Manual* displays what might be called "prototrigonometry."[4] In any case, it seems that Liu's methods were somewhat more advanced mathematically than the methods of the Roman surveyors.

Chinese astronomers did use trigonometric methods involving tables in the eighth century. The Chinese emperors, like rulers elsewhere, had always been interested in problems of the calendar, that is, in predicting various celestial events such as eclipses. Unfortunately, Chinese astronomers were not very successful in predicting eclipses because they did not fully understand the motions of the sun and moon. Indian astronomers, because of Greek influence in the creation of a geometrical model, were more successful. Thus in the eighth century, when Buddhism was strong both in India and China and there were many reciprocal visits of Buddhist monks, the Chinese emperors of the Tang dynasty brought in Indian scholars as well to provide a new expertise. These scholars, led by Chutan Hsita (early eighth century), prepared an astronomical work in Chinese in 718, the *Chiu-chih li* (*Nine Planets* [*sun, moon, five ordinary planets and two invisible ones*]), based on Indian sources. In particular, this work contained a description of the construction of a sine table in steps of 3°45′ using a circle radius of 3438′. (More details will be given in section 6.6.1.)

In 724, the State Astronomical Bureau of the Tang dynasty began an extensive program of field research to determine the length of the shadows cast by a standard gnomon (of length eight feet) at various latitudes, ranging from 29° to 52° along the same meridian (114° E), at the summer and winter solstices and at the equinoxes. These observations were then analyzed by the chief astronomer, Yi Xing (683–727), himself a Buddhist monk (Figure 6.3). Yi Xing's goal was to use these and other observations, as well as various interpolation techniques, to calculate the length of such shadows, the duration of daylight and night, and the occurrence of eclipses, whatever the position of the observer. (Yi Xing was not aware of the sphericity of the earth and therefore could not make use of the classic Greek model.) Among the tables Yi Xing produced for these purposes in his *Ta yen li* was a shadow table based on the sun's zenith distance α rather than on the latitude and date. The Chinese measured this zenith angle in *tu* rather than degrees, where 365.2565 *tu* = 360°. In other words, the *tu* was chosen so that the sun traveled precisely 1 *tu* along the ecliptic each day. Yi Xing's table gave the length of a shadow of a gnomon of 8 feet for each integral value of the zenith angle α from 1 to 79. In modern terms, this is a table of the function $s(\alpha) = 8 \tan \alpha$ and is the earliest recorded version of a tangent table.

It is not known how Yi Xing calculated the table, but a detailed comparison of his work with the standard Indian astronomical works and with the sine table in the *Chiu-chih li* leads one at least to the tentative conclusion that he proceeded by first converting *tu* to degrees, next interpolating in the sine table, and finally calculating the shadow lengths by the formula $\bar{s}(\alpha) = 8\frac{\sin \alpha}{\sin(90 - \alpha)}$.[5] In any case, although the *Ta yen li* and even the *Chiu-chih li* were preserved in Chinese compendia, Yi Xing's tangent table ideas were not

FIGURE 6.3
Yi Xing on a Chinese stamp.

continued in his own country. Trigonometric methods do not appear again in China until after general contact with the West was opened in the seventeenth century. On the other hand, the next appearance of a shadow (tangent) table is in Islamic sources in the ninth century. Whether transmittal of this idea occurred across central Asia during that century is not known.

6.3 INDETERMINATE ANALYSIS

Calendrical problems led the Chinese mathematicians to another mathematical question, that of solving systems of indeterminate linear equations. For example, the Chinese assumed that at a certain point in time, the *Shang yuan,* there occurred simultaneously the beginning of the 60-day cycle used in Chinese dating, the winter solstice, and the new moon. If in a certain other year, the winter solstice occurred r days into a 60-day cycle and s days after the new moon, then that year was N years after *Shang yuan,* where N satisfied the simultaneous congruences

$$aN \equiv r \pmod{60} \qquad aN \equiv s \pmod{b}$$

where a is the number of days in the year and b is the number of days from new moon to new moon. In the extant records of ancient calendars, however, there is no indication given as to how the Chinese astronomers solved such problems.

6.3.1 The Chinese Remainder Problem

Simpler versions of congruence problems occur in mathematical works. The earliest such example is in the *Sunzi suanjing* (*Master Sun's Mathematical Manual*), a work probably written late in the third century which later became part of the required course of study for civil servants. This manual consisted chiefly of methods of arithmetical operation, but it also contained the following example of what is today called the Chinese remainder problem.

"We have things of which we do not know the number; if we count them by threes, the remainder is 2; if we count them by fives, the remainder is 3; if we count them by sevens, the remainder is 2. How many things are there?" In modern notation the problem is to find N which simultaneously satisfies

$$N = 3x + 2 \qquad N = 5y + 3 \qquad N = 7z + 2$$

for integral values x, y, z, or, what amounts to the same thing, which satisfies the congruences

$$N \equiv 2 \pmod{3} \qquad N \equiv 3 \pmod{5} \qquad N \equiv 2 \pmod{7}.$$

Sun Zi gives the answer, 23, as well as his method of solution: "If you count by threes and have the remainder 2, put 140. If you count by fives and have the remainder 3, put 63. If you count by sevens and have the remainder 2, put 30. Add these numbers and you get

233. From this subtract 210 and you get 23." Sun Zi explains further: "For each unity as remainder when counting by threes, put 70. For each unity as remainder when counting by fives, put 21. For each unity as remainder when counting by sevens, put 15. If the sum is 106 or more, subtract 105 from this and you get the result."[6]

In modern notation, Sun Zi has apparently noted that

$$70 \equiv 1 \ (\text{mod } 3) \equiv 0 \ (\text{mod } 5) \equiv 0 \ (\text{mod } 7),$$

$$21 \equiv 1 \ (\text{mod } 5) \equiv 0 \ (\text{mod } 3) \equiv 0 \ (\text{mod } 7),$$

and

$$15 \equiv 1 \ (\text{mod } 7) \equiv 0 \ (\text{mod } 3) \equiv 0 \ (\text{mod } 5).$$

Hence $2 \times 70 + 3 \times 21 + 2 \times 15 = 233$ satisfies the desired congruences. Since any multiple of 105 is divisible by 3, 5, and 7, one subtracts off 105 twice to get the smallest positive value. Because this problem is the only one of its type presented by Sun Zi, it is not known whether he had developed a general method of solving

$$x \equiv 1 \ (\text{mod } n_1), \quad x \equiv 0 \ (\text{mod } n_2), \quad x \equiv 0 \ (\text{mod } n_3), \ldots, x \equiv 0 \ (\text{mod } n_k)$$

for given integers $n_1, n_2, n_3, \ldots, n_k$, the most difficult part of the complete solution. The numbers in this particular problem are easy enough to find by inspection, but note for future reference that $70 = 2 \times \frac{3\cdot5\cdot7}{3}$, $21 = 1 \times \frac{3\cdot5\cdot7}{5}$, and $15 = 1 \times \frac{3\cdot5\cdot7}{7}$.

Perhaps two centuries after Sun Zi there appeared the *Zhang Quijian suanjing* (*Zhang Quijian's Mathematical Manual*) (c. 475), another of the mathematical works adopted during the Tang dynasty as part of the examination system. In this book, which also contains interesting material on progressions and the solving of numerical equations, is the first appearance of the problem of the "hundred fowls," famous because it also occurs in various guises in mathematics texts in India, the Islamic world, and Europe. Zhang's original problem is as follows: "A rooster is worth 5 coins, a hen 3 coins, and 3 chicks 1 coin. With 100 coins we buy 100 of the fowls. How many roosters, hens, and chicks are there?"[7] In modern notation, with the number of roosters set as x, the number of hens y, and the number of chicks z, two equations need to be solved:

$$5x + 3y + \frac{1}{3}z = 100$$

$$x + y + z = 100.$$

Zhang gives three answers: 4 roosters, 18 hens, 78 chicks; 8 roosters, 11 hens, 81 chicks; and 12 roosters, 4 hens, 84 chicks; but he only hints at a method: "Increase the roosters every time by 4, decrease the hens every time by 7, and increase the chicks every time by 3." Namely, he has noted that changing the values this way preserves both the cost and the number of fowls. It is possible to solve this problem by a modification of the "Gaussian elimination" method known from the *Jiuzhang suanshu* and get as a general solution $x = -100 + 4t, y = 200 - 7t, z = 3t$ from which Zhang's description follows. In fact, Zhang's answers are the only ones in which all three values are positive. It is not known, however, if Zhang used this method or some other one.

Several Chinese authors over the next centuries commented on this hundred fowls problem, but none succeeded in giving a reasonable explanation of the method or a way of generalizing it to other problems. No explanation of Sun Zi's remainder problem appeared either, although there is a record of a calendrical computation by Yi Xing in the early eighth century which used indeterminate analysis to relate several astronomical cycles by solving the simultaneous congruences

$$N \equiv 0 \ (\text{mod } 1{,}110{,}343 \times 60)$$

$$N \equiv 44{,}820 \ (\text{mod } 60 \times 3040)$$

$$N \equiv 49{,}107 \ (\text{mod } 89{,}773)$$

The answer is given as $N = 96{,}961{,}740 \times 1{,}110{,}343$.

6.3.2 Qin Jiushao and the *Ta-Yen* Rule

It was not until 1247, during the Song dynasty, that Qin Jiushao (c. 1202–1261) published a general method for solving systems of linear congruences in his *Shushu jiuzhang* (*Mathematical Treatise in Nine Sections*). This work was greatly influenced by the old *Jiuzhang suanshu* as were most Chinese mathematical works through the fifteenth century. Like the earlier work, Qin's *Mathematical Treatise* was a collection of problems with solutions and methods. Many of these problems were similar to those of the old text and were solved by similar methods, but they did tend to be somewhat more difficult. The significant new development, however, was the *ta-yen* rule for solving simultaneous linear congruences, congruences which in modern notation are written $N \equiv r_i \ (\text{mod } m_i)$ for $i = 1, 2, \ldots, n$.

Ten of the problems of the *Mathematical Treatise* are remainder problems of this type. In particular, we consider problem I, 4 which, in modern notation, asks to determine N such that

$$N \equiv 10 \ (\text{mod } 12) \equiv 0 \ (\text{mod } 11) \equiv 0 \ (\text{mod } 10) \equiv 4 \ (\text{mod } 9)$$

$$\equiv 6 \ (\text{mod } 8) \equiv 0 \ (\text{mod } 7) \equiv 4 \ (\text{mod } 6).$$

The first step is to reduce the moduli to ones relatively prime. [Note that in the example from Sun Zi, the moduli were already relatively prime.] Qin gives a detailed procedure for doing this, which amounts to finding the least common multiple of the moduli, $m_1, m_2, \ldots, m_n$, then finding integers $\mu_1, \mu_2, \ldots, \mu_n$, relatively prime in pairs, whose least common multiple is the same and such that μ_i divides m_i for each i. The solution to $N \equiv r_i \ (\text{mod } \mu_i)$ for all i is then the same as that to $N \equiv r_i \ (\text{mod } m_i)$ as long as $r_i - r_j$ is divisible by the greatest common divisor of m_i, m_j for all i, j. (This condition is not noted by Qin although all of his examples actually satisfy it.) In the present case, 11 and 7, being prime, are left as is as are $9 = 3^2$ and $8 = 2^3$. But since $12 = 2^2 \times 3$, $10 = 2 \times 5$, and $6 = 2 \times 3$, and since 2 and 3 already appear to higher powers, the modulus 10 is reduced to 5 and the moduli 12 and 6 both to 1. The new relatively prime moduli, called the *dingshu*, are then $\mu_1 = 1$, $\mu_2 = 11$, $\mu_3 = 5$, $\mu_4 = 9$, $\mu_5 = 8$, $\mu_6 = 7$, and $\mu_7 = 1$. The *yenmu* θ, equal to the product of the *dingshu* or to the least common multiple of the original moduli is, in this case, $\theta = 11 \times 5 \times 9 \times 8 \times 7 = 27{,}720$.

Biography   **Qin Jiushao (1202–1261)**

Qin Jiushao was probably born in Sichuan during the time when the Mongols under Ghenghis Khan were completing their conquest of North China. The Song dynasty's capital at this time was at Hangzhou and it was there that Qin studied in the Board of Astronomy, the agency responsible for calendrical computations. Subsequently, Qin wrote, "I was instructed in mathematics by a recluse scholar. At the time of troubles with the barbarians [the mid-1230s], I spent some years at the distant frontier; without care for my safety among the arrows and stone missiles, I endured danger and unhappiness for ten years." To console himself, he then spent time thinking about mathematics. "I made inquiries among well-versed and capable [persons] and investigated mysterious and vague matters . . . As for the details [of the mathematical problems], I set them out in the form of problems and answers meant for practical use . . . I selected

eighty-one problems and divided them into nine classes; I drew up their methods and their solutions and elucidated them by means of diagrams."[8] The "diagrams" of his *Mathematical Treatise in Nine Sections* are of the positions of the rods on the counting board as solutions to the various problems are described.

Qin served the government later in several offices, but since he "was extravagant and boastful [and] obsessed with his own advancement," he was several times relieved of his duties because of corruption. Nevertheless, he became rich. On a magnificently situated plot of land which he obtained by trickery, he had an enormous house constructed, in the back of which there was a "series of rooms for lodging beautiful female musicians and singers." In fact he developed an impressive reputation in love affairs.[9]

For the second step, Qin divides the *yenmu* by each of the *dingshu* in turn to get what he calls the *yenshu*, which we will designate by M_i. Here $M_1 = \theta \div 1 = 27{,}720$, $M_2 = \theta \div 11 = 2520$, $M_3 = 5544$, $M_4 = 3080$, $M_5 = 3465$, $M_6 = 3960$, and $M_7 = 27{,}720$. Each M_i satisfies $M_i \equiv 0 \pmod{\mu_j}$ for $j \neq i$.

In the third step, Qin subtracts from each of the *yenshu* as many copies of the corresponding *dingshu* as possible, that is, he finds the remainders of M_i modulo μ_i. These remainders, labeled N_i, are $N_1 = 27{,}720 - 27{,}720 \times 1 = 0$, $N_2 = 2520 - 229 \times 11 = 1$, $N_3 = 4$, $N_4 = 2$, $N_5 = 1$, $N_6 = 5$, and $N_7 = 0$. Of course, $N_i \equiv M_i \pmod{\mu_i}$ for each i.

It is finally time to solve congruences, in particular, the congruences $N_i x_i \equiv 1 \pmod{\mu_i}$. Once this is done, one answer to the modified problem with relatively prime moduli is easily seen to be

$$N = \sum_{i=1}^{n} r_i M_i x_i,$$

in analogy with the solution to Sun Zi's problem. Since each μ_i divides θ, any multiple of θ can be subtracted from N to get other solutions.[10]

To solve $N_i x_i \equiv 1 \pmod{\mu_i}$ with N_i and μ_i relatively prime, Qin uses what he calls the "technique of finding one." In effect, this procedure involves the Euclidean algorithm. Qin describes it using diagrams of the counting board. Because the congruences in our

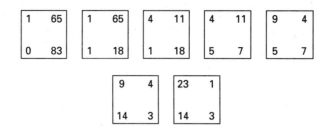

FIGURE 6.4
Counting board diagrams for solving $65x \equiv 1 \pmod{83}$ by method of Qin Jiushao.

example can be solved by inspection, however, we demonstrate the technique by using the example $65x \equiv 1 \pmod{83}$ (Figure 6.4). Qin begins by placing 65 in the upper right of a counting board with four squares, 83 in the lower right, 1 at the upper left, and nothing in the lower left. As he writes, "first divide right bottom by right top, multiply the quotient obtained by the top left and [add it to] the bottom left, [at the same time replacing the bottom right by the remainder of the division]. And then use the right column top and bottom; using the smaller to divide the greater, dividing alternately, immediately multiply by the quotient obtained [and add it] successively . . . into the left column top or bottom until finally the top right is just 1, then stop. Then take the top left result [as the solution]."[11] These diagrams represent the following computations:

$$83 = 1 \cdot 65 + 18 \qquad 1 \cdot 1 + 0 = 1$$
$$65 = 3 \cdot 18 + 11 \qquad 3 \cdot 1 + 1 = 4$$
$$18 = 1 \cdot 11 + 7 \qquad 1 \cdot 4 + 1 = 5$$
$$11 = 1 \cdot 7 + 4 \qquad 1 \cdot 5 + 4 = 9$$
$$7 = 1 \cdot 4 + 3 \qquad 1 \cdot 9 + 5 = 14$$
$$4 = 1 \cdot 3 + 1 \qquad 1 \cdot 14 + 9 = 23$$

The second column can be thought of as representing the absolute values of the successive coefficients of 65 obtained by substitution. Namely, begin with $18 = 83 - 1 \cdot 65$ and substitute this into $11 = 65 - 3 \cdot 18$ to get $11 = 65 - 3 \cdot (83 - 1 \cdot 65) = 4 \cdot 65 - 3 \cdot 83$, where the 4 is the result of the second calculation in the second column. Similarly, $7 = 18 - 1 \cdot 11 = (83 - 1 \cdot 65) - 1 \cdot (4 \cdot 65 - 3 \cdot 83) = 4 \cdot 83 - 5 \cdot 65$. The final result is that $1 = 23 \cdot 65 - 18 \cdot 83$, so that 23 is a solution to the congruence. (Qin always adjusts matters so that the final coefficient is positive.)

In the main example, the solutions x_i to the congruences with $N_i \neq 0$ are $x_2 = 1, x_3 = 4, x_4 = 5, x_5 = 1$, and $x_6 = 3$. These solutions are called the *chenglu*. Taking $x_1 = x_7 = 0$, Qin completes the solution by multiplying each x_i by the corresponding M_i and r_i to get the *tsungshu* $r_i M_i x_i$. In this case, these values are nonzero only for $i = 4$ and $i = 5$: $r_4 M_4 x_4 = 4 \times 3080 \times 5 = 61{,}600$ and $r_5 M_5 x_5 = 6 \times 3465 \times 1 = 20{,}790$. Their sum, 82,390, is then reduced by twice the *yenmu*, or 55,440, to get the final result: $N = 26{,}950$.

The outline given of Qin's procedure is abstracted from the detailed verbal description given in the text itself.[12] In his actual solutions of the ten remainder problems in the *Mathematical Treatise*, Qin often modifies the steps in one way or another, not only to simplify matters but also sometimes to demonstrate that he could deal with very large numbers. Not surprisingly, two of the problems are calendrical problems in which the question is to find the common cycle of several different cyclical events. Although the numbers in these problems are quite large and involve fractions or decimals, Qin was able to modify his method appropriately to find the solutions.

6.4 SOLVING EQUATIONS

In addition to the first description of the procedure for solving systems of linear congruences, the *Shushu jiuzhang* also contains many problems in solving polynomial equations. The methods for solving these equations which appear here and in several other Chinese works can be considered as a generalization of the methods for solving pure quadratic equations ($x^2 = a$) detailed in the *Jiuzhang suanshu* and discussed in Chapter 1. Recall that there is one example in that text of a mixed quadratic equation, but that no method of solution is indicated.

Similar problems appear in other Chinese works through the centuries. But although evidently a method of solving quadratic and even cubic equations existed during the first millenium, it was not until the mid-eleventh century that there is any record of the method. At that time, Jia Xian, in a work now lost, generalized the square and cube root procedures of the *Jiuzhang suanshu* to higher roots by using the array of numbers known today as the Pascal triangle and also extended and improved the method into one usable for solving polynomial equations of any degree. Some of Jia Xian's methods are discussed in a work of Yang Hui written about 1261.

Jia's basic idea stems from the original square and cube root algorithms which made use of the binomial expansions $(r + s)^2 = r^2 + 2rs + s^2$ and $(r + s)^3 = r^3 + 3r^2s + 3rs^2 + s^3$ respectively. For example, consider the solution of the equation $x^3 = 12,812,904$, which a reasonable guess shows is a three-digit number starting with 2. In other words, the closest integer solution can be written as $x = 200 + 10b + c$. Ignoring temporarily the c, we need to find the largest b so that $(200 + 10b)^3 = 200^3 + 3 \cdot 200^2 \cdot 10b + 3 \cdot 200 \cdot (10b)^2 + (10b)^3 \le 12,812,904$, or so that $3 \cdot 200^2 \cdot 10b + 3 \cdot 200 \cdot 100b^2 + 1000b^3 = b(1,200,000 + 60,000b + 1000b^2) \le 4,812,904$. By trying in turn $b = 1, 2, 3, \ldots$, one discovers that $b = 3$ is in fact the largest value satisfying the inequality. Since $3(1,200,000 + 60,000 \cdot 3 + 1000 \cdot 3^2) = 4,167,000$, one next subtracts 4,167,000 from 4,812,904 and derives a similar inequality for c: $c(3 \cdot 230^2 + 3 \cdot 230c + c^2) \le 645,904$. In this case it turns out that $c = 4$ satisfies this as an equality, so the solution to the original equation is $x = 234$.

Jia realized that this solution process could be generalized to nth order roots for $n > 3$ by determining the binomial expansion $(r + s)^n$. In fact, as Yang Hui reports, not only did he write out the Pascal triangle of binomial coefficients through the sixth row (Figure 6.5), but he also developed the usual method of generating the triangle: "Add the numbers in the two places above in order to find the number in the place below."[13] Yang Hui further explained how Jia used the binomial coefficients to find higher order roots by a method analogous to that just described.

Evidently, Jia went even further. He saw that his method could be used to solve arbitrary polynomial equations, especially since these appeared as part of the root extrac-

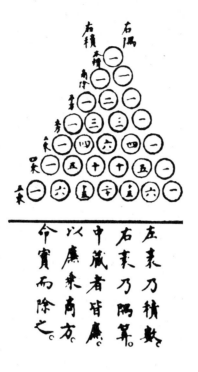

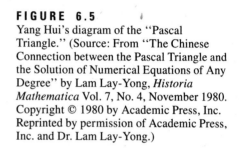

FIGURE 6.5
Yang Hui's diagram of the ''Pascal
Triangle.'' (Source: From ''The Chinese
Connection between the Pascal Triangle and
the Solution of Numerical Equations of Any
Degree'' by Lam Lay-Yong, *Historia
Mathematica* Vol. 7, No. 4, November 1980.
Copyright © 1980 by Academic Press, Inc.
Reprinted by permission of Academic Press,
Inc. and Dr. Lam Lay-Yong.)

tion process, but that it would be simpler on the counting board to generate the various multiples by binomial coefficients step-by-step rather than from the triangle itself.

6.4.1 Qin Jiushao and the Solution of Polynomial Equations

The first detailed account of Jia's method for solving equations, probably somewhat improved, appears in Qin Jiushao's *Shushu jiuzhang*. This method will be presented in the context of a particular equation, $-x^4 + 763,200x^2 - 40,642,560,000 = 0$. (The equation itself came from a geometry problem.) The initial steps in solving such an equation are the same as those in the solving of the pure equation, namely, first, determine the number of decimal digits in the answer, and second, guess the appropriate first digit. In this case it is found, by experience or by trial and error, that the answer will be a three-digit number beginning with 8. Qin's approach, like that of the old cube root algorithm, is, in effect, to set $x = 800 + y$, substitute this value into the equation, and then derive a new equation in y whose solution will be only a two-digit number. One can then guess the first digit of y and repeat the process. Given the decimal nature of the Chinese number system, the Chinese could in theory at least repeat this algorithm as often as desired to approximate the answer to any predetermined level of accuracy. In practice, however, Qin usually states remainders as fractions.

The Chinese did not, of course, use modern algebra techniques to "substitute" $x = 800 + y$ into the original equation as William Horner did in his essentially similar method of 1819. The problem was set up on a counting board with each row standing

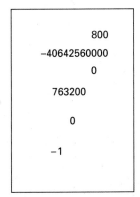

FIGURE 6.6
Initial counting
board configuration
for solution of
$-x^4 + 763{,}200x^2$
$- 40{,}642{,}560{,}000 = 0.$

for a particular power of the unknown (Figure 6.6). For reasons of space, however, we will write the coefficients horizontally. Thus for the problem at hand, the opening configuration is

$$-1 \qquad 0 \qquad 763200 \qquad 0 \qquad -40642560000.$$

Given that the initial approximation to the root was 800, Qin described what is now called the repeated (synthetic) division of the original polynomial by $x - 800 (= y)$. The first step gives:

$$
\begin{array}{r|rrrrr}
800 & -1 & 0 & 763200 & 0 & -40642560000 \\
 & & -800 & -640000 & 98560000 & 78848000000 \\
\hline
 & -1 & -800 & 123200 & 98560000 & \lfloor 38205440000 \\
\end{array}
$$

Qin's description of the counting board process tells exactly what numbers to multiply and add (or subtract) to give the arrangement on the third line. For example, the -1 is multiplied by 800 and the result added to the 0. That result (-800) is then multiplied by 800 and the product subtracted from the 763,200. In algebraic symbolism, this first step shows that the original polynomial has been replaced by

$$(x - 800)(-x^3 - 800x^2 + 123200x + 98560000) + 38205440000$$

$$= y(-x^3 - 800x^2 + 123200x + 98560000) + 38205440000.$$

Qin repeated the procedure three more times, dividing each quotient polynomial by the same $y = x - 800$. The final result is

$$0 = -x^4 + 763200x^2 - 40642560000$$

$$= y\{y[y(-y - 3200) - 3076800] - 826880000\} + 38205440000$$

or

$$-y^4 - 3200y^3 - 3076800y^2 - 826880000y + 38205440000 = 0.$$

Of course, Qin only has numbers on the counting board. His diagrams (one for each step) are combined here into a single large diagram:

$$
\begin{array}{r|rrrrr}
800 & -1 & 0 & 763200 & 0 & -40642560000 \\
 & & -800 & -640000 & 98560000 & 78848000000 \\
800 & -1 & -800 & 123200 & 98560000 & \lfloor 38205440000 \\
 & & -800 & -1280000 & -925440000 & \\
800 & -1 & -1600 & -1156800 & \lfloor -826880000 & \\
 & & -800 & 1920000 & & \\
800 & -1 & -2400 & \lfloor -3076800 & & \\
 & & -800 & & & \\
800 & -1 & \lfloor -3200 & & & \\
 & \lfloor -1 & & & & \\
\end{array}
$$

$$
\begin{array}{r|rrrrr}
40 & -1 & -3200 & -3076800 & -826880000 & 38205440000 \\
 & & -40 & -129600 & -128256000 & -38205440000 \\
\hline
 & -1 & -3240 & -3206400 & -955136000 & \lfloor 0 \\
\end{array}
$$

The third line from the bottom contains the coefficients of Qin's equation for y along with his guess of 4 as the first digit of the two-digit answer. (This came simply from dividing 38205440000 by 826880000.) In the example, as is normally the case in our texts today, the answer "comes out even." The equation for y is exactly divisible by $y - 40$. The solution to the original equation is then $x = 840$.

To see the relationship of Qin's description to Jia's method by the Pascal triangle and how the binomial coefficients are generated step-by-step, consider how the equation $x^3 = 12,812,904$ would be solved using Qin's procedure. The layout of the figures in this case is

200⌋	1	0	0	− 12812904
		200	40000	8000000
200⌋	1	200	40000	⌊−4812904
		200	80000	
200⌋	1	400	⌊120000	
		200		
200⌋	1	⌊600		
	⌊1			

30⌋	1	600	120000	−4812904
		30	18900	4167000
30⌋	1	630	138900	⌊−645904
		30	19800	
30⌋	1	660	⌊158700	
		30		
30⌋	1	⌊690		
	⌊1			

4⌋	1	690	138700	−645904
		4	2776	645904
	1	694	161476	⌊0

One can easily see the binomial coefficients in this table. For example, the ninth line implies that the equation for the second decimal digit $10b$ is $(10b)^3 + 3 \cdot 200 \cdot (10b)^2 + 3 \cdot 200^2 \cdot 10b + (200^3 − 12812904) = 0$, exactly as specified by Jia.

Qin himself gave no theoretical justification of his procedure, nor did he mention the Pascal triangle. But since he solved twenty-six different equations in the *Shushu jiuzhang* by the method and since several of his contemporaries solved similar equations by the same method, it is evident that he and the Chinese mathematical community in general were in possession of a correct algorithm for solving these problems. This algorithm, since it was rediscovered in Europe more than five centuries after Qin's time, deserves a few additional comments.

First, the texts only briefly state how the guessed values for the digits of the root are found. In some cases it is clear that the solver simply made a trial division of the constant term by the coefficient of the first power of the unknown, as is generally done in the square root algorithm itself. Sometimes several trials are indicated and the author picks one that works. But in general, one can only surmise that the Chinese mathematicians possessed extensive tables of powers which could be used to make the various guesses. Second, there is no mention in the texts of multiple roots. Qin's fourth degree equation above, in fact, has another positive root, 240, as well as two negative ones. The root 240 could easily have been found by the same method provided one had guessed 2 for the initial digit. But in this case the real problem from which the equation was derived had only one solution, 840, and Qin did not deal with equations in the abstract. Third, operations with negative numbers were performed as easily as those with positives. Recall that the Chinese used different colored counting rods to represent the two types of numbers and had long before discovered the correct arithmetic algorithms for computations. On the other hand, negative roots do not appear. Fourth, because they could deal with negative numbers, the Chinese generally represented equations in a form equivalent to $f(x) = 0$. This represents a basic difference in approach compared to the ancient Babylonian method or to the medieval Islamic one. Finally, it appears that the Chinese method of solving quadratic equations is totally different from that of the Babylonians. The latter essentially developed a formula which could only be applied to such equations. The Chinese developed a numerical algorithm that they ultimately generalized to equations of any degree.

6.4.2 The Work of Li Ye, Yang Hui, and Zhu Shijie

Qin Jiushao had three contemporaries who also made significant contributions to the mathematics of solving equations, Li Ye (1192–1279), Yang Hui (second half of the thirteenth century), and Zhu Shijie (late thirteenth century). But probably due to the war between the Mongols and the two Chinese dynasties of the Jin and the Southern Song, which lasted most of the century, there is doubt that any of these mathematicians had much influence on the others.

Li Ye wrote two major mathematical works, the *Ceyuan haijing* (*Sea Mirror of Circle Measurements*) in 1248 and the *Yigu yanduan* (*Old Mathematics in Expanded Sections*) in 1259, as well as numerous works in other fields. The *Ceyuan haijing* dealt with the properties of circles inscribed in right triangles, but was chiefly concerned with the setting up and solution of algebraic equations for dealing with these properties. The *Yigu yanduan* similarly dealt with geometric problems on squares, circles, rectangles, and trapezoids, but again its main object was the teaching of methods for setting up the appropriate equations, invariably quadratic, for solving the problem.

We give one example of Li Ye's methods from his *Yigu yanduan*:[14]

> Problem 8: There is a circular pond inside a square field and the area outside the pond is 3300 square feet. The sum of the perimeters of the square and the circle is 300 feet. Find the two perimeters.

Li's discussion is virtually identical to what one would find in a modern text. He sets x to be the diameter of the circle and $3x$ ($\pi = 3$) to be the circumference. Then $300 - 3x$ is the perimeter of the square. Squaring that value, he gets $90,000 - 1800x + 9x^2$ as the area of 16 square fields. Also, since $\frac{3x^2}{4}$ is the area of one circular pond, $12x^2$ is the area of 16 circular ponds. The difference of the two expressions, namely $90,000 - 1800x - 3x^2$ is equal to 16 portions of the area outside the pond, or $16 \times 3300 = 52,800$. The desired equation is then $37,200 - 1800x - 3x^2 = 0$. In contrast to the work of Qin, Li Ye now

---

Biography   Li Ye (1192–1279)

Li Ye was born into a bureaucratic family in Zhending in Hebei province north of the Yellow river. In 1230 he passed the civil service examination and took a government post in the northern kingdom of Jin. But his district, and the entire Jin kingdom, fell to the Mongols within a few years, so Li gave up hope of an official career and devoted the rest of his life to scholarship. After Kublai Khan ascended the throne in 1260, Li was asked to serve in the Mongol government, and did so briefly. He retired for good in 1266 and returned to seclusion in the Mt. Fenglong district of his birth.

---

merely asserts that 20 is the root, or the diameter, and therefore that 60 is the circumference of the circle and 240 that of the square.

It is interesting that Li Ye nearly always follows his algebraic derivation with a geometric derivation (Figure 6.7). Here the side of the large square is 300, the sum of the given perimeters. The shaded areas represent 16×3300. Since $300x$ is the area of each long strip, x^2 the area of each small square, and $12x^2$ the total area of the 16 circular ponds, he derives the equation $300^2 - 16 \times 3300 = 6 \times 300x - 9x^2 + 12x^2 = 1800x + 3x^2$ or $37{,}200 = 1800x + 3x^2$ as before. (Note that the diagram indicates the 3 small squares at the bottom right.)

The text thus provides more evidence for the development of Chinese mathematics. Not only did the solution method originally have a geometric basis, but also the very setting up of the problems did as well. Because the numerical results were recorded and calculated on the counting board, the Chinese scholars ultimately recognized patterns on that board and developed them into numerical algorithms. At the same time they probably began to abstract the geometrical concept of, for instance, square, into simply a position on the counting board and then into the algebraic idea of the square of an unknown numerical quantity. Once the notion of squares of an unknown became abstract, there was

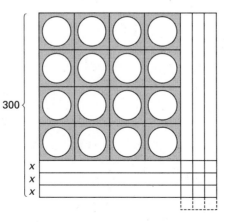

FIGURE 6.7
Problem 8 from Li Ye's *Yigu yanduan.*

no barrier to considering equations of higher degree. Qin Jiushao's equations were based on real, and even geometric, problems, but he had no hesitation about using powers of the unknown which had no geometric meaning whatsoever.

About Yang Hui, whose reports on the work of Jia Xian have been discussed earlier, little is known other than that he lived under the Song dynasty in the south of China. Two major works of his are still extant, the *Xiangjie jiushang suanfa* (*A Detailed Analysis of the Arithmetical Rules in the Nine Sections*) of 1261 and the collection known as *Yang Hui suanfa* (*Yang Hui's Methods of Computation*) of 1275. The latter work, like the work of Li Ye, contained material on quadratic equations. In contrast to Li's work, however, Yang Hui gave a detailed account of his methods. In general, Yang used the same method as Qin, but he also gave alternate methods more reminiscent of the Chinese method of square root extraction described earlier, namely, the explicit use of double the first approximation in deriving the second equation. In addition, Yang presented geometric diagrams consisting of squares and rectangles illustrating the various numerical methods used.

Little is also known about the life of the last of the important thirteenth century Chinese mathematicians, Zhu Shijie. He was probably born near present day Beijing, but spent most of his life as a wandering teacher, that is, as a professional mathematics educator. He wrote two major works, the *Suanxue Qimeng* (*Introduction to Mathematical Studies*) in 1299 and the *Sijuan yujian* (*Precious Mirror of the Four Elements*) in 1303.

One of Zhu's major contributions was his adaptation of Qin's method of solving polynomial equations into a procedure for solving systems of equations. We illustrate his method by considering the first problem of the *Precious Mirror*: "Given that the length of the diameter of a circle inscribed in a right triangle multiplied by the product of the lengths of the two legs equals 24, and the length of the vertical leg added to the length of the hypotenuse equals 9, what is the length of the horizontal leg?"[15] Note here that in the Chinese wording of the problem, single Chinese characters stand for such words as "the length of the diameter of a circle inscribed in a right triangle" or the "product of the lengths of the two legs." Hence the actual wording is virtually equivalent to our symbolic equations and a modern translation remains in the spirit of the Chinese work. Thus, let a stand for the vertical leg, b the horizontal leg, c the hypotenuse, and d the diameter of the circle (Figure 6.8). The problem can be translated into the two equations

$$dab = 24$$

$$a + c = 9.$$

Zhu in addition assumes as known the two equations

$$a^2 + b^2 = c^2$$

$$d = b - (c - a),$$

where the second gives the relationship between the diameter of the inscribed circle and the lengths of the sides of the triangle.

As is unfortunately common to many of these Chinese texts, the method of solution is only briefly indicated and the fifth degree equation satisfied by b is simply written down. Evidently, it was expected that teachers would fill in the missing details. But there are

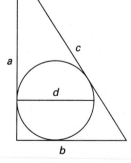

FIGURE 6.8
Problem 1 of the *Sijuan yujian* of Zhu Shijie.

indications that the solution procedure was the following: First, from $b^2 = c^2 - a^2 = (c - a)(c + a)$ and $c + a = 9$, conclude that $b^2 = 9(c - a)$. Next, multiply the equation $(c + a) - (c - a) = 2a$ by 9 to get $9(c + a) - 9(c - a) = 18a$. Thus $81 - b^2 = 18a$ and

$$18ab = 81b - b^3. \qquad\qquad \textbf{(6.1)}$$

Third, multiply $d = b - (c - a)$ by 9 to get $9d = 9b - 9(c - a)$ or

$$9d = 9b - b^2. \qquad\qquad \textbf{(6.2)}$$

Multiplying together equations 6.1 and 6.2 gives

$$162dab = 729b^2 - 81b^3 - 9b^4 + b^5.$$

(This is the first equation actually written out by Zhu.) Since $dab = 24$, Zhu needs to solve the fifth degree equation in b:

$$b^5 - 9b^4 - 81b^3 + 729b^2 - 3888 = 0.$$

Zhu does not, however, show his method of solution, merely writing that $b = 3$.

As can be surmised from its title, Zhu's other text is on a much more elementary level, probably being intended for beginners or for reference in the Office of Mathematics. In general, problems and methods are repeated, or only slightly modified, from the classic *Jiuzhang suanshu*. The first several chapters deal with procedures of arithmetic. These are followed by such items as proportions, interest, taxation, areas, and volumes. In fact, the various formulas used show no improvement over the sometimes erroneous results given in the Han dynasty work. Other chapters deal with such procedures for solving systems of linear equations as the surplus and deficiency method and the Gaussian elimination method. Similarly, the *Yang Hui suanfa* contains many examples of systems of equations. It also contains the early indeterminate problems of Sun Zi as well as the "hundred fowls" problem. Yang also includes some problems involving the summation of series, as does Zhu in his second book.

What can we conclude in general about Chinese mathematics in the medieval period? The Chinese mathematicians were proficient in solving many kinds of algebraic problems. Many of their methods probably stemmed originally from geometric considerations but ultimately were apparently translated into purely algebraic procedures. From the texts we have, it also appears that the Chinese scholars were primarily interested in solving problems of importance to the Chinese bureaucracy. Although there is evidently some development of better techniques over the centuries, to a large extent progress was stifled by the general Chinese reverence for the past. Hence even incorrect methods from such works as the *Jiuzhang suanshu* were repeated down through the centuries. Although the thirteenth century mathematicians exploited the counting board to the fullest, its very use provided limits. Equations remained numerical, and there could be no development of any theory of equations as was to take place several centuries later in the West. In fact, a combination of political circumstances led to a decline in Chinese mathematical activity during the Mongol (Yuan) and the later Ming dynasties, so that even some of the great thirteenth century works were no longer studied. Finally, in the late sixteenth century, with the arrival of the Jesuit priest Matteo Ricci (1552–1610), Western mathematics entered China and the indigenous tradition began to disappear (Figure 6.9).

FIGURE 6.9
Matteo Ricci on a stamp from Taiwan.

6.5 INTRODUCTION TO THE MATHEMATICS OF MEDIEVAL INDIA

A number of rival Aryan states developed in north India during the middle of the first millenium B.C.E. In the sixth century, one, Magadha, gradually gained prominence. The caste system, having originated during the Aryan invasions, became solidified. The brahmins, or priests, were the bearers of the religious traditions that grew into Hinduism. Because the lower castes were believed unworthy of learning, however, the basic traditions were not put into written form but handed down orally from brahmin to brahmin. To assist in memorization of the important ideas, many of the works were put into poetic stanzas. Not only were the strictly religious works treated this way, but also other aspects of culture, including, for example, the basics of astronomy and mathematics. Astronomical knowledge in India as elsewhere was from ancient times part of what kept the ruling class in power. In any case, this lack of written work clouds our knowledge today of the developments in Indian mathematics. Even when the mathematics was committed to writing, the condensed nature of the verses often makes the mathematics difficult to understand.

In 327 B.C.E. Alexander the Great crossed the Hindu Kush mountains into northeastern India and, during the following two years, conquered the small Indian kingdoms of the area. Greek influence began to spread into India. Alexander came with scientists and historians in his entourage, not just as a conqueror interested in plunder but on a mission to "civilize" the East. Naturally, the Indians believed they were already civilized. Each people considered the other barbarians. Alexander's grand designs ended with his premature death in 323. His Indian provinces were soon reconquered by Chandragupta Maurya, who had earlier become king of Magadha. Chandragupta established friendly relations with Seleucus, Alexander's successor in Western Asia, and through this relationship there was evidently some interchange of ideas. Shortly after Chandragupta's death, Ashoka succeeded to the throne. He proceeded to conquer most of India but then converted to Buddhism and sent missionaries both east and west to convert the neighboring kingdoms. Ashoka left records of his reign in edicts carved on pillars throughout his kingdom. These pillars contain some of the earliest written evidence of Indian numerals (Figure 6.10).

During the first century C.E., northern India was conquered by Kushan invaders. The Kushan empire soon became the center of a flourishing trade between the Roman world and the East. Early in the fourth century, northern India was again united under a native dynasty, that of the Guptas. Under their rule, which only lasted about a century and a half, India reached a high point of culture with the flowering of art and medicine and the opening of universities. It was also during this period that Indian colonists spread Hindu culture to various areas of southeast Asia, including Burma, Malaya, and Indochina. Although the Gupta empire came to an end around 480, the dynasty itself continued to rule several eastern provinces of India into the eighth century.

A northern Indian kingdom was revived in 606 by Harsha, a remarkably tolerant and just ruler, but after his death in 647 his empire collapsed and India broke up into many small states. Nevertheless, there was still cultural unity in the Indian subcontinent and so, even after the seventh century one can still speak of Indian mathematics. Beginning in the eighth century there were periodic incursions of Moslem Arabs into the north and major battles between Moslems and Hindus. Finally, toward the end of the twelfth century, northern India was conquered by a Moslem army under Mohammed Ghori, and in 1206

FIGURE 6.10
A pillar of Ashoka is on the right of this Indian stamp.

the Moslem Sultanate of Delhi was established, an empire which was to last over 300 years. The Sultanate even succeeded in conquering parts of the Hindu kingdoms in the south of India, kingdoms which had generally been independent even of the earlier native kingdoms of the north.

Through the various invasions and new kingdoms, it does appear that the study of astronomy was always encouraged. Whoever ruled the country always seemed to need astronomers to help with calendrical questions and, of course, to give astrological advice. Thus most of the Indian mathematics works of this period were parts of astronomical works. Nevertheless, here, as elsewhere, creative mathematicians went beyond the strict requirements of practical problem solving to develop new areas of mathematics which they found of interest. We will consider in the remainder of this chapter first the mathematics of astronomy and surveying exemplified in the Indian texts of the fourth and fifth century, next the work of Brahmagupta in the seventh century and Bhāskara in the twelfth century on the solution of two classes of indeterminate equations, including the so-called Pell equation, and finally the general work in algebra and combinatorics of various Indian mathematicians of the entire medieval period.

6.6 INDIAN TRIGONOMETRY AND SURVEYING

During the first centuries of the common era, in the period of the Kushan empire and that of the Guptas, there is strong evidence of the transmission of Greek astronomical knowledge to India, probably along the Roman trade routes. Curiously, Ptolemy's astronomy and mathematics were not transmitted but the work of some of his predecessors instead, in particular, the work of Hipparchus. Just as the needs of Greek astronomy led to the development of trigonometry, the needs of Indian astronomy led to Indian improvements in this field.

6.6.1 The *Sūrya-Siddhānta*

The earliest Indian work which contains trigonometric material is the *Sūrya-Siddhānta* (system) [of astronomy]. This is one of several similar works written around the fourth century. The *Sūrya-Siddhānta* is written, as are most of the medieval Indian works, in Sanskrit verse, in stanzas of two lines, each line being composed of two halves of eight syllables each. Here are the first three stanzas of the material on sines.

> The eighth part of the minutes of a sign is called the first sine; that increased by the remainder left after subtracting from it the quotient arising from dividing it by itself is the second sine.
>
> Thus, dividing the tabular sines in succession by the first and adding to them, in each case, what is left after subtracting the quotients from the first, the result is twenty-four tabular sines in order as follows.
>
> Two hundred and twenty-five; four hundred and forty-nine; six hundred and seventy-one; eight hundred and ninety; eleven hundred and five; thirteen hundred and fifteen.[16]

These stanzas require explanation. First, the Sanskrit term translated as "sine" is *jya-ardha,* literally chord-half (Side Bar 6.1). The Hindus had decided that the half-chord of double the arc was more useful than Hipparchus' chord (or that of Ptolemy). Recall that in

---

Side Bar 6.1   **The Etymology of "Sine"**

The English word "sine" comes from a series of mistranslations of the Sanskrit *jya-ardha* (chord-half). Āryabhaṭa frequently abbreviated this term to *jya* or *jiva*. When some of the Hindu works were later translated into Arabic, the word was simply transcribed phonetically into an otherwise meaningless Arabic word *jiba*. But since Arabic is written without vowels, later writers interpreted the consonants *jb* as *jaib*, which means bosom or breast. In the twelfth century, when an Arabic trigonometry work was translated into Latin, the translator used the equivalent Latin word *sinus*, which also meant bosom, and by extension, fold (as in a toga over a breast) or a bay or gulf. This Latin word has now become our English "sine."

---

many of Ptolemy's calculations, it was this half-chord that was used. He calculated at each step what the Indian author decided simply to tabulate.

Second, the author of the *Sūrya-Siddhānta* refers to the "eighth part of the minutes of a sign." Recall that an (astrological) "sign" was taken by the Babylonians and then the Greeks as a 30° interval on the ecliptic. Hipparchus used steps of 15° and $7\frac{1}{2}°$ in his calculations. The Indian author here simply takes the division one step further and calculates sines of intervals of $3\frac{3}{4}° = 3°45'$.

Third, the value of the sines, the lengths of the half-chords, are based on the same value for the radius as Hipparchus used, $r = 3438'$ (equal to $360° × 60'$, the number of minutes in the circumference, divided by 2π). The first sine s_1, of the arc $3°45' = 225'$, is taken to be equal to the arc itself, $225'$. (Note that this is only true if the radius is equal to 3438.) The indicated calculation of the remaining sines is through a finite difference procedure. For example, the second sine s_2 (of $7°30'$) is given as $s_1 + \left(s_1 - \frac{s_1}{s_1}\right) = 225' + 224' = 449'$; the third sine s_3 (of $11°15'$) is $s_2 + \left(s_1 - \frac{s_1}{s_1} - \frac{s_2}{s_1}\right) = 449' + 222' = 671'$. It is unlikely, however, that the sines were actually calculated in this way. Rather, they were calculated as Hipparchus did: The sine of 90° is equal to the radius 3438'; the sine of 30° is half the radius, 1719'; the sine of 45° is $3438/\sqrt{2} = 2431'$; and the sines of the other arcs are calculated by use of the Pythagorean theorem and the half-angle formula.

Once the table of sines from $3°45'$ to 90° in steps of $3°45'$ had been constructed, a table of differences and second differences could also have been constructed. If the Indians noticed then that the second differences were proportional to the sines, it would not have been difficult to construct the rule given in the text. Similar sine tables of roughly the same accuracy were produced in India by many authors over the next several hundred years, using various values for the radius, including 120 by Varāhamihira (sixth century), 3270 by Brahmagupta (seventh century), and 3415 by Śrīpati (eleventh century). Varāhamihira tabulated the cosine as well as the sine for his radius and described the standard relationships between these functions. No one, however, until the time of Bhāskara (twelfth century), calculated, or evidently required, any such table for arcs closer together than $3\frac{3}{4}°$.

The *Sūrya-Siddhānta* itself, although it did not contain tables of any other trigonometric function, was probably the source of the Chinese calculation of the tangent function discussed earlier and even hints at the secant. For verses 21–22 of Chapter 3, in discussing the shadow cast by a gnomon, read, "Of [the sun's meridian zenith distance] find the base sine and the perpendicular sine [cosine]. If then the base sine and radius be multiplied respectively by the measure of the gnomon in digits, and divided by the perpendicular sine, the results are the shadow and hypotenuse at mid-day."[17]

6.6.2 Āryabhaṭa and the *Āryabhaṭīya*

The sine table of the *Sūrya-Siddhānta* appears again in Āryabhaṭa's *Āryabhaṭīya,* the earliest Indian mathematics and astronomy work by an identifiable author. Little is known about Āryabhaṭa (b. 476), however, other than that he wrote the book in 499 at Kusuma-pura, near the Gupta capital of Pāṭalipura (modern Patna) on the Ganges in Bihar in northern India. The *Āryabhaṭīya,* basically an astronomical work, was a brief book of four sections and 123 stanzas, the second section of 33 stanzas dealing with mathematics. By no means a detailed working manual, it was only a brief descriptive work perhaps intended to be memorized and surely only intended as a resumé of either a more detailed treatise or simply of lectures given by the author. As such, given that the oral tradition accompanying the work eventually decayed, we today cannot fully comprehend the author's intentions. In any case, samples of some of the stanzas follow, each of which gives a rule of procedure for solving some problem.[18]

Stanza 5 provides a rule for calculating cube roots.

Stanza 5.   *One should divide the second aghana by three times the square of the cube root of the preceding* ghana. *The square (of the quotient) multiplied by three times the* purva *(that part of the cube root already found) is to be subtracted from the first* aghana *and the cube (of the quotient of the above division) is to be subtracted from the* ghana.

The technical terms in the stanza refer to the places in the given number. Counting from right to left, the first, fourth, and so on places are named *ghana* (cubic), the second, fifth, and so on are called the first *aghana* (noncubic), while the third, sixth, and so on are called the second *aghana*. But as the example of calculating the cube root of 12,977,875 shows, certain steps are not spelled out explicitly, probably due to the limitations of Sanskrit verse.

$$
\begin{array}{r}
1\ \ 2\ \ 9\ \ 7\ \ 7\ \ 8\ \ 7\ \ 5\quad)2 \qquad\qquad \text{First digit } 2 \approx \sqrt[3]{12}\\
8 \qquad\qquad\qquad\qquad\qquad 2^3
\end{array}
$$

$$
1\ \ 2\ \ \overline{\begin{array}{l}4\ \ 9\end{array}}\qquad)3 \qquad\qquad 12 = 3\times 2^2
$$

$$
\begin{array}{r}
3\ \ 6 \qquad\qquad 3 = \text{quotient;}\quad (4 \text{ is too large})\\
\overline{1\ \ 3\ \ 7}\\
5\ \ 4 \qquad\qquad 3^2\ \text{multiplied by } 3\times 2\\
\overline{8\ \ 3\ \ 7}\\
2\ \ 7 \qquad\qquad\qquad 3^3
\end{array}
$$

$$
1\ \ 5\ \ 8\ \ 7\ \ \overline{\begin{array}{l}8\ \ 1\ \ 0\ \ 8\end{array}}\qquad)5 \qquad\qquad 1587 = 3\times 23^2
$$

$$
\begin{array}{r}
7\ \ 9\ \ 3\ \ 5 \qquad\qquad 5 = \text{new quotient}\\
\overline{1\ \ 7\ \ 3\ \ 7}\\
1\ \ 7\ \ 2\ \ 5 \qquad\qquad 5^2\ \text{multiplied by } 3\times 23\\
\overline{1\ \ 2\ \ 5}\\
\overline{1\ \ 2\ \ 5} \qquad\qquad\qquad 5^3
\end{array}
$$

Stanza 10. *Add 4 to 100, multiply by 8, and add 62,000. The result is approximately the circumference of a circle of which the diameter is 20,000.*

There is no record of how Āryabhata derived this value of 3.1416 for π.

Stanza 16. *The distance between the ends of the two shadows multiplied by the length of the first shadow and divided by the difference in length of the two shadows gives the* koṭī. *The* koṭī *multiplied by the length of the gnomon and divided by the length of the (first) shadow gives the length of the* bhujā.

This stanza gives a method for finding the height of a pole (*bhujā*) with a light at the top by measuring various shadows. In Figure 6.11, DE and $D'E'$ are two gnomons of length g. The lengths $DF = s_1$ and $D'C = s_2$ of the shadows cast by the *bhujā* AB as well as the distance FC between the shadow ends are known. The lengths of the *bhujā* $AB = x$ and the *koṭī* $AF = y$ are to be found. The formulas of the stanza may be translated as

$$y = \frac{(s_2 + t)s_1}{s_2 - s_1} \quad \text{and} \quad x = \frac{yg}{s_1}.$$

Note that this problem is very similar both in form and solution method to problem 1 in the Chinese *Sea Island Mathematical Manual*.

Stanza 19. *The desired number of terms minus one, halved, . . . multiplied by the common difference between the terms, plus the first term, is the middle term. This multiplied by the number of terms desired is the sum of the desired number of terms. Or the sum of the first and last terms is multiplied by half the number of terms.*

Āryabhata is here providing us with a formula for the sum S_n of an arithmetic progression with initial term a and common difference d. The formula translates to

$$S_n = n\left[\left(\frac{n-1}{2}\right)d + a\right] = \frac{n}{2}[a + (a + (n-1)d)]. \tag{6.3}$$

Stanza 20. *Multiply the sum of the progression by eight times the common difference, add the square of the difference between twice the first term and the common dif-*

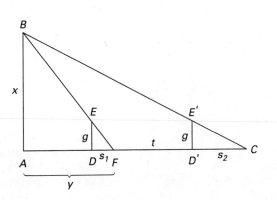

FIGURE 6.11
Measuring distance and height in the *Āryabhatīya*.

ference, take the square root of this, subtract twice the first term, divide by the common difference, add one, divide by two. The result will be the number of terms.

In the same circumstances as above, S_n is given and n is to be found. The formula given is

$$n = \frac{1}{2}\left[\frac{\sqrt{8S_n d + (2a - d)^2} - 2a}{d} + 1 \right]. \qquad (6.4)$$

If equation 6.3 for S_n is rewritten as a quadratic equation in n, namely $dn^2 + (2a - d)n - 2S_n = 0$, then the value for n in equation 6.4 follows from the quadratic formula. Although Āryabhaṭa does not give here a quadratic formula, Brahmagupta, a century and a quarter later, wrote out the formula in the form needed. It appears then that the formula was known to Āryabhaṭa as well.

Stanza 22.   *The sixth part of the product of three quantities consisting of the number of terms, the number of terms plus one, and twice the number of terms plus one is the sum of the squares. The square of the sum of the (original) series is the sum of the cubes.*

These two statements give us formulas for the sums S_n^2, S_n^3 of the first n integral squares and cubes, namely, $S_n^2 = \frac{1}{6}n(n + 1)(2n + 1)$ and $S_n^3 = (1 + 2 + \cdots + n)^2$. The first of these formulas was in essence known to Archimedes. The second formula is almost obvious, at least as a hypothesis, if one tries a few numerical examples. How Āryabhaṭa discovered them, of course, is a matter of conjecture.

The final two stanzas of the mathematics portion of the *Āryabhaṭīya* deal with a method of solving indeterminate equations of the first degree in integers. The conciseness of the expression has led to various conflicting interpretations of exactly what method Āryabhaṭa had in mind. Rather than deal with these here, we will consider the clearer explanations of the same problem in the work of Brahmagupta.

6.7 INDIAN INDETERMINATE ANALYSIS

Brahmagupta was born in 598 in northwestern India. He probably lived most of his life in Bhillamāla (modern Bhinmal in Rajasthan) in the empire of Harsha. Brahmagupta himself, whose major work, the *Brāhmasphuṭasiddhānta* (*Correct Astronomical System of Brahma*) was written when he was 30, is often referred to as Bhillamālacarya, the teacher from Bhillamāla.

Like most other mathematical works of medieval India, the mathematical work of Brahmagupta was imbedded as chapters in an astronomical work. The mathematical techniques described are applied to various astronomical problems. Like Āryabhaṭa, Brahmagupta wrote in verse. His descriptions of procedures are generally fuller than those of his predecessor however, and he gives some examples. Nevertheless, either because of faulty copying over the years or because the oral tradition never required that every step be written down, there are many problems in which Brahmagupta's description of the method simply does not match the steps of his examples. The modern explanations to be presented do, however, convey his main ideas. Brahmagupta himself has nothing a modern reader would consider a proof. He just presents algorithms for solving the problems.

6.7.1 Linear Congruences

Brahmagupta was interested in solving the problem of finding an integer which when divided by each of two positive integers leaves a given remainder. In modern notation, his goal is to find N satisfying $N \equiv a \pmod{r}$ and $N \equiv b \pmod{s}$, or to find x and y such that $N = a + rx = b + sy$, or so that $a + rx = b + sy$, or finally, setting $c = a - b$, so that $rx + c = sy$.

We follow Brahmagupta's rule, the method of *kuṭṭaka*, by using an example from his text: $N \equiv 10 \pmod{137}$ and $N \equiv 0 \pmod{60}$. The problem can be written as the single equation $137x + 10 = 60y$.

> The divisor which yields the greatest remainder is divided by that which yields the least; the residue is reciprocally divided; and the quotients are severally set down one under the other.

Therefore, divide 137 by 60 and continue by dividing the residues. In other words, apply the Euclidean algorithm until the final nonzero remainder is reached:

$$137 = 2 \cdot 60 + 17$$
$$60 = 3 \cdot 17 + 9$$
$$17 = 1 \cdot 9 + 8$$
$$9 = 1 \cdot 8 + 1$$

Then list the quotients one under the other:

$$2$$
$$3$$
$$1$$
$$1$$

Brahmagupta lists 0 for the first quotient, evidently taking the first division as $60 = 0 \cdot 137 + 60$, despite his statement of which divisor is divided into which.

> The [final] residue is multiplied by an assumed number such that the product having added to it (or subtracted from it) the difference of the remainders is exactly divisible. That multiplier is to be set down [underneath] and the quotient last.

The final residue is 1. Multiply that by some number v so that $1 \cdot v \pm 10$ is exactly divisible by the last divisor, in this case 8. Brahmagupta explains that one uses the $+$ when there are an even number of quotients and the $-$ when there are an odd number. Here, since 0 is one of the quotients, the last equation becomes $1v - 10 = 8w$. Choose $v = 18$ and $w = 1$. The new column of numbers is then:

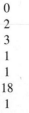

$$0$$
$$2$$
$$3$$
$$1$$
$$1$$
$$18$$
$$1$$

The penultimate term is multiplied by the term next above it, and the product is added to the ultimate term. [The process is continued to the top of the column.] The number at the top is [the *agrante*].

Multiply 18 by 1 and add 1 to get 19. Then replace the term "above," namely 1, by 19, and remove the last term. Continue in this way as in the table until there are only two terms.

0	0	0	0	0	130
2	2	2	2	297	297
3	3	3	130	130	
1	1	37	37		
1	19	19			
18	18				
1					

The top term, the *agrante,* is 130. So $x = 130$, $y = 297$ is a solution to the original equation. Brahmagupta, however, wants a smaller solution, so he first determines N:

[The *agrante*] is divided by the divisor yielding the least remainder; and the residue, multiplied by the divisor yielding the greatest remainder and added to the greater remainder, is a remainder of the product of the divisors.[19]

Therefore, we divide 130 by 60 and obtain a remainder of 10. Multiplying 10 by 137 and adding the product to 10 then gives 1380 as the value for N modulo the product of 137 and 60, or, $N \equiv 1380 \pmod{8220}$. Brahmagupta then solves for y by dividing 1380 by 60 (since $N = 60y$) and calculates a new value for x. Hence $y = 23$, $x = 10$ is a solution to the equation $137x + 10 = 60y$.

Although we do not know how Brahmagupta justified his procedure to his own students, we will present a modern explanation. Begin with the equation $137x + 10 = 60y$ and make step-by-step substitutions in accordance with the successive quotients appearing in the Euclidean algorithm. Note that after each substitution the sign of 10 changes.

$$137x + 10 = 60y \qquad\qquad 60 = 0 \cdot 137 + 60$$
$$137x + 10 = (0 \cdot 137 + 60)y$$
$$137(x - 0y) + 10 = 60y \qquad z = x - 0y$$
$$137z + 10 = 60y$$
$$60y - 10 = 137z \qquad\qquad 137 = 2 \cdot 60 + 17$$
$$60y - 10 = (2 \cdot 60 + 17)z$$
$$60(y - 2z) - 10 = 17z \qquad t = y - 2z$$
$$60t - 10 = 17z$$
$$17z + 10 = 60t \qquad\qquad 60 = 3 \cdot 17 + 9$$
$$17z + 10 = (3 \cdot 17 + 9)t$$
$$17(z - 3t) + 10 = 9t \qquad u = z - 3t$$
$$17u + 10 = 9t$$
$$9t - 10 = 17u \qquad\qquad 17 = 1 \cdot 9 + 8$$
$$9t - 10 = (1 \cdot 9 + 8)u$$
$$9(t - 1u) - 10 = 8u \qquad v = t - 1u$$
$$9v - 10 = 8u$$

$$8u + 10 = 9v \qquad\qquad 9 = 1 \cdot 8 + 1$$
$$8u + 10 = (1 \cdot 8 + 1)v$$
$$8(u - 1v) + 10 = 1v \qquad\qquad w = u - 1v$$
$$8w + 10 = 1v$$
$$1v - 10 = 8w$$

This last equation, already mentioned in Brahmagupta's own explanation is solved by inspection: $v = 18$, $w = 1$. The remaining variables are then found by substitution, the process of which is exactly that described by Brahmagupta.

$$u = 1v + w = 1 \cdot 18 + 1 = 19$$
$$t = 1u + v = 1 \cdot 19 + 18 = 37$$
$$z = 3t + u = 3 \cdot 37 + 19 = 130$$
$$y = 2z + t = 2 \cdot 130 + 37 = 297$$
$$x = 0y + z = 0 \cdot 297 + 130 = 130$$

In modern terminology, both Brahmagupta and Qin Jiushao were interested in solving systems of linear congruences, but a close inspection shows that the two methods were quite different, especially since the Indian author usually dealt with a system of two congruences, while the Chinese author dealt with a larger system. Even when Brahmagupta did deal with a problem similar to a "Chinese remainder problem," such as, "What number, divided by 6, has a remainder of 5; and divided by 5, a remainder of 4, and by 4, a remainder of 3; and by 3, a remainder of 2?," he solved these congruences two at a time. Namely, he first solved $N \equiv 5$ (mod 6) and $N \equiv 4$ (mod 5) to get $N \equiv 29$ (mod 30), then solved $N \equiv 29$ (mod 30) and $N \equiv 3$ (mod 4) and so on. It appears, then, that the only similarity between the Indian and Chinese methods is that both make use of the Euclidean algorithm. A more interesting question, then, unanswerable with current evidence, is whether either culture learned the algorithm from the Greeks, whether all three learned it from an earlier culture, or whether the two Asian cultures simply discovered the algorithm independently.[20]

There is good evidence, however, that Brahmagupta and Āryabhaṭa were interested in congruence problems for the same basic reason as the Chinese, namely, for use in astronomy. The Indian astronomical system of the fifth and sixth century had been heavily influenced by Greek astronomy, especially in the notion that the various planets traveled on epicycles which in turn circled the earth. Therefore Indian astronomers, like their Greek counterparts, needed trigonometry to be able to calculate positions. But a significant idea of Hindu astronomy, similar to one from ancient China but not particularly important in Greece, was that of a large astronomical period at the beginning and end of which all the planets (including the sun and moon) had longitude zero. It was thought that all worldly events would recur with this same period. For Āryabhaṭa, the fundamental period was the *Mahayuga* of 4,320,000 years, the last quarter of which, the *Kaliyuga,* began in 3102 B.C.E. For Brahmagupta, the fundamental period was the *Kalpa* of 1000 *Mahayugas.*

In any case, to do calculations with heavenly bodies one had to know their average motion. Since it was difficult to determine these motions empirically, it became necessary to calculate them from current observations and the fact that all the planets were at approximately the same place at the beginning of the period. These calculations were made by solving linear congruences.[21]

6.7.2 The Pell Equation

An entirely different type of indeterminate equation of interest to Indian mathematicians was the quadratic equation of the form $Dx^2 \pm b = y^2$, whose first occurrence is in the work of Brahmagupta. Today the special case $Dx^2 \pm 1 = y^2$ is usually referred to as Pell's equation (mistakenly named after the seventeenth century Englishman, John Pell). There is some evidence that certain special cases of this equation were solved in Greece, but the earliest extant documentation of an effort to solve this equation in some generality is from India. Brahmagupta, as in the case of the *kuṭṭaka*, introduces rules for dealing with equations of this type, in conjunction with examples. Consider the following.

> [A person who can] make the square of [a number] . . . multiplied by ninety-two . . . with one added to the product an exact square within a year [is] a mathematician.

This equation, $92x^2 + 1 = y^2$, will be solved here in considerably less than a year. Brahmagupta's solution rule begins as follows.

> A root [is set down] two-fold, and [another, deduced] from the assumed square multiplied by the multiplier, and increased or diminished by a quantity assumed.

So set down any value, say 1, and note that if 92 is multiplied by 1^2 and the product added to 8 (the quantity assumed), then the sum is a square, namely 100. Thus, three numbers x_0, b_0, y_0 have been found satisfying the equation $Dx_0^2 + b_0 = y_0^2$. For convenience, we will write that (x_0, y_0) is a solution for additive b_0. In this case, $(1, 10)$ is a solution for additive 8. Brahmagupta next writes this solution in two rows as

$$
\begin{array}{ccc}
x_0 & y_0 & b_0 \\
x_0 & y_0 & b_0
\end{array}
$$

or

$$
\begin{array}{ccc}
1 & 10 & 8 \\
1 & 10 & 8
\end{array}
$$

> The product of the first [pair] taken into the multiplier, with the product of the last [pair] added, is a last root.

Namely, a new value for the "last root" y is found by setting $y_1 = Dx_0^2 + y_0^2$. In this example, $y_1 = 92(1)^2 + 10^2 = 192$.

> The sum of the products of oblique multiplication is a first root. The additive is the product of the like additive or subtractive quantities.[22]

A new value for the "first root" x is determined as $x_1 = x_0 y_0 + x_0 y_0$ or $x_1 = 2x_0 y_0$, while a new additive is $b_1 = b_0^2$. In other words, $(x_1, y_1) = (20, 192)$ is a solution for additive $b_1 = 64$. Brahmagupta also uses the more general result, that if (u_0, v_0) is a solution for additive c_0 and (u_1, v_1) is a solution for additive c_1, then $(u_0 v_1 + u_1 v_0, Du_0 u_1 + v_0 v_1)$ is a solution for additive $c_0 c_1$. This can be verified by considering the identity

$$
D(u_0 v_1 + u_1 v_0)^2 + c_0 c_1 = (Du_0 u_1 + v_0 v_1)^2,
$$

given that $Du_0^2 + c_0 = v_0^2$ and $Du_1^2 + c_1 = v_1^2$. We will call this new solution the **composition** of the solutions (u_0, v_0) and (u_1, v_1). Brahmagupta concludes his basic rule: "The roots [so found] divided by the [original] additive or subtractive quantity, are [roots] for additive unity." In the present example, divide 20 and 192 by 8 to get that (5/2, 24) is a solution for additive 1. Since, however, one of these roots is not an integer, this is not a satisfactory answer for the author. So he composes this solution with itself to get the integral solution for additive 1, (120, 1151). In other words, $92 \cdot 120^2 + 1 = 1151^2$.

This example, as well as illustrating Brahmagupta's method, shows its limitations. The solution for additive 1 in the general case is the pair $\left(\frac{x_1}{b_0}, \frac{y_1}{b_0}\right)$. There is no guarantee that these will be integers or even that one can generate integers by combining this solution with itself. Brahmagupta himself simply gives several more rules and examples, without noting the conditions under which integral solutions exist. First, he notes that composition allows him to get other solutions for any additive provided he knows one solution for this additive as well as a solution for additive 1. In general, the given equation will have infinitely many solutions.

Second, if he has found a solution (u, v) for additive 4, he shows how to find a solution for additive 1. Namely, if v is odd or u is even, then

$$(u_1, v_1) = \left(u\left(\frac{v^2 - 1}{2}\right), v\left(\frac{v^2 - 3}{2}\right)\right)$$

is the desired solution. In the case where v is even and u is odd,

$$(u_1, v_1) = \left(\frac{2uv}{4}, \frac{Du^2 + v^2}{4} = \frac{2v^2 - 4}{4}\right)$$

is an integral solution. As an example of the first case, Brahmagupta solves $3x^2 + 1 = y^2$ by beginning with the solution $u = 2$, $v = 4$ for $3u^2 + 4 = v^2$.

Brahmagupta gives a similar rule for subtractive 4, as well as rules for solving the Pell equation in other special circumstances. Although his methods are always correct, the text has no proofs. One can surmise that proofs were discussed orally. Since Brahmagupta or some predecessor did figure out the method, someone must have had an understanding of why it worked. Why the Indian mathematicians were interested in this problem is, however, a mystery. Some of Brahmagupta's examples use astronomical variables for x and y, but there is no indication that the problems actually came from real-life situations.

In any case, the Pell equation became a tradition in Indian mathematics. It was studied through the next several centuries and was solved completely by the otherwise unknown Acarya Jayadeva (c. 1000). The solution given by Bhāskara (1114–1185), the most famous of all medieval Indian mathematicians, is more easily followed, however.

Bhāskara's goal, in the chapter on algebra in his text, the *Siddhantasiromani,* was to show how any equation of the form $Dx^2 + 1 = y^2$ can be solved in integers. He began by recapitulating Brahmagupta's procedure. In particular, he emphasized that once one has found one solution pair, indefinitely many others can be found by composition. The most important material, however, is his discussion of the so-called cyclic method (*chakravāla*). The basic idea is that by continued appropriate choices of solution pairs for various additives by use of the *kuṭṭaka* method, one eventually reaches one which has the desired additive 1. We present Bhāskara's rule for the general case $Dx^2 + 1 = y^2$ and follow its use in one of his examples, $67x^2 + 1 = y^2$.

Making the first and last roots and additive a dividend, additive and divisor, let the multiplier be thence found.

Biography   **Bhāskara (1114–1185)**

Bhāskara (sometimes referred to as Bhāskara II since there was an earlier mathematician of the same name) was born in the south of India into an old learned family. His father was a renowned Brahmin scholar and astronomer. Bhāskara served much of his adult life as the head of the astronomical observatory at Ujjain and came to be widely acclaimed for his extraordinary skills not only in astronomy and

mathematics, but also in the mechanical arts. His grandson received a royal grant in 1206 to establish a college where the works of Bhāskara were studied. His major work, the *Siddhāntasiromani,* like that of his predecessors, was a treatise on astronomy. The mathematics is found in two chapters, the *Lilavati,* named after Bhāskara's daughter, on arithmetic, and the *Bijaganita,* on algebra.

Begin as before by choosing a solution pair (u, v) for any additive b. In this example take $(1, 8)$ as a solution for additive -3. Next, solve the indeterminate equation $um + v = bn$ for m, here $1m + 8 = -3n$. The result is easily calculated to be $m = 1 + 3t$, $n = -3 - t$ for any integer t.

> The square of that multiplier being subtracted from the given coefficient, or the coefficient being subtracted from the square, (so the remainder is small,) the remainder, divided by the original additive, is a new additive; which is reversed if the subtraction be [of the square] from the coefficient. The quotient corresponding to the multiplier will be the first root; whence the last root may be deduced.

In other words, choose t so that the square of m is as close to D as possible and take $b_1 = \pm\frac{D - m^2}{b}$ (which may be negative) for the new additive. The new first root is $u_1 = \frac{um + v}{b}$ while the new last root is $v_1 = \sqrt{Du_1^2 + b_1}$. In the given example, Bhāskara wants m^2 close to 67, so he chooses $t = 2$ and $m = 7$. Then $D - m^2 = 67 - 49 = 18$ divided by $b = -3$ gives -6. But, because the subtraction is of the square from the coefficient, the new additive is 6. The new first root is then $u_1 = \frac{1 \cdot 7 + 8}{-3} = -5$, but since these roots are always squared, u_1 can be taken as positive. Then $v_1 = \sqrt{67 \cdot 25 + 6} = \sqrt{1681} = 41$, and $(5, 41)$ is a solution for additive 6.

> With these, the operation is repeated, setting aside the former roots and additive. This method mathematicians call that of the *chakravāla.* Thus are integral roots found with four, two, or one additive [or subtractive]; and composition serves to deduce roots for additive unity from those which answer to additives [or subtractives] four and two.[23]

Bhāskara here notes that if the above operation is repeated, eventually a solution for additive or subtractive four, two, or one will be reached. As already noted, from a solution with additive or subtractive 4, a solution for additive 1 can be found. This is also easy to do with additive or subtractive 2 and with subtractive 1. Before continuing with the example, however, we need to discuss two questions, neither of which are addressed by Bhāskara. First, why does the method always give integral values at each stage? Second, why does the repetition of the method eventually give a solution pair for additives ± 4, ± 2, or ± 1?

To answer the first question, note that Bhāskara's method can be derived by composing the first solution (u, v) for additive b with the obvious solution $(1, m)$ for additive $m^2 - D$. It follows that $(u', v') = (mu + v, Du + mv)$ is a solution for additive $b(m^2 - D)$. Dividing the resulting equation by b^2 gives the solution $(u_1, v_1) = \left(\frac{mu + v}{b}, \frac{Du + mv}{b}\right)$ for additive $\frac{m^2 - D}{b}$. It is then clear why m must be found so that $mu + v$ is a multiple of b. It is not difficult to prove, although as usual the text does not have a proof, that if $\frac{mu + v}{b}$ is integral, so are $\frac{m^2 - D}{b}$ and $\frac{Du + mv}{b} = \pm\sqrt{Du_1^2 + b_1}$.[24]

The reason that $m^2 - D$ is chosen "small" is so that the second question can be answered. Unfortunately, the proof that the process eventually reaches additive 1 is quite difficult; the first published version only dates to 1929.[25] It may well be that neither Bhāskara nor Jayadeva proved the result. They may simply have done enough examples to convince themselves of its truth. In fact, one can show that the *chakravāla* method leads to the smallest possible solution of the equation and therefore to every solution.[26]

In any case, we continue with Bhāskara's example. Beginning with $67 \cdot 1^2 - 3 = 8^2$, we have derived $67 \cdot 5^2 + 6 = 41^2$. The next step is to solve $5m + 41 = 6n$, with $|m^2 - 67|$ small. The appropriate choice is $m = 5$. Then $(u_2, v_2) = (11, 90)$ is a solution for additive -7, or $67 \cdot 11^2 - 7 = 90^2$. Again, solve $11m + 90 = -7n$. The value $m = 9$ works and $(u_3, v_3) = (27, 221)$ is a solution for additive -2 or $67 \cdot 27^2 - 2 = 221^2$. At this point, since additive -2 has been reached, it is only necessary to compose $(27, 221)$ with itself. This gives $(u_4, v_4) = (11{,}934, 97{,}684)$ as a solution for additive 4. Dividing by 2, Bhāskara finally gets the desired solution $x = 5967$, $y = 48{,}842$ to the original equation $67x^2 + 1 = y^2$.

6.8 ALGEBRA AND COMBINATORICS

The Indian mathematicians of the medieval period were basically algebraists. Their works are filled with rules for calculating with positive and negative numbers, with fractions, and with algebraic expressions. They teach how to solve linear and quadratic equations in one and several variables. They give rules for summing arithmetic and geometric expressions, as well as squares, cubes, and triangular numbers. They give procedures for dealing with surds. They indicate how to count combinations and permutations. And perhaps most important, all the rules and methods are correct. What is missing in most cases is any indication of how the methods and rules were derived or why they are true.

The preserved texts also give some geometric rules, rules for finding areas and volumes. Here to modern eyes the Indians are less successful. Formulas are indicated which are not strictly true but are only approximations. No mention is made, however, that they are only approximate. These formulas appear to have the same standing as the strictly correct rules appearing on nearby pages. Again, of course, there are no written proofs of any type, no arguments to convince one of the correctness of the results.

The lack of written proofs, however, is probably due to the nature of the surviving texts, generally written in verse. Thus steps were left out, and words were used for their poetic rather than their mathematical value. In addition, we infer that the works were designed easily to be memorized. After all, writing materials were scarce and expensive.

What the students needed to memorize, then as now, were rules for solving problems. What is not known about Indian mathematics is how the mathematicians explained the material to their students. The texts clearly needed interpretation and clarification. Some of the students surely asked why the rules and methods worked. Just as surely, the teachers must have answered them. Unfortunately, their answers, the proofs and reasons, have not been recorded.

We will, nevertheless, consider some of the Indian rules. Brahmagupta, for example, presents the rules for operations on positive and negative numbers.

> The sum of two positive quantities is positive; of two negative is negative; of a positive and a negative is their difference; or, if they are equal, zero. . . . In subtraction, the less is to be taken from the greater, positive from positive; negative from negative. When the greater, however, is subtracted from the less, the difference is reversed. . . . When positive is to be subtracted from negative, and negative from positive, they must be thrown together. The product of a negative quantity and a positive is negative; of two negative, is positive; of two positive, is positive. . . . Positive divided by positive or negative by negative is positive. . . . Positive divided by negative is negative. Negative divided by positive is negative.[27]

Zero was generally treated even in multiplication and division like any other number. Brahmagupta noted that the product of zero with any quantity was zero, but also stated that zero divided by zero was zero and that "a positive or negative, divided by zero, is a fraction with that for denominator."[28] Bhāskara explained further that "this fraction, of which the denominator is zero, is termed an infinite quantity. In this quantity consisting of that which has zero for its divisor, there is no alteration, though many be inserted or extracted; as no change takes place in the infinite and immutable God, at the period of the destruction or creation of worlds, though numerous orders of beings are absorbed or put forth."[29] Nevertheless, multiplication and division by zero could be used when desired, for although "the product of zero [with a definite quantity] is zero, . . . it must be retained as a multiple of zero if any further operation impends. . . . Should zero afterwards become a divisor, the definite quantity must be understood to be unchanged."[30] Thus Bhāskara could set the problem, "What number is it, which multiplied by zero, and added to half itself, and multiplied by three, and divided by zero, amounts to the given number 63?"[31] He was simply thinking of the equation

$$\frac{3\left(0x + \frac{1}{2}0x\right)}{0} = 63,$$

which, by factoring out the zeros in the numerator and "cancelling," becomes $3x + \frac{3}{2}x = 63$, an equation whose solution is 14.

Bhāskara also was quite aware of the distributive law:

> Multiplicand 135. Multiplicator 12. Product (multiplying the digits of the multiplicand successively by the multiplicator)—1620. Or, subdividing the multiplicator into parts, as 8 and 4; and severally multiplying the multiplicand by them; adding the products together, the result is the same—1620. . . . Or the multiplicand being multiplied by the multiplicator less two, 10, and added to twice the multiplicand, the result is the same—1620.[32]

Brahmagupta presented the quadratic formula virtually in the same form as we know it.

Take absolute number on the side opposite to that on which the square and simple unknown are. To the absolute number multiplied by four times the [coefficient] of the square, add the square of the [coefficient of the] unknown; the square root of the same, less the [coefficient of the] unknown, being divided by twice the [coefficient of the] square is the [value of the] unknown.

As an example, he presented the equation *ya v* 1 *ya* − 10 *ru* −9. He here used some symbolization for the unknown: *ya* is an abbreviation for *yavat-tavat* (how much), *ru* for *rupa* (absolute number), *ya v* for *yavat-tavat varga* (square of the unknown). This equation then represents $x^2 - 10x = -9$.

Now to the absolute number [−9] multiplied by four times the [coefficient of the] square [−36], and added to the square [100] of the [coefficient of the] unknown, (making 64), the square root being extracted [8], and lessened by the [coefficient of the] unknown [−10], the remainder 18 divided by twice the [coefficient of the] square [2] yields the value of the unknown 9.[33]

Note here that the given equation actually has a second positive solution, corresponding to taking the negative of the square root less the coefficient of the unknown. Although Brahmagupta makes no mention of this second solution, his words can easily be translated into the formula

$$x = \frac{\sqrt{4ac + b^2} - b}{2a}$$

for finding one solution of the equation $ax^2 + bx = c$.

Bhāskara, on the other hand, does deal with multiple roots. His basic technique for solving quadratic equations is that of completing the square. Namely, he adds an appropriate number to both sides of $ax^2 + bx = c$ so that the left side becomes a perfect square: $(rx - s)^2 = d$. He then solves the equation $rx - s = \sqrt{d}$ for x. But, he notes, "if the root of the absolute side of the equation is less than the number, having the negative sign, comprised in the root of the side involving the unknown, then putting it negative or positive, a two-fold value is to be found of the unknown quantity."[34] In other words, if $\sqrt{d} < s$, then there are two values for x, namely

$$\frac{s + \sqrt{d}}{r} \quad \text{and} \quad \frac{s - \sqrt{d}}{r}.$$

Bhāskara does, however, hedge his bets. As he says, "this [holds] in some cases." We will consider an example to see what he means.

The fifth part of a troop of monkeys less three, squared, had gone to a cave; and one monkey was in sight having climbed on a branch. Say how many they were?

Bhāskara writes the equation as $\left(\frac{1}{5}x - 3\right)^2 + 1 = x$, or $x^2 - 55x = -250$, and finds the two roots 50 and 5. "But the second [root] is in this case not to be taken; for it is incongruous. People do not approve a negative absolute number."[35] Here, the negative number is not from the equation itself but from the problem. One cannot subtract three monkeys from one-fifth of five. In the case of quadratic equations which to us have a positive and a negative root, Bhāskara simply finds the positive root. He never gives examples of quadratic equations having two negative roots or no real roots at all, nor does he give examples of quadratic equations having irrational roots. In every example, the square root in the formula is a rational number.

The Indian mathematicians also easily handled equations in several variables. Thus Mahāvīra (ninth century), a mathematician from Mysore in southern India, presented a version of the hundred fowls problem in his major treatise, the *Ganitasārasangraha*: "Doves are sold at the rate of 5 for 3 coins, cranes at the rate of 7 for 5, swans at the rate of 9 for 7, and peacocks at the rate of 3 for 9. A certain man was told to bring at these rates 100 birds for 100 coins for the amusement of the king's son and was sent to do so. What amount does he give for each?"[36]

Mahāvīra gave a rather complex rule for the solution. Bhāskara, on the other hand, presented the same problem with a procedure showing explicitly why the problem has multiple solutions. He put his unknowns, which we label d, c, s, and p, equal to the number of "sets" of doves, cranes, swans, and peacocks respectively. From the prices and the numbers of birds he derived the two equations

$$3d + 5c + 7s + 9p = 100$$
$$5d + 7c + 9s + 3p = 100$$

and proceeded to solve them. He solved each equation for d, then equated the two expressions and found the equation $c = 50 - 2s - 9p$. Taking an arbitrary value 4 for p, he reduced the equation to the standard indeterminate form $c + 2s = 14$ for which the solution is $s = t, c = 14 - 2t$ with t arbitrary. It follows that $d = t - 2$. Then setting $t = 3$, he calculated that $d = 1, c = 8, s = 3$, and $p = 4$, hence that the number of doves is 5, of cranes 56, of swans 27, and of peacocks 12, their prices being respectively 3, 40, 21, and 36. He noted further that other choices of t gave different values for the solution. Thus, "by means of suppositions, a multitude of answers may be obtained."[37]

Bhāskara, and to some extent earlier mathematicians, were also familiar with permutations and combinations, although again they gave no indication in their written work how they had derived the various formulas. Bhāskara, in particular, only presented the rules and examples in typical poetic style: "Let the figures from one upwards, differing by one, put in the inverse order, be divided by the same in the direct order; and let the subsequent be multiplied by the preceding, and the next following by the foregoing. The several results are the changes, ones, twos, threes, and so on. This is a general rule. It serves in poetry, for those versed therein, to find the variations of meter; in the arts to compute the changes upon apertures; . . . in medicine, the combinations of different savours."[38]

As an example of such a calculation, Bhāskara asks, "In a pleasant, spacious and elegant edifice, with eight doors, constructed by a skillful architect, as a palace for the lord of the land, tell me the combinations of apertures taken one, two, three, and so on." He then computed the results as his rule states. Namely, he first divides 8 by 1, to get 8 ways in which the doors can be opened by ones, then divides 8×7 by 1×2 to get 28 ways in which the doors can be opened by twos, and so on. He concludes by noting that the total number of ways in which the doors may be opened is 255. (He did not include "all doors being closed" as one of the possibilities.)

Although the medieval Indian mathematicians chiefly dealt with algebraic problems, the needs of astronomy always required them to be familiar with the ideas of trigonometry (Side Bar 6.2). It was out of the continuing study of these notions that Indian mathemati-

---

Side Bar 6.2   Indian Reasons for Studying Mathematics

Why were Indian scholars interested in mathematics at all? One can deduce some answers to this question by looking at the types of problems included in their works, though, unlike the case in China, many of the problems there are not at all "practical." A more general answer to this question is found in the introduction to Mahāvīra's *Gaṇitasārasaṅgraha:*

In all those transactions which relate to wordly, Vedic, or . . . religious affairs, calculation is of use. In the science of love, in the science of wealth, in music and in the drama, in the art of cooking, and similarly in

medicine and in things like the knowledge of architecture; in prosody, in poetics and poetry, in logic and grammar and such other things, . . . the science of computation is held in high esteem. In relation to movements of the sun and other heavenly bodies, in connection with eclipses and the conjunction of planets . . . it is utilized. The number, the diameter and the perimeter of islands, oceans, and mountains, the extensive dimensions of the rows of habitations and halls belonging to the inhabitants of the world, . . . all of these are made out by means of computation.[39]

---

cians of the fifteenth and sixteenth century in Kerala, in the south of India, developed ideas on ways of representing the sine (and related functions) by infinite series, ideas which were not seen in Europe until the seventeenth century. But since they are closely related to ideas in calculus, we will postpone a discussion of them until Chapter 12.

6.9 THE HINDU-ARABIC DECIMAL PLACE-VALUE SYSTEM

This chapter concludes with a brief discussion of the origins of our own number system, the decimal place-value system, usually referred to as the Hindu-Arabic system because of its supposed origins in India and its transmission to the West through the Arabs. However, the actual origins of the important components of this system, the digits 1 through 9 themselves, the notion of place value, and the use of 0, are to some extent lost to the historical record. We present here a summary of the most recent scholarship on the beginnings and development of these three ideas.

Symbols for the first nine numbers of our number system have their origins in the Brahmi system of writing in India which dates back to at least the time of King Ashoka (mid-third century B.C.E.). The numbers appear in various decrees of the king inscribed on pillars throughout India. There is a fairly continuous record of the development of these forms. Probably in the eighth century these digits were picked up by the Moslems during the time of the Islamic incursions into northern India and their conquest of much of the Mediterranean world. These digits then appear a century later in Spain, and still later in Italy and the rest of Europe (Figure 6.12).

More important than the form of the number symbols themselves, however, is the notion of place value, and here the evidence is somewhat weaker. The Babylonians had a place-value system, but it was based on 60. Although this system was used continuously for astronomical purposes, it is doubtful that it had much influence on the writing of numbers in other situations. The Chinese from earliest times had a multiplicative system

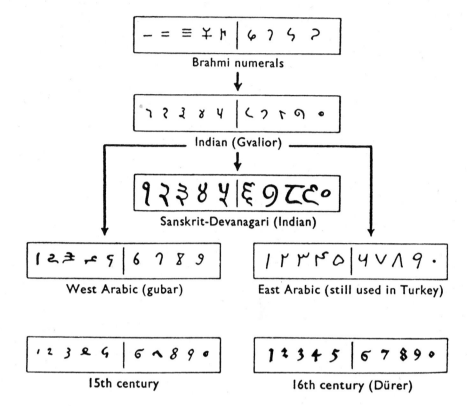

with base 10. This probably was derived from the Chinese counting board, which itself contained columns each representing a different power of 10. In India, although there were number symbols to represent the numbers 1 through 9, there were also symbols to represent 10 through 90. Larger numbers were represented, as in the Chinese written forms, by combining a symbol for 100 or 1000 with a symbol for one of the first 9 numbers. Hence in the early centuries of our era, the Indians, like the Chinese, used a multiplicative system. Āryabhata, in fact, lists names for the various powers of ten in his text: "*dasa* [ten], *sata* [hundred], *sahasra* [thousand], *ayuta* [ten thousand], *niyuta* [hundred thousand] . . ."[40]

Around the year 600, the Indians evidently dropped the symbols for numbers higher than 9 and began to use their symbols for 1 through 9 in our familiar place-value arrangement. The earliest dated reference to this use, however, does not come from India itself. In a fragment of a work of Severus Sebokht, a Syrian priest, dated 662, is the remark that the Hindus have a valuable method of calculation "done by means of nine signs."[41] Severus only speaks of nine signs, and there is no mention of a sign for zero. However, in the Bakhshālī manuscript, a mathematical manuscript in fairly poor condition discovered in 1881 in a village called Bakhshālī in northwestern India, the numbers are written using the place-value system and a dot to represent zero. The best evidence we have is that this manuscript also dates from the seventh century. Perhaps Severus did not consider the dot as a "sign." In other Indian manuscripts from the same period, numbers are generally written in a quasi place-value system to accommodate the poetic nature of the documents.

For example, in the work of Mahāvīra, certain words stand for numbers: moon for 1, eye for 2, fire for 3, and sky for 0. Then the word fire-sky-moon-eye would stand for 2103 and moon-eye-sky-fire for 3021. Note that the place value begins on the left with the units.

Curiously, the earliest dated inscriptions using the decimal place-value system including the zero are found in Cambodia. The earliest one appears in 683, where the 605th year of the Saka era there is represented by three digits with a dot in the middle and the 608th year by three digits with our modern zero in the middle. The dot as symbol for 0 as part of a decimal place-value system also appears in the *Chiu-chih li*, the Chinese astronomical work of 718 compiled by Indian scholars in the employ of the Chinese emperor. Although the actual symbols for the other Indian digits are not known, the author does give details of how the place-value system works: "Using the [Indian] numerals, multiplication and division are carried out. Each numeral is written in one stroke. When a number is counted to ten, it is advanced into the higher place. In each vacant place a dot is always put. Thus the numeral is always denoted in each place. Accordingly there can be no error in determining the place. With the numerals, calculation is easy . . ."[42]

The question remains then as to why the Indians early in the seventh century dropped their own multiplicative system and introduced the place-value system including a symbol for zero. We cannot answer that definitively. It has been suggested, however, that the true origins of the system in India come from the Chinese counting board. The counting board was a portable object. Certainly, Chinese traders who visited India carried these along. In fact, since southeast Asia is the border between Hindu culture and Chinese influence, it may have well been in that area where the interchange took place. What may have happened is that the Indians were impressed with the idea of using only nine symbols. But they naturally took for their symbols the ones they had already been using. They then improved the Chinese system of counting rods by using exactly the same symbols for each place value rather than alternating two types of symbols in the various places. And since they needed to be able to write numbers in some form, rather than just have them on the counting board, they were forced to use a symbol, the dot and later the circle, to represent the blank column of the counting board.[43] If this theory is correct, it is somewhat ironic that Indian scientists then returned the favor and brought this new system back to China early in the eighth century.

In any case, we can certainly put a fully developed decimal place-value system for integers in India by the eighth century, even though the earliest definitively dated decimal place-value inscription there dates to 870. Well before then, though, this system had been transmitted not only to China but also west to Baghdad, the center of the developing Islamic culture. It is well to note, however, that although decimal fractions were used in China, again as places on the counting board, in India itself there is no early evidence of these. It was the Moslems who completed the Indian written decimal place-value system by introducing these decimal fractions.

Exercises

Surveying Problems from China

1. Solve problem 3 of the *Sea Island Mathematical Manual*: To measure the size of a square walled city *ABCD*, we erect two poles 10 feet apart at *F* and *E* (Figure 6.13).

By moving northward 5 feet from *E* to *G* and sighting on *D*, the line of observation intersects the line *EF* at a point *H* such that *HE* = 3 93/120 feet. Moving to point *K* such that *KE* = 13 1/3 feet, the line of sight to

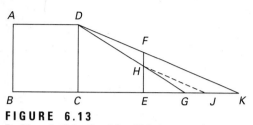

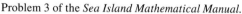

FIGURE 6.13
Problem 3 of the *Sea Island Mathematical Manual.*

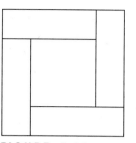

FIGURE 6.14
From the *Yang Hui suanfa.*

D passes through *F*. Find *DC* and *EC*.[44] (Liu Hui gets *DC* = 943 3/4 feet while *EC* = 1245 feet.) *Hint:* Begin by constructing *JH* parallel to *KD*.

2. Solve problem 24 of Chapter 9 of the *Jiuzhang suanshu.* (This is an example of the type of elementary surveying problem which stimulated Liu Hui to write his *Sea Island Mathematical Manual*): A deep well 5 feet in diameter is of unknown depth (to the water level). If a 5 foot post is erected at the edge of the well, the line of sight from the top of the post to the edge of the water surface below will pass through a point 0.4 feet from the lip of the well below the post. What is the depth of the well?[45]

Problems in Indeterminate Analysis from China

3. Check that the answer given to Yi Xing's calendrical congruence problem, namely $N = 96,961,740 \times 1,110,343$, is correct.

4. Solve problem I, 1 from the *Shushu jiuzhang*: $N \equiv 1$ (mod 1), $N \equiv 1$ (mod 2), $N \equiv 3$ (mod 3), $N \equiv 1$ (mod 4).

5. Solve problem I, 5 from the *Shushu jiuzhang*: $N \equiv 32$ (mod 83), $N \equiv 70$ (mod 110), $N \equiv 30$ (mod 135).

6. From *Zhang Qiujian's Mathematical Manual*: Each pint of high quality wine costs 7 coins, each pint of ordinary wine costs 3 coins, and 3 pints of wine dregs cost 1 coin. If 10 coins are used to buy 10 pints in all, find the amount of each type and the total money spent on each.

Problems in Numerical Equation Solving

7. Solve the following equations numerically using Qin Jiushao's procedure. All are taken from his text.

   a. $16x^2 + 192x - 1863.2 = 0$

   b. $-x^4 + 15,245x^2 - 6,262,506.25 = 0$

   c. $x^{10} + 15x^8 + 72x^6 - 864x^4 - 11,664x^2 - 34,992 = 0$

   The following four problems are from the *Yang Hui suanfa*. They are to be solved numerically using Qin's method.

8. The area of a rectangle is 864. The width is less than the length by 12. Find the length.

9. The area of a rectangle is 864. The difference between the length and width is 12. Find the sum of the length and width. In Yang Hui's diagram (Figure 6.14), the given rectangle is put in four times and the central square has side equal to the difference of the length and width.

10. The area of a rectangle is 864. The sum of the length and width is 60. By how much does the length exceed the width? (Use the same diagram as in problem 9.)

11. The area of a rectangle is 864. The sum of the length, twice the width, three times the sum of the length and width, and four times the difference of the length and width is 312. Find the width.

12. Use Qin's method to solve the pure quadratic equation $x^2 = 55,225$. Compare this method to the procedure given in Chapter 1 to solve this same equation.

13. Solve the pure fourth degree equation $y^4 = 279,341$ using Qin's procedure. Show how the fourth order coefficients of the Pascal triangle 4 6 4 1 appear in the solution procedure.

14. This problem is from Zhu Shijie: The sum of the square root of the circumference of a circle and the area of the circle is 114. Find the circumference and the diameter. (Zhu derives a fourth degree equation in order to solve this problem.)

15. The sum of the base, altitude, and hypotenuse of a right triangle divided by the difference of the hypotenuse and altitude is equal to twice the area of the triangle. The sum of the hypotenuse and the difference of the altitude and base divided by the difference of the hypotenuse and base is equal to the base. Find the hypotenuse. (This problem is also from Zhu, who eventually derives a fourth degree polynomial equation for the base *x*.)

Exercises **219**

Problems from Indian Surveying

16. Show that Śrīpati's value of 3415′ for the radius of a circle used as the base in his sine table comes from using $\sqrt{10}$ as an approximation to π.

17. Calculate the fourth, fifth, and sixth sine (of 15°, 18°45′, and 22°30′ respectively) according to the finite difference method given by the *Sūrya-Siddhānta* and compare your values to the given values. Then calculate these same sines using the half-angle formula and compare your results.

18. Calculate the cube root of 13,312,053 using Āryabhaṭa's method.

19. Calculate the shadow (tangent) and hypotenuse (secant) of a gnomon of height 60 if the sun's zenith distance is 18°45′ using the sine value calculated in problem 17 and the *Sūrya Siddhānta's* value of 3256 for the sine of 71°15′.

Problems in Indeterminate Analysis in India

20. Solve the equations $N \equiv 23 \pmod{137}$, $N \equiv 0 \pmod{60}$ and the equations $N \equiv 94 \pmod{137}$, $N \equiv 0 \pmod{30}$, using Brahmagupta's procedure.

21. Solve $1096x + 1 = 3y$ using Brahmagupta's method. Given a solution to this equation (with "additive" 1), it is easy to find solutions to equations with other additives by simply multiplying. For example, solve $1096x + 10 = 3y$.

22. Prove Brahmagupta's rule of *kuṭṭaka*. Begin by noting that the Euclidean algorithm allows one to express the greatest common divisor of two positive integers as a linear combination of these integers. Note further that a condition for the solution procedure to exist is that this greatest common divisor must divide the "additive." Brahmagupta does not mention this, but Bhāskara and others do.

23. Solve the problem $N \equiv 5 \pmod 6 \equiv 4 \pmod 5 \equiv 3 \pmod 4 \equiv 2 \pmod 3$ by the Indian procedure and by the Chinese procedure. Compare the methods.

24. Solve the congruence $N \equiv 10 \pmod{137} \equiv 0 \pmod{60}$ by the Chinese procedure and compare your solution step-by-step with the solution by the *kuṭṭaka* method. How do the two methods compare?

25. Solve the indeterminate equation $17n - 1 = 75m$ by both the Indian and Chinese methods using explicitly the Euclidean algorithm. Compare the solutions.

26. Prove that $D(u_0v_1 + u_1v_0)^2 + c_0c_1 = (Du_0u_1 + v_0v_1)^2$ given that $Du_0^2 + c_0 = v_0^2$ and $Du_1^2 + c_1 = v_1^2$.

27. Solve $83x^2 + 1 = y^2$ by Brahmagupta's method. Begin by noting that $(1, 9)$ is a solution for subtractive 2.

28. Show that if (u, v) is a solution to $Dx^2 - 4 = y^2$, then $(u_1, v_1) = \left(\frac{1}{2}uv(v^2 + 1)(v^2 + 3), (v^2 + 2)\left[\frac{1}{2}(v^2 + 1)(v^2 + 3) - 1\right]\right)$ is a solution to $Dx^2 + 1 = y^2$ and that both u_1 and v_1 are integers regardless of the parity of u or v.

29. Solve $13x^2 + 1 = y^2$ by noting that $(1, 3)$ is a solution for subtractive 4 and applying the method of problem 28.

30. Show that if (u, v) is a solution to $Dx^2 + 2 = y^2$, then $(u_1, v_1) = (uv, v^2 - 1)$ is a solution to $Dx^2 + 1 = y^2$. Deduce a similar rule if (u, v) is a solution to $Dx^2 - 2 = y^2$.

31. Solve $61x^2 + 1 = y^2$ by Bhāskara's *chakravāla* process. The solution is $x = 226,153,980$, $y = 1,766,319,049$.

Algebra Problems from India

32. Solve the following problem from Mahāvīra: "One night, in a month of the spring season, a certain young lady . . . was lovingly happy along with her husband on . . . the floor of a big mansion, white like the moon, and situated in a pleasure garden with trees bent down with the load of the bunches of flowers and fruits, and resonant with the sweet sounds of parrots, cuckoos, and bees which were all intoxicated with the honey obtained from the flowers therein. Then on a love quarrel arising between the husband and the wife, that lady's necklace made up of pearls became sundered and fell on the floor. One-third of that necklace of pearls reached the maid-servant there; one-sixth fell on the bed; then one-half of what remained (and one-half of what remained thereafter and again one-half of what remained thereafter and so on, counting six times in all) fell all of them everywhere; and there were found to remain (unscattered) 1,161 pearls. . . . Give out the (numerical) measure of the pearls (in that necklace)."[46]

33. Solve the following problem from Mahāvīra: There are 4 pipes leading into a well. Among these, each fills the well (in order) in 1/2, 1/3, 1/4, and 1/5 of a day. In how much of a day will all of them together fill the well and each of them to what extent?

34. Solve the following problem from the Bakhshālī manuscript: One person goes 5 yojanas a day. When he has proceeded for seven days, the second person, whose

speed is 9 yojanas a day, departs. In how many days will the second person overtake the first?

35. Rule for permutations of digits when some are alike (from Bhāskara): The permutations found as before, being divided by the permutations separately computed for as many places as are filled by like digits, will be the variations of number. Use this rule to determine the number of permutations of the digits 2, 2, 1, 1 and also the digits 4, 8, 5, 5, 5.

36. Solve the following problem from Bhāskara: Find two numbers such that the sum of the cube of the one and square of the other is a square, and the sum of the numbers themselves is a square.

37. Another problem from Bhāskara: The shadow of a gnomon twelve inches high being lessened by a third part of the hypotenuse became fourteen inches long. How long was the shadow?

38. A problem from Mahāvīra: If 3 peacocks cost 2 coins, 4 pigeons cost 3 coins, 5 swans cost 4 coins, and 6 *sārasa* birds cost 5 coins, and if you buy 72 birds for 56 coins, how many of each type of bird do you have?

39. Another problem from Mahāvīra: Two travelers found a purse containing money. The first said to the second, "By securing half of the money in this purse, I shall become twice as rich as you." The second said to the first, "By securing two-thirds of the money in the purse, I shall, with the money I have on hand, have three times as much money as what you have on hand." How much did each have and how much was in the purse?

FOR DISCUSSION . . .

40. Devise a lesson for a number theory course developing the Chinese remainder theorem in the manner of Qin Jiushao.

41. Devise a lesson for a precalculus course showing how synthetic division can be used to find numerical solutions to polynomial equations.

42. Compare the methods of Liu Hui and Āryabhata for determining the distance and height of a distant object. How are these methods related to the procedures of Heron?

43. Devise a lesson for a number theory course on solving indeterminate equations of the form $rx + c = sy$, using the methods of Brahmagupta.

44. Compare the Chinese and Indian attitudes toward the study of mathematics.

45. Look up the story of Matteo Ricci, the Jesuit who visited China in the late sixteenth century, and write a report on his influence on the study of mathematics in China.

References and Notes

As noted in the references to Chapter 1, the best surveys of Chinese mathematics are J. Needham, *Science and Civilization in China* (Cambridge: Cambridge University Press, 1959), vol. 3, and Li Yan and Du Shiran, *Chinese Mathematics—A Concise History* (translated by John N. Crossley and Anthony W. C. Lun) (Oxford: Clarendon Press, 1987). A somewhat older work is Yoshio Mikami, *The Development of Mathematics in China and Japan* (New York: Chelsea, 1974). A detailed work on aspects of Chinese mathematics in the thirteenth century and its relationship to mathematics at other times and in other countries is Ulrich Libbrecht, *Chinese Mathematics in the Thirteenth Century: The Shushu chiu-chang of Ch'in Chiu-shao* (Cambridge: MIT Press, 1973). A recent guide to the literature on Chinese mathematics is Frank Swetz and Ang Tian Se, "A Brief Chronological and Bibliographic Guide to the History of Chinese Mathematics," *Historia Mathematica* 11 (1984), 39–56. General histories of Indian mathematics, neither of them comprehensive, include B. Datta and A. N. Singh, *History of Hindu Mathematics* (Bombay: Asia Publishing House, 1961) and C. N. Srinivasiengar, *The History of Ancient Indian Mathematics* (Calcutta: The World Press Private Ltd., 1967). The text of Āryabhata is available in an English translation by Walter E. Clark: *The Āryabhaṭīya of Āryabhaṭa* (Chicago: University of Chicago Press, 1930). The major mathematical texts of Bhāskara and Brahmagupta were translated by H. T. Colebrooke in *Algebra with Arithmetic and Mensuration from the Sanskrit of Brahmegupta and Bhāscara* (London: John Murray, 1817).

1. E. L. Konantz, "The precious mirror of the four elements," *China Journal of Arts and Science* 2 (1924), 304–310. This quotation is from the introduction to Zhu Shijie's work by Chien Chiu Shimoju.

2. This translation is adapted from excerpts from "A Chinese Mathematical Classic of the Third Century: THE SEA ISLAND MATHEMATICAL MANUAL OF LIU HUI" by Ang Tian Se and Frank J. Swetz, *Historia*

Mathematica, Vol. 13, No. 2, May 1986, p. 105. Copyright © 1986 by Academic Press, Inc. Reprinted by permission of Academic Press, Inc. and Dr. Frank J. Swetz.

3. Ibid.

4. Ibid., pp. 108ff.

5. This material is adapted from the work of Yabuuti Kiyosi, "Researches on the *Chiu-chih li*—Indian Astronomy under the T'ang Dynasty," *Acta Asiatica* 36 (1979), 7–48, and Christopher Cullen, "An Eighth Century Chinese Table of Tangents," *Chinese Science* 5 (1982), 1–33. The first article contains an English translation and commentary on the *Chiu-chih li*, while details on how and why Yi Xing developed his table of tangents are found in the second article.

6. Libbrecht, *Chinese Mathematics,* p. 269.

7. Ibid., p. 277.

8. Ibid., p. 62.

9. Ibid., p. 31. Here Libbrecht quotes Chou Mi, a contemporary of Qin.

10. The earliest proof that Qin's method works is by V. A. Lebèsgue, *Exercices d'analyse numérique* (Paris, 1859), p. 56. Kurt Mahler in "On the Chinese Remainder Theorem," *Mathematische Nachrichten* 18 (1958), 120–122, gives a modern proof.

11. Li Yan and Du Shiran, *Chinese Mathematics,* p. 163.

12. See Libbrecht, *Chinese Mathematics,* Chapter 17 for more details.

13. Lam Lay Yong, "On the Existing Fragments of Yang Hui's Hsiang Chieh Suan Fa," *Archive for History of Exact Sciences* 5 (1969), 82–86. For more details on the use of the Pascal triangle in solving equations, see Lam Lay Yong, "The Chinese Connection between the Pascal Triangle and the Solution of Numerical Equations of Any Degree," *Historia Mathematica* 7 (1980), 407–424.

14. Lam Lay Yong and Ang Tian Se, "Li Ye and his Yi Gu Yan Duan," *Archive for History of Exact Sciences* 29 (1984), 237–266.

15. Problem quoted from Jock Hoe, *Les systemes d'equations polynomes dans le Siyuan Yujian (1303),* (Paris: Collège de France, Institut des Hautes Études Chinoises, 1977), pp. 94ff.

16. E. Burgess, "Translation of the Sūrya-Siddhānta, a Textbook of Hindu Astronomy," *Journal of the American Oriental Society* 6 (1860), 141–498, p. 196.

17. Ibid., p. 252.

18. Excerpts from pages 24–37 from *The Āryabhaṭīya of Āryabhaṭa* translated with notes by Walter Eugene Clark. Copyright 1930 by the University of Chicago. All rights reserved. Reprinted by permission.

19. Colebrooke, *Algebra with Arithmetic,* pp. 325ff.

20. See the discussion of this in B. L. Van der Waerden, *Geometry and Algebra in Ancient Civilization,* Chapter 5A.

21. Ibid.

22. Colebrooke, *Algebra with Arithmetic,* pp. 364ff.

23. Ibid., pp. 175ff.

24. See the proof in C. N. Srinivasiengar, *History of Ancient,* p. 113.

25. Krishnaswami A. A. Ayyangar, *Journal of the Indian Mathematics Society* 18 (1929), 232–245.

26. A detailed discussion of the entire process of solving the Pell equation is found in C. O. Selenius, "Rationale of the Chakravāla Process of Jayadeva and Bhāskara II," *Historia Mathematica* 2 (1975), 167–184.

27. Colebrooke, *Algebra with Arithmetic,* pp. 339–340.

28. Ibid., p. 340.

29. Ibid., p. 135.

30. Ibid., p. 19.

31. Ibid., p. 40.

32. Ibid., p. 7.

33. Ibid., pp. 346–347.

34. Ibid., pp. 207–208.

35. Ibid., p. 216.

36. Mahāvīra, *Gaṇitāsarasaṅgraha,* M. Rangācārya, ed. and trans. (Madras: Government Press, 1912), p. 134.

37. Colebrooke, *Algebra with Arithmetic,* p. 235.

38. Ibid., p. 49.

39. Mahāvīra, *Gaṇitāsarasaṅgraha,* sec. 1.

40. Clark, *The Āryabhaṭīya,* p. 21.

41. Quoted in D. E. Smith, *History of Mathematics* (New York: Dover, 1958), vol. 1, p. 166. For more details on Indian mathematical notation, see Saradakanta Ganguli, "The Indian Origin of the Modern Place-Value Arithmetical Notation," *American Mathematical Monthly* 39 (1932), 251–256, 389–393, and 40 (1933), 25–31, 154–157.

42. Yabuuti Kiyosi, "Researches on the *Chiu-chih li*," p. 12.

43. For more details of this argument, see Lam Lay Yong, "The Conceptual Origins of our Numeral System and the Symbolic Form of Algebra," *Archive for History of Exact Sciences* 36 (1986), 184–195, and "A Chinese Genesis: Rewriting the History of Our Numeral System," *Archive for History of Exact Sciences* 38 (1988), 101–108.

44. Frank J. Swetz and Ang Tian Se, "A Chinese Mathematical Classic," p. 106.

45. Frank J. Swetz and T. I. Kao, *Was Pythagoras Chinese?* (Reston: National Council of Teachers of Mathematics, 1977), p. 60.

46. Mahāvīra, *Gaṇitāsārasaṅgraha*, p. 73.

Summary of Medieval China and India

Third century	Liu Hui	Mathematics of surveying
Late third century	Sun Zi	Chinese remainder problem
Fourth century	*Sūrya-Siddhānta*	Trigonometric methods
Fifth century	Zhang Quijian	Hundred fowls problem
Late fifth century	Āryabhaṭa	Mathematical methods
Seventh century	Brahmagupta	Indeterminate equations
Early eighth century	Chutan Hsita	Indian sine table in China
683–727	Yi Xing	Tangent table
Ninth century	Mahāvīra	Algebraic problems
Eleventh century	Jia Xian	Pascal triangle
1114–1185	Bhāskara	Pell equation
1192–1279	Li Ye	Algebraic equations for geometry
1202–1261	Qin Jiushao	Linear congruences; equation solving techniques
Late thirteenth century	Yang Hui	Equation solving techniques
Late thirteenth century	Zhu Shijie	Systems of equations
1552–1610	Matteo Ricci	Entry of Western mathematics into China

Chapter 7

The Mathematics of Islam

"You know well . . . for which reason I began searching for a number of demonstrations proving a statement due to the ancient Greeks . . . and which passion I felt for the subject . . . so that you reproached me my preoccupation with these chapters of geometry, not knowing the true essence of these subjects, which consists precisely in going in each matter beyond what is necessary. . . . Whatever way he [the geometer] may go, through exercise will he be lifted from the physical to the divine teachings, which are little accessible because of the difficulty to understand their meaning . . . and because of the circumstance that not everybody is able to have a conception of them, especially not the one who turns away from the art of demonstration." (Preface to the Book on Finding the Chords in the Circle *by al-Bīrūnī, c. 1030)*[1]

*T*here is a story, probably apocryphal, that as a student, Omar Khayyam made a compact with two fellow students, Niẓām al Mulk and Ḥassan ibn Sabbah, to the effect that the one who first achieved a high position and great fortune would help the other two. It was Niẓām who in fact became the grand vizier of the Seljuk Sultan Jalāl al-Dīn Malik-shāh and proceeded to fulfill his promise. Ḥassan received the position of court chamberlain, but after he attempted to supplant his friend in the sultan's favor, he was banished from the court. Omar, on the other hand, declined a high position, accepting instead a modest salary which permitted him to have the leisure to study and write.

In the first half of the seventh century a new civilization came out of Arabia. Under the inspiration of the prophet Muḥammad, the new monotheistic religion of Islam developed and quickly attracted the allegiance of the inhabitants of the Arabian peninsula. In less than a century after Muḥammad's capture of Mecca in 630, the Islamic armies conquered an immense territory as they propagated the new religion first among the previously polytheistic tribes of the Middle East and then among the adherents of other faiths. First Syria and then Egypt were wrested from the Byzantine empire. Persia was conquered by 642, and soon the victorious armies had reached as far as India and parts of central Asia.

In the West, North Africa was quickly overrun, and in 711 the Islamic forces entered Spain. Their forward progress was eventually halted at Tours by the army of Charles Martel in 732. Already, however, the problems of conquest were being replaced by the new problems of governing the immense new empire. Muhammad's successors, the caliphs, originally set up their capital in Damascus, but after about a hundred years of wars, including great victories but also some substantial defeats, the Caliphate split up into several parts. In the eastern segment, under the Abbasid caliphs, the growth of luxury and the cessation of wars of conquest created favorable conditions for the development of a new culture.

In 766 the caliph al-Manṣūr founded his new capital in Baghdad, a city which soon became a flourishing commercial and intellectual center. The initial impulses of Islamic orthodoxy were soon replaced by a more tolerant atmosphere, and the intellectual accomplishments of all residents of the caliphate were welcomed. The caliph Hārūn al-Rashīd, who ruled from 786 to 809, established a library in Baghdad. Manuscripts were collected from various academies in the Near East which had been established by scholars fleeing from the persecutions of the ancient academies in Athens and Alexandria. These manuscripts included many of the classic Greek mathematical and scientific texts. A program of translation into Arabic was soon begun. Hārūn's successor, the caliph al-Ma'mūn (813–833), established a research institute, the *Bayt al-Ḥikma* (House of Wisdom), which was to last over 200 years. Scholars from all parts of the caliphate were invited to this institute to translate Greek and Indian works as well as to conduct original research. By the end of the ninth century, many of the principal works of Euclid, Archimedes, Apollonius, Diophantus, Ptolemy, and other Greek mathematicians had been translated into Arabic and were available for study to the scholars gathered in Baghdad. Islamic scholars also absorbed the ancient mathematical traditions of the Babylonian scribes, still evidently available in the Tigris-Euphrates valley, and in addition learned the trigonometry of the Hindus.

The Islamic scholars did more than just bring these sources together. They amalgamated them into a new whole and, in particular, as the opening quotation indicates, infused their mathematics with what they felt was divine inspiration. Creative mathematicians of the past had always carried investigations well beyond the dictates of immediate necessity, but in Islam many felt that this was a requirement of God. Islamic culture in general regarded "secular knowledge" not as in conflict with "holy knowledge," but as a way to it. Learning was therefore encouraged and those who had demonstrated sparks of creativity were supported by the rulers (usually both secular and religious authorities) so that they could pursue their ideas as far as possible. The mathematicians responded by always invoking the name of God at the beginning and end of their works and even occasionally referring to Divine assistance throughout the texts. Furthermore, since the rulers were naturally interested in the needs of daily life, the Islamic mathematicians, unlike their Greek predecessors, nearly all contributed not only to theory but also to practical applications.[2]

Given the influence of Islam on science in general, and mathematics in particular, the mathematics of this period will be referred to here as "Islamic" rather than "Arabic," even though not all of the mathematicians were themselves Moslems. Nevertheless, it was the Arabic language which was generally in use in the Islamic domains and hence the works to be discussed were all written in that language. A complete history of mathematics of medieval Islam cannot yet be written, since so many of these Arabic manuscripts lie unstudied and even unread in libraries throughout the world. The situation has been

FIGURE 7.1
The Arab contribution to
science on a Tunisian
stamp.

improving recently as more and more texts are being edited and translated, but political
difficulties continue to block access to many important collections. Still, the general outline of mathematics in Islam is known. In particular, Islamic mathematicians fully developed the decimal place-value number system to include decimal fractions, systematized
the study of algebra and began to consider the relationship between algebra and geometry,
studied and made advances on the major Greek geometrical treatises of Euclid, Archimedes, and Apollonius, and made significant improvements in plane and spherical
trigonometry (Figure 7.1).

7.1 DECIMAL ARITHMETIC

The decimal place-value system had spread from India at least to Syria by the mid-seventh
century. It was certainly available in Islamic lands by the time of the founding of the House
of Wisdom. In fact, in 773 an Indian scholar visited the court of al-Mansūr in Baghdad,
bringing with him a copy of a *Siddhānta,* an Indian astronomical text. The caliph ordered
this work translated into Arabic. Besides containing the Indian astronomical system, this
work included at least some indication of the Hindu number system. The Moslems, however, already had a number system with which those who needed to use mathematics were
quite content. In fact, there were two systems in use. The merchants in the marketplace
generally used a form of finger reckoning which had been handed down for generations.
In this system, calculations were generally carried out mentally. Numbers were expressed
in words, and fractions were generally expressed in the Babylonian scale of sixty. When
numbers had to be written, a ciphered system was used in which the letters of the Arabic
alphabet denoted numbers. Many Arabic arithmetic texts in which one or the other of
these systems was discussed were written between the eighth and the thirteenth centuries.

Gradually the knowledge of the Hindu system began to seep into Islamic mathematics.
The earliest available arithmetic text which deals with these Hindu numbers is the *Kitāb
al-jam'wal tafrīq bi ḥisāb al-Hind* (*Book on Addition and Subtraction after the Method
of the Indians*) by Muḥammad ibn-Mūsā al-Khwārizmī (c. 780–850), an early member of
the House of Wisdom (Side Bar 7.1). Unfortunately, there is no extant Arabic manuscript
of this work, only several different Latin translations made in Europe in the twelfth century. In his text al-Khwārizmī introduced nine characters to designate the first nine numbers and, as the Latin translations tell us, a circle to designate zero. He demonstrated how
to write any number using these characters in our familiar place-value notation. He then
described the algorithms of addition, subtraction, multiplication, division, halving, doubling, and determining square roots, and gave examples of their use. The algorithms,
however, were generally set up to be performed on the dust board, a writing surface on
which sand was spread. Thus calculations were generally designed to have figures erased
at each step as one proceeded to the final answer. Although not consistent in this regard,
in general he expressed fractions in the Egyptian mode as sums of unit fractions. It is
important to note that one of the most important features of our place-value system,
decimal fractions, was still missing. Nevertheless, al-Khwārizmī's work was important not
only in the Islamic world but also because it introduced many Europeans to the basics of
the decimal place-value system (Side Bar 7.2).

Side Bar 7.1   Arabic Names

In our initial reference to a particular Islamic mathematician, we give his complete name, although afterwards we abbreviate it for reasons of space. Note that the Arabic name includes not only the given name of the person, but also may include his lineage to one or more generations ("ibn" means "son of"), the place of his or his ancestors' birth, the name of his son ("abū" means "father of"), and one or more appellations indicating some particular characteristic. For example, al-Uqlīdīsī means having to do with Euclid. Namely, the mathematician in question was probably a copyist of Arabic versions of Euclid's works.

Numerous arithmetic works were written in Arabic over the next centuries explaining the Indian methods, both on their own and in connection with the older systems already mentioned. The earliest extant Arabic arithmetic, the *Kitāb al-fuṣūl fī-l-ḥisāb al-Hindī*, (*The Book of Chapters on Hindu Arithmetic*) of Abu l-Ḥasan al-Uqlīdīsī, was written in 952 in Damascus. The author made clear one of the major reasons for what he knew would be the ultimate success of the Indian numbers:

> Most scribes will have to use it [the Indian method] because it is easy, quick, and needs little precaution, little time to get the answer, and little keeping of the heart busy with the working that he has to see between his hands, to the extent that if he talks, that will not spoil his work; and if he leaves it and busies himself with something else, when he turns back to it, he will find it the same and thus proceed, saving the trouble of memorizing it and keeping the heart busy with it. This is not the case in the other (arithmetic) which requires finger bending and other necessaries. Most calculators will have to use it [the Indian method] with numbers that cannot be managed by the hand because they are big.[3]

Al-Uqlīdīsī's text, like that of al-Khwārizmī, dealt with the various algorithms of arithmetic. But there were two major innovations. First, the author showed how to perform arithmetic calculations on paper. As he noted, some think it "ugly to see the [dust board] in the hands of the scribe . . . sitting in the market places [so] . . . we have substituted for it something that will not require [the dust board]." For example, al-Uqlīdīsī gave the following procedure for multiplying 3249 by 2735. He wrote the first number above the second, multiplied each digit of the first by the entire second number, then added the resulting terms together. For example, the first line of the calculation is 6 21 9 15 (= 2 · 3, 7 · 3, 3 · 3, 5 · 3).

$$
\begin{array}{cccccc}
 & & 3249 & & & \\
 & & 2735 & & & \\
6 & 21 & 9 & 15 & & \\
 & 4 & 14 & 6 & 10 & \\
 & & 8 & 28 & 12 & 20 \\
 & & & 18 & 63 & 27 & 45 \\
\end{array}
$$

The result, 8,886,015, is found by careful adding of the columns, keeping track of the various places. Thus the second digit from the right in the answer comes from adding the

---

Side Bar 7.2   **Mathematical Words from Arabic**

Al-Khwārizmī's arithmetic text was probably the source of three English mathematical words. One of the Latin manuscripts of this work begins with the words "Dixit Algorismi," or, "Al-Khwārizmī says." The word "algorismi," through some misunderstanding, soon became a term referring to various arithmetic operations and, ultimately, the English word "algorithm." Our word "zero" probably derives from the Arabic *sifr*, which was Latinized into "zephirum." The word *sifr* itself was an Arabic translation of the Sanskrit word *sūnyā*, meaning "empty." An alternate Medieval translation of *sifr* into "cifra" led to our modern English "cipher."

---

0 and 7 of 20 and 27 to the 4 in 45. The third digit from the right comes from adding the "carry" (1) from the previous addition to the 2 in 20, the 2 in 27, the 0 in 10, the 2 in 12, and the 3 in 63. In any case, all the numbers are written down and preserved so one can check them.

Second, al-Uqlīdīsī treated decimal fractions, the earliest recorded instance of these fractions outside of China. This treatment is in al-Uqlīdīsī's section on halving: "In what is drawn on the principle of numbers, the half of one in any place is 5 before it. Accordingly, if we halve an odd number we set the half as 5 before it, the units place being marked by a sign ′ above it, to denote the place. The units place becomes tens to what is before it. Next, we halve the five as is the custom in halving whole numbers. The units place becomes hundreds in the second time of halving. So it goes always."[4] The central idea of decimal fractions is clear here. In dealing with numbers less than one, one operates on them in exactly the same manner as on whole numbers. It is only after performing the operation that one worries about the decimal place. Al-Uqlīdīsī provided as an example the halving of 19 five times. In order, he gets 9′5, 4′75, 2′375, 1′1875, and 0′59375. He read the latter number as 59,375 of a hundred thousand. Similarly, in a section on increasing numbers he noted that to find one-tenth of a number, one simply repeats it "one place down." So to increase 135 by one-tenth of itself five times, he wrote

$$1 \quad 3 \quad 5$$
$$1 \quad 3 \quad 5$$

The sum is 148′5. One tenth of this is 14′85; the new sum is 163′35. Continuing this process another three times gives the final answer of 217′41885.

Although al-Uqlīdīsī in fact introduced the notion of decimal fractions, essentially as an ad hoc device, their more important use in the context of approximation first appeared in a work of al-Samaw'al ibn Yahyā ibn Yahūda al-Maghribī (c. 1125–1180). For example, when al-Samaw'al divided 210 by 13, he noted that the division does not come out even, but can be carried as far as desired. He wrote the result to five places as 16 plus 1 part of 10, plus 5 parts of 100, plus 3 parts of 1000, plus 8 parts of 10000, plus 4 parts of 100000. That is, unlike his predecessor, he still used words in part rather than the complete place-value system. Nevertheless, the idea of the decimal fraction as an approximation to a rational number made its appearance here. Similarly, he calculated the square root of 10 verbally to be, in our notation, 3.162277. In fact, al-Samaw'al showed how to extract

FIGURE 7.2
Al-Kāshī on an Iranian stamp.

higher roots by essentially the same method used by the Chinese. But he went further than the Chinese in explicitly noting the purpose of the successive steps of the algorithm: "And thus we can determine the root of the cube, the square-square, the square-cube and other [powers]. We are able by this method . . . to obtain answers infinite in number, each of which is more precise and closer to the truth than that which precedes it."[5] Al-Samaw'al evidently realized that, in theory at least, one can calculate an infinite decimal expansion of a number, and that the finite decimals of this expansion "converge" to the exact value, a value not expressible in any finite form.

But even with this important work, the development of the place-value system was not yet complete. It was only the work of Ghiyāth al-Dīn Jamshīd al-Kāshī (d. 1429) in the early fifteenth century that first displayed a total command both of the idea of and a convenient notation for decimal fractions (Figure 7.2). (Al-Kāshī used a vertical line where we use a decimal point.) We can then say that the Hindu-Arabic place-value system was complete. The system, including decimal fractions, also appeared around this time in a Byzantine textbook, with the method described as "Turkish," that is, Islamic. It was not until the end of the following century that the complete system was in use in Europe.

7.2 ALGEBRA

The most important contributions of the Islamic mathematicians lie in the area of algebra. They took the material already developed by the Babylonians, combined it with the classical Greek heritage of geometry, and produced a new algebra, which they proceeded to extend. By the end of the ninth century, the chief Greek mathematical classics were well known in the Islamic world. Islamic scholars studied them and wrote commentaries on them. The most important idea they learned from their study of these Greek works was the notion of proof. They absorbed the idea that one could not consider a mathematical problem solved unless one could demonstrate that the solution was valid. How does one demonstrate this, particularly for an algebra problem? The answer seemed clear. The only real proofs were geometric. After all, it was geometry which was found in Greek texts, not algebra. Hence Islamic scholars generally set themselves the tasks of justifying algebraic rules, either the ancient Babylonian ones or new ones which they themselves discovered, and justifying them through geometry.

7.2.1 The Algebra of al-Khwārizmī and ibn Turk

One of the earliest Islamic algebra texts, written about 825 by al-Khwārizmī, was entitled *Al-kitāb al-muḫtaṣar fī ḥisāb al-jabr wa-l-muqābala* (*The Condensed Book on the Calculation of al-Jabr and al-Muqabala*) and ultimately had even more influence than his arithmetical work. The term *al-jabr* can be translated as "restoring" and refers to the operation of transposing a subtracted quantity on one side of an equation to the other side where it becomes an added quantity. The word *al-muqābala* can be translated as "comparing" and refers to the reduction of a positive term by subtracting equal amounts from both sides of the equation. For example, converting $3x + 2 = 4 - 2x$ to $5x + 2 = 4$ is an example of *al-jabr* while converting the latter to $5x = 2$ is an example of *al-muqābala*. Our own word

> ### *Biography* Muḥammad ibn Mūsā al-Khwārizmī (c. 780–850)
>
> Al-Khwārizmī, or perhaps his ancestors, came from Khwarizm, the region south of the Aral Sea now part of Uzbekistan and Turkmenistan (Figure 7.3). Al-Khwārizmī was one of the first scholars in the House of Wisdom founded by the caliph al-Maʾmūn and also was one of the astronomers called to cast a horoscope for the dying caliph al-Wāthiq in 847. The story is told that although al-Khwārizmī assured the caliph he would live another fifty years, in fact the caliph died ten days later. Perhaps al-Khwārizmī felt it was not good policy to be the bearer of bad news to one's ruler. Besides the contributions to mathematics detailed in the text, al-Khwārizmī wrote a work on geography in which he developed a map of the Islamic world much superior to that known from the work of Ptolemy.

FIGURE 7.3
Al-Khwārizmī on a stamp from the Soviet Union.

"algebra" is a corrupted form of the Arabic *al-jabr*. It came into use when this and other similar treatises were translated into Latin. No translation was made of the word *al-jabr*, which thus came to be taken for the name of this science.

Al-Khwārizmī explained in his introduction why he came to write his text:

> That fondness for science, by which God has distinguished the Imam al-Maʾmūn, the Commander of the Faithful, . . . has encouraged me to compose a short work on calculating by *al-jabr* and *al-muqābala*, confining it to what is easiest and most useful in arithmetic, such as men constantly require in cases of inheritance, legacies, partition, law-suits, and trade, and in all their dealings with one another, or where the measuring of lands, the digging of canals, geometrical computation, and other objects of various sorts and kinds are concerned.[6]

Al-Khwārizmī was interested in writing a practical manual, not a theoretical one. Nevertheless, he had already been sufficiently influenced by the introduction of Greek mathematics into the House of Wisdom that even in such a manual he felt constrained to give geometric proofs of his algebraic procedures. The geometric proofs, however, are not Greek proofs. They appear to be, in fact, very similar to the Babylonian geometric arguments out of which the algebraic algorithms grew. Again, like his oriental predecessors, al-Khwārizmī gave numerous examples and problems, but the Greek influence showed through in his systematic classification of the problems he intended to solve, as well as in the very detailed explanations of his methods.

Al-Khwārizmī began by noting that "what people generally want in calculating . . . is a number."[7] These numbers are the solutions of equations. Thus the text was to be a manual for solving equations. The quantities he dealt with were generally of three kinds, the square [of the unknown], the root of the square [the unknown itself], and the absolute numbers [the constants in the equation]. He then noted that there are six types of equations which can be written using these three kinds:

1. Squares are equal to roots. $(ax^2 = bx)$

2. Squares are equal to numbers. $(ax^2 = c)$

3. Roots are equal to numbers. $(bx = c)$

4. Squares and roots are equal to numbers. $(ax^2 + bx = c)$

5. Squares and numbers are equal to roots. $(ax^2 + c = bx)$
6. Roots and numbers are equal to squares. $(bx + c = ax^2)$

The reason for the six-fold classification is that Islamic mathematicians, unlike the Hindus, did not deal with negative numbers at all. Coefficients, as well as the roots of the equations, must be positive. The types listed are the only types which have positive solutions. Our standard form $ax^2 + bx + c = 0$ would make no sense for al-Khwārizmī, because if the coefficients are all positive, the roots cannot be.

Al-Khwārizmī's solutions to the first three types of equations are straightforward. We only need note that 0 is not considered as a solution to the first type. His rules for the compound types of equations are more interesting. We present his solution to type 4. Because al-Khwārizmī used no symbols, we will follow him in writing everything out in words, including the numbers of his example: "What must be the square which, when increased by ten of its own roots, amounts to thirty-nine? The solution is this: you halve the number of roots, which in the present instance yields five. This you multiply by itself; the product is twenty-five. Add this to thirty-nine; the sum is sixty-four. Now take the root of this which is eight, and subtract from it half the number of the roots, which is five; the remainder is three. This is the root of the square which you sought for."[8]

Al-Khwārizmī's verbal description of his procedure is essentially the same as that of the Babylonian scribes. Namely, in modern notation, the solution of $x^2 + bx = c$ is

$$x = \sqrt{\left(\frac{b}{2}\right)^2 + c} - \frac{b}{2}.$$

Al-Khwārizmī's geometric justification of this procedure also demonstrates his Babylonian heritage. Beginning with a square representing x^2, he adds two rectangles, each of width five ("half the number of roots") (Figure 7.4). The sum of the area of the square and the two rectangles is then $x^2 + 10x = 39$. One now completes the square with a single square of area 25 to make the total area 64. The solution $x = 3$ is then easily found. This geometric description corresponds to the Babylonian description of the solution of $x^2 + \frac{4}{3}x = \frac{11}{12}$. (See Chapter 1, p. 33 and Figure 1.25.)

Although al-Khwārizmī's geometric description of his method appears to have been taken over from Babylonian sources, he or his (unknown) predecessors in this field succeeded in changing the focus of quadratic equation solving away from the actual finding of sides of squares into that of finding numbers satisfying certain conditions. For example, he explains the term "root" not as a side of a square but as "anything composed of units which can be multiplied by itself, or any number greater than unity multiplied by itself, or that which is found to be diminished below unity when multiplied by itself."[9] Also, his procedure for solving quadratic equations of type 4, when the coefficient of the square term is other than one, is the arithmetical method of first multiplying or dividing appropriately to make the initial coefficient one, and then proceeding as before. Al-Khwārizmī even admits somewhat later in his text, when he is discussing the addition of the "polynomials" $100 + x^2 - 20x$ and $50 + 10x - 2x^2$, that "this does not admit of any figure, because there are three different species, i.e., squares and roots and numbers, and nothing corresponding to them by which they might be represented. . . . [Nevertheless], the elucidation by words is easy."[10]

Finally, al-Khwārizmī's presentation of the method and geometric description for type 5, squares and numbers equal to roots, shows that, unlike the Babylonians, he could deal with an equation with two positive roots, at least numerically. In this case, $x^2 + c = bx$, his verbal description of the solution procedure easily translates into our formula

$$x = \frac{b}{2} \pm \sqrt{\left(\frac{b}{2}\right)^2 - c}. \tag{7.1}$$

In fact, he states that one can employ either addition or subtraction to get a root and also notes the condition on the solution: "If the product [of half the number of roots with itself] is less than the number connected with the square, then the instance is impossible; but if the product is equal to the number itself, then the root of the square is equal to half of the number of roots alone, without either addition or subtraction."[11] The geometric demonstration in this case, which reminds us of the Babylonian description for the system $x + y = b$, $xy = c$ into which they would have converted this equation (see Chapter 1, p. 32 and Figure 1.23), only deals with the subtraction in formula 7.1. In Figure 7.5, square $ABCD$ represents x^2 while rectangle $ABNH$ represents c. Therefore HC represents b. Bisect HC at G, extend TG to K so that $GK = GA$, and complete the rectangle $GKMH$. Finally, choose L on KM so that $KL = GK$ and complete the square $KLRG$. It is then clear that rectangle $MLRH$ equals rectangle $GATB$. Since the area of square $MKTN$ is $\left(\frac{b}{2}\right)^2$, while that square less square $KLRG$ equals rectangle $ABNH$ or c, it follows that square $KLRG$ equals $\left(\frac{b}{2}\right)^2 - c$. Since the side of that square is equal to AG, it follows that $x = AC = CG - AG$ is given by formula 7.1 using the minus sign. Although al-Khwārizmī briefly noted that CR could also represent a solution, he did not demonstrate this by a diagram, nor did he deal in his diagram with the special conditions mentioned in his verbal description.

Al-Khwārizmī's text does contain the word "condensed" in the title, thus leading one to believe that there were other books at the time discussing algebraic procedures and their attendant geometric justifications in more detail. There is, however, only a fragment of such a work now extant, the section *Logical Necessities in Mixed Equations* from a longer work *Kitāb al-jabr wa'l muqābala* by 'Abd al-Hamīd ibn Wāsi ibn Turk al-Jīlī, a contemporary of al-Khwārizmī about whom very little is known. The sources even differ as to whether ibn Turk was from Iran, Afghanistan, or Syria.

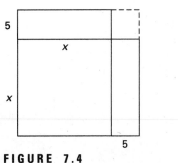

FIGURE 7.4
Al-Khwārizmī's geometric justification for the solution of $x^2 + 10x = 39$.

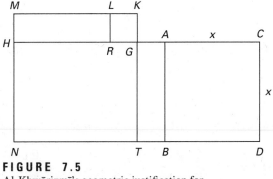

FIGURE 7.5
Al-Khwārizmī's geometric justification for the solution of $x^2 + c = bx$.

In any case, the extant chapter of ibn Turk's book deals with quadratic equations of al-Khwārizmī's types 1, 4, 5, and 6 and includes a much more detailed geometric description of the method of solution than is found in al-Khwārizmī's work. In particular, in the case of type 5, ibn Turk gave geometric versions for all possible cases. His first example is the same as al-Khwārizmī's, namely $x^2 + 21 = 10x$, but he began the geometrical demonstration by noting that G, the midpoint of CH, may be either on the line segment AH, as in al-Khwārizmī's diagram, or on the line segment CA of Figure 7.6. In this case, squares and rectangles are completed, similar in form to those in Figure 7.5, but the solution $x = AC$ is now given as $CG + GA$, thus using the plus sign in formula 7.1. In addition, ibn Turk discussed what he called the "intermediate case," where the root of the square is exactly equal to half the number of roots. His example for this situation is $x^2 + 25 = 10x$; the geometric diagram then simply consists of a rectangle divided into two equal squares.

Ibn Turk further noted that "there is the logical necessity of impossibility in this type of equation when the numerical quantity . . . is greater than [the square of] half the number of roots,"[12] as, for example, in the case $x^2 + 30 = 10x$. Again, he resorted to a geometric argument. Assuming that G is located on the segment AH, we know as before that the rectangle $KMNT$ is greater than the rectangle $HABN$ (Figure 7.7). But the conditions of the problem show that the latter rectangle equals 30 while the former only equals 25. A similar argument works in the case where G is located on CA.

Although the section on quadratic equations of ibn Turk's algebra is the only part still extant, al-Khwārizmī's text contains much else of interest, including an introduction to manipulation with algebraic expressions, explained by reference to similar manipulations with numbers. For example, he notes that if $a \pm b$ is multiplied by $c \pm d$, where the letters stand for numbers, then four multiplications are necessary. Although none of his numbers are negative, he does know the rules for dealing with multiplication and signs. As he states, "I have explained this, that it might serve as an introduction to the multiplication of unknown sums, when numbers are added to them, or when numbers are subtracted from them, or when they are subtracted from numbers."[13]

Al-Khwārizmī's text continues with a large collection of problems, many of which involve these manipulations, and most of which result in a quadratic equation. For exam-

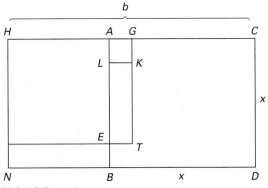

FIGURE 7.6
Ibn Turk's geometric justification for one case of $x^2 + c = bx$.

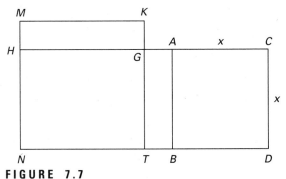

FIGURE 7.7
Ibn Turk's geometric justification of the impossibility of solving $x^2 + 30 = 10x$.

ple, one problem states, "I have divided ten into two parts, and having multiplied each part by itself, I have put them together, and have added to them the difference of the two parts previously to their multiplication, and the amount of all this is fifty-four."[14] It is not difficult to translate this problem into the equation $(10 - x)^2 + x^2 + (10 - x) - x = 54$. The author reduces this to the equation $x^2 + 28 = 11x$ and then uses his rule for this equation of type 5 to get $x = 4$. He ignores here the second root, $x = 7$, for then the sum of the two squares would be 58 and the conditions of the problem could not be met.

Although al-Khwārizmī promised in his preface that he would write about what is useful, very few of his problems leading to quadratic equations deal with any practical ideas. Many of them are similar to the example and begin with "I have divided ten into two parts." There are a few problems concerned with dividing money among a certain number of men, but even these problems are in no sense practical. In fact, one of these reduces to the equation $x^2 + x = 3/4$, where x is the number of men, the solution of which is $x = 1/2$! An entire section of the text is devoted to elementary problems of mensuration, which will be discussed later, and a brief section is devoted to the "rule of three," but neither of these provides any practical uses of quadratic equations either. Finally, the second half of the text is entirely devoted to problems of inheritance. Dozens of complicated situations are presented, for the solution of which one needs to be familiar with Islamic legacy laws. The actual mathematics needed, however, is never more complicated than the solution of linear equations. One can only conclude that although al-Khwārizmī was interested in teaching his readers how to solve mathematical problems, and especially how to deal with quadratic equations, he could not think of any real-life situations that required these equations. Things apparently had not changed in this regard since the time of the Babylonians.

7.2.2 The Algebra of Thābit ibn Qurra and Abū Kāmil

Within 50 years of the works by al-Khwārizmī and ibn Turk, the Islamic mathematicians had decided that the necessary geometric foundations to the algebraic solution of quadratic equations should be based on the work of Euclid rather than on the ancient traditions. Perhaps the earliest of these justifications was given by Thābit ibn Qurra (c. 830–890). Thābit was born in Harran (now in southern Turkey), was discovered there by one of the scholars from the House of Wisdom, and was brought to Baghdad in about 870 where he himself became a great scholar. Among his many writings on mathematical topics is a short work entitled *Qawl fī taṣḥīḥ masā'il al-jabr bi l-barāhīn al-handasīya* (*On the Verification of Problems of Algebra by Geometrical Proofs*). To solve the equation $x^2 + bx = c$, for example, Thābit used Figure 7.8, where AB represents x, square $ABCD$ represents x^2, and BE represents b. It follows that the rectangle $DE = AB \times EA$ represents

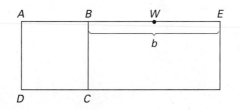

FIGURE 7.8
Thābit ibn Qurra's geometric justification for the solution of $x^2 + bx = c$.

c. If W is the midpoint of BE, Euclid's *Elements* II-6 implies that $EA \times AB + BW^2 = AW^2$. But since $EA \times AB$ and BW^2 are known (equaling respectively c and $(b/2)^2$), it follows that AW^2 and therefore AW are known. Then $x = AB = AW - BW$ is determined. Thābit noted explicitly that the geometric procedure of *Elements* II-6 is completely analogous to the procedure of "the algebraists," that is, the algorithm stated by al-Khwārizmī, and therefore provides the necessary justification. Thābit also showed how to use this same proposition to solve $x^2 = bx + c$ and how to use *Elements* II-5 to solve $x^2 + c = bx$.

Similar justifications of these solutions using *Elements* II were given by the Egyptian mathematician Abū Kāmil ibn Aslam (c. 850–930) in his own algebra text, *Kitāb fī al-jabr wa'l-muqābala*: "I shall explain their rule using geometric figures clarified by wise men of geometry and which are explained in the Book of Euclid."[15] Abū Kāmil, however, unlike Thābit, proved Euclid's results anew in the course of his discussion and also presented numerical examples, in fact the same initial numerical examples as al-Khwārizmī. Like his predecessor, Abū Kāmil followed his discussion of the various forms of quadratic equations by a treatment of various algebraic rules and then a large selection of problems. But he made some advances over the earlier mathematician by dealing with many more complicated identities and more complex problems, including in particular manipulations with surds.

Abū Kāmil was not at all worried about dealing with irrationals. He used them freely in his problems, many of which, like those of al-Khwārizmī, start with "divide 10 into 2 parts." For example, consider problem 37: "If one says that 10 is divided into two parts, and one part is multiplied by itself and the other by the root of 8, and subtract the quantity of the product of one part times the root of 8 from . . . the product of the other part multiplied by itself, it gives 40."[16] The equation in this case is $(10 - x)(10 - x) - x\sqrt{8} = 40$. After rewriting this equation in the form $x^2 + 60 = 20x + \sqrt{8x^2}$ $(=(20 + \sqrt{8})x)$, Abū Kāmil carried out the algorithm for the case squares and numbers equal roots to conclude that $x = 10 + \sqrt{2} - \sqrt{42 + \sqrt{800}}$ and that $10 - x$, the "other part," is equal to $\sqrt{42 + \sqrt{800}} - \sqrt{2}$.

Abū Kāmil even applied substitutions to simplify problems and could deal with equations of degree higher than 2 as long as they were quadratic in form. Problem 45 illustrates both ideas: "One says that 10 is divided into two parts, each of which is divided by the other, and when each of the quotients is multiplied by itself and the smaller is subtracted from the larger, then there remains 2."[17] The equation is

$$\left(\frac{x}{10 - x}\right)^2 - \left(\frac{10 - x}{x}\right)^2 = 2.$$

Abū Kāmil made a new "thing" y equal to $\frac{10 - x}{x}$ and derived the new equation $\frac{1}{y^2} = y^2 + 2$. Multiplying both sides by y^2 gave him the quadratic equation in y^2: $(y^2)^2 + 2y^2 = 1$ for which the solution is $y^2 = \sqrt{2} - 1$. Hence, $y = \sqrt{\sqrt{2} - 1}$. Then,

$$\frac{10 - x}{x} = \sqrt{\sqrt{2} - 1},$$

and Abū Kāmil proceeded to solve for x by first squaring both sides of this equation. The final result is $x = 10 + \sqrt{50} - \sqrt{50 + \sqrt{20,000}} - \sqrt{5000}$.

When considering Abū Kāmil's algebra, remember that it, like all Islamic algebra texts of his era, was written without symbols. Thus the algebraic manipulation which modern symbolism makes almost obvious is carried out completely verbally. More importantly, however, Abū Kāmil is willing to use the algebraic algorithms which had been systematized by the time of al-Khwārizmī with any type of positive number. He makes no distinction between operating with 2 or with $\sqrt{8}$ or even with $\sqrt{\sqrt{2} - 1}$. Since these algorithms came from geometry, on one level that is not surprising. After all, it was the Greek failure to find a "numerical" representation of the diagonal of a square which was one of the causes leading to their use of the geometric algebra of line segments and areas. But in dealing with these quantities, Abū Kāmil interpreted all of them in the same way. It did not matter whether a magnitude was technically a square or a fourth power or a root or a root of a root. For Abū Kāmil, the solution of a quadratic equation was not a line segment, as it would be in the interpretation of the appropriate propositions of the *Elements*. It was a "number," even though Abū Kāmil could not perhaps give a proper definition of that term. There was therefore no compunction about combining the various quantities which appeared in the solutions, using general rules. Abū Kāmil's willingness to handle all of these quantities by the same techniques helped pave the way toward a new understanding of the concept of number fully as important as al-Samaw'al's use of decimal approximations.

7.2.3 Al-Karajī, al-Samaw'al, and the Algebra of Polynomials

The process of relating arithmetic to algebra, begun by al-Khwārizmī and Abū Kāmil, continued in the Islamic world with the work of Abū Bakr al-Karajī (d. 1019) and al-Samaw'al over the next two centuries. These latter mathematicians were instrumental in showing that the techniques of arithmetic could be fruitfully applied in algebra and, reciprocally, that ideas originally developed in algebra could also be important in dealing with numbers.

Little is known of the life of al-Karajī other than that he worked in Baghdad around the year 1000 and wrote many mathematics works as well as works on engineering topics. In the first decade of the eleventh century he composed a major work on algebra entitled *al-Fakhrī* (*The Marvelous*). The aim of *al-Fakhrī,* and of algebra in general according to al-Karajī, was "the determination of unknowns starting from knowns."[18] In pursuit of this aim he made use of all the techniques of arithmetic, converted into techniques of dealing with unknowns. He began by making a systematic study of the algebra of exponents. Although earlier writers, including Diophantus, had considered powers of the unknown greater than the third, al-Karajī was the first to understand fully that these powers can be extended indefinitely. In fact, he developed a method of naming the various powers x^n and their reciprocals $\frac{1}{x^n}$. Each power was defined recursively as x times the previous power. It followed that there was an infinite sequence of proportions

$$1 : x = x : x^2 = x^2 : x^3 = \ldots$$

and a similar one for reciprocals

$$\frac{1}{x} : \frac{1}{x^2} = \frac{1}{x^2} : \frac{1}{x^3} = \frac{1}{x^3} : \frac{1}{x^4} = \ldots.$$

Biography   Al-Samaw'al (1125–1180)

Al-Samaw'al was born in Baghdad to well-educated Jewish parents. His father was in fact a Hebrew poet. Besides giving him a religious education, they encouraged him to study medicine and mathematics. Because the House of Wisdom no longer existed in Baghdad, he had to study mathematics independently and therefore traveled to various other parts of the Middle East. He wrote his major mathematical work *Al-Bāhir* when he was only 19. His interests later turned to medicine, and he became a successful physician and wrote several medical works. The only extant one is entitled *The Companion's Promenade in the Garden of Love*, a treatise on sexology and a collection of erotic stories. When he was about 40, he decided to convert to Islam. To justify his conversion to the world, he wrote an autobiography in 1167 stating his arguments against Judaism, a work which became famous as a source of Islamic polemics against the Jews.

Once the powers were understood, al-Karajī could establish general procedures for adding, subtracting, and multiplying monomials and polynomials. He could only deal, however, with division of monomials by monomials and of polynomials by monomials. This was due partly to his inability to incorporate rules for negative numbers into his theory and partly to his totally verbal means of expression.

Al-Karajī was more successful in continuing the work of Abū Kāmil in applying arithmetic operations to irrational quantities. In particular, he explicitly interpreted the various classes of incommensurables in *Elements* X as classes of "numbers," on which the various operations of arithmetic were defined. Like Abū Kāmil, he gave no definition of "number," but just dealt with the various surd quantities using numerical rather than geometrical techniques. As part of this process, he developed various formulas involving surds, such as,

$$\sqrt{A + B} = \sqrt{\frac{A + \sqrt{A^2 - B^2}}{2}} + \sqrt{\frac{A - \sqrt{A^2 - B^2}}{2}}$$

and

$$\sqrt[3]{A} + \sqrt[3]{B} = \sqrt[3]{3\sqrt[3]{A^2B} + 3\sqrt[3]{AB^2} + A + B}.$$

Further work in dealing with algebraic manipulation was accomplished by al-Samaw'al who, in particular, introduced negative coefficients. He expressed his rules for dealing with these coefficients quite clearly in his algebra text *Al-Bāhir fi'l-ḥisāb* (*The Shining Book of Calculation*):

> If we subtract an additive number from an empty power ($0x^n - ax^n$), the same subtractive number remains; if we subtract the subtractive number from an empty power ($0x^n - (-ax^n)$), the same additive number remains. If we subtract an additive number from a subtractive number, the remainder is their subtractive sum; if we subtract a subtractive number from a greater subtractive number, the result is their subtractive difference; if the number from which one subtracts is smaller than the number subtracted, the result is their additive difference.[19]

Given these rules, al-Samaw'al could easily add and subtract polynomials by combining like terms.

Al-Samaw'al also gave a very clear formulation of the law of exponents. Al-Karajī had in essence used this law, as had Abū Kāmil and others. However, since the product of, for example, a square and a cube was expressed in words as a square-cube, the numerical property of adding exponents could not be seen. Al-Samaw'al decided that this law could best be expressed by using a table consisting of columns, each column representing a different power of either a number or an unknown. In fact, he also saw that he could deal with powers of $1/x$ as easily as with powers of x. In his work the columns are headed by the Arabic letters standing for the numerals, reading both ways from the central column labeled 0. We will simply use the Arabic numerals themselves. Each column then has the name of the particular power or reciprocal power. For example, the column headed by a 2 on the left is named "square," that headed by a 5 on the left is named "square-cube," that headed by a 3 on the right is named "part of cube," and so on. To simplify matters we will just use powers of x. In his initial explanation of the rules, al-Samaw'al also put a particular number under the 1 on the left, such as 2, and then the various powers of 2 in the corresponding columns:

7	6	5	4	3	2	1	0	1	2	3	4	5	6	7
x^7	x^6	x^5	x^4	x^3	x^2	x	1	x^{-1}	x^{-2}	x^{-3}	x^{-4}	x^{-5}	x^{-6}	x^{-7}
128	64	32	16	8	4	2	1	$\frac{1}{2}$	$\frac{1}{4}$	$\frac{1}{8}$	$\frac{1}{16}$	$\frac{1}{32}$	$\frac{1}{64}$	$\frac{1}{128}$

Al-Samaw'al now used the chart to explain what we call the law of exponents, $x^n x^m = x^{m+n}$: "The distance of the order of the product of the two factors from the order of one of the two factors is equal to the distance of the order of the other factor from the unit. If the factors are in different directions then we count (the distance) from the order of the first factor towards the unit; but, if they are in the same direction, we count away from the unit."[20] So, for example, to multiply x^3 by x^4, count four orders to the left of column 3 and get the result as x^7. To multiply x^3 by x^{-2}, count two orders to the right from column 3 and get the answer x^1. Using these rules, al-Samaw'al could easily multiply polynomials in x and $1/x$ as well as divide such polynomials by monomials.

Al-Samaw'al was also able to divide polynomials by polynomials using a similar chart. In this new chart, which reminds us of the Chinese counting board as used in solving polynomial equations, each column again stands for a given power of x or of $1/x$. But now the numbers in each column represent the coefficients of the various polynomials involved in the division process. For example, to divide $20x^2 + 30x$ by $6x^2 + 12$, he first sets the 20 and the 30 in the columns headed by x^2 and x respectively and the 6 and 12 below these in the columns headed by x^2 and 1. Since there is an "empty order" for the divisor in the x column, he places a 0 there. He next divides $20x^2$ by $6x^2$, getting 3 1/3. That number goes in the units column on the answer line. The product of 3 1/3 by $6x^2 + 12$ is $20x^2 + 40$. The next step is subtraction. The remainder in the x^2 column is naturally 0. In the x column the remainder is 30, while in the units column the remainder is -40. Al-Samaw'al now writes a new chart in which the 6, 0, 12 are shifted one place to the right and the directions are given to divide that into $30x - 40$. The initial quotient of $30x$ by $6x^2$ is $5 \cdot 1/x$, so a 5 is placed in the answer line in the column headed by

$1/x$ and the process is continued. We display here al-Samaw'al's first two charts for this division problem.

x^2	x	1	$\frac{1}{x}$	$\frac{1}{x^2}$	$\frac{1}{x^3}$
		$3\frac{1}{3}$			
20	30				
6	0	12			

x^2	x	1	$\frac{1}{x}$	$\frac{1}{x^2}$	$\frac{1}{x^3}$
		$3\frac{1}{3}$	5		
30	-40				
6	0	12			

In this particular example, the division is not exact. Al-Samaw'al continues the process through eight steps to get

$$3\frac{1}{3} + 5\left(\frac{1}{x}\right) - 6\frac{2}{3}\left(\frac{1}{x^2}\right) - 10\left(\frac{1}{x^3}\right) + 13\frac{1}{3}\left(\frac{1}{x^4}\right) + 20\left(\frac{1}{x^5}\right) - 26\frac{2}{3}\left(\frac{1}{x^6}\right) - 40\left(\frac{1}{x^7}\right).$$

To show his fluency with his multiplication procedure, he now checks the answer by multiplying it by the divisor. Since the product differs from the dividend by terms only in $\frac{1}{x^6}$ and $\frac{1}{x^7}$, he calls the result given "the answer approximately." Nevertheless, he also notes that there is a pattern to the coefficients of the quotient. In fact, if a_n represents the coefficient of $\frac{1}{x^n}$, the pattern is given by $a_{n+2} = -2a_n$. He then proudly writes out the next 21 terms of the quotient, ending with $54,613\frac{1}{3}\left(\frac{1}{x^{28}}\right)$.

Given that al-Samaw'al thought of extending division of polynomials into polynomials in $1/x$, and thought of partial results as approximations, it is not surprising that he would divide whole numbers by simply replacing x by 10. As already noted, al-Samaw'al was the first to recognize explicitly that one could approximate fractions more and more closely by calculating more and more decimal places. The work of al-Karajī and al-Samaw'al was thus extremely important in developing the idea that algebraic manipulations and manipulations with numbers are parallel. Virtually any technique that applies to one can be adapted to apply to the other.

7.2.4 Induction, Sums of Powers, and the Pascal Triangle

Another important idea introduced by al-Karajī and continued by al-Samaw'al and others was that of an inductive argument for dealing with certain arithmetic sequences. Thus al-Karajī used such an argument to prove the result on the sums of integral cubes already known to Āryabhaṭa (and even, perhaps, to the Greeks). Al-Karajī did not, however, state a general result for arbitrary n. He stated his theorem for the particular integer 10:

$$1^3 + 2^3 + 3^3 + \cdots + 10^3 = (1 + 2 + 3 + \cdots + 10)^2.$$

His proof, nevertheless, was clearly designed to be extendable to any other integer.

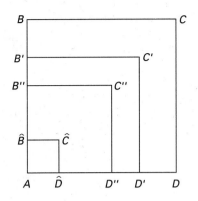

FIGURE 7.9
Al-Karajī's proof of the formula for the sum
of the integral cubes.

Consider the square $ABCD$ with side $1 + 2 + 3 + \cdots + 10$ (Figure 7.9). Setting $BB' = DD' = 10$, and completing the gnomon $BCDD'C'B'$, al-Karajī calculates the area of the gnomon to be

$$2 \cdot 10(1 + 2 + \cdots + 9) + 10^2 = 2 \cdot 10 \cdot \frac{9 \cdot 10}{2} + 10^2 = 9 \cdot 10^2 + 10^2 = 10^3.$$

Since the area of square $ABCD$ is the sum of the areas of square $AB'C'D'$ and the gnomon, it follows that $(1 + 2 + \cdots + 10)^2 = (1 + 2 + \cdots + 9)^2 + 10^3$. A similar argument then shows that $(1 + 2 + \cdots + 9)^2 = (1 + 2 + \cdots + 8)^2 + 9^3$. Continuing in this way to the final square $A\hat{B}\hat{C}\hat{D}$ of area $1 = 1^3$, al-Karajī proves his theorem from the equality of square $ABCD$ to square $A\hat{B}\hat{C}\hat{D}$ plus the sum of the gnomons of areas $2^3, 3^3, \ldots, 10^3$.

Al-Karajī's argument includes in essence the two basic components of a modern argument by induction, namely the truth of the statement for $n = 1$ ($1^2 = 1^3$) and the deriving of the truth for $n = k$ from that for $n = k - 1$. Of course, this second component is not explicit since, in some sense, al-Karajī's argument is in reverse. That is, he starts from $n = 10$ and goes down to 1 rather than proceeding upwards. Nevertheless, his argument in *al-Fakhrī* is the earliest extant proof of the sum formula for integral cubes.

The cube formula itself, however, had appeared somewhat earlier in Islamic mathematics in a treatise of Abū l-Ṣaqr ʿAbd al-ʿAzīz ibn ʿUthmān al-Qabīsī (tenth century) from Baghdad. The discovery of this formula is not difficult if one considers a few examples. What is more remarkable about the work of al-Qabīsī is that it also includes a formula for the sum of fourth powers. Al-Qabīsī, however, does not tell us how he discovered this result nor does he give a proof.

Fortunately, a proof of the sum formula for fourth powers, inductive in nature, appeared early in the eleventh century in a work by the Egyptian mathematician Abū ʿAlī al-Hasan ibn al-Hasan ibn al-Haytham (965–1039). Ibn al-Haytham, in fact, also gave proofs of the formulas for the sums of the integers, their squares, and their cubes, proofs all similar in nature and easily generalizable to the discovery and proof of formulas for the sum of any given powers of the integers. That he did not state any such generalization is probably due to his needing only the formulas for the second and fourth powers in his computation of the volume of a paraboloid, to be discussed in section 7.3.4.

Biography   Ibn al-Haytham (965–1039)

Ibn al-Haytham, known in Europe as Alhazen and one of the most influential of Islamic scientists, was born in Basra, now in Iraq, but spent most of his life in Egypt, after he was invited by the caliph al-Hakim to work on a Nile control project (Figure 7.10). Although the project never came to fruition, ibn al-Haytham did produce in Egypt his most important scientific work, the *Optics* in seven books. The *Optics* was translated into Latin in the early thirteenth century and was studied and commented on in Europe for several centuries thereafter. Ibn al-Haytham's fame as a mathematician chiefly rests on his treatment of "Alhazen's problem." This problem was to find the point or points on some reflecting surface at which the light from one of two points outside that surface is reflected to the other. In the fifth book of the *Optics* he attempted to solve the problem for a variety of surfaces, spherical, cylindrical, and conical, concave, and convex. Although he was not completely successful, his accomplishments showed him to be in full command of both the elementary and advanced geometry of the Greeks. In the final years of his life, ibn al-Haytham earned his living by copying annually, among others, Euclid's *Elements*, Apollonius' *Conics*, and Ptolemy's *Almagest*.

FIGURE 7.10
Ibn al-Haytham's work on optics honored on a stamp from Pakistan.

The central idea in ibn al-Haytham's proof of the sum formulas was the derivation of the equation

$$(n + 1)\sum_{i=1}^{n} i^k = \sum_{i=1}^{n} i^{k+1} + \sum_{p=1}^{n}\left(\sum_{i=1}^{p} i^k\right). \tag{7.2}$$

Naturally, he did not state this result in general form but only for particular integers, namely $n = 4$ and $k = 1, 2, 3$. However, like the proof of al-Karajī, his proof is also immediately generalizable to any values of n and k. Ibn al-Haytham used this equation in the cases $k = 1$ and $k = 2$ to derive the known results for sums of squares and cubes, stated for arbitrary n:

$$\sum_{i=1}^{n} i^2 = \left(\frac{n}{3} + \frac{1}{3}\right)n\left(n + \frac{1}{2}\right)$$

$$\sum_{i=1}^{n} i^3 = \left(\frac{n}{4} + \frac{1}{4}\right)n(n + 1)n = \frac{n^4}{4} + \frac{n^3}{2} + \frac{n^2}{4}.$$

He then derived equation 7.2 for the case $k = 3$ by the following steps:

$$1^3 \cdot 2 = 1^4 + 1^3$$

$$(1^3 + 2^3)3 = 2^4 + 1^4 + (1^3 + 2^3) + 1^3$$

$$(1^3 + 2^3 + 3^3)4 = 3^4 + 2^4 + 1^4 + (1^3 + 2^3 + 3^3) + (1^3 + 2^3) + 1^3$$

$$(1^3 + 2^3 + 3^3 + 4^3)5 = 4^4 + 3^4 + 2^4 + 1^4 + (1^3 + 2^3 + 3^3 + 4^3)$$
$$+ (1^3 + 2^3 + 3^3) + (1^3 + 2^3) + 1^3.$$

The truth of the first line is clear, while the others follow in turn by adding equal quantities to each side. In modern notation, it would be straightforward to formulate this argument by an induction on n. In any case, ibn al-Haytham was able to complete his argument deriving the formula for sums of fourth powers by using the corresponding formulas for cubes and squares:

$$(1^3 + 2^3 + 3^3 + 4^3)5 = 1^4 + 2^4 + 3^4 + 4^4 + \frac{1}{4}(1^4 + 2^4 + 3^4 + 4^4)$$

$$+ \frac{1}{2}(1^3 + 2^3 + 3^3 + 4^3) + \frac{1}{4}(1^2 + 2^2 + 3^2 + 4^2)$$

$$(1^3 + 2^3 + 3^3 + 4^3)5 = \frac{5}{4}(1^4 + 2^4 + 3^4 + 4^4) + \frac{1}{2}(1^3 + 2^3 + 3^3 + 4^3)$$

$$+ \frac{1}{4}(1^2 + 2^2 + 3^2 + 4^2)$$

$$1^4 + 2^4 + 3^4 + 4^4 = \frac{4}{5}(1^3 + 2^3 + 3^3 + 4^3)(4 + \frac{1}{2}) - \frac{1}{5}(1^2 + 2^2 + 3^2 + 4^2)$$

$$1^4 + 2^4 + 3^4 + 4^4 = \frac{4}{5}\left(4 + \frac{1}{2}\right)\left(\frac{4}{4} + \frac{1}{4}\right)4(4 + 1)4 - \frac{1}{5}\left(\frac{4}{3} + \frac{1}{3}\right)4\left(4 + \frac{1}{2}\right)$$

and finally,

$$1^4 + 2^4 + 3^4 + 4^4 = \left(\frac{4}{5} + \frac{1}{5}\right)4\left(4 + \frac{1}{2}\right)\left[(4 + 1)4 - \frac{1}{3}\right].$$

From this result for the case $n = 4$, ibn al-Haytham simply stated his general result:

$$\sum_{i=1}^{n} i^4 = \left(\frac{n}{5} + \frac{1}{5}\right)n\left(n + \frac{1}{2}\right)\left[(n + 1)n - \frac{1}{3}\right].$$

Another inductive argument, this time in relation to the binomial theorem and the Pascal triangle, is found in al-Samaw'al's *Al-Bāhir* where he refers to al-Karajī's treatment of these subjects. Because the particular work of al-Karajī's in which this discussion occurs is no longer extant, we consider al-Samaw'al's version. The binomial theorem is the result

$$(a + b)^n = \sum_{k=0}^{n} C_k^n a^{n-k} b^k$$

where n is a positive integer and the values C_k^n are the binomial coefficients, the entries in the Pascal triangle. Naturally, al-Samaw'al, having no symbolism, writes this formula in words in each individual instance and, in fact, makes no claim that he is stating a general theorem. For him the separate results in each case were probably only techniques for calculating certain products in line with his earlier development of the multiplication of polynomials.[21] For example, in the case $n = 4$ he writes, "For a number divided into two parts, its square-square (fourth power) is equal to the square-square of each of the parts, four times the product of each by the cube of the other, and six times the product of the squares of each of the two parts."[22] Al-Samaw'al then provides a table of binomial coefficients to show how to generalize this rule for greater values of n:

x	x^2	x^3	x^4	x^5	x^6	x^7	x^8	x^9	x^{10}	x^{11}	x^{12}
1	1	1	1	1	1	1	1	1	1	1	1
1	2	3	4	5	6	7	8	9	10	11	12
	1	3	6	10	15	21	28	36	45	55	66
		1	4	10	20	35	56	84	120	165	220
			1	5	15	35	70	126	210	330	495
				1	6	21	56	126	252	462	792
					1	7	28	84	210	462	924
						1	8	36	120	330	792
							1	9	45	165	495
								1	10	55	220
									1	11	66
										1	12
											1

The procedure he provides for constructing this table is the familiar one, that any entry comes from adding the entry to the left of it to the entry just above that one. He then notes that one can use the table to read off the expansion of any power up to the twelfth of "a number divided into two parts."

With this table in mind, let us see how al-Samaw'al demonstrates the quoted result for $n = 4$. Assume the number c is equal to $a + b$. Since $c^4 = cc^3$ and c^3 is already known to be given by $c^3 = (a + b)^3 = a^3 + b^3 + 3ab^2 + 3a^2b$, it follows that $(a + b)^4 = (a + b)(a + b)^3 = (a + b)(a^3 + b^3 + 3ab^2 + 3a^2b)$. By using repeatedly the result $(r + s)t = rt + st$, which al-Samaw'al quotes from Euclid's *Elements* II, he finds that this last quantity equals $(a + b)a^3 + (a + b)b^3 + (a + b)3ab^2 + (a + b)3a^2b = a^4 + a^3b + ab^3 + b^4 + 3a^2b^2 + 3ab^3 + 3a^3b + 3a^2b^2 = a^4 + b^4 + 4ab^3 + 4a^3b + 6a^2b^2$. The coefficients here are the appropriate ones from the table, and the expansion shows that the new coefficients are formed from the old ones exactly as stated in the table construction. Al-Samaw'al next quotes the result for $n = 5$ and then states that the higher powers can be done similarly. Like the proofs of al-Karajī and ibn al-Haytham, al-Samaw'al's argument contains the two basic elements of an inductive proof. He begins with a value for which the result is known, here $n = 2$, and then uses the result for a given integer to derive the result for the next. Since al-Samaw'al did not have any way of stating the general binomial theorem, however, he cannot be said to have proved it, by induction or otherwise. What he had done was provide a method acceptable to his readers for expanding binomials up to the twelfth power, a method which was used in medieval Islam as well as in China to develop an algorithm to calculate roots of numbers (also up to the twelfth).

7.2.5 Omar Khayyam and the Solution of Cubic Equations

There was another strand of development in algebra in the Islamic world alongside of its arithmetization and the development of inductive ideas, namely the application of geometry. By the end of the ninth century Islamic mathematicians, having read the major Greek

---

Biography   Al-Khayyāmī (1048–1131)

Al-Khayyāmī was born in Nishapur, Iran in 1048 shortly after the area was conquered by the Seljuk Turks. He was able during most of his life to enjoy the support of the Seljuk rulers. In fact, he spent many years at the observatory in Isfahan at the head of a group working to reform the calendar. At various times, as ruler replaced ruler, he fell into disfavor, but he was able ultimately to garner enough support to write many mathematical and astronomical works, as well as poetry and philosophical works. He is best known in the West for the collection of poems known as the *Rubaiyat*. In the preface of his great algebra work, he complained how difficult it had been for him to work, but then thanks the ruler who provided him with the necessary support:

> I had not been able to find time to complete this work, or to concentrate my thoughts on it, hindered as I had been by troublesome obstacles. . . . Most of our

contemporaries are pseudo-scientists who mingle truth with falsehood, who are not above deceit and pedantry, and who use the little that they know of the sciences for base material purposes only. When they see a distinguished man intent on seeking the truth, one who prefers honesty and does his best to reject falsehood and lies, avoiding hypocrisy and treachery, they despise him and make fun of him. When God favored me with the intimate friendship of His Excellency, our glorious and unique Lord, the supreme judge, the Imām, Sayid Abū-Ṭāhir . . . after I had despaired of meeting such a man . . . who combined in himself profound power in science with firmness of action . . . my heart was greatly rejoiced to see him. . . . My power was strengthened by his liberality and his favors. In order that I might come nearer to his sublime position I found myself obliged to take up again the work which the vicissitudes of time had caused me to abandon in summarizing what I had verified of the essence of philosophical theories.[23]

---

texts, had noticed that certain geometric problems led to cubic equations, equations which could be solved through finding the intersection of two conic sections. Such problems included the doubling of the cube and Archimedes' splitting of a sphere into two parts whose volumes are in a given ratio. Several Islamic mathematicians during the tenth and eleventh centuries also solved certain cubic equations by taking over this Greek idea of intersecting conics. But it was the mathematician and poet 'Umar ibn Ibrāhīm al-Khayyāmī (1048–1131) (usually known in the West as Omar Khayyam), who first systematically classified and then proceeded to solve all types of cubic equations by this general method.

Al-Khayyāmī's major mathematics text, the *Risāla fi-l-barāhīn 'ala masā'il al-jabr wa'l-muqābala* (*Treatise on Demonstrations of Problems of al-Jabr and al-Muqabala*), is primarily devoted to the solution of cubic equations. As the author makes clear in the preface, the reader of the work must be thoroughly familiar with Euclid's *Elements* and *Data* as well as the first two books of Apollonius' *Conics*, since cubic equations can only be solved geometrically using the properties of conic sections. Nevertheless, the text addresses algebraic, not geometric, problems, and al-Khayyāmī would have liked to have provided algebraic algorithms for solving cubic equations, analogous to al-Khwārizmī's three algorithms for solving quadratic equations. As he wrote, "When, however, the object of the problem is an absolute number, neither we, nor any of those who are concerned with algebra, have been able to solve this equation—perhaps others who follow us will be able to fill the gap."[24] It was not until the sixteenth century that 'Umar's hope was realized.

Al-Khayyāmī began his work, in the style of al-Khwārizmī, by giving a complete classification of equations of degree up to three. Since for al-Khayyāmī, as for his predecessors, all numbers were positive, he had to list separately the various forms that might possess positive roots. Among these were fourteen not reducible to quadratic or linear equations. These were in three groups, one binomial equation, $x^3 = d$, six trinomial equations, $x^3 + cx = d$, $x^3 + d = cx$, $x^3 = cx + d$, $x^3 + bx^2 = d$, $x^3 + d = bx^2$, and $x^3 = bx^2 + d$; and seven tetranomial equations, $x^3 + bx^2 + cx = d$, $x^3 + bx^2 + d = cx$, $x^3 + cx + d = bx^2$, $x^3 = bx^2 + cx + d$, $x^3 + bx^2 = cx + d$, $x^3 + cx = bx^2 + d$, and $x^3 + d = bx^2 + cx$. Each of these equations is analyzed in detail by the author. He describes the conic sections necessary for their solution, proves that his solution is correct, and finally discusses the conditions under which there may be no solutions or more than one solution. We will discuss here al-Khayyāmī's solution of $x^3 + cx = d$ or, as he puts it, the case where "a cube and sides are equal to a number."

Unlike al-Karajī and al-Samaw'al, al-Khayyāmī was very careful to keep to the Greek idea of homogeneity. Namely, he conceives of the cubic equation as an equation between solids. Since x represents a side of a cube, c must represent an area (expressible as a square), so that cx is a solid, while d itself represents a solid. To construct the solution, al-Khayyāmī sets AB equal in length to a side of the square c, or $AB = \sqrt{c}$ (Figure 7.11). He then constructs BC perpendicular to AB so that $BC \cdot AB^2 = d$, or $BC = d/c$. Next, he extends AB in the direction of Z and constructs a parabola with vertex B, axis BZ, and parameter AB. In modern notation, this parabola has the equation $x^2 = \sqrt{c}y$. Similarly, he constructs a semicircle on the line BC. Its equation is

$$\left(x - \frac{d}{2c}\right)^2 + y^2 = \left(\frac{d}{2c}\right)^2 \qquad \text{or} \qquad x\left(\frac{d}{c} - x\right) = y^2.$$

The circle and the parabola intersect at a point D. It is the x coordinate of this point, here represented by the line segment BE, which provides the solution to the equation.

Al-Khayyāmī proved that this solution is correct by using the basic properties of the parabola and the circle. If $BE = DZ = x_0$ and $BZ = ED = y_0$, then first, $x_0^2 = \sqrt{c}y_0$ or $\frac{\sqrt{c}}{x_0} = \frac{x_0}{y_0}$, since D is on the parabola, and second, $x_0\left(\frac{d}{c} - x_0\right) = y_0^2$ or $\frac{x_0}{y_0} = \frac{y_0}{(d/c) - x_0}$, since D is on the circle. It follows that

$$\frac{c}{x_0^2} = \frac{x_0^2}{y_0^2} = \frac{y_0^2}{\left(\frac{d}{c} - x_0\right)^2} = \frac{y_0}{\frac{d}{c} - x_0}\frac{x_0}{y_0} = \frac{x_0}{\frac{d}{c} - x_0}$$

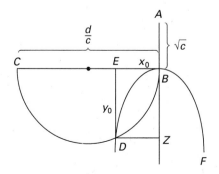

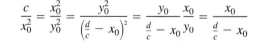

FIGURE 7.11
Al-Khayyāmī's construction for the solution of $x^3 + cx = d$.

and then that $x_0^3 = d - cx_0$, so x_0 is the desired solution. Al-Khayyāmī notes here, without any indication of a proof, that this class of equations always has a single solution. In other words, the parabola and circle always intersect in one point other than the origin. The origin, though, does not provide a solution to the problem. Al-Khayyāmī's remark reflects the modern statement that the equation $x^3 + cx = d$ always has exactly one positive solution.

Al-Khayyāmī treats each of his fourteen cases in the same manner. In those in which a positive solution does not always exist he gives a geometric condition for the existence. Namely, there are zero, one, or two solutions depending on whether the conic sections involved do not intersect or intersect at one or two points with positive coordinates. His one failure in this analysis is in the case of the equation $x^3 + cx = bx^2 + d$, where he does not discover the possibility of three solutions. In general, however, he does not relate the existence of one or two solutions to conditions on the coefficients. Even when he does so, in the case $x^3 + d = bx^2$, it is in only a limited way. In that equation he notes that if $\sqrt[3]{d} = b$, there can be no solution. If x were a solution, then $x^3 + b^3 = bx^2$, so $bx^2 > b^3$ and $x > b$. Since $x^3 < bx^2$, it is also true that $x < b$, a contradiction. Similarly, there can be no solution if $\sqrt[3]{d} > b$. The condition $\sqrt[3]{d} < b$, however, does not guarantee a solution. Al-Khayyāmī notes again that there may be zero, one, or two (positive) solutions, depending on how many times the conics for this problem (a parabola and a hyperbola) intersect.

7.2.6 Sharaf al-Dīn al-Ṭūsī and Cubic Equations

Al-Khayyāmī's methods were improved on by Sharaf al-Dīn al-Ṭūsī (d. 1213), a mathematician born in Tus, Persia. Like his predecessor, he began by classifying the cubic equations into several groups. His groups differed from those of al-Khayyāmī, because he was interested in determining conditions on the coefficients which determine the number of solutions. Therefore his first group consisted of those equations that could be reduced to quadratic ones, plus the equation $x^3 = d$. The second group consisted of the eight cubic equations that always have at least one (positive) solution. The third group consists of those types that may or may not have (positive) solutions, depending on the particular values of the coefficients. These include $x^3 + d = bx^2$, $x^3 + d = cx$, $x^3 + bx^2 + d = cx$, $x^3 + cx + d = bx^2$, and $x^3 + d = bx^2 + cx$.

For the second group of equations, his method of solution is the same as al-Khayyāmī's. He determines the solution by intersecting two appropriately chosen conic sections. Even here he goes beyond al-Khayyāmī by always giving a careful discussion as to why the two conics in fact intersect. It is in the third group, however, that he makes his most original contribution.

Consider Sharaf al-Dīn's analysis of $x^3 + d = bx^2$. He begins by putting the equation in the form $x^2(b - x) = d$. He then notes that the question of whether the equation has a solution depends on whether the "function" $f(x) = x^2(b - x)$ reaches the value d or not. In other words, he needs to consider the question of the maximum value of $x^2(b - x)$ (Figure 7.12). He then claims that the value $x_0 = \frac{2b}{3}$ in fact provides the maximum value for $f(x)$, that is, for any x between 0 and b, $x^2(b - x) \le \left(\frac{2b}{3}\right)^2\left(\frac{b}{3}\right) = \frac{4b^3}{27}$. It is curious that he does not say why he has chosen this particular value for x_0. Perhaps he guessed it by analogy to the fact already known to the Greeks (*Elements* VI-28) that $x = b/2$ provides the maximum value for the expression $x(b - x)$, or by a close study of problem 4 of

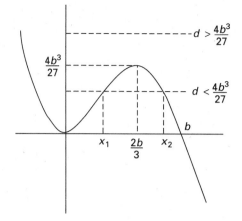

FIGURE 7.12
Modern graphic interpretation of Sharaf al-Dīn al-Ṭūsī's analysis of the cubic equation $x^3 + d = bx^2$.

Archimedes' *On the Sphere and Cylinder II,* a problem which involves a cubic equation of this type. It has also been suggested that he found this maximum by considering the conditions on x under which $f(x) - f(y) > 0$ for both $y < x$ and $y > x$, that is, in essence calculating a zero of the "derivative" of $f(x)$.[25] However he derived it, he does give a perfectly correct geometric proof that this value is in fact the maximum. He was able similarly to complete the same analysis for each of the five equations of his third group.

Given that $\frac{2b}{3}$ provides the maximum, Sharaf al-Dīn then notes that if the maximum value $\frac{4b^3}{27}$ is less than the given d, there can be no solutions to the equation. If $\frac{4b^3}{27}$ equals d, there is only one solution, $x = \frac{2b}{3}$. Finally if $\frac{4b^3}{27}$ is greater than d, there are two solutions, x_1 and x_2, where $0 < x_1 < \frac{2b}{3}$ and $\frac{2b}{3} < x_2 < b$. Knowing that solutions now exist, Sharaf al-Dīn proceeds to solve the equation by reducing it to a form already known, in this case the equation $x^3 + bx^2 = k$ where $k = \frac{4b^3}{27} - d$. He demonstrates that if a solution X to that equation is found geometrically by the intersection of two conic sections, then the larger solution x_2 to the given equation is $x_2 = X + \frac{2b}{3}$. To find the remaining root x_1, the author finds the positive solution Y to the quadratic equation $x^2 + (b-x_2)x = x_2(b-x_2)$ and then demonstrates that $x_1 = Y + b - x_2$ is the other positive root of the original equation. Hence the root of the new polynomial is related to that of the old by this change of variable formula. It is clear, therefore, that Sharaf al-Dīn had a solid understanding of the nature of cubic equations and the relationship of their roots and coefficients. Unlike his predecessors, he was able to see that the various types of cubic equations were related. Solutions of one type could be conveniently used in solving a second type. Note also that although he in effect used the discriminant of a cubic equation, here $\frac{4b^3}{27} - d$, to determine whether positive solutions existed, he was not able to use it algebraically to determine the numerical solutions.

On the other hand, Sharaf al-Dīn was interested in finding numerical solutions to these cubic equations. The example he gave in the case discussed was $x^3 + 14{,}837{,}904 = 465x^2$. By the above method, he first calculated that $\frac{4b^3}{27} = 14{,}895{,}500$ and $k = \frac{4b^3}{27} - d = 57{,}596$. It followed that there were two solutions x_1, x_2, with $0 < x_1 < 310$ and $310 < x_2 < 465$. To find x_2 he needed to solve $x^3 + 465x^2 = 57{,}596$. He found that 11 is a solution and therefore that $x_2 = \frac{2b}{3} + 11 = 310 + 11 = 321$. To find x_1 he needed to solve the quadratic equation $x^2 + 144x = 46,224$. The (positive) solution is an irrational number approximately equal to 154.73, a solution he found by a numerical method related to the Chinese method discussed in Chapter 6. The solution x_1 to the original equation is then 298.73.

The Islamic algebraists evidently made great strides in developing the algebra which they had received from the Babylonians. They incorporated into their work the notion of proof derived from the Greeks and thereby put their methods on a firm footing. It was unfortunate for the future development of mathematics that the work of most of the Islamic algebraists of the eleventh and twelfth centuries was not transmitted to Europe in time to play a role in further developments in that part of the world.

7.3 GEOMETRY

Islamic mathematicians dealt at an early stage with practical geometry, but later worked on various theoretical aspects of the subject, including the parallel postulate of Euclid, the concept of an irrational magnitude, and the exhaustion principle for determining volumes of solids.

7.3.1 Practical Geometry

The earliest extant Arabic geometry, like the earliest algebra, is due to al-Khwārizmī, and occurs as a separate section of his algebra text. A brief reading makes it clear that in his geometry, even more so than in his geometric demonstrations in algebra, al-Khwārizmī was not at all influenced by theoretical Greek mathematics. His text is an elementary compilation of rules for mensuration such as might be needed by surveyors. There are no axioms or proofs (except for a proof of the Pythagorean theorem for isosceles right triangles), and it appears very similar to the older Hebrew geometry, the *Mishnat ha-Middot*, discussed in Chapter 4.

We begin with al-Khwārizmī's rules for the circle:

In any circle, the product of its diameter, multiplied by three and one-seventh, will be equal to the circumference. This is the rule generally followed in practical life, though it is not quite exact. The geometricians have two other methods. One of them is, that you multiply the diameter by itself, then by ten, and hereafter take the root of the product; the root will be the circumference. The other method is used by the astronomers among them. It is this, that you multiply the diameter by sixty-two thousand eight hundred thirty-two and then divide the product by twenty thousand. The quotient is the circumference. Both methods come very nearly to the same effect. . . . The area of any circle will be found by multiplying half of the circumference by half of the diameter, since, in every polygon of equal sides and angles, . . . the area is found by multiplying half of the perimeter by half of the diameter of the middle circle that may be drawn through it. If you multiply the diameter of any circle by itself, and subtract from the product one-seventh and half of one-seventh of the same, then the remainder is equal to the area of the circle.[26]

The first of the approximations for π given here is the Archimedean one, 3 1/7, familiar to Heron as well as included in the *Mishnat ha-Middot*. In fact, al-Khwārizmī

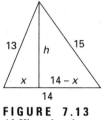

FIGURE 7.13
Al-Khwārizmī's
calculation of the
area of a triangle.

gives the same example as the *Mishnat ha-Middot* when he calculates the area of a circle of diameter 7 by squaring 7 to give 49, then subtracting off one-seventh and half of one-seventh, or 10 1/2, to get 38 1/2. The approximation of π by $\sqrt{10}$, attributed to geometricians, was used in India. Interestingly, however, it is less exact than the "not quite exact" value of 3 1/7. The earliest known occurrence of the third approximation, 3.1416, was also in India, in the work of Āryabhaṭa. The attribution of this value to astronomers is probably connected with its use in the Indian astronomical works which were translated into Arabic. One might assume that al-Khwārizmī's reference to geometricians for these latter approximations meant that he knew of some demonstration for them. He did not give any such in his own geometry work, however.

The formula for the area of a rhombus is given simply: "You may then compute the area [of the rhombus] . . . from both [diagonals]. . . . You multiply the one by the half of the other."[27] The example given here is the same as that of the Hebrew work, namely a rhombus with sides equal to five and diagonals equal to six and eight. The area is then 24.

The discussion of the volume of the frustum of a pyramid is also quite similar to that in the *Mishnat ha-Middot* and the example is again the same. Rather than give a formula directly, as in the Moscow Papyrus, al-Khwārizmī calculates the height to the top of the completed pyramid by using similar triangles, then subtracts the volume of the upper pyramid from that of the lower.

Al-Khwārizmī's work differs from the Hebrew work, however, in his computation of the area of a triangle with sides 13, 14, and 15. Rather than use Heron's formula, he drops a perpendicular to the side of length 14 from the opposite vertex, takes as his unknown the distance x of the base of that perpendicular from one end of that side, and uses the Pythagorean theorem twice to calculate the height h of the triangle (Figure 7.13). Thus, $13^2 - x^2 = 15^2 - (14 - x)^2$, $x = 5$, $h = 12$, and the area of the triangle is 84.

7.3.2 The Parallel Postulate

Al-Khwārizmī's geometry can certainly be classified as practical. Islamic authors were, however, greatly influenced by their reading of the Greek theoretical works and soon became interested in pure geometrical questions stemming from their knowledge of Euclid and other authors. One of the ideas that recurs in Islamic geometry is that of parallel lines and the provability of Euclid's fifth postulate. Even in Greek times, mathematicians were disturbed with this postulate. Many attempts were made to prove it from the others. So too in the Islamic world. One of the attempts to deal with this question was in the work of ibn al-Haytham entitled *Maqāla fī sharḥ muṣādarāt kitāb Uqlīdis* (*Commentary on the Premises of Euclid's Elements*) in which he attempted to reformulate Euclid's theory of parallels. He began by redefining the concept of parallel lines, deciding that Euclid's own definition of parallel lines as two lines which never meet was inadequate. His "more evident" definition included the assumption of the constructibility of such lines. Namely, he wrote that if a straight line moves so that one end always lies on a second straight line and so that it always remains perpendicular to that line, then the other end of the moving line will trace out a straight line parallel to the second line. In effect, this definition characterized parallel lines as lines always equidistant from one another and also introduced the concept of motion into geometry. Later commentators, including al-Khayyāmī, were unhappy with this. As they knew, Euclid had only used motion in generating new

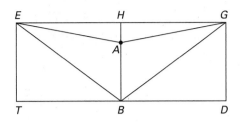

FIGURE 7.14
Ibn al-Haytham's proof of his lemma dealing with the parallel postulate.

objects from old, as a sphere is generated by rotating a semicircle. Nevertheless, ibn al-Haytham used this idea in his "proof" of the fifth postulate.

The crucial step in ibn al-Haytham's proof is the following

Lemma.   *If two straight lines are drawn at right angles to the two endpoints of a fixed straight line, then every perpendicular line dropped from the one line to the other is equal to the fixed line.*

In Figure 7.14, if *GA* and *DB* are perpendicular to *AB* and a perpendicular is dropped from *G* to the line *DB*, it must be proved that *GD* is equal to *AB*. Ibn al-Haytham's proof is by contradiction. He first assumes that *GD* > *AB*. He extends *GA* past *A* so that *AE* = *AG*. Similarly *BD* is extended past *B*. From the point *E* a perpendicular is dropped to the line *DB* extended, meeting it at *T*. Then the lines *GB* and *BE* are drawn. Triangles *EAB* and *GAB* are congruent by side-angle-side. Therefore ∠*GBA* = ∠*EBA*, so ∠*GBD* = ∠*EBT*, and *GB* = *BE*. It follows that the triangles *EBT* and *GBD* are congruent and therefore that *GD* = *ET*. Now, using his concept of motion, ibn al-Haytham imagines line *ET* moving along line *TD* and remaining always perpendicular to it. When *T* coincides with *B*, point *E* will be outside line *AB*, since *ET* > *AB*. We call *ET* at this particular time *HB*. Of course, when *ET* reaches *GD*, the two lines will coincide. It now follows from the definition of parallelism that line *GHE* is a straight line parallel to *DBT*. By construction, *GAE* is also a straight line, so there would be two different straight lines with the same endpoints and therefore two straight lines would enclose a space. This, of course, is impossible. A similar contradiction results from the assumption that *GD* < *AB*. Hence the proof is complete.

Because *GD* = *AB*, it follows easily that ∠*AGD*, like the three other angles of quadrilateral *ABDG*, is a right angle. It is then not difficult to demonstrate Euclid's postulate. Of course, what ibn al-Haytham did not realize was that his original definition of parallel lines already implicitly contained that postulate. In any case, his result makes clear the reciprocal relationship between the parallel postulate and the fact that the angle sum of any quadrilateral is four right angles.

Al-Khayyāmī was also interested in this question of parallelism. In his *Sharh mā ashkala min musādarāt kitāb Uqlīdis* (*Commentary on the Problematic Postulates of the Book of Euclid*), he began with the principle that two convergent straight lines intersect, and it is impossible for them to diverge in the direction of convergence. By convergent lines, he meant lines which approached one another. Given this postulate, al-Khayyāmī proceeded to prove a series of eight propositions culminating in Euclid's fifth postulate. He began by constructing a quadrilateral with two perpendiculars of equal length, *AC* and *BD*, at the two ends of a given line segment *AB* and then connecting the points *C* and *D* (Figure 7.15). He proceeded to prove that the two angles at *C* and *D* were both right angles

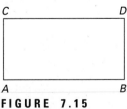

FIGURE 7.15
Al-Khayyāmī's quadrilateral: *AC* = *BD*, *AC* ⊥ *AB*, and *BD* ⊥ *AB*. Are the angles at *C* and *D* acute, obtuse, or right?

Biography   Naṣīr al-Dīn al-Ṭūsī (1201–1274)

Naṣīr al-Dīn, from Tus in Iran, completed his formal education in Nishapur, then a major center of learning, and soon gained a great reputation as a scholar (Figure 7.16). The thirteenth century, however, was a time of great turmoil in Islamic history. The only places of peace in Iran were the forts controlled directly by the Isma'īlī rulers. Fortunately, Naṣīr al-Dīn persuaded one of these rulers to allow him to work at such a fort. After the Mongol leader

Hūlāgū defeated the Isma'īlīs in 1256, al-Ṭūsī was able to transfer his allegiance. He served Hūlāgū as a scientific adviser and gained his approval to construct an observatory at Maragha, a town about 50 miles south of Tabriz. It was here that Naṣīr al-Dīn spent the rest of his life as head of a large group of astronomers and where he computed a new set of very accurate astronomical tables.

FIGURE 7.16
Naṣīr al-Dīn al-Ṭūsī.

by showing that the two other possibilities, that they were both acute or both obtuse, led to contradictions. If they were acute, *CD* would be longer than *AB* while if they were obtuse, *CD* would be shorter than *AB*. In each case he showed that the lines *AC* and *BD* would diverge or converge on both sides of *AB*, and this would contradict his original postulate. Al-Khayyāmī was now able to demonstrate Euclid's fifth postulate. In some sense, his treatment was better than ibn al-Haytham's because he explicitly formulated a new postulate to replace Euclid's rather than have the latter hidden in a new definition.

About a century after al-Khayyāmī, another mathematician, Naṣīr al-Dīn al-Ṭūsī (1201–1274) subjected the works of his predecessor to detailed criticism and then attempted his own proof of the fifth postulate in his book of about 1250 entitled *Al-risāla al-shāfiya 'an al-shakk fi-l-khuṭūt al-mutawāziya* (*Discussion Which Removes Doubt about Parallel Lines*). He considered the same quadrilateral as al-Khayyāmī and also tried to derive a contradiction from the hypotheses of the acute and obtuse angles. But in a manuscript probably written by his son Ṣadr al-Dīn in 1298, based on Naṣīr al-Dīn's later thoughts on the subject, there is a new argument based on another hypothesis, also equivalent to Euclid's, that if a line *GH* is perpendicular to *CD* at *H* and oblique to *AB* at *G*, then the perpendiculars drawn from *AB* to *CD* are greater than *GH* on the side on which *GH* makes an obtuse angle with *AB* and less on the other side (Figure 7.17). The importance

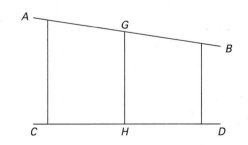

FIGURE 7.17
Naṣīr al-Dīn al-Ṭūsī's hypothesis on parallels and perpendiculars.

of this latter work is that it was published in Rome in 1594 and was studied by European geometers. In particular, it became the starting point for the work of Saccheri and ultimately for the discovery of non-Euclidean geometry.[28]

7.3.3 Incommensurables

Another geometric topic of interest to Islamic mathematicians was that of incommensurables. In fact, many Arabic commentaries were written on the topic of Euclid's *Elements* Book X. Recall that Islamic algebraists early on began to use irrational quantities in their work with equations, ignoring the Euclidean distinction between number and magnitude. There were, however, several commentators who made some attempt to reconcile this use and to put it into a theoretical framework consistent with the Euclidean work.

In the *Risāla fi'l-maqādir al-mushtaraka wa'l-mutabāyana* (*Treatise on Commensurable and Incommensurable Magnitudes*), written sometime around 1000, Abū 'Abdallāh al-Hasan ibn al-Baghdādī attempted to reconcile the operational rules already being used for irrational quantities with the main principles of the *Elements* and thus to prove that the contemporary methods of computation were valid. He was quite aware that these numerical methods of computation were simpler than the geometric modes of Euclid: "It is easier . . . to assume a number and to base oneself on it than to make a similar assumption concerning a magnitude."[29] Since he knew of Aristotle and Euclid's fundamental distinction between number and magnitude, he began by relating the two concepts by establishing a correspondence between numbers and line segments in what appears to be a modern way. Namely, given a unit magnitude a, each "whole number" n corresponds to an appropriate multiple na of the unit magnitude. Parts of this magnitude, such as $(m/n)(a)$, then correspond to parts of a number m/n. Ibn al-Baghdādī considered any magnitude expressible this way as a rational magnitude. He showed that these magnitudes relate to one another as numbers to numbers, as in *Elements* X-5. Magnitudes which are not "parts" are considered irrational magnitudes. In effect, ibn al-Baghdādī attempted to imbed the rational numbers into a number line. But he also wanted to connect irrational magnitudes to "numbers."

Ibn al-Baghdādī made the connection through the idea of a root. The root of a number n was the middle term x in the continuous proportion $n : x = x : 1$. Such a root may or may not exist. He then defined the root of a magnitude na similarly as the mean proportional between the unit magnitude a and the magnitude na. This quantity is always constructible by ruler and compass, so it necessarily exists. It may, of course, be either rational or irrational. Since "rational numbers" correspond to "rational magnitudes," and since the latter always have roots, which may or may not be rational, he could consider roots of the former to continue this correspondence. In particular, he noted that for magnitudes, roots and squares were of the same geometric type. In other words, the root of a magnitude expressed as a line segment was another line segment, just as the square of a line segment could be expressed as a line segment. Ibn al-Baghdādī, like some of his Islamic predecessors, hence moved away from the Greek insistence on homogeneity and toward the notion that all "quantities" can be expressed in the same way, essentially as "numbers."

Ibn al-Baghdādī concluded his treatment by dealing extensively with the various types of irrational magnitudes treated by Euclid in Book X. As a result of that discussion, he was able to prove a result on the "density" of irrational magnitudes, namely, that between any two rational magnitudes there exist infinitely many irrational magnitudes. For example, he considered the magnitudes represented by the consecutive numbers 2 and 3. The squares of these magnitudes are represented by 4 and 9. Between those magnitudes are magnitudes represented by the numbers 5, 6, 7, and 8. Their roots, $\sqrt{5}$, $\sqrt{6}$, $\sqrt{7}$, and $\sqrt{8}$, which ibn al-Baghdādī calls magnitudes of the first order of irrationality, lie between 2 and 3. Similarly, the squares of 4 and 9, namely 16 and 81, also represent magnitudes, as do the squares 25, 36, 49, and 64. Corresponding to the integers 17, 18, ... , 24 are magnitudes of the first order of irrationality $\sqrt{17}$, $\sqrt{18}$, ... , $\sqrt{24}$ as well as magnitudes of the second order of irrationality $\sqrt{\sqrt{17}}$, $\sqrt{\sqrt{18}}$, ... , $\sqrt{\sqrt{24}}$. The latter magnitudes lie between the original magnitudes 2 and 3. Ibn al-Baghdādī noted that one can continue in this way to find as many magnitudes as one wants, of various higher orders of irrationality, between the two original ones. Ibn al-Baghdādī's work thus demonstrated that Islamic authors both understood the arguments of their Greek predecessors in keeping separate the realms of magnitude and number and wanted to break the bonds imposed by this dichotomy so that they could justify their increasing use of "irrationals" in computation.

7.3.4 Volumes and the Method of Exhaustion

One final area of geometry we will discuss also demonstrates that Islamic authors understood the works of the Greeks and wanted to go beyond them, namely, the work in calculating volumes of solids through the method of exhaustion pioneered by Eudoxus and used so extensively by Archimedes. It turned out that although Islamic mathematicians read Archimedes' work *On the Sphere and the Cylinder*, they did not have available his work *On Conoids and Spheroids* in which Archimedes showed how to calculate the volume of the solid formed by revolving a parabola about its axis. Thus Thābit ibn Qurra found his own proof, which was quite long and complicated, and some 75 years later Abū Sahl al-Kūhī (tenth century), from the region south of the Caspian sea, simplified Thābit's method and solved some similar problems on volumes and analogous problems on the centers of gravity. Al-Kūhī in turn was criticized shortly afterwards by ibn al-Haytham for not solving the paraboloid problem in all generality, that is, for not considering the volume of the solid formed by revolving a segment of a parabola about a line perpendicular to its axis. It is this latter problem which ibn al-Haytham proceeded to solve himself.

In modern terminology, ibn al-Haytham proved that the volume of the solid formed by rotating the parabola $x = ky^2$ around the line $x = kb^2$ (which is perpendicular to the axis of the parabola) is 8/15 of the volume of the cylinder of radius kb^2 and height b. His formal argument was a typical exhaustion argument. Namely, he assumed that the desired volume was greater than 8/15 of that of the cylinder and derived a contradiction, then assumed that it was less and derived another contradiction. But the essence of ibn al-Haytham's argument involved "slicing" the cylinder into n disks, each of thickness $h = b/n$, the intersection of each with the paraboloid providing an approximation to the volume of a slice of the paraboloid (Figure 7.18). The ith disk in the paraboloid has radius

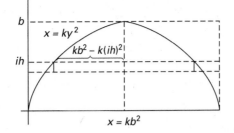

FIGURE 7.18
Revolving a segment of a parabola around a
line perpendicular to its axis.

$kb^2 - k(ih)^2$ and therefore has volume $\pi h(kh^2n^2 - ki^2h^2)^2 = \pi k^2h^5(n^2 - i^2)^2$. The total
volume of the paraboloid is therefore approximated by

$$\pi k^2h^5\sum_{i=1}^{n-1}(n^2 - i^2)^2 = \pi k^2h^5\sum_{i=1}^{n-1}(n^4 - 2n^2i^2 + i^4).$$

But ibn al-Haytham already knew formulas for the sums of integral squares and integral
fourth powers. Using these, he could calculate that

$$\sum_{i=1}^{n-1}(n^4 - 2n^2i^2 + i^4) = \frac{8}{15}(n - 1)n^4 + \frac{1}{30}n^4 - \frac{1}{30}n = \frac{8}{15}n \cdot n^4 - \frac{1}{2}n^4 - \frac{1}{30}n$$

and therefore that

$$\frac{8}{15}(n - 1)n^4 < \sum_{i=1}^{n-1}(n^2 - i^2)^2 < \frac{8}{15}n \cdot n^4.$$

But the volume of a typical slice of the circumscribing cylinder is $\pi h(kb^2)^2 = \pi k^2h^5n^4$,
and therefore the total volume of the cylinder is $\pi k^2h^5n \cdot n^4$, while the volume of the
cylinder less its "top slice" is $\pi k^2h^5(n - 1)n^4$. Therefore the inequality shows that
the volume of the paraboloid is bounded between 8/15 of the cylinder less its top slice
and 8/15 of the entire cylinder. Since the top slice can be made as small as desired by
taking n sufficiently large, it follows that the paraboloid is exactly 8/15 of the cylinder
as asserted.

7.4 TRIGONOMETRY

An Indian *Siddhānta* was brought to Baghdad late in the eighth century and translated
into Arabic. Thus Islamic scholars were made aware of the trigonometrical knowledge of
the Hindus, which had earlier been adapted from the Greek version of Hipparchus. They
were also soon aware of Ptolemy's trigonometry as detailed in his *Almagest* when that
work was translated into Arabic as well. As in other areas of mathematics, the Islamic
mathematicians absorbed what they found from other cultures and gradually infused the
subject with new ideas.

Biography   Al-Bīrūnī (973–1055)

Al-Bīrūnī was born in Khwarizm, near a town now named Biruni in Uzbekistan, and began scientific studies early in life under the guidance of Abū Naṣr Manṣūr ibn 'Irāq, a prominent astronomer from the region (Figure 7.19). Political strife in his homeland compelled him to flee in 995, but two years later he was back in Kāth, the principal city of Khwarizm, to observe a lunar eclipse. He had previously arranged that Abu'l-Wafā' would observe the same eclipse in Baghdad, so that the time difference of the two occurrences would enable him to calculate the difference in longitude of the two sites. In 1017, Khwarizm was conquered by Sultan Maḥmūd of Ghazna, in Afghanistan, who soon ruled an extensive empire which included parts of northern India. Al-Bīrūnī was taken to the sultan's court, from where he traveled to India and where he wrote a major work on all aspects of Indian culture, including such varied topics as the caste system, Hindu religious philosophy, the rules of chess, notions of time, and calendrical procedures.

As in both Greece and India, trigonometry in Islam was intimately tied to astronomy, so in general mathematical texts on trigonometry were written as chapters of more extensive astronomical works. The mathematicians were particularly interested in using trigonometry to solve spherical triangles because Islamic law required that Moslems face the direction of Mecca when they prayed. To determine the appropriate direction at one's own location required an extensive knowledge of the solution of such triangles on the sphere of the earth. The solution of both plane and spherical triangles was also important in the determination of the correct time for prayers. These times were generally defined in relation to the onset of dawn and the end of twilight as well as the length of daylight and the altitude of the sun on a given day, notions which again required spherical trigonometry to determine accurately.

7.4.1 The Trigonometric Functions

FIGURE 7.19
Al-Bīrūnī on a Syrian stamp.

Recall that Ptolemy used only one trigonometric function, the chord, in his trigonometric work, while the Hindus modified that into the more convenient sine. Early in Islamic trigonometry, both the chord and the sine were used concurrently, but eventually the sine won out. (The Islamic sine of an arc, like that of the Hindus, was the length of a particular line in a circle of given radius R.) It is not entirely clear who introduced the other functions, but we do know that Abū 'Abdallāh Muḥammad ibn Jābir al-Battānī (c. 855–929) used the "sine of the complement to 90°" (our cosine) in his astronomical work designed to be an improvement on the *Almagest*. Because he did not use negative numbers, he defined the cosine only for arcs up to 90°. For arcs between 90° and 180°, he used the versine, defined as versin $\alpha = R + R \sin(\alpha - 90°)$. Since al-Battānī did not make use of the tangent, however, his formulas were no less clumsy than those of Ptolemy.

The tangent, cotangent, secant, and cosecant functions made their appearance in Islamic works in the ninth century, perhaps earliest in the work of Aḥmad ibn 'Abdallāh al-Marwazī Habas al-Ḥāsib (c. 770–870), although the tangent function had already been used in China in the eighth century. We consider here, however, the discussion of these

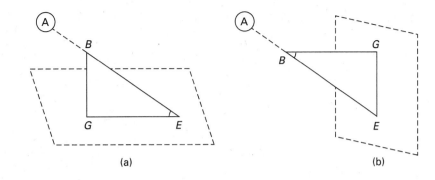

FIGURE 7.20
Al-Bīrūnī's definition of tangent, cotangent, secant, and cosecant. In (a), *GE* is the cotangent of angle *E* and *EB* is the cosecant. In (b), *GE* is the tangent of angle *B* and *BE* is the secant.

functions by Abu l-Rāyhan Muhammad ibn Ahmad al-Bīrūnī (973–1055) in his *Exhaustive Treatise on Shadows*. "An example of the direct shadow [cotangent] is: Let *A* be the body of the sun and *BG* the gnomon perpendicular to *EG*, which is parallel to the horizon plane, and *ABE* the sun's ray passing through the head of the gnomon *BG* (Figure 7.20a). . . . *EG* is that which is called the direct shadow such that its base is *G* and its end *E*. And *EB*, the line joining the two ends of the shadow and the gnomon, is the hypotenuse of the shadow [cosecant]."[30] The tangent and secant are defined similarly by using a gnomon parallel to the horizon plane. In Figure 7.20b, *GE* is called the "reversed shadow" (tangent) while *BE* is called the "hypotenuse of the reversed shadow" (secant).

Al-Bīrūnī demonstrated the various relationships among the trigonometric functions. For example, he showed that the "ratio of the gnomon to the hypotenuse of the shadow is as the ratio of the sine of the altitude to the total sine."[31] By the "total sine," al-Bīrūnī meant the radius *R* of the circle on which one is measuring the arcs. The formula can then be translated as

$$\frac{g}{g \csc \alpha} = \frac{R \sin \alpha}{R}$$

(where *g* is the length of the gnomon) or as

$$\csc \alpha = \frac{1}{\sin \alpha}.$$

Again, he noted that "if we are given the shadow at a certain time, and we want to find the altitude of the sun for that time, we multiply the shadow by its equal and the gnomon by its equal and we take [the square root] of the sum, and it will be the cosecant. Then we divide by it the product of the gnomon by the total sine, and there comes out the sine of the altitude. We find its corresponding arc in the sine table and there comes out the altitude of the sun at the time of that shadow."[32] In modern notation, al-Bīrūnī has used the relationship

$$\sqrt{g^2 \cot^2 \alpha + g^2} = g \csc \alpha \qquad (\text{or } \cot^2 \alpha + 1 = \csc^2 \alpha)$$

and then the previous formula in the form

$$\frac{gR}{g \csc \alpha} = R \sin \alpha$$

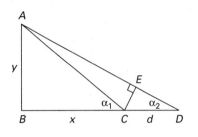

FIGURE 7.21a
Al-Qabīsī's method for determining height and distance by way of two angle determinations.

FIGURE 7.21b
Astrolabe on an Iranian stamp.

to determine the sine function based on the particular value of the radius R used. He then consulted his sine table in reverse to determine α. Al-Bīrūnī similarly gave rules equivalent to $\tan^2 \alpha + 1 = \sec^2 \alpha$ and $\tan \alpha = \frac{\sin \alpha}{\cos \alpha}$ and presented a table for the tangent and cotangent in which he used the relationship $\cot \alpha = \tan(90° - \alpha)$.

It is perhaps surprising that with the wealth of trigonometrical knowledge collected in his text al-Bīrūnī only used it for dealing with astronomical problems. For determining terrestrial heights and distances he described nontrigonometrical methods. For example, to determine the height of a minaret, where the base is accessible, he suggested that, "if surveyed at a time when the altitude of the sun equals an eighth of a revolution [45°], there will be between the end of the shadow and the foot of the vertical a distance equal to [its height]."[33] If the base is not accessible, however, al-Bīrūnī described a procedure similar to the Chinese and Indian procedures discussed in Chapter 6. Unlike his Indian and Chinese predecessors, however, he gave a description in his text of his reasoning, using the idea of similar triangles.

Three quarters of a century before the appearance of *On Shadows*, however, al-Qabīsī had described a trigonometric method, using only the sine, for determining the height and distance of an inaccessible object. One sights the summit A from two locations C, D, and determines, using an astrolabe (an angle-measuring instrument usually used for astronomical purposes), the angles $\alpha_1 = \angle ACB$ and $\alpha_2 = \angle ADB$ (Figure 7.21). If $CD = d$, then the height $y = AB$ and the distance $x = BC$ are given by

$$y = \frac{d \sin \alpha_2}{\sin(90 - \alpha_2) - \frac{\sin(90 - \alpha_1)\sin \alpha_2}{\sin \alpha_1}}, \qquad x = \frac{y \sin(90 - \alpha_1)}{\sin \alpha_1}.$$

7.4.2 Spherical Trigonometry

Although there were a few instances of the use of trigonometry on earth, the major use of the trigonometric functions was to solve spherical triangles arising from astronomical problems. Islamic mathematicians were able to derive simpler methods than those of Ptolemy for dealing with these problems. It appears that the basic results were discovered independently by two contemporaries of al-Bīrūnī, Abū Naṣr Manṣūr ibn 'Irāq (d. 1030), one of al-Bīrūnī's teachers, and Muhammad Abu'l-Wafā' al-Būzjānī (940–997), an important astronomer of Baghdad. The first result is what has become known as "the rule of four quantities":

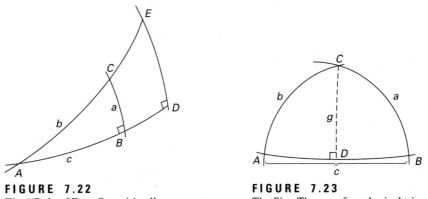

FIGURE 7.22
The "Rule of Four Quantities."

FIGURE 7.23
The Sine Theorem for spherical triangles.

Theorem. *If ABC and ADE are two spherical triangles with right angles at B, D, respectively and a common acute angle at A, then sin BC : sin CA = sin DE : sin EA (Figure 7.22).*

An immediate corollary of this theorem is one of the special cases of Menelaus' theorem discussed in Chapter 4: If *ABC* is a right spherical triangle with right angle at *B*, then sin *A* = sin *a*/sin *b*. To prove this, extend the hypotenuse *AC* to a point *E* such that *AE* is a quadrant of a great circle, drop a perpendicular to *D* on *AB* extended, and apply the theorem. This result was in essence used by Ptolemy in many of his calculations. Abu'l-Wafā also gave proofs of other special cases of the Menelaus theorem, including the results cos *a*/cos *b* = 1/cos *c* and sin *c*/tan *a* = 1/tan *A*. In addition, both Abū Naṣr and Abu'l-Wafā gave a proof of the sine theorem for arbitrary spherical triangles.

Theorem. *In any spherical triangle ABC, $\frac{\sin a}{\sin A} = \frac{\sin b}{\sin B} = \frac{\sin c}{\sin C}$.*

This result is proved by dropping a perpendicular from *C* to *AB* and considering the two right spherical triangles so produced (Figure 7.23).

Given the sine theorem, al-Bīrūnī was able to show how to determine the *qibla*, the direction of Mecca relative to one's own location in which a Moslem must face during prayer. One of al-Bīrūnī's solutions to this problem is outlined here.[34] Assume that *M* is the position of Mecca and that *P* is one's current location (Figure 7.24). Let arc *AB* represent the equator and *T* the north pole, and draw meridians from *T* through *P* and *M* respectively. The *qibla* is then ∠*TPM* on the earth's sphere. Assuming the latitudes α, β and the longitudes γ, δ of *P* and *M* respectively are known, then arcs *TP* and *TM* are known (90° − α, 90° − β respectively), and also ∠*PTM*(= δ − γ) is known. Unfortunately, the sine theorem itself is not sufficient to solve the triangle *PTM* since no single angle and side opposite are known. Al-Bīrūnī, however, used the theorem repeatedly on a series of triangles.

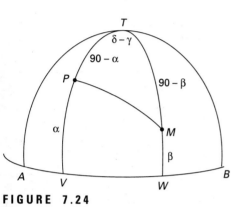

FIGURE 7.24
The problem of the *qibla*.

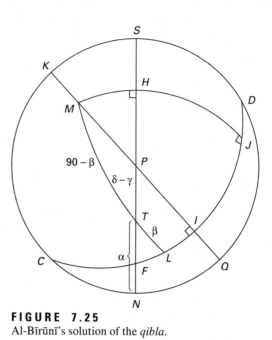

FIGURE 7.25
Al-Bīrūnī's solution of the *qibla*.

We follow al-Bīrūnī's method, taking as example P to be Jerusalem (latitude 31°47′ N, longitude 35°13′ E). Mecca itself has latitude 21°45′ N, longitude 39°49′ E. Let the circle $KSQN$ represent the horizon circle of the point P (or its local zenith) as viewed from above and M the zenith of Mecca (Figure 7.25). If S is the south point of the horizon (P is northwest of M), N the north point, and the arcs PMK and NPS are drawn, the arc NK represents the *qibla*. Let circle CFD represent the horizon circle of Mecca and circle MHJ the horizon circle of F and draw circle MTL through the north celestial pole T. The data of this problem give $TN = \alpha = 31°47′$, $TL = \beta = 21°25′$, $MT = 90° - \beta = 68°35′$, and $\angle MTH = \delta - \gamma = 4°36′$. Since MT, $\angle MTH$ and $\angle THM = 90°$ are known, the sine theorem for triangle MTH shows that

$$\sin MH = \frac{\sin MT \sin \angle MTH}{\sin \angle THM} = .07466.$$

Therefore $MH = 4°17′$ and $HJ = 90° - MH = 85°43′$. Because $\angle TFL = HJ$ and TL and $\angle TLF = 90°$ are known, the sine theorem applied to triangle TFL determines

$$\sin TF = \frac{\sin TL \sin \angle TLF}{\sin \angle TFL} = .36617,$$

so $TF = 21°29′$, and therefore $FN = \alpha - TF = 10°18′$ and $FP = 90° - FN = 79°42′$. Next, apply the rule of four quantities to triangles FPI and FHJ. Again, because FP, $FH = 90°$, and HJ are known, $\sin PI$ is determined as

$$\sin PI = \frac{\sin PF \sin HJ}{\sin FH} = .98114,$$

so $PI = 78°51'$ and $IQ = 90° - PI = 11°9'$. But C is the pole of circle $KMPIQ$. Therefore $\angle FCN(= IQ)$ is also known. Finally, apply the sine theorem to triangle CFN. Again, three quantities are known, namely $\angle FCN$, $\angle CFN(= \angle TFL)$, and FN, so the fourth quantity, CN, is determined. Thus,

$$\sin CN = \frac{\sin \angle CFN \sin FN}{\sin \angle FCN} = .92204,$$

$CN = 67°14'$ and the *qibla* $NK = CK + CN = 90° + CN = 157°14'$.

The sine theorem and the theorem of the four quantities, along with their various corollaries and the tables of the various functions, enabled Islamic mathematicians to solve the spherical triangles important for astronomical and associated religious purposes. It was not until the thirteenth century, however, that a systematic and comprehensive work on spherical and plane trigonometry, independent of astronomy, was written, the *Kitāb al shakl al-qiṭā'* (*Treatise on the Transversal Figure,* usually known as the *Treatise on the Complete Quadrilateral*), by Naṣir al-Dīn al-Ṭūsī. In this work is found for the first time the theorem of sines for plane triangles, to the effect that in any such triangle ABC, always $\frac{a}{\sin A} = \frac{b}{\sin B} = \frac{c}{\sin C}$. Naṣir al-Dīn then used this result systematically to solve plane triangles. For example, if two angles A, B, and one side a are known, the third angle is calculated by $C = 180° - A - B$ and side b is computed from the ratio $\frac{b}{\sin B} = \frac{a}{\sin A}$. Side c is calculated similarly.

Naṣir al-Dīn's work also contained the rules for solving spherical triangles. Not only did he include all four of the special cases of Menelaus' theorem for spherical right triangles mentioned in Chapter 4, but he also included two additional ones: $\cos c = \cot A \cot B$ and $\cos A = \cos a \sin B$. Using these six results systematically, he showed how to solve right spherical triangles as well as nonright triangles. In particular, he provided the first discussion of the solution of the case in which the three angles are known.

7.4.3 Trigonometric Tables

To deal with the problems of astronomy and geography, not only were formulas for solving triangles necessary, but also highly accurate tables were needed. These were gradually developed. Al-Bīrūnī, for example, calculated a table of sines at intervals of $15'$ accurate to four sexagesimal places. As in the calculations of Ptolemy, the accuracy of such a table depended primarily on the accuracy of the calculation of $\sin 1°$. Various methods were used to do this calculation, the most impressive being the method of al-Kāshī in the early fifteenth century. He started with a version of the triple angle formula $\sin 3\theta = 3 \sin \theta - 4 \sin^3 \theta$. Putting $\theta = 1°$ gives a cubic equation for $x = \sin 1°$: $\sin 3° = 3x - 4x^3$. Since al-Kāshī calculated his sine table based on a circle of radius 60, he needed to calculate $y = 60 \sin 1° = 60x$. His equation was therefore $60 \sin 3° = 3y - \frac{4y^3}{60^2}$, or

$$y = \frac{900(60 \sin 3°) + y^3}{45 \cdot 60}.$$

Recall that $\sin 3°$ can be calculated to whatever accuracy needed by use of the difference and half-angle formulas. Al-Kāshī in fact used as his value for $60 \sin 3°$ the sexagesimal 3:8,24,33,59,34,28,15. His equation, written in sexagesimal notation, was therefore

FIGURE 7.26
Ulūgh Beg.

$$y = \frac{47,6;8,29,53,37,3,45 + y^3}{45,0}..$$

He proceeded to solve this equation by an iterative procedure, given that he knew that the solution was a value close to 1. Writing the equation symbolically as $y = \frac{q + y^3}{p}$, and assuming that the solution is given as $y = a + b + c + \ldots$, where the various letters represent successive sexagesimal places, one begins with the first approximation $y_1 = q/p \approx a(= 1)$. To find the next approximation $y_2 = a + b$, solve for b by setting

$$y_2 = \frac{q + y_1^3}{p} \quad \text{or} \quad a + b = \frac{q + a^3}{p}.$$

Then $b \approx \frac{q - ap + a^3}{p}(= 2)$. Similarly, if $y_3 = a + b + c$, set

$$y_3 = \frac{q + y_2^3}{p}$$

and find

$$c \approx \frac{q - (a + b)p + (a + b)^3}{p}(= 49).$$

Al-Kāshī did not justify this iterative approximation procedure, but he evidently knew it converged more rapidly than the older solution procedures for cubic equations used by his predecessors. In this case, he calculated $y = 1;2,49,43,11,14,44,16,26,17$, a result equivalent to a decimal value for sin 1° of 0.017452406437283571, quite a feat for the days before calculators. Al-Kāshī's patron, Ulūgh Beg, himself an astronomer who ruled a domain in central Asia from his capital of Samarkand, used this work to calculate sine and tangent tables for every minute of arc to five sexagesimal places, a total of 5400 entries in each table (Figure 7.26)!

By the time of al-Kāshī, however, Islamic scientific civilization was in a state of decline. There were few other scientists of consequence in the years following. Even before the fifteenth century, though, mathematical activity had resumed in Europe. A central factor of this revival was the work of the translators of the twelfth century who made available to Europeans a portion of the Islamic mathematical corpus. Unfortunately for the future development of mathematics, some of the most significant work of Islamic mathematicians, including that of al-Bīrūnī, al-Samaw'al, al-Khayyāmī, and Sharaf al-Dīn al-Tūsī, and most of that of ibn al-Haytham, never reached the translators. So rather than building on these Islamic contributions, European mathematicians were forced to rediscover much of the same material centuries later.

Exercises

Problems from Decimal Arithmetic

1. Multiply 8023 by 4638 using the method of al-Uqlīdīsī.

2. Use al-Uqlīdīsī's method to increase 135 by one-tenth of itself 5 times, and check that the resulting value is 217.41885, as given in the text.

Problems from Algebra

3. Al-Khwārizmī gives the following rule for his 6th case, $bx + c = x^2$: Halve the number of roots. Multiply this by itself. Add this square to the number. Extract the square root. Add this to the half of the number of roots.

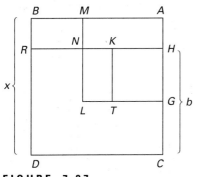

FIGURE 7.27
Al-Khwārizmī's justification for the solution rule for
$bx + c = x^2$.

That is the solution. Translate this rule into a formula. Give a geometric argument for its validity using Figure 7.27 where $x = AB = BD$, $b = HC$, c is represented by rectangle *ABRH*, and G is the midpoint of *HC*.

4. Solve the following problems due to Al-Khwārizmī by applying the appropriate formula:

    a. $\left(\frac{1}{3}x + 1\right)\left(\frac{1}{4}x + 1\right) = 20$

    b. $x^2 + (10 - x)^2 = 58$

    c. $x/3 \cdot x/4 = x + 24$

5. Solve $\frac{1}{2}x^2 + 5x = 28$ by multiplying first by 2 and then using Al-Khwārizmī's procedure. Similarly, solve $2x^2 + 10 = 48$ by first dividing by 2.

6. Solve the problem of Al-Khwārizmī: I have divided ten into two parts, and have divided the first by the second, and the second by the first and the sum of the quotients is 2 1/6. Find the parts.

7. Solve the following problems of Abū Kāmil:

    a. Suppose 10 is divided into two parts and the product of one part by itself equals the product of the other part by the square root of 10. Find the parts.

    b. Suppose 10 is divided into two parts, each one of which is divided by the other, and the sum of the quotients equals the square root of 5. Find the parts. (Abū Kāmil solves this in two ways, once directly for *x*, and a second time by first setting $y = \frac{10 - x}{x}$.)

8. Solve the following problems of Abū Kāmil:

    a. $[x - (2\sqrt{x} + 10)]^2 = 8x$ (First substitute $x = y^2$.)

    b. $\left(x + \sqrt{\frac{1}{2}x}\right)^2 = 4x$ (Abū Kāmil does this three different ways; he first solves directly for *x*, next substitutes $x = y^2$, and finally substitutes $x = 2y^2$.)

    c. $(x + 7)\sqrt{3x} = 10x$ (Abū Kāmil gives two solutions.)

9. Solve the following problem in three variables due to Abū Kāmil: $x < y < z, x^2 + y^2 = z^2, xz = y^2, xy = 10$. (Begin by setting $y = 10/x$, $z = 100/x^3$ and substituting in the first equation.)

10. Complete al-Samaw'al's procedure of dividing $20x^2 + 30x$ by $6x^2 + 12$ to get the result stated in the text. Prove that the coefficients of the quotient satisfy the rule $a_{n+2} = -2a_n$ where a_n is the coefficient of $1/x^n$.

11. Use al-Samaw'al's procedure to divide $20x^6 + 2x^5 + 58x^4 + 75x^3 + 125x^2 + 196x + 94 + 40\frac{1}{x} + 50\frac{1}{x^2} + 90\frac{1}{x^3} + 20\frac{1}{x^4}$ by $2x^3 + 5x + 5 + 10\frac{1}{x}$. (His answer is $10x^3 + x^2 + 11x + 10 + 8\frac{1}{x^2} + 2\frac{1}{x^3}$.)

12. Give a complete inductive proof of the result

$$\sum_{i=1}^{n} i^3 = \left(\sum_{i=1}^{n} i\right)^2$$

and compare this with al-Karajī's proof.

13. Use ibn al-Haytham's procedure to derive the formula for the sum of the fifth powers of the integers:

$$1^5 + 2^5 + \cdots + n^5 = \frac{1}{6}n^6 + \frac{1}{2}n^5 + \frac{5}{12}n^4 - \frac{1}{12}n^2$$

14. Show, using the formulas for sums of fourth powers and squares, that

$$\sum_{i=1}^{n-1} (n^4 - 2n^2i^2 + i^4) = \frac{8}{15}(n - 1)n^4 + \frac{1}{30}n^4 - \frac{1}{30}n$$

$$= \frac{8}{15}n \cdot n^4 - \frac{1}{2}n^4 - \frac{1}{30}n.$$

15. Show that one can solve $x^3 + d = cx$ by intersecting the hyperbola $y^2 - x^2 + \frac{d}{c}x = 0$ with the parabola $x^2 = \sqrt{c}y$. Sketch the two conics. Find sets of values for *c* and *d* for which these conics do not intersect, intersect once, and intersect twice (in the first quadrant).

16. Show that one can solve $x^3 + d = bx^2$ by intersecting the hyperbola $xy = d$ and the parabola $y^2 + dx - db = 0$. Assuming that $\sqrt[3]{d} < b$, determine the conditions on *b* and *d* which give zero, one, or two intersections of these two conics. Compare your answer with Sharaf al-Dīn al-Ṭūsī's analysis of the same problem.

17. Show that $x^3 + cx = bx^2 + d$ is the only one of al-Khayyāmī's cubics which could have three positive solutions. Under what conditions do these three positive solutions exist?

18. Show using calculus that $x_0 = \frac{2b}{3}$ does maximize the function $x^2(b - x)$. Then use calculus to analyze the graph of $y = x^3 - bx^2 + d$ and confirm Sharaf al-Dīn's conclusion on the number of positive solutions to $x^3 + d = bx^2$.

19. Show, as did Sharaf al-Dīn al-Ṭūsī, that if x_2 is the larger positive root to the cubic equation $x^3 + d = bx^2$, and if Y is the positive solution to the equation $x^2 + (b - x_2)x = x_2(b - x_2)$, then $x_1 = Y + b - x_2$ is the smaller positive root of the original cubic.

20. Analyze the possibilities of positive solutions to $x^3 + d = cx$ by first showing that the maximum of the function $x(c - x^2)$ occurs at $x_0 = \sqrt{c}/3$. Use calculus to consider the graph of $y = x^3 - cx + d$ and determine the conditions on the coefficients giving it zero, one, or two positive solutions.

Problems from Geometry

21. Demonstrate Euclid's postulate 5 (the parallel postulate) by assuming the truth of ibn al-Haytham's lemma: If two straight lines are drawn at right angles to the two endpoints of a fixed straight line, then every perpendicular line dropped from the one line to the other is equal to the fixed line.

22. Demonstrate the following equalities, typical examples of material on irrationals occurring in works of Islamic commentators on *Elements* X:

a. $\sqrt{6 \pm \sqrt{20}} = \sqrt{5} \pm 1$.

b. $\sqrt{\sqrt{8} \pm \sqrt{6}} = \sqrt[4]{4\frac{1}{2}} \pm \sqrt[4]{\frac{1}{2}}$.

c. $\sqrt[4]{12} \pm \sqrt[4]{3} = \sqrt{\sqrt{27} \pm \sqrt{24}} = \sqrt{51 \pm \sqrt{2592}}$.

23. Abū Sahl al-Kūhī knew from his own work on centers of gravity and the work of his predecessors that the center of gravity divides the axis of certain plane and solid figures in the following ratios:

Triangle—$\frac{1}{3}$	Tetrahedron—$\frac{1}{4}$
Segment of a parabola—$\frac{2}{5}$	Paraboloid of revolution—$\frac{2}{6}$
	Hemisphere—$\frac{3}{8}$

Noting the pattern, he guessed that the corresponding value for a semicircle was 3/7. Show that al-Kūhī's first five results are correct, but that his guess for the semicir-

cle implies that $\pi = 3\ 1/9$. (Al-Kūhī realized that this value contradicted Archimedes' bounds of 3 10/71 and 3 1/7, but concluded that there was an error in the transmission of Archimedes' work.)

Problems from Trigonometry

24. Derive al-Qabīsī's trigonometric formula for determining the height y of an inaccessible object.

25. Prove the general sine theorem for spherical triangles: In triangle ABC, $\frac{\sin a}{\sin A} = \frac{\sin b}{\sin B} = \frac{\sin c}{\sin C}$. Hint: Drop a perpendicular from CD to AB and apply the right triangle result that $\sin A = \frac{\sin g}{\sin b}$ (see Figure 7.23).

26. Use al-Bīrūnī's procedure to determine the *qibla* for Rome (latitude 41°53′ N, longitude 12°30′ E).

27. Write a computer program using al-Bīrūnī's algorithm which will compute the *qibla* for any location whose latitude and longitude are specified. Use the program to compute the *qibla* for your location.

28. Al-Bīrūnī devised a method for determining the radius r of the earth by sighting the horizon from the top of a mountain of known height h. That is, al-Bīrūnī assumed that one could measure α, the angle of depression from the horizontal at which one sights the apparent horizon (Figure 7.28). Show that r is determined by the formula

$$r = \frac{h \cos \alpha}{1 - \cos \alpha}.$$

Al-Bīrūnī performed this measurement in a particular case, determining that $\alpha = 0°34′$ as measured from the summit of a mountain of height 652;3,18 cubits. Calculate the radius of the earth in cubits. Assuming that a

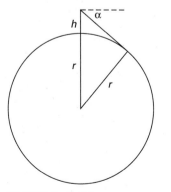

FIGURE 7.28
Al-Bīrūnī's method for calculating the earth's radius.

cubit equals 18″, convert your answer to miles and compare to a modern value. Comment on the efficacy of al-Bīrūnī's procedure.

29. Verify that if

$$y = \frac{900(60 \sin 3°) + y^3}{45 \cdot 60}$$

and if $\sin 3° = 3;8,24,33,59,34,28,15$, then

$$y = \frac{47,6;8,29,53,37,3,45 + y^3}{45,0}.$$

30. Calculate the first four sexagesimal places of the approximation to y (of problem 29) following the method indicated in the text. Your calculation should show why the iteration method works.

FOR DISCUSSION . . .

31. Why did it take many centuries after its introduction for the decimal place-value system to become the system of numeration universally used in the Islamic world?

32. Outline a lesson teaching the quadratic formula using geometric arguments in the style of al-Khwārizmī.

33. Compare and contrast the geometric proofs of the quadratic formulas of al-Khwārizmī and Thābit ibn Qurra. Which method would be easier to explain?

34. Design a lesson for a calculus class using the methods of Sharaf al-Dīn al-Ṭūsī which will demonstrate the number of solutions to various cubic equations.

35. Design a lesson for a trigonometry class showing the application of the rules for solving spherical triangles to various interesting problems.

36. Should spherical trigonometry be taught in schools today? Give arguments for and against, but first determine in which countries this subject is in fact taught.

37. Given ibn al-Haytham's "integration" to determine the volume of a paraboloid of revolution and his general rule for determining the sums of kth powers of integers, why did Islamic mathematicians not discover that the area under the curve $y = x^n$ was $\frac{x^{n+1}}{n+1}$ for an arbitrary positive integer n? What needed to happen in Islamic civilization for Islamic mathematicians to discover calculus?

References and Notes

The best general work on Islamic mathematics is Adolf P. Youschkevitch, *Les Mathématiques Arabes (VIIIᵉ–XVᵉ siècles)* (Paris: J. Vrin, 1976). This is one part of a more general work on medieval mathematics (in Russian) and was translated into French by M. Cazenave and K. Jaouiche. Another excellent work in English which treats various important ideas of Islamic mathematics is J. Lennart Berggren, *Episodes in the Mathematics of Medieval Islam* (New York: Springer Verlag, 1986). This work, although not a history of Islamic mathematics as such, treats certain important mathematical ideas considered by Islamic mathematicians at a level accessible to university mathematics students. A general survey of Islamic work in the sciences, which includes material on mathematics, is Edward S. Kennedy, "The Arabic Heritage in the Exact Sciences," *Al-Abhath* 23 (1970), 327–344. This article, and many other useful articles on Islamic science, can also be found in D. A. King and M. H. Kennedy, eds., *Studies in the Islamic Exact Sciences* (Beirut: American University of Beirut Press, 1983).

1. Jens Høyrup, "The Formation of 'Islamic Mathematics': Sources and Conditions," *Science in Context* 1 (1987), 281–329, pp. 306–307.

2. Ibid. This article contains a detailed development of these ideas. The author contends that there was an Islamic "miracle," comparable to the Greek one, involving the integration of mathematical theory and practice, which was also crucial for the creation of modern science.

3. A. S. Saidan, *The Arithmetic of Al-Uqlīdisī* (Boston: Reidel, 1978), p. 35. This work provides a complete translation of al-Uqlīdisī's work, along with a commentary.

4. Ibid., p. 110.

5. Al-Samaw'al is quoted in Roshdi Rashed, "L'extraction de la racine n'ième et l'invention des fractions décimales (XIᵉ–XIIᵉ siècles)," *Archive for History of Exact Sciences* 18 (1978), 191–243.

6. Frederic Rosen, *The Algebra of Muhammed ben Musa* (London: Oriental Translation Fund, 1831), p. 3. This translation from the Arabic of al-Khwārizmī's *Algebra* is long out of print and difficult to find. A more recent English translation, from a twelfth century Latin translation, is Louis Karpinski, *Robert of Chester's Latin Translation of the Algebra of al-Khowarizmi* (Ann Arbor: University of Michigan Press, 1930).

7. Ibid., p. 5.

8. Ibid., p. 8.

9. Karpinski, *Robert of Chester's,* p. 69.

10. Rosen, *Algebra of,* pp. 33–34.

11. Ibid., p. 12.

12. A. Sayili, *Logical Necessities in Mixed Equations by 'Abd al-Hamīd ibn Turk and the Algebra of his Time* (Ankara: Türk Tarih Kurumu Basimevi, 1962), p. 166.

13. Rosen, *Algebra of,* p. 23.

14. Ibid., p. 43.

15. Martin Levey, *The Algebra of Abū Kāmil, Kitāb fī'l-muqābala, in a Commentary by Mordecai Finzi* (Madison: University of Wisconsin Press, 1966), p. 32. This is an edition and English translation of the fifteenth century Hebrew translation of Abū Kāmil's work, and provides a detailed discussion of the relation of his algebra to previous Greek and Islamic work.

16. Ibid., p. 144.

17. Ibid., p. 156.

18. Franz Woepcke, *Extrait du Fakhrī, traité d'algèbre par Aboù Bekr Mohammed ben Alhaçan al-Karkhī* (Paris: L'imprimerie Impériale, 1853), p. 45. (This translation is available in a 1982 reprint by Georg Olms Verlag, Hildesheim.) Note that al-Karajī's name is here given as al-Karkhī. It is not known which transliteration is correct, since available Arabic manuscripts give both versions.

19. Adel Anbouba, *Al Samaw'al,* in the *Dictionary of Scientific Biography* (New York: Scribners, 1970–1980), vol. XII, 91–95; p. 92.

20. Berggren, *Mathematics of Medieval,* p. 114.

21. See J. Lennart Berggren, "Proof, Pedagogy, and the Practice of Mathematics in Medieval Islam," *Interchange* 21 (1990), 36–48. See also M. Yadegari, "The Binomial Theorem: A Widespread Concept in Medieval Islamic Mathematics," *Historia Mathematica* 7 (1980), 401–406.

22. Roshdi Rashed, "L'induction mathématique: al-Karajī, as-Samaw'al," *Archive for History of Exact Sciences* 9 (1972), 1–21, p. 4. This paper, as well as some of Rashed's other works, all of interest, are collected in *Entre arithmétique et algèbre: recherches sur l'histoire des mathématique arabes* (Paris: Société d'Édition Les Belles Lettres, 1984).

23. Daoud S. Kasir, *The Algebra of Omar Khayyam* (New York: Columbia Teachers College, 1931), p. 44. This is the only available English translation of al-Khayyāmī's work on algebra and includes a detailed discussion of his contributions to the subject. For more on Omar Khayyam, see D. J. Struik, "Omar Khayyam, Mathematician," *Mathematics Teacher* 51 (1958), 280–285 and B. Lumpkin, "A Mathematics Club Project from Omar Khayyam," *Mathematics Teacher* 71 (1978), 740–744.

24. Ibid., p. 49.

25. For more details on Sharaf al-Dīn al-Ṭūsī's work, see J. Lennart Berggren, "Innovation and Tradition in Sharaf al-Dīn al-Ṭūsī's *al Muʿādalāt,*" *Journal of the American Oriental Society* 110 (1990), 304–309, and Jan P. Hogendijk, "Sharaf al-Dīn al-Ṭūsī on the Number of Positive Roots of Cubic Equations," *Historia Mathematica* 16 (1989), 69–85. Sharaf al-Dīn's text has been translated into French by Roshdi Rashed as *Sharaf al-Dīn al-Ṭūsī, oeuvres mathématiques: Algèbre et géométrie au XIIᵉ siècle* (Paris: Société d'Édition Les Belles Lettres, 1986). Rashed gives not only an edited Arabic text and a translation, but also an extensive commentary on the mathematics involved.

26. Rosen, *Algebra of,* pp. 71–72. Another English translation of the *Geometry* of al-Khwārizmī appears in Solomon Gandz, "The Mishnat Ha-Middot and the Geometry of Muhammed ibn Musa al-Khowarizmi," reprinted in Solomon Gandz, *Studies in Hebrew Astronomy and Mathematics* (New York: Ktav, 1970).

27. Ibid., p. 76.

28. For more information on non-Euclidean geometry in the Islamic world, see Chapter 2 of B. A. Rosenfeld, *A History of Non-Euclidean Geometry: Evolution of the Concept of a Geometric Space,* Abe Shenitzer, trans., (New York: Springer Verlag, 1988) and Chapter 3 of Jeremy Gray, *Ideas of Space: Euclidean, Non-Euclidean, and Relativistic,* 2nd ed., (Oxford: Clarendon Press, 1989). See also D. E. Smith, "Euclid, Omar Khayyam, and Saccheri," *Scripta Mathematica* 3 (1935), 5–10.

29. Ibn al-Baghdādī is quoted in Galina Matvievskaya, "The Theory of Quadratic Irrationals in Medieval Oriental Mathematics," in D. A. King and G. Saliba, eds. *From Deferent to Equant: A Volume of Studies in the History of Science in the Ancient and Medieval Near East in Honor of E. S. Kennedy* (New York: New York Academy of Sciences, 1987), 253–277, p. 267.

30. E. S. Kennedy, *The Exhaustive Treatise on Shadows by Abū al-Rayhān al-Bīrūnī* (Aleppo: University of Aleppo, 1976), p. 64. This work provides a translation and detailed commentary on the work of al-Bīrūnī. See also E. S. Kennedy, "An Overview of the History of Trigonometry," in *Historical Topics for the Mathematics Classroom* (Reston, Va.: National Council of Teachers of Mathematics, 1989), 333–359.

31. Ibid., p. 89.

32. Ibid., p. 90.

33. Ibid., p. 255.

34. See Berggren, *Episodes in the Mathematics of Medieval Islam,* for more details.

Summary of Medieval Islam

780–850	Muhammad al-Khwārizmī	Arithmetic, algebra, practical geometry
Ninth century	'Abd al-Hamīd ibn Turk	Quadratic equations
830–890	Thābit ibn Qurra	Euclidean justification of algebra
850–930	Abū Kāmil ibn Aslam	Algebra using irrationals
Mid-tenth century	Abū Sahl al-Kūhī	Centers of gravity
Mid-tenth century	Abu l'Hasan al-Uqlīdīsī	Earliest Arabic arithmetic
Mid-tenth century	'Abd al-'Azīz al-Qabīsī	Sums of integral powers, trigonometric methods
940–997	Muhammad Abu'l-Wafā	Theorems of spherical trigonometry
Early eleventh century	Abū Bakr al-Karajī	Algebra, early use of induction
Early eleventh century	Abū Nasr Mansūr	Theorems of spherical trigonometry
Early eleventh century	Ibn al-Baghdādī	Irrationals
965–1039	Abū 'Ali ibn al-Haytham	Integral powers, paraboloid, parallel postulate
973–1055	Muhammad al-Bīrūnī	Trigonometry and applications
1048–1131	'Umar al-Khayyāmī	Cubic equations, parallel postulate
1125–1180	Ibn Yahyā al-Samaw'al	Decimal fractions, polynomials, binomial theorem
Late twelfth century	Sharaf al-Dīn al-Tūsī	Analysis of cubic equations
1201–1274	Nasīr al-Dīn al-Tūsī	Parallel postulate, trigonometry text
Early fifteenth century	Ghiyāth al-Dīn al-Kāshi	Calculations with decimals

Chapter 8

Mathematics in Medieval Europe

"Who wishes correctly to learn the ways to measure surfaces and to divide them, must necessarily thoroughly understand the general theorems of geometry and arithmetic, on which the teaching of measurement . . . rests. If he has completely mastered these ideas, he . . . can never deviate from the truth."
(Introduction to the Liber embadorum, *Plato of Tivoli's Latin translation of the Hebrew* Treatise on Mensuration and Calculation *by Abraham Bar Hiyya, 1116)*[1]

Leonardo of Pisa writes about meeting the Holy Roman Emperor Frederick II (1194–1250): "After being brought to Pisa by Master Dominick to the feet of your celestial majesty, most glorious prince, I met Master John of Palermo; he proposed to me a question that had occurred to him, pertaining not less to geometry than to arithmetic . . . When I heard recently from a report from Pisa and another from the Imperial Court that your sublime majesty deigned to read the book I composed on numbers, and that it pleased you to listen to several subtleties touching on geometry and numbers, I recalled the question proposed to me at your court by your philosopher. I took upon myself the subject matter and began to compose in your honor this work, which I wish to call *The Book of Squares.* I have come to request indulgence if in any place it contains something more or less than right or necessary, for to remember everything and be mistaken in nothing is divine rather than human; and no one is exempt from fault nor is everywhere circumspect."[2]

The Roman Empire in the West collapsed in 476 under the onslaught of various "barbarian" tribes. Feudal societies were soon organized in parts of the empire and the long process of the development of the European nation-states began. For the next five centuries, however, the general level of culture in Europe was very low. Serfs worked the land, and few of the barons could read or write, let alone understand mathematics. In fact, there was little practical need for the subject, because the feudal estates were relatively self-sufficient and trade was almost nonexistent, especially after the Moslem conquest of the Mediterranean sea routes.

Despite the lack of mathematical activity, the early middle ages had inherited from antiquity the notion of the **quadrivium**—arithmetic, geometry, music, and astronomy—

as part of the requirements for an educated man, even in the evolving Roman Catholic culture. Thus St. Augustine (354–430) had written in his *City of God* that "we must not despise the science of numbers, which, in many passages of Holy Scripture, is found to be of eminent service to the careful interpreter. Neither has it been without reason numbered among God's praises: 'Thou hast ordered all things in number, and measure, and weight'."[3] For the study of the "science of numbers," however, virtually all early medieval Europe had available were Latin versions of Nicomachus' *Arithmetic*, parts of Euclid's *Elements* by the Roman scholar Boethius (480–524), and similar material from Isidore of Seville (560–636) which included some elementary work in music and astronomy.

The only schools in early medieval Europe were those connected to the Catholic Church, and it was there that the only significant mathematical problem of the time was considered, the determination of the calendar. In particular, there was a debate in the Church as to whether Easter should be determined using the Roman solar calendar or the Jewish lunar calendar. The two reckonings could be reconciled, but only by those with some mathematical knowledge. Charlemagne, in fact, even before his coronation in 800 as Holy Roman Emperor, formally recommended that the mathematics necessary for Easter computations be part of the curriculum in church schools. Also dating from this time period is a collection of fifty-three arithmetical problems, entitled *Propositiones ad acuendos juvenes* (*Propositions for Sharpening Youths*) and attributed to Alcuin of York (735–804), Charlemagne's educational adviser. The problems of the collection often require some ingenuity for solving, but do not depend on any mathematical theory or rules of procedure.

In the tenth century, a revival of interest in mathematics began with the work of Gerbert d'Aurillac (945–1003), who became Pope Sylvester II in 999 (Figure 8.1). In his youth, Gerbert studied in Spain, where he probably learned some of the mathematics of the Moslems. Later, under the patronage of Otto II, the Holy Roman Emperor, Gerbert reorganized the cathedral school at Rheims and successfully reintroduced the study of mathematics. Besides teaching basic arithmetic and geometry, Gerbert dealt with the mensuration rules of the Roman surveyors and the basics of astronomy. He also taught the use of a counting board, divided into columns representing the (positive) powers of ten, in each of which he would place a single counter marked with the western Arabic form of one of the numbers 1, 2, 3, . . . , 9. Zero was represented by an empty column. Gerbert's work represents the first appearance in the Christian West of the Hindu-Arabic numerals, although the absence of the zero and the lack of suitable algorithms for calculating with these counters showed that Gerbert did not understand the full significance of the Hindu-Arabic system.

The limited sources available to Europeans at the turn of the millenium did mention that much had been achieved in mathematics by Greek authors, at the time completely inaccessible. This heritage, as well as some of the mathematics developed in the Islamic world, was only brought into Western Europe through the efforts of the translators. European scholars discovered the major Greek scientific works (primarily in Arabic translation) beginning in the twelfth century and started the process of translating these into Latin. Much of this work was accomplished at Toledo in Spain which at the time had only recently been retaken by the Christians from the former Moslem rulers. Here could be found repositories of Islamic scientific manuscripts as well as people straddling the two cultures. In particular, there was a flourishing Jewish community, many of whose members

FIGURE 8.1
Gerbert of Aurillac, Pope Sylvester II.

Side Bar 8.1 **Translators and Their Translations**

Adelard of Bath (flourished 1116–1142)
Astronomical Tables of al-Khwārizmī
Elements of Euclid
Liber ysagogarum Alchorismi, the arithmetical
    work of al-Khwārizmī

John of Seville and Domingo Gundisalvo
(fl. 1135–1153)
Liber alghoarismi de practica arismetrice, an
    elaboration of al-Khwārizmī's *Arithmetic*

Plato of Tivoli (fl. 1134–1145)
Spherica of Theodosius (c. 100 B.C.E.)
De motu stellarum of al-Battānī, which contains
    important material on trigonometry
Measurement of a Circle of Archimedes
Liber embadorum of Abraham bar Ḥiyya

Robert of Chester (fl. 1141–1150)
Algebra of al-Khwārizmī
Revision of al-Khwārizmī's astronomical tables
    for the meridian of London

Gerard of Cremona (fl. 1150–1185)
De sphaera mota of Autolycus
Elements of Euclid
Data of Euclid
Measurement of a Circle of Archimedes
Conics of Apollonius
Spherica of Theodosius
Almagest of Ptolemy
De figuris sphaericis of Menelaus
Algebra of al-Khwārizmī
Algebra of Abū Kāmil
Gebri de astronomia by Jābir ibn Aflaḥ
    (twelfth century), an Islamic work on
    astronomy which contained much
    trigonometry

Wilhelm of Moerbeke (fl. 1260–1280)
On Spirals of Archimedes
On the Equilibrium of Planes of Archimedes
Quadrature of the Parabola of Archimedes
Measurement of a Circle of Archimedes
On the Sphere and Cylinder of Archimedes
On Conoids and Spheroids of Archimedes
On Floating Bodies of Archimedes

---

were fluent in Arabic. The translations then were often made in two stages, first by a Spanish Jew from Arabic into Spanish, and then by a Christian scholar from Spanish into Latin. The list of the translations of major mathematical works (with their dates) is fairly extensive (Side Bar 8.1).

Among the earliest of the translating teams were John of Seville and Domingo Gundisalvo, who were active in the first half of the twelfth century. John was born a Jew, his original name probably being Solomon ben David, but converted to Christianity, while Gundisalvo was a philosopher and Christian theologian. The most important of their mathematical translations was of an elaboration of al-Khwārizmī's work on arithmetic. They also translated a large number of astronomical, medical, and philosophical works.

A contemporary of John of Seville was Adelard of Bath (1075–1164), who was born in Bath and spent much of his early years traveling in France, southern Italy, Sicily, and the Near East, the latter two places in particular having many Arabic treatises available. Adelard was responsible for the first translation, from the Arabic, of Euclid's *Elements*. He also translated the astronomical tables of al-Khwārizmī in 1126. This translation contains the first sine tables available in Latin as well as the first tangent tables, the latter having been added to al-Khwārizmī's work by an eleventh century editor. Another Englishman, Robert of Chester, who lived in Spain for several years, translated the *Algebra* of al-Khwārizmī in 1145, thus introducing to Europe the algebraic algorithms for solving quadratic equations. Interestingly enough, in the same year Plato of Tivoli translated from the Hebrew the *Liber embadorum* (*Book of Areas*) by the Spanish-Jewish scholar Abraham bar Ḥiyya, a work which also contained the Islamic rules for solving quadratic equations.

Perhaps the greatest of all the translators was Gerard of Cremona (1114–1187), an Italian who worked primarily in Toledo and is credited with the translation of more than eighty works. Undoubtedly, not all of these are due to him alone. Many were done by others under his direction, but the names of these assistants have been lost to history. Among Gerard's works was a new translation of Euclid's *Elements* from the Arabic of Thābit ibn Qurra and the first translation of Ptolemy's *Almagest* from the Arabic in 1175.

By the end of the twelfth century, then, many of the major works of Greek mathematics and a few Islamic works were available to scholars in Europe who read Latin. During the next centuries, these works were assimilated and new mathematics began to be created by the Europeans themselves. It is well to note, however, that some Spanish-Jewish scholars had earlier read the Arabic works in the original and had produced works on their own, in Hebrew. During the twelfth century, in fact, the cultural exchange among the three major civilizations of Europe and the Mediterranean basin—the Jewish, Christian, and Islamic—was very intense. The Islamic supremacy of previous centuries was on the wane, and the other two were gaining strength. By the end of the next century, the genius of western Christendom had manifested itself, while various restrictions on the lives of the Jews began to lessen the Jewish contribution.

This chapter will discuss both the Jewish and Christian contributions of the twelfth through the fourteenth centuries. We will first consider geometry and trigonometry, next the initial developments in combinatorics, third, the algebraic developments which grew out of the introduction of Islamic algebra into Europe, and finally, some of the mathematics of kinematics that grew out of the study of Aristotle's works in the medieval universities.

8.1 GEOMETRY AND TRIGONOMETRY

Euclid's *Elements* was translated into Latin early in the twelfth century. Before then, of course, Arabic versions were available in Spain. And so, when Abraham bar Ḥiyya (d. 1136) of Barcelona wrote his *Hibbur ha-Meshihah ve-ha-Tishboret* (*Treatise on Mensuration and Calculation*) in 1116 to help French and Spanish Jews with the measurement of their fields, he began the work with a summary of some important definitions, axioms, and theorems from Euclid. Not much is known of the life of Abraham bar Ḥiyya, but from his Latin title of Savasorda, a corruption of the Arabic words meaning "captain of the bodyguard," it is likely that he had a court position, probably one in which he gave mathematical and astronomical advice to the Christian monarch.

8.1.1 Abraham bar Ḥiyya's *Treatise on Mensuration*

Like most of those who dealt with geometry over the next few centuries, Abraham was not so much interested in the theoretical aspects of Euclid's *Elements* as in the practical application of geometric methods to measurement. Nevertheless, he took over the Islamic tradition of proof, absorbed from the Greeks, and gave geometric justifications of methods for solving the algebraic problems he included as part of his geometrical discussions. In particular, Abraham included in his work the major results of *Elements* II on geometric algebra and used them to demonstrate methods of solving quadratic equations. In fact, Abraham's work was the first in Europe to give the Islamic procedures for solving such equations.

For example, Abraham posed the question, "If from the area of a square one subtracts the sum of the (four) sides and there remains 21, what is the area of the square and what is the length of each of the equal sides?"[4] We can translate Abraham's question into the quadratic equation $x^2 - 4x = 21$, an equation which he solves in the familiar way beginning with the halving of 4 to get 2. Abraham's statement of the problem is not geometrical, in that he writes of subtracting a length (the sum of the sides) from an area. But in his geometric justification he restates the problem to mean the cutting off of a rectangle of sides 4 and x from the original square of unknown side x to leave a rectangle of area 21. He then bisects the side of length 4 and applies *Elements* II-6 to justify the algebraic procedure. Thus Abraham evidently had learned his algebra not from al-Khwārizmī (whose *Algebra* was translated into Latin in the same year as Abraham's work), but from an author such as Thābit ibn Qurra, who used Euclidean justifications. Abraham similarly presented the method and Euclidean proof for examples of the two other Islamic classes of mixed quadratic equations, $x^2 + 4x = 77$, and $4x - x^2 = 3$. In the latter case, he gave both positive solutions. Abraham also solved such quadratic problems as the systems $x^2 + y^2 = 100$, $x - y = 2$, and $xy = 48$, $x + y = 14$.

Abraham's original contribution, however, is found in his section on measurements in circles. He began by giving the standard rules for finding the circumference and area of a circle, first using 3 1/7 for π but then noting that if one wants a more exact value, as in dealing with the stars, one should use $3\frac{8\ 1/2}{60}\left(= 3\frac{17}{120}\right)$. To measure areas of segments of circles, Abraham noted that one must first find the area of the corresponding sector by multiplying the radius by half the length of the arc (Figure 8.2). One then subtracts the area of the triangle formed by the chord of the segment and the two radii at its ends. But how does one calculate the length of the arc, assuming one knows the length of the chord? By the use of a table relating chords and arcs. And so for the first time in Europe there appears what one can call a trigonometric table (Figure 8.3). Unlike the table of sines of al-Khwārizmī, which appeared in Latin translation shortly after Abraham's book and which used degrees to measure arcs and a circle radius of 60, Abraham's table was a table of arcs to given chords using what seemed to Abraham a more convenient measure for the circle. Namely, he used a radius of 14 parts, so the semicircumference would be integral (44), and then gave the arc (in parts, minutes, and seconds) corresponding to each integral value of the chord from 1 to 28. So to determine the length of the arc of a segment of a circle, given the chord s and the distance h from the center of the chord to the circumference, Abraham first determined the diameter d of the circle by the formula

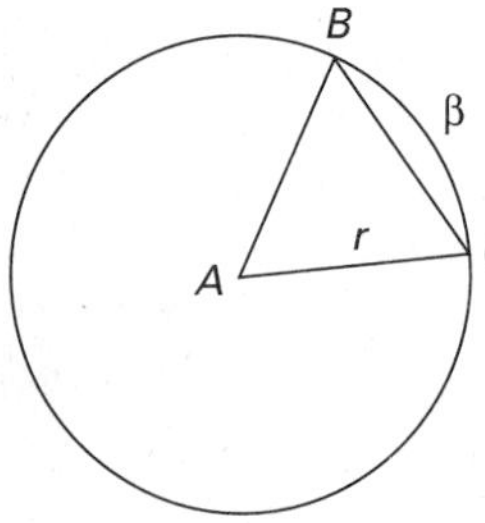

FIGURE 8.2
Area of segment $B\beta C =$
Area of sector $AB\beta C$
$-$ Area of triangle ABC;
Area of sector $= r\frac{\beta}{2}$.

$$d = \frac{s^2}{4h} + h$$

(See Figure 8.4). Then, he multiplied the given chord by 28/d (to convert to a circle of diameter 28), consulted his table to determine the corresponding arc α, and then multiplied

Partes Cordatum	Arcus		
	Partes	Min.	Sec.
1	1	0	2
2	2	0	8
3	3	0	26
4	4	0	55
5	5	1	44
6	6	2	54
7	7	4	42
8	8	7	11
9	9	9	56
10	10	13	42
11	11	18	54
12	12	24	38
13	13	31	9
14	14	40	0
15	15	50	10
16	17	2	16
17	18	16	36
18	19	33	27
19	20	53	26
20	22	17	10
21	23	45	6
22	25	19	24
23	27	0	0
24	28	49	56
25	31	26	37
26	33	20	52
27	36	27	32
28	44	0	0

FIGURE 8.3
Arcchord table of Abraham bar Ḥiyya.

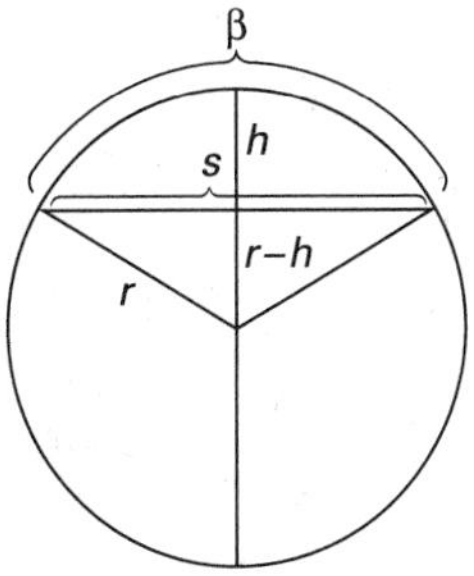

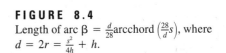

FIGURE 8.4
Length of arc $\beta = \frac{d}{28}$arcchord$\left(\frac{28}{d}s\right)$, where $d = 2r = \frac{s^2}{4h} + h$.

α by $d/28$ to find the actual arc length. Abraham's procedure here provides one of the earliest uses of a trigonometric table to measure quantities on earth, rather than in the heavens. But note that trigonometry is used to measure pieces of circles rather than triangles.

8.1.2 The *Artis Cuiuslibet Consummatio*

Abraham's geometrical text was only the first of many practical geometrical works to appear in medieval Europe. Another early one, which appeared late in the twelfth century and was translated into French from the original Latin about a hundred years later, has come down to the present with no author noted. Since the work also has no title, it is generally known by the first three words of the manuscript, *Artis cuiuslibet consummatio*

(*The perfection of any art*). Even more so than Abraham's treatise, this work has an intensely practical nature, emphasized in the rather poetic introduction:

> The perfection of any art, seen as a whole, depends on two aspects: theory and practice. Anyone deprived of either of these aspects is labeled semiskilled. Truly the modern Latins . . . [by] neglecting the practice fail to reap where they sowed the richest fruits as if picking a spring flower without waiting for its fruit. What is sweeter when once the qualities of numbers have been known through arithmetic than to recognize their infinite dispositions by subtle calculation, the root, origin, and source necessarily available for every science? What is more pleasant when once the proportion of sounds has been known through music than to discern their harmonies by hearing? What is more magnificent when once the sides and angles of surfaces and solids have been proved through geometry than to know and investigate exactly their quantities? What is more glorious or excellent when once the motion of the stars has been known through astronomy than to discover the eclipses and secrets of the art? We prepare for you therefore a pleasant treatise and delightful memoir on the practice of geometry so that we may offer to those who are thirsty what we have drunk from the most sweet source of our master.[5]

To be truly educated, the author seems to be saying, one not only must study the theoretical aspects of the quadrivium, but also must understand how these subjects are used in the real world. *Artis cuiuslibet consummatio* is designed to show, then, the practical aspects of one of the quadrivial subjects, namely, geometry.

The book is divided into four parts: area measurement, height measurement, volume measurement, and calculation with fractions. The last section is designed to help the reader with the computations necessary in the earlier parts. The first part, on areas, begins with the basic procedures for finding the areas of the various types of triangles, as well as rectangles and parallelograms, most of which procedures are justified by some appeal to Euclidean propositions. The author follows this with a section on the areas of various equilateral polygons, all of the formulas for which are incorrect. Instead of being formulas for areas of pentagons, hexagons, heptagons, and so on, of side n, the formulas are always those for the nth pentagonal, hexagonal, heptagonal, number. For example, the procedure given for finding the area of a pentagon of side n amounts to using the formula

$$A = \frac{3n^2 - n}{2}.$$

The author may well have been influenced by works on figurate numbers, similar to that of Nicomachus, which had always been available in Latin.

The part of the book on heights is most interesting because the author demonstrates some knowledge of trigonometry. For example, the procedure for measuring the altitude of the sun using the shadow of a vertical gnomon of length 12 is given: "Let the shadow be multiplied by itself. Let 144 be added to the product. Let the root of the whole sum be taken. And then let the shadow be multiplied by 60. Let the product be divided by the root found. The result will be the sine; let its arc be found. Let the arc be subtracted from 90; the remainder will be . . . the altitude of the sun."[6] Namely, if the shadow is designated by s, the altitude α is given by

$$\alpha = 90 - \arcsin\left(\frac{60s}{\sqrt{s^2 + 144}}\right)$$

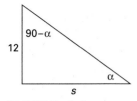

FIGURE 8.5
The calculation of the altitude of the sun given the shadow and conversely from the *Artis cuiuslibet consummatio:*

$$\alpha = 90 - \arcsin\left(\frac{60s}{\sqrt{s^2 + 144}}\right);$$
$$s = \frac{12 \sin(90 - \alpha)}{\sin \alpha}.$$

where, as in most of the Islamic trigonometrical works, the sines are computed using a radius of 60 (Figure 8.5). Similarly, the author calculates the shadow from the altitude by using

$$s = \frac{12 \sin(90 - \alpha)}{\sin \alpha}.$$

These two problems show that the author knew the use of a table of sines but probably did not know of the tables of cosines, tangents, or cotangents, even though these had already been developed in the Islamic world. It was only the earliest of the Hindu and Islamic improvements on Greek work which was available.

To show further that old traditions remain long after one might assume improved methods would replace them, we simply turn to the last section of this part of the *Artis cuiuslibet consummatio*. To measure the height of a tower, the author not only does not use trigonometric methods, but also reverts to probably the oldest (and simplest) method available: "Wait until the altitude of the sun is 45 degrees . . . ; then the shadow lying in the plane of any body will be equal to its body."[7] If the tower is inaccessible, the author uses the ancient methods requiring two sightings similar to those in the Chinese and Indian sources. As in almost all of Indian, Islamic, and medieval European sources, even though trigonometric methods are known, they are applied solely to heavenly triangles, not to earthly ones.

The *Artis cuiuslibet consummatio*, as a typical example of a twelfth century practical geometry, gives us an idea of the state of geometrical knowledge in northern Europe of the time. Some familiarity with Greek geometrical traditions is combined with very practical methods, also dating to ancient times and not all strictly correct, for actually computing geometrical quantities of use in daily life. In southern Europe, however, the Islamic influence was stronger and Euclidean traditions of proof are more in evidence, as in the work of Abraham bar Ḥiyya. Another example is provided by the geometrical work of one of the first Italian mathematicians, Leonardo of Pisa (c. 1170–1240).

8.1.3 Leonardo of Pisa's *Practica Geometriae*

Leonardo's *Practica geometriae* (1220) is more closely related to the work of Abraham bar Ḥiyya than to the *Artis cuiuslibet consummatio*. In fact, some of the sections appear to be taken almost directly from the *Liber embadorum*. Leonardo's work is, however, somewhat more extensive. As in the earlier book, Leonardo begins with a listing of various definitions, axioms, and theorems of Euclid, including especially the propositions of Book II. So in his section on measuring rectangles, in which he includes the standard methods for solving quadratic equations, he is able to quote Euclid in justification of his procedures.

Again, like Abraham, Leonardo calculated areas of segments and sectors of circles. To do this, he too needed a table of arcs and chords. Strangely enough, although he defined the sine of an arc in the standard way, he did not give a table of sines, but one of chords, and in fact reproduced the Ptolemaic procedure for determining the chord of half an arc from that of the whole arc. His chord table, though, was not Ptolemaic. In fact, it may well be original to Leonardo since it is based on a radius of 21. Like the value 14 of Abraham, this was chosen so the semicircumference of the circle is integral, but unlike Abraham's table, this table is a direct chord table (Figure 8.6). For each integral arc from 1 to 66 rods

Arcus pertice	Arcus pertice	Corde pertice	Ar pedes	Cuu vncie	M puncta	Arcus pertice	Arcus pertice	Corde pertice	Ar pedes	CV vncie	VM puncta
1	131	0	5	17	17	34	98	30	2	6	17
2	130	1	5	17	13	35	97	31	0	8	5
3	129	2	5	17	4+	36	96	31	4	8	7
4	128	3	5	17	2	37	95	32	2	5	15
5	127	4	4	12	10	38	94	33	0	1	9
6	126	5	5	16	7+	39	93	34	3	13	0
7	125	6	5	14	5	40	92	35	1	4	15
8	124	7	5	12	9	41	91	35	4	12	10
9	123	8	5	8	16	42	90	36	2	0	0
10	122	9	5	7	8	43	89	36	5	3	5
11	121	10	5	4	2	44	88	37	2	4	6
12	120	11	4	17	18	45	87	37	5	3	2
13	119	12	4	13	6	46	86	38	1	17	15
14	118	13	4	7	16	47	85	38	4	12	13
15	117	14	4	1	0	48	84	38	1	4	0
16	116	15	3	11	18	49	83	39	3	11	15
17	115	16	3	3	12	50	82	39	5	17	2
18	114	17	2	12	8	51	81	40	2	2	1
19	113	18	2	0	15	52	80	40	4	2	10
20	112	19	1	8	12	53	79	40	0	0	11
21	111	20	0	13	18	54	78	40	1	14	5
22	110	21	0	0	0	55	77	41	3	7	8
23	109	21	5	2	16	56	76	41	4	16	2
24	108	22	4	4	5	57	75	41	0	4	12
25	107	23	3	4	8	58	74	41	1	8	1
26	106	24	2	3	2	59	73	41	2	9	0
27	105	25	1	0	6	60	72	41	3	7	14
28	104	25	5	16	2	61	71	41	4	9+	2
29	103	26	4	8	0	62	70	41	4	15	10
30	102	27	3	0	3	63	69	41	5	6	9
31	101	28	1	9	7	64	68	41	5	12	17
32	100	28	5	16	4	65	67	41	5	6	14
33	99	29	4	3	9	66	66	42	0	0	0

FIGURE 8.6
Chord table of Leonardo of Pisa.

(and also from 67 to 131), the table gives the corresponding chord, in the same measure, with fractions of the rods not in sixtieths, but in the Pisan measures of feet (6 to the rod), unciae (18 to the foot), and points (20 to the uncia). Leonardo demonstrated how to use the chord table to calculate arcs to chords in circles of radius other than 21. In particular, he gave a detailed discussion of how to interpolate in the table.

Like Abraham bar Ḥiyya, Leonardo used the table of chords only to calculate areas of circular sectors and segments. When, later in the same chapter, he calculated the lengths of the sides and diagonals of a regular pentagon inscribed in a circle, he did not use what seems to us the obvious method of consulting his table of chords. He returned to Euclid and quoted appropriate theorems from Book XIII relating the sides of a hexagon, pentagon, and decagon to enable him actually to perform the calculations. And toward the end of the book, when he wanted to calculate heights, again he did not use trigonometry. He used the old methods of similar triangles, starting with a pole of known height to help sight the top of the unknown object, then measuring the appropriate distances along the ground.

8.1.4 Trigonometry

That trigonometry in the medieval period was not used to measure earthly triangles is further demonstrated by two fourteenth century trigonometry works, one by the Englishman Richard of Wallingford (1291–1336) and the other by the French Jew Levi ben Gerson (1288–1344).

Richard of Wallingford was a monk who spent the final nine years of his life as the abbot at St. Albans monastery. The *Quadripartitum*, a four-part work on the fundamentals of trigonometry, was written while he was still a student at Oxford, probably around 1320. Its goal, like that of most texts on trigonometry, was to teach the methods required for the solution of problems in spherical trigonometry, which in turn was required for astronomy. Although extant Islamic treatises on trigonometry covered basically the same material, they were in general compiled in the eastern part of the Islamic world and never reached the translators of Toledo. The chief source of the *Quadripartitum*, then, was the *Almagest* of Ptolemy. Thus Richard's work can reasonably be claimed as the first comprehensive trigonometry treatise written in Europe.

Although the chief ideas of the work come from the *Almagest*, Richard from the beginning deals with the Hindu sines rather than Ptolemy's chords. Thus he defines the **sinus rectus** (right sine) of a given arc as half of the chord on double that arc. Using theorems from the *Elements*, Richard proves the basic results of plane trigonometry, including the rules for calculating the sine of half an arc, the sine of the complement of an arc, the sine of double an arc (which is not explicitly found in Ptolemy's work), and finally, using Ptolemy's theorem, the sine of the difference of two arcs. In another slight extension of Ptolemy, he also shows how to determine the chord (not sine) of the sum of three arcs, given the chords of each. Richard concludes the first part of the treatise by showing how to calculate the sines of arcs from 1° to 90° in increments of 1° using the previous theorems and Ptolemy's approximation methods.

The remaining sections of Richard's treatise provide a detailed study of the theorem of Menelaus in both plane and spherical versions. Because this theorem is concerned with ratios among the various sides in the Menelaus configuration, Richard needed first to consider the basics of the theory of proportions. His study of proportions is closely related to the work of several contemporaries in the universities and will be considered in section 8.4.1. Here we only note that in his treatment of Menelaus' theorem, Richard considered all the possible cases of Menelaus' configuration and proved the result anew each time. While modern readers might consider his work tedious, he evidently felt that such detail was necessary for the less mathematically experienced readers for whom he was writing. One also sees here Richard's commitment to strictly Euclidean rigor of argument as he exhausts all the cases. Recall that even though mathematical knowledge was at a low ebb during the early middle ages, the basic notion of a mathematical proof survived and was reinvigorated, as, for example, by Richard, once the need for more mathematics had established itself.

The trigonometrical work of Levi ben Gerson was roughly contemporaneous with the *Quadripartitum*. It formed part of an astronomical treatise which in turn formed part of a major philosophical work, *Sefer milhamot Adonai* (*Wars of the Lord*). Levi's trigonometry, like Richard of Wallingford's, is based chiefly on Ptolemy, though again, like Richard, Levi generally uses sines rather than chords. Levi's main departure from Ptolemy, and also from Richard, is that he gives detailed procedures for solving plane triangles. He first gives the standard methods for solving right triangles and then proceeds to general triangles. In the case where three sides are known, Levi solves the triangle by dropping a perpendicular from one vertex to the opposite side (or opposite side extended), thus creating two right triangles that can be solved by his earlier methods. The same method works also where two sides and the included angle are known. For the case where two sides and the angle opposite one of them are known, Levi uses (with proof) the law of sines. He does not, however, mention the possible ambiguity of this case. Of course in any particular problem, one of the unknown angles is required to be acute or obtuse, so a single solution of the triangle can be found. Finally, Levi notes that the case where two angles and a side are known can also be solved using the law of sines. Although again there is little in Levi's work which could not be found in one of the major Islamic trigonometries, it still provides one of the earliest treatments of trigonometry available in Europe. But as in those Islamic works and the practical geometry texts, the methods Levi presents are used only for solving astronomical triangles, never for solving earthly ones.

8.2 COMBINATORICS

Although there was some interest in combinatorics in medieval India and many correct rules are presented in the Indian mathematical classics, no proofs or motivations are mentioned in the manuscripts. There is evidence from early in the first millennium that Jewish scholars were also interested in combinatorial questions. In contrast to India, however, by the medieval period detailed derivations of various combinatorial rules had been written down.

The earliest Jewish source on this topic seems to be the mystical work *Sefer Yetsirah* (*Book of Creation*), written sometime before the eighth century and perhaps as early as the second century. In it the unknown author calculates the various ways in which the 22 letters of the Hebrew alphabet can be arranged. He was interested in this calculation because the Jewish mystics believed that these letters had magic powers. Suitable combinations could therefore subjugate the powers of nature. Thus, "Two stones [letters] build two houses [words], three build six houses, four build twenty-four houses, five build one hundred and twenty houses, six build seven hundred and twenty houses, seven build five thousand and forty houses."[8] Evidently, the author understood that the number of possible arrangements of n letters was $n!$. He also noted that the number of combinations of the letters taken two at a time was $231 \left(= \frac{22 \cdot 21}{2} \right)$. (This particular rule was also known to Boethius.)

8.2.1   The Work of Abraham ibn Ezra

A more detailed study of combinations was carried out by Rabbi Abraham ben Meir ibn Ezra (1090–1167), a Spanish-Jewish philosopher, astrologer, and biblical commentator. It was in an astrological text that ibn Ezra discussed the number of possible conjunctions of the seven "planets" (including the sun and the moon). It was believed that these conjunctions would have a powerful influence on human life. Ibn Ezra thus calculated C_k^7 for each integer k from two to seven and noted that the total was 120. (The notation C_k^n represents the number of combinations of n elements taken k at a time or the number of k-element subsets of a set of order n.) He began with the simplest case, that the number of binary conjunctions was 21. This number was equal to the sum of the integers from one to six, and could be calculated by the general rule:

$$C_2^n = \sum_{i=1}^{n-1} i = \frac{n(n-1)}{2}.$$

To calculate ternary combinations, ibn Ezra explains, "We begin by putting Saturn with Jupiter and with them one of the others. The number of the others is five; multiply 5 by its half and by half of unity. The result is 15. And these are the conjunctions of Jupiter."[9] Namely, there are five ternary combinations involving Jupiter and Saturn, four involving Jupiter and Mars, but not Saturn, and so on. Hence, there are $C_2^6 = 15 \left(= 5 \cdot \frac{5}{2} + 5 \cdot \frac{1}{2} \right)$ ternary conjunctions involving Jupiter. Similarly, to find the ternary conjunctions involving Saturn but not Jupiter, ibn Ezra needs to calculate the number of choices of two planets from the remaining five: $C_2^5 = 10$. He then finds the ternary conjunctions involving Mars, but neither Jupiter nor Saturn, and finally concludes with the result

$$C_3^7 = C_2^6 + C_2^5 + C_2^4 + C_2^3 + C_2^2 = 15 + 10 + 6 + 3 + 1 = 35.$$

Ibn Ezra next calculates the quaternary conjunctions by analogous methods. The conjunctions involving Jupiter require choosing three planets from the remaining six. Those with Saturn but not Jupiter require choosing three from five. So finally, $C_4^7 = C_3^6 + C_3^5 + C_3^4 + C_3^3 = 20 + 10 + 4 + 1 = 35$. Ibn Ezra then just states the results for the conjunctions involving five, six, and seven planets. Essentially, he has given an argument for the case $n = 7$, easily generalizable to the general combinatorial rule:

$$C_k^n = \sum_{i=k-1}^{n-1} C_{k-1}^i.$$

Besides his work on astrology, ibn Ezra also wrote a work on arithmetic (1146) in which he introduced the Hebrew speaking community to the decimal place-value system. He used the first nine letters of the Hebrew alphabet to represent the first nine numbers and then instructed his readers on the meaning of place value, the use of the zero (which he wrote as a circle), and the various algorithms for calculation in the Hindu-Arabic system.

Ibn Ezra's place-value system extended only to integers. The earliest occurrence of decimal fractions in Europe was apparently in a brief manuscript by Immanuel ben Jacob Bonfils (mid-fourteenth century), a Jewish mathematician and astronomer who worked mainly in Tarascon in southern France. Rabbi Immanuel began: "Know that the unit is divided into ten parts which are called Primes, and each Prime is divided into ten parts which are called Seconds, and so on into infinity. I also want to call to your attention that I am calling the degrees of the tens Prime Integers and the hundreds Second Integers, and on into infinity. The degree of the units, however, I am calling by their name Units, for it is an intermediate between the integers and the fractions."[10] With the basic system set out, Immanuel could now consider the rules for multiplying and dividing integers and decimal fractions, at least with regard to the powers of ten involved. Thus he noted that in multiplication one adds the degrees of the factors, while in division one subtracts. And although he does not explicitly call the degree of a decimal fraction negative, he does give the rules for operating with the degrees equivalent to rules for dealing with negatives. For example, in division, "if the [degree] of the divisor is higher [than that of the dividend], then subtract the [degree] of the dividend from the [degree] of the divisor, and where the number of the remaining [degree] is, there the quotient will fall, in the opposite direction, that is, among the fractions, if both are integers, or among the integers, if both are fractions."[11]

Although Immanuel understood the basic rules for operating with decimal fractions, and the necessity of numbering the places appropriately so that simple rules of exponents could be stated, he did not derive any notation for such fractions. In fact, his work, like the more extensive work of al-Samaw'al in the same area, had little subsequent effect.

8.2.2 Levi ben Gerson and Induction

Perhaps 30 years before the work of Rabbi Immanuel, Levi ben Gerson dealt with various aspects of combinatorics in a major work, the *Maasei Hoshev* (*The Art of the Calculator*) (1321), in which he gave careful, rigorous proofs of every result. Levi's text is divided into two parts, a first theoretical part in which every theorem receives a detailed proof, and a second applied part in which explicit instructions are given for performing various types of calculation. Levi's theoretical first section begins with a quite modern justification for considering the theory at all:

Because the true perfection of a practical occupation consists not only in knowing the actual performance of the occupation but also in its explanation, why the work is done in a particular way, and because the art of calculating is a practical occupation, it is clear that it is pertinent to concern oneself with its theory. There is also a second reason to inquire about the theory in this field. Namely, it is clear that this field contains many types of operations, and each type itself concerns so many different types of material that one could believe that they cannot all belong to the same subject. Therefore, it is only with the greatest difficulty that one can

Biography Levi ben Gerson (1288–1344)

Levi was born probably in the village of Bagnols-sur-Cèze in the south of France and spent most of his life in the nearby town of Orange. He was not only a mathematician, but also an astronomer, philosopher, and biblical commentator. Not much is known of his life, except that he maintained contact with many important Christians, at the request of some of whom he composed a set of astronomical tables. His various works show that he was acquainted with the major Greek philosophical, astronomical, and mathematical writings, probably all in Arabic translations. His best known contribution to astronomy is his invention of the Jacob Staff, which was used for centuries to measure the angular separation between heavenly bodies. In particular, the Jacob Staff was popular with sixteenth century European sailors who used it for navigation purposes (Figure 8.7).

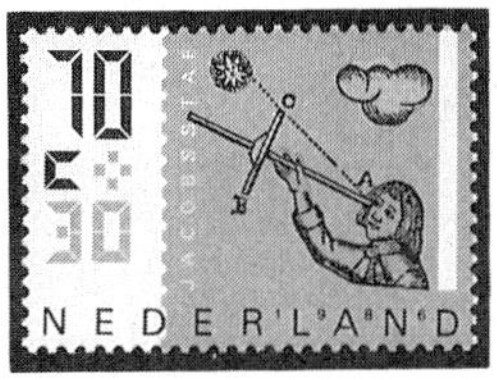

FIGURE 8.7
Jacob Staff, due to Levi ben Gerson.

achieve understanding of the art of calculating, if one does not know the theory. With the knowledge of the theory, however, complete mastery is easy. One who knows it will understand how to apply it in the various cases which depend on the same foundation. If one is ignorant of the theory, one must learn each kind of calculation separately, even if two are really one and the same.[12]

The most important aspects of Levi's work are the combinatorial theorems and their proofs. Before examining them, however, we consider two of the earliest theorems in the book, theorems that deal with associativity and commutativity of multiplication. It is in the proofs of these theorems that Levi introduces, somewhat more explicitly than his Islamic predecessors, the essentials of the method of mathematical induction, what he calls the process of "rising step-by-step without end." In general, when Levi uses such a proof, he first proves the **inductive step,** the step that allows one to move from k to $k + 1$, then notes that the process begins at some small value of k, then finally states the complete result. Nowhere does he state the modern principle of induction, but it does appear that he knows how to use it.

Proposition 9. *If one multiplies a number which is the product of two numbers by a third number, the result is the same as when one multiplies the product of any two of these three numbers by the third.*

Proposition 10. *If one multiplies a number which is the product of three numbers by a fourth number, the result is the same as when one multiplies the product of any three of these four numbers by the fourth.*

In modern notation, the first result states that $a(bc) = b(ac) = c(ab)$ while the second extends that result to four factors. The proof of proposition 9 simply involves counting the number of times the various factors of the product appear in that product. In the proof of proposition 10, Levi notes that $a(bcd)$ contains bcd a times. Since by proposition 9, bcd can be thought of as $b(cd)$, it follows that the product $a(bcd)$ contains acd b times, or, $a(bcd) = b(acd)$, as desired. Levi then generalizes these two results to any number of factors: "By the process of rising step-by-step without end, this is proved; that is, if one multiplies a number which is the product of four numbers by a fifth number, the result is the same as when one multiplies the product of any four of these by the other number. Therefore, the result of multiplying any product of numbers by another number contains any of these numbers as many times as the product of the others."[13] We see here the essence of the principle of mathematical induction. Levi uses the principle again in proving that $(abc)d = (ab)(cd)$ and concludes that one can use the same proof to demonstrate the result without end: Any number contains the product of two of its factors as many times as the product of the remaining factors.

Levi is certainly not consistent about applying his induction principle. The middle of the text contains many theorems dealing with sums of various sequences of integers, theorems that could be proved by induction. But for many of these, Levi uses other methods. But in his proof of the formula for the sum of the first n integral cubes, he does use induction, in a way reminiscent of al-Karajī's proof of the same result. The basic inductive step is

Proposition 41.   *The square of the sum of the natural numbers from 1 up to a given number is equal to the cube of the given number added to the square of the sum of the natural numbers from 1 up to one less than the given number. (In modern notation, the theorem says that* $(1 + 2 + \cdots + n)^2 = n^3 + (1 + 2 + \cdots + (n - 1))^2.$)

We present Levi's proof in modern notation. First, $n^3 = n \cdot n^2$. Also, $n^2 = (1 + 2 + \cdots + n) + (1 + 2 + \cdots + (n - 1))$. (This result is Levi's proposition 30.) Then $n^3 = n[(1 + 2 + \cdots + n) + (1 + 2 + \cdots + (n - 1))] = n^2 + n[2(1 + 2 + \cdots + (n - 1))]$. But $(1 + 2 + \cdots + n)^2 = n^2 + 2n(1 + 2 + \cdots + (n - 1)) + (1 + 2 + \cdots (n - 1))^2$. It follows that $n^3 + (1 + 2 + \cdots + (n - 1))^2 = (1 + 2 + \cdots + n)^2$.

Levi next notes that although 1 has no number preceding it, "its third power is the square of the sum of the natural numbers up to it." In other words, he gives the first step of a proof by induction for the result stated as

Proposition 42.   *The square of the sum of the natural numbers from 1 up to a given number is equal to the sum of the cubes of the numbers from 1 up to the given number.*

Levi's proof is not quite a modern proof by induction, but the basic idea is there. Instead of arguing from n to $n + 1$, he argues, as did al-Karajī, from n to $n - 1$. He notes that, first of all, $(1 + 2 + \cdots + n)^2 = n^3 + (1 + 2 + \cdots + (n - 1))^2$. The final summand is, also by the previous proposition, equal to $(n - 1)^3 + (1 + 2 + \cdots + (n - 2))^2$. Continuing in this way, Levi eventually reaches $1^2 = 1^3$, and the result is proved.

Inductive proofs are also evident in the final section of the theoretical part of the *Maasei Ḥoshev*, that on permutations and combinations. Levi's first result in this section shows that the number of permutations of a given number n of elements is what we call $n!$:

Proposition 63.   *If the number of permutations of a given number of different elements is equal to a given number, then the number of permutations of a set of different elements*

containing one more number equals the product of the former number of permutations and the given next number.

Symbolically, the proposition states that $P_{n+1} = (n + 1)P_n$ (where P_k stands for the number of permutations of a set of k elements). This result provides the inductive step in the proof of the proposition $P_n = n!$, although Levi does not mention that result until the end. His proof of proposition 63 is very detailed. Given a permutation, say *abcde*, of the original n elements and a new element *f*, he notes that *fabcde* is a permutation of the new set. Because there are P_n such permutations of the original set, there are also P_n permutations of the new set beginning with *f*. Also, if one of the original elements, for example *e*, is replaced by the new element *f*, there are P_n permutations of the set *a, b, c, d, f* and therefore also P_n permutations of the new set with *e* in the first place. Because any of the n elements of the original set, as well as the new element, can be put in the first place, it follows that the number of permutations of the new set is $(n + 1)P_n$. Levi finishes the proof of proposition 63 by showing that all of these $(n + 1)P_n$ permutations are different. He then concludes, "Thus it is proved that the number of permutations of a given set of elements is equal to that number formed by multiplying together the natural numbers from 1 up to the number of given elements. For the number of permutations of 2 elements is 2, and that is equal to $1 \cdot 2$, the number of permutations of 3 elements is equal to the product $3 \cdot 2$, which is equal to $1 \cdot 2 \cdot 3$, and so one shows this result further without end."[14] Namely, Levi mentions the beginning step and then notes that with the inductive step already proved, the complete result is also proved.

After proving, using a counting argument, that $P_2^n = n(n - 1)$ (where P_k^n represents the number of permutations of k elements in a set of n), Levi proves that $P_k^n = n(n - 1)(n - 2) \cdots (n - k + 1)$ by induction on k. As before, he states the inductive step as a theorem:

Proposition 65. *If a certain number of elements is given and the number of permutations of order a number different from and less than the given number of elements is a third number, then the number of permutations of order one more in this given set of elements is equal to the number which is the product of the third number and the difference between the first and the second numbers.*

Modern symbolism replaces Levi's convoluted wording with a brief phrase: $P_{j+1}^n = (n - j)P_j^n$. Levi's proof is quite similar to that of proposition 63. At the end, he states the complete result: "It has thus been proved that the permutations of a given order in a given number of elements are equal to that number formed by multiplying together the number of integers in their natural sequence equal to the given order and ending with the number of elements in the set."[15] To clarify this statement, Levi first gives the initial step of the induction by quoting his previous result in the case $n = 7$, that is, the number of permutations of order two in a set of seven is equal to $6 \cdot 7$. Then, the number of permutations of order 3 is equal to $5 \cdot 6 \cdot 7$ (since $5 = 7 - 2$). Similarly, the number of permutations of order 4 is equal to $4 \cdot 5 \cdot 6 \cdot 7$, "and so one proves this for any number."

In the final three propositions of the theoretical part of *Maasei Hoshev*, Levi completes his development of formulas for permutations and combinations. Proposition 66 shows that $P_k^n = C_k^n P_k^k$, while proposition 67 simply rewrites this as $C_k^n = P_k^n/P_k^k$. Since he has already given formulas for both the numerator and denominator of this quotient, Levi thus has demonstrated the standard formula for C_k^n:

$$C_k^n = \frac{n(n-1)\cdots(n-k+1)}{1\cdot 2\cdots k}.$$

After proving in proposition 68 that $C_k^n = C_{n-k}^n$, Levi proceeds in the second part of his work to show how to apply his results in various practical situations.

8.3 MEDIEVAL ALGEBRA

Although the theory of combinatorics appears to have developed in Europe through the Jewish tradition, the writers on algebra in medieval Europe were direct heirs to Islamic work.

8.3.1 Leonardo of Pisa's *Liber Abbaci*

One of the earliest European writers on algebra was Leonardo of Pisa, most famous for his masterpiece, the *Liber abbaci* or *Book of Calculation*. (The word *abbaci* (from *abacus*) does not refer to a computing device, but simply to calculation in general.) The first edition of this work appeared in 1202, while a slightly revised one was published in 1228. The many surviving manuscripts testify to the wide readership the book enjoyed. The sources for the *Liber abbaci* were largely in the Islamic world, which Leonardo visited during many journeys, but he enlarged and arranged the material he collected through his own genius. The book contained not only the rules for computing with the new Hindu-Arabic numerals, but also numerous problems of various sorts in such practical topics as calculation of profits, currency conversions, and measurement, supplemented by the now standard topics of current algebra texts such as mixture problems, motion problems, container problems, the Chinese remainder problem, and, at the end, various forms of problems solvable by use of quadratic equations. Interspersed among the problems is a limited amount of theory, such as methods for summing series and geometric justifications of the quadratic formulas.

Leonardo used a great variety of methods in his solution of problems. Often, in fact, he used special procedures designed to fit a particular problem rather than more general methods. One basic method used often is the old Egyptian method of "false position" in which a convenient, but wrong, answer is given first and then adjusted appropriately to get the correct result. Leonardo also used the methods of al-Khwārizmī for solving quadratic equations. For many of the problems, it is possible to cite Leonardo's sources. He often took problems verbatim from such Islamic mathematicians as al-Khwārizmī, Abū Kāmil, and al-Karajī, many of which he found in Arabic manuscripts discovered in his travels. Some of the problems seem ultimately to have come from China or India, but Leonardo probably learned these in Arabic translations. The majority of the problems, however, are of his own devising and show his creative abilities. A few of Leonardo's problems and solutions should give the flavor of this most influential mathematical work.

Biography   **Leonardo of Pisa (c. 1170–1240)**

Leonardo, often known today by the name Fibonacci (son of Bonaccio) given to him by Baldassarre Boncompagni, the nineteenth century editor of his works, was born around 1170. His father was a Pisan merchant who had extensive commercial dealings in Bugia on the North African coast (now Bejaia, Algeria). Leonardo spent much of his early life there learning Arabic and studying mathematics under Islamic teachers. Later he traveled throughout the Mediterranean, probably on business for his father. At each location, he met with Islamic scholars and absorbed the mathematical knowledge of the Islamic world. After his return to Pisa about 1200, he spent the next 25 years writing works in which he incorporated what he had learned. The ones which have been preserved include the *Liber abbaci* (1202, 1228), the *Practica geometriae* (1220), and the *Liber quadratorum* (1225). Leonardo's importance was recognized both at the court of Frederick II, as noted in the opening story, and also in the city of Pisa, which in 1240 granted him a yearly stipend in thanks for his teaching and other services to the community.

Leonardo began his text by introducing the Hindu-Arabic numerals: "The nine figures of the Indians are 9, 8, 7, 6, 5, 4, 3, 2, 1. With these nine figures, and with the sign 0, which the Arabs call "zephirum" (cipher), can be written every number, as we will demonstrate below."[16] He then shows precisely that, giving the names to the various places in the place-value system (for integers only). Leonardo then deals with various algorithms for adding, subtracting, multiplying, and dividing whole numbers and common fractions. His notation for mixed numbers differs from ours in that he writes the fractional part first, but his algorithms are generally close to the ones we use today. For example, to divide 83 by 5 2/3 (or, as he writes, 2/3 5), Leonardo multiplies 5 by 3 and adds 2, giving 17. He then multiplies 83 by 3, giving 249, and finally divides 249 by 17, giving 14 11/17. Notations aside, Leonardo was able to use his procedures effectively to show his readers how to perform calculations including in particular the intricate ones necessary to convert among the many currencies in use in the Mediterranean basin during his time.

Leonardo presents several versions of the classic problem of buying birds. In the first he asks how to buy 30 birds for 30 coins, if partridges cost 3 coins each, pigeons 2 coins each, and sparrows 2 for 1 coin. He begins by noting that he can buy 5 birds for 5 coins by taking 4 sparrows and 1 partridge. Similarly, 2 sparrows and 1 pigeon will give him 3 birds for 3 coins. By multiplying the first transaction by 3 and the second by 5, he procures 12 sparrows and 3 partridges for 15 coins and 10 sparrows and 5 pigeons also for 15 coins. Adding these two transactions gives the desired answer: 22 sparrows, 5 pigeons, 3 partridges.

Another classic problem is that of the lion in the pit: The pit is 50 feet deep. The lion climbs up 1/7 of a foot each day and then falls back 1/9 of a foot each night. How long will it take him to climb out of the pit? Leonardo here uses a version of false position. He assumes the answer to be 63 days, since 63 is divisible by both 7 and 9. Thus in 63 days the lion will climb up 9 feet and fall down 7, for a net gain of 2 feet. By proportionality, then, to climb 50 feet, the lion will take 1575 days. (By the way, Leonardo's answer is incorrect. At the end of 1571 days the lion will be only 8/63 of a foot from the top. On the next day he will reach the top.)

Problems in later chapters of the book tend to be more explicitly algebraic. For example, suppose that two men each have some money. The first says to the second, "If you give me one denarius, we will each have the same amount." The second says to the first, "If you give me one denarius, I will have ten times as much as you." How much does each have? In modern notation, if x and y represent the amounts held by the first and second man respectively, this problem becomes the system of equations $x + 1 = y - 1, y + 1 = 10(x - 1)$. Leonardo, however, looks at the problem somewhat differently by introducing the new unknown $z = x + y$ (the total sum of money). Then $x + 1 = \frac{1}{2}z$ and $y + 1 = \frac{10}{11}z$. Adding these two equations together gives $z + 2 = \frac{31}{22}z$, from which $z = \frac{44}{9}$, $x = 1\frac{4}{9}$, and $y = 3\frac{4}{9}$.

Leonardo also deals comfortably with determinate and indeterminate problems in more than two unknowns. For example, suppose there are four men such that the first, second, and third together have 27 denarii, the second, third, and fourth together have 31, the third, fourth, and first have 34, while the fourth, first, and second have 37. To determine how much each man has requires solving a system of four equations in four unknowns. Leonardo accomplished this expeditiously by adding the four equations together to determine that four times the total sum of money equals 129 denarii. The individual amounts are then easily calculated.

The most famous problem of the *Liber abbaci* is the rabbit problem: "How many pairs of rabbits can be bred in one year from one pair? A certain person places one pair of rabbits in a certain place surrounded on all sides by a wall. We want to know how many pairs can be bred from that pair in one year, assuming it is their nature that each month they give birth to another pair, and in the second month after birth, each new pair can also breed."[17] Leonardo proceeds to calculate: After the first month there will be two pairs, after the second, three. In the third month, two pairs will produce, so at the end of that month there will be five pairs. In the fourth month, three pairs will produce, so there will be eight. Continuing in this fashion, he shows that there will be 377 pairs by the end of the twelfth month. Listing the sequence 1, 2, 3, 5, 8, 13, 21, 34, 55, 89, 144, 233, 377 in the margin, he notes that each number is found by adding the two previous numbers, and "thus you can do it in order for an infinite number of months." This sequence, calculated recursively, is known today as a Fibonacci sequence. It turns out that it has many interesting properties unsuspected by Leonardo, not the least of which is its connection with the Greek problem of dividing a line in extreme and mean ratio.

In his final chapter, Leonardo demonstrates his complete command of the algebra of his Islamic predecessors as he shows how to solve equations that reduce ultimately to quadratic equations. He discusses in turn each of the six basic types of quadratic equation, as given by al-Khwārizmī, and then gives geometric proofs of the solution procedures for each of the three mixed cases. He follows the proofs with some 50 pages of examples, including the familiar ones beginning with "divide 10 into two parts."

The content of the *Liber abbaci* contained no particular advance over mathematical works then current in the Islamic world. The chief value of the work was that it provided Europe's first comprehensive introduction to the mathematics of those works. Those reading it were afforded a wide variety of methods to solve mathematical problems, methods that provided the starting point from which advances could ultimately be made. On the other hand, Leonardo did make original contributions to mathematics. These occur in a much shorter work, the *Liber quadratorum (Book of Squares)*, which appeared in 1225.

8.3.2 The *Liber Quadratorum*

The *Liber quadratorum* is a book on number theory, in which Leonardo, following in the footsteps of Diophantus, some of whose work was available in al-Karajī's *al-Fakhrī*, discusses the solving in rational numbers of various equations involving squares. The book originated in a question posed to Leonardo by a Master John of Palermo to "find a square number from which, when five is added or subtracted, always arises a square number." As Leonardo reports, he "saw, upon reflection, that this solution itself and many others have origin in the squares and the numbers which fall between the squares."[18] The initial problem, to find x, y, z so that $x^2 + 5 = y^2$ and $x^2 - 5 = z^2$ is solved as the seventeenth of the 24 propositions of the book, but Leonardo first develops various properties of square numbers and sums of square numbers.

To solve Master John's problem, Leonardo introduces what he calls **congruous** numbers, numbers n of the form $ab(a + b)(a - b)$ when $a + b$ is even and $4ab(a + b)$ $(a - b)$ when $a + b$ is odd. He shows that congruous numbers are always divisible by 24 and that integral solutions of $x^2 + n = y^2$ and $x^2 - n = z^2$ can be found only if n is congruous. The original problem is therefore not solvable in integers. Nevertheless, since $720 = 12^2 \cdot 5$ is a congruous number (with $a = 5$ and $b = 4$) and since $41^2 + 720 = 49^2$ and $41^2 - 720 = 31^2$, it follows by dividing both equations by 12^2 that $x = 41/12$, $y = 49/12$, $z = 31/12$ provides a solution in rational numbers to $x^2 + 5 = y^2$, $x^2 - 5 = z^2$.

Leonardo concludes the book with various other propositions involving squares, including a solution to the question proposed by a Master Theodore, Philosopher to the Emperor Frederick II: "To find three numbers which added together with the square of the first number make a square number. Moreover, this square, if added to the square of the second number, yields also a square number. To this square, if the square of the third number is added, a square number similarly results."[19] The numbers 35, 144, 360 form Leonardo's integral solution to the problem.

In respect to the number theory of the *Liber quadratorum*, Leonardo had no successor until Diophantus' *Arithmetica* was again available in Europe several centuries later. In fact, his original contributions and use of various techniques of number theory were not appreciated until modern times. On the other hand, the practical material in the *Liber abbaci* and the *Practica geometriae* was picked up by various Italian surveyors and masters of computation (*maestri d'abbaco*) who were influential in the next several centuries in bringing a renewed sense of mathematics into Italy. It took a full 300 years, however, for this renewed mathematical knowledge to increase to the point where conditions in Italy were ripe enough for new mathematics to be created.

8.3.3 Jordanus de Nemore's *De Numeris Datis*

The fate of being ignored also affected some of the mathematical work of another European mathematician of the early thirteenth century, Jordanus de Nemore. About Jordanus virtually nothing is known, although it is suspected that he taught in Paris around 1220.

Nonbiography   Jordanus de Nemore

Although Jordanus has been recognized as one of the best mathematicians of the Middle Ages, there is virtually no available evidence about his life, other than that he appears to have been connected with the University of Paris in the early decades of the thirteenth century. Some years ago, he was identified with Jordanus de Saxonia, the second Master General of the Dominican order, but recent scholarly work has shown that this identification is impossible. The translator of *De numeris datis*, Barnabas Hughes, concludes that Jordanus is *sine patre, sine matre, sine genealogia*. He also notes in a letter, however, that "the only explanation that appealed to me [as to why no biographical information is extant] was that the name is a pseudonym. But why a *nom de plume*? Could it be that Jordanus was really a woman? Shades of Hypatia! Thirteenth century women were good for writing poems, songs, and prayers; but science?"[20]

His writings include several works on arithmetic, geometry, and mechanics, but here we will only discuss his major work on algebra, *De numeris datis (On Given Numbers)*.

De numeris datis is an analytic work on algebra, based on but differing in spirit from the Islamic algebras that had made their way into Europe by the early thirteenth century. It appears to be modeled on Euclid's *Data*, available to Jordanus in a Latin translation of Gerard of Cremona. It presents problems in which certain quantities are given and then shows that other quantities are therefore also determined. The problems in *De numeris datis*, however, are algebraic rather than geometric. Jordanus' proofs are also algebraic, or, perhaps, arithmetical. In fact, one of his aims is apparently to base the new algebra on arithmetic, the most fundamental of the subjects of the quadrivium, rather than on geometry. He also organized his subject in a logical fashion and, in a major departure from his Euclidean model, provided numerical examples for most of his theoretical results.

Although many of the actual problems and the numerical examples were available in the Islamic algebras, Jordanus adapted them to his own purposes. In particular, he made the major change of using letters to stand for arbitrary numbers. Jordanus' algebra is no longer entirely rhetorical. That is not to say that his symbolism looks modern. He picks his letters in alphabetical order with no distinction between letters representing known quantities and those representing unknowns and uses no symbols for operations. Sometimes a single number is represented by two letters. At other times the pair of letters ab represents the sum of the two numbers a and b. The basic arithmetic operations are always written in words. And Jordanus does not use the new Hindu-Arabic numerals. All of his numbers are written as Roman numerals. Nevertheless, the idea of symbolism, so crucial to any major advance in algebraic technique, is found, at least in embryonic form, in Jordanus' work.

To understand Jordanus' contribution, we consider a few of the text's more than 100 propositions, organized into four books. Like Euclid, Jordanus writes each proposition in a standard form. The general enunciation is followed by a restatement in terms of letters. By use of general rules, the letters representing numbers are manipulated into a canonical form from which the general solution can easily be found. Finally, a numerical example is calculated following the general outlines of the abstract solution. The canonical forms themselves are among the earliest of the propositions.

Proposition I-1.   *If a given number is divided into two parts whose difference is given, then each of the parts is determined.*

Jordanus' proof is straightforward: "Namely, the lesser part and the difference make the greater. Thus the lesser part with itself and the difference make the whole. Subtract therefore the difference from the whole and there will remain double the lesser given number. When divided [by two], the lesser part will be determined; and therefore also the greater part. For example, let 10 be divided in two parts of which the difference is 2. When this is subtracted from 10 there remains 8, whose half is 4, which is thus the lesser part. The other is 6."[21]

In modern symbolism, Jordanus' problem amounts to the solution of the two equations $x + y = a$, $x - y = b$. Jordanus notes first that $y + b = x$, so that $2y + b = a$ and therefore $2y = a - b$. Thus $y = \frac{1}{2}(a - b)$ and $x = a - y$.

Jordanus uses this initial proposition in many of the remaining problems of Book I. For example, consider

Proposition I-3.   *If a given number is divided into two parts, and the product of one by the other is given, then of necessity each of the two parts is determined.*

This proposition presents one of the standard Babylonian problems: $x + y = m$, $xy = n$. Jordanus' method of solution, however, is different from the classic Babylonian solution, and, in addition, he uses symbolism as indicated: Suppose the given number abc is divided into the parts ab and c. Suppose ab multiplied by c is d and abc multiplied by itself is e. Let f be the quadruple of d, and g be the difference of e and f. Then g is the square of the difference between ab and c. Its square root b is then the difference between ab and c. Since now both the sum and difference of ab and c are given, both ab and c are determined according to the first proposition. Jordanus' numerical example has 10 as the sum of the two parts and 21 as the product. He notes that 84 is quadruple 21, that 100 is 10 squared, and that 16 is their difference. Then the square root of 16, namely 4, is the difference of the two parts of 10. By the proof of the first proposition, 4 is subtracted from 10 to get 6. Then 3 is the desired smaller part while 7 is the larger.

Jordanus' solution, translated into modern symbolism, amounts to using the identity $(x - y)^2 = (x + y)^2 - 4xy = m^2 - 4n$ to determine $x - y$ and reduce the problem to proposition I-1. The solution is then $x = m - \frac{1}{2}\left(m - \sqrt{m^2 - 4n}\right)$, $y = \frac{1}{2}\left(m - \sqrt{m^2 - 4n}\right)$. Jordanus' method appears to be new with him, and he continues to use his own methods throughout the work. Thus in proposition I-4, which proposes to solve the system $x + y = m$, $x^2 + y^2 = n$, Jordanus first determines $2xy$ from the identity $2xy = (x + y)^2 - (x^2 + y^2)$. Since $(x - y)^2 = x^2 + y^2 - 2xy$, both the sum and difference of the unknowns are determined and by I-1, x and y can be found. Similar methods are used in propositions I-5 and I-6, the solution of the systems $x - y = m$, $xy = n$ and $x - y = m$, $x^2 + y^2 = n$ respectively.

Proposition I-7 gives us Jordanus' first example of a pure quadratic equation, although in the context of the system $x + b = y$, $x^2 + bx = a$. Jordanus in effect notes first that the second equation can be written as $(x + b)x = a$, thus reducing the problem to the system of proposition I-5. Jordanus' numerical example, $x^2 + 6x = 40$ has the solution $x = 4$, so y, the sum of the two parts x and b, is 10. In fact, every proposition in Book I deals with a number divided into two parts. Although some of the problems are more complex than the ones already discussed, in every numerical example but one the number to be divided is 10. The solution methods may differ somewhat from those in the Islamic texts, but it is clear that al-Khwārizmī's problems live on!

Many of the propositions in the remaining three books of Jordanus' treatise deal with numbers in given proportion. They demonstrate his fluency in dealing with the rules of proportion found in Book V of Euclid's *Elements*. Consider

Proposition II-18. *If a given number is divided into however many parts, whose continued proportions are given, then each of the parts is determined.*

Since Jordanus, like his contemporaries, had no way to express arbitrarily many parts, he deals in his proof with a number divided into three parts: $a = x + y + z$. Then $x : y = b$ and $y : z = c$ are both known ratios. Jordanus notes that the ratio $x : z$ is also known. It follows that the ratio of x to $y + z$ is known and therefore also that of a to x. Since a is known, x and then y and z can be determined. His example enables us to follow his verbal description. The number 60 is divided into three parts, of which the first is double the second and the second is triple the third. That is, $x + y + z = 60$, $x = 2y$, $y = 3z$. Then $x = 6z$ and therefore $y + z = \frac{2}{3}x$. So $60 = 1\frac{2}{3}x$, and $x = 36$, $y = 18$, $z = 6$. One notes that Jordanus easily inverts ratios if necessary and also knows how to combine them.

Among the propositions in Book IV are three giving the three standard forms of the quadratic equation, all presented with algebraic rather than geometric justifications. For these problems, however, Jordanus does use the standard Islamic algorithm. Consider

Proposition IV-9. *If the square of a number added to a given number is equal to the number produced by multiplying the root and another given number, then two values are possible.*

Thus Jordanus asserts that there are two solutions to the equation $x^2 + c = bx$. He then gives the procedure for solving the equation: take half of b, square it to get f, and let g be the difference of x and $\frac{1}{2}b$, that is,

$$g = \pm\left(x - \frac{1}{2}b\right).$$

Then,

$$x^2 + f = x^2 + c + g^2 \qquad \text{and} \qquad f = c + g^2.$$

Jordanus concludes by noting that x may be obtained by either subtracting g from $b/2$ or by adding g to $b/2$. Jordanus' example makes his symbolic procedure clearer. To solve

$x^2 + 8 = 6x$, he squares half of 6, giving 9, and then subtracts 8 from it, leaving 1. The square root of 1 is 1, and this is the difference between x and 3. Hence x can be either 2 or 4.

Among the other quadratic problems Jordanus solves in Book IV are the systems $xy = a, x^2 + y^2 = b$ and $xy = a, x^2 - y^2 = b$. In each case, as in all the previous cases, the given example results in a positive integral answer. While Jordanus often uses fractions as part of his solution, he has carefully arranged matters so that final answers are always whole numbers. If, in fact, he had read Abū Kāmil's *Algebra*, which had been translated by Gerard of Cremona, Jordanus would have seen nonintegral, and even nonrational, solutions to this type of problem. He nevertheless rejected such solutions when he made up his examples. Given his very formal style, however, Jordanus may still have been under the influence of Euclid and have felt that irrational numbers simply did not belong in a work based on arithmetic. Jordanus himself wrote a theoretical arithmetic work modeled after Books VII–IX of the *Elements*, including definitions, postulates, and axioms. This *Arithmetica* was solely concerned with "numbers" in the Euclidean sense. Hence, although *De numeris datis* represents an advance from the Islamic works in the use of analysis and in some symbolization, it returns to the strict Greek separation of number from magnitude, an idea from which Jordanus' Islamic predecessors had already departed.

8.4 THE MATHEMATICS OF KINEMATICS

The algebraic work of Jordanus de Nemore was not developed further in the thirteenth century, even though a group of followers had appeared in Paris by the middle of that century. Perhaps Europe was not then ready to resume the study of pure mathematics. By early in the fourteenth century, however, certain other aspects of mathematics began to develop in the universities of Oxford and Paris out of attempts to clarify certain remarks in Aristotle's physical treatises (Side Bar 8.2).

8.4.1 The Study of Ratios

One of the new mathematical ideas came from the effort to derive a relationship among the force F applied to an object, its resistance R, and its velocity V. A basic postulate of medieval physics was that F must be greater than R for motion to be produced. (The medieval philosophers did not attempt to measure these quantities in any particular units.) The simplest relationship among these quantities implied by Aristotle's own words may be expressed by the statement that F/R is proportional to V. This mathematical relationship, however, quickly leads to a contradiction of the postulate. For if F is left fixed, the continual doubling of R is equivalent to the continual halving of V. Halving a positive velocity keeps it positive, but the doubling of R eventually makes R greater than F, thus contradicting the notion that F must be greater than R for motion to take place.

Thomas Bradwardine (1295–1349) of Merton College, Oxford, in his 1328 *Tractatus de proportionibus velocitatum in motibus* (*Treatise on the Proportions of Velocities in*

Side Bar 8.2   The Medieval Universities

It was during the late twelfth century that Europe saw the beginnings of the institutions which were to have immense influence in the development of science in general and mathematics in particular, the universities. We cannot assign any specific date for the origins of the earliest universities. They were formed as societies, or guilds, of masters and pupils and appeared on the scene when there was enough learning available in western Europe to justify their existence. The earliest of these institutions were in Paris, Oxford, and Bologna. In Paris, the university grew out of the cathedral school of Notre Dame. The masters and students gradually grouped themselves into the four faculties of arts, theology, law, and medicine. Although there is evidence of the existence of the university in the late twelfth century, the first official charter dates from 1200. The University at Oxford emerged not out of a church school but from a group of English students who had returned from Paris. Again, although the university certainly existed in the late twelfth century, the first official document dates from 1214. At Bologna, the university began as a law school, perhaps as early as the eleventh century. The Italian university differed from its northern counterparts, however, since it was a guild of students rather than one of professors that initially constituted the organization. The students elected the professors and other officials. Student control was somewhat weakened, however, because salaries were paid by the Bolognese municipality and the faculty conducted the examinations.

The curriculum in arts at all of the universities was based on the ancient trivium of logic, grammar, and rhetoric and the quadrivium of arithmetic, geometry, music, and astronomy. This study in the faculty of arts provided the student with preparation for the higher faculties of law, medicine, or theology. The centerpiece of the arts curriculum was the study of logic, and the primary texts for this were the logical works of Aristotle, all of which had by this period been translated into Latin. The masters felt that logic was the appropriate first area of study since it taught the methods for all philosophic and scientific inquiry. Gradually, other works of Aristotle were also added to the curriculum. For several centuries, the great philosopher's works were the prime focus of the entire arts curriculum. Other authors were studied insofar as they allowed one a better understanding of this most prolific of the Greek philosophers. In particular, mathematics was studied only as it related to the work of Aristotle in logic or the physical sciences. The mathematical curriculum itself—the quadrivium—usually consisted of arithmetic, taken from such works as Boethius' adaptation of Nicomachus or a medieval text on rules for calculation; geometry, taken from Euclid and one of the practical geometries; music, taken also from a work of Boethius; and astronomy, taken from Ptolemy's *Almagest* and some more recent Latin translations of Islamic astronomical works.

Movements), proposed a solution to this dilemma, that is, a "correct" interpretation of Aristotle's remarks. The rule noted above implies that for two forces F_1, F_2, two resistances R_1, R_2, and two velocities V_1, V_2, the equation

$$\frac{F_2}{R_2} = \frac{V_2}{V_1}\frac{F_1}{R_1}$$

is satisfied. Bradwardine suggested that this should be replaced by the relationship expressed in modern notation as

$$\frac{F_2}{R_2} = \left(\frac{F_1}{R_1}\right)^{\frac{v_2}{v_1}}.$$

In other words, the multiplicative relation should be replaced by an exponential one. This solution does indeed remove the absurdity noted above. Given initially that $F > R$ (or $F/R > 1$), halving the velocity in this situation is equivalent to taking roots of the ratio F/R. Consequently, F/R will remain greater than 1, and R will never be greater than F. Neither Bradwardine nor anyone else in this period, however, attempted to give any experimental justification for this relationship. The scholars at Merton wanted a mathematical explanation of the world, not a physical one. As it turned out, Bradwardine's idea was discarded as a physical principle by the middle of the next century, but the mathematics behind it led to important new ideas. To deal with these required a systematic study of ratios, in particular of the idea of compounding (or multiplying) ratios.

Up to the fourteenth century, compounding was performed in the classical Greek style. Thus, to deal with the ratio compounded of $a : b$ and $c : d$, one needed to find a magnitude e such that $c : d = b : e$. Then the compound ratio would be $a : e$. Gradually, however, the more explicit notion of multiplication of ratios was introduced. For example, Bradwardine's contemporary at Oxford, Richard of Wallingford, defined ratios as well as their compounding and dividing in part II of his *Quadripartitum*:

1. A **ratio** is a mutual relation between two quantities of the same kind.

2. When one of two quantities of the same kind divides the other, what results from the division is called the **denomination** of the ratio of the dividend to the divisor.

3. A ratio [is said to be] **compounded** of ratios when the product of the denominations gives rise to some denomination.

4. A ratio [is said to be] **divided** by a ratio when the quotient of the denominations gives rise to some denomination.[22]

There are several important notions here. First, Richard emphasizes that ratios can be taken only between quantities of the same kind. This Euclidean idea meant that velocity could not be treated as a ratio of distance to time. Second, the word *denomination* in these definitions refers to the "name" of the ratio in "lowest terms" as given in the terminology due to Nicomachus now standard in Europe. For example, the ratio 3 : 1 is called a triple ratio while that of 3 : 2 is called the sesquialter. Finally, definitions 3 and 4 show that for Richard, unlike for Euclid, multiplication (of numbers) was involved in compounding, and the inverse notion of division could also be applied. Thus, although he compounded the ratios 4 : 16, 16 : 2, and 2 : 12 to get 4 : 12, he noted that since the first ratio is a subquadruple (1 : 4), the second an octuple (8 : 1), and the third a subsextuple (1 : 6), the compound ratio can also be formed by first dividing 8 by 4 to get 2, and then dividing 6 by 2 to get 1 : 3 (a subtriple) as the final result. Thus one can actually use the standard algorithm for multiplying fractions to compound ratios.

Nicole Oresme (1320–1382), a French cleric and mathematician associated with the University of Paris, undertook a very detailed study of ratios in his *Algorismus proportionum* (*Algorithm of Ratios*) and his *De proportionibus proportionum* (*On the Ratios of Ratios*) a few years later. In addition to performing compounding in the traditional manner, Oresme noted explicitly that one can also compound ratios by multiplying the antecedents and then multiplying the consequents. Thus $4 : 3$ compounded with $5 : 1$ is $20 : 3$. The connecting link between the two methods is presumably that $a : b$ can be expressed as $ac : bc$, $c : d$ as $bc : bd$, and hence the compound of $a : b$ with $c : d$ as the compound of $ac : bc$ with $bc : bd$ or as $ac : bd$. In any case, given a way of multiplying two ratios, Oresme also noted that one could reverse the procedure and divide two ratios. Thus the quotient of $a : b$ by $c : d$ was the ratio $ad : bc$.

Now that the product of any two ratios had been defined, Oresme discussed the product of a given ratio with itself. Thus $a : b$ compounded with itself n times gives what would now be written as $(a : b)^n$. More importantly, given any ratio, Oresme devised a language for discussing what are now called "roots" of that ratio. Thus since $2 : 1$ is a double ratio, Oresme called that ratio which when compounded twice with itself equals $2 : 1$ half of a double ratio. (In modern terminology, this is the ratio $(2 : 1)^{1/2}$). Similarly, he called $(3 : 1)^{3/4}$ three fourth parts of a triple ratio. Oresme next developed an arithmetic for these ratios. For example, to multiply $(2 : 1)^{1/3}$ by $3 : 2$, Oresme cubed the second ratio to get $27 : 8$, multiplied this by $2 : 1$ to get $27 : 4$ and then took the cube root of the ratio considered as a fraction to get $\left(6\frac{3}{4}\right)^{1/3}$. Similarly to divide $(2 : 1)^{1/2}$ by $4 : 3$, he divided $2 : 1$ by the square of $4 : 3$, namely $16 : 9$, to get $9 : 8$ and then took the square root of that, $(9 : 8)^{1/2}$. In some sense, then, Oresme's works show for the first time operational rules for dealing with exponential expressions with fractional exponents.

Oresme even attempted to deal with what we would call irrational exponents. He felt intuitively that "every ratio is just like a continuous quantity with respect to division," that is, that one could take any possible "part" of such a ratio. So, "there will be some ratio which will be part of a double ratio and yet will not be half of a double nor a third part or fourth part or two-thirds part, etc., but it will be incommensurable to a double and, consequently [incommensurable] to any [ratio] commensurable to this double ratio."[23] Because Oresme had no notation for irrational exponents, he could only convey his sense of them negatively. Namely, he felt that ratios of the form $(2 : 1)^r$ should exist even when r was not a rational number. "And further, by the same reasoning there could be some ratio incommensurable to a double and also to a triple ratio and [consequently incommensurable] to any ratios commensurable to these. . . . And there might be some irrational ratio which is incommensurable to any rational ratio. Now the reason for this seems to be that if some ratio is incommensurable to two [rational ratios] and some ratio is incommensurable to three rational ratios and so on, then there might be some ratio incommensurable to any rational ratio whatever. . . . However, I do not know how to demonstrate this."[24] What Oresme was apparently expressing, in terms of modern ideas, was that since the number line is continuous and since, for example, the fractional powers of 2 do not exhaust all (real) numbers, there must be (nonfractional) powers of 2 equal to the real numbers not already included. In fact, somewhat later in the text he states a theorem to the effect that irrational ratios are much more prevalent than rational ones:

Proposition III-10.   *It is probable that two proposed unknown ratios are incommensurable because if many unknown ratios are proposed it is most probable that any [one] would be incommensurable to any [other].*

Although Oresme had no formal way of proving this result, he noted that if one considers all the integral ratios from $2 : 1$ up to $101 : 1$, there are 4950 ways of comparing these two by two in terms of exponents (always comparing a greater ratio to a smaller), but only 25 ways with rational exponents. For example, $4 : 1 = (2 : 1)^2$ and $8 : 1 = (4 : 1)^{3/2}$. On the other hand, there is no rational exponent r such that $3 : 1 = (2 : 1)^r$. Oresme then used a probability argument to conclude that astrology must be fallacious. His argument is that with great probability the ratio of any two unknown ratios, for example those which represent various celestial motions, will be irrational. Since, therefore, there can be no exact repetitions of planetary conjunctions or oppositions, and since astrology rests on such endless repetitions, the whole basis of that science is false.

8.4.2 Velocity

The efforts to turn Aristotle's ideas on motion into quantitative results also resulted in new mathematics. In particular, these ideas were developed in the early fourteenth century by Bradwardine and another scholar at Merton College, William Heytesbury. Recall that Greek mathematicians, including Autolycus and Strato, had dealt with the notion of uniform velocity and, to some extent, accelerated motion, but never considered velocity or acceleration as independent quantities which could be measured. Velocities were only dealt with by comparing distances and times, and therefore, in essence, only average velocities (over certain time periods) could be compared.

In the fourteenth century, however, there appeared the beginning of the notion of velocity, and in particular instantaneous velocity, as measurable entities. Thus Bradwardine in his *Tractatus de continuo* (*Treatise on the Continuum*) (c. 1330) defined the "grade" of motion as "that part of the matter of motion susceptible to 'more' and 'less'."[25] Bradwardine then showed how to compare velocities: "In the case of two local motions which are continued in the same or equal times, the velocities and distances traversed by these [movements] are proportional, i.e., as one velocity is to the other, so the space traversed by the one is to the space traversed by the other . . . In the case of two local motions traversing the same or equal spaces, the velocities are inversely proportional to the time, i.e., as the first velocity is to the second, so the time of the second velocity is to the time of the first."[26] In other words, if two objects travel at (uniform) velocities v_1, v_2 respectively in times t_1, t_2 and cover distances s_1, s_2, then (1) if $t_1 = t_2$, then $v_1 : v_2 = s_1 : s_2$, and (2) if $s_1 = s_2$, then $v_1 : v_2 = t_2 : t_1$. Bradwardine thus considered uniform velocity itself as a type of magnitude, capable of being compared with other velocities.

Heytesbury, only a few years later in his *Regule solvendi sophismata* (*Rules for Solving Sophisms*) (1335), gave a careful definition of instantaneous velocity for a body whose motion is not uniform: "In nonuniform motion . . . the velocity at any given instant will be measured by the path which would be described by the . . . point if, in a period of time, it were moved uniformly at the same degree of velocity with which it is moved in that given instant, whatever [instant] be assigned."[27]

Heytesbury also dealt with acceleration in this same section: "Any motion whatever is uniformly accelerated if, in each of any equal parts of the time whatsoever, it acquires an equal increment of velocity. . . . But a motion is nonuniformly accelerated . . . when it acquires . . . a greater increment of velocity in one part of the time than in another equal part. . . . And since any degree of velocity whatsoever differs by a finite amount from zero velocity . . . , therefore any mobile body may be uniformly accelerated from rest to any assigned degree of velocity."[28] This statement provides not only a very clear definition of uniform acceleration, but also, in nascent form at least, the notion of velocity changing with time.

How does one determine the distance traveled by a body being uniformly accelerated? The answer, generally known today as the **mean speed rule**, was first stated by Heytesbury in this same work: "When any mobile body is uniformly accelerated from rest to some given degree [of velocity], it will in that time traverse one-half the distance that it would traverse if, in that same time it were moved uniformly at the [final] degree [of velocity] . . . For that motion, as a whole, will correspond to . . . precisely one-half that degree which is its terminal velocity."[29] In modern notation, if a body is accelerated from rest in a time t with a uniform acceleration a, then its final velocity is $v_f = at$. What Heytesbury is saying is that the distance traveled by this body is $s = (1/2)v_f t$. Substituting the first formula in the second gives the standard modern formulation $s = (1/2)at^2$.

Heytesbury gave a proof of the mean speed theorem by an argument from symmetry, taking as his model a body d accelerating uniformly from rest to a velocity of 8 in one hour. (The number 8 does not represent any particular speed, but is just used as the basis for his example.) He then considers three other bodies, a moving uniformly at a speed of 4 throughout the hour, b accelerating uniformly from 4 to 8 in the first half hour, and c decelerating uniformly from 4 to 0 in that same half hour. First, he notes that body d goes as far in the first half hour as does c and as far in the second half hour as does b. Therefore, d travels as far in the whole hour as the total of b and c in the half hour. Second, he argues that since b increases precisely as much as c decreases, together they will traverse as much distance in the half hour as if they were both held at the speed of 4. This latter distance is the same that a travels in the whole hour. It follows that d goes exactly as far as does a in the hour, and the mean speed theorem is demonstrated, at least to Heytesbury's satisfaction. He then proves the easy corollary, that the body d traverses in the second half hour exactly three times the distance it covered in the first half hour.

Other scholars at Merton College in the same time period began to explore the idea of representing velocity, as well as other varying quantities, by line segments. The basic idea seems to come, in effect, from Aristotle, since both velocities and line segments were conceived of as magnitudes in the Greek philosopher's distinction between the two types of quantities. Both were infinitely divisible, and hence it was not unreasonable to attempt to represent the somewhat abstract idea of velocity by the concrete geometric idea of a line segment. Velocities of different "degrees" would thus be represented by line segments of different lengths. Oresme carried this idea to its logical conclusion by introducing a two dimensional representation of velocity changing with respect to time. In fact, in his *Trac-*

FIGURE 8.8
Uniform velocity.

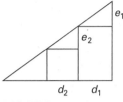

FIGURE 8.9
Uniformly difform
velocity, where
$d_1 : d_2 = e_1 : e_2$.

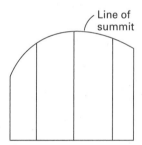

FIGURE 8.10
Difformly difform
velocity, or nonuniform
acceleration.

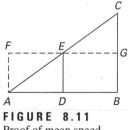

FIGURE 8.11
Proof of mean speed
theorem due to Oresme.

tatus de configurationibus qualitatum et motuum (*Treatise on the Configuration of Qualities and Motions*) of about 1350, Oresme even generalized this idea to other cases where a given quantity varied in intensity over either distance or time.

Oresme began by explaining why one can use lines to represent such quantities as velocity:

> Every measurable thing except numbers is imagined in the manner of continuous quantity. Therefore, for the mensuration of such a thing, it is necessary that points, lines, and surfaces, or their properties, be imagined. For in them, as [Aristotle] has it, measure or ratio is initially found . . . Therefore, every intensity which can be acquired successively ought to be imagined by a straight line perpendicularly erected on some point.[30]

From these straight lines Oresme constructed what he called a **configuration**, a geometrical figure consisting of all the perpendicular lines drawn over the base line. In the case of velocities, the base line represented time, while the perpendiculars represented the velocities at each instant. The entire figure represented the whole distribution of velocities, which Oresme interpreted as representing the total distance traveled by the moving object. Oresme did not use what we call coordinates. There was no particular fixed length by which a given degree of velocity was represented. The important idea was only that "equal intensities are designated by equal lines, a double intensity by a double line, and always in the same way if one proceeds proportionally."[31]

For Oresme, then, a uniform quality, for example, a body moving with uniform velocity, is represented by a rectangle, for at each point the velocity is the same (Figure 8.8). The area of the rectangle represents the total distance traveled. The distance traveled by a body beginning at rest and then moving with constant acceleration, representing what Oresme calls a "uniformly difform" quality, one whose intensity changes uniformly, is the area of a right triangle (Figure 8.9). As Oresme notes, "A quality uniformly difform is one in which if any three points [of the subject line] are taken, the ratio of the distance between the first and the second to the distance between the second and the third is as the ratio of the excess in intensity of the first point over that of the second point to the excess of that of the second point over that of the third point, calling the first of those three points the one of greatest intensity."[32] This equality of ratios naturally defines a straight line, the hypotenuse of the right triangle. Finally, a "difformly difform" quality, such as nonuniform acceleration, is represented by a figure whose "line of summit" is a curve which is not a straight line (Figure 8.10). Namely, Oresme in essence has developed the idea of representing the functional relationship between velocity and time by a curve. In fact, he notes, "the aforesaid differences of intensities cannot be known any better, more clearly, or more easily than by such mental images and relations to figures."[33] In other words, this geometrical representation of varying quantities provides the best way to study them.

Given this representation of the motion of bodies, it was easy for Oresme to give a geometrical proof of the mean speed theorem. For if $\triangle ABC$ represents the configuration of a body moving with uniformly accelerated motion from rest, and if D is the midpoint of the base AB, then the perpendicular DE represents the velocity at the midpoint of the journey and is half the final velocity (Figure 8.11). The total distance traveled, represented

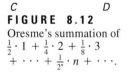

FIGURE 8.12
Oresme's summation of
$\frac{1}{2} \cdot 1 + \frac{1}{4} \cdot 2 + \frac{1}{8} \cdot 3$
$+ \cdots + \frac{1}{2^n} \cdot n + \cdots$.

by triangle ABC, is then equal to the area of the rectangle $ABGF$, precisely as stated by the Mertonians.

Oresme's geometric technique reappeared some 250 years later in the work of Galileo. The difference between the two lay mainly in that Galileo assumed that uniform acceleration from rest was the physical rule obeyed by bodies in free fall, while Oresme was studying the subject only abstractly. This abstraction is evident in Oresme's consideration of cases involving velocities increasing without bound. For example, he considers the case where the velocity of an object during the first half of the time interval AB, taken equal to 1 unit, is equal to 1, that in the next quarter is 2, that in the next eighth is 3, in the next sixteenth 4, and proceeds to calculate the total distance traveled. In effect, he is summing the infinite series

$$\frac{1}{2} \cdot 1 + \frac{1}{4} \cdot 2 + \frac{1}{8} \cdot 3 + \cdots + \frac{1}{2^n} \cdot n + \cdots.$$

His result is that the sum, representing the total distance, is 2, or, as he puts it, "precisely four times what is traversed in the first half of the [time]."[34] His proof, given geometrically, is very elegant. He draws a square of base CD equal to $AB(=1)$ and divides it "to infinity into parts continually proportional according to the ratio 2 to 1" (Figure 8.12). Namely, E represents half of the square, F one quarter, G one eighth, and so on. The rectangle E is placed over the right half of the square on AB, F atop the new configuration over its right quarter, G atop the right eighth, and so on. It is then evident that the total area of the new configuration, which represents the total distance traveled, is not only equal to the sum of the infinite series but also equal to the sum of the areas of the two original squares.

Oresme's idea of representing velocities, as well as other qualities, geometrically, was continued in various works by others over the next century. However, no one was able to extend the representation of distances to situations more complex than Oresme's uniformly difform qualities. Eventually, even this idea was lost. Much the same fate befell the ideas of the other major European mathematicians of the medieval period. Their works were not studied and their new ideas had to be rediscovered centuries later. This lack of progress is evident in the stagnant mathematical curricula at the universities. With the works of Aristotle continuing to be the basis of the curriculum, the only mathematics studied was that which was of use in understanding the works of the great philosopher. Although an Oresme might carry these ideas further, such men were rare. In addition, the ravages of the Black Death and the Hundred Years War caused a marked decline in learning in France and England. It was therefore in Italy and Germany that a few of the ideas of the medieval French and English mathematicians found other mathematicians to study them. It was there that new ideas were generated in the Renaissance.

Exercises

Problems from Alcuin's **Propositiones ad acuendos juvenes**

1. A cask is filled to 100 *metreta* capacity through three pipes. One-third of its capacity plus 6 *modii* flows in through one pipe; one-third of its capacity flows in through another pipe; but only one-sixth of its capacity flows in through the third pipe. How many *sextarii* flow in through each pipe? (Here a *metreta* is 72 *sextarii* and a *modius* is 200 *sextarii*.)[35]

2. A man must ferry across a river a wolf, a goat, and a head of cabbage. The available boat, however, can only carry the man and one other thing. The goat cannot be left alone with the cabbage, nor the wolf with the goat. How should the man ferry his three items across the river?

3. A hare is 150 paces ahead of a hound which is pursuing him. If the hound covers 10 paces each time the hare covers 6, in how many paces will the hound overtake the hare?

Problems from Abraham bar Ḥiyya

4. Given a chord of length 6 in a circle of diameter 10 1/2 find the length of the arc cut off by the chord.

5. Find the area of the circle segment determined by the chord in problem 4.

6. Find the length of the chord which cuts off an arc of length 5 1/2 in a circle of diameter 33.

7. If a chord of length 8 has distance 2 from the circumference, find the diameter of the circle.

8. In a rectangle whose diagonal is 10 and whose length exceeds the width by 2 find the length, the width, and the area.

9. How long is the side of a rhombus, if one diagonal is 16 and the other 12?

Problems in Practical Geometry

10. Show that

$$A = \frac{3n^2 - n}{2}$$

provides a formula for the nth pentagonal number. Calculate the area of regular pentagons with sides of length $n = 1, 2, 3$ and compare to the answer given by the formula. How close an approximation does the given formula provide?

11. Use Leonardo's table of chords to solve the following: Suppose a given chord in a circle of diameter 10 is 8 rods, 3 feet, 16 2/7 unciae. Find the length of the arc cut off by the chord.

12. From Leonardo's *Practica geometriae*: Given the quadrilateral inscribed in a circle with $ab = ag = 10$ and $bg = 12$, find the diameter ad of the circle (Figure 8.13).

13. Using Figure 8.13, suppose now that $ag = 15$, $ab = 13$, and $bg = 14$. Find the diameter ad.

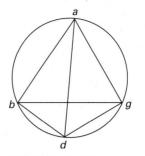

FIGURE 8.13
Determining the diameter of a circle in the work of Leonardo of Pisa.

Problems from Levi ben Gerson

14. Prove this theorem from Levi's *Trigonometry*: If all sides of any triangle whatever are known, its angles are also known. Start by dropping a perpendicular from one vertex to the opposite side (or opposite side extended) and show how one can calculate the angles.

15. Prove the general combinatoric rule by induction:

$$C_k^n = \sum_{i=k-1}^{n-1} C_{k-1}^i$$

16. Prove proposition 30 from the *Maasei Hoshev*: $(1 + 2 + \cdots + n) + (1 + 2 + \cdots + (n - 1)) = n^2$.

17. Prove proposition 32 of the *Maasei Hoshev*:

$$1 + (1 + 2) + (1 + 2 + 3) + \cdots + (1 + 2 + \cdots + n) = \begin{cases} 1^2 + 3^2 + \cdots + n^2 & n \text{ odd}; \\ 2^2 + 4^2 + \cdots + n^2 & n \text{ even}. \end{cases}$$

18. Prove proposition 33 of the *Maasei Hoshev*:

$$(1 + 2 + 3 + \cdots + n) + (2 + 3 + \cdots + n) + (3 + \cdots + n) + \cdots + n = 1^2 + 2^2 + \cdots + n^2$$

19. Prove proposition 34 of the *Maasei Hoshev*:

$$[(1 + 2 + \cdots + n) + (2 + 3 + \cdots + n) + \cdots + n] + [1 + (1 + 2) + \cdots + (1 + 2 + \cdots + (n - 1))]$$
$$= n(1 + 2 + \cdots + n)$$

20. Use the three previous results to prove:

$$1^2 + 2^2 + \cdots + n^2 =$$
$$[n - \frac{1}{3}(n - 1)][1 + 2 + \cdots + n]$$

Problems from Leonardo of Pisa

21. The Fibonacci sequence (the sequence of rabbit pairs) is determined by the recursive rule $F_0 = F_1 = 1$ and $F_n = F_{n-1} + F_{n-2}$. Show that

$$F_{n+1} \cdot F_{n-1} = F_n^2 - (-1)^n$$

and that

$$\lim_{n \to \infty} \frac{F_n}{F_{n-1}} = \frac{1 + \sqrt{5}}{2}.$$

22. Five men each having money find a purse containing a certain amount of money. If the first man took the amount in the purse, he would have double the total of the second and the third. If the second took the amount he would have triple the total of the third and the fourth. Similarly, the third would have four times the total of the fourth and fifth, the fourth five times that of the fifth and first, and the fifth six times that of the first and second. How much money did each man have originally and how much was in the purse?

23. Find a number which when divided by 2 has remainder 1; by 3 has remainder 2; by 4, 3; by 5, 4; by 6, 5; and is evenly divisible by 7. (Leonardo begins by finding the smallest number evenly divisible by 2, 3, 4, 5, 6, namely, 60.)

24. Divide 10 into two parts; divide the larger by the smaller and multiply the quotient by the difference of the parts. The result is 24. Find the parts.

25. Prove that Leonardo's "congruous" numbers are always divisible by 24.

26. From the *Book of Squares*: Find a square number for which the sum of it and its root is a square number and for which the difference of it and its root is similarly a square number. (In modern notation, find x, y, z such that $x^2 + x = z^2$ and $x^2 - x = y^2$. Leonardo begins his solution by using the congruous number 24 in solving $a^2 + 24 = b^2, a^2 - 24 = c^2$; he then divides everything by 24.)

27. From the *Book of Squares*: Find a square number which when twice its root is added or subtracted always makes a square number. (Again, begin with the basic result on congruous numbers from problem 26.)

Problems from Jordanus de Nemore

28. If the sum of the product of the two parts of a given number and of their difference is known, then each of them is determined. Namely, solve the system $x + y = a, xy + x - y = b$. Use Jordanus' example where $a = 9$ and $b = 21$.

29. If a given number is divided into two parts and one is multiplied by a given number, and if this product is divided by one of the parts and the result is given, then each of the parts is determined. The example given by Jordanus is $x + y = 10, 5x/y = 7\ 1/2$.

30. If the sum of two numbers is given together with the product of their squares, then each of them is determined. Jordanus' example is $x + y = 9, x^2y^2 = 324$.

Problems from Oxford and Paris

31. Use Oresme's technique to divide a sesquialterate ratio $(3 : 2)$ by a third part of a double ratio $(2 : 1)^{1/3}$.

32. Show that there are in fact 4950 ways of comparing (by ratio) the 100 integral ratios from $2 : 1$ up to $101 : 1$ and that precisely 25 will have rational exponents.

33. Show that under the assumptions of the mean speed theorem, if one divides the time interval into four equal subintervals, the distances covered in each interval will be in the ratio $1 : 3 : 5 : 7$. Generalize this statement to a division of the time interval into n equal subintervals and prove your result.

34. From Oresme's *Tractatus de configurationibus qualitatum et motuum*: Show geometrically that the sum of the series

$$48 \cdot 1 + 48 \cdot \frac{1}{4} \cdot 2 + 48 \cdot \left(\frac{1}{4}\right)^2 \cdot 4$$
$$+ \cdots + 48\left(\frac{1}{4}\right)^n \cdot 2^n + \cdots$$

is equal to 96.

35. Solve the following problem of Oresme: Divide the line AB of length 1 (representing time) proportionally to infinity in a ratio of $2 : 1$; that is, divide it so the first part is one-half, the second one-quarter, the third one-eighth, and so on. Let there be a given finite velocity (say, 1) in the first interval, a uniformly accelerated velocity (from 1 to 2) in the second, a constant velocity (2) in the third, a uniformly accelerated velocity (from 2 to 4) in the fourth, and so on. (Figure 8.14). Show that the total distance traveled is 7/4.

36. Prove the result of Oresme: $1 + 1/2 + 1/3 + 1/4 + \cdots$ becomes infinite. (This series is usually called the harmonic series.)

FOR DISCUSSION ...

37. Determine what mathematics was necessary to solve the Easter problem. What was the result of the debate in the Church? How is the date of Easter determined today?

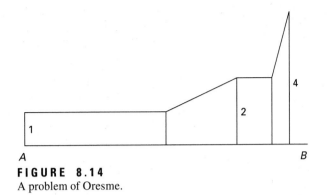

FIGURE 8.14
A problem of Oresme.

38. There were many works on arithmetic written in the medieval period, among which are those due to Abraham bar Ḥiyya and Jordanus de Nemore. Determine the types of algorithms dealt with by the various authors and compare their treatments. One place to begin is Martin Levey, "Abraham Savasorda and his Algorism: A Study in Early European Logistic," *Osiris* 11 (1954), 50–64.

39. Compare Levi ben Gerson's use of "induction" to that of al-Karajī. Should the methods of either be considered "proof by induction?" Discuss.

40. Write a lesson demonstrating proof by induction using some of Levi ben Gerson's examples.

41. Write a lesson developing some of the basic combinatoric rules using the methods of Abraham ibn Ezra and Levi ben Gerson.

42. Write a short class lesson demonstrating geometrically the sums of various infinite series. Start with those of Oresme and generalize.

References and Notes

Among the best general sources on the mathematics of medieval Europe are Marshall Clagett, *Mathematics and Its Applications to Science and Natural Philosophy in the Middle Ages* (Cambridge: Cambridge University Press, 1987) and David C. Lindberg, ed., *Science in the Middle Ages* (Chicago: University of Chicago Press, 1978). In particular, Chapter 5 on Mathematics by Michael S. Mahoney and Chapter 7 on The Science of Motion by John E. Murdoch and Edith D. Sylla in the latter work provide good surveys. An excellent collection of original source materials translated into English is in Edward Grant, ed., *A Source Book in Medieval Science* (Cambridge: Harvard University Press, 1974). Another source book dealing mainly with aspects of mechanics is Marshall Clagett, *The Science of Mechanics in the Middle Ages* (Madison: University of Wisconsin Press, 1961). Basic biographical information on mathematicians of the Middle Ages can be found in George Sarton, *Introduction to the History of Science* (Huntington, New York: Robert E. Krieger, 1975), especially Volumes II and III. Finally, a systematic survey of mathematics in the Middle Ages not only in Europe but also in Islam, China, and India is Adolf P. Juschkewitsch, *Geschichte der Mathematik im Mittelalter* (Leipzig: Teubner, 1964), a German translation of the Russian original of 1961.

1. Maximilian Curtze, "Der Liber Embadorum des Abraham bar Chijja Savasorda in der Übersetzung des Plato von Tivoli," *Abhandlungen zur Geschichte der mathematischen Wissenschaften* 12 (1902), 1–183, p. 11. This article provides a German translation as well as an edition of the Latin.

2. Leonardo Pisano Fibonacci, *The Book of Squares*, edited and translated by L. E. Sigler (Boston: Academic Press, 1987), p. 3. Sigler annotates and explains Leonardo's work and translates much of the mathematics into modern notation.

3. St. Augustine, *City of God*, XI, 30, quoting *Wisdom of Solomon*, 11:20.

4. Curtze, "Liber Embadorum," p. 35.

5. Stephen Victor, *Practical Geometry in the High Middle Ages. Artis Cuiuslibet Consummatio and the Pratike de Geometrie* (Philadelphia: American Philosophical Society, 1979), pp. 109–111. This work contains much general information on the tradition of practical geometry in the Middle Ages as well as the translation of the two texts of the title. Another practical geometry is discussed in H. L. Busard, "The *Practica Geometriae* of Dominicus de Calvasio," *Archive for History of Exact Sciences* 2 (1965), 520–575. Other similar works are discussed in Gillian Evans, "The 'Sub-Euclidean' Geometry of the Earlier Middle Ages, up to the Mid-Twelfth Century," *Archive for History of Exact Sciences* 16 (1976), 105–118.

6. Ibid., p. 221.

7. Ibid., p. 295.

8. Quoted in Nachum L. Rabinovitch, *Probability and Statistical Inference in Ancient and Medieval Jewish Literature* (Toronto: University of Toronto Press, 1973), p.

144. This book provides an interesting look at the beginnings of various central ideas in probability and statistics in the discussions of the rabbis of ancient and medieval times on various points of Jewish law. An older work on Jewish mathematics is M. Steinschneider, *Mathematik bei den Juden* (Hildesheim: Georg Olms, 1964), a reprint of original articles of 1893–1901.

9. J. Ginsburg, "Rabbi ben Ezra on Permutations and Combinations," *The Mathematics Teacher* 15 (1922), 347–356, p. 351.

10. Solomon Gandz, "The invention of the decimal fractions and the application of the exponential calculus by Immanuel Bonfils of Tarascon (c. 1350)," *Isis* 25 (1936), 16–45, p. 39.

11. Ibid., p. 41.

12. Gerson Lange, *Sefer Maasei Choscheb. Die Praxis der Rechners. Ein hebräisch-arithmetisches Werk des Levi ben Gerschom aus dem Jahre 1321* (Frankfurt: Louis Golde, 1909), p. 1. This book has a Hebrew edition and a German translation of Levi ben Gerson's work on combinatorics. The propositions which follow are all translated from the German. For more details on Levi's ideas on induction, see Nachum L. Rabinovitch, "Rabbi Levi ben Gershon and the Origins of Mathematical Induction," *Archive for History of Exact Sciences* 6 (1970), 237–248.

13. Ibid., p. 8.

14. Ibid., p. 49.

15. Ibid., p. 51.

16. B. Boncompagni, ed., *Scritti di Leonardo Pisano* (Rome: Tipografia delle scienze matematiche e fisiche, 1857–1862), vol. 1, p. 2. This work is the standard edition of all of Leonardo's writings. The first volume contains the *Liber abbaci*. The various problems of Leonardo's which we discuss are also taken from this edition.

17. Ibid., p. 283.

18. Leonardo Pisano, *The Book of Squares*, p. 3.

19. Ibid., p. 107.

20. The letter is quoted in Jens Høyrup, "Jordanus de Nemore, 13th Century Mathematical Innovator: an Essay on Intellectual Context, Achievement, and Failure," *Archive for History of Exact Sciences* 38 (1988), 307–363. This article provides a broad overview of the work of Jordanus. The article by Wilbur Knorr, "On a Medieval Circle Quadrature: *De circulo quadrando*," *Historia*

Mathematica 18 (1991), 107–128 makes a new attempt to place Jordanus in the context of early thirteenth century Paris.

21. Barnabas Hughes, *Jordanus de Nemore: De numeris datis* (Berkeley: University of California Press, 1981), p. 57. This work provides a critical Latin edition and an English translation of Jordanus' work on algebra. Hughes also discusses the sources of the work and gives a modern symbolic translation of each proposition. The propositions quoted on the next few pages have been translated directly from the Latin to give a better flavor of Jordanus' words than Hughes' more modern translation.

22. These definitions are taken from John North, *Richard of Wallingford. An Edition of His Writings with Introductions, English Translations and Commentary* (Oxford: Clarendon Press, 1976), p. 59. The three volumes of this set provide critical editions and English translations of much of Richard's mathematical and astronomical work, as well as commentaries.

23. Edward Grant, *Nicole Oresme: De proportionibus proportionum and Ad pauca respicientes* (Madison: University of Wisconsin Press, 1966), p. 161. This edition of Oresme's *On the Ratios of Ratios* provides not only the Latin version and a complete English translation, but also an extensive analysis of the background of Oresme's ideas.

24. Ibid., pp. 161–163.

25. Clagett, *The Science of Mechanics*, p. 230. A detailed introduction to the work of Bradwardine and his contemporaries is found in A. G. Molland, "The Geometrical Background to the 'Merton School': An Exploration into the Application of Mathematics to Natural Philosophy in the Fourteenth Century," *British Journal for the History of Science* 4 (1968), 108–125.

26. Ibid., pp. 230–231.

27. Grant, *A Source Book in Medieval Science*, p. 238. This quotation from William of Heytesbury is also found in Clagett, *The Science of Mechanics in the Middle Ages*, p. 236. A detailed study of the work of Heytesbury is Curtis Wilson, *William Heytesbury: Medieval Logic and the Rise of Mathematical Physics* (Madison: University of Wisconsin Press, 1956).

28. Grant, *Source Book*, p. 238 and Clagett, *The Science of Mechanics*, p. 237.

29. Grant, *Source Book*, p. 239 and Clagett, *The Science of Mechanics*, p. 271.

30. Marshall Clagett, *Nicole Oresme and the Medieval Geometry of Qualities and Motions. A Treatise on the Uniformity and Difformity of Intensities Known as Tractatus de configurationibus qualitatum* (Madison, University of Wisconsin Press, 1968), pp. 165–167. Besides the Latin text and a complete English translation, this edition contains a detailed commentary as well as translations of some related works.

31. Ibid., p. 167. A study of the use of lines to express functions is in Edith Sylla, "Medieval Concepts of the Latitude of Forms: The Oxford Calculators," *Archives d'Histoire Doctrinal et Littérarie du Moyen Âge* 40 (1973), 223–283.

32. Ibid., p. 193.

33. Ibid., p. 193.

34. Ibid., p. 415.

35. David Singmaster, "Some early sources in recreational mathematics," in Cynthia Hay, ed., *Mathematics from Manuscript to Print: 1300–1600* (Oxford: Clarendon Press, 1988), p. 199.

Summary of Mathematics in Medieval Europe

480–524	Boethius	Latin version of Greek texts
560–636	Isidore of Seville	Latin texts for the quadrivium
735–804	Alcuin of York	Arithmetic problems
945–1003	Gerbert d'Aurillac	Counting board
1075–1164	Adelard of Bath	Translations
Early twelfth century	John of Seville	Translations
Early twelfth century	Domingo Gundisalvo	Translations
Early twelfth century	Abraham bar Ḥiyya	Geometry and trigonometry
Early twelfth century	Plato of Tivoli	Translations
1090–1167	Abraham ibn Ezra	Combinatorics
Mid-twelfth century	Robert of Chester	Translations
1114–1187	Gerard of Cremona	Translations
1170–1240	Leonardo of Pisa	Arithmetic, algebra, geometry
Early thirteenth century	Jordanus de Nemore	Algebra
1225–1286	Wilhelm of Moerbeke	Translations
1288–1344	Levi ben Gerson	Trigonometry, combinatorics, induction
1291–1336	Richard of Wallingford	Trigonometry, proportions
1295–1349	Thomas Bradwardine	Kinematics
Early fourteenth century	William Heytesbury	Kinematics
1320–1382	Nicole Oresme	Kinematics, exponentials, graphs
Mid-fourteenth century	Immanuel Bonfils	Decimal fractions

Interchapter

Mathematics Around the World

Having studied in some detail the mathematics of China, India, the Islamic world, and Europe up to about the year 1300, we will at this point attempt a comparison of the mathematics known in these places at that time. And because there were many common ideas in these widely separated cultures, we will also discuss the question of whether these concepts were indigenous or transmitted from one culture to another. Another obvious question is what mathematical ideas were known in other parts of the world at this time. This question is difficult to answer in much detail, given the current state of our knowledge. Nevertheless, we will present in the second half of this chapter an overview of what is known about mathematics in the Americas, in Africa, and in the Pacific.

I.1 MATHEMATICS AT THE TURN OF THE FOURTEENTH CENTURY

We begin with geometry. Practical geometry, that is, the measure of fields, the determination of unknown distances and heights, and the calculation of volumes, was performed by much the same techniques in the four societies studied. All of them knew how to calculate areas and volumes, at least approximately, and all knew and used the Pythagorean theorem when dealing with right triangles. Even the techniques of determining the height of a distant tower were nearly the same.

As far as theoretical geometry was concerned, it was in the world of Islam that the heritage of classical Greek geometry was preserved and studied and in which certain advances were made. It was there that questions were raised and answered about the exact volumes of certain solids and about the locations of centers of gravity, using

both heuristic methods for arriving at answers and the technique of exhaustion for giving proofs. It was there that questions were raised and answers attempted about Euclid's parallel postulate. And it was there that the Greek idea of proof from stated axioms was most fully understood and developed.

Although Europe had always had at least a version of Euclid's *Elements* available, the beginning of the fourteenth century saw only the bare beginnings of a renewed interest in Euclid and other Greek geometers, stimulated by the appearance of a mass of translations of this material in the twelfth and thirteenth centuries. But although the idea of proof survived, there was still no new work in theoretical geometry. Neither India nor China had been exposed to classical Greek geometry, as far as is known, but that is not to say that they had no notion of proof. In the works of the Chinese mathematicians and their numerous commentators, there are always derivations of results, as, for example, in the *Sea Island Classic* of Liu Hui. These derivations are not, however, based on explicitly stated axioms. They are, on the other hand, examples of logical arguments. Less is known about Indian arguments than those from China, but even there, there exist commentaries of later writers who present reasons as to why the various methods are valid.

Related to geometry is the subject of trigonometry, developed in the Hellenic world as a part of the study of astronomy. By the year 1300, trigonometry was in active use in India, Islam, and in Europe, generally for the same purpose of studying the heavens. The subject was modified and extended as it traveled from one country to another, but those interested in the heavens in those three civilizations were all fluent with at least some version of the subject. It appears that only China was lacking trigonometry, even though Indian scholars had introduced at least the elements of the subject in their visits in the eighth century. It is probable that trigonometry was simply not useful to the Chinese in their own astronomical and calendrical calculations.

It was in certain aspects of algebra, on the other hand, that the Chinese first developed techniques that were later used elsewhere. For example, they had from early times developed efficient methods of solving systems of linear equations. And by the fourteenth century, they had developed their early root finding techniques, which involved the use of the Pascal triangle, into a detailed procedure for solving polynomial equations of any degree. They also worked out the basis of what is today called the Chinese remainder theorem, a procedure for solving simultaneous linear congruences.

Linear congruences were also solved in India, but by a method somewhat different from that of the Chinese, the only similarity being the use of the Euclidean algorithm. Indian scholars were even prouder, however, of the techniques they developed for solving the quadratic indeterminate equations known today as the Pell equations. The Indian mathematicians were also familiar with the standard techniques of solving quadratic equations, but since there is no documentation as to how they thought about the method, we do not know whether they developed the technique independently or absorbed it from the ancient Babylonians or from Islam.

For Islam, of course, there is copious documentation of an interest in algebra. Not only did Islamic mathematicians study the quadratic equation in great detail, giving geometric justifications for the various algebraic procedures involved in the solution, but also they studied cubic equations. For these equations, Islamic mathematicians developed a solution method involving conic sections and were able to understand some aspects of the relationship of the roots to the coefficients of these equations. They were also familiar with a method similar to the Chinese one of solving cubic equations numerically.

Islamic mathematicians worked with the Pascal triangle, probably some years earlier than their Chinese counterparts. More importantly, they developed techniques of proof closely resembling our modern proof by induction, techniques they used not only in dealing with this triangle of coefficients but also with various other series of numbers. Such techniques were also developed in Europe by Levi ben Gerson in relation to various com-

binatorial formulas. Furthermore, Islamic algebraists developed in great detail the techniques for manipulating algebraic expressions, especially those involving surds, and thereby began the process of negating the classical Greek separation of number and magnitude.

By the turn of the fourteenth century, algebraic techniques were only beginning their appearance in Europe. Those techniques which were available were clearly based on the Islamic work, although Jordanus de Nemore considered the material from a somewhat different point of view. He also introduced a form of symbolism in his algebraic work, something missing entirely in Islamic algebra but also present, in different form, in India and China. On the other hand, European algebra of this time period, like its Islamic counterpart, did not consider negative numbers at all. India and China, however, were very fluent in the use of negative quantities in calculation, even if they were still hesitant about using them as answers to mathematical problems.

The one mathematical subject present in Europe in this time period which was apparently not considered in the other areas was the complex of ideas surrounding motion. It was apparently only in Europe that mathematicians considered the mathematical question of the meaning of instantaneous velocity and therefore were able to develop the mean speed rule. Thus the seed was planted which ultimately grew into one branch of the subject of calculus nearly three centuries later.

It appears that the level of mathematics in these four areas of the world was comparable at the turn of the fourteenth century. Although there were specific techniques available in each culture which were not available in others, there were many mathematical ideas and methods common to two or more. The question then arises as to whether ideas developed independently in the four areas or whether there was transmission among them.

For certain ideas, the lines of transmission are clear. Thus trigonometry moved from Greece to India to Islam and back to Europe, with each culture modifying the material to meet its own requirements. Also, the decimal place-value system, with its beginnings in China or India (or perhaps on the border between them) moved to Baghdad in the eighth century and then to Europe (via both Italy and Spain) in the eleventh and twelfth centuries.

But for other common ideas, the situation is less clear. For example, in trigonometry the first tabulation of the tangent function was in China in the early eighth century, where Yi Xing developed this idea probably with the aid of Indian computations of the sine. The next appearance of a tabulated tangent function was in Islam. Was this notion carried there by Chinese technicians captured in the Battle of the Talas River in 751 which established Islamic hegemony in western Central Asia? In any case, although tangent tables were brought to Europe early in the twelfth century with the translation of an edited version of al-Khwārizmī's astronomical tables, the tangent function is not found in the early European trigonometry works.

What about the Pascal triangle? It appears in Islam in the early eleventh century and then in China perhaps in the middle of that century. Was there transmission? There was certainly contact in this period between Islam and China along the famous silk route. Recall also that al-Bīrūnī in this time period was at the court of Sultan Mahmūd of Ghazna, where he studied the culture of India, and there was always some contact, particularly through Buddhism, between India and China.

Or consider the methods of determining heights and distances in all four of these cultures. The earliest documented appearance of the basic method of two sightings is in China in the third century. By the thirteenth century, the method was being used in Europe. Similarly, recreational problems such as that of the hundred fowls or that of the faucets emptying (or filling) a tank appear in Chinese, Islamic, and early medieval European work. Did these problems travel, and, if so, how? Again, the silk route comes to mind as a method of transmission. Or, on a more specific level, there was a group of Jewish merchants known as the Radhanites who regularly traveled from southern France to China, via Damascus and India, in the ninth century, carrying eunuchs, fur, and swords to the East and returning with musk, spices, and medicinal plants. Did they learn of any Chinese mathematics and bring it back to Europe or to any of their way-stations, or, conversely, did they take any Islamic or Indian mathematics to China? This question applies with perhaps more relevance to the question of the Jacob Staff, the surveying device first described in Europe by Levi ben Gerson early in the fourteenth century but available in China by the eleventh century. Was this carried from China by Jewish merchants?

Going back even further, we have not even touched the questions of the origins of the Egyptian mathematics

of 1800 B.C.E. or of the Babylonian mathematics of the same time period. Recent studies of more ancient sources in Babylonia have pushed back the origins of the place-value system to the middle of the third millenium B.C.E. and the origins of number symbols even further.[1] Similarly, Egyptian number hieroglyphics date back at least to the time of Menes, the first Pharaoh of Upper and Lower Egypt (c. 3100 B.C.E.). And there is at least some evidence suggesting that certain mathematical ideas present in the Egyptian papyri have their origins in central Africa.[2] There are also records of contacts between Egypt and Mesopotamia as far back as 3300 B.C.E., but it is not known whether any mathematical ideas were transmitted from one culture to the other.

The answers to many of the questions of transmission can at this point only be speculation. Any documentation of such transmission remains to be discovered. But transmission or not, it does appear that the common mathematical ideas were adapted to meet the mathematical needs of each civilization.

I.2 MATHEMATICS IN AMERICA, AFRICA, AND THE PACIFIC

There were mathematical ideas in the world in civilizations different from the four major medieval societies already considered. Unfortunately, most of the other civilizations were nonliterate, and so written documentation is not available. Thus, any description of the mathematics of these societies necessarily comes from artifacts or from the descriptions of ethnologists. Much research on mathematics in various societies has been performed in recent years, but there are still many unanswered questions. We can only present here a brief sketch of what is known about the mathematical ideas of various societies. References to the current literature are provided, however, so that the interested reader can pursue these matters further.

We begin in the Americas with the Mayans, the society in the New World about whose mathematics the most is known, primarily because the Mayans did have a written language. Mayan civilization flourished in southern Mexico, Guatemala, Belize, and Honduras and reached its high point between the third and ninth centuries. Thereafter, the Mayans came under the influence of

FIGURE I.1
Detail from a Mayan ceramic vessel (c. 750 C.E.) depicting two mathematicians. The one on the left is a man, while the one in the upper right corner is a woman. The mathematicians are identified by scrolls with number symbols emerging from their armpits.[3]

other peoples of Mexico and many of their cultural centers fell into ruin. Nevertheless, a strong Mayan culture still existed when the Spaniards arrived in the early sixteenth century.

Like many ancient civilizations, the Mayans had a priestly class which studied mathematics and astronomy and kept the calendar (Figure I.1). The records of the priests were written down and preserved on a bark-paper or else carved into stone monuments. Unfortunately, the Spanish conquerors destroyed most of the documents they found, so there are very few remaining. And because modern day Mayans cannot read the ancient hieroglyphics, it has been a long and tedious process to decipher the few documents which remain, in particular, the Dresden codex (named for the library which owns it), which dates from the twelfth century and deals with aspects of the Mayan calendar (Figure I.2). Nevertheless, scholars today understand the basics of the classic Mayan calendrical and numeration systems. The documents only provide, however, the results of calculations. There is no record of the methods by which the calculations were made. Some of what follows in the description of Mayan mathematics is therefore speculative.

The Mayan numeration system was a mixed system, like the Babylonian. It was a place-value system with base 20 on one level, but for the representation of numbers less than twenty, it was a grouping system with base 5. The Mayans used only two symbols to represent numbers, a dot ($\cdot$) to represent 1 and a line (——) to represent 5. These were grouped in the appropriate way to represent numbers up to 19. Thus $\cdots$ represented 8, and $\equiv$ represented 17. For numbers larger than 19, a place-value system was used. The first place represented the units, the second place the 20s, the third place the 400s, and so on. Unlike the Babylonians, however, the Mayans did have a symbol for 0, namely   , used to designate an "empty" place. Mayan numbers were generally written vertically, with the highest place value at the top, but for convenience we will write them horizontally, using the same conventions as we used to represent the Babylonian numbers. Thus 3,5 will represent $3 \times 20 + 5$, or 65.

For calendrical purposes, the Mayans modified their numeration system slightly, using the third place from the bottom to represent 360s, rather than 400s, with every

FIGURE I.2
Mayan calendar on a Guatemalan stamp.

other place still representing 20 times the place before. It is this system we will use in what follows, because it was in calendrical calculations that the Mayans used numbers most extensively. In this calendrical numeration system, then, 2,3,5 will represent $2 \times 360 + 3 \times 20 + 5$, or 785, while 2,0,12,15 will represent $2 \times 7200 + 12 \times 20 + 15$ or 14,655.

For the Babylonians, there is quite extensive evidence as to the methods of calculation used. The natural question to ask with regard to the Mayans then is how they calculated. Unfortunately, all that exists in the Mayan documents are the results of various computations without any record of the methods themselves. It is surmised that for addition and subtraction the Mayans used some sort of counting board device to collect the dots and lines in each place and move any excess over 20 into the next place. To do multiplication, presumably all one needs to know are three basic facts: $1 \times 1 = 1$, $1 \times 5 = 5$, and $5 \times 5 = 1,5$. To perform any other multiplication, one then just needs the distributive law and a way to keep track of the places, but, naturally, multiplication tables up to 19×19 would make computation easier. There is no record, however, of the use of such tables.

The most important use of computation for the Mayan priests appears to have been for calendrical computations. The Mayans used two different calendars at the same time. First, there was the 260-day almanac which was the product of two cycles, one of length 13 and the other of length 20. We specify a day in the almanac by a pair (t, v), where t is a day number from 1 to 13 and v is one of 20 day names. For example, because the list of 20 day names begins with *Imix* and *Ik*, the day 1 *Imix* will be written as $(1,1)$ while the day 5 *Ik* as $(5,2)$. The second calendar was the 365-day year. This calendar year was divided into 18 months of 20 days and an extra period of 5 days. For our purposes, however, it is sufficient to designate a day in the 365-day year by its number y. Thus, because *Muan* is the fifteenth month of the calendar, the second day of Muan will be designated by $y = 282$. The three cycles of 13 day numbers, 20 day names, and 365 days of the year were traversed independently and thus the complete cycle of triples (t,v,y) was repeated after $13 \cdot 20 \cdot 73 = 18,980$ days, or 52 calendar years, or 73 almanacs. This entire cycle is generally called the "calendar-round."

The two basic calendrical problems the Mayans needed to solve were first, given a date (as a triple) and a specified number of days later, to determine the new date, and second, given two Mayan dates (as triples), to determine the least number of days between them. If the specified number of days is denoted, in base 20 calendrical notation, by m,n,p,q,r, where $0 \le m,n,p,r \le 19$, $0 \le q \le 17$, the first question can be written in modern notation as given an initial date (t_0,v_0,y_0), determine the date (t,v,y) which is m,n,p,q,r days later. One can show that this date is given by

$$t = t_0 - m - 2n - 4p + 7q + r \pmod{13}$$

$$v = v_0 + r \pmod{20}$$

$$y = y_0 + 190m - 100n - 5p + 20q + 4 \pmod{365}$$

(The coefficients in the first equation come from the fact that $20 \equiv 7 \pmod{13}$, $18 \times 20 \equiv -4 \pmod{13}$, $20 \times 18 \times 20 \equiv -2 \pmod{13}$, and $20 \times 20 \times 18 \times 20 \equiv -1 \pmod{13}$.) For example, if the given date is $(4,15,120)$, the new date $0,2,5,11,18$ days later is $(10,13,133)$. The second calendrical problem requires one to determine the smallest intervals between the dates in each of the three component cycles, to combine the first two to determine the smallest interval in the almanac, and then to combine this value with the third to determine the number of days in the calendar-round.

There is no evidence of the exact method used by the Mayan priests to solve either of these problems, because only the results are given in the surviving documents. But it seems virtually certain that they must have gone through a type of computation similar to that indicated by the algebraic formulas above. In any case, the priests were able to use their base 20 notation to solve the problems they needed to keep track of their calendar and thus to provide the Mayan government with the correct days on which to celebrate festivals, make appropriate sacrifices, plant the maize, or accomplish whatever other tasks were necessary to run the Mayan kingdom (Figure I.3)[4].

About 2000 miles south of the Mayan heartland, there was another major civilization of about four million people, the Inca, which flourished in what is now Peru and surrounding areas from about 1400 to 1560. The Incas did not have a written language, but did possess a logical-numbering system of recording in the knots and cords of what are called *quipus*. The *quipus* were the

FIGURE I.3
Mayan observatory, El
Caracol, at Chichen Itza in the
Yucatan peninsula of Mexico.

means by which the Inca leadership monitored its domains. They necessarily received and sent many messages daily, including details of items that were needed in storehouses, taxes which were owed, numbers of workers needed for certain public works projects, and so on. The messages were encoded on the *quipus* and sent to their destination by a series of runners (Figure I.4). Of necessity, the messages had to be concise and compact, so *quipu* makers were trained in Cuzco, the capital, to design and create the *quipus* on which the messages were carried.

A *quipu* is a collection of colored knotted cords, where the colors, the placement of the cords, the knots on the individual cords, the placement of the knots, and the spaces between the knots all contribute to the meaning of the recorded data. Every *quipu* has a main cord, thicker than the others, to which are attached other cords, called pendant cords, to each of which may be attached further cords, called subsidiary cords. Sometimes there is a top cord, a cord placed near the center of several pendant cords and tied so that when the *quipu* lies flat it falls in a direction opposite to the pendant cords. Data is recorded on the cords (other than the main cord) by a system of knots. The knots are clustered together in groups separated by spaces and represent numbers using a base 10 place-value system with the highest value place

closest to the main cord. Thus a cord with three knots near the top and nine knots near the bottom represents the number 39. As additional help for reading the numbers, the knots representing units are generally larger knots than those representing higher powers of ten. The largest number so far discovered on a *quipu* is 97,357. Zeros are generally represented by a particularly wide space (Figure I.5).

The pendant cords on *quipus* are themselves generally clustered in groups, sometimes with each group consisting of the same set of distinct colors. It is assumed that each color refers to a particular class of data which is being recorded on the *quipu*. In addition, one often has a top cord associated to a group of cords on which is recorded the sum of the numbers on the individual cords of the group. Sometimes certain of the pendant cords record sums of the numbers on other such cords. Sometimes it appears that the knots on a particular cord do not represent data at all but are simply labels. In any case, the

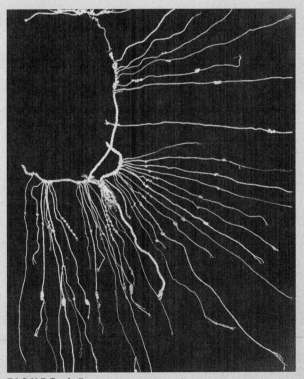

FIGURE I.5
Quipu from the Smithsonian Collection.[5]

FIGURE I.4
Inca runner carrying quipu.   .

quipus are not calculating tools, but only records. The calculations on which these records are based must have been done elsewhere, probably with some sort of counting board.

In modern terminology, a *quipu* can be thought of as a particular type of graph known as a tree. (See Chapter 18.) Certainly the *quipu* makers had to ask themselves the types of questions often associated with the study of trees, including how many different trees can be constructed with a given number of edges. And since the Inca officials associated numbers to each edge of the tree as well as colors, the questions which had to be answered in designing these objects so that they could be useful were not trivial ones.[6]

In the Inca civilization, as in the Mayan, there was a professional class of mathematicians, people who had to deal with the mathematics of the culture on a regular basis to help the civilization maintain itself. But in the other cultures to be discussed next, such a class did not exist. In fact, these peoples had no category in their lives called "mathematics." Nevertheless, there are certain aspects of their culture which we today recognize as being mathematical. The people involved did not distinguish and classify these aspects as we do. The mathematical ideas were simply part of what they needed to conduct their lives, to farm, to build, to worship. This mathematics of a group of people, used on a regular basis, is what today is often called "ethnomathematics," the study of which allows us to see the importance of mathematical ideas to various such groups.[7]

Perhaps the most sophisticated civilization in what is now the United States in pre-Columbian times was that of the Anasazi, who lived in the Four Corners area of the southwest from about 600 B.C.E. to around 1300 C.E. The high point of their civilization was reached in the years after 1000 in which they constructed elaborate pueblos and ceremonial structures at various sites, the most prominent being at Mesa Verde (in southwestern Colorado) and at Chaco Canyon (in northwestern New Mexico). It appears that viewing areas were set up in many of their structures for the same purposes as sightlines at the temple at Stonehenge, namely to determine the occurrences of certain important astronomical events including the summer and winter solstice and even the 18.6 year cycle of moonrise positions.

Although the Anasazi have left us no documented records of their mathematics, we can speculate about what mathematics was necessary for their lives by considering the archaeological remnants of their civilization. For example, one important notion in the Anasazi religion, coming from the myths of their origins, is that of the four cardinal directions. It was evidently important to the Anasazi to align their major buildings in these directions and even to build their roads that way. One of the major roads out of Chaco Canyon was aligned due north and built that way for many miles, irrespective of topographical obstacles. And the great ceremonial structure in Chaco Canyon, Casa Rinconada, is a sixty-three foot diameter circle, the roof of which was originally supported by four pillars forming the corners of a square aligned exactly along the cardinal directions. The question, then, is how the Anasazi determined the direction of true north. One possibility, because they were certainly aware of the daily and yearly motions of the sun, is that they used the same techniques as the Roman surveyors of a millenium earlier, to draw a circle centered on a pole, then record the curve of the endpoints of the pole's shadow throughout the day and determine the two points where the curve intersects the circle. The line connecting those two points is an east-west line to which a perpendicular bisector can be drawn to determine a north-south line. (See Figure 5.1.)[8]

Other North American Indians built carefully aligned structures and even entire urban areas, thus displaying a knowledge of astronomy and geometry. For example, the Cahokian mounds in East St. Louis, Illinois, built by the civilization known as the Mississippian during the period from 900 to 1200, display not only alignments to important celestial events but also evidence of detailed city planning. Similarly, the Bighorn Medicine Wheel, near the summit of Medicine Mountain in the Bighorn Mountains of northern Wyoming, and the Moose Mountain Medicine Wheel in southeastern Saskatchewan were probably constructed by Plains Indians to determine the summer solstice. The Moose Mountain Medicine Wheel probably dates from over two thousand years ago, while the Bighorn Medicine Wheel is of much more recent origin.

Like the North American Indians, most African cultures of the past did not leave written records, so it is not possible to say with any degree of certainty when and how mathematical ideas were created in these cultures. What makes matters even worse for the historian is that few artifacts are even available from which mathematical ideas can be inferred, partly because comparatively little archaeology has been done in Africa south of the Sahara. One major ancient structure which is only now being studied in detail is Great Zimbabwe, a massive stone complex 17 miles south of Nyanda, Zimbabwe, which was probably built in the twelfth century. It is evident that the empire which built this complex required mathematics to deal with the administrative and engineering requirements of the construction as well as with the trade, taxes, and calendars required to keep the empire functioning. Similarly, the bureaucracies of the West African states of the medieval period, including Ghana, Mali, and Songhai, also required mathematics like their colleagues elsewhere in the world. Because the influence of Islam penetrated to much of west Africa, and because an Islamic university was in existence in Timbuktu from the fourteenth century at least until 1600, scholars of that region were probably exposed to some of the mathematics of Islam. However, we have no direct information on mathematics or mathematicians of this time and place.

Until more archaeological finds have been discovered, the most we can do to find out about the mathematics of the peoples of Africa is to consider the reports of the ethnographers who studied these peoples in the nineteenth and twentieth centuries and pull out of their studies what we consider as mathematical ideas. But the sources and the dates of such ideas are virtually impossible to determine.

One mathematical idea which appears in the Bushoong culture in Zaire and also in the Tshokwe culture of northeastern Angola is the graph theoretical idea of tracing out certain figures in a continuous curve without lifting one's finger from the sand. In western mathematics, this idea was first dealt with by Leonhard Euler in 1736. (See Chapter 14.) The Bushoong children, who first showed their diagrams to a European ethnologist in 1905, were evidently not only aware of the conditions which insured that the graph could be drawn continuously but also knew the procedure which permitted its drawing most expeditiously. For the Tshokwe, figure drawing is not a children's game, but part of a storytelling tradition among the elders. As part of their storytelling, dots are used to represent humans and the rather complex curves are drawn including certain dots within the figure and excluding certain others. In fact the procedure for drawing is to set out a rectangular grid of dots on which the curve is superimposed (Figure I.6). Without a special

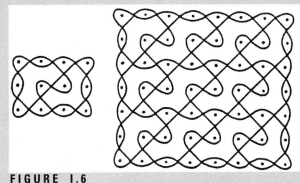

FIGURE I.6
Examples of Tshokwe graphs.[9]

study of the diagrams, it is not easy to determine which dots are inside and which are outside, but the detailed drawing rules which the Tshokwe follow enable them to construct the curves quickly in one continuous motion.[10]

Another mathematical idea which occurs in many African cultures is the idea of a geometric pattern, as used in cloth weaving or decorative metal work. There are numerous examples from all over Africa of patterned strips, using the seven possible strip patterns, as well as most of the 17 other plane patterns.[11] In fact the Bakuba people from Zaire use all seven strip patterns in their cloth, as well as at least 12 of the plane patterns. The artists of Benin (Nigeria) decorate their bronze castings with all of the strip patterns and some of the other plane patterns as well (Figure I.7).

Mathematical games and puzzles occur in Africa too. For example, the board game known variously as

FIGURE I.7
Examples of strip patterns from Benin.[12]

wari, omweso, and *mankala* is played throughout Africa and is quite useful in teaching children counting and strategy. Similarly, the familiar puzzle story of a person attempting to transport three objects, *A, B, C,* across a river, but only being able to take one at a time and not being able to leave either *A* or *C* alone with *B,* occurs in several African cultures. Among the Bamileke (Cameroon), the objects are a tiger, a sheep, and a big spray of reeds. A different problem, where the person can take two objects at a time is found, among other places, in Algeria (where the objects are a jackal, a goat, and a bundle of hay), in Liberia (a cheetah, a fowl, and some rice), and in Zanzibar (a leopard, a goat, and some tree leaves).

Moving now to the South Pacific, we find the idea of tracing figures continuously in the sand also in Malekula, in the Republic of Vanuatu, an island chain some 1200 miles northeast of Australia. The drawing of figures here is imbedded in Malekula religious life. In fact, passage to the Land of the Dead requires being able to draw these figures accurately. The Malekulans devised standard algorithms for tracing their quite complicated figures using symmetry operations on a few basic drawings. Thus, one can analyze the Malekulan figures using some of the language of modern day group theory.

Group theory is also convenient in analyzing the kin relationships in Malekula. In fact, the elders explained these relationships to an anthropologist using diagrams which can easily be transformed into a group table. The basic idea is that the society is divided up into six sections, and men of one section can only marry women of a different section, while their children belong to still another section. If a given male belongs to the section we label as *e* (identity), his mother will belong to section *m* and his father to section *f*. Then the mother of his father will be in section *mf* and the father of his mother in section *fm*. It turns out that the kin rules are such that all the possible "products" of *m* and *f* form the dihedral group of order 6, that is the group of six elements generated by the elements *m, f* with the relations $m^3 = e, f^2 = e,$ and $(mf)(mf) = e.$ Marriage can only take place between *A* and *B* if *B* belongs to the section of the mother of the father of *A,* or, equivalently, if *A* belongs to the section of the mother of the father of *B.*[13]

Mathematical games also occur in the South Pacific, the most famous of which is *mu torere,* popular among the Maori of New Zealand. A description of this game, played between two people with four markers each on a game board in the shape of an eight-pointed star, would take too much space. Suffice it to say that a complete analysis of the game involves many of the common formulas of combinatorics. Played between two players of equal ability, the game involves many careful choices and consideration of multiple opportunities in order to avoid pitfalls and to trap one's opponent.

This brief trip through the world of ethnomathematics shows us that two of the central ideas of mathematics, logical thought and pattern analysis, occur in societies around the world. And although most societies did not have the formal "mathematics" of the literate civilizations of China, India, Islam, or Europe, mathematics was, and is, a force in the lives of people in all parts of the globe.

Exercises

1. Show why the formulas for determining the Mayan date a specified number of days later than a given date are valid.

2. Given the Mayan date (8,10,193), determine the Mayan date which is 0,2,3,5,10 days later.

3. Find an algorithm for deciding the minimum number of days between two Mayan dates (t_0, v_0, y_0) and (t_1, v_1, y_1). It might be easier to first ignore the 365-day cycle altogether and simply determine the minimum number of days between the two almanac dates of (t_0, v_0) and (t_1, y_1). For help with this problem, consult the works of Closs and Lounsbury mentioned in note 4.

4. Show that the minimum number of days between the two Mayan dates of (8,20,13) and (6,18,191) is 1,8,15,18 (= 10,398). Because these two dates are the birth and death dates of Pacal, a Mayan king, and because it is known from other sources that Pacal's age at death was more than 60 and less than 100, determine the number of days of Pacal's life and his age at death. (Recall that the number of days calculated by the algorithm in problem 3 is only determined up to a multiple of one calendar-round, 18,980 days or 52 calendar years.)

5. Work out the group table of the kin structure in Malekulan society. For a woman in each of the six sections, de-

termine the section of her husband, her mother, her father, and her children.

6. Write a report on the seven possible strip patterns and the 17 possible plane patterns of symmetry. Find examples of each in wallpaper patterns or in fabric patterns. Consult D. K. Washburn and D. W. Crowe, *Symmetries of Culture* (Seattle: University of Washington Press, 1988).

7. Learn the game *mankala* and design a lesson for young children using the game to teach various arithmetic concepts. See Laurence Russ, *Mancala Games* (Algonac, Mich.: Reference Publications, 1984), H. J. R. Murray, *A History of Board Games Other than Chess* (Oxford: Clarendon Press, 1952), or M. B. Nsimbi, *Omweso, a Game People Play in Uganda* (Los Angeles: African Studies Center, UCLA, 1968) for details.

8. Read the paper by Anna Sofaer, Rolf M. Sinclair, and Joey B. Donahue, "Solar and Lunar Orientations of the Major Architecture of the Chaco Culture of New Mexico," in *Proceedings of the Colloquio Internazionale Archeologia e Astronomia* (Venice, 1990) and some of the references cited in the paper. Do these articles convince you that the Anasazi used mathematical tools in the orientation and basic design of their buildings? What other types of evidence would be worth considering?

9. Chapter 6 of Marcia Ascher and Robert Ascher, *Code of the Quipu*, cited in note 6, deals with *quipus* in terms of the mathematical structure known as a tree. Read the chapter and do some of the exercises. What kinds of analyses of trees did Inca *quipu* makers have to perform in order to create *quipus* appropriate for various purposes?

References and Notes

1. See Georges Ifrah, *From One to Zero: A Universal History of Numbers*, translated by Lowell Buir (New York: Penguin, 1985), Marvin Powell, "The Antecedents of Old Babylonian Place Notation and the Early History of Babylonian Mathematics," *Historia Mathematica* 3 (1976), 417–439, and D. Schmandt-Besserat, *Before Writing: From Counting to Cuneiform* (Austin: University of Texas Press, 1992).

2. See Théophile Obenga, "Système opératoire négro-africain," in Théophile Obenga, *L'Afrique dans l'Antiquité: Egypte pharaonique—Afrique noire* (Paris: Présence Africaine, 1973) and Paulus Gerdes, *Ethnogeometrie: Kulturanthropologische Beiträge zur Genese und Didaktik der Geometrie* (Bad Salzdetfurth: Franzbecker, 1991).

3. This picture is taken from Persis B. Clarkson, "Classic Maya Pictorial Ceramics: A Survey of Content and Theme," in Raymond Sidrys, ed., *Papers on the Economy and Architecture of the Ancient Maya* (Los Angeles: Institute of Archaeology, UCLA, 1978), 86–141. Clarkson identified the female scribe. Michael Closs identified her as a mathematician because of the number scroll coming from her armpit.

4. See Michael Closs, "The Mathematical Notation of the Ancient Maya," in *Native American Mathematics*, Michael P. Closs, ed. (Austin: University of Texas Press, 1986), 291–369, and Floyd G. Lounsbury, "Maya Numeration, Computation, and Calendrical Astronomy," in *Dictionary of Scientific Biography* (New York: Scribners, 1978), vol. 15, 759–818 for more details on Mayan mathematical techniques.

5. Smithsonian Institution, catalog no. 289613, Department of Anthropology.

6. See Marcia Ascher and Robert Ascher, *Code of the Quipu: A Study in Media, Mathematics, and Culture* (Ann Arbor: University of Michigan Press, 1980) for more details on *quipus*. This work provides a mathematical analysis of various techniques of *quipu* making and also provides exercises to help students learn the relevant mathematical ideas.

7. Claudia Zaslavsky, *Africa Counts: Number and Pattern in African Culture* (Boston: Prindle, Weber, and Schmidt, 1973) is the most complete work to date on the mathematics of African peoples. A more recent work, extending to other areas of the world as well, is Marcia Ascher, *Ethnomathematics: A Multicultural View of Mathematical Ideas* (Pacific Grove, Ca.: Brooks/Cole, 1991). Both of these books provide not only great detail on the mathematical ideas but also numerous references for further reading. A discussion of the general idea of ethnomathematics is found in Marcia Ascher and Robert Ascher, "Ethnomathematics," *History of Science* 24 (1986), 125–144. See also Ubiratan D'Ambrosio, *Etnomatemática: Arte ou técnica de explicar e conhecer* (Sao Paulo: Editora Ática S.A., 1990) for a more philosophical study of the notion of ethnomathematics and the *Newsletter* of the International Study Group on Ethnomathematics for current information in this field.

8. For more details on the Anasazi, consult William Ferguson and Arthur Rohn, *Anasazi Ruins of the Southwest in Color* (Albuquerque: The University of New Mexico Press, 1986). For details on the astronomy of the Anasazi

and other North American Indians, see Ray A. Williamson, *Living the Sky: The Cosmos of the American Indian* (Norman: University of Oklahoma Press, 1987) and E. C. Krupp, ed., *In Search of Ancient Astronomies* (New York: McGraw-Hill, 1978).

9. This graph is sketched in Paulus Gerdes, "On Mathematical Elements in the Tchokwe 'Sona' Tradition," *For the Learning of Mathematics* 10 (1990), 31–34, p. 32.

10. See Ascher, *Ethnomathematics*, Chapter 2 and also Paulus Gerdes, ibid.

11. See Zaslavsky, *Africa Counts*, Chapter 14. This chapter, entitled "Geometric Symmetries in African Art," was written by D. W. Crowe.

12. Source: Fig. 14-23, p. 193, Copyright © 1979 by Claudia Zaslavsky. Reprinted from *Africa Counts: Number and Pattern in African Culture* by Claudia Zaslavsky by permission of the publisher, Lawrence Hill Books (Brooklyn, New York).

13. See Ascher, *Ethnomathematics*, Chapter 3.

Chapter 9

Algebra in the Renaissance

"But of number, cosa *[unknown], and* cubo *[cube of the unknown], however
they are compounded . . . , nobody until now has formed general rules,
because they are not proportional among them. . . . And therefore, until now,
for their equations, one cannot give general rules except that, sometimes, by
trial, . . . in some particular cases. And therefore when in your equations you
find terms with different intervals without proportion, you shall say that the
art, until now, has not given the solution to this case, . . . even if the case may
be possible."*
(*From the* Summa de Arithmetica, Geometrica, Proportioni et Proportionalita
of Luca Pacioli, 1494)[1]

Gerolamo Cardano's account of the discovery of the rule for the algebraic solution of a cubic
equation, as reported in Chapter 11 of his *Ars Magna*: "Scipio Ferro of Bologna well-nigh
thirty years ago [c. 1515] discovered this rule and handed it on to Antonio Maria Fior of
Venice, whose contest with Niccolò Tartaglia of Brescia gave Niccolò occasion to discover it.
He [Tartaglia] gave it to me in response to my entreaties, though withholding the demonstration. Armed with this assistance, I sought out its demonstration in [various] forms. This was
very difficult."[2]

Many changes began to take place in the European economy in the fourteenth century that
eventually had an effect on mathematics. The general cultural movement of the next two
centuries, known as the Renaissance, also had its impact, particularly in Italy, so it is to
that country that we now return.

The Italian merchants of the Middle Ages generally were what one might call venture
capitalists. They traveled to distant places in the East, bought goods which were wanted
back home, then returned to Italy to sell these goods in hopes of making a profit. These
traveling merchants needed very little mathematics other than the ability to determine their
costs and revenues for each voyage. A commercial revolution spurred originally by the
demands of the Crusades had begun by the early fourteenth century to change this system
greatly. New technologies in shipbuilding and greater safety on the shipping lanes helped

to replace the traveling merchants of the Middle Ages with the sedentary merchants of the Renaissance. These "new men" were able to remain at home in Italy and hire others to travel to the various ports, to make the deals, to act as agents, and to arrange for shipping. Thus there grew up international trading companies centered in the major Italian cities, companies that had a need for more sophisticated mathematics than did their predecessors. They had to deal with letters of credit, bills of exchange, promissory notes, and interest calculations. Double-entry bookkeeping began as a way of keeping track of the various transactions. Business was no longer composed of single ventures but of a continuous flow of goods consisting of many shipments from many different ports enroute simultaneously. The medieval economy, based in large part on barter, was gradually being replaced by the money economy of today.

The Italian merchants needed a new facility in mathematics to be able to deal with the new economic circumstances, but the mathematics they needed was not the mathematics of the quadrivium, the mathematics studied in the universities. They needed new tools for calculating and problem solving. To meet this need there appeared in early fourteenth century Italy a new class of "professional" mathematicians, the *maestri d'abbaco* or abacists, who wrote the texts from which they taught the necessary mathematics to the sons of the merchants in new schools created for this purpose.

The first section of this chapter will therefore discuss the mathematics of the abacists in Italy, and, in particular, their algebra. Because the commercial revolution soon spread to other parts of Europe as well, the next section will deal with late fifteenth and early sixteenth century algebra in France, England, Germany, and Portugal. But because the major new discoveries in algebra in this time period took place in Italy, partly in response to Luca Pacioli's statement in the opening quotation that, as of 1494, cubic equations were in general unsolvable algebraically, we will go back to Italy to tell the marvelous story of the ultimate discovery of such a solution in the work of Scipione del Ferro, Niccolò Tartaglia, Gerolamo Cardano, and Rafael Bombelli.

All of the algebraists mentioned based their work on the Islamic algebras first translated into Latin in the twelfth century. But by the middle of the sixteenth century, virtually all of the surviving works of Greek mathematics, newly translated into Latin from the Greek manuscripts which had been stored in Constantinople, were available to European mathematicians. The last section of this chapter is thus devoted to the works of François Viète, who used his understanding of Greek mathematics to revamp entirely the study of algebra, and of Simon Stevin, who once and for all eliminated the Aristotelian distinction between number and magnitude, in effect giving us our current concept of "number."

9.1 THE ITALIAN ABACISTS

The Italian abacists of the fourteenth century were instrumental in teaching the merchants the "new" Hindu-Arabic decimal place-value system and the algorithms for using it. As usual, when a new system replaces an old traditional one, there was great resistance to the change. For many years, account books were still kept in Roman numerals. The current system of writing out the amounts on checks in words dates from this time. It was believed that the Hindu-Arabic numerals were too easily altered to depend on them alone in recording large commercial transactions. The advantages of the new system, however, eventually overcame the initial hesitation. The old counting board system required not only a board but a bag of counters to be carried around, while the new system required only pen and paper and could be used anywhere. In addition, the nature of operating on the counting board required that preliminary steps in the calculation be eliminated as one worked toward the final answer. With the new system, all the steps were available for checking when the calculation was finished. (Of course, these advantages would have meant nothing had there not been recently introduced a steady supply of cheap paper.) The abacists instructed entire generations of middle-class Italian children in the new methods of calculation, and these methods soon spread throughout the continent.

In addition to the algorithms of the Hindu-Arabic number system, the abacists taught their students methods of problem solving using the tools both of arithmetic and of the Islamic algebra. The texts written by the abacists, of which several hundred different ones still exist, are generally large compilations of problems along with their solutions.[3] These include not only genuine business problems of the type the students would have to solve when they joined their fathers' companies, but also recreational problems of the type found in modern elementary algebra texts. There were also sometimes geometrical problems as well as problems dealing with elementary number theory, the calendar, and astrology. The solutions in the texts were written in great detail with every step fully described. In general, there was no reason given for the various steps, nor any indication of the limitations of a particular method. The texts did not tell the reader what to do if one came upon an impossible situation. Whether the teachers discussed such situations in class is unknown. In any case, it seems clear that these abacus texts were designed not only for classroom use, but also to serve as reference manuals for the merchants themselves. The solution of a particular type of problem could easily be found and readily be followed without the necessity of the merchant understanding the theory behind the solution.

The following are examples of the types of problems found in these texts, most of which can be solved by using the ancient methods of the rule of three or false position:

> The gold florin is worth 5 *lire*, 12 *soldi*, 6 *denarii* in Lucca. How much (in terms of gold florins) are 13 *soldi*, 9 *denarii* worth? (One needs to know that 20 *soldi* make up 1 *lira* and 12 *denarii* make 1 *soldo*.)

> The *lira* earns 3 *denarii* a month in interest. How much will 60 *lire* earn in 8 months? (This is a problem in simple interest. Problems in compound interest also appeared where the period of compounding was generally one year.)

> A field is 150 feet long. A dog stands at one corner and a hare at the other. The dog leaps 9 feet in each leap while the hare leaps 7. In how many feet and leaps will the dog catch the hare?

Although these texts were strictly practical, they did have significant influence on the development of mathematics because they instilled in the Italian merchant class a facility with numbers without which no future advances could be made. Furthermore, some of these texts also brought to this middle class the study of Islamic algebra as a basic part of the curriculum. During the fourteenth and fifteenth centuries the abacists extended the Islamic methods by introducing abbreviations and symbolism, developing new methods for dealing with complex algebraic problems, and expanding the rules of algebra into the domain of equations of degree higher than the second. More important than the introduction of a few new techniques, however, was the general teaching of how algebra could be used to solve practical problems. With a growing competence in algebra brought about by the study of these abacus texts, it was only natural that European scholars would attempt to apply these techniques also to solve more theoretical problems arising from the rediscovery of many of the classic Greek mathematical texts. This combination of algebra and Greek geometry was to lead in the seventeenth century to the new analytic techniques which serve as the basis of modern mathematics.

9.1.1 New Algebraic Techniques

Islamic algebra was entirely rhetorical. There were no symbols for the unknown or its powers nor for the operations performed on these quantities. Everything was written out in words. The same was generally true in the works of the early abacists and in the earlier Italian work of Leonardo of Pisa. Early in the fifteenth century, however, some of the abacists began to substitute abbreviations for unknowns. For example, in place of the standard words *cosa* (thing), *censo* (square), *cubo* (cube), and *radice* (root), some authors used the abbreviations *C*, *Ce*, *Cu*, and *R*. Combinations of these abbreviations were used for higher powers. Thus, *ce di ce* stood for *censo di censo* or fourth power. Near the end of the fifteenth century, Luca Pacioli added to these abbreviations the ones $\bar{p}$ and $\bar{m}$ to represent plus and minus (*più* and *meno*). As with other innovations, however, there was no great movement on the part of all the writers to use abbreviations. This change was a slow one. New symbols gradually came into use in the fifteenth and sixteenth centuries, but modern algebraic symbolism was not fully formed until the mid-seventeenth century.

Even without much symbolism, the Italian abacists, like their Islamic predecessors, were competent in handling operations on algebraic expressions. For example, Paolo Gerardi, in his *Libro di ragioni* of 1328, gave the rule for adding the fractions $100/x$ and $100/(x + 5)$:

> You place 100 opposite one *cosa* [x], and then you place 100 opposite one *cosa* and 5. Multiply crosswise as you see indicated, and you say . . . 100 times the one *cosa* that is across from it makes 100 *cose*. And then you say 100 times one *cosa* and 5 makes 100 *cose* and 500 in number. Now you must add one with the other which makes 200 *cose* and 500 in number. Then multiply one *cosa* times 1 *cosa* and 5 in number, making 1 *censo* [x^2] and 5 *cose*. Now you must divide 200 *cose* and 500 in number by one *censo* and 5 *cose* [$(200x + 500)/(x^2 + 5x)$].[4]

Similarly, the rules of signs were also written out in words and even justified, here in a late fourteenth century manuscript by an unknown author:

> Multiplying minus times minus makes plus. If you would prove it, do it thus: You must know that multiplying 3 and 3/4 by itself will be the same as multiplying 4 minus 1/4 [by itself]. Multiplying 3 and 3/4 by 3 and 3/4 gives 14 and 1/16. To multiply 4 minus 1/4 by 4 minus 1/4 . . . , say 4 by 4 is 16; now multiply across and say 4 times minus one quarter makes minus 4 quarters, which is 1, and 4 times minus 1/4 makes minus 1, so you have minus 2.

Subtract this from 16 and it leaves 14. Now take minus 1/4 times minus 1/4. That gives 1/16, so one has the same as the other [multiplication].[5]

Antonio de' Mazzinghi (1353–1383), one of the few abacists about whom any biographical details are known, taught in the *Bottega d'abbaco* at the monastery of S. Trinita in Florence. His algebraic problems survive in several fifteenth century manuscripts. Antonio was expert in devising clever algebraic techniques for solving complex problems. In particular, he explicitly used two different names for the two unknown quantities in many of these problems. For example, consider the following: "Find two numbers such that multiplying one by the other makes 8 and the sum of their squares is 27."[6] The abacist began the solution by supposing that the first number is *un cosa meno la radice d'alchuna quantità* (a thing minus the root of some quantity) while the second number equals *una cosa più la radice d'alchuna quantità* (a thing plus the root of some quantity). The two words *cosa* and *quantità* then serve in his rhetorical explication of the problem as the equivalent of our symbols x and y, that is, the first number is equal to $x - \sqrt{y}$, the second to $x + \sqrt{y}$. Antonio also taught how to divide algebraic expressions. Thus he gave names for the fractions $\frac{1}{x}, \frac{1}{x^2}, \ldots, \frac{1}{x^6}$: 1 *cosa-esemi*, 1 *censo-esemi*, ..., 1 *cubo dicubo-esemi*, and explained that to divide 3 *cosa*, 8 *censi*, and 18 *cubi* by 3 *censi*, the result is 6 *cosa*, 2 2/3, and 1 *cosa-esemi*. That is,

$$(3x + 8x^2 + 18x^3) \div 3x^2 = 6x + 2\frac{2}{3} + \frac{1}{x}.^7$$

9.1.2 Higher Degree Equations

The third major innovation of the Italian abacists was the extension of Islamic quadratic-equation solving techniques to higher degree equations. In general, all of the abacists began their treatments of algebra by presenting al-Khwārizmī's six types of linear and quadratic equations and showing how each can be solved. But Maestro Dardi of Pisa in a work of 1344 extended this list to 198 types of equations of degree up to four, some of which involved radicals.[8] Most of the equations can be solved by a simple reduction to one of the standard forms, but in each case Dardi gave the solution anew, presenting both a numerical example and a recipe for solving the particular type of equation. For example, Dardi noted that the equation $ax^4 = bx^3 + cx^2$ has the solution given by

$$x = \sqrt{\left(\frac{b}{2a}\right)^2 + \frac{c}{a}} + \frac{b}{2a},$$

that is, it has the same solution as the standard equation $ax^2 = bx + c$. (Note that 0 is never considered as a solution.) Again, the equation $n = ax^3 + \sqrt{bx^3}$ can be solved for x^3 by reducing it to a quadratic equation in $\sqrt{x^3}$.

More interesting than these quadratic equations are four examples of irreducible cubic and quartic equations. Dardi's cubic equation is $x^3 + 60x^2 + 1200x = 4000$. His rule tells us to divide 1200 by 60, cube the result (which gives 8000), add 4000 (giving 12,000), take the cube root $\left(\sqrt[3]{12,000}\right)$, and finally subtract the quotient of 1200 by 60. Dardi's answer, which is correct, is that $x = \sqrt[3]{12,000} - 20$. If we write this equation using modern notation and then give Dardi's solution rule, we obtain the solution to the equation $x^3 + bx^2 + cx = d$ in the form

$$x = \sqrt[3]{\left(\frac{c}{b}\right)^3 + d} - \frac{c}{b}.$$

It is easy enough to see that this solution is wrong in general, and Dardi even admits as much. How then did Dardi figure out the correct solution to his particular case? We can answer this question by considering the problem which illustrates the rule, a problem in compound interest: A man lent 100 *lire* to another and after 3 years received back a total of 150 *lire* in principal and interest, where the interest was compounded annually. What was the interest rate? Dardi set the rate for 1 *lira* for 1 month at x *denarii*. Then the annual interest on 1 *lira* is $12x$ *denarii* or $(1/20)x$ *lire*. So the amount owed after 1 year is $100(1 + x/20)$ and after 3 years is $100(1 + x/20)^3$. Dardi's equation therefore is

$$100\left(1 + \frac{x}{20}\right)^3 = 150 \qquad \text{or} \qquad 100 + 15x + \frac{3}{4}x^2 + \frac{1}{80}x^3 = 150$$

or finally

$$x^3 + 60x^2 + 1200x = 4000.$$

Because the left side of this equation comes from a cube, it can be completed to a cube once again by adding an appropriate constant. In general, since $(x + r)^3 = x^3 + 3rx^2 + 3r^2x + r^3$, to complete $x^3 + bx^2 + cx$ to a cube, we must find r satisfying two separate conditions, $3r = b$ and $3r^2 = c$, conditions which cannot always be satisfied. In this particular case, however, where $b = 60$ and $c = 1200$, such an r exists: $r = 20$. For the solution to be valid in the general case, r must equal c/b, while b and c must satisfy the relation $3c = b^2$. If this condition is satisfied, Dardi's formula follows easily.

Cubic equations occur in certain other manuscripts of the *maestri d'abbaco*, sometimes with incorrect answers. On the other hand, one anonymous manuscript suggests that the equation $x^3 + px^2 = q$ can be solved by setting $x = y - p/3$ where y is a solution of $y^3 = 3\left(\frac{p}{3}\right)^2 y + \left[q - 2\left(\frac{p}{3}\right)^3\right]$. This is correct as far as it goes, but the author has only managed to replace one cubic equation by another. In the numerical example presented, he solves the new equation by trial, but this could also have been done with the original. Nevertheless, although the abacists did not manage to give a complete general solution to the cubic equation, they, like their Islamic predecessors, wrestled with the problem and arrived at partial results, as noted in the opening quotation from the work of Luca Pacioli (1445–1517).

Pacioli, one of the last of the abacists, was ordained as a Franciscan friar in the 1470s and taught mathematics at various places in Italy during the remainder of his career. He became so famous as a teacher that there is a painting of him by Jacopo di Barbari now hanging in the Naples Museum which shows him teaching geometry to a young man tentatively identified as Guidobaldo, the son of his patron, the Duke of Urbino (Figure 9.1). As part of his teaching, Pacioli composed three different abacus texts for his students. He regretted what he believed to be the low ebb to which teaching had fallen. Because he felt that one of the problems was the scarcity of available subject material, he gathered mathematical materials for some 20 years and in 1494 completed the most comprehensive mathematics text of the time, and one of the earliest mathematics texts to be printed. This

FIGURE 9.1
Luca Pacioli teaching
geometry in a painting
by Jacopo di Barbari.
(Source: Alinari/Art
Resource, N.Y.)

was the *Summa de Arithmetica, Geometrica, Proportioni et Proportionalita*, a 600-page
work written in the Tuscan dialect rather than in Latin. It contained not only practical
arithmetic but also much of the algebra already discussed, the first published treatment of
double-entry bookkeeping, and a practical geometry similar to that of Leonardo of Pisa.
(In fact, it contains the same chord table.) There was little that was original in this work,
aside from the abbreviations for plus and minus mentioned earlier, but its comprehensiveness and the fact that it was the first such work to be printed made it into a widely
circulated and influential text, extensively studied by sixteenth century Italian mathematicians. It became the common base from which these men were able to extend the range of
algebra. Before considering these advances, however, we first turn to contemporaneous
developments elsewhere in Europe. It is not only from Italy that our algebra comes.

9.2 ALGEBRA IN FRANCE, GERMANY, ENGLAND, AND PORTUGAL

Algebra was developing in Italy in the fourteenth and fifteenth centuries owing in large
measure to a need generated by the developing mercantile economy. The economy was
changing in the remainder of Europe as well, although somewhat later than in Italy. Mathematics texts appeared elsewhere also to meet the new needs of the society. We will
consider here the work of Nicolas Chuquet in France, Christoff Rudolff and Michael Stifel

in Germany, Robert Recorde in England, and Pedro Nuñez in Portugal. There is much similarity among their works in algebra and also similarities between these works and the Italian algebra of the fifteenth century. It is clear that these mathematicians all had some knowledge of the contemporaneous work elsewhere in Europe, even though explicit reference to the work of others is generally limited or lacking entirely. But each of them also seems to have some original material. It appears that the knowledge of Islamic algebra had spread widely in Europe by the fifteenth century. Each person attempting to write new works used this material and works in algebra from elsewhere in Europe, adapted them to fit the circumstances of his own country, and introduced some of his own new ideas. By the late sixteenth century, with the spread of printing, new ideas could circulate more rapidly throughout the continent, and those generally felt to be most important were absorbed into a new European algebra.

9.2.1 France: Nicolas Chuquet

Nicolas Chuquet (d. 1487) was a French physician who wrote his mathematical treatise in Lyon near the end of his life. Lyon in the late fifteenth century was a thriving commercial community with a growing need, as in the Italian cities, for practical mathematics. It was probably to meet this need that Chuquet composed his *Triparty* in 1484, a work on arithmetic and algebra in three parts, followed by three related works containing problems in various fields in which the rules established in the *Triparty* are used.[9] These supplementary problems show many similarities to the problems in Italian abacus works, but the *Triparty* itself is on a somewhat different level in that it is a text in mathematics itself. Most of the mathematics in it was certainly known to the Islamic algebraists and also to Leonardo of Pisa. Nevertheless, since it is the first detailed algebra in fifteenth century France, we will consider some of its important ideas.

The first part of the *Triparty* is concerned with arithmetic. Like the Italian works, it begins with a treatment of the Hindu-Arabic place-value system and details the various algorithms for the basic operations of arithmetic, both with whole numbers and with fractions. One of Chuquet's procedures with fractions is a rule "to find as many numbers intermediate between two neighboring numbers as one desires."[10] His idea is that to find a fraction between two fractions, one simply adds the numerators and adds the denominators. Thus between 1/2 and 1/3 is 2/5, and between 1/2 and 2/5 is 3/7. Chuquet gives no proof that the rule is correct, but he does apply it to deal with finding roots of polynomials. For example, to find the root of $x^2 + x = 39\ 13/81$, Chuquet begins by noting that 5 is too small to be a root while 6 is too large. He then proceeds to find the correct intermediate value by first checking, in turn, 5 1/2, 5 2/3, 5 3/4, and 5 4/5 and determining that the root must be between the two last values. Applying his rule to the fractional parts, he next checks 5 7/9, which turns out to be the correct answer.

In part two of the *Triparty*, Chuquet applies his fraction rule to the calculation of square roots of numbers which are not perfect squares. Noting that 2 is too small and 3 too large to be the square root of 6, he begins the next stage of his approximation procedure by determining that 2 1/3 is too small and 2 1/2 too large. His next several approximations are, in turn, 2 2/5, 2 3/7, 2 4/9, 2 5/11, and 2 9/20. At each stage he calculates the square of the number chosen and, depending on whether it is larger or smaller than 6, determines between which two values to use his rule of intermediates. He notes that "by this manner one may proceed, . . . until one approaches very close to 6, a little more or a little less, and until it is sufficient. And one should know that the more one should continue in this way, the nearer to 6 one would approach. But one would never attain it precisely. And from all this follows the practice, in which the good and sufficient root of 6 is found to be 2 89/198, which root multiplied by itself produces 6 plus 1/39,204."[11] Chuquet is evidently aware of the irrationality of $\sqrt{6}$ and has developed a new recursive algorithm to calculate it to whatever accuracy may be desired. He has therefore taken another step on the road to denying the usefulness of the Greek dichotomy between the discrete and the continuous, the final elimination of which was to occur about a century later.

Chuquet also displays in the second part of his work the standard methods for calculating the square and cube roots of larger integers, one integral place at a time, but as is usual in the discussion of these methods, he does not take the method below the unit. He shows no knowledge of the idea of a decimal fraction. If the standard method does not give an exact root, one has the choice of calculating using common fractions by his method of intermediates or (and this is the method he prefers) of simply not bothering to calculate at all and leaving the answer in the form $R^2 6$ or $R^3 12$, his notation for our $\sqrt{6}$ and $\sqrt[3]{12}$. Chuquet also uses the Italian $\bar{p}$ and $\bar{m}$ for plus and minus, but introduces an underline to indicate grouping. Thus what we would write as $\sqrt{14 + \sqrt{180}}$, Chuquet writes as $R^2 \underline{14\bar{p}R^2 180}$. He proceeds to use this notation with complete understanding through the rest of this second part as he displays a solid knowledge of computations with radical expressions, both simple and compound, including the necessary rules for dealing with positives and negatives in addition, subtraction, multiplication, and division.

The third part of the *Triparty* is more strictly algebraic. Chuquet here shows how to manipulate with polynomials and how to solve various types of equations. As part of his discussion of polynomials, he introduces an exponential notation for the powers of the unknown which makes calculation somewhat easier than the Italian abbreviations. Thus he writes 12^2 for what we would write as $12x^2$ and $\bar{m}12^{\overline{2m}}$ for what we would write as $-12x^{-2}$. He even notes that the exponent 0 is to be used when one is dealing with numbers themselves. These expressions (*diversities*) involving exponents (*denominations*) are to be added, subtracted, multiplied, and divided using our standard rules. Thus "whoever would multiply 8^3 by $7^{\overline{1m}}$ it is first necessary to multiply 8 by 7 coming to 56, then he must add the denominations, that is to say $3\bar{p}$ with $1\bar{m}$ coming to 2. Thus this multiplication comes to 56^2, and so should others be understood."[12] Not only does he give this rule, but he also justifies it. He writes down in two parallel columns the powers of 2 (beginning with $1 = 2^0$ and ending with $1,048,576 = 2^{20}$) and the corresponding *denomination* and then notes that multiplication in the first column corresponds to addition in the second. Since the addition rule of exponents works for numbers, he simply extends it to his *diversities*.

Chuquet also has a few innovations in his equation-solving techniques. First, he generalizes al-Khwārizmī's rules to equations of any degree that are of quadratic type, thus going somewhat further than the Italian abacists. For example, he gives the solution of the equation $cx^m = bx^{m+n} + x^{m+2n}$ to be

$$x = \sqrt[n]{\sqrt{(b/2)^2 + c} - (b/2)}.$$

Second, he notes that a particular system of two equations in three unknowns has multiple solutions. To solve the system $x + y = 3z$, $x + z = 5y$, he first picks 12 for x and then finds $y = 3\ 3/7$ and $z = 5\ 1/7$. Then he picks 8 for y and calculates $x = 28$ and $z = 12$. "Thus," he concludes, "it appears that the number proposed alone determines the varying answer."[13] Finally, although he is not consistent about this, Chuquet is willing under some circumstances to consider negative solutions to equations. For example, he solves the problem $\frac{5}{12}\left(20 - \frac{11}{20}x\right) = 10$ to get $x = -7\frac{3}{11}$. He then checks the result carefully and concludes that the answer is correct. In other problems, however, he rejects negative solutions as "impossible," and he never considers 0 to be a solution.

The three supplements to the *Triparty* contain hundreds of problems in which the techniques of that work are applied. Many of the problems are commercial, of the same type found in the Italian abacus works, while others are geometrical, both practical and theoretical. This work may have been intended as a text, but, unfortunately, it was never printed and exists today only in manuscript form. Some parts of it were incorporated into a work of Estienne de la Roche (probably one of Chuquet's students) in 1520, but neither this work nor Chuquet's itself had much influence.

9.2.2 Germany: Christoff Rudolff and Michael Stifel

Algebra in Germany first appears late in the fifteenth century, probably due to the same reasons which led to its development in Italy somewhat earlier. It is likely, in fact, that many of the actual techniques were also imported from Italy. The very name given to algebra in Germany, the Art of the *Coss*, reveals its Italian origin. *Coss* was simply the German form of the Italian *cosa*, or thing, the name usually given to the unknown in an algebraic equation. Two of the most important Cossists in the first half of the sixteenth century were Christoff Rudolff (first half of the sixteenth century) and Michael Stifel (1487–1567).

Christoff Rudolff wrote his *Coss*, the first comprehensive German algebra, in Vienna in the early 1520s.[14] It was published in Strasbourg in 1525. As usual the book begins with the basics of the place-value system for integers, giving the algorithms for calculation as well as a short multiplication table. In a section dealing with progressions, Rudolff included a list of nonnegative powers of 2 alongside their respective exponents, just as Chuquet had done. He also noted that multiplication in the powers corresponded to addition in the exponents. This idea is then extended to powers of the unknown. Although Rudolff did not have the exponential notation of his French predecessor, he did have a system of abbreviations of the names of these powers (Side Bar 9.1).

To help the reader understand these terms, Rudolff gave as examples the powers of various numbers. He then showed how to add, subtract, multiply, and divide expressions formed from these symbols. Because it is not obvious how to multiply these symbols, unlike the situation in Chuquet's system, Rudolff presented a multiplication table for use with them, which showed, for instance, that ℞ times ⅜ was ℀. To simplify matters, he then included numerical values for his symbols. Thus *radix* was labeled as 1, *zensus* as 2, *cubus* as 3, and so on, and he noted that in multiplying expressions one could simply add the corresponding numbers to find the correct symbol. In this section Rudolff also dealt with binomials, terms connected by an operation sign, and included, for the first time in

---

Side Bar 9.1   Rudolff's System for Powers of the Unknown

dragma	φ	radix	$\mathfrak{x} \leftrightarrow x$
zensus	$\mathfrak{z} \leftrightarrow x^2$	cubus	$\mathfrak{c} \leftrightarrow x^3$
zens de zens	$\mathfrak{zz} \leftrightarrow x^4$	sursolidum	$\int\mathfrak{z} \leftrightarrow x^5$
zensicubus	$\mathfrak{zc} \leftrightarrow x^6$	bissursolidum	$^b\!\int\mathfrak{z} \leftrightarrow x^7$
zenszensdezens	$\mathfrak{zzz} \leftrightarrow x^8$	cubus de cubo	$^c\mathfrak{c} \leftrightarrow x^9$

---

an algebra text, the current symbols of $+$ and $-$ to represent addition and subtraction. The signs had been used earlier in an arithmetic work of 1518 of Heinrich Schreiber (Henricus Grammateus), Rudolff's teacher at the University of Vienna. These signs had appeared even earlier in a work of Johann Widman of 1489. There, however, they represented excess and deficiency rather than operations.

Rudolff also introduced in this work the modern symbol $\sqrt{}$ for square root. He modified this symbol somewhat to indicate cube roots and fourth roots, but did not use modern indices. He did, however, give a detailed treatment of operations on surds, showing how to use conjugates in division as well as how to find the square roots of surd expressions, that is, $\sqrt{27 + \sqrt{200}}$. He also introduced a symbol for "equals," namely a period, as in $1\,\mathfrak{x}\,.2$ ($x = 2$). Often, however, he relied on the German *gleich*.

The second half of Rudolff's *Coss* is devoted to the solving of algebraic equations. Rather than using the standard al-Khwārizmī six-fold classification of such equations, Rudolff uses his own eight-fold classification. The rule for the solution of each type of equation is given in words and then illustrated with examples. Although Rudolff deals with equations of higher degree than two in his classes, like Chuquet he includes only those which can be solved by reduction to a quadratic equation or by simple roots. Thus, for example, one of his classes is that now written as $ax^n + bx^{n-1} = cx^{n-2}$. The solution given is the standard

$$x = \sqrt{\left(\frac{b}{2a}\right)^2 + \frac{c}{a}} - \frac{b}{2a}.$$

His sample equations illustrating this class include $3x^2 + 4x = 20$ and $4x^7 + 8x^6 = 32x^5$, both of which have the solution $x = 2$. Like most other authors, Rudolff does not deal with the negative roots or with zero as a root.

After presenting the rules, Rudolff, as is typical, gives several hundred examples of problems that can be solved using the rules. Many are commercial problems, but those needing a version of the quadratic formula are generally artificial ones, including the ubiquitous "divide 10 into two parts such that . . . ." At the end of the text, Rudolff presents three irreducible cubic equations with their answers but without giving a method of solution. He simply notes that others who come later will continue the algebraic art and teach how to deal with these. Curiously, on the final page there is a drawing of a cube of side $3 + \sqrt{2}$ divided into eight rectangular prisms. Whether Rudolff intended this diagram to be a hint for the solution of the cubic equation is not known.

Biography   Michael Stifel (1487–1567)

Michael Stifel was ordained as a priest in 1511. Reacting to various clerical abuses he became an early follower of Martin Luther. In the 1520s he became interested in what he called *wortrechnung* (word calculus), the interpretation of words through the numerical values of the letters involved. Through interpreting certain Biblical passages using his numerical methods, he finally came to the belief that the world would end on October 18, 1533. He assembled his congregation in the church on that morning, but to his great dismay, nothing happened. He was subsequently discharged from his parish and for a time placed under house arrest. Because he had now been cured of prophesying, however, he was given another parish in 1535 through the intervention of Luther. Subsequently he devoted himself to the study of mathematics at the University of Wittenberg and soon became an expert in algebraic methods, publishing his *Deutsche Arithmetica* in 1545, one year after the *Arithmetica Integra*. Later in life, however, he resumed his *wortrechnung* and wrote two books on the subject.

A new edition of Rudolff's text was brought out in 1553 by Michael Stifel. Stifel earlier had prepared a text of his own, the *Arithmetica Integra* in 1544, a work which included the second European publication of a version of the "Pascal triangle" of binomial coefficients (Table 9.1).[15] (The first publication was on the title page of Peter Apianus' *Arithmetic* of 1527.) Stifel noted that he had discovered these coefficients and the procedure for using them in finding roots only with great difficulty as he had been unable to find any written accounts of them. Thus, although these coefficients had been used for that purpose in China several centuries earlier, the lack of transmission required the rules to be discovered anew in Europe.

Stifel used the same symbols as Rudolff for the powers of the unknowns, but he was more consistent in using the correspondence between these letters and the integral "exponents." In fact, he went further than Rudolff by writing out a table of powers of 2 along with their exponents which included the negative values -1, -2, -3 as corresponding to 1/2, 1/4, and 1/8 respectively.

Although Stifel, like most of his contemporaries, did not accept negative roots to equations, he was the first to compress the three standard forms of the quadratic equation into the single form $x^2 = bx + c$ where b and c were either both positive or were of opposite parity. The solution, expressed in words, was then equivalent to

$$x = \frac{b}{2} \pm \sqrt{\left(\frac{b}{2}\right)^2 + c},$$

where the negative sign was only possible in the case where b was positive and c negative. In that case, as long as $(b/2)^2 + c > 0$, there were two positive solutions. Combining the three cases of the quadratic into one does not seem a major advance, but in the context of the sixteenth century it was significant. It was another step toward the extension of the number concept, although another two centuries were to pass before all algebra texts adopted this procedure.

TABLE 9.1

Apianus' Version

						3		3						
					4		6		4					
				5		10		10		5				
			6		15		20		15		6			
		7		21		35		35		21		7		
	8		28		56		70		56		28		8	
9		36		84		126		126		84		36		9

Stifel's Version

1							
2							
3	3						
4	6						
5	10	10					
6	15	20					
7	21	35	35				
8	28	56	70				
9	36	84	126	126			
10	45	120	210	252			
11	55	165	330	462	462		
12	66	220	495	792	924		
13	78	286	715	1,287	1,716	1,716	
14	91	364	1,001	2,002	3,003	3,432	
15	105	455	1,365	3,003	5,005	6,435	6,435
16	120	560	1,820	4,368	8,008	11,440	12,870
17	136	680	2,380	6,188	12,376	19,448	24,310

9.2.3 England: Robert Recorde

The *Arithmetica Integra* and Stifel's 1553 revision of Rudolff's *Coss* were very important in Germany, influencing textbook writers well into the next century and helping to develop in Germany, as had already been done in Italy, mathematical awareness in the middle classes. They also had influence in England, where they were the major source of the first English algebra, *The Whetstone of Witte*, published in 1557 by the first English author of mathematical works in the Renaissance, Robert Recorde (1510–1558) (see Figure 9.2).

The Whetstone of Witte had little that was original in technique, because it was based on the German sources and even used the German symbols for powers of the unknown,

---

Biography   **Robert Recorde (1510–1558)**

Robert Recorde graduated from Oxford in 1531 and was licensed in medicine soon thereafter. Although he probably practiced medicine in London in the late 1540s, his only known positions were in the civil service, positions in which he was not notably successful. On the other hand, he did write several successful mathematics textbooks besides *The Whetstone of Witte*, including *The Ground of Arts* (1543) on arithmetic, *The Pathway to Knowledge* (1551) on geometry, and *The Castle of Knowledge* (1556) on astronomy. His works show that he was especially interested in pedagogy. In particular, his books were set in the form of a dialogue between master and pupil, in which each step in a particular technique was carefully explained.

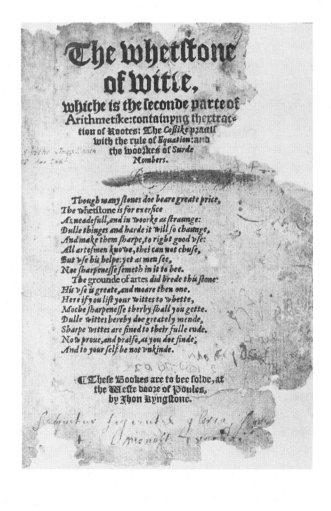

FIGURE 9.2
Title page of Robert Recorde's *The Whetstone of Witte* (1557). (Source: Smithsonian Institution Photo No. 92-338)

but there are a few points of interest in the text, which taught algebra to an entire generation of English scientists. First, Recorde created the modern symbol for equality: "To avoid the tedious repetition of these words—is equal to—I will set as I do often in work use, a pair of parallels, or gemow [twin] lines of one length, thus ====, because no 2 things can be more equal."[16] Second, he modified and extended the German symbolization of powers of the unknown to powers as high as the eightieth, setting the integer of the power next to each symbol and noting that multiplication of these symbols corresponds to addition of the corresponding integers. In fact, he showed how to build the symbol for any power out of the square $\mathcal{z}$, the cube $\mathcal{C}$, and various sursolids (prime powers higher than the third) *$\int\mathcal{z}$ (where * stands for a letter designating the order of the prime). The fifth power is written $\int\mathcal{z}$, the seventh power as $^b\int\mathcal{z}$ (second sursolid), and the eleventh power as $^c\int\mathcal{z}$ (third sursolid). Then, for instance, the ninth power is written $\mathcal{C}$ $\mathcal{C}$ (cube of the cube), the twentieth power as $\mathcal{z}\mathcal{z}\int\mathcal{z}$ (square of the square of the fifth power), and the twenty-first power as $\mathcal{C}^b\int\mathcal{z}$ (cube of the seventh power). Finally, to help students remember the various rules of operation, he gave them in poetic form. His verse giving the procedure for multiplying and dividing expressions of the form ax^n, where the power n is called the "quantity" of the expression, includes the standard rule of signs for those operations as well as the rule of exponents:

> Who that will multiplie,
> Or yet divide trulie:
> Shall like still to have more,
> And mislike lesse in store.
> Their quantities doe kepe soche rate,
> That .M. doeth adde; and .D. abate.

9.2.4 Portugal: Pedro Nuñez

Mathematics was also necessary in Portugal, where already in the fifteenth century navigators were extending the European knowledge of the rest of the world. It was here that Pedro Nuñez (1502–1578) wrote his *Libro de Algebra* in 1532.[17] Nuñez was influenced by his reading of the work of Pacioli. His notation clearly derives from the Italian writers, and it appears that he had no knowledge of his German contemporaries. Thus he uses Italian abbreviations for the various powers of the unknown—*co* for *cosa*, *ce* for *censo*, *cu* for *cubo*—as well as $\bar{p}$ and $\bar{m}$ for plus and minus. In his text, he deals with the procedures for combining algebraic expressions, for solving equations, and for dealing with radicals and proportions. He includes dozens of problems, but unlike those in most of the other algebra texts mentioned, his are all abstract.

To give the flavor of Nuñez's text, we consider how he solved one of the standard problems, to find two numbers whose product and the sum of whose squares is known. The product is given as 10 and the sum of squares as 30. Nuñez solves the problem three different ways to demonstrate various algebraic techniques. First, letting x be the smaller of the two numbers, he takes $10/x$ as the larger, squares each and gets the equation

$$x^2 + \frac{100}{x^2} = 30.$$

Biography   **Pedro Nuñez (1502–1578)**

Pedro Nuñez studied at the University of Salamanca but received a degree in medicine in Lisbon in 1525 (Figure 9.3). He made several contributions to the science of navigation, and became chief cosmographer to the King of Portugal and professor of mathematics at the University of Coimbra, writing his *Libro de Algebra* in 1532. Many of his students later held high positions at court. Although Nuñez was of Jewish origin, he was never persecuted by the Inquisition, probably because one of these students became the Inquisitor General, Cardinal Don Henrique. Nuñez's algebra text was originally written in Portuguese, but because he felt that it would have more influence if it were available in Spanish, he translated it into Spanish some 30 years later and had it printed in the Netherlands in 1567. He wrote his astronomical works chiefly in Latin, however. In addition to his scientific work, Nuñez was also a poet of some note.

FIGURE 9.3
Pedro Nuñez on a Portuguese stamp. Notice the algebra problem in the background.

Multiplying by x^2, he reduces this to a quadratic equation in x^2 whose solution by one of al-Khwārizmī's formulas is $x^2 = 15 \pm \sqrt{125}$. Thus the two desired numbers are $\sqrt{15 - \sqrt{125}}$ and $\sqrt{15 + \sqrt{125}}$. For his second method, he notes that the two numbers cannot be equal, then represents the squares of the two numbers respectively as $15 - x$ and $15 + x$. The numbers themselves being the square roots of these quantities, he derives the equation $\sqrt{15 - x}\sqrt{15 + x} = 10$, which easily reduces to $x = \sqrt{125}$. Therefore, the solution is the same as before. Nuñez's third solution makes use of the identity $(a + b)^2 = a^2 + 2ab + b^2$. The square of the sum of the two numbers is 50, so their sum is $\sqrt{50}$. The two numbers, therefore, are $(1/2)\sqrt{50} - x$ and $(1/2)\sqrt{50} + x$. Multiplying these together gives the equation

$$12\frac{1}{2} - x^2 = 10,$$

whence $x = \sqrt{2\ 1/2}$. The two numbers in this case are then $\sqrt{12\ 1/2} - \sqrt{2\ 1/2}$ and $\sqrt{12\ 1/2} + \sqrt{2\ 1/2}$. Nuñez's problem now is to show that this pair of numbers is the same as the pair found by the first two methods. He does this by comparing their respective squares, even though he realizes that the equality of squares does not necessarily imply the equality of the roots. Although he is not sure how to avoid this difficulty, he is convinced that the solutions are in fact the same, since both pairs do satisfy the original equation.

9.3 THE SOLUTION OF THE CUBIC EQUATION

Fra Luca Pacioli noted in 1494 that there was not yet an algebraic solution to the general cubic equation, but throughout the fifteenth and early sixteenth centuries many mathematicians were working on this problem. Finally, sometime between 1500 and 1515, Scipione del Ferro (1465–1526), a professor at the University of Bologna, discovered an algebraic method of solving the cubic equation $x^3 + cx = d$. Recall from our study of the Islamic solutions to cubic equations that, since most mathematicians still did not deal with negative numbers even as coefficients of equations, there were 13 different types of mixed irreducible cubic equations depending on the relative positions of the (positive) quadratic, linear, and constant terms. So del Ferro had only begun the process of solving the cubic equation with his solution of one of these cases.

In modern academia, professors announce and publish new results as quickly as possible to ensure priority, so it may be surprising to learn that del Ferro did not publish, nor even publicly announce, his major breakthrough. But academic life in sixteenth century Italy was far different from that of today. There was no tenure. University appointments were mostly temporary and subject to periodic renewal by the university senate. One of the ways a professor convinced the senate that he was worthy of continuing in his position was by winning public challenges. Two contenders for a given position would present each other with a list of problems, and, in a public forum some time later, each would present his solutions to the other's problems. Often, considerable amounts of money, aside from the university positions themselves, were dependent on the outcome of such a challenge. As a result, if a professor discovered a new method for solving certain problems, it was to his advantage to keep it secret. He could then pose these problems to his opponents secure in the knowledge that he would prevail.

Before he died del Ferro disclosed his solution to his pupil, Antonio Maria Fiore (first half of the sixteenth century), and to his successor at Bologna, Annibale della Nave (1500–1558). Although neither of these men publicized the solution, word began to circulate among Italian mathematicians that this old problem had been, or soon would be, solved. Another mathematician, Niccolò Tartaglia of Brescia (1499–1557), in fact, boasted that he too had discovered the solution to a form of the cubic, $x^3 + bx^2 = d$. In 1535, Fiore challenged Tartaglia to a public contest, hoping to win on the strength of his knowledge of the earlier case. Each of his 30 submitted problems dealt with that class of cubic equations. For example, one of the problems reads, "A man sells a sapphire for 500 ducats, making a profit of the cube root of his capital. How much is this profit?" [$x^3 + x = 500$].[18] But Tartaglia, the better mathematician, worked long and hard on that case, and, as he later wrote, on the night of February 12, 1535 he discovered the solution. Since Fiore was unable to solve many of Tartaglia's questions covering other areas of mathematics besides the cubic, Tartaglia was declared the winner, in this case of 30 banquets prepared by the loser for the winner and his friends. (Tartaglia, probably wisely, declined the prize, accepting just the honor of the victory.)

Word of the contest and the new solutions of the cubic soon reached Milan, where Gerolamo Cardano (1501–1576) was giving public lectures in mathematics, supported by a grant from the will of the scholar Tommasso Piatti for the instruction of poor youths. Cardano wrote to Tartaglia, asking that Tartaglia show him the solution so it could be included, with full credit, in the arithmetic text Cardano was then writing. Tartaglia initially refused, but after many entreaties and a promise from Cardano to introduce him and his new inventions in artillery to the Milanese court, he finally came to Milan in early 1539. Tartaglia, after extracting an oath from Cardano that he would never publish Tartaglia's discoveries—Tartaglia planned to publish them himself at some later date—divulged the secret of the equation $x^3 + cx = d$ to Cardano in the form of a poem:

When the cube and its things near
Add to a new number, discrete,

Biography   Gerolamo Cardano (1501–1576)

Cardano was trained as a physician but was denied admission to the Milan College of Physicians because of his illegitimate birth. For several years, therefore, he practiced medicine in a small town near Padua, returning to Milan in 1533 where he treated occasional private patients as well as lectured in mathematics and wrote a textbook on arithmetic. Finally, he convinced the College of Physicians to change its mind. Cardano soon became the most prominent physician in Milan and in great demand throughout Europe. His most important patient was probably John Hamilton, the Archbishop of Scotland, who in 1551 requested Cardano's services to help him overcome steadily worsening attacks of asthma. Cardano, after spending a month observing the archbishop's symptoms and habits, decided he had a severe allergy to the feathers in the bed he slept in. Thus Cardano recommended that the bedding be changed to silk and the pillow to linen. The archbishop's health improved immediately, and he remained extremely grateful to Cardano for the rest of his life, offering money and other assistance whenever Cardano might need it. Cardano was not so successful in his attempt to cast a horoscope for the young king Edward VI on his return from Scotland through England. He predicted a long life, but unfortunately, the sixteen-year-old king died shortly thereafter. Cardano's own life was filled with many tragedies, including the death of his wife in 1546 and the execution of his son for the murder of his own wife in 1560. The final blow came in 1570 when he was brought before the Inquisition on an accusation of heresy. Fortunately, the sentence was relatively lenient. Cardano spent his last few years in Rome where he wrote his autobiography *De Propria Vita*.

Determine two new numbers different
By that one; this feat
Will be kept as a rule
Their product always equal, the same,
To the cube of a third
Of the number of things named.
Then, generally speaking,
The remaining amount
Of the cube roots subtracted
Will be your desired count.[19]

9.3.1 Gerolamo Cardano and the *Ars Magna*

Cardano kept his promise not to publish Tartaglia's result in his arithmetic book, which soon appeared. In fact, he sent Tartaglia a copy off the press to show his good faith. Cardano then began to work on the problem himself, probably assisted by his servant and student, Lodovico Ferrari (1522–1565). Over the next several years he worked out the solutions and their justifications for all of the various cases of the cubic. Ferrari himself managed to solve the fourth degree equation as well. Meanwhile, Tartaglia still had not published anything on the cubic. Cardano did not want to break his solemn oath, but he was eager that the solutions should be made available. Acting on rumors of the original discovery by del Ferro, he and Ferrari journeyed to Bologna and called on della Nave. He graciously gave the two permission to inspect del Ferro's papers, and they were able to verify that del Ferro had discovered the solution first. Cardano no longer felt an obligation

HIERONYMI CAR
DANI, PRÆSTANTISSIMI MATHE-
MATICI, PHILOSOPHI, AC MEDICI,
ARTIS MAGNÆ,
SIVE DE REGVLIS ALGEBRAICIS,
Lib.unus. Qui & totius operis de Arithmetica, quod
OPVS PERFECTVM
inscripsit, est in ordine Decimus.

Habes in hoc libro, studiose Lector, Regulas Algebraicas (Itali, de la Cos-
sa uocant) nouis adinuentionibus, ac demonstrationibus ab Authore ita
locupletatas, ut pro pauculis antea uulgò tritis, iam septuaginta euaserint. Ne-
cp solùm , ubi unus numerus alteri, aut duo uni, uerum etiam, ubi duo duobus,
aut tres uni æquales fuerint, nodum explicant.   Hunc aut librum ideo seor-
sim edere placuit, ut hoc abstrusissimo, & planè inexhausto totius Arithmeti-
cæ thesauro in lucem eruto, & quasi in theatro quodam omnibus ad spectan-
dum exposito, Lectores incitarentur, ut reliquos Operis Perfecti libros, qui per
Tomos edentur, tanto auidius amplectantur, ac minore fastidio perdiscant.

FIGURE 9.4
Title page of Gerolamo
Cardano's *Ars Magna*.
(Source: Smithsonian
Institution Photo No.
76-15322)

to Tartaglia. After all, he would not be publishing Tartaglia's solution, but one discovered some 20 years earlier by a man now deceased. So in 1545 Cardano published his most important mathematical work, the *Ars Magna, sive de Regulis Algebraicis* (*The Great Art, or On the Rules of Algebra*), chiefly devoted to the solution of cubic and quartic equations (Figure 9.4). Tartaglia, of course, was furious when Cardano's work appeared. He felt he had been cheated of the rewards of his labor, even though Cardano did mention that Tartaglia was one of the original discoverers of the method. Tartaglia's protests availed him nothing. In an attempt to recoup his prestige, he had another public contest, this time with Ferrari, but was defeated. To this day, the formula providing the solution to the cubic equation is known as Cardano's formula.

It is now time to consider the details of the cubic formula, as presented in poetry by Tartaglia (seen earlier) and in prose by Cardano in the *Ars Magna*, for the equation $x^3 + cx = d$: "Cube one-third the coefficient of the thing; add to it the square of one-half the constant of the equation; and take the square root of the whole. You will duplicate this and to one of the two you add one-half the number you have already squared and from the other you subtract one-half the same. . . . Then subtracting the cube root of the [second] from the cube root of the [first], the remainder [or] that which is left is the value of the thing."[20] Tartaglia's poem tells us (lines 3–4) to find two numbers u, v, such that $u - v = d$ and (lines 6–8) $uv = (c/3)^3$. Then (lines 10–12), $x = \sqrt[3]{u} - \sqrt[3]{v}$. Cardano explains this further. If one solves the system of two equations in u, v by solving the second for v $\left(v = \left(\frac{c}{3}\right)^3 \frac{1}{u}\right)$ and substituting in the first, one gets $u - \left(\frac{c}{3}\right)^3 \frac{1}{u} = d$ or

$$u^2 - \left(\frac{c}{3}\right)^3 = du,$$

a quadratic equation in u. This is easily solved to get

$$u = \sqrt{\left(\frac{d}{2}\right)^2 + \left(\frac{c}{3}\right)^3} + \frac{d}{2}.$$

The solution for v differs only by a sign. One then has precisely the formula enunciated by Cardano:

$$x = \sqrt[3]{\sqrt{\left(\frac{d}{2}\right)^2 + \left(\frac{c}{3}\right)^3} + \frac{d}{2}} - \sqrt[3]{\sqrt{\left(\frac{d}{2}\right)^2 + \left(\frac{c}{3}\right)^3} - \frac{d}{2}}.$$

Cardano proves this result by a geometric argument involving cubes applied to a specific example. The essence of his proof can be seen more easily through the following algebraic argument: If $x = \sqrt[3]{u} - \sqrt[3]{v}$ where $u - v = d$ and $uv = (c/3)^3$, then

$$x^3 + cx = \left(\sqrt[3]{u} - \sqrt[3]{v}\right)^3 + c\left(\sqrt[3]{u} - \sqrt[3]{v}\right)$$

$$= u - 3\sqrt[3]{u^2 v} + 3\sqrt[3]{u v^2} - v + 3\sqrt[3]{uv}\left(\sqrt[3]{u} - \sqrt[3]{v}\right)$$

$$= u - v - 3\sqrt[3]{uv}\left(\sqrt[3]{u} - \sqrt[3]{v}\right) + 3\sqrt[3]{uv}\left(\sqrt[3]{u} - \sqrt[3]{v}\right)$$

$$= d.$$

To clarify his rule, Cardano presents the example $x^3 + 6x = 20$. The formula tells us, since $c/3 = 2$ and $d/2 = 10$, that

$$x = \sqrt[3]{\sqrt{108} + 10} - \sqrt[3]{\sqrt{108} - 10}.$$

There is an obvious question about this answer. It is clear to us, and presumably was also to Cardano, that the solution to the equation $x^3 + 6x = 20$ is $x = 2$. The answer given by the formula is equal to 2, but that is certainly not evident. Cardano, interestingly enough, makes no mention of this difficulty, and goes on to more examples.

Similarly, in the chapter "On the cube equal to the thing and number," that is, $x^3 = cx + d$, Cardano presents and proves a rule differing little from his rule for cube and thing equal to number. In this case, the formula becomes

$$x = \sqrt[3]{\frac{d}{2} + \sqrt{\left(\frac{d}{2}\right)^2 - \left(\frac{c}{3}\right)^3}} + \sqrt[3]{\frac{d}{2} - \sqrt{\left(\frac{d}{2}\right)^2 - \left(\frac{c}{3}\right)^3}}.$$

After presenting as examples the equation $x^3 = 6x + 40$ and $x^3 = 6x + 6$, he notes the difficulty here if $(c/3)^3 > (d/2)^2$. In that case, one could not take the square root. To circumvent the difficulty, Cardano describes other methods for special cases. As we will see later, it was Rafael Bombelli who showed how to deal with the square roots of negative numbers in the Cardano formula.

Cardano's discussion of the solution of the various cases of the cubic are in Chapters 11–23 of the *Ars Magna*. But the text opens with some general results, including a discussion of the number of roots a given equation could have, whether the roots are positive (true) or negative (fictitious), and how the roots of one equation determine the roots of a related equation. For example, Cardano tells us that equations of the form $x^3 + cx = d$ always have one positive solution and no negative ones. Conversely, the number and sign of the roots of the equation $x^3 + d = cx$ depend on the coefficients. If

$$\frac{2c}{3}\sqrt{\frac{c}{3}} = d,$$

then this equation has one positive root $r = \sqrt{c/3}$ and one negative one $-s = -2\sqrt{c/3}$. If

$$\frac{2c}{3}\sqrt{\frac{c}{3}} > d,$$

then there are two positive roots r and s and one negative root $-t$, where $t = r + s$. In addition, t is the positive root of the equation $x^3 = cx + d$. Parenthetically, Cardano notes that in the first case, one could consider the positive root r as two separate roots, for the negative root equals $-2r$. Finally, if

$$\frac{2c}{3}\sqrt{\frac{c}{3}} < d,$$

then there are no positive roots. There is one negative root $-s$, where s is the positive root of $x^3 = cx + d$. Sharaf al-Dīn al-Ṭūsī had given a similar discussion of the roots of this equation (and certain others) some 300 years earlier and had arrived at the same criteria for the existence of positive roots, but whether Cardano used the same method of considering maximums is unknown. Cardano does provide more information than his Islamic predecessor, however, since he considers negative roots. Thus he is also able to understand, if not prove, that when there are three real roots to a cubic equation, their sum is equal to the coefficient of the x^2 term.

Cardano's pupil Lodovico Ferrari succeeded in finding the solution to the fourth degree equation. Cardano presented this solution briefly near the end of the *Ars Magna*, where he listed the 20 different types of quartic equations, outlined a basic procedure, and calculated a few examples. This basic procedure begins with a linear substitution which eliminates the term in x^3, leaving an equation of the form $x^4 + cx^2 + e = dx$, for instance. To solve this equation, second degree and constant terms are added to both sides to turn each side into a perfect square. One then takes square roots and calculates the answer. We illustrate the procedure with one of Cardano's examples: $x^4 + 3 = 12x$. If we add $2bx^2 + b^2 - 3$ to both sides (where b is to be determined), the left side becomes

$x^4 + 2bx^2 + b^2$, a perfect square, while the right side becomes $2bx^2 + 12x + b^2 - 3$. For this latter to be a perfect square, we must have

$$2b(b^2 - 3) = \left(\frac{12}{2}\right)^2 \qquad \text{or} \qquad 2b^3 = 6b + 36.$$

Therefore, we need to solve a cubic equation in b. (This equation is now called the resolvent cubic of the given quartic.) Cardano, of course, has a rule for solving this equation, but in this case it is clear that $b = 3$ is a solution. Thus the added polynomial is $6x^2 + 6$, and the original equation is transformed into $x^4 + 6x^2 + 9 = 6x^2 + 12x + 6$. Taking square roots gives $x^2 + 3 = \sqrt{6}(x + 1)$, the solutions to which are easily found to be

$$x = \sqrt{1\frac{1}{2} \pm \sqrt{\sqrt{6} - 1\frac{1}{2}}}.$$

Are these the only roots of the quartic? One can attempt to find others by taking negative square roots in the equation $x^4 + 6x^2 + 9 = 6x^2 + 12x + 6$, but that leads to complex values for x which Cardano ignores. In other examples, he does use both sets of roots. One could also look for other roots by using a second solution of the resolvent cubic. Cardano evidently considered this possibility but only teases us about what happens: "I need not say whether having found another value for b . . . we would come to two other solutions [for x]. If this operation delights you, you may go ahead and inquire into this for yourself."[21]

Much else of interest is found in Cardano's masterpiece, including the use of negative numbers as solutions to problems and the first appearance of complex numbers, not in connection with cubic equations, but in connection with a quadratic problem. This problem is simply to divide 10 into two parts (here we go again!) such that the product is 40. By standard techniques of solving quadratic equations, Cardano showed that the two parts must be $5 + \sqrt{-15}$ and $5 - \sqrt{-15}$. Although he checked that these answers in fact satisfy the conditions of the problem, he was not entirely happy with the solution, for, as he wrote, "So progresses arithmetic subtlety the end of which, as is said, is as refined as it is useless."[22] Cardano thus left off the discussion and wrote no more about complex numbers.

9.3.2 Rafael Bombelli and Complex Numbers

Cardano's *Ars Magna* was extremely influential, marking the first substantive advance over the Islamic algebra so long studied in Europe. The author himself was quite proud of his work. At the end of the text there appears in large type, "WRITTEN IN FIVE YEARS, MAY IT LAST AS MANY THOUSANDS." Nevertheless, the book itself was difficult to read. Its arguments were often prolix and not easily followed, and its organization left much to be desired. To improve the teaching of the subject, and to clear up some of the difficulties still remaining, Rafael Bombelli (1526–1572) some 15 years later decided to write a more systematic text in Italian to enable students to master the material on their own. Although only the first three of the five parts were published in his lifetime, and

---

Biography **Rafael Bombelli (1526–1572)**

Bombelli was educated as an engineer and spent much of his adult life working on engineering projects in the service of his patron, a Roman nobleman who was a favorite of Pope Paul III. The largest project in which he was involved was the reclamation of the marshes in the Val di Chiana into arable land. Today that valley, extending southeast for about 60 miles between the Arno and the Tiber, is still one of the most fertile in central Italy. Bombelli later served as a consultant on a proposed project for the draining of the Pontine Marshes near Rome. During a lull in the reclamation work caused by a war in the area, he was able to work on his algebra treatise at his patron's villa in Rome sometime between 1557 and 1560. Other professional engagements delayed the printing of it, however, and it did not appear until shortly before his death in 1572.

---

although in the questions concerning multiple roots of cubic equations Bombelli did not achieve as much as Cardano, nevertheless Bombelli's *Algebra* marks the high point of the Italian algebra of the Renaissance.

Bombelli's *Algebra* was more in the tradition of Luca Pacioli's *Summa* and the German *Coss* works than was Cardano's book. Bombelli began the book with elementary material and gradually worked up to the solving of cubic and quartic equations. Like Cardano, he gave a separate treatment to each class of cubics, but he expanded on Cardano's brief treatment of quartics by giving a separate section to each of those classes as well. After dealing with the theoretical material, he presented the student with a multitude of problems using the techniques developed. He had originally intended to include practical problems similar to those of the earlier abacus works, but after studying a copy of Diophantus' *Arithmetica* at the Vatican Library, he decided to replace these with abstract numerical problems taken from Diophantus and other sources.

Algebraic notation was gradually replacing the strictly verbal accounts of the Moslems and of the earliest Italian algebraists, and Bombelli contributed to that change. He used *R.q.* to denote the square root, *R.c.* to denote the cube root, and similar expressions to denote higher roots. He used $\lfloor\ \rfloor$ as parentheses to enclose long expressions, as in *R.c.*$\lfloor 2p\ R.q.21 \rfloor$, but kept the standard Italian abbreviations of *p* for plus and *m* for minus. His major innovation was the use of a semicircle around a number *n* to denote the *n*th power of the unknown. Thus $x^3 + 6x^2 - 3x$ would be written as $1\overset{3}{\smile}p\ 6\overset{2}{\smile}m\ 3\overset{1}{\smile}$. Writing powers numerically rather than in the German form of symbols allowed him to express easily the exponential laws for multiplying and dividing monomials.

Late in the first part of the *Algebra* Bombelli introduced "another sort of cube root much different from the former, which comes from the chapter on the cube equal to the thing and number; ... this sort of root has it own algorithms for various operations and a new name."[23] This root is the one that occurs in the cubic equations of the form $x^3 = cx + d$ when $(d/2)^2 - (c/3)^3$ is negative. Bombelli proposed a new name for these numbers, which are neither positive (*più*) nor negative (*meno*), that is, the modern imaginary numbers. The numbers written today as bi, $-bi$ respectively, Bombelli called *più di meno* (plus of minus) and *meno di meno* (minus of minus). For example, he wrote $2 + 3i$ as 2 *p di m* 3 and $2 - 3i$ as $2\ m\ di\ m$ 3. Bombelli presented the various laws of multiplication for these new (complex) numbers, such as *più di meno* times *più di meno* gives *meno* and *più di meno* times *meno di meno* gives *più* $((bi)(ci) = -bc, bi(-ci) = bc)$.

To illustrate his rules, Bombelli next gave numerous examples of the four arithmetic operations on these new numbers. Thus, to find the product of $\sqrt[3]{2 + \sqrt{-3}}$ and $\sqrt[3]{2 + \sqrt{-3}}$ one first multiplies $\sqrt{-3}$ by itself to get -3, then 2 by itself to get 4, then adds these two to get 1 for the "real" part. Next one multiplies 2 by $\sqrt{-3}$ and doubles the result to get $\sqrt{-48}$. The answer is $\sqrt[3]{1 + \sqrt{-48}}$. To divide 1000 by $2 + 11i$, Bombelli multiplied both numbers by $2 - 11i$, the conjugate (*residue*) of the denominator. He then divided the new denominator, 125, into 1000, giving 8, which in turn he multiplied by $2 - 11i$ to get $16 - 88i$ as the result. The standard rules for adding and subtracting are also presented in detail with examples. Bombelli, although he noted that "the whole matter seems to rest on sophistry rather than on truth,"[24] nevertheless presented here for the first time the rules of operation for complex numbers. It seems clear from his discussion that he developed these rules strictly by analogy to the known rules for dealing with real numbers. Arguing by analogy is a common method of making mathematical progress, even if one is not able to give rigorous proofs. Bombelli of course, because he did not know what these numbers really were, could give no such proofs.

Proofs notwithstanding, with the rules for dealing with complex numbers now available, Bombelli could discuss how to use Cardano's formula for the case $x^3 = cx + d$ whether $(d/2)^2 - (c/3)^3$ is positive or negative. He first considered the example $x^3 = 6x + 40$. The formula easily enough gives $x = \sqrt[3]{20 + \sqrt{392}} + \sqrt[3]{20 - \sqrt{392}}$, even though it is obvious that the answer is $x = 4$. Bombelli showed how one can see that the sum of the two cube roots is in fact 4. He assumed that $20 + \sqrt{392}$ equals the cube of a quantity of the form $a + \sqrt{b}$ for some numbers a and b, or $\sqrt[3]{20 + \sqrt{392}} = a + \sqrt{b}$. This implies that $\sqrt[3]{20 - \sqrt{392}} = a - \sqrt{b}$. Multiplying these two equations together gives $\sqrt[3]{8} = a^2 - b$ or

$$a^2 - b = 2.$$

Furthermore, cubing the first equation and equating the parts without square roots gives

$$a^3 + 3ab = 20.$$

Bombelli did not attempt to solve this system of two equations in two unknowns by a general argument. Rather, he noted that the first equation implies that $a^2 > 2$ and the second shows that $a^3 < 20$. The only integer satisfying both inequalities is $a = 2$. Fortunately, $b = 2$ then provides the other value in each equation, so Bombelli had shown that $\sqrt[3]{20 + \sqrt{392}} = 2 + \sqrt{2}$. It follows that the solution to the cubic equation may be written as $x = (2 + \sqrt{2}) + (2 - \sqrt{2}) = 4$ as desired.

For the equation $x^3 = 15x + 4$, the Cardano formula gives

$$x = \sqrt[3]{2 + \sqrt{-121}} + \sqrt[3]{2 - \sqrt{-121}},$$

although again it is clear that the answer is $x = 4$. Bombelli used his newfound knowledge of complex numbers to apply the same method as above. He first assumed that

$\sqrt[3]{2 + \sqrt{-121}} = a + \sqrt{-b}$. Then $\sqrt[3]{2 - \sqrt{-121}} = a - \sqrt{-b}$, and a short calculation leads to the two equations

$$a^2 + b = 5$$

and

$$a^3 - 3ab = 2.$$

Thus Bombelli needed to find a number a whose square was less than 5 and whose cube was greater than 2. He very carefully showed that neither $a = 1$ nor $a = 3$ would work, and therefore that $a = 2$ was the only possibility. Then $b = 1$ provides the other solution and the desired cube root is $2 + \sqrt{-1}$. It followed that the solution to the cubic equation is $x = (2 + \sqrt{-1}) + (2 - \sqrt{-1})$ or $x = 4$.

Bombelli presented several more examples of the same type, where in each case he was able somehow to calculate the appropriate values of a and b. He did note, however, that this was not in general possible. If one attempts to solve the system in a and b by a general method, such as substitution, one is quickly led back to another cubic equation. Bombelli also showed that complex numbers could be used to solve quadratic equations which previously had been thought to have no solution. For example, he used the standard quadratic formula to show that $x^2 + 20 = 8x$ has the solutions $x = 4 + 2i$ and $x = 4 - 2i$. Although he could not answer all questions about the use of complex numbers, his ability to use them to solve certain problems provided mathematicians with the first hint that there was some sense to dealing with them. Since mathematicians were still not entirely happy with using negative numbers—Cardano called them fictitious and Bombelli did not consider them as roots at all—it is not surprising that it took many years before they were entirely comfortable with using complex numbers.

Bombelli was the last of the Italian algebraists of the Renaissance. His *Algebra*, however, was widely read in other parts of Europe. Two men, one in France and one in the Netherlands, just before the turn of the seventeenth century, used both Bombelli's work and some newly rediscovered Greek mathematical works to take algebra into new directions.

9.4 THE WORK OF VIÈTE AND STEVIN

The European algebraists of the sixteenth century had achieved about as much as possible in their continuation of the Islamic algebra of the Middle Ages. They were now expert in algebraic manipulations, even though their symbolism still left something to be desired, and they knew how to solve any algebraic equation of degree up to four. The solutions, however, were given in the form of rules of procedure. Most of these authors used some symbolization for the unknown and its powers, but there were no symbols for the coefficients. Thus the best that could be done to illustrate the procedure was to use numerical examples. In none of these algebra texts is there a formula written down, like the quadratic formula found in every current algebra textbook. To be able to write down such formulas required a new approach to symbols.

There was another major trend in mathematics during the sixteenth century in Italy besides this continuation of Islamic algebra. As part of the general revival of knowledge of classical antiquity, there developed a great interest in retrieving all of the Greek mathematical works to be found. The basic works of Euclid, Archimedes, and Ptolemy had been translated several centuries earlier. Since the translators were not expert mathematicians, however, their versions were not always completely understandable. In the sixteenth century, however, there was a concerted effort made to retranslate these works as well as translate other Greek mathematical works from the original Greek, these new translations to be prepared by mathematicians. The most important figure in this mathematical renaissance was the Italian geometer Federigo Commandino (1509–1575), who singlehandedly prepared Latin translations of virtually all of the known works of Archimedes, Apollonius, Pappus, Aristarchus, Autolycus, Heron, and others. Commandino's mathematical talents allowed him to conquer many of the obscurities that centuries of copyists had introduced into the Greek manuscripts. Thus he included with each translation extensive mathematical commentaries, clarifying difficulties and providing references from one treatise to other related ones.

With the entire corpus of Greek mathematical works that had survived the destruction of the major libraries in late antiquity now available to Europeans, the question of how the Greeks discovered their theorems began to be addressed in earnest. In particular, with the availability of Pappus' *Mathematical Collection* and especially Book VII, *On the Domain of Analysis*, European mathematicians began to search for the "methods of analysis" used by the ancient Greeks. Most Greek mathematics texts were models of synthetic reasoning, beginning with axioms and proceeding step-by-step to increasingly complex results. The texts generally gave little clue to how the results were found or how one might find similar results. Pappus' work was the only one available to provide some hints of the Greek method of geometrical analysis. Not only did Pappus discuss the basic procedure of analysis, but he also gave in Book VII a guide to the major Greek texts which provided the tools enabling one to use analysis in the discovery of new results or in the solving of new problems. Unfortunately, aside from Euclid's *Data* and Apollonius' *Conics*, the treatises referred to by Pappus were no longer available. With a mixture of curiosity and frustration, the Europeans then studied the available Greek texts to try to ferret out the Greek methods and to reconstruct the lost texts of the *Domain of Analysis* using the hints and descriptions of Pappus.

René Descartes in 1629 best expressed these feelings in Rule IV of his *Rules for the Direction of the Mind*:

> But when I afterwards bethought myself how it could be that the earliest pioneers of Philosophy in bygone ages refused to admit to the study of wisdom any one who was not versed in Mathematics, . . . I was confirmed in my suspicion that they had knowledge of a species of Mathematics very different from that which passes current in our time. I do not indeed imagine that they had a perfect knowledge of it . . . but I am convinced that certain primary germs of truth implanted by nature in human minds . . . had a very great vitality in that rude and unsophisticated age of the ancient world. . . . But my opinion is that these writers then with a sort of low cunning, deplorable indeed, suppressed this knowledge. Possibly they acted just as many inventors are known to have done in the case of their discoveries; i.e., they feared that their method being so easy and simple would become cheapened on being divulged, and they preferred to exhibit in its place certain barren truths, deductively demon-

strated with show enough of ingenuity, as the results of their art, in order to win from us our admiration for these achievements, rather than to disclose to us that method itself which would have wholly annulled the admiration accorded. Finally, there have been certain men of talent who in the present age have tried to revive this same art. For it seems to be precisely that science known by the barbarous name of Algebra, if only we could extricate it from that vast array of numbers and inexplicable figures by which it is overwhelmed, so that it might display the clearness and simplicity which we imagine ought to exist in a genuine Mathematics.[25]

9.4.1 François Viète and *The Analytic Art*

One of the first men of talent who attempted to identify the Greek analysis with the new algebra, and who tried to display this new algebra with clearness and simplicity was François Viète (1540–1603). In the closing years of the sixteenth century, he composed the several treatises collectively known as *The Analytic Art*, in which he effectively reformulated the study of algebra by replacing the search for solutions to equations by the detailed study of the structure of these equations, thus developing the earliest consciously articulated theory of equations.

Viète began his program in his *In Artem Analyticem Isagoge* (*Introduction to the Analytic Art*) of 1591 with an announcement of what he wanted to accomplish:

> There is a certain way of searching for the truth in mathematics that Plato is said first to have discovered. Theon called it analysis. . . . Although the ancients propounded only two kinds of analysis, zetetics and poristics, to which the definition of Theon best applies, I have added a third, which may be called rhetics or exegetics. It is properly zetetics by which one sets up an equation or proportion between a term that is to be found and the given terms, poristics by which the truth of a stated theorem is tested by means of an equation or proportion, and exegetics by which the value of the unknown term in a given equation or proportion is determined. Therefore the whole analytic art, assuming this three-fold function for itself, may be called the science of correct discovery in mathematics.[26]

The two kinds of analysis of the Greeks were called theorematic and problematic in the discussion of Pappus in Chapter 5. Viète entirely altered the meaning of these two terms, while adding a new one. For Viète, **zetetic analysis** is the procedure by which one transforms a problem into an equation linking the unknown and the various knowns; **poristic analysis** is the procedure exploring the truth of a proposed theorem by appropriate symbolic manipulation; and, finally, **exegetics** is the art of transforming the equation found by zetetics to find a value for the unknown. It is not entirely surprising that Viète tried to identify Greek analysis with algebra. The procedure for solving an equation assumes that one can treat the unknown x as if it were known by using the basic rules of operation. At the end of the series of operations, one then has the unknown expressed ($x =$) in terms of the knowns. This is in some sense the same procedure that Pappus described of assuming what is sought as known and then proceeding through its consequences to something already given. But Viète's use of the terms is not the same as that of Pappus. The actual burden of finding the unknown is borne by the new third kind of analysis, exegetics, rather than by the two earlier kinds. In any case, we can understand clearly Viète's goal, as stated in the last sentence of the quotation of the previous paragraph as well as in the final paragraph of the *Introduction*: "Finally, the analytic art endowed

---

Biography   **François Viète (1540–1603)**

Viète was born in Fontenay-le-Comte, a village in western France near the Bay of Biscay. After receiving a law degree from the University of Poitiers, he returned to his native village to begin the practice of that profession. His legal reputation grew through his association with a prominent local family, and he was called to Paris by King Henri III for private advice, confidential negotiations, and finally in 1580 a seat on the privy council. One of his tasks for Henri, after the latter moved the court to Tours in 1589, was to act as a cryptanalyst of intercepted messages between the king's enemies. He was so successful at this that he was denounced by some who thought that the decipherment could only have been made by sorcery. Viète spent much of the remainder of his life in political service to Henri III and his successor Henri IV. Mathematics for Viète could therefore be only an avocation.

---

with its three forms of zetetics, poristics, and exegetics, claims for itself the greatest problem of all, which is *to leave no problem unsolved.*"[27]

In the *Introduction*, Viète presents one of his most important contributions, a new form of symbolization. "Numerical logistic is that which employs numbers; symbolic logistic that which uses symbols, as, say, the letters of the alphabet."[28] Viète will thus manipulate letters as well as numbers. "Given terms are distinguished from unknown by constant, general, and easily recognized symbols, as (say) by designating unknown magnitudes by the letter A and the other vowels $E, I, O, U,$ and Y and given terms by the letters B, G, D and the other consonants."[29] While Viète uses a convention differing from the current one in distinguishing knowns from unknowns, he is now able to manipulate completely with these letters. Furthermore, these letters do not need to stand for numbers only. They may stand for any quantity to which one can apply the basic operations of arithmetic. Viète has, however, not entirely broken away from his predecessors. He continues to use words or abbreviations for powers rather than exponents as suggested by Bombelli and Chuquet. Instead of using A^2, B^3, or C^4, Viète writes *A quadratum, B cubus,* or *C quadrato-quadratum*, the first and third of which he sometimes abbreviates to *A quad* or *C quad-quad*. He therefore must give verbal rules for multiplying and dividing powers, like, for example, *latus* (side) times *quadratum* equals *cubus*, and *quadratum* times itself equals *quadrato-quadratum*. In numerical examples, Viète often uses a different scheme of symbolization, namely *N* stands for *numerus* (number), *Q* for *quadratus* (square), and *C* for *cubus* (cube).

In symbolizing operations, Viète adopted the German forms $+$ and $-$ for addition and subtraction. For multiplication, he generally used the word *in*, while for division, he used the fraction bar. Hence

$$\frac{A \; in \; B}{C \; quadratum}$$

means, in modern notation,

$$\frac{AB}{C^2}.$$

Square roots are written using the symbol L for *latus*: $L64$ means the square root of 64 and $LC64$ stands for the cube root of 64. Viète, like most of his predecessors, insisted on the law of homogeneity, that all terms in a given equation must be of the same degree. So to make sense of the equation we would write as $x^3 + cx = d$, Viète insisted that c be a plane (so that cx is a solid) and that d be a solid. He would write the equation as *A cubus + C plano in A aequetus D solido*. Note that he does not use a symbol for "equals" but a word.

While Viète had come only part way toward modern symbolism, the crucial step of allowing letters to stand for numerical constants enabled him to break away from the style of examples and verbal algorithms of his predecessors. He could now treat general examples, rather than specific ones, and give formulas rather than rules. In addition, the elimination of the possibility of actually carrying out numerical computations involving the symbolic constants made it possible to focus on the procedures of the solution rather than the solution itself. One could see that the solution procedures could be applied to quantities other than numbers, such as, for example, line segments or angles. Further, solving equations symbolically made the structure of the solution more evident. Instead of replacing 5 + 3 by 8, for example, one kept the expression $B + D$ in the displayed formula so that at the end of the argument one could consider its relationship to the original constants. Viète was thus able in some circumstances to discover how the roots of an equation were related to the expressions from which the equation was constructed.

We consider here a few of Viète's problems and methods of solutions in the various treatises that make up *The Analytic Art*, beginning with material from the *Prior Notes on Symbolic Logistic*, probably written at the same time as the *Introduction* but not published until 1631. In these *Notes*, Viète shows how to operate on symbolic quantities. He derives many of the standard algebraic identities, most of which were previously known in verbal form at least but are here written for the first time purely in symbols. For example, Viète notes that $A - B$ times $A + B$ equals $A^2 - B^2$ and that $(A + B)^2 - (A - B)^2 = 4AB$. Viète also writes out the expansion of $(A + B)^n$ for each integer n from two to six, but does not write out the general binomial theorem. The reason that he did not see this generalization is probably related to his using words rather than numbers to designate the various powers. Similarly, Viète writes out the products of $A - B$ with $A^2 + AB + B^2$, $A^3 + A^2B + AB^2 + B^3, \ldots$ to get $A^3 - B^3, A^4 - B^4, \ldots, A^6 - B^6$ but does not attempt to give any general rule.

In another section of the *Prior Notes*, Viète applies his algebra to trigonometry. Using the identity $(BG + DF)^2 + (DG - BF)^2 = (B^2 + D^2)(F^2 + G^2)$, he shows that given two right triangles, one with base D, perpendicular B, and hypotenuse Z, the other with base G, perpendicular F, and hypotenuse X, a new right triangle can be constructed with base $DG - BF$, perpendicular $BG + DF$, and hypotenuse ZX. The angle at the base of this new triangle is then the sum of the angles at the base of the two original triangles (assuming that that sum is less than 90°). It follows that if one starts with two identical triangles with base D, perpendicular B, and hypotenuse Z, the new triangle, with base $D^2 - B^2$, perpendicular $2BD$, and hypotenuse Z^2, has base angle double that of the original. The results here are equivalent to the familiar double-angle formulas of trigonometry. Viète then performs the same construction using the "double-angle" triangle and the original one to get the "triangle of the triple angle." This triangle has base $D^3 - 3B^2D$, perpendicular $3BD^2 - B^3$, and hypotenuse Z^3. These formulas for the base, perpendicular, and hypotenuse are equivalent to the modern triple-angle formulas for cosine and sine. Viète continues his construction to generate formulas for the quadruple and quintuple angles as well.

In the *Five Books of Zetetics* (1591), Viète uses his symbolic methods of calculation to deal with a large number of algebraic problems drawn from a variety of sources, both ancient and contemporary. In particular, many are taken from Diophantus' *Arithmetica*, because Viète believed that this was an ancient work of algebra, which showed to some extent the hidden analysis of the Greeks. In each problem, as promised, he shows how to derive the equation relating the unknown and the known quantities. He begins with the same problem with which both Diophantus and Jordanus de Nemore began their texts: Given the difference between two numbers and their sum, to find the numbers. Viète's procedure is straightforward: Letting B be the difference, D the sum, and A the smaller of the two numbers, he notes that $A + B$ is the greater. Then the sum of the numbers is $2A + B$, which equals D. Hence $2A = D - B$ and $A = (1/2)D - (1/2)B$. The other number is then $E = (1/2)D + (1/2)B$. Having written down the solution in symbols, Viète then restates it in words: "Half the sum of the numbers minus half the difference equals the lesser number, plus that difference, the greater."[30] He concludes with an example: If B is 40 and D is 100, then A is 30 and E is 70. This format is typical of Viète's work. Although he has introduced symbolic methods, he often restates his answers in words as if to convince perhaps skeptical readers that the new symbolic method can always be translated back into the more familiar verbal mode of expression.

It is enlightening to compare the same problem in Diophantus, Jordanus, and Viète to see the differences. Diophantus, although stating the problem generally, in fact solves it only for a particular numerical example, the same one which Viète uses. Jordanus solves it generally but in words: "Subtract the difference from the whole and there will remain double the lesser given number." Viète solves it totally symbolically. This problem exemplifies the change in algebra over 1350 years. Similar comparisons can be made using many other common problems.

The second book of zetetics deals with products of unknowns as well as various powers. Viète shows that if one knows the product of two values and the sum of their squares, or if one knows the product of two values and their sum, or if one knows the sum of two values and the difference between their squares, one can find the unknown values. Several of the results of this book are important in Viète's later treatment of cubic equations. For example, problem 20 asks to find two values given their sum and the sum of their cubes. In this case, Viète sets G equal to the sum of the unknown values, D equal to the sum of the cubes, and A equal to the product of the unknowns. From the formula for the expansion of the cube of a binomial, Viète then derives the result $G^3 - D = 3GA$, or, in modern notation,

$$(r + s)^3 - (r^3 + s^3) = 3(r + s)rs. \tag{9.1}$$

Thus the product A is known and one can find the two unknown values.

9.4.2 Viète's Theory of Equations

The central work in Viète's theory of equations is found in the *Two Treatises on the Recognition and Emendation of Equations*. Here Viète shows how to transform various kinds of equations into a few canonical forms, each of which he then shows how to solve.

For example, rather than give a separate procedure for each of the 13 cases of the cubic, as did Cardano and Bombelli, Viète shows in each case how to transform the cubic into a form containing no term of second degree. He then shows how to solve each of these latter forms. Before dealing with these formulas, however, Viète demonstrates how cubic equations can be constructed, generally in terms of the classical theory of proportions rather than in terms of the roots.

We consider, in particular, the equation $x^3 - cx = d$. In Chapter 4 of the first of the two treatises on equations, Viète shows that this equation, which he here writes as $x^3 - b^2x = b^2d$, partly in keeping with his homogeneity rules, comes from the existence of four continued proportionals, the first of which is b and the second x, while d is the difference between the fourth and the second. That is, if

$$b : x = x : y = y : x + d,$$

then

$$\left(\frac{b}{x}\right)^2 = \frac{x}{y}\frac{y}{x + d} = \frac{x}{x + d}.$$

It follows that $x^3 = b^2(x + d)$ and the given equation is evident. As an example, Viète notes that if $x^3 - 64x = 960$, there are four continued proportionals beginning with 8 such that the difference between the fourth and second is $960/64 = 15$. He concludes that the proportionals are 8, 12, 18, and 27, and the solution is $x = 12$.

In Chapter 6, Viète derives this same equation another way. Here he writes it as $x^3 - 3bx = d$ and refers back to identity 9.1. Thus x is the sum of two values r and s, whose product rs equals b and such that $r^3 + s^3 = d$. As an example, he gives $x^3 - 6x = 9$, where $6/3 = 2$ is the product of two numbers the sum of whose cubes is 9. These numbers are 1 and 2, and it follows that the unknown x is 3, their sum. Viète further notes here a condition for this derivation, namely that $d^2 > 4b^3$. This condition follows from another one of Viète's algebraic identities, namely, $(r^3 - s^3)^2 = (r^3 + s^3)^2 - 4(rs)^3$.

If the inequality is not satisfied, there are two other methods of deriving the equation. First, Viète uses a more complicated identity, $(r + s)^3 - (r^2 + s^2 + rs)(r + s) = rs(r + s)$, to show that $x = r + s$ if $3b = r^2 + s^2 + rs$ and $d = rs(r + s)$. Thus, in the example $x^3 - 21x = 20$, he needs to find r, s so that $r^2 + s^2 + rs = 21$ and $rs(r + s) = 20$. These values are 1 and 4, so the desired root is 5.

Viète displays a second alternate method for analyzing this cubic under the condition $d^2 < 4b^3$ using the trigonometrical identities he had developed earlier, since "the constitution of equations of this sort is more elegantly and clearly brought out by an angular section analysis."[31] He begins by rewriting the equation as

$$x^3 - 3b^2x = b^2d. \tag{9.2}$$

The inequality then becomes $b > d/2$. Viète had already shown that the base of a "triple-angle" triangle could be expressed as $D^3 - 3B^2D$ and the hypotenuse as Z^3 where the original triangle had base D, perpendicular B, and hypotenuse Z. In modern terminology, the formula for the base can be rewritten as $\cos 3\alpha = \cos^3 \alpha - 3 \sin^2 \alpha \cos \alpha = 4 \cos^3 \alpha - 3 \cos \alpha$ or as

$$\cos^3 \alpha - \frac{3}{4} \cos \alpha = \frac{1}{4} \cos 3\alpha. \tag{9.3}$$

By setting $x = r \cos \alpha$ and substituting in equation 9.2, that equation becomes $r^3 \cos^3 \alpha - 3b^2 r \cos \alpha = b^2 d$ or

$$\cos^3 \alpha - \frac{3b^2}{r^2} \cos \alpha = \frac{b^2 d}{r^3}. \tag{9.4}$$

Comparing equations 9.4 and 9.3 shows first that

$$\frac{3b^2}{r^2} = \frac{3}{4} \quad \text{or} \quad r = 2b$$

and second that

$$\frac{1}{4} \cos 3\alpha = \frac{b^2 d}{r^3} = \frac{b^2 d}{8b^3} \quad \text{or} \quad \cos 3\alpha = \frac{d}{2b}.$$

The inequality for the coefficients of equation 9.2 insures that this final equation makes sense. Thus if α satisfies $\cos 3\alpha = d/2b$ and $r = 2b$, then $x = r \cos \alpha$ is a solution to equation 9.2. To solve $x^3 - 300x = 432$, he notes that $b = 10$ and $d = 432/100$. It follows that $\cos 3\alpha = 432/2000$. By consulting tables, one determines that $\cos \alpha = 0.9$ and thus $x = 2b \cos \alpha = 18$.

Given Viète's analysis of the equation, the question still remains of how to solve a cubic equation algebraically. After all, one can only use the identities in relatively simple cases. In the second of the two treatises on equations, Viète shows how to reduce every cubic equation to one of the three types $x^3 + bx = d$, $x^3 - bx = d$, or $bx - x^3 = d$ by a linear substitution of the form $x = y \pm c$ where $3c$ is the coefficient of the x^2 term of the original equation. Thus $x^3 - 3cx^2 = d$ is transformed by use of the substitution $x = y + c$ to $y^3 - 3c^2 y = d + 2c^3$ and $x^3 + 3cx^2 - bx = d$ is transformed by use of the substitution $x = y - c$ into $y^3 - (3c^2 + b)y = d - 2c^3 - bc$. Given these transformations, Viète finally demonstrates the method of solution. For the equation written now as $x^3 - 3bx = 2d$, Viète first notes that b^3 must be less than d^2. (The trigonometric method must suffice if this condition is not satisfied.) Since x must be the sum of two numbers whose product is b, he sets up the equation $y(x - y) = b$, or $xy - y^2 = b$, a quadratic equation in y with two distinct solutions. Solving the equation for x gives

$$x = \frac{b + y^2}{y}.$$

Substituting this expression into $x^3 - 3bx = 2d$ and multiplying all terms by y^3 produces the quadratic equation in y^3,

$$2dy^3 - (y^3)^2 = b^3.$$

The solutions to this are $y^3 = d \pm \sqrt{d^2 - b^3}$, so the reason for the inequality condition is evident. Because the desired root x is the sum of the two values for y, the final result is the formula slightly modified from that of Cardano,

$$x = \sqrt[3]{d + \sqrt{d^2 - b^3}} + \sqrt[3]{d - \sqrt{d^2 - b^3}}.$$

Although Viète did not consider negative or complex roots to equations, he did deal to some extent with the relationship of the roots to the coefficients. For example, Viète was aware that the quadratic equation $bx - x^2 = c$ could have two positive roots. To discover the relationship between these two roots, x_1 and x_2, he equates the two expressions $bx_1 - x_1^2$ and $bx_2 - x_2^2$. Then $x_1^2 - x_2^2 = bx_1 - bx_2$, and, dividing through by $x_1 - x_2$, he finds that $x_1 + x_2 = b$, that is, "b is the sum of the two roots being sought." Substituting $x_1 + x_2$ for b in the equation $bx_1 - x_1^2 = c$, he gets the other relationship $x_1 x_2 = c$, or "c is the product of the two roots being sought."[32]

Viète tries the same device for the cubic equation $bx - x^3 = d$, which he also knows may have two positive roots, x_1, x_2. In this case the results are not so simple. Viète finds that the coefficient b is equal to $x_1^2 + x_2^2 + x_1 x_2$, while the constant d is $x_1^2 x_2 + x_2^2 x_1$. What about cubic equations having three positive roots? Such an equation, of course, must have a term of the second degree. Viète's normal method of solving it would replace the equation by an equation without a second degree term through the use of a linear substitution and the new equation would have at most two positive roots. So Viète would not normally be able to find a third root. In fact, in an example he gave of such a reduction, using the equation $x^3 - 18x^2 + 88x = 80$, he only calculated the integer root of the reduced equation $20y - y^3 = 16$ and thus gave only one root for the original equation. Nevertheless, at the very end of the second treatise on equations, Viète stated four propositions without proof, one for each degree of equation from two through five, expressing the coefficients of the equation as elementary symmetric functions of the roots. Thus, for the third degree, "If $x^3 - x^2(b + d + g) + x(bd + bg + dg) = bdg$, x is explicable by any of the three b, d, or g" and for the fourth, "If $x(bdg + bdh + bgh + dgh) - x^2(bd + bg + bh + dg + dh + gh) + x^3(b + d + g + h) - x^4 = bdgh$, x is explicable by any of the four b, d, g, or h."[33] Viète considered these theorems "elegant and beautiful" and a "crown" of this work, an apt point to conclude the study of this brilliant French algebraist.

9.4.3 Simon Stevin and Decimal Fractions

Another mathematician who made a substantial contribution to a major change in mathematical thinking around the turn of the seventeenth century was Simon Stevin (1548–1620). He was a contemporary of Viète, but lived most of his life in the Netherlands. Stevin's major mathematical contribution was the creation of a well thought out notation for decimal fractions, for the use of which he proved himself a strong advocate. He also played a fundamental role in changing the basic concept of number and in erasing the Aristotelian distinction between number and magnitude. These contributions are set forth in his most important mathematical works, *De Thiende* (*The Art of Tenths*), known in its French version as *La Disme*, and *l'Arithmétique*, a work containing both arithmetic and algebra, both published in 1585.[34]

Decimal fractions were not used in Europe in the late Middle Ages or in the Renaissance. The various arithmetic texts written throughout Europe from the thirteenth through the sixteenth centuries, although they invariably discussed the Hindu-Arabic place-value system, used it only for integers. If fractions were needed, they were written as common fractions or, in many trigonometric works, as sexagesimal fractions. Both Rudolff and Viète displayed hints of the use of decimal fractions in works written during the sixteenth century, but neither they nor any other European mathematician of the time demonstrated a clear understanding of the concept. In the Islamic development some centuries earlier, there were occasional examples of a decimal fraction notation, but these were separated in

---

Biography   Simon Stevin (1548–1620)

Stevin was born into a wealthy family in Bruges, in what is now Belgium, but eventually left the area, then under Spanish rule, for the new Republic of Holland (Figure 9.5). Much of his adult life was spent in the service of the Stadhouders of Holland, William of Orange, who was assassinated in 1584, and his son and successor Maurice of Nassau. Stevin served Maurice as engineer, tutor in mathematics and ballistics, and advisor in various other mathematically dependent fields such as finance and navigation. From 1593 until his death, Stevin served the Dutch government as quartermaster general of the army, responsible for organizing military camps. At the request of Maurice he organized a school of engineering at the University of Leiden, where Dutch, rather than the traditional Latin, was the language of instruction. He was responsible for meeting the growing need of the Dutch nation for technically trained engineers, merchants, surveyors and navigators. In fact, Stevin wrote textbooks in Dutch for several of the subjects taught at Leiden.

FIGURE 9.5
Simon Stevin on a Belgian stamp.

time and place from al-Samaw'al's discussion of the concept. Even after al-Samaw'al, it took more than 200 years before both the concept and the notation of a decimal fraction existed together in the world of Islam.

Stevin, who was probably not influenced by the Islamic development, put both idea and notation together in his *De Thiende*. In the preface Stevin states the purpose of his work: "It teaches (to speak in a word) the easy performance of all reckonings, computations, and accounts, without broken numbers [common fractions], which can happen in man's business, in such sort as that the four principles of arithmetic, namely addition, subtraction, multiplication, and division, by whole numbers may satisfy these effects."[35] Thus Stevin promised to show that all operations using his new system can be performed exactly as if one were using whole numbers. That, of course, is the basic advantage of decimal fractions. Stevin begins *De Thiende* by defining *thiende* as arithmetic based on geometric progression by tens, using the Hindu-Arabic numerals, and by calling a whole number a *commencement* with the notation ⓪. Thus 364 is to be thought of as 364 *commencements* and is written as 364⓪. His major definition, in which he describes his terminology and notation for decimal fractions, is the third: "And each tenth part of the unity of the commencement we call the *prime*, whose sign is ①, and each tenth part of the unity of the prime we call the *second*, whose sign is ② and so of the other; each tenth part of the unity of the precedent sign, always in order one further."[36] To explain this, he gives examples: 3①7②5③9④ means 3 primes, 7 seconds, 5 thirds, and 9 fourths, or 3/10, 7/100, 5/1000, 9/10,000, or, altogether, 3759/10,000. Similarly, 8 937/1000 is written as 8⓪9①3②7③. Stevin makes the point that no fractions are used in his notation and that except in the case of ⓪, there are only single digits to the left of the signs (circled digits). The numbers written according to these rules Stevin names **decimal numbers**.

Stevin proceeds in the second part of this brief pamphlet to show how the basic operations are performed on decimal numbers. The important idea, naturally, is that operations are performed exactly as on whole numbers, with the proviso that one must take into account the appropriate signs. Thus in addition and subtraction the numbers must be lined up with all ①s, for example, under one another. For multiplication, Stevin notes that once the multiplication in integers is performed, the sign of the right-most digit is determined by adding the signs of the right-most digits of the multiplicands. For division, one similarly subtracts the right-most sign of the divisor from the right-most sign of the dividend. He also gives rules for determining the sign when finding square and cube roots. Thus, although his notation is somewhat different from our own, Stevin has clearly set out the basic rules and rationale for using decimal fractions in calculation. The concluding section of *De Thiende* consists of pleas to use his new decimal system for calculations in various trades. He suggests using a known basic unit in each case as the *commencement* and then applying his system for fractions of that unit. His suggestion, however, was not generally carried out until 200 years later, when the French revolutionary government introduced the metric system.

How the idea of decimal fractions in *De Thiende* is connected with a change in the basic concept of number is demonstrated in Stevin's other mathematical work of 1585, *l'Arithmétique*. Certainly, many authors over the centuries had been treating irrational quantities as "numbers," that is, had been dealing with them using the same rules and concepts as with the whole numbers. Gradually, the Euclidean distinction between number and magnitude, between discrete quantity and continuous quantity, had broken down. It was Stevin who first stated this breakdown explicitly. Thus, he began *l'Arithmétique* with two definitions.

1. Arithmetic is the science of numbers.
2. Number is that which explains the quantity of each thing.

Thus, at the very beginning of the work, Stevin makes the point that number represents quantity, any type of quantity at all. Number is no longer to be only a collection of units, as defined by Euclid. Stevin even writes in capitals at the top of the page, THAT UNITY IS A NUMBER. The Greeks had rejected this notion. To them, unity was not a number, but only the generator of number, as the point was the generator of a line. Through the centuries, this idea had been argued. As late as 1547, one of the questions which Ferrari sent to Tartaglia as part of the challenge competition mentioned earlier was whether unity was a number. Tartaglia complained that the question did not have to do with mathematics but with metaphysics. He then hedged his answer by asserting that unity was a number "in potential" but not one "in actuality." Stevin, by contrast, is very sure of himself. His basic philosophical argument is that since the part is of the same matter as the whole, and since unity is a part of a multitude of units (that is, a "number"), then unity must itself be a number. The mathematical argument is simply that one can operate on unity just as on other "numbers." In particular, one can divide unity into as many parts as desired. The Euclidean special role of unity as the basis of "collections of units" and therefore as the basis of the distinction between the discrete and the continuous, no longer makes sense to Stevin. He boldly asserts this particular idea as well, that "number is *not* discontinuous quantity."[37] Any quantity, including the unit, can be divided "continuously." In some sense, this is the basis of the idea of a decimal fraction. One can continue the signs as far as one likes to determine any division of unity, however fine.

Stevin further explains what number should mean by giving several special definitions. For example, "number explaining the value of the geometric quantity is called a geometric number and receives the name conforming to the species of the quantity that it explains." A "square number" represents a square and a "cube number" represents a cube. Stevin points out, however, that any (positive) number is a square number, and thus the root of any square number is also a number: "The part is of the same matter that is the whole. The root of 8 is part of its square 8. Therefore $\sqrt{8}$ is of the same matter that is 8. But the matter of 8 is number. Therefore the matter of $\sqrt{8}$ is number. And, by consequence, $\sqrt{8}$ is a number."[38] The decimal number system of *De Thiende* then enables Stevin to represent $\sqrt{8}$ to any accuracy desired, just as it enables one to represent 8 itself.

Stevin does distinguish between pairs of numbers which are commensurable (have a common measure) and incommensurable (do not have a common measure). But all of these quantities are numbers in his sense. Thus, there is no real point in Euclid's multitude of distinctions of classes of irrational lines in Book X. All of these lines are represented as numbers and can be dealt with by the standard arithmetical operations. Thus, Stevin notes, we can consider more kinds of lines than the ones considered by Euclid by simply taking more roots and combinations of roots. And all of these lines (or numbers) can be calculated using his decimal arithmetic.

From the current vantage point, where the discrete Euclidean "numbers" have long been incorporated into the continuous number line, it is somewhat difficult to understand the fundamental contribution of Stevin. But Euclid had always been the center of the study of mathematics. His ideas always had to be confronted. If one wanted to change his notions, one needed to make strong and continued arguments. Many mathematicians who read Euclid did ignore his distinctions, both in the Islamic world and in Europe. In particular, the algebraists studied in this chapter tended to manipulate with all quantities in the same way. Others more philosophically inclined, however, were somewhat bothered by this generally cavalier attitude toward Euclid's work. These mathematicians needed to be convinced that there was no longer any mathematical necessity for the distinction. Naturally, Stevin alone did not do this. Not until the nineteenth century was the work of imbedding "discrete arithmetic" into "continuous magnitude" completed. Nevertheless, Stevin stood at a watershed of mathematical thinking. Ultimately he was so successful that it is difficult to understand what happened before.

Exercises

Problems from Italian Abacus Texts

1. The gold florin is worth 5 *lire*, 12 *soldi*, 6 *denarii* in Lucca. How much (in florins) are 13 *soldi*, 9 *denarii* worth? (Note that 20 *soldi* make 1 *lira* and 12 *denarii* make 1 *soldo*.)

2. If 8 *braccia* of cloth are worth 11 florins what are 97 *braccia* worth?

3. The *lira* earns 3 *denarii* per month. How much will 60 *lire* earn in 8 months? (As stated this is a simple interest problem. Try it also assuming that the interest is compounded each month.)

4. The next three problems are from the work of Piero della Francesca. Three men enter into a partnership. The first puts in 58 ducats, the second 87; we do not know how much the third puts in. Their profit is 368, of which the

first gets 86. What shares of profit do the second and third receive and how much did the third invest?

5. Of three workmen, the second and third can complete a job in 10 days. The first and third can do it in 12 days, while the first and second can do it in 15 days. In how many days can each of them do the job alone?

6. A fountain has two basins, one above and one below, each of which has three outlets. The first outlet of the top basin fills the lower basin in two hours, the second in three hours, and the third in four hours. When all three outlets are shut, the first outlet of the lower basin empties it in three hours, the second in four hours, and the third in five hours. If all the outlets are opened, how long will it take for the lower basin to fill?

7. Solve this problem from the work of Antonio de' Mazzinghi. Find two numbers such that multiplying one by

the other makes 8 and the sum of their squares is 27. (Put the first number equal to $x + \sqrt{y}$ and the second equal to $x - \sqrt{y}$; then the two equations are $x^2 - y = 8$ and $2x^2 + 2y = 27$.)

8. Divide 10 into two parts such that if one squares the first, subtracts it from 97, and takes its square root, then squares the second, subtracts it from 100, and takes its square root, the sum of the two roots is 17. (This problem is also from the work of Antonio de' Mazzinghi. Antonio set the parts u, v equal to $5 + x$ and $5 - x$ respectively and derived an equation in x.)

9. Master Dardi gives a rule to solve the fourth degree equation $x^4 + bx^3 + cx^2 + dx = e$ as $x = \sqrt[4]{(d/b)^2 + e} - \sqrt{d/b}$. His problem illustrating the rule is the following: A man lent 100 *lire* to another and after 4 years received back 160 *lire* for principal and (annually compounded) interest. What is the interest rate? As in the text's example, set x as the monthly interest rate in *denarii* per *lira*. Show that this problem leads to the equation $x^4 + 80x^3 + 2400x^2 + 32{,}000x = 96{,}000$ and that the solution found by "completing the fourth power" is given by the stated formula.

10. Master Dardi gives another rule to solve the quartic equation $x^4 + dx = bx^3 + cx^2 + e$ in the form $x = \sqrt[4]{(c/4)^2 + e} + (b/4) - \sqrt{d/2b}$. Use this formula to solve one of the problems Dardi gave to illustrate it: Divide 10 into two parts such that their product divided by their difference is $\sqrt{28}$. Show that this formula is not true in general.

11. The equation $6x^3 = 43x^2 + 79x + 30$ is solved in the *Summa* of Luca Pacioli as follows: "Add the number to the *cose* to form a number, and then you get one *cubo* equal to 7 1/6 *censi* plus 18 1/6, after you have reduced to one *cubo* [divided all the terms by 6]. Then divide the *censi* in half and multiply this half by itself, and add it onto the number. It will be 31 1/144 and the *cosa* is equal to the root of this plus 3 7/12, which is half of *censi*."[39] Show that Pacioli's answer is incorrect. What was he thinking of in presenting his rule?

Problems from Chuquet

12. Carry Chuquet's approximation procedure for $\sqrt{6}$ further. That is, since $2\ 4/9 < \sqrt{6}$, $2\ 5/11 > \sqrt{6}$, and $2\ 9/20 > \sqrt{6}$, the next approximation is $2\ 13/29$. Continue the procedure until you reach Chuquet's final value of $2\ 89/198$. Then continue to the value $2\ 881/1960$, found in an earlier work.

13. Use Chuquet's approximation procedure to calculate his values for $\sqrt{5}$ (2 161/682) and $\sqrt{12}$ (3 181/390).

14. Find two numbers in the proportion $5 : 7$ such that the square of the smaller multiplied by the larger gives 40.

15. Find a number which, when multiplied by 20 and then having 7 added to the product, has the sum in the proportion $3 : 10$ with the number formed by multiplying the original number by 30 and subtracting 9. (Chuquet notes that the problem is impossible. Why?)

16. In a vessel full of wine there are three taps such that if one opens the largest it will empty the vessel in 3 hours, if one opens the middle one it will empty it in 4 hours, and if one uses the smallest tap it will empty it in 6 hours. How long would it take to empty the vessel if all three taps are open?

17. A man makes a will and dies leaving his wife pregnant. His will disposes of 100 *écus* such that if his wife has a daughter, the mother should take twice as much as the daughter, but if she has a son, he should have twice as much as the mother. [Sexist problem!] The mother gives birth to twins, a son and a daughter. How should the estate be split, respecting the father's intentions?

Problems from Rudolff's **Coss**

18. Express $\sqrt{27 + \sqrt{200}}$ as $a + \sqrt{b}$.

19. I am owed 3240 *florins*. The debtor pays me 1 *florin* the first day, 2 the second day, 3 the third day, and so on. How many days does it take to pay off the debt?

20. Divide 10 into two parts such that their product is $13 + \sqrt{128}$.

Problems from Recorde's **Whetstone of Witte**

21. There is a certain army composed of Dukes, Earls, and soldiers. Each Duke has under him twice as many Earls as there are Dukes. Each Earl has under him four times as many soldiers as there are Dukes. The two hundredth part of the number of soldiers is 9 times as many as the number of Dukes. How many of each are there?

22. A gentleman, willing to prove the cunning of a bragging Arithmetician, said thus: I have in both hands 8 crowns. But if I count the sum of each hand by itself severally and add to it the squares and the cubes of the both, it will make in number 194. Now tell me, what is in each hand?[40]

Problems from Cardano's **Ars Magna**

23. Show that if r, s are two positive roots of $x^3 + cx = d$, then $t = r + s$ is a root of $x^3 = cx + d$.

24. Show that if $x = t$ is a root of $x^3 = cx + d$, then $r = t/2 + \sqrt{c - 3(t/2)^2}$ and $s = t/2 - \sqrt{c - 3(t/2)^2}$ are both roots of $x^3 + d = cx$. Apply this rule to solve $x^3 + 3 = 8x$.

25. Suppose m, n are numbers such that $c = m + n$ and $d = m\sqrt{n}$. Show that $r = \sqrt{n}$ is a positive solution to $x^3 + d = cx$. Furthermore, show that $s = \sqrt{c - 3(r/2)^2} - r/2$ is a second positive solution. Find the third solution. Apply this rule to the equations $x^3 + 3 = 10x$ and $x^3 + 60 = 46x$.

26. Prove that the equation $x^3 + cx = d$ always has one positive solution and no negative ones.

27. Use Cardano's formula to solve $x^3 + 3x = 10$ and $x^3 + 6x = 2$.

28. Use Cardano's formula to solve $x^3 = 6x + 40$ and $x^3 = 6x + 6$.

29. Consider the equation $x^3 = cx + d$. Show that if $(c/3)^3 > (d/2)^2$ (and thus that Cardano's formula involves imaginary quantities), then there are three real solutions.

30. Solve $x^3 + 21x = 9x^2 + 5$ completely by first using the substitution $x = y + 3$ to eliminate the term in x^2 and then solving the resulting equation in y.

31. Use Ferrari's method to solve the quartic equation $x^4 + 4x + 8 = 10x^2$. Begin by rewriting this as $x^4 = 10x^2 - 4x - 8$ and adding $-2bx^2 + b^2$ to both sides. Determine the cubic equation which b must satisfy so that each side of the resulting equation is a perfect square. For each solution of that cubic, find all solutions for x. How many different solutions to the original equation are there?

32. The dowry of Francis' wife is 100 *aurei* more than Francis' own property and the square of the dowry is 400 more than the square of his property. Find the dowry and the property. (Note the negative answer for Francis' property; Cardano interpreted this as a debt.)

33. I have 12 *aurei* more than Francis and the square of my *aurei* is 128 more than the cube of Francis' *aurei*. How much do we each have? (Again, Cardano interprets the negative answer.)

Challenge Problems from the Contest between Ferrari and Tartaglia

34. Find two numbers x, y with $x > y$ such that $x + y = y^3 + 3yx^2$ and $x^3 + 3xy^2 = x + y + 64$. (Tartaglia's so-

lution is $x = \sqrt[3]{4 + \sqrt{15\frac{215}{216}}} + \sqrt[3]{4 - \sqrt{15\frac{215}{216}}} + 2$ while $y = x - 4$.)

35. Divide 8 into two parts x, y such that $xy(x - y)$ is a maximum. (Note that this was posed in the days before calculus.)

Problems from Bombelli

36. It is obvious that 3 is a root of $x^3 + 3x = 36$. The Cardano formula gives $x = \sqrt[3]{\sqrt{325} + 18} - \sqrt[3]{\sqrt{325} - 18}$. Using Bombelli's methods, show that this number is in fact equal to 3.

37. Express $\sqrt[3]{52 + \sqrt{-2209}}$ in the form $a + b\sqrt{-1}$.

Problems from Viète

38. Given a right triangle with base D, perpendicular B, and hypotenuse Z, and a second right triangle with base G, perpendicular F, and hypotenuse X, show that the right triangle constructed in the text with base $DG - BF$, perpendicular $BG + DF$, and hypotenuse ZX has its base angle equal to the sum of the base angles of the original triangles.

39. Derive the formulas for $\sin 4\alpha$ and $\cos 4\alpha$ in a way analogous to Viète's derivation of formulas for the sine and cosines of 2α and 3α.

40. Given the product of two numbers and their ratio, find the roots. Let A, E, be the two roots, $AE = B$, $A : E = S : R$. Show that $R : S = B : A^2$ and $S : R = B : E^2$. Viète's example has $B = 20$, $R = 1$, $S = 5$. Show that $A = 10$ and $E = 2$. (Jordanus has the same problem but with different numbers.)

41. Given the difference between two numbers and the difference between their cubes, find the numbers. Let E be the sum of the numbers, B the difference between them, and D the difference between the cubes. Show that $E^2 = \frac{4D - B^3}{3B}$. Once E^2 is known, so is E and then the numbers themselves. Find the solution when $B = 6$ and $D = 504$. (Diophantus has the same problem twice, once in Book IV with these numerical values and once in Book B.)

42. Show that if $x^3 + bx = b^2c$, then there are four continued proportionals, the first of which is b, the sum of the second and fourth being c, and the unknown x being the second. Use this result to solve $x^3 + 64x = 2496$.

Problems from Stevin

43. Write 13.395 and 22.8642 in Stevin's notation. Use his rules to multiply the two numbers together and to divide the second by the first.

44. Given the two numbers 237 ⓪ 5 ① 7 ② 8 ③ and 59 ⓪ 7 ① 3 ② 9 ③, subtract the second from the first.

FOR DISCUSSION . . .

45. There were other algebra texts written in the sixteenth century besides the ones discussed. Look up and report on the algebras of John Buteon, Jacques Peletier, Guillaume Gosselin, and/or Nicolas Petri.

46. Why is Cardano's formula no longer generally taught in a college algebra course? Should it be? What insights can it bring to the study of the theory of equations?

47. Outline a lesson introducing the study of complex numbers by way of the problems with Cardano's formula giving a real root as the sum of two complex values. Discuss the merits of such an approach.

48. Compare the various notations for unknowns used by the mathematicians discussed in the text. Write a brief essay on the importance of a good notation for increasing a student's understanding in algebra.

49. The first printed mathematics book is the so-called *Treviso Arithmetic* of 1478, by an unknown author. Write a brief essay on its contents and its importance. Consult Frank J. Swetz, *Capitalism and Arithmetic: The New Math of the 15th Century* (La Salle, Ill.: Open Court, 1987) for a translation of this work.

50. Why was the knowledge of mathematics necessary for the merchants of the Renaissance? Did they really need to know the solutions of cubic equations? What, then, was the purpose of the detailed study of these equations in the works of the late sixteenth century?

51. Compare the symbolism of Jordanus and Viète. In what way is Viète's work an advance on that of Jordanus?

52. Explain why mathematicians of the sixteenth century equated the new algebra with the Greek analysis as described by Pappus.

References and Notes

General works on the material of this chapter include Paul Lawrence Rose, *The Italian Renaissance of Mathematics: Studies on Humanists and Mathematicians from Petrarch to Galileo* (Geneva: Droz, 1975), Warren Van Egmond, "The Commercial Revolution and the Beginnings of Western Mathematics in Renaissance Florence, 1300–1500," Ph.D. diss., University of Indiana, 1976, and R. Franci and L. Toti Rigatelli, "Towards a History of Algebra from Leonardo of Pisa to Luca Pacioli," *Janus* 72 (1985), 17–82. Chapter 2 of B. L. Van der Waerden, *A History of Algebra from al-Khwarizmi to Emmy Noether* (New York: Springer, 1985) also provides a good introduction to the material.

1. Quoted in R. Franci and L. Toti Rigatelli, "Towards a History of Algebra," pp. 64–65.

2. Gerolamo Cardano, *The Great Art, or The Rules of Algebra*, translated and edited by T. Richard Witmer (Cambridge: MIT Press, 1968), p. 96.

3. Warren van Egmond, "Commercial Revolution." This dissertation examines the works of the *maestri d'abbaco*, including both the arithmetic and the algebra contained in them. It also discusses the great importance of these works for reintroducing the basic ideas of mathematics into the general culture.

4. R. Franci and L. Toti Rigatelli, "Towards a History of Algebra," p. 31. This article provides a detailed look at the works of various *maestri d'abbaco* and analyzes their relationship to the works of Leonardo of Pisa and Luca Pacioli.

5. R. Franci and L. Toti Rigatelli, "Fourteenth-century Italian algebra," in Cynthia Hay, ed., *Mathematics from Manuscript to Print: 1300–1600* (Oxford: Clarendon Press, 1988), 11–29, p. 16. This article summarizes the contents of several important fourteenth century abacus manuscripts.

6. Van Egmond, "Commercial Revolution," p. 266.

7. For more details on Mazzinghi, see R. Franci and L. Toti Rigatelli, "Towards a History of Algebra," and B. L. van der Waerden, *History of Algebra.*

8. Warren Van Egmond, "The Algebra of Master Dardi of Pisa," *Historia Mathematica* 10 (1983), 399–421.

9. A translation of much of Chuquet's manuscript by Graham Flegg, Cynthia Hay, and Barbara Moss, has been published as *Nicolas Chuquet, Renaissance Mathematician* (Boston: Reidel, 1985). In addition to the work of Chuquet himself, this volume contains an extensive discussion of what is known of the author and his place in the history of mathematics.

10. Ibid., p. 90.

11. Ibid., p. 105.

12. Ibid., p. 151.

13. Ibid., p. 177.

14. There is no modern edition of Rudolff's *Coss* of 1525. There is an extensive study of all of the German algebra works of the Renaissance by P. Treutlein, "Die Deutsche Coss," *Abhandlungen zur Geschichte der mathematischen Wissenschaften* 2 (1879), 1–124.

15. The work of Michael Stifel is discussed in Joseph Hofmann, *Michael Stifel 1487?–1567: Leben, Wirken und Bedeutung für die Mathematik seiner Zeit* (Wiesbaden: Franz Steiner Verlag, 1968).

16. A photographic reprint of Robert Recorde, *The Whetstone of Witte* (New York: Da Capo Press, 1969) is available. The pages are unnumbered.

17. Information on Nuñez's *Algebra* can be found in H. Bosmans, "Sur le 'Libro de algebra' de Pedro Nuñez," *Bibliotheca Mathematica* (3) 8 (1907), 154–169 and H. Bosmans, "L'Algebre de Pedro Nuñez," *Annaes scientificos da academia politechnica do Porto* 3 (1908), 222–271. A more general treatment of his life and work is by Rodolpho Guimaräes, *Sur la vie et l'oeuvre de Pedro Nuñes* (Coimbra: Imprimerie de l'Université, 1915).

18. John Fauvel and Jeremy Gray, eds., *The History of Mathematics: A Reader* (London: Macmillan, 1987), p. 254. Many of the problems involved in the challenges between Fiore and Tartaglia and between Ferrari and Tartaglia, as well as Ferrari's description of his discussion with Cardano are translated in the Interchapter of this book.

19. Translated from the Italian by my daughter, Sharon Katz, with the assistance of other students at Princeton University.

20. Cardano, *The Great Art*, pp. 98–99. This work will repay a careful reading as there are many gems included in it not discussed in this text. An English language biography of Cardano is Oystein Ore, *Cardano: The Gambling Scholar* (Princeton: Princeton University Press, 1953). Cardano's autobiography is available as *Cardano, The Book of My Life*, translated by J. Stoner (New York: Dover, 1962). See also Richard Feldman, "The Cardano-Tartaglia Dispute," *Mathematics Teacher* 54 (1961), 160–163 and James Bidwell and Bernard Lange, "Girolamo Cardano: A Defense of His Character," *Mathematics Teacher* 64 (1971), 25–31.

21. Ibid., p. 250.

22. Ibid., p. 220.

23. Rafael Bombelli, *Algebra* (Milan: Feltrinelli, 1966), p. 133. This is a modern Italian reprint of the original 1572 edition. There is no English translation available. For more on Bombelli, see three articles of S. A. Jayawardene: "The Influence of Practical Arithmetics on the Algebra of Rafael Bombelli," *Isis* 64 (1973), 510–523; "Unpublished Documents Relating to Rafael Bombelli in the Archives of Bologna," *Isis* 54 (1963), 391–395; and "Rafael Bombelli, Engineer-Architect: Some Unpublished Documents of the Apostolic Camera," *Isis* 56 (1965), 298–306.

24. Bombelli, *Algebra*, p. 133.

25. René Descartes, *Rules for the Direction of the Mind*, translated by Elizabeth S. Haldane and G. R. T. Ross, *Great Books* edition (Chicago: Encyclopedia Britannica, 1952), pp. 6–7.

26. François Viète, *The Analytic Art*, translated and edited by T. Richard Witmer (Kent, Ohio: Kent State University Press, 1983), pp. 11–12. All further quotations are taken from this edition, but sometimes have been amended to give a better sense of the original Latin. There is an alternative translation of the *Two Treatises on the Recognition and Emendation of Equations*, somewhat more faithful to the original—and therefore somewhat harder for a modern reader to understand—by Robert Schmidt, published in *The Early Theory of Equations: On Their Nature and Constitution* (Annapolis: Golden Hind Press, 1986). There is much in Viète's work which we have not considered in the text. Many of his methods could well be adapted to modern use. Unfortunately, there is no English study of these methods.

27. Ibid., p. 32.

28. Ibid., p. 17.

29. Ibid., p. 24.

30. Ibid., p. 84.

31. Ibid., p. 174.

32. Ibid., p. 210.

33. Ibid., p. 310.

34. The second half of Charles Jones' dissertation "On the Concept of One as a Number" (University of Toronto, 1978) is primarily devoted to a study of Stevin's ideas on decimals and what Jones calls "the breakdown of the Greek number concept." I have summarized some of his arguments in the text.

35. Henrietta O. Midonick, ed., *The Treasury of Mathematics* (New York: Philosophical Library, 1965), p. 737. The English translation of *De Thiende* was made by Robert Norton in 1608 and is reprinted in the volume cited as

well as in *The Principal Works of Simon Stevin*, volume II, edited by Dirk J. Struik (Amsterdam: Swets and Zeitlinger, 1958). More on Stevin can be found in Dirk J. Struik, *The Land of Stevin and Huygens: A Sketch of Science and Technology in the Dutch Republic During the Golden Century* (Dordrecht: Reidel, 1981).

36. Midonick, *Treasury*, p. 740.

37. Charles Jones, "Concept of One," p. 239.

38. Ibid., p. 248.

39. R. Franci and L. Toti Rigatelli, "Towards a History of Algebra," p. 65.

40. Robert Recorde, *The Whetstone of Witte*.

Summary of Renaissance Algebra

Mid-fourteenth century	Maestro Dardi of Pisa	Abacus work
1353–1383	Antonio de' Mazzinghi	Abacus work
c. 1430–1487	Nicolas Chuquet	French algebra
1445–1517	Luca Pacioli	Italian arithmetic and algebra
1465–1526	Scipione del Ferro	Cubic equations
Early sixteenth century	Christoff Rudolff	German *Coss*
1487–1567	Michael Stifel	German *Coss*
1499–1557	Niccolò Tartaglia	Cubic equations
1500–1558	Annibale della Nave	Cubic equations
1501–1576	Gerolamo Cardano	Cubic equations
1502–1578	Pedro Nuñez	Portuguese algebra
1509–1575	Federigo Commandino	Translations from the Greek
1510–1558	Robert Recorde	English algebra
1522–1565	Lodovico Ferrari	Quartic equations
1526–1572	Rafael Bombelli	Complex numbers
1540–1603	François Viète	Theory of equations
1548–1620	Simon Stevin	Decimal fractions

Chapter 10

Mathematical Methods in the Renaissance

"Philosophy is written in this grand book, the universe, which stands continually open to our gaze. But the book cannot be understood unless one first learns to comprehend the language and read the letters in which it is composed. It is written in the language of mathematics, and its characters are triangles, circles, and other geometric figures without which it is humanly impossible to understand a single word of it; without these, one wanders about in a dark labyrinth."
(From Galileo Galilei, The Assayer, *Rome, 1623)[1]*

*O*n 26 February, 1616, Galileo was called to a meeting at the home of Cardinal Bellarmine, an advisor to the Pope and a cardinal of the Inquisition. Three months later the cardinal gave Galileo an affidavit certifying what happened at that meeting: "Sig. Galileo . . . was . . . told of the declaration made by his holiness [Pope Paul V] and published by the Congregation of the Index, that [to say] the earth moves around the sun and that the sun stands still in the center of the universe without motion from east to west is contrary to Sacred Scripture and therefore may not be defended or held."[2]

Algebra was not the only mathematical concern of the Renaissance. In fact, Galileo, in the opening quotation, virtually equated mathematics with geometry. As part of the general revival of interest in classical learning, Renaissance scholars studied the Greek geometry texts. First, naturally, they studied Euclid, whose *Elements,* in various Latin versions, was a major part of the mathematics curriculum at the European universities of the time. It was expected that anyone having any pretense of learning whatsoever would be familiar with Euclid's work.

Because there were many who did not know Latin and did not attend the universities, vernacular versions of the *Elements* began to appear in the sixteenth century. Tartaglia prepared an Italian version in 1543; Johann Scheubel and Wilhelm Holzmann (Xylander) translated major portions into German in 1558 and 1562; Pierre Forcadel did the same in French in 1564–1566; while Rodrigo Camorano made a Spanish translation of the first six books in 1576. The most impressive of the vernacular versions, however, was the English

---

Biography   John Dee (1527–1608)

Dee took his B.A. from Cambridge University in 1545 and shortly thereafter journeyed to the continent where he studied with various mathematicians, learned much about such fields as geography, astronomy, and astrology, and lectured on Euclid in Paris. Returning to England, he served as a court astrologer to Queen Elizabeth. His own writings encompassed such varied topics as logic, astronomy, perspective, burning mirrors, and astrology, but later in life he became enamored of the mystical elements in mathematics. Thus he studied and wrote about how various symbols could be combined in certain figures, the proper understanding of which would enable the reader to understand the hidden secrets of the physical world. Like some of his contemporaries, he also involved himself with gematria, the study of the numerical values of words, and alchemy. Ultimately, his mysticism and accusations that he was involved with the practice of "black magic" caused him to lose his royal patronage. He died in poverty.

---

translation of Henry Billingsley in 1570. Its nearly 1000 pages included all 13 original books of the *Elements* as well as three additional books traditionally ascribed to Euclid. It also contained numerous additions and notes from various ancient and modern authors. The printer evidently spared no pains in the production of this work. For example, the discussions of solid geometry from Book XI contain "pop-up" diagrams, pasted onto the relevant pages, which enabled the reader actually to construct the three-dimensional figures (Figure 10.1).

The most noteworthy part of the Billingsley *Euclid,* however, is the *Mathematical Preface* by the sixteenth century English scientist and mystic John Dee (1527–1608). Dee was well qualified to write a preface to the translation of Euclid. He had acquired a wide knowledge of various fields in which mathematics was employed and wanted to convince those about to work their way through this great geometrical work of its value. Thus he gave detailed descriptions of some 30 different fields that need mathematics and the relationships among them, organized into what he called a "groundplat," or chart. Dee's framework offers an overview of "applied mathematics" in the Renaissance, the subject of this chapter (Figure 10.2).

A careful student of Greek mathematics, Dee began his preface by noting that "of Mathematical things are two principal kinds; namely Number, and Magnitude."[3] The science of number is called arithmetic, that of magnitude, geometry. These are the two principal divisions of the mathematical arts. Dee notes that arithmetic originally meant the study of whole numbers, but arithmeticians have "extended their name farther than to numbers whose least part is a unit."[4] Various other kinds of numbers have been introduced, including common fractions, sexagesimal (or astronomical) fractions, and radical numbers (roots). Arithmetic has also been extended into the "Arithmetical Art of Equation," that is, algebra. It is, however, the application of geometry, the "science of magnitude," to which most of his preface is devoted.

Dee gives a brief history of geometry to justify its name (meaning "land measuring") which, he says, is too "base and scant for a science of such dignity and ampleness." The name "has been suffered to remain, that it might carry with it a perpetual memory of the first and notablest benefit, by that science, to common people showed, which was, when

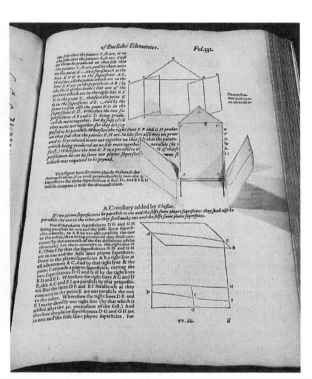

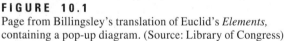

FIGURE 10.1
Page from Billingsley's translation of Euclid's *Elements,*
containing a pop-up diagram. (Source: Library of Congress)

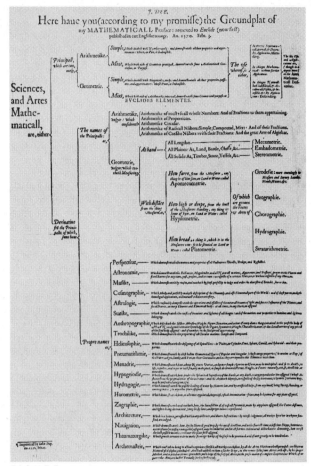

FIGURE 10.2
Dee's "Groundplat" from the preface to the Billingsley
translation of Euclid's *Elements.* (Source: Smithsonian
Institution Photo No. 92-345)

bounds . . . of land and ground were lost and confounded or, that ground bequeathed, were
to be assigned . . . or . . . that commons were distributed into severalties. For, where, upon
these and such like occasions, some by ignorance, some by negligence, some by fraud,
and some by violence, did wrongfully limit, measure, encroach, or challenge . . . those
lands and grounds, great loss, disquietness, murder and war did (full oft) ensue. Till, by
God's mercy and man's industry, the perfect science of lines, planes, and solids . . . gave
unto every man, his own."[5] It is good to know that this science at its origins prevented war
and helped to dispense justice.

Dee divides the applications of geometry into two classes, "vulgar" geometry, which
includes the various sciences of measurement such as stereometry, the measure of solids,
and geography, the study of the methods for creating maps, and the "methodical arts,"
"which, declining from the purity, simplicity, and immateriality of our principal science
of magnitudes, do yet nevertheless use the great aid, direction, and method of the said

principal science."[6] Among these methodical arts are perspective, astronomy, music, astrology, statics, architecture, navigation, trochilike (the study of circular motions), menadry (the study of simple machines), and zography (the study of painting). This chapter will survey some of these fields, discussing both Dee's analysis and the actual work of some of the practitioners in the sixteenth and early seventeenth centuries.

10.1 PERSPECTIVE

According to Dee, "Perspective is an Art Mathematical which demonstrates the manner and properties of all radiations direct, broken and reflected." This art explains why "walls parallel . . . approach afar off" and why "roof and floor parallels, the one to bend downward, the other to rise upward, at a little distance from you."[7] Closely related to perspective is the art of Zography, "which teaches and demonstrates how the intersection of all visual pyramids, made by any plane assigned . . . may be by lines . . . represented."[8] It is these two arts with which a painter must be well acquainted in order that "in winter he can show you the lively view of summer's joy and riches and in summer exhibit the countenance of winter's doleful state and nakedness. Cities, towns, forts, woods, armies, yea whole kingdoms . . . can he, with ease bring with him, home (to any man's judgment) as patterns lively of the things rehearsed."[9]

Although there was some use of perspective in ancient times, it was only in the Renaissance that painters began in earnest to attempt to give visual depth to their works. The earliest painters accomplished this through trial and error, but by the fifteenth century, artists were attempting to derive a mathematical basis for displaying three-dimensional objects on a two-dimensional surface. Clearly, objects which are farther away from the observer must be made smaller to give the picture realism. The question then becomes how small a given object should be. The answer to this question, painters ultimately realized, had to come from geometry. Filippo Brunelleschi (1377–1446) was the first Italian artist to make a serious study of the geometry of perspective, but Leon Battista Alberti (1404–1472) wrote the first text on the subject, the *Della Pittura* of 1435 (Figure 10.3) Alberti noted in this treatise that the first requirement of a painter is to know geometry. Thus he presented a geometrical result showing how to represent a set of squares in the "ground plane" on the plane of the canvas, the "picture plane."

The picture plane may be thought of as pierced by rays of light from the various objects in the picture to the artist's eye, whose position is called the "station point." Hence the picture plane is a section of the projection from the eye to the scene to be pictured (Figure 10.4). The perpendicular from the station point to the picture plane intersects the latter in a point V called the "center of vision" or the "central vanishing point." The horizontal line AV through the central vanishing point is called the "vanishing line" or "horizon line." The words "vanishing point" and "vanishing line" are used because all horizontal lines in the picture perpendicular to the picture plane must be drawn to intersect at the "vanishing point." All other sets of parallel horizontal lines will intersect at some point on the "vanishing line."

To represent a set of squares in the ground plane with sides parallel to and perpendicular to the picture plane, Alberti began by marking off a set of equal distances on the line of intersection BC of the picture plane and the ground plane, the "ground line." He

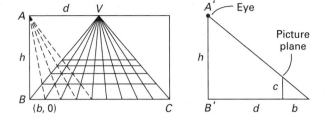

FIGURE 10.4
Alberti's rule for perspective drawing of a set of squares.

connected these to the central vanishing point, thus giving one set of sides. To deal with the set of sides parallel to the ground line Alberti invented the following method, which for simplicity is described using algebra. Suppose that the eye is at a distance d from the picture plane and is situated at a height h above the ground line. If a line in the ground plane is parallel to the ground line and at a distance b behind it, then its position in the picture plane should be at a distance c above the ground line, where c is determined by the proportion $c : b = h : d + b$ derived from the similar triangles in the diagram. Thus $c = (hb)/(d + b)$. To draw in the squares easily, mark the point A on the vanishing line at distance d from the vanishing point V. If AB and the ground line BC are taken as coordinate axes, the equation of the line connecting B and V is

$$y = \frac{h}{d}x.$$

If the line is then drawn from the point with coordinates $(b, 0)$ to $A = (0, h)$, its equation is

$$y = -\frac{h}{b}x + h.$$

The y coordinate of the intersection of the two lines is then $(hb)/(d + b)$, and the construction is correct. One can easily use this construction for as many parallel lines as desired.

    This checkerboard construction is at the heart of the system of "focused perspective" used by artists from the fifteenth century to the present day. Alberti himself did not discuss any more advanced perspective constructions, but Piero della Francesca (1420–1492), in his work *De perspectiva pingendi* (*On Painter's Perspective*), written sometime between 1470 and 1490, gave a detailed discussion of how to draw various two- and three-dimensional geometrical objects in focused perspective. Della Francesca, besides being an artist, was a competent mathematician, problems from whose abacus tract appeared in the exercises to Chapter 9. His text on perspective includes a drawing showing the calculations the artist made in preparing a painting in focused perspective.[10]

    Another artist-mathematician of this same period was the German Albrecht Dürer (1471–1528), who spent several years in Italy studying works on perspective before writing a major treatise of his own (Figure 10.5). The *Underweysung der Messung mit Zirckel und Richtscheyt in Linien, Ebnen, und gantzen Corporen* (*Treatise on Mensuration with the Compass and Ruler in Lines, Planes, and Whole Bodies*), published in 1525, was the first geometric text written in German.[11] Dürer had to create a new German vocabulary for scientific terms, including abstract mathematical concepts. If possible, he used the expression handed down from generation to generation by artisans. For example, "der

FIGURE 10.5
Self-portrait of Albrecht Dürer.

neue Mondschein" (crescent) denoted the intersection of two circles; "Gabellinie" (fork line) meant hyperbola; and "Eierlinie" (egg line) meant ellipse.

Dürer believed that he needed to instruct the German artists in many of the preliminary geometrical ideas involved in drawing before they could approach perspective, dealt with at the end of the *Underweysung der Messung.* In his opinion, German painters were equal to any in practical skill and imagination, but they were well behind the Italians in rational knowledge. "And since geometry is the right foundation of all painting, I have decided to teach its rudiments and principles to all youngsters eager for art."[12] Therefore, the work is eminently practical. Dürer shows how to apply geometric principles to the representation of objects on canvas (Figure 10.6).

The first of the four books of the *Underweysung der Messung* deals with the representation of space curves. Dürer's idea is to project the curve onto both the yz plane and the xy plane in order to determine its nature. As an example, consider his construction of an ellipse from its definition as a section of a right circular cone (Figure 10.7). Dürer first

FIGURE 10.6
St. Jerome in His Study. Here, Dürer illustrates his application of the theory of perspective. (Source: The Nelson-Atkins Museum of Art, Kansas City, Missouri, 58–70/21, gift of Robert B. Fizzell)

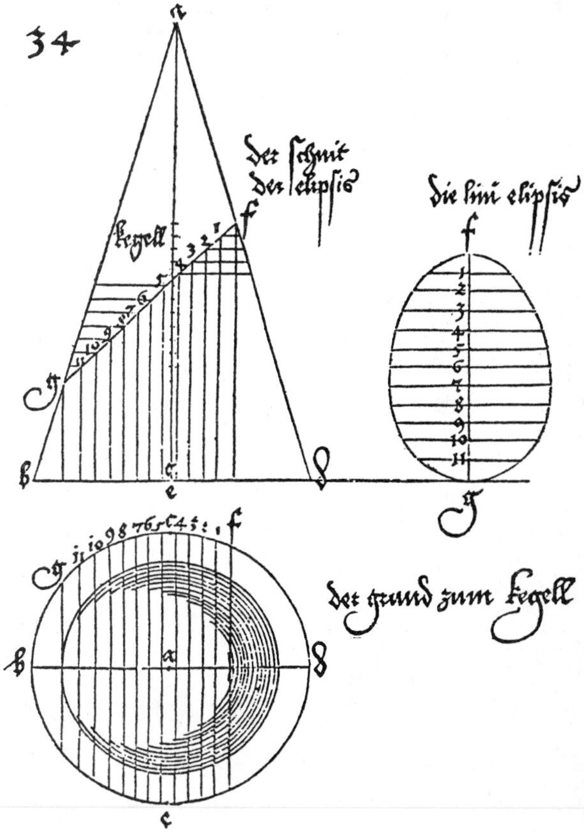

FIGURE 10.7
Dürer's construction of an ellipse by projection.

projects the cone with its cutting plane onto the yz plane. The line segment fg representing the diameter of the ellipse is divided into 12 equal parts, and both vertical and horizontal lines are drawn through the division points. At each of the 11 points i, the horizontal line represents part of the diameter of the circular section C_i made by a horizontal cutting plane. The two points of intersection of this circle with the ellipse are symmetrically located on the ellipse with respect to its diameter and therefore determine the width w_i of the ellipse there. The projection of the cone onto the xy plane then consists of this series of concentric circles C_i. The continuation of each vertical line becomes a chord in the corresponding circle whose length is w_i. Dürer thus has a rough projection of the ellipse. The outline of the ellipse is, however, not symmetric about its minor axis, since this projection is not taken from a direction perpendicular to the plane of the ellipse itself. But when Dürer attempts to draw the ellipse from its projection, he simply transfers the line segment representing the axis of the ellipse to a new vertical line fg, divides it at the same points i, draws horizontal line segments through each of width w_i, and then sketches the curve through the ends of these line segments. Dürer's drawing is therefore in error, because the curve is wider at the bottom than at the top. A possible reason Dürer did not realize that the ellipse should be symmetric about its minor axis is that the centerline of the cone, around the projection of which all the circles are drawn, does not pass through the center of the ellipse. Although one can prove that $w_i = w_{12-i}$ ($i = 1, 2, 3, 4, 5$) by an analytic argument, Dürer probably believed that the ellipse was in fact egg-shaped—he does call it an Eierlinie—because the cone itself widens toward the bottom.[13]

After describing the construction and representation by projections of other space curves, Dürer continued in the second book of the *Underweysung* to describe methods for constructing various regular polygons, both exact ones using the classical tools of straightedge and compass and approximate ones, taken from the tradition of artisans. Thus the work, which was published in Latin several years after its German edition, served both to introduce the artisans to the Greek classics and also to familiarize professional mathematicians with the practical geometry of the workshop. The third book of the text was purely practical, showing how geometry could be applied in such varied fields as architecture and typography. Here Dürer suggested new types of columns and roofs as well as the methods of accurately constructing both Roman and Gothic letters. In the final book Dürer returned to more classical problems and dealt with the geometry of three-dimensional bodies. In particular, he presented a construction of the five regular solids by paper folding, a method still found in texts today, as well as similar procedures for certain semiregular solids. He also presented other problems of construction, including that of doubling the cube, before concluding the work with the basic rules for the perspective drawing of these solid figures.

10.2 GEOGRAPHY AND NAVIGATION

Two related aspects of mathematics discussed by Dee and extremely important to the world of the sixteenth century were geography and navigation. "The art of Navigation demonstrates how, by the shortest good way, by the aptest direction, and in the shortest time, a sufficient ship, between any two places (in passage navigable) assigned, may be conducted; and in all storms and natural disturbances chancing, how to use the best possible means, whereby to recover the place first assigned."[14] In the fifteenth and sixteenth centuries, Europeans were exploring the rest of the world, and methods of navigation were of central importance. The country that could employ new techniques well had great advantages in the quest for new colonies and their attendant natural resources.

The major problem of navigation on the seas was the determination of the ship's latitude and longitude at any given time. The first of these was not too difficult. One's latitude, in the northern hemisphere, was equal to the altitude of the north celestial pole, and this was marked, approximately, by Polaris, the pole star. A good approximation of the latitude was found simply by taking the altitude of that star. An alternate method of finding latitude was by observation of the sun. The zenith distance of the sun at local noon is equal to the latitude minus the sun's declination. Navigators of the fifteenth century had accurate tables of the declination for any day of the year, so they needed only to take a reading of the sun's altitude at noon. This altitude was, of course, the highest altitude of the day and could be determined by finding the shortest shadow of a standard pole.

The determination of longitude was much more difficult. Knowing the difference between the longitudes of two places is equivalent to knowing the difference between their local times, because 15° of longitude is equivalent to one hour. Theoretically, if one had a clock set to the time at a place of known longitude and could determine when, on that clock, local noon occurred at one's current location, the difference in time would enable one to make a determination of longitude. Alternatively, one could compare the known time of an astronomical event, such as an eclipse of the moon, at the place of known longitude with its local time at one's current location. Unfortunately, these methods could not work given the state of time keeping devices in the Renaissance. They were simply not accurate enough, especially if operated on the moving decks of a ship at sea. When Columbus attempted to determine longitude on his second voyage to America in 1494 using an eclipse of the moon, his error was about 18°. Despite rewards offered by various European governments for the invention of accurate methods of determining longitude at sea, the problem remained unsolved until the eighteenth century, when an accurate marine clock was finally devised.

Given these difficulties of finding one's location at sea, it is not surprising that seamen often used methods of "guesstimation" rather than mathematical astronomy. While scholars were aware that a great circle route was the shortest distance between two points, sailors generally preferred to sail to the latitude of their destination as quickly as possible and then head due east or west until they reached land. Whatever the method of navigating, however, the seamen needed accurate maps. Dee called the making of these maps Geography: "Geography teaches ways by which in sundry forms (as spherical, plane, or other) the situation of cities, towns, villages, forts, castles, mountains, woods, havens, rivers, creeks, and such other things, upon the outface of the earthly globe . . . may be described and designed in commensurations analogical to nature and verity, and most aptly to our view, may be represented."[15]

Maps had been drawn since antiquity. Ptolemy in his *Geography* had analyzed some of the problems of mapping the round earth onto flat paper, had exhibited the longitude and latitude of the known localities of the inhabited world, and had included some 26 regional maps as well as a world map. To construct his maps, he needed to use some form of projection, that is, some way of constructing a function from a portion of the earth's spherical surface to a flat piece of paper. Presumably, Ptolemy wanted the resultant maps to represent the shape of the land masses depicted as closely as possible. In any case, a projection is determined by the grid of longitude and latitude lines, generally known as meridians and parallels, respectively. For his regional maps, Ptolemy simply used a rectangular grid for these lines. On the spherical earth, however, because the spacing of the meridians depends on the latitude, he chose a scale in the two directions so that it corresponded approximately to the ratio of the length of one degree of longitude on the middle parallel of the map to one degree of latitude. This ratio is equal to $MN : AB$ (because the length of a degree of longitude at the equator is equal to that of a degree of latitude), which in turn equals $NP : BC$, $NP : NC$, and finally $\cos \phi$ (Figure 10.8). For example, because Ptolemy's map of Europe reaches from latitude 42° to latitude 54°, the given ratio should be approximately $\cos 48° = .6691$, or 2 : 3.

For his world map, which included only what he calculated to be 180° of longitude, stretching from the Strait of Gibralter to China, Ptolemy chose two different methods. In the first, the parallels were represented by concentric circles centered on the north pole while the meridians were straight lines from the pole, which was not included on the map. Ptolemy realized that this projection could not preserve the proper ratio between degrees of longitude and degrees of latitude, except within a small region of a particular parallel, which he took as the parallel of Rhodes, 36°. Thus, as in all projections of a major portion of the earth's surface, some distortion was inevitable. Ptolemy later developed a more natural appearing map by modifying the meridians into circular arcs as well. This map gives a correct representation of distance on three selected parallels through which the circular arcs are drawn, but still has distortion far from the center of the map (Figure 10.9).

Because it is impossible to make an absolutely correct map on a flat piece of paper, the mapmaker always needs to make some choice of the particular qualities of the projection desired. The mapmaker can choose to preserve areas or shapes or directions or distances. The larger the portion of the earth's surface to be represented, the more difficult it is for the map to have several of these qualities, even approximately. In general, the maps used by seamen during the early Renaissance were constructed by using a different criterion, ease of drawing. These "plane charts" used a rectangular grid for parallels and meridians, with the same scale on each. Because the distances between the meridians were the same at all latitudes, and because the true distance depends on the cosine of the latitude, shapes on these maps had the appearance of being elongated in the horizontal direction. Thus shape was not preserved and more important for the sailor, lines of con-

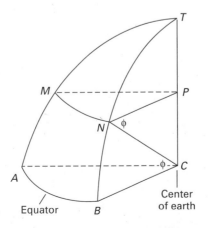

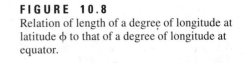

FIGURE 10.8
Relation of length of a degree of longitude at latitude ϕ to that of a degree of longitude at equator.

FIGURE 10.9
Ptolemy's world map for the 1552 Basel edition of his *Geography*. (Source: Smithsonian Institution Photo No. 90-15779)

stant compass bearing, called rhumbs, were not represented by straight lines. When such maps were of relatively small areas, the rhumb lines were straight enough and were often drawn in for each of 8 or 16 compass directions. But as long sea voyages became increasingly common, improvements were required.

One of the first to attempt to apply mathematics to the improvement of mapmaking methods was Pedro Nuñez, in his *Tratado da sphera* of 1537. He discovered that on a sphere a rhumb line or **loxodrome**, as it is now called, becomes a spiral terminating at the pole. Using globes for navigation, however, was inconvenient because they could not be made large enough. Nuñez therefore attempted to develop a map in which loxodromes were straight lines. For accuracy, however, it was necessary that the meridians converge near the poles. Although Nuñez was able to design a device that enabled sailors to measure the number of miles in a degree along each parallel, he was not able to solve the problem he had set.

By 1569, Nuñez's problem was solved from a slightly different point of view by Gerard Mercator (1512–1594), with a new projection known ever since as Mercator's projection (Figure 10.10). Both parallels and meridians were represented by straight lines on this map. To compensate for the "incorrect" spacing of the meridians, therefore, Mercator increased the spacing of the parallels toward the poles. He claimed that on his new map rhumb lines were now straight and a navigator could simply lay a straightedge on his map between his origin and his destination to determine the constant compass bearing to follow. Mercator did not explain the mathematical principle he followed for increasing the distance between the parallels, and some believe that he did it by guesswork alone. Not

FIGURE 10.10
Mercator on a Belgian stamp.

FIGURE 10.11
A world map in Mercator
projection on a Canadian
stamp.

until the work of Edward Wright (1561–1615), *On Certain Errors in Navigation* (1599),
did an explanation of Mercator's methods appear in print (Figure 10.11).

Recall that the ratio of the length of a degree of longitude at latitude ϕ to one at the
equator is equal to cos ϕ. If meridians are straight lines, then, the distances between them
at a latitude ϕ are stretched by a factor of sec ϕ. For loxodromes to be straight on such a
map, then, the vertical distances must also be stretched by the same factor. Because sec ϕ
varies at each point along a meridian, the stretching factor needs to be considered for each
small change of latitude. If we denote by $D(\phi)$ the distance on the map between the equator
and the parallel of latitude ϕ, the change dD in $D(\phi)$ caused by a small change $d\phi$ in ϕ is
determined by $dD = \sec \phi d\phi$. Because the same factor applies horizontally as well, any
"small" region on the globe will be represented on the map by a "small" region of the
same shape. The angle at which a line crosses a meridian on the globe will be transformed
into that same angle on the map and loxodromes will be straight. It follows from this
argument that, in modern terminology, the map distance between the equator and the
parallel at latitude ϕ is given by

$$D(\phi) = \int_0^\phi \sec \phi d\phi, \tag{10.1}$$

where the radius of the globe is taken as 1. Wright, of course, did not use integrals. He
took for his $d\phi$ an angle of $1'$ and computed a table of what he called "meridional parts"
by adding the products sec $\phi d\phi$ for latitudes up to 75°. $D(\phi)$ can be calculated by calculus
techniques as

$$\ln(\sec \phi + \tan \phi) \qquad \text{or} \qquad \ln\left(\tan\left(\frac{\phi}{2} + \frac{\pi}{4}\right)\right).^{[16]}$$

John Dee met Mercator on one of his trips to the continent. He returned with several
of Mercator's globes and probably conferred with Wright concerning the mathematical
details of Mercator's projection. Thus he was involved in the process of making maps
"analogical to nature." Mercator's map, although well suited for navigation, was unfortunately not "analogical to nature" for regions far from the equator. The spacing out of the
parallels greatly increased the relative size of such regions. The popularity of the map led
generations of students to believe, for example, that Greenland is larger than South America. Nevertheless, its simplicity of use made it the prime sea chart during the age of
European exploration.

10.3 ASTRONOMY AND TRIGONOMETRY

According to Dee, "Astronomy is an art mathematical which demonstrates the distance,
magnitudes, and all natural motions, appearances, and passions proper to the planets and
fixed stars, for any time past, present and to come, in respect of a certain horizon, or
without respect of any horizon. By this art we are certified of the distance of the starry
sky and of each planet from the center of the earth, and of the greatness of any fixed star
seen, or planet, in respect of the earth's greatness."[17] Thus, the purpose of astronomy is
to predict the motions of the heavenly bodies as well as to determine their sizes and
distances. A related art is Cosmography, "the whole and perfect description of the heav-

enly, and also elemental part of the world, . . . and mutual collation necessary."[18] As Dee notes further, cosmography explains the relationship of heavenly to earthly events, allowing us to determine "the rising and setting of the sun, the lengths of days and nights . . . with very many other pleasant and necessary uses."

10.3.1 Regiomontanus

Since astronomy and cosmography in the Renaissance, like astronomy in earlier periods, were heavily dependent on trigonometry, we begin with a discussion of the trigonometry text *De Triangulis Omnimodis* (*On Triangles of Every Kind*) of Johannes Müller (1436–1476), generally known as Regiomontanus because he was born near Königsberg in Lower Franconia. (*De Triangulis* was written about 1463 but not published until 70 years later.)

Regiomontanus had made a new translation of Ptolemy's *Almagest* directly from the Greek and, after completing it, realized that there was a need for a compact systematic treatment of the rules governing the relationships of the sides and angles in both plane and spherical triangles which would improve on Ptolemy's seemingly *ad hoc* approach. He considered such a treatment a necessary prerequisite to the study of the *Almagest*: "You, who wish to study great and wonderful things, who wonder about the movement of the stars, must read these theorems about triangles. Knowing these ideas will open the door to all of astronomy and to certain geometric problems."[19]

Regiomontanus presented his material in *On Triangles* in careful geometric fashion, beginning with definitions and axioms. He proved each theorem by using the axioms, results from Euclid's *Elements*, or earlier results in the text. Most theorems are accompanied by diagrams and many are followed by examples illustrating the material. Regiomontanus based his trigonometry on the sine of an arc, defined as the half chord of double the arc, but he did note that one can also consider the sine as depending on the corresponding central angle. Like his European predecessors, he made no use of the tangent function, even though he must have been aware of tables of tangents which had appeared in Europe, mostly taken from Islamic astronomical works. These tables appeared as tables of shadow lengths of a gnomon of particular size and were available not only in texts but also on many astrolabes to make astronomical calculations convenient. Perhaps Regiomontanus did not include these in his book because they were not necessary to his theoretical treatment. He did include a table of tangents, which he called a *tabula fecunda* (fruitful table), in a compilation of astronomical tables he prepared in 1467. In any case, Regiomontanus was able to solve all of the standard problems of trigonometry using just the sine, an extensive table for which, based on a radius (or total sine) of 60,000, he appended to the text.

The first half of Regiomontanus' text deals with plane triangles, the second half with spherical ones. Among his results are various methods for solving triangles. Conceptually, there is nothing particularly new in his methods, but unlike earlier European authors on trigonometry, Regiomontanus often provides clear and explicit examples of his procedures. For example, theorem I-27 shows how to determine the angles of a right triangle if two sides are known, while theorem I-29 shows how to determine the unknown sides of a right triangle, if one of the two acute angles and one side is given. In both cases, Regiomontanus uses his sine table. His example for the second of these theorems assumes that one acute angle is 36° and that the hypotenuse equals 20. Thus the other angle is 54°, and the two sides would be 35,267 and 48,541 respectively, if the hypotenuse were 60,000. Using the rule of three, Regiomontanus calculates that because the hypotenuse is 20, these sides are equal to 11 3/4 and 16 11/60 respectively.

In theorem II-1 Regiomontanus proves the law of sines: "In every rectilinear triangle the ratio of [one] side to [another] side is as that of the right sine of the angle opposite one of [the sides] to the right sine of the angle opposite the other side."[20] Since Regiomontanus' sines are lines in a circle of given radius, his proof of the theorem for the triangle ABG requires circles drawn with centers B and G having equal radii BD and GA respectively (Figure 10.12). Drawing perpendiculars to BG from A and D, intersecting that line at K and H respectively, Regiomontanus then notes that DH is the sine of $\angle ABG$ while AK is the sine of $\angle AGB$, using circles of the same radius. Since $BD = GA$, $\angle ABG$ is opposite side GA, and $\angle AGB$ is opposite side AB, the similarity of triangles ABK and DBH provides the desired result.

In the remainder of Book II of *On Triangles*, Regiomontanus shows how to determine various parts of triangles if certain information is given, such as the ratio of the sides or the length of the perpendicular from a vertex to the opposite side. In two of these theorems, rather than using geometric arguments, he uses arguments from algebra, what he calls "the art of thing and square," because he claims that no "geometric" proof of his result is available. Thus, to find the sides AB, AG of a triangle given that the base $BG = 20$, the perpendicular $AD = 5$, and the ratio $AB : AG = 3 : 5$, Regiomontanus sets segment DE equal to BD and, for algebraic simplicity, sets the unknown segment EG as $2x$ (Figure 10.13). Then $BE = 20 - 2x$, $BD = 10 - x$, and $DG = 10 + x$. Since $AB^2 = BD^2 + AD^2$ and $AG^2 = DG^2 + AD^2$, and since the ratio $AD^2 : AG^2 = 9 : 25$, Regiomontanus concludes that

$$\frac{(10 - x)^2 + 25}{(10 + x)^2 + 25} = \frac{9}{25}.$$

This equation reduces easily to $16x^2 + 2000 = 680x$. Regiomontanus stops his solution here, noting only that "what remains [to be done], the rules of the art [of algebra] show."[21]

Book III of *On Triangles* provides a basic introduction to spherical geometry, including especially many results on great circles. This discussion is preliminary to the standard material on spherical trigonometry contained in the final two books of the text, where Regiomontanus includes the law of sines for both right and arbitrary spherical triangles

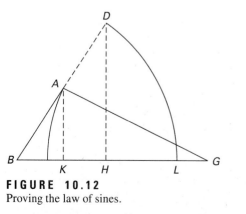

FIGURE 10.12
Proving the law of sines.

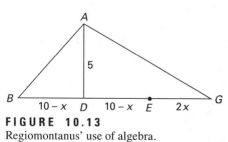

FIGURE 10.13
Regiomontanus' use of algebra.

and the various methods of solving spherical triangles already found in Ptolemy's *Almagest* or in Islamic trigonometries. The Menelaus theorem does not appear explicitly, but the familiar diagram does occur in some of Regiomontanus' proofs. In the discussion of one of the methods of solution of a spherical triangle, Regiomontanus provides a necessary warning in using the law of sines to solve triangles: "I would like you truly to be cautious in selecting the arcs from the given sines, lest [your] parchment be spoiled by [your] repeating the same [thing] a hundred times. For it has often been mentioned that every sine that is less than the whole sine corresponds to two arcs, of which one is greater than a quadrant and the other is smaller. Therefore, when, from a given sine, one wishes to find its arc, it is to be decided whether its arc is greater than a quadrant or less than one."[22]

Regiomontanus' *On Triangles* included examples demonstrating his methods. But all of the examples are purely numerical. Because he was writing a book on trigonometry, not on astronomy, even the theorems on spherical triangles were illustrated only with artificial examples rather than with examples drawn from actual problems in the heavens. Like all of the previous works on trigonometry, this one contained no examples of the application of the plane trigonometrical results to solving triangles on earth.

After Regiomontanus, roughly a score of other works on trigonometry appeared in the last two-thirds of the sixteenth century, many quite similar to his work.[23] Some authors improved his tables and included tables of all of the other trigonometric functions. Generally, these were, like the sine, defined as lengths of certain lines depending on a given arc in a circle of a fixed radius, usually of the form 10^n or 6×10^n. The value of n tended to be larger toward the end of the century as more and more accuracy was required for astronomical calculations. The large radius enabled all values to be given in integers, since decimal fractions were still not in use. George Joachim Rheticus (1514–1574), however, defined the trigonometric functions in his work directly in terms of angles of a right triangle, holding one of the sides fixed at a large numerical value. Rheticus thus called the sine the "perpendiculum" and the cosine the "basis" of the triangle with fixed hypotenuse. Other authors gave other names to the trigonometric functions. The first author to use the modern terms "tangent" and "secant" was Thomas Finck (1561–1656) in his *Geometria rotundi libra XIV* of 1583. He called the three co-functions "sine complement," "tangent complement," and "secant complement." Many of these trigonometry texts gave various numerical examples to illustrate methods of solving plane and spherical triangles, but not until the work of Bartholomew Pitiscus (1561–1613) in 1600 did there appear any problem in such a text explicitly involving the solving of a real plane triangle on earth. Pitiscus, in fact, invented the term "trigonometry." He titled his book *Trigonometriae sive, de dimensione triangulis, Liber* (*Book of Trigonometry, or the Measurement of Triangles*).

Pitiscus intended in the text to show how to measure triangles and, in appendix 2 on Altimetry, he gave trigonometric methods for determining the height BC of a distant tower. In Figure 10.14 a quadrant is used to measure $\angle AKM = \angle ABC = 60°20'$. The distance AC from the observer to the tower is measured as 200 feet. Pitiscus sets up the proportion $\sin 60°20' : AC = \sin 29°40' : BC$ and then calculates that $BC = 113\frac{80,204}{86,892}$, or, approximately, 114 feet. This calculation uses Pitiscus' sine table, calculated to a radius of 100,000. He gives a second procedure using his tangent table, in which the required proportion is $AC : 100,000 = BC : \tan 29°40'$. The major difference between Pitiscus' methods and current ones is that he always has to adjust for the fact that his trigonometric values are lengths of certain lines in a particular circle. The trigonometric ratios in use today had yet to arrive.

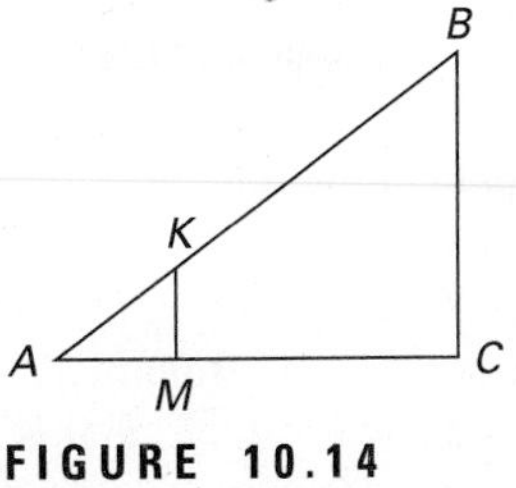

FIGURE 10.14
Measuring the height of a tower.

10.3.2 Nicolaus Copernicus and the Heliocentric System

The trigonometry of the fifteenth century exemplified by Regiomontanus' work, even without trigonometric ratios, provided the mathematics necessary to attack the astronomical problems of the day. Some of these problems, discussed in Regiomontanus' edition of the *Almagest*, were ones involving fundamental questions about Ptolemy's system, still the accepted view of the nature of the universe. Islamic astronomers through the centuries had noted certain discrepancies between Ptolemy's predictions and their own observations and had made minor adjustments to some of Ptolemy's details. But the Christian view of the universe at the beginning of the Renaissance was still based on the views of Aristotle and Ptolemy to the effect that the universe was composed of a system of nested spheres centered on the earth and that it was the rotation of these spheres, to which were attached the planets, which caused the appearances in the heavens. The various additional parts of the model, such as epicycles and eccenters, were all somehow imbedded in the various spheres. This basic view of the universe can perhaps most easily be seen in Dante's *Divine Comedy* (1328), which describes the poet's journey through each of the celestial spheres holding the planets and stars up to the final immovable sphere containing the throne of God.

By the fifteenth century, however, astronomers were having very serious difficulties accepting Ptolemy's system in detail. One type of error was pointed out by Regiomontanus, who noted that Ptolemy's theory of the moon required the observed size to vary considerably more than it really does. More importantly, because even small errors tended to accumulate over the centuries, astronomers found many occasions when Ptolemy's predictions of planetary positions or lunar eclipses were greatly in error. And as European explorers set out on voyages around the globe, they needed improved navigational techniques that could come only through correct astronomical tables. In addition, through these explorations, Europeans found so many parts of the world previously unknown to them that they realized that Ptolemy's geography was also in error. The way was prepared for believing that the fundamentals of his astronomy could be wrong.

The Catholic Church was also aware by the early Renaissance that the Julian calendar, used since the time of the Roman Empire, had serious inadequacies. In particular, since the true solar year was 11 1/4 minutes less than the 365 1/4 days on which that calendar was based, the cumulative errors threatened to change the relationship of the calendar months to the seasons. For instance, according to Church law, Easter was to be celebrated on the first Sunday after the first full moon following the vernal equinox. The equinox was always reckoned as March 21, but by the sixteenth century it actually took place about March 11. Without correction, Easter would eventually arrive in the summer rather than in the spring. When calendar reform became an official Church project, however, the astronomers advised that existing astronomical observations were inadequate and did not yet permit an accurate, mathematically based calendar change.

Among the astronomers who refused an invitation to participate in the reform of the calendar was Nicolaus Copernicus (1473–1543), who, having studied Ptolemy's system in great detail and having become aware of all its inaccuracies, came to the conclusion that it was impossible to patch up the earth-centered approach any longer. "[The astronomers] have not been able to discover or deduce from [their hypotheses] the chief thing, that is the form of the universe, and the clear symmetry of its parts. They are just like someone including in a picture hands, feet, head, and other limbs from different places, well painted

---

Biography   Nicolaus Copernicus (1473–1543)

Copernicus was born in Torun in West Prussia (now Poland) into the family of a wealthy merchant and was sent at the age of eighteen to study at the University of Cracow (Figure 10.15). Upon leaving Cracow, he was appointed to a clerical post through the influence of his uncle, the Bishop of Warmia. He therefore not only received a salary but also was permitted to travel to Italy to study at Bologna and Padua over the next ten years. Finally returning home, he spent the remainder of his life in Warmia, serving as a Canon of Frauenburg Cathedral. The job not being a particularly demanding one, he was generally free to

concentrate on his study of astronomy and was able to complete his manuscript of *De Revolutionibus* by about 1530. He was unwilling, however, to publish the work. In about 1514, he had already written a brief outline of his system, *The Commentariolus*, which was circulated to various scholars. But it was not until George Rheticus, Professor of Mathematics at the University of Wittenberg, arrived in Frauenburg in 1539 to learn first hand about Copernicus' system, that Copernicus was finally persuaded to allow his masterwork to be published.

---

indeed, but not modelled from the same body, and not in the least matching each other, so that a monster would be produced from them rather than a man."[24] To redo the "painting" and eliminate the monster, Copernicus decided to read all the opinions of the ancients to determine whether anyone had proposed a system of the universe different from the earth-centered one. Having discovered that some Greek philosophers had proposed a sun-centered (**heliocentric**) system in which the earth moves, Copernicus explored the consequences of reforming the system under that assumption: "Thus assuming the motions which I attribute to the Earth, . . . I eventually found by long and intensive study that if the motions of the wandering stars are referred to the circular motion of the Earth and calculated according to the revolution of each star, not only do the phenomena agree with the result, but also it links together the arrangement of all the stars and spheres, and their sizes, and the very heaven, so that nothing can be moved in any part of it without upsetting the other parts and the whole universe."[25]

Copernicus' fundamental treatise in which he expounded his system of the universe was *De Revolutionibus Orbium Coelestium (On the Revolutions of the Heavenly Spheres)*, a book that represented the work of a lifetime but that was only published in 1543, the year of his death. This book sets forth the first mathematical description of the motions of the heavens based on the assumption that the earth moves, for, as Copernicus notes in his preface, "Mathematics is written for mathematicians."[26] This work, following very closely the model of Ptolemy's *Almagest*, is a very technical work in which the author uses detailed mathematical calculation, based on the assumption that the sun is at the center of the universe and buttressed by the results of observations taken by Copernicus and his predecessors, to describe the orbits of the moon and the planets and to show how these orbits are reflected in the positions observed in the skies. Copernicus sketched his theory very briefly in the first book of *De Revolutionibus* and presented the simplified diagram of the sun in the center of seven concentric spheres, one each for the six planets, including the earth, and one for the fixed sphere of the stars (Figure 10.16).

It should be emphasized that Copernicus, like his predecessors, conceived of the system of the universe as a series of nested spheres containing the planets, rather than as

FIGURE 10.15
Copernicus and his system on a Hungarian stamp.

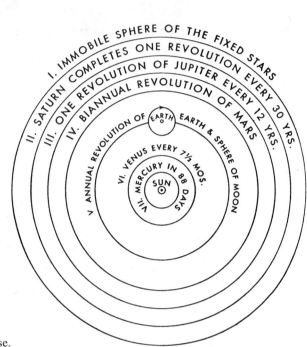

FIGURE 10.16
Copernicus' system of the universe.

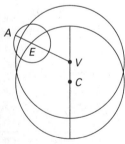

FIGURE 10.17
Ptolemy's equant: The planet *A* moves on the epicycle centered at *E*; *E* travels around the circle centered at *C* in such a way that the radius vector *VE* moves uniformly around *V*, the equant point.

empty space through which the planets travel in circles. Copernicus had no physics to keep the planets in their orbits. Spheres for Copernicus, however, as for Aristotle, had a natural motion which needed no other physical basis: "The movement of a sphere is a revolution in a circle, expressing its shape by the very action, in the simplest of figures, where neither beginning nor end is to be found, nor can the one be distinguished from the other, as it moves always in the same place."[27] In fact, one of Copernicus' aims in his reform of Ptolemy's work was to return to one of the classic principles of astronomy, that all heavenly motion must be composed of uniform motion of circles about their centers. Copernicus believed Ptolemy had violated that principle by accounting for certain aspects of a planet's motion through the use of the equant, a point within the planet's orbit around which the radius vector to the center of the planet's epicycle revolved uniformly (Figure 10.17). The uniform motion in that case was not about the center of the circle on which the epicycle traveled. Islamic astronomers at Maragha, led by Nasir al-Dīn al-Ṭūsī, had also been bothered by this problem. Copernicus adapted their solution in his own work. But the Islamic astronomers had not taken Copernicus' major step of challenging the centrality and immovability of the earth.

Copernicus himself does not—and could not—present any real evidence for either the earth's daily rotation on its axis or its yearly revolution about the sun. For the first motion, he simply argues that it is more reasonable to assume that the relatively small earth rotates rather than the immense sphere of the stars. For the second motion, his argument is in essence that the qualitative behavior of the planets can more easily be understood by attributing part of their motion to the earth's own yearly revolution. Thus, retrogression can be explained in terms of the combined orbital motions of the earth and the planet rather than by an epicycle (Figure 10.18). The observed variation in the planets' distances from the earth also is more easily understood in terms of the two orbits. Copernicus answers the objection to the earth's motion around the sun, that it would cause the fixed stars to appear different at different times of the year (the so-called annual parallax), by assuming that the radius of the earth's orbit is so much smaller than the radius of the sphere of the stars that no such parallax could be observed. Thus one of the effects of Copernicus' theory was vastly to increase the size of the universe.

After his basic introduction to the new system, Copernicus follows his mentor Ptolemy by presenting an outline of the plane and spherical trigonometry necessary to solve the mathematical problems presented by the movements of the celestial bodies. Despite the advances in trigonometric technique now available in Europe following the work of Regiomontanus, Copernicus' own treatment stays very close to that of the second century astronomer, even to the use of chords. He does, however, make two concessions to the 1400 years of work since the time of Ptolemy. First, he uses 100,000 for his circle radius (now that Arabic numerals were in general use) rather than the 60 used by the ancients. Second, his table does not give the chords of the various arcs but instead half the chords of twice the arcs "because the halves come more frequently into use in demonstration and calculation than the whole chords do."[28] Copernicus does not use the now common term "sine." Nor, in his brief outline of methods of solving plane triangles does he mention the law of sines. His basic method of solution involves drawing appropriate perpendiculars and then dealing with right triangles.

Copernicus follows his treatment of plane triangles with a collection of results on spherical triangles. Again, he uses only chords (or half chords). Not only does he fail to give the law of sines for spherical triangles but also he does not give any of the simplifications of Menelaus' results which come from using the other trigonometric relations. Nevertheless, his discussion of how to solve spherical triangles is comprehensive and is sufficient for the astronomical work to follow.

In the remaining books of his treatise, Copernicus uses his new sun-centered model, along with both ancient and modern observations, to calculate the basic parameters of the

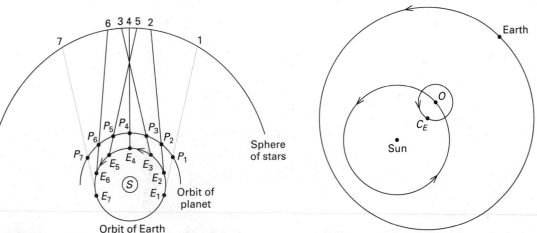

FIGURE 10.18
Retrograde motion for a planet outside the orbit of the earth. The observed positions of the planet against the sphere of the stars are marked, in order, 1, 2, 3, 4, 5, 6, 7. Retrogression takes place between 3 and 5.

FIGURE 10.19
Earth revolves around C_E which rotates on an epicycle centered at O, where O revolves around the sun.

FIGURE 10.20
The four hundredth anniversary of the Gregorian calendar on a Vatican stamp.

orbits of the moon and the planets. It is in reading these later books that one sees that moving the center of the universe away from the earth has not simplified Ptolemy's picture very much. Copernicus found that simply placing the planets on sun-centered spheres did not satisfy the requirements of observation. Thus he, like Ptolemy, had to introduce more complicated devices such as epicycles. For example, Copernicus' calculations showed that the center of the earth's (circular) orbit was not the sun, but a point C_E in space that revolved on an epicycle whose center O revolved about the sun (Figure 10.19). Similarly, the centers of the various planetary orbits were neither located at the sun nor even at the center of the earth's orbit. In the end, the full system as described in *De Revolutionibus* was of the same order of complexity as that of Ptolemy.

The mathematical details of Copernicus' work made it unreadable to all but the best astronomers of his day, its primary audience. Over the next several decades these mathematicians found that calculations of astronomical phenomena were simplified by applying Copernicus' theory and techniques. It was unnecessary to believe in the movement of the earth to use these techniques. Therefore, many people, both astronomers and educated laymen, took Copernicus' work merely as a mathematical hypothesis and not as a physical theory. In fact, the preface to the printed text of *De Revolutionibus*, written by Andreas Osiander, the Lutheran theologian who saw the book through the press, claimed that Copernicus' views on the earth's motion should not be taken as true, but only as a hypothesis for calculation, "since the true laws cannot be reached by the use of reason."[29]

During the latter half of the sixteenth century, however, various churchmen, particularly Protestant clerics deeply involved in the fierce conflict with the Roman Catholic Church, began to express severe opposition to Copernicus' ideas because they explicitly contradicted various Biblical passages asserting the earth's stability. These Protestant leaders believed that the Roman Church had departed greatly from the views expressed in the Bible. They vehemently rejected any doctrines seen as deviating from the literal words of Scripture. During this same period the Catholic Church itself had little to say about Copernicus' work. *De Revolutionibus*, in fact, was taught at various Catholic universities, and the astronomical tables derived from it provided the basis for the reform of the calendar promulgated for the Catholic world by Pope Gregory XIII in 1582 (Figure 10.20). Ironically, it was not until the seventeenth century, after most astronomers were convinced of the earth's movement by new evidence and a better heliocentric theory than that of Copernicus, that the Catholic Church brought its full power to bear against the heresy then seen to be represented by the moving earth.

10.3.3 Tycho Brahe

One astronomer who used Copernicus' work as the basis for astronomical calculations was Erasmus Reinhold (1511–1553). In 1551 he issued the first complete set of astronomical tables prepared in Europe for over three centuries, generally called the Prutenic tables after his patron, the Duke of Prussia. These tables were markedly superior to the older ones, partly because they were based on more and better data. Nevertheless, they were not intrinsically more accurate than tables based on Ptolemy's work. There were still errors of a day or more in the prediction of lunar eclipses.

One way to improve the results no matter how one calculated the tables, however, was to have better observations. Tycho Brahe (1546–1601) was one astronomer who devoted

FIGURE 10.21
Tycho Brahe's observatory, with one of his quadrants and the nova of 1572, on a stamp from Ascension Island.

much of his life to making these observations. To do so obviously required excellent instruments, which in return required funds. It was fortunate that in 1576 he was able to convince King Frederick II of Denmark to allow him the use of the island of Hveen near Copenhagen and to provide him with funds for building a magnificent observatory and also for hiring the assistants necessary to provide year-round observations using the newly constructed instruments (Figure 10.21). Tycho was the first astronomer to realize the necessity for making continuous observations of the various planets. Although he eventually left Denmark and moved to Prague to work for the Austrian emperor Rudolph II, he was able to accumulate enormous amounts of data over a 25 year period, generally to an accuracy of a couple minutes of arc, an accuracy far in excess of the best work of any of the ancients.

Two of Tycho's most important series of observations convinced him that the Ptolemaic system with its Aristotelian philosophy could not be correct. First, beginning in late 1572 he tracked for 16 months a new object which had appeared in the heavens, a nova. Because this object did not move with respect to the sphere of the stars—Tycho demonstrated this by very precise observation—he concluded that it belonged to the region of the fixed stars. Hence, despite Aristotle, change was possible in the heavens, and therefore one distinction between the earth and the heavens was removed. The possibility of change in the heavens was further confirmed by his observation of a comet in 1577. Again, by a comparison of the parallax of the comet with that of the moon and the planets, he concluded that the comet lay beyond the moon and that it revolved around the sun at a distance greater than that of Venus. Since its distance from the sun apparently varied greatly during the course of his observations, Tycho further concluded that the heavens could not be filled by solid spheres carrying the planets. There must in fact be space between the planets in which another heavenly object could travel.

10.3.4 Johannes Kepler and Elliptical Orbits

Tycho was primarily an excellent observer, rather than a theoretician. He did devise a model of the universe "intermediate" between that of Ptolemy and Copernicus in which all of the planets except the earth traveled around the sun, while the whole system revolved around the central immovable earth, but he was not able to elaborate it mathematically. Johannes Kepler (1571–1630), who worked with Tycho for the final two years of his life in Prague, was the astronomer able to use the mass of Tycho's observations to construct a new heliocentric theory which could accurately predict heavenly events without the elaborate machinery of epicycles.

It was perhaps his theological training, combined with a philosophical bent, which provided Kepler with the goal from which he never wavered, of discovering the mathematical rules God used for creating the world. As he stated in his earliest work, the *Mysterium Cosmographicum* (*The Secret of the Universe*) of 1596, "Quantity was created in the beginning along with matter."[30] In a note to the second edition of 1621, Kepler clarified what he meant: "Rather the ideas of quantities are and were coeternal with God, and God himself . . . On this matter the pagan philosophers and the Doctors of the Church agree."[31] Throughout his life, Kepler attempted through both philosophical analysis and prodigious calculation to demonstrate the numerical relationships with which God had created the world. His goal appeared to be nothing less than to reconfirm on a higher level the

---

Biography   Johannes Kepler (1571–1630)

Kepler was born in Weil-der-Stadt in southwest Germany and studied in the University of Tübingen where he became acquainted with Copernicus' theory and convinced himself that in essence it represented the correct system of the world. Although he had originally planned to become a Protestant minister, fate intervened, and he was recommended by the university to fill a job as mathematics professor at the Protestant school in the Austrian town of Graz. When the school was closed several years later and all Protestant officials were exiled, an exception was made in Kepler's case. He was allowed to return and to spend time thinking about mathematics and astronomy.

Kepler knew that to work out in complete detail a correct version of Copernicus' theory, he had to have access to the observations of Tycho Brahe. He therefore began a correspondence with the Dane which finally resulted in his being appointed his assistant in Prague by Emperor Rudolph II. Although Tycho died about 18 months after Kepler's arrival, Kepler had by this time learned enough about Tycho's work to be able to use the material in working on his own major project. He was himself appointed as Imperial Mathematician to succeed Tycho Brahe and spent the next 11 years in Prague (Figure 10.22).

FIGURE 10.22
Kepler and his system of the universe on a Hungarian stamp.

Pythagorean doctrine that the universe is made up of number. Taking as his starting point Copernicus' placing of the sun at the center of the universe, he was able to discover the three laws of planetary motion, today known as Kepler's laws, and also many other relationships that today we tend to dismiss as mystical.

Kepler discusses one of these relationships in great detail in the *Mysterium Cosmographicum*: Why are there precisely six planets? Because "God is always a geometer," the Supreme Mathematician wanted to separate the planets with the regular solids. Euclid had proved that there could be only five such solids, so Kepler took this as the reason that God chose to provide just six planets. He then worked out the idea that between each pair of spheres containing the orbits of adjacent planets there was inscribed one of the regular solids (Figure 10.23). Thus, inside the sphere of Saturn was to be inscribed a cube which in turn circumscribed the orbit of Jupiter. Similarly, between the orbits of Jupiter and Mars was a tetrahedron, between Mars and Earth a dodecahedron, between Earth and Venus an icosahedron, and between Venus and Mercury an octahedron. These solids lay in the interspherical spaces, and their sizes provided a measure of the relationship between the sizes of the various planetary orbits. For example, Kepler noted that the diameter of Jupiter's orbit is triple that of Mars, while the ratio of the diameter of the sphere circumscribed about the tetrahedron is triple that of the sphere inscribed in the tetrahedron. Not all of the values came out exactly correct. There was still some discrepancy. But even this fact did not bother Kepler too much. He gave various reasons why the values could not be expected to be exact, including the fact that even the data from the Prutenic tables was not entirely accurate. He was so convinced of the correctness of his basic views that such discrepancies were of little moment. Kepler's views on this matter were not a mere function of his youth. In fact, he returned time and again to this basic proposition, each time attempting to adduce new reasons for its correctness.

Kepler was well-schooled in music. Thus he would have been very familiar with the Pythagorean ratios of string lengths which give consonant harmonies: a ratio of 1 : 2 is

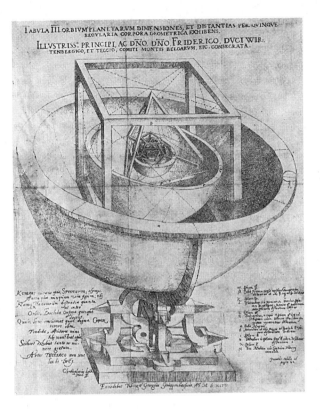

that of an octave, of 2 : 3 a fifth, of 3 : 4 a fourth, and so on. In his *Harmonices Mundi*
(*Harmonies of the World*, 1619), Kepler attempted to assign these harmonic ratios to
various numbers connected with the different planets. First he tried the periods of revolution, but these did not give any harmonic ratios. Next, he tried the volumes of the planets,
the greatest and smallest solar distances, the extreme velocities, and the variations in time
needed by a planet to cover a unit length of its orbit. Nothing appeared to work. Finally,
after a lengthy argument, Kepler hit on the "right" numbers, the apparent daily angular
movements of the planets as seen from the sun. Thus the daily movement of Saturn at
aphelion (the point on its orbit farthest from the sun) is $1'46''$, while its daily movement at
perihelion (the point closest to the sun) is $2'15''$. The ratio between these two values is
approximately 4 : 5, a major third. The corresponding ratio for Mars is $26'14'' : 38'1''$ or
approximately 2 : 3, a fifth. Not only did Kepler find consonances between the extreme
movements of the individual planets, but he also found them between movements of different planets. The ratio of Saturn at perihelion to Jupiter at aphelion turned out to be
1 : 2, an octave. Further, when he transposed a particular set of these relations into a
common key, Kepler found a major scale beginning with the aphelion of Saturn and a
minor one beginning with Saturn's perihelion. Kepler included in his book the various
notes "played" by the planets, both singly and together, concluding with several multipart
harmonies: "Accordingly the movements of the heavens are nothing except a certain ever-
lasting polyphony . . . Hence it is no longer a surprise that man, the ape of his Creator,

should finally have discovered the art of singing polyphonically . . . in order that he might play the everlastingness of all created time in some short part of an hour by means of an artistic concord of many voices and that he might to some extent taste the satisfaction of God the Workman with His own works, in that very sweet sense of delight elicited from this music which imitates God."[32]

One may well wonder if Kepler could indeed be thought of as a scientist, given his propensity toward what we might today call mysticism. The answer, however, is a resounding yes. Kepler was responsible for some of the most important astronomical discoveries of his time. There is a direct line from his three laws of planetary motion to the fundamental work of Newton on the laws of motion.

Kepler announced in his *Mysterium Cosmographicum* that among his goals was to discover the "motion of the circles" of the planets, that is, to determine their orbits. He gave numerous arguments in that work for the basic correctness of the Copernican system, but by the end of the century he realized that Copernicus' mathematical details did not give the complete solution to the problem. To correct Copernicus' work, Kepler knew that he needed better observational data, data which could only come from Tycho Brahe. With these finally in hand by 1601, Kepler could proceed to determine the exact details of the planetary orbits. He began with the case of Mars, because that planet's orbit had always been the most difficult to comprehend. If he could understand Mars' orbit, Kepler believed, he could understand them all.

In his *Astronomia Nova* (*New Astronomy*) of 1609, Kepler takes us through his eight years of detailed calculations, false starts, stupid mistakes, and continued perseverance to calculate the orbit of Mars. He first decided that he needed accurate parameters for the earth's own orbit, because his overriding Copernican theory was that the motion of Mars was viewed from a moving earth. Kepler took the earth's orbit as a circle with radius r centered on a point C, with the sun at a point S making $CS = e = 0.018r$ (Figure 10.24). (The orbit of the earth is very close to circular, so the assumption that it was a circle did not lead Kepler astray.) Furthermore, there was another point A on the diameter CS, with $AC = CS$, such that $\angle EAQ$ varied uniformly with time, where Q is the earth's aphelion. In other words, he reintroduced the equant which Copernicus had rejected. Because the earth moved with a constant angular velocity on its orbit with respect to A, its linear velocity necessarily changed as its distance from the sun changed. Kepler showed that the earth's velocity near aphelion and perihelion varied inversely with its distance from the sun, and then generalized this result to the rest of the orbit. (Unfortunately, that rule turned out to be incorrect. As Kepler understood later, it is the component of the planet's velocity perpendicular to the radius vector which varies inversely with the distance from the sun.)

Unlike Ptolemy or Copernicus, however, Kepler was not interested in only the pure mathematics of the celestial motions, that is, in "saving the appearances," but in the physics as well. He was trying to describe the actual orbit of the earth through space and so wanted to know what caused the earth to move, what kept it in its orbit, and why the velocity changed with the distance to the sun. Having read the work of William Gilbert, *On Magnets* (1601), Kepler settled on the fact that some force emanating from the sun acts on the planet and sweeps it around in its orbit. He could understand this force acting on the earth as it moved around in a circle much better than he could see it acting on a planet moving on an epicycle. It also made sense that, like magnetic force, the sun's force weakened with distance, so that the planet's velocity was smaller at a greater distance. This

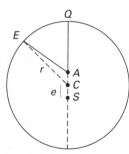

FIGURE 10.24
Kepler's assumption for the earth's orbit.

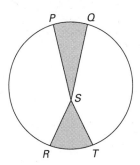

FIGURE 10.25
Kepler's second law: A plant sweeps out equal areas in equal times. The time of the planet's motion from Q to P is equal to that from R to T when area SPQ is equal to area SRT.

change from a mathematical to a physical point of view was one of the reasons Kepler felt comfortable in reintroducing the equant as well as rejecting epicycles.

Returning to the motion of Mars, Kepler started by assuming its orbit was circular because he knew that this assumption provided approximately correct results. His aim was to calculate the relation between the length of arc QP traveled by the planet after aphelion Q and the time it takes to traverse that arc. He was aware that the planet moved more slowly the farther it was from the sun. Since the calculation of the exact relationship between velocity and arc, however, was beyond his capabilities, Kepler resorted to approximation. His assumption, now taken for Mars as well as for the earth, that the planet's velocity varied inversely with the length of the radius vector, implied that the time required to pass over an (infinitesimal) arc was proportional to that vector. The time could therefore be represented by the radius vector, with appropriate choice of units. Kepler then argued that the total time required to pass over a finite arc QP could be thought of as the sum of the radius vectors making up that part of the circle, or as the area swept out by the radius vector (Figure 10.25). Kepler realized that such an infinitesimal argument was not rigorous, but he stated it anyway as a law based on the incorrect circular orbit and the incorrect velocity law: The radius vector sweeps out equal areas in equal times. This law is generally referred to as Kepler's second law, because it is today regarded as a supplement to the first law. Interestingly enough, Kepler made no attempt to prove it differently even when he discovered that the correct planetary orbit was an ellipse.

That the shape of the orbit is an ellipse is the content of Kepler's first law. Kepler informs us fully how he discovered this law as well. Having worked out the orbit of the earth, he made various calculations of the distances of Mars to the supposed center of its orbit and found that they were larger near aphelion and perihelion and smaller during the remainder of the orbit. Thus the circularity of the orbit was impossible. Kepler concluded that the orbit had to be an oval of some sort. It was somewhat strange to reject the comforting circularity of the Greeks and replace it with a rather vaguely shaped oval, because such a curve would seemingly destroy all possibility of the "harmony of the spheres" for which Kepler had been searching. Nevertheless, Kepler began the long process of calculating the exact shape of the oval.

After two years of calculation, the result appeared to Kepler virtually by accident. To aid in certain computations, he had been approximating the oval by an ellipse. He noted that the distance AR between the circumference of the circle and the end of the minor axis of the ellipse was equal to 0.00429 (the radius of the circle being set at 1), which turned out to be $(1/2)e^2$ where $e = CS$ was the distance between the center of the circle and the sun (Figure 10.26). It followed that the ratio

$$CA : CR = 1 : 1 - \frac{e^2}{2} \approx 1 + \frac{e^2}{2} = 1.00429.$$

What struck Kepler about this number was that he had seen it before. It was equal to the secant of 5°18′, the value of the angle ϕ between the directions AC and AS, where A is the point on the circle 90° away from the aphelion point Q. The secant in this case is the ratio of the length of the radius vector to its projection onto a diameter. Realizing that $CA : CR \approx SA : SB \approx SA : CA$, Kepler then had the brilliant inspiration that when the angle between CQ and a direction CP had any value β (and not just 90°), the ratio of the distance SP to that actual sun-Mars distance was also the ratio of SP to its perpendicular projection PT

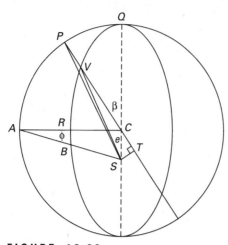

FIGURE 10.26
Kepler's derivation of the elliptical orbit.

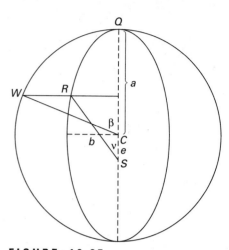

FIGURE 10.27
Kepler's proof that the curve of the orbit is
an ellipse.

on a diameter. In other words, he realized that the actual sun-Mars distance was PT, where
$PT = PC + CT = 1 + e \cos \beta$. The remaining question for Kepler was how to lay off
this distance. Kepler first decided to lay it off with one end at the sun and the other on the
radius vector PC, that is, to make $SV = PT$. Unfortunately, the curve so traced turned out
to be not quite in accord with observation and, in fact, was not exactly an ellipse, which,
for various reasons, Kepler was now convinced was the true orbit.

Kepler finally discovered the correct result, that the distance given by $\rho = 1 + e \cos \beta$
should be laid off from the sun so that the endpoint is on a line perpendicular to the line
CQ, where β is the angle that CQ makes with the line from C to the intersection W of that
perpendicular with the auxiliary circle (Figure 10.27). (One should note that the difference
between Kepler's first idea and this one is extremely small, producing discrepancies of at
most about 5′ of arc.) Kepler was able to demonstrate that the curve he now produced was
an ellipse using an argument summarized here in modern notation. Assume that an ellipse
is centered at C with $a = 1$ and $b = 1 - \frac{e^2}{2}$, where $e = CS$. This ellipse can be thought
of as being formed from the circle of radius 1 by reducing all the ordinates perpendicular
to QC in the proportion b. If v represents the angle at S subtended by the arc RQ, then
$\rho \cos v = e + \cos \beta$ and $\rho \sin v = b \sin \beta$. Squaring the two equations and adding gives

$$\rho^2 = e^2 + 2e \cos \beta + \cos^2 \beta + \left(1 - \frac{e^2}{2} \right)^2 \sin^2 \beta$$

$$= e^2 + 2e \cos \beta + 1 - e^2 \sin^2 \beta + \frac{e^4}{4} \sin^2 \beta.$$

Neglecting the term in e^4 then produces the result

$$\rho^2 = 1 + 2e \cos \beta + e^2 \cos^2 \beta$$

$$= (1 + e \cos \beta)^2.$$

Thus the equation of the ellipse can be written as $\rho = 1 + e \cos \beta$, exactly the same equation as already derived for the curve of the orbit itself. In addition, the distance c of the center of the ellipse from the focus is given by

$$c^2 = 1 - b^2 = 1 - \left(1 - \frac{e^2}{2}\right)^2 = e^2,$$

if again the term in e^4 is ignored. It follows that the sun is at one focus of the ellipse and that e is the eccentricity. We have now derived Kepler's first law of planetary motion: A planet travels in an ellipse around the sun with the sun at one focus. Kepler himself, having derived the law for the case of Mars, merely checked it briefly for the other planets before asserting its general validity.[33]

Kepler's third law appears for the first time in the *Harmonice Mundi*, stated as an empirical fact. In some sense it was a culmination of the work begun earlier in the *Mysterium Cosmographicum* because it provided another answer to the general questions Kepler had asked regarding the size and motions of the orbits: "It is absolutely certain and exact that the ratio which exists between the periodic times of any two planets is precisely the ratio of the $\frac{3}{2}$th power of the mean distances [of the planet to the sun]."[34] Kepler discovered the law by studying more of Tycho Brahe's measurements, but never gave any derivation of it from other principles.

Kepler's three laws of planetary motion had great consequences in the development of astronomical as well as of physical theory. Their discovery provides an excellent example of the procedures used by scientists. They need some theory to begin with, but then must always compare the results of the theory with the results of observation. If they have confidence in their observations, and these do not agree with the predictions of their theory, they must modify the theory. Kepler did this often until he finally reached theoretical results agreeing with his observations. He spent years performing the necessary calculations. Toward the end of his life, however, the invention of logarithms greatly simplified Kepler's calculations as well as those of other astronomers.

10.4 LOGARITHMS

The idea of the logarithm probably had its source in the use of certain trigonometric formulas which transformed multiplication into addition or subtraction. For example, the formula $\sin \alpha \sin \beta = \frac{1}{2}[\cos(\alpha - \beta) - \cos(\alpha + \beta)]$ was employed in the sixteenth century to simplify some of the tedious calculations made by astronomers. If one needed to solve a triangle using the law of sines, a multiplication and a division were required. Because sines were generally calculated to seven or eight digits, these calculations were long and errors were often made. Astronomers realized that it would be simpler and reduce the number of errors if one could replace the multiplications and divisions by additions and subtractions. A second, more obvious, source of the idea of a logarithm was probably found in the work of such algebraists as Stifel and Chuquet, who both displayed tables relating the powers of 2 to the exponents and showed that multiplication in one column corresponded to addition in the other. But because these tables had increasingly large gaps, they could not be used for the necessary calculations. Around the turn of the seventeenth century, however, two men working independently, the Scot John Napier (1550–1617) and the Swiss Jobst Bürgi (1552–1632) came up with the idea of producing an extensive table which would allow one to multiply any desired numbers together (not just powers of 2) by performing additions. Napier published his work first.

10.4.1 The Idea of the Logarithm

Napier's logarithmic tables first appeared in 1614 in a book entitled *Mirifici Logarithmorum Canonis Descriptio* (*Description of the Wonderful Canon of Logarithms*). This work contained only a brief introduction, showing how the tables were to be used. His second work on logarithms, describing the theory behind the construction of the tables, *Mirifici Logarithmorum Canonis Constructio* (*Construction of the Wonderful Canon of Logarithms*) appeared in 1619, two years after his death. In this latter work appears his imaginative idea of using geometry to construct a table for the improvement of arithmetic.

Realizing that astronomers' calculations involved primarily trigonometric functions, especially sines, Napier aimed to construct a table by which multiplications of these sines could be replaced by addition. For the definition of logarithms Napier conceived of two number lines, on one of which an increasing arithmetic sequence, $0, b, 2b, 3b, \ldots$ is represented, and on the other a sequence whose distances from the right endpoint form a decreasing geometric sequence, $ar, a^2r, a^3r, \ldots$, where r is the length of the second line (Figure 10.28). (Napier chose r to be 10,000,000, because that was the radius for his table of sines, and a to be a number smaller than but very close to 1 (Side Bar 10.1).) The points on this line can be marked $0, r - ar, r - a^2r, r - a^3r, \ldots$. For Napier, these points generally represent sines of certain angles.

Napier now considers points P and Q moving to the right on each line as follows: P moves on the upper line "arithmetically" (that is, with constant velocity). Thus P covers each equal interval $[0, b], [b, 2b], [2b, 3b], \ldots$ in the same time. Q moves on the lower line "geometrically." Its velocity changes so that it too covers each (decreasing) interval $[0, r - ar], [r - ar, r - a^2r], [r - a^2r, r - a^3r], \ldots$ in the same time. Napier then shows that a point moving geometrically has its velocity over each interval proportional to the distance of the beginning of that interval from the right end of the line. The distances traveled in each interval form a decreasing geometrical sequence $r(1 - a), ar(1 - a), a^2r(1 - a), \ldots$, each member of the sequence being the same multiple of the distance of the left endpoint of the interval to the right end of the line. Because distances covered in equal times have the same ratios as the velocities, Napier's conclusion follows. It appears that Napier initially thought of the velocity of the lower point as changing abruptly when it passed each marked point, remaining constant in each of the given intervals. In his

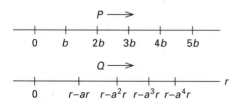

FIGURE 10.28
Napier's moving points.

Napier is primarily responsible for the introduction of our modern notation for decimal fractions. Stevin had detailed the idea, along with a suggestion for a notation. But Napier, near the beginning of the *Constructio*, after noting that accuracy of computation requires the use of large numbers like 10,000,000 as the base for a table of sines, wrote: "In computing tables, these large numbers may again be made still larger by placing a period after the number and adding ciphers. . . . In numbers distinguished thus by a period in their midst, whatever is written after the period is a fraction, the denominator of which is unity with as many ciphers after it as there are figures after the period."[35] For example, he writes, 25.803 is the same as $25\frac{803}{1000}$ and 9999998.0005021 means $9999998\frac{5021}{10,000,000}$. The publication of Napier's tables, in which these decimal fractions appeared, soon resulted in their general use throughout Europe. It had taken about 400 years since the introduction into Europe of the Hindu-Arabic numbers for the complete decimal place-value system to be generally accepted.

---

definition of logarithm, however, Napier smoothed out these changes by considering the second point's velocity as changing continuously (without, naturally, using that terminology). Thus a point moves geometrically if its velocity is always proportional to its distance from the right end of the line. For Napier, "the logarithm of a given sine is that number which has increased arithmetically with the same velocity throughout as that with which radius began to decrease geometrically, and in the same time as radius has decreased to the given [number]."[36] In other words, if the upper point P begins moving from 0 with constant velocity equal to that with which the lower point Q also begins moving (geometrically) from 0, and if P has reached y when Q has reached a point whose distance from the right endpoint (radius) is x, then y is said to be the **logarithm** of x.

In modern calculus notation, Napier's idea is reflected in the differential equations

$$\frac{dx}{dt} = -x, \quad x(0) = r; \qquad \frac{dy}{dt} = r, \quad y(0) = 0.$$

The solution to the first equation is $\ln x = -t + \ln r$, or $t = \ln \frac{r}{x}$. Combining this with the solution $y = rt$ of the second equation shows that Napier's logarithm y (here written as $y = \text{Nlog } x$) may be expressed in terms of the modern natural logarithm as $y = \text{Nlog } x = r \ln \frac{r}{x}$. Napier's logarithm is thus closely related to the natural logarithm. It does not, however, share the common properties of the natural logarithm since, for example, its value decreases when the value of x increases.

10.4.2 The Use of Logarithms

Although Napier's definition is somewhat different from the modern one, he nevertheless is able to derive important properties of logarithms analogous to those of our modern logarithm as well as show how to construct a table of logarithms. He begins by noting that the definition implies immediately that $\text{Nlog } r = 0$, for the upper point will not have moved at all. Napier in fact realized that he could have assigned 0 to be the logarithm of any fixed number, but, he says, "it was best to fit it to the whole sine, that the addition or subtraction of the logarithm which is most frequent in all calculations, might never after be any trouble to us."[37] Similarly, if $\alpha/\beta = \gamma/\delta$, then Nlog α − Nlog β = Nlog γ − Nlog δ. This result also follows from the definition because the geometrical motion of the lower point implies that its time to travel from α to β equals its time to travel from γ to δ. From this result follow rules enabling one to use logarithms in calculation. For example, if $x : y = y : z$, then 2Nlog y = Nlog x + Nlog z, and if $x : y = z : w$, then Nlog x + Nlog w = Nlog y + Nlog z. On the other hand, Napier does not show us how to calculate the logarithm of a product, probably because he is not interested in pure multiplications as such. He constructed his logarithms with trigonometry in mind, and many calculations involved in the solving of triangles require the finding of a fourth proportional, for which his rules indeed apply.

As an example of this type of calculation, consider the right triangle whose hypotenuse c and leg a are known. The problem is to find the angle α opposite the given leg. Napier makes use of the basic trigonometric relation

$$\frac{\sin \alpha}{r} = \frac{a}{c},$$

where $r = 10^7$ is the radius of the circle in which the sines are defined. Napier then uses his table and the rule for proportions given above to calculate Nlog sin α = Nlog a − Nlog c + Nlog r. Because Nlog r = 0, he has now found the logarithm of the sine of α in terms of the logarithms of the sides. Reading his table in reverse gives the desired angle. Although Napier's table is a table of logarithms of sines, he can calculate the logarithms of the numerical lengths needed in this problem by looking in the table for a sine which is close enough to the desired number, making appropriate adjustments for the number of digits in one or the other, and then taking the logarithm of that sine value.

Napier gives many other examples of the use of the table. To solve a plane triangle given two sides a, b and the angle α opposite side a, Napier applies his logarithm laws to the law of sines

$$\frac{\sin \alpha}{a} = \frac{\sin \beta}{b},$$

noting that there are two possible values for β, one less than a right angle and one greater. To solve a triangle given two sides a, b and the included angle γ, Napier does not use the standard method of dropping a perpendicular, because that method is not suited to logarithmic calculation. Instead, he makes use of the **law of tangents**,

$$\frac{a + b}{a - b} = \frac{\tan \frac{1}{2}(\alpha + \beta)}{\tan \frac{1}{2}(\alpha - \beta)}.$$

If γ is given, then $\alpha + \beta$ is known. Applying logarithms to this proportion allows him to find $\tan \frac{1}{2}(\alpha - \beta)$, therefore $\frac{1}{2}(\alpha - \beta)$, and therefore both α and β. How did Napier calculate the logarithm of a tangent from his table of logarithms of sines? To answer this question, we present one line of Napier's actual table, given in seven columns for each minute of arc from 0° to 45°.

| 34°40′ | 5688011 | 5642242 | 3687872 | 1954370 | 8224751 | 55°20′ |

The first column gives the value of an arc (or angle), while the second gives the sine of that arc. The final column gives the arc which is complementary to that in the first column, while the sixth column gives its sine. It follows that the sixth column gives the sine of the complement of the arc of the first column, that is, the cosine of that arc. The third and fifth columns give Napier's logarithms of the sines in the second and sixth columns, respectively or, as Napier also notes, the logarithms of the sine complements of the sixth and second columns, respectively. Finally, the middle column represents the difference of the entries in the third and fifth columns, or, Napier's logarithm of the tangent of the arc of the first column. Because the logarithm of 10,000,000 is 0, logarithms of numbers greater than 10,000,000 must be negative and are defined by simply reversing the directions of the moving points in the original definition. These numbers, of course, cannot represent sines, but can represent tangents or secants. In this case, the negative of the logarithm in the middle column is the logarithm of the tangent of 55°20′, while the negative of the logarithm in the third column is the logarithm of the secant of that same angle.

Although we cannot detail Napier's actual construction of his table of logarithms from his kinematical definition, we note that the construction took him twenty years.[38] And even though this work was done in the era of hand calculations, there were remarkably few errors. Late in his life, however, Napier decided that it would be more convenient to have logarithms whose value was 0 at 1 rather than at 10,000,000. In that case the familiar properties of logarithms, $\log xy = \log x + \log y$ and $\log \frac{x}{y} = \log x - \log y$, would hold. Furthermore, if the logarithm of 10 were set at 1, the logarithm of $a \times 10^n$, where $1 \le a < 10$, would simply be n added to the logarithm of a. Napier died before he could construct a new table based on these principles, but Henry Briggs (1561–1631), who discussed this matter thoroughly with Napier in 1615, began the calculation of such a table. Rather than simply convert Napier's logarithms to these new "common" logarithms by simple arithmetic procedures, however, Briggs worked out the table from scratch. Starting with log 10 = 1, he calculated successive square roots of 10, that is $\sqrt{10}$, $\sqrt{\sqrt{10}}$, $\sqrt{\sqrt{\sqrt{10}}}$, . . . , until after 54 such root extractions he reached a number very close to 1. All of these calculations were carried out to 30 decimal places. Since $\log\sqrt{10} = 0.5000$, $\log\sqrt{\sqrt{10}} = 0.2500$, . . . , $\log(10^{1/2^{54}}) = 1/2^{54}$, he was able to build up a table of logarithms of closely spaced numbers using the laws of logarithms. Briggs' table, completed by Adrian Vlacq in 1628, became the basis for nearly all logarithm tables into the twentieth century. Astronomers very quickly discovered the great advantages of using logarithms for calculations. Logarithms became so important that the eighteenth century French mathematician Pierre-Simon de Laplace was able to assert that the invention of logarithms, "by shortening the labors, doubled the life of the astronomer."

10.5 KINEMATICS

The final mathematical arts of John Dee which we will consider are those which deal with motion. Thus, "Statics is an art mathematical which demonstrates the causes of heaviness and lightness of all things and of motions and properties to heaviness and lightness belonging,"[39] while "Trochilike . . . demonstrates the properties of all circular motions, simple and compound."[40] The man generally considered to be the founder of modern

---

Biography   Galileo Galilei (1564–1642)

Galileo is today probably most famous for his clash with the Catholic Church over his publication of the *Dialogue Concerning the Two Chief World Systems* (1632) in which he presented the arguments for and against both the Ptolemaic and the Copernican theories of the universe. As noted in the opening of the chapter, Galileo had been warned by Church authorities in 1616 that the Church's official position was that the earth did not move and that Galileo must not hold or defend such views. Galileo in his book therefore took some pains to present the Copernican position as a hypothetical one and simply to consider its consequences as well as the failings of the traditional Ptolemaic position. Nevertheless, a careful reading of the text shows that in fact Galileo was convinced of the truth of the earth's motion around the sun—not surprising at this date—and made the defenders of the older position in his *Dialogue* appear foolish. In any case, in 1633 Galileo was brought before the Inquisition in Rome and forced to confess his error. He was then sentenced to house imprisonment for the remainder of his life and forbidden to publish any more books. He did manage, however, to publish his most important work, the *Discourses and Mathematical Demonstrations Concerning Two New Sciences* in 1638 by sending the manuscript beyond the reach of the Inquisition to Leiden in the Netherlands, where it was printed by the publishing house of the Elseviers. Both the *Dialogue* and the *Discourses* were written in Italian rather than in the more scholarly Latin, because Galileo was aiming them not at university scholars but at the general educated public. Given then the immediate popularity of the *Dialogue*, the Church's banning of it had little practical effect. Even though the authorities made attempts to destroy available copies, there were so many in circulation that they had the effect that Galileo desired and the Church feared, of convincing the public that the Copernican system was in fact true and the earth did move, statements in the Bible to the contrary notwithstanding (Figure 10.29).

---

physics, Galileo Galilei (1564–1642), was responsible in large measure for reformulating the laws of motion considered first by the Greeks and later by certain medieval scholars. But like his predecessors, he proposed to use geometry, not algebra, to explicate his ideas. Although he did work in what today is generally called statics, his most important new ideas, dealing with the "natural" accelerated motion of free fall and the "violent" motion of a projectile, were published in 1638 in his *Discourses and Mathematical Demonstrations Concerning Two New Sciences*. Galileo thus applied mathematics to the study of motion on earth much as Copernicus and Kepler had applied it to the study of motion in the heavens.

10.5.1 Accelerated Motion

The *Two New Sciences* is written in the form of a dialogue among three people, whose discussion of motion is carried out around the framework of a formal treatise on motion written in the Euclidean format including a definition, a postulate, and many theorems and proofs. The definition reads, "Motion is equably or uniformly accelerated which, abandoning rest, adds on to itself equal momenta of swiftness in equal times."[41] Galileo, although beginning with the same definition as his medieval predecessors, made two major advances. First he discovered by 1604 that uniformly accelerated motion is precisely that of a freely falling body, and second he worked out numerous mathematical consequences of this fact, some of which he could confirm by experiment.[42]

FIGURE 10.29
Galileo on an Italian
stamp.

At one time Galileo believed that the velocity of a falling body increased in the ratio of the distance fallen rather than in the ratio of the times. In the *Two New Sciences* he gives an argument showing that this possibility is erroneous. First, he notes that if two different velocities of a given body are proportional to the distances covered while the body has each velocity in turn, then the times for the body to cover those distances are equal. This statement is virtually obvious for velocities constant over the given period of time. Galileo then assumes it to be true also for continuously changing velocities. Thus, "if the speeds with which the falling body passed the space of four *braccia* were the doubles of the speeds with which it passed the first two *braccia*, as one space is double the other space, then the times of those passages are equal."[43] Galileo here compares two (infinite) sets of velocities, those at each instant in which the falling body passes a point in the first two *braccia* (an Italian measure of distance) with those at each instant it passes a point in the first four *braccia* twice as far from the point of origin. His statement that the total times are equal results from applying the statement for finite times to infinitesimal times and adding up the entire set of these infinitesimal times. Galileo concludes that it is ridiculous that a given fallen body starting from rest could cover both two *braccia* and four *braccia* in the same time, and thus that it is false that speed increases as the distance traveled.

Galileo's argument by comparing two infinitesimal sets provides one of the first such arguments in mathematical history. Earlier in this particular work, Galileo had considered the notion of one-to-one correspondence in the classic case of comparing integers to squares. It is "obvious," Galileo has one of his characters state, that the set of all numbers is more numerous than the set of squares. On the contrary, however, Galileo's character also notes that there are just as many square numbers as there are roots, "since every square has its root and every root its square, nor is there any square that has more than just one root, or any root that has more than just one square."[44] Because the set of roots is the same as the set of numbers, he concludes that square numbers are as numerous as all numbers, in contradiction to the earlier "obvious" result. Galileo's basic conclusion from this dilemma is that one cannot consider the attributes equal to, less than, or greater than when dealing with infinite sets.

In his treatment of motion, Galileo's first theorem is essentially the same as the medieval mean-speed rule. His proof, however, is somewhat different.

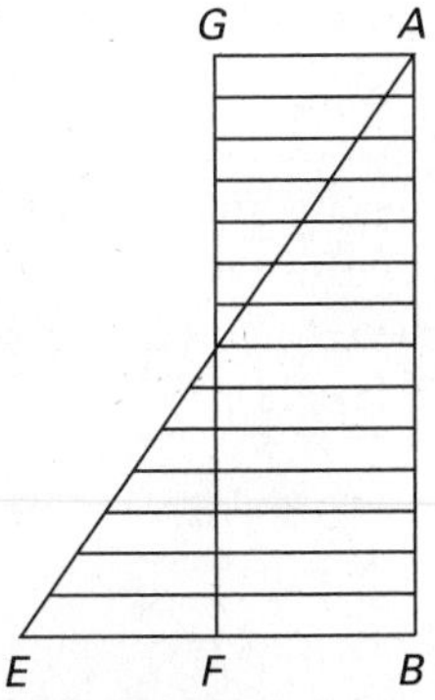

Theorem. *The time in which a certain space is traversed by a moveable in uniformly accelerated movement from rest is equal to the time in which the same space would be traversed by the same moveable carried in uniform motion whose degree of speed is one-half the maximum and final degree of speed of the previous, uniformly accelerated, motion.*[45]

Galileo's proof again uses one-to-one correspondence of infinite sets. With *AB* representing the time of travel, *EB* representing the maximum speed attained by the moveable, and *F* the midpoint of *BE*, Galileo constructs right triangle *ABE* and rectangle *ABFG*, which are of equal areas (Figure 10.30). There are then one-to-one correspondences between the instants of time represented by points of the line *AB* and the parallels in the triangle representing the increasing degrees of speed on the one hand, and those instants and the parallels in the rectangle representing the equal speeds at half the final speed on the other. Galileo concludes that "there are just as many momenta of speed consumed in

FIGURE 10.30
Galileo's proof of the
mean-speed theorem.

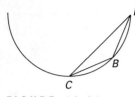

FIGURE 10.31
Galileo and the
brachistochrone problem.

the accelerated motion according to the increasing parallels of triangle *AEB* as in the equable motion according to the parallels of the [rectangle] *GB*,"[46] because the deficit above the halfway point is made up by the surplus below it. Since these "momenta" of speed for each instant of time are proportional to the distances traveled in those instants, it follows that the total distance in each case is the same. As before, Galileo uses an argument with infinitesimals. One may wonder why he believed in such arguments, given that they violated classic geometric concepts. But use them he did.

As a corollary to the theorem, Galileo proves that in the case of a moveable falling from rest, the distances traveled in any times are as the squares of those times. Namely, if the body falls a distance d_1 in time t_1, and d_2 in time t_2, then $d_1 : d_2 = t_1^2 : t_2^2$. Galileo's result in modern notation is written as $d = kt^2$, but Galileo always used Euclidean proportionality concepts instead of modern "function" concepts. To prove this corollary, Galileo first notes that for two bodies both traveling at constant, but unequal, velocities, the distances traveled are in the ratio compounded of the ratios of the speeds and the times, or $d_1 : d_2 = (v_1 : v_2)(t_1 : t_2)$. This result is derived from the facts that for equal times distances are proportional to velocities and for equal velocities distances are proportional to times. By the theorem, the distances traveled by the falling body in the two times are the same as if in each case the body had a constant velocity equal to half its final velocity. These halves of the final velocities are also proportional to the times. It follows that in the compound ratio, the ratio of the velocities can be replaced by the ratio of the times, and the corollary is proved. Galileo's language of proportionality does not enable him to state the proportionality constant $k = 1/2$, but this value is certainly clearly implied in his proof.

Galileo stated and proved some 38 propositions on naturally accelerated motion. He was interested in comparing velocities, times, and distances for motion along inclined planes as well as for free fall. Thus he presented a postulate to the effect that the velocity acquired by an object sliding down an inclined plane (without friction) depends only on the height of the plane and not on the angle of inclination. Using this postulate, he deduced results such as that the times of descent for a given object along two different inclined planes of the same height are to one another as the lengths of the planes, and, conversely, the times of descent over planes of equal lengths are to one another inversely as the square roots of their heights. Galileo also made progress toward solving the brachistochrone problem, that is, of discovering the path by which an object moves in shortest time from one point to another point at a lower level. He showed that in a given vertical circle, the time taken for a body to descend along a chord from any point to the bottom of the circle, say *DC*, is greater than its time to descend along the two chords *DB*, *BC*, the first beginning at the same point as the original chord, the second ending at the same bottom point (Figure 10.31). (Here *DC* must subtend an arc no greater than 90°.) By extending this result to more and more chords, he concluded, erroneously, that the path of swiftest descent is a circular arc. It was not until the end of the century that several mathematicians deduced that this curve was in fact a cycloid.

10.5.2 The Motion of Projectiles

In the final part of the *Two New Sciences*, Galileo discusses the motion of projectiles. These motions are compounded from two movements, the horizontal one being of constant velocity and the vertical one being naturally accelerated. As he wrote,

I mentally conceive of some moveable projected on a horizontal plane, all impediments being put aside. Now it is evident . . . that equable motion on this plane would be perpetual if the plane were of infinite extent; but if we assume it to be ended, and [situated] on high, the moveable . . . , driven to the end of this plane and going on further, adds on to its previous equable and indelible motion that downward tendency which it has from its own heaviness. Thus there emerges a certain motion, compounded from equable horizontal and from naturally accelerated downward [motion], which I call projection.[47]

Galileo here states part of the fundamental law of inertia, that a body moving on a frictionless horizontal plane at constant velocity will not change its motion, because, as he had noted earlier, "there is no cause of acceleration or retardation."[48] Isaac Newton extended this principle into one of his laws of motion by replacing Galileo's "causes" by his own notion of force. Galileo, however, was not so interested in the law itself as in the path of the projectile. Thus he proved

Theorem.   *When a projectile is carried in motion compounded from equable horizontal and from naturally accelerated downward [motions], it describes a semiparabolic line in its movement.*

Galileo discovered this theorem in 1608 in connection with an experiment rolling balls off tables, an experiment which convinced him that the horizontal motion was unaffected by the downward motion due to gravity.[49] His proof in the *Two New Sciences* uses this assumption. Galileo draws a careful graph of the path of the object, noting that in equal times the horizontal distances traveled are equal while in those same times the vertical distances increase in proportion to the squares of the times. Therefore the curve has the property that for any two points on it, say F, H, the ratio of the squares of the horizontal distances, $FG^2 : HL^2$ is the same as that of the vertical distances (to the plane), $BG : BL$ (Figure 10.32). Galileo then concludes from his familiarity with the work of Apollonius that the curve is a parabola as claimed.

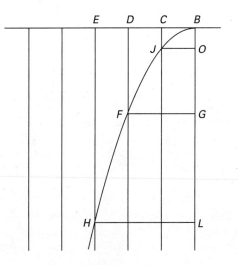

FIGURE 10.32
Galileo and the parabolic motion of a projectile.

Galileo continued this discussion of projectile motion by proving that for objects fired at an angle to the horizontal, as from a cannon, the path would also be parabolic. In fact he calculated several tables giving the height and distance traveled by such a projectile as functions of the initial angle of elevation, showing, for example, that the maximum range is achieved by an initial angle of 45°. Gunners had already observed this fact, but Galileo was able to demonstrate "something that has perhaps not been observed through experiment, and this is . . . that of the other shots, those are equal [in range] to one another whose elevations exceed or fall short of [45°] by equal angles."[50]

More important than the results on projectiles, however, was Galileo's discussion of the basic principles of the application of mathematics to physics. In particular, he realized that the parabolas were only approximations to the true path, since there were various "impediments" to the motion, including, for example, air resistance. Galileo answers these objections as follows: "No firm science can be given of such events of heaviness, speed, and shape which are variable in infinitely many ways. Hence to deal with such matters scientifically, it is necessary to abstract from them. We must find and demonstrate conclusions abstracted from the impediments, in order to make use of them in practice under those limitations that experience will teach us. . . . Indeed, in projectiles that we find practicable, . . . the deviations from exact parabolic paths will be quite insensible."[51] Galileo thus states his firm belief in the application of mathematics to physics. One must always form a mathematical model by considering only the most important ideas in a given situation. Only after deriving mathematically the consequences of one's model and comparing these to experiment can one decide whether adjustments to the model are necessary. Galileo, like Kepler, followed these basic precepts of mathematical modeling of physical phenomena. Kepler, because he was dealing with astronomical phenomena, could only compare his theoretical results to observation. Galileo, on the other hand, conducted experiments to verify (or refute) the results of his reasoning. The detailed explication of this process of mathematical modeling, even more than his actual physical theorems, formed Galileo's most fundamental contribution to the mutual development of mathematics and physics. His ideas came into full flower as the scientific revolution of the seventeenth century reached its climax in the work of Isaac Newton.

Exercises

Problems in Perspective

1. Make a perspective drawing of a checkerboard. First establish a reasonable distance for the vanishing line and vanishing point and then construct the horizontal lines using the rules given in the text.

2. Read Chapter 3 of *The Mathematics of Great Amateurs* by Julian Lowell Coolidge on Piero della Francesca and write a brief outline of Piero's method of drawing a cube in space.

3. Prepare a small collection of copies of Renaissance paintings from some art texts, including some from Piero, Alberti, and Dürer. Mark the vanishing point and vanishing line and indicate some lines in the painting which meet at the vanishing point.

Problems from Geography and Navigation

4. Find out the details of the problem of the determination of longitude in the sixteenth century and how it was related to the problem of timekeeping. Find out how the problem was ultimately solved. Consult Daniel J. Boorstin, *The Discoverers* (New York: Random House, 1985) for more information.

5. Prepare a brief research paper on the problem of map projection. Give a mathematical description of a few of

the most common projections used today and describe their properties. Include in your description examples of stereographic, conical, and cylindrical projections.

6. This problem provides details on constructing a Mercator chart to represent the region between the equator and 30° N. latitude and between 75° and 85° W. longitude. Draw a line 10 cm. long to represent the equator between those meridians. Divide it into intervals of 1 cm and draw the meridians perpendicular to the line of the equator. Then 1° of longitude is taken as 1 cm. To find the distance on the map to the 10° parallel, note first that since 1 cm corresponds to 1° on a great circle, the radius of the corresponding sphere must be 180/π. One must therefore multiply this value by $D(10°)$, computed by formula 10.1. Similarly, to calculate the distance to the parallel at 20° from that at 10°, find $D(20°) - D(10°)$ and multiply by the radius. Also calculate the distance of the 30° parallel from the equator. To make the chart somewhat more precise, determine the distances of the parallels at 5°, 15°, and 25°.

7. Modify the calculation of problem 6 for placing parallels in order to map a region between 80° and 100° W. longitude and between 40° and 60° N. latitude, assuming that 1 cm on the parallel of 40° N. corresponds to 1° of longitude.

Problems from Trigonometry

8. Solve the problem from *On Triangles* discussed in the text of finding two sides AB, AG of triangle ABG given that $BG = 20$, the perpendicular $AD = 5$, and the ratio $AB : AG = 3:5$ (see Figure 10.13, p. 366).

9. In triangle ABC, suppose the ratio $\angle A : \angle B = 10 : 7$ and the ratio $\angle B : \angle C = 7 : 3$. Find the three angles and the ratio of the sides. (This problem is also from *On Triangles*.)

10. Prove theorem II-26 of *On Triangles*: If the area of a triangle is given together with the product of the two sides, then either the angle opposite the base is known or [that angle] together with its known [exterior] angle equals two right angles. Show that this result gives implicitly the trigonometric formula for the area of a triangle.

11. The following problem is from Pitiscus' *Trigonometry*: Find the area of the field $ABCDE$ given the following measurements: $AB = 7$, $BC = 9$, $AC = 13$, $CD = 10$, $CE = 11$, $DE = 4$, and $AE = 17$. Begin by drawing $BF \perp AC$, $CG \perp AE$, and $DH \perp CE$ (Figure 10.33).

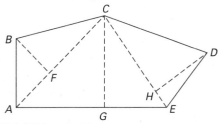

FIGURE 10.33
A problem in area from Pitiscus' *Trigonometry*.

12. This problem is from Copernicus' *De Revolutionibus*. Given the three sides of an isosceles triangle, to find the angles. Circumscribe a circle around the triangle and draw another circle with center A and radius $AD = \frac{1}{2}AB$ (Figure 10.34). Then show that each of the equal sides is to the base as the radius is to the chord subtending the vertex angle. All three angles are then determined. Perform the calculations with $AB = AC = 10$ and $BC = 6$.

13. Prove that

$$\sin \alpha \sin \beta = \frac{1}{2}[\cos(\alpha - \beta) - \cos(\alpha + \beta)].$$

Derive similar formulas involving addition and/or subtraction for $\cos \alpha \cos \beta$ and $\sin \alpha \cos \beta$.

Problems on Logarithms

The next five exercises outline a method of deriving the modern natural logarithm by slightly modifying Napier's definition.

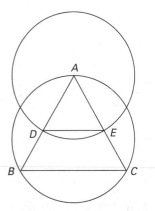

FIGURE 10.34
To find the angles of an isosceles triangle.

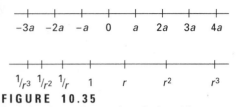

FIGURE 10.35
Deriving the modern notion of a logarithm.

14. Conceive of two number lines, the top one of which is marked with an arithmetic sequence . . . $-4a$, $-3a$, $-2a$, $-a$, 0, a, $2a$, $3a$, $4a$, . . . and the bottom one with a geometric sequence . . . $1/r^3$, $1/r^2$, $1/r$, 1, r, r^2, r^3, . . . where $r > 1$ (Figure 10.35). Consider points P, Q moving on the two lines respectively so that P covers each marked interval $[0, a]$, $[a, 2a]$, $[2a, 3a]$, . . . in the same time, while Q also covers each marked interval $[1, r]$, $[r, r^2]$, $[r^2, r^3]$, . . . in the same time. Consider the velocity of Q constant in each of the marked intervals. Show that this definition of the motion of Q implies that the velocity of Q at any marked point is proportional to the point's distance from the left endpoint, 0.

15. Show that given any two marked intervals on the bottom line $[\alpha, \beta]$, $[\gamma, \delta]$ of whatever length, with $\beta/\alpha = \delta/\gamma$, the time for Q to cover $[\alpha, \beta]$ equals the time for Q to cover $[\gamma, \delta]$.

16. Smooth out the velocity jumps in the lower line somewhat by introducing in that line the point $s = \sqrt{r}$ between 1 and r as well as points representing all integral powers of s. Similarly, introduce on the upper line the point $b = \frac{1}{2}a$ between 0 and a and all its integral multiples. Show that if Q is still required to cover each new marked interval $[1, s]$, $[s, s^2]$, . . . in equal times while P continues at constant velocity, then the results of exercises 14 and 15 still hold.

17. To eliminate all velocity jumps, assume that the velocity of Q increases smoothly so that at any point of the lower line that velocity is proportional to the distance from 0. Assume further that P travels at constant velocity, say v, and that Q starts at 1 with velocity v at the same time that P starts at 0. Suppose now that Q is at the number x at the same time as P is at the number y. Define the **logarithm** of x, written $\log x$, to be y. Show that $\log 1 = 0$ and that if $\beta/\alpha = \gamma/\delta$, then Q covers $[\alpha, \beta]$ in the same times as it covers $[\gamma, \delta]$. Conclude therefore that $\log \beta - \log \alpha = \log \delta - \log \gamma$.

18. From the final relationship of exercise 17, conclude by appropriate specialization that $\log (\delta/\gamma) = \log \delta - \log \gamma$,

that $\log (\beta\gamma) = \log \beta + \log \gamma$, and that for integral n, $\log(\beta^n) = n \log \beta$. Also show that for m a positive integer, $\log(\sqrt[m]{\beta}) = \log \beta/m$ and that for any rational value r, $\log(\beta^r) = r \log \beta$.

19. Show using calculus that the logarithm function defined in problem 17 is the modern natural logarithm function.

20. Using the definition of the function Nlog presented in the text, determine Nlog(xy) and Nlog(x/y) in terms of Nlog x and Nlog y.

21. Find out how the use of logarithms was mechanized in the seventeenth century by the invention of the slide rule. Give examples of the various types of slide rules used. When did the slide rule itself become obsolete and why?

Problems from Galileo's **Two New Sciences**

22. Prove: If the same moveable is carried from rest on an inclined plane and also along a vertical of the same height, the times of movements will be to one another as the lengths of the plane and of the vertical. A corollary is that the times of descent along differently inclined planes of the same height are to one another as the lengths.

23. Prove: The times of motion of a moveable starting from rest over equal planes unequally inclined are to each other inversely as the square root of the ratio of the heights of the planes.

24. Show that a projectile fired upward at an angle α from the horizontal follows a parabolic path.

25. Show that if a projectile fired at an angle α from the horizontal at a given initial speed reaches a distance of 20,000 if $\alpha = 45°$, then with the same initial speed it will reach a distance of 17,318 if $\alpha = 60°$ or $\alpha = 30°$.

26. Show that if a projectile fired at a given initial speed at an angle α to the horizontal reaches a maximum height of 5000 if $\alpha = 45°$, then with the same initial speed it will reach a height of 2499 when $\alpha = 30°$ and a height of 7502 when $\alpha = 60°$.

FOR DISCUSSION . . .

27. Compare Galileo's and Kepler's attitudes toward the interaction of experiment (or observation) and theory in developing a new body of knowledge.

28. Read Kepler's biography in Arthur Koestler's *The Sleepwalkers*, referred to in the references to this chapter. Koestler also discusses the lives of Copernicus and Galileo. How believable is Koestler's ''sleepwalking'' hypothesis to explain the discovery of the new ideas in astronomy? Comment.

29. Look up the eccentricities of the orbits of Mars and the other visible planets. Compare these with the eccentricity of the earth to see why Kepler was able to assume the circularity of the earth's orbit. Considering these eccentricities, why did Kepler study Mars in detail rather than Mercury?

30. Does the treatment of natural logarithms outlined in problems 14 to 18 form a reasonable method of introducing the subject in a precalculus class? Compare this method with the methods found in standard texts. Comment.

31. Are there any living scientists who exhibit some of the mysticism attributed to Dee and Kepler? Find out about them and their contributions to science.

32. Look up a treatment of geometrical perspective in a modern text on techniques of painting. How does it compare to the discussion of Alberti?

33. Look up the review of Galileo's case by the Roman Catholic Church in the 1980s. Has the Church revised its opinion that Galileo disobeyed orders?

References and Notes

There are several books which discuss aspects of the applied mathematics of the Renaissance in greater detail. Julian Lowell Coolidge, *The Mathematics of Great Amateurs*, Second Edition (Oxford: Clarendon Press, 1990) contains three chapters dealing with the mathematics of the artists of the Renaissance as well as a chapter on John Napier. Thomas S. Kuhn, *The Copernican Revolution* (Cambridge: Harvard University Press, 1957) and E. J. Dijksterhuis, *The Mechanization of the World Picture* (Princeton: Princeton University Press, 1986) each contain sections on the developments in astronomy. The latter work, in particular, has mathematical details. Arthur Koestler, *The Sleepwalkers* (New York: Penguin, 1959) provides a lively treatment of the history of astronomy from Greek times to the time of Galileo, including biographies of Copernicus, Brahe, Kepler, and Galileo. Many of his interpretations are controversial. Finally, Ernan McMullin, ed., *Galileo, Man of Science* (Princeton Junction: The Scholar's Bookshelf, 1988) is a collection of essays on various aspects of Galileo's scientific achievements.

1. Galileo, *The Assayer*, in Stillman Drake, trans., *Discoveries and Opinions of Galileo* (Garden City: Doubleday, 1957), pp. 237–238.

2. Stillman Drake, *Galileo at Work* (Chicago: University of Chicago Press, 1978), pp. 347–348. There is an enormous amount of literature on Galileo and his conflict with the Church. A glance at the appropriate shelf in any university library will provide several titles.

3. John Dee, *Mathematical Preface* (New York: Science History Publications, 1975), p. 3. This is a photographic reproduction of the original with an introduction by Allen G. Debus, who discusses Dee's life and influence. The pages are unnumbered just as in the original. A discussion of Dee's philosophy and of the mathematical preface is also found in F. A. Yates, *Theatre of the World* (Chicago: Chicago University Press, 1969).

4. Ibid., p. 5.

5. Ibid., p. 13.

6. Ibid., p. 19.

7. Ibid.

8. Ibid., p. 38.

9. Ibid.

10. For more information on Leon Battista Alberti and Piero della Francesca, see Julian Lowell Coolidge, *Mathematics of Great*, Chapter 3. For Alberti's work on perspective, see J. and P. Green, "Alberti's Perspective: A Mathematical Comment," *Art Bulletin* 64 (1987), 641–645.

11. An English version of this work is available by W. Strauss, trans., *The Painter's Manual* (New York: Abaris, 1977).

12. Erwin Panofsky, "Dürer as a Mathematician," in James R. Newman, ed., *The World of Mathematics* (New York: Simon and Schuster, 1956), vol. 1, 603–621, pp. 611–612. This section is taken from Erwin Panofsky, *Albrecht Dürer* (Princeton: Princeton University Press, 1945), a work that discusses Dürer's life and art in great detail.

13. For more information, see Roger Herz-Fischler, "Dürer's Paradox or Why an Ellipse is not Egg-Shaped," *Mathematics Magazine* 63 (1990), 75–85.

14. Dee, *Mathematical Preface*, p. 42.

15. Ibid., p. 15.

16. More details can be found in V. Frederick Rickey and Philip M. Tuchinsky, "An Application of Geography to

Mathematics: History of the Integral of the Secant,'' *Mathematics Magazine* 53 (1980), 162–166. See also Florian Cajori, ''On an Integration Ante-dating the Integral Calculus,'' *Bibliotheca Mathematica* (3) 14 (1914), 312–319.

17. Dee, *Mathematical Preface*, p. 20.

18. Ibid., p. 23.

19. Barnabas Hughes, *Regiomontanus on Triangles* (Madison: University of Wisconsin Press, 1967), p. 27. This work has the original Latin of Regiomontanus' *De Triangulis Omnimodis* as well as an English translation, an introduction, and extensive notes.

20. Ibid., p. 109.

21. Ibid., p. 119.

22. Ibid., p. 249.

23. For more details on the history of trigonometry see M. C. Zeller, ''The Development of Trigonometry from Regiomontanus to Pitiscus,'' Ph.D. diss., University of Michigan, 1944 (Ann Arbor: Edwards Bros., 1946). This work has an extensive table comparing various aspects of trigonometrical notation and terminology in European texts to the beginning of the seventeenth century. See also J. D. Bond, ''The Development of Trigonometric Methods Down to the Close of the XVth Century,'' *Isis* 4 (1921), 295–323.

24. A. M. Duncan, trans., *Copernicus: On the Revolutions of the Heavenly Spheres* (New York: Barnes and Noble, 1976), p. 25. This is a new translation of Copernicus' *De Revolutionibus*. An older translation, by Charles Glenn Wallis, appears in volume 16 of the *Great Books* (Chicago: Encyclopedia Britannica, 1952).

25. Ibid., p. 26.

26. Ibid., p. 27.

27. Ibid., p. 38.

28. Ibid., p. 60.

29. Ibid., p. 22.

30. A. M. Duncan, trans., *The Secret of the Universe* (New York: Abaris, 1981), p. 67. This is the first English translation of Kepler's *Mysterium Cosmographicum* and contains an introduction and commentary by E. J. Aiton. The translation is of the second edition, so it includes not only the original text of 1596 but also Kepler's notes added in 1621. This modern edition also contains the Latin text and copies of the original diagrams.

31. Ibid., p. 73.

32. Johannes Kepler, *The Harmonies of the World*, Book V, translated by Charles Glenn Wallis in the *Great Books*, vol. 16, p. 1048. It is an interesting exercise to play on a piano the notes which Kepler assigns to the various planets and so to understand his version of the ''Music of the Spheres.''

33. For more details on Kepler's discovery of the elliptical path, see Curtis Wilson, ''How Did Kepler Discover His First Two Laws,'' *Scientific American* 226 (March, 1972), 92–106, Eric Aiton, ''How Kepler Discovered the Elliptical Orbit,'' *Mathematical Gazette* 59 (1975), 250–260, and also Dijksterhuis, *Mechanization of the World*, pp. 303–323.

34. Kepler, *Harmonies*, p. 1020.

35. John Napier, *Constructio*, translated by William R. Mac-Donald (London: Dawsons, 1966), p. 8. This work is a photographic reprint of the original translation. A careful study will show in detail the ingenious interpolation schemes which Napier used to guarantee 8-place accuracy in his tables. Many essays on Napier's work are found in C. G. Knott, ed., *Napier Tercentenary Memorial Volume* (London: Longmans, Green and Co., 1915). A general discussion of logarithms is found in E. M. Bruins, ''On the History of Logarithms: Bürgi, Napier, Briggs, de Decker, Vlacq, Huygens,'' *Janus* 67 (1980), 241–260 and in F. Cajori, ''History of the Exponential and Logarithmic Concepts,'' *American Mathematical Monthly* 20 (1913), 5–14, 35–47, 75–84, 107–117.

36. Napier, *Constructio*, p. 19.

37. John Napier, *Descriptio*, translated by Edward Wright (New York: Da Capo Press, 1969), p. 6. This is a photographic reprint of the original translation and includes many examples of the use of logarithms in solving both plane and spherical triangles as well as Napier's original table.

38. See C. H. Edwards, *The Historical Development of the Calculus* (New York: Springer-Verlag, 1979), Chapter 6 for details.

39. Dee, *Mathematical Preface*, p. 25.

40. Ibid., p. 34.

41. Galileo, *Two New Sciences*, translated by Stillman Drake (Madison: University of Wisconsin Press, 1974), p. 154. Drake also wrote an introduction, extensive notes, and a glossary of Galileo's technical terms. For other essays on

various aspects of Galileo's scientific life, see Stillman Drake, *Galileo Studies: Personality, Tradition and Revolution* (Ann Arbor: University of Michigan Press, 1970).

42. See Stillman Drake, "Galileo's Discovery of the Law of Free Fall," *Scientific American* 228 (May, 1973), 84–92 for extensive details on Galileo's discovery of the law of free fall.

43. Galileo, *Two New Sciences*, p. 160.

44. Ibid., p. 40.

45. Ibid., p. 165.

46. Ibid.

47. Ibid., p. 217.

48. Ibid., p. 196.

49. Stillman Drake, "Galileo's Discovery of the Parabolic Trajectory," *Scientific American* 232 (March, 1975), 102–110, has many details on Galileo's manuscript notes detailing his experiments.

50. Galileo, *Two New Sciences*, p. 246.

51. Ibid., p. 225.

Summary of Renaissance Applied Mathematics

1377–1446	Filippo Brunelleschi	Perspective
1404–1472	Leon Battista Alberti	Perspective
1420–1492	Piero della Francesca	Perspective
1436–1476	Regiomontanus	Trigonometry
1471–1528	Albrecht Dürer	Geometry of perspective
1473–1543	Nicolaus Copernicus	Astronomy
1502–1578	Pedro Nuñez	Map making
1512–1594	Gerard Mercator	Map making
1514–1574	George Rheticus	Trigonometry
1527–1608	John Dee	Mathematical Preface
1546–1601	Tycho Brahe	Astronomy
1550–1617	John Napier	Logarithms
1552–1632	Jobst Bürgi	Logarithms
1561–1613	Bartholomew Pitiscus	Trigonometry
1561–1615	Edward Wright	Map making
1561–1631	Henry Briggs	Logarithms
1561–1656	Thomas Finck	Trigonometry
1564–1642	Galileo Galilei	Kinematics
1571–1630	Johannes Kepler	Astronomy

Chapter 11

Geometry, Algebra, and Probability in the Seventeenth Century

"Whenever two unknown magnitudes appear in a final equation, we have a locus, the extremity of one of the unknown magnitudes describing a straight line or a curve."
(Pierre de Fermat's Introduction to Plane and Solid Loci, *1637)*[1]

*T*he Duke of Roannez had a salon in Paris early in the 1650s, which provided a meeting place for mathematicians among others. It was Roannez who introduced Blaise Pascal to Antoine Gombaud, the chevalier de Méré about 1652. De Méré took the opportunity of his new friendship to put two betting questions to Pascal, one on the number of tosses of two dice necessary to have at least an even chance of getting a double six and one on the equitable division of stakes in a game interrupted before its conclusion. It was out of Pascal's answers to these two questions that the theory of probability grew.

It is in the study of the early seventeenth century that one is first aware of the speeding up of mathematical developments. Printing was now well established, and communication, both through letter and through the printed word, was becoming much more rapid. The ideas of one mathematician were passed on to others, to be criticized, commented upon, and finally extended. In this chapter we survey some of the newly developing areas of mathematics.

Viète's ideas on the use of algebra in analysis were reformulated in the 1630s into the new subject of **analytic geometry**, a subject formed by the merger of algebra and geometry whose basic premise is stated in the opening quotation. The two central figures in the development of analytic geometry, which was to prove vital in the subsequent invention of the calculus, are Pierre de Fermat and René Descartes. Both of these men played central roles in other areas of mathematics as well. Descartes, along with Albert Girard, reworked some of Viète's algebraic ideas into a theory of equations. Fermat was involved, in his correspondence with Blaise Pascal, in the early development of probability theory, the first textbook on which was written by Christian Huygens in 1656. Fermat also was responsible for the first new work in number theory since Leonardo of Pisa, while Pascal, along with Girard Desargues, made some of the earliest contributions to the subject of projective geometry.

11.1 ANALYTIC GEOMETRY

Analytic geometry was born in 1637 of two fathers, René Descartes (1596–1650) and Pierre de Fermat (1601–1665). Naturally, there had been a gestation period, but early in that year Fermat sent to his correspondents in Paris a manuscript entitled *Ad locos planos et solidos isagoge* (*Introduction to Plane and Solid Loci*). At about the same time Descartes was readying for the printer the galley proofs of his *Discours de la méthode pour bien conduire sa raison et chercher la vérité dans les sciences* (*Discourse on the Method for Rightly Directing One's Reason and Searching for Truth in the Sciences*) with its three accompanying essays, among which was *La géométrie* (*The Geometry*). Both Fermat's *Introduction* and Descartes' *Geometry* present the same basic techniques of relating algebra and geometry, the techniques whose further development culminated in the modern subject of analytic geometry. Both men came to the development of these techniques as part of the effort of rediscovering the "lost" Greek techniques of analysis. Both were intimately familiar with the Greek classics, in particular with the *Domain of Analysis* of Pappus, and both tested their new ideas against the four-line-locus problem of Apollonius and its generalizations. But Fermat and Descartes developed distinctly different approaches to their common subject, differences caused by their differing points of view toward mathematics.

11.1.1 Fermat and the *Introduction to Plane and Solid Loci*

Fermat began his study of mathematics with the normal university curriculum at Toulouse, which probably covered little more than an introduction to Euclid's *Elements*. But after completing his baccalaureate degree and before beginning his legal education, he spent several years in Bordeaux studying mathematics with former students of Viète, who during the late 1620s were engaged in the editing and publishing of their teacher's work. Fermat became familiar both with Viète's new ideas for symbolization in algebra and with his program of discovering and elucidating the secret analysis of the Greek mathematicians. In Bordeaux, Fermat began his own project of using Pappus' annotations and lemmas in the *Domain of Analysis* to restore the *Plane Loci* of Apollonius. Fermat tried to reconstruct the original work along with Apollonius' reasoning in the discovery of the various theorems. His study of Viète naturally led Fermat to attempt to replace Apollonius' geometric analysis with an algebraic version. It is this algebraic version of Apollonius' locus theorems which provided the beginnings of Fermat's analytic geometry.

For example, Fermat considered the following result: If, from any number of given points, straight lines are drawn to a point, and if the sum of the squares of the lines is equal to a given area, the point lies on a circumference [circle] given in position. The theorem deals with an indeterminate number of points, but Fermat's treatment of the

---

Biography   Pierre de Fermat (1601–1665)

Fermat was born into an upwardly mobile family in Beaumont-de-Lomagne in the south of France. He received his undergraduate education at the University of Toulouse and took a bachelor of civil law degree in 1631 at Orléans. He then returned to Toulouse, in the vicinity of which he spent the remainder of his life practicing law. Fermat was evidently never a brilliant lawyer, probably because he spent much of his time on his first love, mathematics. Due to the state of his health and the press of his legal work, however, he never traveled far from his home. Thus, all of his mathematical work was communicated to others by way of his extensive correspondence.

Fermat always considered mathematics a hobby, a refuge from the continual disputes with which he had to deal as a jurist. He therefore refused to publish any of his discoveries, because to do so would have forced him to complete every detail and to subject himself to possible controversies in another arena. In many cases it is not known what, if any, proofs Fermat constructed nor is there always a systematic account of certain parts of his work. Fermat often tantalized his correspondents with hints of his new methods for solving certain problems. He would sometimes provide outlines of these methods, but often his promises to fill in gaps "when leisure permits" remained unfulfilled. Nevertheless, a study of his manuscripts, published by his son 14 years after his death, as well as his many letters, enables scholars today to have a reasonably complete picture of Fermat's methods.[2]

---

simplest case, that of two points, contains the germs of the two major ideas of analytic geometry: the correspondence between geometric loci and indeterminate algebraic equations in two or more variables, and the geometric framework for this correspondence, a system of axes along which lengths are measured.

Fermat takes the two given points A, B and bisects the line AB at E. With IE as radius (I yet to be determined) and E as center, he describes a circle (Figure 11.1).[3] He then shows that any point P on this circle satisfies the conditions of the theorem, namely that $AP^2 + BP^2$ equals the given area M, provided that I is chosen so that $2(AE^2 + IE^2) = M$. The important ideas in this proof are that the locus, the circle, is determined by the sum of the squares of two variable quantities, AP and BP, and that the point I is determined in terms of its "coordinate" measured from the "origin" E.

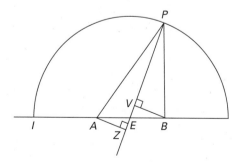

FIGURE 11.1
Fermat's analysis of a special case of a theorem of Apollonius:

$$AP^2 + BP^2 = PZ^2 + AZ^2 + PV^2 + BV^2$$
$$= (PE + EZ)^2 + AE^2 - EZ^2 +$$
$$(PE - EV)^2 + BE^2 - EV^2$$
$$= PE^2 + 2PE \cdot EZ + EZ^2 +$$
$$AE^2 - EZ^2 + PE^2 -$$
$$2PE \cdot EV + EV^2 + BE^2 - EV^2$$
$$= 2PE^2 + 2AE^2 = 2(AE^2 + IE^2).$$

The idea of using an origin from which to determine the (horizontal) coordinate of a point also is apparent in Fermat's treatment of the case of Apollonius' theorem with several noncollinear points. In that situation, he takes a base line such that all the points lie on one side and drops perpendiculars from the given points to the line. Not only does he use the horizontal coordinates GH, GL, GK in his analysis of the problem, but also he uses the vertical coordinates GA, HB, LD, KC measured along perpendiculars to the base line (Figure 11.2). In fact, he shows that the horizontal coordinate GM of the center O of the desired circle is given by $GM = \frac{1}{4}(GH + GL + GK)$ while the vertical coordinate $MO = \frac{1}{4}(GA + HB + LD + KC)$. The radius OP is determined by the equation

$$M = AO^2 + BO^2 + CO^2 + DO^2 + 4OP^2,$$

where M is the given area.

In his treatment of the general case of Apollonius' theorem, however, Fermat did not express the circle by means of an equation, probably because he was trying to write the text as Apollonius would have. But two years after he finished his reconstruction, he set down his new ideas on analytic geometry in his *Introduction to Plane and Solid Loci* with the sentence quoted in the opening of the chapter sounding the central theme. Fermat asserted that if in solving algebraically a geometric problem one ends up with an equation in two unknowns, the resulting solution is a locus, either a straight line or a curve, the points of which are determined by the motion of one endpoint of a variable line segment, the other endpoint of which moves along a fixed straight line.

Fermat's chief assertion in this brief introduction is that if the moving line segment makes a fixed angle with the fixed line, and if neither of the unknown quantities occurs to a power greater than the square, then the resulting locus will be a straight line, a circle, or one of the other conic sections. He proceeds to prove his result by a treatment of each of the various cases which occur. We illustrate his method first in the case of the straight line: "Let NZM be a straight line given in position, with point N fixed. Let NZ be equal to the unknown quantity A, and ZI, the line drawn to form the angle NZI, the other unknown quantity E. If D times A equals B times E, the point I will describe a straight line given in position."[4] Fermat thus begins with a single axis NZM and a linear equation (Figure 11.3).

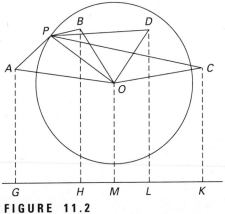

FIGURE 11.2
A second special case of Apollonius' theorem.

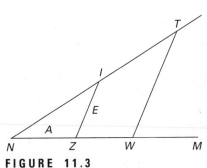

FIGURE 11.3
Fermat's analysis of the equation D times A equals B times E.

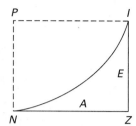

FIGURE 11.4
Fermat's analysis of the equation Aq equals D times E.

(Fermat uses Viète's convention of vowels for unknowns and consonants for knowns.) He wants to show that this equation, which we write as $dx = by$, represents a straight line. Since $D \cdot A = B \cdot E$, also $B : D = A : E$. Because $B : D$ is a known ratio, the ratio $A : E$ is determined, and so also the triangle NZI. Thus the line NI is, as Fermat says, given in position. Fermat dismisses as easy the necessary completion of the argument, to show that any point T on NI determines a triangle TWN with $NW : TW = B : D$.

Although the basic notions of modern analytic geometry are apparent in Fermat's description, Fermat's ideas differ somewhat from current ones. First, Fermat uses only one axis. The curve is thought of, not as made up of points plotted with respect to two axes, but as generated by the motion of the endpoint I of the variable line segment ZI as Z moves along the given axis. Fermat often takes the angle between ZI and ZN as a right angle, although there is no particular necessity for so doing. A second difference is that to Fermat, as to Viète and most others of the time, the only proper solutions to algebraic equations were positive. Thus Fermat's "coordinates" ZN and ZI, solutions to his equation $D \cdot A = B \cdot E$, represent positive numbers. Hence, Fermat only draws the ray emanating from the origin into the first quadrant.

Fermat's restriction to the first quadrant is quite apparent also in his treatment of the parabola: "If Aq equals D times E, point I lies on a parabola."[5] Fermat intended to show that the equation $x^2 = dy$ (in modern notation) determines a parabola. He began with the basic two line segments NZ and ZI, in this case at right angles. Drawing NP parallel to ZI, he then asserted that the parabola with vertex N, axis NP, and latus rectum D is the parabola determined by the given equation (Figure 11.4). In fact, Apollonius' construction of the parabola showed that the rectangle contained by D and NP was equal to the square on PI (or NZ), a statement translated into algebra by the equation $dy = x^2$. Although Fermat knew what a parabola looked like, his diagram only included part of half of it. He did not deal with negative lengths along the axis.

Fermat proceeded to determine the curves represented by five other quadratic equations in two variables. In modern notation, $xy = b$ and $b^2 + x^2 = ay^2$ represent hyperbolas, $b^2 - x^2 = y^2$ represents a circle, $b^2 - x^2 = ay^2$ represents an ellipse, and $x^2 \pm xy = ay^2$ represents a straight line. In each case, Fermat assumed that the reader was familiar with the Greek constructions of the relevant curve. Finally, Fermat sketched a method of reducing any quadratic equation to one of his seven canonical forms, by showing how to change variables. For example, he asserted that any equation containing ax^2 and ay^2 along with bx and/or cy can be reduced to the canonical equation of a circle, provided that the angle between the axis and the line tracing out the curve is a right angle. Thus, the equation

$$p^2 - 2hx - x^2 = y^2 + 2ky$$

is transformed by first adding k^2 to both sides so the right side becomes a square. Setting

$$r^2 = h^2 + k^2 + p^2 \quad \text{or} \quad r^2 - h^2 = k^2 + p^2,$$

Fermat could now rewrite the original equation as

$$r^2 - (x + h)^2 = (y + k)^2,$$

the canonical equation of a circle if $x + h$ is replaced by x' and $y + k$ by y'. Fermat also dealt with equations containing an xy term by an appropriate change of variable.

Fermat was able to determine the locus corresponding to any quadratic equation in two variables and show that it had to be a straight line, a circle, or a conic section. To conclude his *Introduction*, Fermat noted that one could apply his methods to the following generalization of the four-line-locus problem: "If, given any number of lines in position, lines be drawn from one and the same point to each of them at given angles, and if [the sum of] the squares of all of the drawn lines is equal to a given area, the point will lie on a [conic section] given in position."[6] Fermat, however, left the actual solution to the reader.

11.1.2 Descartes and the *Geometry*

Fermat's brief treatise created a stir when it reached Paris. A circle of mathematicians, centered on Marin Mersenne (1588–1648), had been gathering regularly to discuss new ideas in mathematics and physics. Mersenne acted as the recording and corresponding secretary of the group, and as such, received material from various sources, copied it, and distributed it widely. Mersenne thus served as France's "walking scientific journal." Fermat had begun a regular correspondence with Mersenne in 1636, but because many of his manuscripts were brief and lacking in detail, Mersenne often forwarded to Fermat requests to amplify his work. Nevertheless, the reception of Fermat's *Introduction* was positive and established his name as a first-class mathematician. The manuscript had, however, reached Paris and then Descartes just prior to the publication of Descartes' own version of analytic geometry. One can only imagine Descartes' chagrin at seeing material similar to his own appearing before his own work reached its intended audience.

Descartes' analytic geometry was, nevertheless, somewhat different from that of Fermat. To understand it, one must realize that the *Geometry* was written to demonstrate the application to geometry of Descartes' methods of correct reasoning discussed in the *Discourse*, reasoning based on self-evident principles. Like Fermat, Descartes had studied the works of Viète and saw in them the key to understanding the analysis of the Greeks. But rather than dealing with the relationship of algebra to geometry through the study of loci, Descartes was more concerned with demonstrating this relationship through the geometric construction of solutions to algebraic equations. In some sense, then, he was merely following in the ancient tradition, a tradition that had been continued by such Islamic mathematicians as al-Khayyāmī and Sharaf al-Dīn al-Ṭūsī. But Descartes did take the same crucial step as Fermat, a step which his Islamic predecessors failed to take, of using coordinates to study this relationship between geometry and algebra.

The *Geometry* begins, "Any problem in geometry can easily be reduced to such terms that a knowledge of the lengths of certain straight lines is sufficient for its construction."[7] To relate line lengths to numbers most clearly, one needs to pick a given length as a unit. The first of the three books of this work is devoted to showing how to construct geometric solutions to problems requiring only the use of lines and circles, the standard Euclidean curves. But Descartes makes these Euclidean techniques appear modern in his clear use of algebraic techniques. For example, to construct the solution of the quadratic equation

---

Biography René Descartes (1596–1650)

Descartes was born at La Haye (now La Haye-Descartes) near Tours. Because he was sickly throughout his youth, he was permitted during his school years to rise late. He thus developed the habit of spending his mornings in meditation. His thoughts led him to the conclusion that little he had learned in school was certain. In fact, he became so full of doubts that he decided to abandon his studies. As he reported in his *Discourse on Method*, "I used the rest of my youth to travel, to see courts and armies, to frequent people of differing dispositions and conditions, to store up various experiences, to prove myself in the encounters with which fortune confronted me, and everywhere to reflect upon the things that occurred, so that I could derive some profit from them."[8] In 1628 Descartes finally settled in Holland, where he felt able to accomplish his lifelong goal of creating a new philosophy suited to discovering truth about the world. He resolved to accept as true only ideas so clear and distinct that they would cause no doubt and then to follow the model of mathematical

reasoning through simple, logical steps to discern new truths. He soon wrote a major treatise on physics, but at the last minute, having heard of Galileo's condemnation by the Church, decided not to publish it for fear that a small doctrinal error might lead to the banning of his entire philosophy. He was soon persuaded, however, that he should share his new ideas with the world. In 1637 he published his *Discourse on Method*, along with three essays on optics, meteorology, and geometry designed to show the efficacy of the "method" (Figure 11.5).

Descartes' international reputation was enhanced with the publication of several other philosophical works, and, in 1649, he was invited by Queen Christina of Sweden to come to Stockholm to tutor her. He reluctantly accepted. Unfortunately, his health could not withstand the severity of the northern climate, especially since Christina required him, contrary to his long established habits, to rise at an early hour. Descartes soon contracted a lung disease which led to his death in 1650.

FIGURE 11.5
Descartes and his
Discours de la Méthode
on a French stamp.

$z^2 = az + b^2$, he constructs a right triangle *NLM* with $LM = b$ and $LN = \frac{1}{2}a$ (Figure 11.6). Prolonging the hypotenuse to *O*, where $NO = NL$ and constructing the circle centered on *N* with radius *NO*, he concludes that *OM* is the required value *z*, because the value of *z* is given by the standard formula

$$z = \frac{1}{2}a + \sqrt{\frac{1}{4}a^2 + b^2}.$$

Under the same conditions, *MP* is the solution to $z^2 = -az + b^2$, while if *MQR* is drawn parallel to *LM*, then *MQ* and *MR* are the two solutions to $z^2 = az - b^2$.

Descartes notes, however, that "often it is not necessary thus to draw the lines on paper, but it is sufficient to designate each by a single letter."[9] As long as it is known what operations are possible geometrically, it is feasible just to perform the algebraic operations and state the result as a formula. In these algebraic operations Descartes takes another major step. The terms a^2 and a^3—for which he is the first to use modern exponential notation consistently—are represented by Descartes as line segments, rather than as geometric squares and cubes. Thus he can also consider higher powers without worrying about their lack of geometric meaning. Descartes makes only a brief bow to the homogeneity requirements carefully kept by Viète by noting that any algebraic expression could

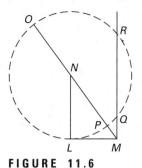

FIGURE 11.6
Descartes' construction
of the solution to a
quadratic equation.

be considered to include as many powers of unity as necessary for this purpose, but in fact he freely adds algebraic expressions whatever the power of the terms. Furthermore, Descartes replaces Viète's vowel-consonant distinction of unknowns and knowns with the current usage of letters near the end of the alphabet for unknowns and those near the beginning for knowns.

Descartes concludes his first book with a detailed discussion of Apollonius' problem of the four lines. It is here that he introduces a coordinate axis to which all the lines as well as the locus of the solution is referred. The problem requires the finding of points from which lines drawn to four given lines at given angles satisfy the condition that the product of two of the line lengths bears a given ratio to the product of the other two. Descartes notes that "since there are always an infinite number of different points satisfying these requirements, it is also required to discover and trace the curve containing all such points."[10]

Using Figure 11.7, Descartes notes that matters are simplified if all lines are referred to two principal ones. Thus he sets x as the length of segment AB along the given line EG and y as the length of segment BC along the line BC to be drawn, where C is one of the points satisfying the requirements of the problem. The lengths of the required line segments, CB, CH, CF, and CD (drawn to the given lines EG, TH, FS, and DR respectively), can each be expressed as a linear function of x and y. For example, because all angles of the triangle ARB are known, the ratio $BR : AB = b$ is also known. It follows that $BR = bx$ and $CR = y + bx$. Because the three angles of triangle DRC are also known, so is the ratio $CD : CR = c$, and therefore $CD = cy + bcx$. Similarly, setting the fixed distances $AE = k$ and $AG = \ell$ and the known ratios $BS : BE = d$, $CF : CS = e$, $BT : BG = f$, and $CH : TC = g$, one shows in turn that $BE = k + x$, $BS = dk + dx$, $CS = y + dk + dx$, $CF = ey + dek + dex$, $BG = \ell - x$, $BT = f\ell - fx$, $CT = y + f\ell - fx$, and finally $CH = gy + fg\ell - fgx$. Because the problem involves comparing the products of certain pairs of the line lengths, it follows that the equation expressing the desired locus is a quadratic equation in x and y. Furthermore, as many points of the locus as desired can be constructed, because if any value of y is given, the value of x is expressed in the form of a determinate quadratic equation whose solution has already been provided. The required curve can then be drawn. In book two of the *Geometry*, Descartes returns to this problem

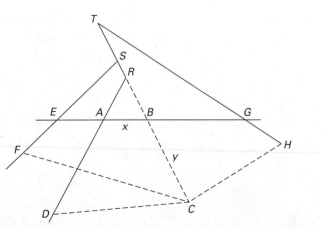

FIGURE 11.7
The problem of the four-
line locus.

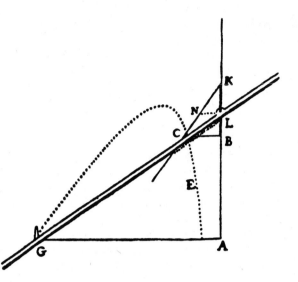

FIGURE 11.8
Descartes' curve-drawing
instrument.

and shows that the curve given by the quadratic equation in two variables is either a circle or one of the conic sections, depending on the values of the various constants involved.

Descartes' main concern in the *Geometry* was the actual construction of points that were the solution of geometric problems. Implicit in this work was the necessity of deciding what types of curves were legitimate in such a construction. Naturally, circles and straight lines, which he used to construct the points on the locus of the four-line problem, were among the legitimate curves and were the simplest of these. These were the curves used by Euclid. Other Greek authors, however, had not hesitated to use conic sections as well as certain other curves.[11]

Descartes decided to base his definition of curves acceptable in geometry on Euclid's Postulates 1 and 3 on drawing straight lines and circles and on the following new one: "Two or more lines can be moved, one upon the other, determining by their intersection other curves."[12] He therefore accepted only curves traced by some continuous motion generated by certain machines. It is not entirely clear today how Descartes would decide which curves fall under his terminology, but he gave several examples of instruments designed to trace such curves. For example, in Figure 11.8, *GL* is a ruler pivoting at *G*. It is linked at *L* with a device *CNKL* which allows *L* to be moved along *AB*, always keeping the line *KN* parallel to itself. The intersection *C* of the two moving lines *GL* and *KN* determines a curve. Descartes found the equation of this curve from simple geometrical considerations. Setting $CB = y$, $BA = x$, and the constants $GA = a$, $KL = b$, and $NL = c$, Descartes calculated that $BK = (b/c)y$, $BL = (b/c)y - b$ and $AL = x + (b/c)y - b$. Since $CB : BL = GA : AL$, Descartes derived the equation

$$\frac{ab}{c}y - ab = xy + \frac{b}{c}y^2 - by,$$

or finally,

$$y^2 = cy - \frac{c}{b}xy + ay - ac.$$

Descartes stated, without proof, that this curve is a hyperbola. Slight modifications of this machine enabled Descartes to produce curves whose equations were of higher degree.

It appears that Descartes' basic reason for defining what he called "geometric" curves was that "all points of those curves . . . must bear a definite relation to all points of a straight line, and that this relation must be expressed by means of a single equation."[13] In other words, any such curve must be expressible as an algebraic equation. It is apparent that Descartes also believed the converse of this statement, that any algebraic equation in two variables determines a curve whose construction could be realized by an appropriate machine. He could not prove such a statement, but he devoted much of the third book of the *Geometry* to showing how to construct points on algebraic curves of degree higher than two. Descartes was convinced that curves for which one could construct arbitrary points were amenable to a tracing by continuous motion by one of his machines.

There are probably several reasons why Descartes defined "geometric" curves by continuous motion rather than directly as curves having an algebraic equation. First, Descartes was interested in reforming the study of geometry. Defining acceptable curves by a completely algebraic criterion would have reduced his work to algebra. Second, because he wanted to be able to construct the solution points of geometric problems, he needed to be able to determine intersections of algebraic curves. It was evident to him that defining curves by continuous motion would explicitly determine intersection points. It was not at all evident that curves defined by algebraic equations had intersection points. In line with his basic philosophy Descartes could not adopt the algebraic definition as an axiom. Finally, Descartes evidently was not convinced that an algebraic equation was the best way to define a curve. Nowhere in the *Geometry* does he begin with an equation. Unlike Fermat, Descartes always describes a curve geometrically and then, if appropriate, derives its equation. An equation for Descartes is thus only a tool in the study of curves and not the defining criterion.

It must also be asked, on the other hand, why Descartes rejected curves not definable geometrically. He was certainly aware of curves without algebraic equations. An ancient example was the quadratrix, defined by a combination of a rotary and a linear motion. (See Figure 3.6, pg. 103.) What bothered Descartes about such a curve, as it had also bothered the ancients, was that the two motions had no exact, measurable relation, because one could not precisely determine the ratio of the circumference of the circle to its radius. As Descartes wrote, "the ratios between straight and curved lines are not known, and I believe cannot be discovered by human minds, and therefore no conclusion based upon such ratios can be accepted as rigorous and exact."[14] Unfortunately for Descartes, the first determination of the exact lengths of various curves in the 1650s as well as the study of areas under his nongeometric (or transcendental) curves soon undermined Descartes' basic distinction between acceptable and nonacceptable curves in geometry.

It is clear that both Fermat and Descartes understood the basic connection between a geometric curve and an algebraic equation in two unknowns. Both used as their basic tool a single axis along which one of the unknowns was measured rather than the two axes used today. Both used as their chief examples the familiar conic sections, although both were also able to construct curves whose equations were of degree higher than two. Finally, although both dealt with the analytic geometry of curves rather than functions, each understood in his own way the idea of a function, that change in one variable determined change in the other.

The two men came at the subject of analytic geometry, however, from different viewpoints. Fermat gave a very clear statement that an equation in two variables determines a curve. He always started with the equation and then described the curve. Descartes, on the other hand, was more interested in geometry. For him, the curves were primary. Given a geometric description of a curve, he was able to come up with the equation. Thus Descartes was forced to deal with algebraic equations considerably more complex than those of Fermat. It was the very complication of Descartes' equations that led him to discover methods of dealing with polynomial equations of high degree, methods to be discussed in section 11.2.2.

Descartes and Fermat emphasized the two different aspects of the relationship between equations and curves. Unfortunately, Fermat never published his work. Although it was presented clearly and circulated through Europe in manuscript, it never had the influence of a published work. Descartes' work, conversely, proved very difficult to read. It was published in French, rather than the customary Latin, and had so many gaps in arguments and complicated equations that few mathematicians could fully understand it. Descartes was actually proud of the gaps. He wrote at the end of the work, "I hope that posterity will judge me kindly, not only as to the things which I have explained, but also as to those which I have intentionally omitted so as to leave to others the pleasure of discovery."[15] But a few years after the publication of the *Geometry*, Descartes changed his mind somewhat. He encouraged other mathematicians to translate the work into Latin and to publish commentaries to explain what he had intended. It was only after the publication of the Latin version by Frans van Schooten (1615–1660), a professor at the engineering school in Leiden, first in 1649 with commentary by van Schooten himself and by Florimond Debeaune (1601–1652) and then with even more extensive commentaries and additions in 1659–1661, that Descartes' work achieved the recognition he desired.

11.1.3 The Work of Jan de Witt

One of the additions to van Schooten's 1659–1661 edition of Descartes' *Geometry* was a treatise on conic sections by Jan de Witt (1623–1672). In his student days, de Witt had studied with van Schooten, who had known Descartes and had studied Fermat's works during a sojourn in Paris. Through van Schooten, de Witt became acquainted with the works of both of the inventors of analytic geometry. In 1646, at the age of 23, he composed the *Elementa curvarum linearum* (*Elements of Curves*) in which he treated the subject of conic sections from both a synthetic and an analytic point of view. The first of the two books of the *Elements* was devoted to developing the properties of the various conic sections using the traditional methods of synthetic geometry. In the second book, the first systematic treatise on conic sections using the new method, de Witt extended Fermat's ideas into a complete algebraic treatment of the conics beginning with equations in two variables. Although the methodology was similar to that of Fermat, de Witt's notation was the modern one of Descartes.

For example, de Witt uses Fermat's methods to show that $y^2 = ax$ represents a parabola. He also shows the graphs of parabolas determined by such equations as $y^2 = ax + b^2$, $y^2 = ax - b^2$, $y^2 = b^2 - ax$ and the equations formed from these by interchanging x

Biography   **Jan de Witt (1623–1672)**

Jan de Witt was a talented mathematician who, because of his family background, could devote little time to mathematics (Figure 11.9). Born into a politically active Dutch family, he became a leader in his hometown of Dort and, after the death of Prince William II of Orange, was appointed in 1653 to the position of grand pensionary of Holland, in effect the prime minister. He guided Holland through difficult times over the next 19 years, successfully balancing the conflicting demands of England and France. When France invaded Holland in 1672, however, the people called William III to return to power. Violent demonstrations ensued against de Witt and he was murdered by an infuriated mob.

FIGURE 11.9
Jan de Witt on a Dutch stamp.

and y. Like his predecessor, however, he only displays that part of the graph for which both x and y are positive. But de Witt also considers the more complicated equation

$$y^2 + \frac{2bxy}{a} + 2cy = bx - \frac{b^2x^2}{a^2} - c^2.$$

Setting $z = y + bx/a + c$ reduces the equation to

$$z^2 = \frac{2bc}{a}x + bx,$$

or with $d = \frac{2bc}{a} + b$, to $z^2 = dx$, an equation which de Witt knows represents a parabola. He then shows how to use the transformation equations to draw the locus. Measuring x along AE and y along ED, he sets $BE = AG = c$ and extends DB to C such that $GB : BC = a : b$ or $BC = bx/a$ (Figure 11.10). It follows that $DC = y + c + bx/a = z$. Also, setting $GB : GC = a : e$ gives $GC = ex/a$, where GC is to be the axis. (In fact, the transformation equations have the effect of converting from the oblique axes along AE and ED to new perpendicular axes along DC and GC.) The length of the latus rectum, the ratio of the square on the ordinate $z = DC$ to the corresponding length GC of the abscissa, is

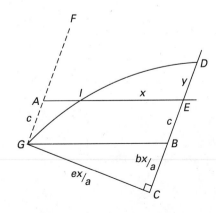

FIGURE 11.10
De Witt's construction of the parabola

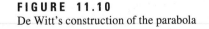

therefore $(dx)/(ex/a) = ad/e$. De Witt then draws the parabola with vertex G, axis GC, and latus rectum of length ad/e. Only the part ID above the axis AE, however, can serve as the desired locus. De Witt completes the proof by noting that given an arbitrary point D on this locus, the basic property of a parabola implies that the square on DC equals the rectangle on $GC(= ex/a)$ and the latus rectum ad/e. Thus $z^2 = dx$, and substituting $z = y + c + bx/a$ gives back the original equation.

De Witt similarly gives detailed treatments of both the ellipse and the hyperbola, presenting standard forms such as

$$\frac{ey^2}{g} = f^2 - x^2 \text{ (ellipse)}$$

and

$$\frac{ey^2}{g} = x^2 - f^2 \qquad \text{and} \qquad xy = f^2 \text{ (hyperbolas)}$$

first and then showing how other equations can be reduced to one of these by appropriate substitutions. Although de Witt does not state the conditions on the original equation which determine whether the locus is a parabola, ellipse, or hyperbola, it is easy enough to discover these by analyzing his examples. De Witt concludes his work by noting that any quadratic equation in two variables can be transformed into one of the standard forms and therefore represents a straight line, circle, or conic section. Although both Fermat and Descartes had sketched this same result, it was de Witt who provided all the details to solve the locus problem for quadratic equations.

11.2 THE THEORY OF EQUATIONS

Algebraic methods for solving cubic and quartic equations were discovered in Italy in the sixteenth century and improved on somewhat by Viète near the turn of the seventeenth century. But Cardano was hampered by a lack of a convenient notation and Viète always restricted himself to positive solutions. Thus even though the former gave various examples of relationships among the roots of a single cubic equation and between roots of related equations and the latter was able to express algebraically the relationship between the coefficients and the solutions of equations of degree up to five, provided all values were positive, the general theory was still incomplete. It was Albert Girard (1595–1632) in his 1629 work *Invention nouvelle en l'algèbre* (*A New Discovery in Algebra*) who gave a complete statement of this relationship, albeit without proof.

11.2.1 Albert Girard and the Fundamental Theorem of Algebra

Girard was probably born in St. Mihiel in the French province of Lorraine, but spent much of his life in the Netherlands, where he studied at Leiden and served as a military engineer in the army of Frederick Henry of Nassau. Although he wrote a work on trigonometry and edited the works of Stevin, his most important contributions are to algebra. In *A New Discovery in Algebra*, Girard clearly introduced the notion of a fractional exponent, "the numerator is the power and the denominator the root"[16] as well as the current notation for higher roots, that is, $\sqrt[3]{}$ for cube root as an alternative to exponent 1/3. Furthermore, he was among the first to note the geometric meaning of a negative solution to an equation: "The minus solution is explicated in geometry by retrograding; the minus goes backward where the plus advances."[17] He even gave an example of a geometric problem whose algebraic translation has two positive and two negative solutions. He noted on the relevant diagram that the negative solutions were to be interpreted as being laid off in the direction opposite that of the positive ones.

Not only did Girard understand the meaning of negative solutions to equations, but also he generalized the work of Viète and considered **factions**, today known as the elementary symmetric functions of n variables: "When several numbers are proposed, the entire sum may be called the first *faction*; the sum of all the products taken two by two may be called the second *faction*; the sum of all the products taken three by three may be called the third *faction*; and always thus to the end, but the product of all the numbers is the last *faction*. Now, there are as many *factions* as proposed numbers."[18] He pointed out that for the numbers 2, 4, 5, the first *faction* is 11, their sum; the second is 38, the sum of all products of pairs; while the third is 40, the product of all three numbers. He also noted that the Pascal triangle of binomial coefficients, which Girard called the "triangle of extraction," tells how many terms each of the *factions* contains. In the case of four numbers the first *faction* contains 4 terms, the second, 6, the third, 4, and the fourth and last, 1.

Girard's basic result in the theory of equations is the following

Theorem. *Every algebraic equation . . . admits of as many solutions as the denomination of the highest quantity indicates. And the first faction of the solutions is equal to the [coefficient of the second highest] quantity, the second faction of them is equal to the [coefficient of the third highest] quantity, the third to the [fourth], and so on, so that the last faction is equal to the [constant term]—all this according to the signs that can be noted in the alternating order.*[19]

What Girard meant by the last statement about signs is that one first needs to arrange the equation so that the degrees alternate on each side of the equation. Thus $x^4 = 4x^3 + 7x^2 - 34x - 24$ should be rewritten as $x^4 - 7x^2 - 24 = 4x^3 - 34x$. The roots of this equation being 1, 2, -3, and 4, the first *faction* is equal to 4, the coefficient of x^3; the second to -7, the coefficient of x^2; the third to -34, the coefficient of x; and the fourth to -24, the constant term. Similarly, the equation $x^3 = 167x - 26$ can be rewritten as $x^3 - 167x = 0x^2 - 26$. Because -13 is one solution, his result implies that the product of the two remaining roots is 2, while their sum is 13. To find these simply requires solving a quadratic equation. The answers are $6\ 1/2 + \sqrt{40\ 1/4}$ and $6\ 1/2 - \sqrt{40\ 1/4}$.

In the first part of the theorem, Girard asserts the truth of the fundamental theorem of algebra, that every polynomial equation has a number of solutions equal to its degree (denomination of the highest quantity). As his examples show, he acknowledged that a given solution could occur with multiplicity greater than one. He also fully realized that in his count of solutions he would have to include imaginary ones (which he called impossible). So in his example $x^4 + 3 = 4x$, he noted that the four *factions* are 0, 0, 4, 3. Because 1 is a solution of multiplicity 2, the two remaining solutions have the property that their product is 3 and their sum is -2. It follows that these are $-1 \pm \sqrt{-2}$. In answer to the anticipated question of the value of these impossible solutions, Girard answered that "they are good for three things: for the certainty of the general rule, for being sure that there are no other solutions, and for its utility."[20]

11.2.2 Descartes and Equation Solving

Girard used his *factions* in certain cases to reduce the degree of an equation once one solution α was known. The standard method of dividing the original polynomial by $x - α$ was first explicated by Descartes in the third book of his *Geometry*. Descartes begins his own study of equations by quoting—almost—Girard's result that "every equation can have as many distinct roots as the number of dimensions of the unknown quantity in the equation."[21] Descartes uses "can have" rather than Girard's "admits of" because he only considers distinct roots and because, at least initially, he does not want to consider imaginary roots. Later on, however, he does note that roots are sometimes imaginary and that "while we can always conceive of as many roots for each equation as I have already assigned [that is, as many as the dimension], yet there is not always a definite quantity corresponding to each root so conceived of."[22] Descartes goes further than Girard by showing how equations are built up from their solutions. Thus if $x = 2$ or $x - 2 = 0$ and if also $x = 3$ or $x - 3 = 0$, Descartes notes that the product of the two equations is $x^2 - 5x + 6 = 0$, an equation of dimension 2 with the two roots 2 and 3. Again, if this latter equation is multiplied by $x - 4 = 0$, there results an equation of dimension 3, $x^3 - 9x^2 + 26x - 24 = 0$, with the three roots 2, 3, and 4. Multiplying further by $x + 5 = 0$, an equation with a "false" root 5, produces a fourth degree equation with four roots, three "true" and one "false."

Descartes concludes that "it is evident from the above that the sum of an equation having several roots [that is, the polynomial itself] is always divisible by a binomial consisting of the unknown quantity diminished by the value of one of the true roots, or plus the value of one of the false roots. In this way, the degree of an equation can be lowered. Conversely, if the sum of the terms of an equation is not divisible by a binomial consisting of the unknown quantity plus or minus some other quantity, then this latter quantity is not a root of the equation."[23] This is the earliest statement of the modern factor theorem. In his usual fashion, Descartes does not give a complete proof. He just says that the result is "evident." Similarly, he also states without proof the result today known as **Descartes' Rule of Signs**: "An equation can have as many true [positive] roots as it contains changes of sign, from + to − or from − to +; and as many false [negative] roots as the number of times two + signs or two − signs are found in succession."[24] As illustration, the equation $x^4 - 4x^3 - 19x^2 + 106x - 120 = 0$ has three changes of sign and one pair of consecutive − signs. Thus it can have up to three positive roots and one negative one. In fact, the roots are 2, 3, 4, and −5.

Descartes is, however, primarily interested in constructing solutions to equations, so toward the end of the third book he demonstrates explicitly some methods for equations of higher degree. In particular, for equations of degree 3 or 4, he uses the intersection of a parabola and a circle, both of which meet Descartes' criteria for constructible curves. Descartes' methods are similar to those of al-Khayyāmī, but, unlike his Islamic predecessor, Descartes realized that certain intersection points represented negative (false) roots of the equation and also that "if the circle neither cuts nor touches the parabola at any point, it is an indication that the equation has neither a true nor a false root, but that all the roots are imaginary."[25]

Descartes showed further how to solve equations of degree higher than the fourth by intersecting a circle with a curve constructed by one of his machines. Although he only briefly sketched his methods and applied them to a few examples, Descartes believed that it was "only necessary to follow the same general method to construct all problems, more and more complex, ad infinitum; for in the case of a mathematical progression, whenever the first two or three terms are given, it is easy to find the rest."[26] Over the remainder of the century, various mathematicians attempted to generalize Descartes' methods to find other geometrical means for constructing the solutions to various types of equations. Geometric methods, however, proved inadequate to gaining a complete understanding of the nature of such solutions. It turned out that algebraic methods as well as the new ideas of calculus were better suited for solving even the kinds of geometrical problems to which Descartes applied his construction techniques.

11.3 ELEMENTARY PROBABILITY

The modern theory of probability is usually considered to begin with the correspondence of Pascal and Fermat in 1654, partially in response to the gambling questions de Méré raised to Pascal which are noted in the opening of the chapter. But because gambling is one of the oldest leisure activities, it would seem that from earliest times people probably had considered the basic ideas of probability, at least on an empirical basis, and, in particular, had some vague conception of how to calculate the odds of the occurrence of any given event in a gambling game. Dice from several ancient cultures have been found. Although it is not always known what the purpose of these objects was, there are strong indications that they were used for predicting the future and for gaming. Unfortunately, no written evidence survives from any of these civilizations about how the various games were played and whether any calculations of odds were made.

11.3.1 The Earliest Beginnings of Probability Theory

In Europe in the late middle ages some elementary probabilistic ideas connected with dice playing are spelled out. For example, there are several documents that calculate the number of different ways two or three dice can fall, 21 ways in the case of two dice and 56 in the case of three. These numbers are correct, assuming one only counts the different sets of dots which can occur, without examining the order in which they happen. Thus, in the case of two dice, there is one way to roll a 2, one way to roll a 3, two ways to roll a 4 (2, 2 and 1, 3), two ways to roll a 5 (1, 4 and 2, 3), and so on. In modern terms, these ways are not "equiprobable" (equally likely) and could not serve as the basis for calculating odds in play. But counting the ways the dice could fall most likely came from the earlier use of dice in divination where it was the actual dice faces showing that determined the future and where odds were not involved. The earliest known comment that the 56 ways three dice fall are not equiprobable occurs in an anonymous Latin poem *De Vetula* written sometime between 1200 and 1400: "If all three [dice] are alike there is only one way for each number; if two are alike and one different there are three ways; and if all are different there are six ways."[27] An analysis of the situation according to the stated rule then shows that the total number of ways for three dice to fall is 216 (Figure 11.11).

By the sixteenth century the idea of equiprobable events was beginning to be understood and thus it became possible for actual probability calculations to be made. The earliest systematic attempt to make these calculations is in the *Liber de Ludo Aleae* (*Book on Games of Chance*) written about 1526 by Cardano, although not published during his lifetime. Besides counting accurately the number of ways two or three dice can fall, Cardano demonstrated an understanding of the basic notions of probability. Thus, having counted that there are 11 different throws of two dice in which a 1 occurs, 9 additional ones in which a 2 occurs, and 7 more in which a 3 occurs, he calculated that for the problem of throwing a 1, 2, or 3 there are 27 favorable occurrences and 9 unfavorable ones, and therefore the odds are 3 to 1. It follows that a fair wager would be 3 coins for the one betting on getting a 1, 2, or 3 versus 1 coin for the player betting against, because in four throws they would expect to come out even.

Cardano also was aware of the multiplication rule of probabilities for independent events, but in his book he recorded his initial confusion as to what exactly should be multiplied. Thus he calculated that the chances of at least one 1 appearing in a toss of three dice is 91 out of 216, so the odds against are 125 to 91. To determine the odds against throwing at least one 1 in two successive rolls, he squared the odds and calculated the result as 15,625 to 8281, or approximately 2 to 1. After consideration of the matter,

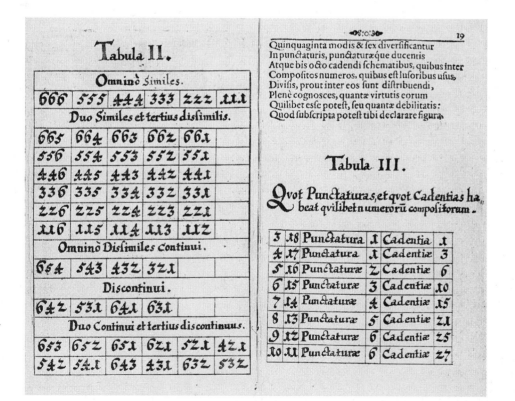

FIGURE 11.11
Manuscript page from *De Vetula* showing all 56 ways three dice can fall. (Source: The Houghton Library, Harvard University)

however, he noted that this reasoning must be false, because if the chances of a given event are even (odds of 1 to 1), the reasoning would imply that the chances would still be even of the given event occurring twice or three times in succession. This, he noted, is "most absurd. For if a player with two dice can with equal chances throw an even and an odd number, it does not follow that he can with equal fortune throw an even number in each of three successive casts."[28] Cardano then proceeded to correct his error. After careful calculation in some easy cases he realized that it is the probabilities which must be multiplied and not the odds. Thus by counting in a case where the odds for success are 3 to 1, or the probability of success is 3/4, he showed that for two successive plays, there are 9 chances of repeated success and 7 otherwise. Therefore, the probability of succeeding twice is 9/16, while the odds are 9 to 7 in favor. He then generalized and noted that for n repeated trials in a situation with f possible outcomes and s successes, the correct odds in favor are s^n to $f^n - s^n$.

Cardano also discussed the problem de Méré had posed to Pascal of determining how many throws must be allowed to provide even odds for attaining two sixes on a pair of dice, a problem which evidently was popular for years. Cardano argued that since there is one chance in 36 of throwing two sixes, on average such a result will occur once every 36 rolls. Therefore, the odds are even that one will occur in half that number of rolls, or 18. He similarly argued that in dealing with one die, there are even odds that a 2 would appear in three rolls. Cardano's reasoning implies that in six rolls of one die a 2 is certain or in 36 rolls of two dice a double 6 is certain, but he did not realize his error.

The problem posed by de Méré to Pascal on the division of the stakes had also been considered earlier in Italy, in particular, in the *Summa* of Luca Pacioli. Pacioli's version of the problem has two players playing a fair game which was to continue until one player had won six rounds. The game actually stops when the first player has won five rounds and the second three. Pacioli's answer to the division of the stakes was that they should be split in the ratio of 5 to 3. Tartaglia, in his *Generale Trattato* written some 60 years later, noted that this answer must be wrong, for the reasoning implied that if the first player had won one round and the second none when the game was suspended, the first player would collect all of the stakes, an obviously unjust result. Tartaglia argued that since the difference between the two scores was two games, one-third of the number needed to win, the first player should take one-third of the second's share of the stake, and therefore the total stake should be divided in the ratio of 2 to 1. Tartaglia, however, was evidently not entirely confident of his answer either, for he concluded that "the resolution of such a question is judicial rather than mathematical, so that in whatever way the division is made there will be cause for litigation."[29]

11.3.2 Blaise Pascal, Probability, and the Pascal Triangle

The ideas of Cardano and Tartaglia on probability, however, were not taken up by others of their time and were forgotten. It was only in the decade surrounding 1660 that probability entered European thought and then in two senses, first as a way of understanding stable frequencies in chance processes, and second as a method of determining reasonable degrees of belief. The work of Blaise Pascal (1623–1662) exemplifies both of these senses. In his mathematical answer to de Méré's division problem Pascal deals with a game of chance, while in his decision-theoretic argument for belief in God there is no concept whatever of chance.

Biography   **Blaise Pascal (1623–1662)**

Pascal was born at Clermont Ferrand, France, and showed his mathematical precocity very early. His father Étienne introduced him as a young man to the circle around Mersenne. Thus the young Pascal was soon acquainted with the major mathematical developments in France, including the work of Fermat. He began his own mathematical and scientific researches before he was 20. Among his accomplishments was the invention of a calculating machine and the investigation of the action of fluids under the pressure of air. After 1654, however, his scientific interests were overshadowed by an increasing interest in religious affairs. Much of the rest of his life was devoted to theological considerations. Never in good health, he died at the age of 39 after a violent illness (Figure 11.12).

FIGURE 11.12
Pascal on a French stamp.

Pascal described his solution to the division problem in several letters to Fermat in 1654 and then in more detail a few years later at the end of his *Traité du triangle arithmétique* (*Treatise on the Arithmetical Triangle*). He began with two basic principles to apply to the division. First, if the position of a given player is such that a certain sum belongs to him whether he wins or loses, he should receive that sum even if the game is halted. Second, if the position of the two players is such that if one wins, a certain sum belongs to him and if he loses, it belongs to the other, and if both players have equally good chances of winning, then they should divide the sum equally if they are unable to play.

Pascal next noted that what determines the split of the stakes is the number of games remaining and the total number that the rules say either player must win to obtain the entire stake. Therefore, if they are playing for a set of two games with a score of 1 to 0 or for a set of 11 games with the score 10 to 9, the results of the division of the stakes at the time of interruption should be the same. In both cases, the first player needs to win 1 more game, while the second player needs 2.

As an example of Pascal's principles, suppose that the total stake in the contest is 80 dollars. First, if each player needs one game to win and the contest is stopped, simply divide the 80 dollars in half, so each gets 40. Second, suppose that the first player needs one game to win and the second player two. If the first player wins the next game, he will win the 80 dollars. If he loses, then both players will need one game, so by the first case, the first player will win 40 dollars. If they stop the contest now, the first player is therefore entitled to the 40 dollars he would win in any case plus half of the remaining 40, that is, to 60 dollars, the mean of the two possible amounts he could win. Similarly, if the first player needs one game to win while the second player needs three games, there are two possibilities for the next game. If the first player triumphs, he wins the 80 dollars, while if he loses, the situation is the same as in the second case, in which he is entitled to 60 dollars. It follows that if that next game is not played, the first player should receive 60 dollars plus half of the remaining 20, that is, 70 dollars, the mean of his two possible winnings.

The general solution to the division problem, it turns out, requires some of the properties of Pascal's triangle. Before considering Pascal's solution, therefore, we must first consider his construction and use of what he called the **arithmetical triangle**, the triangle of numbers which had been used in various parts of the world already for more than 500 years. Pascal's *Treatise on the Arithmetical Triangle*, famous also for its explicit statement

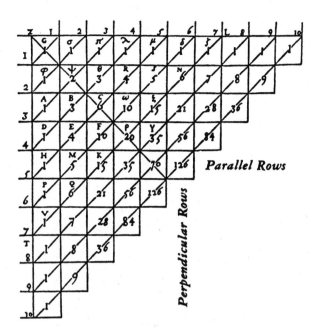

FIGURE 11.13
Pascal's version of the arithmetical triangle.

of the principle of mathematical induction, begins with his construction of the triangle starting with a 1 in the upper left hand corner and then using the rule that each number is found by adding together the number above it and the number to its left (Figure 11.13). In the discussion of Pascal's results, however, it will be clearer to use the modern table and modern notation to identify the various entries in the triangle. The standard binomial symbol $\binom{n}{k}$ will be used to name the kth entry in the nth row (where the initial column and initial row are each numbered 0). The basic construction principle is then that

$$\binom{n}{k} = \binom{n-1}{k} + \binom{n-1}{k-1}.$$

Row/Column	0	1	2	3	4	5	6	7	8
0	1								
1	1	1							
2	1	2	1						
3	1	3	3	1					
4	1	4	6	4	1				
5	1	5	10	10	5	1			
6	1	6	15	20	15	6	1		
7	1	7	21	35	35	21	7	1	
8	1	8	28	56	70	56	28	8	1

Pascal begins his study by considering how various entries are related to sums of others. His proofs are usually by the method of "generalizable example," because, like his forebears, he had no good way of symbolizing general terms. For example, Pascal's "third consequence" (of the definition of the triangle) states that any entry is the sum of all the elements in the preceding column up to the preceding row:

$$\binom{n}{k} = \sum_{j=k-1}^{n-1} \binom{j}{k-1}.$$

In this case, Pascal takes as his example the particular entry $\binom{4}{2}$, which, by the method of construction, is equal to $\binom{3}{1} + \binom{3}{2}$. Because $\binom{3}{2} = \binom{2}{1} + \binom{2}{2}$ and $\binom{2}{2} = \binom{1}{1}$, the result follows. Pascal gives a similar proof of the eighth consequence, that the sum of the elements in the nth row is equal to 2 raised to the nth power:

$$\sum_{k=0}^{n} \binom{n}{k} = 2^n.$$

It is in the proof of the twelfth consequence that Pascal states and uses the principle of mathematical induction, but in the context of the specific result to be proved,

$$\binom{n}{k} : \binom{n}{k+1} = k+1 : n-k.$$

Pascal notes that "although this proposition has an infinity of cases, I shall demonstrate it very briefly by supposing two lemmas," namely, the two basic parts of an induction argument. "The first, which is self evident, [is] that this proportion is found in the [first row], for it is perfectly obvious that $\left[\binom{1}{0} : \binom{1}{1}\right] = $]1 : 1. The second [is] that if this proportion is found in any [row], it will necessarily be found in the following [row]. Whence it is apparent that it is necessarily in all the [rows]. For it is in the second [row] by the first lemma; therefore by the second lemma it is in the third [row], therefore in the fourth, and to infinity."[30] Although this is a clear statement of the induction principle for the specific case at hand, and of the reason for its use in demonstrating a general result, Pascal does not prove the second lemma generally. As before, he only shows, as a generalizable example, that the truth of the lemma in the third row implies its truth in the fourth. Thus, to demonstrate that $\binom{4}{1} : \binom{4}{2} = 2 : 3$, he first notes that $\binom{3}{0} : \binom{3}{1} = 1 : 3$ and therefore that $\binom{4}{1} : \binom{3}{1} = \left(\binom{3}{1} + \binom{3}{0}\right) : \binom{3}{1} = 4 : 3$. Next, since $\binom{3}{1} : \binom{3}{2} = 2 : 2$, it follows that $\binom{4}{2} : \binom{3}{1} = \left(\binom{3}{2} + \binom{3}{1}\right) : \binom{3}{1} = 4 : 2$. The desired result comes from dividing the first of these two proportions by the second. Pascal is aware that this proof is not general, for he completes it by noting that "the proof is the same for all other [rows], since it requires only that the proportion be found in the preceding [row], and that each [entry] be equal to the [entry above it and the entry to the left of that one], which is everywhere the case."[31] In any case, this twelfth consequence does enable Pascal to demonstrate easily, by compounding ratios, that

$$\binom{n}{k} : \binom{n}{0} = (n-k+1)(n-k+2) \cdots n : k(k-1) \cdots 1$$

or, since $\binom{n}{0} = 1$, that

$$\binom{n}{k} = \frac{n(n-1) \cdots (n-k+1)}{k!}.$$

Having set out the basic properties of the arithmetical triangle, Pascal shows how to apply it. For example, he demonstrates, using an argument by induction, that $\binom{n}{k}$ equals the number of combinations of k elements in a set of n elements. Also, he shows that the row entries in the triangle are the binomial coefficients, that is, that the numbers in row n are the coefficients of the powers of a in the expansion $(a + 1)^n$. But one of the most important applications of the triangle, Pascal believed, was to the problem of the division of stakes. He solved this in the following

Theorem. *Suppose that the first player lacks r games of winning the set while the second player lacks s games where both r and s are at least 1. If the set of games is interrupted at this point, the stakes should be divided so that the first player gets that proportion of the total as $\sum_{k=0}^{s-1} \binom{n}{k}$ is to 2^n, where $n = r + s - 1$ (the maximum number of games left).*

The theorem asserts that the probability of the first player winning is the ratio of the sum of the first s terms of the binomial expansion of $(1 + 1)^n$ to the total 2^n. One can consider the first term of the expansion as giving the number of chances for the first player to win n points, the second the number of chances to win $n - 1$, and so on, while the sth term gives the number of chances to win $n - (s - 1) = r$ points. Since one may as well assume that in fact exactly n more games must be played, these coefficients give all of the ways the first player can win.

Pascal proves the theorem by induction, beginning with the case where $n = 1$, or $r = s = 1$, the case where the stakes should be evenly split. The assertion of the theorem is that they should be divided so that the first player gets the proportion $\binom{1}{0}$ to 2, or 1/2, and therefore the result is true for $n = 1$. The next step is to assume that the result is true when the maximum number of games left is m and prove it for the case where the maximum number of games left is $m + 1$, where the first player lacks r games and the second s games. As before, Pascal's proof of this inductive step is by a generalizable example, taking $m = 3$. But we will provide the complete proof, using modern notation. Consider the two possibilities if the players were to play one more game. If the first player wins, he would then lack $r - 1$ games while the second player would still lack s games. Since $r - 1 + s - 1 = m$, the induction hypothesis shows that the first player should get that proportion of the stakes that $\sum_{k=0}^{s-1} \binom{m}{k}$ is to 2^m. On the other hand, if the first player loses the next game, the induction hypothesis shows that he should be awarded that proportion of the stakes that $\sum_{k=0}^{s-2} \binom{m}{k}$ is to 2^m. Thus, by Pascal's basic principles, the award to the first player if that next game is not played should be the mean of those two values, namely that proportion of the stakes which

$$\sum_{k=0}^{s-1} \binom{m}{k} + \sum_{k=0}^{s-2} \binom{m}{k}$$

is to $2 \cdot 2^m$. The sum of binomial coefficients can be rewritten as

$$\binom{m}{0} + \sum_{k=1}^{s-1} \binom{m}{k} + \sum_{k=1}^{s-1} \binom{m}{k-1}.$$

By the rule for construction of the arithmetic triangle, and because $\binom{m}{0} = \binom{m+1}{0}$, this sum is in turn equal to

$$\sum_{k=0}^{s-1} \binom{m+1}{k}.$$

Since $2 \cdot 2^m = 2^{m+1}$, the award to the first player is precisely as asserted by the theorem for the case $n = m + 1$, and the proof is complete.

Pascal had thus answered completely de Méré's problem of division. In his correspondence with Fermat, the two men discussed the same problem when there were more than two players and found themselves in agreement on the solution. Pascal also mentioned briefly the other problem, of determining the number of throws of two dice for which there are even odds that a pair of sixes will occur. He noted that in the analogous problem for one die, the odds for throwing a six in four throws are 671 to 625, but did not show his method of calculating the result. De Méré evidently believed that since four throws were sufficient to guarantee at least even odds in the case of one die (where there are six possible outcomes), the same ratio 4 : 6 would hold no matter how many dice were thrown. Because there were 36 possibilities in tossing two dice, he thought that the correct value should be 24. He probably posed the question to Pascal because this value did not seem to be empirically correct. Pascal noted that the odds are against success in 24 throws, but did not detail in his letters or in any other work the theory behind this statement.

Pascal's decision-theoretic argument in favor of belief in God demonstrates the second side of probabilistic reasoning, a method of coming to a "reasonable" decision. Either God is or God is not, according to Pascal. One has no choice but to "wager" on which of these statements is true, where the wager is in terms of one's actions. In other words, a person may act either with complete indifference to God or in a way compatible with the (Christian) notion of God. Which way should one act? If God is not, it does not matter much. If God is, however, wagering that there is no God will bring damnation while wagering that God exists will bring salvation. Because the latter outcome is infinitely more desirable than the former, the outcome of the decision problem is clear, even if one believes that the probability of God's existence is small: the "reasonable" person will act as if God exists.

11.3.3 Christian Huygens and the Earliest Probability Text

FIGURE 11.14
Christian Huygens on a Dutch stamp.

Pascal's argument in favor of belief in God is certainly valid, given his premises. (Whether one accepts the premises is a different matter.) In fact, his notion of somehow calculating the "value" of a particular action became the basis for the first systematic treatise on probability, written in 1656 by Christian Huygens (1629–1695), a student of van Schooten (Figure 11.14). Huygens became interested in the question of probability during a visit to Paris in 1655 and wrote a brief book on the subject, the *De Ratiociniis in Aleae Ludo* (*On the Calculations in Games of Chance*), which appeared in print in 1657.

Huygens' work contained only 14 propositions and concluded with five exercises for the reader. The propositions included ones dealing with both of de Méré's problems, but Huygens also gave detailed discussions of the reasoning behind the solutions, in particular how to calculate in a game of chance: "Although in a pure game of chance the results are uncertain, the chance that one player has to win or to lose depends on a determined value."[32] Huygens' "value" is similar to Pascal's notion in his wager, but in the case of games of chance, Huygens could calculate it explicitly. In modern terms, the "value" of a chance is the **expectation**, the average amount that one would win if one played the game many times. It is this amount that a player would presumably pay to have the privilege of playing an equitable game. For example, Huygens' first proposition is: "To have equal chances of winning [amounts] a or b is worth $(a+b)/2$ to me."[33] This proposition is the same as one of the principles Pascal stated in solving the division problem. Huygens, however, gave a proof. He postulated two players each putting in a stake of $(a+b)/2$ with each player having the same chance of winning. If the first wins, he receives a and his opponent b. If the second wins, the payoffs are reversed. Huygens considered this an equitable game. In modern terminology, since the probability of winning each of a or b is 1/2, the expectation for each player is $(1/2)a + (1/2)b$, Huygens' "value" of the chance.

Huygens generalized this result in his third proposition: "To have p chances to win a and q chances to win b, the chances being equivalent, is worth $(pa + qb)/(p + q)$ to me."[34] In other words, if $p + q = r$, if the probability of winning a is p/r, and if the probability of winning b is q/r, then the expectation is given by $(p/r)a + (q/r)b$. Huygens proved this result by imbedding the problem in a symmetric game played by $p + q$ players arranged in a circle, each of whom puts in the same stake x and each of whom has the same chance of winning.[35] If a given player wins, he takes the entire stake, pays b to each of the $q - 1$ players to his left and a to each of the p players to his right, retaining the remainder. To make this remainder equal to b, it must be true that

$$(p + q)x - (q - 1)b - pa = b \qquad \text{or} \qquad x = \frac{pa + qb}{p + q}.$$

But now it is clear that each player has q chances of winning b and p chances of winning a, so the game is equitable, and each player should be willing to risk the stated stake.

Huygens took as an axiom that each player in an equitable game would be willing to risk the calculated fair stake and would not be willing to risk more. In fact, however, as the history of gambling shows, that assumption is, at the very least, debatable. It is not at all clear that the fair stake defined by Huygens is the most a given person is willing to pay for the chance to participate in a game. The success of state run lotteries, not to mention the gambling palaces in Las Vegas and Atlantic City, testifies to precisely the opposite. Nevertheless, Huygens based the remainder of his treatise on the results of his third proposition, and even today the concept of expectation is considered a useful one.

Huygens' discussion of de Méré's problem of division was similar to Pascal's, but he gave a more extensive analysis of the problem of the dice in proposition 11. He showed how to determine the number of times two dice should be thrown, so that one would be willing to wager $(1/2)a$ in order to win a if two sixes appear in that many plays. Huygens proceeded in stages. Supposing that one wins a when two sixes turn up, he argued that on the first throw the player has 1 chance of winning a and 35 chances of winning 0, so the value of a chance on one throw is $(1/36)a$. If the player fails on the first throw, he takes a second, whose value is naturally the same $(1/36)a$. Hence for the first throw the player has 1 chance of winning a and 35 chances of taking the second throw, which is worth $(1/36)a$. The value of his chance of throwing a double six on the two throws is, by the third proposition,

$$\frac{1a + 35(1/36)a}{1 + 35} \qquad \text{or} \qquad \frac{71}{1296}a.$$

Huygens next moved to the case of four throws. If the player gets a double six on one of the first two plays, he wins a; if not, he has a second pair of chances, the value of which is $(71/1296)a$. Since there are 71 chances of winning a on the first pair of plays and therefore 1225 chances of not winning (out of 1296), there are 1225 chances of reaching the second pair of plays, whose value is $(71/1296)a$. Again by the third proposition, the value of the player's chance on a double six in four throws is

$$\frac{71a + 1225(71/1296)a}{1296} \quad \text{or} \quad \frac{178991}{1679616}a.$$

Since this value is still considerably less than the desired $(1/2)a$, Huygens had to continue the process. Although he did not present any further calculations, he noted that one next considers 8 throws, then 16, and then 24 and 25. The results show that in 24 throws the player is at a very slight disadvantage on the bet of $(1/2)a$ while for 25 he has a very slight advantage.[36]

At the conclusion of his little treatise, Huygens presented as exercises some problems of drawing different colored balls from urns, problems of the type which today appear in every elementary probability text. These problems were discussed by many mathematicians over the next decades, especially since Huygens' text was the only introduction available to the theory of probability until the early eighteenth century. Even then, its influence continued, because Jakob Bernoulli incorporated it into his own more extensive work on probability, the *Ars Conjectandi* of 1713.

11.4 NUMBER THEORY

Fermat, involved in the beginnings of analytic geometry and probability, also made contributions to number theory, contributions which were virtually ignored during his lifetime and indeed until the middle of the next century. One of the reasons for this was probably his deep secrecy about his methods. Thus, although many of his results are known, because he announced them proudly in letters to his various correspondents and presented them with challenges to solve similar problems, there is virtually no record of any of his proofs and only vague sketches of some of his methods.

Fermat's earliest interest in number theory grew out of the classical concept of a perfect number, one equal to the sum of all of its proper divisors. Book IX of Euclid's *Elements* contains a proof that if $2^n - 1$ is prime, then $2^{n-1}(2^n - 1)$ is perfect. The Greeks had, however, only been able to discover four perfect numbers, 6, 28, 496, and 8128, because it was difficult to determine the values of n for which $2^n - 1$ is prime. Fermat discovered three propositions which could help in this regard, propositions which he communicated to Mersenne in a letter of June, 1640. The first of these results was that if n is not itself prime, then $2^n - 1$ cannot be prime. The proof of this result just exhibited the factors: If $n = rs$, then $2^n - 1 = 2^{rs} - 1 = (2^r - 1)(2^{r(s-1)} + 2^{r(s-2)} + \cdots + 2^r + 1)$. The basic question therefore reduced to asking for which primes p is $2^p - 1$ prime.

Such primes are today called Mersenne primes in honor of Fermat's favorite correspondent.

Fermat's second proposition was that if p is an odd prime, then $2p$ divides $2^p - 2$, or p divides $2^{p-1} - 1$. His third was that, with the same hypothesis, the only possible divisors of $2^p - 1$ are of the form $2pk + 1$. Fermat indicated no proofs of these results in his letter, but only gave a few numerical examples. He confirmed that $2^{37} - 1$ was composite by testing its divisibility by numbers of the form $74k + 1$ until he found the factor $223 = 74 \cdot 3 + 1$. But in a letter written a few months later to Bernard Frenicle de Bessy (1612–1675) he stated a more general theorem of which these two propositions are easy corollaries. This theorem, today known as Fermat's Little Theorem, is, in modern terminology, that if p is any prime and a any positive integer, then p divides $a^p - a$. (The theorem is often written in the form $a^p \equiv a \pmod{p}$, or, adding the condition that a and p are relatively prime, in the form $a^{p-1} \equiv 1 \pmod{p}$. It follows that if n is the smallest positive integer such that p divides $a^n - 1$, then n divides $p - 1$ and, in addition, that all powers k such that p divides $a^k - 1$ are multiples of n.) Fermat gave no indication in any of his writings how he discovered or proved this result. In any case, the second of the propositions in the letter to Mersenne is simply the case $a = 2$ of the theorem (where $p > 2$). The third proposition requires only a bit more work. Suppose q is a prime divisor of $2^p - 1$. The theorem then implies that p divides $q - 1$ or that $q - 1 = hp$ for some integer h. Since $q - 1$ is even, 2 must divide hp and therefore must divide h. It follows that $h = 2k$ and $q = 2kp + 1$ as asserted.

Fermat's Little Theorem turned out to be an extremely important result in number theory with many applications. But his work on another aspect of primality showed that even Fermat could be mistaken. In his correspondence he repeatedly asserted that the so-called Fermat numbers, those of the form $2^{2^n} + 1$, were all prime. As late as 1659, he wrote that he had found a proof. It is not difficult to show that the numbers of this form for $n = 0, 1, 2, 3,$ and 4 are prime. But Leonhard Euler discovered in 1732 that 641 was a factor of $2^{2^5} + 1$, and, in fact, no prime Fermat numbers have been found beyond $2^{2^4} + 1$. How did Fermat make such an error? It is quite likely that his attempted proof was of the type he outlined in another area of his number theoretic work, the method of infinite descent, and that he simply believed that the methods used for integers up to 4 would work for larger ones.

The method of infinite descent is involved in the only number theoretic proof which Fermat actually wrote out in detail, the problem of finding an integral right triangle whose area is a given number. Fermat commented that the area of such a triangle cannot be a square, that is, it is impossible to find integers x, y, z, w such that $x^2 + y^2 = z^2$ and $(1/2)xy = w^2$. The method of infinite descent, by which Fermat proved this result, demonstrates the nonexistence of positive integers having certain properties by showing that the assumption that one integer has such a property implies that a smaller one has the same property. By continuing the argument, one gets an infinite decreasing sequence of positive integers, an impossibility.

One can pull out of Fermat's proof of this result another argument by infinite descent showing that one cannot find three positive integers a, b, c such that $a^4 - b^4 = c^2$. It follows that one also cannot express a fourth power as a sum of two other fourth powers. It is the generalization of this result, to the effect that "one cannot split a cube into two cubes, nor a fourth power into two fourth powers, nor in general any power beyond the square *in infinitum* into two powers of the same name,"[37] which Fermat wrote as a marginal note to Diophantus' problem II, 8 in his copy of the 1621 Latin edition of the *Arithmetica*, that constitutes the content of what has become known as Fermat's Last Theorem. (See Figure 5.4, pg. 167.) In modern terms, this conjecture asserts that there do not exist nonzero integers a, b, c, and $n > 2$, such that $a^n + b^n = c^n$. This result, of which Fermat claimed he had "a truly marvelous demonstration . . . which this margin is too narrow to contain," has provided mathematicians since the seventeenth century with a so far unsolved challenge. Attempts at its proof have generated much new mathematics, but no one as of this writing has found a complete proof or a counterexample. Because so many people have worked on this problem over the past 300 years, most historians believe that Fermat erred in his own claim of a proof, probably because he erroneously assumed that the method of infinite descent which works easily in the cases $n = 3$ and $n = 4$ would generalize to larger values of n.

Although Fermat's claim in the case of the Fermat numbers was wrong and his assertion of the truth of Fermat's Last Theorem was premature, most of his claims of results in number theory announced in his correspondence or scribbled in the margins of his copy of Diophantus have proved true. But although he tried on many occasions to stimulate other European mathematicians to work on his various number theoretic problems, his pleas fell on deaf ears. It took until the next century before a successor could be found to continue the work in number theory begun by the French lawyer.

11.5 PROJECTIVE GEOMETRY

The fate of being ignored also befell Girard Desargues (1591–1661), a French engineer and architect whose most original contributions to mathematics were in the field of projective geometry. As part of his professional interests, he wanted to continue the study of perspective begun by the Renaissance artists. Having mastered the geometrical works of the Greeks, especially that of Apollonius, he proposed to unify the various methods, not by algebraicizing them as did Fermat, but by subsuming them under new synthetic techniques of projection. In particular, he attempted in his *Brouillon projet d'une atteinte aux événemens des rencontres d'un cone avec un plan* (*Rough Draft of an Attempt to Deal with the Outcomes of the Meetings of a Cone with a Plane*) of 1639 to unify the study of conics by use of these techniques. It was well known, for example, that a circle viewed obliquely appears as an ellipse. Because viewing obliquely is equivalent to projecting the circle from a certain point not in its plane onto another plane, Desargues wanted to study those properties of conics that are invariant under projections.

As part of his study, Desargues had to consider points at infinity, the points, like the vanishing point in a drawing in perspective, where parallel lines meet. "Every straight line is, if necessary, taken to be produced to infinity in both directions." When several straight lines are either parallel or intersect at the same point, Desargues says that they belong to the same **ordinance**. "Thus any two lines in the same plane belong to the same ordinance, whose *butt* [intersection point] is at a finite or infinite distance."[38] The collection of all the points at infinity makes the line at infinity. It follows that every plane must be considered to extend to infinity in all directions. In addition, because the cylinder could be considered a cone with vertex at infinity, Desargues treated cones and cylinders simultaneously. Thus, two plane sections of a cone are related by a projective tranformation, projection from the vertex, and two plane sections of a cylinder are related by projection from the point at infinity. Since the circle is a plane section of a cone (or cylinder), Desargues was able to regard all conics as projectively equivalent to a circle. The ellipses are those projections which do not meet the line of infinity in their plane; the parabolas are those which just touch it; while the hyperbolas are those projections of circles which cut the line at infinity. Thus any property of the circle invariant under projections could easily be proved to be a property of all conics.

Desargues' most famous result, however, occurs not in the *Rough Draft*, but in the appendix to a practical work by a friend, Abraham Bosse (1602–1676), *Maniére univer-selle de M. Desargues pour practiquer la perspective* (*Universal Method of M. Desargues for Using Perspective*): "When the lines *HDa*, *HEb*, *CED*, ℓga, ℓfb, HℓK, *DgK*, *EfK*, in different planes or in the same plane, having any order or direction whatsoever, meet in like points, the points *c*, *f*, *g* lie in one line *cfg*."[39] In modern terminology, Desargues is considering two triangles, *KED* and *abℓ*, which are related by a projection from the "like" point *H* (Figure 11.15). In other words, the lines joining pairs of corresponding vertices meet at *H*. The conclusion is then that the intersection points *g*, *f*, *c* of pairs of corresponding sides, here *DK*, *aℓ*; *EK*, *bℓ*; and *DE*, *ab*, all lie on the same line. Desargues proved the result by applying Menelaus' theorem.

Desargues' work was not well received, partly because he invented and used so many new technical terms that few could follow it and partly because mathematicians were just beginning to appreciate Descartes' analytic unification of geometry and were not ready to consider a new synthetic version. Apparently the only contemporary mathematician to appreciate his work was Pascal, who published in 1640 a brief *Essay on Conics* in which he credited Desargues for introducing him to projective methods. This work contains a version of the theorem known ever since by Pascal's name.

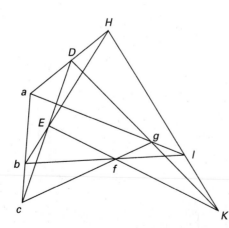

FIGURE 11.15
Desargues' Theorem.

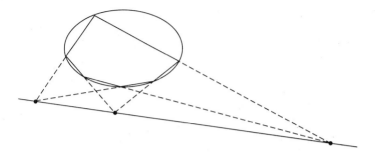

FIGURE 11.16
Pascal's Hexagon
Theorem.

Theorem.   *If a hexagon is inscribed in a conic, then the opposite sides intersect in three collinear points (Figure 11.16).*

Because this theorem is meant to be a statement in projective geometry, among the possible cases are those where some of the pairs of opposite sides are parallel and have their intersection point at infinity. This is naturally the case when the hexagon is a regular one inscribed in a circle. Pascal gave no proof of his theorem in his brief essay. He merely claimed its truth first for circles and then for arbitrary conics. Presumably, he meant to prove the general result by following Desargues' outline. Pascal promised to reveal more of his results along with his methods in a more complete work on conics, a work which he wrote in the mid-1650s. Unfortunately, this larger work was never published, and all manuscript copies have subsequently disappeared. In fact, projective methods in geometry were effectively ignored until early in the nineteenth century.

Exercises

Problems from Fermat's **Introduction**

1. Assuming that $xy = c$ represents a hyperbola with asymptotes the x and y axes, show that $xy + c = rx + sy$ also represents a hyperbola. Find its asymptotes. Do the same for the equation $x^2 + b^2 = ay^2$.

2. Determine the locus of the equation $b^2 - 2x^2 = 2xy + y^2$. (*Hint:* Add x^2 to both sides.)

3. Show that $b^2 + x^2 = ay$ represents a parabola. Draw that portion lying in the first quadrant.

4. Determine the equation of the circle which solves the problem from Apollonius' *Plane Loci* for the case of two points.

5. Determine the equation of the circle which solves the problem from Apollonius' *Plane Loci* for the case of four noncollinear points (x_i, y_i) ($i = 1, 2, 3, 4$).

Problems from Descartes' **Geometry**

6. Using the various constants mentioned in the text, determine the equation of the locus which solves the four-line problem in the special case where the product of the first two lines equals the product of the second two.

7. To solve the fourth degree equation $x^4 - px^2 - qx - r = 0$, Descartes considers the cubic equation in y^2: $y^6 - 2py^4 + (p^2 + 4r)y^2 - q^2 = 0$. If y is a solution, show that the original polynomial factors into two quadratics: $r_1(x) = x^2 - yx + \frac{1}{2}y^2 - \frac{1}{2}p - \frac{q}{2y}$, $r_2(x) = x^2 + yx + \frac{1}{2}y^2 - \frac{1}{2}p + \frac{q}{2y}$, each of which can be solved. Apply this method to solve the equation $x^4 - 17x^2 - 20x - 6 = 0$. Note that the corresponding equation in y, $y^6 - 34y^4 + 313y^2 - 400 = 0$ has the solution $y^2 = 16$.

8. Solve the equation $x^3 - \sqrt{3}x^2 + \frac{26}{27}x - \frac{8}{27\sqrt{3}} = 0$ by first

substituting $y = \sqrt{3}x$ and then $z = 3y$ to get an equation in z with integral coefficients.

9. Construct Descartes' graphical solution to the cubic equation $x^3 = -4x + 16$, which uses the parabola $x^2 = y$ and the circle $(x - 8)^2 + \left(y - \frac{3}{2}\right)^2 = \left(\frac{3}{2}\right)^2 + 8^2$. Construct al-Khayyāmī's graphical solution to the same equation. Which method is simpler? Give reasons for and against introducing such graphical solution techniques into a modern classroom.

10. Read the section from the third book of Descartes' *Geometry* that gives details on the graphical method of constructing solutions to fourth degree polynomial equations. Outline the method and use it to solve $x^4 = x^2 + 5x + 2$.

Problems from de Witt

11. Show that de Witt's substitution $z = y + \frac{b}{a}x + c$ in the equation

$$y^2 + \frac{2bxy}{a} + 2cy = bx - \frac{b^2x^2}{a^2} - c^2$$

has the effect of converting from the oblique x, y axes to new perpendicular axes. (To simplify matters, first set $c = 0$.)

12. Show that de Witt's equation

$$y^2 + \frac{2bxy}{a} + 2cy = \frac{fx^2}{a} + ex + d$$

represents a hyperbola. (Use the substitution $z = y + \frac{b}{a}x + c$ and show that this substitution converts the original oblique axes into perpendicular axes.) Sketch the curve. Find a similar equation which provides the general form of the equation of an ellipse, making sure that all coefficients are positive.

Problems from Girard

13. Solve $x^3 = 300x + 432$ using Girard's technique, given that $x = 18$ is one solution.

14. Solve $x^3 = 6x^2 - 9x + 4$ using Girard's technique. First, determine one solution by inspection.

15. Show that in the equation $x^4 + Bx^2 + D = Ax^3 + Cx$, A is the sum of the roots, $A^2 - 2B$ is the sum of the squares of the roots, $A^3 - 3AB + 3C$ is the sum of the cubes of the roots, and $A^4 - 4A^2B + 4AC + 2B^2 - 4D$ is the sum of the fourth powers of the roots.

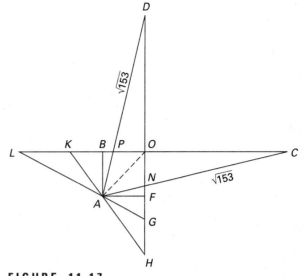

FIGURE 11.17
A problem from Girard.

16. This problem illustrates Girard's geometric interpretation of negative solutions to polynomial equations. Let two straight lines DG, BC intersect at right angles at O (Figure 11.17). Determine A on the line bisecting the right angle at O so that $ABOF$ is a square of side 4. Draw ANC as in the diagram so that $NC = \sqrt{153}$. Find the length FN. (Girard notes that if $x = FN$, then $x^4 = 8x^3 + 121x^2 + 128x - 256$ and so there are four possible solutions, each of which can be calculated. The two positive solutions are represented by FN and FD, while the two negative ones are represented by FG and FH, the latter two taken in the opposite direction from the former two.)

Problems from Pascal

17. Prove by induction on n that

$$\binom{n}{k} = \sum_{j=k-1}^{n-1} \binom{j}{k-1}$$

for all k less than n.

18. Prove by induction that

$$\sum_{k=0}^{n} \binom{n}{k} = 2^n.$$

19. Prove that

$$\binom{n}{k} : \binom{n}{k+1} = (k+1):(n-k).$$

20. Prove that

$$\binom{n}{k} : \binom{n-1}{k} = n : (n-k).$$

21. Prove that

$$\sum_{k=0}^{j}\binom{n}{k} = \sum_{k=0}^{j}\binom{n-1}{k} + \sum_{k=0}^{j-1}\binom{n-1}{k}$$

   for any j less than n.

22. Pascal stated that the odds in favor of throwing a six in four throws of a single die are 671 to 625. Show why this is true.

23. Show that the odds against at least one 1 appearing in a throw of three dice is $125 : 91$. (This answer was stated by Cardano.)

24. Calculate the odds against throwing one six in two throws of one die and also in three throws.

25. Determine the appropriate division of the stakes in a game between two players if the first player is lacking 3 games to win and the second 4. Answer the same question if the first player is lacking 3 games and the second 5.

26. Suppose three players play a fair series of games under the condition that the first player to win 3 games wins the stakes. If they stop play when the first player needs one game while the second and third players each need two games, find the fair division of the stakes. (This problem was discussed in the correspondence of Pascal with Fermat.)

27. For a roll of three dice, show that both a 9 and a 10 can be achieved in 6 different ways. Nevertheless, show that the probability of rolling a 10 is higher than that of rolling a 9. (A discussion of this idea is found in a fragment of a work of Galileo.)

28. Construct a tangent to a point P on a conic section by using Pascal's hexagon theorem. Consider the tangent line as passing through two neighboring points at P. Then pick four other points on the conic and apply the theorem.

Problems from Huygens

29. If two players play a game with two dice with the condition that the first player wins if the sum thrown is 7, the second wins if the sum is 6, and the stakes are split if

there is any other sum, find the expectation (value of the chance) of each player.

30. If I play with another player throwing two dice alternately under the condition that I win when I have thrown a 7 and he wins when he throws a 6, and if he throws first, what is the ratio of my chance to his?

31. There are 12 balls in an urn, 4 of which are white and 8 black. Three blindfolded players, A, B, C draw a ball in turn, first A, then B, then C. The winner is the one who first draws a white ball. Assuming that each (black) ball is replaced after being drawn, find the ratio of the chances of the three players.

32. There are 40 cards, 10 from each suit. A wagers B that he will draw four cards and get one of each suit. What are the fair amounts of the wagers of each?

Problems from Fermat's Number Theory

33. Prove the following: Every prime number of the form $3n + 1$ is equal to a sum of the form $3r^2 + s^2$. (For example, $7 = 3 \cdot 1^2 + 2^2$ and $37 = 3 \cdot 2^2 + 5^2$.)

34. Prove that if p is prime, then $2^p \equiv 2 \pmod{p}$ by writing $2^p = (1+1)^p$, expanding by the binomial theorem, and noting that all of the binomial coefficients $\binom{p}{k}$ for $1 \le k \le p-1$ are divisible by p. Prove $a^p \equiv a \pmod{p}$ by induction on a, using this result and the fact that $(a+1)^p \equiv a^p + 1 \pmod{p}$.

35. For a proof of the Fermat Little Theorem in the case where a and p are relatively prime, consider the remainders of the numbers $1, a, a^2, \ldots$ on division by p. These remainders must ultimately repeat (why?), and so $a^{n+r} \equiv a^r \pmod{p}$ or $a^r(a^n - 1) \equiv 0 \pmod{p}$ or $a^n \equiv 1 \pmod{p}$. (Justify each of these alternatives.) Take n as the smallest positive integer satisfying the last congruence. By applying the division algorithm, show that n divides $p - 1$.

FOR DISCUSSION . . .

36. The best known quotation from Descartes is, "I think, therefore I am," from the *Discourse on Method*. The context is Descartes' resolve only to accept those ideas which are self-evidently true. There is a well known joke based on this quote: Descartes goes into a restaurant. The waiter asks him, "Would you like tonight's special?" He replies, "I think not," and disappears.[40] Comment on the logical validity of this joke.

37. Compare the analytic geometries of Descartes, Fermat, and de Witt. Adapt the formulation of one of these au-

thors to give a presentation of the subject to a precalculus class.

38. What advances in technique and/or understanding enabled Girard to state that every algebraic equation has as many solutions as its degree? Why were Cardano and Viète unable to do so?

39. Outline a lesson in the theory of equations using Descartes' algebraic techniques to teach such results as the factor theorem and the methods of solving polynomial equations of degree higher than two.

40. Outline a lesson in elementary probability theory using the ideas of Cardano. Include material on justification of the various rules involved and on the possible mistakes one can make.

41. Outline a lesson on the principle of mathematical induction using material from Pascal's *Treatise on the Arithmetical Triangle*.

42. Compare Pascal's use of mathematical induction to the use of it by ibn al-Haytham, al-Samaw'al, and Levi ben Gerson. (For more information on the European origin of induction, see W. H. Bussey, "The Origin of Mathematical Induction," *American Mathematical Monthly* 24 (1917), 199–207.)

43. Find out who first used the term "mathematical induction" to describe the technique of Pascal and others. Why was this term chosen? (See Florian Cajori, "The Origin of the Name 'Mathematical Induction'," *American Mathematical Monthly* 25 (1918), 197–201.)

44. Find out about the recent (1988) "near-proof" of Fermat's Last Theorem. What were the problems with the supposed proof?

References and Notes

The only general history of analytic geometry is Carl Boyer, *History of Analytic Geometry* (New York: Scripta Mathematica, 1956). This work covers the subject from its beginnings in ancient times up through the nineteenth century. There is a recent reprint of this work published by The Scholar's Bookshelf. A brief survey of the work of Fermat and Descartes is in Carl Boyer, "Analytic Geometry: The Discovery of Fermat and Descartes," *Mathematics Teacher* 37 (1944), 99–105 and "The Invention of Analytic Geometry," *Scientific American* 180 (Jan, 1949), 40–45. The best general histories of probability which deal to some extent with the seventeenth century and earlier are F. N. David, *Games, Gods, and Gambling* (New York: Hafner, 1962) and Ian Hacking, *The Emergence of Probability* (Cambridge: Cambridge University Press, 1975). The latter book is more philosophical, while the former discusses the relevant texts in more detail. A general history of number theory from a historical point of view is André Weil, *Number Theory: An Approach through History from Hammurapi to Legendre* (Boston: Birkhäuser, 1983). The work of Desargues and Pascal on projective geometry is discussed in J. V. Field and J. J. Gray, *The Geometrical Work of Girard Desargues* (New York: Springer, 1987).

1. David Eugene Smith, *A Source Book in Mathematics* (New York: Dover, 1959), vol. 2, p. 389.

2. The best study of the life and mathematical work of Fermat is Michael S. Mahoney, *The Mathematical Career of Pierre de Fermat 1601–1665* (Princeton: Princeton University Press, 1973). This book contains a detailed analysis of Fermat's work, not only in analytic geometry but also in the various aspects of the calculus to be treated in Chapter 12.

3. Ibid., p. 102–103.

4. Smith, *Source Book*, vol. 2, p. 390.

5. Ibid., p. 392.

6. Mahoney, *Mathematical Career*, p. 91.

7. David Eugene Smith and Marcia L. Latham, trans., *The Geometry of René Descartes* (New York: Dover, 1954), p. 2. This is the standard modern version of the *Geometry*. It contains the original French and the English translation on facing pages as well as Descartes' original notation and diagrams.

8. René Descartes, *Discourse on Method, Optics, Geometry, and Meteorology* translated by Paul J. Olscamp (Indianapolis: Bobbs-Merrill Co., 1965), p. 9. This version contains not only the *Geometry*, but also the other two essays. The *Optics* in particular contains more of Descartes' mathematics as he demonstrates how to construct certain curves, lenses in the shape of which will have prescribed optical properties. For a general study of Descartes' work in mathematics and physics, see J. F. Scott, *The Scientific Work of René Descartes* (London: Taylor and Francis, 1952).

9. Smith and Latham, *Geometry*, p. 5.

10. Ibid., p. 22.

11. See the very interesting paper of H. J. M. Bos, "On the Representation of Curves in Descartes' Géométrie," *Archive for History of Exact Sciences* 24 (1981), 295–338. Bos discusses Descartes' general program for geometry which is outlined in the *Geometry*. See also A. Molland, "Shifting the Foundations: Descartes' Transformation of Ancient Geometry," *Historia Mathematica* 3 (1976), 21–49. See E. G. Forbes, "Descartes and the Birth of Analytic Geometry," *Historia Mathematica* 4 (1977), 141–161 for an argument that analytic geometry should be credited to Marino Ghetaldi (1566–1627), a pupil of Viète.

12. Smith and Latham, *Geometry*, p. 43.

13. Ibid., p. 48.

14. Ibid., p. 91.

15. Ibid., p. 240.

16. *The Early Theory of Equations: On Their Nature and Constitution* (Annapolis: Golden Hind Press, 1986), p. 107. This book contains Ellen Black's translation of Girard's *A New Discovery in Algebra* as well as treatises by Viète and Debeaune.

17. Ibid., p. 145.

18. Ibid., p. 138.

19. Ibid., p. 139.

20. Ibid., p. 141.

21. Smith and Latham, *Geometry*, p. 159.

22. Ibid., p. 175.

23. Ibid., p. 159.

24. Ibid., p. 160.

25. Ibid., p. 200.

26. Ibid., p. 240.

27. Quoted in David, *Games, Gods*, p. 33.

28. Oystein Ore, *Cardano, the Gambling Scholar* (Princeton: Princeton University Press, 1953), p. 203. Gerolamo Cardano's *The Book on Games of Chance* appears in translation by Sydney Henry Gould at the end of this biographical work. Ore discusses Cardano's work in detail and argues more strongly than either David or Hacking for crediting Cardano with originating many of the central ideas of probability.

29. Quoted in Oystein Ore, "Pascal and the Invention of Probability Theory," *American Mathematical Monthly* 67 (1960), 409–419, p. 414.

30. Blaise Pascal, *Treatise on the Arithmetical Triangle*, Richard Scofield, trans., *Great Books of the Western World* (Chicago: Encylopedia Britannica, 1952), vol. 33, p. 452. This edition also contains the letters between Pascal and Fermat on probability. For more on Pascal, see Harold Bacon, "The Young Pascal," *Mathematics Teacher* 30 (1937), 180–185, Morris Bishop, *Pascal, The Life of Genius* (New York: Reynal and Hitchcock, 1936), and Jean Mesnard, *Pascal, His Life and Works* (New York: Philosophical Library, 1952).

31. Pascal, *Treatise*, p. 452.

32. Christian Huygens, *On the Calculations in Games of Chance* in *Oeuvres Completes* (The Hague: 1888–1950), vol. 14, p. 61.

33. Ibid., p. 62.

34. Ibid., p. 64.

35. Huygens' version of his game is incomplete. This improvement is found in Olav Reiersol, "Notes on Some Propositions of Huygens in the Calculus of Probability," *Nordisk Matematisk Tidskrift* 16 (1968), 88–91.

36. For a modern discussion of de Méré's dice problem, see Janet B. Pomeranz, "The Dice Problem—Then and Now," *College Mathematics Journal* 15 (1984), 229–237.

37. Quoted in Mahoney, *Mathematical Career*, p. 344. For more on Fermat's method of infinite descent, see Howard Eves, "Fermat's Method of Infinite Descent," *Mathematics Teacher* 53 (1960), 195–196. See also Michael Mahoney, "Fermat's Mathematics: Proofs and Conjectures," *Science* 178 (1972), 30–36.

38. Field and Gray, *Geometrical Work*, pp. 69–70. For more on Desargues, see N. A. Court, "Desargues and his Strange Theorem," *Scripta Mathematica* 20 (1954), 5–13, 155–164.

39. Smith, *Source Book*, vol. 2, pp. 307 – 308.

40. Told to me by my son Ari.

Summary of Seventeenth Century Geometry, Algebra, and Probability

1501–1576	Gerolamo Cardano	Probability
1581–1638	Claude Gaspard Bachet	Translation of Diophantus
1588–1648	Marin Mersenne	Walking scientific journal
1591–1661	Girard Desargues	Projective geometry
1595–1632	Albert Girard	Theory of equations
1596–1650	René Descartes	Analytic geometry, theory of equations
1601–1665	Pierre de Fermat	Analytic geometry, probability, number theory
1607–1684	Antoine Gombaud	Probability
1615–1660	Frans van Schooten	Analytic geometry
1623–1662	Blaise Pascal	Probability, projective geometry
1623–1672	Jan de Witt	Analytic geometry
1629–1695	Christian Huygens	Probability

Chapter 12

The Beginnings of Calculus

"The prime occasion from which arose my discovery of the method of the Characteristic Triangle, and other things of the same sort, happened at a time when I had studied geometry for not more than six months. . . . At that time I was quite ignorant of Cartesian algebra and also of the method of indivisibles; indeed I did not know the correct definition of the center of gravity. For, when by chance I spoke of it to Huygens, I let him know that I thought that a straight line drawn through the center of gravity always cut a figure into two equal parts . . . Huygens laughed when he heard this, and told me that nothing was farther from the truth. So I, excited by this stimulus, began to apply myself to the study of the more intricate geometry."
(From a 1680 letter from Gottfried Leibniz to Ehrenfried Walter von Tschirnhaus (1651–1708))[1]

On October 24, 1676, Isaac Newton wrote his second (and last) letter for Leibniz (the *Epistola posterior*), giving details of some of his work, and sent it by way of their mutual correspondent Henry Oldenburg. But Newton concealed the basic goal of his version of the calculus by means of the anagram 6accdæ13eff7i3l9n4o4qrr4s8t12ux of the Latin phrase "Data æquatione quotcunque fluentes quantitates involvente, fluxiones invenire; et vice versa" (given an equation involving any number of fluent quantities, to find the fluxions, and vice versa).[2] Although Leibniz responded enthusiastically to Newton's letter, as he had also responded to the *Epistola prior*, giving details of his own work on the calculus and encouraging further dialogue, Newton never replied.

Building on the work of many mathematicians over the centuries who considered the problems of determining the areas of regions bounded by curves and of finding the maximum or minimum values of certain functions, two geniuses of the last half of the seventeenth century, Isaac Newton and Gottfried Leibniz, created the machinery of the calculus, the foundation of modern mathematical analysis and the source of application to an increasing number of other disciplines. The maximum-minimum problem and the area problem, along with the related problems of finding tangents and determining volumes, had been attacked and solved for various special cases over the years. But in virtually every case solved, either in pre-Greek times, or by the Greeks themselves, or by their Islamic

FIGURE 12.1
Quantities connected
with a curve: x is the
abscissa, y the ordinate,
s the arclength, t the
subtangent, τ the tangent,
n the normal, and v the
subnormal.

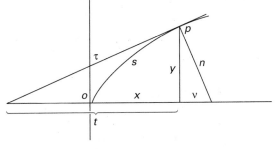

successors, the solution had required an ingenious construction. No one had developed any algorithm which would enable these problems to be solved easily in new situations.

New situations did not often occur in either the Greek or Islamic setting, since these mathematicians had few ways of describing new curves or solids for which to calculate tangents, areas, or volumes. But with the advent of analytic geometry in the first half of the seventeenth century, the possibility suddenly opened up of constructing all sorts of new curves and solids. After all, any algebraic equation determined a curve, and a new solid could be formed, for example, by rotating a curve around any line in its plane. With an infinity of new examples to deal with, mathematicians of the seventeenth century sought for and discovered new ways of finding maximums, constructing tangents, and calculating areas and volumes. These mathematicians were not, however, concerned with functions. They were concerned with curves, defined by some relation between two variables. And in the process of finding tangents they often considered other geometric aspects of the curves. Figure 12.1 illustrates some of the quantities connected with a point on a given curve: the abscissa x, the ordinate y, the arc length s, the subtangent t, the tangent τ, the normal n, and the subnormal v.

In this chapter we will explore first the various methods used to construct tangents and find extrema, next the methods developed to determine areas and volumes, third the idea of a power series which was to prove extremely useful in extending the range of curves to which area techniques could be applied, and fourth the ways of accomplishing what Descartes said could not be done, determining the lengths of curves. We will then consider the work of the mathematicians who were first able to combine these various ideas into an organized whole, chiefly Newton and Leibniz, and conclude with a study of the contents of the earliest textbooks in the new field of calculus.

12.1 TANGENTS AND EXTREMA

In 1615, Kepler wrote his *Nova Stereometria Doliorum Vinariorum* (*New Solid Geometry of Wine Barrels*), in which he showed that Austrian wine merchants had a rather accurate way of determining how much wine remained in a given barrel. As part of this study of various solid shapes, he proved that the largest parallelepiped which can be inscribed in a given sphere is a cube. In fact, he actually tabulated the volumes of parallelepipeds inscribed in a sphere of radius 10 for all integral altitudes from 1 to 20. It was therefore clear to him that near the maximum value of approximately 3080, the volume changed little with small changes in the altitude: "Near the maximum, the decrements on both sides are initially imperceptible."[3]

12.1.1 Fermat's Method of *Adequality*

Fermat, in the late 1620s, was able to turn Kepler's idea into an algorithm, but he was stimulated to consider the question by a study of Viète's work relating the coefficients to the roots of a polynomial: "While I was pondering Viète's method . . . and was exploring more accurately its use in discovering the structure of . . . equations, there came to mind a new method to be derived from it for finding maxima and minima."[4]

Viète had shown that the sum of the two roots x_1, x_2 of $bx - x^2 = c$ was b by equating $bx_1 - x_1^2$ and $bx_2 - x_2^2$ and dividing through by $x_1 - x_2$. The equation $bx - x^2 = c$ comes from the geometric problem of dividing a line of length b into two parts whose product is c. Fermat knew from Euclid that the maximum possible value of c was $b^2/4$ and also that for any number less than the maximum, there were two possible values for x whose sum was b. But what happened as c approached its maximum value? The geometrical situation made it clear to Fermat that even for this maximum value, the equation had two solutions, each of the same value: $x_1 = b/2$ and $x_2 = b - x_1 = b - b/2 = b/2$. This insight gave Fermat his method for maximizing a polynomial $p(x)$: Set $p(x_1) = p(x_2)$. Then divide through by $x_1 - x_2$ to find the relationship between the coefficients and any two roots of the polynomial. Finally, set the two roots equal to one another and solve.

From $bx_1 - x_1^2 = bx_2 - x_2^2$, Fermat derived the fact that $b = x_1 + x_2$, an equation holding for any two roots. Setting $x_1 = x_2 (= x)$ gives $b = 2x$. Thus the maximum occurs when $x = b/2$. Similarly, to maximize $bx^2 - x^3$, Fermat set $bx_1^2 - x_1^3 = bx_2^2 - x_2^3$ and derived $b(x_1^2 - x_2^2) = x_1^3 - x_2^3$ and $bx_1 + bx_2 = x_1^2 + x_1x_2 + x_2^2$. He then set $x_1 = x_2 (= x)$ and determined that $2bx = 3x^2$, from which he concluded that $x = 2b/3$ provides the maximum value. Fermat's procedure brings up two obvious questions. Why did Fermat discard the solution $x = 0$, and how did he know that the solution he chose gave a maximum and not a minimum? For Fermat the answers were simple. The problem was a geometrical one dealing with volumes. Hence the answer $x = 0$ had no meaning, and the positive answer obviously provided the maximum. But more generally, Fermat failed to consider what happens when there are two or more solutions to the final equation his method provides. In any given case, he simply appealed to the geometry of the situation. Fermat's method raises one other question. How can one divide through by $x_1 - x_2$ and then set that value equal to 0? For Fermat, the geometric situation showed that the roots were even distinguishable when their difference was 0. Thus, he never felt he was dividing by 0. He simply assumed that the relationships worked out using Viète's methods were perfectly general (for example, $x_1 + x_2 = b$) and thus held for any particular values of the variables, even those at the maximum.

Fermat did realize, however, that if the polynomial $p(x)$ were somewhat complicated, the division by $x_1 - x_2$ might be rather difficult. Thus he modified his method to avoid this. Instead of considering the two roots as x_1 and x_2, he wrote them as x and $x + e$. Then, after equating $p(x)$ with $p(x + e)$—Fermat actually used the term *adequate* which he had read in Diophantus—he had only to divide by e or one of its powers. In the resulting expression he then removed any term that contained e to get an equation enabling the maximum to be found. Thus, using his original example of $p(x) = bx - x^2$, Fermat

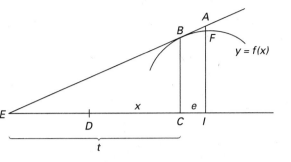

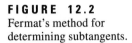

FIGURE 12.2
Fermat's method for
determining subtangents.

adequated $bx - x^2$ with $b(x + e) - (x + e)^2 = bx - x^2 + be - 2ex - e^2$. (We will write this as $bx - x^2 \sim bx - x^2 + be - 2ex - e^2$.) Cancelling common terms gave him $be \sim 2ex + e^2$. On dividing by e, the resulting *adequality* is $b \sim 2x + e$. Finally, removing the term which contains e, Fermat found his known result: $x = b/2$. Fermat noted in his description of this method, which was probably written before 1630 but only reached Paris in late 1636, that "we can hardly expect a more general method."[5]

In this same document, Fermat showed how the method of *adequality* can be adapted to determine a tangent to a curve, in particular to a parabola. Because Fermat discovered this method before he invented analytic geometry, he used a geometric description of the parabola. In 1638, however, once the possibility opened up of defining curves by algebraic equations rather than through geometric properties, Fermat could explain his method more easily. (Descartes, in fact, had strongly criticized his geometric explanation.) To draw a tangent line at B to a curve represented in modern notation by $y = f(x)$, pick an arbitrary point A on the tangent line and drop perpendiculars AI and BC to the axis (Figure 12.2). Fermat's idea is then to *adequate* FI/BC with EI/CE, where F is the intersection of AI with the curve. If $CI = e$, $CD = x$, and $CE =$ the subtangent t, this *adequality* can be written

$$\frac{f(x + e)}{f(x)} \sim \frac{t + e}{t}$$

or $tf(x + e) \sim (t + e)f(x)$. By applying his rules of cancelling common terms, dividing through by e, and then removing any remaining terms containing e, Fermat could calculate the relation between t and x that determines the tangent line. Fermat modified the method to deal with curves expressed in the form $f(x,y) = 0$ and, in fact, justified his method to Descartes by showing him how it could be used to find the tangent to the curve Descartes proposed to him, $x^3 + y^3 = pxy$.

12.1.2 Descartes and the Method of Normals

One of the reasons Descartes was critical of Fermat was that Fermat had discovered the same new mathematics as Descartes, independently of the great philosopher. And Descartes was immensely proud of his own discovery of a method of drawing a normal to a curve at any point, from which, naturally, one could easily determine the tangent as well. As Descartes wrote in his *Geometry*, "I dare say that this is not only the most useful and most general problem in geometry that I know, but even that I have ever desired to know."[6]

Descartes derived his idea for drawing a normal from the realization that a radius of a circle is always normal to the circumference. Thus the radius of a circle tangent to a given

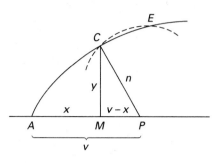

FIGURE 12.3
Descartes' method for finding normals.

curve at the given point will be normal to that curve as well. To construct a circle tangent to a curve required an idea similar to that of Fermat, namely, that the two intersection points of a circle with the curve near the given point will become one if the circle is in fact tangent. To carry out this procedure at a point C of a curve given by $y = f(x)$, assume that P is the center of the required circle, take an arbitrary point A on the axis through P, and set $CP = n$ and $PA = v$ (Figure 12.3). If $C = (x, y)$, then $PM = v - x$, and the equation of the circle is $n^2 = y^2 + (v - x)^2$ or $n^2 = [f(x)]^2 + v^2 - 2vx + x^2$. Descartes then used this equation to determine v, which in turn determines the point P. As he noted, "if the point P fulfills the required conditions, the circle about P as center and passing through the point C will touch but not cut the curve CE; but if this point P be ever so little nearer to or farther from A than it should be, this circle must cut the curve not only at C but also in another point. Now if this circle cuts [the curve also at E], the equation . . . must have two unequal roots. . . . The nearer together the point C and E are taken, however, the less difference there is between the roots; and when the points coincide, the roots are exactly equal."[7] In other words, for P to be the center of a tangent circle, the equation $[f(x)]^2 + v^2 - 2vx + x^2 - n^2 = 0$ must have a double root. As Descartes knew from his study of roots of equations, this meant that the polynomial has a factor of $(x - x_0)^2$ where x_0 is the double root. Setting then $[f(x)]^2 + v^2 - 2vx + x^2 - n^2 = (x - x_0)^2 q(x)$ and equating the coefficients of like powers of x, Descartes could solve for v in terms of x_0.

As usual in the *Geometry*, Descartes provided quite difficult examples of his procedure. We therefore present a simple example to clarify Descartes' method, namely, that of determining a normal to the parabola $y = x^2$ at the point (x_0, x_0^2). In this case, the polynomial which has a double root is $(x^2)^2 + v^2 - 2vx + x^2 - n^2$. Because this is a fourth degree polynomial, it must be equated to $(x - x_0)^2 q(x)$, where $q(x)$ has degree two. Thus,

$$x^4 + x^2 - 2vx + v^2 - n^2 = (x - x_0)^2(x^2 + ax + b)$$

or

$$x^4 + x^2 - 2vx + v^2 - n^2$$
$$= x^4 + (a - 2x_0)x^3 + (b - 2x_0 a + x_0^2)x^2 + (ax_0^2 - 2bx_0)x + bx_0^2.$$

Equating coefficients gives

$$a - 2x_0 = 0$$

$$b - 2x_0 a + x_0^2 = 1$$

$$ax_0^2 - 2bx_0 = -2v$$

$$bx_0^2 = v^2 - n^2$$

Solving the first three equations for v by setting $a = 2x_0$ and $b = 2ax_0 - x_0^2 + 1$ gives $v = 2x_0^3 + x_0$ as the (horizontal) coordinate of the desired point P. (Since v determines n, the fourth equation is not necessary.) Because Descartes was interested just in constructing the normal, he stopped the procedure here with the point P determined. But we note further that the slope of the normal line is

$$\frac{-y_0}{v - x_0} = \frac{-x_0^2}{2x_0^3} = \frac{-1}{2x_0}$$

and therefore the slope of the tangent line is $2x_0$, a familiar result.

12.1.3 The Algorithms of Hudde and Sluse

By the late 1630s, Gilles Persone de Roberval (1602–1675) had discovered a kinematic method of determining tangents by considering a curve to be generated by a moving point. But his method depended on the geometric description of the curve and thus could not meet the need for a simple algebraic algorithm to determine tangents. The procedures of Fermat, and especially of Descartes, lead to such complicated algebra that these methods also could not provide the desired ease of calculation. But a study of these methods led two other mathematicians, Johann Hudde (1628–1704) and René François de Sluse (1622–1685), to discover simpler algorithms in the 1650s.

Hudde was one of van Schooten's students, who, like de Witt, became active in political life in the Netherlands. His contributions to mathematics were made in the late 1650s, when two of his papers appeared in van Schooten's 1659 edition of Descartes' *Geometry*. In *De maximis et minimis* (*On maximums and minimums*) Hudde described his algorithm for simplifying the calculations necessary to determine a double root to a polynomial equation, necessary for Descartes' method of finding normals. Hudde's rule, for which he only sketched a proof, states that if a polynomial $f(x) = a_0 + a_1 x + a_2 x^2 + \cdots + a_n x^n$ has a double root $x = \alpha$, and if $p, p + b, p + 2b, \ldots p + nb$ is an arithmetic progression, then the polynomial $pa_0 + (p + b)a_1 x + (p + 2b)a_2 x^2 + \cdots + (p + nb)a_n x^n$ also has the root $x = \alpha$. In modern terminology, the new polynomial can be expressed as $pf(x) + bxf'(x)$. Hudde's result follows immediately, because if $f(x)$ has a double root, then $f'(x)$ has the same root. Although his rule permitted the arbitrary choice of an arithmetic progression, Hudde most often used the progression with $p = 0, b = 1$. In this case the new polynomial is $xf'(x)$, a result which helped to bring out the computational importance of what we now call the derivative.

As a first example of the rule, consider the problem of determining the normal to the parabola $y = x^2$, where it is necessary to find the relationship between the coefficient v and the double root x_0 of the polynomial $x^4 + x^2 - 2vx + v^2 - n^2$. Using Hudde's rule with $p = 0$ and $b = 1$ gives the new polynomial equation $4x^4 + 2x^2 - 2vx = 0$ or $4x^3 + 2x - 2v = 0$. Because x_0 is a solution of this equation, it follows as before that $v = 2x_0^3 + x_0$ and therefore that the slope of the tangent line is $(v - x_0)/x_0^2 = 2x_0$. An easy generalization of this example makes it possible to show that the slope of the tangent line to $y = x^n$ at (x_0, x_0^n) is nx_0^{n-1}, a result extremely difficult to find using Descartes' procedure.

Hudde also applied his rule to the determination of extreme values, using Fermat's idea that if a polynomial $f(x)$ has an extreme value M, then the polynomial $g(x) = f(x) - M$ has a double root. Thus to maximize $x^2(b - x)$, use the rule with $p = 0, b = 1$ on the polynomial $-x^3 + bx^2 - M$. The new polynomial equation is $-3x^3 + 2bx^2 = 0$, the nonzero root of which, $x = 2b/3$, gives the desired maximum. Hudde was also able to use the rule to find the tangents to curves determined by equations of the form $f(x,y) = 0$, but Sluse gave an even simpler algorithm for this case.

Sluse was born and spent most of his life in Liège in what is now Belgium and, like Hudde, had little time for mathematics. He nevertheless carried on an extensive correspondence with mathematicians all over Europe. His algorithm for determining the subtangent t (and of course the tangent) to a curve given by a polynomial equation $f(x,y) = 0$ was probably discovered in the 1650s but only appeared in print in a letter to Henry Oldenburg (1615–1677) in England in 1673. The algorithm begins with the elimination of constant terms. One then leaves all terms with x on the left and transfers all terms with y to the right with appropriate change of sign. Thus any term containing both x and y will now appear on each side of the equation. Next, one multiplies each term on the right by its exponent of y and each term on the left by its exponent of x. Finally, one replaces one x in each term on the left by t and solves the resultant equation for t. For example, given the equation $x^5 + bx^4 - 2q^2y^3 + x^2y^3 - b^2 = 0$, one eliminates the constant term and transfers all the terms in y to get $x^5 + bx^4 + x^2y^3 = 2q^2y^3 - x^2y^3$. Multiplying by the appropriate exponents and replacing one x in each term on the left by t gives $5x^4t + 4bx^3t + 2txy^3 = 6q^2y^3 - 3x^2y^3$. The subtangent t is therefore given by

$$t = \frac{6q^2y^3 - 3x^2y^3}{5x^4 + 4bx^3 + 2xy^3}$$

and the slope of the tangent by

$$\frac{y}{t} = \frac{5x^4 + 4bx^3 + 2xy^3}{6q^2y^2 - 3x^2y^2}.$$

In modern terms, it is easy enough to see that Sluse has calculated

$$t = -\frac{yf_y(x,y)}{f_x(x,y)} \quad \text{or} \quad \frac{dy}{dx} = -\frac{f_x(x,y)}{f_y(x,y)}.$$

Sluse, however, gave no written justification of his method or hint how he discovered it. The best guess is that he generalized it from a study of many examples. In any case, the importance of the rules of Hudde and Sluse was that they provided general algorithms by which one could routinely construct tangents to curves given by polynomial equations. It was no longer necessary to develop a special technique for each particular curve. Anyone could now determine the tangent.

12.2 AREAS AND VOLUMES

Both Greek and Islamic mathematicians had been able to determine areas and volumes of certain regions bounded by curved lines or surfaces. The texts available, however, generally gave only the result with a proof based on the method of exhaustion. Such results gave seventeenth century mathematicians few clues as to how to determine the areas bounded by the many new curves now available for study or the volumes of solid regions generated by revolving these curves around lines in the plane. The only clear idea passed down from Greek times was that somehow the given region needed to be broken up into very small regions, whose individual areas or volumes were known.

12.2.1 Kepler and Cavalieri: Infinitesimals and Indivisibles

Recall that Kepler used the procedure of adding up small regions in his discovery of the laws of planetary motion. And in his *Nova stereometria*, he calculated the area of a circle of radius AB by first noting that "the circumference . . . has as many parts as points, namely, an infinite number; each of these can be regarded as the base of an isosceles triangle with equal sides AB so that there are an infinite number of triangles in the area of the circle, all having their vertices at the center A."[8] Kepler then stretched the circumference of the circle out into a straight line, upon each point of which, "arranged one next to the other," he placed triangles equal to the ones in the circle, all having the altitude AB (Figure 12.4). It follows that the area of the triangle ABC "consisting of all those triangles, will be equal to all the sectors of the circle and therefore equal to the area of the circle which consists of all of them." Therefore, the area of the circle is one-half of the radius multiplied by the circumference, or, as Kepler put it, the area of the circle is to the square on the diameter as 11 to 14. Similarly, Kepler calculated the volume of a ring (torus) by slicing it into "an infinite number of very thin disks,"[9] each of which is thinner toward the center and thicker toward the outside. Kepler never claimed, however, that his method was rigorous, noting only that "we could obtain absolute and in all respects perfect demonstrations from these books of Archimedes themselves, were we not repelled by the thorny reading thereof."[10]

Bonaventura Cavalieri (1598–1647), a disciple of Galileo, in his *Geometria indivisibilibus continuorum nova quadam ratione promota* (*Geometry, advanced in a new way by the indivisibles of the continua*) of 1635 and his *Exercitationes geometricae sex* (*Six geometrical exercises*) of 1647 attempted rigorous proofs of a similar method of finding areas and volumes by dividing areas up into lines and volumes up into planes (Side Bar 12.1). The central concept of Cavalieri's work was that of *omnes lineae*, or "all the lines" of a plane figure F, to be written as $\mathbb{O}_F(\ell)$. By this Cavalieri meant the collection of intersections of the plane figure with a perpendicular plane moving parallel to itself from one side of the given figure to the other. These intersections are lines, and it is the collection of such lines, thought of as a single magnitude, that Cavalieri dealt with throughout his work. Cavalieri's lines in some sense made up the given figure, but he was careful to distinguish $\mathbb{O}_F(\ell)$ from F itself. He was also able to generalize the idea by considering

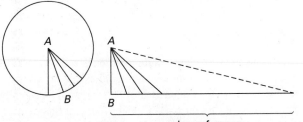

FIGURE 12.4
Kepler's method of determining the area of a circle.

circumference

Biography   **Bonaventura Cavalieri (1598–1647)**

Cavalieri began his study of mathematics in Pisa while a member of a small religious order, and there began a correspondence with Galileo which lasted nearly until Galileo's death. Probably through the latter's influence he obtained a professorship at Bologna in 1629 and succeeded in having the appointment renewed every three years until his own death. Besides the works mentioned in the text, Cavalieri published many other books on mathematics, including a work on astrology, and also investigated lenses and mirrors. His fame rests, however, on the method of indivisibles discussed in the *Geometria*, a work which was widely known although, due to its difficulty, probably little studied.

Side Bar 12.1   **Indivisibles and Infinitesimals**

Seventeenth century mathematicians used two methods to determine areas and volumes, the **method of indivisibles** and the **method of infinitesimals**. In the first method, developed in detail by Cavalieri, but whose roots go back at least to Archimedes, a given geometric object is considered to be made up of objects of dimension one lower. In other words, plane regions are considered to be made up of lines and solid figures of surfaces. In the second method, which was used by Kepler, a figure was made up of "very small" (or infinitesimal) figures of the same dimension. Both of these concepts were alien to classical Greek mathematics. The Greeks certainly considered indivisible points, but never believed that they could "make up" an object of higher dimension. Similarly, infinitesimals did not satisfy definition 4 of *Elements* V, because no matter how one multiplied such a quantity, the result never exceeded any given (noninfinitesimal) magnitude. Despite these problems, the concepts were used, at least heuristically, by all of those concerned with questions of area and volume, and even of tangents. The concept of infinitesimal, in particular, turned out to be the basis for Leibniz's calculus, if not perhaps for him, then certainly for his immediate followers. Newton too used infinitesimals, although he was apparently uncomfortable with the idea and attempted to replace their use by arguments involving limits. The mathematicians of the nineteenth century succeeded in eliminating the use of infinitesimals from calculus, but, more recently, these "very small" quantities have reappeared in a formal axiomatic manner with the work of Abraham Robinson and his colleagues on "nonstandard" analysis.

higher dimensional objects such as "all the squares" or "all the cubes" of a given figure. One can think of "all the squares" of a triangle, for example, as representing a pyramid, each of whose cross sections is a square of side the length of a particular line in the triangle.

The basis for Cavalieri's computations was a result to this day known as Cavalieri's principle: "If two plane figures have equal altitudes and if sections made by lines parallel to the bases and at equal distances from them are always in the same ratio, then the plane figures are also in this ratio."[11] Cavalieri proved this result by an argument using superposition. It followed that if there were a fixed ratio between corresponding lines of the two

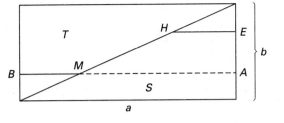

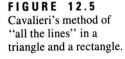

FIGURE 12.5
Cavalieri's method of
"all the lines" in a
triangle and a rectangle.

figures F and G, then $\mathbb{O}_F(\ell) : \mathbb{O}_G(\ell) = F : G$. For example, suppose the rectangle F of length a and width b is divided by its diagonal into two triangles T, S (Figure 12.5). Since each line segment BM in triangle T corresponds to one and only one equal line segment HE in triangle S, then $\mathbb{O}_T(\ell) = \mathbb{O}_S(\ell)$. On the other hand, since every line segment BA of the rectangle is made up of one segment from triangle S and one from triangle T, $\mathbb{O}_F(\ell) = \mathbb{O}_T(\ell) + \mathbb{O}_S(\ell)$. It follows that $\mathbb{O}_F(\ell) = 2\mathbb{O}_T(\ell)$, or, all the lines of the square are double all the lines of the triangle. In modern notation, this result is equivalent to

$$ab = 2\int_0^b \frac{a}{b}t\, dt$$

or, more simply, to

$$b^2 = 2\int_0^b t\, dt.$$

Cavalieri was similarly able to demonstrate that "all the squares" of the rectangle F are triple of "all the squares" of each triangle, or, in modern notation, that

$$a^2 b = 3\int_0^b \frac{a^2}{b^2}t^2\, dt \qquad \text{or} \qquad b^3 = 3\int_0^b t^2\, dt.$$

By 1647, he had demonstrated analogous results for certain higher powers and was able to infer that the area under the "higher parabola" $y = x^k$ inscribed in a rectangle is $\frac{1}{k+1}$ times the area of the rectangle, or that

$$\int_0^b x^k\, dx = \frac{1}{k+1}b^{k+1}.$$

This result had also been discovered by Fermat, Pascal, Roberval, and Evangelista Torricelli in the same time period.

12.2.2 Fermat and the Area Under Parabolas and Hyperbolas

In a letter to Roberval of September 22, 1636 Fermat claimed that he had been able to square "infinitely many figures composed of curved lines," in particular, that he too could calculate the area of a region under any higher parabola. He noted further that "I have had to follow a path other than that of Archimedes in the quadrature of the parabola and that I would never have solved it by the latter means."[12] Fermat was here recalling that Archimedes used triangles in his quadrature. Fermat himself would use simpler figures.

Roberval, writing back in October, claimed that he too had found the same result, using a formula for the sums of powers of the natural numbers: "The sum of the square numbers is always greater than the third part of the cube which has for its root the root of the greatest square, and the same sum of the squares with the greatest square removed is less than the third part of the same cube; the sum of the cubes is greater than the fourth part of the [fourth power] and with the greatest cube removed, less than the fourth part, etc."[13] In other words, finding the area of the region bounded by the parabola $y = px^k$, the x-axis, and a given vertical line depends on the formula

$$\sum_{i=1}^{N-1} i^k < \frac{N^{k+1}}{k+1} < \sum_{i=1}^{N} i^k.$$

It is easy enough to see why this formula is fundamental, by considering the graph of $y = px^k$ over the interval $[0, x_0]$. Divide the base interval into N equal subintervals, each of length $\frac{x_0}{N}$, and erect over each subinterval a rectangle whose height is the y-coordinate of the right endpoint (Figure 12.6). The sum of the areas of these N circumscribed rectangles is then

$$p\frac{x_0^k}{N^k}\frac{x_0}{N} + p\frac{(2x_0)^k}{N^k}\frac{x_0}{N} + \cdots + p\frac{(Nx_0)^k}{N^k}\frac{x_0}{N} = \frac{px_0^{k+1}}{N^{k+1}}(1^k + 2^k + \cdots + N^k).$$

Similarly, one can calculate the sum of the areas of the inscribed rectangles, those whose height is the y-coordinate of the left endpoint of the corresponding subinterval. If A is the area under the curve between 0 and x_0, then

$$\frac{px_0^{k+1}}{N^{k+1}}(1^k + 2^k + \cdots + (N-1)^k) < A < \frac{px_0^{k+1}}{N^{k+1}}(1^k + 2^k + \cdots + N^k).$$

The difference between the outer expressions of this inequality is simply the area of the rightmost circumscribed rectangle. Because x_0 and $y_0 = px_0^k$ are fixed, this difference may

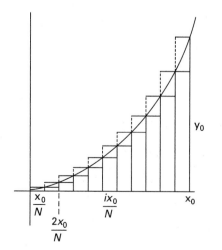

FIGURE 12.6
The area under $y = px^k$ according to Fermat and Roberval.

be made less than any assigned value simply by taking N sufficiently large. It follows from the inequality cited by Roberval that both the area A and the value

$$\frac{px_0^{k+1}}{k+1} = \frac{x_0y_0}{k+1}$$

are "squeezed" between two values whose difference approaches 0. Thus Fermat (and Roberval) found that

$$A = \frac{x_0y_0}{k+1}.$$

The obvious question then is how either of these two men discovered the formula for the sums of powers, a formula which was in essence known to ibn al-Haytham 600 years earlier. Fermat claimed that he had a "precise demonstration" and doubted that Roberval had one. In fact, as is typical in Fermat's work, all we have is his own general statement in terms of triangular numbers, pyramidal numbers, and the other numbers which occur as columns in Pascal's triangle: "The last side multiplied by the next greater makes twice the triangle. The last side multiplied by the triangle of the next greater side makes three times the pyramid. The last side multiplied by the pyramid of the next greater side makes four times the triangulotriangle. And so on by the same progression in infinitum."[14] Fermat's statement, which we write as

$$N\binom{N+k}{k} = (k+1)\binom{N+k}{k+1},$$

is equivalent to Pascal's twelfth consequence. Using the properties of Pascal's triangle, it is then not difficult to derive for each k in turn (beginning with $k = 1$) an explicit formula for the sum of the kth powers. This formula will be of the form

$$\sum_{i=1}^{N} i^k = \frac{N^{k+1}}{k+1} + \frac{N^k}{2} + p(N)$$

where $p(N)$ is a polynomial in N of degree less than k. Roberval's inequality then follows immediately.

It is not known whether Fermat actually proved the general result indicated or merely tried a few values of k and assumed it would be true for any value. After all, the exchange of letters with Roberval occurred 18 years before Pascal wrote his treatise on the triangle with its inductive proofs. Interestingly enough, Pascal himself stated an equivalent result on sums of powers in 1654, noting that "those who are even a little familiar with the doctrine of indivisibles will not fail to see that one may use this result for the determination of curvilinear areas. This result permits one immediately to square all types of parabolas and an infinity of other curves."[15]

In any case, Fermat was not completely satisfied with his method because it only worked for higher parabolas. He could not see how to adapt it for curves of the form $y^m = px^k$ or for "higher hyperbolas" of the form $y^m x^k = p$. In modern terms, this method for finding areas under $y = px^k$ only worked if k were a positive integer. Fermat wanted a method which would work if k were any rational number, positive or negative. Although

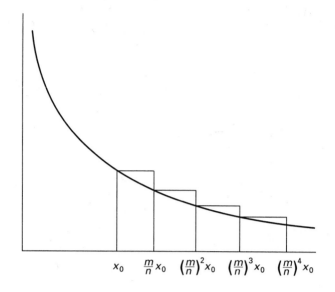

FIGURE 12.7
Fermat's procedure for determining the area under $y = px^{-k}$.

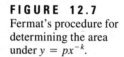

he only announced such a method in his *Treatise on Quadrature* of about 1658, it seems clear that he discovered this new procedure in the 1640s.

To apply his earlier method to the question of determining the area under $y = px^{-k}$ to the right of $x = x_0$ required dividing either the x-axis or the line segment $x = x_0$ from 0 to $y_0 = px_0^{-k}$ into finitely many intervals and summing the areas of the inscribed and circumscribed rectangles. Using the latter procedure, however, would give Fermat an infinite rectangle as the difference between his circumscribed and inscribed rectangles, one for which it was not at all clear that the area could be made as small as desired. On the other hand, there was no way of dividing the (infinite) x-axis into finitely many intervals ultimately to be made as small as one wishes. Fermat's solution to his dilemma was to divide the x-axis into infinitely many intervals, whose lengths were not equal but formed a geometric progression, and then to use the known formula for summing such a progression to add up the areas of the infinitely many rectangles.

Fermat began by partitioning the infinite interval to the right of x_0 at the points $a_0 = x_0, a_1 = \frac{m}{n}x_0, a_2 = \left(\frac{m}{n}\right)^2 x_0, \ldots, a_i = \left(\frac{m}{n}\right)^i x_0, \ldots$ where m and n ($m > n$) are positive integers (Figure 12.7). The intervals $[a_{i-1}, a_i]$ will ultimately be made as small as desired by taking m/n sufficiently close to 1. Fermat next circumscribed rectangles above the curve over each small interval. The first circumscribed rectangle has area

$$R_1 = \left(\frac{m}{n}x_0 - x_0\right)y_0 = \left(\frac{m}{n} - 1\right)x_0\frac{p}{x_0^k} = \left(\frac{m}{n} - 1\right)\frac{p}{x_0^{k-1}}.$$

The second rectangle has area

$$R_2 = \left[\left(\frac{m}{n}\right)^2 x_0 - \left(\frac{m}{n}\right)x_0\right]\frac{p}{\left(\frac{m}{n}x_0\right)^k} = \left(\frac{m}{n}\right)\left(\frac{m}{n} - 1\right)x_0\left(\frac{n}{m}\right)^k\frac{p}{x_0^k} = \left(\frac{n}{m}\right)^{k-1}R_1.$$

Side Bar 12.2   Did Fermat Invent the Calculus?

By the mid-1640s Fermat had determined the area under any curve of the form $y = x^k$ (except, of course, $y = x^{-1}$, a curve for which Fermat realized his method did not apply) and also had been able to construct the tangent to such a curve. Since he had solved the two major problems of the calculus, at least in these significant special cases, why should he not be considered as the inventor of the calculus? The answer must be that Fermat did not realize the inverse relationship between the two problems and therefore could not apply results on tangents to calculate results on areas. A student today, seeing that the derivative of $y = x^k$ was $y' = kx^{k-1}$ and also that the area under $y = x^k$ from 0 to x was $\frac{x^{k+1}}{k+1}$ would probably immediately recognize this inverse property. Fermat did not, because he was not asking the questions which would lead him to it. For Fermat, construction of a tangent meant exactly that: find the length of the subtangent

and then draw the line from the point on the curve to the appropriate point on the axis. Thus, he did not generally consider the slope of the tangent line, what we call the derivative. In dealing with $y = x^k$, he would find that the subtangent t equaled x/k rather than that the slope of the tangent equaled kx^{k-1}. Similarly, to find an area under a curve meant for Fermat to find a suitable rectangle equal in area to the given curvilinear region. In other words, the area under $y = x^k$ from 0 to x_0 equaled the area of the rectangle whose width was x_0 and whose height was $\frac{1}{k+1}y_0$. He never considered the area from a fixed coordinate to a variable one as determining a function, expressible as a new curve. Thus, although Fermat was able to solve the two basic problems of the calculus in many instances, he did not ask the "right" questions. It was others who were able to see what Fermat missed.

Similarly, the third rectangle has area

$$R_3 = \left(\frac{n}{m}\right)^{2(k-1)} R_1.$$

It follows that the sum of all the circumscribed rectangles is

$$R = R_1 + \left(\frac{n}{m}\right)^{k-1} R_1 + \left(\frac{n}{m}\right)^{2(k-1)} R_1 + \cdots$$

$$= R_1\left[1 + \left(\frac{n}{m}\right)^{k-1} + \left(\frac{n}{m}\right)^{2(k-1)} + \cdots\right]$$

or, using the formula for the sum of a geometric series,

$$R = \frac{1}{1 - \left(\frac{n}{m}\right)^{k-1}} R_1 = \frac{1}{1 - \left(\frac{n}{m}\right)^{k-1}} \left(\frac{m}{n} - 1\right) \frac{p}{x_0^{k-1}} = \frac{1}{\frac{n}{m} + \left(\frac{n}{m}\right)^2 + \cdots + \left(\frac{n}{m}\right)^{k-1}} \frac{p}{x_0^{k-1}}.$$

Fermat could have made a similar calculation for the inscribed rectangles, but decided it wasn't necessary. He let the area of the first rectangle "go to nothing," or, in modern terminology, found the limiting value of his sum, by letting n/m approach 1. The value of R then approaches $\frac{1}{k-1}\frac{p}{x_0^{k-1}}$, and therefore, the desired area A is given by

---

Biography   Evangelista Torricelli (1608–1647)

Torricelli studied mathematics in Rome with Benedetto Castelli (1578–1643), a pupil of Galileo and, in 1641, was able to study with Galileo himself at his house in Arcetri. He stayed there until Galileo's death and was soon thereafter appointed to Galileo's old position of mathematician and philosopher to the Grand Duke of Tuscany. Torricelli remained in Florence for the rest of his life, continuing Galileo's work on motion and grinding lenses for more powerful telescopes. He is probably most famous for his discovery of the principle of the barometer in 1643. He died of typhoid fever in 1647 (Figure 12.8).

$$A = \frac{1}{k-1}x_0 y_0.$$

Fermat quickly noticed that this division of the axis into infinite intervals could also be applied to find the known area under the parabolas $y = px^k$ from $x = 0$ to $x = x_0$. He simply divided this finite interval $[0, x_0]$ into an infinite set of subintervals by beginning from the right: $a_0 = x_0$, $a_1 = \frac{n}{m}x_0$, $a_2 = \left(\frac{n}{m}\right)^2 x_0, \ldots, a_i = \left(\frac{n}{m}\right)^i x_0, \ldots$ where here $n < m$, and proceeded as above to show that this area is equal to $\frac{1}{k+1}x_0 y_0$. In the other cases Fermat wanted to solve, namely, the areas under the curves $x^k y^m = p$ and $y^m = px^k$, the method had to be modified slightly to avoid having the geometric series involve fractional powers (Side Bar 12.2). But Fermat did succeed in showing that the area under the "hyperbola" $x^k y^m = p$ to the right of $x = x_0$ is $\frac{m}{k-m}x_0 y_0$ while that under the "parabola" $y^m = px^k$ from 0 to x_0 is $\frac{m}{k+m}x_0 y_0$.

12.2.3 Torricelli and the Infinitely Long Solid

FIGURE 12.8
Torricelli on an Italian stamp.

Evangelista Torricelli (1608–1647), a disciple of Galileo, solved Fermat's problem of determining areas under higher parabolas and hyperbolas by a different argument, although he again inscribed certain rectangles under the curve. He then gave complete classical proofs of his results by *reductio ad absurdum* arguments. His most surprising discovery, however, announced in 1643, was his determination that the volume of the infinitely long solid formed by rotating the hyperbola $xy = k^2$ around the y-axis from $y = a$ to $y = \infty$ was finite and in fact equal to the volume of the cylinder of altitude k^2/a and radius equal to the semidiameter $AS = \sqrt{2}k$ of the hyperbola (Figure 12.9). Torricelli used a method similar to the cylindrical shell method taught today, but he expressed it in terms of curved indivisibles, analogous to the lines of his friend Cavalieri. First, he showed that the lateral surface area of any cylinder inscribed in his infinite hyperbolic solid, such as *POMN*, was equal to the area of the circle of radius AS. (In modern terms, this is simply that $2\pi x\frac{k^2}{x} = \pi(\sqrt{2}k)^2$.) Next, he noted that the infinite solid can be considered to be composed of all these cylindrical surfaces, to each of which there corresponds one of the circles making up the cylinder *ACHI*. It follows that the infinite solid is equal to the cylinder.

Torricelli wrote that "it may seem incredible that although this solid has an infinite length, nevertheless none of the cylindrical surfaces we considered has an infinite length,

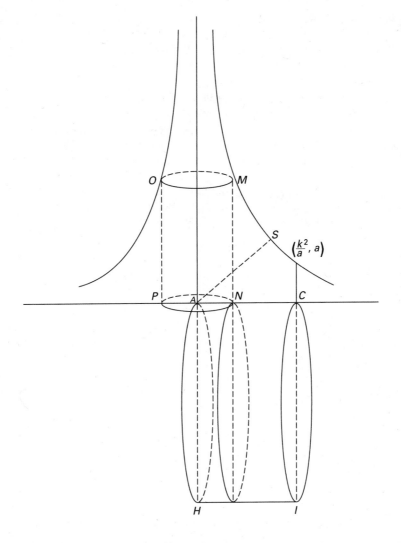

FIGURE 12.9
Torricelli's infinite
hyperbolic solid.

but all of them are finite."[16] Because he believed that this result was "incredible," however, he decided to present a second proof, this one by exhaustion, to lend more strength
to this result.

12.2.4 Wallis and Fractional Exponents

Another mathematician who used Cavalieri's indivisibles was John Wallis (1616–1703).
Wallis, the first mathematician actually to explain fractional exponents and use them
consistently, took an approach somewhat different from Cavalieri's, however, in his discussion of the area under $y = x^k$ found in the *Arithmetica infinitorum* (*Arithmetic of
Infinites*) of 1655. To determine the ratio of the area under $y = x^2$ between $x = 0$ and
$x = x_0$ to the circumscribed rectangle whose area was $x_0 y_0$, he noted that the ratio of the
corresponding line segments over a given abscissa x was $x^2 : x_0^2$. But since there were

---

Biography **John Wallis (1616–1703)**

Although Wallis studied mathematics in his university days in Cambridge, much of his early life was spent in preparing for an ecclesiastical career. Nevertheless, his interest in various scientific questions led him to be involved in the first informal meetings in the 1640s in London of that group of men who formed the Royal Society in 1662. These weekly meetings were devoted to the discussion of "Philosophical Inquiries," including matters of anatomy, geometry, astronomy, and mechanics which were currently undergoing detailed investigations in England as well as on the continent. Wallis' early interest in mathematics revived about 1647. Two years later he was appointed to the vacancy in the Savilian chair of mathematics at Oxford caused by the incumbent finding himself on the wrong side in the English civil war. It was at Oxford that Wallis wrote his mathematical works, which included, besides the *Arithmetica infinitorum*, tracts on algebra, conic sections, and mechanics.

---

infinitely many such abscissas, Wallis needed to calculate the ratio of the sum of the infinitely many antecedents to the sum of the infinitely many consequents. Taking his abscissas in arithmetic progression $0, 1, 2, \ldots$, Wallis wanted to determine what in modern terminology would be

$$\lim_{n \to \infty} \frac{0^2 + 1^2 + 2^2 + \cdots + n^2}{n^2 + n^2 + n^2 + \cdots + n^2}.$$

To calculate this ratio he tried various cases:

$$\frac{0 + 1}{1 + 1} = \frac{1}{2} = \frac{1}{3} + \frac{1}{6}$$

$$\frac{0 + 1 + 4}{4 + 4 + 4} = \frac{5}{12} = \frac{1}{3} + \frac{1}{12}$$

$$\frac{0 + 1 + 4 + 9}{9 + 9 + 9 + 9} = \frac{14}{36} = \frac{1}{3} + \frac{1}{18}$$

and, in general,

$$\frac{0^2 + 1^2 + 2^2 + \cdots + n^2}{n^2 + n^2 + n^2 + \cdots + n^2} = \frac{1}{3} + \frac{1}{6n}.$$

Wallis concluded that if the number of terms was infinite, that is, if the lines "filled up" the desired areas, the ratio would be exactly 1/3. After calculating the analogous ratio for cubes to be 1/4, Wallis took the leap by what he called "induction" to the conclusion that for any positive integer k,

$$\frac{0^k + 1^k + 2^k + \cdots + n^k}{n^k + n^k + n^k + \cdots + n^k} = \frac{1}{k+1}$$

if there were an infinite number of terms.

Wallis' next step was to generalize this result to other powers by using analogy. Thus he noted that given any arithmetic sequence of powers, say 2, 4, 6, . . . , the consequents of the corresponding ratio of areas was also an arithmetic sequence, namely 3, 5, 7, ... It followed that if the consequent of the ratio was 1, the sequence of powers must have index 0, that is, that m^0 must be 1 for every m. Furthermore, he noted that the sequence of second powers, with consequent 3, was composed of the square roots of the sequence of fourth powers, with consequent 5, and that 3 is the arithmetic mean between the 1 and 5, the consequents of the series of powers 0 and 4. Wallis then made another bold generalization. Taking the series of terms $\sqrt{0}, \sqrt{1}, \sqrt{2}, . . .$, whose terms are the square roots of the series 0, 1, 2, . . . , he decided that the consequent of the corresponding ratio should be the arithmetic mean between 1 and 2, the consequents of the series of powers 0 and 1. In other words, the ratio

$$\frac{\sqrt{0} + \sqrt{1} + \sqrt{2} + \cdots + \sqrt{n}}{\sqrt{n} + \sqrt{n} + \sqrt{n} + \cdots + \sqrt{n}}$$

must ultimately be equal to $\frac{1}{1\,1/2} = \frac{2}{3}$. In addition, the power of this series should be the arithmetic mean between 0 and 1, namely 1/2, or, as Wallis put it, the index of $\sqrt{x}$ is 1/2. Wallis similarly concluded that the index of $\sqrt[3]{x}$ must be 1/3 and that of $\sqrt[3]{x^2}$ must be 2/3, while the consequents of their corresponding ratios must be the two arithmetic means between 1 and 2, namely 1 1/3 and 1 2/3. Then, defining a fractional power for an arbitrary positive fraction p/q as the index of the qth root of the pth power, Wallis brought all these generalizations together into a theorem: "If we take an infinite series of quantities, beginning with a point or 0, continuously increasing in the ratio of any power, an integer or a rational fraction, then the ratio of the whole to the series of as many numbers equal to the highest number is 1 divided by the index of this power plus 1."[17]

Although Wallis applied this result to solve the area problem for a curve of the form $y = x^{p/q}$, that is, he found

$$\int_0^1 x^{p/q}\, dx = \frac{1}{\frac{p}{q} + 1},$$

he did not prove that his answer was correct except in the case where the index was $1/q$. He was, however, a firm believer in the power of analogy. Thus, he generalized his ideas of indexes to both negative and irrational numbers and showed that these indexes obeyed our familiar laws of exponents. His methods fell short when he attempted to generalize his theorem and the solution of the area problem to curves of the form $y = x^{-k}$. His basic rule told him that the corresponding ratio in the case of exponent -1 should be $1/(-1 + 1) = 1/0$ while in the case of exponent -2 the ratio should be $1/-1$. It was reasonable to assume that the area under the hyperbola $y = 1/x$ was in some sense 1/0 or infinity, but what did it mean that the area under the curve $y = 1/x^2$ was $1/-1$? Since for indices 3, 2, 1, 0, the corresponding ratios were 1/4, 1/3, 1/2, and 1/1, and these values formed an increasing sequence, he assumed that the ratio $1/-1$ for index -2 should be greater than the ratio 1/0 for index -1. But what it meant for $1/-1$ to be greater than infinity, Wallis could never quite figure out.

Passing over this problem, but also realizing that his method could be applied to finding areas under curves given by sums of terms of the form $ax^{p/q}$, Wallis next attempted to generalize his methods to the more complicated problem of determining arithmetically the area of a circle of radius 1, namely, of finding the area under the curve $y = \sqrt{1 - x^2} = (1 - x^2)^{1/2}$. To use his technique of arguing by analogy, he actually attacked a more general problem, to find the ratio of the area of the unit square to the area enclosed in the first quadrant by the curve $y = (1 - x^{1/p})^n$. The case $p = 1/2, n = 1/2$ is the case of the circle, where the ratio is $4/\pi$. It was easy enough for Wallis to calculate by his known methods the ratios in the cases where p and n were integral. For example, if $p = 2$ and $n = 3$, the area under $y = (1 - x^{1/2})^3$ from 0 to 1 is that under $y = 1 - 3x^{1/2} + 3x - x^{3/2}$, that is, $1 - 2 + 3/2 - 2/5 = 1/10$. Since the area of the unit square is 1, the ratio here is $1 : 1/10 = 10$. Wallis thus constructed the following table of these ratios where, for $p = 0$, he simply used the area under $y = 1^n$:

$p \backslash n$	0	1	2	3	4	5	6	7	...
0	1	1	1	1	1	1	1	1	...
1	1	2	3	4	5	6	7	8	...
2	1	3	6	10	15	21	28	36	...
3	1	4	10	20	35	56	84	120	...
⋮	⋮	⋮	⋮	⋮	⋮	⋮	⋮	⋮	...

Wallis clearly recognized Pascal's arithmetical triangle in his table. What he wanted was to be able to interpolate rows corresponding to $p = 1/2, p = 3/2, \ldots$ and columns corresponding to $n = 1/2, n = 3/2 \ldots$ from which he could find the desired value, which he wrote as □, when both parameters equaled 1/2. From his knowledge of Pascal's triangle, Wallis realized that in his table the relationship $a_{p,n} = \frac{p+n}{n} a_{p,n-1}$ holds, where $a_{p,n} = \binom{p+n}{n}$ designates the entry in row p, column n. Using this same rule for the row $p = 1/2$, he noted first that $a_{1/2,0} = 1$, because all other entries in column 0 were equal to 1. It followed that $a_{1/2,1} = \left(\frac{1/2+1}{1}\right) \cdot 1 = 3/2$, $a_{1/2,2} = \left(\frac{1/2+2}{2}\right) \cdot \frac{3}{2} = \frac{5}{4} \cdot \frac{3}{2} = \frac{15}{8}$, $a_{1/2,3} = \frac{7}{6} \cdot \frac{5}{4} \cdot \frac{3}{2} = \frac{105}{48}$, ... Similarly, since $a_{1/2,1/2} = $ □, he had $a_{1/2,3/2} = \left(\frac{1/2+3/2}{3/2}\right)$□ $= \frac{4}{3}$□, $a_{1/2,5/2} = \frac{6}{5} \cdot \frac{4}{3}$□, ... and the row $p = \frac{1}{2}$ was

$$1 \quad \square \quad \frac{3}{2} \quad \frac{4}{3}\square \quad \frac{15}{8} \quad \frac{8}{5}\square \quad \ldots$$

Wallis was able to fill in the remainder of the table analogously, but since he was interested in calculating □, he considered the ratios in this row. Because it was evident that the ratios of alternate terms continually decreased, that is, $a_{1/2,k+2} : a_{1/2,k} > a_{1/2,k+4} : a_{1/2,k+2}$ for all k, he made the assumption that this was true for ratios of adjoining terms as well. It followed that □ $: 1 > 3/2 :$ □, so □ $> \sqrt{3/2}$; that $3/2 :$ □ $> 4/3$□ $: 3/2$, so □ $< 3/2\sqrt{3/4} = [(3 \times 3)/(2 \times 4)]\sqrt{4/3}$; and, similarly, that □ $> [(3 \times 3)/(2 \times 4)]\sqrt{5/4}$, □ $< [(3 \times 3 \times 5 \times 5)/(2 \times 4 \times 4 \times 6)]\sqrt{6/5}$, ... Wallis was thus able to assert that □ (or $4/\pi$) could be calculated as an infinite product:

$$\square = \frac{4}{\pi} = \frac{3 \times 3 \times 5 \times 5 \times 7 \times 7 \times \cdots}{2 \times 4 \times 4 \times 6 \times 6 \times 8 \times \cdots}.$$

12.2.5 Roberval and the Cycloid

Although an infinite product was not perhaps the kind of area result Wallis had hoped for, other mathematicians of the period also had to be satisfied with answers not strictly arithmetical in their consideration of curves other than the power curves. Roberval, for example, around 1637 determined the area under a **cycloid,** the curve traced by a point attached to the rim of a wheel rolling along a line. Roberval defined this curve as follows: "Let the diameter AB of the circle AGB move along the tangent AC, always remaining parallel to its original position, until it takes the position CD, and let AC be equal to the semicircle AGB [Figure 12.10]. At the same time, let the point A move on the semicircle AGB, in such a way that the speed of AB along AC may be equal to the speed of A along the semicircle AGB. Then, when AB has reached the position CD, the point A will have reached the position D. The point A is carried along by two motions—its own on the semicircle AGB, and that of the diameter along AC."[18]

Roberval began his calculation by dividing the axis AC and the semicircle AGB into infinitely many equal parts. Along the semicircle these parts are $AE = EF = FG = \cdots$ while along the axis they are $AM = MN = NO = \cdots$. Furthermore, since the motion which generates the cycloid is composed of equal motions along the semicircle and the axis, Roberval set $AE = AM$, $EF = MN$, ... Because the point A will then be at E when the base of the diameter is at M, the point M_2, whose horizontal distance from M is the same as that of E from the point E_1 on the axis, is a point on the cycloid. Similarly, the point N_2 whose horizontal distance from N is the same as the distance of F from the point F_1 on the axis, is also a point on the curve, as are the points $O_2, P_2, \ldots$ indicated in Figure 12.10. Roberval then constructed a new curve, the companion of the cycloid, through the points M_1, N_1, O_1, . . . where M_1 has the same x-coordinate as M and the same y-coordinate as E, and so on. In modern notation, this curve is given by $x(t) = at$, $y(t) = a(1 - \cos t)$, or, in nonparametric form, as $y = a\left(1 - \cos \frac{x}{a}\right)$, where a is the radius of the circle. (The cycloid itself is given by $x(t) = a(t - \sin t)$, $y(t) = a(1 - \cos t)$.)

To determine the area under half of one arch of the cycloid, Roberval first determined that the area between the cycloid and its companion is equal to that of half the generating circle. This follows from Cavalieri's principle, since $M_1M_2 = EE_1$, $N_1N_2 = FF_1$, . . . and the corresponding pairs of lines are each at the same altitude. To finish his calculation, Roberval again used the principle to show that the companion curve to the cycloid bisects

FIGURE 12.10
Roberval's determination
of the area bounded by a
cycloid.

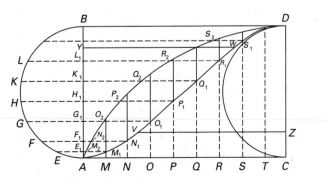

the rectangle $ABCD$. This follows from the fact that to each line VZ in region $ACDM_1$, there corresponds an equal line WY in AM_1DB. Because the area of the rectangle is equal to the product of half the circumference of the circle with the diameter (or $2\pi a^2$), the area under the companion curve is equal to that of the generating circle (πa^2). It follows that the area under half of one arch of the cycloid is equal to 3/2 that of the generating circle or that the area under an entire arch is three times that of the circle.

Roberval's companion to the cycloid was in effect a cosine curve, although Roberval did not identify it as such. But in the same work he did draw, probably for the first time, a curve identified as a curve of sines, although it only consisted of the sines of one quadrant of the circle. In addition, Roberval was able to determine that the area under this curve was equal to the square of the radius defining the particular sines. Pascal, some 20 years later, in a small treatise entitled *Traité des sinus du quart de cercle* (*Treatise on the sines of a quadrant of a circle*) was able to find the area under any portion of that curve. Consider the quadrant ABC of the circle and let D be any point from which the sine DI is drawn to the radius AC (Figure 12.11). Pascal then draws a "small" tangent EDE' and perpendiculars ER, $E'R'$ to the radius. His claim is that "the sum of the sines of any arc of a quadrant is equal to the portion of the base between the extreme sines, multiplied by the radius."[19] By the "sum of the sines," Pascal means the sum of the infinitesimal rectangles formed by multiplying each sine by the infinitesimal arc represented by the tangent EE'. Hence, Pascal's theorem in modern terms is that

$$\int_\alpha^\beta r \sin \theta \, d(r\theta) = r(r \cos \alpha - r \cos \beta).$$

For his proof, Pascal notes that triangles EKE' and DIA are similar, hence $DI : DA = E'K : EE' = RR' : EE'$ and therefore $DI \cdot EE' = DA \cdot RR'$. In other words, the rectangle formed from the sine and the infinitesimal arc (or tangent) is equal to the rectangle formed by the radius and the part of the axis between the ends of the arc, or, $r \sin \theta \, d(r\theta) = r(r \cos(\theta + d\theta) - r \cos(\theta)) = r(d(r \cos \theta))$. Adding up these rectangles between the two given angles produces the result cited. Although this result proved important, and although Pascal generalized it immediately to give formulas for the integrals of powers of the sine, the most significant aspect of Pascal's work was the appearance of the "differential triangle" EKE'. Leibniz's study of this particular work of Pascal was instrumental in his own realization of the connection between the area problem and the tangent problem.

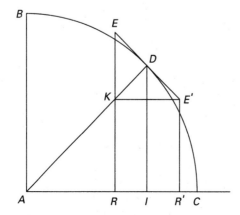

FIGURE 12.11
Pascal's area under the sine curve.

12.2.6 The Area Under the Rectangular Hyperbola

Our final example of a mid-seventeenth century solution to the area problem is the work of the Belgian mathematician Gregory of St. Vincent (1584–1667) on the area under the hyperbola $xy = 1$ (Figure 12.12). In his *Opus geometricum* (*Geometrical Work*) of 1647, Gregory showed that if (x_i, y_i) for $i = 1, 2, 3, 4$ are four points on this hyperbola such that $x_2 : x_1 = x_4 : x_3$, then the area under the hyperbola over $[x_1, x_2]$ equals that over $[x_3, x_4]$ (Figure 12.13). To prove this, divide the interval $[x_1, x_2]$ into subintervals at the points a_i, $i = 0, \ldots, n$. Because $x_2 : x_1 = x_4 : x_3$, it follows that $x_3 : x_1 = x_4 : x_2 = v$ or $x_3 = vx_1$, $x_4 = vx_2$. One can therefore conveniently subdivide the interval $[x_3, x_4]$ at the points $b_i = va_i$, $i = 0, \ldots, n$. If rectangles are then inscribed in and circumscribed about the

FIGURE 12.12

The Frontispiece of Gregory of St. Vincent's *Opus Geometricum*. Gregory claimed that he had squared the circle. (Source: Special Collections Division, USMA Library, West Point, New York)

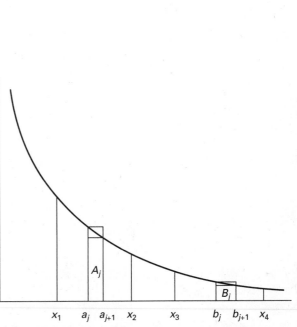

FIGURE 12.13

Gregory of St. Vincent and the area under the hyperbola $xy = 1$.

hyperbolic areas A_j over $[a_j, a_{j+1}]$ and B_j over $[b_j, b_{j+1}]$, it is straightforward to calculate the corresponding inequalities:

$$(a_{j+1} - a_j)\frac{1}{a_{j+1}} < A_j < (a_{j+1} - a_j)\frac{1}{a_j} \quad \text{and} \quad (b_{j+1} - b_j)\frac{1}{b_{j+1}} < B_j < (b_{j+1} - b_j)\frac{1}{b_j}.$$

Substituting the values $b_j = va_j$ into the second set of inequalities gives

$$(a_{j+1} - a_j)\frac{1}{a_{j+1}} < B_j < (a_{j+1} - a_j)\frac{1}{a_j}.$$

Thus both hyperbolic regions are squeezed between rectangles of the same areas. Because both intervals can be divided into subintervals as small as desired, it follows that the two hyperbolic areas are equal.

When the Belgian Jesuit Alfonso Antonio de Sarasa (1618–1667) read Gregory's work in 1649, he immediately noticed that this calculation implied that the area $A(x)$ under the hyperbola from 1 to x had the logarithmic property $A(\alpha\beta) = A(\alpha) + A(\beta)$. (Because the ratio $\beta : 1$ equals the ratio $\alpha\beta : \alpha$, the area from 1 to β equals the area from α to $\alpha\beta$. Because the area from 1 to $\alpha\beta$ is the sum of the areas from 1 to α and from α to $\alpha\beta$, the logarithmic property is immediate.) Thus if one could calculate the area under the hyperbola $xy = 1$, one could calculate logarithms. The search for means of calculating these areas led to the power series methods of Newton and others in the 1660s, methods which were instrumental in Newton's version of the calculus.

12.3 POWER SERIES

In 1668, Nicolaus Mercator (1620–1687) published his *Logarithmotechnica* (*Logarithmic Teachings*), in which appeared the power series expansion for the logarithm. Mercator, having read the hint of de Sarasa that the logarithm was related to the area under a hyperbola and having learned from Wallis how to calculate certain ratios of infinite sums of powers, decided to calculate $\log(1 + x)$ (the area A under the hyperbola $y = 1/(1 + x)$ from 0 to x) by using such infinite sums. He divided the interval $[0, x]$ into n subintervals of length x/n and approximated A by the sum

$$\frac{x}{n} + \frac{x}{n}\left(\frac{1}{1 + \frac{x}{n}}\right) + \frac{x}{n}\left(\frac{1}{1 + \frac{2x}{n}}\right) + \cdots + \frac{x}{n}\left(\frac{1}{1 + \frac{(n-1)x}{n}}\right).$$

Since each term $\frac{1}{1 + (kx/n)}$ is the sum of the geometric series $\sum_{j=0}^{\infty}(-1)^j\left(\frac{kx}{n}\right)^j$, it follows that

$$A \approx \frac{x}{n} + \frac{x}{n}\sum_{j=0}^{\infty}(-1)^j\left(\frac{x}{n}\right)^j + \frac{x}{n}\sum_{j=0}^{\infty}(-1)^j\left(\frac{2x}{n}\right)^j + \cdots + \frac{x}{n}\sum_{j=0}^{\infty}(-1)^j\left(\frac{(n-1)x}{n}\right)^j$$

$$= n\frac{x}{n} - \frac{x^2}{n^2}\sum_{i=1}^{n-1}i + \frac{x^3}{n^3}\sum_{i=1}^{n-1}i^2 + \cdots + (-1)^j\frac{x^{j+1}}{n^{j+1}}\sum_{i=1}^{n-1}i^j + \cdots$$

$$= x - \frac{\sum_{i=1}^{n-1}i}{n \cdot n}x^2 + \frac{\sum_{i=1}^{n-1}i^2}{n \cdot n^2}x^3 + \cdots + (-1)^j\frac{\sum_{i=1}^{n-1}i^j}{n \cdot n^j}x^{j+1} + \cdots.$$

By Wallis' results, the coefficient of x^{k+1} in this expression is equal to $1/(k+1)$ if n is infinite. Therefore,

$$\log(1 + x) = x - \frac{x^2}{2} + \frac{x^3}{3} - \frac{x^4}{4} + \cdots,$$

a power series in x which enabled actual values of the logarithm to be calculated easily.

Power series for other transcendental functions were discovered by James Gregory (1638–1675) in Scotland around 1670 and communicated to John Collins (1625–1683), the secretary of the Royal Society, without any indication of how they were discovered. For example, in a letter of December 19, 1670, Gregory wrote that the arc whose sine is B (where the radius of the circle is R) is expressible as

$$B + \frac{B^3}{6R^2} + \frac{3B^5}{40R^4} + \frac{5B^7}{112R^6} + \frac{35B^9}{1152R^8} + \cdots. [20]$$

In modern terminology, Gregory's series is the series for $1/R \arcsin B/R$, which, if $R = 1$, can be written

$$\arcsin x = x + \frac{x^3}{6} + \frac{3x^5}{40} + \frac{5x^7}{112} + \frac{35x^9}{1152} + \cdots.$$

Similarly, in a letter of February 15, 1671, Gregory included, among others, the series for the arc y given the tangent x and vice versa, written in modern notation as:

$$\arctan x = x - \frac{x^3}{3} + \frac{x^5}{5} - \frac{x^7}{7} + \frac{x^9}{9} - \cdots$$

$$\tan y = y + \frac{y^3}{3} + \frac{2y^5}{15} + \frac{17y^7}{315} + \frac{3233y^9}{181440} + \cdots. [21]$$

However Gregory derived these series, it turns out that the arctangent series, as well as series for the sine and cosine which Newton found in the mid-1660s, had been discovered in southern India perhaps 200 years earlier. These series appear in Sanskrit verse in the *Tantrasaṅgraha-vyākhyā* (c. 1530), a commentary on a work by Kerala Gargya Nīlakaṇṭha (1445–1545) of some 30 years earlier. Unlike the situation for many results of Indian mathematics, a detailed proof of these results exists, in the *Yuktibhāṣa*, a work in Malayalam, the language of Kerala, the southwestern region of India. The *Yuktibhāṣa*, written by Jyesthadeva (1500–1610), credits the arctangent series to the earlier mathematician Madhava (1340–1425), who lived near Cochin.

The Sanskrit verse giving the arctangent series may be translated as follows:

The product of the given sine and the radius, divided by the cosine, is the first result. From the first [and the second, third, etc.] results, obtain [successively] a sequence of results by taking repeatedly the square of the sine as the multiplier and the square of the cosine as the divisor. Divide the above results in order by the odd numbers one, three, etc. [to get the full sequence of terms]. From the sum of the odd terms subtract the sum of the even terms. The result becomes the arc. In this connection . . . the sine of the arc or that of its complement, whichever is smaller, should be taken here [as the given sine]; otherwise the terms obtained by the [above] repeated process will not tend to the vanishing magnitude. [22]

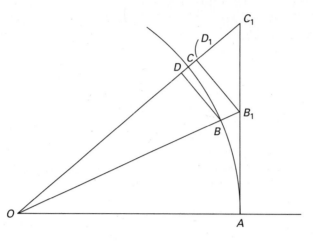

FIGURE 12.14
Jyesthadeva's derivation
of the arctangent series.

It is not difficult to translate these words into the modern symbolism of the same arctangent series which Gregory found, noting that the author has realized that convergence only occurs when $\tan \theta \leq 1$.

Jyesthadeva's proof of the validity of the arctangent series begins with the following lemma where, for simplicity, the radius of the circle is set equal to 1:

Lemma. *Let BC be a small arc of a circle with center O. If OB, OC meet the tangent at any point A of the circle in the points B_1, C_1 respectively, then arc BC is given approximately by arc $BC \approx B_1C_1/(1 + AB_1^2)$ (Figure 12.14).*

If perpendiculars BD, B_1D_1 are drawn to OC, it follows by similarity that $\frac{BD}{B_1D_1} = \frac{OB}{OB_1} = \frac{1}{OB_1}$ and $\frac{B_1D_1}{B_1C_1} = \frac{OA}{OC_1} = \frac{1}{OC_1}$ and therefore that $BD = \frac{B_1C_1}{OB_1 \cdot OC_1}$. When arc BC is very small, $OB_1 \approx OC_1$ and therefore arc $BC \approx BD = \frac{B_1C_1}{OB_1^2} = \frac{B_1C_1}{1 + AB_1^2}$.

Dividing the tangent $t = AC_1$ to arc AC into n equal parts, applying the lemma to each in turn, and then letting n get indefinitely large gives

$$\arctan t = \lim_{n \to \infty} \sum_{r=0}^{n-1} \frac{t/n}{1 + \left(\frac{rt}{n}\right)^2}$$

$$= \lim_{n \to \infty} \sum_{r=0}^{n-1} \frac{t}{n}\left[1 - \left(\frac{rt}{n}\right)^2 + \left(\frac{rt}{n}\right)^4 - \cdots + (-1)^k\left(\frac{rt}{n}\right)^{2k} + \cdots\right]$$

$$= \lim_{n \to \infty} \left[\frac{t}{n} + \frac{t}{n}\left(1 - \frac{t^2}{n^2} + \frac{t^4}{n^4} - \cdots\right) + \frac{t}{n}\left(1 - \frac{2^2t^2}{n^2} + \frac{2^4t^4}{n^4} - \cdots\right)\right.$$

$$\left. + \frac{t}{n}\left(1 - \frac{3^2t^2}{n^2} + \frac{3^4t^4}{n^4} - \cdots\right) + \cdots + \frac{t}{n}\left(1 - \frac{(n-1)^2t^2}{n^2} + \frac{(n-1)^4t^4}{n^4} - \cdots\right)\right]$$

$$= \lim_{n \to \infty} \left[t - \frac{t^3}{n^3}(1^2 + 2^2 + \cdots + (n-1)^2) + \frac{t^5}{n^5}(1^4 + 2^4 + \cdots + (n-1)^4) - \cdots\right].$$

To complete the derivation, Jyesthadeva needed to deal with sums of integral powers. Like ibn al-Haytham, he showed that

$$n\sum_{j=1}^{n} j^{p-1} = \sum_{j=1}^{n} j^p + \sum_{k=1}^{n-1}\sum_{j=1}^{k} j^{p-1},$$

a result which implied Wallis' theorem that

$$\lim_{n\to\infty}\frac{1}{n^{p+1}}\sum_{j=1}^{n} j^p = \frac{1}{p+1}.$$

Substituting that into the earlier formula gave him the final result that

$$\arctan t = t - \frac{t^3}{3} + \frac{t^5}{5} - \frac{t^7}{7} + \cdots.$$

Why were Hindu authors interested in this series? Their main goal seems to have been the calculation of lengths of circular arcs, values of which were necessary for astronomical purposes. This series permitted that calculation. For example, direct substitution of $t = 1$ gives $\pi/4 = 1 - 1/3 + 1/5 - 1/7 + \cdots$. Because this series converges very slowly, however, it was necessary to make various modifications. Thus, the *Tantrasaṅgraha-vyākhyā* contains other series whose convergence is considerably more rapid, including

$$\frac{\pi}{4} = \frac{3}{4} + \frac{1}{3^3 - 3} - \frac{1}{5^3 - 5} + \frac{1}{7^3 - 7} - \cdots.$$

Interestingly enough, it was the same question of determining arc length of a curve which brought European authors to the realization that the tangent problem and the area problem were related.

12.4 RECTIFICATION OF CURVES AND THE FUNDAMENTAL THEOREM

Descartes stated in his *Geometry* that the human mind could discover no rigorous and exact method of determining the ratio between curved and straight lines, that is, of determining exactly the length of a curve. Only two decades after Descartes wrote those words, however, several human minds proved him wrong. Probably the first rectification of a curve was that of the semicubical parabola $y^2 = x^3$ by the Englishman William Neile (1637–1670) in 1657 acting on a suggestion of Wallis. This was followed within the next two years by the rectification of the cycloid by Christopher Wren (1632–1723), the architect of St. Paul's Cathedral and much else in London, and the reduction of the rectification of the parabola to finding the area under a hyperbola by Huygens. The most general procedure, however, was that by Hendrick van Heuraet (1634–1660(?)), which appeared in van Schooten's 1659 Latin edition of Descartes' *Geometry*.

12.4.1 The Work of van Heuraet

Van Heuraet began his paper *De transmutatione curvarum linearum in rectas (On the transformation of curves into straight lines)* by showing that the problem of constructing

---

Biography **Hendrick van Heuraet (1634–1660)**

Van Heuraet was born in Haarlem in the Netherlands and went to Leiden to study mathematics under van Schooten in 1653.[23] His inheritance from his cloth merchant father upon the latter's death the previous year made him rather wealthy, so he could afford to study and travel without worrying about his means of support. His early mathematical work showed such great promise that van Schooten published not only his treatise on rectification but also his work on inflection points. As far as is known, however, van Heuraet died at an early age. There is no extant record of his activities after early 1660.

---

a line segment equal in length to a given arc is equivalent to finding the area under a certain curve. Let P be an arbitrary point on the arc MN of the curve α (Figure 12.15). The length PS of the normal line from P to the axis can be determined by Descartes' method. Taking an arbitrary line segment σ, van Heuraet defines a new curve α' by the ratio $P'R : \sigma = PS : PR$, where P' is the point on α' associated to P. (The σ is included so that both ratios are ratios of lines.) Drawing the differential triangle ACB with AC tangent to α at P, he notes that $PS : PR = AC : AB$. In modern notation, if $AC = ds$ and $AB = dx$, then van Heuraet's ratios yield $P'R : \sigma = ds : dx$ or $\sigma ds = P'R\, dx$. Because the sum of the infinitesimal tangents, or equivalently, the infinitesimal pieces of the arc, over the curve MN gives the length of MN, van Heuraet concludes that $\sigma \cdot$ (length of MN) $=$ area under the curve α' between M' and N'. Thus, if it is possible to derive the equation of α' from that of α and to calculate the area under it, the length of MN can also be calculated. Again using modern terminology, with $z = P'R = \sigma\frac{ds}{dx} = \sigma\sqrt{1 + \left(\frac{dy}{dx}\right)^2}$, van Heuraet's procedure can be written as

$$\sigma \cdot (\text{length of } MN) = \int_a^b z\, dx = \int_a^b \sigma \sqrt{1 + \left(\frac{dy}{dx}\right)^2}\, dx,$$

where a and b represent the x-coordinates of M and N, essentially the modern arc-length formula.

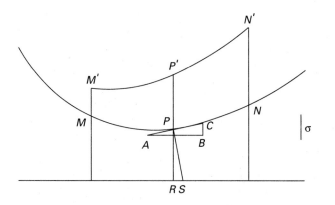

FIGURE 12.15
Van Heuraet's
rectification of a curve.

Van Heuraet illustrated his procedure for one of the few curves for which the area under the associated curve can actually be calculated, the semicubical parabola $y^2 = x^3$ which had been considered by Neile somewhat earlier. Using Descartes' normal method, he calculated that the equation which must have a double root is $x^3 + x^2 - 2vx + v^2 - n^2 = 0$. Using Hudde's rule for finding the double root, he multiplied the terms of this equation by 3, 2, 1, 0 to get $3x^3 + 2x^2 - 2vx = 0$. Therefore $v - x = \frac{3x^2}{2}$, and $PS = \sqrt{(9/4)x^4 + x^3}$. Setting $\sigma = 1/3$, van Heuraet defined the new curve α' by taking

$$z = P'R = \sigma \cdot \frac{PS}{PR} = \frac{1}{3} \frac{\sqrt{\frac{9}{4}x^4 + x^3}}{\sqrt{x^3}} = \sqrt{\frac{1}{4}x + \frac{1}{9}}$$

or, equivalently, $z^2 = 1/4x + 1/9$. Van Heuraet easily identified this curve as a parabola, the area under which he knew how to calculate. The length of the semicubical parabola from $x = 0$ to $x = b$ then equals this area divided by σ, that is,

$$\sqrt{\left(b + \frac{4}{9}\right)^3} - \frac{8}{27}.$$

After remarking that one can similarly explicitly determine the lengths of the curves $y^4 = x^5$, $y^6 = x^7$, $y^8 = x^9 \ldots$, van Heuraet concluded the paper with the more difficult rectification of an arc of a parabola $y = x^2$, a length that depends on the determination of the area under the hyperbola $z = \sqrt{4x^2 + 1}$. That problem, in 1659, had not yet been satisfactorily solved. Nevertheless, van Heuraet's methods soon became widely known. In particular, the use of the differential triangle and the association of a new curve to the given curve helped to lead others to the ideas relating the tangent problem to the area problem.

12.4.2 Gregory and the Fundamental Theorem

Among the mathematicians who related the tangent problem to the area problem were Isaac Barrow (1630–1677) and James Gregory, both of whom decided to organize the material relating to tangents, areas, and rectification gathered in their travels through France, Italy, and the Netherlands, and to present it systematically. Not surprisingly, then, the *Lectiones Geometricae* (*Geometrical Lectures*, 1670) of Barrow and the *Geometriae pars universalis* (*Universal Part of Geometry*, 1668) of Gregory contained much of the same material presented in similar ways. In effect, both of these works were treatises on material today identified as calculus, but with presentations in the geometrical style each author had learned in his university study. Neither was able to translate the material into a method of computation useful for solving problems.

As an example, consider how Gregory presented the fundamental theorem of calculus, the result linking the ideas of area and tangent. This result was the natural outcome of Gregory's study of the general problem of arc length as he found it in the work of van Heuraet. Thus he considers a monotonically increasing curve $y = y(x)$ and constructs two curves associated to it, the normal curve $n(x) = y\sqrt{1 + (dy/dx)^2}$ and $u(x) = \frac{cn}{y} = c\sqrt{1 + (dy/dx)^2}$, where c is a given constant. Now constructing the differential

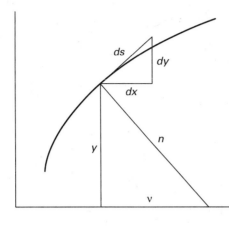

FIGURE 12.16
Gregory's differential triangle.

triangle dx, dy, ds at a given point, he argues from its similarity with the triangle formed by the ordinate y, the subnormal v, and the normal n that $y : n = dx : ds = c : u$ and thus that both $u\,dx = c\,ds$ and $n\,dx = y\,ds$ (Figure 12.16). Summing the first equation over the curve shows Gregory, as it did van Heuraet, that the arc length $\int ds$ can be expressed in terms of the area under the curve $\frac{1}{c}u(x)$. The sum of the second equation enables Gregory to show that the area under $n = n(x)$ is equal, up to a constant multiple, to the area of the surface formed by rotating the original curve around the x-axis. Gregory proves both of these results by a careful Archimedean argument using inscribed and circumscribed rectangles and a double *reductio ad absurdum*.

Having now shown that arc length can be found by an area, Gregory makes a fundamental advance by asking the converse question. Can one find a curve $u(x)$ whose arc length s has a constant ratio to the area under a given curve $y(x)$? In modern notation, Gregory is asking to determine u such that

$$c\int_0^x \sqrt{1 + \left(\frac{du}{dx}\right)^2}\,dx = \int_0^x y\,dx.$$

But this means that $c^2\!\left(1 + \left(\frac{du}{dx}\right)^2\right) = y^2$ or that $\frac{du}{dx} = \frac{1}{c}\sqrt{y^2 - c^2}$. In other words, Gregory must determine a curve u, the slope of whose tangent is equal to a given function. Letting $z = \sqrt{y^2 - c^2}$, Gregory simply defines $u(x)$ to be the area under the curve z/c from the origin to x. His task is then to show that the slope of the tangent to this curve is given by z/c. What he in fact demonstrates, again by a *reductio* argument, is that the line connecting a point K on the u curve to the point on the axis at a distance cu/z from the x-coordinate of K is tangent to the curve at K.

Gregory's crucial advance, then, was to abstract the idea of area under a specific curve between two given x-values into the idea of area as a function of a variable. In other words, he constructed a new curve whose ordinate at any value x was equal to the area under the original curve from a fixed point up to x. Once this idea was conceived, it turned out that it was not difficult to construct the tangent to this new curve and show that its slope at x was always equal to the original ordinate there.

Biography **Isaac Barrow (1630–1677)**

Barrow entered Trinity College, Cambridge in 1643 and received his B.A. in 1648 and his M.A. in 1652. Because he had royalist sympathies, he was ousted from the university in 1655 and prevented from assuming a professorship. He took the opportunity to tour the continent for four years and learn mathematics in France, Italy, and the Netherlands. He returned to Cambridge at the time of the restoration, took holy orders, and became the Regius professor of Greek. In 1662, he accepted concurrently the Gresham profes-sorship of geometry in London and the following year became the first Lucasian professor of mathematics at Cambridge. After presenting several courses of lectures over the next few years, in elementary mathematics, geometry, and optics, he resigned his position in 1669 to become the royal chaplain in London. In 1673, he returned to Trinity College as master and two years later was appointed vice-chancellor of the University, but he died in 1677, probably due to an overdose of drugs.

12.4.3 Barrow and the Fundamental Theorem

Gregory had the idea of constructing a new curve only in dealing with arc length. Isaac Barrow, on the other hand, stated a more general version of part of the fundamental theorem as proposition 11 of lecture X of his *Geometrical Lectures.*

Theorem. *Let ZGE be any curve of which the axis is AD and let ordinates applied to this axis, AZ, PG, DE, continually increase from the initial ordinate AZ. Also let AIF be a curve such that if any straight line EDF is drawn perpendicular to AD, cutting the curves in the points E, F, and AD in D, the rectangle contained by DF and a given length R is equal to the intercepted space ADEZ. Also let DE : DF = R : DT and join FT. Then TF will be tangent to AIF (Figure 12.17).*[24]

Like Gregory, Barrow began with a curve *ZGE*, which we write as $y = f(x)$, and constructed a new curve $AIF = g(x)$ such that $Rg(x)$ is always equal to the area bounded by $f(x)$ between a fixed point and the variable point x. In modern notation,

$$Rg(x) = \int_a^x f(x)\, dx.$$

Barrow then proved that the length $t(x)$ of the subtangent to $g(x)$ is given by $Rg(x)/f(x)$, or that

$$g'(x) = \frac{g(x)}{t(x)} = \frac{f(x)}{R} \qquad \text{or} \qquad \frac{d}{dx}\int_a^x f(x)\, dx = f(x).$$

Barrow proved this result by showing that the line *TF* always lies outside of the curve. If *I* is any point on the curve $g(x)$ on the side of *F* toward *A*, and if *IG* is drawn parallel to *AZ* and *KL* parallel to *AD*, the nature of the curve shows that $LF : LK = DF : DT = DE : R$ or $R \cdot LF = LK \cdot DE$. Because $R \cdot IP$ equals the area of *APZG*, it follows that $R \cdot LF$

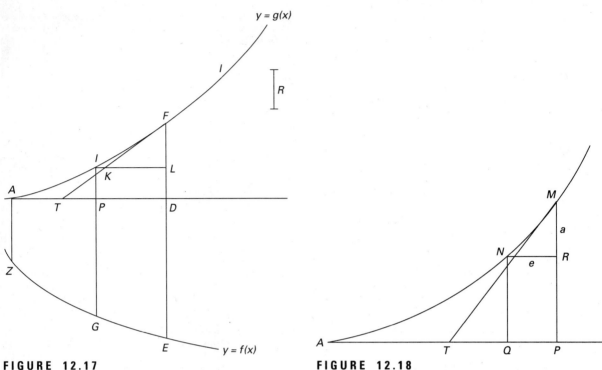

FIGURE 12.17
Barrow's version of the fundamental theorem.

FIGURE 12.18
Barrow's differential triangle.

equals the area of *PDEG*. Therefore $LK \cdot DE = $ area $PDEG < PD \cdot DE$. Hence $LK < PD$ or $LK < LI$ and the tangent line is below the curve at *I*. A similar argument applies for a point *I* on the side of *F* away from *A*.

In proposition 19 of lecture XI, Barrow proved the second part of the fundamental theorem, namely, that

$$\int_a^b Rf'(x)\, dx = R(f(b) - f(a)),$$

by showing a correspondence between infinitesimal rectangles in the region under the curve $Rf'(x)$ and those in the (large) rectangle $R(f(b) - f(a))$.

How did Barrow discover the inverse relationship of the tangent and the area problems? Barrow does not tell us explicitly, but a careful reading of the early parts of the *Geometrical Lectures* shows that he often thought of curves as being generated by the motion of a moving point. Thus he demonstrates that the slope of the tangent line at a point *P* to a curve so generated is equal to the velocity of the moving point at *P*. Furthermore, he also represents the varying velocities of a point by the varying ordinates of a curve whose axis represents time. "Hence, if through all points of a line representing time are drawn straight lines so disposed that no one coincides with another (i.e., parallel lines), the plane surface that results as the aggregate of the parallel straight lines, when each

---

Side Bar 12.3   **Did Barrow Invent the Calculus?**

Given that Barrow knew the algebraic procedures for calculating tangents and areas and was also aware of the fundamental theorem, should he be considered one of the inventors of the calculus? The answer must be no. Barrow presented all of his work in a classic geometric form. It does not appear that he was aware of the fundamental nature of the two theorems presented in the text. Barrow does not mention that they are particularly important. They are presented as two among many geometrical results dealing with tangents and areas. And Barrow never uses them to calculate areas. Perhaps if Newton had not come along, Barrow would have seen the uses to which these theorems could be put. But because he realized that Newton's abilities outshone his own, and because he was more concerned with pursuing theological interests, Barrow abandoned the study of mathematics to his younger colleague and left to him the invention of the calculus.

---

represents the degree of velocity corresponding to the point through which it is drawn, exactly corresponds to the aggregate of the degrees of velocity, and thus most conveniently can be adapted to represent the space traversed also."[25] This idea of representing distance as the area under the velocity curve goes back to Galileo and Oresme, but, combined with the notion of velocity as the slope of a tangent, it could have easily led to Barrow's understanding of the inverse relationship of the differential and integral processes.

During the years just preceding the publication of his *Geometrical Lectures* Barrow was the Lucasian professor of mathematics at Cambridge University. It is not known whether Isaac Newton ever attended any of Barrow's lectures, but he may well have been influenced by Barrow's idea of curves being generated by motion. Newton in fact suggested a few improvements to Barrow's book, in particular, that Barrow include an algebraic method of calculating tangents based on the differential triangle. This method consists of drawing the differential triangle NMR at a point M on a given curve and calculating the ratio of $MP = y$ to $PT = t$ by using the corresponding ratio of $MR = a$ to $NR = e$ in the infinitesimal triangle (Figure 12.18). Thus, if the curve is $y^2 = x^3$, Barrow replaces y by $y + a$, x by $x + e$ and gets $(y + a)^2 = (x + e)^3$ or $y^2 + 2ay + a^2 = x^3 + 3x^2e + 3xe^2 + e^3$. He then removes all terms containing a power of a or e or a product of the two, "for these terms have no value," and gets $y^2 + 2ay = x^3 + 3x^2e$. Next, "rejecting all terms consisting of letters denoting known or determined quantities . . . for these terms, brought over to one side of the equation, will always be equal to zero," he is left with $2ay = 3x^2e$. In the final step he substitutes y for a and t for e to get the ratio $y : t$. In this case, the result is $y : t = 3x^2 : 2y$. Barrow further notes that "if any indefinitely small arc of the curve enters the calculation, an indefinitely small part of the tangent or of any straight line equivalent to it is substituted for the arc."[26] Barrow has thus modified Fermat's method by ignoring completely the idea of *adequality*. He makes no attempt to justify the method but only notes that he has frequently used it in his own calculations (Side Bar 12.3).

The work of Barrow and Gregory can be thought of as a culmination of all of the seventeenth century methods of area and tangent calculations. But neither of these men in 1670 could mold these methods into a true computational and problem solving tool. In the five years before that date, however, Isaac Newton, communicating with practically no one from his rooms at Cambridge, was already using his intense powers of concentration to consolidate and extend the work of all his predecessors into the subject we today call the calculus.

12.5 ISAAC NEWTON

Isaac Newton, according to his most recent biographer Richard Westfall, was "one of the tiny handful of supreme geniuses who have shaped the categories of the human intellect, a man not finally reducible to the criteria by which we comprehend our fellow beings."[27] Because calculus was only one of the many areas in which he made major contributions to our understanding of the world around us, and because his collected mathematical papers, newly edited by Derek Whiteside, fill up eight thick volumes, we can only present here a brief glimpse of the reasons why he is considered such a "supreme genius." But what will be apparent in the next several pages is that over the course of a brief few years in the 1660s, Newton succeeded in consolidating and generalizing all the material on tangents and areas developed by his seventeenth century predecessors into the magnificent problem solving tool exhibited in the thousand page calculus textbooks of our own day.

Having mastered the entire achievement of seventeenth century mathematics by self-study, Newton spent the two years from late 1664 to late 1666 working out his basic ideas on calculus, partly in his room at Cambridge and partly back at his home in Woolsthorpe. More work followed in the next several years, although there were breaks in his mathematical study during that time which he devoted to other topics including optics, mechanics, and alchemy. On at least three occasions, Newton wrote up his researches into a form suitable for publication. Unfortunately, for various reasons, Newton never published any of these three papers on calculus. Nevertheless, the so-called October, 1666 tract on fluxions, the *De analysi per aequationes numero terminorum infinitas* (*On Analysis by Equations with Infinitely Many Terms*) of 1669, and the *Tractatus de methodis serierum et fluxionum* (*A Treatise on the Methods of Series and Fluxions*) of 1671 all circulated to some extent in manuscript in the English mathematical community and demonstrated the great power of Newton's new methods. Because the latter treatise summarizes and deepens

FIGURE 12.19
Newton on a stamp from the Soviet Union.

Biography   **Isaac Newton (1642–1727)**

Newton was born on December 25, 1642 at Woolsthorpe, near Grantham, some 100 miles north of London, to a mother newly widowed in October. When he was 3 years old, his mother remarried and left young Isaac in the care of his grandmother until she ultimately returned to Woolsthorpe in 1653 upon the death of her second husband. In 1655, Newton was sent to Grantham to attend the local grammar school. It was here that he mastered Latin, the mainstay of the classical school curriculum, and also was introduced to the study of mathematics by the somewhat unusual schoolmaster Henry Stokes. Not only did he learn basic arithmetic, but he also studied such advanced topics as plane trigonometry and geometric constructions, thus putting him far ahead of his fellow students on his matriculation at Trinity College, Cambridge in 1661.

Mathematics, however, was not generally part of the course of study at Cambridge. In fact, the university had few requirements at all. If one stayed in residence for four years and paid one's fees, one received a bachelor's degree. On the other hand, because in 1663 Newton started to explore mathematics on his own, it was to his advantage that the university did not particularly care what he studied. He mastered Euclid so that he could understand trigonometry, then the *Clavis Mathematicae* (*Key to Mathematics*) of William Oughtred (1574–1660), a popular book containing the essentials of arithmetic and algebra, then Descartes' *Geometry* in van Schooten's Latin edition along with the hundreds of pages of commentary, Viète's collected works, and finally Wallis' *Arithmetica Infinitorum*. To devote himself fully to research,

however, Newton needed the security of university financial support. This was assured through a scholarship in 1664, a fellowship in 1667, and the appointment as Lucasian professor in 1669.

Apparently one of the central reasons for Newton's success in his development not only of the calculus but also of the basic principles of optics and mechanics was his intense facility for concentration. As John Maynard Keynes wrote, "I believe that the clue to his mind is to be found in his unusual powers of continuous concentrated introspection. . . . His peculiar gift was the power of holding in his mind a purely mental problem until he had seen straight through it. . . . I believe that Newton could hold a problem in his mind for hours and days and weeks until it surrendered to him its secret."[28] Newton's powers of concentration are exemplified by many stories told of him, similar to stories about Archimedes. For example, "when he had friends to entertain at his chamber, if he stept into his study for a bottle of wine, and a thought came into his head, he would sit down to paper and forget his friends."[29] In fact, "thinking all hours lost, that were not spent in his studies, . . . he seldom left his chamber, unless at Term Time, when he read in the schools, as being Lucasian professor." But when he lectured, "so few went to hear him, and fewer that understood him, that ofttimes he did in a manner, for want of hearers, read to the walls."[30] Perhaps Newton was not a success as a professor, but as the central figure in the Scientific Revolution, his works continue to exert their influence on our lives (Figure 12.19).

---

the results of the two earlier ones, we will use it as the framework for the study of Newton's calculus, referring to the others as necessary.

12.5.1 Power Series

It is with the topic of power series that the 1671 treatise begins. The central idea for Newton is the analogy between the infinite decimals of arithmetic and the infinite degree "polynomials" which we call power series:

Since the operations of computing in numbers and with variables are closely similar . . . I am amazed that it has occurred to no one (if you except N. Mercator with his quadrature of the hyperbola) to fit the doctrine recently established for decimal numbers in similar fashion to variables, especially since the way is then open to more striking consequences. For since this doctrine in species has the same relationship to Algebra that the doctrine in decimal numbers has to common Arithmetic, its operations of Addition, Subtraction, Multiplication, Division and Root-extraction may easily be learnt from the latter's provided the reader be skilled in each, both Arithmetic and Algebra, and appreciate the correspondence between decimal numbers and algebraic terms continued to infinity. . . . And just as the advantage of decimals consists in this, that when all fractions and roots have been reduced to them they take on in a certain measure the nature of integers, so it is the advantage of infinite variable-sequences that classes of more complicated terms (such as fractions whose denominators are complex quantities, the roots of complex quantities and the roots of affected equations) may be reduced to the class of simple ones: that is, to infinite series of fractions having simple numerators and denominators and without the all but insuperable encumbrances which beset the others.[31]

Newton proceeds to show by example the advantage of infinite variable-sequences, or power series, which he considers simply as generalized polynomials with which he can operate just as with ordinary polynomials. Thus, for example, the fraction $1/(1+x)$ can be written as the series

$$1 - x + x^2 - x^3 + x^4 - x^5 + \cdots$$

by simply using long division to divide $1 + x$ into 1. Similarly, one can use the standard arithmetic algorithm for determining square roots to calculate the roots of polynomials as power series. Applying this method to $\sqrt{1 + x^2}$, Newton easily calculates the result as

$$1 + \frac{x^2}{2} - \frac{x^4}{8} + \frac{x^6}{16} - \frac{5x^8}{128} + \frac{7x^{10}}{256} - \cdots.$$

The reduction of "affected equations," that is, the solving of an equation $f(x, y) = 0$ for y in terms of a power series in x, is somewhat more difficult, Newton believed, because the method of solving equations $f(y) = 0$ numerically was not completely familiar. Thus Newton explains his method of solving such equations in terms of the example $y^3 - 2y - 5 = 0$. He notes first that the integer 2 can be taken as an initial approximation to a root. He then sets $y = 2 + p$ and substitutes in the original equation to get the new equation $p^3 + 6p^2 + 10p - 1 = 0$. Because p is small, Newton can neglect p^3 and $6p^2$ and solve $10p - 1 = 0$ to get $p = 0.1$. It follows that $y = 2.1$ is the second approximation to the root. The next step is to set $p = 0.1 + q$ and substitute that into the equation for p. In the resulting equation, $q^3 + 6.3q^2 + 11.23q + 0.061 = 0$, the two highest degree terms are again neglected. The linear equation is then solved to get $q = -0.0054$, yielding a new approximation for y as 2.0946. One can continue this method as far as desired. Newton himself stops after one more step with the value $y = 2.09455148$. He then adapts the numerical equation solving method to algebra and calculates several examples. Thus, the solution of $y^3 + a^2y + axy - 2a^3 - x^3 = 0$ is given as

$$y = a - \frac{x}{4} + \frac{x^2}{64a} + \frac{131x^3}{512a^2} + \frac{509x^4}{16384a^3} + \cdots.$$

12.5.2 The Binomial Theorem

Newton's discovery of power series came out of his reading of Wallis' *Arithmetica infini-torum*, especially the section on determining the area of a circle. In fact, he got out of Wallis' work more than Wallis had put in. In considering areas, Wallis had always looked for a specific numerical value, or the ratio of two such values, because he wanted to determine the area under a curve between two fixed values, say, 0 and 1. Newton realized that one could see further patterns if one calculated areas from 0 to an arbitrary value x. Thus in looking at the same problem as Wallis of calculating the area of a circle, he began by considering a sequence of curves similar to those considered by Wallis, namely the curves $y = (1 - x^2)^n$. But Newton began by tabulating the values under these curves in terms of the variable x. For example, using modern notation,

$$\int_0^x (1 - x^2)^0 \, dx = x$$

$$\int_0^x (1 - x^2)^1 \, dx = x - \frac{1}{3}x^3$$

$$\int_0^x (1 - x^2)^2 \, dx = x - \frac{2}{3}x^3 + \frac{1}{5}x^5$$

$$\int_0^x (1 - x^2)^3 \, dx = x - \frac{3}{3}x^3 + \frac{3}{5}x^5 - \frac{1}{7}x^7$$

$$\int_0^x (1 - x^2)^4 \, dx = x - \frac{4}{3}x^3 + \frac{6}{5}x^5 - \frac{4}{7}x^7 + \frac{1}{9}x^9$$

Newton then tabulated not numerical areas, but the coefficients of the various powers of x:

$n = 0$	$n = 1$	$n = 2$	$n = 3$	$n = 4$	...	times
1	1	1	1	1	...	x
0	1	2	3	4	...	$-\frac{x^3}{3}$
0	0	1	3	6	...	$\frac{x^5}{5}$
0	0	0	1	4	...	$-\frac{x^7}{7}$
0	0	0	0	1	...	$\frac{x^9}{9}$

Like Wallis, Newton realized that he had here Pascal's triangle and also attempted to interpolate. In fact, to solve the problem of the area of the circle he needed the values in the column corresponding to $n = 1/2$. To find these values, he rediscovered Pascal's formula $\binom{n}{k} = \frac{n(n-1)(n-2)\cdots(n-k+1)}{k!}$ for positive integer values of n and decided to use the same formula even when n was not a positive integer. Thus the entries in the column $n = 1/2$ would be

$$\binom{\frac{1}{2}}{0} = 1, \quad \binom{\frac{1}{2}}{1} = \frac{1}{2}, \quad \binom{\frac{1}{2}}{2} = \frac{\frac{1}{2}(\frac{1}{2} - 1)}{2} = -\frac{1}{8}, \quad \binom{\frac{1}{2}}{3} = \frac{\frac{1}{2}(\frac{1}{2} - 1)(\frac{1}{2} - 2)}{6} = \frac{1}{16}, \dots.$$

Newton could now fill in the table for columns corresponding to $n = k/2$ for any positive integral k. He realized further that in the original table each entry was the sum of the number to its left and the one above that. If, in his table with extra columns interpolated, he revised that rule slightly to read that each entry should be the sum of the number two columns to its left and the one above that, the new entries found by the binomial coefficient formula satisfied that rule as well. Not only did this give Newton confidence that his interpolation was correct, but it also convinced him to add columns to the left corresponding to negative values of n. The sum rule made it clear to him that in the column $n = -1$ the first number had to be 1, while the next number had to be -1, since $1 + (-1) = 0$, and 0 was the second entry in the column $n = 0$. Similarly, the third number in the $n = -1$ column was 1, the fourth, -1, and so on. Of course the binomial coefficient formula gave these same alternating values of 1 and -1 as well. Newton's interpolated table for calculating the area under $y = (1 - x^2)^n$ from 0 to x was then the following:

$n = -1$	$n = -\frac{1}{2}$	$n = 0$	$n = \frac{1}{2}$	$n = 1$	$n = \frac{3}{2}$	$n = 2$	$n = \frac{5}{2}$	$\ldots$	times
1	1	1	1	1	1	1	1	$\ldots$	x
-1	$-\frac{1}{2}$	0	$\frac{1}{2}$	1	$\frac{3}{2}$	2	$\frac{5}{2}$	$\ldots$	$-\frac{x^3}{3}$
1	$\frac{3}{8}$	0	$-\frac{1}{8}$	0	$\frac{3}{8}$	1	$\frac{15}{8}$	$\ldots$	$\frac{x^5}{5}$
-1	$-\frac{5}{16}$	0	$\frac{3}{48}$	0	$-\frac{1}{16}$	0	$\frac{5}{16}$	$\ldots$	$-\frac{x^7}{7}$
1	$\frac{35}{128}$	0	$-\frac{15}{384}$	0	$\frac{3}{128}$	0	$-\frac{5}{128}$	$\ldots$	$\frac{x^9}{9}$
$\vdots$	$\vdots$	$\vdots$	$\vdots$	$\vdots$	$\vdots$	$\vdots$	$\vdots$	$\ddots$	$\vdots$

Newton soon realized that, first, there was no necessity of dealing only with fractions with denominator 2. The same rule would apply for any fractional value of n, positive or negative. Second, he realized that the terms $(1 - x^2)^n$ for n integral "could be interpolated in the same way as the areas generated by them; and that nothing else was required for this purpose but to omit the denominators 1, 3, 5, 7, etc., which are in the terms expressing the areas"[32] (and, of course, reduce the corresponding powers by 1). Finally, there was no reason to limit himself to binomials of the form $1 - x^2$. With appropriate modification, the coefficients of the power series for $(a + bx)^n$ for any value of n could be calculated using the formula for the binomial coefficients. Thus Newton had discovered, although hardly proved, the general binomial theorem. He was, however, completely convinced of its correctness because it provided him in several cases with the same answer that he had derived in other ways. For example, Newton noted that the series derived from $1/(1 + x)$ by division was the same as that derived from the binomial theorem using exponent -1:

$$(1 + x)^{-1} = 1 + (-1)x + \frac{(-1)(-2)}{2!}x^2 + \frac{(-1)(-2)(-3)}{3!}x^3 + \cdots$$

$$= 1 - x + x^2 - x^3 + \cdots.$$

Using his knowledge that the area under $y = 1/(1 + x)$ was the logarithm of $1 + x$, Newton found the power series for $\log(1 + x)$ by integrating the above series term by term and then proceeded to calculate the logarithms of 1 ± 0.1, 1 ± 0.2, 1 ± 0.01, and 1 ± 0.02 to over 50 decimal places. Using appropriate identities, such as $2 = (1.2 \times 1.2)/(0.8 \times 0.9)$ and $3 = (1.2 \times 2)/0.8$, as well as the basic properties of logarithms, Newton was able to calculate the logarithms of many small positive integers.

The knowledge of the binomial theorem let Newton deal with many other interesting series. For example, he worked out the series for $y = \arcsin x$ using a geometrical argument: Suppose the circle AEC has radius 1 and $BE = x$ is the sine of the arc $y = AE$, or $y = \arcsin x$ (Figure 12.20). The area of the circular sector APE is known to be $\frac{1}{2}y = \frac{1}{2} \arcsin x$. On the other hand, it is also equal to the area under $y = \sqrt{1 - x^2}$ from 0 to x less $\frac{1}{2}x\sqrt{1 - x^2}$. By his earlier calculation, Newton knew that

$$\sqrt{1 - x^2} = 1 - \frac{1}{2}x^2 - \frac{1}{8}x^4 - \frac{1}{16}x^6 - \cdots.$$

It follows by integrating term by term and multiplying the above series by x that

$$y = \arcsin x = 2\int_0^x \sqrt{1 - x^2}\, dx - x\sqrt{1 - x^2} = x + \frac{1}{6}x^3 + \frac{3}{40}x^5 + \frac{5}{112}x^7 + \cdots.$$

Newton could then solve this "equation" for $x = \sin y$ by his method for affected equations. Thus in *De analysi* there occurs for the first time in European mathematics the series

$$x = \sin y = y - \frac{1}{6}y^3 + \frac{1}{120}y^5 - \frac{1}{5040}y^7 + \cdots$$

as well as the series for $x = \cos y$ which Newton derived by calculating $\sqrt{1 - (\sin y)^2}$.

In dealing with power series today, one always considers the question of convergence. It would seem that Newton did not worry much about this problem. Near the end of *De analysi* he wrote, "Whatever common analysis performs by equations made up of a finite number of terms, . . . , this method [of series] may always perform by infinite equations. To be sure, deductions in the latter are no less certain than in the other, nor its equations less exact, even though we, mere men possessed only of finite intelligence, can neither designate all their terms nor so grasp them as to ascertain exactly the quantities we desire from them."[33] Nevertheless, Newton clearly realized the limitations of his methods, at least intuitively. For example, in the course of the calculations which gave him the area

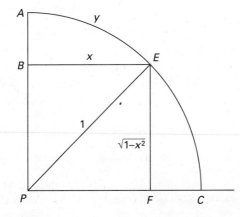

FIGURE 12.20
Newton's power series for $y = \arcsin x$.

under the hyperbola $y = 1/(1 + x)$, he noted that the first few terms of this logarithm series "will be of some use and sufficiently exact provided x be considerably less than [1]."[34]

12.5.3 Algorithms for Calculating Fluxions

Series were of fundamental importance to Newton's calculus. He used them in dealing with every algebraic or transcendental relation not expressible as a finite polynomial in one variable. But there was naturally much more in his *Treatise on Methods*, beginning with the problems which he only indicated in his October, 1676 letter to Leibniz using an anagram, the problems he considered the two basic aspects of calculus: "1. Given the length of the space continuously [that is, at every time], to find the speed of motion at any time proposed. 2. Given the speed of motion continuously, to find the length of the space described at any time proposed."[35] For Newton the basic ideas of calculus had to do with motion. Every variable in an equation was to be considered, at least implicitly, as a distance dependent on time. Of course, this idea was not new with Newton, but he did make the idea of motion fundamental: "I consider quantities as though they were generated by continuous increase in the manner of a space over which a moving object describes its course."[36] The constant increase of time itself Newton considered virtually an axiom, for he gave no definition of time. What he did define was the concept of fluxion: The **fluxion** $\dot{x}$ of a quantity x dependent on time (called a **fluent**) was the speed with which x increased through its generating motion. In his early works Newton did not attempt any further definition of speed. The concept of continuously varying motion was, Newton believed, completely intuitive.

Newton solved problem 1 by a perfectly straightforward algorithm that determined the relationship of the fluxions $\dot{x}$ and $\dot{y}$ of two fluents x and y related by an equation of the form $f(x,y) = 0$: "Arrange the equation by which the given relation is expressed according to the dimensions of some fluent quantity, say x, and multiply its terms by any arithmetical progression and then by $\dot{x}/x$. Carry out this operation separately for each one of the fluent quantities and then put the sum of all the products equal to nothing, and you have the desired equation."[37] As an example, Newton presented the equation $x^3 - ax^2 + axy - y^3 = 0$. First considering this as a polynomial of degree 3 in x, Newton multiplied using the progression 3, 2, 1, 0 to get $3x^2\dot{x} - 2ax\dot{x} + ay\dot{x}$. Next, considering the equation as a polynomial of degree 3 in y and using the same progression, he calculated $axy - 3y^2\dot{y}$. Putting the sum equal to nothing gave the desired relationship $3x^2\dot{x} - 2ax\dot{x} + ay\dot{x} + ax\dot{y} - 3y^2\dot{y} = 0$. In terms of a ratio, this result is $\dot{x} : \dot{y} = (3y^2 - ax) : (3x^2 - 2ax + ay)$.

There are several important ideas to note in Newton's rule for calculating fluxions. First, Newton is not calculating derivatives, for he does not in general start with a function. What he does calculate is the differential equation satisfied by the curve determined by the given equation. In other words, given $f(x,y) = 0$ with x and y both functions of t, Newton's procedure produces what is today written as

$$\frac{\partial f}{\partial x}\frac{dx}{dt} + \frac{\partial f}{\partial y}\frac{dy}{dt} = 0.$$

Second, Newton uses Hudde's notion of multiplying by an arbitrary arithmetic progression. In practice, however, Newton generally uses the progression starting with the highest power of the fluent. Third, if x and y are considered as functions of t, the modern product rule for derivatives is built into Newton's algorithm. Any term containing both x and y is multiplied twice and the two terms added.

Newton presents a justification for his rule, in effect, through infinitesimals. He first defines the **moment** of a fluent quantity to be the amount by which it increases in an "infinitely small" period of time. Thus, the increase of x in an infinitesimal time o is the product of the speed of x by o, or $\dot{x}o$. It follows that after this time interval, x will become $x + \dot{x}o$ and similarly y will become $y + \dot{y}o$. "Consequently, an equation which expresses a relationship of fluent quantities without variance at all times will express that relationship equally between $x + \dot{x}o$ and $y + \dot{y}o$ as between x and y; and so $x + \dot{x}o$ and $y + \dot{y}o$ may be substituted in place of the latter quantities, x and y, in the said equation."[38]

Newton continues by showing how the method applies to the example $x^3 - ax^2 + axy - y^3 = 0$ given earlier. Substituting $x + \dot{x}o$ for x and $y + \dot{y}o$ for y, the new equation becomes

$$(x^3 + 3x^2\dot{x}o + 3x\dot{x}^2o^2 + \dot{x}^3o^3) - (ax^2 + 2ax\dot{x}o + a\dot{x}^2o^2)$$
$$+ (axy + ay\dot{x}o + ax\dot{y}o + a\dot{x}\dot{y}o^2) - (y^3 + 3y^2\dot{y}o + 3y\dot{y}^2o^2 + \dot{y}^3o^3) = 0.$$

"Now by hypothesis $x^3 - ax^2 + axy - y^3 = 0$, and when these terms are erased and the rest divided by o there will remain

$$3x^2\dot{x} + 3x\dot{x}^2o + \dot{x}^3o^2 - 2ax\dot{x} - a\dot{x}^2o + ay\dot{x} + ax\dot{y} + a\dot{x}\dot{y}o - 3y^2\dot{y} - 3y\dot{y}^2o - \dot{y}^3o^2 = 0.$$

But further, since o is supposed to be infinitely small so that it [will] be able to express the moments of quantities, terms which have it as a factor will be equivalent to nothing in respect of the others. I therefore cast them out and there remains $3x^2\dot{x} - 2ax\dot{x} + ay\dot{x} + ax\dot{y} - 3y^2\dot{y} = 0$, as . . . above."[39]

Although this calculation is only an example and not a proof, Newton notes that it is immediately generalizable: "It is accordingly to be observed that terms not multiplied by o will always vanish, as also those multiplied by o of more than one dimension; and that the remaining terms after division by o will always take on the form they should have according to the rule. This is what I wanted to show."[40] In other words, Newton assumes that the reader realizes that the coefficient of $x^{n-1}\dot{x}o$ in the expansion of $(x + \dot{x}o)^n$ is n itself. But note also that Newton's only justification of his step of "casting out" any terms in which o appears is that they are "equivalent to nothing in respect of the others." There is no limit argument here. There is only the intuitive notion of the properties of these infinitesimal increments of time.

The product rule for derivatives is essentially built into Newton's algorithm. Newton's approach to the modern chain rule is through substitution. For example, to determine the relationship of the fluxions in the equation $y - \sqrt{a^2 - x^2} = 0$ he puts z for the square root and deals with the two equations $y - z = 0$ and $z^2 - a^2 + x^2 = 0$. The first gives $\dot{y} - \dot{z} = 0$ while the second gives $2z\dot{z} + 2x\dot{x} = 0$. It follows that $\dot{z} = -x\dot{x}/z$ and that the relationship between the fluxions of x and y is

$$\dot{y} + \frac{x\dot{x}}{\sqrt{a^2 - x^2}} = 0.$$

A similar approach works in dealing with quotients.

To deal with the problem of finding the relationship of quantities given the relationship of their fluxions, Newton simply reverses the above procedure if possible: "Since this problem is the converse of the preceding, it ought to be resolved the contrary way: namely by arranging the terms multiplied by $\dot{x}$ according to the dimensions of x and dividing by $\dot{x}/x$ and then by the number of dimensions, . . . by carrying out the same operation in the terms multiplied by . . . $\dot{y}$, and, with redundant terms rejected, setting the total of the resulting terms equal to nothing."[41] But since this simple "antiderivative" approach often does not work, Newton generally uses the method of power series. Since the fluent equation determined by the fluxional equation $\dot{y} = x^n\dot{x}$ or $\dot{y}/\dot{x} = x^n$ is $y = x^n$, he suggests that when $\dot{y}/\dot{x}$ depends only on x, one should express that ratio by a power series and apply the previous rule to each term. For example, the equation $\dot{y}^2 = \dot{x}\dot{y} + x^2\dot{x}^2$ can be rewritten as $\dot{y}^2/\dot{x}^2 = \dot{y}/\dot{x} + x^2$. This quadratic equation in $\dot{y}/\dot{x}$ can be solved to give $\dot{y}/\dot{x} = 1/2 \pm \sqrt{1/4 + x^2}$. By applying the binomial theorem, one gets the two series

$$\frac{\dot{y}}{\dot{x}} = 1 + x^2 - x^4 + 2x^6 - 5x^8 + \cdots \quad \text{and} \quad \frac{\dot{y}}{\dot{x}} = -x^2 + x^4 - 2x^6 + 5x^8 + \cdots.$$

The solutions to the original problem are then easily found to be

$$y = x + \frac{1}{3}x^3 - \frac{1}{5}x^5 + \frac{2}{7}x^7 + \cdots \quad \text{and} \quad y = -\frac{1}{3}x^3 + \frac{1}{5}x^5 - \frac{2}{7}x^7 + \cdots.$$

The solution method is more complicated if $\dot{y}/\dot{x}$ is given by an equation in both x and y, but even then Newton's basic idea is to express the given equation in terms of a power series.

12.5.4 Applications of Fluxions

With the calculation of fluxions accomplished, Newton uses them to solve various problems. Maxima and minima are found by setting the relevant fluxion equal to zero. For "when a quantity is greatest or least, at that moment its flow neither increases nor decreases; for if it increases, that proves that it was less and will at once be greater than it now is, and conversely so if it decreases."[42] Again he uses the equation $x^3 - ax^2 + axy - y^3 = 0$ as his example for determining the greatest value of x. Setting $\dot{x} = 0$ in the equation involving the fluxions, he gets $-3y^2\dot{y} + ax\dot{y} = 0$ or $3y^2 = ax$. This equation must then be solved simultaneously with the original one to find the desired value for x. Similarly, to find the maximum value of y, one sets $\dot{y} = 0$ and uses the resulting equation $3x^2 - 2ax + ay = 0$. Newton's discussion of this method is brief, however, and he gives no criteria for determining whether the values found are maxima or minima.

Newton's central idea for drawing tangents is to use Barrow's differential triangle. Thus, if x changes to $x + \dot{x}o$ while y changes to $y + \dot{y}o$, then the ratio $\dot{y}o : \dot{x}o = \dot{y} : \dot{x}$ of the sides of this triangle is the slope of the tangent line, thought of as the direction of instantaneous motion of the particle describing the curve. This ratio is in turn equal to that of the ordinate y to the subtangent t. Since to draw the tangent means to find the subtangent, Newton simply notes that $t = y(\dot{x}/\dot{y})$. As a slight simplification in this calculation and others, Newton sometimes sets $\dot{x} = 1$. This is equivalent to considering x as flowing uniformly, or as itself representing time.

A final example of Newton's use of fluxions is his calculation of the curvature of a curve, a problem which "has the mark of exceptional elegance and of being preeminently useful in the science of curves."[43] Newton defines curvature in terms of a circle, that is, he notes that a circle has the same curvature everywhere and that the curvatures of two circles are inversely proportional to their radii. In modern terminology, the curvature of a circle of radius r is generally set as $\kappa = 1/r$. For an arbitrary curve, Newton defines the curvature at a point to be the curvature of the circle tangent to the curve at the point which has the further property that no other tangent circle can be drawn between the curve and the circle. The definition means that this **osculating circle** at a point D also passes through any point d infinitely close to D. To find the curvature at D, one simply needs to find the radius of this circle, that is, the distance DC to the intersection point of the normals to the curve at D and d respectively (Figure 12.21). Drawing the tangent line dDT to the curve through D and d, completing the rectangle $DGCH$, taking g on GC so that $Cg = 1$, and setting $AB = x$, $BD = y$ and $g\delta = z$, Newton concludes from the similarity of triangle DBT to triangle $Cg\delta$ that $Cg : g\delta = TB : BD$ or that $1 : z = \dot{x} : \dot{y}$. Since d is infinitely close to D, it follows that $\delta f = \dot{z}o$, $DE = \dot{x}o$, and $dE = \dot{y}o$. Furthermore, since one can consider DdF a right triangle, $DE : dE = dE : EF$ and therefore $DF = DE + EF = DE + (dE^2/DE) = \dot{x}o + (\dot{y}^2o/\dot{x})$. Thus, $Cg : CG = \delta f : DF = \dot{z}o : (\dot{x}o + \dot{y}^2o/\dot{x})$ and $CG = (\dot{x}^2 + \dot{y}^2)/\dot{x}\dot{z}$. Assuming that $\dot{x} = 1$, Newton concludes that $\dot{y} = z$, $CG = (1 + z^2)/\dot{z}$, $DG = (CG \cdot BD)/BT = z \cdot CG = (z + z^3)/\dot{z}$, and finally that

$$DC = \sqrt{CG^2 + DG^2} = \frac{(1 + z^2)^{3/2}}{\dot{z}}.$$

Newton's result is naturally equivalent to the modern version, that the curvature of $y = f(x)$ equals

$$\frac{y''}{(1 + y'^2)^{3/2}},$$

since his z is our y' and his $\dot{z}$ our y''.

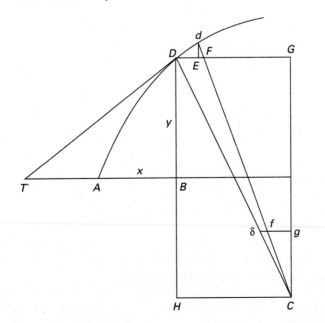

FIGURE 12.21
Newton's method for finding curvature.

12.5.5 Procedures for Finding Areas

Problem 2 of the *Treatise on Methods* asks to find the distance, given the velocity. Newton realized very early in his researches that this problem is equivalent to finding the area under a curve from its equation. From his reading of Wallis, he also knew how to find that area for curves whose equations were finite sums of terms of the form ax^n ($n \neq -1$) and had extended this basic idea to infinite series by his method of power series. Newton also, however, discovered and used the fundamental theorem of calculus to solve area problems. For him, this theorem was virtually self-evident. Because he thought of the curve *AFD* as being generated by the motions of x and y, it followed that the area *AFDB* was generated by the motion of the moving ordinate *BD* (Figure 12.22). It was therefore obvious that the fluxion of the area was in fact the ordinate itself. That is, if z represents the area under the curve, then $\dot{z} = y$, or, taking $\dot{x} = 1$ as before, $\dot{z}/\dot{x} = y$. This equation translates immediately into part of the modern fundamental theorem, that if $A(x)$ represents the area under $y = f(x)$ from 0 to x, then $dA/dx = f(x)$. Newton notes that the area z can be found explicitly from the equation $\dot{z}/\dot{x} = y$ by using the techniques already discussed of finding fluents. But as he writes a few pages later, "hitherto we have exposed the quadrature of curves defined by less simple equations by the technique of reducing them to equations consisting of infinitely many simple terms. However, curves of this kind may sometimes be squared by means of finite equations also."[44]

To square curves by finite equations (that is, to find the area under them), one needs a table of integrals. Newton provides quite an extensive one. The first entry in his table is the simple one—the area under $y = ax^{n-1}$ is $\frac{a}{n}x^n$—but the others are considerably more complex. In this brief excerpt from Newton's table, the function z on the right represents the area under the function y on the left:

$$y = \frac{ax^{n-1}}{(b + cx^n)^2} \qquad z = \frac{\left(\frac{a}{nb}\right)x^n}{b + cx^n}$$

$$y = ax^{n-1}\sqrt{b + cx^n} \qquad z = \frac{2a}{3nc}(b + cx^n)^{3/2}$$

$$y = ax^{2n-1}\sqrt{b + cx^n} \qquad z = \frac{2a}{nc}\left(-\frac{2}{15}\frac{b}{c} + \frac{1}{5}x^n\right)(b + cx^n)^{3/2}$$

$$y = \frac{ax^{2n-1}}{\sqrt{b + cx^n}} \qquad z = \frac{2a}{nc}\left(-\frac{2}{3}\frac{b}{c} + \frac{1}{3}x^n\right)\sqrt{b + cx^n}$$

Comparing Newton's complete table to a modern table of integrals, one notes that there are no transcendental functions listed, no sines or cosines or even logarithms. Although Newton knew the power series of these functions, he never treated them on an equal basis with algebraic functions. He did not operate with the sine, cosine, or logarithm algebraically, by combining them with polynomials and other algebraic expressions. Newton did, however, extend his table to functions whose integrals today would be expressed in terms of transcendental functions by expressing the integrals in terms of areas bounded by certain conic sections, areas which could be calculated by the use of power series techniques. For example, given $y = x^{n-1}/(a + bx^n)$, he wrote that the area z can be expressed as $1/n$ times the area under the hyperbola $v = 1/(a + bu)$, where $u = x^n$.

There is much else in Newton's *Treatise on Methods*, including techniques equivalent to the modern rules of substitution (as in the previous integral example) and integration by

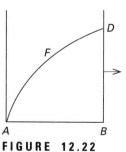

FIGURE 12.22
Newton and the
fundamental theorem.

parts as well as the method of determining arc length. This never published text thus includes virtually all of the important ideas found in the first several chapters of any modern calculus text, as well as some which are thought too advanced for such. One missing idea, however, is that of a limit. There is no clear explanation given for the use of the infinitesimal quantity *o*. In Newton's great work on mechanics, the *Philosophiae Naturalis Principia Mathematica* (*Mathematical Principles of Natural Philosophy*) of 1687, however, this idea is discussed (Figure 12.23).

12.5.6 The *Principia* and the Concept of a Limit

The *Principia*, in which Newton stated his laws of motion and his law of universal gravitation and derived Kepler's three laws of planetary motion mathematically, is written in a classical geometrical style. Although the ideas of Newton's calculus are present, the algebraic machinery he developed earlier is not. But the idea of a limit occurs from the beginning, in terms of Newton's concept of the "ultimate ratio of evanescent quantities." For example, in lemma 7 of section I of Book One he showed that the ultimate ratio of the arc *AB*, chord *AB*, and tangent *AD* as the points *A* and *B* approach one another (or as the length of each "evanesces") is the ratio of equality. Newton knew intuitively what an ultimate ratio ought to be. In effect, he had used this concept in his work with fluxions 20 years before the writing of the *Principia* in his notion of instantaneous speed, for that was a ratio of distance to time as both quantities evanesced. But Newton also knew that those steeped in Greek geometry would object to this concept. In the Scholium to section I he attempted to answer his critics:

> Perhaps it may be objected, that there is no ultimate proportion of evanescent quantities; because the proportion, before the quantities have vanished, is not the ultimate, and when they are vanished, is none. But by the same argument it may be alleged that a body arriving at a certain place, and there stopping, has no ultimate velocity; because the velocity, before the body comes to the place, is not its ultimate velocity; when it has arrived, there is none. But the answer is easy; for by the ultimate velocity is meant that with which the body is moved, neither before it arrived at its last place and the motion ceases, nor after, but at the very instant it arrives. . . . And in like manner, by the ultimate ratio of evanescent quantities is to be understood the ratio of the quantities not before they vanish, nor afterwards, but with which they vanish. . . . There is a limit which the velocity at the end of the motion may attain, but not exceed. This is the ultimate velocity. And there is the like limit in all quantities and proportions that begin and cease to be. . . . Those ultimate ratios with which quantities vanish are not truly the ratios of ultimate quantities [that is, there are no indivisibles] but limits towards which the ratios of quantities decreasing without limit do always converge, and to which they approach nearer than by any given difference, but never go beyond, nor in effect attain to, till the quantities are diminished *in infinitum*.[45]

A translation of Newton's words into an algebraic statement would give a definition of limit close to, but not identical with, the modern one. Newton never made such a translation. Nevertheless, it seems clear that Newton intuitively knew what he was doing in using "limits" to calculate fluxions. He knew that he could deal with the ratios of quantities involving his vanishing *o* and at the end let that *o* be zero. His answers were correct, and in the *Principia* he was able to apply these ideas to the basic principles of physics.

Newton's extremely influential *Principia*, arguably the most important text of the Scientific Revolution, was the work that defined the study of physics for the next 200

FIGURE 12.23
The three hundredth anniversary of Newton's *Principia* honored by a British stamp.

Side Bar 12.4   **Newton, Leibniz, and the Invention of the Calculus**

Newton and Leibniz are considered the inventors of the calculus, rather than Fermat or Barrow or someone else, because they accomplished four tasks. They each developed general concepts, for Newton the fluxion and fluent, for Leibniz the differential and integral, which were related to the two basic problems of calculus, extrema and area. They developed notations and algorithms which allowed the easy use of these concepts. They understood and applied the inverse relationship of their two concepts. Finally, they used these two concepts in the solution of many difficult and previously unsolvable problems. What neither did, however, was establish their methods with the rigor of classical Greek geometry, because both in fact used infinitesimal quantities.

---

years. It secured Newton's reputation and ultimately led to his becoming Master of the Mint in 1696 and President of the Royal Society in 1703. On the other hand, Newton's calculus had relatively little influence because only parts appeared in print and even those many years after they were written. In fact, it was work accomplished some 8 to 10 years after Newton's own discoveries that constituted the basis of the first publication of the ideas of the calculus, work done by the coinventor of the calculus, Gottfried Wilhelm Leibniz (1646–1716) (Side Bar 12.4).

12.6 GOTTFRIED WILHELM LEIBNIZ

As indicated in the opening of the chapter, Christian Huygens brought Leibniz to the frontiers of mathematical research during his stay in Paris from 1672 to 1676 by encouraging him to read such material as van Schooten's edition of Descartes' *Geometry* and the works of Pascal which included the differential triangle. Leibniz was then able to begin the investigations which led to his own invention of the differential and integral calculus toward the end of that period. It was only about ten years later, however, that he began to publish his results in short notes in the *Acta Eruditorum*, the German scientific journal which he helped to found. The presentation here of Leibniz's calculus will be taken from those notes, a work entitled *Historia et origo calculi differentialis* (*History and Origin of the Differential Calculus*) which Leibniz wrote in 1714 as a response to the assertion by English mathematicians that he had stolen his methods from Newton, and the manuscripts of his early work in Paris in which he kept virtually a running log of his thoughts on the new calculus.[46]

12.6.1 Sums and Differences

Leibniz's idea out of which his calculus grew was the inverse relationship of sums and differences in the case of sequences of numbers. He noted that if A, B, C, D, E was an increasing sequence of numbers, and L, M, N, P was the sequence of differences, then $E - A = L + M + N + P$, that is, "the sums of the differences between successive

---

Biography   Gottfried Wilhelm Leibniz (1646–1716)

The second inventor of the calculus, Gottfried Wilhelm Leibniz, was born in Leipzig to the third wife of the vice chairman of the faculty of philosophy at the University of Leipzig. Although his father died when he was only six, the young Leibniz had already been inculcated with a desire to read and study. In 1661 he entered the University of Leipzig where he spent most of his time studying philosophy. Leibniz received his bachelor's degree in 1663 and his master's degree in 1664, but although he prepared a dissertation for the degree of Doctor of Law, the university refused to award it to him, probably because of some political problems in the faculty. Leibniz thus left Leipzig and received his degree in 1667 from the University of Altdorf in Nuremberg.

Meanwhile, Leibniz had been introduced to advanced mathematics during a brief stay at the University of Jena in 1663 and began to work out the details of what he hoped would be his most original contribution to philosophy, the development of an alphabet of human thought, a way of representing all fundamental concepts symbolically and a method of combining these symbols to represent more complex thoughts. Although Leibniz never completed this project, his initial ideas are contained in his *Dissertatio de arte combinatoria* (*Dissertation on the combinatorial art*) of 1666 in which he worked out for himself Pascal's arithmetic triangle as well as the various relations among the quantities included. This interest in finding appropriate symbols to represent thoughts and ways of combining these, however, ultimately led him to the invention of the symbols for calculus we use today.

Soon after Leibniz finished his university studies, he entered upon a career first in diplomacy for the Elector of Mainz and during much of his later life as a Counsellor to the Duke of Hanover. Although there were various periods of his life when his job kept him extremely busy, he was nevertheless able to find time to pursue his ideas on mathematics and to carry on a lively correspondence on the subject with colleagues all over Europe (Figure 12.24).

FIGURE 12.24
Leibniz on a German stamp.

terms, no matter how great their number, will be equal to the difference between the terms at the beginning and the end of the series."[47] It followed that difference sequences were easily summed. Thus Leibniz considered not only the arithmetical triangle of Pascal, in which each column consists of the sums of elements of the preceding column, or conversely, each column consists of the differences of the succeeding column, but also a new triangle of fractions with similar properties, which he called his "harmonic triangle":

$$\frac{1}{1}$$

$$\frac{1}{2} \quad \frac{1}{2}$$

$$\frac{1}{3} \quad \frac{1}{6} \quad \frac{1}{3}$$

$$\frac{1}{4} \quad \frac{1}{12} \quad \frac{1}{12} \quad \frac{1}{4}$$

$$\frac{1}{5} \quad \frac{1}{20} \quad \frac{1}{30} \quad \frac{1}{20} \quad \frac{1}{5}$$

$$\frac{1}{6} \quad \frac{1}{30} \quad \frac{1}{60} \quad \frac{1}{60} \quad \frac{1}{30} \quad \frac{1}{6}$$

$$\frac{1}{7} \quad \frac{1}{42} \quad \frac{1}{105} \quad \frac{1}{140} \quad \frac{1}{105} \quad \frac{1}{42} \quad \frac{1}{7}$$

Each column in this harmonic triangle is formed by taking quotients of the first column with the corresponding columns of the arithmetical triangle. For example, the elements 1/3, 1/12, 1/30, . . . of the third column arise from dividing 1/1, 1/2, 1/3, . . . by 3, 6, 10, . . . , the elements in the third column of Pascal's triangle. Because each column consists of differences of the elements in the column to its left, it follows that the sum of the elements in each column up to a certain value can be found by Leibniz's principle as the difference of the first and last values in the column immediately preceding. Thus, 1/2 + 1/6 + 1/12 = 1/1 − 1/4. Leibniz noted in addition that this rule could be extended to infinite sums because the more terms taken, the smaller the last value of the preceding sequence became. He was therefore able to derive such results as

$$\frac{1}{3} + \frac{1}{12} + \frac{1}{30} + \cdots + \frac{1}{n(n+1)(n+2)/2} + \cdots = \frac{1}{2}.$$

By multiplying this sequence by 3, Leibniz was able to rewrite it as the sum of reciprocals of the pyramidal numbers:

$$\frac{1}{1} + \frac{1}{4} + \frac{1}{10} + \cdots = \frac{3}{2}.$$

Leibniz's actual results here were not new. Their importance lay in what the possibility of summing difference sequences implied when the idea was transfered to geometry. Thus Leibniz considered a curve defined over an interval divided into subintervals and erected ordinates y_i over each point x_i in the division. If one forms the sequence $\{\delta y_i\}$ of differences of these ordinates, its sum, $\Sigma_i \delta y_i$, is equal to the difference $y_n − y_0$ of the final and initial ordinates. Similarly, if one forms the sequence $\{\Sigma y_i\}$, where $\Sigma y_i = y_0 + y_1 + \cdots + y_i$, the difference sequence $\{\delta \Sigma y_i\}$ is equal to the original sequence of the ordinates. Leibniz extrapolated these two rules to handle the situation where there were infinitely many ordinates. He considered the curve as a polygon with infinitely many sides, at each intersection point of which an ordinate y is drawn to the axis. If the infinitesimal difference in ordinates is designated by dy, and if the sum of infinitely many ordinates is designated by $\int y$, the first rule translates into $\int dy = y$ while the second gives $d\int y = y$. Geometrically, the first means simply that the sum of the **differentials** (infinitesimal differences) in a segment equals the segment. (Leibniz assumed here that the initial ordinate equals 0.) The second rule does not have an obvious geometric interpretation, because the sum of infinitely many finite terms is infinite. So Leibniz replaced the finite ordinate y with an infinitesimal area $y\,dx$, where dx was the infinitesimal part of the x-axis determined by the intersection points of the sides of the infinite-sided polygon. Thus, $\int y\,dx$ could be interpreted as the area under the curve and the rule $d\int y\,dx = y\,dx$ simply meant that the differences between the terms of the sequence of areas $\int y\,dx$ are the terms $y\,dx$ themselves.

As part of his quest for the appropriate notation to represent ideas, Leibniz introduced the two notations d and $\int$ to represent his generalization of the idea of difference and sum. The latter is simply an elongated form of the letter S, the first letter of the Latin *summa*, while the former is the first letter of the Latin *differentia*. For Leibniz, both dy and $\int y$ were variables. In other words, d and $\int$ were operators which assigned an infinitely small variable and an infinitely large variable respectively to the finite variable y. But dy is always thought of as an actual difference, that between two neighboring values of the variable y, while $\int y$ is conceived of as an actual sum of all values of the variable y from a certain fixed value to the given one. Since dy is a variable, it too can be operated on by d to give a second order differential, written as $d\,dy$, or even one of higher order. It is perhaps difficult for a modern reader to conceive of these infinitesimal differences and infinite sums, but Leibniz and his followers became extremely adept at using these concepts in developing methods for solving many types of problems.

12.6.2 The Differential Triangle and the Transmutation Theorem

One of the earliest applications Leibniz made of the concept of a differential was to the idea of the differential triangle, a version of which he had seen in his reading of Pascal and, perhaps, of Barrow. The differential triangle, the infinitesimal right triangle the hypotenuse ds of which connects two neighboring vertices of the infinite-sided polygon representing a given curve, is similar to the triangle composed of the ordinate y, the tangent τ, and the subtangent t, so $ds : dy : dx = \tau : y : t$ (Figure 12.25). Since ratios are involved in the idea of a tangent, Leibniz generally made one of these three differentials a constant. In other words, in choosing how to represent a curve as a polynomial with infinitely many sides, he could either make the polygon have equal sides (ds constant or $d\,ds = 0$), the projection of the sides on the x-axis equal (dx constant or $d\,dx = 0$), or the projection of the sides on the y-axis equal (dy constant or $d\,dy = 0$). In some sense, the variable chosen to have a constant differential can be thought of as the independent variable. In any case, it was through manipulations of the differentials in the differential triangle, using his basic rules for manipulating with differentials, that Leibniz found the central techniques for his version of calculus.

Pascal had used the differential triangle in a circle of radius r to show that, in Leibniz's language, $y\,ds = r\,dx$. Leibniz realized that this rule could be generalized to any curve if one replaced the radius by the normal line n, because the triangle made up of the ordinate, normal, and subnormal v was similar to the differential triangle. Therefore, $y : dx = n : ds$ or $y\,ds = n\,dx$. Because $2\pi y\,ds$ can be interpreted as the surface area of the surface formed by rotating ds around the x-axis, this formula replaced a surface area calculation with an area calculation. Similarly, Leibniz noted that $dx : dy = y : v$ or $y\,dy = v\,dx$. Because he realized that $\int y\,dy$ represented a triangle whose area was $(1/2)b^2$ where b was the final value of the ordinate y, he had the result that $\int v\,dx = (1/2)b^2$. Therefore, to find the area under a curve with ordinate z, it was sufficient to find a curve y whose subnormal

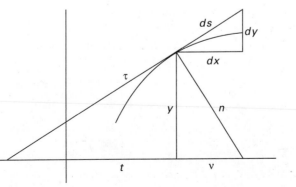

FIGURE 12.25
Leibniz's differential triangle.

v was equal to z. But since $v = y(dy/dx)$, this was equivalent to solving the equation $y(dy/dx) = z$. In other words, an area problem was reduced to what Leibniz called an inverse problem of tangents.

Although these particular rules did not lead Leibniz to any previously unknown result, a generalization of this method gave him his **transmutation theorem** and led him to his arithmetical quadrature of the circle, a series expression for $\pi/4$. In the curve $OPQD$, where P and Q are infinitesimally close, he constructed the triangle OPQ. Extending $PQ = ds$ into the tangent to the curve, drawing OW perpendicular to the tangent, and setting h and z as in Figure 12.26, he showed, using the similarity of triangle TWO to the differential triangle, that $dx : h = ds : z$ or that $z\,dx = h\,ds$. The left side of the second equation is the area under the rectangle $UVRS$ while the right side is twice the area of the triangle OPQ. It follows that the sum of all the triangles, namely the area bounded by the curve $OPQD$ and the line OD, equals half the area under the curve whose ordinate is z, or $(1/2)\int z\,dx = \int y\,dx - (1/2)OG \cdot GD$. Denoting OG by x_0 and GD by y_0, Leibniz's transmutation theorem can now be stated as

$$\int y\,dx = \frac{1}{2}(x_0 y_0 + \int z\,dx).$$

Because $z = y - PU = y - x(dy/dx)$, and because Leibniz could calculate tangents by using the rules of Hudde or Sluse, the transmutation theorem enabled him to find the area under the original curve, provided that $\int z\,dx$ was simpler to compute than $\int y\,dx$. Leibniz applied this result to calculate the area of a quarter of the circle of radius 1 given by $y^2 = 2x - x^2$. In this case

$$z = y - x\left(\frac{1-x}{y}\right) = \frac{x}{y} = \sqrt{\frac{x}{2-x}}$$

or

$$z^2 = \frac{x}{2-x}$$

or, finally,

$$x = \frac{2z^2}{1+z^2}.$$

By Leibniz' transmutation theorem, $\int y\,dx$ (or $\pi/4$) is equal to $(1/2)(1 + \int z\,dx)$. Since it is clear from Figure 12.27 that $\int z\,dx = 1 - \int x\,dz$, Leibniz concluded that $\int y\,dx$ is equal to $1 - \int \frac{z^2}{1+z^2}dz$. By an argument analogous to Mercator's, he showed that

$$\frac{z^2}{1+z^2} = z^2(1 - z^2 + z^4 - z^6 + \cdots)$$

and hence that

$$\int y\,dx = 1 - \frac{1}{3}z^3 + \frac{1}{5}z^5 - \frac{1}{7}z^7 + \cdots.$$

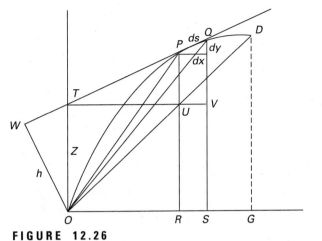

FIGURE 12.26
Leibniz's transmutation theorem.

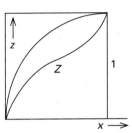

FIGURE 12.27
The transmutation function for the circle:

$$z^2 = \frac{x}{2-x} \qquad \text{or} \qquad x = \frac{2z^2}{1+z^2}.$$

Leibniz' formula for arithmetical quadrature, that $\pi/4 = 1 - 1/3 + 1/5 - 1/7 + \cdots$, followed immediately.

12.6.3 The Calculus of Differentials

Leibniz discovered his transmutation theorem and the arithmetical quadrature of the circle in 1674. During the next two years he discovered all of the basic ideas of his calculus of differentials. He only first published some of these results in "A New Method for Maxima and Minima as well as Tangents, which is neither impeded by fractional nor irrational Quantities, and a remarkable Type of Calculus for them," a brief article appearing in 1684 in the *Acta Eruditorum*. In this paper Leibniz was reluctant to define his differentials dx as infinitesimals because he believed there would be great criticism of these quantities which had not been rigorously defined. Thus he introduced dx as an arbitrary finite line segment. If y was the ordinate of a curve for which x was the abscissa, and if τ was the tangent to the curve at a point with t the subtangent, then dy was defined to be that line such that $dy : dx = y : t$. He then stated some basic rules of operation. If a is a constant, then $da = 0$; $d(v \pm y) = dv \pm dy$; $d(vw) = v\,dw + w\,dv$; and $d(v/y) = (\pm v\,dy \mp y\,dv)/y^2$. (The signs in the quotient rule depend, according to Leibniz, on whether the slope of the tangent line is positive or negative.)

Leibniz had discovered the product and quotient rules in 1675. In fact, in a manuscript of November 11 of that year he writes, "Let us now examine whether $dx\,dy$ is the same thing as $d(xy)$, and whether dx/dy is the same thing as $d(x/y)$."[48] To check his conjecture for the product rule he does an example, where $y = z^2 + bz$ and $x = cz + d$. First he calculates dy as the difference of the y values at $z + dz$ and at z. So $dy = (z + dz)^2 + b(z + dz) - z^2 - bz = (2z + b)dz + (dz)^2$. Since $(dz)^2$ is infinitely less than dz, he discards that term and concludes that $dy = (2z + b)dz$. Similarly, $dx = c\,dz$ and $dx\,dy = (2z + b)c(dz)^2$. He then notes, "but you get the same thing if you work out $d(xy)$ in a straightforward manner." Unfortunately, Leibniz does not work it out here. Later in the manuscript, however, he realized his error in another example by showing that $d(x^2)$ is not the same as $(dx)^2$. Ten days later, he writes the correct version of the product rule, later giving a simple proof by a difference argument: "$d(xy)$ is the same thing as the difference between two successive xy's; let one of these be xy, and the other $(x + dx)(y + dy)$; then we have $d(xy) = (x + dx)(y + dy) - xy = x\,dy + y\,dx + dx\,dy$. The omission of the quantity $dx\,dy$, which is infinitely small in comparison with the rest, . . . will leave $x\,dy + y\,dx$."[49] The quotient rule is proved similarly.

In the 1684 paper, Leibniz continues by giving, also without proof, the power rule $d(x^n) = nx^{n-1}dx$ and the rule for roots $d\sqrt[b]{x^a} = \frac{a}{b}\sqrt[b]{x^{a-b}}dx$, noting that the first law includes the second if a root is written as a fractional power. The chain rule is almost obvious using Leibniz's notation. For example, to calculate the differential of $z = \sqrt{g^2 + y^2}$, where g is a constant, Leibniz sets $r = g^2 + y^2$ and notes that $dr = 2y\,dy$ and $dz = d\sqrt{r} = \frac{dr}{2\sqrt{r}}$. Substituting the first equation into the second, he concludes that

$$dz = \frac{2y\,dy}{2z} = \frac{y\,dy}{z}.$$

To demonstrate the usefulness of his new calculus, Leibniz discusses how to determine maxima and minima. Thus he notes that dv will be positive when v is increasing and negative when v is decreasing, since the ratio of dv to the always positive dx gives the slope of the tangent line. It follows that $dv = 0$ when v is neither increasing nor decreasing. At that place the ordinate will be a maximum (if the curve is concave down) or a minimum (if it is concave up). The tangent there will be horizontal. The question of concavity, Leibniz notes further, depends on the second differences $d\,dv$: "When with increasing ordinates v its increments or differences dv also increase (that is, when dv is positive, $d\,dv$, the difference of the differences, is also positive, and when dv is negative, $d\,dv$ is also negative), then the curve is [concave up], in the other case [concave down]. Where the increment is maximum or minimum, or where the increments from decreasing turn into increasing, or the opposite, there is a *point of inflection*,"[50] that is, when $d\,dv = 0$.

In the final problem of this 1684 paper, an example of one of "the most difficult and most beautiful problems of applied mathematics, which without our differential calculus or something similar no one could attack with any such ease,"[51] Leibniz shows how to solve a problem which had been posed by Debeaune to Descartes in 1639 to find a curve whose subtangent is a given constant a. If y is the ordinate of the proposed curve, the differential equation of the curve is

$$y\frac{dx}{dy} = a \qquad \text{or} \qquad a\,dy = y\,dx.$$

Leibniz sets dx as constant, equivalent to having the abscissas form an arithmetical progression. The equation then can be written as $y = k\,dy$ where k is constant. It follows that the ordinates y are proportional to their increments dy, or that the y's form a geometric progression. Since the relationship of a geometric progression in y to an arithmetic progression in x is as numbers are to their logarithms, Leibniz concludes that the desired curve will be a "logarithmic" curve. (We call it an exponential curve, but, after all, our exponential and logarithmic curves are the same curves referred to different axes.) It follows from Leibniz's discussion that, since $x = \log y$, $d(\log y) = a(dy/y)$ where the constant a determines the particular logarithm used.

Leibniz did not consider the logarithm further in 1684, but after discussions with Johann Bernoulli (1667–1748) some years later, returned in 1695 to the question of the differential not only of the logarithm but also of the exponential function. In a paper of that year, he responded to criticism by Bernard Nieuwentijdt (1654–1718) that his methods would not suffice to calculate the differential of the exponential expression $z = y^x$ (where x and y are both variables).[52] A direct calculation of the differential gives $dz = (y + dy)^{x+dx} - y^x$. Applying the binomial theorem and discarding powers of dy higher than the first as well as multiples $dx\,dy$ produces the equation $dz = y^{x+dx} + xy^{x+dx-1}\,dy - y^x$, a differential equation which is not homogeneous and cannot apparently be simplified further, even in the special case where $y = b$ is constant and $dz = b^{x+dx} - b^x$. To circumvent this difficulty, Leibniz, following a suggestion of Bernoulli of 1694, attacked the problem differently by taking logarithms of both sides of the equation $z = y^x$ to get $\log z = x \log y$. The differential of this equation is then

$$a\frac{dz}{z} = xa\frac{dy}{y} + \log y\,dx.$$

It follows that

$$dz = \frac{xz}{y}dy + \frac{z \log y}{a}dx \quad \text{or} \quad d(y^x) = xy^{x-1}\,dy + \frac{y^x \log y}{a}dx.$$

If $x = r$ is constant, Leibniz noted, this rule reduces to the power rule $d(y^r) = ry^{r-1}\,dy$.

Two years later, Johann Bernoulli published a paper entitled "Principles of the Exponential Calculus," in which he generalized Leibniz's results to find relationships of the differentials in such equations as $y = x^x$, $x^x + x^c = x^y + y$, and $z = x^{y^v}$. He also stated explicitly the standard result on differentials of logarithms in the case where $a = 1$ that "the differential of the logarithm, however composed, is equal to the differential of the [function] divided by the [function]."[53] For example, he wrote,

$$d(\log \sqrt{x^2 + y^2}) = \frac{x\,dx + y\,dy}{x^2 + y^2}.$$

12.6.4 The Fundamental Theorem and Differential Equations

Recall that Leibniz began his researches into what became his calculus with the idea that sums and differences are inverse operations. It followed that the fundamental theorem of calculus was completely obvious. He amplified this idea, however, in a manuscript of about 1680, in which he noted explicitly first, that "I represent the area of a figure by the sum of all the rectangles contained by the ordinates and the differences of the abscissae," or as $\int y\,dx$, and second, that I "obtain the area of a figure by finding the figure of its summatrix or quadratrix; and of this indeed the ordinates are to the ordinates of the given figure in the ratio of sums to differences."[54] That is, to find the area under a curve with ordinates y, one needs to find a curve with ordinates z such that $y = dz$. Leibniz made this idea more explicit in a 1693 paper in the *Acta Eruditorum* where he showed "that the general problem of quadratures can be reduced to the finding of a curve that has a given law of tangency."[55] As he demonstrates, if, given the curve with ordinates y, one can find a curve z such that $dz/dx = y$ (a curve with a given law of tangency), then $\int y\, dx = z$, or, more explicitly in modern notation, assuming that $z(0) = 0$, that

$$\int_0^b y\, dx = z(b).$$

But Leibniz, like Newton, was not so much interested in finding areas as in solving differential equations, especially since it turned out that important physical problems could be expressed in terms of such equations. And Leibniz, also like Newton, used power series methods to solve such equations. His technique, however, was different. For example, consider the equation expressing the relationship between the arc y and its sine x in a circle of radius 1 as discussed by Leibniz in 1693.[56] The differential triangle with sides dy, dt, and dx is similar to the large triangle with corresponding sides 1, x, $\sqrt{1 - x^2}$, so $dt = \frac{x\,dx}{\sqrt{1 - x^2}}$ (Figure 12.28). By the Pythagorean theorem, $dx^2 + dt^2 = dy^2$. Substituting into this the value of dt and simplifying gives Leibniz the differential equation relating the arc and the sine: $dx^2 + x^2 dy^2 = dy^2$. Considering dy as constant, he applies his operator d to this equation and concludes that $d(dx^2 + x^2 dy^2) = 0$ or, using the product rule, that $2dx(d\,dx) + 2x\,dx\,dy^2 = 0$. Leibniz simplifies this into the second order differential equation

$$d^2x + x\,dy^2 = 0 \qquad \text{or} \qquad \frac{d^2x}{dy^2} = -x,$$

the familiar differential equation of the sine. (Note that Leibniz's method of manipulating with second order differentials explains our seemingly strange placement of the 2s in the modern notation for second derivative.)

Given the differential equation, Leibniz next assumed that x could be written as a power series in y: $x = by + cy^3 + ey^5 + fy^7 + gy^9 + \cdots$, with the coefficients to be determined. It was obvious to him that there could be no even degree terms and that, since $\sin 0 = 0$, the constant term was also 0. Differentiating this series twice gives $\frac{d^2x}{dy^2} = 2 \cdot 3cy + 4 \cdot 5ey^3 + 6 \cdot 7fy^5 + 8 \cdot 9gy^7 + \cdots$, a power series to be equated to

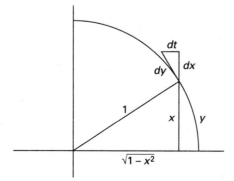

FIGURE 12.28
Leibniz's derivation of the differential equation for the sine.

the power series expressing $-x$. The identity of the coefficients then gives a series of simple equations:

$$2 \cdot 3c = -b$$

$$4 \cdot 5e = -c$$

$$6 \cdot 7f = -e$$

$$8 \cdot 9g = -f$$

$$\cdots = .$$

Setting $b = 1$ as the second initial condition, Leibniz solves these easily to get $c = -\frac{1}{3!}$, $e = \frac{1}{5!}$, $f = -\frac{1}{7!}$, $g = \frac{1}{9!}$, ... and thus derives the sine series, a series he had discovered by 1676:

$$x = \sin y = y - \frac{1}{3!}y^3 + \frac{1}{5!}y^5 - \frac{1}{7!}y^7 + \frac{1}{9!}y^9 + \cdots.$$

Leibniz had by the early 1690s discovered most of the ideas present in current calculus texts, but had never written out a complete, coherent treatment of the material. Nevertheless, like Newton, he too was somewhat bothered by his use of infinitesimals. In some of his later writings, therefore, he attempted to justify their use. Although he never made clear whether he believed infinitesimals actually exist, he justified his procedures with two different approaches. First, he attempted to relate infinitesimals to Archimedean exhaustion: "For instead of the infinite or the infinitely small, one takes quantities as large, or as small, as necessary in order that the error be smaller than the given error, so that one differs from Archimedes' style only in the expression, which are more direct in our method and conform more to the art of invention."[57] Thus, he seemed to think, like Kepler, that any argument using infinitesimals can be replaced by a perfectly rigorous argument in the style of the Greeks. But if one always had to give those arguments, one would never be able to gain new insights. Leibniz's second approach was by use of a law of continuity: "If any continuous transition is proposed terminating in a certain limit, then it is possible to form a general reasoning, which covers also the final limit."[58] In other words, if one determined that a particular ratio is true in general, when, for example, the quantities dx, dy are finite, the same ratio will be true in the limiting case, when these quantities are themselves equal to 0. In any case, the question as to whether infinitesimals exist became for Leibniz an irrelevant one. The basic calculations gave correct answers and, in the end, all infinitesimals were replaced by finite quantities whose ratios are the same as the infinitesimals. Nevertheless, the technique of manipulating with these infinitesimals became a very useful one, particularly for Leibniz's immediate followers, Johann Bernoulli and Jakob Bernoulli (1655–1705). They seemed to accept infinitesimals as actual mathematical entities and through their use achieved many important results both in calculus itself and in its applications to physical problems.

A few words about the priority controversy between Leibniz and Newton are in order here.[59] It should be clear that although the two men discovered essentially the same rules and procedures which today are collectively called the calculus, their approaches to the subject were entirely different. Newton's approach was through the ideas of velocity and distance while Leibniz's was through those of differences and sums. Nevertheless, since

Newton's work was not published until the early eighteenth century, although it was well known in England much earlier, the successes of Leibniz and the Bernoulli brothers in applying their version caused certain English mathematicians to accuse Leibniz of plagiarism, particularly since he had read some of Newton's material during his brief visits to London in the 1670s and had received two letters from Newton through Henry Oldenburg, the secretary of the Royal Society, in which Newton himself discussed some of his results. Conversely, precisely because Newton had not published, the Bernoullis accused Newton of plagiarism from Leibniz. In 1711, the Royal Society, of which Newton was then the president, appointed a commission to look into the charges. Naturally, the commission found Leibniz guilty as charged. The unfortunate result of the controversy was that the English and Continental mathematicians virtually ceased the interchange of ideas. As far as the calculus was concerned, the English adopted Newton's methods and notation, while on the continent, mathematicians used those of Leibniz. It turned out that Leibniz's notation and his calculus of differentials proved easier to work with. Thus progress in analysis was faster on the continent. To its ultimate detriment, the English mathematical community deprived itself of the great progress made there for nearly the entire eighteenth century.

12.7 FIRST CALCULUS TEXTS

The differences between the English and continental approaches appear vividly in the first calculus texts to appear, those of the Marquis de l'Hospital (1661–1704) in France in 1696 and those of Charles Hayes (1678–1760) and Humphry Ditton (1675–1715) in England in 1704 and 1706 respectively. We will conclude this chapter with a brief study of certain aspects of these early texts to give the reader an idea of what an early student of the calculus would need to master.

12.7.1 L'Hospital's *Analyse des Infiniment Petits*

Guillaume François l'Hospital was born into a family of the nobility and served in his youth as an army officer. About 1690 he became interested in the new analysis which was just then beginning to appear in journal articles by Leibniz as well as the Bernoulli brothers. Because Johann Bernoulli was spending time in Paris in 1691, l'Hospital asked him to provide, for a good fee, lectures on the new subject. Bernoulli agreed and some of the lectures were given. After about a year, Bernoulli left Paris to become a professor at the University of Groningen in the Netherlands. Because l'Hospital wanted the instruction to continue, they came to an agreement that for a large monthly salary, Bernoulli would not only continue sending l'Hospital material on the calculus including any new discoveries he might make, but also would give no one else access to them. In effect, Bernoulli was working for l'Hospital. By 1696 l'Hospital decided that he understood differential calculus well enough to publish a text on it, and since he had paid well for Bernoulli's work, he felt no compunction about using much of the latter's organization and discoveries in the new mathematics. Although Bernoulli was somewhat unhappy that his work was being published by another with only a bare acknowledgment, he kept silent on the matter. Since l'Hospital died before he could publish a work on integral calculus, Bernoulli eventually published his own lectures on that material.

L'Hospital begins in this first, extremely successful calculus text, entitled *Analyse des infiniment petits pour l'intelligence des lignes courbes* (*Analysis of infinitely small quantities for the understanding of curves*), by defining variable quantities as those which continually increase or decrease and then gives his fundamental definition of a differential: "The infinitely small part by which a variable quantity increases or decreases continually is called the differential of that quantity." He then presents two postulates which will govern his use of these differentials.

1. Grant that two quantities, whose difference is an infinitely small quantity, may be taken (or used) indifferently for each other; or (which is the same thing) that a quantity which is increased or decreased only by an infinitely small quantity may be considered as remaining the same.

2. Grant that a curve may be considered as the assemblage of an infinite number of infinitely small straight lines; or (which is the same thing) as a polygon of an infinite number of sides, each infinitely small, which determine the curvature of the curve by the angles they make with each other.[60]

For l'Hospital then, there is no question about the existence of infinitesimals. They exist; they can be represented by elements of the differential triangle; and calculations can be made using the various rules which he presents. The rules are generally stated and proved the same way that Leibniz did originally, but although Leibniz and the Bernoullis had begun to consider transcendental curves in their own work, l'Hospital deals virtually exclusively with algebraic curves. He only mentions briefly the logarithmic curve, defined as one whose subtangent $y(dx/dy)$ is constant, and deals not at all with anything resembling a trigonometric curve.

L'Hospital's treatment of maxima and minima is slightly more general than that of Leibniz. He notes that the differential dy will be positive if the ordinates are increasing and negative if they are decreasing, but shows further that dy can change from positive to negative, and the ordinates from increasing to decreasing, in two possible ways, if dy passes through 0 or through infinity. Thus l'Hospital determines that maxima or minima can appear only when dy is either 0 or infinity. As part of this discussion, he presents diagrams illustrating four possibilities, two where the tangent line is horizontal and two where there are cusps and the tangent line is vertical, as well as examples illustrating these possibilities. Thus, to find the maximum of $y - a = a^{1/3}(a - x)^{2/3}$, he calculates

$$dy = -\frac{2\sqrt[3]{a}\,dx}{3\sqrt[3]{a - x}}.$$

Since $dy = 0$ is impossible, he sets dy equal to infinity. This implies that $3\sqrt[3]{a - x} = 0$ or that $x = a$. L'Hospital gives no particular method for distinguishing between maxima and minima, but the nature of the extremum is generally clear from the conditions of the problem. For example, consider the now standard problem of finding among all rectangular parallelepipeds with a given volume a^3 and with one side equal to a given line b, the one with the least surface area. Since the sides of the parallelepiped are b, x, and a^3/bx, the problem reduces to finding the minimum of $y = bx + a^3/x + a^3/b$. L'Hospital concludes that this minimum occurs when $x = \sqrt{a^3/b}$.

L'Hospital naturally discusses second order differences and concludes like Leibniz that points of inflection occur when $d^2y = 0$, assuming dx is taken as constant. He also develops the formula

$$r = \frac{(dx^2 + dy^2)^{3/2}}{dxd^2y}$$

for determining the radius of curvature of a given curve by a method similar to Newton's. But l'Hospital's *Analyse* is probably most famous as the source of l'Hospital's rule—which should probably be renamed Bernoulli's rule—for calculating limits of quotients in the case where the limits of both numerator and denominator are zero.

Proposition.   *Let AMD be a curve (AP = x, PM = y, AB = a) such that the value of the ordinate y is expressed by a fraction, of which the numerator and denominator each become 0 when x = a, that is to say, when the point P corresponds to the given point B. It is required to find what will then be the value of the ordinate BD (Figure 12.29).*[61]

Supposing that $y = p/q$, l'Hospital simply notes that for an abscissa b infinitely close to B, the value of the ordinate y will be given by

$$y + dy = \frac{p + dp}{q + dq}.$$

But since this ordinate is infinitely close to y, and since at B both p and q are 0, l'Hospital simply equates y with dp/dq. In other words, "if the differential of the numerator be found, and that be divided by the differential of the denominator, after having made $x = a \dots$,

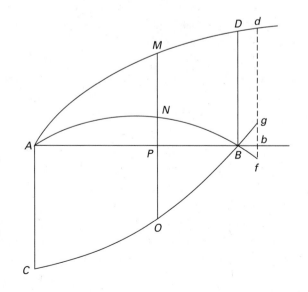

FIGURE 12.29
L'Hospital's diagram illustrating l'Hospital's rule. Notice that the function g is drawn below the x-axis, but the quotient function, represented by curve *AMD*, is above the x-axis. Think of all values of the functions involved as representing positive quantities.

we shall have the value of the ordinate . . . sought." (Note that no limits are involved in the statement or proof of this theorem.) L'Hospital does not believe in trivial examples. His first, communicated to him some years earlier by Bernoulli, is the function

$$y = \frac{\sqrt{2a^3x - x^4} - a\sqrt[3]{a^2x}}{a - \sqrt[4]{ax^3}},$$

where the value is to be found when $x = a$. A straightforward calculation of differentials gives him the answer $y = (16/9)a$.

12.7.2 The Works of Ditton and Hayes

Turning to the English writers, we find in Ditton's *An Institution of Fluxions* and Hayes' *A Treatise of Fluxions* a somewhat different type of text. These two authors were not well known. They had both studied the new calculus, however, and were teaching it to their own students, for whom they felt that an English language text would be useful. Although they had both read the continental authors, they naturally preferred the fluxional approach of Newton to the differential approach of Leibniz. Thus Ditton wrote that quantities are not to be imagined as "the aggregates or sums total of an infinite number of little constituent elements but as the result of a regular flux, proceeding incessantly, from the first moment of its beginning to that of perfect rest. A line is described not by the apposition of little lines or parts, but by the continual motion of a point. . . . The fundamental principle upon which the method of fluxions is built is more accurate, clear, and convincing than those of differential calculus."[62] Ditton then attempted to convince the reader that the discarding of certain terms because they are "nothing" in the differential calculus was not as valid as the removal of terms in the fluxional calculus because they were multiplied by a quantity which "does at last really vanish." Whether these philosophical arguments convinced the students or not, both Hayes and Ditton gave a clear treatment, similar to that of Newton, of the basics of both branches of the calculus.

It is in the detailed calculus of the logarithm and exponential functions, taken presumably from Bernoulli's paper, as well as in the treatment of the integral calculus, that these books differ in content from l'Hospital's work. Thus, both proved Bernoulli's theorem that the fluxion of the logarithm of any quantity is equal to the fluxion of the quantity divided by the quantity. (In fluxional notation, this result is $\dot{\ell}(x) = \dot{x}/x$, where $\ell(x)$ represents the logarithm.) Ditton gave a proof using power series: Since $\ell(1 + x) = x - \frac{1}{2}x^2 + \frac{1}{3}x^3 - \frac{1}{4}x^4 + \cdots$, it follows that $\dot{\ell}(1 + x) = \dot{x} - x\dot{x} + x^2\dot{x} - x^3\dot{x} + \cdots = \dot{x}(1 - x + x^2 - x^3 + \cdots) = \dot{x}/(1 + x)$, a result equivalent to the desired theorem.

With the logarithm taken care of, both authors turned to the exponential function $y = a^x$, treating it in a way directly translated from Leibniz's procedure. To calculate the fluxion of y, the authors note that $\ell(y) = x\ell(a)$ and, taking fluxions of both sides, calculate that

$$\frac{\dot{y}}{y} = \dot{x}\ell(a) + x\dot{\ell}(a).$$

Because a is constant, the fluxion of its logarithm is 0. It follows that $\dot{y} = y\dot{x}\ell(a) = a^x\ell(a)\dot{x}$.

Hayes also deals with the curve determined by the exponential function. Recall that the logarithmic curve is the curve whose ordinates are in geometric progression when its abscissas are in arithmetic proportion or as the curve whose subtangent is constant. Because the subtangent of any curve y is given by $y(\dot{x}/\dot{y})$, and because the subtangent of the curve $y = a^x$ is given by

$$y\frac{\dot{x}}{y\dot{x}\ell(a)} = \frac{1}{\ell(a)},$$

the curve defined by $y = a^x$ must be the logarithmic curve. Furthermore, Hayes calculates the area under it by first noting that, in general, the fluxion of area is $y\dot{x}$. Because in this case the subtangent to the curve is a constant c, it follows that

$$y\frac{\dot{x}}{\dot{y}} = c \qquad \text{or} \qquad y\dot{x} = c\dot{y}.$$

Therefore, the fluxion of the area is $c\dot{y}$ and the area itself must be cy. Hayes' conclusion is that for the logarithmic curve, the area between any two abscissas is proportional to the difference between the corresponding ordinates, a result not explicit in the work of either Leibniz or Bernoulli.

Ditton treated other aspects of the integral calculus in detail, including rectification of curves, areas of curved surfaces, volumes of solids, and centers of gravity. But his text, like those of Hayes and l'Hospital, had no treatment of the calculus of the sine or cosine. There was an occasional mention of these trigonometric relations as part of certain problems, but there is nowhere at the turn of the eighteenth century any treatment of the calculus of these functions. This was not to come until the work of Leonhard Euler in the 1730s.

Exercises

Problems on Tangents and Extrema

1. Show that the largest parallelepiped which can be inscribed in a sphere is a cube. Determine the dimensions of the cube and its volume if the sphere has radius 10.

2. Show that the largest circular cylinder which can be inscribed in a sphere is one in which the ratio of diameter to altitude is $\sqrt{2} : 1$. (Kepler)

3. Show that Fermat's two methods of determining a maximum or minimum of $p(x)$ are both equivalent to solving $p'(x) = 0$.

4. Use one of Fermat's methods to find the maximum of $bx - x^3$. How would Fermat decide which of the two solutions to choose as his maximum?

5. Justify Fermat's first method of determining maxima and minima by showing that if M is a maximum of $p(x)$, then the polynomial $p(x) - M$ always has a factor $(x - a)^2$ where a is the value of x giving the maximum.

6. Determine what quantity Fermat's tangent method maximizes or minimizes. Use the curve $y = x^2$ as an example.

7. Use Fermat's tangent method to determine the relation between the abscissa x of a point B and the subtangent t which gives the tangent line to $y = x^3$.

8. Modify Fermat's tangent method to be able to apply it to curves given by equations of the form $f(x,y) = c$. Begin by noting that if $(x + e, \bar{y})$ is a point on the tangent line near to (x,y), then $\bar{y} = \frac{t+e}{t}y$. Then *adequate* $f(x,y)$ to $f(x + e, \frac{t+e}{t}y)$. Apply this method to determining the subtangent to the curve $x^3 + y^3 = pxy$.

9. Show that in modern notation, Fermat's method of finding the subtangent t to $y = f(x)$ determines t as $t = \frac{f(x)}{f'(x)}$.

Show similarly that the modified method of problem 8 is equivalent in modern terms to determining t as

$$t = -\frac{y\frac{\partial f}{\partial y}}{\frac{\partial f}{\partial x}}.$$

10. Use Fermat's method to determine the subtangent to the ellipse $\frac{x^2}{a^2} + \frac{y^2}{b^2} = 1$. Compare your answer with that of Apollonius in Chapter 3.

11. Use Descartes' circle method to determine the subnormal and the slope of the tangent line to $y = x^{3/2}$ and to $y^2 = x$.

12. Use Hudde's rule applied to Descartes' method to show that the slope of the tangent line $y = x^n$ at (x_0, x_0^n) is nx_0^{n-1}.

13. Maximize $3ax^3 - bx^3 - \frac{2b^2a}{3c}x + a^2b$ using Hudde's rule. (This example is taken from Hudde's *De maximis et minimis*.)

14. Show that Sluse's rule for determining the subtangent t to a curve given by $f(x,y) = 0$ is equivalent to finding

$$t = -\frac{y\frac{\partial f}{\partial y}}{\frac{\partial f}{\partial x}}.$$

15. Apply Sluse's rule to Fermat's equation $x^3 + y^3 - pxy = 0$ and to the circle $x^2 + y^2 = bx$.

16. Derive Sluse's rule for the special case $f(x,y) = g(x) - y$ from Fermat's rule for determining the subtangent to $y = g(x)$. Derive Sluse's general rule from the modification of Fermat's rule discussed in problem 8.

Problems on Areas and Volumes

17. Show that Fermat's rule

$$N\binom{N+k}{k} = (k+1)\binom{N+k}{k+1}$$

is equivalent to

$$N\sum_{j=k-1}^{N+k-1}\binom{j}{k-1} = (k+1)\frac{N(N+1)\cdots(N+k)}{(k+1)!}$$

and also to

$$\sum_{j=1}^{N}\frac{j(j+1)\cdots(j+k-1)}{k!} = \frac{N(N+1)\cdots(N+k)}{(k+1)!}.$$

Setting $k = 3$, derive the formula for the sums of cubes from this last result and the known formulas for the sum of the integers and the sums of the squares.

18. Discover the 5th degree polynomial formula for $\sum_{j=1}^{N} j^4$ by using the formulas of exercise 17.

19. Prove by induction on k that

$$\sum_{j=1}^{N} j^k = \frac{N^{k+1}}{k+1} + \frac{N^k}{2} + p_{k-1}(N)$$

where $p_{k-1}(N)$ is a polynomial of degree less than k.

20. Fermat included the following result in a letter to Roberval of August 23, 1636: If the parabola with vertex A and axis AD is rotated around the line BD, the volume of this solid has the ratio 8:5 to the volume of the cone of the same base and vertex (Figure 12.30). Prove that Fermat is correct and show that this result is equivalent to the result on the volume of this same solid discovered by ibn al-Haytham and discussed in Chapter 7.

21. Determine the area under the curve $y = px^k$ from $x = 0$ to $x = x_0$ by dividing the interval $[0, x_0]$ into an infinite set of subintervals, beginning from the right with the points $a_0 = x_0$, $a_1 = (n/m)x_0$, $a_2 = (n/m)^2 x_0, \ldots$, where $n < m$, and proceeding as in Fermat's derivation of the area under the hyperbola.

22. Show using indivisibles that Wallis' results on the ratio of sums imply that

$$\int_0^{x_0} x^{p/q}\, dx = \frac{q}{p+q}x_0^{p+q/q}.$$

23. Using Wallis' method, interpolate the row and column corresponding to $p = 3/2$ and $n = 3/2$ into his ratio table.

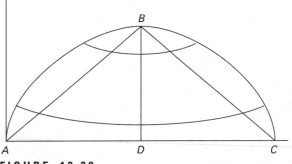

FIGURE 12.30
Fermat's problem of revolving a parabola around a line perpendicular to its axis.

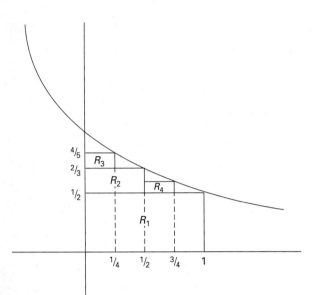

FIGURE 12.31
Determining the area under the hyperbola $y = \frac{1}{x+1}$ by a method of Brouncker.

24. Calculate an infinite series for log 2 by approximating the area under the hyperbola $y = \frac{1}{x+1}$ between $x = 0$ and $x = 1$ by using the method of William, Viscount Brouncker (1620–1684) in 1667 of dividing that area into rectangles (Figure 12.31). Show that $R_1 = \frac{1}{1 \times 2}$, $R_2 = \frac{1}{3 \times 4}$, $R_3 = \frac{1}{5 \times 6}$, $R_4 = \frac{1}{7 \times 8}$. Then find the next four terms of this series. Finally conclude with the general rule for continuation of this series and deduce that $\log 2 = \frac{1}{1 \times 2} + \frac{1}{3 \times 4} + \frac{1}{5 \times 6} + \frac{1}{7 \times 8} + \cdots$.

Problems from van Heuraet and Barrow

25. Find the length of the arc of the curve $y^4 = x^5$ from $x = 0$ to $x = b$ using van Heuraet's method.

26. Show that to find the length of an arc of the parabola $y = x^2$ one needs to determine the area under the hyperbola $y^2 - 4x^2 = 1$.

27. Use Barrow's a, e method to determine the slope of the tangent line to the curve $x^3 + y^3 = a^3$.

28. Barrow was perhaps the first to calculate the slope of the tangent to the curve $y = \tan x$ using his a, e method. Suppose DEB is a quadrant of a circle of radius 1, BX the tangent line at B (Figure 12.32). The tangent curve AMO is defined to be the curve such that if AP is equal to arc BE, then PM is equal to BG, the tangent of arc BE. Use

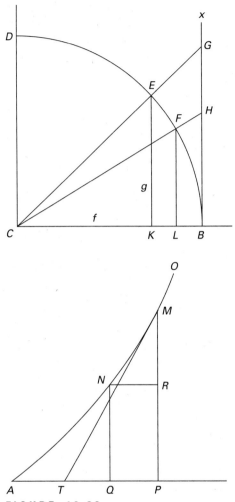

FIGURE 12.32
Barrow's calculation of the tangent line to the tangent function.

the differential triangle to calculate the slope of the tangent to curve AMO as follows: Let $CK = f$, $KE = g$. Since $CE : EK = \text{arc } EF : LK = PQ : LK$, it follows that $1 : g = e : LK$ or $LK = ge$ and $CL = f + ge$. Then $LF = \sqrt{1 - f^2} - 2fge = \sqrt{g^2 - 2fge}$. Since $CL : LF = CB : BH$, one can transfer the ratio in the circle to that on the tangent curve. Demonstrate finally that $PT = t = \frac{BF \cdot CB^2}{CG^2} = \frac{BG \cdot CK^2}{CE^2}$ and show that this result can be translated into the familiar formula

$$\frac{d(\tan x)}{dx} = \sec^2 x.$$

Given this result, can one say that Barrow has differentiated a trigonometric function? Why or why not?

Problems from Newton

29. Calculate a power series for $\sqrt{1 + x}$ by applying the square root algorithm to $1 + x$.

30. Calculate a power series for $1/(1-x^2)$ by using long division.

31. Use Newton's method to solve the equation $x^2 - 2 = 0$ to a result accurate to eight decimal places. How many steps does this take? Compare the efficacy of this method with that of the Chinese square root algorithm.

32. Solve $y^3 + y - 2 + xy - x^3 = 0$ for y as a power series in x. Begin by finding the value of y when $x = 0$, that is, by solving $y^3 + y - 2 = 0$. Since $y = 1$ is a solution, assume that $y = 1 + p$ is a solution to the original equation. Substitute this value for y and get $1 + 3p + 3p^2 + p^3 + 1 + p - 2 + x + px - x^3 = 0$. Removing all terms of degree higher than one in x and p, solve $4p + x = 0$ to get $p = -(1/4)x$. Thus $y = 1 - (1/4)x$ form the first two terms of the desired series. To go further, substitute $p = -(1/4)x + q$ in the equation for p and continue as before. Show that the next two terms in the series are $(1/64)x^2$ and $(131/512)x^3$.

33. Use Newton's method of exercise 32 to solve the equation $(1/5)y^5 - (1/4)y^4 + (1/3)y^3 - (1/2)y^2 + y - z = 0$ for y. Begin with the first approximation $y = z$. Next, substitute $y = z + p$ into the series, delete nonlinear terms in p and solve to get $y = z + (1/2)z^2$ as the second approximation. Continue in this way to get two more terms, $(1/6)z^3$ and $(1/24)z^4$, of this series.

34. Add two more rows to Newton's table of coefficients giving the area under $y = (1 - x^2)^n$ as a power series in x.

35. Square Newton's power series for $(1 - x^2)^{1/2}$ to show that it gives $1 - x^2$.

36. Calculate, using the power series for $\log(1 + x)$, the values of the logarithm of 1 ± 0.1, 1 ± 0.2, 1 ± 0.01, 1 ± 0.02 to eight decimal places. Using the identities presented in the text and others of your own devising, calculate a logarithm table of the integers from one to ten accurate to eight decimal places.

37. Check the details of Newton's calculation of the power series for the arcsine, and show that $2\int\sqrt{1 - x^2}\,dx - x\sqrt{1 - x^2} = x + \frac{1}{6}x^3 + \frac{3}{40}x^5 + \frac{5}{112}x^7 + \cdots$.

38. Calculate the relationship of the fluxions in the equation $x^3 - ax^2 + axy - y^3 = 0$ using multiplication by the progression 4, 3, 2, 1. What do you notice? What would happen if you used a different progression?

39. Find the relationship of the fluxions using Newton's rules for the equations $y^2 - a^2 - x\sqrt{a^2 - x^2} = 0$ and $x^3 - ay^2 + \frac{by^3}{a + y} - x^2\sqrt{ay + x^2} = 0$. In the first equation put $z = x\sqrt{a^2 - x^2}$ and in the second put $z = \frac{by^3}{a + y}$ and $v = x^2\sqrt{ay + x^2}$.

40. Solve the fluxional equation $\dot{y}/\dot{x} = 2/x + 3 - x^2$ by first replacing x by $x + 1$ and then using power series techniques.

Problems from Leibniz

41. Construct Leibniz's harmonic triangle by beginning with the harmonic series $1/1$, $1/2$, $1/3$, $1/4$ ... and taking differences. Develop a formula for the elements in this triangle.

42. Given the curve $y^q = x^p$ ($q > p > 0$), show using the transmutation theorem that

$$\int_0^{x_0} y\,dx = \frac{qx_0y_0}{p + q}.$$

Note that from $y^q = x^p$, it follows that $\frac{q\,dy}{y} = \frac{p\,dx}{x}$, and therefore that $z = y - x\frac{dy}{dx} = \frac{q - p}{q}y$.

43. Prove the quotient rule $d\left(\frac{x}{y}\right) = \frac{y\,dx - x\,dy}{y^2}$ by an argument using differentials.

44. Derive the power series for the logarithm by beginning with the differential equation $dy = \frac{1}{x + 1}dx$, assuming that y is a power series in x with undetermined coefficients, and solving simple equations to determine each coefficient in turn.

45. Derive the power series which determines the number $x + 1$ given its logarithm y, as Leibniz puts it, that is, the power series for the exponential function, by the method of undetermined coefficients. Begin with the differential equation $x + 1 = dx/dy$.

Problems from Early Calculus Textbooks

46. Apply l'Hospital's rule to his example

$$y = \frac{\sqrt{2a^3x - x^4} - a\sqrt[3]{a^2x}}{a - \sqrt[4]{ax^3}}$$

to find the value when $x = a$.

47. Use the method of Hayes and Ditton to calculate the fluxion of $y = x^x$.

FOR DISCUSSION . . .

48. Compare and contrast the "calculuses" of Newton and Leibniz in terms of their notation, their ease of use, and their foundations.

49. Compare the efficacy of the tangent method of Fermat and the circle method of Descartes to determine the slope of the tangent line to the curve $y = x^n$.

50. Outline a lesson introducing the concept of integration through the method of Fermat applied to curves whose equations are of the form $y = x^n$ for n a positive integer.

51. Outline a series of lessons on power series using the ideas of Newton. Is it useful to introduce such series early in a calculus course? Why or why not?

52. Could one structure a calculus course along the lines of Newton's *Treatise on Methods*? How would this differ from the normal organization of a calculus course?

53. Outline a lesson introducing the determination of arc length using the method of van Heuraet. How does this differ from the method normally presented in calculus texts?

54. Modify Newton's method for determining the curvature of a plane curve into a method usable to present in a calculus class today.

55. Is the notion of a differential as an infinitesimal a useful idea to present in a modern calculus class? Would it make the derivation of the basic rules of calculus easier? Why or why not?

56. Outline a lesson on the general binomial theorem following Newton's argument from analogy.

57. Why are Newton and Leibniz considered the inventors of the calculus rather than some of the earlier mathematicians considered in this chapter?

References and Notes

There are three current works on the history of calculus which together provide a fairly complete treatment of the material. The earliest is Carl Boyer, *The History of the Calculus and its Conceptual Development* (New York: Dover, 1959). This work deals primarily with the central concepts underlying the calculus. Although it generally considers ideas only as they prefigure modern ones rather than considering them as part of their own time, the book is still an excellent treatment which covers most of the ideas studied not only in this chapter but also, insofar as they are relevant to calculus, those of several earlier and later chapters. Margaret E. Baron, *The Origins of the Infinitesimal Calculus* (Oxford: Pergamon Press, 1969) traces more of the methods actually used in the calculus up to the time of Newton and Leibniz. It is very strong on the first part of the seventeenth century in particular and provides many examples which are useful in understanding how various mathematicians actually solved the problems they encountered. The third book, C. H. Edwards, *The Historical Development of the Calculus* (New York: Springer, 1979) also is devoted to showing exactly how mathematicians calculated, but unlike the previous work it covers in detail the contributions of Newton and Leibniz as well as the work of their eighteenth and nineteenth century successors. A brief overview of the history of calculus is found in Arthur Rosenthal, "The History of Calculus," *American Mathematical Monthly* 58 (1951), 75–86. A more general discussion of seventeenth century mathematics is found in D. T. Whiteside, "Patterns of Mathematical Thought in the Later Seventeenth Century," *Archive for History of Exact Sciences* 1 (1960–1962), 179–388. Further information on the ideas of calculus as they were developed through the early twentieth century is in Ivor Grattan-Guinness, ed. *From the Calculus to Set Theory, 1630–1910: An Introductory History* (London: Duckworth, 1980).

1. J. M. Child, *The Early Mathematical Manuscripts of Leibniz* (Chicago: Open Court, 1920), p. 215. This work contains the edited translations of many of Leibniz's mathematical manuscripts up to about 1680. The commentaries must be read with care, because Child seems to be most interested in showing that much of Leibniz's work is derived from that of Isaac Barrow.

2. H. W. Turnbull, ed., *The Correspondence of Isaac Newton* (Cambridge: Cambridge University Press, 1960), vol. II, p. 115 and p. 153. This seven volume set contains virtually all the extant letters to and from Newton, as well as other related material. In interpreting the anagram, one must keep in mind that in Latin, the letters "u" and "v" are interchangeable.

3. Johann Kepler, *Gesammelte Werke*, edited by M. Caspar (Munich: Beck, 1960), vol. IX, p. 85.

4. Quoted in Michael Mahoney, *The Mathematical Career of Pierre de Fermat, 1601–1665* (Princeton: Princeton University Press, 1973), p. 148.

5. D. J. Struik, ed., *A Source Book in Mathematics, 1200–1800* (Cambridge: Harvard University Press, 1969), p. 223.

6. David Eugene Smith and Marcia L. Latham, trans., *The Geometry of René Descartes* (New York: Dover, 1954), p. 95.

7. Ibid., pp. 100–104.

8. Struik, *Source Book*, p. 194.

9. Ibid., p. 196.

10. Quoted in Edwards, *Historical Development*, p. 103.

11. Kirsti Andersen, "Cavalieri's Method of Indivisibles," *Archive for History of Exact Sciences* 31 (1985), 291–367, p. 316.

12. Mahoney, *Mathematical Career*, p. 220.

13. Ibid., p. 221.

14. Ibid., p. 230.

15. Blaise Pascal, *Oeuvres*, edited by L. Brunschvicg and P. Boutroux (Paris: Hachette, 1909), vol. III, p. 365.

16. Paolo Mancosu and Ezio Vailati, "Torricelli's Infinitely Long Solid and Its Philosophical Reception in the Seventeenth Century," *Isis* 82 (1991), 50–70, p. 54.

17. Struik, *Source Book*, p. 247. For more on the work of John Wallis, see J. F. Scott, *The Mathematical Work of John Wallis* (New York: Chelsea, 1981).

18. Evelyn Walker, *A Study of the Traité des Indivisibles of Gilles Persone de Roberval* (New York: Columbia University, 1932), p. 174.

19. Struik, *Source Book*, p. 239.

20. H. W. Turnbull, ed., *James Gregory Tercentenary Memorial Volume* (London: G. Bell and Sons, 1939), p. 148.

21. Ibid., p. 170.

22. R. C. Gupta, "The Madhava-Gregory Series," *The Mathematics Educator (India)* 7 (1973), 67–70, p. 68. The earliest twentieth century reports on the Indian power series were in K. Mukunda Marur and C. T. Rajagopal, "On the Hindu Quadrature of the Circle," *Journal of the Bombay Branch of the Royal Asiatic Society* 20 (1944), 65–82 and in C. T. Rajagopal and A. Venkataraman, "The Sine and Cosine Power Series in Hindu Mathematics," *Journal of the Royal Asiatic Society of Bengal, Science* 15 (1949), 1–13. A commentary on the proof of the arctangent series is included in C. T. Rajagopal, "A Neglected Chapter of Hindu Mathematics," *Scripta Mathematica* 15 (1949), 201–209 and C. T. Rajagopal and T. V. Vedamurthi Aiyar, "On the Hindu Proof of Gregory's Series," *Scripta Mathematica* 15 (1951), 65–74. More detailed descriptions and analyses of the series are in C. T. Rajagopal and M. S. Rangachari, "On an Untapped Source of Medieval Keralese Mathematics," *Archive for History of Exact Sciences* 18 (1978), 89–101 and "On Medieval Kerala Mathematics," *Archive for History of Exact Sciences* 35 (1986), 91–99.

23. The best treatment of van Heuraet is in J. A. van Maanen, "Hendrick van Heuraet (1634–1660?): His Life and Mathematical Work," *Centaurus*, 27 (1984), 218–279.

24. J. M. Child, *The Geometrical Lectures of Isaac Barrow* (Chicago: Open Court, 1916), p. 117. Child, in his commentary, seems to credit Barrow with most of the invention of the calculus by translating his geometrical work into modern analytical terms, a step Barrow himself never took. Nevertheless, there is much of interest in Barrow's lectures. See also Florian Cajori, "Who Was the First Inventor of the Calculus," *American Mathematical Monthly* 26 (1919), 15–20 for a review of this book.

25. Ibid., p. 39.

26. Ibid., p. 120.

27. Richard Westfall, *Never at Rest* (Cambridge: Cambridge University Press, 1980), p. x. This biography covers in stimulating detail not only Newton's mathematical achievements but also his work in the various other areas of science. An excellent summary of Newton's mathematical achievements is V. Frederick Rickey, "Isaac Newton: Man, Myth, and Mathematics," *College Mathematics Journal* 18 (1987), 362–389. A summary of Newton's early work is found in two articles of Derek T. Whiteside: "Isaac Newton: Birth of a Mathematician," *Notes and Records of the Royal Society* 19 (1964), 53–62 and "Newton's Marvellous Years: 1666 and All That," *Notes and Records of the Royal Society* 21 (1966), 32–41.

28. John Fauvel, Raymond Flood, Michael Shortland, and Robin Wilson, eds., *Let Newton Be! A new perspective on his life and works* (Oxford: Oxford University Press, 1988), p. 15.

29. Westfall, *Never at Rest*, p. 191.

30. Ibid., pp. 192 and 209.

31. Derek T. Whiteside, *The Mathematical Papers of Isaac Newton* (Cambridge: Cambridge University Press, 1967–

1981), vol. III, pp. 33–35. All of Newton's surviving mathematical manuscripts have been gathered together, translated if appropriate, and edited in this eight volume set. These volumes repay careful browsing. A brief presentation of Newton's calculus is in Philip Kitcher, "Fluxions, Limits and Infinite Littleness: A Study of Newton's Presentation of the Calculus," *Isis* 64 (1973), 33–49.

32. Turnbull, *Correspondence*, vol. II, p. 131. This letter was sent by Newton to Henry Oldenburg, who in turn forwarded it to Leibniz. It, as well as the *epistola prior* of June 13, 1676, were later quoted as central pieces of evidence in the Royal Society report convicting Leibniz of plagiarism.

33. Whiteside, *Mathematical Papers*, II, pp. 241–243.

34. Ibid., II, p. 213.

35. Ibid., III, p. 71.

36. Ibid., III, p. 73.

37. Ibid., III, p. 75.

38. Ibid., III, p. 81.

39. Ibid.

40. Ibid.

41. Ibid., III, p. 83.

42. Ibid., III, p. 117.

43. Ibid., III, p. 151. For more on curvature, see Julian L. Coolidge, "The Unsatisfactory Story of Curvature," *American Mathematical Monthly* 59 (1952), 375–379.

44. Ibid., III, p. 237.

45. Isaac Newton, *Principia*, translated by A. Motte and revised by F. Cajori (Berkeley: University of California Press, 1966), pp. 38–39.

46. The best recent works on Leibniz include Eric Aiton, *Leibniz, A Biography* (Bristol: Adam Hilger Ltd, 1985), a general work covering his entire scientific career, and Joseph E. Hofmann, *Leibniz in Paris, 1672–1676* (Cambridge: Cambridge University Press, 1974), covering in great detail the years in which Leibniz invented his version of the calculus.

47. Child, *Early Mathematical Manuscripts*, pp. 30–31. This is part of the "Historia et Origo."

48. Ibid., p. 100.

49. Ibid., p. 143.

50. Struik, *Source Book*, p. 275.

51. Ibid., p. 279.

52. G. W. Leibniz, *Mathematische Schriften*, C. I. Gerhardt, ed., (Hildesheim: Georg Olms Verlag, 1971), vol. V, 320–328.

53. Johann Bernoulli, *Opera Omnia* (Hildesheim: Georg Olms Verlag, 1968), vol. I, 179–187, p. 183.

54. Child, *Early Mathematical Manuscripts*, p. 138.

55. G. W. Leibniz, *Mathematische Schriften*, vol. V, 294–301. English version is in Struik, *Source Book*, p. 282.

56. G. W. Leibniz, *Mathematische Schriften*, vol. V, 285–288.

57. H. J. M. Bos, "Differentials, Higher-Order Differentials and the Derivative in the Leibnizian Calculus," *Archive for History of Exact Sciences* 14 (1974), pp. 1–90, p. 56. The original paper from which this quotation is taken, may be consulted in Leibniz, *Mathematische Schriften*, vol. V, p. 350. Bos' paper is an excellent study of the general idea of a differential.

58. Ibid.

59. The controversy is discussed in great detail in A. R. Hall, *Philosophers at War: The Quarrel Between Newton and Leibniz* (Cambridge: Cambridge University Press, 1980).

60. Struik, *Source Book*, p. 313. Struik presents a translation of some sections of l'Hospital's *Analyse*. For more details on l'Hospital's work, see Carl Boyer, "The First Calculus Textbook," *Mathematics Teacher* 39 (1946), 159–167.

61. Ibid., pp. 315–316. See also Dirk Struik, "The Origin of l'Hospital's Rule," *Mathematics Teacher* 56 (1963), 257–260.

62. Humphry Ditton, *An Institution of Fluxions* (London: Botham, 1706), p. 1.

Summary of Calculus

1340–1425	Madhava	Power series
1445–1545	Kerala Gargya Nīlakantha	Power series
1500–1610	Jyesthadeva	Power series derivations
1571–1630	Johann Kepler	Maxima, areas, volumes
1584–1667	Gregory of St. Vincent	Area under hyperbola
1596–1650	René Descartes	Normals
1598–1647	Bonaventura Cavalieri	Areas and volumes
1601–1665	Pierre de Fermat	Extrema, tangents, areas
1602–1675	Gilles Persone de Roberval	Tangents, areas
1608–1647	Evangelista Torricelli	Areas and volumes
1615–1677	Henry Oldenburg	Correspondent for calculus
1616–1703	John Wallis	Areas
1618–1667	Alfonso Antonio de Sarasa	Logarithms and areas
1620–1687	Nicolaus Mercator	Power series for logarithms
1622–1685	René François de Sluse	Algorithm for derivatives
1623–1662	Blaise Pascal	Areas
1625–1683	John Collins	Correspondent for calculus
1628–1704	Johann Hudde	Algorithm for derivatives
1630–1677	Isaac Barrow	Areas, tangents, arc length
1634–1660	Henrick van Heuraet	Arc length
1638–1675	James Gregory	Areas, series
1642–1727	Isaac Newton	Series, fluxions, fundamental theorem
1646–1716	Gottfried Wilhelm Leibniz	Differentials, fundamental theorem
1661–1704	Guillaume François l'Hospital	Text on differential calculus
1667–1748	Johann Bernoulli	Calculus of exponentials
1675–1715	Humphry Ditton	Calculus text
1678–1760	Charles Hayes	Calculus text

Chapter 13

Analysis in the Eighteenth Century

*"Jean Bernoulli, public professor of mathematics, pays his best respects to
the most acute mathematicians of the entire world. Since it is known with
certainty that there is scarcely anything which more greatly excites noble
and ingenious spirits to labors which lead to the increase of knowledge than
to propose difficult and at the same time useful problems through the solution
of which . . . they may attain to fame and build for themselves eternal
monuments among posterity, so I should expect to deserve the thanks of the
mathematical world if . . . I should bring before the leading analysts of this
age some problem upon which . . . they could test their methods, exert their
powers, and, in case they brought anything to light, could communicate
with us in order that everyone might publicly receive his deserved praise
from us."(Proclamation made public at Gröningen, the Netherlands,
January, 1697)*[1]

$\mathcal{T}$he calculus of the trigonometric functions was invented by Leonhard Euler in the early
months of 1739. It was then that Euler realized the necessity of using the sine and cosine
functions as solutions to differential equations coming from the theory of vibrations. He made
his invention known through letters to various mathematicians and finally published the
material in his *Introductio* in 1748.

The driving force in the continued development of calculus in the eighteenth century was
the desire to solve physical problems, the mathematical formulation of which was generally in terms of differential equations. Both Newton and Leibniz had solved differential
equations in their study of curves, but the mathematicians who followed them gradually
shifted their emphasis from the study of curves and the geometric variables associated
with them to the study of analytical expressions involving one or more variable quantities
as well as certain constants, that is, functions of one or several variables. The relationship
between the differentials of these variables and the variable dependent on them, deter-

mined by some physical situation often arising from an application of Newton's laws of motion, led to a differential equation whose solution explicitly determined the desired function. In fact, new classes of functions were discovered and analyzed through the differential equations which they satisfied.

The major figure in the development of analysis in the eighteenth century was the most prolific mathematician in history, Leonhard Euler. Much of this chapter will be devoted to his work in the theory of differential equations and multivariable calculus as well as to his three influential textbooks in analysis. The chapter will begin, however, with some of the challenge problems set by the Bernoullis for the mathematicians of Europe, problems whose solutions helped to establish new ideas in mathematics that were later developed by Euler and others. Because influential ideas and techniques also appeared in the works of Thomas Simpson and Colin Maclaurin in England and that of Maria Gaetana Agnesi in Italy, these texts will be considered as well. In particular, because Maclaurin wrote his text partly to answer the criticisms of George Berkeley regarding the foundations of calculus, his response to this criticism will be discussed. The chapter will conclude with the attempt by Joseph Louis Lagrange to eliminate all reference to infinitesimal quantities or even to limits and to base the calculus on the notion of a power series.

13.1 DIFFERENTIAL EQUATIONS

It was the brothers Jakob and Johann Bernoulli (often known as Jacques and Jean, or James and John) who were among the first in Europe to understand the new techniques of Leibniz and to apply them to solve new problems. For example, in 1659 Huygens had discovered using infinitesimals and then proved geometrically that the curve along which an object descending under the influence of gravity would take the same amount of time to reach the bottom, from whichever point on the curve the descent began, was a cycloid. Huygens then used this idea in his invention of a pendulum clock. He realized that if the pendulum were constrained to move in a cycloidal arc, it would keep time perfectly, whatever the size of the oscillation. Jakob Bernoulli in 1690 was able to prove Huygens' result analytically by setting up the differential equation for this curve of equal time, the isochrone.

Succeeding in the isochrone problem, Jakob then proposed a new one, to determine the shape of the catenary, the curve assumed by a flexible but inelastic cord hung freely between two fixed points. Galileo had thought that this curve was a parabola. Jakob himself was unable to solve the problem, but in the *Acta Eruditorum* for June 1691 there appeared solutions by Leibniz, Huygens, and Johann Bernoulli. Johann was immensely proud that he had surpassed his older brother, reporting that the solution had cost him a night of sleep. His solution, which later appeared with more details in the lectures he gave to l'Hospital, began with the differential equation $dy/dx = s/a$ derived from an analysis of the forces acting to keep the cord in position, where s represents arc length. Because $ds^2 = dx^2 + dy^2$, squaring the original equation gives

$$ ds^2 = \frac{s^2\, dy^2 + a^2\, dy^2}{s^2} \qquad \text{or} \qquad ds = \frac{\sqrt{s^2 + a^2}\, dy}{s} $$

or, finally,

$$ dy = \frac{s\, ds}{\sqrt{s^2 + a^2}}. $$

An integration then shows that $y = \sqrt{s^2 + a^2}$ or that $s = \sqrt{y^2 - a^2}$. Bernoulli concluded that

$$ dx = \frac{a\, dy}{s} = \frac{a\, dy}{\sqrt{y^2 - a^2}}. $$

He was not able to express this integral in closed form, but was able to construct the desired curve by making use of certain conic sections. In modern terminology, this equation can be solved in the form $x = a \ln(y + \sqrt{y^2 - a^2})$ or in the form $y = a \cosh \frac{x}{a}$. For Bernoulli in 1691, however, as for his contemporaries, an answer in terms of areas under, or lengths of, known curves was sufficient.

Over the next several years both brothers posed other problems involving differential equations and, along with Leibniz, made much progress in developing methods of solution. In particular, in 1691 Leibniz found the technique of separating variables, that is, of rewriting a differential equation in the form $f(x)dx = g(y)dy$ and then integrating both sides to give the solution. He also developed the technique for solving the homogeneous equation $dy = f(y/x)dx$ by substituting $y = vx$ and then separating variables. By 1694, Leibniz had in addition solved the general first-order linear differential equation $m\, dx + ny\, dx + dy = 0$ where m and n are both functions of x. (In modern notation, this is the equation $dy/dx + ny = -m$.) He defined p by the equation $dp/p = n\, dx$ and substituted to get $pm\, dx + y\, dp + p\, dy = 0$. Because the last two terms on the left side are equal to $d(py)$, an integration gives $\int pm\,dx + py = 0$. This equation, giving the answer in terms of an area, provided Leibniz with the desired solution.

13.1.1 The Brachistochrone Problem

Probably the most significant of the problems proposed by Johann Bernoulli, in terms of its ultimate consequences for mathematics, was that of the **brachistochrone**. He first proposed it in the June 1696 issue of the *Acta Eruditorum* as a "New Problem Which Mathematicians are Invited to Solve: If two points *A* and *B* are given in a vertical plane, to assign to a mobile particle *M* the path *AMB* along which, descending under its own weight, it passes from the point *A* to the point *B* in the briefest time."[2] Bernoulli noted that the required curve was not a straight line, but a curve "well known to geometers." He had requested the solutions by the end of 1696, but in early January, 1697, acting on a suggestion of Leibniz, he extended the deadline to Easter and sent the problem, as mentioned in the opening quotation, to those who had not seen the note in the *Acta*. He offered a prize "neither of gold nor silver . . . Rather, since virtue itself is its own most desirable reward and fame is a powerful incentive, we offer the prize . . . compounded of honor, praise, and approbation; thus we shall crown, honor, and extol, publicly and privately, in letter and by word of mouth, the perspicacity of our great Apollo."[3] Among those to whom the challenge was sent was Newton, who, Bernoulli believed, had stolen Leibniz's methods and would not be able himself to solve this problem.

When Newton received the letter from Bernoulli at about 4:00 P.M. on January 29, he was very tired after a difficult day at the mint. Nevertheless, he stayed up until he had solved the problem by 4:00 the next morning. Bernoulli was forced to acknowledge Newton's talents. Leibniz was sufficiently embarrassed by the incident that he wrote to the Royal Society denying that he had been involved in it. In any case, Newton's solution was published in the May 1697 issue of the *Acta* along with the solutions of Leibniz, Jakob Bernoulli, and Johann himself.

Johann Bernoulli used two physical principles to solve the problem. First, he noted that, according to Galileo, the velocity acquired by a falling body is proportional to the square root of the distance fallen. Second, he recalled Snell's law, that when a light ray passes from a thinner to a denser medium, the ray is bent so that the sine of the angle of incidence is to the sine of the angle of refraction inversely as the densities of the media and therefore directly as the velocities in those media. Bernoulli then assumed that the vertical plane of the problem was composed of infinitesimally thick layers whose densities varied. The brachistochrone, the path of least time, was thus the curved path of a light ray whose direction changed continually as it passed from one layer to the next. At every point the sine of the angle between the tangent to the curve and the vertical axis was proportional to the velocity, and the velocity was in turn proportional to the square root of the distance fallen.

Now, denoting the desired brachistochrone curve by *AMB* and the curve representing the velocity at each point by *AHE*, Bernoulli set x and y to be the vertical and horizontal coordinates respectively of the point *M* measured from the origin *A* and t to be the horizontal coordinate of the corresponding velocity point *H* (Figure 13.1). With m a point infinitesimally close to *M*, he represented the infinitesimals *Cc*, *Mm*, and *nm* by dx, ds, and dy respectively. From the fact that the sine of the angle of refraction (angle *nMm*) is

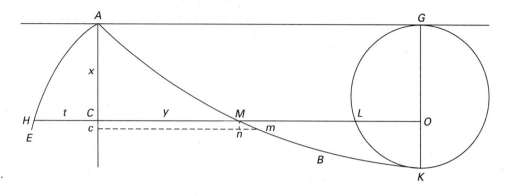

FIGURE 13.1
Johann Bernoulli's
brachistochrone problem.

$dy : ds$, which is in turn proportional to the velocity t, Bernoulli derived the equation $dy : t = ds : a$, or $a\,dy = t\,ds$, or $a^2\,dy^2 = t^2\,ds^2 = t^2\,dx^2 + t^2\,dy^2$, or finally,

$$dy = \frac{t\,dx}{\sqrt{a^2 - t^2}}.$$

Because the curve AHE is a parabola with equation $t^2 = ax$ or $t = \sqrt{ax}$, substitution of this value for t produces the differential equation of curve AMB:

$$dy = dx\sqrt{\frac{x}{a - x}}.$$

Bernoulli immediately recognized this equation as defining a cycloid. To prove this analytically he noted that

$$dx\sqrt{\frac{x}{a - x}} = \frac{a\,dx}{2\sqrt{ax - x^2}} - \frac{(a - 2x)\,dx}{2\sqrt{ax - x^2}}.$$

Given that $y^2 = ax - x^2$ is the equation of a circle GLK, and that the first term on the right is the differential of arc length along this circle, an integration of the equation gives $CM = $ arc $GL - LO$. Because $MO = CO - CM = CO - $ arc $GL + LO$, and with the assumption that CO is equal to half the circumference of the circle, it follows that $MO = $ arc $LK + LO$ or that $ML = $ arc LK. It is then immediate that the curve AMK is a cycloid as asserted. Bernoulli expressed a pleased amazement that this curve was the same as the isochrone. He noted that this was true only because velocity is proportional to the square root of the distance and not to any other power. "We may conjecture," he continued, "that nature wanted it to be thus. For, as nature is accustomed to proceed always in the simplest fashion, so here she accomplished two different services through one and the same curve."[4]

Johann's solution to this problem led to an investigation of the properties of families of curves, which in turn led to some fundamental new concepts in the theory of functions of several variables. Given that one could construct a family $\{C_\alpha\}$ of cycloids, each being the curve of fastest descent from a given point A to a point B_α, Johann Bernoulli posed and solved the problem of finding a new family $\{D_\beta\}$ of curves, called **synchrones**, each point of any one of which was the place reached by particles in a given time t_β descending from A along the various cycloids C_α (Figure 13.2). In physical terms, if the cycloids represent

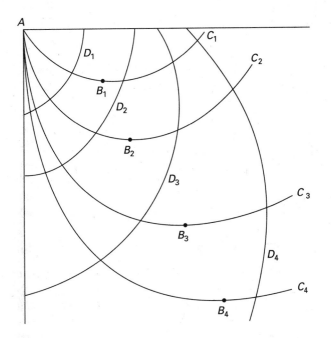

FIGURE 13.2
Two orthogonal families
of curves.

light rays, then the synchrones represent the wave fronts, the simultaneous positions of the various light pulses emitted from A at the same instant. Bernoulli realized that optical theory (developed by Huygens) predicted that the synchrones would intersect the brachistochrone cycloids in right angles. Thus his geometric problem was to find a family $\{D_\beta\}$ of curves orthogonal to a family $\{C_\alpha\}$ of cycloids with a given vertex. Johann was able to construct the synchrones with little difficulty, but he challenged others to solve the general problem of finding orthogonal trajectories to a given family of transcendental curves.

13.1.2 The Differential Calculus of Functions of Two Variables

It was in terms of families $\{C_\alpha\}$ of curves, not, as one might expect from a modern perspective, in terms of surfaces defined by functions of two variables, that the earliest notion of a partial derivative evolved. Such families had been initially considered in the early 1690s by Leibniz. The basic situation is one of two infinitesimally close curves from a given family, C_α and $C_{\alpha+d\alpha}$, intersected by a third curve D defined geometrically in terms of that family. For example, D may be orthogonal to all members of the family. In such a situation, to find the differential equation of D or construct its tangent, it was necessary to consider three different differentials of the ordinate y. Consider the points P, P' on C_α and Q, Q' on $C_{\alpha+d\alpha}$ (Figure 13.3). One differential of y is that between two points on a single curve of the family, say $y(P') - y(P)$. This is the differential of y with x variable and α constant, designated by d_xy. A second differential is that between the y values of two corresponding points on the neighboring curves, say $y(Q) - y(P)$. This is the differential of y with α variable and x constant, designated by $d_\alpha y$. Differentiation using this differential was referred to as differentiation from curve to curve. Finally, there is a third differential, $y(Q') - y(P)$, denoted dy, which is the differential along the curve D.

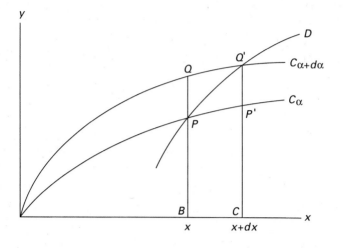

FIGURE 13.3
Partial differentiation in a family of curves.

For Leibniz there was no difficulty in calculating either of the first two differentials provided the curves were given algebraically. He could use his standard rules with either α or x being treated as a constant, in effect taking the partial derivatives with respect to x or α respectively. Unfortunately, many interesting curves were not given algebraically, but in terms of an integral. For example, the family of brachistochrone cycloids above was given by

$$C_\alpha = y(x,\alpha) = \int_0^x \sqrt{\frac{x}{\alpha - x}}\, dx.$$

(The modern notation of limits on the integral sign is being used here for clarity. Neither Leibniz nor the Bernoullis used such symbolism.) The question for Leibniz, brought home to him in a letter from Johann Bernoulli in July 1697, was how to perform the differentiation, with α variable, in a case like this.

Leibniz's solution of the problem goes back to his own basic ideas of the calculus, namely that the differential is the extrapolation of a finite difference and the integral that of a finite sum. Since for two sets of finite quantities, the sum of the differences of the parts is equal to the difference of the sums of the parts, Leibniz discovered what is called the **interchangeability theorem** for differentiation and integration, namely

$$d_\alpha \int_b^x f(x,\alpha)\, dx = \int_b^x d_\alpha f(x,\alpha)\, dx.$$

(The modern formulation of this theorem is in terms of derivatives with respect to α rather than differentials.) Leibniz was then able to apply this result to determining the tangent to the curve D which cuts off equal arcs on the family of logarithmic curves $y(x,\alpha) = \alpha \log x$.

Leibniz sent his solution to Bernoulli, who was certain that Leibniz's result would enable him to make significant progress in solving the problem of orthogonal trajectories for curves defined by integrals involving the parameter. He was in fact able to show that the orthogonal trajectories to the family of curves given by

$$y = \int_0^x p(x,\alpha)\, dx$$

were determined by the differential equation $(1 + p^2)dx + pq\, d\alpha = 0$, where q is defined by $d_\alpha y = q\, d\alpha$. In principle, q could be determined by Leibniz's theorem. Unfortunately, there were many technical difficulties involved in integrating this differential equation, and often there were better methods in particular cases. In the end, although he worked on the problem at various times during the next two decades, Bernoulli was never able to develop a truly general method using Leibniz's result. Nevertheless, the interchangeability theorem became one of the foundations of the differential calculus of functions of two variables.

Two other important aspects of the calculus of functions of two variables, the notion of a total differential and the equality of the mixed second-order differentials, were discovered by Nicolaus I Bernoulli (1687–1759), a nephew of both Johann and Jakob, who lived in the shadow of his more famous uncles and chose to publish very little in mathematics. In an article in the *Acta Eruditorum* of June, 1719, Nicolaus discussed the orthogonal trajectory problem, but he did not provide the demonstrations until years later and then only in manuscript form. In that manuscript, Nicolaus asserted that the differential dy associated with a family of curves with parameter α is given by $dy = p\, dx + q\, d\alpha$, an expression he called the **complete differential equation** of the family, today known as the **total differential**. In this equation, p and q are functions of x, y, and α. Although Bernoulli did not say how he derived this result, one possible method was by way of a geometric argument. The differential dy along the curve D crossing the given family can be expressed by

$$dy = y(Q') - y(P) = [y(Q') - y(Q)] + [y(Q) - y(P)]$$
$$= d_x(y + d_\alpha y) + d_\alpha y = d_x y + d_x d_\alpha y + d_\alpha y.$$

Since the middle term, a second-order differential, is infinitely small compared to the other terms, the result is that $dy = d_x y + d_\alpha y$, or $dy = p\, dx + q\, d\alpha$ (see Figure 13.3).

The equality of mixed second-order differentials follows from another geometrical argument based on Figure 13.3, one in which the line segment $P'Q'$ is found in two different ways. On the one hand, $P'Q' = PQ + d_x PQ = d_\alpha y + d_x d_\alpha y$. On the other, $P'Q' = d_\alpha CP'$, which is in turn equal to $d_\alpha(BP + d_x BP) = d_\alpha BP + d_\alpha d_x BP = d_\alpha y + d_\alpha d_x y$. Comparing the two expressions yields the result $d_x d_\alpha y = d_\alpha d_x y$. Interestingly enough, Nicolaus Bernoulli did not consider this argument as a proof, but simply as an illustration of a result "which I thought to be obvious to anybody from the mere notion of differentials."[5] He used this result in his own work on orthogonal trajectories, but because his main arguments did not appear in his published work, they had little influence. The theorem on equality of mixed partial differentials was thus proved anew about ten years later by Leonhard Euler (1707–1783), who noted that this result and the interchangeability theorem, which both he and Nicolaus Bernoulli derived from the equality theorem, provided the basis of a theory of partial derivatives. Euler used it to give his own solutions to the same problems discussed by his predecessors of finding new curves defined geometrically in terms of given families of curves.

Another important idea in the solution of differential equations, developed around 1740 by Euler, was found independently by Alexis-Claude Clairaut (1713–1765) in 1739, namely the condition for solvability of the homogeneous linear differential equation $P\, dx + Q\, dy = 0$, where P and Q are functions of x and y. If $P\, dx + Q\, dy$ is the total differential of a function $f(x,y)$, then the equality of the mixed second-order differentials

---

Biography **Alexis Clairaut (1713–1765)**

Clairaut, born in Paris, was a child genius who mastered l'Hospital's *Analyse des infiniment petits* by the age of ten and read a paper at the Paris Academy of Sciences two years later. The research on curves which led to his book of 1731 was begun when he was 13 and ultimately led to his election to the Academy at the age of 18. Clairaut soon turned to celestial mechanics and later to pedagogy. His five major works in the former field proved quite influential and his two texts on geometry and algebra were attempts to introduce a historical, or "natural," approach to the teaching of these subjects.

---

shows that $\partial P/\partial y = \partial Q/\partial x$. More importantly, however, Clairaut demonstrated that this condition was sufficient for $P \, dx + Q \, dy$ to be a total differential. He asserted, in fact, that under that condition, the function $f(x,y)$ was given by $\int P \, dx + r(y)$, where r was a function of y to be determined. The differential of $\int P \, dx + r(y)$ was, by Leibniz's result, equal to $P \, dx + dy \int \frac{\partial P}{\partial y} dx + dr$. But since

$$\frac{\partial P}{\partial y} = \frac{\partial Q}{\partial x} \qquad \text{and} \qquad \int \frac{\partial Q}{\partial x} dx = Q + s(y),$$

the differential can be rewritten as $P \, dx + Q \, dy + dr + s \, dy$. Therefore, if r is chosen so that $dr = -s \, dy$, the differential becomes $P \, dx + Q \, dy$ as desired. Clairaut had thus reduced the original two-variable problem to an ordinary differential equation in one variable, an equation he assumed to be solvable. He also easily extended this "exactness" result to homogeneous linear equations in more than two variables.

13.1.3 Differential Equations and the Trigonometric Functions

Although the ideas involving functions of several variables proved important in the solution of differential equations, other techniques were also developed by Euler in the 1730s, techniques which led to his development of the modern notion of the sine and cosine functions. Recall that in 1693 Leibniz had derived the differential equation for the sine by a geometrical argument and then had solved the equation using undetermined coefficients to get the power series representation of the sine. In the early years of the eighteenth century physical problems which led to such equations were typically solved geometrically. For example, the equation

$$dt = \frac{c \, ds}{\sqrt{c^2 - s^2}}$$

would be solved for t as $t = c \arcsin \frac{s}{c}$ by consideration of the geometrical situation, rather than as $s = c \sin \frac{t}{c}$. The sine was still considered as a line in a circle of given radius rather than as a function in modern terms.

In contrast to the "missing" sine function, the exponential function was well known to Euler by 1730. In fact, in a manuscript on differential calculus which he wrote for use as a text shortly after arriving in St. Petersburg, Euler noted that there were two classes of functions, algebraic and transcendental. The latter class consisted solely of the exponential and logarithmic functions, the properties of which he proceeded to discuss. Thus he knew

---

Biography   **Leonhard Euler (1707–1783)**

Born in Basel, Switzerland, Euler showed his brilliance early, graduating with honors from the University of Basel when he was 15. Although his father preferred that he prepare for the ministry, Euler managed to convince Johann Bernoulli to tutor him privately in mathematics. The latter soon recognized his student's genius and persuaded Euler's father to allow him to concentrate on mathematics. In 1726 Euler was turned down for a position at the university, partly because of his youth. A few years earlier, however, Peter the Great of Russia, on the urging of Leibniz, had decided to create the St. Petersburg Academy of Sciences as part of his efforts to modernize the Russian state. Among the earliest members of the Academy, appointed in 1725, were Nicolaus II (1695–1726) and Daniel Bernoulli (1700–1782), two of Johann's sons with whom Euler had developed a friendship. Although there was no position in mathematics available in St. Petersburg in 1726, they nevertheless recommended him for the vacancy in medicine and physiology, a position Euler immediately accepted. (He had studied these fields during his time at Basel.) In 1733, due to Nicolaus' death and Daniel's return to Switzerland, Euler was appointed the Academy's chief mathematician. Late in the same year he married Catherine Gsell with whom he subsequently had 13 children. The life of a foreign scientist was not always carefree in Russia at the time. Nevertheless, Euler was able generally to steer clear of controversies until the problems surrounding the succession to the Russian throne in 1741 convinced him to accept the invitation of Frederick II of Prussia to join the Berlin Academy of Sciences, founded by Frederick I also on the advice of Leibniz. He soon became director of the academy's mathematics section, and with the publication of his texts in analysis as well as numerous mathematical articles, became recognized as the premier mathematician of Europe. In 1755 the Paris Academy of Sciences named him a foreign member, partly in recognition of his winning their biennial prize competition 12 times. Ultimately, however, Frederick tired of Euler's lack of philosophical sophistication. When the two could not agree on financial arrangements or on academic freedom, Euler returned to Russia in 1766 at the invitation of Empress Catherine the Great, whose succession to the throne marked Russia's return to the westernizing policies of Peter the Great. With the financial security of his family now assured, Euler continued his mathematical activities even though he became almost totally blind in 1771. His prodigious memory enabled him to perform detailed calculations in his head. Thus he was able to dictate his articles and letters to his sons and others virtually until the day of his sudden death in 1783 while playing with one of his grandchildren (Figure 13.4).

FIGURE 13.4
Euler on a Swiss stamp.

that the differential equation satisfied by $y = e^{ax}$ was $dy = ae^{ax}dx$, or, to put this the other way around, that the solution to $dy = ay\,dx$ was $y = e^{ax}$. In various papers in the early 1730s, Euler made use of this property of the exponential to solve other differential equations. For example, he noted that the equation

$$dz - 2zdv + \frac{z\,dv}{v} = \frac{dv}{v}$$

could be solved by multiplying through by the "integrating factor" $e^{-2v}v$ to give $e^{-2v}v\,dz - 2e^{-2v}zv\,dv + e^{-2v}z\,dv = e^{-2v}\,dv$. Because the left side is the differential of $e^{-2v}vz$, the solution of the equation is

$$e^{-2v}vz = C - \frac{1}{2}e^{-2v} \qquad \text{or} \qquad 2vz + 1 = Ce^{2v}.$$

Higher-order equations could also be solved by exponential functions, but by the mid-1730s Euler realized that these functions were not sufficient. For example, in 1735 Daniel Bernoulli wrote to Euler to discuss a problem on the vibrations of an elastic band. The problem led to the fourth-order equation $k^4\left(\frac{d^4y}{dx^4}\right) = y$. Both Bernoulli and Euler realized that $e^{x/k}$ was a solution, but Bernoulli wrote that this solution is "not general enough for the present business."[6] Euler was able to solve the equation using power-series methods, but did not recognize a sine or cosine in the resulting answer.

It was not until 1739 that Euler realized that the sine function would enable closed-form solutions of such higher order equations to be given. On March 30 of that year Euler presented a paper to the St. Petersburg Academy of Sciences in which he solved the differential equation of motion of a sinusoidally driven harmonic oscillator, that is, of an object acted on by two forces, one proportional to the distance and one varying sinusoidally with the time. The very statement of the problem is perhaps the earliest use of the sine as a function of time, and the resulting differential equation

$$2a\, d^2s + \frac{s\, dt^2}{b} + \frac{a\, dt^2}{g}\sin\frac{t}{a} = 0$$

(where s represents position and t time) is the earliest use of that function in such an equation. There are two aspects of Euler's solution which are of interest. First, as a special case, he deleted the sine term and solved the equation $2a\, d^2s + \frac{s\, dt^2}{b} = 0$, or, after multiplying through by $b\, ds$, the equation $2ab\, ds\, d^2s = -s\, ds\, dt^2$. An integration with respect to s then gave $2ab\, ds^2 = (C^2 - s^2)\, dt^2$, or

$$dt = \frac{\pm\sqrt{2ab}\, ds}{\sqrt{C^2 - s^2}},$$

the differential equation for the arcsine (with the positive sign) or the arccosine (with the negative sign). Euler, since he was interested in the motion rather than the time, solved the arccosine equation for s instead of t: $s = C\cos(t/\sqrt{2ab})$, the first such explicit analytic solution on record. Second, to solve the general case, Euler postulated a solution of the form $s = u\cos(t/\sqrt{2ab})$, where u is a new variable. He then substituted that solution into the equation and solved for u. This manipulation shows that Euler was already familiar with the basic differentiation rules for the sine and cosine. These had, in effect, been known to Leibniz and, as will be noted below, had already appeared in several printed sources.

There is more to the story of the sine and cosine. On May 5, 1739 Euler wrote to Johann Bernoulli noting that he had solved in finite terms the third order equation $a^3 d^3y = y dx^3$: "Although it appears difficult to integrate, needing a triple integration and requiring the quadrature of the circle and hyperbola, it may be reduced to a finite equation; the equation of the integral is

$$y = be^{x/a} + ce^{-(x/2a)}\sin\frac{(f + x)\sqrt{3}}{2a}$$

... where b, c, f are arbitrary constants arising from the triple integration."[7] Euler did not reveal how he discovered this solution, but a good guess would be that he used the known exponential solution $y = e^{x/a}$ to reduce the order of the equation. In this technique, which Euler had used earlier, one multiplies the original equation $a^3 d^3y - y dx^3 = 0$ by $e^{-(x/a)}$

and assumes that this is the differential of $e^{-(x/a)}(A\, d^2y + B\, dy\, dx + Cy\, dx^2)$. It is then straightforward to show that a new solution of the original equation also satisfies the second-order equation $a^2\, d^2y + a\, dy\, dx + y\, dx^2 = 0$. To solve this latter equation requires a different Eulerian technique. Namely, one guesses a solution to be of the form $y = ue^{\alpha x}$ and substitutes this for y in the equation. Again, a bit of manipulation shows that the term $du\, dx$ can be eliminated by setting $\alpha = -1/2a$. The equation then reduces to $a^2\, d^2u + (3/4)u\, dx^2 = 0$, an equation of the same form as the one Euler solved in March. In this case, the solution is $u = C \sin((x + f)\sqrt{3}/2a)$, from which the general solution to the original third-order solution follows. Note that this reconstruction uses both the quadrature of the hyperbola (the exponential function, related naturally to the logarithm function defined as area under a hyperbola) and the quadrature of the circle (the sine function, related to the arcsine function whose definition involves the area of the circle), as well as three integrations.

Since the sine and exponential functions had been used in the solution of the same differential equation, it was clear that Euler now considered the sine, and by extension, the other trigonometric functions, as functions in the same sense as the exponential function. But even more interesting, it was the very introduction of these functions into calculus which led Euler to the solution method for the class of linear differential equations with constant coefficients, that is, equations of the form

$$y + a\frac{dy}{dx} + b\frac{d^2y}{dx^2} + c\frac{d^3y}{dx^3} + \cdots = 0.$$

Euler noted in a letter to Johann Bernoulli on September 15, 1739 that "after treating this problem in many ways, I happened on my solution entirely unexpectedly; before that I had no suspicion that the solution of algebraic equations had so much importance in this matter."[8] Euler's "unexpected" solution was to replace the given differential equation by the algebraic equation

$$1 + ap + bp^2 + cp^3 + \cdots = 0,$$

and factor this "characteristic polynomial" into its irreducible real linear and quadratic factors. For each linear factor $1 - \alpha p$, one takes as solution $y = Ae^{x/\alpha}$, while for each irreducible quadratic factor $1 + \alpha p + \beta p^2$ one takes as solution

$$e^{-\alpha x/2\beta}\left(C \sin \frac{x\sqrt{4\beta - \alpha^2}}{2\beta} + D \cos \frac{x\sqrt{4\beta - \alpha^2}}{2\beta} \right).$$

The general solution is then a sum of the solutions corresponding to each factor. As an example, Euler solved the equation proposed by Daniel Bernoulli some four years earlier:

$$y - k^4\frac{d^4y}{dx^4} = 0.$$

The corresponding algebraic equation $1 - k^4p^4$ factors as $(1 - kp)(1 + kp)(1 + k^2p^2)$. Thus the solution is

$$y = Ae^{-(x/k)} + Be^{x/k} + C \sin \frac{x}{k} + D \cos \frac{x}{k}.$$

Euler did not say how he arrived at his algebraic solution method. But since several months earlier he had discovered that trigonometric functions were involved in the solution to the equation $y - a^3\frac{d^3y}{dx^3} = 0$, one can surmise that he merely generalized that method. For given one solution of the equation, of the form $y = e^{x/a}$, the reduction procedure indicated there provides in essence a factorization of the characteristic polynomial as $1 - a^3p^3 = (1 - ap)(1 + ap + a^2p^2)$. The general factorization method indicated in Euler's September letter would then have followed easily. In particular, it would have been clear that the sine and cosine terms come from the irreducible quadratic factors.

Johann Bernoulli was somewhat bothered by Euler's solution. He noted that the irreducible quadratic factors of the characteristic polynomial could be factored over the complex numbers, and thus that Euler's method led to complex roots of this polynomial being related to real solutions involving sines and cosines. Euler finally convinced Bernoulli that $2\cos x$ and $e^{ix} + e^{-ix}$ were identical, because they satisfied the same differential equation, and therefore that using imaginary exponentials amounted to the same thing as using sines and cosines. It also followed that complex exponential functions were related to sines and cosines by the relationships

$$e^{ix} = \cos x + i\sin x \qquad \text{and} \qquad e^{-ix} = \cos x - i\sin x.$$

The formulas for e^{ix} and e^{-ix} helped to clear up a controversy over the status of logarithms of negative numbers. If such were to be defined, it would appear, for example, that $2\log(-1) = \log(-1)^2 = \log 1$, or, more generally, that $\log(-x)^2 = \log(+x)^2$, hence that $2\log(-x) = 2\log(+x)$, and $\log(-x) = \log(+x)$. In fact, Johann Bernoulli argued precisely this, that the logarithmic curve must be symmetric around the vertical axis, a position seemingly confirmed by the equality

$$\frac{d(\log(-x))}{dx} = \frac{1}{x} = \frac{d(\log(+x))}{dx}.$$

But Euler's formula shows that $ix = \log(\cos x + i\sin x)$ and thus that a given (complex) number has infinitely many logarithms. For example, $\log 1 = \log(\cos 2n\pi) = 2n\pi i$ where n can be any integer. Because, similarly, $\log(-1) = \log(\cos(2n + 1)\pi) = (2n + 1)\pi i$, it is in fact true that $2\log(-1) = \log 1$ in the sense that doubles of logarithms from one set are equal to logarithms from the other set. Thus, as Euler wrote, his methods made "all the difficulties and contradictions disappear"[9] of dealing with logarithms of negative and complex numbers.

13.2 CALCULUS TEXTS

Although there were a few calculus texts written around the turn of the eighteenth century, including those discussed in Chapter 12, the middle third of the century saw many more, both in England and on the European continent. These included both texts in the vernacular, for the education of the layman, and an important series of texts in Latin, designed for those with a university education.

13.2.1 Thomas Simpson's *Treatise of Fluxions*

In England a growing demand for mathematical knowledge by the middle class was met in part by a group of private teachers who wrote texts to supplement their instruction.

Biography   **Thomas Simpson (1710–1761)**

Born in the village of Market Bosworth, not far from Birmingham, Simpson was raised by his father to become a weaver. Thomas' thirst for a better education, however, led to a rift with his father, and he was forced to leave home. By the age of 25 he was able to learn mathematics on his own and move to London, where he joined the Mathematical Society at Spitalfields, a weaving community. The rules of this society made it the duty of every member "if he be asked any mathematical or philosophical question by another member, to instruct him in the plainest and easiest manner he is able."[10] Through his activity in the society, Simpson became a teacher of mathematics and was soon able to give up weaving and bring his family to London. His reputation in mathematics, enhanced by the publication of several textbooks, finally enabled him in 1743 to secure a position as professor of mathematics at the Royal Military Academy at Woolwich, a school founded to provide military cadets with sufficient mathematical education to succeed as engineers. Shortly thereafter Simpson was elected to the Royal Society.

Thomas Simpson (1710–1761) was one such teacher, whose earliest text, *A New Treatise of Fluxions*, was published in 1737 by subscription of his private students.

Simpson's *Treatise* is basically Newtonian in approach, making much use of infinite series to solve, in particular, problems in integration. It is replete with problems, many of which have become familiar to today's students. Thus, in an early section on maxima and minima, Simpson showed how to find the greatest parallelogram inscribed in a triangle, the smallest isosceles triangle which circumscribes a given circle, and the cone of least surface area with a given volume. Also included in this section is perhaps the earliest solution to the problem of determining the maximum of a function of several variables, obtained by determining when the partial derivatives with respect to each variable vanish simultaneously. The problem is to find x, y, z such that $(b^3 - x^3)(x^2z - z^3)(xy - y^2)$ is maximal. Simpson did not use the language of partial derivatives. He did, however, calculate the fluxion of each of x, y, and z, holding the other two variables constant, before setting each equal to 0 and solving the resulting equations simultaneously. In a later section of the book, buried in a problem on navigation, is probably the first publication in a text of the rule for differentiating the sine. In Simpson's words, "the fluxion of any circular arch is to the fluxion of its sine, as radius to the cosine."[11] The proof, given over 20 years earlier in a paper by Roger Cotes (1682–1716), the editor of the second edition of Newton's *Principia*, uses similar triangles. If z denotes the arc of a circle of radius An centered on A, $x = Ab$ the sine of z, and bn its cosine, then the differential triangle nrm, whose hypotenuse nr represents $\dot{z}$ (the fluxion of the arc) and whose side mr represents $\dot{x}$ (the fluxion of the sine), is similar to triangle Anb (Figure 13.5). It follows that $\dot{z} : \dot{x} = An : bn$, the result quoted. In modern notation, taking $An = 1$, this can be written as $d(\sin z)/dz = \cos z$.

Simpson is most famous today for the rule for numerical integration by parabolic approximation which bears his name. This rule appears not in his calculus text but in his *Mathematical Dissertations on a Variety of Physical and Analytical Subjects* of 1743. It is, however, not original to Simpson, having appeared in the works of other authors even in the seventeenth century.

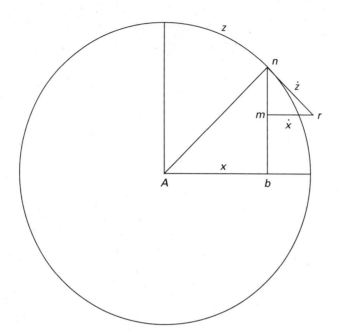

FIGURE 13.5
The fluxion of the sine.

13.2.2 Colin Maclaurin's *Treatise of Fluxions*

The name of another eighteenth century British mathematician, Colin Maclaurin (1698–1746), is also known to today's students from a concept in the calculus text not original to him, the Maclaurin series. The series is found in Maclaurin's *A Treatise of Fluxions*, which appeared in 1742 partly in response to the criticism of the foundations of the theory of fluxions voiced by George Berkeley (1685–1753) eight years earlier. (A discussion of this criticism will be found in section 13.5.1.) Book I of this work treated the foundations of the Newtonian calculus from a geometric point of view. But in Book II, Maclaurin had a different agenda, to demonstrate the rules of fluxions and their applications in an algebraic and algorithmic manner. Maclaurin thus provided details of the entire range of problems to which the calculus was being applied. He discussed maxima and minima and points of inflection; he found tangent lines and asymptotes; he determined curvature; and he gave a complete account of the brachistochrone problem. Maclaurin calculated areas under curves given by y in terms of x by showing that the fluxion of this area was $y\dot{x}$ and then using one of several methods to determine the fluent of this expression. Similarly, he calculated volumes and surface areas of solids of revolution by first determining their fluxions. He used an elementary form of multiple integration to study the gravitational attraction of ellipsoids. Finally, in dealing with logarithms, Maclaurin began with Napier's original definition in terms of motion. It was then easy for him to determine the standard fluxional properties of the logarithm and to use these properties to calculate the fluxions of exponential functions.

Maclaurin's work contained somewhat more about trigonometric functions than Simpson's or any earlier work. For example, in addition to Simpson's theorem on sines, he showed geometrically that the fluxion of an arc is to the fluxion of its tangent in the

Biography **Colin Maclaurin (1698–1746)**

Maclaurin, unlike Simpson, had both a university education and a university career. Born in Kilmodan, a village in western Scotland, he entered the University of Glasgow at the age of 11 and soon mastered the university mathematics curriculum. At the age of 19, Maclaurin was appointed to a chair of mathematics at the University of Aberdeen, but shortly thereafter left for a three year tour of Europe as tutor to the son of a wealthy Lord. The authorities at Aberdeen were not particularly happy with his absence and soon after his return forced him to resign. Meanwhile, Newton had recommended him for a position at Edinburgh, where he remained, teaching subjects ranging from Euclid and elementary algebra to fluxions and Newton's *Principia*, for the rest of his life. In 1745, Maclaurin helped to fortify Edinburgh against the forces of Bonnie Prince Charlie, but when the city fell, he left for York. He fell ill there and never recovered, dying at the age of 48.

duplicate ratio of the radius to the cosine and that the fluxion of an arc is to the fluxion of its secant as the square on the radius is to the rectangle determined by the secant and the tangent. Although these results can be translated into modern calculus theorems $\left(\frac{d}{dx} \tan x = \frac{1}{\cos^2 x}; \frac{d}{dx} \sec x = \sec x \tan x \right)$, for Maclaurin they only gave ratios of fluxions in relation to those of the line segments which represented the trigonometric functions. These results were not applied analytically as were those on the calculus of logarithmic and exponential functions.

The series named for Maclaurin occurs in Book II: Suppose that y is expressible as a series in z, say $y = A + Bz + Cz^2 + Dz^3 + \cdots$. If $E, \dot{E}, \ddot{E}, \ldots$ are the values of y and its fluxions of various orders when z vanishes, then the series can be expressed in the form

$$y = E + \dot{E}z + \frac{\ddot{E}z^2}{1 \times 2} + \frac{\dddot{E}z^3}{1 \times 2 \times 3} + \cdots$$

(with the assumption that $\dot{z} = 1$). Maclaurin's proof is easy, given his assumption that y can be written in a power series. Namely, he first sets $z = 0$ to get $A = E$. Next, he takes the fluxion of the series and again sets $z = 0$. It follows that $B = \dot{E}/\dot{z} = \dot{E}$. He continues to take fluxions and set $z = 0$ to complete the result. Maclaurin noted that this theorem had already been discovered by Brook Taylor (1685–1731) and published in his *Methodus Incrementorum* (*Method of Increments*) in 1715.

Maclaurin worked out many examples of these series, including the series for the sine and cosine in a circle of radius a. For example, if $y = \cos z$ (in a circle of radius a), then $\frac{\dot{y}}{\dot{z}} = \sqrt{a^2 - y^2}/a$. It follows that

$$\frac{\dot{y}^2}{\dot{z}^2} = \frac{a^2 - y^2}{a^2}$$

and that

$$\frac{2\dot{y}\ddot{y}}{\dot{z}^2} = -\frac{2y\dot{y}}{a^2} \quad \text{or} \quad \frac{\ddot{y}}{\dot{z}^2} = -\frac{y}{a^2}.$$

Therefore, since $y = a$ when $z = 0$, it follows that $E = a$, $\dot{E} = 0$, and $\ddot{E} = -\frac{1}{a}$. The first three terms of the series for $y = \cos z$ are then $y = a + 0z - \frac{1}{2a}z^2$. More terms are easily found without any necessity for calculating derivatives of sines and cosines.

Maclaurin also used his series for developing the standard derivative tests for determining maxima and minima: "When the first fluxion of the ordinate vanishes, if at the same time its second fluxion is positive, the ordinate is then a minimum, but is a maximum if its second fluxion is then negative."[12] If the ordinate $AF = E$ and two values of the abscissa, one to the right of A (designated x) and one the same distance to the left (designated $-x$), are given (Figure 13.6), the Maclaurin series shows that the corresponding ordinates are

$$PM = E + \dot{E}x + \frac{\ddot{E}x^2}{2} + \frac{\dddot{E}x^3}{6} + \cdots$$

and

$$pm = E - \dot{E}x + \frac{\ddot{E}x^2}{2} - \frac{\dddot{E}x^3}{6} + \cdots.$$

Assuming that $\dot{E} = 0$ and that x is small enough, Maclaurin concluded that both of these ordinates will exceed the ordinate $AF = E$ when $\ddot{E}$ is positive (so that AF is a minimum) and both will be less than AF when $\ddot{E}$ is negative (so that AF is a maximum). Furthermore, Maclaurin concluded that if $\ddot{E}$ also vanishes, and if $\dddot{E}$ does not, then either $PM > AF$ and $pm < AF$ or vice versa and therefore AF is neither a maximum nor a minimum.

Maclaurin ended his text with probably the earliest analytic proof of part of the fundamental theorem of calculus, at least for the special case of power functions. (He had given a more general geometric proof earlier in the text.) "Supposing n to be any [positive] integer, . . . if the area upon the base AP or x is always equal to x^n, then the ordinate PM or y shall be always equal to nx^{n-1}."[13] Maclaurin began by taking an increment $o = Pp$ of the base x and noting that $PM \times Pp = yo < $ area $PMmp = (x + o)^n - x^n$ (Figure 13.7). By an algebraic result he had proved earlier, it followed that $(x + o)^n$

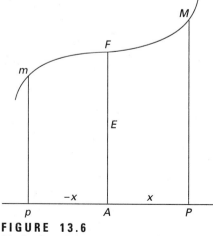

FIGURE 13.6
Maclaurin and the second derivative test.

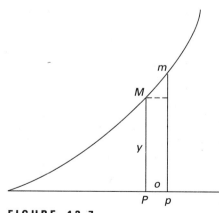

FIGURE 13.7
Maclaurin and the fundamental theorem of calculus.

Biography   Maria Agnesi (1718–1799)

Agnesi was the eldest child of a professor of mathematics at the University of Bologna, who encouraged his daughter to pursue scientific interests by hiring various distinguished professors to tutor her. By the age of 11 she was fluent in seven languages and in her teens was able to dispute important matters in such fields as mechanics, logic, zoology, and mineralogy with the best scholars of the day. Having studied the major mathematical works of the time, she began to instruct her younger brothers in the subject. She soon decided that her work, including material on algebra as well as a complete treatise on the differential and integral calculus, should be published to benefit all Italian youth. This text was so clearly written that a committee of the French Academy, in authorizing its translation into French in 1749, noted that "there is no other book, in any language, which would enable a reader to penetrate as deeply, or as rapidly, into the fundamental concepts of analysis."[14] And John Colson, the Lucasian professor of mathematics at Cambridge in the middle of the eighteenth century, was so impressed with the book that he learned Italian for the sole purpose of translating the work into English so that British youth would have the same benefits as those of Italy.

The Pope, too, recognized Agnesi's talents and appointed her to the chair of mathematics at the University of Bologna. However, it is not clear whether she actually taught there, other than to give a few lectures during her father's final illness. Soon after his death in 1752 she withdrew from all scientific pursuits and spent the rest of her life in religious studies and social work among the poor.

$- x^n < n(x + o)^{n-1}o$ and therefore that $yo < n(x + o)^{n-1}o$. Similarly, using a value of the abscissa to the left of P, Maclaurin found that $yo > n(x - o)^{n-1}o$ and therefore that $n(x - o)^{n-1} < y < n(x + o)^{n-1}$. Rather than use a modern limit argument to show that $y = nx^{n-1}$, Maclaurin used a *reductio* argument. If $y > nx^{n-1}$, then $y = nx^{n-1} + r < n(x + o)^{n-1}$ for any increment o. But if o is chosen to be $\left(x^{n-1} + \frac{r}{n}\right)^{1/(n-1)} - x$, a brief calculation shows that $y = n(x + o)^{n-1}$, a contradiction. A similar contradiction occurs if y is assumed less than nx^{n-1} and Maclaurin's proof is complete.

13.2.3 Maria Agnesi's *Instituzioni Analitiche*

Maclaurin's *Treatise of Fluxions* was read on the European continent, especially after it was translated into French in 1749. The year before, however, the first important successor to l'Hospital's text appeared in Europe, the *Instituzioni analitiche ad uso della gioventu italiana* (*Foundations of Analysis for the use of Italian youth*), by Maria Gaetana Agnesi (1718–1799). Agnesi's text, not surprisingly, showed the influence of Leibniz and his followers rather than Newton. Thus it was written in the language of differentials and infinitesimals rather than that of fluxions. (Interestingly enough, the English translator replaced all of the dx's by $\dot{x}$'s, although he often kept the word "differential" rather than replace it by "fluxion.") The text explained concepts clearly and provided numerous examples. Thus, in her section on maxima and minima, Agnesi presented such problems as that of cutting a line at a point so that the product of the length of one segment and the square of the other is maximal and that of finding the line of minimum length which passes through one vertex of a rectangle and intersects the extensions of both of the opposite sides. She even showed how to find the point of maximal curvature on the logarithmic curve, that defined by $a(dy/y) = dx$.

For Agnesi, the integral calculus is the inverse of the differential calculus, that is, the method of determining, from a given differential expression, the quantity of which that expression is the differential. The symbol $\int y\, dx$ means an antiderivative. But the symbolism of the integrand, that $y\, dx$ represents the area of an infinitesimal rectangle, leads her, virtually as an afterthought, to note that areas under curves can be calculated by this same inverse process.

Agnesi is especially thorough in her treatment of the logarithmic (exponential) curve. She notes that the ordinary rule for integration leads from $dx = ay^{-1}dy$ to

$$x = \frac{ay^{-1+1}}{-1+1} \quad \text{or} \quad \frac{ay^0}{0}$$

and that this "teaches us nothing." Thus she deals with this curve in other ways. She shows first that the curve whose ordinates increase geometrically while the abscissas increase arithmetically has the differential equation $dx = ay^{-1}dy$ and then that one can make computations by using appropriate infinite series. She also shows how to find the area under this curve, both over a finite interval and over an infinite one stretching to the left from a fixed abscissa x. She calculates this "improper integral" (today written as $\int_{-\infty}^{x} e^{t/a}\, dt$) to be ay, where y is the ordinate corresponding to x, and shows how to calculate the volume of the solid generated by revolving the curve around the x-axis. But as in the rest of the books of the first half of the eighteenth century, there is little concerning trigonometric functions.

Curiously, Agnesi's name, like those of the two other textbook authors discussed above, is attached to a small item in her book not even original to her. As an example in analytic geometry, she described geometrically a curve whose equation she determined to be $y = \frac{a\sqrt{a-x}}{\sqrt{x}}$, a curve which had earlier been named *la versiera*, derived from the Latin meaning "to turn." Unfortunately, the word *versiera* was also the abbreviation for the Italian word *avversiera*, meaning "wife of the devil." Because the English translator rendered this word as "witch," the curve has ever since been referred to as the "witch of Agnesi."

13.2.4 Euler's *Introductio*

The three texts discussed so far were all written in the vernacular. It was a series of Latin texts, however, which proved more important for the future. These were the works of Euler, the two volumes of the *Introductio in Analysin Infinitorum* (*Introduction to Analysis of the Infinite*, 1748), the *Institutiones Calculi Differentialis* (*Methods of the Differential Calculus*, 1755), and the three volumes of the *Institutiones Calculi Integralis* (*Methods of the Integral Calculus*, 1768–1770).

The *Introductio*, Euler's "precalculus" text, was an attempt to develop those topics "which are absolutely required for analysis" so that the reader "almost imperceptibly becomes acquainted with the idea of the infinite."[15] And because analysis is concerned with functions, Euler begins his work with their definition: "A **function** of a variable quantity is an analytic expression composed in any way whatsoever of the variable quantity and numbers or constant quantities."[16] The first point to note about Euler's definition is that the word "function" means an "analytic expression," that is, a formula. The second is that his statement as to how these formulas are to be formed, "in any way whatsoever," can only be understood by considering his further discussion. For Euler, there were two basic classes of functions, algebraic and transcendental. The former were formed from the variables and constants by addition, subtraction, multiplication, division, raising to a power, extraction of roots, and the solution of an equation. The latter were those defined by exponentials, logarithms, and, more generally, by integrals. Because integrals could not be discussed in a precalculus work, the transcendental functions discussed in the *Introductio* are limited to the special cases of trigonometric, exponential, and logarithmic functions.

An important tool in Euler's discussion of functions is that of a power series. Euler was convinced that any function could be expressed, except perhaps at isolated points, by a power series, but gave no proof. Rather, he attempted to convince the reader of this truth by showing how to expand any algebraic function as well as various transcendental functions into such a series. His methods for algebraic functions were not new, being a combination of Newton's methods using division (in the case of rational functions) and the binomial theorem (for functions expressible in terms of any powers). And there is no discussion of convergence.

The sections of the *Introductio* which were to have the most influence, however, dealt with the exponential, logarithmic, and trigonometric functions, for it is there that Euler introduced the notations and concepts which were to make obsolete all the discussions of such functions in earlier texts. All modern treatments of these functions are in some sense derived from those of Euler. Thus Euler defined exponential functions as powers in which exponents are variable and then—and this is a first—logarithms in terms of these. Namely, if $a^z = y$, Euler defined z to be the logarithm of y with base a. The basic properties of the logarithm function are then derived from those of the exponential.

Euler developed power series for the exponential and logarithmic functions to arbitrary base a by use of the binomial theorem. His technique made important use of both "infinitely small" and "infinitely large" numbers, concepts whose use today is frowned upon. Nevertheless, Euler rarely erred. For example, Euler noted that since $a^0 = 1$, it follows that $a^\omega = 1 + \psi$ where both ω and ψ are infinitely small. Therefore, ψ must be some multiple of ω, depending on a, and

$$a^\omega = 1 + k\omega \qquad \text{or} \qquad \omega = \log_a(1 + k\omega).$$

Euler noted next that for any j, $a^{j\omega} = (1 + k\omega)^j$, and, expanding the right side by the binomial theorem, that

$$a^{j\omega} = 1 + \frac{j}{1}k\omega + \frac{j(j-1)}{1 \cdot 2}k^2\omega^2 + \frac{j(j-1)(j-2)}{1 \cdot 2 \cdot 3}k^3\omega^3 + \cdots.$$

If j is taken to equal z/ω, where z is finite, then j is infinitely large and $\omega = z/j$. The series now becomes

$$a^z = 1 + \frac{1}{1}kz + \frac{1(j-1)}{1 \cdot 2j}k^2z^2 + \frac{1(j-1)(j-2)}{1 \cdot 2j \cdot 3j}k^3z^3 + \cdots.$$

Because j is infinitely large, $(j-n)/j = 1$ for any positive integer n. The expansion then reduces to the series

$$a^z = 1 + \frac{kz}{1} + \frac{k^2 z^2}{1 \cdot 2} + \frac{k^3 z^3}{1 \cdot 2 \cdot 3} + \cdots$$

where k depends on the base a. Euler also noted that the equation $\omega = \log_a(1 + k\omega)$ implies that if $(1 + k\omega)^j = 1 + x$, then $\log_a(1 + x) = j\omega$. Since then $k\omega = (1 + x)^{1/j} - 1$, it follows that

$$\log_a(1 + x) = \frac{j}{k}(1 + x)^{1/j} - \frac{j}{k}.$$

Another clever use of the binomial theorem finally allows him to derive the series

$$\log_a(1 + x) = \frac{1}{k}\left(\frac{x}{1} - \frac{x^2}{2} + \frac{x^3}{3} + \cdots\right).$$

The choice of $k = 1$, or equivalently, $a = e$, gives the standard power series for e^z and $\ln z$.

Euler's treatment of "transcendental quantities which arise from the circle" is the first textbook discussion of the trigonometric functions which deals with these quantities as functions having numerical values, rather than as lines in a circle of a certain radius. Euler does not, in fact, give any new definition of the sine and cosine. He merely notes that he will always consider the sine and cosine of an arc z to be defined in terms of a circle of radius 1. All basic properties of the sine and cosine, including the addition and periodicity properties, are assumed known, although Euler does derive some relatively complicated identities. More importantly, he derives the power series for the sine and cosine through use of the binomial theorem and complex numbers. Thus from the easily derived identity $(\cos z \pm i \sin z)^n = \cos nz \pm i \sin nz$, Euler concludes that

$$\cos nz = \frac{(\cos z + i \sin z)^n + (\cos z - i \sin z)^n}{2}$$

and, by expanding the right side, that

$$\cos nz = (\cos z)^n - \frac{n(n-1)}{1 \cdot 2}(\cos z)^{n-2}(\sin z)^2$$

$$+ \frac{n(n-1)(n-2)(n-3)}{1 \cdot 2 \cdot 3 \cdot 4}(\cos z)^{n-4}(\sin z)^4 - \cdots.$$

Again letting z be infinitely small, n infinitely large, and $nz = v$ finite, it follows from $\sin z = z$ and $\cos z = 1$ that

$$\cos v = 1 - \frac{v^2}{1 \cdot 2} + \frac{v^4}{1 \cdot 2 \cdot 3 \cdot 4} - \cdots.$$

The remainder of volume one of the *Introductio* includes much else about infinite processes, including infinite products as well as infinite series. For example, Euler considers the hyperbolic functions, although they are not named. He shows that

$$\frac{e^x - e^{-x}}{2} = \frac{x}{1} + \frac{x^3}{1 \cdot 2 \cdot 3} + \frac{x^5}{1 \cdot 2 \cdot 3 \cdot 4 \cdot 5} + \cdots$$

may be factored as

$$\frac{e^x - e^{-x}}{2} = x\left(1 + \frac{x^2}{\pi^2}\right)\left(1 + \frac{x^2}{4\pi^2}\right)\left(1 + \frac{x^2}{9\pi^2}\right)\cdots$$

and similarly

$$\frac{e^x + e^{-x}}{2} = 1 + \frac{x^2}{1\cdot 2} + \frac{x^4}{1\cdot 2\cdot 3\cdot 4} + \cdots = \left(\frac{1 + 4x^2}{\pi^2}\right)\left(\frac{1 + 4x^2}{9\pi^2}\right)\left(\frac{1 + 4x^2}{25\pi^2}\right)\cdots.$$

It was Johann Heinrich Lambert (1728–1777) who introduced the names hyperbolic sine (sinh x) and hyperbolic cosine (cosh x) for these functions in 1768 and worked out the analogies between these functions and the ordinary sine and cosine.

13.2.5   Euler's *Differential Calculus*

Although volume one of the *Introductio* was largely concerned with series—volume two, on analytic geometry, will be dealt with in Chapter 14—Euler considered this material as the algebra necessary for the calculus. He discussed the calculus itself first in his *Institutiones Calculi Differentialis* of 1755. The introduction contains his definition of the differential calculus: "[It] is the method of determining the ratios of the evanescent increments which functions receive to those of the evanescent increments of the variable quantities, of which they are functions."[17] Euler had already given a definition of "function" in the *Introductio*, but here he generalized it somewhat: "When, however, quantities depend on others in such a way that [the former] undergo changes themselves when [the latter] change, then [the former] are called functions of [the latter]; this is a very comprehensive idea which includes in itself all the ways in which one quantity can be determined by others."[18] Thus Euler no longer required a function to be an "analytic expression." The reason for this change is perhaps connected to the controversy over the vibrating string problem to be discussed in section 13.4. And although Euler was well aware of the many applications of the differential calculus to geometry, he insisted that this work was a work of pure analysis. As such, there were to be no diagrams whatsoever.

Because calculus has to do with ratios of "evanescent increments," Euler begins the book with a discussion of increments in general, that is, with finite differences. Given a sequence of values of the variable, say $x, x + \omega, x + 2\omega, \ldots$ and the corresponding values of the function $y, y', y'', \ldots$, Euler considers various sequences of finite differences. The first differences are $\Delta y = y' - y$, $\Delta y' = y'' - y'$, $\Delta y'' = y''' - y''$, $\ldots$; the second differences are $\Delta\Delta y = \Delta y' - \Delta y$, $\Delta\Delta y' = \Delta y'' - \Delta y'$, $\ldots$; third and higher differences are defined analogously. For example, if $y = x^2$, then $y' = (x + \omega)^2$ and $\Delta y = 2\omega x + \omega^2$, $\Delta\Delta y = 2\omega^2$, while the third and higher differences are all 0. Using various techniques, including expansion in series, Euler is able to calculate the differences for all of the standard elementary functions. Furthermore, using the sum Σ to denote the inverse of the Δ operation, he derives various formulas for that operation as well. Thus, since $\Delta x = \omega$, it follows that $\Sigma\omega = x$ and that $\Sigma 1 = x/\omega$. Similarly, since $\Delta x^2 = 2\omega x + \omega^2$, it follows that $\Sigma(2\omega x + \omega^2) = x^2$ and that

$$\sum x = \frac{x^2}{2\omega} - \sum\frac{\omega}{2} = \frac{x^2}{2\omega} - \frac{x}{2}.$$

Euler then easily develops rules for Σ from the corresponding rules for Δ. Rather than discuss the rules for finite differences, however, it will be more useful to discuss Euler's rules for differentials.

"The analysis of infinitesimals . . . is nothing but a particular case of the method of differences . . . which arises when the differences, which were previously assumed to be finite, are taken as infinitely small."[19] Euler's rules for calculating with these infinitely small quantities, the differentials, produce the standard formulas of the differential calculus. For example, if $y = x^n$, then $y' = (x + dx)^n = x^n + nx^{n-1}dx + \frac{n(n-1)}{1 \cdot 2}x^{n-2}dx^2 + \cdots$. Thus $dy = y' - y = nx^{n-1}dx + \frac{n(n-1)}{1 \cdot 2}x^{n-2}dx^2 + \cdots$. "But in this expression the second term with the rest of the sequence will vanish in comparison with the first."[20] Thus $d(x^n) = nx^{n-1}dx$. It should be noted here that Euler intended his argument to apply not just to positive integral powers of x, but to arbitrary powers. The binomial theorem, after all, applies to all powers. Thus the expansion of $(x + dx)^n$ represents not a finite sum, but in fact an infinite series. Euler therefore notes immediately that $d\left(\frac{1}{x^m}\right) = -\frac{m\,dx}{x^{m+1}}$ and, more generally, that $d(x^{\mu/\nu}) = (\mu/\nu)x^{(\mu-\nu)/\nu}dx$.

Euler does not give an explicit statement of the modern chain rule, but does deal with special cases as the need arises. Thus if p is a function of x whose differential is dp, then $d(p^n) = np^{n-1}dp$. Euler's derivation of the product rule is virtually identical to that of Leibniz, but his derivation of the quotient rule is more original. He expands $1/(q + dq)$ into the power series

$$\frac{1}{q + dq} = \frac{1}{q}\left(1 - \frac{dq}{q} + \frac{dq^2}{q^2} - \cdots\right),$$

neglects the higher order terms, and then writes

$$\frac{p + dp}{q + dq} = (p + dp)\left(\frac{1}{q} - \frac{dq}{q^2}\right) = \frac{p}{q} - \frac{p\,dq}{q^2} + \frac{dp}{q} - \frac{dp\,dq}{q^2}.$$

It follows, since the second-order differential $dp\,dq$ vanishes with respect to the first-order ones, that

$$d\left(\frac{p}{q}\right) = \frac{p + dp}{q + dq} - \frac{p}{q} = \frac{dp}{q} - \frac{p\,dq}{q^2} = \frac{q\,dp - p\,dq}{q^2}.$$

The differential of the logarithm requires the power series derived in the *Introductio*. If $y = \ln x$, then

$$dy = \ln(x + dx) - \ln(x) = \ln\left(1 + \frac{dx}{x}\right) = \frac{dx}{x} - \frac{dx^2}{2x^2} + \frac{dx^3}{3x^3} - \cdots.$$

Dispensing with the higher order differentials immediately gives Euler the formula $d(\ln x) = dx/x$. The approach to the arcsine function is through complex numbers. Substituting $y = \arcsin x$ into the formula $e^{iy} = \cos y + i \sin y$ gives $e^{iy} = \sqrt{1 - x^2} + i\,x$. It follows that $y = \frac{1}{i}\ln(\sqrt{1 - x^2} + i\,x)$ and therefore that

$$dy = d(\arcsin x) = \frac{1}{i}\frac{1}{\sqrt{1 - x^2} + i\,x}\left(\frac{-x}{\sqrt{1 - x^2}} + i\right)dx = \frac{dx}{\sqrt{1 - x^2}}.$$

Finally, Euler's derivation of the differential of the sine function begins with the calculation $d(\sin x) = \sin(x + dx) - \sin x = \sin x \cos dx + \cos x \sin dx - \sin x$. Euler then recalls his series expansions of the sine and cosine and, again rejecting higher order terms, notes that $\cos dx = 1$ and $\sin dx = dx$. It follows that $d(\sin x) = \cos x\, dx$ as desired.

Euler's text has many other features, including a discussion of differentials of functions of two variables, an introduction to differential equations, in which he shows how to generate these from a given equation in two variables, a discussion of the Taylor series, a chapter on various methods of converting functions to power series, and an extensive discussion on finding the sums of various series. The remainder of the discussion here, however, will center on Euler's material on determining maxima and minima.

Euler derives the basic criteria for a function to have a maximum or minimum value at $x = \alpha$, in terms of the first and second derivatives, by the use of the Maclaurin series, in a way virtually identical to that of Maclaurin himself. But Euler provides many more examples than his Scottish predecessor and often seeks to generalize. Thus, after considering maxima and minima for several specific polynomials, he discusses in some detail the case of an arbitrary polynomial $y = x^n + Ax^{n-1} + Bx^{n-2} + \cdots + D$. After dealing with several cases of rational functions, he considers the more general rational function

$$\frac{(\alpha + \beta x)^m}{(\gamma + \delta x)^n}.$$

After discussion of the lack of a power series for $x^{2/3}$ around 0, and therefore, the necessity of formulating some different criteria for a maximum or minimum, he deals with the more general case $x^{2p/(2q-1)}$. Most of Euler's examples are of algebraic functions, but he concludes with a few examples using transcendental functions, including the functions $x^{1/x}$ and $x \sin x$, both of which require detailed numerical work to arrive at an exact solution for an extreme value.

For functions V of two variables, Euler begins by considering the special case of functions of the form $X + Y$ where X is a function solely of x and Y of y. In that case, a pair of values (x_0,y_0) such that x_0 is a maximum for X and y_0 a maximum for Y clearly gives a maximum for $X + Y$. For the more general case, Euler realizes, by holding each variable constant in turn, that an extreme value of V can only occur when the differential $dV = P\, dx + Q\, dy = 0$, therefore only when both $P = 0$ and $Q = 0$, that is, when $\frac{\partial V}{\partial x} = 0$ and $\frac{\partial V}{\partial y} = 0$. The question of determining whether a point (x_0,y_0) where both first partial derivatives vanish produces a maximum, a minimum, or neither is more difficult and, in fact, Euler fails to give complete results. He claims, in fact, that if $\frac{\partial^2 V}{\partial x^2}$ and $\frac{\partial^2 V}{\partial y^2}$ are both positive at (x_0,y_0), then the function V has a minimum there, and if they are both negative, there is a maximum. Euler gives several examples illustrating the method, including $V = x^3 + y^2 - 3xy + (3/2)x$. His criteria imply that V has a minimum both when $x = 1$, $y = 3/2$ and when $x = 1/2$, $y = 3/4$. Unfortunately, the latter point is not a minimum, but a saddle point.

13.2.6 Euler's *Integral Calculus*

The final part of Euler's trilogy in analysis, the *Institutiones Calculi Integralis*, begins with a definition of integral calculus. It is the method of finding, from a given relation of differentials of certain quantities, the quantities themselves. Namely, for Euler as it was for Agnesi (and also for Johann Bernoulli), integration is the inverse of differentiation rather than the determination of an area. Thus the first part of the work deals with techniques for integrating (finding antiderivatives of) functions of various types while the remainder of the text deals with the solutions of differential equations. Euler begins with such results as

$$\int ax^n \, dx = \frac{a}{n+1} x^{n+1} + C$$

for $n \neq -1$ and

$$\int \frac{a \, dx}{x} = a \ln x + C = \ln cx^a,$$

and then gives a detailed discussion of partial fraction techniques which enable one to integrate rational functions. He deals with various methods of substitution to integrate functions involving square roots, although not with the modern trigonometric substitutions. One chapter deals with integration by the use of infinite series, Newton's favorite technique, while another treats integration by parts, particularly in the case of functions involving logarithms and exponentials, and a third considers reduction formulas for integration of powers of trigonometric functions. Euler even uses the substitution $\cos \phi = \frac{1 - x^2}{1 + x^2}$, $\sin \phi = \frac{2x}{1 + x^2}$ to convert rational functions involving sines and cosines to ordinary rational functions.

The bulk of the text deals with methods of solving differential equations. Euler solves the general first-order linear equation $dy + Py \, dx = Q \, dx$ (or, in modern terms, $y' + Py = Q$) by separation of variables to get the general solution

$$y = e^{-\int P \, dx} \int e^{\int P \, dx} Q \, dx,$$

a result he had proved in 1734. He shows how to integrate $P \, dx + Q \, dy$ in the "exact" case where $\partial P/\partial y = \partial Q/\partial x$, using Clairaut's idea of 1739. He discusses how to find integrating factors in the case where $P \, dx + Q \, dy$ is not exact, methods which he and Clairaut had also found earlier. He deals with various cases of second and higher order differential equations, including the linear case with constant coefficients. Finally, Euler concludes the book with a discussion of partial differential equations, a prime example of which will be discussed in section 13.4. But even though the original motivation for consideration of differential equations came largely from physical problems, Euler mentions nothing of these in the text.

The *Integral Calculus*, like the *Differential Calculus* and the *Introductio*, is a text in pure analysis, so much so that Euler does not even deal with applications to geometry. This fact perhaps explains the significant differences between Euler's works and a modern calculus text. Thus in the *Differential Calculus* there are no tangent lines or normal lines, no tangent planes, no study of curvature—all topics with which Euler was fully conversant in 1740 but which only appear in some of his geometrical works. Even more surprisingly, there is no calculation of areas in the *Integral Calculus*, nor any material on lengths of curves, or volumes, or surface areas of solids. It follows that the fundamental theorem of calculus, central to modern works, does not appear. There is not even a calculation of a definite integral. Euler was certainly familiar with using antiderivatives to calculate area and in fact used such ideas in various papers. On the other hand, since there does not appear in his work any clear notion of the area under a curve as a function, he did not consider the derivative of such an area function.

Despite the gaps a modern reader might find in Euler's calculus texts, they proved influential to the end of the century in presenting an organization and a clear explanation of the material which Euler and his predecessors had developed. All the mathematicians of the second half of the eighteenth century made constant use of Euler's works. By the early years of the next century, however, the needs of students began to change. It was the new students entering the sciences after the upheavals of the French Revolution who inspired the writing of many new texts, texts which replaced those of Euler and were the direct ancestors of the texts of today.

13.3 MULTIPLE INTEGRATION

Although Euler did not discuss multiple integration in his *Integral Calculus*, he was involved in developing the important ideas of the subject, a subject which had its beginnings in Leibniz's solution to a challenge problem of Vincenzo Viviani (1622–1703) of April 4, 1692. Viviani, hiding behind the anagram D. Pio Lisci Pusillo Geometra (which stood for *Postremo Galileo Discipulo* (the last student of Galileo)), proposed the problem of determining four equal "windows" on a hemispherical surface such that the remainder of the surface was equal in area to a region constructible by straightedge and compass. Leibniz solved the problem on May 27, 1692, the same day he received it. In doing so, he had to calculate areas of various regions on a hemisphere, for which he integrated expressions involving products of two differentials by integrating first with respect to one variable, with the second held constant, and then with respect to the second.

This problem and similar ones were solved somewhat later by the Bernoullis and l'Hospital, among others, but it was not until 1731 that a systematic attempt to calculate volumes of certain regions as well as the areas of their bounding surfaces was published by Clairaut in his *Recherches sur les courbes a double courbure* (*Research on Curves of Double Curvature*). Although Clairaut demonstrated that surfaces could in general be represented by a single equation in three variables, he most often considered cylindrical surfaces generated by a curve in one of the coordinate planes. Thus to calculate the volume of a region between two cylinders given by $y = f(x)$, $z = g(y)$, he showed that the element of volume was given by $dx \int z \, dy$, then used his equations to rewrite z and dy in terms of x so that he could integrate $z \, dy$. With the volume element now given entirely in terms of x, he was able to integrate again to calculate the desired volume. Similarly, he expressed an element of surface area by $dx \int \sqrt{dx^2 + dy^2}$ and performed analogous calculations.

In a work on the calculus of variations in 1760, Joseph-Louis Lagrange (1736–1813) also had to deal with volumes and surface areas. Lagrange simply wrote $\int \int z \, dx \, dy$ for the volume and $\int \int dx \, dy \sqrt{1 + P^2 + Q^2}$ for the surface area, where the equation of the surface is given by $z = f(x,y)$ and where $dz = P \, dx + Q \, dy$. These notations occur without much discussion in both his letters of the 1750s to Euler and in his early papers, although he does note that the double integral signs indicate that two integrations must be performed successively.

13.3.1 The Concept of a Double Integral

It was only in a paper of 1769 that Euler gave the first detailed explanation of the concept of a double integral. Euler began by generalizing his notion of an integral as an antiderivative. Thus $\int \int Z\, dx\, dy$ was to mean a function of two variables which when twice differentiated, first with respect to x alone, second with respect to y alone, gave $Z\, dx\, dy$ as differential. For example, Euler noted that $\int \int a\, dx\, dy = axy + X + Y$ where X is a function of x and Y a function of y. A more complicated example is provided by

$$\int \int \frac{dx\, dy}{x^2 + y^2}.$$

The first integration can be performed with respect to either variable. Integrating each way, Euler found the values

$$\int \frac{dx}{x} \arctan \frac{y}{x} + X \qquad \text{and} \qquad \int \frac{dy}{y} \arctan \frac{x}{y} + Y.$$

Because the only way to perform the second integration, in either case, is by writing the integrand as a power series, Euler did so and showed that both integrations lead to the same final result

$$\int \int \frac{dx\, dy}{x^2 + y^2} = X + Y - \frac{y}{x} + \frac{y^3}{9x^3} - \frac{y^5}{25x^5} + \cdots.$$

Given then the idea of a double integral as double antiderivative, Euler generalized the concept of finding area through a single integration to that of finding volumes using this double integration. His basic idea, like that of Leibniz, was to integrate with respect to one variable first, keeping the other constant, and then deal with the second one. His first example is to find the volume under one octant of the sphere of radius a whose equation is $z = \sqrt{a^2 - x^2 - y^2}$. Taking an element of area $dx\, dy$ in the first quadrant of the circle in the x-y plane, Euler noted that the volume of the solid column above that infinitesimal rectangle is $dx\, dy\sqrt{a^2 - x^2 - y^2}$ (Figure 13.8). It is that function which must be integrated. Holding x constant, Euler integrated with respect to y to get

$$\left[\frac{1}{2} y\sqrt{a^2 - x^2 - y^2} + \frac{1}{2}(a^2 - x^2) \arcsin \frac{y}{\sqrt{a^2 - x^2}} \right] dx$$

as the volume under that piece of the sphere over the rectangle whose width is dx and whose length is y. Replacing y by $\sqrt{a^2 - x^2}$, Euler calculated the volume of the same piece up to that value of y to be $\frac{\pi}{4}(a^2 - x^2)\, dx$. Integrating with respect to x then gives $\frac{\pi}{4}(a^2 x - \frac{1}{3}x^3)$ as the volume from the y-axis to x, and replacing x by a gives the total volume of the octant as $\frac{\pi}{6}a^3$ and that of the whole sphere as $\frac{4\pi}{3}a^3$.

After showing further how to calculate volumes of solid regions bounded above by the sphere and below by various areas of the plane, Euler noted that a double integral may also be used to calculate surface area. He gave, without much discussion, the element of surface area of the sphere as

$$\frac{a\, dx\, dy}{\sqrt{a^2 - x^2 - y^2}},$$

presumably knowing the general formula given earlier by Lagrange. Furthermore, he also noted that $\int\int dx\, dy$ over a region A was precisely the area of A.

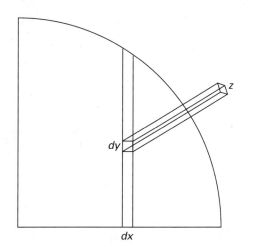

dy

dx

FIGURE 13.8
The volume of a sphere.

13.3.2 Change of Variables in Multiple Integration

The most interesting part of Euler's paper on double integrals was his discussion of what happens to a double integral if the variables are changed. In other words, if x and y are given as functions of two new variables t and v, Euler wanted to determine how to transform the integral $Z\,dx\,dy$ into a new integral with the area element $dt\,dv$.

Euler realized that if the given transformation of variables is a translation followed by a rotation, that is, if

$$x = a + mt + v\sqrt{1 - m^2}$$
$$y = b + t\sqrt{1 - m^2} - mv$$

where m is the cosine of the angle θ of the rotation, then the area elements $dx\,dy$ and $dt\,dv$ should be equal. But when he performed the obvious formal calculation

$$dx = m\,dt + dv\sqrt{1 - m^2}$$
$$dy = dt\sqrt{1 - m^2} - m\,dv$$

and multiplied the two equations together, he arrived at

$$dx\,dy = m\sqrt{1 - m^2}dt^2 + (1 - 2m^2)dt\,dv - m\sqrt{1 - m^2}dv^2,$$

a result which he noted was obviously wrong. Clearly, a similar calculation would also be wrong if t, v were related to x, y by a more complicated transformation. What Euler wanted to do, then, was to develop a method which in the above situation gave $dx\,dy = dt\,dv$ and in the more general case $dx\,dy = W\,dt\,dv$, where W is some function of t and v.

Euler's idea was to deal with one variable at a time, just as was done in double integration. Thus, he first introduced the new variable v such that y was a function of x and v. Then $dy = P\,dx + Q\,dv$. Assuming x fixed, Euler obtained $dy = Q\,dv$ and $\int \int dx\,dy = \int \int Q\,dx\,dv = \int dv \int Q\,dx$. Similarly, now letting x be a function of t and v so that $dx = R\,dt + S\,dv$ and holding v constant, he calculated that $\int dv \int Q\,dx = \int dv \int QR\,dt = \int \int QR\,dt\,dv$. Thus his initial solution to his problem was that $dx\,dy =$

$QR\,dt\,dv$. This answer was not completely satisfactory, however, because Q may depend on x and because the method is not symmetric. Euler therefore continued by considering y as a function of t and v and calculating dy anew:

$$dy = P\,dx + Q\,dv = P(R\,dt + S\,dv) + Q\,dv = PR\,dt + (PS + Q)\,dv.$$

Because it was also true that $dy = T\,dt + V\,dv$, it followed that $PR = T$ and $PS + Q = V$ or that $QR = VR - ST$. Euler's final answer was that $dx\,dy = (VR - ST)dt\,dv$. He noted further that one must, in fact, use the absolute value of $VR - ST$, because area is a positive quantity. In modern notation Euler's result, expressed for double integrals of functions, is that

$$\int\int f(x,y)dx\,dy = \int\int f(x(t,v),y(t,v))\left|\frac{\partial x}{\partial t}\frac{\partial y}{\partial v} - \frac{\partial x}{\partial v}\frac{\partial y}{\partial t}\right|dt\,dv,$$

where the domains over which the integrals are taken are related by the given functional relationship between (x,y) and (t,v).

As is typical of eighteenth century mathematical proofs, Euler's argument is formal. There is no notion of a limit or an infinitesimal approximation or even worries as to points at which the relevant derivatives might not exist. But although Euler was immensely successful in developing new mathematics through the use of such arguments, the lack of an axiomatic foundation on the model of Greek geometry bothered some of his contemporaries. Thus a debate developed as to the proper basis of the chief concepts of the calculus, a debate we will consider in section 13.5.

13.4 PARTIAL DIFFERENTIAL EQUATIONS: THE WAVE EQUATION

The beginnings of the theory of partial differential equations took place in the middle of the eighteenth century with the work not only of Euler, ultimately recorded in his *Integral Calculus*, but also of Jean Le Rond d'Alembert (1717–1783) and Daniel Bernoulli. We will here only discuss one particular type of partial differential equation, the wave equation, because the debate over the subject of vibrating strings, from which the equation was derived, led not only to certain methods of solution but also to a new understanding of the notion of a function.

13.4.1 The Work of d'Alembert

The discussion on the subject of vibrating strings began with a paper of d'Alembert written in 1747, in which he proposed a solution to the problem of the shape of a taut string placed in vibration. Because the position of a point on the string varies both with its abscissa and with time, this shape is determined by a function $y = y(t, x)$ of two variables. D'Alembert considered the string to be composed of an infinite number of infinitesimal masses and then used Newton's laws to derive the partial differential equation for y, now called the **wave equation** and given in modern notation as

$$\frac{\partial^2 y}{\partial t^2} = c^2 \frac{\partial^2 y}{\partial x^2}.$$

Biography   Jean d'Alembert (1717–1783)

D'Alembert was abandoned as an infant on the steps of a Parisian church by his mother, who had just renounced her nun's vows and feared retribution. He was soon adopted into a poor family, but his wealthy natural father provided him with a substantial annuity and helped him gain admission to the Collège Maza-rin, where he received a classical education. Although he became a lawyer in 1738, his true interest was in mathematics, a subject he learned on his own. After

publishing several papers, particularly in the area of differential equations, he was admitted to the Paris Academy in 1741 and soon became one of the leading mathematicians of Europe. His works included not only major treatises on dynamics and fluid mechanics, but also, after 1750, many sections of the French *Encyclopédie*, the 28 volume work which aimed to set forth the basic principles of all the arts and sciences (Figure 13.9).

FIGURE 13.9
D'Alembert on a French stamp.

D'Alembert then solved the equation for the special case $c^2 = 1$ in the form $y = \Psi(t + x) + \Gamma(t - x)$, where Ψ and Γ are arbitrary functions. In other words, any function y of that form will satisfy the wave equation. As d'Alembert pointed out, "this equation includes an infinity of curves."[21] The elaboration of that statement led to much controversy.

D'Alembert himself first discussed the case where $y = 0$ when $t = 0$ for every x, that is, where the string is in the equilibrium position for $t = 0$. He further required that $y = 0$ for $x = 0$ and $x = l$ for all t, that is, that the string be held fixed at the two ends of the interval $[0,l]$. The first requirement shows that $\Psi(x) + \Gamma(-x) = 0$, while the second gives the results $\Psi(t) + \Gamma(t) = 0$ and $\Psi(t + l) + \Gamma(t - l) = 0$. It follows that $\Gamma(t - x) = -\Psi(t - x)$ or that $y = \Psi(t + x) - \Psi(t - x)$; that $\Psi(-x) = -\Gamma(x) = \Psi(x)$ or that Ψ is an even function; and that $\Psi(t + l) = \Psi(t - l)$ or that Ψ is periodic of period $2l$. Furthermore, because the initial velocity is given by $\partial y/\partial t$ at $t = 0$, that is, by $v = \Psi'(x) - \Psi'(-x)$, and because the derivative of an even function is odd, d'Alembert concluded that "the expression for the initial velocity . . . must be such that when reduced to a series it includes only odd powers of x. Otherwise . . . the problem would be impossible."[22] In a paper written shortly afterwards, d'Alembert generalized the solution to the case where the initial position of the string was given by $y(0,x) = f(x)$ and the initial velocity by $v(0,x) = g(x)$. In this case, he obtained the result that the solution is only possible if $f(x)$ and $g(x)$ are odd functions of period $2l$ and, in order that one could operate with these, that each function is given by a single analytic expression which is twice differenti-able. For d'Alembert, then, a function was exactly what Euler defined in his *Introductio*. Thus even though $f(x)$ and $g(x)$ are given just on $[0,l]$, the function $y = \Psi(u)$ determining them must itself be given as an "equation" defined for all values of u. No other type of function could occur as a solution to a physical problem. As d'Alembert concluded in another paper three years later, "in any other case, the problem cannot be solved, at least by my method, and I am not certain whether it will not surpass the power of known analysis. . . . Under this supposition we find the solution of the problem only for the cases in which all the different figures of the vibrating string can be comprehended by one and the same equation. It seems to me that in all other cases it will be impossible to give y a general form."[23]

13.4.2 Euler and Continuous Functions

Two years after d'Alembert's initial paper, Euler published his own solution to the same problem, getting the same formal result as d'Alembert although with a somewhat different derivation. But Euler differed from d'Alembert in what kinds of initial position functions f could be permitted. First, he announced that f could be any curve defined on the interval $[0,l]$, even one which was not determined by an analytic expression. It could be a curve drawn by hand. Thus it was not necessary for the function to be differentiable at every point. Second, it was only the definition on the initial interval which was important. One could make the curve odd and periodic by simply defining it on $[-l,0]$ by $f(-x) = -f(x)$ and then extending it to the entire real line by using $f(x \pm 2l) = f(x)$. After all, Euler reasoned contrary to d'Alembert, as far as the physical situation was concerned, the initial shape of the string could be arbitrary. Even if there were isolated points where the function was not differentiable, one could still consider the curve a solution to the differential equation, because behavior at isolated points was not, Euler believed, relevant to the general behavior of a function over an interval. Euler had already defined, in the second volume of his *Introductio*, the notions of continuous and discontinuous curves: "A **continuous** curve is one such that its nature can be expressed by a single function of x. If the curve is of such a nature that for its various parts . . . different functions of x are required for its expression, that is, after one part . . . is defined by one function of x, then another function is required to express the [next] part . . . , then we call such a curve **discontinuous**. . . . This is because such a curve cannot be expressed by one constant law, but is formed from several continuous parts."[24]. Thus, in this debate, d'Alembert insisted that only "continuous" curves could be used for the initial conditions, while Euler responded that the various parts of the curve need not be connected by any such law of continuity. Ultimately, Euler changed his definition of "function" to reflect his new view of the situation and somewhat later wrote to d'Alembert that the consideration of such functions "opens to us an entirely new field of analysis."[25] D'Alembert, however, continued to maintain that such "discontinuous" curves lay outside the purview of analysis.

13.4.3 Daniel Bernoulli and Physical Strings

Both d'Alembert and Euler, although they carried on their analyses in terms of general "functions," always had in mind the examples of the sine and cosine. The former was odd and periodic while the latter was even and periodic. In fact, in 1750 d'Alembert derived the solution $y = (A \cos Nt)(B \sin Nx)$ to the wave equation by the technique of separation of variables, that is, by assuming $y = f(t)g(x)$ and then differentiating. Nevertheless, it was the third participant in the debate, Daniel Bernoulli, who explicitly referred to combinations of sines and cosines in his attempt to bring the debate back to the reality of physical strings. Bernoulli, whose positions at the University of Basel encompassed medicine, metaphysics, and natural philosophy, and whose chief work was in hydrodynamics and elasticity, wrote in a paper of 1753, "The calculations of Messrs. d'Alembert and Euler, which certainly contain all that analysis can have at its deepest and most sublime, . . . show at the same time that an abstract analysis which is accepted without any synthetic examination of the question under discussion is liable to surprise rather than enlighten us. It seems to me that we have only to pay attention to the nature of the simple vibrations of the strings to foresee without any calculation all that these two great geometers have found by the most thorny and abstract calculations that the analytical mind can perform."[26]

Bernoulli's more physical solution of the problem was to explore the idea that a vibrating string potentially represents an infinity of tones, each superimposed upon the others, and each being separately represented as a sine curve. It followed, although Bernoulli did not write out the result in this generality, that the movement of a vibrating string can be represented by the function

$$y = \alpha \sin \frac{\pi x}{l} \cos \frac{\pi t}{l} + \beta \sin \frac{2\pi x}{l} \cos \frac{2\pi t}{l} + \gamma \sin \frac{3\pi x}{l} \cos \frac{3\pi t}{l} + \cdots$$

where the sum is infinite. The initial position function, over which Euler and d'Alembert quarreled, is then represented by the infinite sum

$$y(0,x) = \alpha \sin \frac{\pi x}{l} + \beta \sin \frac{2\pi x}{l} + \gamma \sin \frac{3\pi x}{l} + \cdots .$$

Interestingly enough, Euler had written these series in 1750, probably intending only a finite sum, as an example of a possible solution to the equation. Bernoulli believed that the latter series could represent any arbitrary initial position function $f(x)$ with the appropriate choice of constants $\alpha, \beta, \gamma, \ldots$ but could give no mathematical argument for the correctness of his view, only writing somewhat later that his representation provides an infinite number of constants which can be used for adjusting the curve to pass through an infinite number of specified points. His view was challenged by Euler, who not only could not see any way of determining these coefficients but also realized that for a function to be represented by such a trigonometric series it had to be periodic. By this argument, though, Euler showed himself to be caught between his original view of what a function was and the more modern view which he was instrumental in helping evolve. Euler, after all, was willing to allow the arbitrary curve $f(x)$ defined over the interval $[0,l]$ to be extended by periodicity to the entire real line. But this was an example of what may be called geometric periodicity. It took no account of the algebraic expression by which f may be expressed. On the other hand, Euler's argument against Bernoulli was based on the algebraic periodicity of the trigonometric functions themselves on the whole real line. Euler had only an inkling of the modern notion of the domain of a function with the concomitant possibility that functions can be represented by different expressions on various parts of their domain.

The debate over the kinds of functions acceptable as solutions to the wave equation was continued through the next decades by these three mathematicians without anyone being convinced by any of the others. Although other mathematicians also entered the debate, it was not until the early years of the nineteenth century that a resolution of the problem was worked out through a complete analysis of the nature of trigonometric series.

13.5 THE FOUNDATIONS OF CALCULUS

The eighteenth century saw extensive development of the techniques of the calculus. Many texts were written explaining the calculus to those who wanted to learn while multitudes of papers appeared in which new procedures and methods of solving various kinds of physical problems were demonstrated. But in the minds of some there was a nagging doubt as to the foundations of the subject. Most mathematicians had read Euclid's *Elements* and regarded it as a model of how mathematics should be done. Yet there was no logical basis for the central procedures of calculus. In general, the practitioners themselves did not worry much about this. Newton, Leibniz, Euler, and the others had a strong intuitive feel for the subject and knew when what they were doing was correct. Even in the light of modern standards, these great mathematicians rarely made errors. Nevertheless, the explanations they themselves gave of the foundations for their procedures left something to be desired.

13.5.1 George Berkeley's *The Analyst*

The most important criticism of both infinitesimals and fluxions was made by the Irish philosopher Bishop George Berkeley (1685–1753) in a 1734 tract entitled *The Analyst* (Figure 13.10). The work was addressed "to an infidel mathematician," generally supposed to be the astronomer Edmund Halley (1656–1741) who financed the publication of Newton's *Principia* and helped see it through the press. Berkeley presumably considered him an infidel because he had persuaded a mutual friend that the doctrines of Christianity were inconceivable. Berkeley's aim in *The Analyst* was not to deny the utility of the calculus or the validity of its many new results, but to show that mathematicians had no valid arguments for the procedures they invoked.

Thus "the Method of Fluxions is the general key by help whereof the modern mathematicians unlock the secrets of Geometry, and consequently of Nature." The fluxions themselves, however, "are said to be nearly as the increments of the flowing quantities [moments], generated in the least equal particles of time; and to be accurately in the first proportion of the nascent, or in the last of the evanescent increments. . . . By moments we are not to understand finite particles . . . [but] only the nascent principles of finite quantities."[27] What are these "nascent principles"? Berkeley notes that even though "the minutest errors are not to be neglected in mathematics,"—a quotation from Newton himself—the actual finding of these fluxions determined by the nascent principles involves precisely that kind of neglect.

Berkeley demonstrates his point by analyzing the calculation of the fluxion of x^n:

In the same time that x by flowing becomes $x + o$, the power x^n becomes $(x + o)^n$, i.e., by the method of infinite series

$$x^n + nox^{n-1} + \frac{n^2 - n}{2}o^2 x^{n-2} + \cdots,$$

and the increments o and $nox^{n-1} + \frac{n^2 - n}{2}o^2 x^{n-2} + \cdots$ are one to another as 1 to nx^{n-1} $+ \frac{n^2 - n}{2}ox^{n-2} + \cdots$. Let now the increments vanish, and their last proportion will be 1 to nx^{n-1}. But it should seem that this reasoning is not fair or conclusive. For when it is said, let the increments vanish, i.e., let the increments be nothing, or let there be no increments, the former supposition that the increments were something, or that there were increments, is destroyed, and yet a consequence of that supposition, i.e., an expression got by virtue thereof, is retained.[28]

FIGURE 13.10
Bishop Berkeley on an Irish stamp.

Berkeley thus questions how one can take a nonzero increment, do calculations with it, and then in the end set it equal to zero. He notes further that the continental mathematicians do things differently. Rather than considering "flowing quantities and their fluxions, they consider the variable finite quantities as increasing or diminishing by the continual addition or subtraction of infinitely small quantities." And these lead to exactly the same kinds of problems. In particular, Berkeley claims he cannot conceive of infinitely small quantities. "But to conceive a part of such infinitely small quantity that shall be still infinitely less than it, and consequently though multiplied infinitely shall never equal the minutest finite quantity, is, I suspect, an infinite difficulty to any man whatsoever."[29] Thus second-order differentials, and similarly, fluxions of fluxions form an "obscure mystery. The incipient celerity of an incipient celerity, the nascent augment of a nascent augment, i.e., of a thing which hath no magnitude—take it in what light you please, the clear conception of it will, if I mistake not, be found impossible."[30] Since Halley could not comprehend the arguments of theology, Berkeley counterattacked by noting that "he who can digest a second or third fluxion, a second or third difference, need not, methinks, be squeamish about any point in divinity."[31]

13.5.2 Maclaurin's Response to Berkeley

Berkeley's criticisms of the foundations of calculus were valid. The question of when a value was zero and when it was not zero extended back even to the work of Fermat, and neither Newton nor Leibniz was ever quite able to resolve it. Nevertheless, several English mathematicians sprang to Newton's defense under Berkeley's attack. The most important response was that of Maclaurin in his *Treatise of Fluxions*. As he noted in his preface, Maclaurin wanted to "deduce those Elements [of the theory of Fluxions] after the Manner of the Ancients from a few unexceptionable Principles by Demonstrations of the strictest Form."[32] He noted that he would not use any indivisible or infinitely small part of time or space as part of the demonstration, "the supposition of an infinitely little magnitude being too bold a Postulate for such a Science as Geometry."[33] Maclaurin therefore had to consider finite lengths and times as his basic elements, for "no quantities are more clearly conceived than the limited parts of space and time."[34] These spaces and times then determine (average) velocity. But because it is instantaneous velocity which is the basic concept necessary to the theory of fluxions, Maclaurin attempted a definition of this as well: "The velocity of a variable motion at any given term of time is not to be measured by the space that is actually described after that term in a given time, but by the space that would have been described if the motion had continued uniformly from that term."[35] This definition is reminiscent of that given by Heytesbury in the fourteenth century. From a modern point of view, however, Maclaurin, by giving such a definition, missed the fundamental idea of instantaneous velocity as a limit of average velocities as the time interval approaches zero. In any case, given the definition of variable velocity, Maclaurin presented axioms for the use of this definition and then proceeded to prove numerous theorems in the "manner of the ancients," using each time a double *reductio ad absurdum*.

In particular, because one of Maclaurin's aims was to show that "infinitesimals" in the arguments of Newton can always be replaced by finite quantities, he showed that even the differential triangle can be derived rigorously.

Proposition. *Let ET be the tangent of the curve FE at E and EI being parallel to the base AD, let IT be parallel to the ordinate DE. Then the fluxions of the base, ordinate, and curve shall be measured by the lines EI, IT, and ET respectively (Figure 13.11).*[36]

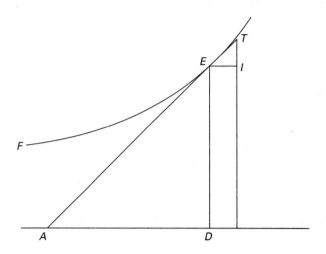

FIGURE 13.11
Maclaurin's differential
triangle.

Maclaurin proved this result by first noting that if the curve is concave up, then the increase of the ordinate in the time of a given increase of the base is greater than that which would have been generated had the motion of the ordinate been uniform. To show that this latter increase, proportional to the fluxion of the ordinate *DE*, is precisely equal to *IT*, he made the initial assumption that it was greater than *IT* and showed by use of his axioms on velocity that a contradiction results. A similar contradiction followed from the contrary assumption. The entire proof was then repeated in the case where the curve was concave down. From a modern point of view, the problem with Maclaurin's proof lies in his definition of a tangent, a concept generally accepted by his contemporaries as "self-evident." True to his belief in the methods of the ancients, however, Maclaurin presented the ancient definition, that a tangent to a curve is a straight line which touches the curve in such a way that no other straight line can be inserted between the curve and the line.

But despite his definition taking instantaneous velocity as the basic concept and his geometrical definition of a tangent, Maclaurin was well aware of Newton's use of the notion of "ultimate proportion of evanescent quantities" or "limits." Thus, he wrote about the ratio which is the limit of the various proportions that finite simultaneous increments of two variable quantities bear to one another as the two increments decrease until they vanish. He noted that to discover this limit one must first determine the ratio of increments in general and then reduce to the simplest terms so that a part of the result might be independent of the value of the increments themselves. The desired limit then readily appears if one supposes the increments to "decrease until they vanish."[37] For example, to find the ratio of the fluxion of x^2 to the fluxion of ax, Maclaurin calculated the ratio of the increments (as x increases to $x + o$) to be $2xo + o^2 : ao$ or $2x + o : a$. "This ratio of $2x + o$ to a continually decreases while o decreases and is always greater than the ratio of $2x$ to a while o is any real increment, but it is manifest that it continually approaches to the ratio of $2x$ to a as its limit."[38]

Maclaurin vehemently denied Berkeley's contention that the method of first supposing a finite increment and then letting that increment vanish is contradictory. In fact, he noted, this method allows one to determine the ratio of the increments when the increments are finite and determine how the ratio varies with the increment. One can then easily determine what limit the ratio approaches as the increments are diminished. As a final response to Berkeley, he even defined the tangent as a limit: "The tangent . . . is the . . . line that limits the position of all the secants that can pass through the point of contact, though strictly speaking it be no secant, [just as] a ratio may limit the variable ratios of the increments, though it cannot be said to be the ratio of any real increments."[39]

The problem with Maclaurin's treatment of the calculus, as noted by many of his contemporaries, was not that he failed rigorously to derive the rules, but that he really did this in the "manner of the ancients." In particular, he used the method of exhaustion and its accompanying *reductio ad absurdum* argument. The use of such a method imposed a heavy toll on the reader. For example, the first 590 pages of this 754 page work do not contain any notation of fluxions. Every new idea is derived geometrically with great verbosity. And in the eighteenth century, few were willing to read through these detailed arguments. Maclaurin himself realized that the advantages of the new calculus were that it enabled old problems to be solved in expeditious fashion and new discoveries to be made with ease. "But when the principles and strict method of the ancients, which had hitherto preserved the evidence of this science entire, were so far abandoned, it was difficult for the Geometricians to determine where they should stop."[40] Nevertheless, although Maclaurin's great efforts answered Berkeley's objections, they were not appreciated by most eighteenth century mathematicians, people who saw themselves as breaking new ground rather than extending the methods of the ancients.

13.5.3 Euler and d'Alembert

Even on the European continent, however, some justification of the procedures was necessary. In his *Differential Calculus*, Euler developed the idea that the ratios involved in the calculation of derivatives were in fact simply versions of the ratio $0 : 0$. For Euler, infinitely small quantities were quantities actually equal to 0, because the latter is the only quantity smaller than any given quantity. "Hence there are not so many mysteries hidden in this concept as there are usually believed to be."[41] But although two zeros are equal in such a way that their difference is always zero, Euler insists that the ratio of two zeros, which depends on the origin of the quantities which are becoming zero, must be calculated in each specific case. As an example, he notes that $0 : 0 = 2 : 1$ is a correct statement because the first quantity on each side of the equal sign is double the second quantity. In fact, then, the ratio $0 : 0$ may be equal to any finite ratio at all. Therefore, "the calculus of the infinitely small is . . . nothing but the investigation of the geometric ratio of different infinitely small quantities."[42]

Interestingly enough, d'Alembert, in the article "Différentiel," which he wrote for the *Encyclopédie* in 1754, combined the ideas of both Euler and Maclaurin. He agreed with Euler that there was no absurdity in considering the ratio $0 : 0$ because it may in fact be equal to any quantity at all. But the central idea of the differential calculus is that dy/dx is the limit of a certain ratio as the quantities involved approach 0. The "most precise and neatest possible definition of the differential calculus" is that it "consists in algebraically determining the limit of a ratio, for which we already have the expression in terms of lines, and in equating those two expressions."[43] As an example of what he meant, d'Alembert calculated the slope of the tangent line to the parabola $y^2 = ax$ by first determining the slope of a secant through the two points (x,y) and $(x + u, y + z)$. This slope, the ratio $z : u$, is easily seen to be equal to $a : 2y + z$. "This ratio is always smaller than $a : 2y$; but the smaller z is, the greater the ratio will be and, since one may choose z as small as one pleases, the ratio $a : 2y + z$ can be brought as close to the ratio $a : 2y$ as we like. Consequently $a : 2y$ is the limit of the ratio $a : 2y + z$."[44] It follows that $dy/dx = a/2y$. D'Alembert's wording is virtually identical to that of Maclaurin. He went somewhat further, however, by giving an explicit definition of the term "limit" in his *Encyclopédie* article on that notion: "One magnitude is said to be the **limit** of another magnitude when the second may approach the first within any given magnitude, however small, though the first magnitude may never exceed the magnitude it approaches."[45] His idea, although apparently geometric rather than arithmetic, was not followed up by his eighteenth century successors. Through the remainder of the century, most of the works on calculus attempted to explain the basis of the subject in terms of infinitesimals, fluxions, or the ratios of zeros.

13.5.4 Lagrange and Power Series

It was Lagrange who near the end of the eighteenth century attempted to give a precise definition of the derivative by eliminating all reference to infinitesimals, fluxions, zeros, and even limits, all of which he believed lacked proper definitions. He sketched his new ideas about derivatives in a paper of 1772 and then developed them in full in his text of 1797, the full title of which expressed what he intended to do: *The Theory of Analytic Functions, containing the principles of the differential calculus, released from every consideration of the infinitely small or the evanescent, of limits or of fluxions, and reduced to the algebraic analysis of finite quantities.* How could Lagrange accomplish the reduction of calculus purely to algebraic analysis? He did so by formalizing the idea which most of his predecessors used without question, the idea that any function can be represented as a power series. For Lagrange, if $y = f(x)$ is any function, then $f(x + i)$, where i is an indeterminate, can "by the theory of series" be expanded into a series in i:

$$f(x + i) = f(x) + pi + qi^2 + ri^3 + \cdots$$

where $p, q, r \ldots$ are new functions of x independent of i. Lagrange then showed that the ratio dy/dx can be identified with the coefficient $p(x)$ of the first power of i in this expansion. He therefore had a new definition of this basic concept of the calculus. Since the function p is "derived" from the original function f, Lagrange named it a *fonction dérivée* (from which comes the English word *derivative*) and used the notation $f'(x)$. Similarly, the derivative of f' is written f'', that of f'' is written f''', and so on. Lagrange easily showed that $q = (1/2)f''$, $r = (1/6)f'''$, ...

To consider the obvious question of why Lagrange believed that every function could be expanded into a power series, one must first consider Lagrange's definition of a function, appearing at the very beginning of his text: "One names a **function** of one or several quantities any mathematical expression in which the quantities enter in any manner whatever, connected or not with other quantities which one regards as having given and constant values, whereas the quantities of the function may take any possible values."[46] In other words, Lagrange has returned in essence to Euler's first definition of function, leaving somewhat vague the notion of "mathematical expression" and "in any manner whatever." Lagrange's experience with functions told him that algebraic expressions can always be expanded in power series except perhaps at particular values of x: "This supposition is in

---

Lagrange was born in Turin into a family of French descent. His father wanted him to study law, but he was attracted to mathematics in school and at the age of 19 became a professor of mathematics at the Royal Artillery School in Turin. At about the same time, having read an article of Euler's on the calculus of variations, he wrote to the latter explaining a better method he had recently discovered. Euler praised Lagrange greatly and arranged to present his paper to the Berlin Academy. Frederick II was also impressed with Lagrange's work, and when Euler left Berlin to return to St. Petersburg, he appointed him to fill Euler's post at the Academy of Sciences. After Frederick's death, Lagrange accepted the invitation of Louis XVI to come to Paris, where he spent the rest of his life. In 1788 he published his most important work, *Analytical Mechanics*, a work which extended the mechanics of Newton, the Bernoullis, and Euler, and emphasized the fact that problems in mechanics can generally be solved by reducing them to the theory of ordinary and partial differential equations. In 1792 he married the 17 year old Renée Le Monnier, who brought renewed joy to his life. Because of his generally introverted personality, he was able to survive the excesses of the French Revolution, in fact being treated with honor, but the death of several of his colleagues disturbed him greatly. After the Terror, he took an active role in improving university education in France. He was ultimately even honored by Napoleon for his life's work (Figure 13.12).

FIGURE 13.12
Lagrange on a French stamp.

fact verified by the expansion of different known functions; but no one that I know has sought to demonstrate it *a priori*."[47]

Lagrange's argument for the expansion of a function f begins with the assertion that $f(x + i) = f(x) + iP$, where $P(x,i)$ is defined by

$$P(x,i) = \frac{f(x + i) - f(x)}{i}.$$

Lagrange assumes further that one can separate out from P that part p which does not vanish at $i = 0$. Namely, $p(x)$ is defined as $P(x,0)$ and then

$$Q(x,i) = \frac{P(x,i) - p(x)}{i}$$

or $P = p + iQ$. It follows that $f(x + i) = f(x) + ip + i^2Q$. Repeating the argument for Q, he writes $Q = q + iR$ and substitutes again. As an example of the procedure, Lagrange takes $f(x)$ to be $1/x$. Since $f(x + i) = 1/(x + i)$, he calculates

$$P = \frac{1}{i}\left(\frac{1}{x + i} - \frac{1}{x}\right) = -\frac{1}{x(x + i)} \quad p = -\frac{1}{x^2}$$

$$Q = \frac{1}{i}\left(-\frac{1}{x(x + i)} + \frac{1}{x^2}\right) = \frac{1}{x^2(x + i)} \quad q = \frac{1}{x^3}$$

$$\cdots$$

Thus the series becomes

$$\frac{1}{x + i} = \frac{1}{x} - \frac{i}{x^2} + \frac{i^2}{x^3} - \frac{i^3}{x^4} + \cdots.$$

At each stage of the expansion, the terms iP, i^2Q, ... can be considered as the error terms resulting from representing $f(x + i)$ by terms up to that point. Furthermore, Lagrange claims, the value of i can always be taken so small that any given term of this series is greater than the sum of the remaining terms, that is, that the remainders are always sufficiently small so that in fact the function is represented by the series. In fact, this result is what Lagrange uses often later on. He also uses a somewhat different form of his expansion result containing what is now called the Lagrange form of the remainder in the Taylor series. Namely, he shows that for any given positive integer n, one can write

$$f(x + i) = f(x) + if'(x) + \frac{i^2 f''(x)}{2} + \cdots + \frac{i^n f^{(n)}(x)}{n!} + \frac{i^{n+1} f^{(n+1)}(x + j)}{(n + 1)!}$$

for some value j between 0 and i. Although this new form is, perhaps, no more convincing to the modern reader than his earlier one, Lagrange himself was satisfied that his principle of a power series representation for every function was correct. After all, he claimed, it enabled him to derive anew all of the basic results of the calculus without any consideration of infinitesimals, fluxions, or limits.

One of these basic results is part of what is known today as the fundamental theorem of calculus, that if $F(x)$ represents the area under the curve $y = f(x)$ from a fixed ordinate, then $F'(x) = f(x)$. (It should be noted that Lagrange had no definition of area. He simply assumed that the area under a curve $y = f(x)$ is a well-determined quantity.) Lagrange began his proof, reminiscent of Maclaurin's proof of the same result for power functions, by noting that $F(x + i) - F(x)$ represents that portion of the area between the abscissas x and $x + i$. Keeping to Euler's dictum that in a text on analysis one should not include diagrams, Lagrange nevertheless wrote that even without a figure one can easily convince oneself that if $f(x)$ is monotonically increasing, then $if(x) < F(x + i) - F(x) < if(x + i)$, with the inequalities reversed if $f(x)$ is monotonically decreasing (Figure 13.13).

Now expanding both $f(x + i)$ and $F(x + i)$, Lagrange determined that

$$f(x + i) = f(x) + if'(x + j)$$

and

$$F(x + i) = F(x) + iF'(x) + \frac{i^2}{2}F''(x + j)$$

where $0 < j < i$ (although the value of j may not be the same in both expansions). It follows that $if(x) < iF'(x) + \frac{i^2}{2}F''(x + j) < if(x) + i^2 f(x + j)$ and therefore that

$$\left| i[F'(x) - f(x)] + \frac{i^2}{2}F''(x + j) \right| < i^2 f'(x + j),$$

where the absolute value sign is necessary to take care of both the increasing and decreasing cases. Lagrange concluded that because the inequality holds no matter how small i is taken to be, it must be true that $F'(x) = f(x)$. He even calculated that if the conclusion were not true, the inequality would fail for

$$i < \frac{F'(x) - f(x)}{f'(x + j) - \frac{1}{2}F''(x + j)}.$$

To finish his proof Lagrange removed the condition that $f(x)$ be monotonic on the original interval $[x, x + i]$. For if it is not, there is a maximum or minimum of f on that interval

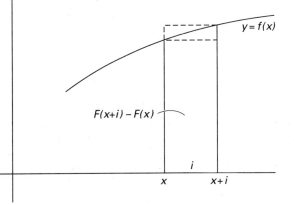

FIGURE 13.13
Lagrange and the
fundamental theorem of
calculus: $if(x) < F(x + i)$
$- F(x) < if(x + i)$.

and i can be chosen small enough so that the extreme value falls outside of the new interval $[x, x + i]$.

It is curious, of course, that despite Lagrange's claim that this work will only use "algebraic analysis of finite quantities," he has in fact in this very important proof, as well as in his remainder arguments, used the notion of a limit. In other sections of the work, where he finds tangent lines, curvature, and maxima and minima, among other geometric quantities, Lagrange uses limits in the same way, along with his central concept of the expansion of the functions involved in a power series. And, in fact, these very arguments were used in the nineteenth century treatments of calculus which used limits explicitly.

Most of the early objections to Lagrange's new foundation for the calculus were aimed at his new notations and the length of some of his calculations rather than at his assertion that any function can be expanded in a power series. Mathematicians in general continued to use the earlier differential methods, especially since Lagrange's book assured them that because there was a correct basis to the entire subject, any method which worked would be legitimate. Even Lagrange in some of his other work continued to employ the notation of differentials rather than that of derivatives. It was not until the second decade of the nineteenth century that various mathematicians pointed out that there existed differentiable functions which do not have a power series representation and thus that Lagrange's basic concept was not tenable. The story of the new attempts to supply a foundation to the ideas of calculus will therefore be continued in Chapter 16.

Exercises

Problems in Differential Equations

Note: The problems in this section generally require a basic knowledge of the subject of differential equations.

1. Derive the equation of the catenary in closed form from Johann Bernoulli's differential equation

$$dy = \frac{a\,dx}{\sqrt{x^2 - a^2}}.$$

2. Determine a procedure for finding the differential equation of a family of orthogonal trajectories to a given family $f(x,y,\alpha) = 0$, assuming that the given function f is algebraic. (Use the fact that orthogonal lines have negative reciprocal slopes.) Use your procedure to find the family orthogonal to the family of hyperbolas $x^2 - y^2 = a^2$.

3. Write the differential equation for the family of syn-

chrones, the family orthogonal to the family of brachis-tochrones. Solve the differential equation.

4. Suppose $A = a_1 + a_2 + \cdots + a_n$ and $B = b_1 + b_2 + \cdots + b_n$. Show that $\Sigma(b_i - a_i) = B - A$, or, that the sum of the differences of the parts is equal to the difference of the sums of the parts.

5. Given the family of curves

$$y = \int_0^x p(x,\alpha)\, dx,$$

show that the orthogonal family is determined by the solution to the differential equation $(1 + p^2)\, dx + pq\, d\alpha = 0$ where q is defined by $d_\alpha y = q\, d\alpha$.

6. Translate Leibniz's solution of $m\, dx + ny\, dx + dy = 0$ into modern terms by noting that $dp/p = n\, dx$ is equivalent to $\ln p = \int n\, dx$ or to $p = e^{\int n\, dx}$. Solve $-3x\, dx + (1/x)y\, dx + dy = 0$ by using Leibniz's procedure.

7. Show that in the homogeneous equation $dy = f(y/x)\, dx$ the variables can be separated by using the substitution $y = vx$. Apply this technique to solve the equation $x^2\, dy = (y^2 + 2xy)\, dx$.

8. Given the differential equation $a\, d^3y - y\, dx^3 = 0$ with one solution $y = e^{x/a}$, show that if one assumes that the product $e^{-(x/a)}(a\, d^3y - y\, dx^3)$ is the differential of $e^{-(x/a)}(A\, d^2y + B\, dy\, dx + Cy\, dx^2)$, then a new solution of the original equation must also satisfy $a^2\, d^2y + a\, dy\, dx + y\, dx^2 = 0$. (*Hint:* Calculate the differential and equate the two expressions. It may be easier if you rewrite the equations in modern notation using derivatives.)

9. Given that $y = e^x$ is a solution of $y''' - 6y'' + 11y' - 6y = 0$, show by a method analogous to that of problem 8 that any other solution must satisfy $y'' - 5y' + 6y = 0$.

10. Solve the previous problem using Euler's procedure of factoring the characteristic polynomial.

11. Show that if $y = ue^{\alpha x}$ is assumed to be a solution of $a^2\, d^2y + a\, dy\, dx + y\, dx^2 = 0$, then if $\alpha = -1/2a$, u is a solution to $a^2\, d^2u + (3/4)u\, dx^2 = 0$.

12. Solve $a^2\, d^2u + (3/4)u\, dx^2 = 0$. First multiply by du and integrate once to get $4a^2\, du^2 = (K^2 - 3u^2)dx^2$ or

$$dx = \frac{2a}{\sqrt{K^2 - 3u^2}}\, du.$$

Integrate a second time to get

$$x = \frac{2a}{\sqrt{3}} \arcsin \frac{\sqrt{3}u}{K} - f.$$

Rewrite this equation for u in terms of x as

$$u = K \sin\left(\frac{(x + f)\sqrt{3}}{2a}\right).$$

13. Since any complex number z can be written as $z = re^{i\theta} = r(\cos\theta + i\sin\theta)$ with $r > 0$, one can define log $z = \log re^{i\theta}$ to be log $r + i\theta$, where log r is the real logarithm of the positive number r. Show therefore that a complex number has infinitely many logarithms. Under what circumstances is any logarithm of a complex number real?

14. Solve the differential equation $(2xy^3 + 6x^2y^2 + 8x)\, dx + (3x^2y^2 + 4x^3y + 3)\, dy = 0$ using Clairaut's method.

Problems from Calculus Textbooks

15. Find the isosceles triangle of smallest area which circumscribes a circle of radius 1. (Simpson)

16. Find the cone of least surface area with given volume V. (Simpson)

17. Show that $(b^3 - x^3)(x^2z - z^3)(xy - y^2)$ has a maximum when $x = \frac{1}{2}b^3\sqrt{5}$, $y = \frac{1}{4}b^3\sqrt{5}$, and $z = \frac{b^3\sqrt{5}}{2\sqrt{3}}$. (Simpson)

18. Calculate the first four terms of the power series for $y = \cos z$ using Maclaurin's technique without explicitly using the derivatives of the cosine or sine.

19. Show that if $E > F$, then $nF^{n-1}(E - F) < E^n - F^n < nE^{n-1}(E - F)$. (Maclaurin used this result in his proof of the special case of the fundamental theorem of calculus.)

20. Find the point of maximal curvature on the curve defined by the differential equation $\frac{a\, dy}{y} = dx$. (Agnesi)

21. A certain man borrowed 400,000 florins at the "usurous" rate of 5% annual interest. Suppose he pays back 25,000 florins each year. How long will it take to pay off the loan? (Euler)

22. Show that k in the series given in the text for a^z and $\log(1 + x)$ is given by $k = \ln a$. (Euler)

23. Derive the power series for $\log(1 + x)$ from the equation

$$\log(1 + x) = \frac{j}{k}(1 + x)^{1/j} - \frac{j}{k}$$

by using the binomial theorem and assuming that j is infinitely large. (Euler)

24. From the formula

$$\log(1 + x) = \frac{j}{k}(1 + x)^{1/j} - \frac{j}{k}$$

with j infinitely large, derive the expression $\ln p = \frac{p^\omega - 1}{p}$ where ω is infinitely small (and $k = 1$). Then show that $d(\ln p) = \frac{dp}{p}$. (Euler)

25. If $y = \arctan x$, show that $\sin y = x/\sqrt{1 + x^2}$ and $\cos y = 1/\sqrt{1 + x^2}$. Then, if $p = x/\sqrt{1 + x^2}$, show that $\sqrt{1 - p^2} = 1/\sqrt{1 + x^2}$. Since $y = \arcsin p$, it follows that $dy = dp/\sqrt{1 - p^2}$ and $dp = \frac{dx}{(1 + x^2)^{3/2}}$. Conclude that

$$dy = \frac{dx}{1 + x^2}. \quad \text{(Euler)}$$

26. Determine all the relative extrema for $V = x^3 + y^2 - 3xy + (3/2)x$, and for each one determine whether it is a maximum or minimum. Compare your answer with that of Euler.

27. Calculate dy for $y = a^x$ by noting that $dy = a^{x+dx} - a^x = a^x(a^{dx} - 1)$ and then expanding $a^{dx} - 1$ into the power series $\ln a \, dx + \frac{(\ln a)^2 dx^2}{2} + \cdots$. (Euler)

28. Calculate dy for $y = \tan x$ by using the addition formula

$$\tan(x + dx) = \frac{\tan x + \tan dx}{1 - \tan x \tan dx}. \quad \text{(Euler)}$$

29. Given that the roots of $z^n - a^n$ are $a(\cos \frac{2k\pi}{n} + i \sin \frac{2k\pi}{n})$, $k = 0, 1, \ldots, n - 1$ and that the complex roots always occur in conjugate pairs, show that $z^2 - 2az \cos \frac{2k\pi}{n} + a^2$ is a real quadratic factor of $z^n - a^n$ for $1 \le k \le \frac{n}{2}$. (Euler)

30. Recall that $a^{j\omega} = (1 + k\omega)^j$ for ω infinitely small and j infinitely large. Set $j\omega = x$ (a finite quantity), $a = e$, and $k = 1$. Then $e^x = (1 + \frac{x}{j})^j$ and similarly $e^{-x} = (1 - \frac{x}{j})^j$. Show from problem 29 that the quadratic factors of $e^x - e^{-x}$ are all of the form

$$\frac{4x^2}{j^2} - \frac{4k^2\pi^2}{j^2} - \frac{4k^2\pi^2 x^2}{j^4}$$

for $k = 1, 2, \ldots$. (Hint: Because j is infinitely large, use only the first two terms in the power series expansion for $\cos \frac{2k\pi}{j}$.) (Euler)

31. Show from problem 30 that, since the third term in the trinomial factor is infinitely small with respect to the other two terms, $\frac{e^x - e^{-x}}{2}$ is divisible by $1 + \frac{x^2}{k^2\pi^2}$ for $k = 1, 2, \ldots$. Conclude that

$$\frac{e^x - e^{-x}}{2} = x\left(1 + \frac{x^2}{\pi^2}\right)\left(1 + \frac{x^2}{4\pi^2}\right)\left(1 + \frac{x^2}{9\pi^2}\right)\cdots.$$

(Euler)

32. Replace x by ix in problem 31 to conclude that

$$\sin x = \frac{e^{ix} - e^{-ix}}{2i}$$

$$= x\left(1 - \frac{x^2}{\pi^2}\right)\left(1 - \frac{x^2}{4\pi^2}\right)\left(1 - \frac{x^2}{9\pi^2}\right)\cdots.$$

Note that this product representation shows that $\sin x$ has zeroes at $0, \pm\pi, \pm 2\pi, \pm 3\pi, \ldots$. (Euler)

33. Since

$$\sin x = x - \frac{x^3}{3!} + \frac{x^5}{5!} + \cdots = x\left(1 - \frac{x^2}{6} + \frac{x^4}{120} - \cdots\right),$$

conclude from the product representation in problem 32 that

$$\sum_{k=1}^{\infty} \frac{1}{k^2\pi^2} = \frac{1}{6} \quad \text{or that} \quad \sum_{k=1}^{\infty} \frac{1}{k^2} = \frac{\pi^2}{6}.$$

This result, sought by both Jakob and Johann Bernoulli, was first demonstrated by Euler in 1735.

Double Integrals

34. Suppose that x and y are given in terms of t and u by the functions

$$x = \frac{t}{\sqrt{1 + u^2}}, \qquad y = \frac{tu}{\sqrt{1 + u^2}}.$$

Show that the change of variable formula in this case is given by

$$dx \, dy = \frac{t \, dt \, du}{1 + u^2}.$$

35. Use Clairaut's technique to calculate the volume of the solid bounded by the cylinders $ax = y^2$, $by = z^2$ and the coordinate planes. First determine the volume element $dx \int z \, dy$ by converting the integrand to a function of x and integrating. Then integrate the volume element with appropriate limits. Compare this method to the standard modern method.

The Wave Equation

36. Suppose that the solution to the wave equation $\frac{\partial^2 y}{\partial t^2} = \frac{\partial^2 y}{\partial x^2}$ is given by $y = \Psi(t + x) - \Psi(t - x)$. Show that the initial conditions $y(0,x) = f(x)$, $y'(0,x) = g(x)$ and the condition $y(t,l) = 0$ for all t lead to the requirements that $f(x)$ and $g(x)$ are odd functions of period $2l$. (D'Alembert)

37. Suppose that $y = F(t)G(x) = \Psi(t + x) - \Psi(t - x)$ is a solution to the wave equation $\frac{\partial^2 y}{\partial t^2} = \frac{\partial^2 y}{\partial x^2}$. Show by differentiating twice that $\frac{F''}{F} = \frac{G''}{G} = C$, where C is some constant, and therefore that $F = ce^{t\sqrt{C}} + de^{-t\sqrt{C}}$ and $G = c'e^{x\sqrt{C}} + d'e^{-x\sqrt{C}}$. Apply the condition $y(0,t) = y(l,t) = 0$ to show that C must be negative and hence

derive the solution $F(t) = A \cos Nt$, $G(x) = B \sin Nx$ for appropriate choice of A, B, and N. (D'Alembert)

Foundations of the Calculus

38. Translate D'Alembert's definition of a limit into algebraic language and compare it with the modern definition of a limit.

39. Use Lagrange's technique to calculate the various quantities $P, p, Q, q, \ldots$ for the power series representation of the function $f(x) = \sqrt{x}$.

40. Show why Lagrange's power series representation fails for the case $f(x) = e^{-(1/x^2)}$.

41. Given that $f(x + i) = f(x) + pi + qi^2 + ri^3 + \cdots$, show that $p = f'(x), q = f''(x)/2!, r = f'''(x)/3!, \ldots$.

FOR DISCUSSION . . .

42. Do eighteenth century mathematicians prove or use the fundamental theorem of calculus in the sense it is used today? What concepts must be defined before one can even consider this theorem? How are these concepts dealt with by eighteenth century mathematicians? Did these mathematicians consider the fundamental theorem as "fundamental"?

43. Develop a lesson to enhance students' understanding of the fundamental theorem of calculus by using the work of Maclaurin, Agnesi, and Lagrange.

44. Develop a lesson for teaching the change of variable theorem in multiple integration using the technique of Euler.

45. Compare Euler's trilogy of precalculus and calculus texts to a modern series. What items are common? What does Euler have that is missing in today's texts and conversely? Could one use Euler's texts today?

46. Trace the development of the concept of the limit from Newton through Maclaurin to D'Alembert. How do their formulations agree? How do they compare with the modern formulation of this concept?

47. Develop several lessons teaching the basic methods for solving various classes of differential equations using the formulations of Leibniz, Clairaut, and Euler.

References and Notes

Although there is no single work treating in detail all of the topics of this chapter, several books treat certain aspects of this material. These include S. B. Engelsman, *Families of Curves and the Origins of Partial Differentiation* (Amsterdam: Elsevier, 1984), Umberto Bottazzini, *The Higher Calculus: A History of Real and Complex Analysis from Euler to Weierstrass* (New York: Springer-Verlag, 1986), and Ivor Grattan-Guinness, *The Development of the Foundations of Mathematical Analysis from Euler to Riemann* (Cambridge: MIT Press, 1970). Two good survey articles dealing with this material are H. J. M. Bos, "Calculus in the Eighteenth Century—The Role of Applications," *Bulletin of the Institute of Mathematics and Its Applications* 13 (1977), 221–227 and Craig Fraser, "The Calculus as Algebraic Analysis: Some Observations on Mathematical Analysis in the 18th Century," *Archive for History of Exact Sciences* 39 (1989), 317–336.

1. David Eugene Smith, *A Source Book in Mathematics* (New York: Dover, 1959), vol. 2, p. 646.

2. Ibid., p. 645. The original problem as well as Johann Bernoulli's solution are presented here.

3. Ibid., p. 647.

4. Ibid., p. 654.

5. Engelsman, *Families of Curves*, p. 106. This work is the best presentation of the history of partial differentiation and the source of much of the material of this section.

6. C. Truesdell, "The Rational Mechanics of Flexible or Elastic Bodies: 1638–1788," in Leonhard Euler, *Opera Omnia* (Leipzig, Berlin, and Zurich: Societas Scientarum Naturalium Helveticae, 1911–), (2) 11, part 2, p. 166. This article contains many details on aspects of Euler's work. It is an introduction to a section of Euler's collected works, now totaling over 70 volumes, which has been appearing in four series for over 80 years and is still not complete. All references to papers of Euler will include a reference to their location in this set. A more detailed treatment of the invention of the calculus of the sine and cosine functions is found in Victor J. Katz, "The Calculus of the Trigonometric Functions," *Historia Mathematica* 14 (1987), 311–324. Various articles on the work of Euler are found in a special journal issue on the bicentenary of his death: *Mathematics Magazine* 56 (5) (1983). Also, see E. A. Fellmann, ed., *Leonhard Euler 1707–1783: Beiträge zu Leben und Werk* (Boston: Birkhäuser, 1983).

7. G. Eneström, "Der Briefwechsel zwischen Leonhard Euler und Johann I Bernoulli," *Bibliotheca Mathematica* (3) 6 (1905), 16–87, p. 31.

8. Ibid., p. 46.

9. John Fauvel and Jeremy Gray, *The History of Mathematics: A Reader* (London: Macmillan, 1987), p. 452. This quotation is from a translation of part of the original paper of Euler, "De la Controverse entre Mrs Leibniz et Bernoulli sur les Logarithmes des Nombres Negatifs et Imaginaires," *Mem. Acad. Sci. Berlin* (1749) = *Opera Omnia* (1) vol. 17, 195–232.

10. Frances Marguerite Clarke, *Thomas Simpson and his Times* (New York: Columbia Univ. Press, 1929), p. 16. This work is the only biography of Thomas Simpson and provides many details of his life and work.

11. Thomas Simpson, *A New Treatise of Fluxions* (London: Gardner, 1737), p. 179.

12. Colin Maclaurin, *A Treatise of Fluxions* (Edinburgh: Ruddimans, 1742), p. 694.

13. Ibid., p. 753. These sections were pointed out by Judith Grabiner in a lecture at a meeting of the Canadian Society for History and Philosophy of Mathematics.

14. Edna Kramer, "Maria Agnesi," *Dictionary of Scientific Biography*, vol. 1, p. 76. A more recent work on Agnesi is C. Truesdell, "Maria Gaetana Agnesi," *Archive for History of Exact Sciences* 40 (1989), 113–147.

15. Euler, *Introduction to Analysis of the Infinite, Book I*, translated by John D. Blanton (New York: Springer-Verlag, 1988), p. v. This work is the first volume of a recent English translation of Euler's *Introductio*. (For the second volume, see note 24.) It repays a careful reading as there is much in there that could not be summarized in the text. See also Carl Boyer, "The Foremost Textbook of Modern Times (Euler's *Introductio in analysin infinitorum*)," *American Mathematical Monthly* 58 (1951), 223–226.

16. Euler, *Introduction*, p. 3. For more on the notion of functions, see A. P. Youshkevitch, "The Concept of Function up to the Middle of the Nineteenth Century," *Archive for History of Exact Sciences* 16 (1982), 37–85.

17. Euler, *Institutiones Calculi Differentialis* in *Opera Omnia* (1) vol. 10, p. 5.

18. Ibid., p. 4.

19. Ibid., p. 84.

20. Ibid., p. 99.

21. D. J. Struik, *A Source Book in Mathematics, 1200–1800* (Cambridge: Harvard Univ. Press, 1969), p. 355. Struik provides translations of the important parts of the papers involved in the controversy over vibrating strings. A dis-cussion of the entire matter can be found in Jerome R. Ravetz, "Vibrating Strings and Arbitrary Functions," a chapter in *The Logic of Personal Knowledge: Essays Presented to Michael Polanyi on his Seventieth Birthday, 11 March 1961* (London: Routledge and Paul, 1961), 71–88, as well as in Chapter 1 of Bottazzini, *Higher Calculus* and in C. Truesdell, "Rational Mechanics." A comprehensive study of the work of d'Alembert is Thomas Hankins, *Jean d'Alembert: Science and the Enlightenment* (Oxford: Clarendon Press, 1970).

22. Quoted in Truesdell, "Rational Mechanics," p. 239.

23. Quoted in Struik, *Source Book*, p. 361.

24. Euler, *Introduction to Analysis of the Infinite, Book II*, translated by John D. Blanton (New York: Springer-Verlag, 1990), p. 6.

25. Quoted in Bottazzini, *Higher Calculus*, p. 27.

26. Quoted in Struik, *Source Book*, p. 361.

27. George Berkeley, *The Analyst*, in James Newman, *The World of Mathematics* (New York: Simon and Schuster, 1956), vol. 1, 288–293, pp. 288-289. Selections of Berkeley's work are reprinted here and also in Struik, *Source Book*, pp. 333–338.

28. Newman, *World of Mathematics*, pp. 291–292.

29. Struik, *Source Book*, p. 335.

30. Newman, *World of Mathematics*, p. 289.

31. Ibid., p. 290.

32. Maclaurin, *Treatise of Fluxions*, p. 1. A brief biography of Colin Maclaurin appears in H. W. Turnbull, *Bicentenary of the Death of Colin Maclaurin* (Aberdeen, University Press, 1951). See also C. Tweedie, "A Study of the Life and Writings of Colin Maclaurin," *Mathematical Gazette* 8 (1915), 132–151 and 9 (1916), 303–305 and H. W. Turnbull, "Colin Maclaurin," *American Mathematical Monthly* 54 (1947), 318–322.

33. Ibid., p. 4.

34. Ibid., p. 53.

35. Ibid., p. 55.

36. Ibid., p. 181.

37. Ibid., p. 420.

38. Ibid., p. 421.

39. Ibid., p. 423.

40. Ibid., p. 38.

41. Struik, *Source Book*, p. 384. Struik has a translation of a

few pages of Euler's discussion of the metaphysics of the calculus from his *Institutiones Calculi Differentialis*.

42. Ibid.

43. Struik, *Source Book*, p. 345. This selection is from the article 'Différentiel' by D'Alembert in the *Encyclopédie*.

44. Ibid., pp. 343–344.

45. Quoted in Bottazzini, *Higher Calculus*, p. 49.

46. Lagrange, *Theory of Analytic Functions*, in *Oeuvres de Lagrange* (Paris: Gauthier-Villars, 1881), vol. 9, p. 15. A brief selection is translated in Struik, *Source Book*, pp. 388–391.

47. Ibid., p. 22.

Summary of Eighteenth Century Analysis

1622–1703	Vincenzo Viviani	Multiple integration problem
1646–1716	Gottfried Leibniz	Partial derivatives, Differential equations
1654–1705	Jakob Bernoulli	Brachistochrone problem
1667–1748	Johann Bernoulli	Brachistochrone problem, Differential equations
1685–1731	Brook Taylor	Taylor series
1685–1753	George Berkeley	Criticism of foundations of calculus
1687–1759	Nicolaus I Bernoulli	Rules for partial derivatives
1698–1746	Colin Maclaurin	Calculus text
1700–1782	Daniel Bernoulli	Vibrating string problem
1707–1783	Leonhard Euler	Differential equations, textbooks
1710–1761	Thomas Simpson	Calculus text
1713–1765	Alexis Clairaut	Differential equations
1717–1783	Jean d'Alembert	Vibrating string problem
1718–1799	Maria Agnesi	Calculus text
1736–1813	Joseph Lagrange	Calculus by way of power series

Chapter 14

Probability, Algebra, and Geometry in the Eighteenth Century

> *"It seems that to make a correct conjecture about any event whatever, it is necessary only to calculate exactly the number of possible cases and then to determine how much more likely it is that one case will occur than another. But here at once our main difficulty arises, for this procedure is applicable to only a very few phenomena. . . . What mortal, I ask, could ascertain the number of diseases, counting all possible cases . . . and say how much more likely one disease is to be fatal than another . . . ? Or who could enumerate the countless changes that the atmosphere undergoes every day and from that predict what the weather will be a month or even a year from now?"*
> *(Jakob Bernoulli's* Ars conjectandi, *1713)*[1]

*O*n September 28, 1794 the National Convention of France passed a law founding the École Centrale des Travaux Publiques, soon renamed the École Polytechnique. All of the best mathematicians of France taught there during the next decades, and many of them wrote texts for use there. The École Polytechnique soon became the model for colleges of engineering throughout Europe and the United States.

Although the development of analysis and its applications in various fields formed the central aspect of the history of mathematics in the eighteenth century, there was also important work in other areas. Jakob Bernoulli took up Huygens' work in probability and extended it, showing along the way a new method of calculating the sums of the powers of the integers and ultimately proving what is today known as the Law of Large Numbers. Abraham De Moivre carried this work even further by applying his knowledge of series and ultimately worked out the curve of the normal distribution as well as some of its properties. Finally, Thomas Bayes and Pierre-Simon de Laplace showed how to determine probability from a consideration of certain empirical data.

There were several major texts in algebra published during the century, including works of Maclaurin and Euler. These books included some systematization of earlier material. Thus Maclaurin's text included a new method of solving systems of linear equations, usually referred to as Cramer's rule, while Euler's book included some details on various methods in number theory. A central goal of algebra, however, was the extension of the equation-solving techniques of Cardano and Ferrari to polynomial equations of fifth degree and higher. No one was successful in this endeavor, but Lagrange, toward the end of the century, produced a detailed review of the methods for solving cubic and quartic equations and developed ideas which he felt would be essential in dealing with higher degree equations.

There were two major areas in geometry which also drew attention. First, there were renewed attempts to prove Euclid's parallel postulate by Girolamo Saccheri and Johann Heinrich Lambert, building on similar work by Islamic mathematicians. The second, and more important, thrust of eighteenth century geometry was the application of analysis to geometric questions. In particular, Euler, Clairaut, and Gaspard Monge contributed many new ideas in analytic and differential geometry. Euler also considered two special questions whose answers provided the seeds out of which the field of topology was to grow at the end of the next century.

The most important political event of the eighteenth century, the French Revolution, had an effect on mathematics and its education as indicated in the opener, so the chapter will conclude with a brief look at the new mathematics curriculum instituted at the École Polytechnique. We will also briefly consider the beginning of the development of mathematics in the colonies of the New World.

14.1 PROBABILITY

The early work on probability discussed in Chapter 11 was chiefly concerned with the question of determining probabilities, or expectations, in cases arising from various types of games or other gambling questions. The idea was to work out efficient ways of counting successes and failures of a particular experiment and thus to determine the probability *a priori*. Jakob Bernoulli's aim, in his study of the subject over some 20 years, was somewhat different. As indicated in the opening quotation, he wanted to be able to calculate probabilities of events in cases where it was impossible to enumerate all possibilities. To do this, he proposed to ascertain probabilities *a posteriori* by looking at the results observed in many similar instances. "For example, if we have observed that out of 300 persons of the same age and with the same constitution as a certain Titius, 200 died within ten years while the rest survived, we can with reasonable certainty conclude that there are twice as many chances that Titius also will have to pay his debt to nature within the ensuing decade as there are chances that he will live beyond that time."[2]

14.1.1 Jakob Bernoulli and the *Ars Conjectandi*

It seemed reasonably obvious to Bernoulli that the more observations one made of a given situation, the better one would be able to predict future occurrences. But he wanted to give a "scientific proof" of this principle, which would show not only that increases in the number of observations would enable the actual probability of the event to be estimated to within any desired degree of accuracy, but also how to calculate exactly how many observations were necessary to ensure that the result was within a predetermined interval around the true answer. By the time of his death in 1705, Bernoulli had provided this scientific proof in his Law of Large Numbers. This was included in his important text on probability, the *Ars conjectandi* (*Art of Conjecturing*), a work not published until 1713.

The Law of Large Numbers appears in the fourth and last part of the *Ars conjectandi*. The first three parts are more in the spirit of earlier work on probability. In fact, part one is essentially a reprint of Huygens' 1657 work, with added commentary. Part two develops anew various laws of combinations, most of which were known in previous centuries, while part three applies these laws to solve more problems about games. There are two original aspects of Bernoulli's work which bear mention, however. First, he generalized Pascal's ideas on the division of stakes in an interrupted game from Pascal's case where the chances of each player winning a given point were equal to the case where the chances of the two players are not equal, or, more generally, to the case of an experiment in which the chances of success or failure are not equal. Bernoulli showed that if the chance of success is a while that of failure is b (out of $a + b$ trials), then the probability of r successes in n trials is the ratio of $\binom{n}{n-r}a^r b^{n-r}$ to $(a + b)^n$. Similarly, the probability of at least r successes in n trials is the ratio of $\sum_{j=0}^{n-r} \binom{n}{j} a^{n-j}b^j$ to $(a + b)^n$.

A second contribution, which also made use of Pascal's arithmetic triangle, was Bernoulli's calculation of the sums of integral powers. Bernoulli surpassed ibn al-Haytham and Jyesthadeva in this regard, not only by writing out formulas for the sums of the integral powers up to order 10, but also by noting a pattern which gave him a general result for any power c:

$$\sum_{j=1}^{n} j^c = \frac{1}{c+1}n^{c+1} + \frac{1}{2}n^c + \frac{c}{2}B_2 n^{c-1} + \frac{c(c-1)(c-2)}{2 \cdot 3 \cdot 4}B_4 n^{c-3}$$

$$+ \frac{c(c-1)(c-2)(c-3)(c-4)}{2 \cdot 3 \cdot 4 \cdot 5 \cdot 6}B_6 n^{c-5} + \cdots,$$

where the series ends at the last positive power of n and where $B_2 = \frac{1}{6}$, $B_4 = -\frac{1}{30}$, $B_6 = \frac{1}{42}$, ... These latter quantities, today called the **Bernoulli numbers**, may be calculated by noting that on the first sum in which a given one occurs, it is that number which "completes to unity" the sum of the previous coefficients of the powers of n. Thus, because $\Sigma j^4 = \frac{1}{5}n^5 + \frac{1}{2}n^4 + \frac{1}{3}n^3 + B_4 n$, $B_4 = 1 - \frac{1}{5} - \frac{1}{2} - \frac{1}{3} = -\frac{1}{30}$.[3]

Part IV of the *Ars conjectandi* is entitled *On the Use and Applications of the Doctrine in Politics, Ethics, and Economics*. Although Bernoulli does not in fact discuss any practical applications, he does discuss various kinds of evidence seen in real life and how these pieces of evidence might be combined into a single probability statement. Realizing that in most real world situations, absolute certainty (or probability equal to 1) is impossible to achieve, Bernoulli introduced the idea of **moral certainty**. He decided that for an outcome to be morally certain, it should have a probability no less than 0.999. Conversely, an outcome with probability no greater than 0.001 he considered to be morally impossible. It was to determine the moral certainty of the true probability of an event that Bernoulli formulated his theorem, the Law of Large Numbers.

To understand the discussion of the theorem, one should keep in mind one of Bernoulli's examples. Suppose there is an urn containing 3,000 white and 2,000 black pebbles, although that number is unknown to the observer. The observer wants to determine the proportion of white to black by taking out, in turn, a certain number of pebbles and recording the outcome, at each step always replacing each pebble before taking out the next. Thus, in what follows, an observation is the removal of one pebble and a success is that the pebble is white. Assume then, in general, that N observations are made, that X of these are successes, and that $p = r/(r + s)$ is the (unknown) probability of a success. (Here r is the total of successful cases and s the total of unsuccessful ones. In the example, $p = 3/5$.) Bernoulli's theorem, in modern terminology, states that given any small fraction ϵ (which Bernoulli always took in the form $1/(r + s)$) and any large positive number c, the number $N = N(c)$ may be found so that the probability that X/N differs from p by no more than ϵ is greater than c times the probability that X/N differs from p by more than ϵ. In symbols, this result can be written as

$$P\left(\left|\frac{X}{N} - p\right| \le \epsilon\right) > cP\left(\left|\frac{X}{N} - p\right| > \epsilon\right).$$

In other words, the probability that X/N is "close" to p is very much greater than the probability that it is not "close." In modern texts, the theorem is usually stated as follows: Given any $\epsilon > 0$ and any positive number c, there exists an N such that

$$P\left(\left|\frac{X}{N} - p\right| > \epsilon\right) < \frac{1}{c + 1}.$$

Because the calculation of probabilities involved the summation of certain terms of the binomial expansion $(r + s)^N$, Bernoulli undertook a detailed analysis of the terms of that expansion. This gave him not only a proof of the theorem, but also a way of determining $N(c)$. In particular, he showed that $N(c)$ could be taken to be any integer greater than the larger of

$$mt + \frac{st(m - 1)}{r + 1} \qquad \text{and} \qquad nt + \frac{rt(n - 1)}{s + 1}$$

where m, n are integers such that

$$m \ge \frac{\log c(s - 1)}{\log(r + 1) - \log r} \qquad \text{and} \qquad n \ge \frac{\log c(r - 1)}{\log(s + 1) - \log s}.$$

In his example, Bernoulli calculated that for $r = 30$ and $s = 20$, the second expression was larger and therefore $N = 25,550$ when $c = 1000$. In other words, Bernoulli's result enabled him to know that 25,550 observations would be sufficient for moral certainty that the relative frequency found would be within 1/50 of the true proportion 3/5. Bernoulli's text ended with this calculation and similar ones for other values of c, perhaps because Bernoulli was unhappy with this result. For the early 1700s, 25,550 was an enormous number, larger than the entire population of Basel, for example. What the result seemed to say was that nothing reliable could be learned in a reasonable number of experiments. Bernoulli may have felt that he had failed in his quest to quantify the measure of uncertainty, especially since his intuition told him that 25,550 was much larger than necessary.[4] He therefore did not include the promised applications of his method to politics and economics. Nevertheless, Bernoulli pointed the way toward a more successful attack on the problem by his slightly younger contemporary, Abraham De Moivre (1667–1754).

14.1.2 De Moivre and *The Doctrine of Chances*

De Moivre's major mathematical work was *The Doctrine of Chances,* first published in 1718, with new editions in 1738 and 1756. This probability text is much more detailed than the work of Huygens, partly because of the general advances in mathematics since 1657. De Moivre gives not only general rules but also detailed applications of these rules, often to the playing of various games common in his time. As an example, consider the dice problem of de Méré, solved as part of a more comprehensive problem.

Problem III.   *To find in how many trials an event will probably happen, or how many trials will be necessary to make it indifferent to lay on its happening or failing, supposing that a is the number of chances for its happening and b the number of chances for its failing in a + b trials.*[5]

De Moivre begins his solution by noting that if there are x trials, then $\frac{b^x}{(a+b)^x}$ is the probability for the event failing x consecutive times. Since there are to be even odds as to whether the event happens at least once in x trials, this probability must equal 1/2, that is, x must satisfy the equation

$$\frac{b^x}{(a+b)^x} = \frac{1}{2} \quad \text{or} \quad (a+b)^x = 2b^x.$$

De Moivre easily solves this equation by taking logarithms:

$$x = \frac{\log 2}{\log(a+b) - \log b}.$$

Furthermore, he notes that if $a : b = 1 : q$, so that the odds against a success are q to 1, then the equation can be rewritten in the form

$$\left(1 + \frac{1}{q}\right)^x = 2 \quad \text{or} \quad x \log\left(1 + \frac{1}{q}\right) = \log 2.$$

By expanding $\log\left(1 + \frac{1}{q}\right)$ in a power series, De Moivre concludes that if "q is infinite, or pretty large in respect to unity,"[6] then the first term $1/q$ of the series is sufficient and the solution can be written as $x = q \log 2$ or $x \approx 0.7q$. Thus to solve de Méré's problem of finding how many throws of two dice are necessary to give even odds of throwing two ones, De Moivre simply notes that $q = 35$, so $x = 24.5$. The required number of throws is therefore between 24 and 25, the same answer as Huygens found by a much more detailed calculation.

De Moivre often uses infinite series to perform his probability calculations. But more important than these calculations themselves is his detailed discussion of approximating the sum of the terms of the binomial $(a + b)^n$, printed as an appendix to *The Doctrine of Chances* in its second and third editions, although first written in 1733, and in which appears for the first time the so-called normal approximation to the binomial distribution. De Moivre's aim in his discussion was to estimate probability by means of experiment: "Supposing for instance that an event might as easily happen as not happen, whether after three thousand experiments it may not be possible it should have happened two thousand times and failed a thousand; and that therefore the odds against so great a variation from equality should be assigned, whereby the mind would be the better disposed in the conclusions derived from the experiments."[7] For De Moivre, as for Bernoulli, the method of calculating the relevant probabilities lay in the calculation of certain binomial coefficients. He initially restricted himself to equally likely occurrences and sought to find the probability of $n/2$ successes in n trials, that is, the ratio which the middle term of $(1 + 1)^n$ has to the sum of all the terms, 2^n, for n large and even. He determined that this ratio $\binom{n}{n/2}/2^n$ approached $\frac{2T(n-1)^n}{n^n\sqrt{n-1}}$ as n got large, where

$$\log T = \frac{1}{12} - \frac{1}{360} + \frac{1}{1260} - \frac{1}{1680} + \cdots = \frac{B_2}{1 \cdot 2} + \frac{B_4}{3 \cdot 4} + \frac{B_6}{5 \cdot 6} + \frac{B_8}{7 \cdot 8} + \cdots,$$

the B_i being the Bernoulli numbers.

De Moivre's derivation of this result shows his great familiarity with infinite series and with logarithms. He began by noting that the middle term $M \binom{n}{n/2} = n!/\left(\frac{n}{2}\right)!^2$, where $n = 2m$, can be written as

$$\frac{(m + 1)(m + 2) \cdots (m + (m - 1))(m + m)}{(m - 1)(m - 2) \cdots (m - (m - 1))m}.$$

It follows that $\log M$ can be written as a sum of logarithms of quotients of the factors. Each of these logarithms can then be expanded in a power series in $1/m$. Thus,

$$\log M = \log \frac{m + 1}{m - 1} + \log \frac{m + 2}{m - 2} + \cdots + \log \frac{m + (m - 1)}{m - (m - 1)} + \log 2,$$

and, for example,

$$\log \frac{m+1}{m-1} = \log \frac{1 + \dfrac{1}{m}}{1 - \dfrac{1}{m}} = 2\left(\frac{1}{m} + \frac{1}{3m^3} + \frac{1}{5m^5} + \cdots\right)$$

and

$$\log \frac{m+2}{m-2} = \log \frac{1 + \dfrac{2}{m}}{1 - \dfrac{2}{m}} = 2\left(\frac{2}{m} + \frac{8}{3m^3} + \frac{32}{5m^5} + \cdots\right).$$

De Moivre then cleverly noted that the sum of these power series can be determined by adding vertically instead of horizontally. Thus, except for the term log 2, this sum can be expressed as the sum of the following columns:

$$\text{col. } 1 = \frac{2}{m}(1 + 2 + \cdots + s)$$

$$\text{col. } 2 = \frac{2}{3m^3}(1^3 + 2^3 + \cdots + s^3)$$

$$\text{col. } 3 = \frac{2}{5m^5}(1^5 + 2^5 + \cdots + s^5)$$

$$\cdots = \cdots$$

where $s = m - 1$. The columns, because they involve sums of integral powers, can be calculated by using Bernoulli's formulas to write each sum as a polynomial in s. De Moivre added the highest degree terms of each polynomial together, getting a power series which he could express as $(2m - 1)\log(2m - 1) - 2m\log m$. Similarly, the sum of the second highest degree terms of each polynomial forms the power series expressing the function $(1/2)\log(2m - 1)$. The sums of third, fourth, ..., highest degree terms are more difficult to determine, but De Moivre showed that in the limit as m approaches infinity, these become $\frac{1}{12}, -\frac{1}{360}, \ldots$. Remembering the extra term log 2, De Moivre concluded that the logarithm of M is

$$\left(2m - \frac{1}{2}\right)\log(2m - 1) - 2m \log m + \log 2 + \frac{1}{12} - \frac{1}{360} + \cdots$$

and, subtracting off $\log 2^n = \log 2^{2m} = 2m \log 2$, that the logarithm of the ratio is

$$n\log(n - 1) - \frac{1}{2}\log(n - 1) - n \log n + \log 2 + \frac{1}{12} - \frac{1}{360} + \cdots,$$

so that the ratio itself is as stated.

Because De Moivre wanted to be able to calculate with this ratio, he showed by use of the series for log T that $2T$ is approximately $2.168 = 2\ 21/125$. He also determined that

$$\sum_{k=1}^{m} \log k = \left(m + \frac{1}{2}\right)\log m - m + \log B \qquad \text{or} \qquad m! = Bm^{m+1/2}e^{-m},$$

where $\log B = 1 - \log T$, a formula today named for James Stirling (1692–1770). Stirling was responsible for calculating that $B = \sqrt{2\pi}$, probably by an argument similar to De Moivre's but starting from Wallis' product for π. It followed then that $\log T = 1 - \frac{1}{2}\log 2\pi$, or that, in modern notation, $T = e/\sqrt{2\pi}$. Since De Moivre knew that if n is large, then $\frac{(n-1)^n}{n^n} = \left(1 - \frac{1}{n}\right)^n$ approximates e^{-1}, he concluded that the ratio of the middle term of $(1 + 1)^n$ to the sum 2^n is equal to $2/\sqrt{2\pi n}$.

To deal with terms other than the middle, De Moivre generalized his method somewhat and showed that if Q is a term of the binomial expansion $(1 + 1)^n$ at a distance t from the middle term M, then

$$\log \frac{M}{Q} = \left(m + t - \frac{1}{2}\right)\log(m + t - 1) + \left(m - t + \frac{1}{2}\right)\log(m - t + 1)$$
$$- 2m \log m + \log \frac{m + t}{m}$$

where $m = \frac{1}{2}n$. He concluded, again approximating the logarithms by power series, that for n large,

$$\log\left(\frac{Q}{M}\right) \approx -\frac{2t^2}{n} \qquad \text{or} \qquad Q \approx M e^{-(2t^2/n)}.$$

In modern notation, this means that

$$P\left(X = \frac{n}{2} + t\right) \approx P\left(X = \frac{n}{2}\right)e^{-(2t^2/n)} = \frac{2}{\sqrt{2\pi n}}e^{-(2t^2/n)}.$$

De Moivre thought of the various values of $Q = P(X = \frac{n}{2} + t)$ as forming a curve: "If the terms of the binomial are thought of as set upright, equally spaced at right angles to and above a straight line, the extremities of the terms follow a curve. The curve so described has two inflection points, one on each side of the maximal term."[8] He calculated that the inflection points of this curve, today known as the normal curve, occurred at a distance $\frac{1}{2}\sqrt{n}$ from the maximum term.

With his approximation to the individual terms of the binomial expansion, De Moivre was able to calculate the sums of large numbers of such terms and thus improve considerably on Bernoulli's quantification of uncertainty. To find

$$\sum_{t=0}^{k} P(X = \frac{n}{2} + t),$$

he approximated this sum by

$$\frac{2}{\sqrt{2\pi n}}\int_0^k e^{-(2t^2/n)}\,dt$$

and evaluated the integral by writing the integrand as a power series and integrating term by term. For $k = \frac{1}{2}\sqrt{n}$, the series converged rapidly enough for him to conclude that the sum was equal to 0.341344 and therefore that "if it was possible to take an infinite number of experiments, the probability that an event which has an equal number of chances to happen or fail shall neither appear more frequently than $\frac{1}{2}n + \frac{1}{2}\sqrt{n}$ nor more rarely than

$\frac{1}{2}n - \frac{1}{2}\sqrt{n}$ times will be expressed by the double sum of the number exhibited . . . , that is, by 0.682688."[9] In modern terminology, De Moivre had shown that for n large, the probability that the number of occurrences of a symmetric binomial experiment would fall within $\frac{1}{2}\sqrt{n}$ of the middle value $\frac{1}{2}n$ was 0.682688. De Moivre then calculated the corresponding values for various other multiples of $\sqrt{n}$. Thus, "to apply this to particular examples, it will be necessary to estimate the frequency of an event's happening or failing by the square root of the number which denotes how many experiments have been, or are designed to be taken; and this square root, . . . will be as it were the *modulus* by which we are to regulate our estimation."[10] For De Moivre, $\sqrt{n}$ was the unit by which distances from the center were to be measured. Thus the accuracy of a probability estimate increases as the square root of the number of experiments.

The discussion above only applies to cases where the chances of an event happening or failing are equal. But De Moivre did sketch a generalization of his method to the general case by showing how to approximate terms in $(a + b)^n$, where $a \neq b$. Using these general methods, one can calculate that far fewer experiments are necessary to achieve the accuracy demanded in Bernoulli's example. In fact, in the case where Bernoulli required 25,550 trials, De Moivre's method only requires 6498. De Moivre himself, however, only gave examples in the equiprobable case. Thus he showed, for example, that 3600 experiments will suffice to give the probability 0.682688 that an event will occur at least 1770 times and no more than 1830 times or the probability 0.99874 that an event will occur at least 1710 times and no more than 1890 times. Unfortunately, although De Moivre's results were in fact more precise than those of Bernoulli, he was not able to apply them. Apparently, he did not even recognize the importance of the curve he had developed other than having the $\sqrt{n}$ serve as a measure for estimating the accuracy of an experiment. Nevertheless, his work was to have profound influence on later developments in the century.

14.1.3 Bayes and Statistical Inference

One reason that De Moivre's work was not immediately applied was that it did not directly answer the question necessary for applications, the question of statistical inference: Given empirical evidence that a particular event happened a certain number of times in a given number of trials, what is the probability of this event happening in general? The first person to attempt a direct answer to this question was Thomas Bayes (1702–1761), in his *An Essay towards solving a Problem in the Doctrine of Chances,* written toward the end of his life and not published until three years after his death.

Bayes began his essay with a statement of the basic problem: "Given the number of times in which an unknown event [i.e., an event of unknown probability] has happened and failed. Required the chance that the probability of its happening in a single trial lies somewhere between any two degrees of probability that can be named."[11] In modern notation, if X is the number of times the event has happened in n trials, x the probability of its happening in a single trial, and r and s the two given probabilities, Bayes' aim is to calculate $P(r < x < s|X)$, that is, the probability that x is between r and s, given X. Bayes proceeded to develop axiomatically, from a definition of probability, the two basic results that he would need. Proposition 3 states that "the probability that two subsequent events will both happen is a ratio compounded of the probability of the 1st, and the probability of the 2nd on supposition the 1st happens." Proposition 5 is: "If there be two subsequent events, the probability of the 2nd b/N and the probability of both together P/N, and it being first discovered that the 2nd event has happened, from hence I guess that the 1st event has also happened, the probability I am in the right is P/b."[12] In modern notation, letting E be the first event and F the second, proposition 3 can be written as $P(E \cap F) = P(E)P(F|E)$, that is, the probability of both happening is the product of the probability of E with the probability of F given E, while proposition 5 can be written as $P(E|F) = P(E \cap F)/P(F)$, that is, the probability of E given that F has happened is the quotient of the probability of both happening divided by the probability of F alone. Bayes' basic problem can be interpreted as the calculation of $P(E|F)$, where E is the event "$r < x < s$" and F is the event "X successes in n trials." To apply proposition 5 to this calculation, he therefore needed a way of calculating the two probabilities $P(E \cap F)$ and $P(F)$.

Bayes naturally knew Bernoulli's result that if the probability of a success were a and that of a failure were b, then the probability of p successes and q failures in $n = p + q$ trials was $\binom{n}{q}a^p b^q$. But whereas Bernoulli could only give a rough approximation to the sum of these terms and De Moivre chiefly considered the equiprobable case where $a = b$, Bayes used De Moivre's approach through area to attack the problem directly. Thus he began by modeling the probabilities by a certain area:

"I suppose the square table . . . *ABCD* (Figure 14.1) to be so made and levelled, that if either of the balls *O* or *W* be thrown upon it, there shall be the same probability that it rests upon any one equal part of the plane as another. . . . I suppose that the ball *W* shall be first thrown, and through the point where it rests a line *ot* shall be drawn parallel to *AD*, and

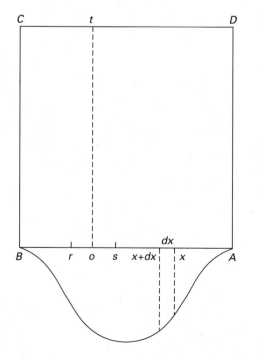

FIGURE 14.1
Bayes' Theorem.

meeting *CD* and *AB* in *t* and *o*; and that afterwards the ball *O* shall be thrown $p + q$ or n times, and that its resting between *AD* and *ot* after a single throw be called the happening of the event *M* in a single trial."[13]

(For simplicity in what follows, the length of *AB* will be 1.) In terms of the basic problem, the position of *W* determines the probability *x*. Bayes notes that the probability of the point *o* falling between any two points *r* and *s* is simply the length *rs*. Similarly, the probability of the event *M* given that *W* has been thrown is the length of *Ao*.

To calculate $P(E \cap F)$, Bayes uses proposition 3. Any given probability range for the point *o* is represented by an interval on the axis *AB*, say $[x, x + dx]$, measured from *A*. Because a particular *x* represents the probability of the ball landing to the right of *ot*, $1 - x$ represents the probability of it landing to the left. The probability that the ball will land to the right *p* times in $p + q = n$ throws is therefore given by $y = \binom{n}{q}x^p(1 - x)^q = \binom{n}{p}x^p(1 - x)^{n-p}$. Bayes draws the curve given by this function below the axis *AB* and uses proposition 3 to conclude that the probability of *W* lying above the coordinate interval $[x, x + dx]$ and the ball landing *p* times to the right of *W* is represented by the area under $[x, x + dx]$ and above the curve. It follows that $P(E \cap F) = P((r < x < s) \cap (X = p))$ is represented by the total area under the interval $[r, s]$ and above the curve, or, in modern notation, by

$$\int_r^s \binom{n}{p}x^p(1 - x)^{n-p}\, dx.$$

Because $P(F) = P(X = p)$ can be thought of as $P((0 < x < 1) \cap (X = p))$, it follows from the above argument that $P(X = p)$ is represented by the entire area under the axis *AB* and above the curve, or by

$$\int_0^1 \binom{n}{p}x^p(1 - x)^{n-p}\, dx.$$

Proposition 5 then implies that

$$P(E|F) = P((r < x < s)|(X = p)) = \frac{\int_r^s \binom{n}{p}x^p(1 - x)^{n-p}dx}{\int_0^1 \binom{n}{p}x^p(1 - x)^{n-p}\, dx}.$$

Bayes thus concludes "that in the case of such an event as I there call *M*, from the number of times it happens and fails in a certain number of trials, without knowing anything more concerning it, one may give a guess whereabouts its probability is, and, by the usual methods computing the magnitudes of the areas there mentioned, see the chance that the guess is right."[14]

Although Bayes' problem was, in fact, formally solved, there were two obstacles to be overcome before one could consider the solution as a practical one. First, does Bayes' physical analogy of rolling balls on a table truly mirror the actual problems to which the theory would be applied? Could nature's choice of an unknown probability *x* really be the same as the rolling of a ball across a level table? Bayes answered this question by, in effect, restricting the application of the rule to just those circumstances in which for any given number n of trials, all possible outcomes $X = 0$, $X = 1$, $X = 2, \ldots$ are equally likely, that is, for events concerning which "I have no reason to think that, in a certain number of trials, it should rather happen any one possible number of times than another."[15] But this particular statement of Bayes has generated wide debate to the present time concerning the nature of the circumstances to which he referred. Is ignorance about the probabilities in a given situation equivalent to all possible outcomes being equally likely?

Second, can one actually calculate the integrals in Bayes' formula? Bayes attempted to do so by expanding the integrands in power series. The integral in the denominator turned out to be $\frac{1}{n+1}$. The integral in the numerator, while not difficult to approximate when either p or $n - p$ is small, turned out to be very difficult otherwise. Richard Price (1723–1791), the friend of Bayes who submitted his paper to the Royal Society, worked out a few special cases when p is close to n. For example, if $p = n$, then the relevant quotient is

$$\frac{\int_r^s x^n \, dx}{\int_0^1 x^n \, dx} = s^{n+1} - r^{n+1}.$$

So suppose nothing is known about an event M except that it has happened once. The chance that the unknown probability x of M is greater than 1/2, that is, between 1/2 and 1, is then $1^2 - (1/2)^2 = 3/4$. Similarly, if M has happened twice, the probability that x is greater than 1/2 is 7/8; in other words, the odds are 7 to 1 that there is more than an even chance of it happening. In this same situation, the odds are still better than even that the probability of x is greater than 2/3.

14.1.4 The Calculations of Laplace

Bayes' formula does provide a start in answering the basic question of statistical inference. Further progress was made a few years later by Pierre-Simon de Laplace (1749–1827). In 1774, Laplace, using principles similar to those of Bayes, derived essentially the same result involving integrals for determining probability, given empirical evidence. Returning to the question of drawing tickets from an urn, he supposed that p white and q black tickets had been drawn from an urn containing an unknown ratio x of white tickets. Given then any guessed value for x, Laplace showed how to calculate the probability that x differed from $\frac{p}{p+q}$ by as small a value ϵ as one wished. He was in fact able to demonstrate that

$$P\left(\left| x - \frac{p}{p+q} \right| \le \epsilon \Big| X = p \right) \cong \frac{2(p+q)^{3/2}}{\sqrt{2\pi}\sqrt{pq}} \int_0^\epsilon e^{-\{[(p+q)^3/2pq]z^2\}} \, dz \cong \frac{2}{\sqrt{2\pi}} \int_0^{\epsilon/\sigma} e^{-(u^2/2)} \, du$$

where $\sigma^2 = pq/(p+q)^3$. To show that this probability approached 1 as $p + q$ became large, whatever the value of ϵ, Laplace had to integrate $\int_0^\infty e^{-(u^2/2)} \, du$. Using a result of Euler's, he was in fact able to show that this integral equaled $\sqrt{\pi/2}$ and therefore established his result. (Note that, in effect, De Moivre had already proved this.)

Biography   Pierre de Laplace (1749–1827)

Born in Normandy, Laplace entered the University of Caen in 1766 to begin preparation for a career in the church. He discovered there his mathematical talents, however, and in 1768 left for Paris to continue his studies. He met with d'Alembert, who was so impressed with him that he secured for Laplace a position in mathematics at the École Militaire, where he taught elementary mathematics to aspiring cadets. Legend has it that he examined, and passed, Napoleon there in 1785. A steady stream of mathematical papers soon began to flow from his pen, winning him election to the Academy of Sciences in 1773.

Laplace's most important accomplishments were in the field of celestial mechanics. Over the period

from 1799 to 1825 he produced his five volume *Traité de Mécanique céleste* (*Treatise on Celestial Mechanics*), in which he successfully applied calculus to the motions of the heavenly bodies and showed, among much else, why Newton's law of gravitation implied the long term stability of the solar system. Laplace also contributed heavily to the field of probability, producing his *Théorie analytique des probabilités* (*Analytic Theory of Probability*) in 1812. Although he was honored by Napoleon, he voted against him in 1814 as a member of the Senate, supporting Louis XVIII instead. Laplace was rewarded with the title of marquis. At his death he was eulogized as "the Newton of France" (Figure 14.2).

FIGURE 14.2
Laplace on a French stamp.

To go further in calculating, naturally, Laplace had to evaluate the integral $\int_0^T e^{-(u^2/2)}\,du$ for arbitrary T. This he did in 1785 by deriving two different series for this integral, one which converged rapidly for small T and one for large T. He then applied his results to a genuine problem in statistical inference. During the 26 year period from 1745 to 1770, 251,527 boys and 241,945 girls had been born in Paris. Setting x as the probability of a male birth, he made a straightforward calculation using his analysis and demonstrated that the probability that $x \le 1/2$ was 1.15×10^{-42}. He therefore concluded that it was "morally certain" that $x > 1/2$. He then extended his analysis using similar data from London to show that it was also morally certain that the probability of a male birth in London was greater than that in Paris.

With a real problem in statistical inference now solved, Laplace turned his attention back to astronomy. Some of his further contributions to probability, in part derived from his analysis of errors of observation in astronomy, will be discussed in Chapter 16.

14.2 ALGEBRA AND NUMBER THEORY

There were few major new developments in algebra in the eighteenth century, in contrast to the work in other fields. The major effort, accomplished by mathematicians whose chief influence was felt elsewhere, was a systematization of earlier material. Thus two major algebra texts will be considered here, one by Maclaurin (published in 1748 although probably written in the 1730s) and one by Euler (published in 1767), both of which served to introduce students to the field and set the basis for future work. Maclaurin's text went through six editions, the last appearing in 1796, while Euler's work appeared in print at least 30 times in six languages during the 50 years after its initial publication. These two texts show what was considered important in algebra in the eighteenth century.

14.2.1 Maclaurin's *Treatise of Algebra*

Maclaurin's work, *A Treatise of Algebra in Three Parts*, defined algebra as "a general method of computation by certain signs and symbols which have been contrived for this purpose and found convenient. It is called an Universal Arithmetic and proceeds by operations and rules similar to those in common arithmetic, founded upon the same principles."[16] In other words, for Maclaurin algebra is not "abstract" but simply generalized arithmetic. For example, he shows how the basic operations can apply to negative numbers. He notes that any quantity can enter algebraic computation as either an increment or a decrement. As examples of these two forms, he includes such concepts as excess and deficit, value of money due to a man and due by him, a line drawn to the right and one to the left, and elevation above horizon and depression below. He notes that one can subtract a greater quantity from a lesser of the same kind, the remainder in that case always being opposite in kind, but one can only do this if it makes sense. For example, one cannot subtract a greater quantity of matter from a lesser. Nevertheless, Maclaurin always considers a negative quantity to be no less real than a positive one. He thus demonstrates how to calculate with positive and negative quantities. In particular, to show the reason for the rule of signs in multiplying such quantities, he notes that since $+a - a = 0$, also $n(+a - a) = 0$. But the first term of this product, $+na$, is positive. The second term must therefore be negative. Therefore $-a$ multiplied by $+n$ is negative. Similarly, since $-n(+a - a) = 0$ and the first term of this product is negative, the second term, $(-n)(-a)$, must be positive and equal to $+na$.

Maclaurin continues in the first part of his work to deal with such topics as manipulation with fractions, powers of binomials, roots of polynomials, and sums of progressions. He shows how to solve linear and quadratic equations, including a fair number of "word problems" as examples. In the case of linear equations in more than one unknown, he shows in the cases of two and three equations in the same number of unknowns that the solution can be found by solving for one unknown in terms of the others and then substituting. He notes that if there are more unknowns than equations, there may be an infinite number of solutions, while in the opposite case, there may be no solutions at all, but he does not give any examples of either situation.

He does, however, present what he calls a "general theorem" for eliminating unknowns in a system of equations, the method known today as **Cramer's rule**, named after the Swiss mathematician Gabriel Cramer (1704–1752) who used it in a book on curves in 1750. If

$$ax + by = c$$
$$dx + ey = f$$

then solving the first equation for x and substituting gives

$$y = \frac{af - dc}{ae - db},$$

and a similar answer for x. The system of three equations

$$ax + by + cz = m$$

$$dx + ey + fz = n$$

$$gx + hy + kz = p$$

is dealt with by first solving each equation for x, thus reducing the problem to one in two unknowns, and then using the earlier rule to find

$$z = \frac{aep - ahn + dhm - dbp + gbn - gem}{aek - ahf + dhc - dbk + gbf - gec}.$$

In addition to giving the answer, Maclaurin notes the general rule that the numerator consists of the various products of the coefficients of x and y as well as the constant terms, each product consisting of one coefficient from each equation, while the denominator consists of products of the coefficients of all three unknowns. He also explains how to determine the sign of each term. Furthermore, he solves for y and x and shows that the general rule determining each of these values is analogous to the one for z. In particular, each of the three expressions has the same denominator. Maclaurin even extends the rule to systems of four equations in four unknowns, but mentions nothing about any further generalization.

The numerators and denominators involved in Maclaurin's solution are, of course, what are known today as **determinants**. But the use of such combinations of coefficients as tools for solving systems of linear equations had appeared somewhat earlier. Leibniz had suggested a similar idea in a letter to l'Hospital in 1693 and had even devised a way of indexing the coefficients of the system by the use of numbers. And halfway around the world, the Japanese mathematician Seki Kowa (1642–1708) described the use of determinants in a manuscript of 1683, carefully showing by use of diagrams how to decide whether a given term is to be positive or negative.

The second part of Maclaurin's work is a treatise on the solving of polynomial equations which presents all that had been discovered up to his time in well-organized form. Thus, Maclaurin includes not only Cardano's rule for solving cubics and Ferrari's rule for quartics, but also Descartes' rule of signs and Newton's methods for approximating numerically the solution to an equation. He notes that the procedure by which equations are generated—multiplying together equations such as $x - a = 0$ or other equations of degree smaller than the given one—shows that no equation can have more roots than the degree of the highest power. Furthermore, "roots become impossible [complex] in pairs" and therefore "an equation of an odd dimension has always one real root."[17] He then discusses the general procedure for finding integral roots of monic polynomials: check all divisors of the constant term as possible roots and, if one such root α is found, divide the polynomial by $x - \alpha$ to reduce the degree.

Maclaurin concludes his text with a discussion of the application of algebraic techniques to geometric problems and, conversely, of the use of geometrical procedures to solve equations. The major difference between the uses of algebra and geometry, he notes, is that in the former, one can express even impossible roots explicitly, but in the latter such quantities do not appear at all. Although there was little new mathematically in the text,

Maclaurin's work provided the students of the eighteenth century with a solid introduction to the algebra of the day.

14.2.2 Euler's *Introduction to Algebra*

An even better introduction to algebra, perhaps, was provided by Euler's *Vollständige Anleitung zur Algebra (Complete Introduction to Algebra)*. Euler, like Maclaurin, begins his text by providing a definition of the subject: "The foundation of all the mathematical sciences must be laid in a complete treatise on the science of numbers, and in an accurate examination of the different possible methods of calculation. This fundamental part of mathematics is called Analysis, or Algebra. In Algebra, then, we consider only numbers, which represent quantities, without regarding the different kinds of quantity."[18] Later on in the text, he makes the definition somewhat more specific: Algebra is "the science which teaches how to determine unknown quantities by means of those that are known."[19] He notes that even ordinary addition of two known quantities can be thought of as fitting under this definition and so the second, perhaps more common, definition really includes the first.

Euler starts with a discussion of the algebra of positive and negative quantities. His discussion of multiplication is somewhat less formal than that of Maclaurin: "Let us begin by multiplying $-a$ by $+3$. Now, since $-a$ may be considered as a debt, it is evident that if we take that debt three times, it must thus become three times greater, and consequently the required product is $-3a$."[20] Euler then notes the obvious generalization that $-a$ times b will be $-ba$ or $-ab$ and continues on to the case of the product of two negatives. Here he simply says that $-a$ times $-b$ cannot be the same as $-a$ times b, or $-ab$, and therefore must be equal to $+ab$.

After discussing various other operations, Euler introduces the concept of an imaginary number:

> Since all numbers which it is possible to conceive are either greater or less than 0, or are 0 itself, it is evident that we cannot rank the square root of a negative number amongst possible numbers, and we must therefore say that it is an impossible quantity. In this manner we are led to the idea of numbers, which from their nature are impossible; and therefore they are usually called *imaginary quantities*, because they exist merely in the imagination. All such expressions as $\sqrt{-1}$, $\sqrt{-2}$ . . . are consequently impossible, or imaginary numbers, since they represent roots of negative quantities; . . . But notwithstanding this these numbers present themselves to the mind; they exist in our imagination, and we still have a sufficient idea of them; since we know that by $\sqrt{-4}$ is meant a number which, multiplied by itself, produces -4; for this reason also, nothing prevents us from making use of these imaginary numbers, and employing them in calculation.[21]

Curiously, Euler does not realize that there may be problems in these calculations. For although he has noted that $\sqrt{-4} \times \sqrt{-4} = -4$, somewhat later he writes that the general rule for multiplying square roots implies that $\sqrt{-1} \times \sqrt{-4} = \sqrt{4} = 2$.

Euler continues the text by discussing logarithms, infinite series, and the binomial theorem. He defines logarithms as he did in the *Introductio*: If $a^b = c$, then b is the logarithm of c with base a. Logarithms are then applied in a chapter on calculation of compound interest. Infinite series are introduced in terms of division, with the first example being $\frac{1}{1-a} = 1 + a + a^2 + a^3 + \cdots$. Although Euler does not discuss convergence as such, he asserts that "there are sufficient grounds to maintain that the value of this infinite series is the same as that of the fraction."[22] He then deals with some examples so that this statement will be "easily understood." Thus, if $a = 1$, the fraction is equal to $1/0$, "a number infinitely great," while the series becomes $1 + 1 + 1 + \cdots$, also infinite, confirming the assertion. But, Euler concludes, "the whole becomes more intelligible" if values for a less than 1 are taken. In that case, "the more terms we take, the less the difference [between the fraction and the series] becomes; and consequently, if we continue the series to infinity, there will be no difference at all between its sum and the value of the fraction."[23]

Over 100 pages of Euler's text are devoted to solving equations of various sorts. His rules, which include the standard ones as well as a new method for solving fourth degree equations, are accompanied by a wide variety of word problems. And although Euler in essence presents the determinant form of the solution for a system of two equations in two unknowns, his standard method for larger systems is that of solving for one variable in terms of the others, thereby reducing the system to one of fewer equations in fewer unknowns. He notes, however, that in many cases there are various tricks which can be used: "When a person is a little accustomed to such calculations, he easily perceives what is most proper to be done."[24]

The final part of Euler's text is devoted to a subject not found at all in the work of Maclaurin, the solution of indeterminate equations. Many of the problems solved in this part are, in fact, the problems of Diophantus' *Arithmetica*. But Euler, like Fermat a century earlier, always gives general solutions to the problems rather than the single solution typical of the Greek algebraist. As an example, consider the following problem, virtually the same as Diophantus' problem II–11.

Question 2.   *To find such a number x, that if we add to it any two numbers, for example, 4 and 7, we obtain in both cases a square.*[25]

Diophantus solved this problem by the method of the double equation. Euler uses a different technique. Setting $x + 4 = p^2$, he concludes that $x + 7 = p^2 + 3$ is a square whose root is $p + q$. Setting $q = r/s$, it follows that $p^2 + 3 = p^2 + 2pq + q^2$ or that $p = (3 - q^2)/2q$ or finally that

$$x = p^2 - 4 = \frac{9 - 22q^2 + q^4}{4q^2} = \frac{9s^4 - 22r^2s^2 + r^4}{4r^2s^2}.$$

Euler then notes that any choice of integers for r and s gives a solution for x.

Much of this section on indeterminate equations, however, is devoted to some general methods rather than specific problems. Euler deals especially with techniques for finding solutions, in either rational numbers or integers, to equations of the form $p(x) = y^2$ where $p(x)$ is a polynomial of degree 2, 3, or 4. As a special case, he considers the solution in integers of the equation $Dx^2 + 1 = y^2$ discussed in Chapter 6, the equation whose solution Euler incorrectly attributed to the English mathematician John Pell (1610–1685) and which Fermat claimed to have had. Rather than presenting a general method of solution, Euler demonstrates a procedure to be applied in each case separately. He then concludes his discussion by presenting a table in which solutions to the equation are listed for values of D from 2 to 100. Although Euler did not prove that solutions exist for every D, such a proof was given by Lagrange in 1766 and included as an appendix in later editions of the *Algebra*.

14.2.3 Euler and Number Theory

Euler worked on and solved many interesting number-theoretical problems during his life, some of which had been suggested by Fermat or grew out of problems Fermat solved. Thus in 1749 Euler proved Fermat's claim that every prime of the form $4n + 1$ can be written as the sum of two squares and in 1773, after working on the problem for many years, gave a proof that every integer can be expressed as a sum of no more than four squares. (Lagrange had proved this result three years before; Euler's proof was a generalization of his earlier proof for two squares.) We will, however, only discuss Euler's detailed study of congruences, his generalization of Fermat's Little Theorem, and his discovery of the law of quadratic reciprocity.

Probably around 1750 Euler began to write an elementary treatise on number theory, but after completing 16 chapters he set it aside. The manuscript was discovered after his death and eventually published in 1849 under the title *Tractatus de numerorum doctrina* (*Treatise on the Doctrine of Numbers*). The early chapters contain the calculation of such number theoretic functions as $\sigma(n)$, the number of divisors of an integer n, and $\phi(n)$, the number of integers prime to n and less than n. The most important part of the treatise, however, beginning in the fifth chapter, is Euler's treatment of the concept of congruence with respect to a given number d, now called the **modulus**. Euler defines the residue of a with respect to d as the remainder r on the division of a by d: $a = md + r$. He notes that there are d possible remainders, and that therefore all the integers are divided into d classes, each class consisting of those numbers having the given remainder. For example, division by 4 divides the integers into four classes, numbers of the form $4m$, $4m + 1$, $4m + 2$, and $4m + 3$. All numbers in a given class he regards as "equivalent." Euler further shows that if A and B are in the class of residues α and β respectively, then $A + B$, $A - B$, nA, and AB are in the class of residues $\alpha + \beta$, $\alpha - \beta$, $n\alpha$, and $\alpha\beta$ respectively. Euler thus demonstrates, in modern terminology, that the function assigning an integer to its "residue class" is a ring homomorphism. In fact, it was out of such ideas that the theory of rings eventually developed.

Similarly, basic ideas of group theory are evident in Euler's discussion of residues of a series in arithmetic progression $0, b, 2b, \ldots$. Euler shows that if the modulus d and the number b are relatively prime, then this series contains d different residues. Therefore, b has an "inverse" with respect to d, a number p such that the residue of pb equals 1. On the other hand, if the greatest common divisor of d and b is g, then there are only d/g different residues and such an inverse does not exist. For example, there are 9 different residues of multiples of 2 with respect to modulus 9, and 5 is the inverse of 2, while there are only 3 distinct residues of multiples of 3 with respect to 9 and no inverse exists for 3.

Euler continues this line of investigation by considering the residues of a geometric series $1, b, b^2, b^3, \ldots$ where b is prime to d. The number n of distinct residues of this series can be no more than $\mu = \phi(d)$. Euler notes that this number n is the smallest number greater than 1 such that b^n has residue 1, because once this power is reached, all subsequent powers simply repeat the same remainders. To show that n is a factor of μ, he uses an argument later to be standard in group theory by, in effect, considering the cosets of the subgroup of powers of b in the multiplicative group of residues of d relatively prime to d and showing that the order of the subgroup divides the order of the group. Euler first demonstrates that if r and s are residues, say of b^ρ and b^σ, then rs is also a residue, of $b^{\rho+\sigma}$. Similarly, r/s is a residue. Thus if r is a residue and $x < d$ is a nonresidue (a number prime to d not a residue of the series of powers), xr must also be a nonresidue. Therefore, if 1, α, β, . . . form the entire set of n residues, then x, $x\alpha$, $x\beta$, . . . form a set of n distinct nonresidues. Because any nonresidue not included in this latter list also leads to a set of n nonresidues, all distinct from the first list, Euler concludes that $\mu = mn$ for some integer m. It follows that $b^\mu = b^{mn}$ has remainder 1 on division by d, or that $b^\mu - 1$ is divisible by d. A special case of this theorem, when d is a prime p, is Fermat's Little Theorem.

The question of residues of powers was of interest to Euler over a long period of his life, and he often did computations using them. Thus the end of the seventh chapter of the manuscript provides a list of tables of powers of numbers and their residues for all moduli d from 2 to 13. Euler also did extensive computing of prime divisors of expressions of the form $x^2 + ny^2$ and attempted to determine which primes can be written in that form. He published a paper in 1751 which simply contained tables of such results for 16 positive and 18 negative values of n. These calculations led Euler by 1783 to a statement of a theorem equivalent to the quadratic reciprocity theorem.

Euler called $p \neq 0$ a **quadratic residue** with respect to a prime q if there exist a and n such that $p = a^2 + nq$, that is, if $x^2 \equiv p \pmod{q}$ has a solution. Note that the condition of being a quadratic residue with respect to q depends only on the residue class of p with respect to q. For example, 1, 4, 9, $5 \equiv 4^2$, and $3 \equiv 5^2$ are quadratic residues with respect to 11, while 2, 6, 7, 8, and 10 are nonresidues. In his paper of 1783 Euler first proved that if $q = 2m + 1$ is an odd prime, then there are exactly m quadratic residues and therefore m nonresidues. Furthermore, he showed that the product and the quotient of two quadratic residues are again quadratic residues. He then determined that -1 is residue with respect to q if q is of the form $4n + 1$ while it is a nonresidue if q is of the form $4n + 3$. At the end of the paper, however, after considering more examples, he wrote down four conjectures relating, for two different odd primes q and s, conditions under which each may or may not be a quadratic residue with respect to the other, conditions which may be written as follows.

1. If $q \equiv 1 \pmod{4}$ and q is a quadratic residue with respect to s, then s and $-s$ are both quadratic residues with respect to q.

2. If $q \equiv 3 \pmod{4}$ and $-q$ is a quadratic residue with respect to s, then s is a quadratic residue and $-s$ is not with respect to q.

3. If $q \equiv 1 \pmod{4}$ and q is not a quadratic residue with respect to s, then s and $-s$ are both nonresidues with respect to q.

4. If $q \equiv 3 \pmod{4}$ and $-q$ is not a quadratic residue with respect to s, then $-s$ is a quadratic residue and s is a nonresidue with respect to q.

Euler was not able to prove these results in 1783. They were restated in a somewhat different form by Adrien-Marie Legendre (1752–1833) in a paper of 1785 and in his textbook *Essai sur la théorie des nombres* of 1798, both times, however, with an incomplete proof. The first complete proof was given by Carl Friedrich Gauss in 1801 in his great work *Disquisitiones Arithmeticae*, to be discussed in Chapter 15.

14.2.4 Lagrange and the Solution of Polynomial Equations

Another aspect of algebra considered in the eighteenth century was the solution of polynomial equations. Many mathematicians, in fact, attempted to generalize the methods of Cardano and Ferrari to solve algebraically polynomial equations of degree five and higher, but without success. Lagrange, in his *Réflexions sur la théorie algébrique des équations* (*Reflections on the Algebraic Theory of Equations*) of 1770, began a new phase in this work by undertaking a detailed review of these earlier solutions to determine why the methods for cubics and quartics worked. He was not able to find analogous methods for higher degree equations, but was able to sketch a new set of principles for dealing with these equations which he hoped might ultimately succeed.

Lagrange began with a systematic study of the methods of solution of the cubic equation $x^3 + nx + p = 0$, starting essentially with the procedure of Cardano. Setting $x = y - (n/3y)$ transforms this equation into the sixth degree equation $y^6 + py^3 - (n^3/27) = 0$ which, putting $r = y^3$, reduces in turn to the quadratic equation $r^2 + pr - (n^3/27) = 0$. This latter equation has two roots, r_1, and $r_2 = -\left(\frac{n}{3}\right)^3 \frac{1}{r_1}$. But whereas Cardano took the sum of the real cube roots of r_1 and r_2 as his solution, Lagrange knew that each equation $y^3 = r_1$ and $y^3 = r_2$ had three roots. Thus there were six possible values for y, namely, $\sqrt[3]{r_1}$, $\alpha\sqrt[3]{r_1}$, $\alpha^2\sqrt[3]{r_1}$, $\sqrt[3]{r_2}$, $\alpha\sqrt[3]{r_2}$, and $\alpha^2\sqrt[3]{r_2}$, where α is a complex root of $x^3 - 1 = 0$, or of $x^2 + x + 1 = 0$. Lagrange could then show that the three distinct roots of the original equation were given by

$$x_1 = \sqrt[3]{r_1} + \sqrt[3]{r_2}$$

$$x_2 = \alpha\sqrt[3]{r_1} + \alpha^2\sqrt[3]{r_2}$$

$$x_3 = \alpha^2\sqrt[3]{r_1} + \alpha\sqrt[3]{r_2}$$

Lagrange next noted that rather than consider x as a function of y, one could reverse the procedure, because the equation for y, which he called the *réduite* or *reduced* equation, was the one whose solutions enabled the original equation to be solved. The idea was to express those solutions in terms of the original ones. Thus Lagrange noted that any of the six values for y could be expressed in the form $y = \frac{1}{3}(x' + \alpha x'' + \alpha^2 x''')$ where (x', x'', x''') was some permutation of (x_1, x_2, x_3). It was this introduction of the permutations of the roots of an equation which provided the cornerstone not only for Lagrange's method but for the methods others were to use in the next century.

In the case of the cubic, there are several important ideas to note. First, the six permutations of the x_i lead to the six possible values for y and thus show that y satisfies an equation of degree six. Second, the permutations of the expression for y can be divided into two sets, one consisting of the identity permutation and the two permutations which interchange all three of the x_i, and the second consisting of the three permutations which just interchange two of the x_i. (In modern terminology, the group of permutations of a set of three elements has been divided into two cosets.) For example, if $y_1 = \frac{1}{3}(x_1 + \alpha x_2 + \alpha^2 x_3)$, then the two nonidentity permutations in the first set change y_1 to $y_2 = \frac{1}{3}(x_2$

$+ \alpha x_3 + \alpha^2 x_1)$ and $y_3 = \frac{1}{3}(x_3 + \alpha x_1 + \alpha^2 x_2)$ respectively. But then $\alpha y_2 = \alpha^2 y_3 = y_1$ and $y_1^3 = y_2^3 = y_3^3$. Similarly, if the results of the permutations of the second set are y_4, y_5, and y_6, it follows that $y_4^3 = y_5^3 = y_6^3$. Thus, because there are only two possible values for $y^3 = \frac{1}{27}(x' + \alpha x'' + \alpha^2 x''')^3$, the equation for y^3 is of degree 2. Finally, the sixth degree equation satisfied by y has coefficients which are rational in the coefficients of the original equation. Lagrange considered several other methods of solution of the cubic equation, but found in each case the same underlying idea. Each led to a rational expression in the three roots which took on only two values under the six possible permutations, thus showing that the expression satisfied a quadratic equation.

Lagrange next considered the solutions of the quartic equation. Ferrari's method of solving $x^4 + nx^2 + px + q = 0$ was to add $2yx^2 + y^2$ to each side, rearrange, and then determine a value for y such that the right side of the new equation

$$x^4 + 2yx^2 + y^2 = (2y - n)x^2 - px + y^2 - q$$

was a perfect square. After taking square roots of each side, he could then solve the resulting quadratic equations. The condition that the right side be a perfect square is that

$$(2y - n)(y^2 - q) = \left(\frac{p}{2}\right)^2 \quad \text{or} \quad y^3 - \frac{n}{2}y^2 - qy + \frac{4nq - p^2}{8} = 0.$$

Therefore, the *réduite* is a cubic, which can, of course, be solved. Given the three solutions for y, Lagrange then showed, as in the previous case, that each is a permutation of a rational function of the four roots x_1, x_2, x_3, x_4 of the original equation. In fact, it turned out that $y_1 = \frac{1}{2}(x_1 x_2 + x_3 x_4)$ and that the 24 possible permutations of the x_i lead to only three different values for that expression, namely y_1, $y_2 = \frac{1}{2}(x_1 x_3 + x_2 x_4)$, and $y_3 = \frac{1}{2}(x_1 x_4 + x_2 x_3)$. The expression must therefore satisfy a third degree equation, again one with coefficients rational in the coefficients of the original equation.

Having studied the methods for solving cubics and quartics, Lagrange was ready to generalize. First, as was clear from the discussion of cubic equations, the study of the roots of equations of the form $x^n - 1 = 0$ was important. For the case of odd n, Lagrange could show that all the roots could be expressed as powers of one of them. In particular, if n is prime and $\alpha \neq 1$ is one of the roots, then α^m for any $m < n$ can serve as a generator of all of the roots. Second, however, Lagrange realized that to attack the problem of equations of degree n, he needed a way of determining a *réduite* of degree $k < n$. Such an equation must be satisfied by certain functions of the roots of the original equation, functions which take on only k values when the roots are permuted by all $n!$ possible permutations. Because relatively simple functions of the roots did not work, Lagrange attempted to find some general rules for determining such functions and the degree of the equation which they would satisfy.

Lagrange noted that if the values of the roots of the *réduite* are $f_1, f_2, \ldots, f_k$, where each f_i is a function of the n roots of the original equation, then the *réduite* is given by $(t - f_1)(t - f_2) \cdots (t - f_k) = 0$. Although he could not prove that the degree of this equation in general is less than $n!$, he was able to show that its degree k, the number of different values taken by f under the permutations of the variables, always divided $n!$. One can read into this statement Lagrange's theorem to the effect that the order of any subgroup of a group divides the order of the group, but Lagrange never treated permutations as a "group" of operations. Lagrange went on, however, to attempt to relate various functions of the roots. He defined two such functions u and v to be **similar** if all permutations of the roots which leave u unchanged also leave v unchanged, and conversely. He then proved that in this case v can be expressed as a rational function of u and the coefficients of the original equation. Furthermore, if u is unchanged by permutations which do change v, and if v takes on r different values for each one taken on by u, then v can be found in terms of u by an equation of degree r in v. For example, in the cubic equation $x^3 + nx + p = 0$, the expression $v = (x_1 + \alpha x_2 + \alpha^2 x_3)^3$ takes on two values under the six permutations of the roots while $u = x_1 + x_2 + x_3$ is unchanged under those permutations. Then $v^2 + \frac{1}{27}pv - \frac{n^3}{27^3} = 0$ is the equation satisfied by v. (Note here that $u = 0$.)

Lagrange presumably hoped to solve the general polynomial equation of degree n by use of this theorem. Namely, he would start with a symmetric function of the roots, say $x_1 + x_2 + \cdots + x_n$, which was unchanged under all $n!$ permutations, then find a function v which takes on r different values under these permutations. Thus v would be a root of an equation of degree r with coefficients rational in the original coefficients (because the given symmetric function was one of those coefficients). If that equation could be solved, then he could find a new function w which takes on, say, s values under the permutations which leave v unchanged. Thus w would satisfy an equation of degree s. He would continue in this way until the function x_1 is reached. Unfortunately, Lagrange was unable to find a general method of determining these intermediate functions such that they were of a form which could be solved by known methods. He was thus forced to abandon his quest. Nevertheless, his work did form the foundation on which all nineteenth century work on the algebraic solution of equations was based. The story will be continued in Chapter 15.

14.3 GEOMETRY

Geometry in the eighteenth century was connected both to algebra, through the relationships codified under the term analytic geometry, and to calculus, through the application of infinitesimal techniques to the study of curves and surfaces. But there was also considerable interest in the continuing problem of attempting to derive Euclid's parallel postulate rigorously from the remaining axioms and postulates and thus show that it was unnecessary for Euclid to have assumed his non-self-evident fifth postulate. In particular, we will consider the work of Girolamo Saccheri (1667–1733) and Johann Lambert in this regard and that of Clairaut, Euler, and Gaspard Monge (1746–1818) in analytic geometry.

14.3.1 Saccheri and the Parallel Postulate

Saccheri entered the Jesuit order in 1685 and subsequently taught philosophy in Genoa, Milan, Turin, and then at the University in Pavia, near Milan, where he held the chair of mathematics until his death. In 1697, he published a work in logic containing a study of certain types of false reasoning in which one begins with hypotheses which are incompatible with one another. Ultimately, he was led to the consideration of Euclid's postulates and the study of whether an alternative to Euclid's parallel postulate would be compatible or incompatible with the remaining axioms and postulates. It was this study which Saccheri finally published in 1733 in his *Euclides ab omni naevo vindicatus* (*Euclid Freed of all Blemish*). Saccheri treats the "blemish" of the parallel postulate in the first part of this work. In the second part he deals with what he considered two other blemishes, one dealing with the existence of fourth proportionals, the other with the compounding of ratios.

Saccheri's aim in Part One, the only part to be considered here, is "clearly to demonstrate the disputed Euclidean axiom"[26] by assuming that it is false and then deriving it as a logical consequence. Saccheri begins with a consideration of the quadrilateral *ABCD* with two equal sides *CA* and *DB*, both perpendicular to the base *AD*, the same quadrilateral considered some 600 years earlier by al-Khayyāmī (Figure 14.3). Using only Euclidean propositions not requiring the parallel postulate, Saccheri easily demonstrates that the angles at *C* and *D* are equal. There are then three possibilities for these angles, that they are both right, both obtuse, or both acute. Saccheri called these possibilities the hypothesis of the right angle, the hypothesis of the obtuse angle, and the hypothesis of the acute angle, respectively. He then shows that these hypotheses are equivalent respectively to the line segment *CD* being equal to, less than, or greater than the line segment *AB*. It was obvious to Saccheri, as it was to all who considered the question in earlier times, that the only "true" possibility was the hypothesis of the right angle, since it is, in fact, implied by the parallel postulate. The other two hypotheses come from the assumption that the parallel postulate is false. Saccheri intended to derive the parallel postulate from each of these two "false" hypotheses, using only the "self-evident" axioms of Euclid. Thus each of these two possibilities would lead to a contradiction, and the "blemish" of an unnecessary postulate would be removed from Euclid's work.

Saccheri began by proving that if either of the hypotheses is true for one quadrilateral, then it is true for all. He then continued with

Proposition VIII. *Given any triangle ABD, right-angled at B; extend DA to any point X, and through A erect HAC perpendicular to AB, the point H being within the angle XAB. I say the external angle XAH will be equal to, or less, or greater than the internal and opposite ADB, according to whether the hypothesis of the right angle, or obtuse angle, or acute angle is true, and conversely.*

Saccheri's proof made use of various propositions of Euclid's *Elements*, Book I. He began by assuming that *AC* is equal to *BD* and connecting *CD*, thus forming a Saccheri quadrilateral *ABCD* (Figure 14.4). By the hypothesis of the right angle, *CD* = *AB*. It

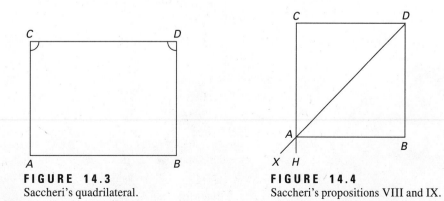

FIGURE 14.3
Saccheri's quadrilateral.

FIGURE 14.4
Saccheri's propositions VIII and IX.

follows that $\angle ADB = \angle DAC = \angle XAH$, and the first case is proved. Under the hypothesis of the obtuse angle, $CD < AB$. Then $\angle XAH = \angle DAC < \angle ADB$, and the second case is proved. Similarly, under the hypothesis of the acute angle, $CD > AB$, and the statement about angles follows. The converse is proved by arguments nearly as brief. This proposition leads to a more important one.

Proposition IX.   *In any right triangle, the two acute angles remaining are, taken together, equal to one right angle, in the hypothesis of the right angle; greater than one right angle, in the hypothesis of the obtuse angle; but less in the hypothesis of the acute angle.*

Because under either hypothesis, angles XAH and HAD together equal two right angles, while angle HAB is a right angle, angles XAH and DAB together equal one right angle. The result is then immediate from proposition VIII. There is, unfortunately, a problem with this theorem which Saccheri evidently did not realize. His statement of the theorem says that both of the nonright angles of the triangle are acute. In fact, this follows from *Elements* I-17 to the effect that any two angles of a triangle are together less than two right angles. That theorem in turn depends on an assumption used by Euclid but never explicitly stated, that a straight line can be extended to any given length, an assumption which turns out not to be valid under the hypothesis of the obtuse angle.

Although Saccheri was unaware of the straight line result, he did prove somewhat later that the hypothesis of the obtuse angle leads to a contradiction of *Elements* I-17 by first demonstrating, in the case of either the hypothesis of the right angle (Proposition XI) or that of the obtuse angle (Proposition XII), that if a line AP intersects PL at right angles and AD at an acute angle, then AD will ultimately intersect PL (Figure 14.5). The central idea of the proof is that if points $M_1, M_2, M_3, \ldots$ are taken along AD with $AM_1 = M_1M_2 = M_2M_3 = \cdots$ and if N_i is the foot of the perpendicular from M_i to AP for each i, then $AN_1 \leq N_1N_2 \leq N_2N_3 \leq \cdots$. It follows that some N_i will lie beyond the point P and therefore that AD intersects PL at some point between M_{i-1} and M_i. Saccheri could now prove Euclid's parallel postulate under these hypotheses.

Proposition XIII.   *If the straight line XA (of given length however great) meets two straight lines AD, XL, making with them on the same side internal angles XAD, AXL less than two right angles, I say that these two, even if neither angle is right, will meet in some point on the side of those angles, and indeed at a finite distance, if either hypothesis holds, of the right angle or of the obtuse angle.*

Again, the proof depends on *Elements* I-17. Because one of the angles, say AXL, is acute, one can drop a perpendicular AP on XL, which, by that proposition, falls on the side of the acute angle AXL (Figure 14.6). Because in either hypothesis the two acute angles PAX and PXA are not together less than a right angle, if these are subtracted from the sum of the given angles XAD and AXL, the remaining angle DAP will be less than a right angle. Propositions XI and XII then allow Saccheri to conclude that the two lines will intersect.

But now, because the acute angles of triangle APX are, under the hypothesis of the obtuse angle, greater than one right angle, Saccheri could choose an acute angle PAD which, together with those two angles, make up two right angles. By proposition XII, the

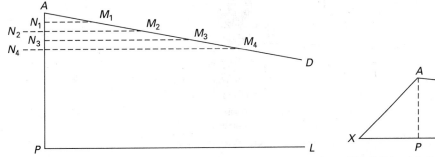

FIGURE 14.5
Saccheri's propositions XI and XII that *AD* and *PL* will
ultimately intersect.

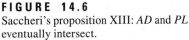

FIGURE 14.6
Saccheri's proposition XIII: *AD* and *PL*
eventually intersect.

line *AD* eventually intersects *XP* extended at, say, *L*. It then follows that two angles of triangle *XAL* themselves sum to two right angles, contradicting I-17. Also, of course, because the parallel postulate has been proved, Saccheri can use it to prove, as Euclid did in *Elements* Book I, that the three angles of any triangle are together equal to two right angles, contradicting, by way of Proposition IX, the hypothesis of the obtuse angle itself. He states it as follows.

Proposition XIV. *The hypothesis of the obtuse angle is absolutely false, because it destroys itself.*

Saccheri next showed that the hypotheses of the right, obtuse, and acute angles are equivalent respectively to the results that the sum of the angles of any triangle is equal to, greater than, or less than two right angles and that the sum of the angles of a quadrilateral is equal to, greater than, or less than four right angles. He then proceeded to investigate in more detail the consequences of the hypothesis of the acute angle. Here, however, he was not able to derive the parallel postulate as a consequence. He did, however, derive other intriguing results. For example, consider

Proposition XVII. *If the straight line AH is at right angles to any straight line AB however small, I say that under the hypothesis of the acute angle it cannot be true that every straight line BD intersecting AB in an acute angle will ultimately meet AH produced.*

Suppose *BM* is also perpendicular to *AB*. Drop a perpendicular from *M* to *AH* intersecting that line at *H* (Figure 14.7). Because the sum of the angles of a quadrilateral is less than four right angles, it follows that angle *BMH* is acute. Similarly, if *BX* is drawn from *B* perpendicular to *HM*, intersecting that line at *D*, then angle *XBA* is also acute. But now *BD* extended cannot intersect *AH* extended, because, the angles at *H* and *D* both being right, this would contradict *Elements* I-17.

Because proposition XVII implies that there are two straight lines in the plane which do not meet, Saccheri could show in proposition XXIII that for such lines, either they

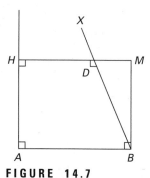

FIGURE 14.7
Saccheri's proposition XVII: *BD* and *AH* do
not ultimately intersect.

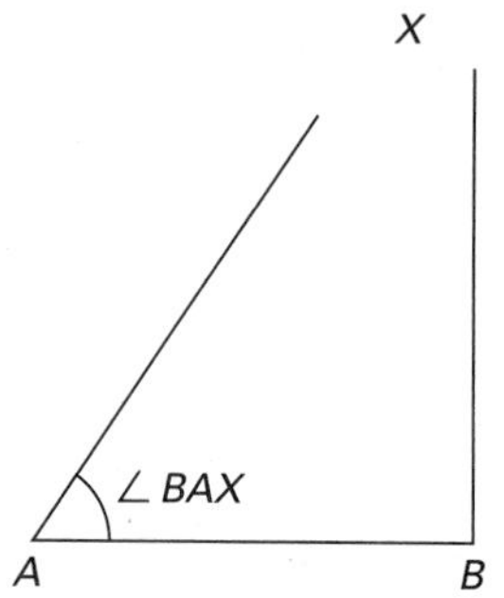

FIGURE 14.8
Saccheri's "angle of parallelism": *AX* and
BX only meet at an "infinite" point *X*.

have a common perpendicular or else "they mutually approach ever more toward each other."[27] Furthermore, in the latter case, the distance between the lines becomes smaller than any assigned length, that is, the lines are asymptotic. Saccheri was then able to prove in proposition XXXII that given a line *BX* perpendicular to a line segment *AB*, there is a certain acute angle *BAX* such that the line *AX* "only at an infinite distance meets *BX*,"[28] while lines making smaller acute angles with *BA* intersect *BX* and those making larger ones all have a common perpendicular with *BX* (Figure 14.8). Saccheri then concluded as follows.

Proposition XXXIII. *The hypothesis of the acute angle is absolutely false, because it is repugnant to the nature of the straight line.*

Saccheri hardly gave a "proof" of this result. It appears, in fact, that he only ended his quest with this proposition because he had faith that the parallel postulate must be true. He merely wrote that, given the hypothesis of the acute angle, there must exist two straight lines which eventually run together "into one and the same straight line, truly receiving, at one and the same infinitely distant point a common perpendicular in the same plane with them."[29] But then, apparently having second thoughts on the matter, he spent the next 30 pages attempting a further justification of his result, showing along the way that two straight lines cannot enclose a space, that two straight lines cannot have a segment in common, and that there is a unique perpendicular to a given line at a given point, all of which ideas relate to finite straight lines and none of which have much to do with his two straight lines which both join together and have a common perpendicular at infinity. Nevertheless, Saccheri believed that he had accomplished his aim.

14.3.2 Lambert and the Parallel Postulate

Johann Lambert, having studied at least a summary of Saccheri's work, attempted to improve upon it. But his work on the parallel postulate, the *Theorie der Parallellinien* (*Theory of Parallel Lines*), finished by 1766, was never published, perhaps because Lam-

---

Lambert was a self-educated mathematician and philosopher, who had mastered mathematics while assisting his father as a tailor in Alsace. In 1748 he moved to Switzerland as a private tutor for a wealthy family and later toured Europe on an educational journey with his pupils. During this period, Lambert was able to study in the family library and carry out both theoretical and experimental investigations. He never, however, adopted conventional bourgeois attitudes. He was finally proposed for a position at the Prussian Academy of Sciences in Berlin. When he arrived there early in 1764, he was welcomed by Euler, but his strange appearance and behavior delayed his appointment for a year. Eventually overcoming the initial hostility of Frederick II, Lambert produced over 150 works before his untimely death at the age of 49.

---

bert was not finally happy with the conclusions. In the book, he considered a quadrilateral with three right angles and made three hypotheses as to the nature of the fourth angle, essentially the same three hypotheses as had Saccheri, that it could be right, obtuse, or acute. Again using the principle that a straight line can be of arbitrarily great length, Lambert rejected the second hypothesis. But he had great difficulty in rejecting the third hypothesis. As he noted, "this hypothesis would not destroy itself at all easily."[30]

Like Saccheri, Lambert began to deduce various consequences from that hypothesis. The most surprising was that in his fundamental quadrilateral the difference between 360° and the sum of the angles was proportional to the area of the quadrilateral, that is, the larger the quadrilateral, the smaller the angle sum. It follows that "if the third hypothesis holds we would have an absolute measure of length for each line, for the content of each surface, and each bodily space."[31] In other words, because Lambert's result implied that a quadrilateral $ABCD$ with right angles at A, B, and C and with $AB = AD$ has a determined acute angle D, an angle which can fit no other such quadrilateral, the measure of D may be taken as the absolute measure of the quadrilateral. Lambert was not able to deduce this absolute measure, that is, he could not determine what the angle D would be if $AB = AD = 1$ foot, but he did realize that this hypothesis destroyed entirely the notion of similar figures. It also implied that the difference between 180° and the sum of the angles of a triangle, the **defect** of the triangle, was proportional to the area of the triangle. Lambert realized that a similar result is true under the second hypothesis, with the defect being replaced by the excess of the angle sum over 180°. But he also knew that spherical triangles had this same property that their angle sum was greater than 180° and that the excess was proportional to the area. He then argued by analogy: "I should almost therefore put forward the proposal that the third hypothesis holds on the surface of an imaginary sphere."[32]

Lambert abandoned his study of Euclid's parallel postulate once he felt that he could not successfully refute the hypothesis of the acute angle, even though it seems he was convinced that Euclid's geometry was true of space. Nevertheless, he believed that because the geometry of the obtuse angle hypothesis was reflected in geometry on the sphere, the sphere of imaginary radius would perform the same function for the acute angle hypothesis. Although by 1770 he had introduced the hyperbolic functions as complex analogues of the circular ones, in the sense that $\cosh ix = \cos x$ and $\sinh ix = i \sin x$, he was not able to apply these functions to develop a geometry on the imaginary sphere based on the hypothesis of the acute angle, nor could he give a construction in three dimensional space of this imaginary sphere. It was only in the early nineteenth century, when analysis of this type could be brought to bear on the alternatives to the parallel postulate, that what is today called non-Euclidean geometry was developed. That story will be related in Chapter 17.

14.3.3 Clairaut and Space Curves

The central thrust of eighteenth century geometry was its connection to analysis. Thus the second volume of Euler's *Introductio* provided a clear organization of the topic of plane curves. Euler began with a classification of quadratic and cubic curves and then gave a treatment of quartics which included 146 different forms. He next dealt with the various properties of curves in general, without the use of calculus, including asymptotes, curvature, and singular points. He even included a chapter on transcendental curves, in which he considered such curves as $2y = x^i + x^{-i}$ (or $y = \cos(\ln x)$) and $y = x^x$, and sketched the earliest graph of $y = \arcsin x$. For certain curves, such as the Archimedean spiral, he made use of polar coordinates, described in a modern fashion. Thus, if s represents the polar angle and z the length of the radius, then the equation of that spiral is $z = as$. Similarly, the equation $z = ae^{s/n}$ represents the logarithmic spiral, whose graph he also displayed.

The earliest published work on curves in space was the 1731 *Recherches sur les courbes á double courbure* (*Researches on Curves of Double Curvature*) by Clairaut. For Clairaut, a curve in space could only be defined as an intersection of certain surfaces. Thus he began his study by dealing with various simple cases of surfaces. He showed from the geometric definition that a sphere has equation $x^2 + y^2 + z^2 = a^2$, that a paraboloid is $y^2 + z^2 = ax$ and that, in general, the equation of a surface of revolution formed by revolving a curve $f(x, u) = k$ around the x-axis is found by replacing u by $\sqrt{y^2 + z^2}$. He showed that every equation in three variables in which each term is of the same degree must be a conic surface. And he proved the general result that an equation in three variables always defines a surface whose properties are determined by the equation.

Clairaut applied the techniques of differential calculus to find tangents and perpendiculars to curves in space and, therefore, considered such curves as being composed of "an infinity of small sides."[33] To determine the tangent line to a curve at a point N is to determine the point t where the extension of the line segment Nn, which connects N to an infinitely close point n on the curve, intersects the xy-plane, or, if M is the projection of N onto that plane, to determine the length Mt (Figure 14.9). Of course, this goal of finding the subtangent Mt is the direct analogue of the standard seventeenth century method of determining tangents to plane curves. The result is the direct analogue of the result in two dimensions as well, although the procedure in three space is complicated somewhat by the necessity of keeping all relevant lines in the same plane. Clairaut takes Mm as the projection in the xy-plane of the infinitesimal side Nn of the curve and extends it to meet the intersection point t. Then the triangle NtM defines the plane in which the tangent line lies. Clairaut only makes use of one axis, the x-axis. So if AP is taken to represent the

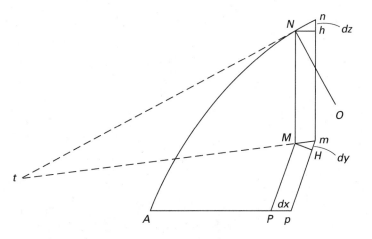

FIGURE 14.9
Clairaut and tangents to
space curves.

x-coordinate of N, the z and y coordinates are represented by the length MN of the perpendicular from N to the xy-plane and the length MP of the perpendicular from there to the axis, respectively. If Ap, nm, pm, are the corresponding coordinates of n, and further, if Nh is drawn parallel to Mm and MH parallel to Ap, then Pp represents dx, nh represents dz, mH represents dy, and $\sqrt{dx^2 + dy^2}$ represents Mm. Because triangles nNh and NMt are similar, Clairaut derives the proportion $nh : Nh = MN : Mt$. Because $Nh = Mm$, it follows that

$$\frac{dz}{\sqrt{dx^2 + dy^2}} = \frac{z}{Mt} \qquad \text{or} \qquad Mt = \frac{z\sqrt{dx^2 + dy^2}}{dz}.$$

The tangent Nt itself is then given by

$$Nt = \sqrt{MN^2 + Mt^2} = \frac{z\sqrt{dx^2 + dy^2 + dz^2}}{dz}.$$

Furthermore, the perpendicular NO from the curve to the xz-plane which is also perpendicular to the plane of triangle NtM is given by

$$NO = \frac{z(dx^2 + dy^2 + dz^2)}{\sqrt{dx^2 + dy^2}}.$$

Clairaut gives several examples of this computation, including the curve determined by the intersection of the two parabolic cylinders $ax = y^2$ and $by = z^2$. In this case $a\,dx = 2y\,dy$ and $b\,dy = 2z\,dz$. It follows that

$$dy = \frac{a\,dx}{2\sqrt{ax}} \qquad dz = \frac{b\,dy}{2\sqrt{by}} = \frac{ab\,dx}{4\sqrt[4]{b^2a^3x^3}} \qquad \text{and} \qquad \sqrt{dx^2 + dy^2} = \frac{dx\sqrt{4x + a}}{4x}.$$

Since $z = \sqrt[4]{ab^2x}$, the subtangent $Mt = \left(z\sqrt{dx^2 + dy^2}\right)/dz$ is given by $Mt = \sqrt{4x + a}$. Similarly, the tangent Nt is found to be $Nt = \sqrt{by + 4x + a}$. An analogous calculation can be made for the perpendicular NO.

14.3.4 Euler and Space Curves and Surfaces

It was not until 1775 that Euler took up the subject of space curves, this time expressing them parametrically through the arc length s.[34] Thus, a curve was given by three equations $x = x(s)$, $y = y(s)$, $z = z(s)$. Taking differentials of each led Euler to the expressions $dx = p\,ds$, $dy = q\,ds$, $dz = r\,ds$ from which he derived the result $p^2 + q^2 + r^2 = 1$. The functions p, q, r, the derivatives of the coordinate functions with respect to arc length, are the components of the unit tangent vector to the curve. These components are also called the **direction cosines** of the tangent line (or of the curve itself) at the specified point. Furthermore, Euler used the unit sphere centered at a point $(x(s), y(s), z(s))$ to define the curvature of the curve, generalizing Newton's earlier concept of the curvature of a plane curve. If the "unit vectors" (p, q, r) at the neighboring parameter values s and $s + ds$, both considered to be emanating from that center, differ by an arc on the sphere equal to ds', then the **curvature** κ at that point was defined as $\left|\frac{ds'}{ds}\right|$, a value measuring how the curve at any point differs from a great circle on the sphere. Because the vector ds' is given by

$$\left(\frac{dx}{ds}(s + ds) - \frac{dx}{ds}(s), \frac{dy}{ds}(s + ds) - \frac{dy}{ds}(s), \frac{dz}{ds}(s + ds) - \frac{dz}{ds}(s)\right) = \left(\frac{d^2x}{ds^2}ds, \frac{d^2y}{ds^2}ds, \frac{d^2z}{ds^2}ds\right),$$

it follows that

$$\kappa = \left|\frac{ds'}{ds}\right| = \sqrt{\left(\frac{d^2x}{ds^2}\right)^2 + \left(\frac{d^2y}{ds^2}\right)^2 + \left(\frac{d^2z}{ds^2}\right)^2}.$$

Euler next defined the **radius of curvature** ρ to be the reciprocal of the curvature. Thus

$$\rho = \frac{ds^2}{\sqrt{(d^2x)^2 + (d^2y)^2 + (d^2z)^2}}.$$

It turned out, although it was not proved until the nineteenth century, that curvature is one of the two essential properties of a space curve. The other quantity is the torsion, which measures the rate at which the curve deviates from being a plane curve. If the curvature and torsion are given as functions of arc length along a curve, then the curve is completely determined up to its actual position in space.

Besides dealing with space curves, Euler also elaborated on Clairaut's treatment of surfaces, both in the second volume of his *Introductio* and in an important paper some years later. In the *Introductio*, he systematized the subject of quadric surfaces. Like Clairaut, Euler used a single coordinate plane, with only one axis defined on it, and represented the third coordinate by the perpendicular distance from a point to that plane. But he did remark that it was possible to use three coordinate planes and often described a surface by means of its trace in various such planes. He gave the equation for a plane in three space as $\alpha x + \beta y + \gamma z = a$ but described the meaning of the coefficients only in terms of the cosine of the angle θ between that plane and the xy-plane: $\cos\theta = \gamma/\sqrt{\alpha^2 + \beta^2 + \gamma^2}$. In his discussion of quadric surfaces themselves, Euler began by noting that the general second degree equation in three variables can be reduced by a change of coordinates to one of the forms $Ax^2 + By^2 + Cz^2 = a^2$, $Ax^2 + By^2 = Cz$, or $Ax^2 = By$. He was then able to classify the quadric surfaces into classes, including the ellipsoid, the elliptic and hyperbolic paraboloids, the elliptic and hyperbolic hyperboloids, the cone, and the parabolic cylinder.

Because the *Introductio* is a precalculus work, Euler made no attempt to deal there with such ideas as tangent planes and normal lines to surfaces in three-space. It was his paper of 1760, *Recherches sur la courbure des surfaces* (*Research on the Curvature of Surfaces*), which one might call the beginning of the differential geometry of surfaces.[35] In that work Euler noted that although the method of finding the curvature of a plane curve at a given point was well known, even to define the curvature of a surface in space at a point was difficult. Each section of a surface by a plane through the given point gives a different curve, and the curvatures of each of these sections may well be different, even if one restricts oneself to plane sections which are perpendicular to the surface. In the paper, Euler calculated these various curvatures and established some relationships among them. First, however, he needed to characterize planes perpendicular to the surface, that is, planes which pass through the normal line to the surface at the given point P. He showed that the plane with equation $z = \alpha y - \beta x + \gamma$ is perpendicular to the surface defined by $z = f(x, y)$ if $\beta \frac{\partial z}{\partial x} - \alpha \frac{\partial z}{\partial y} = 1$. Defining the principal plane to be the plane through P perpendicular both to the surface and to the xy-plane, Euler then demonstrated that if a given plane perpendicular to the surface makes an angle ϕ with the principal plane, the curvature of the section formed by that plane is given by $\kappa_\phi = L + M \cos 2\phi + N \sin 2\phi$ where L, M, N depend solely on the partial derivatives of z at P. Taking the derivative of this expression with respect to ϕ, Euler found that the maximum and minimum curvatures occur when $-2M \sin 2\phi + 2N \cos 2\phi = 0$ or when $\tan 2\phi = N/M$. But since $\tan(2\phi + 180°) = \tan 2\phi$, Euler concluded that if a maximum curvature occurs for a given value of ϕ, the minimum occurs at $\phi + 90°$. He was finally able to show that if κ_1 is the maximum curvature and κ_2 the minimum, and if the minimum curvature occurs at the principal plane, then the curvature of any section made by a plane at angle ϕ to the principal plane is given by $\kappa = \frac{1}{2}(\kappa_1 + \kappa_2) - \frac{1}{2}(\kappa_1 - \kappa_2)\cos 2\phi$.

14.3.5 The Work of Monge

Gaspard Monge systematized the basic results of both analytic and differential geometry and added much new material in several papers beginning in 1771 and finally in two textbooks written for his students at the École Polytechnique at the end of the century. For example, in a paper published in 1784 Monge presented for the first time the point-slope form of the equation of a line: "If one wishes to express the fact that this line [with slope-intercept equation $y = ax + b$] passes through the point M of which the coordinates are x' and y', which determines the quantity b, the equation becomes $y - y' = a(x - x')$, in which a is the tangent of the angle which the straight line makes with the line of x's."[36] Monge's text *Géométrie descriptive* of 1799, on the other hand, did not deal with algebra at all, but relied on the basic ideas of pure geometry. Monge outlined many techniques for representing three dimensional objects in two dimensions. He systematically used projections and other transformations in space to draw in two dimensions various aspects of space figures. He described in detail such concepts as the tangent plane to a surface, the intersection of two surfaces, the notion of a developable surface (a surface which can be flattened out to a plane without distortion), and the curvature of a surface.

> ### 𝐵𝑖𝑜𝑔𝑟𝑎𝑝ℎ𝑦 Gaspard Monge (1746–1818)
>
> Monge was born in Beaune, a town about 150 miles southeast of Paris. He was a brilliant student in Lyon and, after preparing a plan of his native town, was invited to the Royal Engineering School at Mézières, where he soon had an opportunity to display his abilities. He was asked to develop a plan for a particular type of fortification. Instead of using the traditional complex method, he employed a new graphical method, a method he subsequently enlarged into the subject of descriptive geometry. He was therefore promoted to a teaching position from which he was able to influence the scientific training of French military engineers. He was elected to the Academy of Sciences in 1780 and thereafter held many positions of responsibility under the royal, the revolutionary, and finally the imperial governments of France over the next 35 years (Figure 14.10).

FIGURE 14.10

Monge on a French stamp.

In his second text, the *Application de l'analyse á la géométrie* of 1807, which grew out of lecture notes dating from 1795, Monge showed how to apply analysis to geometry. The first part of this work, which used only algebra, contained the earliest detailed presentation of the analytic geometry of lines in two- and three-dimensional space as well as planes in three-dimensional space. Thus Monge indicated that points in space are to be determined by considering perpendiculars to each of three coordinate planes. A line in space is determined by its projection onto two of these three planes, the equations of the projections onto the xy-plane, for example, being given in the slope-intercept form or in the point-slope form. Monge showed how to find the intersection of two lines as well as how to find a line parallel to a given line through a given point and a line through two given points. He also noted that the lines in the plane with equations $y = ax + \alpha$ and $y = a'x + \alpha'$ are perpendicular if $aa' = -1$.

Monge wrote the equation of a plane both in the form $z = ax + by + c$, where a and b are the slopes of the lines of intersection of this plane with the xz-plane and the yz-plane respectively, and in the symmetric form $Ax + By + Cz + D = 0$, where the coefficients A, B, and C determine the direction cosines of the angles between the plane and the coordinate planes. He then proceeded to discuss all of the familiar problems dealing with points, lines, and planes, such as finding the normal line to a plane passing through a given point, finding the shortest distance between two lines, and finding the angle between two lines or between a line and a plane.

The second part of Monge's text was devoted to the study of surfaces. Here he used the entire machinery of calculus to develop analytically all of the topics he had considered in his *Géométrie descriptive*. Thus he considered in detail how to determine, from various types of descriptions, the partial differential equation which represents a given surface as well as how, in certain cases, to integrate that equation. To develop the equations of the tangent plane and normal line to a surface, Monge began by noting that the differential equation which represents the surface $z = f(x, y)$ near a point (x', y', z') is

$$dz = \frac{\partial z}{\partial x}dx + \frac{\partial z}{\partial y}dy,$$

where the partial derivatives are evaluated at x' and y'. On the other hand, the equation of any plane through (x', y', z') can be written as $A(x - x') + B(y - y') + C(z - z') = 0$. For this plane to be a tangent plane, any point on it infinitely near the given point must also be on the surface, that is, must satisfy the differential equation of the surface. So, taking $x - x'$ as dx, $y - y'$ as dy, $z - z'$ as dz, Monge noted that the equation $A\,dx + B\,dy + C\,dz = 0$ must be identical to $dz = \frac{\partial z}{\partial x}dx + \frac{\partial z}{\partial y}dy$. It follows that $A/C = -\partial z/\partial x$, $B/C = -\partial z/\partial y$, and that the equation of the tangent plane is

$$z - z' = (x - x')\frac{\partial z}{\partial x} + (y - y')\frac{\partial z}{\partial y}.$$

The equations of the normal line to the surface, that is, the normal line to the tangent plane, are then calculated to be

$$x - x' + (z - z')\frac{\partial z}{\partial x} = 0; \qquad y - y' + (z - z')\frac{\partial z}{\partial y} = 0.$$

Monge's general idea of connecting partial differential equations with the geometry of space has had great influence through the years. Perhaps even more importantly, his teaching at the École Polytechnique influenced an entire generation of French engineers, mathematicians, and scientists, a matter to which we will return in section 14.4.

14.3.6 Euler and the Beginnings of Topology

It was in the mid-1730s that Euler became aware of a little problem coming out of the town of Königsberg, in East Prussia (now in Russia). In the middle of the river Pregel, which ran through the town, there were two islands. The islands and the two banks of the river were connected by seven bridges. The question asked by the townspeople was whether it was possible to plan a stroll which passes over each bridge exactly once. Euler, as usual, instead of considering this problem in isolation, attacked and solved the general problem of the existence of such a path, whatever the number of regions and bridges. In a paper published in 1736, he noted first that if one labeled the regions by letters A, B, C, $D, \ldots$, one could then label a path by a series of letters representing the successive regions passed through (Figure 14.11). Thus $ABDA$ would represent a path leading from region A to region B and then to D and back to A, regardless of the particular bridges crossed. It followed immediately that a complete path satisfying the desired conditions must contain

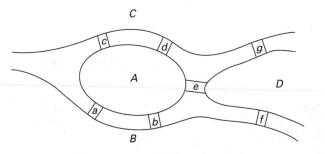

FIGURE 14.11
The seven bridges of
Königsberg.

one more letter than the number of bridges. In the Königsberg case, that number must be eight.

Next, Euler realized that if the number k of bridges leading into a given region is odd, then the letter representing that region must occur $\frac{1}{2}(k + 1)$ times. Thus, if there is one bridge leading to region A, then A will occur once; if there are three bridges, A will occur twice; and so on. It does not matter in this case whether the path starts in the region A or in some other region. On the other hand, if k is even, then the letter representing that region will occur $\frac{1}{2}k$ times if the path starts outside of the region and $\frac{1}{2}k + 1$ times if the path begins in the region. For example, if there are four bridges leading into region A, then A will occur twice if the route begins outside of A and three times if it begins in A. From the two different determinations of the number of letters which occurs in his representation of a particular path, Euler could determine whether a path passing over each bridge exactly once is possible:

> Every route must, of course, start in some one region. . . . If the sum of the resulting numbers [calculated above for each region] is equal to the actual number of bridges plus one, the journey can be accomplished, though it must start in a region approached by an odd number of bridges. But if the sum is one less than the number of bridges plus one, the journey is feasible if its starting point is a region approached by an even number of bridges, for in that case the sum is again increased by one.[37]

In the Königsberg case, there is an odd number of bridges leading into each of the regions A, B, C, D, namely, 5, 3, 3, and 3, respectively. The sum of the corresponding numbers, 3, 2, 2, and 2, is 9, which is more than "the actual number of bridges plus one." It follows that the desired path is impossible. In general, Euler noted that such a path will always be impossible if there are more than two regions approached by an odd number of bridges. If there are exactly two such regions, then the path is possible as long as it starts in one of those regions. Finally, if all the regions are approached by an even number of bridges, then a path crossing each bridge exactly once is always possible. Because once one knows whether a path is possible, the actual construction of it is straightforward, Euler had completely solved the problem he set.

This particular problem, as well as another solved by Euler, that given any simple polyhedron with V vertices, E edges, and F faces, then $V - E + F = 2$, were merely isolated facts in the eighteenth century (Figure 14.12). Euler did note, nevertheless, that they are apparently part of a branch of geometry in which the relations depend on position alone and not at all on magnitudes. It was not until the late nineteenth and early twentieth centuries, however, that these facts and certain others were systematically studied and finally turned into the subject of topology.

FIGURE 14.12
Euler and the polyhedron formula on a stamp from the D.D.R.

14.4 THE FRENCH REVOLUTION AND MATHEMATICS EDUCATION

The major mathematicians of the eighteenth century were associated, not with universities, but with academies founded under the patronage of various monarchs to gain prestige for their countries as well as to provide a ready source of scientific assistance in both military and civilian projects necessary for the advancement of the nation. The universities in general did not provide an advanced education in mathematics, since even in the eighteenth century they were primarily dominated by philosophers. In France in particular there had been no first rate mathematicians associated with the University of Paris since the fourteenth century. The only schools which provided a mathematical and scientific education were the military schools, one of whose major functions was to produce military engineers. Thus early in his career Monge taught at the military school at Mézières, where he developed his earliest ideas on descriptive geometry in connection with drafting plans for military fortifications. Similarly, Laplace and Legendre taught for a time at the École Militaire in Paris.

Because the military schools and the universities were centers of Royalist support during the French Revolution, most were closed by 1794 when the revolution reached its most radical phase. Nevertheless, with the attacks on France by the armies of her neighbors as well as the fleeing of some of her best educated citizens, it was necessary to have schools in which students of nonnoble background, who showed a "constant love of liberty and equality and a hatred of tyrants,"[38] could be trained to serve in both military and civilian capacities as engineers and scientists. It was for this purpose that the National Convention on September 28, 1794 founded the École Centrale des Travaux Publiques, soon renamed the École Polytechnique. The school was to be more than just an engineering school, however. It was also to develop well-educated citizens and, in particular, to stimulate the talent necessary to advance science in general.

Monge, who before the revolution had helped to reform the teaching of science in the naval schools, was appointed to the commission responsible for organizing the École Polytechnique. He was therefore instrumental in developing the "revolutionary course," the three-month concentrated survey of the sciences which the first students began in December, and which was to serve as a preview of the two- or three-year course of study they would ultimately pursue. The students were to study four basic areas, descriptive geometry, chemistry, analysis and mechanics, and physics, as well as have a course in engineering drawing. The latter course met for three hours every evening, while the first three each had an hour lecture scheduled every morning followed by an hour of directed study. The physics course, however, only met for four hours in each ten day *décade*. (The revolutionary calendar divided the month into three ten-day periods, *décades*, rather than into seven-day weeks.)

The first month of the descriptive geometry course taught by Monge himself was to cover essentially the material described in section 14.3.5. Thus the students were to study general methods of projection, the determination of tangents and normals to curves and surfaces, the construction of intersections of surfaces, and the notion of a developable surface. They would also consider applications of these ideas to various questions in such fields as building construction and mapmaking. The second month of the course covered architecture and public works, while the third month dealt with fortifications. The course in analysis, also to be taught by Monge, was to start the first month with the solution of polynomial equations up through the fourth degree and then to continue with algebraic and geometric methods for solving systems of equations and a study of the curves and surfaces these equations represent. The second month was to deal with the theory of series, exponential and logarithmic functions, elementary probability, and differential calculus with applications to geometry. The final month would then consider the integral calculus, including the finding of lengths, areas, and volumes, and the solution of differential equations.

The syllabus was ambitious. Unfortunately, it could not and did not work. Monge was ill when the school opened for classes on December 21, 1794. His course in descriptive geometry was therefore postponed, while C. J. Ferry taught his analysis course, and C. Griffet-Labaume repeated the course for those students who wanted it during the free hour. Unfortunately, it was clear before the first *décade* was finished that most of the students could not comprehend the course at all. It was soon decided by Lagrange, who was on the governing board of the school, to let Griffet-Labaume teach an elementary course in algebra instead of his repetitions. But even though this new course did not go further in the first month than the representation of plane curves by equations in two variables, fewer than a third of the students remained to the end. Things improved somewhat in the second month, with the continuation of the elementary algebra course, the addition of a course in trigonometry, and Monge's return to teach his course in descriptive geometry, but it was still evident that there was a great gap between the plans and the reality. There were several reasons for the poor beginning of the school, among which was the severe Paris winter, aggravated by a food shortage. But the primary reason was the poor preparation of the students. Students had been examined in their hometowns before being admitted, chiefly for "political correctness," but a political test for entrance could not make up for the lack of any consistent academic standards.

Despite the poor beginning, Monge and others were soon able to make vast improvements at the École Polytechnique, which became the model for colleges of engineering throughout Europe and the United States. National standards for education were established, partly through the creation in 1795 of a new national school for teachers, the École Normale Superieure. But even before these standards were established, the École Polytechnique itself sent examiners to the provinces to insure that admitted students were well prepared. The course offerings were made somewhat more realistic. The "revolutionary course" was not taught again, only the regular courses, spread over the three-year program. Finally, France's best mathematicians all taught at the school, including Lagrange, Laplace, and Sylvestre Lacroix (1765–1843), and several wrote elementary textbooks for use there. Lacroix, in particular, wrote texts on arithmetic, trigonometry, analytic geometry, synthetic geometry, and differential and integral calculus. Most of these works went through numerous editions and were translated into several languages. In fact, Lacroix's calculus text, translated into English in 1816, was influential in bringing continental methods to both England and the United States.

Besides changing the nature of technical education in France, the revolutionary government was also responsible for the standardization of weights and measures in France and the introduction of the metric system (Figure 14.13). The Constituent Assembly passed an initial law requiring the standardization in May of 1790, and the Academy of Sciences then formed a committee, including Laplace, Lagrange, and Monge, to consider the subject. The initial recommendation was that the unit of length should be that of a pendulum which beats in seconds, but by March of 1791 the committee decided that the standard length should be the ten-millionth part of a quadrant of a great circle on the earth, since that would be more "natural" than using time. The Assembly then enacted a law providing for a new geodesic survey of the meridian of Paris so this length could be accurately determined. With the unit of length defined—it was named the "meter" a year later on a suggestion of Laplace—it was decided that all subdivisions and multiples would be decimal. Furthermore, measures for area and volume were to be defined in terms of the

FIGURE 14.13
A French stamp commemorating the introduction of the metric system.

measure for length. Thus, a basic unit for area, the square on a side of 100 meters, was to be called an "are." Similarly, a basic unit of mass, the gram, was defined to be the mass of a cubic centimeter of water at a given temperature (Figure 14.14).

The members of the committee went even further, devising decimal systems for money, related to weight through the value of gold and silver, and for angles, by dividing the quadrant of a circle into 100 equal parts, now called grads. Finally, they designed the revolutionary calendar to extend the metric system to the realm of history. Laplace went along with this division of the month into 3 ten-day *décades*, with five extra holidays at the end of the year, even though he realized that this decimalization of the calendar would cause more problems than it would solve, given the incommensurability between the day and the year. Interestingly enough, although the decimalization of weights and measures was accepted over the next century by virtually the entire world, neither the decimalization of angles nor of the calendar lasted more than a dozen years.

Napoleon, having taken control of France in 1799, restored the Gregorian calendar in 1806. Meanwhile, he had been able to gain the support of France's important scientists. Monge, in particular, having traveled to Egypt with Bonaparte in 1798, became a strong supporter of the emperor. In return, he was named a senator for life and later a grand

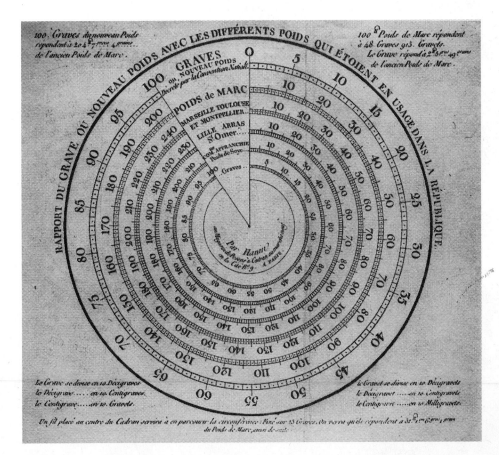

FIGURE 14.14
Plate from French work of the Commission temporaire des poids et mesures (1793) giving conversions from old standards of weight to the new metric standard. Note that the basic unit of weight was called a *grave* or a *nouveau poid*, rather than a kilogram. (Source: Smithsonian Institution Photo No. 89-8736)

officer of the Legion of Honor and count of Péluse. Legendre, Lagrange, and Laplace also received honors, the first being named a Chevalier de l'Empire, and the others counts. With the fall of Napoleon in 1815, Monge lost all of his positions, spending the remaining three years of his life in intellectual exile. Lagrange had died in 1813, but Laplace and Legendre were able to make peace with the restored monarchy and continue their work unabated.

14.5 MATHEMATICS IN THE AMERICAS

The political upheavals of late eighteenth century France were to some extent inspired by the American Revolution of 1776. Mathematics in America, however, was nowhere near the level of that in France. By the time of the Revolution, there were nine colleges in what was to become the United States. Because most were founded with the primary purpose of training clergymen, many of the professors of mathematics were themselves in the clergy and had their chief interest in theology rather than mathematics. Thus the level of instruction in the subject was even lower than the corresponding instruction in European universities of the time. There were generally courses available in arithmetic, algebra through the solution of simple equations, basic geometry, and trigonometry through the solution of plane triangles by use of logarithms, with application to surveying and astronomy. By the middle of the eighteenth century, courses in fluxions began to be offered, at least at Harvard and Yale. There was also private instruction in mathematics available. Thus by 1727 Isaac Greenwood (1702–1745), Hollis Professor of Mathematics at Harvard from 1728 to 1738, was available for teaching both the method of fluxions and the Leibnizian differential calculus, as well as navigation, surveying, mechanics, optics, and astronomy. In 1729 he authored the first published arithmetic work written by an American. In general, however, Americans used textbooks imported from England.

Two scientific societies were founded in America during the century, modeled on the similar societies in Europe. These were the American Philosophical Society, founded in 1743, and the American Academy of Arts and Sciences, organized in 1780. But even the journal of the latter society was pessimistic about American achievement in mathematics at the time. An editorial in the first volume of its *Memoirs* noted that the mathematical papers appearing would be of interest to very few readers and that, in any case, they contained little in the way of research, being chiefly practical.

Although there were no research mathematicians in America during the eighteenth century, there were astronomers and surveyors of note who at least appreciated the importance of the subject. John Winthrop (1714–1779), who served for more than 40 years as Hollis Professor of Mathematics at Harvard, taught mathematics in the Newtonian tradition, including such topics as "the elements of Geometry, together with the doctrine of Proportion, the principles of Algebra, Conic Sections, Plane and Spherical Trigonometry, with general principles of Mensuration of Planes and Solids, the use of globes, [and] the calculations of the motions and phenomena of the heavenly bodies according to the different hypotheses of Ptolemy, Tycho Brahe, and Copernicus."[39] David Rittenhouse (1732–1796), an astronomer, clock maker, and surveyor, who performed many detailed astronomical observations and assisted in the establishment of the Mason-Dixon line, published a few mathematical papers, including one dealing with the integration of powers of the sine.

Biography   Benjamin Banneker (1731–1806)

Benjamin Banneker's father was a freed slave while his mother was the daughter of a former indentured servant from England who had married her own black slave, himself perhaps the son of an African chieftain. His grandmother taught him to read and write and arranged for him to attend a small country school during the winter months. His technical abilities surfaced early in his life, although his circumstances as a free black farmer in Maryland did not allow him to develop his talents. Nevertheless, by the age of 22 he had constructed an accurate clock, mostly out of wood, a clock which continued to operate until it was destroyed in a fire shortly after his death. Banneker was, however, fortunate in that his neighbors included the Ellicott family, a family of businessmen and surveyors from whom he was able to borrow

technical books and a few scientific instruments. Having taught himself the principles of mathematics, surveying, and astronomy, he was invited by Andrew Ellicott to assist him in the survey of the boundaries of the District of Columbia, a task to which Ellicott had been appointed by President Washington in 1791. After his return from this task, Banneker continued his studies to the point that he was able to compile an almanac for 1792 which included daily positions of the sun, moon, and planets, times and descriptions of solar and lunar eclipses, times of rising and setting of the sun, moon, and certain bright stars, and local tide tables. Banneker's almanac proved popular, and he was able to have similar almanacs published yearly through 1797 (Figure 14.15).

FIGURE 14.15
Benjamin Banneker honored on a U.S. stamp.

And Benjamin Banneker (1731–1806), the first American black to achieve distinction in science, taught himself sufficient mathematics and astronomy to publish a series of almanacs in the 1790s.

There were also two major public figures who provided some encouragement to the study of mathematics in the late eighteenth century. Benjamin Franklin (1706–1790), although not himself a mathematician, encouraged its study through essays and through his influence in founding the University of Pennsylvania. Because he was also well-known in Europe and a member of the Royal Society, he was able to provide the beginnings of an interaction between the scientific communities of Europe and the nascent one in the United States. Finally, Thomas Jefferson (1743–1826), who had studied some mathematics at The College of William and Mary but was largely self-educated in the field, developed during his time as ambassador to France a renewed interest in mathematics and its applications, publishing numerous articles in such fields as surveying, astronomy, spherical trigonometry, and the new metric system. In a letter of 1799 written from his home in Monticello, he expressed his views on the value of mathematics:

> Trigonometry . . . is most valuable to every man. There is scarcely a day in which he will not resort to it for some of the purposes of common life. The science of calculation also is indispensible as far as the extraction of the square and cube roots. Algebra as far as the quadratic equation and the use of logarithms are often of value in ordinary cases. But all beyond these is but a luxury, a delicious luxury indeed, but not to be indulged in by one who is to have a profession to follow for his subsistence. In this light I view the conic sections, curves of the higher orders, perhaps even spherical trigonometry, algebraical operations beyond the second dimension, and fluxions.[40]

At the end of the eighteenth century, and indeed well into the nineteenth, the United States had no need of the "luxuries" of mathematics.

In other parts of the Americas, the only concern for mathematics was also a practical one. For example, Father J. P. DeBonnécamps taught applied mathematics, including hydrography, surveying, and astronomy, at Quebec. And south of the Rio Grande, the decision of the Pope to divide that part of the Americas between Spain and Portugal led to the necessity of competent surveyors in South America. The first mathematics work in colonial America was written in Mexico in 1560 by Juan Diez Freyle, the *Sumario Compendioso de las Cuentas* (*Brief Summary of Reckoning*), a book which contained extensive tables relating to gold and silver exchange, among other monetary affairs, as well as arithmetic problems relating to the tables and some elementary algebra. In general, the earliest mathematics texts elsewhere in Latin America also contained very practical material, especially including material useful for military purposes. Unlike the situation to the north, however, there were no universities founded in Latin America during the colonial period. It was not until independence in the nineteenth century that there was a chance for mathematics to develop there. Ultimately, both north and south of the Rio Grande, enough students mastered the basics of mathematics that the mathematics of Europe could be further developed in the New World.

Exercises

Problems from Bernoulli's Ars conjectandi

1. Calculate the Bernoulli numbers B_8, B_{10}, and B_{12}. The sequence of Bernoulli numbers is usually completed by setting $B_0 = 1$, $B_1 = -(1/2)$, and $B_k = 0$ for k odd and greater than 1.

2. Write out explicitly, using Bernoulli's techniques, the formulas for the sums of the first n fourth, fifth, and tenth powers. Then show that the sum of the tenth powers of the first 1000 positive integers is

    91,409,924,241,424,243,242,241,924,242,500.

    Bernoulli claimed that he calculated this value in "less than half of a quarter of an hour" (without a calculator).

3. Show that if one defines the Bernoulli numbers B_i by setting

    $$\frac{x}{e^x - 1} = \sum_{i=0}^{\infty} \frac{B_i}{i!} x^i,$$

    then the values of B_i for $i = 2, 4, 6, 8, 10, 12$ are the same as those calculated in the text and in problem 1.

4. Complete Bernoulli's calculation of his example for the Law of Large Numbers by showing that if $r = 30$ and $s = 20$ (so $t = 50$) and if $c = 1000$, then

    $$nt + \frac{rt(n-1)}{s+1} > mt + \frac{st(m-1)}{r+1}$$

    where m, n are integers such that

    $$m \geq \frac{\log c(s-1)}{\log(r+1) - \log r}$$

    and

    $$n \geq \frac{\log c(r-1)}{\log(s+1) - \log s}.$$

    Conclude that in this case the necessary number of trials is $N = 25,550$.

5. Use Bernoulli's formula to show that if greater certainty is wanted, say $c = 10,000$, then the number of trials necessary is $N = 31,258$. Calculate also the value of N if $c = 100,000$ and if $c = 100$.

6. Suppose that a is the probability of success in an experiment and $b = 1 - a$ is the probability of failure. If the experiment is repeated three times, show that the probabilities of the number of successes being 3, 2, 1, 0 respectively are given by $P(S = 3) = 1a^3$, $P(S = 2) = 3a^2b$, $P(S = 1) = 3ab^2$, $P(S = 0) = 1b^3$.

7. Generalize problem 6 to the case of n trials. Show that the probability of r successes is $P(S = r) = \binom{n}{n-r} a^r b^{n-r}$.

8. Using the results of problem 7, with $a = 1/3$, $b = 2/3$ and $n = 10$, calculate $P(4 \leq S \leq 6)$.

Problems from De Moivre

9. Suppose that the probability of success in an experiment is 1/10. How many trials of the experiment are necessary to insure even odds on it happening at least once? Calculate this both by De Moivre's exact method and his approximation.

10. Solve De Moivre's equation $\left(1 + \frac{1}{q}\right)^x = 2$ for q large by modern methods. Recall that $\left(1 + \frac{1}{q}\right)^q \approx e$ for q large.

11. How many throws of three dice are necessary to insure even odds that three ones will occur at least once?

12. In a lottery in which the ratio of the number of losing tickets to the number of winning tickets is $39 : 1$, how many tickets should one buy to give oneself even odds of winning a prize?

13. Generalize De Moivre's procedure in Problem III (of his text) to solve Problem IV: To find how many trials are necessary to make it equally probable that an event will happen twice, supposing that a is the number of chances for its happening in any one trial and b the number of chances for its failing. (*Hint:* Note that $b^x + xab^{x-1}$ is the number of chances in which the event may succeed no more than once, while $(a + b)^x$ is the total number of chances. Approximate the solution for the case where $a : b = 1 : q$, with q large, and show that $x \approx 1.678q$.)

14. Show that the sums labeled col. 1, col. 2, and col. 3 in De Moivre's derivation of the ratio $\binom{n}{n/2} : 2^n$ may be written explicitly as

$$\text{col. 1} = \frac{s^2 + s}{m}$$

$$\text{col. 2} = \frac{\frac{1}{2}s^4 + s^3 + \frac{1}{2}s^2}{3m^3}$$

$$\text{col. 3} = \frac{\frac{1}{3}s^6 + s^5 + \frac{5}{6}s^4 - \frac{1}{6}s^2}{5m^5}$$

Determine the corresponding value for col. 4.

15. Add the highest degree terms of the columns from problem 14 to get

$$s\left(\frac{s}{m} + \frac{1}{2 \cdot 3}\frac{s^3}{m^3} + \frac{1}{3 \cdot 5}\frac{s^5}{m^5} + \frac{1}{4 \cdot 7}\frac{s^7}{m^7} + \cdots\right)$$

which, setting $x = s/m$, is equal to

$$s\left(\frac{2x}{1 \cdot 2} + \frac{2x^3}{3 \cdot 4} + \frac{2x^5}{5 \cdot 6} + \frac{2x^7}{7 \cdot 8} + \cdots\right).$$

Show that the series in the parentheses can be expressed in finite terms as

$$\log\left(\frac{1 + x}{1 - x}\right) + \frac{1}{x}\log(1 - x^2)$$

and therefore that the original series is

$$mx \log\left(\frac{1 + x}{1 - x}\right) + m \log(1 - x^2).$$

Since $s = m - 1$ (or $mx = m - 1$), show therefore that the sum of the highest degree terms of the columns of problem 14 is equal to

$$(m - 1) \log\left(\frac{1 + \frac{m-1}{m}}{1 - \frac{m-1}{m}}\right)$$

$$+ m \log\left[\left(1 + \frac{m - 1}{m}\right)\left(1 - \frac{m - 1}{m}\right)\right],$$

which in turn is equal to $(2m - 1)\log(2m - 1) - 2m \log m$.

16. Show that the sum of the second highest degree terms of each column from problem 14 is

$$\frac{s}{m} + \frac{s^3}{3m^3} + \frac{s^5}{5m^5} + \frac{s^7}{7m^7} + \cdots$$

which, since $s = m - 1$, is equal to

$$\frac{1}{2}\log\left(\frac{1 + \frac{s}{m}}{1 - \frac{s}{m}}\right) \quad \text{or} \quad \frac{1}{2}\log(2m - 1).$$

17. Use Wallis' product representation for π to derive Stirling's result that

$$1 - \frac{1}{12} + \frac{1}{360} - \frac{1}{1260} + \frac{1}{1680} - \cdots = \frac{1}{2}\log 2\pi.$$

18. Derive De Moivre's result

$$\log\left(\frac{Q}{M}\right) \approx -\frac{2t^2}{n} \quad \text{or equivalently} \quad \log\left(\frac{M}{Q}\right) \approx \frac{2t^2}{n}.$$

(*Hint:* Divide the arguments of the first two logarithm terms in the expression in the text by m. Then simplify and replace the remaining logarithm terms by the first two terms of their respective power series.)

Problems from Bayes

19. Calculate $P(r < x < s | X = n - 1)$ explicitly, using Bayes' theorem. In particular, suppose that you have drawn 10 white and 1 black ball from an urn containing an unknown proportion of white to black balls. If you now guess that this unknown proportion is greater than 7/10, what is the probability that your guess is correct?

20. Show that if an event of unknown probability happens n times in succession, the odds are $2^{n+1} - 1$ to 1 for more than an even chance of its happening again.

21. Evaluate the integral in the denominator of Bayes' formula to get

$$\int_0^1 \binom{n}{p} x^p (1 - x)^{n-p} \, dx = \frac{1}{n + 1}.$$

Problems from Maclaurin's **A Treatise of Algebra**

22. Suppose that the distance between London and Edinburgh is 360 miles and that a courier for London sets out from the Scottish city running at 10 mph at the same time that one sets out from the English capital for Edinburgh at 8 mph. Where will the couriers meet?

23. Derive Cramer's rule for three equations in three unknowns from the rule for two equations in two unknowns: Given the system

$$ax + by + cz = m$$

$$dx + ey + fz = n$$

$$gx + hy + kz = p$$

solve each equation for x in terms of y and z, then form two equations in those variables and solve for z. Finally, determine y and x by substitution.

24. A company dining together find that the bill amounts to $175. Two were not allowed to pay. The rest found that their shares amounted to $10 per person more than if all had paid. How many were in the company?

Problems from Euler's **Complete Introduction to Algebra**

25. How can one reconcile the algebraic rule $\sqrt{a}\sqrt{b} = \sqrt{ab}$ with the computation $\sqrt{-1}\sqrt{-4} = -2 \neq \sqrt{(-1)(-4)}$? Why do you think Euler erred in his discussion of this matter?

26. Calculate, using logarithms, the new principal after 100 years if one earns 5% compounded annually on an initial principal of $1000.

27. Twenty persons, men and women, dine at a tavern. The share of the bill for one man is $8, for one woman $7, and the entire bill amounts to $145. Required, the number of men and women separately.

28. A horse-dealer bought a horse for a certain number of crowns and sold it again for 119 crowns, by which means his profit was as much per cent as the horse cost him. What was the purchase price?

29. Some officers being quartered in a country, each commands three times as many horsemen, and twenty times as many foot soldiers, as there are officers. Also, a horseman's monthly pay amounts to as many florins as there are officers, and each foot soldier receives half that pay. The whole monthly expense is 13,000 florins. Required, the number of officers.

30. Calculate the distinct residues $1, \alpha, \beta, \ldots$ of $1, 5, 5^2, \ldots$ modulo 13. Then pick a nonresidue x of the sequence and determine the coset $x, x\alpha, x\beta, \ldots$. Continue to pick nonresidues and determine the cosets until you have divided the group of all 12 nonzero residues modulo 13 into nonoverlapping subsets, the cosets of the group of powers of 5.

31. Determine the quadratic residues modulo 13 and those modulo 23.

32. Prove that -1 is a quadratic residue with respect to a prime q if and only if $q \equiv 1 \pmod{4}$.

Problems from Lagrange

33. Show that if n is prime, then the roots of $x^n - 1 = 0$ can all be expressed as powers of any such root $\alpha \neq 1$.

34. Let x_1, x_2 be the two roots of the quadratic equation $x^2 + bx + c = 0$. Since $t = x_1 + x_2$ is invariant under the two permutations of the two roots, while $v = x_1 - x_2$ takes on two distinct values, v must satisfy an equation of degree 2 in t. Find the equation. Similarly, x_1 is invariant under the same permutations as $x_1 - x_2$. Thus x_1 can be expressed rationally in terms of $x_1 - x_2$. Find such a rational expression. Use the answers to the above two questions to "solve" the original quadratic equation.

35. Determine the three roots x_1, x_2, x_3 of $x^3 - 6x - 9 = 0$. Use Lagrange's procedure to find the sixth degree equation satisfied by y, where $x = y + 2/y$. Determine all six solutions of this equation and express each explicitly as $\frac{1}{3}(x' + \alpha x'' + \alpha^2 x''')$, where (x', x'', x''') is a permutation of (x_1, x_2, x_3) and α is a complex root of $x^3 - 1 = 0$.

Problems from Clairaut, Euler, and Monge

36. Use Clairaut's methods to calculate the subtangent and the tangent to the curve defined as the intersection of the cylinders $x^2 - a^2 = y^2$, $y^2 - a^2 = z^2$.

37. Calculate the length of the perpendicular from a point P on the curve defined by $ax = y^2$, $by = z^2$ to the xz-plane, where the perpendicular is also perpendicular to the plane defined by the tangent and subtangent to that curve.

38. Prove that the angle θ between the plane $\alpha x + \beta y + \gamma z = a$ and the xy-plane is given by $\cos \theta = \gamma/\sqrt{\alpha^2 + \beta^2 + \gamma^2}$. Determine the cosine of the angles this plane makes with the other two coordinate planes.

39. Show that the plane $z = \alpha y - \beta x + \gamma$ is perpendicular to the surface $z = f(x,y)$ if $\beta\frac{\partial z}{\partial x} - \alpha\frac{\partial z}{\partial y} = 1$. (Show that the plane contains a normal line to the surface.)

40. Find the normal line to the plane $Ax + By + Cz + D = 0$ which passes through the point (x_0, y_0, z_0).

41. Convert Monge's form of the equations of the normal line to the surface $z = f(x, y)$ into the modern vector equation of the line.

42. Show that an "Euler path" over a series of bridges connecting certain regions (a path which crosses each bridge exactly once) is always possible if there are either two or no regions which are approached by an odd number of bridges.

43. Construct Euler paths in the situations of Figure 14.16.

Problems from Banneker's Notebook

44. Benjamin Banneker was fond of solving mathematical puzzles and recorded many in his notebook, including his own version of the old hundred fowls problem: A gentleman sent his servant with £100 to buy 100 cattle, with orders to give £5 for each bullock, 20 shillings for each cow, and one shilling for each sheep. (Recall that 20 shillings equals £1.) What number of each sort of animal did he bring back to his master?[41]

45. Divide 60 into four parts such that the first increased by 4, the second decreased by 4, the third multiplied by 4, and the fourth divided by 4 shall each equal the same number.

46. Suppose a ladder 60 feet long is placed in a street so as to reach a window on one side 37 feet high, and without moving it at the bottom, to reach a window on the other side 23 feet high. How wide is the street?

FOR DISCUSSION . . .

47. The so-called St. Petersburg Paradox was a topic of debate among those mathematicians involved in probability in the eighteenth century. The paradox involves the following game between two players. Player A flips a coin until a tail appears. If it appears on his first flip, player B pays him 1 ruble. If it appears on the second flip, B pays 2 rubles, on the third, 4 rubles, . . . , on the nth flip, 2^{n-1} rubles. What amount should A be willing to pay B for the privilege of playing? Show first that A's expectation, namely, the sum of the probabilities for each possible outcome of the game multiplied by the payoff for each outcome, is

$$\sum_{i=0}^{\infty} \frac{1}{2^i} 2^{i-1}$$

and then that this sum is infinite. Next, play the game 10 times and calculate the average payoff. What would you be willing to pay to play? Why does the concept of expectation seem to break down in this instance?

48. Read parts of a few recent statistics texts and write a brief explanation of the current meaning of Bayesian and non-Bayesian statistics. How is Bayes' original formulation applied today?

49. Outline a lesson for a statistics course deriving Bayes' theorem and discussing its usefulness.

50. Outline a lesson for an algebra course teaching the principles of Cramer's rule using the technique of Maclaurin.

51. Prepare a report on the discovery of determinants by Seki Kowa, the Japanese mathematician.

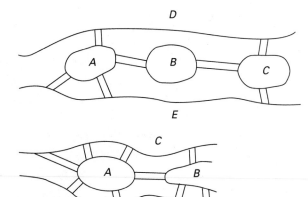

FIGURE 14.16
Bridge path problems.

52. Compare the treatments of multiplication of signed numbers in the texts of Maclaurin and Euler. What laws of arithmetic does each tacitly assume?

53. Develop a lesson in a course in three dimensional analytic geometry which uses the work of Monge to derive the equation of the tangent plane to a surface.

References and Notes

There are several good books on the early history of probability and statistics, including the works of Hacking and David cited in Chapter 11. An earlier book which gives great detail on the mathematics of the early workers in probability is I. Todhunter, *A History of the Mathematical Theory of Probability from the Time of Pascal to that of Laplace* (New York: Chelsea, 1965). An excellent treatment of the early history of statistics, which naturally also includes work on probability, is Stephen M. Stigler, *The History of Statistics* (Cambridge: Harvard University Press, 1986). A more philosophical treatment, concentrating on the ideas behind the notion of probability, is Lorraine Daston, *Classical Probability in the Enlightenment* (Princeton: Princeton University Press, 1988). Three recent books on the history of algebra have sections dealing with the eighteenth century. These are Luboš Nový, *Origins of Modern Algebra* (Prague: Academia Publishing House, 1973), Hans Wussing, *The Genesis of the Abstract Group Concept* (Cambridge: MIT Press, 1984), and B. L. van der Waerden, *A History of Algebra* (New York: Springer-Verlag, 1985). The standard work on the history of non-Euclidean geometry is Roberto Bonola, *Non-Euclidean Geometry, A Critical and Historical Study of its Development*, translated by H. S. Carslaw (New York: Dover, 1955). More recent works, both of which contain material on Saccheri and Lambert, include Jeremy Gray, *Ideas of Space: Euclidean, Non-Euclidean and Relativistic* (Oxford: Clarendon Press, 1989) and Boris A. Rosenfeld, *A History of Non-Euclidean Geometry: Evolution of the Concept of a Geometric Space*, translated by Abe Shenitzer (New York: Springer-Verlag, 1988). A briefer look at the subject is Jeremy Gray, "Non-Euclidean Geometry—A Re-interpretation," *Historia Mathematica* 6 (1979), 236–258. Finally, a general survey of the history of differential geometry is found in Dirk J. Struik, "Outline of a History of Differential Geometry," *Isis* 19 (1933), 92–120, while the history of analytic geometry, including material on the eighteenth century, is detailed in Carl Boyer, *History of Analytic Geometry* (New York: Scripta Mathematica, 1956).

1. From Jakob Bernoulli, *Ars conjectandi*, Part IV, in James Newman, ed., *The World of Mathematics* (New York: Simon and Schuster, 1956), vol. 3, 1452–1455, p. 1452.

2. Ibid., p. 1453.

3. These formulas and related material can be found in the section of Part II of Bernoulli's *Ars conjectandi* in D. J. Struik, *A Source Book in Mathematics, 1200–1800* (Cambridge: Harvard University Press, 1969), 316–320. A brief look at the entire *Ars conjectandi* is in Ian Hacking, "Jacques Bernoulli's Art of Conjecturing," *British Journal for the History of Science* 22 (1971), 209–229.

4. See Stigler, *History of Statistics*, pp. 66–70, for more details. See also O. B. Sheynin, "On the Early History of the Law of Large Numbers," *Biometrika* 55 (1968), 459–467, reprinted in E. S. Pearson and M. G. Kendall, *Studies in the History of Statistics and Probability* (London: Griffin, 1970), 231–240 and Karl Pearson, "James Bernoulli's Theorem," *Biometrika* 17 (1925), 201–210.

5. Abraham De Moivre, *The Doctrine of Chances*, 3rd ed., (New York: Chelsea, 1967), p. 36. This reprint edition of the entire work is well worth perusing for its many problems and examples. More information on De Moivre can be found in H. M. Walker, "Abraham de Moivre," *Scripta Mathematica* 2 (1934), 316–333 and in Ivo Schneider, "Der Mathematiker Abraham de Moivre (1667–1754)," *Archive for History of Exact Sciences* 5 (1968), 177–317.

6. Ibid., p. 37.

7. Ibid., p. 242.

8. Quoted in Stigler, *History of Statistics*, p. 76.

9. De Moivre, *Doctrine of Chances*, p. 246.

10. Ibid., p. 248.

11. Thomas Bayes, "An Essay towards solving a Problem in the Doctrine of Chances," reprinted with a biographical note by G. A. Barnard in E. S. Pearson and M. G. Kendall, *History of Statistics and Probability*, 131–154, p. 136.

12. Ibid., p. 139.

13. Ibid., p. 140.

14. Ibid., p. 143.

15. Ibid. This quotation has led to numerous discussions in the literature as to the types of situations to which Bayes'

theorem applies. Two recent analyses of this matter are in Stigler, *History of Statistics*, pp. 122–131 and in Donald A. Gillies, "Was Bayes a Bayesian?," *Historia Mathematica* 14 (1987), 325–346.

16. Colin Maclaurin, *A Treatise of Algebra in Three Parts*, 2nd ed. (London: Millar and Nourse, 1756), p. 1.

17. Ibid., p. 132.

18. Leonhard Euler, *Elements of Algebra*, translated from the *Vollständige Anleitung zur Algebra* by John Hewlett (New York: Springer-Verlag, 1984), p. 2. This Springer edition, a reprint of the original 1840 English translation, contains an introduction by C. Truesdell. Its many problems and ingenious methods of solution are worth a careful reading.

19. Ibid., p. 186.

20. Ibid., p. 7.

21. Ibid., p. 43.

22. Ibid., p. 91.

23. Ibid., pp. 91–92.

24. Ibid., p. 211.

25. Ibid., p. 413.

26. Girolamo Saccheri, *Euclides Vindicatus*, translated by G. B. Halstead (New York: Chelsea, 1986), p. 9. This Chelsea edition is a reprint of the first English translation of 1920 with added notes by Paul Stäckel and Friedrich Engel. For a brief treatment of Saccheri's ideas, see Louis Kattsoff, "The Saccheri Quadrilateral," *Mathematics Teacher* 55 (1962), 630–636.

27. Ibid., p. 117.

28. Ibid., p. 169.

29. Ibid., p. 173.

30. John Fauvel and Jeremy Gray, eds., *The History of Mathematics: A Reader* (London: Macmillan, 1987), 517–520, p. 517. This source book contains an English selection from Lambert's *Theorie der Parallellinien*. For more on

Lambert, see J. J. Gray and L. Trilling, "Johann Heinrich Lambert, Mathematician and Scientist," *Historia Mathematica* 5 (1978), 13–41.

31. Ibid.

32. Ibid., p. 520.

33. Alexis-Claude Clairaut, *Recherches sur les courbes á double courbure* (Paris: Quillau, 1731), p. 39.

34. Leonhard Euler, "Methodus facilis omnia symptomata linearum curvarum non in eodem plano sitarum investigandi," *Acta Acad. Sci. Petrop.* 1 (1782), 19–57 = *Opera* (1) 28, 348–381.

35. Leonhard Euler, "Recherches sur la courbure des surfaces," *Mem. de l'Academie des Sciences de Berlin* 16 (1760), 119–143 = *Opera* (1) 28, 1–22.

36. Quoted in Boyer, *Analytic Geometry*, pp. 205–206.

37. Newman, *World of Mathematics*, vol. 1, p. 577. This volume contains an English translation of Euler's article on the problem of the seven bridges of Königsberg.

38. Janis Langins, *La République avait besoin de savants* (Paris: Belin, 1987), p. 123. This book provides a detailed study of the first year of the École Polytechnique and includes copies of many of the relevant documents.

39. Quoted in Dirk J. Struik, "Mathematics in Colonial America," in Dalton Tarwater, ed., *The Bicentennial Tribute to American Mathematics: 1776–1976* (Washington: Mathematical Association of America, 1976), 1–7, p. 3.

40. Quoted in David Eugene Smith and Jekuthiel Ginsburg, *A History of Mathematics in America before 1900* (Chicago: Mathematical Association of America, 1934), p. 62.

41. See Silvio A. Bedini, *The Life of Benjamin Banneker* (New York: Scribner's, 1972) for details on Banneker's life and work as well as copies of various relevant documents, including the problems included here.

Summary of Eighteenth Century Probability, Algebra, and Geometry

1642–1708	Seki Kowa	Determinants
1654–1705	Jakob Bernoulli	Combinatorics and probability
1667–1733	Girolamo Saccheri	Non-Euclidean geometry
1667–1754	Abraham De Moivre	Probability, Normal curve
1698–1746	Colin Maclaurin	Algebra text, Cramer's Rule
1702–1745	Isaac Greenwood	Professor at Harvard
1702–1761	Thomas Bayes	Probability, Bayes' Theorem
1704–1752	Gabriel Cramer	Cramer's Rule
1707–1783	Leonhard Euler	Algebra, Number theory, geometry, topology
1713–1765	Alexis Clairaut	Space curves
1714–1779	John Winthrop	Professor at Harvard
1728–1777	Johann Lambert	Non-Euclidean geometry
1731–1806	Benjamin Banneker	Surveying, almanacs
1732–1796	David Rittenhouse	Astronomy, surveying, clockmaking
1736–1813	Joseph Lagrange	Theory of equations
1746–1818	Gaspard Monge	Analytic and differential geometry
1749–1827	Pierre de Laplace	Statistical inference
1752–1833	Adrien Legendre	Number theory
1765–1843	Sylvestre Lacroix	Textbooks

Chapter 15

Algebra in the Nineteenth Century

"It is greatly to be lamented that this virtue of the real integers that they can be decomposed into prime factors which are always the same for a given integer does not belong to the complex integers [of arbitrary cyclotomic number fields], for were this the case, the entire theory, which is still laboring under many difficulties, could be easily resolved and brought to a conclusion. For this reason, the complex integers we are considering are seen to be imperfect, and there arises the question whether other types of complex numbers can be found . . . which would preserve the analogy with the real integers with respect to this fundamental property."

(Ernst Kummer, in "De numeris complexis, qui radicibus unitatis et numeris integris realibus constant," 1847)[1]

---

*I*n a letter to his son, William Rowan Hamilton described his discovery of quaternions: "On the 16th day of [October, 1843]—which happened to be a Monday, and a Council day of the Royal Irish Academy—I was walking in to attend and preside, and your mother was walking with me, along the Royal Canal; . . . and although she talked with me now and then, yet an *under-current* of thought was going on in my mind, which gave at last a *result*, whereof it is not too much to say that I felt *at once* the importance. An *electric* circuit seemed to close; and a spark flashed forth . . . Nor could I resist the impulse—unphilosophical as it may have been—to cut with a knife on a stone of Brougham Bridge . . . the fundamental formula with the symbols i, j, k, namely,

$$i^2 = j^2 = k^2 = ijk = -1,$$

which contains the *Solution* of the *Problem*."[2]

Algebra in 1800 meant the solving of equations. By 1900, the term encompassed the study of various mathematical structures, that is, sets of elements with well-defined operations, satisfying certain specified axioms. It is this change in the notion of algebra that will be explored in this chapter.

The nineteenth century opened with the appearance of the *Disquisitiones Arithmeticae* of Carl Friedrich Gauss, in which the "Prince of Mathematicians" discussed the basics of number theory, not only proving the law of quadratic reciprocity, but also introducing various new concepts which provided early examples of groups and matrices. Gauss' study of higher reciprocity laws soon led to his study of the so-called Gaussian integers, complex numbers of the form $a + bi$ where a and b are ordinary integers. Attempts to generalize the properties of these integers to integers in other number fields led Ernst Kummer to the realization that some of the most important of these properties, such as unique factorization, fail to hold. To recover this property, along with a reasonable new meaning of the term "prime," Kummer created what he called "ideal complex numbers" by 1846, the study of which led Richard Dedekind in the 1870s to define "ideals" in rings of algebraic integers, these ideals having the property of unique factorization into primes.

Gauss' study of the solutions of cyclotomic equations in the *Disquisitiones* as well as the detailed study of permutations by Augustin-Louis Cauchy in 1815 helped with a new attack on the problem of solving algebraic equations of degree higher than four. It was Niels Henrik Abel who finally proved the impossibility of the solution of a general equation of degree five or higher in terms of radicals (in 1827). Soon thereafter, Evariste Galois sketched the relationship between algebraic equations and groups of permutations of the roots, a relationship whose complete development transformed the question of the solvability of equations to one of the study of subgroups and factor groups of the group of the equation. Galois' work was not published until 1846 nor completely understood until somewhat later. In 1854 Arthur Cayley gave the earliest definition of an abstract group, but his work, too, was somewhat in advance of its time and was not fully developed until the late 1870s by Cayley himself and then a few years later by Walther von Dyck and Heinrich Weber. Meanwhile, the study of the "numbers" determined by the solutions of algebraic equations led to the definition of a field of numbers both by Leopold Kronecker and Richard Dedekind and soon after to an abstract definition of a field by Weber.

English mathematicians of the first third of the nineteenth century, including George Peacock and Augustus De Morgan, attempted to axiomatize the basic ideas of algebra and to determine exactly how much one can generalize the properties of the integers to other types of quantities. Such study led eventually to the discovery of quaternions by William Rowan Hamilton in 1843, partly in an attempt to determine a physically meaningful algebra in three-dimensional space. Quaternions, however, were four-dimensional objects, and so the physicists could only use the three-dimensional part of the quaternions in any algebraic manipulations. After a debate lasting nearly to the end of the century, the algebra of vectors developed by Oliver Heaviside and Josiah Willard Gibbs won out over the algebra of quaternions to become the language of the physicists. Meanwhile, the algebraic freedom to determine laws of operation, exploited by Hamilton for use in physics, was applied by George Boole to the study of logic. Boole's work turned out to be important in computer design a century later.

Another aspect of what is today called modern algebra, the theory of matrices, also developed in the mid-nineteenth century. Determinants had been used as early as the seventeenth century, but it was only in 1850 that James Joseph Sylvester coined the term **matrix** to refer to a rectangular array of numbers. Soon thereafter, Cayley developed the algebra of matrices. The study of eigenvalues was begun by Cauchy early in the century in his work on quadratic forms and then fully developed by, among others, Camille Jordan and Georg Frobenius. The latter in particular organized the theory of matrices into essentially the form it has today.

15.1 NUMBER THEORY

Legendre had published his work on number theory in 1798, but at the same time a young man in Brunswick, a city in northern Germany, was putting the finishing touches on a number theoretic work which would ultimately have far more influence than that of Legendre. Carl Friedrich Gauss (1777–1855) noted in the preface to his *Disquisitiones Arithmeticae* (*Investigations in Arithmetic*), published in 1801, that he had written most of it before studying the works of some of his contemporaries. But although some of what he thought he had discovered was already known to Euler or Lagrange or Legendre, Gauss' work contains many new discoveries in the theory of numbers.

15.1.1 Gauss and Congruences

Gauss began in Chapter 1 by presenting the modern definition and notation for congruence: The integers b and c are **congruent** relative to the **modulus** a if a divides the difference of b and c. Gauss wrote this as $b \equiv c \pmod{a}$, noting that he adopted the symbol $\equiv$ "because of the analogy between equality and congruence,"[3] and called b and c each a **residue** of the other. He then discussed the elementary properties of congruence. For example, Gauss showed how to solve both the linear congruence $ax + b \equiv c \pmod{m}$ and the Chinese remainder problem as well as how to calculate the Euler function $\phi(n)$ giving the number of integers less than n which are relatively prime to n. In Chapter 3, Gauss, like Euler before him, considered residues of powers. Noting that if p is prime and a any number less than p, then the smallest exponent m such that $a^m \equiv 1 \pmod{p}$ is a divisor of $p - 1$, Gauss went on to show that in fact "there always exist numbers with the property that no power less than the $(p - 1)$st is congruent to unity."[4] A number a satisfying this property is called a **primitive root** modulo p. If a is a primitive root, then the powers $a, a^2, a^3, \ldots, a^{p-1}$ all have different residues modulo p and thus exhaust all of the numbers $1, 2, \ldots, p - 1$. In particular, $a^{(p-1)/2} \equiv -1 \pmod{p}$, a property crucial to Gauss' proof of

Wilson's Theorem. *$(p - 1)! \equiv -1 \pmod{p}$.*

Let a be a primitive root modulo p. Then

$$(p - 1)! = a^1 a^2 \cdots a^{p-1} = a^{1+2+\cdots+p-1} = a^{p(p-1)/2}.$$

Because $\frac{p(p-1)}{2} \equiv \frac{p-1}{2} \pmod{p - 1}$, it follows that $(p - 1)! \equiv a^{(p-1)/2} \equiv -1 \pmod{p}$. (This theorem had first been stated by John Wilson (1741–1793) and first proved by Lagrange in 1773.)

The central topic of Chapter 4 of the *Disquisitiones* is the law of quadratic reciprocity. Recall that Euler had stated, but not proved, this law stating the conditions under which two odd primes are quadratic residues of each other. Gauss considered this theorem so important that he gave six different proofs, the first of which he found in the spring of 1796 after much effort. In the *Disquisitiones*, Gauss gave numerous examples and special cases derived from calculations before presenting his second proof. He showed first, following Euler, that -1 is a quadratic residue of primes of the form $4n + 1$ and a nonresidue

Biography   Carl Friedrich Gauss (1777–1855)

Gauss was born into a family which, like many others of the time, had recently moved into town, hoping to improve its lot from that of impoverished farm workers. One of the benefits of living in Brunswick was that the young Carl could attend school. There are many stories told about Gauss' early developing genius. At the beginning of the school year, to keep his 100 pupils occupied, the teacher, J. G. Büttner, assigned them the task of summing the first 100 integers. He had barely finished explaining the assignment when the 9 year old Gauss wrote the single number 5050 on his slate and deposited it on the teacher's desk. Gauss had noticed that the sum in question was simply 50 times the sum 101 of the various pairs 1 and 100, 2 and 99, 3 and 98, . . . and had performed the required multiplication in his head. Impressed by his young student, Büttner arranged for Gauss to have special textbooks, to have tutoring by his assistant Martin Bartels (1769–1836), who himself later became a professor of mathematics in Russia, and to be admitted to a secondary school where he mastered the classical curriculum.

In 1791, the Duke of Brunswick granted Gauss a stipend which enabled him first to attend the Collegium Carolinium, a new science oriented academy funded by the Brunswick government to train bureaucrats and military officers, then to matriculate at the University of Göttingen in neighboring Hanover, which already had a reputation in the sciences, and

finally to continue his research back in Brunswick while receiving a Ph.D. from the local University of Helmstedt. Not only did Gauss publish his researches in number theory in 1801, with the book being dedicated to his patron the Duke, but he also developed at the same time a new method for calculating orbits which enabled several asteroids to be discovered. The Duke's patronage lasted until the Duke was killed in battle against France in 1806 and the duchy was occupied by the French army. Fortunately for science, the French general had been given explicit orders to look out for Gauss' welfare. Thus Gauss was able to stay in Brunswick until he accepted a position at Göttingen the following year as professor of astronomy and director of the observatory. Gauss remained at Göttingen for the remainder of his life, doing research in pure and applied mathematics as well as astronomy and geodesy.

Gauss was never particularly happy with teaching classes, because most of the students were uninterested in, and not well prepared for, mathematics, but he was willing to work privately with any actively interested student who approached him. Compared to his predecessor Euler and to his French contemporary Cauchy, Gauss ultimately published little, his collected works occupying only (!) 12 volumes. Nevertheless, his mathematical papers in various fields are of such profundity that they have influenced the progress of the subject to the present (Figure 15.1).[5]

FIGURE 15.1
Gauss on a German stamp.

of primes of the form $4n + 3$. He next dealt with 2 and -2 and concluded that for primes of the form $8n + 1$, both 2 and -2 are quadratic residues; for primes of the form $8n + 3$, -2 is a quadratic residue and 2 is not; for primes of the form $8n + 5$, both 2 and -2 are nonresidues; and for primes of the form $8n + 7$, 2 is a quadratic residue and -2 is not. The demonstrations of these properties are not difficult. For example, to prove the first result, Gauss chose a primitive root a for the prime $8n + 1$ and noted that $a^{4n} \equiv -1 \pmod{8n + 1}$. This congruence can be rewritten in either of the two forms

$$(a^{2n} + 1)^2 \equiv 2a^{2n} \pmod{8n + 1} \qquad \text{or} \qquad (a^{2n} - 1)^2 \equiv -2a^{2n} \pmod{8n + 1}.$$

It then follows that both $2a^{2n}$ and $-2a^{2n}$ are squares modulo $8n + 1$. But because a^{2n} is a square, it follows that so are 2 and -2.

Gauss went on to characterize the primes for which 3 and -3 are quadratic residues as well as those for 5 and -5 and 7 and -7. He could then state the general result.

Quadratic Reciprocity Theorem.   *If p is a prime number of the form 4n + 1, +p will be a [quadratic] residue or nonresidue of any prime number which taken positively is a residue or nonresidue of p. If p is of the form 4n + 3, −p will have the same property.*

The proof is much too long and detailed to be discussed here. Note, however, that Gauss was unaware of Legendre's suggestive notation for quadratic residues, a notation which enables the result to be stated succinctly. Legendre defined the symbol $\left(\frac{p}{q}\right)$ to be equal to 1 if p is a quadratic residue modulo q and -1 if not, where q is an odd prime. The theorem can then be stated in the following elegant form:

$$\left(\frac{p}{q}\right)\left(\frac{q}{p}\right) = (-1)^{\frac{p-1}{2}\frac{q-1}{2}}.$$

Similar formulas can be written expressing quadratic residue properties of -1 and ± 2 modulo a prime p. Given the product rule for residues, that $\left(\frac{a}{p}\right)\left(\frac{b}{p}\right) = \left(\frac{ab}{p}\right)$, where a, b are prime to p, as well as rules for determining the residue situation when two numbers have common factors, Gauss could then determine for any two positive numbers P and Q whether Q is a residue or nonresidue of P. For example, to decide whether 453 is a quadratic residue modulo 1236 ($= 4 \cdot 3 \cdot 103$), he noted first that 453 is a quadratic residue both of 4 and of 3. By the Chinese remainder theorem, it follows that the question is reduced to determining $\left(\frac{453}{103}\right)$. Using Legendre's notation, we have

$$\left(\frac{453}{103}\right) = \left(\frac{41}{103}\right) = \left(\frac{103}{41}\right)(-1)^{\frac{41-1}{2}\frac{103-1}{2}} = \left(\frac{103}{41}\right) = \left(\frac{-20}{41}\right) = \left(\frac{-1}{41}\right)\left(\frac{2^2}{41}\right)\left(\frac{5}{41}\right).$$

Because each of the three factors on the right is equal to 1, it follows that 453 is a quadratic residue modulo 103 and also modulo 1236. Gauss showed, in fact, that $453 \equiv 297^2$ (mod 1236).

Over the next several decades, in attempting to generalize the law of quadratic reciprocity to laws of cubic and quartic reciprocity, Gauss was led to consider the Gaussian integers, the complex numbers $a + bi$ where a and b are ordinary integers. His investigations of these numbers are contained in a paper published in 1832, in which he establishes certain analogies between the Gaussian integers and ordinary integers. After noting that there are four units (invertible elements) among the Gaussian integers, 1, -1, i, and $-i$, he defines the **norm** of an integer $a + bi$ to be the product $a^2 + b^2$ of the integer with its complex conjugate $a - bi$. He then calls an integer **prime** if it cannot be expressed as the product of two other integers, neither of them units. His next task is to determine which Gaussian integers are prime. Because an odd real prime p can be expressed as $p = a^2 + b^2$ if and only if it is of the form $4n + 1$, it follows that such primes, considered as Gaussian integers, are composite: $p = (a + bi)(a - bi)$. Conversely, primes of the form $4n + 3$ are still prime as Gaussian integers. Because $2 = (1 + i)(1 - i)$, 2 is also composite. To determine which other Gaussian integers $a + bi$ are prime, Gauss uses the norm and shows that $a + bi$ is prime if and only if its norm is a real prime, which can only be 2 or of the form $4n + 1$. In other words, 2 and primes of the form $4n + 1$ split as the product of two Gaussian primes in the domain of Gaussian integers, while primes of the form $4n + 3$ remain prime there.

With primes defined in the domain of Gaussian integers, Gauss easily showed that any integer can be factored as the product of such primes. To complete the analogy with ordinary integers, he then proved that this factorization is unique, at least up to unit factors, by first demonstrating, using his description of primes, that if any Gaussian prime p divides the product $qrs \ldots$ of Gaussian primes, then p must itself be equal to one of those primes, or one of them multiplied by a unit. Having established the unique factorization property of the Gaussian integers, Gauss considered congruence modulo Gaussian integers and then quartic reciprocity, stated in terms not of ordinary integers but of Gaussian integers. He was furthermore aware that a law of cubic reciprocity would involve complex numbers of the form $a + b\omega$, where $\omega^3 = 1$ and a, b are ordinary integers. Thus primes and factorization of numbers of that type would need to be investigated, a task Gauss never completed.

15.1.2 Fermat's Last Theorem and Unique Factorization

This entire question of factorization in various domains turned out to be related not only to reciprocity but also to the continued attempts to prove Fermat's Last Theorem, the conjecture that $x^n + y^n = z^n$ has no nontrivial integral solutions if $n > 2$. It was Euler who in 1753 first announced a proof of the case $n = 3$ of the theorem using Fermat's method of infinite descent, publishing it in his *Algebra* of 1770.

The next mathematician to make any progress on the theorem was Sophie Germain (1776–1831). In the early 1820s, she showed that $x^n + y^n = z^n$ has no solution where xyz is not divisible by n for n any odd prime less than 100. Unfortunately, her technique did not help to determine the truth of Fermat's Last Theorem in the case where one of x, y, z is divisible by n. In 1825, however, Legendre succeeded in proving the complete result for $n = 5$. Seven years later, Peter Lejeune-Dirichlet (1805–1859) proved the theorem for $n = 14$, while it took an additional seven years before the result was proved for $n = 7$ by Gabriel Lamé (1795–1870). These latter proofs were all very long, involved difficult manipulations, and did not appear capable of generalization. If the theorem were to be proved, it seemed that an entirely new approach would be necessary.

Such a new approach to proving Fermat's Last Theorem was announced with great fanfare by Lamé at the Paris Academy meeting of March 1, 1847.[6] Lamé claimed he had solved this long outstanding problem and gave a brief sketch of the proof. The basic idea began with the factorization of the expression $x^n + y^n$ over the complex numbers as

$$x^n + y^n = (x + y)(x + \alpha y)(x + \alpha^2 y) \cdots (x + \alpha^{n-1}y)$$

where α is a primitive root of $x^n - 1 = 0$. Lamé next planned to show that if x and y are such that the factors in this expression are all relatively prime and if also $x^n + y^n = z^n$, then each of the factors must itself be an nth power. He would then use Fermat's technique of infinite descent to find a solution in smaller numbers. On the other hand, if the factors were not relatively prime, he hoped to show that they had a common factor. Dividing by this factor would then reduce the problem to the first case. When Lamé finished his presentation, Joseph Liouville (1809–1882) took the floor and cast some serious doubt on

---

Forced to study in private due to the turmoil of the French Revolution and the opposition of her parents, Germain nevertheless mastered mathematics through calculus on her own. She wanted to continue her studies at the École Polytechnique when it opened in 1794, but women were not admitted as students. Nevertheless, she diligently collected and studied the lecture notes from various mathematics classes and even submitted a paper of her own to Lagrange. Mastering Gauss' *Disquisitiones* soon after it was published, she also began a correspondence with him, using the pseudonym of M. Le Blanc. Germain was, in fact, responsible for suggesting to the French general leading the army occupying Brunswick in 1807

that he insure Gauss' safety. Gauss, naturally, did not then know the name of Sophie Germain, but the misunderstanding was cleared up with an exchange of letters. Perhaps surprisingly for a mathematician brought up in Germany, Gauss was pleased to learn that his correspondent and protector was a woman. As he wrote, "When a person of the sex which, according to our customs and prejudices, must encounter infinitely more difficulties than men to familiarize herself with these thorny researches, succeeds nevertheless in surmounting these obstacles and penetrating the most obscure parts of them, then without doubt she must have the noblest courage, quite extraordinary talents, and a superior genius."[7]

Lamé's proposal. Basically, he noted that Lamé's idea to conclude that each factor was an nth power because the factors were relatively prime and their product was an nth power depended on the theorem that any integer can be uniquely factored into a product of primes. It was by no means obvious, he concluded, that such a result was true for complex integers of the form $x + \alpha^j y$.

Over the next several weeks, Lamé tried without success to overcome Liouville's objection. But on May 24, Liouville read into the proceedings a letter from Ernst Kummer (1810–1893) which effectively ended the discussion. Namely, Kummer noted not only that unique factorization fails in the domains in question, but that three years earlier he had published an article, admittedly in a somewhat obscure publication, in which he had demonstrated this failure. Kummer's article of 1744 was related to the question of higher reciprocity laws, but contained a general study of complex numbers involving roots of unity.

15.1.3 Kummer and Ideal Numbers

The complex numbers studied by Kummer, important in connection with Fermat's Last Theorem and with general reciprocity laws, are called the **cyclotomic integers**. They are the complex numbers of the form

$$f(\alpha) = a_0 + a_1\alpha + a_2\alpha^2 + \cdots + a_{n-1}\alpha^{n-1}$$

where $\alpha \neq 1$ is a root of $x^n - 1 = 0$ and each a_i is an ordinary integer. In particular, we will assume here that n is prime. The numbers formed by replacing α by the other solutions $\alpha^i \neq 1$ of the equation $x^n - 1 = 0$, written $f(\alpha^2)$, $f(\alpha^3)$, . . . , $f(\alpha^{n-1})$, are called the **conjugates** of the given number $f(\alpha)$, while the product of all the conjugates is called the

Kummer, born in Sorau, Germany (now Zary, Poland), a town halfway between Berlin and Wrocław (Breslau), entered the University of Halle in 1828 to study theology. He soon changed his specialty to mathematics and, after receiving his doctorate in 1831, taught mathematics and physics in a Gymnasium in Liegnitz (now Legnica) for ten years before receiving an appointment to the University of Breslau in 1842. It was at Breslau that Kummer devoted himself to research in number theory. In 1855, after Dirichlet left Berlin to succeed Gauss at Göttingen, Kummer was appointed to the vacant position at Berlin. In 1861, Kummer and Karl Weierstrass established at Berlin Germany's first ongoing seminar in pure mathematics, a seminar which soon attracted many mathematicians from throughout the world and helped to make Berlin one of the most important world centers of mathematics in the late nineteenth and early twentieth centuries.

norm of $f(\alpha)$, written $N f(\alpha)$, and is an ordinary integer. The norm satisfies the relationship $N[f(\alpha)g(\alpha)] = N f(\alpha) N g(\alpha)$. This multiplication property of the norm became one of Kummer's primary tools to deal with factorization of cyclotomic integers, because a factorization of such an integer $h(\alpha)$ implies a factorization of the ordinary integer $N h(\alpha)$.

To deal with Kummer's argument, we need two definitions. A complex integer in a particular domain is **irreducible** if it cannot be factored into two other integers in that domain, neither of which is a unit. It is **prime** if whenever it divides a product it divides one of the factors. (Note that this definition is different from Gauss'.) It is not difficult to show that a prime is irreducible. Kummer found, however, that in the domain Γ of cyclotomic integers generated by α, a 23rd root of unity, there were irreducible integers which are not prime. He was then able to demonstrate the failure of unique factorization in Γ.

First, Kummer showed that 47 and 139 were not the norms of any integer in Γ. On the other hand, $47 \cdot 139$ was the norm of $\beta = 1 - \alpha + \alpha^{-2}$. It follows that β divides $47 \cdot 139$. If β were prime, it would have to divide either 47 or 139. But this is impossible since $N(\beta)$ divides neither $N(47) = 47^{22}$ nor $N(139) = 139^{22}$. On the other hand, if β could be factored, the norm of one of these factors would have to be 47, contradicting the result that 47 is not a norm. It follows that β is irreducible but not prime. An explicit decomposition of $47 \cdot 139$ into two different sets of irreducible factors is then feasible. First, $47 \cdot 139 = N(\beta)$, giving a factorization into 22 irreducible factors each of norm $47 \cdot 139$. Second, if $h(\alpha) = \alpha^{10} + \alpha^{-10} + \alpha^8 + \alpha^{-8} + \alpha^7 + \alpha^{-7}$ and $g(\alpha) = \alpha^{10} + \alpha^{-10} + \alpha^8 + \alpha^{-8} + \alpha^4 + \alpha^{-4}$, then $N h(\alpha) = 47^2$ and $N g(\alpha) = 139^2$, so $h(\alpha)$, $g(\alpha)$ and all of their conjugates are irreducible. Further, setting $f(\alpha) = h(\alpha)g(\alpha)$, one can show that

$$47 \cdot 139 = f(\alpha)f(\alpha^4)f(\alpha^{-7})f(\alpha^{-5})f(\alpha^3)f(\alpha^{-11})f(\alpha^2)f(\alpha^8)f(\alpha^9)f(\alpha^{-10})f(\alpha^6)$$

where the factors are those conjugates generated by the transformation of α to α^4. This new factorization of $47 \cdot 139$ into 22 irreducibles, half with norm 47^2 and half with norm 139^2, is clearly different from the original one. Interestingly enough, Liouville, before he had heard from Kummer in May of 1847, had written in his notebook a much simpler example of nonunique factorization, in a domain generated by a root of $x^2 = -17$ rather than $x^n = 1$:

$$169 = 13 \cdot 13 = \left(4 + 3\sqrt{-17}\right)\left(4 - 3\sqrt{-17}\right).$$

Having discovered nonunique factorization, Kummer spent the next several years fashioning an answer to the question posed in the opening quotation of this chapter. He devised a new type of complex numbers, "ideal complex numbers," which would uniquely factor into "ideal" prime factors. As an example, consider the domain of the 23rd roots of unity, in which neither 47 nor 139 have prime factors. In terms of "ideal" prime factors, however, because $N(1 - \alpha + \alpha^{-2}) = 47 \cdot 139$, it should be true that each of the 22 irreducible factors on the left would be divisible by two ideal prime factors on the right, one a factor of 47 and the other a factor of 139. To describe such a prime factor, Kummer defined what it meant to be divisible by it. Thus, let P be the prime factor of 47 which divides $\beta = 1 - \alpha + \alpha^{-2}$ and ψ the product of the 21 conjugates of β. Then ψ will be divisible by all of the ideal prime factors of 47 except P, so that $\gamma\psi$ will be divisible by 47 if and only if γ is divisible by P. Hence an integer γ in the domain of the 23rd roots of unity is **divisible** by the ideal prime factor P if $\gamma\psi$ is divisible by 47. Similarly, γ is divisible m times by P if $\gamma\psi^m$ is divisible by 47^m. Kummer generalized this idea to arbitrary domains of cyclotomic integers and, because an arbitrary ideal number was defined to be the (formal) product of prime ideal numbers, succeeded, virtually by definition, in restoring unique factorization into primes to the ideal numbers. By further detailed study of these ideal numbers, Kummer was also able to establish certain conditions on a prime integer n under which Fermat's Last Theorem was true for that n. It is still not known, however, how to prove the theorem in general for primes not meeting these conditions.[8]

In his original paper on ideal complex numbers, Kummer wrote that he intended to generalize his work to domains other than the cyclotomic integers, domains generated by a root of $x^2 - D = 0$, where D is an integer. Kummer never published any such generalization, partly because he could not find the "correct" generalization of the concept of integer to those domains. The obvious generalization appears to be that an integer of that domain would be a complex number of the form $a + b\sqrt{D}$, where a and b are ordinary integers. It turns out, however, that this definition leads to problems. For example, consider the domain of numbers of the form $a + b\sqrt{-3}$, and let $\beta = 1 + \sqrt{-3}$. Then $\beta^3 = -8$. Since 2 does not divide β in this domain, there must be an ideal prime factor P of 2 which divides 2 with a greater multiplicity than it divides β. For simplicity, we write this as $\mu_P(2) > \mu_P(\beta)$. On the other hand, 2^k divides $8\beta^k$ for all k. It follows that $k\mu_P(2) \leq \mu_P(8) + k\mu_P(\beta)$ or $k(\mu_P(2) - \mu_P(\beta)) \leq \mu_P(8)$ for all k. This implies that $\mu_P(2) \leq \mu_P(\beta)$, a contradiction.

15.1.4 Dedekind and Ideals

It was Richard Dedekind (1831–1916) who by 1871 solved this problem of defining the integers. Furthermore, since he was unhappy that "Kummer never defined the ideal numbers themselves but only divisibility by them,"[9] Dedekind created a new concept to restore unique factorization to his newly defined domains. Dedekind began by defining an

---

Biography   **Richard Dedekind (1831–1916)**

Dedekind, like Gauss, was born in Brunswick and studied both at the Collegium Carolinium and then at the University of Göttingen. In 1858, he became a professor at the Polytechnikum in Zurich and four years later returned to Brunswick to teach at the Polytechnikum there, the successor to Collegium Carolinium. Although at various times he could have received an appointment to a major German university, Dedekind chose to remain in Brunswick where he felt he had sufficient freedom to pursue his mathematical research. It was his work as editor of Dirichlet's *Vorlesungen über Zahlentheorie* (*Lectures on Number Theory*, 1863 and later editions) which convinced him to publish his own ideas on the subject, ideas which he had been developing in his lectures over the years. Thus in the second edition of Dirichlet's work (1871), he provided a supplement in which he established the theory of algebraic integers, a theory he expanded in later editions (Figure 15.2).

FIGURE 15.2
Dedekind and ideal factorization on a stamp from the D.D.R.

algebraic number as a complex number θ which satisfies any algebraic equation over the rational numbers, that is, an equation of the form

$$\theta^n + a_1\theta^{n-1} + a_2\theta^{n-2} + \cdots a_{n-1}\theta + a_n = 0$$

where the a_i are rational numbers. He then defined an **algebraic integer** to be an algebraic number which satisfied such an equation with all coefficients being ordinary integers. For example, $\theta = \frac{1}{2} + \frac{1}{2}\sqrt{-3}$ is an algebraic integer because it satisfies the equation $\theta^2 - \theta + 1 = 0$, even though it is not of the "obvious" form $a + b\sqrt{-3}$ with a, b integers. Dedekind then showed that the sum, difference, and product of algebraic integers are also algebraic integers. He defined divisibility in the standard way: an algebraic integer α is **divisible** by an algebraic integer β if $\alpha = \beta\gamma$ for some algebraic integer γ. To develop general laws of divisibility, however, Dedekind needed to restrict himself to only a part of the domain of all algebraic integers. Thus, given any algebraic number θ which satisfies an irreducible equation of degree n, he defined the system of algebraic integers Γ_θ corresponding to θ to be the set of those algebraic integers of the form $x_0 + x_1\theta + x_2\theta^2 + \cdots x_{n-1}\theta^{n-1}$, where the x_i are rational numbers. In any domain Γ_θ, Dedekind could now define, like Kummer, what it meant for an integer to be prime or irreducible.

Dedekind noted that the Gaussian integers Γ_i are a system of algebraic integers and that Gauss had proved that unique factorization into primes is true in this system. But, following the work of Dirichlet, he gave a proof different from that of Gauss. Namely, he first showed that the Euclidean division algorithm is true in this domain in the sense that given any two nonzero Gaussian integers z and m, there always exist two other Gaussian integers q and r such that $z = qm + r$ with $N(r) < N(m)$. (It turns out that such a division algorithm does not generally exist in domains of algebraic integers; a domain in which it is true is called a **Euclidean domain**.) Dedekind then showed, exactly as in the case of ordinary integers, that the repeated use of the division algorithm determines for any two Gaussian integers z and m a greatest common divisor d which can be written in the form $d = az + bm$. In particular, if an irreducible Gaussian integer p divides a product rs of two Gaussian integers, then it must divide one of the factors. For if p does not divide r, then $1 = ap + br$ or $s = aps + brs$ and p divides s. Unique factorization follows immediately, and Dedekind, like Gauss, could determine all the primes of Γ_i.

Dedekind noted further that if θ is a root of any of the equations $x^2 + x + 1 = 0$, $x^2 + x + 2 = 0$, $x^2 + 2 = 0$, $x^2 - 2 = 0$, or $x^2 - 3 = 0$, then a similar division algorithm is valid in Γ_θ and unique factorization follows. On the other hand, he showed that the algorithm does not apply to the domain determined by a root θ of $x^2 + 5 = 0$, where the norm of any integer $\omega = x + y\theta$ is $N(\omega) = x^2 + 5y^2$. Dedekind considered the integers $a = 2, b = 3, b_1 = -2 + \theta, b_2 = -2 - \theta, d_1 = 1 + \theta$, and $d_2 = 1 - \theta$ and proved, by use of the norm, that each of these integers is irreducible. Furthermore, easy multiplications showed that $ab = d_1 d_2$, $b^2 = b_1 b_2$, and $ab_1 = d_1^2$. It followed that unique factorization does not hold. Dedekind went further, however. "Imagine for an instant that the . . . preceding numbers are rational integers."[10] Using the general laws of divisibility and assuming that a and b are relatively prime, and also b_1 and b_2, Dedekind deduced that there would exist integers α, γ, and δ such that $a = \alpha^2, b = \gamma\delta, b_1 = \gamma^2, b_2 = \delta^2$, $d_1 = \alpha\gamma$, and $d_2 = \alpha\delta$. Then, for example, $ab = \alpha^2\gamma\delta = d_1 d_2$. These integers, however, do not exist in the given domain. After all, the original integers are all irreducible. To create substitutes for these new integers and thereby restore unique factorization was Dedekind's goal in the creation of his new concept, the ideal, a concept he believed easier to understand than Kummer's ideal numbers.

Dedekind decided that "there was no more need for any new creation, like that of the ideal number of Kummer, and it sufficed completely to consider a system of numbers already existing."[11] Because Kummer had only defined divisibility by an ideal number, Dedekind took for his "system of numbers already existing" the set I of all those integers in Γ_θ which are divisible by the given ideal number, a set he named an **ideal**. Because for any numbers α and β divisible by an ideal number, the sum $\alpha + \beta$ is also divisible, and because $\omega\alpha$ is divisible for any ω in Γ_θ, these conditions are necessary for a set to be an ideal. But because any set satisfying those conditions was the set divisible by some ideal number, Dedekind could take those conditions as his definition of an ideal I in a domain Γ_θ of algebraic integers. Furthermore, he defined a **principal ideal** (α) to be the set of all multiples of a given integer α.

Dedekind's next task was to define divisibility of ideals. He noted that if α is divisible by β, or $\alpha = \mu\beta$, the principal ideal generated by α is contained in that generated by β. Conversely, if every multiple of α is also a multiple of β, then α is divisible by β. Dedekind therefore extended this definition to arbitrary ideals: An ideal I is **divisible** by an ideal J if every number in I is contained in J. An ideal P, different from Γ_θ, is said to be **prime**, if it has no divisor other than itself and Γ_θ, that is, if it is contained in no other ideal except Γ_θ itself.

Dedekind noted further that there was a natural definition of a **product** of two ideals, namely, that IJ consists of all sums of products of the form $\alpha\beta$ where α is in I and β is in J. It is then obvious that IJ is divisible by both I and J. To complete the relationship between the two notions of product and divisibility, however, he had to prove two further theorems.

Theorem 1. *If the ideal C is divisible by an ideal I, then there exists a unique ideal J such that the product IJ is identical with C.*

Theorem 2. *Every ideal different from Γ_θ is either a prime ideal or may be represented uniquely in the form of a product of prime ideals.*

It is these two theorems which provided Dedekind his new way of restoring unique factorization to any domain of algebraic integers. Thus, in the domain of algebraic integers of the form $x + y\sqrt{-5}$, the (nonexistent) prime factors α, γ, δ were replaced by prime ideals A, G, D so that the principal ideals (a), (b), (d_1), (d_2) factored as $(a) = A^2$, $(b) = GD$, $(d_1) = AG$, and $(d_2) = AD$. The nonunique integer factorization $ab = d_1 d_2$ could then be replaced by the unique factorization into prime ideals: $(a)(b) = A^2 GD = (d_1)(d_2)$.

15.2 SOLVING ALGEBRAIC EQUATIONS

Because the solving of equations was the central concern of algebra before the nineteenth century, it is not surprising that major features of the new forms which algebra took in that century grew out of new approaches to this problem. In fact, some of the central ideas of group theory grew out of these new approaches.

15.2.1 Cyclotomic Equations

Lagrange had studied in detail the solvability of equations of degree less than five and had indicated possible means of attack for equations of higher degree. In the final chapter of his *Disquisitiones Arithmeticae*, Gauss discussed the solution of cyclotomic equations, equations of the form $x^n - 1 = 0$, and the application of these solutions to the construction of regular polygons. Naturally, Gauss knew the solutions of this equation in the form $\cos \frac{2\pi k}{n} + i \sin \frac{2\pi k}{n}$ for $k = 0, 1, 2, \ldots, n - 1$, but his aim in this chapter was to determine these solutions algebraically. Because the solution of the equation for composite integers follows immediately from that for primes, Gauss restricted his attention to the case where n is prime, and because $x^n - 1$ factors as $(x - 1)(x^{n-1} + x^{n-2} + \cdots + x + 1)$, it was the equation

$$x^{n-1} + x^{n-2} + \cdots + x + 1 = 0$$

which provided the focus for his work.

Gauss' plan for the solution of this $(n - 1)$st degree equation was to solve a series of auxiliary equations, each of degree a prime factor of $n - 1$, with the coefficients of each in turn being determined by the roots of the previous equation. Thus for $n = 17$, where $n - 1 = 2 \cdot 2 \cdot 2 \cdot 2$, he wanted to determine four quadratic equations, while for $n = 73$ he needed three quadratics and two cubics. Gauss knew that the roots of $x^{n-1} + x^{n-2} + \cdots + x + 1$ could be expressed as powers r^i $(i = 1, 2, \ldots n - 1)$ of any fixed root r. Furthermore, he realized that if g is any primitive root modulo n, the powers $1, g, g^2, \ldots, g^{n-2}$ include all the nonzero residues modulo n. It follows that the $n - 1$ roots of the equation can be expressed as $r, r^g, r^{g^2}, \ldots, r^{g^{n-2}}$ or even as $r^\lambda, r^{\lambda g}, r^{\lambda g^2}, \ldots, r^{\lambda g^{n-2}}$ for any λ less than n. His method of determining the auxiliary equations involved constructing

periods, certain sums of the roots r^j, which in turn were the roots of the auxiliary equations. An analysis of the particular example $n = 19$ will give the flavor of Gauss' work.

For $n = 19$, the factors of $n - 1$ are 3, 3, and 2. Gauss begins by determining three periods of six terms each, each period to be the root of an equation of degree 3. The periods are found by choosing a primitive root modulo 19, here 2, setting $h = 2^3$, and computing

$$\alpha_i = \sum_{j=0}^{5} r^{ihj} \qquad \text{for} \quad i = 1, 2, 4.$$

In modern terminology, the permutations of the 18 roots of $x^{18} + x^{17} + \cdots + x + 1$ form a cyclic group G determined by the mapping $r \to r^2$, where r is any fixed root. The periods here are the sums which are invariant under the subgroup H of G generated by the mapping $r \to r^h$. These sums contain six elements because $h^6 = 2^{18} \equiv 1 \pmod{19}$, that is, H is a subgroup of G of order 6. Furthermore, for $i = 1, 2, 4$ those mappings of the form $r \to r^{ihj}$ for $j = 0, 1, \ldots, 5$ are precisely the three cosets of H in the group G. For example, because $H = \{r \to r, r^8, r^{64}, r^{512}, r^{4096}, r^{32,768}\}$, it follows that

$$\alpha_1 = r + r^8 + r^7 + r^{18} + r^{11} + r^{12}$$

where the powers are reduced modulo 19. Similarly,

$$\alpha_2 = r^2 + r^{16} + r^{14} + r^{17} + r^3 + r^5 \qquad \text{and} \qquad \alpha_4 = r^4 + r^{13} + r^9 + r^{15} + r^6 + r^{10}.$$

Gauss then shows that $\alpha_1, \alpha_2, \alpha_4$ are roots of the cubic equation $x^3 + x^2 - 6x - 7 = 0$.

The next step is to divide each of the three periods into three further periods of two terms each, where again the new periods will satisfy an equation of degree 3. These periods,

$$\beta_i = \sum_{j=0}^{1} r^{ikj} \qquad \text{for} \quad i = 1, 2, 4, 8, 16, 13, 7, 14, 9$$

where $k = 2^9$, are invariant under the subgroup K generated by the mapping $r \to r^k$. Because $k^2 \equiv 1 \pmod{19}$, K has order 2 and has nine cosets corresponding to the values of i given. For example,

$$\alpha_1 = \beta_1 + \beta_8 + \beta_7 = (r + r^{18}) + (r^8 + r^{11}) + (r^7 + r^{12}).$$

Given these new periods of length 2, Gauss shows that β_1, β_8, and β_7 are all roots of the cubic equation $x^3 - \alpha_1 x^2 + (\alpha_2 + \alpha_4)x - 2 - \alpha_2 = 0$. It turns out that each of the other β_i can be expressed as a polynomial in β_1. Finally, Gauss breaks up each of the periods with two terms into the individual terms of which it is formed and shows that, for example, r and r^{18} are the two roots of $x^2 - \beta_1 x + 1 = 0$. The sixteen remaining roots of the original equation are then simply powers of r or can be found by solving eight other similar equations of degree 2.

Because the equations of degree 3 involved in the above example are solvable by the use of radicals, as is the equation of degree 2, Gauss has demonstrated that the roots of $x^{19} - 1 = 0$ are all expressible in terms of radicals. His more general result, applicable to any equation $x^n - 1 = 0$, only shows that a series of equations can be discovered, each of prime degree less than $n - 1$, whose solutions would determine the solution of the original equation. But, Gauss continues,

> everyone knows that the most eminent geometers have been unsuccessful in the search for a general solution of equations higher than the fourth degree, or (to define the search more accurately) for the reduction of mixed equations to pure equations. [Pure equations are those of the form $x^m - A = 0$, which can be solved by taking an mth root once the solutions of $x^m - 1 = 0$ are known.] And there is little doubt that this problem is not merely beyond the powers of contemporary analysis but proposes the impossible. . . . Nevertheless, it is certain that there are innumerable mixed equations of every degree which admit a reduction to pure equations, and we trust that geometers will find it gratifying if we show that our equations are always of this kind.[12]

Gauss then sketches a proof, although one with a minor gap, that the auxiliary equations involved in his solution of $x^n - 1 = 0$ for n prime can always be reduced to pure equations. He therefore demonstrates, by induction, that these equations are always solvable in radicals. Naturally, if $n - 1$ is a power of 2, all of the auxiliary equations are quadratic and no special proof is necessary. In this case, however, Gauss notes further that the solutions can be constructed geometrically by Euclidean techniques. Because the roots of $x^n - 1 = 0$ can be considered as the vertices of a regular n-gon (in the complex plane), Gauss has proved that such a polygon can be constructed whenever $n - 1$ is a power of 2. The only such primes known to Gauss, and even to us today, are 3, 5, 17, 257, and 65,537. In fact, the story is told that Gauss' discovery of the construction of the regular 17-gon was instrumental in his decision to pursue a career in mathematics. Gauss concludes with a warning: "Whenever $n - 1$ involves prime factors other than 2, we are always led to equations of higher degree. . . . We can show with all rigor that these higher-degree equations cannot be avoided in any way nor can they be reduced to lower degree equations. . . . We issue this warning lest anyone attempt to achieve geometric constructions for sections [of the circle] other than the ones suggested by our theory (e.g. sections into 7, 11, 13, 19, etc. parts) and so spend his time uselessly."[13]

Interestingly, although Gauss made this assertion, he did not in fact prove that regular n-gons, where $n = 7, 11, 13, 19$, and so on cannot be constructed. This gap was filled in 1837 by Pierre Wantzel (1814–1848). Wantzel also gave the final resolution of two classical Greek construction problems by showing that any construction problem which does not lead to an irreducible polynomial equation with degree a power of two and with constructible coefficients cannot be accomplished using straightedge and compass. For example, because the problem of doubling a cube of side a requires the solution of the irreducible cubic $x^3 - 2a^3 = 0$, it is impossible to construct this solution with Euclidean tools. Similarly, the problem of trisecting an angle α requires the solution of the irreducible cubic equation expressing $x = \sin(\alpha/3)$ in terms of the known value $a = \sin \alpha$: $4x^3 - 3x + a = 0$. (See section 9.4.1.) Again, Wantzel's result shows that this construction is impossible with Euclidean tools.

The other important construction problem unsolved by the Greeks, that of squaring the circle, was also shown to be impossible. Algebraically, this problem is equivalent to solving $x^2 - \pi = 0$, a quadratic equation. Unfortunately, one coefficient of this quadratic is π, and the Greeks found no way of constructing, using Euclidean tools, a line segment with that length. By the nineteenth century, it had long been suspected that π could not be expressed as the root of any algebraic equation with rational coefficients, that is, that it was a **transcendental** number rather than an algebraic number. It was Liouville who in 1844 was the first actually to display a transcendental number:

$$\frac{1}{10} + \frac{1}{10^{2!}} + \frac{1}{10^{3!}} + \cdots + \frac{1}{10^{n!}} + \cdots = 0.110001000000000000000000100\ldots.$$

He then attempted without success to show that both e and π were also transcendental. A protegé of Liouville, Charles Hermite (1822–1901), finally demonstrated that e was transcendental in 1873 and, on the basis of Hermite's ideas, Ferdinand Lindemann (1852–1939) showed that π was transcendental nine years later. It followed immediately that it was impossible to square the circle with Euclidean tools.

15.2.2 The Theory of Permutations

It is clear that Gauss was convinced that the general equation of degree higher than four could not be solved by radicals. Recall that Lagrange had attempted some years earlier to find a solution by considering permutations of the roots. To consider the question of higher degree equations in detail, therefore, it was necessary to understand the theory of permutations. Substantial work on this concept was accomplished early in the nineteenth century by Augustin-Louis Cauchy (1789–1857).

Up to Cauchy's time, the term "permutation" generally referred to an arrangement of a certain number of objects, say letters. It was Cauchy who first considered the importance of the action of changing from one arrangement to another. He used the word **substitution** to refer to such an action, what one would today call a permutation, that is, a one-to-one function from the given (finite) set of letters to itself. In a series of papers on the subject nearly 30 years after his initial efforts, Cauchy used the words "substitutions" and "permutations" interchangeably to refer to such functions. To avoid confusion, it is the latter word which will generally be used, in its modern sense, in what follows.

Besides focusing on the functional aspect of a permutation, Cauchy used a single letter, say S, to denote a given permutation and defined the product of two such permutations S, T, written ST, to be the permutation determined by first applying S to a given arrangement and then applying T to the resulting one. He named the permutation which leaves a given arrangement fixed the identity permutation, noted that the powers S, S^2, $S^3, \ldots$ of a given permutation must ultimately result in the identity, and then defined the **degree** of a permutation S to be the smallest power n such that S^n is equal to the identity. Cauchy even defined what he called a **circular** (cyclic) permutation on the letters a_1, $a_2, \ldots, a_n$ to be one which takes a_1 to a_2, a_2 to $a_3, \ldots, a_{n-1}$ to a_n, and a_n to a_1. In 1844, Cauchy introduced the notation $(a_1 a_2 \cdots a_n)$ for such a permutation. At that time he also defined the inverse of a permutation S in the obvious way, using the notation S^{-1}, and introduced the notation 1 for the identity. Further, given any set of permutations on n letters, he defined what he called the **system of conjugate permutations** determined by these, today called the **subgroup** generated by the given set, as the collection of all permutations formed from the original ones by taking all possible products. Finally, he showed that the order of this system (the number of elements in the collection) always divides the order $n!$ of the complete system of permutations on n letters.

Biography   Niels Henrik Abel (1802–1829)

Abel, born near Stavanger in Norway, unfortunately enjoyed but a brief life. His native abilities in mathematics were discovered by his instructor at the Cathedral School in Oslo, who encouraged Abel to read various advanced mathematics treatises available in the university. Becoming interested in the problem of the fifth degree equation, he believed that he was able, in fact, to solve it using radicals. Because no one in Norway could understand his arguments, he had the paper forwarded to Denmark. Before it could be published, Abel was asked to provide some numerical examples. In searching for these, he realized that his method was incorrect. Although he then proceeded to do research in other areas, in particular the theory of elliptic functions, he continued to work on the solvability question over the next several years, while studying at the University of Oslo, until he managed to prove its impossibility to his own satisfaction. He published the result in a small pamphlet at his own expense in 1824, but the brevity caused by his attempt to save money prevented most mathematicians from understanding it. Thus two years later, during his travels through Europe to visit various mathematicians and better prepare himself for a scientific career, he wrote an expanded version that was published in the first volume of the new German mathematics journal, *Journal für die reine und angewandte Mathematik* (*Journal for Pure and Applied Mathematics*), edited by August Crelle, who soon became one of Abel's best friends. When Abel returned to Norway in 1827, he found that there were no positions available to him, the only mathematics professorship at the university having recently been awarded to his former secondary school teacher. Abel struggled to make a living by tutoring and substituting at the university, meanwhile preparing a large number of new mathematical papers. But in January, 1829 he suffered an attack of tuberculosis from which he was not able to recover. He died in April, two days before Crelle wrote to him with the news that an appointment had been secured for him in Berlin (Figure 15.3).[14]

15.2.3 The Unsolvability of the Quintic

The first proposed proof that the general fifth degree equation could not be solved using radicals appeared in a privately printed treatise by the Italian Paolo Ruffini (1765–1822) in 1798, a treatise whose purported proof no contemporary could understand. In the mid-1820s, however, Niels Henrik Abel (1802–1829) finally gave a complete proof of the impossibility of such a solution.

FIGURE 15.3
Abel honored on a Norwegian stamp.

Abel's unsolvability proof involved applying results on permutations to the set of the roots of the equation, but it is too lengthy to present here.[15] It is well to note, however, that after proving his unsolvability result, Abel continued his research to attempt to solve the following problems: "1. To find all equations of any given degree that are algebraically solvable. 2. To decide whether a given equation is algebraically solvable or not."[16] Although he was not able in what remained of his life to solve either of these questions in its entirety, he did make progress with a particular type of equation. In a paper published in Crelle's *Journal* in 1829, Abel generalized Gauss' solution method for the equations $x^n - 1 = 0$. For that equation, every root is expressible as a power of one of them. Abel was able to show that "if the roots of an equation of any degree are related so that all of them are rationally expressible in terms of one, which we designate as x, and if, furthermore, for any two of the roots θx and $\theta_1 x$ [where θ and θ_1 are rational functions], we have $\theta\theta_1 x = \theta_1\theta x$, then the equation is algebraically solvable."[17] He demonstrated this result by showing that in this situation, as in the cyclotomic case, one could always reduce the solution to that of auxiliary equations of prime degree. It is because of this result that commutative groups today are often referred to as Abelian.

15.2.4 The Work of Galois

Although Abel could not complete his research program, this was largely accomplished by another genius who died young, Evariste Galois (1811–1832). Galois' thoughts on the subject of solvability of algebraic equations by radicals are outlined in the manuscript he submitted to the French Academy in 1831. In that manuscript, he begins by clarifying the idea of rationality. Since an equation has coefficients in a certain domain, for example the set of ordinary rational numbers, to say that an equation is solvable by radicals means that one can express any root using the four basic arithmetic operations and the operation of root extraction, all applied to elements of this original domain. It turns out, however, that it is usually convenient to solve an equation in steps, as Gauss did in the cyclotomic case. Therefore, once one has solved $x^n = \alpha$, for instance, one has available as coefficients in the next step these solutions, expressible as $\sqrt[n]{\alpha}, r\sqrt[n]{\alpha}, r^2\sqrt[n]{\alpha}, \ldots$, where r is an nth root of unity. Galois notes that such quantities are adjoined to the original domain and that any quantity expressible by the four basic operations in terms of these new quantities and the original ones can then also be considered as **rational**.

Galois' main result is expressed as follows.

Proposition I. *Let an equation be given of which a, b, c, . . . are the m roots.* [Galois tacitly assumes that this equation is irreducible and that all the roots are distinct.] *There will always be a group of permutations of the letters a, b, c, . . . which has the following property: 1. that every function of the roots, invariant under the permutations of the group, is rationally known; 2. conversely, that every function of the roots which is rationally known is invariant under the permutations.*[18]

Galois called this group of permutations the **group of the equation**. (Galois' use of the word "group" is somewhat ambiguous. Here it refers to a set of permutations closed under composition.) In modern usage, one normally considers the group of the equation as a group of automorphisms acting on the entire field created by adjoining the roots of the equation to the original field of coefficients. Galois' result is then that the group of the equation is that group of automorphisms of the extension field which leaves invariant precisely the elements of the original field (the elements "rationally known"). Besides giving a brief proof of his result, Galois presented Gauss' example of the cyclotomic polynomial $\frac{x^n-1}{x-1}$ for n prime. In that case, supposing that r is one root and g a primitive root modulo n, the roots can be expressed in the form $a = r, b = r^g, c = r^{g^2}, \ldots$ and the group of the equation is the cyclic group of $n - 1$ permutations generated by the cycle $(abc \ldots k)$. On the other hand, the group of the general equation of degree n, that is, of the equation with literal coefficients, is the group of all $n!$ permutations of n letters.

Biography Evariste Galois (1811–1832)

Galois' tragically brief life has been the subject of a fictionalized biography which included speculation that his death in a duel was engineered by government agents because of his radical political views. The known facts, however, do not support this contention.[19] Galois was born in Bourg-la-Reine, a town not far from Paris in which his father was elected mayor in 1815. He had mixed success in the preparatory school of Louis-le-Grand, especially after discovering his talents in mathematics. Although he published a short paper before he turned 18 and submitted a memoir on the solvability of equations of prime degree to the French Academy at the same time, he nevertheless twice failed the entrance examinations for the École Polytechnique, the first time probably because he had not mastered the basics and the second time perhaps because his father had committed suicide a few days earlier due to a scandal concocted by a reactionary priest. Galois was forced to enroll at the École Normale, whose director locked the students into the building so they could not participate in the political activities leading to the July revolution of 1830. After Galois attacked the director in a December letter for favoring "legitimacy" over "liberty," he was expelled from school and joined a heavily republican division of the National Guard, a division which was soon dissolved because of its perceived threat to the throne occupied by the "bourgeois" King Louis-Phillipe. Now heavily involved in political activity, Galois nevertheless continued his mathematical researches, submitting a revised version of his memoir on solvability of equations to the Academy in January, 1831.

The referee rejected the manuscript some six months later because he could not understand the proofs. He suggested that Galois complete and clarify his theory and resubmit it.

Meanwhile, however, Galois had been arrested twice, the first time for threatening the life of the king and the second time for wearing the uniform of the dissolved National Guard division. For the second offense, he was convicted and sentenced to six months in prison, during which time his hatred of the Academy for their failure to appreciate his work grew to such a degree that he lashed out at France's "official scientists" in a vicious diatribe intended as a preface to the private publication of his work. Before the publication could take place, however, Galois became involved with "an infamous coquette and her two dupes"[20] and, although the exact circumstances have never been clarified, was forced (or chose) to fight a duel in which he was killed, five months before his twenty-first birthday. On the night before the duel, fearing the worst, he wrote a letter to his friend Auguste Chevalier amplifying and annotating some of his earlier manuscripts. He concluded with the following: "I have often dared in my life to state propositions of which I was not certain. But all that I have written here has been clear in my head for more than a year, and it would not be in my interest . . . to announce theorems of which I do not have complete proof. Publicly beseech Jacobi or Gauss to give their opinion not of the truth, but of the importance of these theorems. After that, I hope, there will be men who will find it profitable to decipher all this mess"[21] (Figure 15.4).

FIGURE 15.4
Galois on a French stamp.

With the main theorem stated, Galois explores its application to the solvability question. His second proposition shows what happens when one adjoins to the original field one or all of the roots of some auxiliary equation (or of the original equation). Because any automorphism leaving the new field invariant certainly leaves the original field invariant, the group H of the equation over the new field is a subgroup of the group G over the original field. In fact, G can be decomposed either as $G = H + HS + HS' + \cdots$ or as $G = H + TH + T'H + \cdots$ where $S, S', T, T', \ldots$ are appropriately chosen permutations. Galois explains this entire procedure in his letter to Chevalier and notes that ordinarily these two decompositions do not coincide. When they do, however, and this always happens when all the roots of an auxiliary equation are adjoined, he calls the decomposition **proper**. "If the group of an equation has a proper decomposition so that it is divided into m groups of n permutations, one may solve the given equation by means of two equations, one having a group of m permutations, the other one of n permutations."[22] In modern terminology, a proper decomposition occurs when the subgroup H is **normal**, that is, when the right cosets $\{HS\}$ coincide with the left cosets $\{TH\}$. In these circumstances, the question of solvability reduces to the solvability of two equations each having groups of order less than the original one.

Gauss had already shown that the roots of the polynomial $x^p - 1$ with p prime can be expressed in terms of radicals. It follows that if the pth roots of unity are assumed to be in the original field, then the adjunction of one root of $x^p - \alpha$ amounts to the adjunction of all of the roots. If G is the group of an equation, this adjunction therefore leads to a normal subgroup H of the group G such that the index of H in G (the quotient of the order of G by that of H) is p. Galois also proves the converse, that if the group G of an equation has a normal subgroup of index p, then one can find an element α of the original field (assuming that the pth roots of unity are in that field) such that adjoining $\sqrt[p]{\alpha}$ reduces the group of the equation to H. Galois concludes, both in his manuscript and in his letter, that an equation is solvable by radicals as long as one can continue the process of finding normal subgroups until all of the resulting indices are prime. Galois gave the details of this procedure in the case of the general equation of degree four, showing that the group of the equation of order 24 has a normal subgroup of order 12, which in turn contains one of order 4, which contains one of order 2, which contains the identity. It follows that the solution can be obtained by first adjoining a square root, then a cube root, and then two more square roots. Galois notes that the standard solution in this case amounts to precisely those steps.

15.2.5   Jordan and the Theory of Groups of Substitutions

With Galois' death, his manuscripts lay unread until they were finally published in 1846 by Liouville in his *Journal des mathématiques*. Within the next few years, several mathematicians included Galois' material in university lectures or published commentaries on the work. It was not until 1866, however, that Galois theory was included in a text, the third edition of the *Cours d'algebre* of Paul Serret (1827–1898). Four years later, Camille Jordan (1838–1922) published his monumental *Traité des substitutions et des équations algébriques* (*Treatise on Substitutions and Algebraic Equations*) which contained a somewhat revised version of Galois theory.

It is in Jordan's text, and in some of his papers of the preceding decade which are essentially incorporated in it, that many modern notions of group theory first appear, although always in the context of groups of permutations (substitutions). Thus Jordan defines a **group** to be a system of permutations of a finite set with the condition that the product (composition) of any two such permutations belongs to the system. He is then able to show that every group contains a unit element 1 and, for every permutation a another permutation a^{-1} such that $aa^{-1} = 1$. Jordan defines the **transform** of a permutation a by a permutation b to be the permutation $b^{-1}ab$ and the transform of the group $A = \{a_1, a_2, \ldots, a_n\}$ by b to be the group $B = \{b^{-1}a_1b, b^{-1}a_2b, \ldots, b^{-1}a_nb\}$ consisting of all the transforms. If B coincides with A, then A is said to be **permutable** with b. Although Jordan does not explicitly define a normal subgroup of a group, he does define a **simple** group to be one which contains no subgroup (other than the identity) which is permutable with all elements of the group. For a nonsimple group G, there must then exist a **composition series**, a sequence of groups $G = H_0, H_1, H_2, \ldots, \{1\}$ such that each group is contained in the previous one and is permutable with all its elements (that is, is normal), and that no other such group can be interposed in this sequence. Jordan further proves that if the order of G is n and the orders of the subgroups are successively $\frac{n}{\lambda}, \frac{n}{\lambda\mu}, \frac{n}{\lambda\mu\nu}, \ldots$, then the integers $\lambda, \mu, \nu, \ldots$, are unique up to order, that is, that any other such sequence has the same **composition factors**.[23]

Jordan then restates some of Galois' results in group theoretic language. He defines a **solvable group** to be one which belongs to an equation solvable by radicals. It follows that a solvable group is one which contains a composition series with all composition factors prime. Because a commutative group always has prime composition factors, Jordan could show that an **Abelian equation**, one "of which the group contains only permutations which are interchangeable among themselves,"[24] is always solvable by radicals. On the other hand, because the alternating group on n letters, which has order $\frac{n!}{2}$, is simple for $n > 4$, it follows immediately that the general equation of degree n is not solvable by radicals. With Jordan's work clarifying that of Galois, it was clear to all that the theory of permutation groups was intimately connected with the solvability of equations.

15.3 GROUPS AND FIELDS—THE BEGINNING OF STRUCTURE

Certain concepts of group theory were implicit in the early nineteenth century developments in number theory and the solvability of equations by radicals, in both of which areas Gauss played a significant role. Gauss' work in the theory of quadratic forms was also important in bringing to the fore ideas which were ultimately to be part of the abstract theory of groups.

15.3.1 Gauss and Quadratic Forms

Gauss discussed the theory of quadratic forms, that is, functions of two variables x, y of the form $ax^2 + 2bxy + cy^2$, with a, b, c integers, in Chapter 5 of his *Disquisitiones*. Gauss' aim in his discussion of forms was to determine whether a given integer can be represented by a particular form. As a tool in the solution of this problem, he defined equivalence of two forms. A form $f = ax^2 + 2bxy + cy^2$ is **equivalent** to a form $f' = a'x'^2 + 2b'x'y' + cy'^2$ if there exists a linear substitution $x = \alpha x' + \beta y', y = \gamma x' + \delta y'$ with $\alpha\delta - \beta\gamma = 1$ which transforms f into f'. An easy calculation shows that any two equivalent forms have the same discriminant $b^2 - ac$. On the other hand, two forms with the same discriminant are not necessarily equivalent. Gauss was able to show that for any given value D of the discriminant there were finitely many classes of equivalent forms. In particular, there was a distinguished class, the principal class, consisting of those forms equivalent to the form $x^2 - Dy^2$.

To investigate these classes, Gauss presented a rule of composition for forms. In other words, given forms f, f', of the same discriminant, Gauss defined a new form F composed of f, f' (written $F = f + f'$) which had certain desirable properties. First, Gauss showed that if f and g are equivalent and if f' and g' are equivalent, then $f + f'$ is equivalent to $g + g'$. Therefore, the composition operation is an operation on classes. Gauss next showed that the operation of composition is both commutative and associative. Finally, Gauss showed that "if any class K is composed with the principal class, the result will be the class K itself," that for any class K there is a class L (opposite to K) such that the composite of the two is the principal class, and that "given any two classes K, L of the same [discriminant], . . . we can always find a class M with the same [discriminant] such that L is composed of M and K." Given that composition enjoys the basic properties of addition, Gauss noted that "it is convenient to denote composition of classes by the addition sign, $+$, and identity of classes by the equality sign."[25]

With the addition sign as the sign of operation, Gauss designated the composite of a class C with itself by $2C$. Gauss then proved that for any class C, there is a smallest multiple mC which is equal to the principal class and that, if the total number of classes is n, then m is a factor of n. Naturally, this result reminded him of earlier material in the *Disquisitiones*. "The demonstration of the preceding theorem is quite analogous to the demonstrations [on powers of residue classes] and, in fact, the theory of the [composition] of classes has a great affinity in every way with the subject treated [earlier]."[26] He could therefore assert, without proof, various other results which came from this analogy, in terms of what is now the theory of Abelian groups.

15.3.2 Kronecker and the Structure of Abelian Groups

Gauss, although he recognized the analogy between his two treatments, did not attempt to develop an abstract theory of groups. This development took many years. In the mid-1840s, Kummer, in working out his theory of ideal complex numbers, noted that it was analogous in many respects to Gauss' theory of forms. In particular, Kummer defined an equivalence of ideal complex numbers which partitioned them into classes whose properties were analogous to those of Gauss' classes of forms. But it was Kummer's student, Leopold Kronecker (1823–1891), who finally saw that an abstract theory could be developed out of these analogies.

In a paper of 1870 in which he developed certain properties of the number of classes of Kummer's ideal complex numbers, Kronecker recalled Gauss' work on quadratic forms:

> The very simple principles on which Gauss' method rests are applied not only in the given context but also frequently elsewhere, in particular in the elementary parts of number theory. This circumstance shows, and it is easy to convince oneself otherwise, that these principles belong to a more general, abstract realm of ideas. It is therefore appropriate to free their development from all unimportant restrictions, so that one can spare oneself from the necessity of repeating the same argument in different cases. This advantage already appears in the development itself, and the presentation gains in simplicity, if it is given in the most general admissible manner, since the most important features stand out with clarity.[27]

> ### *Biography*   Leopold Kronecker (1823–1891)
>
> Kronecker, attempting to get the best mathematics education possible, studied at the Universities of Berlin, Bonn, and Breslau, before receiving his doctorate in 1845 at Berlin. For several years thereafter he managed the family business, ultimately becoming financially independent. Having carried on mathematical research as a hobby, he was in 1861 elected to the Berlin Academy and permitted to lecture at the university. In 1880, he took over the editorship of the journal founded by Crelle, the *Journal für die reine und angewandte Mathematik* (*Journal for Pure and Applied Mathematics*) and three years later, on Kummer's retirement, became professor of mathematics in Berlin and, with Karl Weierstrass, codirector of the influential mathematics seminar there.

Kronecker thus begins to develop the simple principles: "Let θ', θ'', θ''', ... be finitely many elements such that to each pair there is associated a third by means of a definite procedure." Kronecker goes on to require that this association, which he first writes as $f(\theta', \theta'') = \theta'''$ and later as $\theta' \cdot \theta'' = \theta'''$, be commutative and associative and that, if $\theta'' \neq \theta'''$, then $\theta'\theta'' \neq \theta'\theta'''$. From the finiteness assumption, Kronecker then deduces the existence of a unit element 1 and, for any element θ, the existence of a smallest power n_θ such that $\theta^{n_\theta} = 1$.

Finally, Kronecker develops the fundamental theorem of Abelian groups, that there exists a finite set of elements θ_1, θ_2, ... , θ_m such that every element θ can be expressed uniquely as a product of the form $\theta_1^{h_1}\theta_2^{h_2} \cdots \theta_m^{h_m}$ where for each i, $0 \leq h_i < n_{\theta_i}$. Furthermore, the θ_i can be arranged so that each n_θ is divisible by its successor and the product of these numbers is precisely the number of elements in the system. With the abstract theorem proved, Kronecker interprets the elements in various ways, noting that analogous results in each case had been proved previously by others.

15.3.3 Cayley and the Definition of a Group

Interestingly enough, Kronecker did not give a name to the system he defined, nor did he interpret it in terms of the permutation groups arising from Galois theory. Kronecker was also probably unaware that, 16 years earlier, Arthur Cayley (1821–1895) had developed a similar abstract theory based on the notion of groups of substitutions.

Cayley, in his "On the Theory of Groups," noted that the idea of a group of permutations was due to Galois and immediately proceeded to generalize it to any set of operations, or functions, on a set of quantities. He used the symbol 1 to represent the function which leaves all quantities unchanged and noted that for functions, there is a well-defined notion of composition which is associative, although not, in general, commutative. But then Cayley abstracted the basic ideas out of the concrete notion of operations and defined a **group** to be a "set of symbols, 1, α, β, ... , all of them different, and such that the product of any two of them (no matter in what order), or the product of any one of them into itself, belongs to the set ... It follows that if the entire group is multiplied by any one of the symbols, either [on the right or the left], the effect is simply to reproduce the

---

***Biography*   Arthur Cayley (1821–1895)**

Cayley studied mathematics at Trinity College, Cambridge, graduating as Senior Wrangler, but because there was no suitable teaching job available, decided to become a lawyer and was called to the bar in 1849. Although he became skilled in legal work, he regarded the law just as the means of providing him with an income and always reserved a substantial portion of his time for mathematics. In fact, in his 14 years as an attorney he produced close to 300 mathematical papers. In 1863, he was elected to the newly created Sadlerian professorship of mathematics at Cambridge, a position he accepted with eagerness, even though it meant a substantial cut in his earnings.

The duties of the Sadlerian professor were to "teach the principles of pure mathematics" and also "to apply himself to the advancement of that science." In regard to the first duty, Cayley was not very successful. His lectures at the university generally attracted few students, partly because he usually spoke about his latest research. On the other hand, his contributions to mathematics were enormous, comprising nearly 1000 papers in various fields. In addition, he served as a referee for hundreds of papers by others and took great pleasure in encouraging young men just beginning their research.

---

group."[28] (Cayley tacitly assumed in his definition that the number of symbols was finite and that the multiplication was associative.) Cayley then introduced the group table

	1	α	β	$\cdots$
1	1	α	β	$\cdots$
α	α	α^2	$\beta\alpha$	$\cdots$
β	β	$\alpha\beta$	β^2	$\cdots$
$\cdots$	$\cdots$	$\cdots$	$\cdots$	$\cdots$

and noted that each row and each column will contain all the symbols of the group. Furthermore, if there are n elements in the group, each element θ satisfies the symbolic equation $\theta^n = 1$.

Cayley showed by a familiar argument that if n is prime, then the group is necessarily of the form $1, \alpha, \alpha^2, \ldots, \alpha^{n-1}$. If n is not prime, there are other possibilities. In particular, he displayed the group tables of the two possible groups of four elements and the two possible groups of six elements. In a further paper of 1859, he described all five groups of order eight by giving a list of the elements and the defining relations as well as the smallest power (index) of each element which equals 1. For example, one of the groups contains the elements $1, \alpha, \beta, \beta\alpha, \gamma, \gamma\alpha, \gamma\beta, \gamma\beta\alpha$ with the relations $\alpha^2 = 1$, $\beta^2 = 1$, $\gamma^2 = 1$, $\alpha\beta = \beta\alpha$, $\alpha\gamma = \gamma\alpha$, and $\beta\gamma = \gamma\beta$. Each element in this group, except the identity, has index 2.

Although Cayley wrote an article for the *English Cyclopedia* in 1860 in which he explained the term "group," among others, no continental mathematician over the next few years paid attention to this (nearly) abstract definition. No one noticed either that Kronecker's definition of 1870 defined, albeit with the extra condition of commutativity, the identical structure. In 1878, however, Cayley published four new papers on the same subject in which he repeated his definition and results of 1854. In particular, he wrote that "a group is defined by means of the laws of combinations of its symbols."[29] Further, "although the theory [of groups] as above stated is a general one, including as a particular case the theory of substitutions, yet the general problem of finding all the groups of a given order n is really identical with the apparently less general problem of finding all the groups of the same order n which can be formed with the substitutions upon n letters."[30] Cayley took this result, today known as Cayley's theorem, as nearly obvious, noting merely that any element of a group may be thought of as acting upon all the elements of the group by the group operation, such operation inducing a permutation of the group elements. However, Cayley noted, this "does not in any wise show that the best or the easiest mode of treating the general problem is thus to regard it as a problem of substitutions; and it seems clear that the better course is to consider the general problem in itself, and to deduce from it the theory of groups of substitutions."[31] Thus Cayley, like Kronecker, realized that problems in group theory could often best be attacked by considering groups in the abstract, rather than in their concrete realizations. In fact, it is often only by dealing with the abstractions that one can make further progress.

15.3.4 The Axiomatization of the Group Concept

Events proceeded rapidly after 1878 in the recognition that the definitions of Kronecker and Cayley could be combined into a single abstract group concept. In particular, there were two publications in 1882 which could be said to mark the complete axiomatization of the group concept. First, Walther von Dyck (1856–1934) published his "Gruppentheoretische Studien" in which he formulated the basic problem: "To define a group of discrete operations, which are applied to a certain object, while one ignores any special form of representation of the individual operations, regarding these only as given by the properties essential for forming the group."[32] Dyck, although he alluded to the associative and inverse properties, did not give these as defining properties of a group. Instead, he showed how to construct a group by use of generators and relations. Namely, he began with a finite number of operations, $A_1, A_2, \ldots, A_m$, then built the "most general" group G on these elements by considering all possible products of powers of these elements and their inverses. This group, today called the **free group** on $\{A_i\}$, automatically satisfies the modern group axioms. Dyck next specialized to other groups by assuming various relations of the form $F(A_1, A_2, \ldots, A_m) = 1$. In fact, he showed that if the group $\overline{G}$ is formed from operations $\overline{A}_1, \overline{A}_2, \ldots, \overline{A}_m$ which satisfy the given relations, then "all these infinitely many operations of the group G, which are equal to the identity in $\overline{G}$, form a [sub]group H and this . . . commutes with all operations $S, S', \ldots$ of the group G."[33] Dyck then proved that the mapping $A_i \rightarrow \overline{A}_i$ defined what he called an **isomorphism** from G onto $\overline{G}$. In modern terminology, Dyck proved that the subgroup H is normal in G and that $\overline{G}$ is isomorphic to the factor group G/H.

A second paper of the same year, this one by Heinrich Weber (1842–1913) on quadratic forms, was the first to give a complete axiomatic description of a finite group without any reference to the nature of the elements composing it:

A system G of h elements of any sort $\theta_1, \theta_2, \ldots, \theta_h$ is called a group of order h if it satisfies the following conditions:

I. Through some rule, which is called composition or multiplication, one derives from any two elements of the system a new element of the same system. In signs, $\theta_r\theta_s = \theta_t$.

II. Always $(\theta_r\theta_s)\theta_t = \theta_r(\theta_s\theta_t) = \theta_r\theta_s\theta_t$.

III. From $\theta\theta_r = \theta\theta_s$ and from $\theta_r\theta = \theta_s\theta$, there follows $\theta_r = \theta_s$."[34]

From the given axioms and the finiteness of the group, Weber derived the existence of a unique unit element and, for each element, the existence of a unique inverse. He further defined a group to be an **Abelian group** if the multiplication is commutative and then proved the fundamental theorem of Abelian groups by essentially the same method used by Kronecker.

Although the use of the abstract group concept became more common over the next several years, it was not until 1893 that Weber published a definition which included infinite groups. He repeated his three conditions of 1882 and noted that if the group is finite these suffice to insure that if any two of the three group elements A, B, C are known there is a unique solution to the equation $AB = C$. On the other hand, this conclusion is no longer valid for infinite groups. In that case, one must assume the existence of unique solutions to $AB = C$ as a fourth axiom. This fourth axiom, even without finiteness, implies a unique identity and unique inverses for every element of the group.

After defining the modern notion of isomorphism of groups, Weber illuminated the basis for his abstract approach: "One can combine all groups isomorphic to one another into a class of groups, which itself is again a group, whose elements are the generic characters which one obtains if one combines the corresponding elements of the individual isomorphic groups into a general concept. The individual isomorphic groups are then to be considered as different representatives of the generic concept, and it is irrelevant which representative one uses to study the properties of the group."[35] Weber produced many examples of groups, including the additive group of vectors in the plane, the group of permutations of a finite set, the additive group of residue classes modulo m, the multiplicative group of residue classes modulo m relatively prime to m, and the group of classes of binary quadratic forms of a given discriminant under Gauss' law of composition. With the publication of this material, and its incorporation in Weber's 1895 algebra text, *Lehrbuch der Algebra*, the abstract concept of a group can be considered to have become part of the mathematical mainstream.

15.3.5 The Concept of a Field

The story of field theory is much simpler to tell than that of group theory. The notion of field is certainly implicit in Galois' work around 1830. Recall that Galois discussed what it meant for quantities to be rational and how to adjoin a new element to a given set of rational quantities. For Galois, the notions of the rational number field Q and of an extension field $Q(\alpha)$ generated by either a transcendental quantity or a root of a given equation were intuitively obvious, and there was no need to name this concept. It was Kronecker, beginning in the 1850s, who tried to be more specific in actually constructing these fields. Kronecker believed that algebra and analysis could be put on a more rigorous basis by basing all concepts on constructions beginning with the whole numbers: "God Himself made the whole numbers—everything else is the work of men."[36] Thus he felt that irrational quantities like $\sqrt{2}$ made no sense unless one could find a definite way of constructing them out of the whole numbers. In terms of fields, then, he wanted to find a method of constructing extension fields of the rational numbers, or indeed of any already determined field, which would not depend on the prior existence of irrational numbers.

Kronecker began with the idea of a domain of rationality determined by certain elements R', R'', ... This domain included all the quantities which were rational functions of R', R'', . . . with integer coefficients. Thus he assumed the existence of the integers and, therefore, the rational numbers. He was then able to solve his problem of adjoining $\sqrt{2}$ to a domain of rationality in which $x^2 - 2$ had no root by considering the remainders of polynomials with rational coefficients on division by $x^2 - 2$. Because two polynomials with the same remainder would be considered equal, it was straightforward to define the basic operations on this set of remainders and thereby construct a new domain of rationality. Another way of looking at this construction is simply to consider the new domain of rationality as containing a new element α as well as all rational functions of α, with the condition that α^2 is always replaced by 2.

Dedekind, also beginning in the 1850s, was more concerned with the set of elements itself than with the process of adjunction. Recall that Dedekind was interested in the arithmetic of algebraic integers, complex numbers which could be expressed as roots of algebraic equations. Thus Dedekind gave the following definition, in his supplement to the second edition of Dirichlet's *Vorlesungen* (1871): "A system A of real or complex numbers α is called a *field* if the sum, difference, product and quotient of every pair of these numbers α belongs to the same system A."[37] (He noted that 0 cannot be a denominator in any such quotient and that a field must contain at least one number other than 0.) The smallest such system, of course, is the field of rational numbers, which is contained in every field, while the largest such system is the field of complex numbers, which contains every field. Thus for Dedekind, unlike Kronecker, the adjunction of an algebraic element to a field always takes place in the field of complex numbers. In fact, given any set K of complex numbers, Dedekind defines the field $Q(K)$ to be the smallest field which contains all the elements K.

For both Dedekind and Kronecker, every field contained the field of rational numbers. Neither attempted to extend his definition to other types of fields, even though as far back as 1830 Galois had published a brief paper which in essence described finite fields. Galois' aim in that paper was to generalize the ideas of Gauss in solving congruences of the form $x^2 \equiv a \pmod{p}$. Galois asked what would happen if, when a solution did not exist, one created a solution, exactly as one created the solution i to $x^2 + 1 = 0$. Thus, designating a solution to an arbitrary congruence $F(x) \equiv 0 \pmod{p}$, by the symbol i (where $F(x)$ is of degree n and no residue modulo p is itself a solution), Galois considered the collection of p^n expressions $a_0 + a_1 i + a_2 i^2 + \cdots + a_{n-1} i^{n-1}$ with $0 \le a_j < p$ and noted that these expressions can be added, subtracted, multiplied, and divided in the obvious manner. Galois next noted that if α is any of the nonzero elements of his set, some smallest power n of α must be equal to 1 and, by arguments analogous to those of Gauss in the case of residues modulo p, showed that all such elements satisfy $\alpha^{p^n - 1} \equiv 1$ and that there is a primitive root β such that every nonzero element is a power of β. Galois concluded the paper by showing that for every prime power p^n one can find an irreducible nth degree congruence modulo p, a root of which generates what is today called the Galois field of order p^n. The simplest way to find such a polynomial, Galois remarked, is by trial and error. As an example, he showed that $x^3 - 2$ is irreducible modulo 7 and therefore that the set of elements $\{a_0 + a_1 i + a_2 i^2\}$, with i a zero of that polynomial and $0 \leq a_j < 7$ for $j = 0, 1, 2$, forms the field of order 7^3.

It was Heinrich Weber who combined the Dedekind–Kronecker version of a field with the finite systems of Galois into an abstract definition of a field in the same paper of 1893 in which he gave an abstract definition of a group. In fact, he used the notion of group in his definition. Thus a **field** was a set with two forms of composition, addition and multiplication, under the first of which it was a commutative group and under the second of which the set of nonzero elements formed a commutative group. Furthermore, the two forms of composition are related by the following rules: $a(-b) = -ab$; $a(b + c) = ab + ac$; $(-a)(-b) = ab$; and $a \cdot 0 = 0$. Weber further noted that in a field a product can only be zero when one of the factors is zero. He then gave several examples of fields, including the rational numbers, the finite fields (of which he only cited the residue classes modulo a prime), and the "form fields," the fields of rational functions in one or more variables over a given field F. As in the case of groups, the notion of a field was incorporated into Weber's algebra text, thus enabling a new generation of students to understand this abstract concept as well.

15.4 SYMBOLIC ALGEBRA

In the nineteenth century, algebra in England was characterized by a new interest in symbolic manipulation and its relation to mathematical truth (Side Bars 15.1 and 15.2). One of the leaders in this new movement in algebra and, in general, a man interested in the reform of mathematical study in England, was George Peacock (1791–1858). Peacock explained his new symbolical approach to algebra in his *Treatise on Algebra* of 1830, which he extensively enlarged and revised in 1842–1845. Peacock's interest in reform can be traced back to questions on the meaning of negative and imaginary numbers, questions which had been raised by several English mathematicians of the late eighteenth century. Negatives and imaginaries were freely used in the eighteenth century (and earlier) and were considered necessary in obtaining all sorts of algebraic results. But mathematicians were unable to explain their meanings in any way other than by various physical analogies. It was this lack of an adequate foundation for these concepts which led Francis Maseres (1731–1824) and William Frend (1757–1841) to write algebra texts specifically renouncing their use. It was clear, however, that this was too radical a step to be generally accepted, given the practical value of negatives and imaginaries in the study of solutions of equations.

15.4.1 Peacock's *Treatise on Algebra*

Peacock took it upon himself to rescue negatives and imaginaries by distinguishing two types of algebra, what he called "arithmetical algebra" and "symbolical algebra." Arithmetical algebra was universal arithmetic, that is, a means of developing the basic principles of the arithmetic of the nonnegative real numbers by use of letters rather than the numbers themselves. Thus in arithmetical algebra one can write that $a - (b - c) = a + c - b$ but only under the conditions that $c < b$ and $b - c < a$, so that the subtractions can in

Side Bar 15.1   **Mathematics at Cambridge**

By the mid-eighteenth century, mathematics had become central in the system of Cambridge studies, so much so that the most important examination at Cambridge, the Senate House examination, usually called the *tripos* after the three-legged stool on which originally candidates had sat during the questioning, was primarily devoted to mathematics. After all, the study of mathematics was presumed to develop the mind and thus help to prepare the English gentleman to assume his place in the leadership of the Church or state. The mathematics necessary to pass the tripos exam included the synthetic mathematics of Euclid and Apollonius, together with algebra, trigonometry, fluxions, and the elements of physics as presented in Book I of Newton's *Principia*. The more serious students, however, who hoped to become *wranglers*, that is, to finish at the top of the honors list, studied more advanced mathematics on their own. This material included the remainder of the *Principia* and, increasingly in the early nineteenth century, the work of such French mathematicians as Lagrange, Lacroix, and Laplace. Becoming a wrangler virtually guaranteed one a college fellowship at Cambridge and thus a beginning on a career, especially important to any student not of independent means.

The traditional mode of mathematics instruction throughout the eighteenth century was the synthetic geometric approach, the approach even Newton followed in his *Principia*. It was therefore not easy for Cambridge students to understand the analytical methods being practiced with such success on the continent. To remedy this situation, several Cambridge undergraduates decided in 1812 to form a new society, the Analytical Society, whose purpose was to advance continental analytic mathematics in Britain and, in particular, to bring this mathematics into the regular curriculum at Cambridge. Although the Analytical Society only lasted about a year, many of the original members, including George Peacock and Charles Babbage (1792–1871), were influential in the conversion of Cambridge to the new analytic style by the mid-1820s. One of the effects of their work was ultimately to change the role of mathematics at Cambridge from that of the mainstay of a liberal education to that of a profession in its own right, one whose goal was the development of new mathematics as part of the general advancement of knowledge.

fact be performed. In symbolical algebra, on the other hand, the symbols (letters) need not have any particular interpretation. Manipulations with the symbols are to be derived from analogous manipulations in arithmetic, but in symbolical algebra it is not necessary to limit their range of applicability. For example, the equation above is universally valid in symbolical algebra.

Peacock's answer to the question what *is* a negative number is that it is simply a symbol of the form $-a$. One operates with these symbols in the way derived from arithmetic. Since $(a - b)(c - d) = ac - ad - bc + bd$ in arithmetic, provided that $a > b$ and $c > d$, the same rule applies in symbolical algebra without that restriction. It follows then, setting $a = c = 0$, that $(-b)(-d) = bd$ and, setting $a = d = 0$, that $(-b)c = -bc$. Similarly, $\sqrt{-1}$ is simply a symbol which obeys the same rules that the square root symbols obey in arithmetic. Therefore $\sqrt{-1}\sqrt{-1} = -1$. These are examples of what Peacock called the principle of the permanence of equivalent forms: "Whatever algebraical forms are equivalent, when the symbols are general in form but specific

---

Side Bar 15.2   The Tripos Examination for 1785

The question paper was introduced by a memorandum telling the candidates to write distinctly and to observe that "at least as much will depend upon the clearness and precision of the answers as upon the quantity of them."[38]

1. To prove how many regular Solids there are, what are those Solids called, and why there are no more.

2. To prove the Asymptotes of an Hyperbola always external to the Curve.

3. Suppose a body thrown from an Eminence upon the Earth, what must be the Velocity of Projection, to make it become a secondary planet to the Earth?

4. To prove in all the conic sections generally that the force tending to the focus varies inversely as the square of the Distance.

5. Supposing the periodical times in different Ellipses round the same center of force, to vary in the sesquiplicate ratio of the mean distances, to prove the forces in those mean distances to be inversely as the square of the distance.

6. What is the relation between the 3rd and 7th Sections of Newton, and how are the principles of the 3rd applied to the 7th?

7. To reduce the biquadratic equation $x^4 + qx^2 + rx + s = 0$ to a cubic one.

8. To find the fluent of $\dot{x} \times \sqrt{a^2 - x^2}$.

9. To find a number from which if you take its square, there shall remain the greatest difference possible.

10. To rectify [an arbitrary] arc DB of the circle $DBRS$.

Biography   George Peacock (1791–1858)

Peacock was born in Denton, a town in Lincolnshire only a few miles from Newton's birthplace. In 1809 he entered Trinity College, Cambridge and four years later graduated as a second wrangler, becoming in turn a fellow at Trinity, a college lecturer, a tutor, and, in 1837, professor of astronomy and geometry. It was his position as moderator of the tripos exam from 1817 to 1819 which enabled him to introduce continental mathematics into that exam. Some years later, he participated in the commission which rewrote Cambridge's statutes to remove religious tests as prerequisites for receiving a degree.

---

in value, will be equivalent likewise when the symbols are general in value as well as in form."[39] In other words, any law of arithmetic, expressible as an equation, determines a law of symbolical algebra by the removal of any limitations on the interpretation of the symbols involved. As an example not involving negatives or imaginaries, Peacock noted that since

$$(1 + x)^m = 1 + mx + m(m - 1)\frac{x^2}{1 \cdot 2} + \cdots$$

is true in arithmetic if m is rational and $0 < x < 1$, the same equation holds in symbolical algebra, no matter what the values of x and m.

In the first version of his *Algebra* in 1830, Peacock defined symbolical algebra as "the science which treats of the combinations of arbitrary signs and symbols by means of defined though arbitrary laws."[40] He thus began a shift in the focus of algebra away from the meaning of the symbols to the laws of operation on these symbols. Peacock, however, did not avail himself of the "arbitrary laws" of combination he advocated, either in 1830 or in 1845. All the laws in his symbolical algebra were in fact derived by the principle of permanence from the corresponding laws of arithmetic for the same operations. In fact, in 1845 he wrote that "I believe that no views of the nature of Symbolical Algebra can be correct or philosophical which made the selection of its rules of combination arbitrary and independent of arithmetic."[41] But despite his failure to use his asserted freedom to create laws for symbolical algebra, Peacock's statement that the results of this algebra "may be said to exist by convention only"[42] marks the beginning of a new meaning for the entire subject of algebra, a meaning which was soon to be exploited by other English mathematicians.

15.4.2 De Morgan and the Laws of Algebra

Augustus De Morgan (1806–1871) was influenced by his reading of Peacock's treatises, but recognized more clearly than his predecessor that the laws of algebra could be created without using those of arithmetic as a suggestive device. Rather than beginning with the laws of arithmetic, De Morgan believed that one could create an algebraic system by beginning with arbitrary symbols and creating (somehow) a set of laws under which these symbols are operated on. Only afterwards would one provide interpretations of these laws. He gave a simple example of such a creation in 1849:

> Given symbols M, N, $+$, and one sole relation of combination, namely that $M + N$ is the same as $N + M$. Here is a symbolic calculus: how can it be made a significant one? In the following ways, among others. 1. M and N may be magnitudes, and $+$ the sign of addition of the second to the first. 2. M and N may be numbers, and $+$ the sign of multiplying the first by the second. 3. M and N may be lines, and $+$ the direction to make a rectangle with the antecedent for a base, and the consequent for an altitude. 4. M and N may be men and $+$ the assertion that the antecedent is the brother of the consequent. 5. M and N may be nations, and $+$ the sign of the consequent having fought a battle with the antecedent.[43]

Although De Morgan asserted the freedom to create algebraic axioms for his symbols and even realized that the symbols could represent things other than "quantities" or "magnitudes," he, like Peacock, did not attempt to create any new system which obeyed laws different from those obeyed by the numbers of arithmetic. In fact, in 1841 he set out what he believed were the rules which were "essential to algebraical process."[44] These rules include the substitution principle (that two expressions connected by an $=$ sign can be substituted for one another), the inverse principle for both addition and multiplication (that $+$ and $-$ as well as $\times$ and $\div$ are "opposite in effect"), the commutative principle for addition and multiplication, the distributive laws (of multiplication over both addition and subtraction), and the exponential laws $a^b a^c = a^{b+c}$ and $(a^b)^c = a^{bc}$. Having presented the laws, which he believed to be "neither insufficient nor redundant," he commented that

Augustus De Morgan, born in India into the family of an English army officer, studied at Cambridge in the 1820s after some of Peacock's reforms had gone into effect. Thus he was introduced to the continental analytic mathematics from the start. Because he graduated in 1827 only as a fourth wrangler, partly because other interests interfered with the "cramming" generally necessary to do well in the tripos, he felt that this showing was too poor for him to attempt a career in mathematics and so prepared for the bar.

Nevertheless, in 1828 he was selected for the chair in mathematics at the newly created London University, a position he held for most of the rest of his life. De Morgan was a dedicated teacher, regularly giving four courses in each semester, courses ranging from elementary arithmetic to the calculus of variations. He spent much of his creative talent on devising better ways of instruction and wrote not only various mathematics texts but also articles and books on the teaching of mathematics.[45]

the "most remarkable point . . . is that the laws of operation prescribe much less of connexion between the successive symbols $a + b$, ab and a^b than a person who has deduced these laws from arithmetical explanation would at first think sufficient."[46] In other words, there is no necessity of deriving the meaning of multiplication from that of addition nor the meaning of exponentiation from that of multiplication. Nevertheless, although one can certainly derive all sorts of algebraic results by using just these principles, such an algebra would have no more meaning than the putting together of a jigsaw puzzle by using the backs of the pieces. True mathematics, De Morgan believed, must have real content. Laying out the axiomatic structure of a system was far less important than the task of interpretation. It was only the interpretation, which De Morgan recognized was outside of the logical framework established by axioms, which gave a mathematical system its meaning and significance.

15.4.3 Hamilton: Complex Numbers and Quaternions

It was the Irish mathematician and physicist William Rowan Hamilton (1805–1865) who was able finally to create a new algebraic system having a genuine interpretation but not conforming to all of the axioms set out by De Morgan. Like Peacock and De Morgan, Hamilton wanted to be able to justify the use of negatives and imaginaries in algebra, concepts he agreed had poor foundations. As he wrote in his fundamental paper of 1837, "Theory of Conjugate Functions, or Algebraic Couples; with a Preliminary and Elementary Essay on Algebra as the Science of Pure Time,"

It requires no peculiar scepticism to doubt, or even to disbelieve, the doctrine of Negatives and Imaginaries, when set forth (as it has commonly been) with principles like these: that a *greater magnitude may be subtracted from a less*, and that the remainder is *less than nothing*; that *two negative numbers* or numbers denoting magnitudes each less than nothing, may be *multiplied* . . . and that the product will be a *positive* number . . . and that although the *square* of a number . . . is therefore *always positive*, whether the number be positive or negative, yet that numbers, called *imaginary* can be found or conceived or determined, and operated on by

---

Biography   William Rowan Hamilton (1805–1865)

Hamilton was born in Dublin, but educated in the town of Trim, some 30 miles to the northwest, under the tutelage of his uncle, a scholar of the classics. Because Hamilton showed signs of genius at an early age, his uncle proceeded to turn this precocity to the study of languages. By the time he was ten, William was fluent not only in Latin, Greek and the modern European languages, but also in Hebrew, Persian, Arabic, and Sanskrit, among others. From an early age, Hamilton also learned arithmetic, using methods of computation of his own devising, but his mathematical interest was spurred by his contact with an American calculating prodigy who was consistently able to best Hamilton in competition. Soon afterwards, he discovered Euclidean geometry as well as more modern areas of mathematics. By the time of his entrance into Trinity College, Dublin, in 1823, he was prepared to

deal with the continental analytic mathematics taught at Trinity in line with the reforms instituted at Cambridge. Hamilton swiftly moved beyond the prescribed curriculum and soon mastered the mathematical texts in use at the École Polytechnique. His first important original work was in optics, rather than in pure mathematics, and, in fact, he is today more famous for his work in dynamics than for his mathematics. It was this work in physics which led to Hamilton's appointment, even before he received his degree, as Astronomer Royal of Ireland in 1827, a position which he held for the remainder of his life. His contributions to mathematics and physics led to his being named in 1865 the first foreign associate of the newly created National Academy of Sciences of the United States (Figure 15.5).

FIGURE 15.5
Hamilton on an Irish stamp.

all the rules of positive and negative numbers, as if they were subject to those rules, *although they have negative squares* and must therefore be supposed to be themselves neither positive nor negative, nor yet null numbers, so that the magnitudes which they are supposed to denote can neither be greater than nothing, nor less than nothing, nor even equal to nothing.[47]

To place algebra, like geometry, on a firm foundation required creating for it certain intuitive principles, and these Hamilton felt could come from the intuition of pure time. In other words, Hamilton assumed that there was a set M of "moments" which were ordered by a relation $<$ such that for all A, B in M, either $A = B$ or $A < B$ or $A > B$. Hamilton then defined an equivalence relation on the set of pairs of moments by setting (A,B) equivalent to (C,D) if the following conditions are satisfied: "If the moment B be identical with A, then D must be identical with C; if B be later than A, then D must be later than C, and exactly so much later; and if B be earlier than A, then D must be earlier than C, and exactly so much earlier."[48] To avoid confusion later, it should be noted that Hamilton did not use the modern pair notation in discussing this equivalence. He represented the equivalence class defined by this relation first by the suggestive notation $B - A$, and later by a single symbol a, a symbol one can think of as denoting the **time step** from A to B.

It is the time steps which provide the basis of Hamilton's construction of negatives. Namely, if a represents the pair $B - A$, then Θa (Hamilton's notation for $-a$) represents the pair $A - B$. (Hamilton created this particular notation from the letter O, the initial letter of the Latin *oppositio* (*opposite*).) Taking a given step a as a unit, and using a natural definition of the sum of two steps, Hamilton then proceeded to construct the set of rational numbers. Positive integers are determined by multiples (successive sums) of a with itself while negatives come from multiples of Θa. Rational numbers are defined through the comparison of two integral multiples of the step a. Hamilton then demonstrated the standard rules for the arithmetic operations on these (positive and negative) multiples. For example, the product of two negative multiples of a must be positive since such a product involves reversing the direction of the step a twice. Having to his own satisfaction answered the objections to negative numbers indicated above, without resorting to quantities "less than nothing," Hamilton attempted next to construct the real numbers from the rationals. Not only was this attempt a failure from a modern perspective, but it also had little effect on the arithmetization of analysis carried out in Germany later in the century. On the other hand, his construction of the complex numbers from the reals in the final part of this same essay is the one often used in textbooks today.

In this construction, Hamilton considered couples, or pairs, of moments, time steps, and numbers. Thus two pairs (A_1, B_1), (A_2, B_2) of moments determine a pair $(a, b) = (B_1 - A_1, B_2 - A_2)$ of time steps, while the ratio α of the two pairs of steps $(\alpha a, \alpha b)$, (a, b) led Hamilton to conceive that any two pairs of steps would have a ratio expressible as a pair of numbers (α, β). (The original ratio α would then be replaced by the pair $(\alpha, 0)$.) It was clear that addition and subtraction of these pairs should be defined by

$$(\alpha, \beta) \pm (\gamma, \delta) = (\alpha \pm \gamma, \beta \pm \delta).$$

Assuming the distributive law for multiplication, Hamilton then argued that a general rule for multiplication can be given by

$$(\alpha, \beta)(\gamma, \delta) = (\alpha\gamma - \beta\delta, \beta\gamma + \alpha\delta).$$

It followed that division should be defined as

$$\frac{(\alpha, \beta)}{(\gamma, \delta)} = \left(\frac{\alpha\gamma + \beta\delta}{\gamma^2 + \delta^2}, \frac{\beta\gamma - \alpha\delta}{\gamma^2 + \delta^2} \right).$$

As Hamilton wrote, "these definitions are really *not arbitrarily chosen*, and that though others might have been assumed, no others would be equally proper,"[49] because from them follow the known laws of operation on complex numbers. For example, $(0,1)(0,1) = (-1,0)$ and therefore, with the identification of $(\alpha,0)$ with the number α, $\sqrt{-1}$ can be identified with the pair $(0,1)$. The complex number $\alpha + \beta\sqrt{-1}$ can then simply be *defined* to be the number pair (α,β). Hamilton thus succeeded in constructing the complex numbers from the reals, bypassing any appeal to "imaginary" numbers, thereby answering the question of what complex numbers *really* are.

Hamilton concluded his essay with the statement that he hoped to develop a "Theory of Triplets and Sets of Moments, Steps and Numbers, which includes this Theory of Couples."[50] He had already learned that the operations on complex numbers have a geometrical interpretation in the two-dimensional plane. But because much of physics took place in three-dimensional space, a system of operations (that is, an algebra) of triplets would prove immensely useful. As he wrote to De Morgan in 1841, "if my view of Algebra be just, it *must* be possible, in *some* way or other, to introduce not only triplets but *polyplets*, so as in some sense to satisfy the symbolical equation $a = (a_1, a_2, \ldots, a_n)$; a being here one symbol, as indicative of one (complex) thought; and $a_1, a_2, \ldots, a_n$ denoting n real numbers positive or negative."[51] The struggle for Hamilton, of course,

FIGURE 15.6
The rules for quaternions
inscribed on an Irish
stamp in commemoration
of their discovery.

was not in the addition of his triplets—that was easy—but in the multiplication. Knowing the basic laws for his couples, he wanted his triplets similarly to satisfy the associative and commutative properties of multiplication as well as the distributive law. He wanted division to be always possible (except by 0) and to always have a unique result. He wanted the moduli to multiply, that is, if $(a_1, a_2, a_3)(b_1, b_2, b_3) = (c_1, c_2, c_3)$, then $(a_1^2 + a_2^2 + a_3^2)(b_1^2 + b_2^2 + b_3^2) = c_1^2 + c_2^2 + c_3^2$. Finally, he wanted the various operations to have a reasonable interpretation in three-dimensional space. Hamilton had begun his search for a multiplicative law for triplets as early as 1830. After 13 years of considering the problem, he finally solved it, but not in the way he had hoped, in an experience described in the opening of this chapter.

Hamilton's solution was not to consider triplets at all, but quadruplets, (a, b, c, d), which he wrote, in analogy with the standard notation for complex numbers, as $a + bi + cj + dk$. The basic multiplication laws $i^2 = j^2 = k^2 = ijk = -1$ and the derived rules $ij = k, ji = -k, jk = i, kj = -i, ki = j$, and $ik = -j$, when extended by the distributive law to all quadruplets, or **quaternions**, gave this system all of the properties Hamilton sought, with the sole exception of the commutative law of multiplication (Figure 15.6). In modern terminology, the set of quaternions forms a noncommutative division algebra over the real numbers. Hamilton's system was the first significant system of "quantities" which did not obey all of the standard laws which Peacock and De Morgan had set down. As such, its creation broke a barrier to the consideration of systems violating these laws, and, soon, the freedom of creation advocated by Peacock became a reality.

Hamilton himself was so taken with his discovery that he spent the remainder of his life writing several tomes on the theory of quaternions and strongly advocating their use in various areas of physics. Although he converted few physicists to his position, his work marked the beginning of today's common use of vector terminology in physical theories. In fact, Hamilton himself noted the convenience of writing a quaternion $Q = a + bi + cj + dk$ in two parts, the real part a and imaginary part $bi + cj + dk$. He named the former the **scalar** part, because all values it can attain are on "one *scale* of progression of number from negative to positive infinity," while he named the latter the **vector** part, because it can be geometrically constructed in three-dimensional space as a "straight line or radius vector."[52] (The word "radius vector" had been part of mathematical vocabulary since the early eighteenth century. It was Hamilton, however, who first used the word "vector" in today's more general sense.) Thus Hamilton wrote $Q = S.Q + V.Q$, where $S.Q$ is the scalar part and $V.Q$ the vector part. In particular, if we consider the product

$$(ai + bj + ck)(xi + yj + zk) = $$
$$-(ax + by + cz) + (bz - cy)i + (cx - az)j + (ay - bx)k$$

of two quaternions α, β with zero scalar parts, then $S.\alpha\beta$ is the negative of the modern dot product of the vectors α and β, while $V.\alpha\beta$ is the modern cross product.

15.4.4 Quaternions and Vectors

Hamilton's successors in the advocacy of quaternion concepts for use in physics were Peter Guthrie Tait (1831–1901) and James Clerk Maxwell (1831–1879) (Figure 15.7), friends and fellow students at both the University of Edinburgh and Cambridge University. Tait in fact wrote an *Elementary Treatise on Quaternions* in 1867 in which he advocated quater-

FIGURE 15.7
Maxwell honored
on a Mexican stamp
for his work on
electromagnetism.

nion methods in physics. Tait's treatise contained equivalents of virtually all the modern laws of operation of the dot and cross product of vectors, although written in quaternion notation. In particular, he showed that $S.\alpha\beta = -T\alpha T\beta \cos\theta$, where $T\alpha$ is the length of α and θ is the angle between α and β, and that $V.\alpha\beta = T\alpha T\beta \sin\theta \cdot \eta$ where η is a unit vector perpendicular to both α and β.

Maxwell, in his *Treatise on Electricity and Magnetism*, also advocated Hamilton's ideas. His main purpose, however, as stated in his opening chapter, was "to avoid explicitly introducing the Cartesian coordinates, and to fix the mind at once on a point of space instead of its three coordinates, and on the magnitude and direction of a force instead of its three components."[53] Thus quaternions and the associated vectors were to be used to represent physical quantities in a more conceptual way than the usual coordinate form. In the treatise itself Maxwell generally expressed his physical results both ways, in coordinate form and in quaternion form.

It was, however, Josiah Willard Gibbs (1839–1903) at Yale University and Oliver Heaviside (1850–1925) in England who independently realized, after their reading of Tait and Maxwell, that the full algebra of quaternions was not necessary for discussing the various physical concepts. All that was necessary were the two types of products of vectors, the dot product and the cross product. Gibbs published his version of vector analysis privately in 1881 and 1884 and lectured on the subject for many years at Yale, while Heaviside first published his method in papers on electricity in 1882 and 1883. It is to the former, however, that our modern notations of $A \cdot B$ and $A \times B$, for the dot product and the cross product respectively, are due. With the formal publication of Gibbs' *Vector Analysis* in a 1901 work derived from his lectures, it was clear to the physics community that the use of vectors in describing physical concepts was all that was needed. Although quaternions were to remain important mathematically, their use in physics soon died a quiet death.

15.4.5 Boole and Logic

The algebraic freedom advocated by Peacock and De Morgan was exploited in a different way by the self-taught English logician George Boole (1815–1864). In 1847, Boole published a small book, *The Mathematical Analysis of Logic*, and, seven years later, expanded it into *An Investigation of the Laws of Thought*. Boole's aim in these books was to "investigate the fundamental laws of those operations of the mind by which reasoning is performed; to give expression to them in the symbolical language of a Calculus, and upon this foundation to establish the science of Logic and construct its method."[54]

Because algebra is studied by means of signs, Boole put into his opening proposition the basic signs by which logic would be analyzed.

Proposition 1. *All the operations of Language, as an instrument of reasoning, may be conducted by a system of signs composed of the following elements, viz.:*

1st. Literal symbols, as x, y, etc., representing things as subjects of our conceptions.

2nd. Signs of operation, as $+$, $-$, $\times$, standing for those operations of the mind by which the conceptions of things are combined or resolved so as to form new conceptions involving the same elements.

3rd. The sign of identity, $=$.

And these symbols of Logic are in their use subject to definite laws, partly agreeing with and partly differing from the laws of the corresponding symbols in the science of Algebra.[55]

Boole next defined the laws of his symbols of language. A letter was to represent a class, or set, of objects. Thus x could stand for the class of "men" and y for the set of "good things." The combination xy would then stand for the class of things to which both x and y are applicable, that is, the class of "good men." It was obvious to Boole that the commutative law holds for his multiplication: $xy = yx$. Among other laws for multiplication which Boole derived are that $x^2 = x$, since the class to which x and x are applicable is simply that of x, and $xy = x$ in the case where the class represented by x is contained in that represented by y.

Addition for Boole, written $x + y$, represented the conjunction of the two classes represented by x and y, while subtraction, written $x - y$, stood for the class of those things represented by x with the exception of those represented by y. The commutative law of addition then holds as well as the distributive laws $z(x \pm y) = zx \pm zy$. With 0 used to represent the empty class and 1 the universal class, Boole similarly derived the familiar laws $0y = 0$ and $1y = y$ as well as the not so familiar one $x(1 - x) = 0$, which "affirms that it is impossible for any being to possess a quality, and at the same time not to possess it."[56]

Because the stated laws agree with the laws of numbers restricted to just the values 0 and 1, Boole decided that his algebra of logic would deal with variables which only take on these values. In particular, he considered functions of one or several logical variables, $f(x), f(x,y), \ldots$, in which the variables can only take on the values 0 and 1. A function $f(x)$ can be expanded in the form $f(x) = f(1)x + f(0)(1 - x)$, or, putting $\bar{x} = 1 - x$, in the form $f(x) = f(1)x + f(0)\bar{x}$. Similarly, $f(x,y) = f(1,1)xy + f(1,0)x\bar{y} + f(0,1)\bar{x}y + f(0,0)\bar{x}\bar{y}$. For example, the function $1 - x + xy$ can be expanded as $1xy + 0x\bar{y} + 1\bar{x}y + 1\bar{x}\bar{y} = xy + \bar{x}y + \bar{x}\bar{y}$.

Boole then proved that if V is some function, one can interpret the equation $V = 0$ by expanding V according to the above rules and equating to 0 every constituent whose coefficient does not vanish. As an instance of this procedure, Boole considered the definition of "clean beasts" from Jewish law: Clean beasts are those which both divide the hoof and chew the cud. With x, y, z representing clean beasts, beasts which divide the hoof, and beasts which chew the cud respectively, the definition of clean beasts is given by the equation $x = yz$ or $V = x - yz = 0$. Expanding $x - yz$, Boole found that

$$V = 0xyz + xy\bar{z} + x\bar{y}z + x\bar{y}\bar{z} - \bar{x}yz + 0\bar{x}y\bar{z} + 0\bar{x}\bar{y}z + 0\bar{x}\bar{y}\bar{z}.$$

Equating each nonvanishing term to 0 then gives

$$xy\bar{z} = 0 \qquad x\bar{y}z = 0 \qquad x\bar{y}\bar{z} = 0 \qquad \bar{x}yz = 0.$$

The interpretation of these is the assertion of the nonexistence of certain classes of animals. For example, the first equation asserts that there are no beasts which are clean and divide the hoof, but do not chew the cud.

The algebra of classes developed by Boole, seemingly dormant for many years after Boole published it, is today known as Boolean algebra and has resurfaced as central in the study of the algebra of circuit design, the algebra by which the logic behind modern calculators and computers is developed. Boole would probably be pleased that his calculus of the laws of thought is in fact used in nearly the way he forecast over a century ago.

15.5 THE THEORY OF MATRICES

The idea of a matrix has a long history, dating at least from its use by Chinese scholars of the Han period for solving systems of linear equations. In the eighteenth century, and even somewhat earlier, determinants of square arrays of numbers were calculated and used, even though the square arrays themselves were not singled out for attention. Other work in the nineteenth century led to more formal computations with such arrays and by mid-century to a definition of a matrix and the development of the algebra of matrices. But alongside this formal work, there was a deeper side to the development of the theory of matrices, namely the work growing out of Gauss' study of quadratic forms which ultimately led to the concepts of similarity, eigenvalues, diagonalization, and finally the classification of matrices through canonical forms.

15.5.1 Basic Ideas of Matrices

Recall that Gauss, in his theory of binary quadratic forms, dealt with the idea of a linear substitution which transforms one form into another. Namely, if $F = ax^2 + 2bxy + cy^2$, then the substitution

$$x = \alpha x' + \beta y'$$
$$y = \gamma x' + \delta y'$$

converts F into a new form F' whose coefficients depend on the coefficients of F and those of the substitution. Gauss noted that if F' is transformed into F'' by a second linear substitution

$$x' = \epsilon x'' + \zeta y''$$
$$y' = \eta x'' + \theta y''$$

then the composition of the two substitutions gives a new substitution transforming F into F'':

$$x = (\alpha\epsilon + \beta\eta)x'' + (\alpha\zeta + \beta\theta)y''$$
$$y = (\gamma\epsilon + \delta\eta)x'' + (\gamma\eta + \delta\theta)y''$$

The coefficient "matrix" of the new substitution is the product of the coefficient matrices of the two original substitutions. Gauss performed an analogous computation in his study of ternary quadratic forms $Ax^2 + 2Bxy + Cy^2 + 2Dxz + 2Eyz + Fz^2$, which in effect gave the rule for multiplying two 3×3 matrices. But although he wrote the coefficients of the substitution in a rectangular array and even used a single letter S to refer to a particular substitution, Gauss did not explicitly refer to this idea of composition as a "multiplication."

In 1815 Cauchy published a fundamental memoir on the theory of determinants, in which he not only introduced the name "determinant" to replace several older terms, but also used the abbreviation $(a_{1,n})$ to stand for what he called the "symmetric system"

$$
\begin{array}{cccc}
a_{1,1} & a_{1,2} & \cdots & a_{1,n} \\
a_{2,1} & a_{2,2} & \cdots & a_{2,n} \\
\vdots & \vdots & \ddots & \vdots \\
a_{n,1} & a_{n,2} & \cdots & a_{n,n}
\end{array}
$$

to which the determinant is associated. Although many of the basic results on calculating determinants had been known earlier, Cauchy gave the first complete treatment of these in this memoir, including such ideas as the array of minors associated to a given array (the adjoint) and the procedure for calculating a determinant by expanding on any row or column. In addition, he followed Gauss in explicitly recognizing the idea of composing two systems $(\alpha_{1,n})$ and $(a_{1,n})$ to get a new system $(m_{1,n})$, where the latter is defined by the familiar law of multiplication

$$
m_{i,j} = \sum_{k=1}^{n} \alpha_{i,k} a_{k,j}.
$$

He then showed that the determinant of the new system was the product of those of the two original ones.

Ferdinand Gotthold Eisenstein (1823–1852), a student of Gauss who visited Hamilton in Ireland in 1843, introduced the explicit notation $S \times T$ to denote the substitution composed of S and T in his discussion of ternary quadratic forms in a paper of 1844, perhaps because of Cauchy's product theorem for determinants. About this notation Eisenstein wrote, "Incidentally, an algorithm for calculation can be based on this; it consists in applying the usual rules for the operations of multiplication, division and exponentiation to symbolical equations between linear systems; correct symbolical equations are always obtained, the sole consideration being that the order of the factors, i.e., the order of the composing systems, may not be altered."[57] It is interesting, but probably futile, to speculate whether Eisenstein's discussions with Hamilton in 1843 stimulated either to realize the possibility of an algebraic system with a noncommutative multiplication.

15.5.2 Matrix Operations

Eisenstein never developed fully his idea of an algebra of substitutions because of his untimely death at the age of 29. That development was carried out in England by Arthur Cayley and James Joseph Sylvester (1814–1897) in the 1850s.

In 1850 Sylvester coined the term **matrix** to denote "an oblong arrangement of terms consisting, suppose, of m lines and n columns" because out of that arrangement "we may form various systems of determinants."[58] (The English word "matrix" meant "the place from which something else originates.") Sylvester himself made no use of the term at the time. It was his friend Cayley who put the terminology to use in papers of 1855 and 1858.

Biography   James Joseph Sylvester (1814–1897)

Sylvester, who was born into a Jewish family in London and studied for several years at Cambridge, was not permitted to take his degree there for religious reasons. Therefore, he received his degree from Trinity College, Dublin. In 1841, he accepted a professorship at the University of Virginia but remained there only a short time. His horror of slavery and an altercation with a student who did not show him the respect he felt he deserved led to his resignation in 1843.

After his return to England, he spent 10 years as an attorney and 15 years as professor of mathematics at the Royal Military Academy at Woolwich, before accepting in 1871 the chair of mathematics at the newly opened Johns Hopkins University in Baltimore. While at Hopkins, Sylvester founded the *American Journal of Mathematics* and helped develop a tradition of graduate education in mathematics in the United States.

In the former, Cayley noted that the use of matrices is very convenient for the theory of linear equations. Thus he wrote

$$(\xi, \eta, \zeta, \ldots) = \begin{pmatrix} \alpha, & \beta, & \gamma, & \cdots \\ \alpha', & \beta', & \gamma', & \cdots \\ \alpha'', & \beta'', & \gamma'', & \cdots \\ \vdots & \vdots & \vdots & \vdots \end{pmatrix} (x, y, z, \ldots)$$

to represent the system of equations

$$\xi = \alpha x + \beta y + \gamma z + \ldots$$

$$\eta = \alpha' x + \beta' y + \gamma' z + \ldots$$

$$\zeta = \alpha'' x + \beta'' y + \gamma'' z + \ldots$$

He then determined the solution of this system using what he called the inverse of the matrix:

$$(x, y, z, \ldots) = \begin{pmatrix} \alpha, & \beta, & \gamma, & \cdots \\ \alpha', & \beta', & \gamma', & \cdots \\ \alpha'', & \beta'', & \gamma'', & \cdots \\ \vdots & \vdots & \vdots & \vdots \end{pmatrix}^{-1} (\xi, \eta, \zeta, \ldots).$$

This representation came from the basic analogy of the matrix equation to a simple linear equation in one variable. Cayley, however, knowing Cramer's rule, then described the entries of the inverse matrix in terms of fractions involving the appropriate determinants.

In 1858, Cayley introduced single letter notation for matrices and showed not only how to multiply them but also how to add and subtract:

> It will be seen that matrices (attending only to those of the same order) comport themselves as single quantities; they may be added, multiplied or compounded together, &c.: the law of the addition of matrices is precisely similar to that for the addition of ordinary algebraical quantities; as regards their multiplication (or composition), there is the peculiarity that

matrices are not in general convertible [commutative]; it is nevertheless possible to form the powers (positive or negative, integral or fractional) of a matrix, and thence to arrive at the notion of a rational and integral function, or generally of any algebraical function, of a matrix.[59]

Cayley then exploited his idea, making constant use of the analogy between ordinary algebraic manipulations and those with matrices, but carefully noting where this analogy fails. Thus, using the formula for the inverse of a 3 × 3 matrix, he wrote that "the notion of the inverse . . . matrix fails altogether when the determinant vanishes; the matrix is in this case said to be indeterminate . . . It may be added that the matrix zero is indeterminate; and that the product of two matrices may be zero, without either of the factors being zero, if only the matrices are one or both of them indeterminate."[60]

It was Cayley's use of the notational convention of single letters for matrices which perhaps suggested to him the result known as the Cayley-Hamilton theorem. For the case of a 2 × 2 matrix

$$M = \begin{pmatrix} a & b \\ c & d \end{pmatrix},$$

Cayley stated this result as

$$\det\begin{pmatrix} a - M & b \\ c & d - M \end{pmatrix} = 0.$$

Cayley first communicated this "very remarkable" theorem in a letter to Sylvester in November of 1857. In 1858, he proved it by simply showing that the determinant $M^2 - (a + d)M^1 + (ad - bc)M^0$ equaled 0 (where M^0 is the identity matrix). Stating the general version in essentially the modern form that M satisfies the equation in λ, $\det(M - \lambda I) = 0$, Cayley noted that he had "verified" the theorem in the 3 × 3 case, but wrote further that "I have not thought it necessary to undertake the labour of a formal proof of the theorem in the general case of a matrix of any degree."[61] A complete proof was first given some 20 years later by Frobenius.

Cayley's motivation in stating the Cayley-Hamilton theorem was to show that "any matrix whatever satisfies an algebraical equation of its own order" and therefore that "every rational and integral function . . . of a matrix can be expressed as a rational and integral function of an order at most equal to that of the matrix, less unity."[62] Cayley went on to show that one can adapt this result even for irrational functions. In particular, he showed how to calculate $L = \sqrt{M}$, where M is the 2 × 2 matrix given above. The result is given in the form

$$L = \begin{pmatrix} \dfrac{a + Y}{X} & \dfrac{b}{X} \\ \dfrac{c}{X} & \dfrac{d + Y}{X} \end{pmatrix}$$

where $X = \sqrt{a + d + 2\sqrt{ad - bc}}$ and $Y = \sqrt{ad - bc}$. Cayley failed, however, to give conditions under which this result holds. A similar argument, again dependent on manipulation of symbols without any consideration of special cases in which the manipulation fails, enabled Cayley to come to a false characterization of all the matrices L which commute with M. In fact, it was that very question which led Camille Jordan 10 years later to develop a fundamental classification of matrices by means of what today is called the Jordan Canonical Form.

15.5.3 The Notion of an Eigenvalue

Jordan's classification depends not on formal manipulation of matrices but on spectral theory, the results surrounding the concept of an eigenvalue. In modern terminology, an **eigenvalue** of a matrix is a solution λ of either the matrix equation $AX = \lambda X$, where A is an $n \times n$ matrix and X is an $n \times 1$ matrix or of $XA = \lambda X$, where A is $n \times n$ and X is $1 \times n$. This concept, in its origins and its later development, was independent of matrix theory *per se*; it grew out of a study of various ideas which ultimately were included in that theory. Thus the context within which the earliest eigenvalue problems arose during the eighteenth century was that of the solution of systems of linear differential equations with constant coefficients. D'Alembert, in works dating from 1743 to 1758, and motivated by the consideration of the motion of a string loaded with a finite number of masses (here restricted for simplicity to three), had to consider the system

$$\frac{d^2y_i}{dt^2} + \sum_{k=1}^{3} a_{ik}y_k = 0 \quad i = 1, 2, 3.$$

To solve this system he decided to multiply the ith equation by a constant v_i for each i and add the equations together to obtain

$$\sum_{i=1}^{3} v_i \frac{d^2y_i}{dt^2} + \sum_{i,k=1}^{3} v_i a_{ik} y_k = 0.$$

If the v_i are then chosen so that $\sum_{i=1}^{3} v_i a_{ik} + \lambda v_k = 0$ for $k = 1, 2, 3$, that is, if (v_1, v_2, v_3) is an eigenvector corresponding to the eigenvalue $-\lambda$ for the matrix $A = (a_{ik})$, the substitution $u = v_1 y_1 + v_2 y_2 + v_3 y_3$ reduces the original system to the single differential equation

$$\frac{d^2u}{dt^2} + \lambda u = 0,$$

an equation which, after Euler's work on differential equations, was easily solved and led to solutions for the three y_i. A study of the three equations in which it appears shows that λ is determined by a cubic equation which has three roots. D'Alembert realized that for the solutions to make physical sense they had to be bounded as $t \to \infty$. This, in turn, would only be true provided that the three values of λ were distinct, real, and positive.

It was Cauchy who first solved the problem of determining in a special case the nature of the eigenvalues from the nature of the matrix (a_{ik}) itself. In all probability, he was not influenced by d'Alembert's work on differential equations, but by the study of quadric surfaces, a study necessary as part of the analytic geometry which Cauchy was teaching from 1815 at the École Polytechnique. A quadric surface (centered at the origin) is given by an equation $f(x, y, z) = K$, where f is a ternary quadratic form. To classify such surfaces Cauchy, like Euler earlier, needed to find a transformation of coordinates under which f is converted to a sum or difference of squares. In geometric terms, this problem amounts to finding a new set of orthogonal axes in three-dimensional space by which to express the surface. But Cauchy then generalized the problem to quadratic forms in n variables, the coefficients of which can be written as a symmetric matrix. For example, the binary quadratic form $ax^2 + 2bxy + cy^2$ determines the symmetric 2×2 matrix

$$\begin{pmatrix} a & b \\ b & c \end{pmatrix}.$$

Cauchy's goal then was to find a linear substitution on the variables such that the matrix resulting from this substitution was diagonal, a goal he achieved in a paper of 1829. Because the details in the general case are somewhat involved and because the essence of Cauchy's proof is apparent in the two-variable case, it is that case which will be dealt with here.

To find a linear substitution which converts the binary quadratic from $f(x, y) = ax^2 + 2bxy + cy^2$ into a sum of squares it is necessary to find the maximum and minimum of $f(x, y)$ subject to the condition that $x^2 + y^2 = 1$. The point at which such an extreme value of f occurs is then a point on the unit circle which also lies on the end of one axis of the family of ellipses (or hyperbolas) described by the quadratic form. If one takes the line from the origin to that point as one of the axes and the perpendicular to that line as the other, the equation in relation to those axes will only contain the squares of the variables. By the principle of Lagrange multipliers, the extreme value occurs when the ratios $f_x/2x$ and $f_y/2y$ are equal. Setting each of these equal to λ gives the two equations

$$\frac{ax + by}{x} = \lambda \quad \text{and} \quad \frac{bx + cy}{y} = \lambda,$$

which can be rewritten as the system

$$(a - \lambda)x + by = 0 \quad \text{and} \quad bx + (c - \lambda)y = 0.$$

Cauchy knew that this system has nontrivial solutions only if its determinant equals 0, that is if $(a - \lambda)(c - \lambda) - b^2 = 0$. In matrix terminology, this equation is the characteristic equation $\det(A - \lambda I) = 0$, the equation dealt with by Cayley some 30 years later.

To see how the roots of this equation allow one to diagonalize the matrix, let λ_1 and λ_2 be the roots of the characteristic equation and (x_1, y_1), (x_2, y_2) the corresponding solutions for x and y. Thus

$$(a - \lambda_1)x_1 + by_1 = 0 \quad \text{and} \quad (a - \lambda_2)x_2 + by_2 = 0.$$

If one multiplies the first of these equations by x_2, the second by x_1, and subtracts, the result is the equation

$$(\lambda_2 - \lambda_1)x_1x_2 + b(y_1x_2 - x_1y_2) = 0.$$

Similarly, starting with the two equations involving $c - \lambda_i$, one arrives at the equation

$$b(y_2x_1 - y_1x_2) + (\lambda_2 - \lambda_1)y_1y_2 = 0.$$

Adding these two equations gives $(\lambda_2 - \lambda_1)(x_1 x_2 + y_1 y_2) = 0$. Therefore, if $\lambda_1 \neq \lambda_2$—and this is surely true in the case being considered, unless the original form is already diagonal—then $x_1 x_2 + y_1 y_2 = 0$. Because $(x_1, y_1), (x_2, y_2)$ are only determined up to a constant multiple, one can arrange to have $x_1^2 + y_1^2 = 1$ and $x_2^2 + y_2^2 = 1$. In modern terminology, the linear substitution

$$x = x_1 u + x_2 v$$

$$y = y_1 u + y_2 v$$

is orthogonal. One easily computes that the new quadratic form arising from this substitution is $\lambda_1 u^2 + \lambda_2 v^2$ as desired. That λ_1 and λ_2 are real follows from assuming, on the contrary, that they are complex conjugates of one another. In that case, x_1 would be the conjugate of x_2 and y_1 that of y_2, and $x_1 x_2 + y_1 y_2$ could not be 0. Cauchy had therefore shown that all eigenvalues of a symmetric matrix are real and, at least in the case where they are all distinct, that the matrix can be diagonalized by use of an orthogonal substitution.

15.5.4 Canonical Forms

The basic arguments of Cauchy's paper provided the beginnings to an extensive theory dealing with the eigenvalues of various types of matrices and with canonical forms. In general, however, throughout the middle of the nineteenth century, these results were all written in terms of forms, not in terms of matrices. Quadratic forms lead to symmetric matrices. The more general case of bilinear forms, functions of $2n$ variables of the form

$$\sum_{i,j=1}^{n} a_{ij} x_i y_j,$$

lead to general square matrices.

The most influential part of the theory of forms was worked out by Camille Jordan in his *Traité des substitutions*. Jordan came to the problem of classification, not through the study of bilinear forms, but through the study of the linear substitutions themselves. He had made a detailed study of Galois' work on solutions of algebraic equations and especially of his work on solving equations of prime power degree. These solutions involved the study of linear substitutions on these roots, substitutions whose coefficients could be considered to be elements of a finite field of order p. Such a substitution on the roots $x_1, x_2, \ldots x_n$ could be expressed in terms of a matrix A. In other words, if X represents the $n \times 1$ matrix of the roots x_i, then the substitution can be written as $X' \equiv AX \pmod{p}$. Jordan's aim was to find what he called a "transformation of indices" so that the substitution could be expressed in as simple terms as possible. In matrix notation, that means that he wanted to find an $n \times n$ invertible matrix P so that $PA \equiv DP$ where D is the "simplest possible" matrix. Thus if $Y \equiv PX$, then $PAP^{-1}Y \equiv PAX \equiv DPX \equiv DY$ and the substitution on Y is "simple." Using the characteristic polynomial for A, Jordan noted that if all of the roots of $\det(A - \lambda I) \equiv 0$ are distinct, then D could be taken to be diagonal, with the diagonal elements being the eigenvalues. On the other hand, if there are multiple roots, Jordan showed that a substitution can be found so that the resulting D is in block form

$$\begin{pmatrix} D_1 & 0 & 0 & \cdots & 0 \\ 0 & D_2 & 0 & \cdots & 0 \\ \cdots & \cdots & \cdots & \ddots & \cdots \\ 0 & 0 & 0 & \cdots & D_m \end{pmatrix}$$

where each block D_i is a matrix of the form

$$\begin{pmatrix} \lambda_i & 0 & 0 & \cdots & 0 & 0 \\ \lambda_i & \lambda_i & 0 & \cdots & 0 & 0 \\ \vdots & \vdots & \vdots & \ddots & \vdots & \vdots \\ 0 & 0 & 0 & \cdots & \lambda_i & \lambda_i \end{pmatrix}$$

and $\lambda_i \not\equiv 0 \pmod{p}$ is a root of the characteristic polynomial. The canonical form known today as the Jordan canonical form, where the values λ_i off of the main diagonal of the matrix are all replaced by 1's, was introduced by Jordan in 1871 when he realized that his method could be applied to the solution of systems of linear differential equations, whose coefficients, instead of being taken from a field of p elements, were either real or complex numbers. Thus Jordan returned, over a hundred years after the work of d'Alembert, to the origins of the entire complex of ideas associated with the eigenvalues of a matrix.

Jordan, however, did not use Cayley's single letter notation to represent linear substitutions. It was Frobenius who in 1878 combined the ideas of his various predecessors into the first complete monograph on the theory of matrices. In particular, Frobenius dealt with various types of relations among matrices. For example, he defined two matrices A and B to be **similar** if there were an invertible matrix P such that $B = P^{-1}AP$ and **congruent** if a P existed with $B = P^tAP$, where P^t is the transpose of P. He showed that when two symmetric matrices were similar, the transforming matrix P could be taken to be **orthogonal**, that is, one whose inverse equaled its transpose. Frobenius then made a detailed study of orthogonal matrices and showed, among other things, that their eigenvalues were complex numbers of absolute value 1. Frobenius concluded his paper by showing the relationship between his symbolical matrix theory and the theory of quaternions. Namely, he determined four 2×2 matrices whose algebra was precisely that of the quantities $1, i, j$, and k of quaternion algebra.

Frobenius and others made further contributions to matrix theory over the next decades. But it was not until the beginning of the twentieth century that textbooks appeared in which all of the above material was organized in the terminology of matrices. And it was not until the fourth decade of the century that the fundamental relationship of matrices to linear transformations of vector spaces was explicitly recognized. For that to happen, it was necessary for the abstract idea of a vector space to be made explicit. Because that development grew out of certain geometric ideas, its discussion will be postponed until Chapter 17.

Exercises

Problems from Gauss' Number Theory

1. Prove that if p is prime and $0 < a < p$, then the smallest exponent m such that $a^m \equiv 1 \pmod{p}$ is a divisor of $p - 1$.

2. For the prime $p = 7$, calculate for each integer a with $1 < a < 7$ the smallest exponent m such that $a^m \equiv 1 \pmod{7}$. Show that the theorem in problem 1 holds for all a.

3. Determine the primitive roots of $p = 7$, $p = 13$, and $p = 19$, that is, determine numbers a such that $p - 1$ is the smallest exponent such that $a^{p-1} \equiv 1 \pmod{p}$.

4. Prove that for primes of the form $8n + 3$, -2 is a quadratic residue and 2 is not.

5. Prove that for primes of the form $8n + 7$, 2 is a quadratic residue and -2 is not.

6. Complete Gauss' determination that 453 is a quadratic residue modulo 1236 by showing that

   a. If $x^2 \equiv 453 \pmod 4$, $x^2 \equiv 453 \pmod 3$, and $x^2 \equiv 453 \pmod{103}$ are all solvable, then so is $x^2 \equiv 453 \pmod{4 \cdot 3 \cdot 103}$.

   b. 453 is a quadratic residue modulo both 4 and 3.

   c. $\left(\frac{453}{103}\right) = \left(\frac{41}{103}\right)$.

   d. $\left(\frac{5}{41}\right) = 1$.

7. Show that the Gaussian integer $a + bi$ is prime if and only if the norm $a^2 + b^2$ is an ordinary prime.

8. Show that if any Gaussian prime p divides the product $abc \cdots$ of Gaussian primes, then p must equal one of those primes, or one of them multiplied by a unit. (*Hint:* Take norms of both sides.)

9. Factor $3 + 5i$ as a product of Gaussian primes.

Problems from Algebraic Number Theory

10. Show that a prime complex integer is irreducible.

11. Show that in the domain of integers of the form $a + b\sqrt{-17}$, Liouville's factorization $169 = 13 \cdot 13 = (4 + 3\sqrt{-17})(4 - 3\sqrt{-17})$ in fact demonstrates that unique factorization into primes fails in that domain. (*Hint:* Use norms to show that each of the four factors is irreducible.)

12. Show that the Gaussian integers form a Euclidean domain. That is, show that, given two Gaussian integers z, m, there exist two others q, r such that $z = qm + r$ and $N(r) < N(m)$.

13. Show that the domain of complex integers Γ_θ, where θ is a root of $x^2 - 2 = 0$, is Euclidean. First, determine explicitly the analogue of the Euclidean algorithm in this domain.

14. Show that in the domain of numbers of the form $a + b\sqrt{-5}$, the integers 2, 3, $-2 + \sqrt{-5}$, $-2 - \sqrt{-5}$, $1 + \sqrt{-5}$, $1 - \sqrt{-5}$ are all irreducible.

Problems in Cyclotomic Equations

15. Determine the cosets of the cyclic subgroup of order 6 of the cyclic group of order 18.

16. Use Gauss' method to solve the cyclotomic equations $x^4 + x^3 + x^2 + x + 1 = 0$ and $x^6 + x^5 + x^4 + x^3 + x^2 + x + 1 = 0$.

17. In the example dealing with Gauss' solution to $x^{19} - 1 = 0$, show that β_1, β_8, and β_7 are roots of the cubic equation $x^3 - \alpha_1 x^2 + (\alpha_2 + \alpha_4)x - 2 - \alpha_2 = 0$ where the α's and β's are as in the text.

18. In the example dealing with Gauss' solution to $x^{19} - 1 = 0$, show that r and r^{18} are both roots of $x^2 - \beta_1 x + 1 = 0$, where r and β_1 are as in the text.

Problems from Group Theory and Field Theory

Note: Some of the problems in this section require knowledge of group theory and Galois theory.

19. Show that two equivalent quadratic forms have the same discriminant.

20. Use Cardano's formula to calculate the group G of the equation $x^3 + 6x = 6$ over the rational numbers. Show that this group has a normal subgroup H such that both H and the index of H in G are primes.

21. Use Ferrari's method to calculate the group G of the equation $x^4 + 3 = 12x$ over the rational numbers. Show explicitly that this group has a composition series of the type indicated in the text.

22. Describe the five distinct groups of order 8.

23. Create fields of orders 5^3 and 3^5 by finding appropriate irreducible congruences modulo 5 and 3 respectively.

24. Compare Weber's definition of a group with the standard modern definition. Show that they are equivalent.

25. Compare Weber's definition of a field with the standard modern definition. Can some of Weber's axioms be proved from other ones?

Problems in Symbolic Algebra

26. Show that Hamilton's laws of operation on number couples (α, β) mirror the analogous laws of operation on complex numbers $\alpha + \beta i$.

27. Try to create a multiplication for number triples, written, say, in the form $\alpha + \beta i + \gamma j$, which satisfies Hamilton's critera for a reasonable multiplication. Namely, the multiplication must satisfy the commutative and associative laws, must be distributive over addition, must allow unique division, and must satisfy the modulus multiplication rule.

28. Let $\alpha = 3 + 4i + 7j + k$ and $\beta = 2 - 3i + j - k$ be quaternions. Calculate $\alpha + \beta$, $\alpha\beta$, α/β.

29. Define the modulus $|\alpha|$ of a quaternion $a + bi + cj + dk$ by $|\alpha| = a^2 + b^2 + c^2 + d^2$. Show that $|\alpha\beta| = |\alpha||\beta|$ and $\left|\frac{\alpha}{\beta}\right| = \frac{|\alpha|}{|\beta|}$.

30. Determine the general form of the expansion of a function $f(x, y, z)$ of three logical variables into a polynomial in terms of the form $x'y'z'$, where x', for example, represents either x or $\bar{x}$. Use this expansion to expand the function $V = x - yz$.

31. Interpret the remaining three of Boole's equations $x\bar{y}z = 0$, $x\bar{y}\bar{z} = 0$, $\bar{x}yz = 0$ in the case in the text, where x stands for clean beasts, y for beasts which divide the hoof, and z for beasts which chew the cud.

Problems from Matrix Theory

32. Show that if the substitution $x = \alpha x' + \beta y'$, $y = \gamma x' + \delta y'$ with $\alpha\delta - \beta\gamma = 1$ transforms the quadratic form $F = ax^2 + 2bxy + cy^2$ into the form $F' = a'x'^2 + 2b'x'y' + c'y'^2$, then there is an "inverse" substitution of the same form which transforms F' into F.

33. Prove that if the product of two matrices is the zero matrix, then at least one of the factors has determinant 0.

34. Show explicitly the truth of the Cayley-Hamilton theorem that $\det(M - \lambda I) = 0$ in the cases where M is a 2×2 and a 3×3 matrix.

35. Show that the matrix

$$L = \begin{pmatrix} \dfrac{a + Y}{X} & \dfrac{b}{X} \\ \dfrac{c}{X} & \dfrac{d + Y}{X} \end{pmatrix},$$

where

$$X = \sqrt{a + d + 2\sqrt{ad - bc}}$$

and

$$Y = \sqrt{ad - bc},$$

is the square root of the matrix

$$M = \begin{pmatrix} a & b \\ c & d \end{pmatrix}.$$

36. Determine the conditions on the 2×2 matrix M of problem 35 so that a square root exists. How many square roots are there?

37. Determine explicitly the square root L of an arbitrary 3×3 matrix M. Use the fact that $L^2 = M$ and that M satisfies its characteristic equation.

38. Determine explicitly the matrices which commute with the 2×2 matrix

$$M = \begin{pmatrix} a & b \\ c & d \end{pmatrix}.$$

Show that these depend on the number of roots of the characteristic polynomial of M.

39. Using d'Alembert's method, determine explicitly the solution to the system of differential equations

$$\frac{d^2 y_i}{dt^2} + \sum_{k=1}^{3} a_{ik} y_k = 0 \qquad \text{for} \qquad i = 1, 2, 3.$$

Show why the three eigenvalues of the matrix (a_{ik}) must be distinct, real, and positive for the solution to make physical sense.

40. Use Cauchy's technique to find an orthogonal substitution which converts the quadratic form $2x^2 + 6xy + 5y^2$ into a sum or difference of squares.

41. Find four 2×2 matrices over the complex numbers whose multiplication mimics the multiplication of the units $1, i, j, k$ of quaternion theory.

FOR DISCUSSION . . .

42. Find a copy of Weber's *Lehrbuch der Algebra* (1895) and compare it to a standard modern algebra text.

43. Design a lesson introducing the concept of a group through

   a. the notion of a permutation of a finite set;

   b. the notion of composition of quadratic forms;

   c. the notion of the residue classes modulo a prime *p*.

44. Compare the advantages and disadvantages of introducing a class to algebraic number fields by Kronecker's method of construction to Dedekind's method of considering subfields of the complex numbers.

45. Compare De Morgan's version of the laws of algebra with Weber's axioms for a field.

46. Design a lesson explaining negative numbers using

   a. Peacock's principle of permanence of equivalent forms;

   b. Hamilton's formulation using pairs of positive numbers.

   Which formulation would work better in a classroom? Why?

47. How do students in high school "understand" negative numbers? Do they understand why a negative times a negative is positive? Is such an understanding necessary?

48. Design a lesson explaining complex numbers using Hamilton's ordered pairs.

49. Outline a lesson on manipulation of matrices following Cayley's 1858 treatment of the subject.

References and Notes

General references on the history of algebra include the three works mentioned in the notes to Chapter 14, the books of Nový, Wussing, and van der Waerden. Other useful works include Michael J. Crowe, *A History of Vector Analysis: The Evolution of the Idea of a Vectorial System* (Notre Dame: University of Notre Dame Press, 1967) for vector algebra and two books by Harold Edwards, *Fermat's Last Theorem. A Genetic Introduction to Number Theory* (New York: Springer-Verlag, 1977) and *Galois Theory* (New York: Springer-Verlag, 1984). An excellent survey of the history of group theory is Israel Kleiner, "The Evolution of Group Theory: A Brief Survey," *Mathematics Magazine* 59 (1986), 198–215. There have been several excellent studies of British algebra, including Joan Richards, "The Art and Science of British Algebra: A Study in the Perception of Mathematical Truth," *Historia Mathematica* 7 (1980), 343–365; Helena Pycior, "George Peacock and the British Origins of Symbolical Algebra," *Historia Mathematica* 8 (1981), 23–45, and Ernest Nagel, " 'Impossible Numbers': A Chapter in the History of Modern Logic," *History of Ideas* 3 (1935), 429–474. These articles contain a much fuller picture of British algebra than could be presented in the text. The history of the theory of matrices is presented in three articles by Thomas Hawkins: "Cauchy and the Spectral Theory of Matrices," *Historia Mathematica* 2 (1975), 1–29; "Another Look at Cayley and the Theory of Matrices," *Archives Internationales d'Histoire des Sciences* 26 (1977), 82–112; and "Weierstrass and the Theory of Matrices," *Archive for History of Exact Sciences* 17 (1977), 119–163. A reading of these articles gives a very detailed picture of the development of matrix theory and related areas of mathematics. A briefer and more elementary survey of the history of matrix theory is R. W. Feldmann, "History of Elementary Matrix Theory," *Mathematics Teacher* 55 (1962), 482–484, 589–590, 657–659 and 56 (1963), 37–38, 101–102, 163–164.

1. Ernst Kummer, "De numeris complexis, qui radicibus unitatis et numeris integris realibus constant," *Journal de mathématiques pures et appliquées* 12 (1847), 185–212, reprinted in Kummer's *Collected Papers* (New York: Springer-Verlag, 1975), vol. 1, 165–192, p. 182.

2. Michael J. Crowe, *Vector Analysis*, p. 29.

3. Carl Friedrich Gauss, *Disquisitiones Arithmeticae*, translated by Arthur A. Clarke (New York: Springer-Verlag, 1986), p. 1. A perusal of the entire book is well worth the trouble.

4. Ibid., p. 35.

5. A recent biography of Gauss, which includes a bibliography of his works and a survey of the secondary literature, is W. K. Bühler, *Gauss: A Biographical Study* (New York: Springer-Verlag, 1981).

6. The story of the Paris Academy meeting of March 1, 1847 is told in more detail in Edwards, *Fermat's Last Theorem*, pp. 76–80.

7. Quoted in Edwards, *Fermat's Last Theorem*, p. 61. A

recent article on Sophie Germain is J. H. Sampson, "Sophie Germain and the Theory of Numbers," *Archive for History of Exact Sciences* 41 (1991), 157–161.

8. For a more detailed discussion of divisors and of Kummer's work on Fermat's Last Theorem, consult Chapter 4 of Edwards, *Fermat's Last Theorem*. See also H. M. Edwards, "The Genesis of Ideal Theory," *Archive for History of Exact Sciences* 23 (1980), 321–378.

9. Richard Dedekind, "Sur la Théorie des Nombres entiers algébriques," *Bulletin des Sciences mathématiques et astronomiques* 11 (1877) 1–121, part of which is reprinted in his *Mathematische Werke*, vol. 3, 262–296, p. 268.

10. Ibid., p. 280.

11. Ibid., p. 268. For more on the creation of ideals, see H. M. Edwards, "Dedekind's Invention of Ideals," in Esther Phillips, ed., *Studies in the History of Mathematics* (Washington: M.A.A., 1987), 8–20.

12. Gauss, *Disquisitiones*, p. 445.

13. Ibid., p. 459.

14. For a biography of Abel, see Oystein Ore, *Niels Henrik Abel: Mathematician Extraordinary* (New York: Chelsea, 1974).

15. See Van der Waerden, *History of Algebra*, pp. 85–88 for details of the proof.

16. Quoted in Wussing, *Abstract Group Concept*, p. 98.

17. Ibid., p. 100.

18. A detailed discussion of Galois' propositions and their proofs is in Edwards, *Galois Theory*. A further clarification is in H. M. Edwards, " A Note on Galois Theory," *Archive for History of Exact Sciences* 41 (1991), 163–169. B. Melvin Kiernan, "The Development of Galois Theory from Lagrange to Artin," *Archive for History of Exact Sciences* 8 (1971), 40–154, provides the best history of the entire subject of Galois theory.

19. The best recent article on the life of Galois is T. Rothman, "Genius and Biographers: the Fictionalization of Evariste Galois," *American Mathematical Monthly* 89 (1982), 84–106. The fictionalized biography is Leopold Infeld, *Whom the Gods Love: The Story of Evariste Galois* (New York: Whittlesey House, 1948). Another biography is found in E. T. Bell, *Men of Mathematics* (New York: Simon and Schuster, 1937).

20. Quoted in Rothman, "Fictionalization of Evariste Galois," p. 97.

21. R. Bourgne and J. P. Azra, eds., *Ecrits et Mémoires Mathématiques d'Evariste Galois* (Paris: Gauthier-Villars, 1962), p. 185.

22. Ibid., p. 175.

23. Camille Jordan, *Traité des substitutions et des équations algébriques* (Paris: Gauthier-Villars, 1870), sec. 54.

24. Jordan, *Traité des substitutions*, sec. 402.

25. Gauss, *Disquisitiones*, pp. 264–265.

26. Ibid., p. 366.

27. Quoted in Wussing, *Abstract Group Concept*, p. 64.

28. Arthur Cayley, "On the Theory of Groups, as depending on the Symbolic Equation $\theta^n = 1$," *Philosophical Magazine* (4) 7, 40–47, p. 41. This paper is also found in Cayley, *The Collected Mathematical Papers* (Cambridge: Cambridge University Press, 1889–1897), vol. 2, 123–130.

29. Arthur Cayley, "On the Theory of Groups," *Proceedings of the London Mathematical Society* 9 (1878), 126–133, p. 127. This paper is reprinted in Cayley, *Collected Mathematical Papers*, vol. 10, 324–330.

30. Arthur Cayley, "The Theory of Groups," *American Journal of Mathematics* 1 (1878), 50–52, p. 52. This paper is reprinted in Cayley, *Collected Mathematical Papers*, vol. 10, 401–403.

31. Ibid.

32. Walther von Dyck, "Gruppentheoretische Studien," *Mathematische Annalen* 20 (1882), 1–44, p. 1. This paper is discussed further in Wussing, *Abstract Group Concept*, p. 240 and in van der Waerden, *History of Algebra*, p. 152.

33. Ibid., p. 12.

34. Heinrich Weber, "Beweis des Satzes, dass jede eigentlich primitive quadratische Form unendlich viele Primzahlen fähig ist," *Mathematische Annalen* 20 (1882), 301–329, p. 302. Weber's work is also discussed in the works of Wussing and van der Waerden.

35. Heinrich Weber, "Die allgemeinen Grundlagen des Galois'schen Gleichungstheorie," *Mathematische Annalen* 43 (1893), 521–549, p. 524.

36. Quoted in Kurt-R. Biermann, "Kronecker," *Dictionary of Scientific Biography* (New York: Scribners, 1970–1980), vol. 7, 505–509.

37. Richard Dedekind, Supplement XI to Dirichlet, *Vorlesungen über Zahlentheorie*, (Braunschweig: Vieweg und Sohn, 1893) p. 452. The quoted edition is the fourth one,

but the material also occurs in the second and third editions.

38. William Rouse Ball, *The Origin and History of the Mathematical Tripos* (Cambridge: Cambridge University Press, 1880), p. 195.

39. George Peacock, *A Treatise on Algebra*, reprint edition, (New York: Scripta Mathematica, 1940), vol. 2, p. 59.

40. Quoted in Pycior, "George Peacock," p. 35.

41. Peacock, *Treatise on Algebra*, p. 453.

42. Ibid., p. 449.

43. Augustus De Morgan, *Trigonometry and Double Algebra* (London: Taylor, 1849), pp. 92–93. This is quoted in Nagel," 'Impossible Numbers'," pp. 185–186.

44. Augustus De Morgan, "On the Foundation of Algebra, No. II," *Transactions of the Cambridge Philosophical Society* 7 (1839–1842), 287–300, p. 287.

45. See Abraham Arcavi and Maxim Bruckheimer, "The didactical De Morgan: a selection of Augustus de Morgan's thoughts on teaching and learning mathematics," *For the Learning of Mathematics* 9 (1989), 34–39. This article is made up of numerous quotations from De Morgan's works illustrating his commitment to better teaching of mathematics.

46. De Morgan, "On the Foundation of Algebra, No. II," pp. 288–289.

47. William Rowan Hamilton, "Theory of Conjugate Functions, or Algebraic Couples; with a Preliminary and Elementary Essay on Algebra as the Science of Pure Time," *Transactions of the Royal Irish Academy* 17 (1837), 293–422, reprinted in Hamilton, *Mathematical Papers* (Cambridge: Cambridge University Press, 1967), vol. 3, 3–96, p. 4. For more information on Hamilton's work, see Jerold Mathews, "William Rowan Hamilton's Paper of 1837 on the Arithmetization of Analysis," *Archive for History of Exact Sciences* 19 (1978), 177–200 and Thomas Hankins, "Algebra of Pure Time: William Rowan Hamilton and the Foundations of Algebra," in P. J. Mackamer and R. G. Turnbull, eds., *Motion and Time, Space and Matter: Interrelations in the History and Philosophy of Science* (Columbus: Ohio State University Press, 1976), 327–359.

48. Hamilton, *Mathematical Papers*, p. 10.

49. Ibid., p. 83.

50. Ibid., p. 96.

51. Quoted in Crowe, *Vector Analysis*, p. 27.

52. Ibid., p. 32.

53. Clerk Maxwell, *Treatise on Electricity and Magnetism* (London: Oxford University Press, 1873), pp. 9–10.

54. George Boole, *An Investigation of the Laws of Thought* (New York: Dover, 1958), p. 1. More on Boole is found in D. MacHale, *George Boole: His Life and Work* (Dublin: Boole Press, 1985).

55. Ibid., p. 27.

56. Ibid., p. 49.

57. Quoted in Hawkins, "Another Look at Cayley," p. 86.

58. James Joseph Sylvester, "On a New Class of Theorems," *Philosophical Magazine* (3) 37 (1850), 363–370, reprinted in *Collected Mathematical Papers* (Cambridge: Cambridge University Press, 1904–1912), vol. 1, 145–151, p. 150.

59. Arthur Cayley, "A Memoir on the Theory of Matrices," *Philosophical Transactions of the Royal Society of London* 148 (1858), reprinted in Cayley, *Collected Mathematical Papers*, vol. 2, 475–496, p. 476.

60. Ibid., p. 481.

61. Ibid., p. 483.

62. Ibid.

Summary of Nineteenth Century Algebra

1765–1822	Paolo Ruffini	Fifth degree polynomial equations
1776–1831	Sophie Germain	Number theory
1777–1855	Carl Friedrich Gauss	Number theory, cyclotomic polynomials
1789–1857	Augustin-Louis Cauchy	Permutations, determinants, eigenvalues
1791–1858	George Peacock	Symbolical algebra
1802–1829	Niels Henrik Abel	Fifth degree polynomial equations
1805–1859	Peter Lejeune-Dirichlet	Number theory
1805–1865	William Rowan Hamilton	Complex numbers, quaternions
1806–1871	Augustus De Morgan	Symbolical algebra
1809–1882	Joseph Liouville	Number theory
1810–1893	Ernst Kummer	Factorization, divisors
1811–1832	Evariste Galois	Groups of equations, finite fields
1814–1897	James Joseph Sylvester	Matrices
1814–1848	Pierre Wantzel	Construction problems
1815–1864	George Boole	Logic
1821–1895	Arthur Cayley	Abstract groups, matrices
1823–1852	Ferdinand Gotthold Eisenstein	Forms, matrices
1823–1891	Leopold Kronecker	Abstract groups, fields
1831–1879	James Clerk Maxwell	Quaternions, vectors
1831–1901	Peter Guthrie Tait	Quaternions
1831–1916	Richard Dedekind	Ideals, fields
1838–1922	Camille Jordan	Groups, linear substitutions
1839–1903	Josiah Willard Gibbs	Vectors
1842–1913	Heinrich Weber	Groups, fields
1849–1917	Georg Frobenius	Theory of matrices
1850–1925	Oliver Heaviside	Vectors
1852–1939	Ferdinand Lindemann	Transcendance of π
1856–1934	Walther von Dyck	Group theory

Chapter 16

Analysis in the Nineteenth Century

"In the above mentioned work of M. Cauchy [Cours d'Analyse de l'École Polytechnique] . . . one finds the following theorem: 'If the different terms of the series $u_0 + u_1 + u_2 + u_3 + \cdots$ are functions of one and same variable quantity x, and indeed are continuous functions with respect to that variable in the neighborhood of a particular value, for which the series converges, then the sum s of the series, in the neighborhood of this particular value, is also a continuous function of x.' But it appears to me that this theorem admits exceptions. For example, the series $\sin x - \frac{1}{2}\sin 2x + \frac{1}{3}\sin 3x - \cdots$ is discontinuous for each value $(2m + 1)\pi$ of x, where m is an integer. It is well known that there are many series with similar properties."
(Niels Henrik Abel, in "Untersuchungen über die Reihe $1 + \frac{m}{1}x + \frac{m}{1}\frac{m-1}{2}x^2 + \frac{m}{1}\frac{m-1}{2}\frac{m-2}{3}x^3 + \cdots$," 1826)[1]

In the autumn of 1858, Richard Dedekind, professor in the Polytechnic School in Zurich, had to lecture for the first time on the elements of the differential calculus. In preparing for these lectures, he decided that although the traditional geometric approach to the idea of a limit had pedagogic value in an introductory course, there was still no truly "scientific" foundation for the subject. He decided therefore to concentrate his energies on providing the basis for an arithmetic definition of the concept of the real numbers. On November 24, 1858, Dedekind reached his goal and shortly thereafter communicated the results to a friend and to some of his best students. But because he did not feel completely at ease with his presentation, he did not publish his idea of "Dedekind cuts" until 1872.

Toward the end of the eighteenth century, with the French Revolution's restructuring of mathematics education throughout the European continent and with the growing necessity for mathematicians to teach rather than just do research, there was an increasing concern with how mathematical ideas should be presented to students and a concomitant increase in concern for "rigor." Recall that Lagrange attempted to base all of calculus on the notion of a power series. And although Lacroix wrote his calculus texts using Lagrange's method, among others, it was soon discovered that not all functions could be expressed by such series.

It was Augustin-Louis Cauchy, the most prolific mathematician of the nineteenth century, who first established the calculus on the basis of the limit concept so familiar today. Although the notion of limits had been discussed much earlier, even by Newton, Cauchy was the first to translate the somewhat vague notion of a function approaching a particular value into arithmetic terms by means of which one could actually prove the existence of limits. Cauchy used his notion of limit in defining continuity (in the modern sense) and convergence of sequences, both of numbers and of functions. Cauchy's notion of convergence, first published in 1821, was also developed in essence by the Czech mathematician Bernhard Bolzano in 1817 and by the Portuguese mathematician José Anastácio da Cunha as early as 1782. Unfortunately, the works of these latter two appeared in the far corners of Europe and were not appreciated, nor even read, in the mathematical centers of France and Germany. Thus it was out of Cauchy's work that today's notions developed.

One of Cauchy's important results, that the sum of an infinite series of continuous functions is continuous, assuming this sum exists, turned out not to be true. Counterexamples were discovered as early as 1826 in connection with the series of sine and cosine functions now known as Fourier series. These series, although considered briefly by Daniel Bernoulli in the middle of the eighteenth century, were first studied in detail by Joseph Fourier in his work on heat conduction in the early nineteenth century. Fourier's works stimulated Peter Lejeune-Dirichlet to study in more detail the notion of a function and Bernhard Riemann to develop the concept known today as the Riemann integral.

Some unresolved questions in the work of Cauchy and Bolzano as well as the study of points of discontinuity growing out of Cauchy's wrong theorem led several mathematicians in the second half of the century to consider the structure of the real number system. In particular, Richard Dedekind and Georg Cantor each developed methods of constructing the real numbers from the rational numbers and, in this connection, began the detailed study of infinite sets.

In his calculus texts, Cauchy defined the integral as a limit of a sum rather than as an antiderivative, as had been common in the eighteenth century. His extension of this notion of the integral to the domain of complex numbers led him by the 1840s to develop many of the important concepts studied today in a first course in complex variables, including the idea of a residue. These ideas were further developed and extended by Riemann in the following decade.

Because integration in the complex domain can be thought of as integration over a real two-dimensional plane, Cauchy was also able to state the theorem today known as Green's theorem relating integration around a closed curve to double integration over the region it bounds. Similar theorems relating integrals over a region to integrals over the boundary of the region were discovered by Mikhail Ostrogradsky and William Thomson. These theorems, today known as the divergence theorem and Stokes' theorem, were soon applied in physics to such areas as electricity and magnetism.

A final aspect of analysis to be dealt with in this chapter is the early nineteenth century progress in probability and statistics. In particular, we will consider the work of Legendre and Gauss in the development of the method of least squares and Gauss' derivation of the normal curve of errors. The chapter will conclude with a brief look at Laplace's synthesis of all earlier work in the field.

16.1 RIGOR IN ANALYSIS

Silvestre-Francois Lacroix replaced Lagrange at the École Polytechnique in 1799. Two years earlier the first of the three volumes of his *Traité du calcul différentiel et du calcul intégral* had appeared. Lacroix intended this work to be a survey of the methods of calculus developed since the time of Newton and Leibniz. Thus Lacroix presented not only Lagrange's view of the derivative of a function $f(x)$ as the coefficient of the first-order term in the Taylor series for f, but also dealt with the definition of dy/dx as a limit in the style of d'Alembert and as a ratio of infinitesimals in the manner of Euler. Lacroix was proud of the book's comprehensiveness and hoped that the true metaphysics of the subject would be found in what the various methods had in common.

For his teaching in Paris, however, Lacroix wrote a shortened one volume version of the text entitled *Traité élémentaire du calcul différentiel et du calcul intégral*, a text whose continued popularity is attested by its appearance in nine editions between 1802 and 1881. In this work, Lacroix decided to base the differential calculus initially on the notion of a limit, defined in the process of his determination of the limit of a differential quotient. Thus he showed that if $u = ax^2$ and $u_1 = a(x + h)^2$, then $2ax$ is "the **limit** of the ratio $(u_1 - u)/h$, or, is the value towards which this ratio tends in proportion as the quantity h diminishes, and to which it may approach as near as we choose to make it."[2] After calculating several other limits of ratios, Lacroix explained that in fact "the differential calculus is the finding of the limit of the ratios of the simultaneous increments of a function and of the variables on which it depends."[3] Thus Lacroix, following his predecessors Euler and Lagrange, made no attempt at the beginning of the book to motivate the differential calculus in terms of slopes of tangent lines. The differential calculus was part of "analysis" and required no geometrical motivation or diagrams. Tangent lines were simply an application of the calculus and as such were discussed in section 7 of the text, "On the Application of the Differential Calculus to the Theory of Curves."

Despite Lacroix's decision in the *Traité élémentaire* to begin differential calculus with limits, he rapidly moved to establish the Taylor series of a function. Like Lagrange, he believed that all functions could be expressed as series except perhaps at isolated points. He then proceeded to use the Taylor series representation to develop the differentiation formulas for various transcendental functions and the methods for determining maxima and minima, even for functions of several variables. As part of this latter discussion, he corrected Euler's error in giving the conditions for a function of two variables to have an extreme value. In fact, he showed that a sufficient condition for $u(x, y)$ to have an extreme value at a point where both first partial derivatives vanish is that

$$\frac{\partial^2 u}{\partial x^2}\frac{\partial^2 u}{\partial y^2} > \left[\frac{\partial^2 u}{\partial x \partial y}\right]^2$$

at that point.

Lagrange's method of power series also appealed to the reform minded members of the Cambridge Analytical Society, including George Peacock, Charles Babbage, and John Herschel (1792–1871), who translated Lacroix's *Traité élémentaire* into English in 1816 to provide an analytic text for use at Cambridge. The translators were so disappointed that

---

Biography   Augustin-Louis Cauchy (1789–1857)

Although Cauchy was the most prolific mathematician of the nineteenth century, he was never easy to deal with. As Abel wrote in a letter to a friend in 1826 during his visit to Paris, "there is no way to get along with him, although he is at present the mathematician who knows best how mathematics ought to be treated. . . . Cauchy is immoderately Catholic and bigoted, a very strange thing for a mathematician. Otherwise he is the only one [in Paris] who at present works in pure mathematics."[4] Born in the capital in the year the French Revolution began, he studied engineering at the École Polytechnique from 1805 to 1807. While working as an engineer from 1810 to 1813 on various military projects, he showed such a strong interest in pure mathematics that he was encouraged by Laplace and Lagrange to leave engineering. With their help he secured a teaching position at the École Polytechnique and several years later also

at the Collége de France. Upon the appearance of his texts in analysis, he became one of the most respected members of the French mathematical community. He wrote so many mathematical papers that the journal of the Paris Academy was forced to limit the contributions of any one person. Cauchy got around these restrictions by establishing his own journal.

When the July Revolution of 1830 led to the overthrow of the last Bourbon king, Cauchy as an ardent conservative refused to take the oath of allegiance to the new king and went into a self-imposed exile in Italy and then in Prague. He returned to Paris in 1838 but did not return to his teaching posts until the Revolution of 1848 led to the removal of the requirement of an oath of allegiance. It was perhaps because of his political views that the French government did not honor him with a postage stamp until the two hundredth anniversary of his birth (Figure 16.1).

FIGURE 16.1
Cauchy honored on a recent French stamp.

Lacroix had "substituted the method of limits of d'Alembert in the place of the more correct and natural method of Lagrange,"[5] that they provided notes so that the reader could use Lagrange's method rather than that of limits.

In spite of the appeal of Lagrange's method in England, Cauchy, back in France, found that this method was lacking in "rigor." Cauchy in fact was not satisfied with what he believed were unfounded manipulations of algebraic expressions, especially infinite ones. These expressions were only true for certain values, those values for which the infinite series was convergent. In particular, Cauchy discovered that the Taylor series for the function

$$f(x) = e^{-x^2} + e^{-(1/x^2)}$$

does not converge to the function. Thus, because from 1813 he was teaching at the École Polytechnique, Cauchy began to rethink the basis of the calculus entirely. In 1821, at the urging of several of his colleagues, he published his *Cours d'Analyse de l'École Royale Polytechnique* in which he introduced new methods into the foundations of the calculus. We will study Cauchy's ideas on limits, continuity, convergence, derivatives, and integrals in the context of an analysis of this text as well as its sequel of 1823, *Résumé des Leçons donnees a l'École Royale Polytechnique sur le Calcul Infinitesimal*, for it is these texts, used in Paris, which were to provide the model for calculus texts for the remainder of the century.

Side Bar 16.1   What Is a Limit?

Leibniz (1684): If any continuous transition is proposed terminating in a certain limit, then it is possible to form a general reasoning, which covers also the final limit.

Newton (1687): The ultimate ratio of evanescent quantities . . . [are] limits towards which the ratios of quantities decreasing without limit do always converge; and to which they approach nearer than by any given difference, but never go beyond, nor in effect attain to, till the quantities are diminished *in infinitum*.

Maclaurin (1742): The ratio of $2x + o$ to a continually decreases while o decreases and is always greater than the ratio of $2x$ to a while o is any real increment, but it is manifest that it continually approaches to the ratio of $2x$ to a as its limit.

D'Alembert (1754): This ratio $[a : 2y + z]$ is always smaller than $a : 2y$, but the smaller z is, the greater the ratio will be and, since one may choose z as small as one pleases, the ratio $a : 2y + z$ can be brought as close to the ratio $a : 2y$ as we like. Consequently, $a : 2y$ is the limit of the ratio $a : 2y + z$.

Lacroix (1806): The limit of the ratio $(u_1 - u)/h$ . . . is the value towards which this ratio tends in proportion as the quantity h diminishes, and to which it may approach as near as we choose to make it.

Cauchy (1821): If the successive values attributed to the same variable approach indefinitely a fixed value, such that they finally differ from it by as little as one wishes, this latter is called the limit of all the others.

16.1.1 Limits

Cauchy's definition of limit occurs near the beginning of his *Cours d'Analyse*: "If the successive values attributed to the same variable approach indefinitely a fixed value, such that finally they differ from it by as little as one wishes, this latter is called the **limit** of all the others."[6] As an example, Cauchy noted that an irrational number is the limit of the various fractions which approach it (Side Bar 16.1). He also defined an **infinitely small quantity** to be a variable whose limit is zero. Note that Cauchy is not defining the modern concept $\lim_{x \to a} f(x) = b$, for that concept involves two different variables. He seems to suppress the role of the independent variable entirely. Furthermore, it may appear that Cauchy's definition of a limit is little different from that of d'Alembert. To see, however, what Cauchy meant by his verbal definition and to discover the difference between it and the definitions of his predecessors, it is necessary to consider his use of the definition to prove certain specific results on limits. In fact, Cauchy not only deals with both the dependent and independent variables, but also translates his statement arithmetically by use of the language of inequalities. As an example, consider Cauchy's proof of the following

Theorem.   *If, for increasing values of x, the difference f(x + 1) − f(x) converges to a certain limit k, the fraction f(x)/x converges at the same time to the same limit.*[7]

Cauchy begins by translating the hypothesis of the theorem into an arithmetic state-ment: Given any value ϵ, as small as one wants, one can find a number h such that if $x \geq h$, then $k - \epsilon < f(x + 1) - f(x) < k + \epsilon$. He now proceeds to use this translation in his proof. Because each of the differences $f(h + i) - f(h + i - 1)$ for $i = 1, 2, \ldots n$ satisfies the inequality, so does their arithmetic mean

$$\frac{f(h + n) - f(h)}{n}.$$

It follows that

$$\frac{f(h + n) - f(h)}{n} = k + \alpha$$

where $-\epsilon < \alpha < \epsilon$ or, setting $x = h + n$, that

$$\frac{f(x) - f(h)}{x - h} = k + \alpha.$$

But then $f(x) = f(h) + (x - h)(k + \alpha)$ or

$$\frac{f(x)}{x} = \frac{f(h)}{x} + \left(1 - \frac{h}{x}\right)(k + \alpha).$$

Because h is fixed, Cauchy concludes that as x gets large, $\frac{f(x)}{x}$ approaches $k + \alpha$ where $-\epsilon < \alpha < \epsilon$. Because ϵ is arbitrary, the conclusion of the theorem holds. Cauchy also proves the theorem for the cases $k = \infty$ and $k = -\infty$ and then uses the result to conclude, for example, that as x gets large $\frac{\log x}{x}$ converges to 0 and $\frac{a^x}{x}$ $(a > 1)$ has limit ∞.

16.1.2 Continuity

Given the definition of a limit, Cauchy could now define the crucial concept of continuity. Recall that for Euler a continuous function was one expressed by a single expression while a discontinuous one was expressed in different parts of its domain by different expressions. Cauchy realized that such a definition was contradictory. For example, the function $f(x)$ which is equal to x for positive values and $-x$ for negative values would appear to be discontinuous according to Euler's definition. On the other hand, one can write this same function using the single analytic expression

$$f(x) = \frac{2}{\pi} \int_0^\infty \frac{x^2\, dt}{t^2 + x^2},$$

so $f(x)$ is continuous. The geometric notion of a continuous curve, one without any breaks, was generally understood, but Cauchy sought to find an analytic definition expressing this idea for functions. Lagrange had attempted such a definition earlier in the specific case of a function "continuous at 0" and having value 0 there: "We can always find an abscissa h corresponding to an ordinate less than any given quantity; and then all smaller values of h correspond also to ordinates less than the given quantity."[8]

Having read Lagrange's work, Cauchy generalized Lagrange's idea and gave his own new definition: "The function $f(x)$ will be, between two assigned values of the variable x,

Side Bar 16.2 Definitions of Continuity

Euler (1748): A continuous curve is one such that its nature can be expressed by a single function of x. If a curve is of such a nature that for its various parts . . . different functions of x are required for its expression, . . . , then we call such a curve discontinuous.

Bolzano (1817): A function $f(x)$ varies according to the law of continuity for all values of x inside or outside certain limits . . . if [when] x is some such value, the difference $f(x + \omega) - f(x)$ can be made smaller than any given quantity provided ω can be taken as small as we please.

Cauchy (1821): The function $f(x)$ will be, between two assigned values of the variable x, a continuous function of this variable if for each value of x between these limits, the [absolute] value of the difference $f(x + \alpha) - f(x)$ decreases indefinitely with α.

Dirichlet (1837): One thinks of a and b as two fixed values and of x as a variable quantity that can progressively take all values lying between a and b. Now if to every x there corresponds a single, finite y in such a way that, as x continuously passes through the interval from a to b, $y = f(x)$ also gradually changes, then y is called a continuous function of x in this interval.

Heine (1872): A function $f(x)$ is continuous at the particular value $x = X$ if for every given quantity ϵ, however small, there exists a positive number η_0 with the property that for no positive quantity η which is smaller than η_0 does the absolute value of $f(X \pm \eta) - f(X)$ exceed ϵ. A function $f(x)$ is continuous from $x = a$ to $x = b$ if for every single value $x = X$ between $x = a$ and $x = b$, including $x = a$ and $x = b$, it is continuous.

a **continuous function** of this variable if for each value of x between these limits, the numerical [absolute] value of the difference $f(x + \alpha) - f(x)$ decreases indefinitely with α (Side Bar 16.2). In other words, the function $f(x)$ will remain continuous with respect to x between the given values if, between these values, an infinitely small increment of the variable always produces an infinitely small increment of the function itself."[9] Note that Cauchy presents both an arithmetic definition and one using the more familiar language of infinitely small quantities. But because Cauchy had already defined such quantities in terms of limits, the two definitions mean the same thing. Cauchy demonstrated his clear understanding of his definition by showing, for example, that $\sin x$ is continuous (on any interval). For $\sin(x + \alpha) - \sin x = 2 \sin \frac{1}{2}\alpha \cos(x + \frac{1}{2}\alpha)$, and the right side clearly decreases indefinitely with α.

It is interesting that Bernhard Bolzano (1781–1848), a Czech mathematician also familiar with the work of Lagrange, had some four years earlier given a definition of continuity virtually identical to Cauchy's. As part of his plan to prove rigorously the intermediate value theorem, "that between any two values of the unknown quantity which give results of opposite sign [when substituted in a continuous function $f(x)$] there must always lie at least one real root of the equation $[f(x) = 0]$,"[10] Bolzano needed to give a clear definition of the type of functions for which the theorem would hold. Noting that others had given definitions of continuity in terms of nonmathematical concepts such as time and motion, Bolzano gave what he called a "correct definition": "A function $f(x)$

Biography   Bernhard Bolzano (1781–1848)

Bolzano studied mathematics, philosophy, and physics in the university of his hometown of Prague. In 1805 he was appointed there to a chair in the philosophy of religion, a position established by order of Emperor Franz I of Austria to counter the new trends of enlightenment being spread through Europe in the wake of the French Revolution. Bolzano, however, was not a particular sympathizer of the Catholic restoration and expressed his own enlightened views on religion in his lectures. He was finally dismissed from his post in 1819 and put under police supervision on suspicion of heresy. Meanwhile, however, his philosophical training had attracted him to questions about the foundations of analysis, questions which he was able to resolve to his satisfaction with new definitions and proofs related to the intuitive ideas of limit and continuity (Figure 16.2).

---

varies according to the law of continuity for all values of x inside or outside certain limits . . . if [when] x is some such value, the difference $f(x + \omega) - f(x)$ can be made smaller than any given quantity provided ω can be taken as small as we please."[11] The similarity of this definition to that of Cauchy is clear. In modern terms, both defined continuity over an interval, not at a point. But there is no convincing evidence that Cauchy had read Bolzano's work when he produced his own definition. Both men, being interested in giving a definition from which certain "obvious" results could be proved, came up with essentially the same idea.

16.1.3 Convergence

Cauchy's concept of the convergence of series also had predecessors. But it was his definition, accompanied by explicit criteria which could be used actually to test for convergence, that has lasted to the present. Cauchy's definition appears in Chapter 6 of his *Cours d'Analyse*: "Let $s_n = u_0 + u_1 + u_2 + \cdots + u_{n-1}$ be the sum of the first n terms [of a series], n designating an arbitrary integer. If, for increasing values of n, the sum s_n approaches indefinitely a certain limit s, the series will be called **convergent**, and the limit in question will be called the **sum** of the series. On the contrary, if, as n increases indefinitely, the sum s_n does not approach any fixed limit, the series will be **divergent** and will not have a sum."[12] To clarify the definition, Cauchy stated what has become known as the "Cauchy criterion" for convergence. He realized that for a series to be convergent, it was necessary that the individual terms u_n must decrease to zero. This condition, however, was not sufficient. Convergence could only be assured if the various sums $u_n + u_{n+1}, u_n + u_{n+1} + u_{n+2}, u_n + u_{n+1} + u_{n+2} + u_{n+3}, \cdots$ "taken, from the first, in whatever number one wishes, finish by constantly having an absolute value less than any assignable limit."[13] Cauchy offered no proof of this sufficiency condition, because without some arithmetical definition of the real numbers no such proof is possible. He did, however, offer examples. Thus he showed that the geometric series $1, x, x^2, x^3, \ldots$, with $|x| < 1$, converges, because the sums $x^n + x^{n+1}, x^n + x^{n+1} + x^{n+2}, \ldots$, respectively equal to $x^n \frac{1-x^2}{1-x}, x^n \frac{1-x^3}{1-x}, \ldots$, are always between x^n and $\frac{x^n}{1-x}$ and the latter values both converge to zero with increasing n.

FIGURE 16.2
Bolzano on a Czechoslovakian stamp.

Biography José Anastácio da Cunha (1744–1787)

Da Cunha was educated in Lisbon and became a military officer during the French and Spanish invasion of Portugal in 1762. Pursuing the study of ballistics, he wrote a memoir on the subject in 1769, analyzing various manuals used for instruction on the subject. His work brought him to the attention of the Marquis of Pombal, the powerful minister of King José I who had been able to reduce the powers of the Inquisition and remove the Jesuits from their positions of power. Pombal arranged for da Cunha to be appointed to the chair of geometry at the newly reorganized University of Coimbra in 1773. Upon the death of the king in 1777, however, Pombal lost power and his protegé da Cunha, having gained a reputation as a free thinker, was arrested by the Inquisition and convicted of heterodox religious opinions. He was pardoned in 1781 and spent the remainder of his life as a professor of mathematics in a Lisbon school organized for the education of poor children.

Cauchy had been preceded in his statement of the Cauchy criterion not only by Bolzano, but also by the Portuguese scholar José Anastácio da Cunha (1744–1787), who included this material in his *Principios Mathematicos*, a comprehensive text which stretched from basic arithmetic to the calculus of variations and was published in sections beginning in 1782. Da Cunha's version of convergence and the Cauchy criterion is as follows: "A **convergent** series, so Mathematicians say, is one whose terms are similarly determined, each one by the number of the preceding terms, in such a way that the series can always be continued, and finally it need not matter whether it does or does not because one may neglect without considerable error the sum of any number of terms one might wish to add to those already written or indicated."[14] Da Cunha even used the criterion to demonstrate, like Cauchy, the convergence of the geometric series. Unfortunately, although da Cunha's work was translated into French in 1811, it was apparently little noticed and had little influence.

Bolzano's work, published as well in a distant corner of Europe, also had little contemporary influence. But in this work Bolzano showed that the Cauchy criterion implied the least upper bound principle, eventually seen to be one of the defining properties of the real number system. Bolzano's convergence definition and his statement of the Cauchy criterion (applied to a series of functions rather than constants) are contained in a

Theorem. *If a series of quantities $F_1(x)$, $F_2(x)$, $F_3(x)$, . . . , $F_n(x)$, . . . [where each $F_i(x)$ can be thought of as representing the sum of the first i terms of a series], has the property that the difference between its nth term $F_n(x)$ and every later term $F_{n+r}(x)$, however far from the former, remains smaller than any given quantity if n has been taken large enough, then there is always a certain constant quantity, and indeed only one, which the terms of this series approach, and to which they can come as close as desired if the series is continued far enough.*[15]

Bolzano's proof of uniqueness of the limit is straightforward, but his proof of the existence for each x of a number $X(x)$ to which the series converges is faulty because Bolzano, like Cauchy, had no way of defining an arbitrary real number X. Nevertheless, he did show how to determine the X to within any degree of accuracy d. If n is taken sufficiently large so that $F_{n+r}(x)$ differs from $F_n(x)$ by less than d for every r, then $F_n(x)$ is the desired approximation to X.

Bolzano could now also prove the least upper bound property of the real numbers.

Theorem.   *If a property M does not belong to all values of a variable x, but does belong to all values which are less than a certain u, then there is always a quantity U which is the greatest of those of which it can be asserted that all smaller x have property M.*

Bolzano's proof of the existence of this least upper bound U of all numbers having the property M involves the creation of a series to which the convergence criterion can be applied. Because M is not valid for all x, there must exist a quantity $V = u + D$ such that M is not valid for all x smaller than V. Now Bolzano considers the quantities $V_m = u + D/2^m$ for each positive integer m. If M is valid for all x less than every such quantity, then u itself must be the desired least upper bound. On the other hand, suppose that M is valid for all x less than $u + D/2^m$ but not for all x less than $u + D/2^{m-1}$. The difference between those two quantities is $D/2^m$, so Bolzano next applies this bisection technique to the interval $[u + D/2^m, u + D/2^{m-1}]$ and determines the smallest integer n so that M is valid for all x less than $u + D/2^m + D/2^{m+n}$ but not for all x less than $u + D/2^m + D/2^{m+n-1}$. Continuing this procedure, Bolzano constructs a sequence u, $u + D/2^m$, $u + D/2^m + D/2^{m+n}$, . . . which satisfies his Cauchy criterion and therefore must converge to a value U, a value which he can easily prove satisfies the conditions of the theorem. (Bolzano's proof was modified slightly by Weierstrass in the 1860s to demonstrate what is now called the Bolzano-Weierstrass theorem, that given any bounded infinite set S of real numbers, there exists a real number r in every neighborhood of which there are other points of S.)

The least upper bound principle easily implies the intermediate value theorem. For suppose $f(\alpha) < 0$ and $f(\beta) > 0$. Without loss of generality, we can also assume that $f(x) < 0$ for all $x < \alpha$. Then the property M that $f(x) < 0$ is not satisfied by all x but is satisfied by all x smaller than a certain $u = \alpha + \omega$, where $\omega < \beta - \alpha$ (because f is assumed continuous). It follows that there is a value U which is the largest such that $f(x) < 0$ for all $x < U$. It is straightforward to show that $f(U)$ can neither be positive nor negative. Thus $f(U) = 0$ and the theorem is proved.

Cauchy presented a slightly different proof of the same result in an appendix to his text. In the main body of the text, however, he only gave a geometric argument, noting that the curves $y = f(x)$ and $y = 0$ must cross each other. One can only speculate as to which method Cauchy actually presented in his classes.

Given the Cauchy criterion, Cauchy proceeded to develop in his text various tests by which one could demonstrate convergence in particular cases, beginning with tests for series of positive terms, say $u_0, u_1, u_2, \ldots$. The comparison test, that if a given series is term-by-term bounded by a convergent series, then the given series is itself convergent, was used by Cauchy without any particular comment. His most common comparison was to a geometric series with ratio less than 1, a series whose convergence Cauchy proved initially by use of the Cauchy criterion. In fact, he used the comparison test to demonstrate the validity of many of his other tests. For example, Cauchy proved the root test, that if the greatest of the limits of the values $\sqrt[n]{u_n}$ is a number k less than 1, then the series converges. Choosing a number U such that $k < U < 1$, Cauchy noted that for n sufficiently large, $\sqrt[n]{u_n} < U$ or $u_n < U^n$. It then follows by comparison with the convergent geometric series

$1 + U + U^2 + U^3 + \cdots$ that the given series also converges. Similarly, if the limit of the roots is a number greater than 1, then the series diverges.

Cauchy used a similar proof to demonstrate the ratio test: "If, for increasing values of n, the ratio u_{n+1}/u_n converges to a fixed limit k, the series u_n will be convergent if one has $k < 1$ and divergent if one has $k > 1$."

For series involving positive and negative terms, Cauchy deals with the idea of absolute convergence (although not with that terminology), adapts the root and ratio test to that case, proves the alternating series test, and shows how to calculate the sum and product of two convergent series. He also adapts his various tests to sequences of functions. In particular, he shows how to find the interval of convergence of a power series. Although some of the individual results on series had been known previously, Cauchy was the first to organize them into a coherent theory which allowed him and others to generalize to the case of series of complex numbers and functions.

There is one significant result in the *Cours d'Analyse*, however, which, as Abel noted in 1826, is incorrect as stated:

Theorem 6-1-1. *When the different terms of the series* $[\sum_{n=0}^{\infty} u_n]$ *are functions of the same variable x, continuous with respect to that variable in the neighborhood of a particular value for which the series is convergent, the sum s of the series is also, in the neighborhood of this particular value, a continuous function of x.*

Cauchy's "proof" of this result is quite simple. We will present his argument in words as Cauchy did and then translate the words into modern symbols. Writing s_n as the sum of the first n terms of the series and s as the sum of the entire series, Cauchy denotes by r_n the remainder $s - s_n$. (Here s, s_n, and r_n are all functions of x.) To prove continuity of s he needs to show that an infinitely small increment in x leads to an infinitely small increment in $s(x)$, that is, given $\epsilon > 0$,

$$\exists \delta \quad \text{such that} \quad \forall a, |a| < \delta \Rightarrow |s(x + a) - s(x)| < \epsilon. \tag{16.1}$$

Although in certain earlier theorems, Cauchy actually calculated appropriate values for δ, in this case he just attempted an argument using arbitrary infinitely small quantities. Thus, he wrote that an infinitely small increment α of x leads to an infinitely small increment of $s_n(x)$ because the latter is continuous for every n, or

$$\exists \delta \quad \text{such that} \quad \forall a, |a| < \delta \Rightarrow |s_n(x + a) - s_n(x)| < \epsilon. \tag{16.2}$$

Next, because the series converges for any x, r_n will itself be infinitely small for n large enough and will remain so for an infinitely small increment of x, or

$$\exists N \quad \text{such that} \quad \forall n, n > N \Rightarrow |r_n(x)| < \epsilon \text{ and } |r_n(x + a)| < \epsilon. \tag{16.3}$$

Because the increment of s is the sum of those of s_n and r_n, Cauchy concludes that this increment is also infinitely small and therefore that s is itself continuous. We would write

$$|s(x + a) - s(x)| = |s_n(x + a) + r_n(x + a) - s_n(x) - r_n(x)|$$

$$\leq |s_n(x + a) - s_n(x)| + |r_n(x + a)| + |r_n(x)| \leq \epsilon + \epsilon + \epsilon = 3\epsilon. \tag{16.4}$$

What Cauchy failed to notice was that the δ in equation 16.2 depends on ϵ, x, and n, while the N in equation 16.3 depends on ϵ, x, and a. Unless we know that there is some value M

which will work in equation 16.3 for all a with $|a| < \delta$, we cannot assert the truth of equation 16.4 (or equation 16.1). Cauchy's arguments with infinitely small increments obscured the needed relationships among the various quantities involved. For the proof to be valid, a notion of uniform convergence was necessary, that is, one needs the additional hypothesis that the number M can be chosen independently of x, at least in some fixed interval. We will return to this question in section 16.1.8.[16]

16.1.4 Derivatives

Cauchy's *Cours d'Analyse* provided a treatment of the basic ideas of functions and series. In his 1823 text *Résumé des Leçons donnees a L'École Royale Polytechnique sur le Calcul Infinitésimal*, Cauchy applied his new ideas on limits to the study of the derivative and the integral, the two basic concepts of the infinitesimal calculus.

After beginning this text with the same definition of continuity as in the earlier one, Cauchy proceeded in Lesson 3 to define the derivative of a function, what he called the *fonction dérivée*, as the limit of $[f(x + i) - f(x)]/i$ as i approaches the limit of 0, as long as this limit exists. Just as he did in his definition of continuity, Cauchy defined the concept of a derivative over an interval, in fact an interval in which the function f is continuous. He noted that this limit will have a definite value for each value of x, and therefore is a new function of that variable, a function for which he used Lagrange's notation $f'(x)$. The definition itself, although it can be thought of as expressing the quotient of infinitesimal differences as in Euler's work, is more directly taken from that section of Lagrange's *Analytic Functions* in which Lagrange, as part of his power series expansion of f, showed that $f(x + i) = f(x) + if'(x) + iV$ where V is some function which goes to zero with i. Cauchy was able to translate this theorem about derivatives into a definition of the derivative. He then calculated the derivative of several elementary functions. For example, if $f(x) = \sin x$, then the quotient of the definition reduces to

$$\frac{\sin \frac{1}{2}i}{\frac{1}{2}i} \cos(x + \frac{1}{2}i),$$

whose limit $f'(x)$ is seen to be $\cos x$.

There is, of course, nothing new about Cauchy's calculations of derivatives. Nor is there anything particularly new about the theorems Cauchy was able to prove about derivatives. Lagrange had derived the same results from his own definition of the derivative. But because Lagrange's definition of a derivative rested on the false assumption that any function could be expanded into a power series, the significance of Cauchy's work lies in his explicit use of the modern definition of a derivative, translated into the language of inequalities through Cauchy's definition of limit, to prove theorems. The most important of these results, in terms of its later use, was in Lesson 7.

Theorem.   *If the function $f(x)$ is continuous between the values $x = x_0$ and $x = X$, and if we let A be the smallest, B the largest value of the derivative $f'(x)$ in that interval, then the ratio of the finite differences*

$$\frac{f(X) - f(x_0)}{X - x_0}$$

must be between A and B.[17]

Cauchy's proof of this theorem is the first to use the δ and ε so familiar to today's students. Cauchy began by choosing δ and ε so that for all values of i with $|i| < δ$ and for any value of x in the interval $[x_0, X]$, it is true that

$$f'(x) - ε < \frac{f(x + i) - f(x)}{i} < f'(x) + ε.$$

That such values exist follows from Cauchy's definition of the derivative as a limit. Note, however, that Cauchy used the fact, implicit in his definition of derivative on an interval rather than at a point, that, given ε, the same δ works for every x in the interval. In any case, Cauchy next interposed $n - 1$ new values $x_1 < x_2 < \cdots < x_{n-1}$ between x_0 and $x_n = X$ with the property that $x_i - x_{i-1} < δ$ for each i and applied the above inequality to the subintervals determined by each successive pair of values. It follows that for $i = 1, 2, \ldots, n$,

$$A - ε < \frac{f(x_i) - f(x_{i-1})}{x_i - x_{i-1}} < B + ε.$$

Cauchy then used an algebraic result to conclude that the sum of the numerators divided by the sum of the denominators also must satisfy the same inequality:

$$A - ε < \frac{f(X) - f(x_0)}{X - x_0} < B + ε.$$

Because this result is true for every ε, the conclusion of the theorem follows.

As an immediate consequence of this theorem, Cauchy derived the mean value theorem for derivatives. Assuming that $f'(x)$ is continuous in the given interval, an assumption which of course justifies the assumption that it has a largest and smallest value, Cauchy used the intermediate value theorem to conclude that

$$\frac{f(X) - f(x_0)}{X - x_0} = f'(x_0 + θ(X - x_0))$$

for some value θ between 0 and 1. Using this mean value theorem, Cauchy then proved that a function with positive derivative on an interval is increasing there, one with negative derivative is decreasing, and one with zero derivative is constant.

16.1.5 Integrals

Cauchy's treatment of the derivative, although using his new definition of limits, was closely related to the treatments in the works of Euler and Lagrange. Cauchy's treatment of the integral, on the other hand, broke entirely new ground. Recall that the standard eighteenth century definition of integration was simply the inverse of differentiation. Even Lacroix wrote that "the integral calculus is the inverse of the differential calculus, its object being to ascend from the differential coefficients to the function from which they are derived."[18] Although Leibniz had developed his notation to remind one of the integral as an infinite sum of infinitesimal areas, the problems inherent in the use of infinities convinced eighteenth century mathematicians to take the notion of the indefinite integral, or antiderivative, as their basic notion for the theory of integration. It was recognized, naturally, that one could evaluate areas not only by use of antiderivatives but also by various approximation techniques. But it was Cauchy who first took these techniques as fundamental and proceeded to construct a theory of definite integrals upon them.

There are probably several reasons why Cauchy felt compelled to define the integral as the limit of a sum rather than in terms of antiderivatives. First, there were many situations where it was clear that an area under a curve made sense even though it could not be calculated by evaluating an antiderivative at the endpoints of an interval; such was the case in particular for certain piecewise continuous functions which showed up in the work of Fourier on series of trigonometric functions. A second reason may well have developed out of Cauchy's work, to be discussed in section 16.3.2, in developing a theory of integrals of complex functions. Finally, Cauchy may have realized in the course of organizing his material for lectures at the École Polytechnique that there is no guarantee that an antiderivative exists for every function. Cauchy's own explanation of his reason for choosing to define an integral in terms of a sum, however, was that it works. As he wrote in an 1823 article, "it seems to me that this manner of conceiving a definite integral [as the sum of the infinitely small values of the differential expression placed under the integral sign] ought to be adopted in preference . . . because it is equally suitable to all cases, even to those in which we cannot pass generally from the function placed under the ∫ sign to the primitive function."[19] Furthermore, he noted, "when we adopt this manner of conceiving definite integrals, we easily demonstrate that such an integral has a unique and finite value, whenever, the two limits of the variable being finite, the function under the ∫ sign itself remains finite and continuous in the entire interval included between these values."[20]

In the second part of his *Résumé*, Cauchy presented the details of a rigorous definition of the integral using sums. Cauchy probably took his definition from the work on approximations of definite integrals by Euler and by Lacroix. But rather than consider this method a way of approximating an area, presumably understood intuitively to exist, Cauchy made the approximation into a definition. Thus, supposing that $f(x)$ is continuous on $[x_0, X]$, he took $n - 1$ new intermediate values $x_1 < x_2 < \cdots < x_{n-1}$ between x_0 and $x_n = X$ and formed the sum

$$S = (x_1 - x_0)f(x_0) + (x_2 - x_1)f(x_1) + \cdots + (X - x_{n-1})f(x_{n-1}).$$

Cauchy noted that S depends both on n and on the particular values x_i selected. But, he wrote, "it is important to observe that if the numerical values of the elements $[x_{i+1} - x_i]$ become very small and the number n very large, the method of division will have only an insensible influence on the value of S."[21]

To prove this result, Cauchy noted that if one chose a new subdivision of the interval by subdividing each of the original subintervals, the corresponding sum S' could be re-written in the form

$$S' = (x_1 - x_0)f(x_0 + \theta_0(x_1 - x_0)) + (x_2 - x_1)f(x_1 + \theta_1(x_2 - x_1)) + \cdots$$
$$+ (X - x_{n-1})f(x_{n-1} + \theta_{n-1}(X - x_{n-1})),$$

where each θ_i is between 0 and 1. By the definition of continuity, this expression can be rewritten as

$$S' = (x_1 - x_0)[f(x_0) + \epsilon_0] + (x_2 - x_1)[f(x_1) + \epsilon_1] + \cdots + (X - x_{n-1})[f(x_{n-1}) + \epsilon_{n-1}]$$

$$= S + (x_1 - x_0)\epsilon_1 + (x_2 - x_1)\epsilon_2 + \cdots + (X - x_{n-1})\epsilon_{n-1}$$

$$= S + (X - x_0)\epsilon'$$

where ϵ' is a value between the smallest and largest of the ϵ_i. Cauchy then argued that if the subintervals are sufficiently small, the ϵ_i and consequently ϵ' will be very close to zero, so that the taking of a subpartition does not change the value of the sum appreciably. Given any two sufficiently small subdivisions, one can take a third that subdivides each. The value of the sum for this third is then arbitrarily close to the values for each of the original two. It follows that "if we let the numerical values of [the lengths of the subdivisions] decrease indefinitely by increasing their number, the value of S ultimately becomes sensibly constant or, in other words, it will end by attaining a certain limit that will depend uniquely on the form of the function $f(x)$ and the extreme values x_0, X attributed to the variable x. This limit is what we call the definite integral [written $\int_{x_0}^{X} f(x)\, dx$]."[22] This definition thus uses a generalization of Cauchy's criterion for convergence to sequences not necessarily indexed by the natural numbers.

With the integral now defined in terms of a limit of sums, it was not difficult for Cauchy to prove the mean value theorem for integrals, that

$$\int_{x_0}^{X} f(x)\, dx = (X - x_0)f[x_0 + \theta(X - x_0)]$$

where $0 \le \theta \le 1$, and also the additivity theorem for integrals over intervals. He then easily demonstrated the

Fundamental Theorem of Calculus.   *If $f(x)$ is continuous in $[x_0, X]$, if $x \in [x_0, X]$, and if*

$$F(x) = \int_{x_0}^{x} f(x)\, dx,$$

then $F'(x) = f(x)$.

To prove the theorem, Cauchy used the mean value theorem and additivity to get

$$F(x + \alpha) - F(x) = \int_{x}^{x+\alpha} f(x)\, dx = \alpha f(x + \theta\alpha).$$

If one divides both sides by α and passes to the limit, the conclusion follows from the continuity of $f(x)$. This version of the fundamental theorem can be considered the first one meeting modern standards of rigor, because it was the first in which $F(x)$ was clearly defined through an existence proof for the definite integral.

Although one of Cauchy's original hypotheses for the existence of a definite integral of $f(x)$ was that $f(x)$ be continuous on the interval of integration, Cauchy also realized that his definition made sense even if that condition were relaxed somewhat. Hence he showed that if $f(x)$ had finitely many discontinuities in the given interval, the integral could still be defined by breaking the interval into subintervals at the points of discontinuity and defining the integral by a further limit argument. Thus, for example, if $f(x)$ is continuous on $(a, b]$, Cauchy made the definition

$$\int_a^b f(x) \, dx = \lim_{\epsilon \to 0} \int_{a+\epsilon}^b f(x) \, dx.$$

Cauchy made similar definitions for integrals of functions over infinite intervals.

Cauchy used a different generalization of his definition of an integral in his presentation of a new method of integrating the differential equation $A \, dx + B \, dy$ in the case where $\partial A / \partial y = \partial B / \partial x$. Cauchy showed that the desired integral $f(x, y)$ could be defined by taking definite integrals in the plane from a fixed point (x_0, y_0):

$$f(x, y) = \int_{x_0}^x A(x, y) \, dx + \int_{y_0}^y B(x_0, y) \, dy.$$

Because this definition implies that

$$\frac{\partial f}{\partial x} = A(x, y)$$

and

$$\frac{\partial f}{\partial y} = \int_{x_0}^x \frac{\partial A}{\partial y}(x, y) \, dx + B(x_0, y) = \int_{x_0}^x \frac{\partial B}{\partial x}(x, y) \, dx + B(x_0, y) =$$

$$B(x, y) - B(x_0, y) + B(x_0, y) = B(x, y),$$

this function indeed solves the problem.

The *Cours d'Analyse* and the *Résumé* were the bases of Cauchy's first year course at the École Polytechnique. In the second year, he gave a detailed introduction to the theory of differential equations. Much of that course was concerned with the standard techniques for solving such equations already developed in the eighteenth century, including, of course, the theorem just mentioned. But with his new concern for rigor in analysis, Cauchy wanted to determine the conditions under which one could prove the existence of a solution to $y' = f(x, y)$ satisfying a prescribed initial condition. The approximation technique he used in his proof of beginning with the given initial point (x_0, y_0) and constructing small straight line segments to approximate the desired curve had in essence been used in the eighteenth century. It was Cauchy who was able to demonstrate, however, by use of a version of the Cauchy criterion, that if both $f(x, y)$ and $\partial f(x, y)/\partial y$ are finite, continuous, and bounded in a region of the plane containing (x_0, y_0), then the approximation method produces polygons which converge to a solution curve to the differential equation in some neighborhood of (x_0, y_0) contained in the original region.

There is a curious story connected with Cauchy's treatment of differential equations. Cauchy never published an account of this second year course, and it is only recently that proof sheets for the first 13 lectures of the course have come to light. It is not clear why these notes stop at this point, but there is evidence that Cauchy was reproached by the directors of the school. He was told that, because the École Polytechnique was basically an engineering school, he should use class time to teach applications of differential equations rather than to deal with questions of rigor. Cauchy was forced to conform and

Orphaned at the age of 9, Fourier was placed by the bishop of his home town of Auxerre, 90 miles southwest of Paris, in the local military school, where he soon displayed a talent for mathematics. Although he hoped to become an army engineer, such a career was not available to him at the time because he was not of noble birth. He therefore took up a teaching position. With the outbreak of the Revolution, Fourier became prominent in local affairs. His defense of victims of the Terror in 1794, however, led to his arrest. Fortunately, after the death of Robespierre he was released and was appointed in 1795 as an assistant to Lagrange and Monge at the École Polytechnique.

Three years later, during Napoleon's Egyptian campaign, he became secretary of the Institut d'Égypte, in which position he was able to conduct extensive research on Egyptian antiquities. On his return to France, he served for 12 years as prefect of the department of Isére in southeastern France and succeeded in accomplishing many public improvements. Fortunately, even with the fall of Napoleon, Fourier was able to be elected to the reconstituted Académie des Sciences and, in 1822, became its perpetual secretary, that is, in modern terms, its executive director, a position he held until his death.

announced that he would no longer give completely rigorous demonstrations. He evidently then felt that he could not publish his lectures on the material, because they did not reflect his own conception of how the subject should be handled.

16.1.6 Fourier Series and the Notion of a Function

Abel's counterexample to Cauchy's false result on convergence was a Fourier series, a series of trigonometric functions of the type Euler and Daniel Bernoulli argued about in the middle of the eighteenth century. It was Joseph Fourier (1768–1830) who made a detailed study of these series in connection with his investigation of heat diffusion early in the nineteenth century. He first presented his work to the French Academy in 1807 and later reworked and expanded it into his *Théorie analytique de la chaleur* (*Analytic theory of heat*) in 1822. Fourier began by considering the special case of the temperature distribution $v(t, x, y)$ at time t in a rectangular lamina infinite in the positive x direction, of width 2 in the y direction, with the edge $x = 0$ being maintained at a constant temperature 1 and the edges $y = \pm 1$ being kept at temperature 0. By making certain assumptions about the flow of heat, Fourier was able to show that v satisfies the partial differential equation

$$\frac{\partial v}{\partial t} = \frac{\partial^2 v}{\partial x^2} + \frac{\partial^2 v}{\partial y^2}.$$

Fourier then solved this equation under the condition that the temperature of the lamina had reached equilibrium, that is, that $\partial v/\partial t = 0$. Assuming that $v = \phi(x)\psi(y)$ (the method of separation of variables), he differentiated twice with respect to each variable to get $\phi''(x)\psi(y) + \phi(x)\psi''(y) = 0$ or

$$\frac{\phi(x)}{\phi''(x)} = -\frac{\psi(y)}{\psi''(y)} = A$$

for some constant A. The obvious solutions to these equations are $\phi(x) = \alpha e^{mx}$, $\psi(y) = \beta \cos ny$, where $m^2 = n^2 = 1/A$. Physical reasoning dictates that m be negative (for otherwise the temperature would tend toward infinity for x large), so Fourier concluded that the general solution of the original partial differential equation is a sum of functions of the type $v = ae^{-nx} \cos ny$. By using the boundary conditions $v = 0$ when $y = \pm 1$, Fourier showed that n must be an odd multiple of $\pi/2$ and therefore that the general solution is given by the infinite series

$$v = a_1 e^{-(\pi x/2)} \cos\left(\frac{\pi y}{2}\right) + a_2 e^{-(3\pi x/2)} \cos\left(\frac{3\pi y}{2}\right) + a_3 e^{-(5\pi x/2)} \cos\left(\frac{5\pi y}{2}\right) + \cdots.$$

To determine the coefficients a_i, Fourier used the additional boundary condition that $v = 1$ when $x = 0$. With $u = \pi y/2$, that implied that the a_i satisfied the equation

$$1 = a_1 \cos u + a_2 \cos 3u + a_3 \cos 5u + \cdots,$$

a single equation for infinitely many unknowns which Fourier turned into infinitely many equations by differentiating it infinitely often and each time setting $u = 0$. By noting the patterns determined by solving the first several of these equations, Fourier was able to determine that $a_1 = 4/\pi$, $a_2 = -(4/3\pi)$, $a_3 = 4/5\pi$, ... and therefore solve the partial differential equation. But in the usual spirit of mathematicians, once the original problem was solved he began to consider the mathematical ramifications of his new type of solution. First, he noted that his values for the coefficients implied that

$$\cos u - \frac{1}{3} \cos 3u + \frac{1}{5} \cos 5u - \cdots = \frac{\pi}{4}$$

with $u \in \left(-\frac{\pi}{2}, \frac{\pi}{2}\right)$. But the same series clearly represents 0 for $u = \frac{\pi}{2}$ and $-\frac{\pi}{4}$ for $u \in \left(\frac{\pi}{2}, \frac{3\pi}{2}\right)$. Fourier realized that this result would not be immediately believable to his readers, however: "As these results appear to depart from the ordinary consequences of the calculus, it is necessary to examine them with care and to interpret them in their true sense. One considers the equation $y = \cos u - \frac{1}{3} \cos 3u + \frac{1}{5} \cos 5u - \frac{1}{7} \cos 7u + \cdots$ as that of a line of which u is the abscissa and y the ordinate. One sees ... that this line must be composed of separate parts aa, bb, cc, dd, ... of which each is parallel to the axis and equal to $[\pi]$. These parallels are placed alternatively above and below the axis at a distance of $[\pi/4]$ and are joined by the perpendiculars ab, cb, cd, ed, ... which are themselves part of the line."[23] In other words, Fourier considered the infinite cosine series to represent the "square wave" of Figure 16.3. To a modern reader, it is not clear why Fourier drew in the

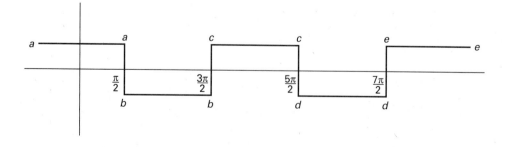

FIGURE 16.3
Fourier's square wave:
$y = \cos u$
$- \frac{1}{3} \cos 3u$
$+ \frac{1}{5} \cos 5u$
$- \frac{1}{7} \cos 7u + \cdots.$

line segments perpendicular to the abscissa, for the value of the series at $u = k\pi/2$ for k odd is always 0. Fourier, however, realizing that the partial sums of this series could always be represented by a curve without breaks, thought that the infinite sum too should be represented by such a curve. The questions of whether this curve represented a "function" in the modern sense or whether the series represented a "continuous" function in Cauchy's sense were not relevant to Fourier's work. He was interested in a physical problem and probably conceived of this solution in geometrical terms, where he could draw a "continuous" curve without worrying about whether it represented a "function."

Fourier next investigated the representation of various functions by series of trigonometric functions, the most general being the series of the form $c_0 + c_1 \cos x + c_2 \cos 2x + \cdots + d_1 \sin x + d_2 \sin 2x + \cdots$. Usually, however, Fourier limited himself to either a sine series or a cosine series. Furthermore, since he wanted to convince his readers of the validity of his methods, he determined anew the coefficients of his original series and others by methods different from his original method. For example, he showed that if one writes the function $(1/2)\pi\phi(x)$ as a sine series

$$\frac{1}{2}\pi\phi(x) = a_1 \sin x + a_2 \sin 2x + a_3 \sin 3x + \cdots,$$

one can multiply both sides by $\sin nx \, dx$ for each integer n in turn and integrate over the interval $[0, \pi]$. Because

$$\int_0^\pi \sin mx \sin nx \, dx = \begin{cases} 0, & \text{if } m \neq n; \\ \pi/2, & \text{if } m = n, \end{cases}$$

it follows that

$$a_k = \int_0^\pi \phi(x) \sin kx \, dx,$$

as long as the integrals representing the coefficients make sense, that is, as long as the area under $\phi(x) \sin kx$ is well-defined.

To clarify what he meant by a function, Fourier gave a definition: "In general, the function $f(x)$ represents a succession of values or ordinates each of which is arbitrary. An infinity of values being given to the abscissa x, there is an equal number of ordinates $f(x)$. All have actual numerical values, either positive or negative or null. We do not suppose these ordinates to be subject to a common law; they succeed each other in any manner whatever, and each of them is given as if it were a single quantity."[24] Despite this modern sounding definition, Fourier himself never considered what would today be called "arbitrary functions." His examples show that he only intended to consider piecewise continuous functions. And, of course, Fourier only asserted that the series represented the given arbitrary function on the interior of a particular finite interval, such as $[0, \pi]$. The value of the series at the endpoints could easily be calculated separately, while the periodicity properties of the sine function enabled one to extend the original function geometrically to the entire real line.

Fourier attempted in certain cases to prove that his expansion actually represented the function by using trigonometric identities to rewrite the partial sum of the first n terms in closed form and then considering the limit as n increased. In general, however, he believed that his explicit calculation of the coefficients in his proposed expansion of an arbitrary

Because the level of mathematics instruction in his native Germany was generally low during his youth, Dirichlet went to Paris in 1822 to study at the Collége de France. He became a tutor to the children of a famous French general and was thus able to meet many of the most prominent French intellectual figures, including Joseph Fourier, who ultimately proved the strongest influence on his mathematics. Dirichlet returned to Germany in 1825 and three years later was appointed to the faculty of the University of Berlin, a position he held for 27 years. He generally had a very heavy teaching load in Berlin, partly because he taught at the military academy as well as the university. He was therefore willing to accept an invitation to move to Göttingen upon Gauss' death in 1855, where he had increased time for research. Unfortunately, his time there lasted only three and a half years until his death in 1859.

function in terms of integrals which represented real areas was a convincing enough argument that the expansion was valid. Thus, for example, he calculated the expansion given later by Abel:

$$\frac{1}{2}x = \sin x - \frac{1}{2}\sin 2x + \frac{1}{3}\sin 3x - \frac{1}{4}\sin 4x + \cdots.$$

This series represented $(1/2)x$ on $[0, \pi/2]$ and the function of Figure 16.4 over the entire real line. Abel realized that not only did this function violate Cauchy's result on the sum of a series of continuous functions, but also that Fourier's attempts at a proof that the Fourier series converged to the original function were not sufficient.

A few years later, Dirichlet considered the same question, but rather than try to show that the Fourier series for an "arbitrary" function converged, he lowered his sights drastically and found sufficient conditions on the function which would assure this convergence. In particular, he showed that if a function $f(x)$ defined on $[-\pi, \pi]$ was continuous and bounded on that interval, except perhaps for a finite number of finite discontinuities, and in addition had only a finite number of turning values in that interval, then the Fourier

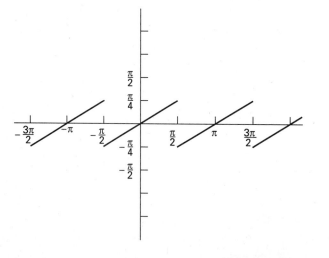

FIGURE 16.4
Fourier's graph of $y = \sin x - \frac{1}{2}\sin 2x + \frac{1}{3}\sin 3x - \frac{1}{4}\sin 4x + \cdots$.

series converged at each x in $(-\pi, \pi)$ to the value $\lim_{\epsilon \to 0} \frac{1}{2}[f(x - \epsilon) + f(x + \epsilon)]$. (This value is equal to $f(x)$ if f is continuous at x.) The result held at the endpoints as well if, like Fourier, one interpreted $f(x)$ as being geometrically periodic outside the given interval. Dirichlet chose his conditions in order to be able to integrate products of the given function with trigonometric functions over certain intervals. And Cauchy's new formulation of the definite integral only guaranteed the existence of the integral for functions with finitely many discontinuities. Dirichlet realized the difficulties of extending this result to functions with infinitely many discontinuities in a given interval. In fact, he provided an example of a function which did not satisfy his original conditions, a function which is continuous nowhere: "$f(x)$ equals a determined constant c when the variable x takes a rational value, and equals another constant d, when this variable is irrational. The function thus defined has finite and determined values for every value of x, and meanwhile one cannot substitute it into the series, seeing that the different integrals which enter into that series lose all significance in this case."[25]

16.1.7 The Riemann Integral

In 1853, Georg Bernhard Riemann (1826–1866) attempted to generalize Dirichlet's result by first determining precisely which functions were integrable according to Cauchy's definition of the integral $\int_a^b f(x)\, dx$. He began, in fact, by changing the definition somewhat. Like Cauchy, he divided the interval $[a, b]$ into n subintervals $[x_{i-1}, x_i]$ for $i = 1, 2, \ldots, n$. Setting $\delta_i = x_i - x_{i-1}$, he now considered the sum

$$S = \sum_{i=1}^{n} \delta_i f(x_{i-1} + \epsilon_i \delta_i)$$

where each ϵ_i is between 0 and 1. This sum is more general than Cauchy's because Riemann allowed the argument of the function f to take any value in the relevant subinterval. He then defined the integral to be the limit to which S tends, provided it exists, no matter how the δ_i and ϵ_i are taken. Riemann now asked a question which Cauchy had not: In what cases is a function integrable and in what cases not? Cauchy himself had only shown that a certain class of functions was integrable, but had not tried to find all such functions. Riemann, on the other hand, formulated a necessary and sufficient condition for a finite function $f(x)$ to be integrable: "If, with the infinite decrease of all the quantities δ, the total size s of the intervals in which the variations of the function $f(x)$ are greater than a given quantity σ always becomes infinitely small in the end, then the sum S converges when all the δ become infinitely small"[26] and conversely. (The **variation** of a function in an interval is the difference between the maximum and minimum values of the function on that interval.) As an example of a function defined on $[0, 1]$ which does not satisfy Cauchy's criterion for integrability but is Riemann-integrable, Riemann gave

$$f(x) = \sum_{n=1}^{\infty} \frac{\phi(nx)}{n^2},$$

where $\phi(x)$ is defined to be x minus the nearest integer, or, if there are two equally close integers, is defined to be equal to 0. It turns out that f is continuous everywhere except at the infinitely many points $x = p/2n$ with p and n relatively prime. But, because the variation of f near such a point is equal to $\pi^2/8n^2$, the points near which the variation is

---

Biography   Karl Weierstrass (1815–1897)

Born in Westphalia, Germany, Weierstrass entered the University of Bonn in 1834 at his father's urging to study public finance and administration with the aim of a career in the Prussian civil service. His natural bent for mathematics, combined with his interest in the cameraderie of the Bonn taverns, prevented him from doing well in his intended field. He left Bonn in 1838 with no degree. To earn a living, therefore, he studied for a teaching certificate and, beginning in 1841, taught such subjects as mathematics, physics, German, botany, geography, history, gymnastics, and calligraphy at various gymnasia for the next 14 years. After a series of brilliant mathematics papers in Crelle's *Journal*, he was awarded an honorary doctorate by the University of Königsberg in 1854 and finally received an appointment as an *extraordinarius* (associate professor) at the University of Berlin in 1856. Although health problems caused him to teach while seated, with an advanced student writing on the blackboard, his clear lectures soon won him a European wide reputation, and his classes were attended each year by hundreds of students. In 1861, he, along with Kummer, introduced the mathematics seminar to the University of Berlin, a factor in making the university the premier university in the world for pure mathematics during the late nineteenth century.

---

greater than any $\sigma > 0$ are finite in number and the function satisfies Riemann's integrability criterion.

With a new class of functions now proved to be integrable, Riemann was able to extend Dirichlet's result on convergence of Fourier series. But rather than determine sufficient conditions on a function which insure the convergence of its series, Riemann attacked the problem in reverse by beginning with a function which is representable by a trigonometric series and attempting to determine what consequences this representation has for the behavior of the function. Riemann was able in this way to find many types of functions which were representable by trigonometric series, but never answered the entire question to his own satisfaction. Probably this was the reason that Riemann never published the manuscript in which this material appears.

16.1.8 Uniform Convergence

The work of Dirichlet and Riemann made it absolutely clear that a Fourier series can represent a discontinuous function and that therefore Cauchy's theorem on the sum of a series of continuous functions had to be modified. In 1847, two mathematicians, Philip Seidel (1821–1896) and George Stokes (1819–1903), independently discovered critical properties of convergence which allowed the sum of a series of continuous functions to be discontinuous at a point. It was, however, Karl Weierstrass who realized how to insure that the sum function was continuous over an entire interval as Cauchy had originally stated his theorem. In the course of his lectures at the University of Berlin beginning in the 1850s, Weierstrass made a careful distinction between convergence of a series of numbers and that of a series of functions, a distinction which Cauchy had glossed over. He was then able to identify a crucial property of convergence of functions, the property of uniform convergence over an interval: An infinite series $\Sigma u_n(x)$, is **uniformly convergent** in $[a, b]$ when, given any $\epsilon > 0$, there exists an N (dependent on ϵ) such that $|r_n(x)| < \epsilon$ for every $n > N$ and for every x in the interval $[a, b]$.

Biography   Sofia Kovalevskaya (1850–1891)

While some little girls awoke to delicate flowers on their nursery walls, Sofia Vasilevna Korvin's room was papered with the lecture notes from Mikhail Ostrogradsky's course in calculus. Her father, an army officer, had liked mathematics and allowed Sofia to study with a tutor. She grew to like mathematics too, but could not pursue her studies because she was a woman. Russian universities did not yet permit women to attend officially, and her family was not about to let her go off alone to a European university. Sofia solved this problem by contracting a "marriage of convenience" with Vladimir Kovalevsky, a publisher of scientific and political works and himself an aspiring scientist.

With a husband, Sofia was able to go abroad and study mathematics, first at the University of Heidelberg and then in Berlin with Weierstrass. Even though Weierstrass and others made a strong case for her to the faculty senate, the University of Berlin, unlike that of Heidelberg, refused to admit a woman officially. Sofia studied privately with Weierstrass and, after writing several publishable mathematics papers, the most significant being on the theory of partial differential equations, received her doctorate in 1874 from the University of Göttingen, a university which was willing to grant doctorates *in absentia*. Sofia Kovalevskaya was the first European woman since the Renaissance to earn a Ph.D. in mathematics.

Returning to Russia, the Kovalevskys became connubial enough to produce a daughter in 1878. A few years later, Sofia resumed the mathematical research she had briefly set aside in favor of domestic and societal concerns. Upon the death of her husband in 1883, Sofia was able to secure a position as a professor at the University of Stockholm, another first for a woman in modern times. She soon became an active member of the European mathematical community, serving as an editor of the Swedish journal *Acta Mathematica* and receiving the Prix Bordin of the French Academy of Sciences in 1888 for her work on the revolution of a solid body about a fixed point.

Life was difficult for Sofia Kovalevskaya as a single mother. As she wrote in a letter to a friend, "All these stupid but unpostponable practical affairs are a serious test of my patience, and I begin to understand why men treasure good, practical housewives so highly. Were I a man, I'd choose myself a beautiful little housewife who would free me from all this."[27] Unfortunately, Kovalevskaya had little time to fulfill her mathematical promise. In early 1891, she contracted pneumonia on a trip to France and Germany and died a few days after her return to Sweden. For years after her untimely death, her grave in Stockholm remained a place of pilgrimage, not only for mathematicians but also for supporters of women's rights (Figure 16.5).

FIGURE 16.5
Sofia Kovalevskaya on a Russian stamp.

Given Weierstrass' definition, it was simple enough to correct Cauchy's proof for the case where the convergence of the series was uniform. But this definition also had a deeper influence. Not only did Weierstrass make absolutely clear how certain quantities in his definition depended on other quantities, but he also completed the transformation away from the use of terms such as "infinitely small." Henceforth, all definitions involving such ideas were given arithmetically. For example, Eduard Heine (1821–1881), a student of Weierstrass, not only gave a definition of continuity at a point in a paper of 1872, but also reworked Cauchy's definition of continuity over an interval into the following: "A function $f(x)$ is called . . . **uniformly continuous** from $x = a$ to $x = b$ if for any given quantity ϵ, however small, there exists a positive quantity η_0 such that for all positive values η which are less than η_0, $f(x \pm \eta) - f(x)$ remains less than ϵ. Whichever value one may give to x, assuming only that x and $x \pm \eta$ belong to the interval from a to b, the same η_0 must effect the required [property]."[28] Heine then went on to prove that a function continuous over a closed interval is uniformly continuous in that interval, making implicit use of what is today called the Heine-Borel theorem, and also that a continuous function on a closed interval attains a maximum and a minimum.

Because Weierstrass himself did not publish many of his new ideas, it was through the efforts of Heine and his other students, including the first woman to earn a doctorate in mathematics, Sofia Kovalevskaya (1850–1891), that Weierstrass' concepts became the standard in mathematical analysis, a standard still in place today.

16.2 THE ARITHMETIZATION OF ANALYSIS

Even with the new definitions of continuity and convergence by Bolzano and Cauchy, it was apparent by the middle of the nineteenth century that there was a crucial step missing in their proofs of such results as the intermediate value theorem and the existence of a limit of a bounded increasing sequence. Although the new definitions enabled mathematicians to show that a certain sequence of numbers satisfied the Cauchy criterion, there was no way to assert the existence of a limit if one could not specify in advance what type of "number" this limit would be. Cauchy, among others, understood intuitively what the real numbers were. He had even asserted that an irrational number can be considered as a limit of a certain sequence of rational numbers. But he was thereby asserting the *a priori* existence of such a number, without any argument as to how that assertion could be justified.

16.2.1 Dedekind Cuts

By the middle of the century, several mathematicians were actively considering this matter of exactly what an irrational number is. They were no longer content to assume the existence of such objects as their eighteenth century predecessors had, particularly because similar such "obvious" assumptions were leading to incorrect conclusions. For example, Weierstrass, in the course of teaching the basics of analysis in Berlin, constructed a function which was everywhere continuous but nowhere differentiable, a function which no one in the previous century would have believed possible. Thus Weierstrass and Dedekind began the detailed consideration of this question of the meaning of the irrationals. As Dedekind wrote in the introduction to his brief work *Stetigkeit und irrationale Zahlen* (*Continuity and Irrational Numbers*), first worked out in 1858 but only published in 1872,

> As professor in the Polytechnic School in Zürich I found myself for the first time obliged to lecture upon the elements of the differential calculus and felt more keenly than ever before the lack of a really scientific foundation for arithmetic. In discussing the notion of the approach of a variable magnitude to a fixed limiting value . . . I had recourse to geometric evidences. Even now such resort to geometric intuition in a first presentation of the differential calculus, I regard as exceedingly useful, from the didactic standpoint, and indeed indispensible if one does not wish to lose too much time. But that this form of introduction into the differential calculus can make no claim to being scientific, no one will deny. . . . The statement is so frequently made that the differential calculus deals with continuous magnitude, and yet an explanation of this continuity is nowhere given; even the most rigorous

expositions of the differential calculus . . . depend upon theorems which are never established in a purely arithmetic manner. . . . It then only remained to discover the true origin [of these theorems] in the elements of arithmetic and thus at the same time to secure a real definition of the essence of continuity.[29]

Dedekind's program of research led to what is now called the **arithmetization of analysis**, the deduction of the theorems of analysis first from the basic postulates defining the integers and then from the principles of the theory of sets.

To secure his definition of the "essence of continuity" in a "purely arithmetic manner," Dedekind began by considering the order properties of the set R of rational numbers. The ones he considered most important were first, that if $a > b$ and $b > c$, then $a > c$; second, that if $a \neq c$, then there are infinitely many rational numbers lying between a and c; and third, that any rational number a divides the entire set R into two classes A_1 and A_2, with A_1 consisting of those numbers less than a and A_2 consisting of those numbers greater than a, the number a itself being assigned to either one of the two classes. These two classes have the obvious property that every number in A_1 is less than every number in A_2. In making the correspondence between rational numbers and points on the line, Dedekind noted that it was clear even to the Greeks that "the straight line L is infinitely richer in point-individuals than the domain R of rational numbers in number-individuals." But one cannot use the geometrical line to define numbers arithmetically. Dedekind's aim, therefore, was "the creation of new numbers such that the domain of numbers shall gain the same completeness, or . . . the same continuity, as the straight line."[30]

To create these new numbers Dedekind decided to transfer to the domain of number the property which he considered the essence of continuity of the straight line, namely, that "if all points of the straight line fall into two classes such that every point of the first class lies to the left of every point of the second class, then there exists one and only one point which produces this division."[31] Thus Dedekind generalized his third property of the rationals into a new definition: "If any separation of the system R into two classes A_1, A_2 is given which possesses only this characteristic property that every number a_1 in A_1 is less than every number a_2 in A_2, then . . . we shall call such a separation a **cut** and designate it by (A_1, A_2)."[32] Every rational number a determines a cut for which a is either the largest of the numbers in A_1 or the smallest of those in A_2, but there are certainly cuts not produced by rational numbers, for example the cut determined by defining A_2 to be the set of all positive rational numbers whose square is greater than 2 and A_1 to be the set of all other rational numbers. "In this property that not all cuts are produced by rational numbers consists the incompleteness or discontinuity of the domain R of all rational numbers. Whenever, then, we have to do with a cut (A_1, A_2) produced by no rational number, we create a new, an *irrational* number α, which we regard as completely defined by this cut (A_1, A_2); we shall say that the number α corresponds to this cut, or that it produces this cut."[33] Dedekind thus considered α to be a new creation of the mind which corresponded to the cut. Other mathematicians, however, felt that it was better to define the real number α to *be* the cut.

In any case, the collection of all such cuts determines the system $\mathfrak{R}$ of real numbers. Dedekind was able to show that this system possessed a natural ordering $<$ which satisfies the three properties of order satisfied by the rational numbers. He then proved that the system $\mathfrak{R}$ also possessed the attribute of continuity, that if it is broken up into two classes

$\mathscr{A}_1$, $\mathscr{A}_2$ such that every number α_1 in $\mathscr{A}_1$ is less than every number α_2 in $\mathscr{A}_2$, then there exists one and only one real number α which is either the greatest number in $\mathscr{A}_1$ or the smallest number in $\mathscr{A}_2$. In fact, α is the real number corresponding to the cut (A_1, A_2) where A_1 consists of all rational numbers in $\mathscr{A}_1$ and A_2 consists of all rational numbers in $\mathscr{A}_2$. It is straightforward to show that α possesses the requisite properties.

Dedekind completed his essay by defining the standard arithmetic operations in his new system $\mathscr{R}$. He was thus able to "arrive at real proofs of theorems (as, for example, $\sqrt{2}\sqrt{3} = \sqrt{6}$), which to the best of my knowledge have never been established before."[34] One of these theorems was that every bounded increasing sequence $\{\beta_i\}$ of real numbers has a limit. Letting $\mathscr{A}_2$ be the set of all numbers γ such that $\beta_i < \gamma$ for all i, and $\mathscr{A}_1$ all remaining numbers, Dedekind easily showed that the cut $(\mathscr{A}_1, \mathscr{A}_2)$ determines the number α which is the least number in $\mathscr{A}_2$ and which is the required limit.

16.2.2 Cantor and Fundamental Sequences

Dedekind's work on cuts appeared in print in 1872. Interestingly enough, the question of defining the real numbers arithmetically was so much in the air at the time that at least four other works accomplishing the same goal were published around that same time, all based on the general idea of defining an irrational number in terms of a sequence satisfying Cauchy's convergence criterion. Charles Meray (1835–1911) was the first to publish (in 1869), followed, in 1872, by Ernst Kossak (1839–1902), who gave an account of Weierstrass' method, Georg Cantor (1845–1918), and Eduard Heine, who used Cantor's method to derive the basic theorems of the arithmetic of the real numbers. We will just discuss Cantor's method here, because it had far-reaching implications for the theory of sets.

Cantor came at the problem of creating the real numbers from a point of view different from that of Dedekind. He was interested in the old problem of the convergence of Fourier series and took up the question of whether a trigonometric series which represents a given function is necessarily unique. In 1870 he managed to prove uniqueness under the assumption that the trigonometric series converged for all values of x. But then he succeeded in weakening the conditions. First, in 1871, he showed that the theorem was still true if either the convergence or the representation failed to hold at a finite number of points in the given interval. Second, in the following year, he was able to prove uniqueness even if the number of these exceptional points was infinite, provided that the points were distributed in a specified way. To describe accurately this distribution of points, Cantor realized that he needed a new way of describing the real numbers.

Beginning like Dedekind with the set of rational numbers, Cantor introduced the notion of a **fundamental sequence**, a sequence $a_1, a_2, \ldots, a_n, \ldots$ with the property that "for any positive rational value ϵ there exists an integer n_1 such that $|a_{m+n} - a_n| < \epsilon$ for $n \geq n_1$, for any positive integer m."[35] Such a sequence, now called a Cauchy sequence, satisfies the criterion Cauchy set out in 1821. For Cauchy, it was obvious that such a sequence converged to a real number b. Cantor, on the other hand, realized that to say this was to commit a logical error, for that statement presupposed the existence of such a real number. Therefore, Cantor used the fundamental sequence to *define* a real number b. In other words, Cantor associated a real number to every fundamental sequence of rational numbers. The rational number r was itself associated to a sequence, the sequence $r, r, \ldots$,

---

Biography   **Georg Cantor (1845–1918)**

Georg Cantor might have followed in the footsteps of his mother's family of musicians as a violinist and sometimes wondered why he did not. Instead, he became interested in mathematics during his school years in St. Petersburg and studied it at the University of Zurich beginning in 1862 and then in Berlin under Weierstrass a year later. It took him only ten years to become a full professor at the University of Halle, but he aspired to a better paying and more prestigious position in Berlin. Kronecker, who opposed Cantor's theory of infinite sets, managed to keep him from the capital. Despite mental illness during his later years, Cantor was able to organize the Association of German Mathematicians in 1890 and was chiefly responsible for setting up the first International Congress of Mathematicians, held in Zurich in 1897.

$r, \ldots$, but there were also sequences which were not associated to rationals. For example, the sequence 1, 1.4, 1.41, 1.414, $\ldots$, generated by a familiar algorithm for calculating $\sqrt{2}$, was such a fundamental sequence.

Realizing that two fundamental sequences could well converge to the same real number, Cantor went on to define an equivalence relation on the set of such sequences. Thus, the number b associated to the sequence $\{a_i\}$ was said to be equal to the number b' associated to the sequence $\{a_i'\}$ if for any $\epsilon > 0$, there exists an n_1 such that $|a_n - a_n'| < \epsilon$ for $n > n_1$. The set B of real numbers was then the set of equivalence classes of fundamental sequences. It was not difficult to define an order relationship on these sequences as well as to establish the basic arithmetic operations. But Cantor wanted to show that the set he had defined was in some sense the same as the number line. It was clear to Cantor that to every point on the line there corresponds a fundamental sequence, but he realized that the converse required an axiom, namely that to every real number (equivalence class of fundamental sequences) there corresponds a definite point on the line.

Having now defined real numbers Cantor returned to his original question in the theory of trigonometric series. By using his identification of the real numbers with the points on the line, he defined the **limit point** of a point set P to be "a point of the line so placed that in every neighborhood of it we can find infinitely many points of P. . . . By the neighborhood of a point should here be understood every interval which has the point in its interior. Thereafter it is easy to prove that a [bounded] point set consisting of an infinite number of points always has at least one limit point."[36] Cantor denoted the set of these limit points by P', calling this set the first **derived set** of P. Similarly, if P' is infinite, Cantor defined the second derived set P'' to be the set of the limit points of P'. (If P' is finite, the set of its limit points is empty.) Continuing in this way, Cantor defined derived sets of any finite order. He then distinguished two types of bounded point sets. Those of the first species were ones for which the derived set $P^{(n)}$ was empty for some value of n, while those of the second species were those which did not satisfy this condition. For example, in the interval [0, 1] the point set 1, 1/2, 1/3, . . . has derived set $\{0\}$ and is therefore of the first species, while the set of rational numbers in that interval has the entire interval as the derived set and is therefore of the second species. Cantor was able to show that point sets of the first species existed for which $P^{(n)}$ was finite for any given n and was then able to demonstrate that the trigonometric series corresponding to a function was unique provided that either the convergence or the representation failed only at a point set of the first species.

16.2.3 Infinite Sets

There were many new questions which occurred to Cantor in dealing with these point sets. For example, he knew that the rational numbers were dense, but they were not continuous. It would seem, therefore, that in some sense there should be "more" real numbers than rational numbers. In November, 1873 he posed this question in a letter to Dedekind: "Take the collection of all positive whole numbers n and denote it by (n); then think of the collection of all real numbers x and denote it by (x); the question is simply whether (n) and (x) may be corresponded so that each individual of one collection corresponds to one and only one of the other? . . . As much as I am inclined to the opinion that (n) and (x) permit no such unique correspondence, I cannot find the reason."[37]

Dedekind could not answer Cantor's question, but only a month later Cantor was able to show that such a correspondence was impossible. His proof was by contradiction. If the real numbers in the interval (a, b) could be put into one-to-one correspondence with the natural numbers, then one could list these real numbers sequentially: $r_1, r_2, r_3, \ldots,$ $r_n, \ldots$. Cantor then proceeded to find a real number in the interval which was not included in the list. He picked the first two numbers a', b' from the sequence such that $a' < b'$. Similarly, he picked the first two numbers a'', b'' in (a', b') such that $a'' < b''$. Continuing in this way he determined a nested sequence of intervals $(a, b), (a', b'), (a'', b''), \ldots$. There were then two possibilities. First, the number of such intervals could be finite. In that case, there was certainly a real number in the smallest interval $(a^{(n)}, b^{(n)})$ which was not in the original list. Second, if the number of such intervals were infinite, they determined two bounded monotonic sequences $\{a^{(i)}\}, \{b^{(i)}\}$, which had the limits $\bar{a}, \bar{b}$, respectively. If $\bar{a} \neq \bar{b}$, then the interval $(\bar{a}, \bar{b})$ surely contained a real number not in the original list. Finally, if $\bar{a} = \bar{b} = \eta$, then η cannot be in the list either. (If it were equal to r_k for some k, it would not be in the intervals past a certain index, while from its definition as a limit it must be in all of the intervals.)

In the paper of 1874 which contained the above proof, Cantor also included a proof that the set of all algebraic numbers can be put into one-to-one correspondence with the set of natural numbers. It followed that there were an infinite number of transcendental numbers. More importantly, however, Cantor had established a technique of counting infinite collections and had determined a clear difference in the size (or cardinality) of the continuum of real numbers on the one hand and the set of rational or algebraic numbers on the other. Shortly afterwards, in another letter to Dedekind, he asked whether it might be possible to find a one-to-one correspondence between a square and an interval. The obvious answer here was "no" and, in fact, some of Cantor's colleagues felt that the question was ridiculous. But within three years Cantor discovered that the answer was "yes." He constructed the correspondence by mapping the pair (x, y) represented by the infinite decimal expansions $x = a_1a_2a_3 \ldots, y = b_1b_2b_3 \ldots$ to the point z represented by the expansion $z = a_1b_1a_2b_2a_3b_3 \ldots$. There was a slight problem with this correspondence, related to the fact that 0.19999 . . . and 0.20000 . . . represented the same number, but

Cantor soon corrected his proof and established that a one-to-one correspondence existed. Although the result surprised the mathematical community, Dedekind pointed out that Cantor's mapping was discontinuous and therefore had little to do with the geometric meaning of dimension. In fact, several mathematicians soon offered proof that no such continuous map was possible and that, therefore, dimension was invariant under continuous, one-to-one correspondences.

16.2.4 The Theory of Sets

Cantor soon realized that his concept of a one-to-one correspondence could be placed at the foundation of a new theory of sets. In 1879, he used it to begin the study of the cardinality of an infinite set. Two sets A and B were defined to be of the **same power** if there was a one-to-one correspondence between the elements of A and the elements of B. Cantor initially singled out two special cases, those sets of the same power as the set N of natural numbers—these were called denumerable sets—and those of the power of the set of real numbers. In his further attempt to understand the properties of the continuum, Cantor was led to establish over the next two decades a detailed theory of infinite sets, of which only the beginnings can be discussed here. Much of this theory was outlined in two papers published in 1895 and 1897 and collectively titled *Beiträge zur Begründung der transfiniten Mengenlehre* (*Contributions to the Founding of Transfinite Set Theory*).

The *Beiträge* began with a definition of a set as "any collection into a whole M of definite and separate objects m of our intuition or our thought."[38] "Every set M has a definite 'power,' which we will also call its 'cardinal number,' " he continued. "We will call by the name 'power' or 'cardinal number' of M the general concept which, by means of our active faculty of thought, arises from the set M when we make abstraction of the nature of its various elements m and of the order in which they are given."[39] By this "abstraction" Cantor meant that the cardinality of an infinite set was the generalization of the concept of "number of elements" for a finite set. Thus the set of natural numbers and the set of real numbers have different cardinal numbers. That of the set of natural numbers, the set of smallest transfinite cardinality, Cantor called "aleph-zero," written $\aleph_0$, while that of the real numbers is denoted $\mathscr{C}$. Two sets are equivalent, or have the same cardinality, if there is a one-to-one correspondence between them. Cantor also defined the notion of $<$ for transfinite cardinals: The cardinality $\overline{\overline{M}}$ of a set M is less than that of a set N if there is no part of M which is equivalent to N while there is a part of N which is equivalent to M. It is then clear that for two sets M and N no more than one of the relations $\overline{\overline{M}} = \overline{\overline{N}}$, $\overline{\overline{M}} < \overline{\overline{N}}$, or $\overline{\overline{N}} < \overline{\overline{M}}$ can occur.

Because $\aleph_0 < \mathscr{C}$, Cantor posed the question whether any other cardinalities were possible for subsets of the real numbers. In 1878, he thought he had answered this question in the negative: "Through a process of induction, which we do not describe further at this point, one is led to the theorem that the number of classes of linear point-sets yielded by this [equivalence] is finite and, indeed, that it is equal to *two*."[40] This conjecture that every subset of the real numbers has cardinality either $\aleph_0$ or $\mathscr{C}$ is called the **Continuum Hypothesis**. Although Cantor many times believed he had a proof of the result, and at least once believed that he had a proof of its negation, neither he nor anyone else was able to prove or disprove the hypothesis. In fact, the Continuum Hypothesis was eventually shown to be unprovable by use of any reasonable collection of axioms for the theory of sets.

Although Cantor was unable to answer all of the questions he raised in connection with the theory of infinite sets, his conception of these sets soon received both wide acceptance and strong criticism. In particular, Leopold Kronecker believed that any mathematical construct must be capable of being completed in a finite number of operations. Because some of Cantor's constructions did not meet Kronecker's criterion, Kronecker, as an editor of Crelle's *Journal*, held up the publication of one of Cantor's articles for so long that Cantor refused to publish again in the *Journal*, even though it was the most influential of the mathematics journals of the time. Nevertheless, although Kronecker and others continued to oppose Cantor's transfinite methods, there were also a growing number of mathematicians who supported his new approach to set theory. The conflict between the two groups, however, continues to this day.

16.2.5 Dedekind and Axioms for the Natural Numbers

Cantor had developed some of the more advanced ideas of set theory and had shown, along with Dedekind, how to construct the real numbers starting from the rational numbers. But it was Dedekind who completed the process of arithmetizing analysis by characterizing the natural numbers, and therefore the rational numbers, in terms of sets. In the work in which he accomplished this task, *Was sind und was sollen die Zahlen? (What are the (natural) numbers and what do they mean?)*, developed over a 15 year period but only published in 1888, he also provided an introduction to the basic notions of set theory.

To characterize the natural numbers, Dedekind began with the notion that the natural numbers form a set of *things*, or "objects of our thought." Therefore, Dedekind defined the term *systeme*, here translated as *set*: "It very frequently happens that different things, a, b, c, . . . for some reason can be considered from a common point of view, can be associated in the mind, and we say that they form a set S. . . . Such a set S as an object of our thought is likewise a thing; it is completely determined when with respect to every thing it is determined whether it is an element of S or not."[41] Given this necessarily somewhat vague definition, Dedekind proceeded to describe various simple relations involving sets. For example, a set A is a **part** of a set S when every element of A is also an element of S. Also, the set **compounded** out of any sets A, B, C, . . . , denoted $\mathcal{M}(A, B, C, . . .)$ consists of those elements which are in at least one of the sets A, B, C, . . . , while the set of elements common to A, B, C, . . . , is denoted $\mathcal{G}(A, B, C, . . .)$. In modern terminology, Dedekind's "part" has become our "subset," $\mathcal{M}(A, B, C, . . .)$ has become the "union" of the sets A, B, C, . . . , and $\mathcal{G}(A, B, C, . . .)$ is their "intersection."

A fundamental property of the natural numbers is the notion that to each number there is a unique successor. In other words, there is a function ψ from the set N of natural numbers to itself given by $\psi(n) = n + 1$. In general, Dedekind defined a function ϕ on S to be a "law according to which to every determinate element s of S there belongs a determinate thing which is called the **transform** of s and denoted by $\phi(s)$."[42] Because different elements of N have different successors, Dedekind was led to the notion of an *ähnlich* (**similar** or **injective**) transformation, one for which "to different elements a, b of the set S there always correspond different transforms $a' = \phi(a)$, $b' = \phi(b)$."[43] In this case there is an inverse transformation $\overline{\phi}$ of the system $S' = \phi(S)$, defined by assigning to every element s' of S' the unique element s which was transformed into it by ϕ. Two sets

R and S are then said to be **similar** to one another if there exists an injective transformation ϕ defined on R such that $S = \phi(R)$.

The natural numbers also have the property that the image of N under the successor transformation is a proper subset of N itself, with the only element not belonging to that image being the element 1. It is, in fact, in the property that the image is a proper subset that the infinitude of the set N resided: "A set S is said to be infinite when it is similar to a proper part of itself; in the contrary case S is said to be a finite set."[44] But do there exist infinite sets at all? Dedekind was hesitant to prove results about such sets without an argument that they exist, so he gave one: "The totality S of all things which can be objects of my thought is infinite. For if s signifies an element of S, then is the thought s', that s can be object of my thought, itself an element of S."[45] Given the transformation from S to itself defined by $s \rightarrow s'$, for which it is clear that the image is not all of S and that the transformation is injective, Dedekind concluded that the set S does satisfy the requirements of his definition.

Dedekind realized that the properties that N was a set possessing an injective successor function whose image was a proper subset of N did not characterize N uniquely. There may well be extraneous elements in any set S which satisfied these properties, elements which are not natural numbers. For example, the set of positive rational numbers satisfies all of the properties. So Dedekind added one more property, that an element belongs to N if and only if it is an element of every subset K of S having the property that 1 belongs to K and that the successor of every element of K is also in K. In other words, N is characterized by being the intersection of all sets satisfying the original properties. N thus contains a base element 1, the successor $\phi(1)$ of 1, the successor $\phi(\phi(1))$ of that element, and so forth, but no other elements.

From Dedekind's characterization of the natural numbers, he was able to derive the principle of mathematical induction as well as to give a definition and derive the properties of the order relationship on N and the operations of addition and multiplication. Two other mathematicians, Giuseppe Peano (1858–1932) and Gottlob Frege (1848–1925), also considered the same question of the construction of the natural numbers and the derivation of their important properties in the 1880s. Frege's work appeared in print in 1884 and Peano's in 1889. It was this work, together with the work of Weierstrass and his school, which enabled calculus to be placed on a firm foundation beginning with the basic notions of set theory. It also showed that calculus has an existence independent of the physical world of motion and curves, the world used by Newton to create the subject in the first place.

16.3 COMPLEX ANALYSIS

Recall that William Rowan Hamilton had by 1837 developed the theory of complex numbers as ordered pairs of real numbers, thus giving one answer to the question of what this mysterious square root of -1 really was. But mathematicians had been using complex numbers since the sixteenth century and even after Hamilton's work did not generally conceive of them in this abstract form. It was the geometrical representation of these numbers, first published by the Norwegian surveyor Caspar Wessel (1745–1818) in an essay in 1797, which ultimately became the basis for a new way of thinking about complex quantities, a way which soon convinced mathematicians that they could use these numbers without undue worry.

16.3.1 Geometrical Representation of Complex Numbers

Wessel's aim in his *On the Analytical Representation of Direction* was not initially related to complex numbers as such. He felt that certain geometrical concepts could be more clearly understood if there was a way to represent both the length and direction of a line segment in the plane by a single algebraic expression. Wessel made clear that these expressions had to be capable of being manipulated algebraically. In particular, he wanted a way of algebraically expressing an arbitrary change of direction more general than the simple use of a negative sign to indicate the opposite direction.

Wessel began by dealing with addition: "Two straight lines are added if we unite them in such a way that the second line begins where the first one ends and then pass a straight line from the first to the last point of the united lines. This line is the sum of the united lines."[46] Thus, whatever the algebraic expression of a line segment was to be, the addition of two had to satisfy this obvious property drawn from Wessel's conception of motion. In other words, he conceived of line segments as representing vectors. It was multiplication, however, which provided Wessel with the basic answer to his question of the representation of direction. To derive this multiplication, he established a number of properties which he felt were essential. First, the product of two lines in the plane had to remain in the plane. Second, the length of the product line had to be the product of the lengths of the two factor lines. Finally, if all directions were measured from the positive unit line, which he called 1, the angle of direction of the product was to be the sum of the angles of direction of the two factors. Designating by ϵ the line of unit length perpendicular to the line 1, he easily showed that his desired properties implied that $\epsilon^2 = (-\epsilon)^2 = -1$ or that $\epsilon = \sqrt{-1}$. A line of unit length making an angle θ with the positive unit line could now be designated by $\cos \theta + \epsilon \sin \theta$ and, in general, a line of length A and angle θ by $A(\cos \theta + \epsilon \sin \theta) = a + \epsilon b$ where a and b are chosen appropriately (Figure 16.6). Thus from Wessel's algebraic interpretation of a geometrical line segment arose the geometrical interpretation of the

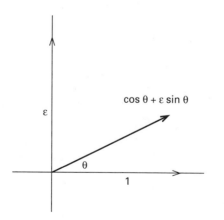

FIGURE 16.6
Wessel's geometric interpretation of complex numbers.

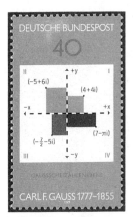

FIGURE 16.7
The Gaussian complex
plane on a German
stamp.

complex numbers. The obvious algebraic rule for addition satisfied Wessel's requirements
for that operation, while the multiplication $(a + \epsilon b)(c + \epsilon d) = ac - bd + \epsilon(ad + bc)$
satisfied his axioms for multiplication. Wessel also easily derived from his definitions the
standard rules for division and root extraction of complex numbers.

Unfortunately, Wessel's essay remained unread in most of Europe for many years after
its publication. The same fate awaited the similar geometric interpretation of the complex
numbers put forth by the Swiss bookkeeper Jean-Robert Argand (1768–1822) in a small
book published in 1806. It was only because Gauss used the same geometric interpretation
of the complex numbers in his proofs of the fundamental theorem of algebra and in his
study of quartic residues that this interpretation gained acceptance in the mathematical
community (Figure 16.7). Gauss was so intrigued with the fundamental theorem, that
every polynomial $p(x)$ with real coefficients has a real or complex root, that he published
four different proofs of it, in 1799, 1815, 1816, and 1848. Each proof used in some form
or other the geometric interpretation of complex numbers, although in the first three Gauss
hid this notion by considering the real and imaginary parts of the numbers separately. It
was only in 1848 that Gauss believed mathematicians would be comfortable enough with
the geometric interpretation of complex numbers so that he could use it explicitly in a
proof. In fact, in that proof, similar to his first one, he even permitted the coefficients of
the polynomial to be complex.

16.3.2   Complex Integration

By the second decade of the century, Gauss, with his clear understanding of the meaning
of complex numbers, began the development of the theory of complex functions. In a letter
of 1811 to his friend Friedrich Wilhelm Bessel (1784–1846), Gauss not only discussed the
geometric interpretation of the complex numbers but also discussed the meaning of
$\int_{\mu}^{v} \phi(x)\, dx$ where the variable x is complex:

> We must assume that x passes through infinitely small increments (each of the form $\alpha + \beta i$)
> from the value for which the integral is 0 to $x = a + bi$, and then sum all the $\phi(x)\, dx$. In this
> way the meaning is completely established. But the passage can occur in infinitely many
> ways; just as one can think of the entire domain of all real magnitudes as an infinite straight
> line, so one can make the entire domain of all magnitudes, real and imaginary, meaningful as
> an infinite plane, wherein each point determined by abscissa $= a$ and ordinate $= b$ represents
> the magnitude $a + bi$ as it were. The continuous passage from one value of x to another
> $a + bi$ accordingly occurs along a line and is consequently possible in infinitely many ways.[47]

Gauss went on to assert the "very beautiful theorem" that as long as $\phi(x)$ is never
infinite within the region enclosed by two different curves connecting the starting and
ending points of this integral, then the value of the integral is the same along both curves.
Although he did not express himself in those terms, Gauss was considering $\phi(x)$ as an
analytic function. In any case, he never published a proof of this result. Such a proof was
published in 1825 by Cauchy, however, so the theorem is generally called Cauchy's integral
theorem.

Cauchy first considered the question of integration in the complex domain in a memoir
written in 1814 but not published until 1827. In this work he was mainly interested in the
evaluation of definite integrals where one or both of the limits of integration is infinite. To

perform such an evaluation, he attempted to make rigorous various procedures developed by Euler and Laplace involving moving the paths of integration into the complex plane. In particular, he used an idea of Euler's to derive the Cauchy-Riemann equations. Euler, in a paper written about 1777, asserted that the most important theorem about complex functions was that every function $Z(z) = Z(x + iy)$ which can be written as the sum $M(x, y) + iN(x, y)$ has the property that $Z(x - iy) = M - iN$. In this case it follows that if

$$V = \int Z\, dz = \int (M + iN)(dx + i\, dy) = \int (M\, dx - N\, dy) + i \int (N\, dx + M\, dy) = P + iQ,$$

then

$$P - iQ = \int (M\, dx - N\, dy) - i \int (N\, dx + M\, dy).$$

Therefore $P = \int (M\, dx - N\, dy)$ and $Q = \int (N\, dx + M\, dy)$, where, as usual for Euler, the integral signs stand for antidifferentiation. Because P is the integral of the differential $M\, dx - N\, dy$ it follows that

$$\frac{\partial M}{\partial y} = -\frac{\partial N}{\partial x}.$$

Similarly, the expression for Q shows that

$$\frac{\partial M}{\partial x} = \frac{\partial N}{\partial y}.$$

These two equations, the Cauchy-Riemann equations, ultimately became the characteristic property of complex functions.

In his 1821 *Cours d'Analyse*, Cauchy dealt with complex quantities, as had Euler, by considering separately the real and imaginary parts. Thus he considered the "symbolic expressions" $a + ib$ and multiplied them together using normal algebraic rules "as if $\sqrt{-1}$ was a real quantity whose square was equal to -1."[48] He defined a function of a complex variable in terms of two real functions of two real variables and showed what is meant by the various standard transcendental functions in the complex domain. He then generalized most of his results on convergence of series to complex numbers, by using the modulus $\sqrt{a^2 + b^2}$ of the quantity $z = a + ib$ as the analogue of the absolute value of a real number. He also defined continuity for a complex function in terms of the continuity of its two constituent functions.

It was not until 1825, however, having discovered his new definition of a definite integral, that Cauchy was able to deal with complex functions in their own right. In his *Mémoire sur les intégrales définies prises entre des limites imaginaires* (*Memoir on definite integrals taken between imaginary limits*), he explicitly defined the definite complex integral

$$\int_{a+ib}^{c+id} f(z)\, dz$$

to be the "limit or one of the limits to which the sum of products of the form $[(x_1 - a) + i(y_1 - b)]f(a + ib)$, $[(x_2 - x_1) + i(y_2 - y_1)]f(x_1 + iy_1)$, . . . $[(c - x_{n-1}) + i(d - y_{n-1})]f(x_{n-1} + iy_{n-1})$ converges when each of the two sequences $a, x_1, x_2, \ldots, x_{n-1}, c$ and $b, y_1, y_2, \ldots, y_{n-1}, d$ consist of terms that increase or decrease from the first to the last and approach one another indefinitely as their number increases without limit."[49] In other words, Cauchy directly generalized his definition of a real definite integral by simply taking partitions of the two intervals $[a, b]$ and $[c, d]$. Cauchy realized, however, as had Gauss, that there were infinitely many different paths of integration beginning at $a + ib$ and ending at $c + id$. It was therefore not clear that this definition made sense. To demonstrate his integral theorem, which in effect stated that the definition did make sense, he began by considering a path determined by the parametric equations $x = \phi(t)$, $y = \psi(t)$, where ϕ and ψ are monotonic differentiable functions of t in the interval $[\alpha, \beta]$, with $\phi(\alpha) = a$, $\phi(\beta) = c$, $\psi(\alpha) = b$, and $\psi(\beta) = d$. The two sequences $\{x_j\}$ and $\{y_j\}$ are then determined by taking a single sequence $\alpha, t_1, t_2, \ldots, t_{n-1}, \beta$ and calculating the values of this sequence under ϕ and ψ respectively. Assuming that the lengths of the various subintervals determined by the t_j are small, Cauchy noted that $x_j - x_{j-1} \approx (t_j - t_{j-1})\phi'(t_j)$ and $y_j - y_{j-1} \approx (t_j - t_{j-1})\psi'(t)$. It follows that the definite integral is the limit of sums of terms of the form $(t_j - t_{j-1})[\phi'(t_j) + i\psi'(t_j)]f[\phi(t_j) + i\psi(t_j)]$ and therefore can be rewritten in the form

$$\int_{a+ib}^{c+id} f(z)\, dz = \int_{\alpha}^{\beta} [\phi'(t) + i\psi'(t)]f[\phi(t) + i\psi(t)]\, dt,$$

or, setting $x' = \phi'(t)$, $y' = \psi'(t)$, as

$$\int_{\alpha}^{\beta} (x' + iy')f(x + iy)\, dt.$$

"Now suppose that the function $f(x + iy)$ remains bounded and continuous as long as x stays between the limits a and c, and y between the limits b and d. In this special case one easily proves that the value of the integral . . . is independent of the nature of the functions $x = \phi(t)$, $y = \psi(t)$."[50] Cauchy's proof of this statement, which requires the existence and continuity of $f'(z)$—and Cauchy had not explicitly defined what was meant by the derivative of a complex function—was based on the calculus of variations. Cauchy varied the curve infinitesimally by replacing the functions ϕ and ψ by $\phi + \epsilon u$, $\psi + \epsilon v$, where ϵ is "an infinitesimal of the first order," and u, v both vanish at $t = \alpha$ and $t = \beta$, and expanded the corresponding change in the integral in a power series in ϵ. Using an integration by parts, Cauchy demonstrated that the coefficient of ϵ in this series is 0 and therefore that an infinitesimal change in the path of integration produces an infinitesimal change in the integral of the order of ϵ^2. Cauchy concluded that a finite change in the path, that is, a change from one path of integration to a second such path, can produce but an infinitesimal change in the integral, that is, no change at all. The integral theorem was therefore proved according to Cauchy's, if not modern, standards.

Cauchy next considered the case where f becomes infinite at some value $z_1 = r + is$ in the rectangle $a \leq x \leq c$, $b \leq y \leq d$. The integrals along two paths which together enclose z_1 are no longer the same. Defining R to be $\lim_{z \to z_1}(z - z_1)f(z)$, Cauchy calculated the difference in the integrals along two paths infinitely close to each other and to the point z_1 to be $2\pi Ri$. For example, if $f(z) = 1/(1 + z^2)$, then f becomes infinite at $z = i$. Because

$$\lim_{z \to i}\frac{z - i}{1 + z^2} = \lim_{z \to i}\frac{z - i}{(z - i)(z + i)} = \frac{1}{2i},$$

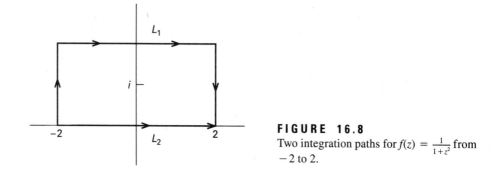

FIGURE 16.8

Two integration paths for $f(z) = \frac{1}{1+z^2}$ from -2 to 2.

it follows that the difference in the values of the integrals of this function over the two paths L_1 and L_2 from -2 to 2 in Figure 16.8 is

$$2\pi \frac{1}{2i} i = \pi.$$

In a paper written in 1826, Cauchy generalized his integral theorem somewhat. Given a value z_1 for which $f(z)$ is infinite, Cauchy noted that the expansion of $f(z_1 + \epsilon)$ in powers of ϵ will begin with negative powers. The coefficient of $1/\epsilon$ in this expansion is what Cauchy terms the **residue** of $f(z)$ at z_1, denoted by $R(f, z_1)$. Thus, if $(z - z_1)f(z) = g(z)$ is bounded near z_1, then

$$f(z_1 + \epsilon) = \frac{g(z_1 + \epsilon)}{\epsilon} = \frac{1}{\epsilon} g(z_1) + g'(z_1 + \theta\epsilon)$$

for θ a number between 0 and 1. It follows that the residue of $f(z)$ at z_1 is $g(z_1)$, the same value denoted earlier as R.

Cauchy noted that his theory of residues had applications to such problems as the splitting of rational fractions, the determination of the values of certain definite integrals, and the solution of certain types of equations. For example, he demonstrated that

$$\int_{-\infty}^{\infty} \frac{\cos x}{1 + x^2}\, dx = \pi e^{-1}$$

by extending the interval of integration to a closed path in the complex plane containing the value i for which the integrand becomes infinite. The central idea in this calculation is that the integral over the path consisting of a half circle and an interval on the real line can be calculated by means of residues, but as the radius of the half circle (and the length of the interval) get larger, the part of the integral taken over the half circle approaches 0.

16.3.3 Complex Functions and Line Integrals

There are many other standard results in complex function theory for which Cauchy was at least partially responsible, most being applications of his integral theorem or his calculus of residues. But the discussion of his work will be concluded with a brief analysis of a paper of 1846 which, although it did not mention complex functions at all, led to a new way of proving the integral theorem and also provided the beginning of some fundamental ideas in both vector analysis and topology. This short paper, *Sur les intégrales qui s'éten-dent a tous les points d'une courbe fermée (On the integrals which extend to all the points of a closed curve)* contained the bare statement of several theorems, without proofs. Cau-chy promised to provide the proofs later, but apparently did not do so. The theorems deal with a function k of several variables x, y, z, . . . which is to be integrated along the boundary curve Γ of a surface S lying in a space of an unspecified number of dimensions. The most important results are collected in the following

Theorem.   *Suppose*

$$k = X\frac{dx}{ds} + Y\frac{dy}{ds} + Z\frac{dz}{ds} + \cdots$$

where $X\,dx + Y\,dy + Z\,dz + \cdots$ is an exact differential. (To say that this differential is exact is to say that $\partial X/\partial y = \partial Y/\partial x$, $\partial X/\partial z = \partial Z/\partial x$, $\partial Y/\partial z = \partial Z/\partial y$, ...) Suppose that the function k is finite and continuous everywhere on S except at finitely many points P, P', P'', . . . in its interior. If α, β, γ, . . . are closed curves in S surrounding these points respectively, then

$$\int_\Gamma k\,ds = \int_\alpha k\,ds + \int_\beta k\,ds + \int_\gamma k\,ds + \cdots.$$

In particular, if there are no such singular points, then

$$\int_\Gamma k\,ds = 0.$$

In the two dimensional case, where S is a region of the plane and k is an arbitrary differential, then

$$\int_\Gamma k\,ds = \pm \int\int_S \left(\frac{\partial X}{\partial y} - \frac{\partial Y}{\partial x}\right) dx\,dy.$$

If k is an exact differential, then $\partial X/\partial y = \partial Y/\partial x$, so the right side, and therefore the left, vanish.

The Cauchy integral theorem follows from the last statement. A complex function $f(z) = f(x + iy)$ can be expressed as $f(x,y) = u(x,y) + iv(x,y)$ and, therefore, since $dz = dx + idy$,

$$\int f(z)\,dz = \int (u\,dx - v\,dy) + i\int (v\,dx + u\,dy).$$

The Cauchy-Riemann equations then imply that both integrands are exact differentials and therefore that the integral theorem holds.

More interesting than the integral theorem, however, is the appearance in Cauchy's paper both of the concept of a line integral in n-dimensional space (and of the matter-of-fact occurrence of a space of dimension higher than three) and of the statement (in the next to the last sentence) of the theorem today generally known as Green's theorem. In

---

Biography   **Georg Bernhard Riemann (1826–1866)**

Riemann needed his father's permission to switch from the study of theology and philology to the study of mathematics in 1846 when he enrolled at the University of Göttingen. He had started life in the village of Breselenz, about 60 miles southeast of Hamburg, and now he would journey to Berlin because mathematics education was not particularly strong at Göttingen. In Berlin he met Dirichlet, who became his mentor. He returned to Göttingen a few years later to study with Gauss and received his Ph.D. in 1851.

For two years he researched and prepared his lectures for his *Habilitation* to qualify to teach at Göttingen. In 1857 he was appointed as an associate professor and two years later, on the death of Dirichlet, who had in the meantime come to Göttingen, as full professor. His mathematical work was brilliant, but tuberculosis cut his work short when it claimed his life in the summer of 1866 during one of his several trips to Italy to find a cure.

---

fact, results somewhat akin to that theorem appear in an 1828 paper of George Green (1793–1841) dealing with electricity and magnetism, but Cauchy's version is the first printed statement of the result so named in today's textbooks. Finally, the expression of the line integral around the boundary of the surface as a sum of line integrals around isolated singular points, whose values are called **periods**, marked the beginning of the study of the relationships of integrals to surfaces over which they are not defined everywhere. Since Cauchy never published the proof of his 1846 theorem, one can only speculate as to how far he carried all of these new concepts. It was Riemann, however, who restated Cauchy's results a few years later, with full proofs, and extended the result on periods far beyond Cauchy's conception.

16.3.4 Riemann and Complex Functions

Riemann's dissertation, *Grundlagen für eine allgemeine Theorie der Functionen einer veränderlichen complexen Grösse* (*Foundations for a general theory of functions of one complex variable*), began with a discussion of an important distinction between real and complex functions. Although the definition of function, "to every one of [the] values [of a variable quantity z] there corresponds a single value of the indeterminate quantity w,"[51] can be applied both to the real and complex case, Riemann realized that in the latter case, where $z = x + iy$ and $w = u + iv$, the limit of the ratio dw/dz defining the derivative could well depend on how dz approaches 0. Because for functions defined algebraically one could calculate the derivative formally and not have this problem, Riemann decided to make this existence of the derivative the basis for the concept of a complex function: "The complex variable w is called a function of another complex variable z when its variation is such that the value of the derivative dw/dz is independent of the value of dz."[52] Cauchy, of course, had essentially used this notion in his entire discussion of complex functions but had only made it explicit toward the end of his career.

As a first application of this definition, Riemann showed that such a complex function considered as a mapping from the z-plane to the w-plane preserves angles. For suppose

p' and p'' are infinitely close to the origin P in the z-plane, with their images q', q'' infinitely close to the image Q of P. Writing the infinitesimal distance from p' to P both as $dx' + i\,dy'$ and as $\epsilon'e^{i\phi'}$, and that from q' to Q as both $du' + i\,dv'$ and $\eta'e^{i\psi'}$, with similar notations for the other infinitesimal distances, Riemann noted that his condition on the function implies that

$$\frac{du' + i\,dv'}{dx' + i\,dy'} = \frac{du'' + i\,dv''}{dx'' + i\,dy''}$$

or that

$$\frac{du' + i\,dv'}{du'' + i\,dv''} = \frac{\eta'}{\eta''}e^{i(\psi'-\psi'')} = \frac{dx' + i\,dy'}{dx'' + i\,dy''} = \frac{\epsilon'}{\epsilon''}e^{i(\phi'-\phi'')}.$$

It follows that $\eta'/\eta'' = \epsilon'/\epsilon''$ and that $\psi' - \psi'' = \phi' - \phi''$, or, in other words, that the infinitesimal triangles $p'Pp''$ and $q'Qq''$ are similar. Such an angle preserving mapping is called a **conformal** mapping. In some sense, both Euler and Gauss knew that analytic complex functions had this property, but it was Riemann who gave this argument and who, in addition, was able to demonstrate the Riemann mapping theorem, that any two simply connected regions in the complex plane can be mapped conformally on each other by means of a suitably chosen complex function.

Riemann's first important application of his definition of a complex function was his derivation of the Cauchy-Riemann equations by determining what the existence of the derivative means in terms of the two functions u and v:

$$\frac{dw}{dz} = \frac{du + i\,dv}{dx + i\,dy} = \frac{\frac{\partial u}{\partial x}dx + \frac{\partial u}{\partial y}dy + i\left(\frac{\partial v}{\partial x}dx + \frac{\partial v}{\partial y}dy\right)}{dx + i\,dy}$$

$$= \frac{\left(\frac{\partial u}{\partial x} + i\frac{\partial v}{\partial x}\right)dx + \left(\frac{\partial v}{\partial y} - i\frac{\partial u}{\partial y}\right)i\,dy}{dx + i\,dy}.$$

If this value is independent of how dz approaches 0, then setting dx and dy in turn equal to zero and equating the real and imaginary parts of the two resulting expressions shows that

$$\frac{\partial u}{\partial x} = \frac{\partial v}{\partial y} \quad \text{and} \quad \frac{\partial v}{\partial x} = -\frac{\partial u}{\partial y}.$$

Conversely, if those Cauchy-Riemann equations are satisfied, then the desired derivative is easily calculated to be $\partial u/\partial x + i\,\partial v/\partial x$, a value independent of dz. Riemann made these equations the center of his theory of complex functions, along with the second set of partial differential equations easily derived from them:

$$\frac{\partial^2 u}{\partial x^2} + \frac{\partial^2 u}{\partial y^2} = 0 \quad \text{and} \quad \frac{\partial^2 v}{\partial x^2} + \frac{\partial^2 v}{\partial y^2} = 0.$$

As an example, Riemann gave a detailed proof of the Cauchy integral theorem following the outline provided by Cauchy in 1846. The important idea was Green's theorem, which Riemann stated in the following form.

Theorem.   *Let X and Y be two functions of x and y continuous in a finite region T with infinitesimal area element designated by dT. Then*

$$\int_T \left(\frac{\partial X}{\partial x} + \frac{\partial Y}{\partial y} \right) dT = -\int_S (X \cos \xi + Y \cos \eta)\, ds$$

where the latter integral is taken over the boundary curve S of T, ξ, η designating the angles the inward pointing normal line to the curve makes with the x- and y-axis respectively.

Riemann proved this by using the fundamental theorem of calculus to integrate $\partial X/\partial x$ along lines parallel to the x-axis, getting values of X where the lines cross the boundary of the region. Since $dy = \cos \xi\, ds$ at each of those points, he could integrate with respect to y to get

$$\int \left[\int \frac{\partial X}{\partial x}\, dx \right] dy = -\int X\, dy = -\int X \cos \xi\, ds.$$

The other half of the theorem is proved similarly. Riemann then noted that

$$\frac{dx}{ds} = \pm \cos \eta \qquad \text{and} \qquad \frac{dy}{ds} = \mp \cos \xi$$

where the sign depends on whether one gets from the tangent line to the inward normal line by traveling counterclockwise or clockwise. It follows that Green's theorem can be rewritten as

$$\int_T \left(\frac{\partial X}{\partial x} + \frac{\partial Y}{\partial y} \right) dT = \int_S \left(X \frac{dy}{ds} - Y \frac{dx}{ds} \right) ds,$$

from which the Cauchy integral theorem follows easily.

Much of Riemann's dissertation involved the introduction of an entirely new concept in the study of complex functions, the idea of a Riemann surface. In the case of functions of a real variable, it is possible to picture the function by a curve in two-dimensional space. Such a representation is no longer possible for complex functions, because the graph would need to be in a space of four real dimensions. An alternative way of picturing complex functions, then, is to trace the independent variable z along a curve in one plane and consider the curve generated by the dependent variable w in another plane. Riemann realized from the fact that a complex function always had a power series representation that "a function of $x + iy$ defined in a region of the (x,y) plane can be continued analytically in only one way." It follows that once one knows the values in a certain region, one can continue the function and even return to the same z value by, say, a continuous curve. There are then two possibilities. "Depending on the nature of the function to be continued, either this function will always assume the same value for the same value of [z], no matter how it is continued, or it will not."[53] In the first case, Riemann called the function single-valued, while in the second it is multiple-valued. As a simple example of the latter, one can take $w = z^{1/2}$. To study such functions effectively, it was not possible simply to use two planes as indicated above, for one would not know which value the function had for a given point on the first plane. Thus Riemann came up with a new idea, to use a multiple plane, a covering of the *z* plane by as many sheets as the function has values. These sheets are attached along a line, say the negative real axis, in such a way that whenever one moves in a curve across that line one changes from one sheet to another. In this way the multiple-valued function has only one value defined at each point of this Riemann surface. Since it may happen that after several circuits (two in the example above) one returns to a former value, the top sheet of this covering must be attached to the bottom one. It follows that it is not in general possible to construct a physical model of a Riemann surface in three-dimensional space. Nevertheless, the study of Riemann surfaces, initiated by Riemann to deal with multiple-valued complex functions, soon led Riemann and others into the realm of what is today called **topology**. The connection of topology with integration along curves and surfaces, barely touched by Cauchy in 1846, was explored in great detail in the second half of the nineteenth century and the early years of the twentieth.

16.4   VECTOR ANALYSIS

Green's theorem was stated by Riemann in 1851 in terms of the equality of a double integral with an integral along a curve taken with respect to the curve element *ds*. It was the use in physics of integrals over curves to represent work done along the curves that seems to have inspired a change in notation which occurred in the 1850s in which the curve integral was replaced by a line integral, an integral of the form $\int p\, dx + q\, dy$. Although this notation had been used in complex integration, the physicists converted it into an expression involving vectors. Other physical concepts involving vectors were incorporated in other integral theorems during the nineteenth century.

16.4.1 Line Integrals and Multiple Connectivity

Clerk Maxwell noted in 1855 that if α, β, and γ are the components of the "intensity of electric action" ϵ parallel to the *x*, *y*, and *z* axes respectively and if ℓ, *m*, *n* are the corresponding direction cosines of the tangent to the curve (the cosines the tangent makes with the three coordinate axes), then ϵ can be written in the form $\ell\alpha + m\beta + n\gamma$. Because $\ell\, ds = dx$, $m\, ds = dy$, and $n\, ds = dz$, Maxwell wrote $\int \epsilon\, ds = \int \alpha\, dx + \beta\, dy + \gamma\, dz$. The following year this notation appeared in a physics text by Charles Delaunay (1816–1872). Delaunay was somewhat clearer than Maxwell, noting that if *F* is a force and F_1 its tangential component along a curve, then the work done by the force acting along the curve could be represented by $\int F_1\, ds$. Again, if the rectangular components of *F* are *X*, *Y*, *Z*, this latter integral can be rewritten as $\int X\, dx + Y\, dy + Z\, dz$.

The line integral notation quickly became standard in physics and was adopted by Riemann in a paper of 1857 in which he studied the (Riemann) surfaces *R* on which were described the curves over which these line integrals were taken. Riemann began by observing that the integral of an exact differential $X\, dx + Y\, dy$ is zero when taken over the perimeter of a region in this surface:

> Hence the integral $\int (X\, dx + Y\, dy)$ has the same value when taken between two fixed points along two different paths, provided the two paths together form the entire boundary of a region of *R*. Thus, if every closed curve in the interior of *R* bounds a region of *R*, then the

integral always has the same value when taken from a fixed initial point to one and the same endpoint, and is a continuous function of the position of the endpoint which is independent of the path of integration. This gives rise to a distinction among surfaces: simply connected ones, in which every closed curve bounds a region of the surface—as, for example, a disk— and multiply connected ones, for which this does not happen—as, for example, an annulus bounded by two concentric circles.[54]

Riemann proceeded to refine the notion of multiple connectedness: "A surface F is said to be $(n + 1)$-ply connected when n closed curves $A_1, \ldots, A_n$ can be drawn on it which neither individually nor in combination bound a region of F, while if augmented by any other closed curve A_{n+1}, the set bounds some region of F."[55] Riemann noted further that an $(n + 1)$-ply connected surface can be changed into an n-ply connected one by means of a cut, a curve going from one boundary point through the interior, to another boundary point. For example, an annulus, which is doubly connected, can be reduced to a simply connected region by any cut q which does not disconnect it. A double annulus needs two cuts to be reduced to a simply connected region.

Using the idea of cuts, Riemann was able to describe exactly what happened when one integrated an exact differential on an $(n + 1)$-ply connected surface R. If one removes n cuts from this surface, there remains a simply connected surface R'. Integration of the exact differential $X\,dx + Y\,dy$ from a fixed starting point over any curve in R' then determines, as before, a single-valued continuous function Z of position on this surface. However, whenever the path of integration crosses a cut, the value jumps by a fixed number dependent on the cut. In fact, there are n such numbers, one for each cut. This notion of multiple connectedness turned out to be important in physics, particularly in fluid dynamics and electromagnetism, and so it was extended to regions of three dimensional space by such physicists as Hermann von Helmholtz (1821–1894), William Thomson (1824–1907), and Maxwell.

16.4.2 Surface Integrals and the Divergence Theorem

Physicists were interested not only in line integrals but also in surface integrals, integrals of functions and vector fields over two-dimensional regions. Recall that as early as 1760 Lagrange had given an explicit expression for the element of surface dS in the process of calculating surface areas. It was not until 1811, however, in the second edition of his *Mécanique analytique*, that Lagrange introduced the general notion of a surface integral. He noted that if the tangent plane at dS makes an angle γ with the xy-plane, then simple trigonometry allows one to rewrite $dx\,dy$ as $\cos\gamma\,dS$. It followed that if A is a function of three variables, then $\int A\,dx\,dy = \int A\cos\gamma\,dS$, the second integral being taken over a region in the surface, the first over the projection of that region in the plane. Similarly, if β is the angle the tangent plane makes with the xz-plane and α that with the yz-plane, then $dx\,dz = \cos\beta\,dS$ and $dy\,dz = \cos\alpha\,dS$. Lagrange noted that α, β, and γ could also be considered as the angles which a normal to the surface element makes with the x, y, and z axes respectively.

Lagrange used surface integrals in dealing with fluid dynamics. Gauss in 1813 used the same concept in considering the gravitational attraction of an elliptical spheroid. But Gauss went further than Lagrange in showing how to calculate an integral with respect to

---

$Biography$ Mikhail Ostrogradsky (1801–1861)

Mikhail Ostrogradsky found his way into mathematics through a desire to be an army officer. Since he had been born into a family of modest means in Ukraine, he could not manage the expensive lifestyle of an officer without an independent income. To support his future career, he enrolled in the University of Kharkov in 1816. He became interested in mathematics and physics and passed the exam for the degree in 1820. He did not actually receive the degree because the minister of religious affairs and national education decided to punish Ostrogradsky's teacher, T. F. Osipovsky, the rector of the university, for failing to

instill the proper religious and pro-Czarist attitudes in his students.

Ostrogradsky left Russia to study for several years in Paris, where he produced some of his most important mathematical work. In 1828 he returned to St. Petersburg and was elected a member of the Academy of Sciences. He connected with his original military ambition by teaching mathematics at military academies. In 1847 he became responsible for all mathematics education in these schools and later wrote several important texts for use there (Figure 16.9).

FIGURE 16.9
Ostrogradsky on a Russian stamp.

dS in the case where the surface S is given parametrically by three functions $x = x(p, q)$, $y = y(p, q), z = z(p, q)$. Using a geometrical argument, he demonstrated that

$$dS = \left[\left(\frac{\partial(y, z)}{\partial(p, q)} \right)^2 + \left(\frac{\partial(z, x)}{\partial(p, q)} \right)^2 + \left(\frac{\partial(x, y)}{\partial(p, q)} \right)^2 \right]^{1/2} dp \, dq$$

and hence that any integral with respect to dS can be reduced to an integral of the form $\int f \, dp \, dq$, where f is either explicitly or implicitly a function of the two variables p, q.

Gauss used his study of integrals over surfaces to prove certain special cases of what is today known as the divergence theorem. The general case of this theorem was, however, first stated and proved in 1826 by Mikhail Ostrogradsky (1801–1861), a Russian mathematician who was studying in Paris in the 1820s.[56] In his paper entitled "Proof of a Theorem in Integral Calculus," which came out of his study of the theory of heat, Ostrogradsky considered a surface with surface element ϵ bounding a solid region with volume element ω. With p, q, and r being three differentiable functions of x, y, z and with the angles α, β, and γ as defined above, Ostrogradsky stated the divergence theorem in the form

$$\int \left(\frac{\partial p}{\partial x} + \frac{\partial q}{\partial y} + \frac{\partial r}{\partial z} \right) \omega = \int (p \cos \alpha + q \cos \beta + r \cos \gamma) \epsilon,$$

where the left hand integral is taken over the solid V, the right hand one over the boundary surface S. Today the theorem is generally written, by use of Lagrange's idea, in the form

$$\iiint_V \left(\frac{\partial p}{\partial x} + \frac{\partial q}{\partial y} + \frac{\partial r}{\partial z} \right) dx \, dy \, dz = \iint_S p \, dy \, dz + q \, dz \, dx + r \, dx \, dy.$$

This result, like Green's theorem, is a generalization of the fundamental theorem of calculus, so Ostrogradsky's proof uses that theorem. To integrate $(\partial p / \partial x) \omega$ over a "narrow

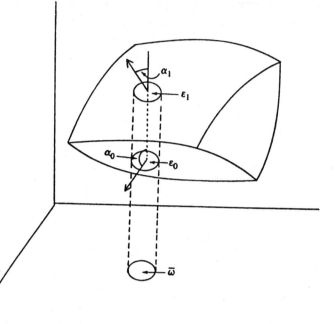

FIGURE 16.10
Ostrogradsky's proof of
the divergence theorem.

cylinder" going through the solid in the x-direction with cross-sectional area $\overline{\omega}$, he used
the fundamental theorem to express the integral as $\int (p_1 - p_0)\overline{\omega}$, where p_0 and p_1 are the
values of p on the pieces of surface where the cylinder intersects the solid (Figure 16.10).
Because $\overline{\omega} = \epsilon_1 \cos \alpha_1$ on one section of the surface and $\overline{\omega} = -\epsilon_0 \cos \alpha_0$ on the other,
where α_1 and α_0 are the angles made by the normal at the surface elements ϵ_1, ϵ_0 respectively, Ostrogradsky had demonstrated that

$$\frac{\partial p}{\partial x}\omega = \int p_1 \epsilon_1 \cos \alpha_1 + \int p_0 \epsilon_0 \cos \alpha_0 = \int (p \cos \alpha)\epsilon$$

where the left integral is over the cylinder and the right ones over the two pieces of surface.
Adding up the integrals over all such cylinders gives one-third of the desired result, the
other two-thirds being done similarly. Interestingly enough, Ostrogradsky generalized his
result to n dimensions in 1836, thus giving one of the earliest statements of a result in
dimension greater than three.

16.4.3 Stokes' Theorem

The divergence theorem relates an integral over a solid to one over the bounding surface,
while Green's theorem relates an integral over a region in the plane to one over the boundary curve. A similar result comparing an integral over a surface in three dimensions to one
around the boundary curve, a result now known as Stokes' theorem, first appeared in print
in 1854. George Stokes had for several years been setting the Smith's Prize Exam at
Cambridge University and, in the February 1854 examination, asked the following.

Biography   **George Stokes (1819–1903)**

Although his three brothers followed his father in an ecclesiastical career in their native Ireland, Stokes was drawn to mathematics through the influence of a teacher. He graduated from Cambridge in 1841 as senior wrangler and eight years later was appointed to the Lucasian chair of mathematics, a position he held until his death. His theoretical and experimental studies during his career spanned much of natural philosophy, including such areas as hydrodynamics, elasticity, and the diffraction of light. His various excursions into pure mathematics were caused by his need to develop methods to solve particular physical problems or to justify the validity of the mathematical techniques he was already using. Stokes served the scientific community in various official posts. In particular, he was the Secretary of the Royal Society from 1854 to 1885, its president from 1885 to 1890, and the representative of the University of Cambridge in Parliament from 1887 to 1891.

Problem 8.   *If X, Y, Z be functions of the rectangular coordinates x, y, z; dS an element of any limited surface; ℓ, m, n the cosines of the inclinations of the normal at dS to the axes; ds an element of the boundary line, shew that*

$$\iint \left[\ell\left(\frac{\partial Z}{\partial y} - \frac{\partial Y}{\partial z}\right) + m\left(\frac{\partial X}{\partial z} - \frac{\partial Z}{\partial x}\right) + n\left(\frac{\partial Y}{\partial x} - \frac{\partial X}{\partial y}\right) \right] dS =$$

$$\int \left(X\frac{dx}{ds} + Y\frac{dy}{ds} + Z\frac{dz}{ds} \right) ds$$

. . . the single integral being taken all around the perimeter of the surface.[57]

It does not seem to be known whether any of the students proved the theorem. However, the theorem had already appeared in a letter of William Thomson to Stokes on July 2, 1850, and the integrand on the left side had appeared in two earlier works of Stokes where it represented the angular velocity of a certain fluid. The first published proof of the result appeared in a monograph of Hermann Hankel (1839–1873) in 1861, at least for the case where the surface is given explicitly as a function $z = z(x, y)$. Hankel substituted the value of z and $dz = (\partial z/\partial x)\, dx + (\partial z/\partial y)\, dy$ into the right hand integral, thus reducing it to an integral in two variables, then used Green's theorem to convert it to a double integral easily seen to be equal to the surface integral on the left.

Stokes himself proved another important result dealing with the integrand appearing on the left in Stokes' theorem. It was clear that

$$\frac{\partial}{\partial x}\left(\frac{\partial Z}{\partial y} - \frac{\partial Y}{\partial z}\right) + \frac{\partial}{\partial y}\left(\frac{\partial X}{\partial z} - \frac{\partial Z}{\partial x}\right) + \frac{\partial}{\partial z}\left(\frac{\partial Y}{\partial x} - \frac{\partial X}{\partial y}\right) = 0.$$

In 1849, Stokes proved what amounted to the converse, namely, that if A, B, C are functions satisfying

$$\frac{\partial A}{\partial x} + \frac{\partial B}{\partial y} + \frac{\partial C}{\partial z} = 0,$$

then there exist functions X, Y, Z such that

$$A = \frac{\partial Z}{\partial y} - \frac{\partial Y}{\partial z} \qquad B = \frac{\partial X}{\partial z} - \frac{\partial Z}{\partial x} \qquad C = \frac{\partial Y}{\partial x} - \frac{\partial X}{\partial y}.$$

This result, like Clairaut's result in two dimensions that an exact differential is the differential of a function, is only valid in certain simple domains. Neither Stokes nor Thomson, who gave a different proof in 1851, dealt with that limitation. Their proofs required that certain differential equations be solvable, and they simply assumed that the solutions could be found without worrying about the specific conditions which would assure this. In any case, their result, combined with Stokes' theorem itself, shows that under the conditions stated on A, B, and C, the surface integral $\int \int (\ell A + mB + nC) \, dS$ (or, in more modern notation, $\int \int A \, dy \, dz + B \, dz \, dx + C \, dx \, dy$) does not depend on the surface but only on the boundary curve.

Both Stokes' theorem and the divergence theorem appear in the opening chapter of Maxwell's *Treatise on Electricity and Magnetism* and are used often in the remainder of the work. Because Maxwell was an advocate of quaternion notation in physics, he wrote out these theorems in quaternion form, using the fact that if the vector operator $\nabla = (\partial/\partial x)i + (\partial/\partial y)j + (\partial/\partial z)k$ is applied to the vector $\sigma = Xi + Yj + Zk$, the resulting quaternion can be written

$$\nabla\sigma = -\left(\frac{\partial X}{\partial x} + \frac{\partial Y}{\partial y} + \frac{\partial Z}{\partial z}\right) + \left(\frac{\partial Z}{\partial y} - \frac{\partial Y}{\partial z}\right)i + \left(\frac{\partial X}{\partial z} - \frac{\partial Z}{\partial x}\right)j + \left(\frac{\partial Y}{\partial x} - \frac{\partial X}{\partial y}\right)k.$$

Maxwell named the scalar and vector parts of $\nabla\sigma$ the **convergence** of σ and the **curl** of σ respectively, because of his interpretation of their physical meaning. Maxwell's convergence is the negative of what is today called the **divergence** of σ.

The vector form of the divergence theorem and Stokes' theorem finally appeared in the work of Gibbs near the end of the century. Setting $dV = dx \, dy \, dz$ to be the element of volume, $d\mathbf{a} = dy \, dz \, i + dz \, dx \, j + dx \, dy \, k$ the element of surface area, and $d\mathbf{r} = i \, dx + j \, dy + k \, dz$ the element of length, Gibbs wrote the divergence theorem in the form

$$\int \int \int \nabla \cdot \sigma \, dV = \int \int \sigma \cdot d\mathbf{a}$$

and Stokes' theorem as

$$\int \int (\nabla \times \sigma) \cdot d\mathbf{a} = \int \sigma \cdot d\mathbf{r}.$$

Even in vector form it is not obvious that Green's theorem, Stokes' theorem, and the divergence theorem can be united into a single result. But Vito Volterra (1860–1940) in 1889 was able to unite them in the course of a study of hypersurfaces in n-dimensional space. (The study of n-dimensional space was new in 1836, but 50 years later, it was already commonplace. How this change occurred will be considered in Chapter 17.) Not only did Volterra state a result using a plethora of indices, of which the three theorems mentioned were all low-dimensional special cases, but also he, along with Henri Poincaré (1854–1912), generalized to higher dimensions the result of Stokes and Thomson on what Poincaré called the integrability conditions. These integrability conditions were the conditions on line integrals, surface integrals, and their higher dimensional analogues which insured that the integrals did not depend on the curve, surface, or hypersurface over which they were being integrated, but only on the boundary of that geometric object. It was this generalization, now known as Poincaré's lemma, which helped to provide the tools for Poincaré's extension of Riemann's ideas on the relationship of such multiple integrals to the topology of the domains of integration.

16.5 PROBABILITY AND STATISTICS

The nineteenth century saw extensive application of statistical methods in various fields, particularly in the social sciences. The discussion of this development, however, would take us too far afield, so this section will be limited to the beginnings of certain important statistical methods up to the time of the first detailed mathematical treatment of the entire theory of probability and statistics by Pierre Laplace.

16.5.1 Legendre and the Method of Least Squares

Perhaps the most important statistical method of the nineteenth century was that of least squares. This method provided one of the primary tools for what was called the "combination of observations," a way of collecting numerous observational measurements of a particular event into a single "best" result. For example, suppose one knows that a particular physical relationship is expressed by a linear function $y = a + bx$. One performs several observations of the phenomenon in question and finds the data points (x_1, y_1), (x_2, y_2), . . . , (x_k, y_k). Replacing x and y in the equation by these k pairs in turn gives k equations for the two unknown coefficients a, b. The system is thus overdetermined and, in general, has no exact solution. The idea then is somehow to determine the "best" approximation to a solution. In geometrical terms, the problem is to find that straight line which is "closest," in some sense, to passing through the k observed points.

This problem of the combination of observations was discussed by various mathematicians in the eighteenth century, primarily in regard to astronomical observations. But the method which has come down to the present as the best method for solving an overdetermined system of linear equations is the method of least squares, developed by Legendre in 1805. This method appears as an appendix to a work on the determination of cometary orbits.

Legendre began his discussion by giving a reason for introducing the method of least squares: "In the majority of investigations in which the problem is to get from measures given by observation the most exact results which they can furnish, there almost always arises a system of equations of the form $E = a + bx + cy + fz + \cdots$ in which $a, b, c, f, \ldots$ are the known coefficients which vary from one equation to another, and $x, y, z, \ldots$ are the unknowns which must be determined in accordance with the condition that the value of E shall for each equation reduce to a quantity which is either zero or very small."[58] In more modern terminology, one has a system of m equations $V_j(\{x_i\}) = a_{j0} + a_{j1}x_1 + a_{j2}x_2 + \cdots + a_{jn}x_n = 0$ $(j = 1, 2, \ldots, m)$ in n unknowns $(m > n)$ for which one wants to find the "best" approximate solutions $\bar{x}_1, \bar{x}_2, \ldots, \bar{x}_n$. For each equation, the value $V_j(\{\bar{x}_i\}) = E_j$ is the error associated with that solution. Legendre's aim was to make all the

E_i small: "Of all the principles which can be proposed for that purpose, I think there is none more general, more exact, and more easy of application, than that of which we have made use in the preceding researches, and which consists of rendering the sum of the squares of the errors a minimum. By this means there is established among the errors a sort of equilibrium which, preventing the extremes from exerting an undue influence, is very well fitted to reveal that state of the system which most nearly approaches the truth."[59]

To determine the minimum of the squares of the errors, Legendre applied the tools of calculus. Namely, for the sum of the squares $E_1^2 + E_2^2 + \cdots + E_m^2$ to have a minimum when x_1 varies, its partial derivative with respect to x_1 must be zero:

$$\sum_{j=1}^{m} a_{j1}a_{j0} + x_1 \sum_{j=1}^{m} a_{j1}^2 + x_2 \sum_{j=1}^{m} a_{j1}a_{j2} + \cdots + x_n \sum_{j=1}^{m} a_{j1}a_{jn} = 0.$$

Because there are analogous equations for $i = 2, 3, \ldots, n$, Legendre noted that he now had n equations in the n unknowns x_i and therefore that the system could be solved by "established methods." Although he offered no derivation of the method from first principles, he did observe that his method was a generalization of the method of finding the ordinary mean of a set of observations of a single quantity. For in that case (the special case where $n = 1$ and $a_{j1} = -1$ for each j), if we set $b_j = a_{j0}$, then the sum of the squares of the errors is $(b_1 - x)^2 + (b_2 - x)^2 + \cdots + (b_m - x)^2$. The equation for making that sum a minimum is $(b_1 - x) + (b_2 - x) + \cdots + (b_m - x) = 0$, so that the solution

$$x = \frac{b_1 + b_2 + \cdots + b_m}{m}$$

is just the ordinary mean of the m observations.

16.5.2 Gauss and the Derivation of the Method of Least Squares

Within ten years of Legendre's publication, the method of least squares was a standard method in solving astronomical and geodetical problems throughout the European continent. In particular, it appeared in 1809 in Gauss' *Theoria motus corporum celestium* (*Theory of Motion of the Heavenly Bodies*). Gauss, however, did not quote Legendre. In fact, Gauss claimed that he had been using the principle himself since 1795. Gauss' statement led to a pained reaction from Legendre, who noted that priority in scientific discoveries could only be established by publication. The quarrel between the two lasted for a number of years. As late as 1827, Legendre was still berating Gauss for appropriating the discoveries of others.[60]

The priority dispute notwithstanding, Gauss went further with the method than Legendre. First, he realized that it was not enough to say that one can use "established methods" to solve the system of n equations in n unknowns which the method of least squares produces. In real applications, there are often many equations and the coefficients are not integers but real numbers calculated to several decimal places. Cramer's method in these cases would require enormous amounts of calculation. Gauss therefore devised a systematic method of elimination for systems of equations, a method of multiplying the equations by appropriately chosen values and then adding these new equations together.

The procedure, now known as the method of Gaussian elimination and virtually identical with the Han Chinese method of 1800 years earlier, results in a triangular system of equations, that is, a system in which the first equation involves but one unknown, the second two, and so on. Thus the first equation can be easily solved for its only unknown, the solution substituted in the second to get the value for the second unknown, and so on until the system is completely solved. Gauss' procedure was improved somewhat later in the century by the German geodesist Wilhelm Jordan (1842–1899), who used the method of least squares to deal with surveying problems. Jordan devised a method of substitution, once the triangular system had been found, to further reduce the system to a diagonal one in which each equation only involved one unknown. This Gauss-Jordan method is the one typically taught in modern linear algebra courses as the standard method for solving systems of linear equations.[61]

Second, Gauss developed a much better justification for the method of least squares than the somewhat vague "general principle" of Legendre. Namely, he derived the method from his prior discovery of a suitable function $\phi(x)$ describing the probability of an error of magnitude x in the determination of an observable quantity. Earlier attempts to discover a reasonable error function had been made by Thomas Simpson in 1755 and by Laplace in the 1770s. The former, in a paper read to the Royal Society, attempted to show that the error in observations would be diminished by taking the mean of several observations. He did this by assuming, for example, that the probability of errors in seconds of sizes -5, -4, -3, -2, -1, 0, 1, 2, 3, 4, 5 in a particular astronomical measurement was respectively proportional to $1, 2, 3, 4, 5, 6, 5, 4, 3, 2, 1$. Thus the probability that a single error does not exceed 1 second is $16/36 = 0.444$ and that it does not exceed 2 seconds is $24/36 = 0.667$. On the other hand, he calculated that the probability that the mean of six errors does not exceed 1 second is 0.725 and that it does not exceed 2 seconds is 0.967, thus showing the advantage of taking means.

Some twenty years later, Laplace tried to derive a reasonable error function amenable to more careful analysis by making explicit assumptions on the conditions such a function should meet. These conditions were first, that $\phi(x)$ should be symmetric about zero, assuming explicitly that the instruments were such that it is equally probable that an observation is too big as that it is too small; second, that the curve must be asymptotic to the real axis in both directions, because the probability of an infinite error is zero; and third, that the total area under $\phi(x)$ should be 1, since the area under that curve between any two values represents the probability that the observation has error between those values. Unfortunately, there were many curves that satisfied Laplace's requirements. Through the use of various other arguments, Laplace settled on the curve $y = (m/2)e^{-m|x|}$ for some positive value m. Laplace soon found out, however, that calculations based on this error function led to great difficulties. Although he then tried other functions, it was Gauss who in 1809 succeeded in deriving a new answer to Laplace's question.

Gauss began with the same basic criteria as Laplace. He joined these, however, to the original problem of Legendre of determining the values of m linear functions $V_1, V_2, \ldots,$ V_m of n unknowns $x_1, x_2, \ldots, x_n$. Supposing that the observed values of these were $M_1, M_2, \ldots, M_m$, with corresponding errors $\Delta_1, \Delta_2, \ldots, \Delta_m$, Gauss noted that because the various observations were all independent, the probability of all these errors occurring was $\Omega = \phi(\Delta_1)\phi(\Delta_2) \cdots \phi(\Delta_m)$. To find the most probable set of values meant maximizing Ω, but to do this required a better knowledge of ϕ. Thus Gauss made the further assumption that "if any quantity has been determined by several direct observations, made under the same circumstances and with equal care, the arithmetical mean of the observed values affords the most probable value."[62] Taking each V_i as the simplest linear function of one variable, namely $V_i = x_1$, he determined ϕ by supposing that $x_1 = (1/m)(M_1 + M_2 + \cdots + M_m)$, the mean of the observations, gives the maximum value of Ω. That maximum occurs when $\frac{\partial \Omega}{\partial x_i} = 0$ for all i. Because Ω is a product, Gauss replaced these equations with $\frac{\partial}{\partial x_i}(\log \Omega) = 0$, or

$$\frac{\frac{\partial}{\partial x_i}(\phi(\Delta_1))}{\phi(\Delta_1)} + \frac{\frac{\partial}{\partial x_i}(\phi(\Delta_2))}{\phi(\Delta_2)} + \cdots + \frac{\frac{\partial}{\partial x_i}(\phi(\Delta_m))}{\phi(\Delta_m)} = 0. \tag{16.5}$$

For each j, $\frac{\partial}{\partial x_i}(\phi(\Delta_j)) = \frac{\partial \phi}{\partial \Delta_j} \frac{\partial \Delta_j}{\partial x_i}$. Because $\frac{\partial \Delta_j}{\partial x_i} = 0$ for $i > 1$ and $\frac{\partial \Delta_j}{\partial x_1} = -1$ for all j, the n equations (16.5) all reduce to the single equation

$$\frac{\phi'(\Delta_1)}{\phi(\Delta_1)} + \frac{\phi'(\Delta_2)}{\phi(\Delta_2)} + \cdots + \frac{\phi'(\Delta_m)}{\phi(\Delta_m)} = 0, \tag{16.6}$$

where we have written $\phi'(\Delta_j)$ in place of $\frac{\partial \phi}{\partial \Delta_j}$. To simplify further, Gauss supposed that each of the observations $M_2, M_3, \ldots, M_m$ was equal to $M_1 - mN$ for some value N. It followed that $x_1 = M_1 - (m-1)N$, that $\Delta_1 = M_1 - x_1 = (m-1)N$, and that $\Delta_i = -N$ for $i > 1$. Substituting these values in equation (16.6) gave Gauss the relation

$$\frac{\phi'[(m-1)N]}{\phi[(m-1)N]} = (1-m)\frac{\phi'(-N)}{\phi(-N)}.$$

Because this is true for every positive integer m, Gauss concluded that

$$\frac{\phi'(\Delta)}{\Delta\phi(\Delta)} = k$$

for some constant k and therefore that $\log(\phi(\Delta)) = \frac{1}{2}k\Delta^2 + C$ or that $\phi(\Delta) = Ae^{(1/2)k\Delta^2}$. The Laplacian conditions on ϕ enabled Gauss to conclude that k is negative, say $k = -h^2$, and then finally that

$$\phi(\Delta) = \frac{h}{\sqrt{\pi}}e^{-h^2\Delta^2}.$$

That this was the "correct" error function followed for Gauss because he was able easily to derive from it the method of least squares. After all, given this function ϕ, the product Ω in the general case was given by

$$\Omega = h^m \pi^{-(1/2)m} e^{-h^2(\Delta_1^2 + \Delta_2^2 + \cdots + \Delta_m^2)}.$$

To maximize Ω, therefore, it is necessary to minimize the $\Sigma\Delta_i^2$, that is, to minimize the sum of the squares of the errors, the very procedure that Legendre had developed.

That the distribution of errors is "normal," that is, is determined by Gauss' function, gained even more credence because it was soon supported by much empirical evidence. In particular, Friedrich Bessel made three sets of measurements of star positions for several hundred stars and compared the theoretical number of errors between given limits, according to the normal law, with the actual values. The comparison showed very close agreement. Meanwhile, Laplace gave a new theoretical derivation of the normal law in a paper of 1810 and established the function $y = Ae^{-kx^2}$ as that representing error distributions and, in general, probability distributions, in a wide variety of situations.

Laplace included the normal law in his major work, the *Théorie analytique des probabilités (Analytic Theory of Probability)*, published in 1812. In that book Laplace collected all the material so far developed in probability theory, beginning with the definition of probability of an event as "the ratio of the number of cases favorable to it, to the number of possible cases, when there is nothing to make us believe that one case should occur rather than any other."[63] One major result in the book was the central limit theorem, Laplace's generalization of De Moivre's calculations of the previous century involving the terms of the binomial theorem. Laplace used this result to prove the normal law and also applied it to the question of the inclinations of the orbits of comets, a problem which he had considered for many years. Furthermore, he dealt with the applications of the theory of probability to such topics as insurance, demographics, decision theory, and the credibility of witnesses. In fact, it was Laplace's view that through the theory of probability mathematics could be brought to bear on the social sciences, just as calculus was the major tool in mathematizing the physical sciences. It took the rest of the century, however, for social scientists to develop the statistical tools necessary to provide a conceptual framework for using mathematics successfully. For the details of this fascinating story, the reader is referred to several recent books.[64]

Exercises

Problems from Cauchy

1. Show that the arithmetic mean of $f(h + i) - f(h + i - 1)$ for $i = 1, 2, \ldots, n$ is

$$\frac{f(h + n) - f(h)}{n}.$$

2. Prove the theorem of Cauchy:

If $\lim\limits_{x\to\infty} f(x + 1) - f(x) = \infty$, then $\lim\limits_{x\to\infty} f(x)/x = \infty$.

3. Use the theorem of problem 2 to show that

$$\lim\limits_{x\to\infty} \frac{\log x}{x} = 0 \quad \text{and} \quad \lim\limits_{x\to\infty} \frac{a^x}{x} = \infty.$$

4. Show that

$$\frac{2}{\pi} \int_0^\infty \frac{x^2\, dt}{t^2 + x^2} = |x|.$$

5. Use the modern definition of continuity and Cauchy's trigonometric identity

$$\sin(x + \alpha) - \sin x = 2 \sin \tfrac{1}{2}\alpha \cos(x + \tfrac{1}{2}\alpha)$$

to show that $\sin x$ is continuous at any value of x.

6. Use the Cauchy criterion to show that the series

$$1 + \frac{1}{1!} + \frac{1}{2!} + \frac{1}{3!} + \cdots$$

converges.

7. Show that if the sequence $\{a_i\}$ converges to a and if f is continuous, then the sequence $\{f(a_i)\}$ converges to $f(a)$.

8. Show that the series $\{u_k(x)\}$, where $u_1(x) = x$ and $u_k(x) = x^k - x^{k-1}$ for $k > 1$, satisfies the hypotheses of Cauchy's theorem 6-1-1 in a neighborhood of $x = 1$ but not the conclusion. Analyze Cauchy's proof for this case to see where it fails.

9. Use the trigonometric formula of problem 5 to prove that the derivative of the sine function is the cosine.

10. By putting $a^i = 1 + \beta$, use Cauchy's definition of derivative to show that the derivative of $y = a^x$ is $y' = a^x/\log_a(e)$.

11. Prove the algebraic result used by Cauchy in his main theorem about derivatives:

If $\quad A < \dfrac{a_i}{b_i} < B \quad$ for $\quad i = 1, 2, \ldots, n, \quad$ then

$$A < \frac{\sum_{i=1}^{n} a_i}{\sum_{i=1}^{n} b_i} < B.$$

12. Show that if $f(x)$ is continuous on $[a, b]$ and if $a = x_0 < x_1 < \cdots < x_n = b$ is a partition of $[a, b]$ into subintervals, then the sum

$$f(x_0)(x_1 - x_0) + f(x_1)(x_2 - x_1)$$
$$+ \cdots + f(x_{n-1})(x_n - x_{n-1})$$

is equal to $(b - a)f(x_0 + \theta(b - a))$ for some θ between 0 and 1.

13. Let $f(x) = x^2 + 3x$ on $[1, 3]$. Partition $[1, 3]$ into eight subintervals and determine the θ which satisifes the property of problem 12.

Problems from Bolzano

14. Complete Bolzano's proof of the least upper bound criterion by showing that the value U to which the constructed sequence converges is the least upper bound of all numbers having the property M.

15. Let A be the set of numbers in $(3/5, 2/3)$ that have decimal expansions containing only finitely many zeros and sixes after the decimal point and no other integer. Find the least upper bound of A.

Problems from Fourier

16. Show that $\phi(x) = \alpha e^{mx}$, $\psi(y) = \beta \cos ny$, with $m^2 = n^2 = 1/A$, are solutions to

$$\frac{\phi(x)}{\phi''(x)} = -\frac{\psi(y)}{\psi''(y)} = A.$$

Conclude that $v = ae^{-nx} \cos ny$ is a solution to

$$\frac{\partial^2 v}{\partial x^2} + \frac{\partial^2 v}{\partial y^2} = 0.$$

17. Show that

$$\int_0^{\pi} \sin mx \sin nx \, dx = \begin{cases} 0, & \text{if } m \neq n; \\ \pi/2, & \text{if } m = n. \end{cases}$$

18. Calculate the coefficients b_i in the Fourier cosine series of a function $\phi(x)$. That is, determine b_i if

$$\frac{1}{2}\pi\phi(x) = b_0 + b_1 \cos x + b_2 \cos 2x + b_3 \cos 3x + \cdots.$$

19. Calculate the Fourier series for

$$\phi(x) = \begin{cases} 1, & \text{if } 0 < x < \pi; \\ -1, & \text{if } \pi < x < 2\pi. \end{cases}$$

Problems on Continuity

20. Consider the Riemann function

$$f(x) = \sum_{n=1}^{\infty} \frac{\phi(nx)}{n^2}$$

defined on $[0, 1]$, where $\phi(x)$ is equal to x minus the nearest integer, or, if there are two equally near integers, equal to 0. Show that f is continuous except at the infinitely many points $x = p/2n$ with p and n relatively prime.

21. Prove Cauchy's theorem on the continuity of the sum of a series of continuous functions under the additional assumption that the series converges uniformly.

Problems from the Arithmetization of Analysis

22. Let $\mathcal{R}$ be the set of all real numbers as defined by Dedekind through his cuts. Show that this set possesses the basic attribute of continuity. Namely, show that if $\mathcal{R}$ is split into two classes $\mathcal{A}_1, \mathcal{A}_2$ such that every real number in $\mathcal{A}_1$ is less than every real number in $\mathcal{A}_2$, then there exists exactly one real number α which is either the greatest number in $\mathcal{A}_1$ or the smallest number in $\mathcal{A}_2$.

23. Define a natural ordering $<$ on Dedekind's set of real numbers $\mathcal{R}$ defined by the notion of cuts. That is, given two cuts $\alpha = (\mathcal{A}_1, \mathcal{A}_2)$, and $\beta = (\mathcal{B}_1, \mathcal{B}_2)$, define what it means for $\alpha < \beta$. Show that this ordering $<$ satisfies the same basic properties on $\mathcal{R}$ as it satisfies on the set of rational numbers.

24. Define an addition on Dedekind's cuts. Show that $\alpha + \beta = \beta + \alpha$ for any two cuts α and β.

25. What did Dedekind mean by a "proof" of the result $\sqrt{2}\sqrt{3} = \sqrt{6}$? How can this result be proved using his definition of a real number? (*Hint:* How would one define multiplication on cuts?)

26. Prove using Dedekind's techniques the theorem that every bounded increasing sequence of real numbers has a limit number. Prove the same result using Cantor's method. Which proof is easier?

27. Show that if $\{a_i\}$ and $\{b_i\}$ are fundamental sequences with $\{b_i\}$ not defining the limit 0, then $\{a_i/b_i\}$ is also a fundamental sequence.

28. Define the product AB of two fundamental sequences $A = \{a_i\}$ and $B = \{b_i\}$ as the sequence consisting of the products $\{a_i b_i\}$. Show that this definition makes sense and that if $AB = C$, then $B = C/A$, where division is defined as in problem 27.

29. Determine explicitly a point set P whose first and second derived sets, P', and P'', are different from P and from each other.

30. Cantor in 1890 gave a second proof that the real numbers of the interval $(0, 1)$ could not be placed in one-to-one correspondence with the natural numbers. Suppose that these numbers were in one-to-one correspondence with the natural numbers. Then there is a listing $r_1, r_2, r_3, \ldots$ of the real numbers in the interval. Write each such number in its infinite decimal form:

$$r_1 = 0.a_{11}a_{12}a_{13} \ldots$$

$$r_2 = 0.a_{21}a_{22}a_{23} \ldots$$

$$r_3 = 0.a_{31}a_{32}a_{33} \ldots$$

Now write, as did Cantor, a number b not in the list by choosing $b = 0.b_1 b_2 b_3 \ldots$ where $b_1 \neq a_{11}$, $b_2 \neq a_{22}$, $b_3 \neq a_{33}, \ldots$. Show that b cannot be in the original list.

31. Show that the rational numbers and the algebraic numbers can be put in one-to-one correspondence with the natural numbers.

Problems from Complex Analysis

32. Show explicitly using residues that

$$\int_{-\infty}^{\infty} \frac{\cos x}{1 + x^2}\, dx = \frac{\pi}{e}.$$

33. Show using residues that

$$\int_0^\infty \frac{dx}{1 + x^6} = \frac{2\pi}{3}.$$

34. Let the complex function $w(z)$ be given as the sum $u(x, y) + iv(x, y)$. Suppose that the Cauchy-Riemann equations are satisfied, that is, that $\partial u/\partial x = \partial v/\partial y$ and $\partial v/\partial x = -\partial u/\partial y$. Show that the derivative dw/dz is equal to $\partial u/\partial x + i\partial v/\partial x$.

Problems from Vector Analysis

35. Suppose a surface S is defined by three parametric equations $x = x(p, q)$, $y = y(p, q)$, $z = z(p, q)$. Show geomet-

rically that the element of surface dS can be written in the form

$$dS = \left[\left(\frac{\partial(y, z)}{\partial(p, q)} \right)^2 + \left(\frac{\partial(z, y)}{\partial(p, q)} \right)^2 + \left(\frac{\partial(x, y)}{\partial(p, q)} \right)^2 \right]^{1/2} dp\, dq.$$

36. Show that if $\sigma = Ai + Bj + Ck$ is a vector field with $\text{div } \sigma = 0$, then $\sigma = \text{curl } \tau$ for some vector field τ.

37. Use Stokes' theorem to show that if $\text{curl } \sigma = 0$, then $\int_C \sigma \cdot d\mathbf{r}$ is independent of the curve C but depends only on its endpoints. Similarly, if $\text{div } \sigma = 0$, then $\int_S \sigma \cdot d\mathbf{a}$ is independent of the surface C but depends only on its boundary curve.

FOR DISCUSSION . . .

38. What does Cauchy mean by his statement that an irrational number is the limit of the various fractions which approach it? What does Cauchy understand by the term "irrational number" or, even by the term "number"?

39. Explain the differences between Cauchy's definition of continuity on an interval and the usual modern definition of continuity at a point. Does a function which satisfies Cauchy's definition satisfy the modern one for every point in the interval? Does a function which satisfies the modern definition for every point in an interval satisfy Cauchy's definition?

40. Look up the construction of the natural numbers by both Peano and Frege and compare their work with the construction of Dedekind.

41. Develop a lesson plan for teaching the concept of uniform convergence by beginning with Cauchy's incorrect theorem and proof.

42. Read Heine's paper of 1872 "Die Elemente der Functionenlehre" (see note 28) and develop an outline of the beginning of a course in real analysis using the paper as a basis.

43. Note that Cauchy did not use Rolle's theorem to prove the mean value theorem as most current calculus books do today. Discuss the advantages and disadvantages of Cauchy's approach vis à vis the standard approach.

44. Compare the Eulerian and the Riemannian derivations of the Cauchy-Riemann equations. Which makes a better introduction for a course in complex analysis?

References and Notes

Among the best recent general works on the history of analysis in the nineteenth century are Ivor Grattan-Guinness, ed., *From the Calculus to Set Theory* (London: Duckworth, 1980), a collection of essays by experts on the various topics covered, Umberto Bottazzini, *The Higher Calculus: A History of Real and Complex Analysis from Euler to Weierstrass* (New York: Springer, 1986), and Ivor Grattan-Guinness, *The Development of the Foundations of Mathematical Analysis from Euler to Riemann* (Cambridge: MIT Press, 1970).

1. Abel, "Investigation of the series $1 + \frac{m}{1}x + \frac{m(m-1)}{1 \cdot 2}x^2 + \cdots$," Crelle's *Journal* 1 (1826), 311–339, p. 316. A part of this paper is translated in G. Birkhoff, *A Source Book in Classical Analysis* (Cambridge: Harvard University Press, 1973), pp. 68–70. The Birkhoff *Source Book* contains many of the most important papers in various areas of analysis during the nineteenth century.

2. Sylvestre Lacroix, *An Elementary Treatise on the Differential and Integral Calculus*, translated by Charles Babbage, George Peacock, and John Herschel (Cambridge: J. Deighton, 1816), p. 2.

3. Ibid., p. 5.

4. Oystein Ore, *Niels Henrik Abel: Mathematician Extraordinary* (New York: Chelsea, 1974), p. 147.

5. Lacroix, *Elementary Treatise,* Preface.

6. Cauchy, *Cours d'analyse de l'école royale polytechnique*, reprinted in Cauchy, *Oeuvres complète d'Augustin Cauchy* (Paris: Gauthier-Villars, 1882–) (2), 3, p. 19. (All further page references to the *Cours d'analyse* will be from this edition. All other references to Cauchy will also be to the *Oeuvres.*) The best treatment of Cauchy's work on calculus is Judith V. Grabiner, *The Origins of Cauchy's Rigorous Calculus* (Cambridge: MIT Press, 1981). Much of the detail in the first section of this chapter is adapted from this book. For a brief overview of the work of Cauchy and others on the notion of continuity, see Judith V. Grabiner, "Who gave you the epsilon? Cauchy and the origins of rigorous calculus," *American Mathematical Monthly* 90 (1983), 185–194. For more details on the development of the concept of the derivative, see Judith V. Grabiner, "The changing concept of change: The derivative from Fermat to Weierstrass," *Mathematics Magazine* 56 (1983), 195–203.

7. Cauchy, *Cours d'analyse*, p. 54.

8. Lagrange, *Théorie des fonctions analytique*, in *Oeuvres de Lagrange* (Paris: Gauthier-Villars, 1867–1892), vol. 9, p. 28. Quoted in Grabiner, *Cauchy's Rigorous Calculus*, p. 95.

9. Cauchy, *Cours d'analyse*, p. 43.

10. S. B. Russ, "A Translation of Bolzano's Paper on the Intermediate Value Theorem," *Historia Mathematica* 7 (1980), 156–185, p. 159. For more discussion of the work of Bolzano, see I. Grattan-Guinness, "Bolzano, Cauchy and the 'New Analysis' of the Early Nineteenth Century," *Archive for History of Exact Sciences* 6 (1970), 372–400. Grattan-Guinness claims that Cauchy took the central ideas of his definitions of continuity and convergence from Bolzano. But see also H. Freudenthal, "Did Cauchy Plagiarize Bolzano?" *Archive for History of Exact Sciences* 7 (1971), 375–392 for a contrary view.

11. Ibid., p. 162.

12. Cauchy, *Cours d'analyse*, p. 114.

13. Ibid., p. 116.

14. Quoted in A. J. Franco de Oliveira, "Anastácio da Cunha and the Concept of Convergent Series," *Archive for History of Exact Sciences* 39 (1988), 1–12, p. 4. For more on da Cunha, see João Filipe Queiró, "José Anastácio da Cunha: A Forgotten Forerunner," *The Mathematical Intelligencer* 10 (1988), 38–43 and A. P. Youschkevitch, "J. A. da Cunha et les fondements de l'analyse infinitésimale," *Revue d'Histoire des Sciences* 26 (1973), 3–22.

15. Russ, "Translation of Bolzano's Paper," p. 171.

16. This treatment of Cauchy's erroneous proof is adapted from that found in V. Frederick Rickey, *Using History in Teaching Calculus* (Washington: MAA, 1993). This very useful volume in the MAA Notes series contains numerous examples of how the history of calculus can be used in teaching the subject and, in particular, shows how to use the "proof" in teaching uniform convergence.

17. Cauchy, *Résumé des leçons données a l'école polytechnique sur le calcul infinitésimal*, in *Oeuvres* (2), 4, p. 44.

18. Lacroix, *Elementary Treatise*, p. 179.

19. Cauchy, "Mémoire sur l'intégration des équations lineares aux différentielles partielles a coefficients constantes," in *Oeuvres*, (2), 1, 275–357, p. 354.

20. Ibid., p. 334.

21. Cauchy, *Résumé*, pp. 122–123.

22. Ibid., p. 125.

23. Quoted in Grattan-Guinness, *From the Calculus to Set Theory*, p. 158.

24. Ibid., p. 153.

25. Quoted in Grattan-Guinness, *Foundations of Mathematical Analysis*, p. 104.

26. Quoted in Bottazzini, *Higher Calculus*, p. 244.

27. Quoted in Ann Hibner Koblitz, *Sofia Kovalevskaia: Scientist, Writer, Revolutionary* (Boston: Birkhäuser, 1983), p. 197. This book provides a detailed nontechnical biography of Kovalevskaya. Her mathematical work is treated in Roger Cooke, *The Mathematics of Sonya Kovalevskaya* (New York: Springer-Verlag, 1984).

28. E. Heine, "Die Elemente der Functionenlehre," *Journal für die Reine und Angewandte Mathematik* 74 (1872), 172–188, p. 184.

29. Dedekind, *Continuity and Irrational Numbers*, translated by Wooster Beman, in Dedekind, *Essays on the Theory of Numbers* (La Salle, Ill.: Open Court, 1948), pp. 1–2.

30. Ibid., p. 9.

31. Ibid., p. 11.

32. Ibid., pp. 12–13.

33. Ibid., p. 15.

34. Ibid., p. 22.

35. Quoted in Bottazzini, *Higher Calculus*, p. 277.

36. Ibid., p. 278.

37. Quoted in Joseph Dauben, *Georg Cantor: His Mathematics and Philosophy of the Infinite* (Princeton: Princeton University Press, 1979), p. 49. Dauben's biography provides a detailed study of Cantor's work and how it relates to the mathematics and philosophy of his day.

38. Cantor, *Contributions to the Founding of the Theory of Transfinite Numbers*, translated by Philip Jourdain (Chicago: Open Court, 1915), p. 85.

39. Ibid., p. 86.

40. Quoted in Gregory H. Moore, "Towards a History of Cantor's Continuum Problem," in David Rowe and John McCleary, eds., *The History of Modern Mathematics* (San Diego: Academic Press, 1989), vol. 1, 79–121, p. 82. The two volume work of Rowe and McCleary presents the proceedings of a conference on nineteenth century mathematics held in 1988. The papers are well worth reading.

41. Dedekind, *The Nature and Meaning of Numbers*, translated by Wooster Beman, in Dedekind, *Theory of Numbers*, p. 45.

42. Ibid., p. 50.

43. Ibid., p. 53.

44. Ibid., p. 63.

45. Ibid., p. 64.

46. Wessel, "On the Analytical Representation of Direction," translated in David Smith, *A Source Book in Mathematics* (New York: Dover, 1959), 55–66, p. 58.

47. Quoted in Bottazzini, *Higher Calculus*, p. 156, from the letter published in Gauss, *Werke* (Göttingen, 1866), vol. 8, pp. 90–92.

48. Cauchy, *Cours d'analyse*, p. 154.

49. Cauchy, *Mémoire sur les intégrales définies prises entre des limites imaginaires* (Paris, 1825), pp. 42–43, translated in Birkhoff, *Source Book*, p. 33.

50. Ibid., p. 44; p. 34.

51. Riemann, *Grundlagen für eine allgemeine Theorie der Functionen einer veränderlichen complexen Grösse*, parts of which are translated in Birkhoff, *Source Book*, 48–50, p. 48.

52. Ibid., p. 49.

53. Riemann, "Theorie der Abel'sche Funktionen," *Crelle's Journal* 54 (1857), parts of which are translated in Birkhoff, *Source Book*, 50–55, p. 51.

54. Ibid., p. 52–53.

55. Ibid.

56. More information on these theorems can be found in Victor J. Katz, "The History of Stokes' Theorem," *Mathematics Magazine* 52 (1979), 146–156.

57. Stokes, *Mathematical and Physical Papers* (Cambridge: Cambridge University Press, 1905), vol. 5, p. 320.

58. Legendre, "Sur la Méthode des moindres quarrés," in Legendre, *Nouvelles Méthodes pour la détermination des orbites des cometes* (Paris, 1805), translated in Smith, *Source Book*, 576–579, p. 576.

59. Ibid., p. 577.

60. For more details, see R. L. Plackett, "The discovery of the method of least squares," *Biometrika* 59 (1972), 239–251. This paper is reprinted in M. G. Kendall and R. L. Plackett, eds., *Studies in the History of Statistics and Probability*, vol. 2 (New York: Macmillan, 1977), 279–291. This volume and its companion, E. S. Pearson and M. G. Kendall, eds., *Studies in the History of Statistics and Probability*, vol. 1 (Darien, Conn.: Hafner, 1970), provide a valuable collection of essays on the history of probability and statistics, most of which originally appeared in *Biometrika*.

61. For more details, see Steven C. Althoen and Renate McLaughlin, "Gauss-Jordan Reduction: A Brief History," *The American Mathematical Monthly* 94 (1987), 130–142 and Victor J. Katz, "Who is the Jordan of Gauss-Jordan?" *Mathematics Magazine* 61 (1988), 99–100.

62. Gauss, *Theoria motus corporum celestium* (Hamburg: Perthes et Besser, 1809), translated by C. H. Davis as *Theory of Motion of the Heavenly Bodies Moving about the Sun in Conic Sections* (Boston: Little, Brown, 1857), p. 258. For more information on Gauss and the method of least squares, see O. B. Sheynin, "C. F. Gauss and the Theory of Errors," *Archive for History of Exact Sciences* 20 (1979), 21–72 and William C. Waterhouse, "Gauss's First Argument for Least Squares," *Archive for History of Exact Sciences* 41 (1991), 41–52. For a discussion of the work of Robert Adrain (1775–1843), an early American mathematician who also contributed to the method of least squares, see Jacques Dutka, "Robert Adrain and the Method of Least Squares," *Archive for History of Exact Sciences* 41 (1991), 171–184.

63. Laplace, *Théorie analytique des probabilités* (Paris: Courcier, 1812), p. 181.

64. The new books include Stigler, *History of Statistics*, Theodore M. Porter, *The Rise of Statistical Thinking* (Princeton: Princeton University Press, 1986), and Lorraine Daston, *Classical Probability in the Enlightenment* (Princeton: Princeton University Press, 1988).

Summary of Nineteenth Century Analysis

1736–1813	Joseph Louis Lagrange	Surface integrals
1744–1787	José Anastácio da Cunha	Definition of convergence
1745–1818	Caspar Wessel	Geometric representation of complex numbers
1749–1827	Pierre Simon Laplace	Analytic probability
1752–1833	Adrien Marie Legendre	Least squares
1765–1843	Sylvestre-François Lacroix	Calculus texts
1768–1830	Joseph Fourier	Fourier series
1777–1855	Carl Friedrich Gauss	Complex variables, normal distribution
1781–1848	Bernhard Bolzano	Continuity and convergence
1789–1857	Augustin-Louis Cauchy	Rigorization of calculus, complex analysis
1801–1861	Mikhail Ostrogradsky	Divergence theorem
1802–1829	Niels Henrik Abel	New ideas on convergence
1805–1859	Peter Lejeune-Dirichlet	Convergence of Fourier series
1815–1897	Karl Weierstrass	Uniform convergence
1819–1903	George Stokes	Vector analysis
1821–1881	Eduard Heine	New ideas in continuity
1826–1866	Georg Bernhard Riemann	Integration; beginning of topology
1831–1879	James Clerk Maxwell	Line integrals
1831–1916	Richard Dedekind	Dedekind cuts; axioms for natural numbers
1845–1918	Georg Cantor	Set theory
1850–1891	Sofia Kovalevskaya	Partial differential equations

Chapter 17
Geometry in the Nineteenth Century

"I am ever more convinced that the necessity of our geometry cannot be proved—at least not by human *reason for human reason. It is possible that in another lifetime we will arrive at other conclusions on the nature of space that we now have no access to. In the meantime we must not put geometry on a par with arithmetic that exists purely* a priori *but rather with mechanics." (Gauss, in a letter to Heinrich Olbers (1758–1840), 1817)*[1]

As part of his qualification to become a lecturer at the University of Göttingen, Bernhard Riemann needed to present an inaugural lecture to the members of the philosophical faculty, a lecture designed to show that he could apply his mathematical research to more general intellectual issues. He submitted three possible topics, the first two being closely related to some already completed research on complex functions and trigonometric series.

Gauss, however, acting on behalf of the faculty, picked Riemann's third topic, "On the Hypotheses which Lie at the Foundation of Geometry." On June 10, 1854, Riemann made his presentation. The lecture had few mathematical details but was packed with so many ideas about what geometry means that mathematicians have been studying it for well over a century.

The importance of analysis in the late eighteenth century, spurred in large extent by the work of Euler and continued in the early nineteenth century by Cauchy, among others, tended to lessen the importance of pure geometry during that period. But the applications of analysis to geometry led to many important new geometrical ideas.

Although Gauss early in his career had considered various aspects of what is today called differential geometry, it was only during his work on a detailed geodetic survey of the kingdom of Hanover, required of him as the Director of the Göttingen Astronomical Observatory, that he finally clarified his ideas on the subject of the theory of surfaces. He published these ideas in 1827 in a brief but densely packed paper in which he carried forward the introductory work of Euler on the theory of surfaces, applying the techniques of calculus to show that some of the basic notions of surface theory, including the notion of curvature, were intrinsic to the surface and did not depend on how the surface was situated in three-dimensional space.

In his work on surfaces, Gauss established a relationship between curvature and the sum of the angles of a triangle on the surface. This relationship turned out to be closely related to the old question of the parallel postulate, whose truth implied that the sum of the angles of a plane triangle was equal to two right angles. Toward the end of his life, Gauss noted that he had long been convinced that the parallel postulate could not be proved and that the acceptance of alternatives could well lead to new and interesting geometries, the "truth" of which for the physical world could only be established by experiment. Nevertheless, Gauss never published any of his ideas on the subject. It was therefore left for Nikolai Lobachevsky and János Bolyai in the 1820s to publish, independently of each other, the first full treatments of a non-Euclidean geometry.

It took nearly 40 years, however, for the ideas of non-Euclidean geometry to make an impression in the mathematical community. It was only with the work of Riemann in 1854 and Hermann von Helmholtz in 1868 on the general notion of a geometrical manifold of arbitrary dimension that the meaning of these new ideas for the study of geometry took hold. Shortly thereafter, various models of non-Euclidean geometries in Euclidean space were introduced, thus convincing the mathematical public that the non-Euclidean geometries were as valid as the Euclidean one from a logical standpoint and that the question of the "truth" of Euclidean geometry for the world in which we live no longer had an obvious answer.

There were also advances in the subject of projective geometry over the early work of Pascal and Desargues, accomplished by such mathematicians as Jean-Victor Poncelet, Michel Chasles, and Julius Plücker. In 1871, Felix Klein showed a connection between projective and non-Euclidean geometry by way of the idea of a metric. The following year he gave a new definition of a geometry in terms of transformations, a definition which demonstrated the relationship of projective to Euclidean geometry and also the connection of geometry with the emerging theory of groups.

Geometry as studied to the middle of the nineteenth century dealt with objects of dimension no greater than three. But with the increasing use of analytical and algebraic methods, it became clear that for many geometric ideas there was no particular reason to limit the number of dimensions to just the number which could be physically realized. Thus, various mathematicians generalized their formulas and theorems to n dimensions, where n could be any positive integer. It was Hermann Grassmann, however, who, beginning in 1844, first attempted a detailed study of n-dimensional vector spaces from a geometric point of view. Unfortunately, Grassmann's work, like that of Bolyai and Lobachevsky, was not appreciated until much later. It was finally Giuseppe Peano who was able to provide axioms for a finite-dimensional vector space and thus provide a basis for the study of higher dimensional geometry.

With the creation of the various new geometries, many mathematicians toward the end of the century felt that it was time to redo the foundations of the entire subject, just as was being done in analysis. Flaws had been discovered in Euclid's reasoning, and these flaws led to certain problems in developing the non-Euclidean geometries. Thus both Peano and David Hilbert brought out new sets of axioms for Euclidean geometry which helped to remove the various flaws and clarify exactly what had to be assumed in order to develop both the old and the new geometries.

17.1 DIFFERENTIAL GEOMETRY

Having led the survey of Hanover from 1820 to 1825, and having introduced various new methods establishing geodesy as a recognized science, Gauss was finally able by 1827 to put on paper the results of his thoughts of over a quarter century on the subject of curved surfaces. Gauss noted in the abstract of his paper, "Disquisitiones Generales circa Superficies Curvas" (*General Investigations of Curved Surfaces*), that:

> although geometers have given much attention to general investigations of curved surfaces and their results cover a significant portion of the domain of higher geometry, this subject is still so far from being exhausted, that it can well be said that, up to this time, but a small portion of an exceedingly fruitful field has been cultivated. Through the solution of the problem, to find all representations of a given surface upon another in which the smallest elements remain unchanged, the author sought some years ago to give a new phase to this study. The purpose of the present discussion is further to open up other new points of view and to develop some of the new truths which thus become accessible.[2]

Gauss had already solved his problem of establishing the conditions for mapping one surface conformally onto another (so that "the smallest elements remain unchanged") by 1822 in a challenge question which he had suggested the Copenhagen Scientific Society pose. The answer was that the function had to be representable as a complex analytic one in the parameters representing the two surfaces. (Gauss did not, however, use complex function theory at the time.) But in the course of developing this answer, Gauss realized that a central idea involved in the study of surfaces was that of curvature, and, in particular, he realized how curvature could be calculated in terms of an analytic description of the surface in question.

17.1.1 The Definition of Curvature

Gauss began his "Disquisitiones Generales" with the notion of the curvature of a curved surface. He decided that he would only deal with surfaces, or parts of surfaces, with "continuous curvature," that is, surfaces (or parts) which possess tangent planes at all points. Thus, the vertex of a cone would not be considered, since there is no tangent plane, and no curvature, at that point. Because the sphere was Gauss' "model" surface, a surface with constant curvature analogous to the constant curvature of the circle in the plane, Gauss decided to define the curvature at a point on a surface by comparing a region around that point to a corresponding region around a point on the unit sphere. Curvature is a local property on a surface S. However it is to be defined, it is clear that the curvature may vary from point to point. On the unit sphere, however, the curvature at every point is set at 1. To make his comparison, therefore, Gauss created a mapping n (today called the Gauss normal map) from S to the unit sphere defined so that the vector from the center to $q = n(p)$ is parallel to the normal vector to S at p (that is, to the normal vector to the tangent plane to S at p) (Figure 17.1). This mapping takes a bounded region A of the surface S to

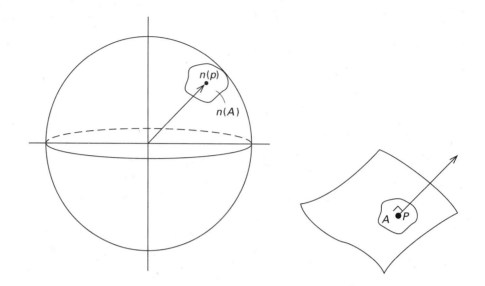

FIGURE 17.1
Gauss' normal map.

a bounded region $n(A)$ of the sphere. Gauss then defined the **total curvature** of A to be the area of $n(A)$, while he defined the more important concept of the **measure of curvature** at a point as the "quotient when the [total] curvature of the surface element about a point is divided by the area of the element itself; and hence it denotes the ratio of the infinitely small areas which correspond to one another on the curved surface and on the sphere."[3] In more modern terminology, Gauss defined the curvature at a point p to be

$$k(p) = \lim_{A \to p} \frac{\text{area of } n(A)}{\text{area of } A}$$

where the limit is taken as the region A around p shrinks to the point p itself. It follows that the total curvature of a region A is equal to $\int k \, d\sigma$ where $d\sigma$ is the element of area on the surface and the integral is taken over the region A.

A modern reader might note at least two problems with Gauss' definition of the measure of curvature. First, how is the area on an arbitrary curved surface to be defined and, second, assuming this is done, how do we know that the limit, if it exists, is independent of the way the shrinking of the region is done? Gauss did not address these problems. In fact, he didn't really define the curvature as a limit but merely as the ratio of infinitesimals. He then used his geometric intuition to assure himself that the definition made sense. For example, if the surface S is a sphere of radius r, then the area of $n(A)$ is equal to $1/r^2$ times the area of A (for any region A) and therefore the curvature at every point is equal to $1/r^2$. Similarly, if S is a plane, then $n(A)$ is equal to a point for any region A. Its area is therefore 0, as is the curvature. A somewhat more surprising result, but one which convinced Gauss that his definition is correct, occurs when S is a circular cylinder. In this case, the image $n(A)$ of a region A is simply a curve on the sphere and thus again has area 0. It follows that the curvature of the cylinder, like that of the plane, is also 0. Why this result is reasonable will be discussed shortly.

17.1.2 Curvature and the *Theorema Egregium*

Gauss was able to use his definition to calculate the curvature in terms of the equation of the given surface. Because the tangent plane to S at p is parallel to the tangent plane to the unit sphere at $n(p)$, the ratio of the area of $n(A)$ to that of A is equal to the ratio of the area of the projection of $n(A)$ onto the xy-plane to that of the projection of A on that same plane. Considering, therefore, a triangular region A whose projection has the three vertices (x, y), $(x + dx, y + dy)$, $(x + \delta x, y + \delta y)$, Gauss noted that the area of the triangle is $\frac{1}{2}(dx\,\delta y - dy\,\delta x)$. Similarly, if the functions $X(x, y)$, $Y(x, y)$ represent the composition of the projection with the normal function n, then the corresponding triangle for $n(A)$ has area $\frac{1}{2}(dX\,\delta Y - dy\,\delta X)$. It follows that

$$k = \frac{dX\,\delta Y - dY\,\delta X}{dx\,\delta y - dy\,\delta x}.$$

It now remained for Gauss to determine the value of this fraction if the surface is defined either by an equation $z = z(x, y)$, by an equation $W(x, y, z) = 0$, or by the parametric equations $x = x(u, v)$, $y = y(u, v)$, $z = z(u, v)$. For the first method of representation, since

$$dX = \frac{\partial X}{\partial x}\,dx + \frac{\partial X}{\partial y}\,dy, \quad dY = \frac{\partial Y}{\partial x}\,dx + \frac{\partial Y}{\partial y}\,dy,$$

$$\delta X = \frac{\partial X}{\partial x}\,\delta x + \frac{\partial X}{\partial y}\,\delta y, \quad \delta Y = \frac{\partial Y}{\partial x}\,\delta x + \frac{\partial Y}{\partial y}\,\delta y,$$

Gauss determined that

$$k = \frac{\partial X}{\partial x}\frac{\partial Y}{\partial y} - \frac{\partial X}{\partial y}\frac{\partial Y}{\partial x}.$$

It is then a straightforward, although messy, calculation to rewrite this expression in terms of the partial derivatives of z to get

$$k = \frac{z_{xx}z_{yy} - z_{xy}^2}{(1 + z_x^2 + z_y^2)^2}.$$

Gauss calculated similar expressions for k when the surface is represented by an equation in three variables and by parametric equations. Gauss was then able to derive a series of beautiful theorems.

First, he showed that the measure of curvature at a point p was expressible in terms of the curvatures of two specific sections of the surface through p. By use of a suitable choice of axes, he rewrote the equation $z = z(x, y)$ in a form where $p = (0, 0, 0)$ and $z_x(0, 0) = z_y(0, 0) = z_{xy}(0, 0) = 0$. It then followed that the maximum and minimum curvatures of all curves formed by normal sections through p are $z_{xx}(0, 0)$ and $z_{yy}(0, 0)$. Thus, the measure of curvature $k(p)$ at p is $z_{xx}z_{yy}$, the product of the two extreme curvatures of the normal sections.

Second, Gauss proved his *theorema egregium* to the effect that the measure of curvature was an isometric invariant of the surface, that is, it did not change if the surface was transformed by a distance-preserving transformation. To accomplish this, he took the parametric form of representation of the surface, $x = x(u, v)$, $y = y(u, v)$, $z = z(u, v)$, set $E = x_u^2 + y_u^2 + z_u^2$, $F = x_u x_v + y_u y_v + z_u z_v$, and $G = x_v^2 + y_v^2 + z_v^2$, and derived a formula for the curvature expressed solely in terms of E, F, G and their partial derivatives of first and second order with respect to u and v. But because E, F, G are unchanged by a distance-preserving function, the curvature is also unchanged. As Gauss stated his "remarkable" theorem, if one surface is "developed" onto another, that is, if there is a one-to-one function from one surface to another which preserves the element of length, then the measures of curvature at corresponding points of the two surfaces are always equal. For example, because the plane can be developed onto a cylinder, the curvature of the cylinder equals that of the plane, namely 0. Gauss emphasized that his result was only the beginning of a new and important method of studying a surface,

> not as the boundary of a solid, but as a flexible, though not extensible, solid, one dimension of which is supposed to vanish. [In this way] the properties of the surface depend in part upon the form to which we can suppose it reduced, and in part are absolute and remain invariable, whatever may be the form into which the surface is bent. To these latter properties, the study of which opens to geometry a new and fertile field, belong the measure of curvature and the integral curvature. . . . From this point of view, a plane surface and a surface developable on a plane, *e.g.* cylindrical surfaces, conical surfaces, etc., are to be regarded as essentially identical.[4]

Finally, Gauss demonstrated the important relationship between the total curvature of a triangle formed by geodesic arcs on the surface (arcs of shortest length) and the sum of the angles in that triangle. In fact, he calculated that the total curvature $\int k \, d\sigma$ over a geodesic triangle was equal to $A + B + C - \pi$, where A, B, C are the measures of the three angles of that triangle. For example, on a surface of constant positive curvature every geodesic triangle has angle sum greater than π, while on a surface of constant negative curvature every such triangle has angle sum less than π. Gauss' result was a generalization of the well-known result that on a unit sphere, where the total curvature of a region equals its area, the angle sum of a triangle composed of great circle arcs (geodesics) was greater than π by a value equal to the area of the triangle.

Gauss' treatise on the differential geometry of surfaces was not only significant in and of itself but also had great consequences for future work. In particular, it turned out that the relationship of angle sum of a triangle to the intrinsic geometry on the surface helped lead to the solution of the question of the validity of Euclid's parallel postulate. Furthermore, Gauss' characterization of a surface in terms of its length element proved to be the beginning of the general theory of n-dimensional manifolds, many important aspects of which were developed in the work of Riemann some 30 years later.

17.2 NON-EUCLIDEAN GEOMETRY

Recall that in the eighteenth century both Saccheri and Lambert attempted to prove Euclid's parallel postulate by assuming that it was false and trying to derive a contradiction. Saccheri believed that he had succeeded in this endeavor, but Lambert realized that his attempt was a failure. Both attacked the problem through synthethic means, trying to use the methodology of Euclid to show that he had assumed an unnecessary postulate. The nineteenth century, however, with its increasing use of analysis to solve all sorts of problems, provided a new approach to this one as well. And interestingly enough, it was the hyperbolic functions of Lambert which were called into service to make the connection between analysis and a new geometry, a connection that Lambert himself had missed.

17.2.1 Taurinus and Log-Spherical Geometry

Lambert had noted that the hypothesis of the acute angle would seem to hold on the surface of a sphere of imaginary radius, but it was Franz Taurinus (1794–1874), a man of independent means who pursued mathematics as a hobby, who actually made this connection explicit in a work of 1826. Taurinus began with a formula of spherical trigonometry connecting the sides and an angle of an arbitrary spherical triangle on a sphere of radius K:

$$\cos\frac{a}{K} = \cos\frac{b}{K}\cos\frac{c}{K} + \sin\frac{b}{K}\sin\frac{c}{K}\cos A,$$

where the triangle has sides a, b, c and opposite angles A, B, C (Figure 17.2). Replacing K by iK, that is, making the radius of the sphere imaginary (whatever that means), and recalling that $\cos ix = \cosh x$ and $\sin ix = i \sinh x$, Taurinus derived the new formula

$$\cosh\frac{a}{K} = \cosh\frac{b}{K}\cosh\frac{c}{K} - \sinh\frac{b}{K}\sinh\frac{c}{K}\cos A. \tag{17.1}$$

Taurinus called the geometry defined by this formula "log-spherical geometry," but he realized that this geometry was not possible in the plane. Exploring the consequences of the formula gives some idea of the properties of this geometry, however. For example, if the triangle is equilateral ($a = b = c$), the formula becomes

$$\cosh\frac{a}{K} = \cosh^2\frac{a}{K} - \sinh^2\frac{a}{K}\cos A$$

or

$$\cos A = \frac{\cosh^2\frac{a}{K} - \cosh\frac{a}{K}}{\sinh^2\frac{a}{K}} = \frac{\cosh\frac{a}{K}\left(\cosh\frac{a}{K} - 1\right)}{\cosh^2\frac{a}{K} - 1} = \frac{\cosh\frac{a}{K}}{\cosh\frac{a}{K} + 1}.$$

Because $\cosh\frac{a}{K} > 1$, it follows that $\cos A > 1/2$ and therefore that $A < 60°$. In other words, the sum of the angles of an equilateral triangle in this geometry is less than 180°. On the other hand, it is easy to see that as either the sides get smaller or the radius K gets larger, the angle A approaches 60° and the geometry approaches Euclid's geometry. In fact, one can also show (by using appropriate power series expansions) that in the limit as K approaches ∞, Taurinus' formula (17.1) reduces to the Euclidean law of cosines $a^2 = b^2 + c^2 - 2bc\cos A$.

A second important formula of spherical trigonometry, which connects the angles and a side of a spherical triangle, is

$$\cos A = -\cos B \cos C + \sin B \sin C \cos\frac{a}{K}.$$

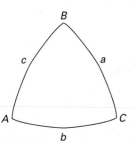

FIGURE 17.2
Spherical triangle on sphere of radius K.

On replacing K by iK, this formula becomes a formula of log-spherical geometry:

$$\cos A = -\cos B \cos C + \sin B \sin C \cosh \frac{a}{K}. \tag{17.2}$$

For the special case where $A = 0°$ and $C = 90°$, formula (17.2) reduces to

$$\cosh \frac{a}{K} = \frac{1}{\sin B}.$$

Naturally, a triangle with right angle at C and an angle of zero degrees at A does not exist in Euclid's geometry. Recall, however, that Saccheri had realized that the hypothesis of the acute angle led to the concept of asymptotic straight lines. Thus Taurinus' triangle must be thought of as one in which two sides are asymptotic (Figure 17.3). The angle B then depends on the length a of the third side through the formula $\sin B = \text{sech } \frac{a}{K}$. One can rewrite this formula in the form

$$\tan \frac{B}{2} = e^{-(a/K)},$$

a formula which was to become fundamental in the work of Lobachevsky.

Taurinus derived other analytic consequences of his log-spherical geometry. Thus, he showed that if one constructed an isosceles right triangle with the two legs both asymptotic to the hypotenuse, divided into two right triangles with angle $B = 45°$, then a is the maximum possible altitude h of any isosceles right triangle (Figure 17.4). In that case $\cosh \frac{h}{K} = \sqrt{2}$, and

$$K = \frac{h}{\ln(1 + \sqrt{2})}.$$

Furthermore, the area of a triangle is proportional to its defect (as Lambert had already discovered), the length of the circumference of a circle of radius r is $2\pi K \sinh \frac{r}{K}$, and the area of a circle of radius r is $2\pi K^2(\cosh \frac{r}{K} - 1)$. It is noteworthy that these latter results, as well as much of the work of Lobachevsky and Bolyai to be discussed shortly, had all been worked out by Gauss somewhat earlier in his private papers. Gauss, however, perhaps because he did not feel that he had proved all the various results to his own high standards, never published anything on the subject directly. On the other hand, his work

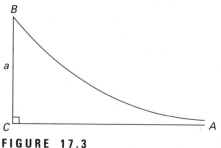

FIGURE 17.3
Triangle in which $\angle C = 90°$ and $\angle A = 0°$.

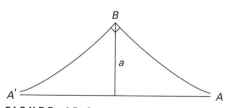

FIGURE 17.4
Isosceles right triangle with both base angles equal to $0°$.

relating curvature of a surface to the defect or excess of a triangle must have somehow been stimulated by his thoughts on this new geometry.

17.2.2 The Non-Euclidean Geometry of Lobachevsky and Bolyai

Despite his analytic results, Taurinus was not convinced that his geometry on an imaginary sphere applied in any "real" situation. The formulas were simply a collection of pretty results with no real content. But because neither Saccheri nor Lambert had succeeded in their attempts to refute the hypothesis of the acute angle, other mathematicians began to believe that a plane geometry in which that hypothesis was valid could exist. And that geometry would have as its analytic basis the formulas of Taurinus. The two mathematicians who first had confidence enough in their new ideas to publish them were the Russian Nikolai Ivanovich Lobachevsky (1792–1856) and the Hungarian János Bolyai (1802–1860). Both began work on the problem of parallels determined to find a correct refutation of the hypothesis of the acute angle. And both gradually changed their minds.

In 1826, Lobachevsky gave a lecture at Kazan University in which he outlined a geometry having more than one parallel to a given line through a given point. Three years later, he published an extended version of his lecture in his Russian text *On the Principles of Geometry*. Over the next decade, he published several other versions of his new geometrical researches, including a detailed summary in 1840 in German entitled *Geometrische Untersuchungen zur Theorie der Parallellinien* (*Geometrical Investigations on the Theory of Parallel Lines*). Bolyai, also doing most of his creative work in the 1820s, published his material (in Latin) in 1831 as an appendix to a geometric work of his father Farkas Bolyai, called *Appendix exhibiting the absolutely true science of space, independent of the truth or falsity of Axiom XI [the Parallel Postulate] of Euclid, that can never be decided* a priori. Because the ideas of Lobachevsky and Bolyai turned out to be remarkably similar, we will concentrate on the work of the former, as detailed in his *Geometrical Investigations*.

FIGURE 17.5
Lobachevsky on a Russian stamp.

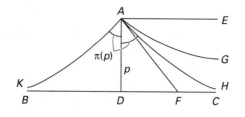

FIGURE 17.6
Lobachevsky's angle of parallelism.

Lobachevsky began with a summary of certain geometrical results which were true independent of the parallel postulate. He then stated clearly his new definition of parallels: "All straight lines which in a plane go out from a point can, with reference to a given straight line in the same plane, be divided into two classes—*cutting* and *not-cutting*. The *boundary lines* of the one and other class of those lines will be called **parallel** to the given line."[5] Thus if *BC* is a line, *A* a point not on the line, and *AD* the perpendicular from *A* to *BC*, one can first draw a line *AE* perpendicular to *AD* (Figure 17.6). The line *AE* does not meet *BC*. Lobachevsky then assumed that there may be other lines through *A*, such as *AG*, which also do not meet *BC*, however far they are prolonged. In passing from a cutting line, such as *AF*, to a not-cutting line, such as *AG*, there must be a line *AH* which is the boundary between these two sets. It is *AH* which is the parallel to *BC*. The angle *HAD* between *AH* and the perpendicular *AD*, an angle dependent on the length *p* of *AB*, is what Lobachevsky calls the **angle of parallelism**, written $\Pi(p)$. If $\Pi(p) = 90°$, then there is only one line through *A* parallel to *BC* and the situation is the Euclidean one. If, however, $\Pi(p) < 90°$, then there will be a corresponding line *AK* on the other side of *AD* from *AH* which also makes the same angle $\Pi(p)$ with *AD*. It is thus always necessary in this non-Euclidean case to distinguish two different sides in parallelism. In any case, on each side of *AD*, under the non-Euclidean assumption, there are infinitely many lines through *A* which do not meet *BC*.

From the non-Euclidean assumption, Lobachevsky derived many results, some of which were in essence known to Saccheri and/or Lambert. For example, he showed that the property that $\Pi(p) < 90°$ for any *p* is equivalent to the property that the angle sum of every triangle is less than 180°. In that case, not only is the equation $\Pi(p) = \alpha$ solvable for any α less than a right angle, but also parallel lines are asymptotic to one another. To define the nature of parallel lines more precisely, Lobachevsky defined a new curve: "We call the **boundary line** (or **horocycle**) that curve lying in a plane for which all perpendiculars erected at the mid-points of chords are parallel to each other."[6] In other words, given a line *AB* with *A* to be on the horocycle, any other point *C* is on the horocycle if *AC* makes an angle $\Pi(AC/2)$ with the line *AB*, for in that case the perpendicular *DE* to *AC* at its midpoint will be parallel to *AB* (Figure 17.7). In fact, it turns out that all perpendiculars to the horocycle will be parallel. One can think of this curve as a circle of infinitely great radius, which, under the Euclidean assumption, would be a straight line. Letting *A*, *B* be two points on a horocycle separated by a distance *s* and drawing two perpendiculars *AA'*, *BB'* to the horocycle such that $AA' = BB' = x$, Lobachevsky constructed a new horocycle through *A'*, *B'* with the distance *A'B'* set equal to *s'* (Figure 17.8). He then showed that s'/s depends only on the distance *x*, that is, $s'/s = f(x)$. If a new horocycle *A"B"* is constructed at a distance *x* from *A'B'* in the same manner with $A"B" = s"$, then it turns out also that $s"/s' = f(x)$ and therefore that $f(2x) = s"/s = f(x)^2$. Similarly, $f(nx) = f(x)^n$, and

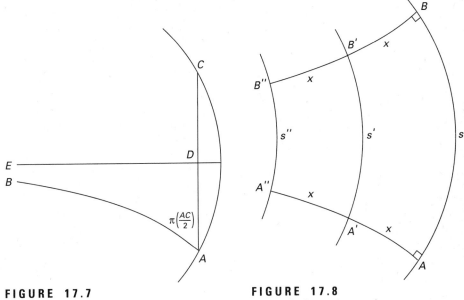

FIGURE 17.7
Lobachevsky's horocycle.

FIGURE 17.8
Perpendiculars to horocycles.

Lobachevsky could conclude that $s' = sa^{-x}$ for some constant a. Because the units of measure are arbitrary, he took a to be e. The distance between the parallel lines AA', BB' is then given by the function $s' = se^{-x}$, where x is measured from A and/or B away from the horocycle. It follows that the parallel lines are indeed asymptotic.

Lobachevsky's most interesting results, and ones which were unknown to Saccheri and Lambert, involved the trigonometry of the non-Euclidean plane. Through a complex argument involving spherical triangles and triangles in the non-Euclidean plane, he was able to evaluate explicitly the function $\Pi(x)$ in the form

$$\tan \frac{1}{2}\Pi(x) = e^{-x},$$

essentially the same result obtained by Taurinus. It followed that $\Pi(0) = \pi/2$ (or that for small values of x, the geometry is close to the Euclidean one) and that $\lim_{x \to \infty} \Pi(x) = 0$. Lobachevsky could then derive new relationships among the sides a, b, c and the opposite angles A, B, C of an arbitrary non-Euclidean triangle:

$$\sin A \cot \Pi(b) = \sin B \cot \Pi(a), \tag{17.3}$$

$$\cos A \cos \Pi(b) \cos \Pi(c) + \frac{\sin \Pi(b) \sin \Pi(c)}{\sin \Pi(a)} = 1, \tag{17.4}$$

$$\cot A \sin C \sin \Pi(b) + \cos C = \frac{\cos \Pi(b)}{\cos \Pi(a)}, \tag{17.5}$$

$$\cos A + \cos B \cos C = \frac{\sin B \sin C}{\sin \Pi(a)}. \tag{17.6}$$

Lobachevsky's formulas imply standard Euclidean formulas when the sides of the triangle are small. For in that case, using the explicit formula for $\Pi(x)$, one can first show that

$$\cot \Pi(x) = \sinh x \qquad \cos \Pi(x) = \tanh x \qquad \sin \Pi(x) = \frac{1}{\cosh x}$$

and then approximate the values of the hyperbolic functions by the terms of their power series up to degree 2: $\cot \Pi(x) = x$, $\cos \Pi(x) = x$, and $\sin \Pi(x) = 1 - \frac{1}{2}x^2$. Substituting these approximations into the four equations and neglecting terms of degree higher than 2 gives the results

$$b \sin A = a \sin B,$$

$$a^2 = b^2 + c^2 - 2bc \cos A,$$

$$a \sin(A + C) = b \sin A,$$

$$\cos A + \cos(B + C) = 0.$$

The first two results are the familiar laws of sines and cosines, respectively, while the last two, when combined with the first two, are equivalent to the result that $A + B + C = \pi$. It also follows that if one replaces the sides a, b, c of the triangle by ia, ib, ic respectively, Lobachevsky's results transform into standard results of spherical trigonometry. Thus Lobachevsky's geometry is essentially the same as Taurinus' log-spherical geometry on the sphere of imaginary radius.

Lobachevsky probably never read Taurinus' work. So what he saw in his trigonometric formulas was simply "a sufficient foundation for considering the assumption of [non-Euclidean] geometry as possible. Hence," he concluded, "there is no means, other than astronomical observations, to use for judging of the exactitude which pertains to the calculations of the ordinary geometry."[7] Gauss, too, realized that the creation of a new and apparently valid geometry in which Euclid's parallel postulate did not hold showed that there was no "necessity" to Euclid's geometry and that one could not automatically conclude that Euclid's geometry held in the world in which we live. It was necessary to experiment to decide whether the geometry of the physical universe was Euclidean or not. Lobachevsky in fact attempted such an experiment, using data on star positions, but the results were inconclusive.

Bolyai also commented that it is not decided whether Euclidean geometry or a non-Euclidean geometry represents "reality." In fact, he claimed, "it remains to demonstrate the impossibility (apart from any supposition) of deciding *a priori* whether [Euclidean geometry] or some [non-Euclidean] geometry (and which one) exist. This, however, is reserved for a more suitable occasion."[8] Although the "more suitable occasion" never occurred, Bolyai did derive most of the same mathematical results as Lobachevsky. He was, however, more explicit in dealing with **absolute geometry**, the collection of theorems which were true independent of the parallel postulate. For example, he proved that "in any rectilinear triangle, the [circumferences of the] circles with radii equal to its sides are as the sines of the opposite angles."[9] In Euclidean geometry, where the circumference of a circle of radius r is $2\pi r$, this result is simply that $a : b : c = \sin A : \sin B : \sin C$, the law of sines. In non-Euclidean geometry, where the corresponding circumference is

---

Born in Kolozsvar, Hungary (now Cluj, Romania), Bolyai received his early education in Maros-Vásárhely (now Tirgu Mures), where his father, Farkas Bolyai, a friend of Gauss, was a professor. At the age of 16, he entered the imperial military academy in Vienna and became a military officer, serving in that capacity in Temesvar (Timisoara), Arad, and Ľvov (Lemberg). He had to retire from the service in 1833, however, due to his physical condition. Meanwhile, his father, always being interested in the question of parallels, had corresponded with Gauss on the question over several years, without any resolution. He ultimately wearied of the matter and warned his son too about attacking this subject. Nevertheless, János persisted and, in 1823, told his father that he had made "wonderful discoveries" in the theory of parallels. He finally published the discoveries in 1831. Although János continued to develop his theory of space, his great disappointment at Gauss' response that he (Gauss) had already discovered the basic ideas of non-Euclidean geometry caused him to give up any further thought of publishing his own ideas (Figure 17.9).

FIGURE 17.9
Bolyai on a Hungarian stamp.

$2\pi K \sinh\frac{r}{K}$, for some constant K (each such constant determining for Bolyai a different geometry), the theorem translates into

$$\sinh\frac{a}{K} : \sinh\frac{b}{K} : \sinh\frac{c}{K} = \sin A : \sin B : \sin C,$$

a result equivalent to Lobachevsky's equation (17.3).

The work of Bolyai and Lobachevsky, although responding to an age-old question about the parallel postulate, drew very little response from the mathematical community before the 1860s. There are several reasons for this, ranging from the fact that some (though not all) of their articles were published in somewhat obscure sources and not in the major languages of the day, to the general difficulties of the acceptance of an entirely new idea into mathematics. But it would appear that the most important reason that the discoveries of the Hungarian and the Russian did not immediately become part of the mainstream of mathematics was that few could really understand what a non-Euclidean plane was. Although the arguments of the founders were correct and logically coherent, and although they displayed what appeared to be reasonable mathematical formulas involving known functions, the "reality" of this new geometry simply was not accepted. Until non-Euclidean geometry could be seen as part of a more general system of geometry and be connected through this system to Euclidean geometry, it was not to be anything more than a curious sidelight.

17.2.3 Riemann's Hypotheses for Geometry

The first mathematician to create a new general system of geometry was Riemann. He presented his ideas to the professors of the philosophical faculty at Göttingen in his inaugural lecture entitled "Über die Hypothesen welche der Geometrie zu Grunde liegen" (On the Hypotheses which lie at the Foundation of Geometry).

In the lecture, Riemann gave few mathematical details, but devoted much time to explaining what geometry ought to be about. Thus he began with "the task of constructing the concept of a multiply-extended quantity from general notions of quantity. It will be shown that a multiply-extended quantity is susceptible of various metric relations, so that space constitutes only a special case of a triply-extended quantity. From this, however, it is a necessary consequence that the theorems of geometry cannot be deduced from general notions of quantity, but that those properties which distinguish space from other conceivable triply-extended quantities can only be deduced from experience."[10] In other words, for Riemann the most general geometrical notion was what is today called a manifold. On a manifold one can establish various metric relations, ways of determining distance. The usual "space" dealt with by (three-dimensional) Euclidean geometry and which had generally been assumed to be the physical space in which we live, is then a special case of a three-dimensional manifold, having attached to it the Euclidean metric, expressed infinitesimally as $ds^2 = dx^2 + dy^2 + dz^2$. Agreeing with Gauss, he also intended to argue that the precise nature of physical space could not be determined *a priori* but only by "experience."

In part one of his lecture, Riemann dealt with the idea of a manifold of n-dimensions. He constructed such a manifold inductively, beginning with the idea of a one-dimensional manifold, or curve, "whose essential characteristic is, that from any point in it a continuous movement is possible in only two directions, forwards and backwards."[11] A two-dimensional manifold is created by having one one-dimensional manifold pass continuously into a second one; the two-dimensional manifold then consists of all the points formed by this passage. Similarly, a three-dimensional manifold is formed by the continuous passage of one two-dimensional manifold into a second and so on for ever higher dimensions. The central idea for Riemann appears to be that the introduction of each higher dimension results in the addition of one further direction in which one can go at a point or, in more modern terminology, the addition of a new dimension to the tangent space to the manifold at a point. Riemann noted, further, that one can reverse the procedure in some sense and define an $(n - 1)$-dimensional manifold as the zeroes of a function defined on an n-dimensional one. One can take these functions to be what are today called **coordinate functions**. It then follows that each point on the manifold is determined by n numerical quantities, that is, n coordinates.

Part two of Riemann's lecture is devoted to the idea of a metric relation on the manifold, a way of determining the length of a curve on the manifold independent of its position. This is the only section of the talk which involved any mathematical formulas, but even here Riemann just presented them without any derivations. Riemann's basic assumption, based on Gauss' earlier work, was that the expression for the length ds of an infinitesimal element of a curve is the square root of a homogeneous positive definite quadratic function of the dx_i, that is,

$$ds^2 = \sum_{i=1}^{n} \sum_{j=1}^{n} g_{ij}\, dx_i\, dx_j$$

where $g_{ij} = g_{ji}$ and all of the g_{ij} are continuous functions on the manifold. Ordinary (Euclidean) space has the simplest case of this metric, namely $ds^2 = \Sigma dx_i^2$. Riemann showed that one cannot in general transform any given metric into another. Therefore, the manifolds having this simplest metric form a special class, a class Riemann named "flat."

To deal with curved (or nonflat) manifolds, Riemann constructed a special set of coordinates, today called Riemannian normal coordinates. Through the use of these coordinates, Riemann defined a notion of curvature which generalizes the idea of Gauss and showed that it too is intrinsic to the manifold and depends only on the coefficients g_{ij}. All the properties of the geometry of the manifold can then be described in terms of the metric and the coordinate grid. For example, one could use asymptotic parallel lines and their associated horocycles as the coordinate grid on a two-dimensional surface and, using an appropriate metric, develop all the properties of Lobachevsky's non-Euclidean geometry without making any assumption about the existence of parallels to a given line. Riemann noted that manifolds with constant curvature have the important property that "figures can be moved in them without stretching." In fact, "since the metric properties of the manifold are completely determined by the curvature, they are therefore exactly the same in all the directions around any one point as in the directions around any other . . . Consequently, in the manifolds with constant curvature figures may be given any arbitrary position."[12]

The final section of Riemann's lecture deals with the relationship of his ideas to our usual concept of three-dimensional (Euclidean) space. Riemann presented three sets of conditions, each of which he claimed is sufficient to determine whether a three-dimensional manifold is flat. One of these sets of conditions is that bodies be free to move and turn and that all triangles have the same angle sum. Riemann noted that it is difficult to determine whether these conditions hold because of the necessity of extending our observations to the immeasurably large and immeasurably small. He did state, however, that one can assume that physical space forms a three-dimensional unbounded manifold. The unboundedness, however, does not imply that space is infinite, because if the curvature is constant but positive, space would necessarily be finite. As to the immeasurably small, Riemann concluded that the metric relations do not necessarily follow from those in the large and that, in fact, the curvature may vary from point to point as long as the total curvature of every measurable portion of space is close to zero. The accurate determination of the curvature and the associated metric, however, is a matter for physics and not mathematics.

17.2.4   The Systems of Geometry of Helmholtz and Clifford

Riemann made no effort to publish his lecture, perhaps because he had not initially desired to give this talk on geometry and was therefore working on several other projects at the time. Thus, although Gauss was quite impressed with it, its new ideas had very little effect elsewhere until it was published in 1868 after Riemann's untimely death. Once this occurred, however, Riemann's work met with widespread acclaim. In particular, Hermann von Helmholtz (1821–1894) in Germany (Figure 17.10) and William Clifford (1845–1879) in England both were influenced by Riemann's work and published their own interpretations and extensions.

Helmholtz, in a paper appearing shortly after the publication of Riemann's lecture and with a title remarkably similar, "Über die Thatsachen die der Geometrie zu Grunde liegen" (On the Facts which Lie at the Foundations of Geometry), attempted to list a set of hypotheses which would provide the basis for any reasonable study of geometry. First, like Riemann, he assumed that a space of n dimensions is a manifold. His definition was somewhat more explicit than Riemann's, however, in that he assumed the existence of n

FIGURE 17.10
Helmholtz on a German stamp.

independent coordinates near a point, at least one of which varies continuously as the point moves. Helmholtz's second axiom was that rigid bodies exist. This assumption permits one to equate two different spatial objects by superposition. Third, Helmholtz asserted that rigid bodies can move freely. In other words, any point in such a body can be moved to any other point in space; other points in the body will be carried by this motion to other points whose coordinates are related to the first by a particular set of equations. With the further assumption that $n = 3$ these hypotheses led Helmholtz to his own concept of the physical space in which we live, namely that of a three-dimensional manifold of constant curvature. It follows that there are three possibilities for physical space: its curvature could be positive, negative, or zero. The third option leads to Euclidean geometry. Contrariwise, "if the measure of curvature is positive we have **spherical** space, in which straightest lines return upon themselves and there are no parallels. Such a space would, like the surface of a sphere, be unlimited but not infinitely great. A constant negative measure of curvature on the other hand gives **pseudospherical** space, in which straightest lines run out to infinity, and a pencil of straightest lines may be drawn, in any flattest surface, through any point, which does not intersect another given straightest line in that surface."[13] Thus Helmholtz succeeded in placing Lobachevsky's non-Euclidean geometry in the context of Riemann's work. Furthermore, spherical geometry also turned out to be a non-Euclidean geometry, one in which no parallel lines exist. Thus the two possible negations of Euclid's parallel postulate could both lead to possible geometries of our physical space.

In a series of lectures in England in the early 1870s, William Clifford also attempted to determine the postulates of physical space. He noted, more specifically than Helmholtz, that one way to distinguish between Euclidean and non-Euclidean spaces was by a postulate of similarity, "that any figure may be magnified or diminished in any degree without altering its shape."[14] This postulate turns out to be equivalent to the assumption of zero curvature and is therefore not true for Lobachevsky's geometry. Clifford was particularly impressed, however, by the revolution wrought by Lobachevsky's ideas, comparing the effect of these in respect to Euclid to that of Copernicus' ideas with respect to Ptolemy. In both cases, humanity's view of the universe was fundamentally altered. In particular, with non-Euclidean geometry of either positive or negative curvature a possibility for physical space, it turned out that man's knowledge of that space, particularly in its far reaches, was limited to the distances his powers of observation could reach.

Clifford also noted in a brief paper of 1876 that in fact, although finite portions of space do have curvature zero, to within the limits of our experimental accuracy, we do not really know whether all the axioms of space apply for very small portions of space. In fact, Clifford provided some new speculations which contradicted Helmholtz's concept of constant curvature. Explaining these ideas, which in more recent times have come to the forefront of research in cosmology, Clifford wrote,

I hold in fact

1. That small portions of space *are* in fact of a nature analogous to little hills on a surface which is on the average flat; namely, that the ordinary laws of geometry are not valid in them.

2. That this property of being curved or distorted is continually being passed on from one portion of space to another after the manner of a wave.

3. That this variation of the curvature of space is what really happens in that phenomenon which we call the *motion of matter*.

4. That in the physical world nothing else takes place but this variation, subject (possibly) to the law of continuity.[15]

Clifford's speculations about the physical world thus made Riemann's ideas on the theory of manifolds into an important research tool in physics. In fact, they turned out to be central in the revolutionary developments in physics related to the theory of relativity which occurred early in the twentieth century.

17.2.5 Models of Non-Euclidean Geometry

Because Lobachevsky's geometry appeared to be valid on a surface of constant negative curvature, Eugenio Beltrami (1835–1900), an Italian mathematician who held chairs in mathematics in Bologna, Pisa, Pavia, and finally Rome, attempted to construct such a surface, the so-called **pseudosphere**. It turned out that one could construct only a portion of the surface in Euclidean three-dimensional space. Nevertheless, Beltrami succeeded in determining the appropriate metric on this surface and showing the connection between this metric and Lobachevsky's trigonometric laws for non-Euclidean space. He began by parametrizing the sphere of radius k (and curvature $1/k^2$) situated in Euclidean three-dimensional space by

$$x = \frac{uk}{\sqrt{a^2 + u^2 + v^2}}, \qquad y = \frac{vk}{\sqrt{a^2 + u^2 + v^2}}, \qquad z = \frac{ak}{\sqrt{a^2 + u^2 + v^2}}$$

for some value a. It is then straightforward to calculate the metric form ds^2 on the sphere by substitution into the Euclidean form $ds^2 = dx^2 + dy^2 + dz^2$:

$$ds^2 = k^2 \frac{(a^2 + v^2)du^2 - 2uv\, du\, dv + (a^2 + u^2)dv^2}{(a^2 + u^2 + v^2)^2}.$$

To transform this result into one on a pseudosphere of curvature $-1/k^2$, Beltrami simply replaced u by iu and v by iv. The resulting metric

$$ds^2 = k^2 \frac{(a^2 - v^2)du^2 + 2uv\, du\, dv + (a^2 - u^2)dv^2}{(a^2 - u^2 - v^2)^2}$$

turned out to have the required properties.

On the pseudosphere, the curves $u = c$ and $v = c$ are geodesics orthogonal to $v = 0$ and $u = 0$ respectively, for any constant $c < a$. Thus Beltrami could consider a right triangle with one vertex at the origin, one leg along the curve $v = 0$, one leg along a curve $u = c$ and the hypotenuse along a geodesic through the origin which makes an angle θ with $v = 0$ (Figure 17.11). He calculated the lengths of these three sides by integration of the appropriate metric forms. For the hypotenuse, set $u = r \cos \theta$ and $v = r \sin \theta$. It follows that

$$ds = \frac{ka\, dr}{a^2 - r^2}.$$

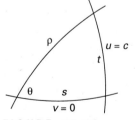

FIGURE 17.11
Beltrami's calculations on the pseudosphere:

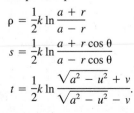

$$\rho = \frac{1}{2}k \ln \frac{a + r}{a - r}$$

$$s = \frac{1}{2}k \ln \frac{a + r \cos \theta}{a - r \cos \theta}$$

$$t = \frac{1}{2}k \ln \frac{\sqrt{a^2 - u^2} + v}{\sqrt{a^2 - u^2} - v}.$$

This element of arc is easily integrated from 0 to r to get the length ρ of the hypotenuse:

$$\rho = \frac{1}{2}k \ln\frac{a + r}{a - r}.$$

Similarly, along the curve $v = 0$,

$$ds = \frac{a\, du}{a^2 - u^2},$$

and the length s of this leg of the triangle up to a given u is

$$s = \frac{1}{2}k \ln\frac{a + u}{a - u} = \frac{1}{2}k \ln\frac{a + r\cos\theta}{a - r\cos\theta}.$$

Finally, the metric along $u = c$ is given by

$$ds = \frac{k\sqrt{a^2 - u^2}}{a^2 - u^2 - v^2}\, dv,$$

a differential whose integral up to a particular v is

$$t = \frac{1}{2}k \ln\frac{\sqrt{a^2 - u^2} + v}{\sqrt{a^2 - u^2} - v}.$$

With a bit of algebraic manipulation on the values for ρ, s, and t, Beltrami showed that

$$\frac{r}{a} = \tanh\frac{\rho}{k} \qquad \frac{r}{a}\cos\theta = \tanh\frac{s}{k} \qquad \frac{v}{\sqrt{a^2 - u^2}} = \tanh\frac{t}{k}.$$

It then follows that

$$\cosh\frac{s}{k}\cosh\frac{t}{k} = \cosh\frac{\rho}{k}.$$

This result, identical with Taurinus' equation (17.1) and Lobachevsky's equation (17.4) for the case of a right triangle, shows that Beltrami's surface with its associated metric gives the same geometry as Lobachevsky's non-Euclidean plane. In other words, Beltrami's calculations showed that the apparently mysterious use of a sphere of imaginary radius by Taurinus was equivalent to the introduction of a new metric on an appropriate two-dimensional manifold.

A different way of looking at Lobachevsky's geometry is simply to consider the imaginary sphere to be projected onto the interior of the circle $u^2 + v^2 = a^2$, where u and v are the parameters given above. It then turns out that straight lines in the Lobachevskian plane are represented by chords in the circle (Figure 17.12). Parallel straight lines are those whose intersection is at the circumference of the circle, with the circumference itself representing points at "infinity." Chords that do not intersect inside the circle represent lines which do not intersect at all in the Lobachevskian plane. Beltrami did not explicitly calculate distances between points in this model, but this gap was filled in in 1872 by Felix Klein (1849–1925), who used some concepts from projective geometry to be discussed in section 17.3.3. A similar model of Lobachevskian geometry in the interior of a circle was

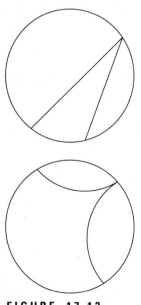

FIGURE 17.12
Klein's and Poincaré's models of the Lobachevskian plane.

developed by Henri Poincaré (1854–1912) in 1882. In this model, straight lines are represented by arcs of circles which are orthogonal to the boundary circle. Parallel lines are then represented by circular arcs which intersect at the boundary. This model has the advantage that angles between circles are measured in the Euclidean way. Figure 17.12 then shows why the angle sum of a triangle is less than π.

It was the use of models of Lobachevskian geometry as subsets of the ordinary Euclidean plane which helped to convince mathematicians by the end of the century that non-Euclidean geometry was as valid as Euclid's. Any contradiction in the former geometry would, by translation to the model, lead to a contradiction in the latter. Saccheri's attempts to "vindicate" Euclid had failed. With the work of Lobachevsky, Bolyai, Beltrami, Klein, and Poincaré, it was now clear that Euclid was truly vindicated. He had been completely correct in his decision 2200 years earlier to take the parallel postulate as a postulate. Because the Lobachevskian alternative to Euclid's parallel postulate led to a geometry as valid as Euclid's, it was impossible to prove that postulate as a theorem.

17.3 PROJECTIVE GEOMETRY

The work of Monge on descriptive geometry late in the eighteenth century, particularly his work on representing three-dimensional objects in two dimensions by various types of projection, led in the early nineteenth century to a renewed interest in the study of geometry from a synthetic point of view. In particular, his work led to a renewed interest in the study of projective geometry, those aspects of geometric figures which were invariant under various types of projections. Certain aspects of projective geometry had been studied by artists in the Renaissance as part of their effort to master the theory of perspective, and Desargues and Pascal in the mid-seventeenth century had worked out the beginnings of the theory of projective geometry. But it was only in the nineteenth century that mathematicians expanded the scope of this study.

17.3.1 Poncelet and Duality

It was a student of Monge, Jean-Victor Poncelet (1788–1867) who composed the first text in synthetic projective geometry in 1822, *Traité des propriétés projectives des figures* (*Projective Properties of Figures*). Poncelet started with the theory of polars in conic sections. Given a conic section *C*, one can associate to any point *p* its **polar** π, the straight line joining the points of contact of the tangents drawn from *p* to *C*. Similarly, to any line π' crossing the conic, one can associate a point *p'*, its **pole**, the intersection point of the tangent lines to the conic at the points where π' meets *C*. Poncelet saw that these concepts were reciprocal, that is, that if *p'* lies on π, then π', the polar of *p'*, goes through *p*, the pole of π (Figure 17.13).

Out of this duality of pole and polar, Poncelet developed a more general notion of duality between points and lines. He saw that, in general, a true proposition about "points" and "lines" remains true if we interchange the two words. For example, the statement, "Two distinct points determine exactly one line on which they both lie," becomes "Two distinct lines determine exactly one point through which they both pass." (Note that the

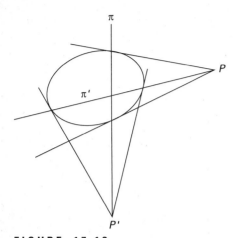

FIGURE 17.13
Reciprocal relationship of poles and polars.

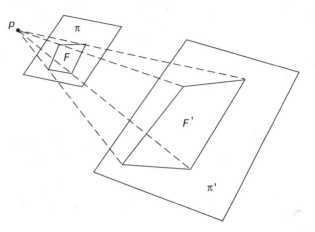

FIGURE 17.14
Central projection of F onto F'.

latter statement is not true in ordinary Euclidean geometry, but becomes true when points at infinity are added to the plane to provide intersections of parallel lines. This idea will be discussed below.) As a more complicated example, recall Pascal's theorem: "If the six vertices of a hexagon lie on a conic, then the points of intersection of the three pairs of opposite sides lie on a line." The dual theorem is: "If the six sides of a hexagon are tangent to a conic, then the lines joining the three pairs of opposite vertices intersect in a point." Although Poncelet did not establish the principle of duality as a theorem, he did use it as a valuable tool of discovery.

The results of Poncelet's discoveries, and the primary objective of his text, however, were the properties of central projection. Given a figure F in a plane π and a point P outside the plane, a central projection of F onto another plane π' is the figure consisting of the points of intersection with π' of all lines from P through the points of F (Figure 17.14). For example, the projection of a square in π is a quadrilateral in π', not necessarily a square, while the projection of a circle is a conic section. Poncelet's aim was to determine which properties of figures are invariant under such a projection. Clearly, the length of a line segment is not a projective invariant, but the property that a straight line can intersect a conic in at most two points is such an invariant. Poncelet noted that because a projection may transform parallel lines into intersecting lines, and because projections preserve intersections, it was necessary to introduce the points at infinity to be used as the intersection points of ordinary parallel lines. It is then useful to assume that all the points at infinity in a given plane together make up a line at infinity. Thus Poncelet's ideas led to the consideration of a new object, the projective plane, consisting of the ordinary points on the plane as well as the points at infinity.

To deal with points at infinity it proved necessary to develop a coordinate system for the projective plane. This was accomplished by Julius Plücker (1801–1868) with his introduction in 1831 of homogeneous coordinates. A point P in the plane with rectangular coordinates (X, Y) has the homogeneous coordinates (x, y, t), if $x = Xt$ and $y = Yt$. With this definition, a point does not have a unique set of coordinates; any two sets differ by a constant multiple. Nevertheless, the use of these coordinates means that any polynomial equation $f(X, Y) = 0$ (in rectangular coordinates) is rewritten in the form $g(x, y, t) = 0$ where all terms of g have the same degree (thus the name "homogeneous"). In addition, the points at infinity of the projective plane have homogeneous coordinates $(x, y, 0)$. Plücker noted that any straight line in the projective plane has the equation $ax + by + ct = 0$ in these coordinates. Namely, given constants (a, b, c), the set of points $\{(x, y, t)\}$ satisfying the equation all lie in a particular line. But surprisingly, if one takes (x, y, t) as the constants, this equation also characterizes the set of lines $\{(a, b, c)\}$ which all pass through the given point (x, y, t). Thus Poncelet's interchange of "point" and "line" in the principle of duality is justified algebraically by the interchange of "constant" and "variable" in the equation $ax + by + ct = 0$.

17.3.2 The Cross Ratio

Poncelet did not discover the most important projective invariant, the cross ratio of four points on a line. This concept was, however, thoroughly investigated by his younger countryman Michel Chasles (1793–1880) under the name of the anharmonic ratio. Recall that if a segment AB on a line p is projected onto a segment $A'B'$ on a line p' by a central projection from a point S, then the length $A'B'$ in general differs from the length AB. Similarly, if C is a point on segment AB and C' the corresponding point on $A'B'$, the ratio $AC : CB$ differs from the ratio $A'C' : C'B'$. But if C and D are two points on the segment AB whose corresponding points on segment $A'B'$ are C', D' respectively, then the **cross ratio**

$$\frac{AC}{CB} : \frac{AD}{DB}$$

is preserved by the projection, that is,

$$\frac{AC}{CB} : \frac{AD}{DB} = \frac{A'C'}{C'B'} : \frac{A'D'}{D'B'}.$$

The proof of this result is not difficult. Draw segments A_1B and AB_2 parallel to the line p' and determine the projections of C, D onto those two lines (Figure 17.15). From the similarity of triangles ACC_2 and BCC_1 and of triangles ADD_2 and BDD_1, one gets that

$$\frac{AC}{CB} = \frac{AC_2}{C_1B} \quad \text{and} \quad \frac{AD}{DB} = \frac{AD_2}{D_1B}.$$

But $D_1B/C_1B = D_2B_2/C_2B_2$. It follows that

$$\frac{AC}{CB} : \frac{AD}{DB} = \frac{AC_2}{C_2B_2} : \frac{AD_2}{D_2B_2},$$

or, that the cross ratio on line p equals the cross ratio on the line determined by segment AB_2. That the cross ratio of the points on the latter line segment is equal to the cross ratio on line p' follows easily from basic principles of similarity.

The standard notation for the cross ratio in the previous paragraph is (AB, CD). By permuting the four letters, one can calculate 24 cross ratios among these four points. (In

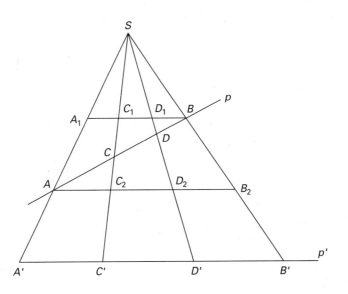

FIGURE 17.15
Cross ratios are preserved
under central projection.

this context, one considers a segment such as AB positive if A is to the left of B and negative in the contrary case.) Chasles noted that among the 24 apparent cross ratios there are really only six different ones and even these are closely related. Thus, for example, $(AB, CD) = 1/(AB, DC)$ and $(AB, CD) = 1 - (AC, BD)$.

Interestingly, although projective geometry aimed to study properties of figures not dependent on such concepts as that of length, the very basis of the definition of a cross ratio was in fact the length of a line segment. It was Christian von Staudt (1798–1867) who was able to correct this problem in 1847 by outlining an axiomatic system for projective geometry, based on the notion of a projective mapping as one which preserved harmonic tetrads. A harmonic tetrad is a set of four points A, B, C, D such that $(AB, CD) = -1$. Although this definition again seems to require lengths, von Staudt showed in fact that given three collinear points A, B, C, one could find the "fourth harmonic" D, that is, the point such that $(AB, CD) = -1$, by a simple projective construction. Von Staudt's work in projective geometry was central in making the subject into a clearly defined area of study and set the stage for the idea of defining a notion of distance in a nonmetric geometry.

17.3.3 Projective Metrics and Non-Euclidean Geometry

It was Cayley who in 1859 first provided a definition for a metric in the projective plane. Given a conic section C, he gave a rather complex definition of a function $d_C(P_1, P_2)$, depending on C, which satisfied the basic property of a distance, namely, that if P_1, P_2, P_3 lie on the same line, then $d(P_1, P_2) + d(P_2, P_3) = d(P_1, P_3)$. Twelve years later, Klein noticed that if Cayley's conic section is a circle, then the part of the projective plane inside that circle can be considered as a model for Lobachevskian geometry and Cayley's metric can be transformed into a distance function for that geometry.

Klein defined his modified metric for the non-Euclidean plane in terms of the cross ratio. Consider the Lobachevskian plane as the interior of the circle $u^2 + v^2 = 1$. Given

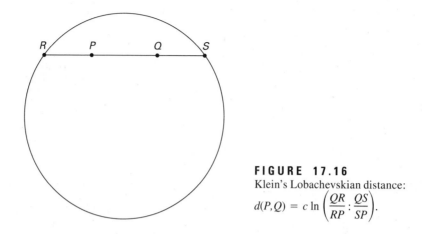

FIGURE 17.16
Klein's Lobachevskian distance:
$$d(P,Q) = c \ln \left(\frac{QR}{RP} : \frac{QS}{SP} \right).$$

two points P, Q in the interior of this circle, join them by a straight line which intersects the circle at the points R, S (Figure 17.16). The (directed) distance from P to Q is then

$$d(P,Q) = c \ln (QP,RS) = c \ln \left(\frac{QR}{RP} : \frac{QS}{SP} \right) = c \ln \left(\frac{QR \cdot SP}{RP \cdot QS} \right)$$

for some constant c which determines the unit of length. It is straightforward to show that if P, Q, Q' are three points on the line, then $d(PQ) + d(QQ') = d(PQ')$, so that the function d satisfies the major property of a distance function. By putting $P = R$, for example, it is also simple to show that the length of the entire chord is infinite, that is, that the circle itself represents the points at infinity. Klein's metric on the Lobachevskian plane is equivalent to the one derived by Beltrami on the pseudosphere. To see this, take P to be the origin and Q to be a point at Euclidean distance $r < 1$ to the right of P. Then

$$d(P, Q) = c \ln \left(\frac{-(1 + r) \cdot (-1)}{1 \cdot (1 - r)} \right) = c \ln \left(\frac{1 + r}{1 - r} \right)$$

which, for appropriate choice of c, is the same distance value Beltrami had calculated.

Klein also used the cross ratio to give a definition of the angle between two lines in the plane. This definition is, however, easier to understand in a slightly different form. Suppose that the two lines are given in projective coordinates by the triples (a_1, b_1, c_1), (a_2, b_2, c_2), that is, by the equations $a_1x + b_1y + c_1t = 0$ and $a_2x + b_2y + c_2t = 0$. These lines intersect in the point $x_0 = b_1c_2 - c_1b_2$, $y_0 = c_1a_2 - a_1c_2$, $t_0 = a_1b_2 - b_1a_2$. Their angle of intersection α is then given by

$$\alpha = \arcsin \frac{\sqrt{t_0^2 - x_0^2 - y_0^2}}{\sqrt{(a_1^2 + b_1^2 - c_1^2)(a_2^2 + b_2^2 - c_2^2)}}.$$

Notice that this formula implies that the angle of intersection is 0 if the point of intersection of the lines is on the boundary circle $x^2 + y^2 = t^2$, that is,

$$\left(\frac{x}{t} \right)^2 + \left(\frac{y}{t} \right)^2 = 1 \qquad \text{or} \qquad u^2 + v^2 = 1.$$

Felix Klein was primarily responsible for creating the mathematical institute at the University of Göttingen, an institute which transformed that small university into the mathematical center of the world for the first third of the twentieth century. After a brilliant early career, in which he contributed many beautiful ideas to the study of geometry, he suffered a nervous breakdown in the mid-1880s, following which he devoted his professional life to teaching, writing, and organizational activities. He became the editor of *Mathematische Annalen*, one of the leading mathematics journals of the time, and wrote several books consisting of edited versions of his lectures on various subjects. Many of the books were specifically aimed at mathematics teachers and dealt with the central ideas behind the mathematics taught in secondary schools. He also wrote a major work on the history of nineteenth century mathematics and directed the publication of the *Enzyklopädie der Mathematischen Wissenschaften*, an encyclopedia whose aim, ultimately impossible to fulfill, was to collect all the results and methods in mathematics obtained up to that time.

Klein was further able to show that, with somewhat different choices of boundary curve, an analogous definition of distance would lead to either Euclidean geometry or to the non-Euclidean geometry of the sphere. (Klein named the geometry of the sphere **elliptic geometry**, Euclidean geometry, **parabolic geometry**, and Lobachevsky's geometry, **hyperbolic geometry**.) In fact, Klein modified the non-Euclidean geometry of the sphere, because as it stood, two points did not always determine a unique straight line. The modification lay in identifying diametrically opposite points on the sphere. The new "half-sphere" could no longer exist in ordinary three-dimensional space (because opposite points along the equator were identified), but a distance function could be defined on it by multiplying the logarithm of a particular cross ratio by an imaginary constant. Klein's unification of Euclidean and non-Euclidean geometry by use of similar distance functions was an additional factor in convincing mathematicians that non-Euclidean geometry was as consistent as the Euclidean version.

17.3.4 Klein's *Erlanger Programm*

In 1872, Klein made another important contribution to the study of geometry in his *Erlanger Programm*. This paper detailed the notion that the various geometrical studies of the nineteenth century could all be unified and classified by viewing geometry in general as the study of those properties of figures which remained invariant under the action of a particular group of transformations on the underlying space (or manifold). In fact, Klein's demonstration of the relationship of geometry to groups of transformations helped to provide the impetus for the development of the abstract notion of a group by the end of the century.

Klein's starting point may have been his realization that any projective transformation of the projective plane into itself which preserved the boundary circle of his model of the Lobachevskian plane left the cross ratio unaltered and hence preserved both distance and angle. It is these transformations which are the rigid motions on his model of the Lobachevskian plane. Klein realized further that because the composition of any two transformations in this set belonged to the set and the inverse of any transformation in the set again belonged to it, the set of all the projective transformations preserving the boundary curve formed a group of transformations. Moreover, because the basic properties of figures in this plane are invariant under this group, one could consider the geometry of this non-Euclidean plane to be precisely the study of those invariant properties. Thus in his paper of 1872, Klein defined in general what a "geometry" was to mean: "Given a manifold and a group of transformations of the same; to investigate the configurations belonging to the manifold with regard to such properties as are not altered by the tranformations of the group."[16]

Klein provided several examples of geometries and their associated groups. Ordinary Euclidean geometry in two dimensions corresponded to what Klein called the **principal group**, the group composed of all rigid motions of the plane along with similarity transformations and reflections. Projective geometry consists of the study of those figures which are left unchanged by projections, or, what is the same thing, by collineations, those transformations which take lines into lines. These transformations can be expressed analytically in the form

$$x' = \frac{a_{11}x + a_{12}y + a_{13}}{a_{31}x + a_{32}y + a_{33}}, \qquad y' = \frac{a_{21}x + a_{22}y + a_{23}}{a_{31}x + a_{32}y + a_{33}},$$

where $\det(a_{ij}) \neq 0$. Because the principal group can be expressed analytically as the set of transformations of the form

$$x' = ax - by + c, \qquad y' = bex + aey + d$$

with $a^2 + b^2 \neq 0$ and $e = \pm 1$, it is clear that it is a subgroup of the projective group. It then follows that there are fewer invariants of the latter than of the former and so any theorem of projective geometry remains a theorem in Euclidean geometry, but not conversely.

Klein's *Erlanger Programm* paper, which contained very little detailed analysis, was intended as the basis of a research program for the study of geometry. It was, however, little noticed until it was translated into Italian, French, and English in the early 1890s. After that, however, Klein's idea that the invariants of a group of transformations were the important object of study in any field of geometry became a central facet in geometrical research well into the twentieth century.

17.4 GEOMETRY IN *N* DIMENSIONS

We have already seen examples of how the Greek limitations of geometry to three dimensions had been breached by various mathematicians in the early nineteenth century. For example, Ostrogradsky in the 1830s had generalized his divergence theorem to *n* dimensions almost casually (by simply adding three dots to the end of various formulas). The actual term "geometry of n dimensions" seems to have first appeared in the title of a paper of Cayley in 1843. The article itself, however, was purely algebraic and only touched on geometry in passing.

17.4.1 Grassmann and the *Ausdehnungslehre*

The first mathematician to present a detailed theory of spaces of dimension greater than three was Hermann Grassmann (1809–1877), a German mathematician whose brilliant work was unfortunately not recognized during his lifetime. Grassmann's aim in his *Die lineale Ausdehnungslehre* (*The science of linear extension*) of 1844 and in its reworking in 1862 was to develop a systematic method of expressing geometrical ideas symbolically, beginning with the notion of geometrical multiplication.

Four years before he wrote his first detailed work discussing *n*-dimensional spaces, Grassmann was already able to deal with the multiplication of vectors in two- and three-dimensional spaces in a paper devoted to a new explanation of the theory of the tides. He defined the **geometrical product of two vectors** to be "the surface content of the parallelogram determined by these vectors" and the **geometrical product of three vectors** to be the "solid (a parallelepiped) formed from them."[17] Defining in an appropriate way the sign of such products, he was able to show that the geometrical product of two vectors is distributive and anticommutative and that the geometrical product of three vectors all lying in the same plane is zero. Because the area of the parallelogram which is the geometrical product of two vectors is equal to the product of the lengths of the two vectors and the sine of the angle between them, this product is identical in numerical value to the modern cross product. Its difference, of course, is in the geometrical nature of the object produced by multiplying. Rather than the product being a new vector, it is a two-dimensional object. But because there is a one-to-one correspondence between Grassmann's parallelograms (considering two as equal if they have the same area and lie in the same or parallel planes) and modern vectors in three-dimensional space, determined by associating to each parallelogram the normal vector whose length equals the parallelogram's area, the two multiplications are essentially identical. The advantage of Grassmann's, however, is that it, unlike the cross product, is generalizable to higher dimensions. It is that generalization which is the basis of Grassmann's major works of 1844 and 1862.

Grassmann began the discussion in his text, particularly in the clearer 1862 version, with the notion of a vector as a straight line segment with fixed length and direction. Two vectors are to be added in the standard way by joining the beginning point of the second vector to the end point of the first. Subtraction is simply the addition of the negative, that is, the vector of the same length and opposite direction. Vectors are the simplest examples of what Grassmann calls an **extensive** quantity. In general, such a quantity is defined abstractly as "any expression derived by means of numbers from a system of units."[18] Grassmann means here that one begins with a set $\epsilon_1, \epsilon_2, \ldots, \epsilon_n$ of linearly independent quantities and then takes as an extensive quantity any linear combination of these "units." Addition of extensive quantities is by the obvious method: $\Sigma\alpha_i\epsilon_i + \Sigma\beta_i\epsilon_i = \Sigma(\alpha_i + \beta_i)\epsilon_i$. Similarly, one can multiply an extensive quantity by a scalar. Grassmann notes that the basic laws of algebra hold for his extensive quantities and defines the **space** of the quantities $\{\epsilon_i\}$ to be the set of all linear combinations of them.

Grassmann next defines multiplication of extensive quantities by use of the distributive law: $(\Sigma\alpha_i\epsilon_i)(\Sigma\beta_j\epsilon_j) = \Sigma\alpha_i\beta_j[\epsilon_i\epsilon_j]$ where each quantity $[\epsilon_i\epsilon_j]$ is called a quantity of the second order. Because this new sum must be an extensive quantity, it too must be expressible as a linear combination of units. Thus Grassmann needs to define second-order units.

He demonstrates that, assuming that the multiplication rules defined on units extend to the same rules on any extensive quantities, there are only four basic possibilities for defining second-order units. First, all of the quantities $[\epsilon_i\epsilon_j]$ could be independent. Second, one could have $[\epsilon_i\epsilon_j] = [\epsilon_j\epsilon_i]$. (This multiplication satisfies all of the ordinary algebraic multiplication rules.) Third, one could have $[\epsilon_i\epsilon_j] = -[\epsilon_j\epsilon_i]$. (This implies that $[\epsilon_i\epsilon_i] = 0$ for all *i*.) Finally, one could have all products $[\epsilon_i\epsilon_j] = 0$. It is the third form of multiplication, called **combinatory** multiplication, with which Grassmann primarily deals in the remainder of his work. According to his condition, then, for any first-order extensive quantities *A* and *B*, the multiplication rule $[AB] = -[BA]$ holds.

With the combinatory product of two first-order units defined, it is straightforward for Grassmann to define products of three or more first-order units using the same basic rules. For example, if there are three first-order units ϵ_1, ϵ_2, ϵ_3, then there are three second-order units $[\epsilon_1\epsilon_2]$, $[\epsilon_2\epsilon_3]$, $[\epsilon_3\epsilon_1]$ and one third-order unit $[\epsilon_1\epsilon_2\epsilon_3]$. (Any other product of three first-order units would have two factors in common and would therefore be equal to 0.) If there are four first-order units, then there are six second-order units, four third-order units, and one fourth-order unit. Grassmann notes further that the product of *n* linear combinations of *n* first-order units, $(\Sigma\alpha_{1i}\epsilon_i)(\Sigma\alpha_{2i}\epsilon_i) \cdots (\Sigma\alpha_{ni}\epsilon_i)$, is equal to $\det(\alpha_{ij})[\epsilon_1\epsilon_2 \cdots \epsilon_n]$, where the bracketed expression is the single unit of *n*th order.

Grassmann's combinatory product of his *Ausdehnungslehre* determines, in modern terminology, the exterior algebra of a vector space. Its ideas came from Grassmann's desire to express symbolically various geometrical concepts. So, in particular, he thought of his second-order quantities as parallelograms and his third-order quantities as parallelepipeds. But even though there was no specifically geometric interpretation of higher order quantities, Grassmann saw that the symbolic manipulations did not require the limitation to any particular number of dimensions. Not only did he construct the exterior algebra, however, but he also developed many of the important ideas relating to vector spaces, that is, to the space of all linear combinations of *n* units. As early as 1840 he had developed the notion of the inner product of two vectors as the "algebraic product of one vector multiplied by the perpendicular projection of the second onto it"[19] and showed that in coordinate form, the inner product was given by $(\Sigma\alpha_i\epsilon_i)(\Sigma\beta_j\epsilon_j) = \Sigma\alpha_i\beta_i$. A few years later Grassmann developed the notion of linear independence and showed that an orthogonal system of quantities is linearly independent (where two vectors are orthogonal if their inner product is zero). The ideas of subspace and dimension also appear in his work, and he proved the well-known result that for two subspaces *U*, *W* of a space *V*,

$$\dim(U + W) = \dim U + \dim W - \dim(U \cap W).$$

Grassmann's work was not appreciated during his lifetime. But although his geometric ideas were entirely forgotten, his ideas in linear algebra and the exterior algebra were rediscovered late in the nineteenth century and applied to many new areas of mathematics. In fact, it was Grassmann's exterior multiplication which provided the clue for the development of the theory of differential forms by Elie Cartan (1869–1951) at the turn of the twentieth century, a development which eventually led to the field of differential topology. On the other hand, the idea of a vector space gradually seeped into the mathematical mainstream in the late nineteenth century when it became clear that such basic notions as linear independence and linear combination were already being used in various parts of mathematics.

17.4.2 Vector Spaces

The first mathematician to give an abstract definition of a vector space was Giuseppe Peano in his *Calcolo geometrico* of 1888. Peano's aim in the book, as the title indicates, was the same as Grassmann's, namely to develop a geometric calculus. Thus much of the book consists of various calculations dealing with points, lines, planes, and solid figures. But in Chapter IX Peano gives a definition of what he calls a **linear system**. Such a system consists of quantities provided with operations of addition and scalar multiplication. The addition must satisfy the commutative and associative laws, while the scalar multiplication satisfies two distributive laws, an associative law, and the law that $1v = v$ for every quantity v. In addition, Peano includes as part of his axiom system the existence of a zero quantity satisfying $v + 0 = v$ for any v as well as $v + (-1)v = 0$. Peano also defined the **dimension** of a linear system as the maximum number of linearly independent quantities in the system. In connection with this idea, Peano noted that the set of polynomial functions in one variable forms a linear system, but that there is no such maximum number of linearly independent quantities and therefore the dimension of this system must be infinite.

Peano's work, like that of Grassmann, had no immediate effect on the mathematical world. His definition was forgotten, although mathematicians continued to use the basic concepts involved. The definition only became a part of standard mathematical terminology in 1918 when Hermann Weyl (1885–1955) repeated Peano's axioms in his book *Raum-Zeit-Materie* (*Space-Time-Matter*). Besides giving several examples satisfying the definitions, Weyl also gave a philosophic reason for adopting such a definition:

> Not only in geometry, but to a still more astonishing degree in physics, has it become more and more evident that as soon as we have succeeded in unravelling fully the natural laws which govern reality, we find them to be expressible by mathematical relations of surpassing simplicity and architectonic perfection. It seems to me to be one of the chief objects of mathematical instruction to develop the faculty of perceiving this simplicity and harmony, which we cannot fail to observe in the theoretical physics of the present day. It gives us deep satisfaction in our quest for knowledge. Analytical geometry [the axiom system he has presented] . . . conveys an idea even if inadequate, of this perfection of form.[20]

A better reason for determining systems of axioms for mathematical constructs can hardly be found.

17.5 THE FOUNDATIONS OF GEOMETRY

The late nineteenth century saw the appearance of axiom systems for various types of mathematical structures. The notions of group and field were axiomatized as was the notion of a vector space. Similarly, axioms were developed for the set of positive integers and a great deal of effort went into the precise definition of the idea of a real number. Of course, the oldest axiom system in existence was that of Euclid for the study of geometry. In fact, that system provided the model for the creation of the various axiom systems in this period. There were, however, several flaws in Euclid's system. In particular, various mathematicians through the ages noticed that Euclid had made assumptions in some of his proofs which were not explicitly mentioned in his list of axioms and postulates. With the new developments in non-Euclidean geometry causing mathematicians to reexamine

---

Biography   David Hilbert (1862–1943)

One of the last of the universal mathematicians, who contributed greatly to many areas of mathematics, Hilbert spent the first 33 years of his life in and around Königsberg, then capital of East Prussia, now in Russia. He attended the university there and, after receiving his doctorate, joined the faculty in 1885. He only rose to prominence, however, after he was called by Felix Klein to Göttingen, where he soon became one of the major reasons for that university's becoming the preeminent university for mathematics in Germany, and probably the world, through the first third of the twentieth century. Hilbert began his career with the study of algebraic forms, then turned to algebraic number theory, the foundations of geometry, integral

equations, theoretical physics, and finally the foundations of mathematics. He is probably most famous for his lecture at the International Congress of Mathematicians in Paris in 1900, where he presented a list of 23 problems which he felt would be of central importance for mathematics in the twentieth century. Hilbert firmly believed that it was problems which drove mathematical progress and was always confident that, *"wir mussen wissen, wir werden wissen* (we must know, we will know).'' After the Nazi seizure of power, Hilbert was forced to witness the demise of the Göttingen he knew and loved and died a lonely man during the Second World War.

---

the nature of the various axioms, it is not surprising that a concerted effort was made by several mathematicians to rectify the situation with regard to Euclid's work and thus put Euclid's geometry on as strong a foundation as possible.

17.5.1 Hilbert's Axioms

Of the attempts to set up a complete set of axioms from which Euclidean geometry could be derived, the most successful was that by probably the premier mathematician of the late nineteenth and early twentieth centuries, David Hilbert (1862–1943). In 1899, Hilbert published his *Grundlagen der Geometrie* (*Foundations of Geometry*), essentially a record of his lectures on Euclidean geometry presented at the University of Göttingen in the winter semester of 1898–1899. His aim in this work was "to choose for geometry a simple and complete set of independent axioms and to deduce from these the most important geometrical theorems in such a manner as to bring out as clearly as possible the significance of the different groups of axioms and the scope of the conclusions to be derived from the individual axioms.''[21]

Hilbert's idea was to begin with three undefined terms, point, straight line, and plane, and to define their mutual relations by means of the axioms. As Hilbert noted, it is only the axioms which define the relationships. One should not have to use any geometrical intuition in proving results. In fact, one could easily replace the three notions by other ones—Hilbert suggested chair, table, and beer-mug—as long as they satisfy the axioms. One sees, therefore, that Hilbert's idea of an axiom system was somewhat different from those of Euclid and Aristotle. The Greek thinkers had attempted merely to state certain "obvious" facts about concepts they already understood intuitively. Hilbert, on the other hand, like those who stated the group axioms, was determined to abstract the desired properties away from any concrete interpretation. Thus, any object could be a "point" or a "line" as long as the "points" and "lines" satisfied the axioms of the geometry, just as any set of objects could be a group as long as there was a law of "multiplication" of these objects which satisfied the axioms of a group.

Hilbert divided his axioms into five sets: the axioms of connection, of order, of parallels, of congruence, and of continuity and completeness. The first group of seven axioms established the connections among his three fundamental concepts of point, line, and plane. Thus not only do two points determine a line (axiom I, 1), but they determine precisely one line (axiom I, 2). Similarly, three points not on the same line determine one (I, 3) and only one (I, 4) plane. The fifth axiom asserts that if two points of a straight line lie in a given plane, then so does the entire line. The sixth axiom says that any two planes which have a common point have at least a second common point. Finally, the seventh axiom of the first group asserts the existence of at least two points on every straight line, three noncollinear points on each plane, and four noncoplanar points in space. After listing these axioms, Hilbert noted as consequence that two straight lines in a plane either have one point or no points in common and that two planes either have no point or a straight line in common.

Hilbert's second group of axioms enabled him to define the idea of a line segment AB as the set of points lying between the two points A and B. Euclid himself had assumed the properties of "betweenness" implicitly, probably because of the "obviousness" of the diagrams in which they occurred. Hilbert made Euclid's assumptions explicit by axiomatizing this idea of "between." For example, axiom II, 3 asserts that of any three points on a straight line, there is always one and only one which lies between the other two, while axiom II, 5 asserts that any line passing through a point of one side of a triangle and not passing through any of the vertices must pass through a point of one of the other two sides. With these axioms, Hilbert was able to deduce the important theorem that any simple polygon divides the plane into two disjoint regions, an interior and an exterior, and that the line joining any point in the interior with any point in the exterior must have a point in common with the polygon.

The third group of axioms consists solely of Hilbert's version of the parallel axiom: "In a plane α there can be drawn through any point A, lying outside of a straight line a, one and only one straight line which does not intersect the line a."[22] The fourth group asserts the basic ideas of congruence. Recall that Euclid proved his first triangle congruence theorem by "placing" one triangle on the second. Many questioned the validity of this method of superposition, and it is for this reason that Hilbert listed six axioms concerning the undefined term congruence. For example, axiom IV, 1 asserts that given a segment AB and a point A', there is always a point B' such that segment AB is congruent to segment $A'B'$, while axiom IV, 2 states in essence that congruence is an equivalence relation. The final axiom of this group almost asserts *Elements* I-4: If two triangles have two sides and the included angle congruent, then the remaining angles will also be congruent. Hilbert does not assert the congruence of the third sides, but proves it by an argument by contradiction. He also proves the other two triangle congruence theorems, Euclid's postulate that all right angles are equal to one another, the alternate-interior angle theorem, and the theorem that the sum of the angles of a triangle is equal to two right angles.

Hilbert's final set of axioms contains two which concern the basic idea of continuity. First, there is the axiom of Archimedes: Suppose A, B, C, D are four distinct points. Then on the ray AB there is a finite set of distinct points, $A_1, A_2, \ldots, A_n$ such that each segment A_iA_{i+1} is congruent to the segment CD and such that B is between A and A_n. In other words, given any line segment and any measure, there is an integer n such that n units of measure yield a line segment greater than the given line segment. Among the consequences of this axiom, when used in conjunction with the earlier axioms, is that there is no limit to the length of a line. Thus this tacit assumption of Euclid, important in Saccheri's and Lambert's rejection of the hypothesis of the obtuse angle, is now made explicit. Hilbert's final axiom states in essence that the points on the line are in one-to-one correspondence with the real numbers. In other words, there are no "holes" in the line. This axiom answers the objection made to Euclid's construction of an equilateral triangle in *Elements* I-1 that there is no guarantee that the two constructed circles actually intersect. According to Hilbert's axiom, no points can be added to the two circles and therefore they cannot simply "pass through" one another.

17.5.2 Consistency, Independence, and Completeness

Having stated the axioms, Hilbert proceeded to show that they were **consistent**, that is, that one could not deduce any contradiction from them, at least under the assumption that arithmetic had no contradictions. His idea was to construct a geometry, using only arithmetic operations, which satisfied the axioms. For example, beginning with a certain set Ω of algebraic numbers, Hilbert defined a point p to be an ordered pair (a, b) of numbers in Ω and a line L to be a ratio $(u : v : w)$ of three numbers in Ω, where u, v are not both 0. Then p lies on L if $ua + vb + w = 0$. With every geometric concept interpreted arithmetically in this fashion and with all the axioms satisfied in the interpretation, Hilbert had created an arithmetic model of his axioms for geometry. If the axioms led to a contradiction in geometry, there would be an analogous contradiction in arithmetic. Therefore, assuming the axioms of arithmetic are consistent, so are the axioms of geometry.

Another important characteristic of an axiom system is **independence**, that is, that no axiom can be deduced from the remaining ones. Although Hilbert did not demonstrate independence completely, he did show that the various groups of axioms were independent by constructing several interesting models in which one group was not satisfied. Hilbert did not deal with a further characteristic of an axiom system, that of **completeness**, that any statement which can be formulated within the system can be shown to be either true or false. It is virtually certain, however, that Hilbert believed that his system was complete. In fact, several mathematicians soon showed that all of the theorems of Euclidean geometry could be proved using Hilbert's axioms.

The importance of Hilbert's work lay not so much in his answering the various objections to parts of Euclid's deductive scheme, but in reinforcing the notion that any mathematical field must begin with undefined terms and axioms specifying the relationships among the terms. As we have discussed, there were many axiom schemes developed in the late nineteenth century to clarify various areas of mathematics. Hilbert's work can be considered the culmination of this process, because he was able to take the oldest such scheme and show that, with a bit of tinkering, it had stood the test of time. Thus the mathematical ideas of Euclid and Aristotle were reconfirmed at the end of the nineteenth century as still the model for pure mathematics. Nearly 100 years later, these ideas continue to prevail.

Exercises

Problems from Gauss' Differential Geometry

1. Show that the area of an (infinitesimal) triangle with vertices (x, y), $(x + dx, y + dy)$, $(x + \delta x, y + \delta y)$ is equal to $\frac{1}{2}(dx\, \delta y - dy\, \delta x)$.

2. Show that if a surface is given in the form $z = z(x,y)$, then the measure of curvature k can be expressed as

$$k = \frac{z_{xx}z_{yy} - z_{xy}^2}{(1 + z_x^2 + z_y^2)^2}.$$

   Hint: Show first that if X, Y, Z are coordinates on the unit sphere corresponding to the point $(x, y, z(x, y))$ on the given surface, then

$$X = \frac{-z_x}{\sqrt{1 + z_x^2 + z_y^2}}, \qquad Y = \frac{-z_y}{\sqrt{1 + z_x^2 + z_y^2}},$$

$$Z = \frac{1}{\sqrt{1 + z_x^2 + z_y^2}}.$$

3. Calculate the curvature function k of the paraboloid $z = x^2 + y^2$.

4. Investigate the formulas for the measure of curvature k in the cases where the surface is given by an equation $w(x, y, z) = 0$ and parametrically by $x = x(u, v)$, $y = y(u, v)$, $z = z(u, v)$.

5. Calculate the curvature function of the ellipsoid $x^2 + \frac{y^2}{4} + \frac{z^2}{9} - 1 = 0$.

6. If $x = x(p, q)$, $y = y(p, q)$, $z = z(p, q)$ are the parametric equations of a surface and if $E = x_p^2 + y_p^2 + z_p^2$, $F = x_p x_q + y_p y_q + z_p z_q$, and $G = x_q^2 + y_q^2 + z_q^2$, show that

$$dx^2 + dy^2 + dz^2 = E\, dp^2 + 2F\, dp\, dq + G\, dq^2.$$

7. Calculate E, F, G on the unit sphere parametrized by $x = \cos u \cos v$, $y = \cos u \sin v$, $z = \sin u$, and show that $ds^2 = du^2 + \cos^2 u\, dv^2$.

Problems from Non-Euclidean Geometry

8. Derive the formula $\cos a = \cos b \cos c + \sin b \sin c \cos A$ for an arbitrary spherical triangle with sides a, b, c and opposite angles A, B, C on a sphere of radius 1 by dividing the triangle into two right triangles and applying the formulas of Chapter 4.

9. Show that the formula in problem 8 changes to

$$\cos\frac{a}{K} = \cos\frac{b}{K}\cos\frac{c}{K} + \sin\frac{b}{K}\sin\frac{c}{K}\cos A$$

   if the sphere has radius K.

10. By using power series, show that Taurinus' "log-spherical" formula

$$\cosh\frac{a}{K} = \cosh\frac{b}{K}\cosh\frac{c}{K} - \sinh\frac{b}{K}\sinh\frac{c}{K}\cos A$$

   reduces to the law of cosines as $K \to \infty$.

11. Show that Taurinus' formula for an asymptotic right triangle on a sphere of imaginary radius i, namely $\sin B = 1/\cosh x$, is equivalent to Lobachevsky's formula for the angle of parallelism, $\tan\frac{B}{2} = e^{-x}$.

12. Show that the circumference of a circle of radius r on the sphere of imaginary radius iK is $2\pi K \sinh\frac{r}{K}$. Show that this value approaches $2\pi r$ as $K \to \infty$. (*Hint:* First determine the circumference of a circle on an ordinary sphere of radius K.)

13. Given that $\tan\frac{1}{2}\Pi(x) = e^{-x}$ where $\Pi(x)$ is Lobachevsky's angle of parallelism, derive the formulas

$$\sin\Pi(x) = \frac{1}{\cosh x} \qquad \text{and} \qquad \cos\Pi(x) = \tanh x$$

   and show that their power series expansions up to degree 2 are $\sin\Pi(x) = 1 - \frac{1}{2}x^2$ and $\cos\Pi(x) = x$ respectively.

14. Substitute the results of problem 13 into Lobachevsky's formulas

$$\sin A \tan\Pi(a) = \sin B \tan\Pi(b)$$

$$\cos A \cos\Pi(b)\cos\Pi(c) + \frac{\sin\Pi(b)\sin\Pi(c)}{\sin\Pi(a)} = 1$$

   to derive the laws of sines and cosines when the sides a, b, c of the non-Euclidean triangle are "small."

15. Show that if ABC is an arbitrary triangle with sides a, b, c, then the formulas $a\sin(A + C) = b\sin A$ and $\cos A + \cos(B + C) = 0$, along with the law of sines, imply that $A + B + C = \pi$.

16. Show that Lobachevsky's basic triangle formulas (17.3), (17.4), (17.5), and (17.6) transform into standard formulas of spherical trigonometry if one replaces the sides a, b, c of the triangle by ia, ib, ic, respectively. (For simplicity, assume that angle C is a right angle.)

17. Describe geometrically Beltrami's parametrization of the sphere of radius k given by

$$x = \frac{uk}{\sqrt{a^2 + u^2 + v^2}} \qquad y = \frac{vk}{\sqrt{a^2 + u^2 + v^2}}$$

$$z = \frac{ak}{\sqrt{a^2 + u^2 + v^2}}.$$

18. Show that replacing u, v by iu, iv respectively transforms the sphere of problem 17 with curvature $1/k^2$ to a pseudosphere with curvature $-1/k^2$.

19. Show that Beltrami's formulas for the lengths ρ, s, t of the sides of a right triangle on his pseudosphere transform into

$$\frac{r}{a} = \tanh \frac{\rho}{k} \qquad \frac{r}{a}\cos\theta = \tanh\frac{s}{k} \qquad \frac{v}{\sqrt{a^2 - u^2}} = \tanh\frac{t}{k}$$

and then show that

$$\cosh\frac{s}{k}\cosh\frac{t}{k} = \cosh\frac{\rho}{k}.$$

Problems from Projective Geometry

20. Demonstrate how a central projection can transform parallel lines into intersecting lines.

21. Show the following inequalities for the cross ratio:

$$(AB, CD) = 1 - (AC, BD) \qquad (AB, CD) = \frac{1}{(AB, DC)}.$$

22. Denote the cross ratios (AB, CD), (AC, DB), (AD, BC) by λ, μ, ν, respectively. Show that

$$\lambda + \frac{1}{\mu} = \mu + \frac{1}{\nu} = \nu + \frac{1}{\lambda} = -\lambda\mu\nu = 1.$$

23. Show that if the point p' lies on the polar π of a point p with respect to a conic C, then π', the polar of p', goes through p. (*Hint:* Assume first that C is a circle.)

24. Determine homogeneous coordinates of the points $(3, 4)$ and $(-1, 7)$. Write the homogeneous coordinates of the point at infinity on the line $2x - y = 0$.

25. Determine rectangular coordinates of the points $(3, 1, 1)$ and $(4, -2, 2)$ given in homogeneous coordinates. Determine the equation (in rectangular coordinates) of a line which passes through the point at infinity $(2, 1, 0)$.

26. Show that every circle in the plane passes through the two points at infinity $(1, i, 0)$ and $(1, -i, 0)$.

27. Given three collinear points A, B, P, show that the point Q determined by the construction in Figure 17.17 makes A, B, P, Q into a harmonic tetrad, that is, that $(AB, PQ) = -1$.

28. Using Klein's definition of distance d in the interior of the circle representing the Lobachevskian plane, show that if P, Q, Q' are three points on a line, then $d(P, Q) + d(Q, Q') = d(P, Q')$.

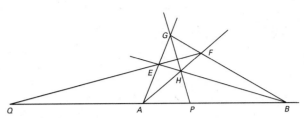

FIGURE 17.17

Given points A, B, P, to determine Q so that $(AB, PQ) = -1$. Draw any two lines through A. Draw a line through P intersecting those lines at G and H. Connect G and H to B. The line through the intersection points F and E (of BG with AF and of BE with AG, respectively) intersects lines APB at the desired point Q.

Problems from Grassmann

29. Letting i, j, k be first-order units in three-dimensional space, determine the combinatory product of $2i + 3j - 4k$, $3i - j + k$, $i + 2j - k$.

30. Show that in Grassmann's combinatory multiplication,

$$\left(\sum \alpha_{1i}\epsilon_i\right)\left(\sum \alpha_{2i}\epsilon_i\right) \cdots \left(\sum \alpha_{ni}\epsilon_i\right) = \det(\alpha_{ij})[\epsilon_1\epsilon_2 \cdots \epsilon_n],$$

where each linear combination is of a given set of n first-order units and where $[\epsilon_1\epsilon_2 \cdots \epsilon_n]$ is the single unit of nth order.

FOR DISCUSSION . . .

31. Study several new high school geometry texts. Do they follow Euclid's axioms or Hilbert's axioms or some combination? Comment on the usefulness of using Hilbert's reformulation in teaching a high school geometry class.

32. Is the analytic form of non-Euclidean geometry as presented by Taurinus, Lobachevsky, and Beltrami a better way of presenting the subject than the synthetic form? How can one make sense of a sphere of imaginary radius?

33. Why were the two first publishers of non-Euclidean geometry from countries not in the mainstream of nineteenth century mathematics? Is this by chance or are there substantive reasons?

34. Read a complete version of Riemann's lecture "On the Hypotheses which lie at the Foundation of Geometry." Describe Riemann's major new ideas and comment on how they have been followed up in the twentieth century. In particular, comment on the oft-repeated statement that Riemann's work was a precursor of Einstein's general theory of relativity.

References and Notes

Some of the important works on various aspects of the history of geometry in the nineteenth century include Jeremy Gray, *Ideas of Space: Euclidean, Non-Euclidean and Relativistic* (Oxford: Clarendon Press, 1989), B. A. Rosenfeld, *A History of Non-Euclidean Geometry* (New York: Springer Verlag, 1988), Roberto Bonola, *Non-Euclidean Geometry: A Critical and Historical Study of Its Development* (New York: Dover, 1955), Julian Lowell Coolidge, *A History of Geometrical Methods* (New York: Dover, 1963), Michael J. Crowe, *A History of Vector Analysis* (New York: Dover, 1985), and, for aspects of the history related to group theory, Hans Wussing, *The Genesis of the Abstract Group Concept* (Cambridge: MIT Press, 1984).

1. Gauss, letter to Olbers, quoted in Rosenfeld, *History of Non-Euclidean Geometry*, p. 215.

2. Gauss, *Abstract of the Disquisitiones Generales Circa Superficies Curvas*, presented to the Royal Society of Göttingen, included in *General Investigations of Curved Surfaces*, translated with notes by Adam Hiltebeitel and James Morehead (Hewlett, NY: Raven Press, 1965), p. 45.

3. Gauss, *General Investigations of Curved Surfaces*, p. 10. This paper, a somewhat earlier version, and the historical background are all included in the Raven Press edition (see note 2) and are well worth reading. For those with experience in differential geometry, it is worth attempting to translate Gauss' ideas into modern language. A discussion of this matter is found in Michael Spivak, *A Comprehensive Introduction to Differential Geometry* (Waltham, MA: Brandeis University, 1970), vol. II. Gauss' work is also discussed in D. J. Struik, "Outline of a History of Differential Geometry," *Isis* 19 (1933), 92–120 and *Isis* 20 (1933), 161–191.

4. Ibid., p. 21.

5. Lobachevsky, *Geometrical Investigations on the Theory of Parallel Lines*, translated by G. B. Halsted, in Bonola, *Non-Euclidean Geometry*, p. 13. This work by Bonola contains not only a translation of this fundamental paper of Lobachevsky, but also a translation of Bolyai's *Appendix*. The two treatments of non-Euclidean geometry can therefore be easily compared. For more on Lobachevsky, see A. Vucinich, "Nikolai Ivanovich Lobachevski, The Man Behind the First Non-Euclidean Geometry," *Isis* 53 (1962), 465–481 and V. Kagan, *N. Lobachevski and his Contribution to Science* (Moscow: Foreign Language Publishing House, 1957).

6. Ibid., p. 30.

7. Ibid., pp. 44–45.

8. Bolyai, *Appendix exhibiting the absolutely true science of space*, translated by G. B. Halsted, in Bonola, *Non-Euclidean Geometry*, p. 48.

9. Ibid., p. 20.

10. Riemann, "On the Hypotheses which lie at the Foundation of Geometry," translated by Spivak in his *A Comprehensive Introduction to Differential Geometry*, vol. II, pp. 4A-4–4A-20. In addition to translating Riemann's lecture, Spivak also discusses the work in detail.

11. Ibid., p. 4A-7.

12. Ibid., p. 4A-15.

13. Helmholtz, "On the Origin and Significance of Geometrical Axioms," in James Newman, ed., *The World of Mathematics* (New York: Simon and Schuster, 1956), vol. 1, 646–668, p. 657.

14. Clifford, "The Postulates of the Science of Space," in Newman, *World of Mathematics*, 552–567, p. 564.

15. Clifford, "On the Space Theory of Matter," in Newman, *World of Mathematics*, 568–569.

16. Klein, "Vergleichende Betrachtungen über neuere geometrische Forschungen," (Erlangen: Deichert, 1872) and *Mathematische Annalen* 43 (1893), 63–100, translated by M. W. Haskell as "A Comparative Review of Recent Researches in Geometry," *Bulletin of the New York Mathematical Society*, 2 (1893), 215–249, p. 218. There have been several recent articles on Klein's seminal work, including T. Hawkins, "The Erlanger Programm of Felix Klein: Reflections on its Place in the History of Mathematics," *Historia Mathematica* 11 (1984), 442–470, and David Rowe, "A Forgotten Chapter in the History of Felix Klein's Erlanger Programm," *Historia Mathematica* 10 (1983), 448–454.

17. Grassmann, "Theorie der Ebbe und Flut," quoted in Crowe, *History of Vector Analysis*, p. 61. Crowe's work provides a detailed study of various aspects of the history of vector analysis, in particular the conflict between the adherents of quaternions and those of vectors in the last half of the nineteenth century.

18. Grassmann, *Ausdehnungslehre*, quoted in David Eugene Smith, *A Source Book in Mathematics* (New York: Dover, 1959), p. 685. The *Ausdehnungslehre* is discussed in detail in J. V. Collins, "An Elementary Exposition of

Grassmann's *Ausdehnungslehre*, or Theory of Extension,'' *American Mathematical Monthly* 6 (1899), 193–198, 261–266, 297–301, and 7 (1900), 31–35, 163–166, 181–187, 207–214, 253–258. A recent work on Grassmann's creation of linear algebra is Desmond Fearnley-Sander, ''Hermann Grassmann and the Creation of Linear Algebra,'' *American Mathematical Monthly* 86 (1979), 809–817.

19. Grassmann, ''Theorie der Ebbe und Flut,'' quoted in Crowe, *History of Vector Analysis*, p. 63.

20. Weyl, *Space-Time-Matter* (New York: Dover, 1952), p. 23.

21. Hilbert, *The Foundations of Geometry* (La Salle, Il.: Open Court, 1902), p. 1. See also Michael Toepell, ''Origins of David Hilbert's *Grundlagen der Geometrie*,'' *Archive for History of Exact Sciences* 35 (1986), 329–344.

22. Ibid., p. 12.

Summary of Nineteenth Century Geometry

1777–1855	Carl Friedrich Gauss	Differential geometry
1788–1867	Jean-Victor Poncelet	Synthetic projective geometry
1792–1856	Nikolai Ivanovich Lobachevsky	Non-Euclidean geometry
1793–1880	Michel Chasles	Cross ratio
1794–1874	Franz Taurinus	Geometry on imaginary sphere
1798–1867	Christian von Staudt	Projective geometry
1801–1868	Julius Plücker	Duality; homogeneous coordinates
1802–1860	János Bolyai	Non-Euclidean geometry
1809–1877	Hermann Grassmann	Geometry in n dimensions
1821–1894	Hermann von Helmholtz	Foundations of n-dimensional geometry
1821–1866	Georg Bernhard Riemann	Hypotheses of geometry
1835–1900	Eugenio Beltrami	Surface of negative curvature
1845–1879	William Clifford	Postulates of physical space
1849–1925	Felix Klein	Erlanger Programm; metric non-Euclidean geometry
1854–1912	Henri Poincaré	Model for non-Euclidean geometry
1858–1932	Giuseppe Peano	Vector space axioms
1862–1943	David Hilbert	Foundations of geometry
1885–1955	Hermann Weyl	Vector space axioms

Chapter 18

Aspects of the Twentieth Century

"[Emmy Noether] had great stimulating power, and many of her suggestions took final shape only in the works of her pupils or co-workers . . . Hasse acknowledges that he owed the suggestion for his beautiful papers on the connection between hypercomplex quantities and the theory of class fields to casual remarks by Emmy Noether. She could just utter a far-seeing remark like this, 'Norm rest symbol is nothing else than cyclic algebra,' in her prophetic lapidary manner, out of her mighty imagination that hit the mark most of the time and gained in strength in the course of years; and such a remark could then become a signpost to point the way for difficult future work. . . . She originated above all a new and epoch-making style of thinking in algebra."

(From an address in memory of Emmy Noether by Hermann Weyl, 1935)[1]

About July 24, 1976, Kenneth Appel and Wolfgang Haken, with the help of a computer, completed a proof of the four-color theorem by showing that the last of the 1936 unavoidable configurations was reducible. This theorem, that four colors suffice to color any map, had first been proposed in 1852. On July 26, the two men submitted a report of their work to the American Mathematical Society, a report which appeared in the September issue of the *Bulletin of the A. M. S.* Interestingly enough, in mid-June, Haken had submitted an abstract to the A. M. S. of a paper he planned to present at the summer meeting in August. The paper was entitled, "Why is the Four Color Problem Difficult?"

A quick glance at any research library in mathematics shows that the mathematical output of the twentieth century far exceeds that of all previous centuries put together. There are shelves upon shelves of journals, most of which were started in this century. And the twentieth century segments of even the strongest of the journals of the nineteenth century, such as *Crelle's* or *Liouville's*, dominate their designated spaces. For better or for worse, however, most of the mathematics taught to undergraduates dates from the nineteenth century or earlier. So in a text designed to be read by undergraduates, one can barely scratch the surface of the mathematics of the twentieth century. We will therefore concentrate in this chapter on only four selected areas of twentieth century mathematics, areas which are to some extent covered in a typical undergraduate curriculum. For those who want to explore more extensively the mathematics of the twentieth century, the library shelves are open and there are many resources available as a guide.

We begin with a consideration of the problems in the foundations of mathematics at the beginning of the century. Cantor's work in the theory of infinite sets, which caused immediate controversy on its inception late in the nineteenth century, continued to cause problems early in the twentieth. Cantor himself, as well as numerous other mathematicians, attempted to prove seemingly "obvious" results for infinite sets with only limited success. For example, Cantor's attempt to prove the trichotomy law for infinite cardinals, the law that for any two cardinal numbers A, B, exactly one of the properties $A = B$, $A < B$, or $A > B$ held, ran into unexpected trouble, as did his attempt to prove that the real numbers could be well-ordered. The key to solving these problems turned out to be a new axiom for set theory, the axiom of choice, an axiom which was in fact used implicitly for many years until it was explicitly stated by Ernst Zermelo in 1904. Zermelo's statement of this new axiom, however, caused new controversy, in answer to which Zermelo produced an axiomatization of set theory. This axiomatization also helped to resolve various paradoxes in set theory which had been discovered around the turn of the century by Bertrand Russell. Zermelo's hope, and that of others, was that once a sound axiomatization of set theory was found, the theory of arithmetic could be based on it and mathematics in general given a secure foundation. Things turned out differently from what most mathematicians expected, however, when in 1931 Kurt Gödel established his Incompleteness Theorems, to the effect that any theory in which the arithmetic of natural numbers could be expressed had true results which could not be proved from the axioms of that theory. Gödel's result thus closed, in some sense, the axiomatic phase of mathematics, the attempt to give complete and consistent sets of axioms on which to base the various parts of the subject.

A second aspect of twentieth century mathematics to be discussed is the growth of topology, both point set topology and combinatorial topology, subjects barely begun in the previous century but destined to become one of the "growth" areas in mathematics in this century. The roots of point set topology lie in the work of Cantor on the theory of sets of real numbers, extended by various mathematicians into the consideration of other kinds of sets. The roots of combinatorial topology are in Riemann's attempts to integrate complex functions in regions with "holes." The beginnings of a theory of such regions, however, was in the work of Henri Poincaré in the 1890s, particularly with his definition of homology. Early in the twentieth century various mathematicians modified his definition by the use of simplexes, but it was not until the 1920s that the connections of combinatorial topology with algebra were recognized at Göttingen by the group of mathematicians led by Emmy Noether.

The subsequent algebraization of topology is part of the third aspect of twentieth century mathematics to be dealt with, the growth of algebraic techniques in all areas of mathematics. This growth included new ideas in the theory of fields with the work of Kurt Hensel and Ernst Steinitz and the intensive study of a new structure, that of a ring, centered again in the work of Noether. The growth of abstraction in algebra continued and perhaps culminated in the theory of categories and functors, introduced by Samuel Eilenberg and Saunders MacLane in 1945.

Algebra turned out also to be important in the growth of machine computation, including the development of the electronic computer during and immediately after World

War II. The final section of this chapter will be devoted to the study of various aspects of mathematics important in this development as well as a look at the implications computers have had in the practice of mathematics. We will therefore consider the early attempts at developing a computer in the work of Charles Babbage and Ada Byron, the theoretical basis for the programmable computer in the work of Alan Turing, and some aspects of the construction of the Institute for Advanced Study computer under John von Neumann. We will conclude with a brief look at two parts of mathematics which have been impacted by the advent of the computer, the theory of error correcting codes and the theory of graphs.

18.1 SET THEORY: PROBLEMS AND PARADOXES

Georg Cantor raised many questions about the theory of infinite sets in his work of the late nineteenth century, questions he attempted over the years to answer, in many cases, unsuccessfully. Other mathematicians also attacked these questions around the turn of the twentieth century.

18.1.1 Trichotomy and Well-Ordering

Cantor had shown that for two sets M and N, with cardinality $\overline{\overline{M}}, \overline{\overline{N}}$ respectively, no more than one of the relations $\overline{\overline{M}} = \overline{\overline{N}}, \overline{\overline{M}} < \overline{\overline{N}}$, or $\overline{\overline{N}} < \overline{\overline{M}}$ can occur. It seemed obvious to him as far back as 1878, furthermore, that exactly one of these relations should hold, namely, that if the two sets did not have the same cardinality, then one set must have the same cardinality as a subset of the other. It was only later that he realized that it was not a trivial matter to deny the existence of two sets, neither one of which was equivalent to a subset of the other. In fact, in his *Beiträge* of 1895, he made explicit mention of this trichotomy principle and stressed that not only did he not have a proof of it but that its proof must surely be difficult. He therefore carefully avoided using this principle in other proofs in his theory.

Cantor also realized that this question of trichotomy was closely related to another principle, that every set can be well-ordered. This principle states that for any set A, there exists an order-relation $<$ such that each nonempty subset B of A contains a least element, an element c such that $c < b$ for every other b in B. The natural numbers are well-ordered under their natural ordering. The real numbers, on the other hand, are not well-ordered under their natural ordering, but Cantor in 1883 thought it nearly self-evident that a well-ordering existed. By the mid-1890s, however, he began to realize that this result too needed a proof, a proof he believed he found in 1897, but soon realized was incomplete.

The well-ordering principle was thought to be so important by David Hilbert that in his address to the International Congress of Mathematicians in Paris in 1900, he presented the question of whether the set of real numbers can be well-ordered as part of the first of the important problems he suggested for mathematicians to consider in the twentieth century (Side Bar 18.1). As Hilbert put it, "It appears to me most desirable to obtain a direct proof of this remarkable statement of Cantor's, perhaps by actually giving an arrangement of numbers such that in every partial system a first number can be pointed out."[2] In other words, Hilbert wanted someone actually to construct an explicit well-ordering of the natural numbers.

Another troubling aspect of set theory at the beginning of the century was the appearance of a number of seeming paradoxes. One of the earliest is today called Russell's paradox, because it was published by Bertrand Russell (1872–1970) in 1903 (Figure 18.1). However, it appears that the paradox, which involves sets containing themselves as elements, was discovered a year or two earlier by Ernst Zermelo (1871–1953), as reported in a note of his friend Edmund Husserl (1859–1938):

Theorem. *A set M, which contains each of its subsets m, m', . . . as elements, is an inconsistent set, i.e., such a set . . . leads to contradictions.*

FIGURE 18.1
Russell on a stamp from Grenada.

Proof: "We consider those subsets m which do not contain themselves as elements. . . . These constitute in their totality a set M_0 . . . and now I prove of M_0, (1) that it does not contain itself as an element, (2) that it contains itself as an element. Concerning (1): M_0, being a subset of M, is itself an element of M, but not an element of M_0. For otherwise M_0 would contain as an element a subset of M (namely, M_0 itself) which contains itself as an element, and that would contradict the notion of M_0. Concerning (2): Hence M_0 itself is a subset of M which does not contain itself as an element. Thus it must be an element of M_0."[3]

Biography   **Ernst Zermelo (1871–1953)**

Zermelo grew up in Berlin as the son of a professor. He attended the universities in Berlin, Halle, and Freiburg, finally receiving his doctorate from the University of Berlin with a dissertation on the calculus of variations. He taught for several years at Göttingen, but in 1910 moved to Zurich. Ill health forced him to resign in 1916, but David Hilbert arranged for him to receive funds in recognition of his important work on set theory so that he was able to move to the Black Forest for a long period of recuperation. He returned to teaching in 1926 at Freiburg, but resigned in 1935 because he could not accept the new Nazi policies at the university. After World War II, he returned to Freiburg, where he finished his career.

Russell himself published several other versions of this paradox, the simplest being the barber paradox: A barber in a certain town has stated that he will cut the hair of all those persons and only those persons in the town who do not cut their own hair. Does the barber cut his own hair?

By early in the twentieth century it was therefore clear that Cantor's approach to set theory, although fruitful in the development of many new concepts, had flaws which needed correcting. Not only did some seemingly obvious results resist any proof, but also some of his intuitive ideas apparently led to contradictions. Interestingly, although many other fields of mathematics were being axiomatized during this time period, Cantor himself did not attempt to base his set theory on any collection of axioms. His definitions and arguments grew out of his intuition. Recall, in particular, that he had defined a set as any collection of objects of thought. It was this extremely broad definition of set which was at the heart of the above paradoxes. And it was the lack of appropriate axioms which were what prevented a solution to the problems of trichotomy and well-ordering.

18.1.2 The Axiom of Choice

The twin objections to Cantor's set theory were cleared away by Zermelo in the period from 1904 to 1908. In 1904, Zermelo published a proof of the well-ordering theorem, basing it on a principle which had appeared implicitly in various arguments for many years, the principle known today as the Axiom of Choice, which Zermelo first made explicit: "Imagine that with every subset M' [of an arbitrary nonempty set M] there is associated an arbitrary element m', that occurs in M' itself; let m' be called the distinguished element of M'."[4] Thus Zermelo asserted as an axiom that there always existed a "choice" function, a function $\gamma : S \rightarrow M$ (where S is the set of all subsets of M) such that $\gamma(M') \in M'$ for every M' in S. In other words, one can always somehow "choose" an element from each subset of a given set.

Zermelo realized that the axiom he had introduced was an important principle. In fact, he noted, "this logical principle cannot . . . be reduced to a still simpler one, but it is applied without hesitation everywhere in mathematical deduction. For example, the validity of the proposition that the number of parts into which a set decomposes is less than or equal to the number of all of its elements cannot be proved except by associating with each of the parts in question one of its elements."[5] (Here, "number" means "cardinality.") As Zermelo knew, this result had been employed by Cantor in the 1880s. In fact, the axiom of choice had been used, although not stated, even earlier. Strictly speaking, for choices in which a rule for the choice is specified, the axiom is not necessary. But as early as 1871, both Cantor and Heine, in their work publicizing the unpublished research of Weierstrass, used the axiom of choice implicitly in the case where no choice rule can be specified to prove the result that a real function sequentially continuous (see section 18.2.2) at a point p is also continuous at p. (A sequence of numbers had to be chosen, each in a specified interval.) Dedekind too had implicitly used the axiom in picking representatives of certain equivalence classes with respect to an ideal. Notwithstanding the fact that the axiom had been used for the past 30 years, its publication by Zermelo soon raised a storm of controversy.

The essence of the controversy was whether the use of infinitely many arbitrary choices was a legitimate procedure in mathematics. This question soon became part of the broader question as to what methods were permissible in mathematics at all, and if all methods must be constructive. And then of course the question came up of what constituted a construction or of what it meant to say that a mathematical object existed. Mathematicians had rarely debated such points before, but the use of the seemingly innocuous principle of making choices now led to the proof of a result, the well-ordering theorem, of which many mathematicians were skeptical. There was therefore a wide diversity of opinion in the mathematical community as to the validity of Zermelo's results. Some accepted them fully and went on to use the axiom of choice explicitly in their own research, while others denied the validity of both the axiom and Zermelo's proof. One group of mathematicians, of which Russell was a prominent member, steered a middle ground by making a careful attempt to determine what results already accepted in set theory had proofs dependent on the axiom of choice, so as to know explicitly what a rejection of that axiom would mean. Unfortunately, even Russell was at the time unable to prove that a particular result required the axiom of choice for its proof.

18.1.3 The Axiomatization of Set Theory

One of the problems with this debate within the mathematical community was that there was no accepted collection of axioms for set theory by which one could decide what methods were acceptable. There were many principles in print, especially after Zermelo's proof of the well-ordering theorem, which some mathematicians accepted while others denied. Hence, Zermelo decided that to solidify his proof and to clarify the terms of the argument surrounding it, an axiomatization of the theory of sets would be appropriate in which his proof would be imbedded. This axiomatization would include not only the axiom of choice, but also an axiom designed to eliminate some of the paradoxes caused by Cantor's overly broad definition of a set. Zermelo's method of axiomatization, influenced by Hilbert's axioms for geometry, began with a collection of unspecified objects and a relation among them which was defined by the axioms. In other words, Zermelo started with a domain $\mathcal{B}$ of objects and a relation $\in$ of membership between some pairs of these objects. An object is called a set if it contains another object (except as specified by axiom 2). To say that $A \subseteq B$ meant that if $a \in A$, then also $a \in B$. Zermelo's seven axioms were as follows (with the names he gave them).

1. *Axiom of Extensionality:* If, for the sets S and T, $S \subseteq T$ and $T \subseteq S$, then $S = T$.

2. *Axiom of Elementary Sets:* There is a set with no elements, called the empty set, and for any objects a and b in $\mathcal{B}$, there exist sets $\{a\}$ and $\{a, b\}$.

3. *Axiom of Separation:* If a propositional function $P(x)$ is definite (see below) for a set S, then there is a set T containing precisely those elements x of S for which $P(x)$ is true.

4. *Power Set Axiom:* If S is a set, then the power set $\mathcal{P}(S)$ of S is a set. (The power set of S is the set of all subsets of S.)

5. *Axiom of Union:* If S is a set, then the union of S is a set. (The union of S is the set of all elements of the elements of S.)

6. *Axiom of Choice:* If S is a disjoint set of nonempty sets, then there is a subset T of the union of S which has exactly one element in common with each member of S.

7. *Axiom of Infinity:* There is a set Z containing the empty set and such that for any object a, if $a \in Z$, then $\{a\} \in Z$.[6]

Zermelo never discussed exactly why he chose the particular axioms he did. But one can surmise the reasons for most of them. The first axiom merely asserts that a set is determined by its members, while the second was probably motivated by Zermelo's desire to have the empty set as a legitimate set and also to distinguish between an element and the set consisting solely of that element. Similarly, the power set axiom and the axiom of union were designed to make clear the existence of certain types of sets constructed from others, types which were used in many arguments. The axiom of separation is Zermelo's method of correcting Cantor's definition of a set as defined by any property, thereby eliminating Russell's paradoxes. Namely, by this axiom, there must first be a given set S to which the function describing the property applies, and second a definite propositional function, a function defined in such a way that the membership relation on $\mathcal{B}$ and the laws of logic always determine whether $P(x)$ holds for any particular x in S. Finally, the axiom of infinity was designed by Zermelo to clarify Dedekind's argument as to the existence of infinite sets. That argument had been met with disapproval by many mathematicians, partly because it seemed to be a psychological rather than a mathematical argument. Zermelo thus proposed his own axiom which asserts that an infinite set can be constructed.

The reaction to Zermelo's axiomatization was mixed. First, Zermelo was criticized for not proving his axioms to be consistent. After all, Hilbert had done so for his geometrical axioms by relying on the consistency of the real numbers. Zermelo admitted that he could not prove consistency, but felt that it could be done eventually. He was convinced, however, that his system was complete in the sense that from it all of Cantorian set theory could be derived. Second, Zermelo was criticized for the specific axioms he included and those he left out. There was certainly no consensus for the correct basis of an axiomatic system, and because Zermelo could not show that his system had no flaws, it was difficult to convince the community at large that this set of axioms would accomplish the desired goal.

To gain any consensus, two changes had to take place in Zermelo's system. First, the axioms themselves needed to be somewhat modified. On the suggestion of several mathematicians, Zermelo himself in 1930 introduced a new system, now called Zermelo-Fraenkel set theory, (after Abraham Fraenkel (1891–1965)). The major change from

Zermelo's original system was the introduction of a new axiom, the axiom of replacement, intended to insure that the set $\{\mathcal{N}, \mathcal{P}(\mathcal{N}), \mathcal{P}(\mathcal{P}(\mathcal{N})), \ldots\}$ exists in the Zermelo theory, where $\mathcal{N}$ is the set of natural numbers. Fraenkel's original formulation of this axiom was: "If M is a set and if M' is obtained by replacing each member of M with some object of the domain [$\mathcal{B}$], then M' is also a set."[7] As a second change, the nature of a "definite" propositional function had to be clarified, since this was essential to the axiom of separation. It turned out that this clarification had more to do with logic than with set theory, and ultimately it became the accepted view that axiomatic set theory needed to be imbedded in the field of logic. For various reasons which we will not discuss here, there are even today certain schools of mathematicians who do not accept one or more of Zermelo's axioms. But it is fair to say that the successes achieved within mathematics on the basis of these axioms have convinced the great majority of working mathematicians that they form a workable basis for the theory of sets.[8]

The axiom of choice itself, although probably the most controversial of Zermelo's axioms, turned out to have numerous applications. It was applied not only in analysis but also in algebra. For example, it was used to prove that every vector space has a basis and that in a commutative ring every proper ideal can be extended to a maximal ideal. It was used repeatedly in the new discipline of topology, where, among many other results, it provided the basis for the proof that the product of any family of compact spaces was compact. It also proved essential in the study of mathematical logic.

One of the most important mathematical tools derived from the axiom of choice was a maximal principle, usually known as Zorn's lemma, which was ultimately proved equivalent to the axiom. Although there were many precursors to the maximal principle, we state it in the form given by Max Zorn in 1935: If $\mathcal{A}$ is a family of sets which contains the union of every chain $\mathcal{B}$ contained in it, then there is a set A^* in $\mathcal{A}$ which is not a proper subset of any other $A \in \mathcal{A}$. (By a chain $\mathcal{B}$ is meant a set of sets such that for every two sets B_1, B_2 in $\mathcal{B}$, either $B_1 \subseteq B_2$ or $B_2 \subseteq B_1$.) Zorn's aim in stating this axiom was, in fact, to replace the well-ordering theorem in various proofs in algebra. He claimed that the latter, although equivalent to his own axiom, did not belong in algebraic proofs because it was somehow a transcendental principle. In any case, Zorn's lemma soon became an essential part of the mathematician's toolbox and was used extensively both in algebra and topology. In fact, it was published in 1939 in the first volume of Nicolas Bourbaki's monumental *Eléments de mathématique: Les structures fondamentales de l'analyse* and used consistently throughout the work (Side Bar 18.2).

Even though the axiom of choice proved useful, however, some of its consequences were unsettling and totally unexpected. Among the most surprising of these results was the Banach-Tarski paradox first noted by Stefan Banach (1892–1945) (Figure 18.2) and Alfred Tarski (1901–1983) in 1924. They proved, using the axiom, that any two spheres of different radii are equivalent under finite decomposition. One can take a sphere A of radius one inch and a sphere B the size of the earth and partition each into the same number of pieces $A_1, A_2, \ldots A_m$ and $B_1, B_2, \ldots B_m$, respectively, such that A_i is congruent to B_i for each i. With results such as this provable using the axiom of choice, there was great interest in clarifying its exact status with regard to the other axioms of set theory. It was certainly not clear that the axiom could not lead to a contradiction.

Zermelo realized that a proof of the consistency of his axioms would be extremely difficult. Although he and others worked on this problem during the next two decades, it was not until 1931 that Kurt Gödel (1906–1978), an Austrian mathematician who spent

FIGURE 18.2
Stefan Banach on a
Polish stamp.

Side Bar 18.2   **Nicolas Bourbaki**

Nicolas Bourbaki is the collective name of a group of primarily French mathematicians who, since the mid-1930s, have been writing texts devoted to what they deem the fundamental structures of analysis. The main part of their work is organized into six sections: Set Theory, Algebra, General Topology, Functions of a Real Variable, Topological Vector Spaces, and Integration. Material in other fields, including Commutative Algebra, Differentiable Manifolds, and Lie Groups, has also appeared, although in recent years it seems that Bourbaki's output has slowed considerably from his pace in the 1950s and 1960s.

Bourbaki was founded by Henri Cartan, André Weil, and others who were participating in the Mathematics Seminar led by Gaston Julia at the Institut Henri Poincaré. His original aim was to write a modern version of the classic French *Cours d'Analyse*, the text used (in various versions) in French universities since the early nineteenth century. Taking his cue from van der Waerden's *Modern Algebra*, Bourbaki aimed to develop the fundamental theorems in a given area, basing them on a reasonable set of axioms. He wanted to present theorems, whether recent or ancient, which led to significant applications and concepts of proven importance.

Bourbaki generally takes 10 to 12 years to write a chapter or group of chapters. One member is assigned the task of writing a preliminary version of the work. A year or so later, the work is brought before the Bourbaki meeting and subjected to detailed and merciless criticism. Once this version has been torn apart, someone else is chosen to revise it, and the following year his version is also torn to shreds. Eventually, however, Bourbaki comes to unanimous agreement on the contents and the book is published.

Bourbaki's works are known for their strict rigor and their devotion to broad general concepts. Their usual format is to begin with these general ideas and to derive as much as possible from them, only later considering the particular examples which, historically, often stimulated the discovery of the general theory. Bourbaki's influence was pervasive in graduate schools of mathematics in the United States, particularly in the third quarter of the twentieth century, and even had its effect on the teaching of mathematics at lower levels. In recent years, however, even French educators have swept away Bourbaki's influence on the secondary curriculum.

---

most of his life at the Institute for Advanced Study in Princeton, showed in essence that there could be no such proof. In fact, he showed that in any system which contains the axioms for the natural numbers—Dedekind's axioms, for example, could be proved in Zermelo-Fraenkel set theory—it was impossible to prove the consistency of the axioms within that system. Gödel also showed that this system was incomplete, that is, that there are propositions expressible in the system for which neither they nor their negations are provable. Given these results, the only hope for dealing with the axiom of choice was to prove that it was relatively consistent, that is, that its addition to the set of axioms did not lead to any contradictions that would not already have been implied without it. Gödel was able to give such a proof by the fall of 1935. Within the next three years he also succeeded in showing that the continuum hypothesis was relatively consistent with Zermelo-Fraenkel set theory.

A final result in the determination of the relationship of the axiom of choice to Zermelo-Fraenkel set theory was completed by Paul Cohen (1934–) in 1963. Cohen, using entirely new methods, was able to show that both the axiom of choice and the continuum hypothesis are independent of Zermelo-Fraenkel set theory (without the axiom of choice). In other words, it is not possible to prove or to disprove either of those axioms within set theory, and, furthermore, one is free to assume the negation of either one without fear of introducing any new contradictions to the theory. With these and other more recent results, it seems to be the case in set theory, as had already been shown in geometry, that there is not one version but many different possible versions, depending on one's choice of axioms. Whether this will be good or bad for the progress of mathematics is a matter for the next century to decide.

18.2 TOPOLOGY

Topology, that part of geometry concerned with the properties of figures invariant under transformations which are continuous and have continuous inverses, grew from various roots in the nineteenth century to become a full-fledged division of mathematics by early in the twentieth. The two branches of the subject to be considered here are point set topology, concerned with the properties of sets of points of some abstract "space," and combinatorial topology, concerned with how geometrical objects are built up out of certain well-defined "building blocks."

Point set topology grew out of the studies of Cantor of sets of real numbers. As such, its central aim is to provide an appropriate context for generalizing such properties of the real numbers as the Bolzano-Weierstrass property and the Heine-Borel property. Both of these are closely related to the important modern notion of compactness, a notion central in the generalization of the theorem on the existence of a maximum of a continuous function on a closed interval. The Bolzano-Weierstrass property is the property that every bounded infinite set of real numbers contains at least one **point of accumulation**, a point in every open interval of which there is another point of the set. In other words, this property asserts the "completeness" of the real numbers. We considered Bolzano's proof of this result in section 16.1.3.

The Heine-Borel property, formulated by Emile Borel (1871–1956) in 1894, states that if an infinite set $\mathcal{A}$ of intervals covers a finite closed interval B of real numbers, in the sense that every number of B lies in the interior of at least one element of $\mathcal{A}$, then there is a finite subset of $\mathcal{A}$ which has the same property. (Although Heine had used this result implicitly in the 1870s, it was Borel who first stated and proved this theorem with respect to a countable set $\mathcal{A}$. Henri Lebesgue (1875–1941) generalized it to arbitrary infinite collections $\mathcal{A}$ in 1904.) Borel's proof is by contradiction. Let $\mathcal{A} = \{A_1, A_2, \ldots, A_m, \ldots\}$. If the conclusion is not true, then for every n there is a point $b_n \in B$ such that b_n is not in A_i for every $i \leq n$. If one bisects the interval B, the same statement is true for at least one of the halves. If one continues this bisection process, one obtains a decreasing nested sequence of closed intervals B_i, each of which has the same property as B itself. But $\cap B_i$ contains a point p. By hypothesis, p is in the interior of A_k for some k, so A_k must contain one of the intervals B_i, a contradiction.

The key to Borel's proof is what is often called the nested interval property, that the intersection of a nested family of closed intervals contains a point. It was this result which was later to be abstracted into the earliest definition of compactness. It was also this result

---

Biography   Grace Chisholm Young (1868–1944)

Born Grace Chisholm at Haslemere, near London, she was educated at home and then entered Girton College, Cambridge, the first English institution where women could receive a university education. Having attained a superior score on the Cambridge Tripos exam in 1892, she decided to go to Göttingen to continue her studies because there was no possibility of advanced study in England. Felix Klein was willing to accept women students, but only after he had assured himself through a personal interview that they would succeed. (There were other members of the faculty who objected to admitting women under any conditions.) In any case, Grace Chisholm earned her Ph.D. in 1895, being the first woman to receive a German doctorate in mathematics through the regular procedure. In 1896, she married William Young, an English mathematician who had been her tutor at Girton.

The Youngs spent the next 44 years in a partnership fruitful both in mathematics and in children (six). Although most of the more than 200 mathematical papers and books which ensued were in William Young's name, Grace had a major role in their production. As William noted in an article of 1914, he had discussed the major idea of the work with his wife, and Grace had elaborated the argument and put the paper into publishable form. Their daughter wrote that her father could only generate ideas when he was stimulated by a sympathetic audience. Not only did Grace provide this audience, but she also had the initiative and stamina to complete the various undertakings her husband proposed.

William died at their home in Switzerland after the outbreak of the second world war left Grace in England. She died in 1944, just before she was to receive an honorary degree from the Fellows of Girton College. Two of their sons as well as one of their grandchildren also became mathematicians.

---

which was among the earliest presented in the first systematic exposition of set theory as a whole, the 1906 *The Theory of Sets of Points* by William Young (1863–1942) and Grace Chisholm Young (1868–1944).

18.2.1   The Youngs and *The Theory of Sets of Points*

The text of the Youngs deals with sets of points on the real line or in the real plane and gives explicit definitions of various fundamental concepts which were later to be generalized. For example, a point x which belongs to an interval "unclosed" at both ends is said to be an **internal point** of the corresponding closed interval. A point L is said to be a **limit point** of a given set of real numbers if inside every interval containing L as an internal point there is a point of the set other than L. A set which contains all its limit points is said to be **closed**, while one which does not is said to be **unclosed** or **open**. (Note that this is not the definition of "open" in use today.) The Youngs then reformulated the Bolzano-Weierstrass and Heine-Borel theorems in terms of these definitions and presented proofs.

The Youngs next generalized the notion of "interval" in the line to that of "region" in the plane by considering the latter to be generated by a set of triangles contained in it and then generalized the notion of limit point by replacing "interval" by "region." They noted further that, in analogy with the properties of an interval, a region divides the plane into three disjoint sets, internal points (those internal to at least one generating triangle), boundary points (those other than internal points which are limit points of internal points), and external points (those which are neither internal points nor boundary points). They then easily stated and proved the generalizations of the Bolzano-Weierstrass and Heine-Borel theorems to the plane.

Another fundamental notion in modern topology, the idea of connectedness, stems from some considerations of Cantor. Cantor, as part of his work in dealing with the "continuum," the entire set of real numbers, tried to characterize this set. As part of this attempt, he found it necessary to define what it meant for the set to be in one piece. Because he saw this idea in terms of the minimal distance between points of the set being 0, he defined "connected" in terms of distance: A set T is **connected** if given any two points p and q in it and any positive number ϵ, there is a finite number of points $t_1, t_2, \ldots,$ t_n in T such that the distances $pt_1, t_1t_2, \ldots, t_nq$ are all less than ϵ. The Youngs, among others, realized that it would be better to give a definition not using distance, but purely in set-theoretic terms, so they translated this notion as follows: "A set of points such that, describing a region in any manner round each point and each limiting point of the set as internal point, these regions always generate a single region, is said to be a **connected set** provided it contains more than one point."[9] Using this definition, the authors then proved that a set is connected if and only if it cannot be divided into two closed components without common points.

18.2.2 Fréchet and Function Spaces

In the same year as the Youngs' text appeared, Maurice Fréchet (1878–1973) began the process of generalizing results on points in the plane to more general contexts. In his dissertation dealing with the theory of functionals, functions operating on certain sets of functions, he had to be able to decide when two functions were "close" to each other and therefore decided "to establish systematically certain fundamental principles of the Functional Calculus, and then to apply them to certain concrete examples." By doing so, "one often gains thereby from seeing more clearly that which was essential in the demonstration, . . . from the simplification, and in the freeing [of the proofs] from that which only depends on the particular nature of the elements considered."[10] In other words, Fréchet decided to reconsider the basic notions of the topology of the real line in terms of arbitrary sets and then apply these notions to his particular sets of interest. He realized that he could prove various results on topology in a general context and then apply them in concrete instances rather than prove the same result over and over again. He especially wanted to answer questions about convergence of functionals to limits and determine the circumstances under which the limit functional has the same properties of the functionals of which it is the limit.

Fréchet thus began, not by defining "limit," but by characterizing the notion by means of axioms. That is, a set E belongs to L, the class for which limits are defined, if given any infinite subset $\{a_i\}$ of E it is possible to determine whether or not a unique element a exists having appropriate properties. This element is to be the limit of $\{a_i\}$. The limit element a is subject to the conditions that if each $a_i = a$, then the limit is a, and if a is the limit of $\{a_i\}$, then it is also the limit of any subsequence of $\{a_i\}$. With this abstract notion of a limit, Fréchet stated various definitions, some of which had already been considered by Cantor: The **derived set** E' of E is the set of its limit elements. The set E is **closed** if $E' \subseteq E$ and **perfect** if $E' = E$. The point a is an **interior point** of E if a is not the limit of any sequence not in E.

As part of his study of functionals, Fréchet wanted to generalize the Heine-Weierstrass result on the existence of a maximum and minimum of a continuous function on a closed interval so that it could apply to function spaces. To accomplish this generalization, he took the central idea of the proof of the Heine-Borel theorem as a definition in a brief note of 1904: "We call a **compact** set every set E such that there always exists at least one element common to every infinite sequence of subsets $E_1, E_2, \ldots, E_n, \ldots$ of E, if these (possessing at least one element each) are closed and each is contained in the preceding one."[11] Fréchet then proved that a necessary and sufficient condition for a set E to be compact was that every infinite subset F had at least one limit element in E, an element which is the limit of some sequence of distinct points of F. Noting that compact sets by his definition had properties analogous to those of closed and bounded sets in space, Fréchet was able to generalize the Weierstrass result, using for his definition of continuity the property which today is generally known as sequential continuity: A function f is **continuous** at a in the closed set E if $\lim_{n\to\infty} f(a_n) = f(a)$ for all sequences $\{a_n\}$ in E which converge to a. Interestingly, in his dissertation of 1906, Fréchet took for his definition of "compact" the necessary and sufficient condition above rather than the intersection property, adding the statement that any finite set will also be considered compact.

Having shown that the concept of a distance was not necessary to develop various familiar notions, Fréchet proceeded to reintroduce that notion in a more general setting. Namely, he considered a subclass $\mathcal{E}$ of the class L consisting of members E in which can be defined a **metric**, a real valued function (a, b) satisfying (1) $(a, b) = (b, a) \geq 0$; (2) $(a, b) = 0$ if and only if $a = b$; and (3) $(a, b) \leq (a, c) + (c, b)$ for any three elements a, b, c of E. Using the metric, Fréchet defined a **Cauchy sequence** to be any sequence $\{a_n\}$ such that for every $\epsilon > 0$ there is an m so that for all $p > 0$, $(a_m, a_{m+p}) < \epsilon$. Limit elements and the concepts associated with them are then defined in terms of the metric and Cauchy sequences. In particular, Fréchet considered the subclass of **normal** sets consisting of sets which are perfect, separable (contain a countable dense subset), and in which every Cauchy sequence has a limit (complete, in modern terminology). It is for sets of this type that he could prove a generalization of the Heine-Borel theorem: A normal set E is compact if and only if for every collection G of sets $\{I\}$ such that every element of E is in the interior of at least one member of that collection, there exists a finite subset of G with the same property.

Fréchet gave a number of examples of spaces with metrics for which his general theorems held, including examples of sets of functions. One such example was the set of real functions continuous on a given closed interval. Here the metric was the **maximum norm**

$$(f, g) = \max_{x \in I} |f(x) - g(x)|,$$

under which Fréchet proved that the set was normal. A second example was the set of all sequences of real numbers $x = \{x_1, x_2, \ldots,\}$ with the metric

$$(x, y) = \sum_{p=1}^{\infty} \frac{1}{p!} \frac{|x_p - y_p|}{1 + |x_p - y_p|}.$$

Fréchet noted that this metric is better than a more standard one $(x, y) = \max_p |x_p - y_p|$ in the sense that the former always gives a finite distance while the latter does not. Again,

Fréchet showed that this set is normal and, furthermore, that the same is true for the set of real functions defined on it with an appropriate metric.

18.2.3 Hausdorff and Topological Spaces

It was Felix Hausdorff (1868–1942) who was able to give a full axiomatization of the notion of a topological space derived from standard properties of sets of real numbers. He described this axiomatization in his 1914 text, *Grundzüge der Mengenlehre* (*Foundations of Set Theory*). Although his axioms and definitions have been modified since 1914 and numerous subsidiary definitions and concepts have been introduced, the basics of point set topology as currently taught are all to be found in this fundamental work.

Hausdorff noted that there are three basic concepts by means of which one can base a general theory of topological spaces, the notions of distance, neighborhoods, and limits. Fréchet had in effect used both the first and the third. Hausdorff noted that if one begins with the concept of distance, one can derive the other two, while if one begins with that of neighborhood, one can define the notion of limits. In general, however, one cannot reverse these procedures. Although the method one chooses, according to Hausdorff, is a matter of taste, he decided to begin with the concept of neighborhood to define the notion of a **topological space**. Such a space, today known as a **Hausdorff space**, is a set E to each element x of which is associated a collection of subsets $\{U_x\}$ of E, called neighborhoods, which satisfy the following axioms.

1. Each point x belongs to at least one neighborhood U_x and every neighborhood U_x contains x.
2. If U_x, V_x are two neighborhoods of the same point x, then their intersection also contains a neighborhood of that point.
3. If $y \in U_x$, then there is a neighborhood U_y such that $U_y \subseteq U_x$.
4. For two different points x, y there are two neighborhoods U_x, U_y whose intersection is empty.[12]

After showing that a metric space, with neighborhoods defined by $U_x = \{y \,|\, (y, x) < \rho\}$ for every real number ρ, satisfies the axioms given, Hausdorff developed the same basic theory as had Fréchet. The major change is that the central concept now is the domain or, in more modern terms, the open set. For Hausdorff, a **domain** A is a subset of E containing only interior points, where the latter are defined as points x which have a neighborhood U_x contained in A. That the entire set E and each of the neighborhoods U_x are domains follows from axioms 1 and 3. Hausdorff also showed that the union of arbitrarily many domains and the intersection of finitely many are domains. A **closed** set, on the other hand, is one which contains all of its accumulation points. Hausdorff then showed that the closed sets are precisely the complements of the domains.

Using Fréchet's limit definition for compact sets, Hausdorff proved the intersection property of nested, closed, compact sets and then a generalization of the Heine-Borel theorem, modifying the original proof only slightly. He also gave a new definition of limit point and convergence general enough to apply in any topological space: The point x is a **limit** of the infinite set A, if every neighborhood U_x of x contains all but a finite number of points of A. Further, because any set A either has a single limit x or none at all, one writes in the first case that $x = \lim A$ or that A **converges** to x.[13] Among the other new definitions was one for connectedness: "A set A differing from the empty set is **connected** if it cannot be divided into two disjoint, nonnull, sets both closed relative to A [that is, such that each set is the intersection with A of a closed set in the ambient space E]."[14] Hausdorff noted that because closed sets are the complements of domains, he could equally well have specified in the definition that the components needed to be domains.

Because the idea of a neighborhood was the point of departure for Hausdorff's development, it is not surprising that he defined continuity using that concept. Namely, he took the standard $\epsilon - \delta$ definition of continuity for functions of a real variable, noted that this definition used neighborhoods on the real line, and then translated it into a general definition for topological spaces: "The function $y = f(x)$ is called **continuous** at the point a, if for each neighborhood V_b of the point $b = f(a)$ there exists a neighborhood U_a of the point a, whose image lies in V_b; i.e., $f(U_a) \subseteq V_b$."[15] Hausdorff then generated an equivalent definition: The function $f : A \rightarrow B$ is continuous at a if and only if, for each subset Q of B which has $b = f(a)$ as an interior point, the inverse image $f^{-1}(Q)$ also contains the point a as an interior point. Using this definition, it is easy to prove that a continuous function preserves connectedness and compactness. Hausdorff naturally noted that the first of these facts implies the intermediate value theorem while the second implies the existence of a maximum and minimum value for a continuous function on a closed interval. In other words, Hausdorff showed that a topological space is the natural setting for these classical results about functions of a real variable.

18.2.4 Combinatorial Topology

Combinatorial topology grew out of the study of the idea of multiple connectivity of a surface in space, an idea developed by Riemann in his work on the integration of differentials. This idea was further refined by the physicists of the mid-nineteenth century because it turned out to be important in fields such as fluid dynamics and electromagnetism. It was Enrico Betti (1823–1892), however, who in 1871 generalized the idea of multiple connectedness to n-dimensional spaces by using hypersurfaces without boundaries as the analogs of Riemann's closed curves. Betti furthermore applied this idea to study the integration of differential forms over spaces of dimension n. The differences in connectivity of two surfaces were a way of telling that the surfaces were essentially "different," that there could be no continuous invertible function from one to the other. To deal with this method of distinguishing surfaces, Poincaré developed the new idea of homology.

Poincaré's idea was to determine how hypersurfaces (or, in more modern terminology, manifolds) could be built out of simpler ones. During the first quarter of the twentieth century, various mathematicians developed his ideas into the idea of triangulation, that is, of considering p-dimensional manifolds as being built from continuous images of p-dimensional "triangles." The appropriate definitions were completely worked out by James W. Alexander (1888–1971) by 1926 when he defined a p-**simplex** to be the p-dimensional analogue of a triangle and a **complex** to be a finite set of simplexes such that no two had an interior point in common and such that every face of a simplex of the set was also a ·simplex of the set. An elementary i-chain of a complex was defined to be an expression of the form $\pm V_0 V_1 \ldots V_i$ where the V's are vertices of an i-simplex. The expression changes sign upon any transposition of the V's, thus giving each chain an orientation. An elemen-

---

Biography   Henri Poincaré (1854–1912)

Poincaré, like Hilbert, was a universal mathematician, one who contributed to virtually every area of mathematics, including physics and theoretical astronomy. He was born into an upper middle-class family, many members of which performed various services to the French government. Poincaré displayed a strong interest in mathematics from an early age and won a first prize in the mathematics competition for students in all of the French lycées. In 1873, he entered the École Polytechnique and, after receiving his doctorate in 1879, entered a university career, first at the

University of Caen and then, in 1881, at the University of Paris. Toward the end of his life, he turned to popularization and wrote several books emphasizing the importance of science and mathematics, including such works as *Science and Hypothesis, The Value of Science,* and *Science and Method.* It was in the latter work that Poincaré described the psychology of discovery in mathematics, stressing the subconscious as a central factor in mathematical creativity (Figure 18.3).

FIGURE 18.3
Poincaré on a French stamp.

tary i-chain was then an i-dimensional "face" of a p-simplex, while an arbitrary i-chain was a linear combination of elementary i-chains with integer coefficients. As an example, the tetrahedron with vertices V_0, V_1, V_2, V_3 is a 3-simplex while it together with its four faces (each a 2-simplex), its four edges (each a 1-simplex), and its four vertices (each a 0-simplex), form a complex. The face $V_0V_1V_2$ is then an elementary 2-chain of the 3-simplex. Alexander next defined the boundary of the elementary i-chain $K = V_0V_i \ldots V_i$ to be the $(i - 1)$-chain $K' = \Sigma(-1)^s V_0 \ldots \hat{V}_s \ldots V_i$ and extended this to arbitrary i-chains by linearity. Thus the boundary of $V_0V_1V_2$ is $V_1V_2 - V_0V_2 + V_0V_1$ (Figure 18.4). An easy calculation with this example shows that the boundary of this boundary is zero, and one can show that this result is true in general.

Alexander gave his definition of homology applied to **closed chains** (cycles), chains whose boundary is zero. Namely, a closed chain K is homologous to zero, $K \sim 0$, if it is the boundary of a chain L. Two chains K and K^* are homologous, $K \sim K^*$, if $K - K^*$ is

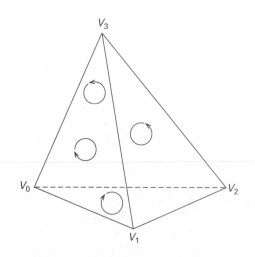

FIGURE 18.4
The boundary of a tetrahedron.

homologous to zero. The pth Betti number of a complex is then the maximum number of closed p-chains which are linearly independent with respect to boundary, that is, such that no linear combination is homologous to zero.

With a commutative operation ("addition") having an inverse being considered on the set of closed chains, it should be clear to modern readers that there is a group hiding among Alexander's definitions. Mathematicians of the 1920s saw this as well. But before discussing the applications of group theory to topology, we need first to consider the growth of algebra in general in the early part of the twentieth century.

18.3 NEW IDEAS IN ALGEBRA

Hilbert had closed out the nineteenth century by developing a new set of axioms for geometry and had shown their consistency, at least in relation to the consistency of arithmetic. As the twentieth century began, various mathematicians, particularly in the United States, attempted to develop sets of axioms for certain algebraic constructs. Since Eliakim H. Moore (1862–1932) showed in 1902 that Hilbert's axioms were not independent, that is, that one of them could in fact be deduced from the others, a strong effort was made to develop sets of independent axioms in algebra. For example, in 1903 Leonard Eugene Dickson (1874–1954) developed a new set of axioms for a field, a set which he believed was an improvement on the work of Weber some ten years earlier. Dickson defined a field to be a set with two rules of combination, designated by $+$ and $\times$, which satisfy the following nine axioms.

1. If a and b belong to the set, then so does $a + b$.
2. $a + b = b + a$.
3. $(a + b) + c = a + (b + c)$.
4. For any two elements a and b of the set, there exists an element x in the set such that $(a + x) + b = b$.
5. If a and b belong to the set, then so does $a \times b$.
6. $a \times b = b \times a$.
7. $(a \times b) \times c = a \times (b \times c)$.
8. For any two elements a and b of the set, such that $c \times a \neq a$ for at least one element c of the set, there exists an element x in the set such that $(a \times x) \times b = b$.
9. $a \times (b + c) = (a \times b) + (a \times c)$.

Naturally, among the axioms are the familiar closure, commutative, and associative laws of addition and multiplication, as well as the distributive law. Dickson's innovation was to replace the identity and inverse laws for both addition and multiplication by the new axioms 4 and 8. He then proceeded to prove independence by creating, for each axiom, a system with two rules of combination which did not satisfy that axiom but did satisfy each of the other eight axioms. For example, a system satisfying every axiom except for the second one consists of all positive rational numbers with ordinary multiplication, but

---

Dickson, born in Independence, Iowa, was the first recipient of a doctorate in mathematics at the University of Chicago. After further study in Leipzig and Paris and a year teaching in Texas, he returned to Chicago in 1900 for the remainder of his career. Dickson was a prolific mathematician, writing hundreds of articles and some 18 books. Among the most important of the latter was his monumental three

volume *History of the Theory of Numbers*, a work in which he traced the evolution of every important concept in that field. Dickson served as editor of the *Transactions of the American Mathematical Society* from 1911 to 1916 and then as President for the following two years. He was elected to the National Academy of Sciences in 1913.

---

with the new addition rule, $a + b = b$. Similarly, the set of all rational numbers with ordinary multiplication but with addition given by $a + b = -a - b$ satisfies all axioms but the third. In this case, axiom 4 is satisfied by taking $x = 2b - a$.

18.3.1 The *p*-adic Numbers

As a result of many papers similar to Dickson's, dealing with such constructs as groups, the algebra of logic, and linear associative algebras (to be discussed in section 18.3.3), the axiomatic approach to algebra gained favor. Once it was clear that a particular construct was important, the attempt was made to discover an independent set of axioms and then derive as much as possible about the system without any appeal to a particular concrete manifestation of the axiom set. Nevertheless, concrete examples were always needed in order to develop new conjectures to prove. Thus a new type of field, the field of ***p*-adic numbers**, was discovered in the opening years of the twentieth century by Kurt Hensel (1861–1941), a field different from the nineteenth century examples known to Weber.

Hensel started by noting that given any prime p, a positive integer A can be written uniquely in the form

$$A = a_0 + a_1 p + a_2 p^2 + \cdots + a_r p^r$$

where each a_i satisfies $0 \leq a_i \leq p - 1$. In this new representation, two numbers are to be thought of as "close" relative to p if they are congruent modulo high powers of p. To develop his theory conveniently, Hensel used the analogy between this representation and the ordinary decimal representation of a fraction to write A in the form $A = a_0.a_1 a_2 \cdots a_r$, where the coefficient of p^k is in the kth place after the period. Then the two numbers $3 + 2 \cdot 5$ and $3 + 2 \cdot 5 + 4 \cdot 5^{10}$, represented respectively by the numbers 3.2 and 3.2000000004 which are close if thought of as decimal fractions, were also to be thought of as "close" relative to the prime 5 because they were congruent modulo each power of 5 up through the tenth. More specifically, Hensel called the numbers $A_k = a_0.a_1 a_2 \cdots a_k$ for $k < r$ the approximate values of A relative to p, since each A_k was congruent to A modulo a higher power of p than A_{k-1}.

To turn the positive integers in this new representation into a field, one needed to be able to perform the usual operations on them. Guided by the analogy with finite (and infinite) decimal fractions, Hensel showed how one must do this. Addition and multiplication are performed in nearly the standard way, by use of the "decimal" analogy, except that one works from left to right and carries forward an appropriate multiple of p when any sum or product is greater than or equal to p. For example, using again $p = 5$, one has the sum

$$
\begin{array}{r}
2.3042134 \\
+\,3.2413123 \\
\hline
0.10113031
\end{array}
$$

As would be expected, it is the operations of subtraction and division which force Hensel to introduce new elements to this set of positive integers, elements which are represented in a surprising way. Consider the subtraction

$$
\begin{array}{r}
3.131312 \\
-\,4.424322 \\
\hline
\end{array}
$$

Beginning at the left and borrowing as necessary from the right, one gets 4.10243 for the first six digits of the answer. In other words, $3.131312 - 4.424322 \equiv 4.10243 \pmod{5^j}$ for $j \leq 5$. Can one, however, get an exact answer rather than just an approximation? After all, if one translates these representations back into the ordinary decimal representation, the answer is a negative integer. To introduce these "negatives," Hensel simply permitted the number of places to the right of the period to increase without limit. In this example, one can see that by placing indefinitely many 0's to the right in both the minuend and the subtrahend, the answer becomes $4.102434444 \cdots$ in the sense that if one cuts off this number at the nth place, the result will be congruent to the actual difference modulo 5^n. Hensel showed further that the use of indefinitely many digits after the period was also necessary to be able to perform divisions A/B in the special case where $B \not\equiv 0 \pmod{p}$ and that in both subtraction and this case of division, the infinite p-adic expansion was periodic.

Having rewritten the elements of a subfield of the field of rational numbers (those quotients A/B such that $B \not\equiv 0, \pmod{p}$) in a periodic p-adic expansion, Hensel extended the analogy with ordinary decimals by introducing a new set, the set of all power series of the form

$$
A = a_0 + a_1 p + a_2 p^2 + \cdots + a_n p^n + \cdots = a_0.a_1 a_2 \ldots a_n \ldots ,
$$

where each coefficient a_i satisfies $0 \leq a_i < p$. As before, the finite series $A_k = a_0.a_1 a_2 \ldots a_k, k = 1, 2, \ldots ,$ are each approximations to A, because A_k is congruent to A modulo p^{k+1}. But now, because there are infinitely many such approximations, each better than the previous one, Hensel turns to analysis and writes

$$
A = \lim_{k \to \infty} A_k.
$$

The set of power series so defined is not, however, a field. Although one can perform addition, subtraction, and multiplication by using the earlier rules on the various approximands and taking limits, the attempt to divide arbitrary power series forces one to generalize these series to those which include a finite number of terms with negative exponents.

(As a very simple case, $6 \div 5$ is written as $1 \cdot 5^{-1} + 1$.) Thus, Hensel included such terms and then showed easily that the set of all series of the form

$$A = a_m p^m + a_{m+1} p^{m+1} + \cdots \qquad (m \text{ any integer})$$

is indeed a field, the field today called the field $\mathbf{Q}_p$ of p-adic numbers.

Having defined a new set of fields, one for each prime p, Hensel was able to apply various concepts from the general theory of fields to these particular ones. Thus the integers of the field are those whose smallest power of p is nonnegative while the units are those whose smallest power of p is the zero power. (The units are precisely those integers whose multiplicative inverse is again an integer.) Hensel dealt with polynomials whose coefficients are in this field and applied the usual constructions to them, including the adjunction of roots of such polynomials to form extension fields to which Galois theory can be applied. But because the field of p-adic numbers has a natural topology, with neighborhoods of a point defined in terms of congruence to a given power of p, one can apply analytic concepts as well as the algebraic ones. For example, the notion of a Cauchy sequence of p-adic numbers can be defined as well as the notion of a limit, and it can be shown that every such Cauchy sequence has a limit.

18.3.2 The Classification of Fields

It was this creation of new fields by Hensel which influenced Ernst Steinitz (1871–1928) to undertake a new detailed study of the entire subject of fields, a study which appeared in 1910 under the title *Algebraische Theorie der Körper* (*Algebraic Theory of Fields*). Steinitz's goal was "to produce a survey of all possible types of fields and to establish the relationships among them."[16] The first distinction among fields involves the concept of the prime field and the characteristic. The **prime field** of any field K is its smallest subfield, that is, the intersection of all of its subfields. If ϵ is the multiplicative unit of this prime field, then there are two possibilities for the set I of all integral multiples $m\epsilon$ of ϵ. First, they could all be distinct. In this case, I is isomorphic to the set of positive integers and the prime field is isomorphic to the field of rational numbers. K is then said to be of **characteristic 0**. The second possibility for I is that there is a smallest natural number p, of necessity prime, such that $p\epsilon = 0$. In this case I is isomorphic to the set of residue classes modulo p. Thus I is itself a field, the finite field of p elements, and K is said to be of **characteristic p**.

Steinitz continued his work by discussing various types of field extensions. In particular, an extension L of a field K is **algebraic** if each element of L is a root of a polynomial equation with coefficients in K while it is a **transcendental** extension otherwise. An algebraic extension is said to be **finite**, if there exists a finite basis for L as a vector space over K and **infinite** otherwise. Infinite algebraic extensions include, for example, the **algebraic closure** of a prime field K, an extension field in which every polynomial over K factors into linear factors. Steinitz showed how to construct such an algebraic closure for the prime fields of characteristic p, but noted that for the field of rational numbers, this construction involved the use of the axiom of choice. Not surprisingly then, his general proof of the existence of an algebraic closure for an arbitrary field also involved that axiom. Finally, Steinitz discussed transcendental extensions. Defining a purely transcendental extension of a field K to be one formed by the adjunction of finitely or infinitely many unknowns, Steinitz was able, using the well-ordering theorem, to demonstrate that every extension of a field K can be formed by taking a purely transcendental extension and then an algebraic extension.

18.3.3 The Theory of Rings

With the work of Steinitz having exhausted the classification of fields, many algebraists of the time concentrated on the study of a different algebraic structure, that of a ring. The name **ring** was first used in the late nineteenth century by Hilbert to denote a set R having two operations, addition and multiplication, such that R is a commutative group with respect to addition, such that multiplication is associative, and such that the distributive law holds. The first detailed study of such objects, under the name of "linear associative algebras," was carried out by the American mathematician Benjamin Peirce (1809–1880) around 1870. By a linear associative algebra, today just called an algebra, Peirce meant a ring which is at the same time a finite dimensional vector space over a field F. (Peirce limited the field of coefficients to the field of real numbers.) Peirce's chief aim in his work was to describe all possible algebras of dimensions 1 through 5 and some of dimension 6, by considering the possible multiplication tables for the basis elements. In the course of this work, however, which turned out to be incomplete, he introduced two important definitions: A nonzero element a of a ring is **nilpotent** if some power a^n is zero, while it is **idempotent** if $a^2 = a$. Peirce was then able to demonstrate the following

Theorem. *In every algebra there is at least one idempotent or at least one nilpotent element.*

The proof is not difficult. Because the algebra is finite-dimensional, any nonzero element A of the algebra must satisfy an equation of the form

$$\sum_{i=1}^{n} a_i A^i = 0$$

for some m. This equation can be rewritten in the form

$$BA + a_1 A = 0 \qquad \text{or} \qquad (B + a_1)A = 0,$$

where B is a linear combination of powers of A. It follows that $(B + a_1)A^k = 0$ for every $k > 0$ and therefore that $(B + a_1)B = 0$ or $B^2 + a_1 B = 0$. It is immediate from the last equation that if $a_1 \neq 0$, then

$$\left(-\frac{B}{a_1}\right)^2 = -\frac{B}{a_1}$$

and $-B/a_1$ is an idempotent. If $a_1 = 0$, $B^2 = 0$, and B is a nilpotent element.

Several other mathematicians in the last quarter of the nineteenth century studied special algebras, in particular, **simple** ones, those having no nontrivial two-sided ideals. (A **two-sided ideal** in an algebra, or in any ring R, is a subset I such that if α and β belong to I so does $\alpha + \beta$, $r\alpha$, and αr for any r in R. This notion generalizes the definition of Dedekind for ideals in a ring of algebraic integers. Naturally, a two-sided ideal can be thought of as a subalgebra.)

Joseph Henry Maclagan Wedderburn (1882–1948), in fact, produced a detailed study of the structure of algebras over any field in a paper of 1907 entitled *On Hyper-complex Numbers* and showed that any simple algebra is a matrix algebra, not necessarily over a field, but over a division algebra. (A **division algebra** is an algebra with multiplicative identity such that every nonzero element has a multiplicative inverse.)

To further classify algebras, it was necessary to classify division algebras. It had already been proved by Frobenius that over the field of real numbers there were only three division algebras, the real numbers, the complex numbers, and the quaternions. Wedderburn himself proved in 1909 that the only finite division algebras were the finite fields themselves, and these were well known. Division algebras over the field of p-adic numbers were classified by Helmut Hasse (1898–1979) in 1931 and over any algebraic number field by Hasse, Richard Brauer (1901–1977), and Emmy Noether (1882–1935) in 1932. Because these classifications involve many advanced concepts from algebraic number theory, including class field theory, they will not be discussed here.

18.3.4 Noetherian Rings

Much work on the theory of rings was accomplished by Noether and others at Göttingen in the 1920s. In particular, Noether was able to develop a decomposition theory for ideals analogous to Dedekind's prime factorization but applicable to rings more general than the rings of integers in algebraic number fields. These more general rings are now called **Noetherian** rings, commutative rings with identity which satisfy the ascending chain condition, that every chain $I_1, I_2, \ldots, I_k, \ldots$ of ideals in the ring such that $I_k \subset I_{k+1}$ breaks off after a finite number of terms.

Noether was also able to characterize those rings R for which the entire Dedekind theory of prime factorization of ideals holds by a set of axioms:

1. R satisfies the ascending chain condition.

2. The residue class ring R/A satisfies the descending chain condition for every nonzero ideal A. (The descending chain is the same as the ascending chain condition with the $\subset$ replaced by $\supset$.)

3. R has a multiplicative identity element.

Biography   **Emmy Noether (1882–1935)**

Emmy Noether received the normal upbringing of an upper middle-class German-Jewish girl, attending finishing school, studying the piano, and taking dance lessons. It was not until 1900, after further study of French and English, that she passed the Bavarian state examinations to qualify to teach in the schools. But about this same time her interest shifted from languages to mathematics, and she spent the next three years auditing mathematics courses at the University of Erlangen, where her father was a professor of mathematics. In fact, in her first semester she was one of only two women allowed even to audit courses. When in 1904 the university officially permitted women to register, she became a regular student and, four years later, received her doctorate with a dissertation on invariants of ternary biquadratic forms.

Noether remained at Erlangen for several more years, until in 1915 David Hilbert called her to Göttingen to assist him in his study of general relativity. Although as a woman she was not permitted to teach officially, or to receive a salary, Hilbert arranged for her to teach courses that were given under his own name. He in fact argued unsuccessfully in the University Senate on her behalf: "I do not see that the sex of the candidate is an argument against her admission as *Privatdozent*. After all, the Senate is not a bathhouse."[17] It was not until after the changes in Germany at the end of World War I that she was able to receive an official position at the university and, after 1922, even a modest salary. During the next ten years at Göttingen, her influence was felt the most, both in Germany and, through her visit in 1928–1929 to Moscow, in the Soviet Union as well. In 1932, she was the only woman mathematician invited to give a plenary lecture at the International Congress of Mathematicians in Zurich.

Her world, as well as the world of many of Germany's mathematicians, changed suddenly in early 1933 on the coming of the Nazis to power. As a Jew, she lost her teaching position at Göttingen, and, along with many of her colleagues, was forced to take refuge abroad. A position was found for her at Bryn Mawr College, near Philadelphia, beginning in fall, 1933, a location which was close enough to Princeton for her to participate regularly in activities at the Institute for Advanced Study. She died suddenly, in April, 1935, after seemingly successful surgery for removal of a tumor.

4. R is an integral domain, that is, it has no zero divisors.

5. R is integrally closed in its quotient field. In other words, every element of the field which satisfies a polynomial equation over the ring is itself in the ring.

She then showed that if R satisfies these five axioms, then its integral closure in a finite separable extension of its quotient field also satisfies them. In particular, because any **principal ideal domain** (integral domain in which every ideal is principal) satisfies these axioms, her result shows not only that all domains of integers in finite algebraic number fields have unique factorization of ideals, but also that the integrally closed subrings in finite algebraic extensions of the fields of algebraic functions in one variable have this property.

As indicated in the opening of this chapter, Noether had great influence on her co-workers, especially in emphasizing the structural rather than the computational aspects of algebra. In fact, the second volume of probably the most important algebra text of the first half of the twentieth century, *Modern Algebra*, by B. L. van der Waerden (1903–), is largely

---

Side Bar 18.3   Women in Mathematics

The attentive reader will have noticed that there are very few female mathematicians discussed in this text. The reason, of course, is that up until recently, very few women have participated in the discipline of mathematics. There were probably women whose names have been lost to history who made contributions to mathematics in ancient times, but in recorded history, both in western and nonwestern cultures, women have in general not been permitted an education which would permit them to achieve success in mathematics. A consideration of the biographies of those women who are included in this book shows that, for the most part, they had a close family member who was willing to teach them mathematics or who, at least, encouraged them to study the subject. Without such a supportive background, evidently, women could not enter the field. And even those who managed to achieve a reasonable knowledge of mathematics were often not able to participate in the mathematical community. Women simply were not supposed to engage in such intense intellectual activities.

Over the past several decades, however, the picture has been changing. Even though there are still significant obstacles for a woman to overcome, particularly the attitudes of those teaching them in school,

it is now possible for women who want to be mathematicians to achieve that aim, even without a family member as a role model. In fact, in recent years approximately 20 percent of the new Ph.D.s in mathematics granted in the United States have been granted to women. And women are gradually entering positions of influence in the mathematical community. The American Mathematical Society had its first female president, Julia Robinson, in the 1980s, and the Mathematical Association of America has had several recent female presidents as well as a female executive director. The Association for Women in Mathematics has over the past 20 years actively sought to increase opportunities for women. It has, for example, worked to find financial support for female graduate students and new doctorates and has sponsored lectures by prominent female mathematicians at major mathematics meetings. On the international level, there has been an increase in the number of female speakers at recent International Congresses, although that number is still absurdly low. In any case, it does appear that progress has been made in providing opportunities for women to enter the mathematics profession. When a history of mathematics text is written at the end of the twenty-first century, there will be far more female mathematicians discussed than in the current text.

---

due to her. A comparison of this text to texts only a few years earlier shows the great changes she initiated (Side Bar 18.3).

18.3.5   Algebraic Topology

One of Noether's suggestions which was to create a whole new field of study developed out of her attending lectures in topology given by Pavel Sergeiivich Aleksandrov (1896–1982) in Göttingen in 1926 and 1927. As Aleksandrov put it in his address in her memory, "When in the course of our lectures she first became acquainted with a systematic construction of combinatorial topology, she immediately observed that it would be worthwhile to study directly the groups of algebraic complexes and cycles of a given polyhedron and the subgroup of the cycle group consisting of cycles homologous to zero; instead of the usual definition of Betti numbers and torsion coefficients, she suggested immediately defining the Betti group as the [quotient] group of the group of all cycles by the subgroup of cycles homologous to zero."[18] With Noether's remarks and the subsequent publications of Leopold Vietoris (1891–) and Heinz Hopf (1894–1971), the subject of algebraic topology began in earnest. Vietoris in 1927 defined the **homology group** $H(A)$ of a complex A to be the quotient group of cycles modulo boundaries, as Noether recommended. About the same time, Hopf defined several other Abelian groups, namely the groups L^p, Z^p, R^p, and $\overline{R}^p$ generated by the p-simplexes, the p-cycles, the p-boundaries (those chains which were the boundary of some chain), and the p-boundary-divisors (those chains for which a multiple was a boundary) respectively. Then for Hopf, the factor group $B_p = Z^p/\overline{R}^p$ was a free group (a group none of whose elements had a multiple equal to 0) whose rank (the number of basis elements) turned out to be the pth Betti number of the complex.

Matters progressed so quickly in this new field that just a year later Walther Mayer (1887–1948) pubished an axiom system for defining homology groups. Namely, Mayer was no longer concerned with the topological complexes themselves, but solely with the algebraic operations which were defined on them. Thus a complex ring Σ was a collection of elements (complexes) $K^{(p)}$, to each of which was attached a dimension p. The p-dimensional elements formed a finitely generated free Abelian group K^p. For each p, a homomorphism $R_p : K^p \to K^{p-1}$ is defined such that $R_{p-1}(R_p(K^p)) = 0$. (R_p is called the pth boundary operator. Often, one just uses R, without subscripts, and then writes the last equation in the form $R^2 = 0$.) Given these axioms, Mayer defined the group of p-cycles C^p to be those elements K of K^p for which $R(K) = 0$ and the group of p-boundaries to be $R(K^{p+1})$. Modifying Hopf's definition slightly, he defined the pth homology group of Σ to be the factor group $H_p(\Sigma) = C^p/R(K^{p+1})$.

It turns out that the assignment of the homology groups to the space carries over to the assignment of functions between spaces and the corresponding groups. That is, if $f : A \to B$ is a continuous function between two manifolds A and B, considered as simplicial complexes, and if $H_k(A)$, $H_k(B)$ are the kth homology groups of A and B respectively, then there is a well-defined group homomorphism $H_k(f) : H_k(A) \to H_k(B)$. In fact, $H_k(f)$ is defined on a k-chain by $H_k(f)(V_0V_1 \ldots V_k) = f(V_0)f(V_1) \ldots f(V_k)$. One can prove that this assignment makes sense in terms of the homology groups, that is, that cycles are taken into cycles and boundaries into boundaries, and therefore that it defines a homomorphism of the appropriate quotient groups.

It was the consideration of relationships between functions defined on members of one collection of objects and new functions defined on related members of a different collection which led Samuel Eilenberg (1913–) and Saunders MacLane (1909–) to the creation of an even more abstract algebraic structure, that of a category, in a paper of 1945. Generalizing in some sense the ideas of Klein's *Erlanger Programm*, they realized that it was always appropriate whenever a new collection of mathematical objects was defined, that a definition be given of mappings between these objects. Thus a **category** C was defined to be a dual collection $\{A,\alpha\}$ consisting of "an aggregate of abstract elements A (for example, groups), called the **objects** of the category, and abstract elements α (for example, homomorphisms), called **mappings** of the category."[19] These mappings were subject to certain axioms, including the existence of an appropriate product mapping which satisfies the associativity property and of an identity mapping corresponding to each object A. Examples of categories besides that of groups and homomorphisms include topological spaces and continuous maps, sets and functions, and vector spaces and linear transformations.

Following their own dictum, Eilenberg and MacLane further introduced the concept of a functor, a function between categories. Namely, if $C = \{A, \alpha\}$ and $\mathcal{D} = \{B, \beta\}$ are two categories, a (covariant) **functor** T from C to $\mathcal{D}$ is a pair of functions, both designated by the same letter T, an object function and a mapping function. The object function assigns to each A in C an object $T(A)$ in $\mathcal{D}$, while the mapping function assigns to each mapping $\alpha : A \to A'$ in C a mapping $T(\alpha) : T(A) \to T(A')$ in $\mathcal{D}$. This pair must further take identity mappings into identity mappings and must satisfy the condition $T(\alpha\alpha') = T(\alpha)T(\alpha')$ whenever the product $\alpha\alpha'$ exists in C. (For a contravariant functor, the mapping function is reversed, that is, $T(\alpha) : T(A') \to T(A)$ and $T(\alpha\alpha') = T(\alpha')T(\alpha)$.) For example, homology is a covariant functor from the category of manifolds and continuous transformations to the category of Abelian groups and homomorphisms. As a second example, the assignment to each finite dimensional vector space V of the vector space $T(V)$ of all real-valued linear functions on V induces a contravariant functor from the category of vector spaces and linear transformations to itself.

The study of categories and functors, what might be considered an abstraction of an abstraction, has actually proved to be important in various recent developments in algebra and also in differential and algebraic geometry. On the other hand, there have been many critics in recent years of a seeming overemphasis in modern algebra on abstractions for abstractions' sake.[20] The arguments in this regard will be left for the students' reading and discussion.

18.4 COMPUTERS AND APPLICATIONS

When a nontechnically educated person thinks of mathematics in the late twentieth century, the most obvious aspect of the subject which comes to mind is that of computers. Mathematicians themselves have to a large degree not yet accepted the entrance of machine computation into their subject. For many mathematicians, pencil and paper are still the most important tools. Yet the rapid advances in computing power since the 1950s have brought the computer into the mainstream of mathematics, and a growing number of mathematicians now make use of it not only in generating examples but also in constructing proofs. Interestingly enough, many aspects of theoretical mathematics that had lain dormant for years have received renewed attention because of their applications to the general field of computer science. Although there is not enough space here to give a detailed history of the development of the computer, we will conclude this chapter with a sketch of the most important aspects of that history and of a few areas of mathematics which are closely related to it.

18.4.1 The Prehistory of Computers

The dream of mechanical calculation must have occurred as far back as Greek times when Ptolemy was probably forced to use a large number of human "computers" to generate the various tables which appear in his *Almagest*. Some Islamic scientists in the middle ages in fact did use certain instruments to help in their own calculations, particularly in those related to astronomy. The calculation of astronomical tables was important in Europe as well by the early seventeenth century, and logarithms were invented in part to help in this

FIGURE 18.5
Schickard's computing machine.

regard. In short order two Englishmen, Richard Delamain (first half of the seventeenth century) and William Oughtred (1574–1660), independently created a physical version of a logarithm table in the form of a slide rule, a circular (later rectilinear) arrangement of movable numerical scales which enabled multiplications and divisions, as well as computations involving trigonometric functions, to be performed easily.

About the same time, Wilhelm Schickard (1592–1635), a professor of astronomy and mathematics at the University of Tübingen, designed and built a machine which performed addition and subtraction automatically, as well as multiplication and division semiautomatically (Figure 18.5). Schickard described the machine in letters to Kepler in 1623 and 1624, but the machine which he intended to build for Kepler's own use was destroyed in a fire before it was completed. The remaining copies of the machine, as well as the designer, perished in the Thirty Years War, so Schickard's device had no influence on later work. Some 20 years later, Pascal constructed a mechanical adding and subtracting machine (Figure 18.6), while in 1671 Leibniz constructed a machine which also did multiplication and division. Leibniz was quite sure that his machine would be of great practical use:

> We may say that it will be desirable to all who are engaged in computations which, it is well known, are the managers of financial affairs, the administrators of others' estates, merchants, surveyors, geographers, navigators, astronomers, and [those connected with] any of the crafts that use mathematics. But limiting ourselves to scientific uses, the old geometric and astronomic tables could be corrected and new ones constructed by the help of which we could measure all kinds of curves and figures . . . Also, the astronomers surely will not have to continue to exercise the patience which is required for computation . . . For it is unworthy of excellent men to lose hours like slaves in the labor of calculation, which could be safely relegated to anyone else if the machine were used.[21]

18.4.2 Babbage's Difference Engine and Analytical Engine

Unfortunately, neither Leibniz's machine nor the various improved models built by others during the following century and a half were actually used to any extent in the way Leibniz envisaged. The mathematical practitioners themselves continued to do calculations by hand, probably because the machines, operated manually, provided little advantage in speed. For complicated calculations, naturally, tables were used, particularly of logarithms

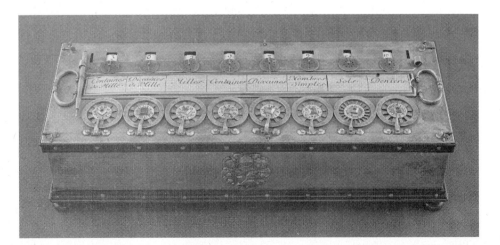

FIGURE 18.6
A replica of Pascal's mechanical calculating device. (Source: Neuhart Donges Neuhart Designers, Inc.)

FIGURE 18.7
Babbage and his
"computer" honored on
a British stamp.

and trigonometric functions, even though these tables, originally calculated by hand, frequently contained errors. It was not until the Industrial Revolution was in full swing in England and the steam engine had been invented that another brilliant mind, that of Charles Babbage, conceived around 1821 the idea of using this new technology to drive a machine which would increase the speed as well as the accuracy of numerical computation (Figure 18.7).

Babbage realized that the calculation of the values of a polynomial function of degree n could be effected by using the fact that the nth order differences were always constant. To take a simple example, consider the following short table for the function $f(x) = x^2$:

x	$f(x)$	**First Difference**	**Second Difference**
1	1		
2	4	3	
3	9	5	2
4	16	7	2
5	25	9	2
6	36	11	2

Note that in this case the second differences, that is, the differences of the first differences of the values, are all 2. Thus to calculate the values of $f(x)$, it is only necessary to perform additions, working backwards from the second difference column to the first difference column to the desired tabular values. (Naturally, one must begin with certain given values, say $2^2 = 4$ and the first first difference 3.) This idea was the principle behind Babbage's original machine, his Difference Engine (Figure 18.8). The plans for the machine called for seven axes, representing the tabular values and the first six differences, each axis containing wheels which could be set to represent numbers of up to 20 decimal digits. The axes would be interconnected so that the constant set up in one of the difference axes would add to the number set up in the next lower difference axis and so on, until the tabular axis was reached. By repeating the process continually, the desired tabular values for polynomial functions of degree up to six could be calculated for as many values of the variable as desired. Babbage realized too that any continuous function could be approximated in an appropriate interval by a polynomial and therefore that the machine could be used to calculate tables for virtually any function of interest to scientists of the day. His aim, in fact, was to attach the machine to a device for making printing plates so that the tables could be printed without any new source of error being introduced. Unfortunately, although Babbage succeeded in convincing the British government to provide him with a grant to help in the building of the Difference Engine, a complete model was never constructed, because of various difficulties in developing machine parts of sufficient accuracy, because ultimately the government lost interest in the project, and because Babbage himself became interested in a new project, the development of a general purpose calculating machine, his Analytical Engine.

Babbage began his new project in 1833 and had elaborated the basic design by 1838. His new machine contained many of the features of today's computers. Constructed again of numerous toothed wheels on axes as hardware, it was to consist of two basic parts, the **store** and the **mill**. The store was the section in which numerical variables were kept until they were to be processed and where the results of the operations were held, while the mill

was the section in which the various operations were performed. To control the operations, Babbage took an idea of Joseph Jacquard (1752–1834), who had automated the weaving industry in France through the introduction of punched cards which described the desired pattern for the loom (Figure 18.9). Babbage thus devised his own system of punched cards which were to contain both the numerical values and the instructions for the machine. Although Babbage never wrote out a complete description of his Analytical Engine and, in fact, never had the financial resources actually to construct it, he did leave for posterity some 300 sheets of engineering drawings, each about 2 by 3 feet, and many thousands of pages of detailed notes on his ideas. Modern scholars have concluded by examining these papers that the technology of the time was probably sufficient to build the engine, but because there was insufficient interest in the British government to finance such a massive project, the engine remained only a theoretical construct.[22]

In 1840, Babbage gave a series of seminars on the workings of the Analytical Engine to a group of Italian scientists assembled in Turin, one of whom summarized the seminars in a published article. The 17 page article was translated into English in 1843 and supplemented by an additional 40 pages of notes by Ada Byron King, Countess of Lovelace (1815–1852). In her notes, Lovelace not only expanded on various parts of the article about the detailed functioning of the engine, but also gave explicit descriptions of how it would solve specific problems. Thus she described, for the first time in print, what would today be called a computer program, in her case a program for computing the Bernoulli numbers. She began from a description of the Bernoulli numbers as the coefficients B_i in the expansion

$$\frac{x}{e^x - 1} = 1 - \frac{x}{2} + B_2\frac{x^2}{2} + B_4\frac{x^4}{4!} + B_6\frac{x^6}{6!} + \cdots.$$

FIGURE 18.9
Joseph Jacquard on a
French stamp.

Biography   Ada Byron King (1815–1852)

Augusta Ada Byron King Lovelace was the child of George Gordon, the sixth Lord Byron, who left England five weeks after his daughter's birth and never saw her again. She was raised by her mother, Anna Isabella Millbanke, a student of mathematics herself, so she received considerably more mathematics education than was usual for girls of her time. Although she never attended any university, she was tutored privately by, and was able to consult with, well-known mathematicians, including William Frend and Augustus De Morgan. In 1833 she met Charles Babbage and soon became interested in his Difference Engine. Her husband, the Earl of Lovelace, was made a Fellow of the Royal Society in 1840 and through this connection, Ada was able to gain access to the books and papers she needed to continue her mathematical studies. Her major mathematical work, discussed in the text, is a heavily annotated translation of a paper by the Italian mathematician L. F. Menabrea dealing with Babbage's Analytical Engine, which contained the first detailed discussion of programming and, in particular, the detailed instructions for programming a computer to compute Bernoulli numbers. Interestingly, only her initials, A.A.L., were used to publish the paper. It was evidently not considered proper in mid-nineteenth century England for a woman of her class to publish a mathematical work.

(See Chapter 14, exercise 3.) By a bit of algebraic manipulation using the power series expansion for e^x, Lovelace rewrote this equation in the form

$$0 = -\frac{1}{2}\frac{2n-1}{2n+1} + B_2\left(\frac{2n}{2!}\right) + B_4\left(\frac{2n(2n-1)(2n-2)}{4!}\right)$$
$$+ B_6\left(\frac{2n(2n-1)\cdots(2n-4)}{6!}\right) + \cdots + B_{2n},$$

a form from which the various B_i can be calculated recursively. Thus to calculate B_{2n}, one needs three numerical values 1, 2, n, as well as the values B_i for $i < 2n$, values which presumably have already been calculated. Instruction cards then are needed to multiply n by 2, subtract 1 from that result, add 1 to that result, divide the two last results, multiply the result by $-(1/2)$, divide $2n$ by 2, multiply the result by B_2, and so on. The results of certain of these calculations, such as $2n - 1$, are used several times during the calculations and therefore need to be moved to various registers where the calculations will take place. At certain stages in the calculation, the machine is instructed to subtract an integer from $2n$ and then decide on the next step depending on whether the result is positive or 0. If it is 0, the equation for B_{2n} is complete and the machine easily solves it; if it is positive, the machine repeats many of the preceding steps. It is not difficult to see that some of the basic concepts of modern day programming, including loops and decision steps, are included in Lovelace's description. Furthermore, she had printed with her notes a detailed diagram of the above program, perhaps the first "flow chart" ever constructed (Figure 18.10).

Besides discussing the basic functioning of the analytical engine, Lovelace described what kinds of jobs it could do and noted explicitly that it could perform symbolic algebraic operations as well as arithmetic ones. But, she noted,

the Analytical Engine has no pretensions whatever to *originate* anything. It can do whatever we *know how to order it* to perform. It can *follow* analysis; but it has no power of *anticipating*

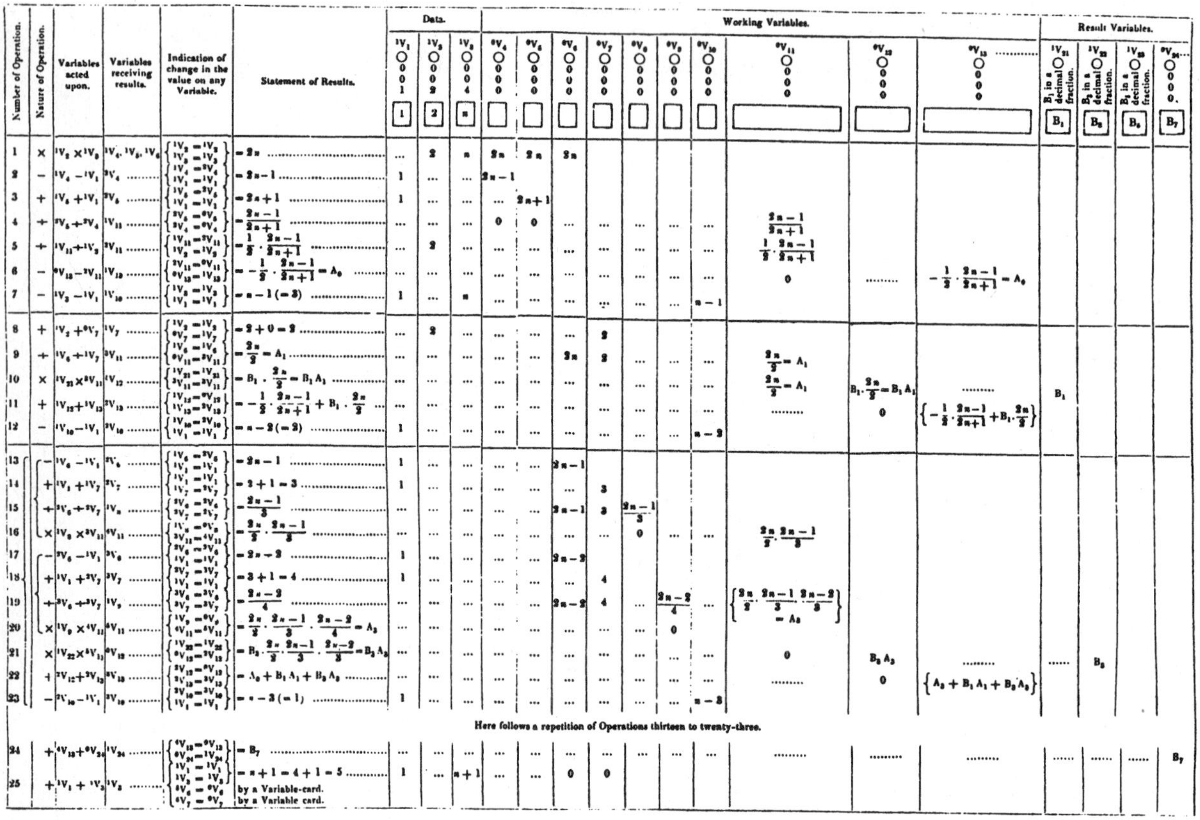

FIGURE 18.10
Ada King's flowchart for calculating Bernoulli numbers.

any analytical relations or truths. Its province is to assist us in making *available* what we are already acquainted with. . . . But it is likely to exert an *indirect* and reciprocal influence on science itself in another manner. For, in so distributing and combining the truths and the formulas of analysis, that they may become most easily and rapidly amenable to the mechanical combinations of the engine, the relations and the nature of many subjects in that science are necessarily thrown into new lights, and more profoundly investigated. . . . It is however pretty evident, on general principles, that in devising for mathematical truths a new form in which to record and throw themselves out for actual use, views are likely to be induced, which should again react on the more theoretical phase of the subject.[23]

A better description of the computer's limitations and its implications for the development of mathematics could hardly be written today.

18.4.3 Turing and Computability

One of the reasons that Babbage's ideas were not brought to fruition with an actual Analytical Engine was that even in mid-nineteenth century England, there was no perceived societal need for it that made it worth the enormous resources which would have

Biography   Alan Turing (1912–1954)

Alan Turing's father, an officer in the British administration in India, decided that his son would be raised in England. Thus Alan's parents saw him only rarely during his formative years. Turing entered King's College, Cambridge in 1931 to study mathematics and received his M.A. degree four years later with a dissertation dealing with Gaussian error functions. Shortly thereafter, however, he began to work in earnest on a major new problem, Hilbert's decision problem, and resolved it in the paper in which he invented the concept of the Turing machine. At about the same time, however, Alonzo Church in Princeton published another solution to the problem, so Turing decided to go to Princeton to work with Church. Returning to England and King's College in 1938, he was called, on the outbreak of World War II, to serve at the Government Code and Cypher School in Bletchley Park in Buckinghamshire. It was there, during the next few years, that Turing led the successful effort to crack the German "Enigma" code, an effort which turned out to be central to the defeat of Nazi Germany.

After the war, Turing continued his interest in automatic computing machines and so joined the National Physical Laboratory to work on the design of a computer. He continued this work at the University of Manchester after 1948. Turing's promising career, however, came to a grinding halt when he was arrested in 1952 for "gross indecency." He was, in fact, a homosexual, and at the time, overt homosexual acts were against the law in England. The penalty for this crime was submission to psychoanalysis and to hormone treatments designed to "cure" this disease. Unfortunately, the cure proved worse than the disease, and in a fit of depression, Turing committed suicide in June of 1954 by eating a cyanide-poisoned apple.

been necessary for its construction. And although various computational devices and analog computers were devised in the century after Babbage's design work, devices generally adapted to solving specific mathematical problems which otherwise would require enormous amounts of manual computation, it was military necessities during the two world wars, especially the second, which were to lead to the actual construction of the first electronic computers, in many essentials based on Babbage's ideas. Still, there were other theoretical ideas worked out in the years immediately before the second world war which were to be fundamental in the development of these computers. One of these was the idea of computability in the work of Alan Turing (1912–1954).

Turing was interested in the problem of computatibility, of determining a reasonable but precise answer to the questions of what a computation is and whether a given computation can in fact be carried out. To answer these questions, Turing extracted from the ordinary process of computation the essential parts and formulated these in terms of a theoretical machine, now known as a Turing machine. Furthermore, he showed that there is a "universal" Turing machine, a machine which can calculate any number or function which can be calculated by any special machine, provided it is given the appropriate instructions.

Turing's machine, presented in a major paper of 1936, was formed from three basic concepts: a finite set of states, or configurations, $\{q_1, q_2, \ldots, q_k\}$; a finite set of symbols $\{a_0, a_1, a_2, \ldots, a_n\}$ which are to be read and/or written by the machine (where a_0 is taken as a blank symbol); and a process of changing both the states and the symbols to be read. To accomplish its job, the machine is supplied with instructions by means of an (infinite) tape running through it, divided into squares, with a finite number of these squares bearing a nonblank symbol. At any given time, there is just one square, say the rth, in the machine, bearing a symbol S_r. To simplify matters further, the possible instructions given by a symbol are limited to replacing the symbol on the square by a new one, moving the tape one square to either the right or the left, and changing the state of the machine. Thus at any given moment, the pair (q_i, S_r) will determine the behavior of the machine according to the particular functional relationship which defines it. If the function is not defined on a particular pair (q_i, S_r), the machine simply halts. It is the symbols printed by the machine, or at least a determined subset of them, which is to represent the number to be computed. Turing's contention, backed up by many arguments in his paper, is that the operations above are all those necessary actually to compute a number.

As an example, Turing constructed a machine to compute the sequence $010101 \cdots$. This machine is to have four states, q_1, q_2, q_3, q_4, and is capable of printing two symbols, 0 and 1. The tape for this machine is entirely blank initially and the machine begins in state q_1. The instructions that the machine uses are as follows.

1. If it is in state q_1 and reads a blank square, it prints 0, moves one square to the right, and changes to state q_2.

2. If it is in state q_2 and reads a blank square, it moves one square to the right and changes to state q_3.

3. If it is in state q_3 and reads a blank square, it prints a 1, moves one square to the right, and changes to state q_4.

4. If it is in state q_4 and reads a blank square, it moves one square to the right and changes to state q_1.

It is easy enough to see that this machine does accomplish what is desired, although Turing for technical reasons arranged the printing so that there are figures only on alternate squares. And although this example does not demonstrate it, the reason for the motion of the tape in either direction is to give the machine a memory. Thus the machine can reread a particular square and act in different ways depending on its particular state at the time. In this way the machine can "remember" numbers written earlier and use them in subsequent computations.

It is the possibility of memory which leads to perhaps the most surprising part of Turing's paper, his proof of the existence of a single machine which can compute any computable number. Turing's idea for this was to take the set of instructions for any given machine, like those written above, and turn that set systematically into a series of symbols, called the standard description of the machine. The "universal" machine is then supplied with a tape containing this standard description followed by the symbols on the input originally supplied to that machine. Turing was in fact able to give a rather explicit description of the behavior of this universal machine in terms of a functional relationship as described earlier. The main idea is that the machine acts in cycles, each cycle representing first, a look at the standard description of the particular machine, second, a look at one square of that machine's input, and finally, a corresponding action. Although Turing did not at the time attempt the physical construction of a machine with the capabilities he proved could exist, it was his idea which led directly to the concept of an all-purpose computer which could be programmed to do any desired computation. Naturally, there are physical limits to the size of a machine and the length of a program, limits which did not exist in Turing's theoretical model with its infinite tape. Modern technology, however, seems to extend these physical limits so often that today's computers are better and better approximations to Turing's universal machine with each passing year.

18.4.4 Shannon and the Algebra of Switching Circuits

A more direct application of mathematical ideas to the construction of a computer was developed by Claude Shannon (1916–) in 1938 as part of his master's thesis at M.I.T. In this work, Shannon applied the algebra of logic developed by Boole a century earlier to the construction of switching circuits which would have desired properties. It is these circuits which form the basis for the internal construction of computing machines. Shannon in fact realized that any circuit can be represented by a set of equations and that the calculus necessary for manipulating these equations is precisely the Boolean algebra of logic. Thus given the desired characteristics of a circuit that one wanted to construct, one used this calculus to manipulate the equations into the simplest possible form, from which the construction of the circuit was then immediate. One could also perform analysis of circuits by this method, by applying the calculus to the equations of a complex circuit and thereby reducing it to a simpler form.

Shannon began his work by dealing simply with switches which may be open or closed. The open ones were represented by 1, the closed ones by 0. Placing two switches in series was represented by the Boolean operation $+$, while placing them in parallel was represented by $\cdot$ (Figure 18.11). Shannon noted the following postulates which these two operations satisfied, with their corresponding circuit interpretations, the truth of which made the analogy of switching circuits with Boolean algebra possible.

1. $0 \cdot 0 = 0; 1 + 1 = 1$. A closed circuit in parallel with a closed circuit is closed while an open circuit in series with an open circuit is open.

2. $1 + 0 = 0 + 1 = 1; 0 \cdot 1 = 1 \cdot 0 = 0$. An open circuit in series with a closed circuit is open while a closed circuit in parallel with an open circuit is closed.

3. $0 + 0 = 0; 1 \cdot 1 = 1$. A closed circuit in series with a closed circuit is closed while an open circuit in parallel with an open circuit is open.

Representing a switch in a circuit by X, Shannon noted that X could take on just the two values 0 and 1. It followed that the laws of Boolean algebra, including the two commutative laws, the two associative laws, and the distributive laws both of multiplication

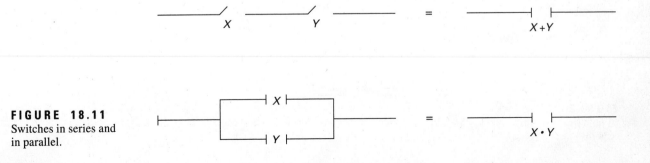

FIGURE 18.11
Switches in series and in parallel.

over addition and of addition over multiplication, could be proved by simply checking each possible case. He also introduced the negation of a variable X, written X', to be that variable which is 1 when X is 0 and 0 when X is 1, and demonstrated some additional laws. For example, $X + X' = 1$, $X \cdot X' = 0$, $(X + Y)' = X' \cdot Y'$, and $(X \cdot Y)' = X' + Y'$. Shannon then recalled Boole's expansion of functions and dualized it by interchanging multiplication and addition. Thus, for example,

$$f(X, Y, Z, \ldots) = [f(0, Y, Z, \ldots) + X][f(1, Y, Z, \ldots) + X'].$$

A very useful rule can then be established by adding X to both sides of this equation (bearing in mind the distributive laws):

$$X + f(X, Y, Z, \ldots) = X + f(0, Y, Z, \ldots).$$

With the various laws of Boolean algebra now established for switching circuits, Shannon was able both to analyze and synthesize circuits.

For example, Shannon presented the circuit (Figure 18.12) whose algebraic representation was

$$W + W'(X + Y) + (X + Z)(S + W' + Z)(Z' + Y + S'V).$$

Using various laws of Boolean algebra, including the special law of the previous paragraph applied three times, he reduced this formula first to

$$W + X + Y + (X + Z)(S + 1 + Z)(Z' + Y + S'V),$$

then to

$$W + X + Y + Z(Z' + S'V),$$

and finally to

$$W + X + Y + ZS'V.$$

This latter formula had a much simpler circuit representation than the original.

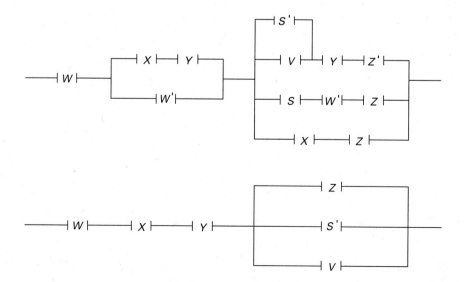

FIGURE 18.12
Simplifying a circuit using Boolean algebra.

Born in Budapest into the family of a well-to-do Jewish banker, von Neumann received his doctorate from the University of Budapest. He taught in Berlin and Hamburg before being invited to Princeton in 1930. Three years later he was chosen as one of the charter members of the Institute for Advanced Study, a position he held for the remainder of his life. Von Neumann was one of the last mathematicians equally at home in pure and applied work. Over the years he produced a steady stream of papers in both areas.

In pure mathematics, he was especially proficient in analysis and combinatorics. He had a great ability to see into complex situations and pull out appropriate axioms which would enable the subject to be treated mathematically. His talents in applied mathematics were in particular demand during the second world war and shortly afterward, when he led the effort to develop the modern computer. He was a member of the Atomic Energy Commission from 1954 until his untimely death from cancer in 1957.

As an example of the synthesis of a circuit having given characteristics, Shannon showed how to construct one which would add two numbers given in binary representation. If the two numbers are represented by $a_n a_{n-1} \cdots a_1 a_0$ and $b_n b_{n-1} \cdots b_1 b_0$, and their sum by $s_{n+1} s_n \cdots s_1 s_0$, then s_0 is equal to 1 if $a_0 = 1$ and $b_0 = 0$ or if $a_0 = 0$ and $b_0 = 1$ and equal to 0 otherwise. There is also a carry digit c_1 which is equal to 1 if both a_0 and b_0 are 1 and 0 otherwise. Thus s_0 is represented by the equation $s_0 = a_0 b'_0 + a'_0 b_0$ and c_1 is represented by $c_1 = a_0 b_0$. Each s_j for $j \geq 1$ requires the addition not only of a_j and b_j but also the carry digit c_j. Thus the formula for s_j is

$$s_j = (a_j b'_j + a'_j b_j)c'_j + (a_j b'_j + a'_j b_j)'c_j$$

while that for the next carry digit c_{j+1} is

$$c_{j+1} = a_j b_j + c_j(a_j b'_j + a'_j b_j).$$

The circuit construction of an adder determined by these equations is the basis for the design of addition methods in modern day calculators and computers.

18.4.5 Von Neumann's Computer

The work of Turing and Shannon were only two facets of the many theoretical and applied problems which had to be solved before the modern computer could be constructed. There were numerous people who worked on these problems, particularly during the 1940s. But the man most responsible for the shape of the ultimate result was probably John von Neumann (1903–1957), who immediately after the second world war gathered a brilliant group of scientists and engineers at the Institute for Advanced Study in Princeton. Their task was to take the experience developed during the war years in the development of two early computers, the ENIAC and the EDVAC, and combine it with recently developed theoretical knowledge to develop what one of its backers called "the most complex research instrument now in existence. . . . Scholars have already expressed great interest in the possibilities of such an instrument and its construction would make possible solutions of which man at the present time can only dream."[24]

The group under von Neumann decided to organize the computer under four main sections, an arithmetic unit, a memory, a control, and an input-output device, the first two being quite analogous to Babbage's mill and store respectively. The arithmetic unit, now generally called the central processing unit, is the place where the machine performs the elementary operations, those operations which it is decided should not be reduced any further. These elementary operations are essentially wired into the machine, like, for example, the addition described above, while any other operation is built out of the elementary ones by a set of instructions. Recall that the number system of Babbage's analytical engine was decimal. But with the advent of electronic, rather than mechanical, devices for representing numbers, it turned out that it was simpler to represent numbers in binary, so that any particular device holding a digit would only need to have two states, on and off, to represent the two possibilities of 1 and 0. Von Neumann was in fact instrumental in designing efficient sets of decimal-binary and binary-decimal conversion instructions so that the operator could enter numbers in the normal decimal mode and receive answers in that mode as well, without compromising the speed and ease of construction of the machine.

The memory unit of the machine needed to be able to take care of two different tasks, storing the numbers which were to be used in the calculations and storing the instructions by which the calculations were to be made. But because instructions themselves can be stored in appropriate numerical code, the machine only needed to be able to distinguish between the actual numbers and coded instructions. Moreover, in order to compromise between the "infinite" memory desired by the user of the machine and the finite memory constructible by the engineer, it was decided to organize the memory in hierarchies, such that some limited amount of memory was immediately accessible while a much larger amount could be accessed at a somewhat slower rate. It was also decided that in order to achieve a sufficiently large memory in a reasonable physical space, the units which stored an individual digit needed to be microscopic parts of some large piece.

The control unit was the section where the instructions to the machine resided, the orders which the machine could actually obey. Again, compromises had to be worked out between the desire for simplicity of the equipment and the usefulness for the sake of speed of a large number of different types of orders. In any case, one of the more important aspects of the control procedure, an aspect of which even Lady Lovelace was aware, was the ability of the machine to use a given sequence of instructions repeatedly. But because the machine must be made aware of when the repetition should end, it was also necessary to design a type of order which lets the machine decide when a particular iteration is complete. Furthermore, the control unit needed to have a set of instructions which integrated the input and output devices into the machine. Von Neumann was particularly interested, in fact, in assuring that the latter devices would allow for both printed and graphical outputs, because he realized that some of the more important results of a particular computation may best be explored graphically.

The computer eventually constructed at the Institute for Advanced Study, based on von Neumann's design and finished in 1951, proved to be the model for the more advanced computers built in succeeding years. The technological achievements in regard to computers over the next four decades have both increased the capacity and decreased the size by factors probably undreamed of by members of the working group of the late 1940s. Computers have now become so much a part of everyday life that one can scarcely imagine how we would accomplish many common tasks without them. Although we will not deal with any particular applications of computers, we will conclude the chapter with brief discussions of two aspects of mathematics which have impacted on and have been impacted by the omnipresence of computers in our time.

18.4.6 Error Detecting and Correcting Codes

When large amounts of information must be sent from one place to another electronically, one is usually concerned with the information being received with as few errors as possible. As Richard Hamming (1915 –), the author of the first major paper on error correcting and detecting codes (1950), noted, one especially needs to have such capability when there is a long period of unattended operation with a minimum of standby equipment, in large, closely interrelated systems where one failure could destroy the entire operation and where there may be noise in the transmission, as, for example, in the case of "jamming."

Hamming's interest in the question of coding had been stimulated in 1947 when he was using an early Bell System relay computer on weekends only. During the weekends the machine was unattended and, in that mode, would dump any work in which it discovered an error and proceed to the next problem. Hamming realized that it would be worthwhile for the machine to be able not only to detect an error but also to correct it, so that his jobs would in fact be completed. He then proceeded to develop his error correcting codes in 1948, although because of patent considerations he was not able to publish them until 1950.

In his paper, Hamming assumed that a code word consisted of exactly n binary digits, where m digits carry the information and the remaining $k = n - m$ check digits are used for error correction and detection. His aim was to determine a single-error correcting code with as few check digits as possible. Each check digit would be a "parity" check of certain code digits, that is, it would be chosen to be 0 or 1 if the total number of 1s in the code digits to be checked was even or odd respectively. The idea was that on the reception of a given code word, the sequence of 0s and 1s of the check digits would be read as a binary number and would be required to give the position of any single error. Therefore, the k check digits must be able to describe $m + k + 1 = n + 1$ possibilities, because an error could be in any one of the n digits or there might be no errors at all. It follows that the total number of possible strings of k digits, namely 2^k, must be at least as great as $n + 1$. To put it another way, the relationship between m and n was given by the inequality

$$2^m \leq \frac{2^n}{n + 1}.$$

For example, a code word with 7 positions would need to use 3 of these as check digits and 4 for actual information carrying. Namely, in using words 7 positions long, of which there are $2^7 = 128$ possibilities, only $2^4 = 16$ of these would be possible to use for actual code symbols.

Hamming developed a geometric model of his method by considering an n-digit code word to be a vertex of the unit cube in the n-dimensional vector space over the field of two elements. He further introduced a metric in the space such that the distance $D(x, y)$

between two such vertices is the number of coordinates in which they differ. For example, if $n = 3$, the distance $D(x, y)$ between $x = 001$ and $y = 111$ is 2. Hamming noted that if all the code points are at least distance 2 from each other, a single error will be detectable, because such an error would change a code point into a point at distance 1 from some other one. If all code points are at least distance 3 from each other, then a single error would still leave the new point closest to exactly one true code point, and thus the error would also be correctable. Thus it turns out that the problem of determining error correcting codes is the same as that of finding subsets of points in the space which are at least a certain minimum distance apart. In particular, if this minimum distance is 3, each point can be surrounded by a "sphere" of radius 1 such that no two spheres have a common point. (A sphere of radius r centered on x is the set of points at distance less than or equal to r from x.) Hamming could now use the geometric model to answer the question as to how many actual code words there could be. Because each sphere of radius 1 has n points on its "surface," as well as a center point, it contains exactly $n + 1$ points. Because the entire space consists of 2^n points, there are at most $\frac{2^n}{n+1}$ spheres, so the number of code points 2^m can be no greater than this number, just as had been noted earlier.

Hamming gave in his paper a particular code of length 7 digits satisfying the requirements for the check digits, a code of a type today known as a Hamming code. It turned out, and Hamming vaguely realized it, that the set of actual code words of 4 digits in this code was a group or, more specifically, a 4-dimensional vector subspace of the 7-dimensional space of all 7-digit binary strings. Ultimately, the Hamming codes could in fact be described as the kernels of certain linear transformations over appropriate vector spaces. The study of codes, begun by Hamming, was to lead to various other areas of mathematics, but the story of this is readily available in a recent book.[25]

18.4.7 Graph Theory

A **graph** in modern terminology consists of a nonempty set V, whose elements are called vertices, and a set E whose elements are edges, where each edge consists of a pair of vertices. In geometrical terminology, the edges of the graph are arcs which join pairs of vertices. Graph theory in the West has its origins in Euler's 1736 solution of the problem of the seven bridges of Königsberg, discussed in section 14.3.6. In his solution to that problem, Euler replaced the regions of the city by vertices and the bridges by edges in order that he could discuss the problem in terms of a graph. Another interesting problem which later became part of graph theory was found by William Rowan Hamilton in 1856. In fact, Hamilton turned the problem into a game which was marketed in 1859. This *Icosian* game consisted of a graph with 20 vertices on which pieces were to be placed in accordance with various conditions, the overriding consideration being that a piece was always placed at the second vertex of an edge on which the previous piece had been placed (Figure 18.13). The first set of extra conditions Hamilton proposed was, given pieces placed on five initial points, to cover the board with the remaining pieces in succession such that the last piece placed is adjacent to the first. In more modern terminology, Hamilton's problem was to discover a cyclic path which passed through each vertex exactly once. Hamilton gave several examples of ways in which this could be accomplished but gave no general method for determining in cases other than his special graph whether or not such a path could be constructed.

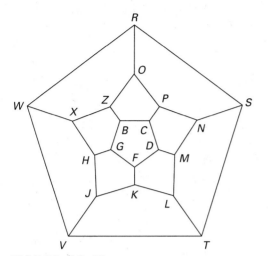

FIGURE 18.13
The Icosian game of William Rowan
Hamilton.

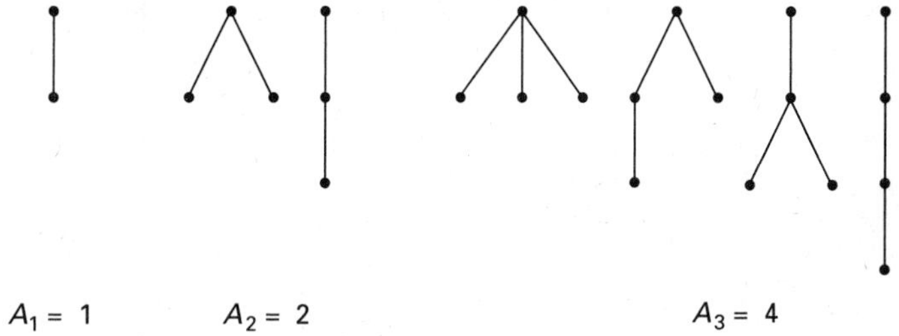

$A_1 = 1$ $A_2 = 2$ $A_3 = 4$

FIGURE 18.14
Different trees with r branches, for
$r = 1, 2, 3$.

The earliest purely mathematical consideration of a special class of graphs appeared in an article of Arthur Cayley in 1857. Cayley, inspired by a consideration of possible combinations of differential operators, defined and analyzed the general notion of a **tree**, a connected graph which has no cyclic paths and therefore in which the number of edges is one fewer than the number of vertices. In particular, Cayley dealt with the notion of a **rooted tree**, a tree in which one particular vertex is designated as the root. He exhibited the possible rooted trees with two, three, and four vertices (which Cayley called **knots**), or, equivalently, with one, two, or three edges (**branches**, to continue the botanical analogy). By a clever combinatorial argument, Cayley then developed a recursive formula for determining the number A_r of different trees with r branches (where "different" is appropriately defined) (Figure 18.14). Thus he showed, for example, that $A_1 = 1$, $A_2 = 2$, $A_3 = 4$, $A_4 = 9$, and $A_5 = 20$. In 1874, Cayley applied his results to the study of chemical isomers and a few years later succeeded in developing a formula for counting the number of unrooted trees with a given number of vertices.

The problem which has played the greatest role in the development of modern day graph theory, a problem which was attacked by many mathematicians from its first formulation in 1852, was the four-color problem. The problem was described in a letter written by De Morgan to Hamilton on October 23 of that year: "A student of mine [Frederick Guthrie] asked me today to give him a reason for a fact which I did not know was a fact—and do not yet. He says that if a figure be anyhow divided and the compartments differently coloured so that figures with any portion of common boundary line are differently coloured—four colours may be wanted, but not more. . . . My pupil says he guessed it in colouring a map of England. The more I think of it, the more evident it seems."[26] De Morgan was not able to think of a case of a map where five colors were required but, although he thought the sufficiency of four colors was "evident," could give no proof of that either. Hamilton was not interested in the problem, but in the following two decades both Cayley and Charles S. Peirce (1839–1914) spent much time in a futile search for a proof. In 1879, Alfred Kempe (1849–1922) published a proof which was accepted as correct and the following year Peter Tait published another one. By 1890, however, both proofs were revealed to contain fallacies. Percy Heawood (1861–1955) demonstrated Kempe's failure, but as part of his discussion was able to prove that five colors are always sufficient for coloring a map.

Many new efforts to solve the problem were made in the early twentieth century, but it was the connection of the map coloring problem with graph theory, made by Hassler Whitney (1907–1989), which both stimulated interest in the latter subject and proved ultimately to be the key to solving the original problem. Whitney defined the notion of the **dual graph** of a map in 1931 to be the graph whose vertices were a set in one-to-one correspondence with the regions of the map and such that two vertices were joined by an edge if and only if the two corresponding regions had a common boundary arc. If we define a coloring of a graph to be an assignment of colors to the vertices so that no two vertices which lie on a common edge have the same color, then the four-color theorem for maps is equivalent to a four-color theorem for graphs. It was thus possible to apply many of the results of the emerging study of graphs to this old problem.

A proof of the four-color theorem for graphs was presented by Kenneth Appel (1932–) and Wolfgang Haken (1928–) in 1976, as mentioned in the chapter opening. Their method of proof involved the reduction of the theorem to a large number of special cases of subgraphs which could occur in a particular graph and then a consideration of the coloring possibilities for each of these subgraphs. For the work to be accomplished in a feasible amount of time, Appel and Haken were forced to use a computer in a nontrivial way. Namely, the number of individual checks which needed to be performed was so large that no human, or collection of humans, could possibly have completed the proof by hand. In fact, Appel and Haken used some six months of computer time to accomplish their task. Although the proof has subsequently been simplified somewhat, there does not seem to be any possibility on the horizon of giving a noncomputer based proof.

A proof by an essential use of the computer is an entirely new phenomenon in mathematics. Computers, since their introduction, have been used to help mathematicians make conjectures, but the proof of the four-color theorem is the first in which the computer was actually used for the detailed work of constructing a formal proof. As was to be expected, this proof has generated much controversy since its appearance. Many mathematicians still do not accept the proof as valid, because the general standard of acceptance of a proof has always been the checking of it by many members of the mathematical community. And although the computer program itself can be checked, there is no way for mathematicians to check the various details of the work actually performed by the computer. How this debate over the use of the computer will be resolved is not at all clear. It is possible that other important results will also be resolved with the assistance of computers, or even that the four-color theorem will one day be proved in the traditional way, but in any case, a new debate over what constitutes a mathematical proof has begun. The question of how one convinces others that a mathematical result is "true," a question which has been asked from as far back as records exist, thus appears to be one which will continue to be asked far into the future.

Exercises

Problems from Set Theory

1. The following is the Richard paradox (named after its originator, Jules Richard (1862–1956)): Arrange all two-letter combinations in alphabetic order, then all three-letter combinations and so on, and then eliminate all combinations which do not define a real number. (For example, "six" defines a real number, while "sx" does not.) Then the set of real numbers definable in a finite number of letters forms a denumerable, well-ordered set $E = \{p_1, p_2, \ldots\}$. Now define the real number $s = .a_1a_2\ldots$ between 0 and 1 by requiring a_n to be one more than the nth decimal of p_n if this decimal is not 8 or 9 and equal to 1 otherwise. Although s is defined by a finite number of letters, it is not in E, a contradiction. How can one resolve this paradox? How is this paradox related to the barber paradox or to Zermelo's original paradox?

2. Show that the trichotomy law follows from Zermelo's well-ordering theorem.

3. Show that Zermelo's axiom of separation resolves Russell's barber paradox as well as the Richard paradox (problem 1) in the sense that certain "sets" are now excluded from discussion.

Problems in Point Set Topology

Note: The problems in this section assume some familiarity with the basic concepts of topology.

4. Formulate and prove the Heine-Borel theorem in the plane.

5. Prove that a connected set A, in the sense of the Youngs, cannot be expressed as $A = B \cup C$, where B and C are closed and $B \cap C = \varnothing$.

6. Show that the set of rational numbers in $[0, 1]$ is not connected.

7. Assume that a set E is compact according to Fréchet's definition in terms of nested sets. Show that every infinite subset E_1 of E has at least one limit element in E.

8. Show by use of examples that the nested set property on the real line depends on the subsets being both closed and bounded.

9. Show that a real function sequentially continuous on a closed and compact set E (in Fréchet's definition) is bounded there and attains its upper bound at least once.

10. Show that if E is closed and compact according to Fréchet's definition, and if $\{E_n\}$ is a nested sequence of closed subsets E, then the intersection $\cap_n E_n$ is not empty.

11. Prove that the space of real functions continuous on $[a, b]$ under the maximum norm metric is "normal" in the sense of Fréchet.

12. Show that the space of all infinite sequences of real numbers $\{x = \{x_1, x_2, \ldots\}\}$ with the metric

$$(x, y) = \sum_{p=1}^{\infty} \frac{1}{p!} \frac{|x_p - y_p|}{1 + |x_p - y_p|}$$

is normal.

13. Show that in the metric space defined in problem 12, there is a number α such that $(x, y) < \alpha$ for all x, y in the space.

14. Show that if E is a topological space and A a closed subset, that is, one containing all its accumulation points, then $E - A$ is a domain (open set) in the notion of Hausdorff. Conversely, if the subset B of E is an open set, show that $E - B$ is closed, using Hausdorff's definition.

15. Show that under Hausdorff's definition of a limit point, a given infinite set A can have no more than one.

16. Show that Hausdorff's two definitions of continuity at a point are equivalent.

17. Use Hausdorff's neighborhood definition of continuity to show that a continuous function preserves connectedness and compactness.

Problems in Combinatorial Topology

18. Determine the boundary of the tetrahedron $V_0V_1V_2V_3$ indicated in Figure 18.4. Show that the boundary of the boundary is 0.

19. Given the face $V_1V_0V_3$ of the tetrahedron of the previous problem, calculate its boundary and show that the boundary of the boundary is 0.

Problems in Algebra

20. Using Dickson's axioms for a field, derive the following theorems:

   a. For any two elements a, b of the set, there exists in the set one element y such that $a + y = b$.

b. If $a + z = a$ for a particular element a, then $b + z = b$ for every element b.

c. If $a + b = a + b'$, then $b = b'$.

21. Prove the same results as in problem 20 with $\times$ instead of $+$ (under the extra condition that $c \times a \neq a$ for at least one element c).

22. Show that the set $\{0, 1, -1\}$ with ordinary addition and multiplication satisfies each of Dickson's axioms except the first.

23. Show that the set $S = \{r\sqrt{2}|r$ rational$\}$ with ordinary addition and multiplication satisfies each of Dickson's axioms except axiom 5. Find the x in this situation which satisfies $(a \times x) \times b = b$.

24. Show that the product 4.324×3.403 is equal to 2.2312242 in Hensel's multiplication relative to $p = 5$.

25. Show that the quotient $3.12 \div 4.21$ is equal in Hensel's arithmetic relative to $p = 5$ to the periodic "decimal" $2.42204220 \ldots$ by actually performing the long division.

26. Show that $3.12 \div 0.2 = 4 \cdot 5^{-1} + 0 + 1 \cdot 5$ in Hensel's field of 5-adic numbers.

27. Show for Hensel's p-adic numbers that the multiplicative inverse of a unit (a number whose smallest power of p is the zero power) is again a unit.

28. Let x be a p-adic number. Define the r-neighborhood of x, $U_r(x)$, where r is an integer, to be $U_r(x) = \{y|y \equiv x \pmod{p^r}\}$. Show that this choice of neighborhoods of x makes the field $\mathbf{Q}_p$ into a topological space in the sense of Hausdorff.

29. Any number x in $\mathbf{Q}_p$ may be written uniquely as $x = p^\alpha e$ where e is a unit. The integer α is called the order of x relative to p, written $v_p(x)$. For any x, y in $\mathbf{Q}_p$ define (x, y) to be $(1/p)^{v_p(x-y)}$. Show that (x, y) defines a metric on $\mathbf{Q}_p$ in the sense of Fréchet.

30. Using the metric of problem 29, define the notion of Cauchy sequence in $\mathbf{Q}_p$ as usual. Show that every Cauchy sequence in $\mathbf{Q}_p$ converges to a limit.

31. Show that each of the following multiplication tables of two basis elements i, j determines an associative algebra of degree 2 over the real numbers. Are there any other ones?

	i	j
i	i	j
j	j	0

	i	j
i	i	j
j	0	0

	i	j
i	j	0
j	0	0

32. Give several examples of categories different from those mentioned in the text. Give several examples of functors using the categories you listed as well as the categories in the text.

Problems from Computers

33. Show that for a polynomial of degree n, the nth order differences are always constant.

34. Construct a difference table which would enable Babbage's Difference Engine to calculate the pyramidal numbers, the numbers which are the sums of the triangular numbers and can be considered as representing, say, the number of cannon balls contained in triangular pyramids of given height.

35. Show that the defining equation for the Bernoulli numbers in problem 3 of Chapter 14 can be transformed into the equation used by Ada Lovelace to write a program for calculating these numbers. (*Hint:* Use the power series for e^x.)

36. Use Ada Lovelace's "flow chart" to program your computer to compute the Bernoulli numbers.

37. Consider a Turing machine with two states q_1, q_2 capable of printing two symbols 0 and 1. Suppose it is defined by the following instructions:

a. If the machine is in state q_1 and reads a 1, it prints 1, moves one square to the right, and remains in state q_1.

b. If the machine is in state q_1 and reads a blank, it prints 1, moves one square to the right, and changes to state q_2.

(Note that there is no instruction for the machine when it is in state q_2.) Suppose the machine begins in state q_1 with a tape whose first square is blank, whose next two squares to the right have 1's, and all of the rest of whose squares to the right are blank, and suppose further that the leftmost 1 is the initial square to be read. Show that the final configuration of the tape will be the same as the initial one except that it will have three 1's instead of two. In general, interpreting a tape with n ones as representing the number $n + 1$, show that this Turing machine will calculate the function $f(n) = n + 1$ for n any nonnegative integer.

38. Determine a Turing machine which computes the function $f(n) = 2n$.

39. Prove the following expansion theorem of Boole: $f(x, y) = [f(0, y) + x][f(1, y) + x']$ where f is a Boolean function of the two Boolean variables x, y.

40. Use the distributive law of addition over multiplication, $a + (bc) = (a + b)(a + c)$ and the theorem of problem 39 to establish the following law for Boolean functions: $x + f(x, y) = x + f(0, y)$.

41. Prove the Boolean expansion theorem $f(x, y) = xf(1, y) + x'f(0, y)$ and the theorem $xf(x, y) = xf(1, y)$. Note that these are the duals of the results in problems 39 and 40.

42. Construct a circuit representing the addition of binary numbers as outlined in the text.

FOR DISCUSSION . . .

43. Look up the Banach-Tarski paradox in a text on set theory and discuss its meaning and implications. Do you believe the result? How does this relate to your belief in the truth of the axiom of choice?

44. Outline a series of lessons developing Hensel's p-adic numbers for a class in modern algebra. Show how one can develop both the algebra and the analysis on such a field.

45. Could one use Hausdorff's text of 1914 in a topology course today? Get a copy and compare it with a current text.

46. Find a copy of an early edition of Van der Waerden's *Modern Algebra* text. Compare it with your current algebra text.

47. Read one or more of the references on the Appel-Haken proof of the four-color theorem. Discuss the implications of this proof. Do you believe that the argument proves the theorem?

References and Notes

Although there is no easily accessible general history of twentieth century mathematics, each of the topics covered in this chapter is dealt with by one or more excellent works. Gregory H. Moore, *Zermelo's Axiom of Choice: Its Origins, Development, and Influence* (New York: Springer, 1982) deals with the material of the first part of this chapter. The original texts of many of the articles referred to can be found in Jean van Heijenoort, ed., *From Frege to Gödel: A Source Book in Mathematical Logic, 1879–1931* (Cambridge: Harvard University Press, 1967). A general history of point set topology is Jerome Manheim, *The Genesis of Point Set Topology* (New York: Macmillan, 1964). The history of algebraic topology is discussed in great detail in Jean Dieudonné, *A History of Algebraic and Differential Topology, 1900–1960* (Boston: Birkhäuser, 1989). Histories of two important concepts of point set topology are Raymond Wilder, "Evolution of the Topological Concept of 'Connected'," *American Mathematical Monthly* 85 (1978), 720–726 and J.-P. Pier, "Historique de la notion de compacité," *Historia Mathematica* 7 (1980), 425–443. A good part of twentieth century algebra, including particularly the theory of algebras, is covered in part three of B. L. Van der Waerden, *A History of Algebra from al-Khwarizmi to Emmy Noether* (New York: Springer, 1985). Among the books which provide histories of computing in one facet or another are B. V. Bowden, ed., *Faster Than Thought* (London: Pitman, 1953), which includes a copy of Lady Lovelace's translation and notes as well as material on British computers of the immediate postwar period; Herman Goldstine, *The Computer from Pascal to von Neumann* (Princeton: Princeton University Press, 1972), which deals primarily with work in which the author was involved in the 1940s and 1950s; and N. Metropolis, *et al.*, eds., *A History of Computing in the Twentieth Century* (New York: Academic Press, 1980), which contains the proceedings of a 1976 conference in which presentations were made by many of the pioneers of computing.

1. Auguste Dick, *Emmy Noether, 1882–1935* (Boston: Birkhäuser, 1981), p. 130.

2. Felix E. Browder, ed., *Mathematical Developments Arising from Hilbert's Problems* (Providence: American Mathematical Society, 1976), p. 9. This book contains Hilbert's address as well as essays discussing the progress on each of Hilbert's problems from 1900 to the time of writing.

3. Quoted in B. Rang and W. Thomas, "Zermelo's Discovery of the 'Russell Paradox'," *Historia Mathematica* 8 (1981), 15–22, pp. 16–17.

4. Zermelo, "Beweis, dass jede Menge wohlgeordnet werden kann," *Mathematische Annalen*, 59 (1904), 514–516, translated in van Heijenoort, *From Frege to Gödel*, 139–141, pp. 139–140.

5. Ibid., p. 141.

6. Zermelo, "Untersuchungen über die Grundlagen der Mengenlehre I," *Mathematische Annalen* 65 (1908), 261–281, translated in van Heijenoort, *From Frege to Gödel*, 199–215, pp. 201–204.

7. Quoted in Moore, *Zermelo's Axiom of Choice*, p. 263.

8. Moore, *Zermelo's Axiom of Choice*, has a good discussion of this point.

9. Young and Young, *The Theory of Sets of Points* (New York: Chelsea, 1972), p. 204. This is a recent reprint of the original work.

10. Quoted in Michael Bernkopf, "The Development of Function Spaces with Particular Reference to Their Origins in Integral Equation Theory," *Archive for History of Exact Sciences* 3 (1966/67), 1–96, p. 37.

11. Fréchet, "Generalisation d'un théoreme de Weierstrass," *Comptes Rendus* 139 (1904), 848–850.

12. Hausdorff, *Grundzüge der Mengenlehre* (New York: Chelsea, 1949), p. 213. This Chelsea reprint gives the full text of the original work.

13. Ibid., p. 232.

14. Ibid., p. 244.

15. Ibid., p. 359.

16. Steinitz, "Algebraische Theorie der Körper," *Journal für die Reine und Angewandte Mathematik* 137 (1910), 167–310, p. 167.

17. Quoted in Constance Reid, *Hilbert* (New York: Springer-Verlag, 1970), p. 143. More information on Noether's mathematics and the developments it inspired is found in James W. Brewer and Martha K. Smith, eds., *Emmy Noether: A Tribute to Her Life and Work* (New York: Marcel Dekker, 1981).

18. Dick, *Emmy Noether*, pp. 173–174. Aleksandrov's entire talk, as well as a memorial talk by van der Waerden, are found in this volume.

19. Eilenberg and MacLane, "General Theory of Natural Equivalences," *Transactions of the American Mathematical Society* 58 (1945), 231–294, p. 237.

20. See, for example, Morris Kline, *Mathematics: The Loss of Certainty* (New York: Oxford University Press, 1980), especially the last few chapters.

21. Leibniz, in David Eugene Smith, *A Source Book in Mathematics* (New York: Dover, 1959), pp. 180–181.

22. See Allan G. Bromley, "Charles Babbage's Analytical Engine, 1838," *Annals of the History of Computing* 4 (1982), 196–217, for more details.

23. Ada Lovelace, "Note G" to her translation of L. F. Menabrea, "Sketch of the Analytical Engine Invented by Charles Babbage," in Philip and Emily Morrison, eds., *Charles Babbage: On the Principles and Development of the Calculator and Other Seminal Writings by Charles Babbage and Others* (New York: Dover, 1961), 225–297, p. 284.

24. Quoted in Goldstine, *The Computer*, pp. 243–244.

25. A history of the subject of error-correcting codes and the mathematical developments which followed from their study is found in Thomas Thompson, *From Error Correcting Codes Through Sphere Packings to Simple Groups* (Washington: Mathematical Association of America, 1983).

26. This letter is quoted in Norman Biggs, E. Keith Lloyd, and Robin J. Wilson, *Graph Theory, 1736–1936* (Oxford: Clarendon, 1976), pp. 90–91. This book contains many of the original papers on aspects of graph theory, including work on the four-color problem. The more recent history of that problem is found in Thomas L. Saaty and Paul C. Kainen, *The Four-Color Problem: Assaults and Conquest* (New York: Dover, 1986). For a discussion of the philosophical implications of the computer assisted proof of the four-color theorem, see two articles by Thomas Tymoczko: "Computers, Proofs and Mathematicians: A Philosophical Investigation of the Four-Color Proof," *Mathematics Magazine* 53 (1980), 131–138, and "The Four-Color Problem and its Philosophical Significance," *Journal of Philosophy* 76 (1979), 57–83.

1792–1871	Charles Babbage	Analytical engine
1809–1880	Benjamin Peirce	Linear associative algebras
1815–1852	Ada Byron King	Computer programming
1821–1895	Arthur Cayley	Trees
1845–1918	Georg Cantor	Well-ordering principle
1854–1912	Henri Poincaré	Algebraic topology
1861–1941	Kurt Hensel	p-adic numbers
1863–1942	William Young	Text on set theory
1868–1942	Felix Hausdorff	Hausdorff spaces
1868–1944	Grace Chisholm Young	Text on set theory
1871–1928	Ernst Steinitz	Field theory
1871–1953	Ernst Zermelo	Axioms for set theory
1871–1956	Emile Borel	Heine-Borel property
1872–1970	Bertrand Russell	Russell's paradox
1874–1954	Leonard Eugene Dickson	Axioms for a field
1878–1973	Maurice Fréchet	Basic notions of topology
1882–1935	Emmy Noether	Ring theory
1882–1948	Joseph H. M. Wedderburn	Algebras over arbitrary fields
1887–1948	Walther Mayer	Axioms for homology
1888–1971	James Alexander	Simplexes
1891–1965	Abraham Fraenkel	Axioms for set theory
1892–1945	Stefan Banach	Banach-Tarski paradox
1894–1971	Heinz Hopf	Homology groups
1896–1982	Pavel Sergeiivich Aleksandrov	Algebraic topology
1901–1983	Alfred Tarski	Banach-Tarski paradox
1903–1957	John von Neumann	Computer design
1906–1978	Kurt Gödel	Impossibility of consistency proofs
1907–1989	Hassler Whitney	Dual graphs
1909–	Saunders MacLane	Category theory
1912–1954	Alan Turing	Turing machines
1913–	Samuel Eilenberg	Category theory
1915–	Richard Hamming	Error-correcting codes
1916–	Claude Shannon	Switching circuits
1928–	Wolfgang Haken	Four-color theorem
1932–	Kenneth Appel	Four-color theorem
1934–	Paul Cohen	Independence of axiom of choice

Answers to Selected Problems

Chapter 1

2. ‖∩∩૧; ૪૪ ૪૪૪

3. ρκε, ξβ, ͵δωκα, $M^{β}$,γωνε

4. ⳡↈↈ, ⊗ⳡↈ≡, ↈↈ≡, ⩠⊗)(═

6. $99\,\overline{2}\,\overline{4}$

7. $2 \div 11$:

1	11	$2 \div 23$:	1	23
$\overline{3}$	$7\,\overline{3}$		$\overline{3}$	$15\,\overline{3}$
$\overline{\overline{3}}$	$3\,\overline{\overline{3}}$		$\overline{\overline{3}}$	$7\,\overline{3}$
$\overline{6}$	$1\,\overline{3}\,\overline{6}\,'$		$\overline{6}$	$3\,\overline{2}\,\overline{3}$
$\overline{66}$	$\overline{6}\,'$		$\overline{12}$	$1\,\overline{2}\,\overline{4}\,\overline{6}\,'$
$\overline{6\,66}$	2		$\overline{276}$	$\overline{12}\,'$
			$\overline{12\,276}$	2

8. $5 \div 13 = \overline{4}\,\overline{13}\,\overline{26}\,\overline{52}$; $6 \div 13 = \overline{4}\,\overline{26}\,\overline{52} = \overline{4}\,\overline{8}\,\overline{13}\,\overline{104}$

11. $18 \leftrightarrow 3,20$; $27 \leftrightarrow 2,13,20$; $32 \leftrightarrow 1,52,30$; $54 \leftrightarrow 1,6,40$; $1,4 \leftrightarrow 56,15$; $1,48 \leftrightarrow 33,20$

12. $25 \times 1,4 = 26,40$; $18 \times 1,21 = 24,18$; $50 \div 16 = 50 \times 3,45 = 3;7,30$; $1,21 \div 25 = 1,21 \times 2,24 = 3;14,24$

13. $16\,\overline{2}\,\overline{8} = 16\frac{5}{8}$

16. $51\frac{41}{109}$, $32\frac{12}{109}$, $16\frac{56}{109}$

17. $\frac{9}{25}$, $\frac{7}{25}$, $\frac{11}{25}$

18. There were 7 people and the price was 53

19. $\frac{15}{24}$ of a day

20. 10.9375 pounds

21. 90° angle: $\frac{1}{4}$, $\frac{\pi}{4} - \frac{1}{2}$

28. $2:0,0,0,0,1,40 = 2.00000000214$

32. 45.2°, 45.7°, 46.2°, 46.7°, 47.9°, 48.5°, 49.7°, 50.2°, 51.3°, 52.6°, 53.1°, 55.0°, 56.1°, 56.7°, 58.1°

33. Line 6: $v + u = 2\frac{2}{9} = 2;13,20$; Line 13: $v + u = 1\frac{7}{8} = 1;52,30$

34. 50.5

36. 30, 25

38. Side of square is 250 pu

41. $x = 3\frac{1}{2}$, $y = 2\frac{1}{3}$

42. $x = 1;0,45$ $y = 0;59,15,33,20$

Chapter 2

4. $n^2 = \frac{(n-1)n}{2} + \frac{n(n+1)}{2}$

5. $8 \cdot \frac{n(n+1)}{2} + 1 = 4n^2 + 4n + 1 = (2n+1)^2$

9. Examples: (a) (3,4,5), (5,12,13), (7,24,25), (9,40,41), (11,60,61); (b) (8,15,17), (12,35,37), (16,63,65), (20,99,101), (24,143,145)

15. $4ax + (a-x)^2 = (a+x)^2$

23. 9; 1

41. Circumference = 250,000 stades = 129,175,000 ft. = 24,465 miles; Diameter = 79577.5 stades = 41,117,680 ft. = 7787 miles

Chapter 3

1. $4\frac{1}{6}$ m from the heavier weight

2. Toward the heavier weight

17. $x^2 = 4ay, y(3a - x) = ab$

20. Focus of $y^2 = px$ is at $\left(\frac{p}{4}, 0\right)$. Length of latus rectum is therefore $2\sqrt{p \cdot \frac{p}{4}} = 2 \cdot \frac{p}{2} = p$.

Chapter 4

1. crd $120° = 5955' = 99, 15$; crd $30° = 1779' = 29, 39$; crd $150° = 6642' = 110, 42$; crd $15° = 897' = 14, 57$; crd $165° = 6817' = 113, 37$

9. Vernal equinox: At latitude $40°$, shadow length is $50;21$. At latitude $23\frac{1}{2}°$, shadow length is $26;5$. Summer solstice: At latitude $36°$, shadow length is $13;19$. At latitude $40°$, shadow length is $17;46$. At latitude $23\frac{1}{2}°$, shadow length is 0. Winter solstice: At latitude $36°$, shadow length is $101;52$. At latitude $40°$, shadow length is $120;20$. At latitude $23\frac{1}{2}°$, shadow length is $64;21$.

11. $45° : \delta = 16°37'$; $315° : \delta = -16°37'$; $90° : \delta = 23°51'$; $270° : \delta = -23°51'$; $120° : \delta = 20°30'$; $240° : \delta = -20°30'$

13. $\lambda = 60°, \rho = 35°47'$; $\lambda = 90°, \rho = 63°45'$; $\lambda = 120°, \rho = 120°19'$; $\lambda = 240°, \rho = 259°41'$; $\lambda = 270°, \rho = 296°15'$; $\lambda = 300°, \rho = 324°13'$

15. At latitude $36°$, length of day is $211°32' = 14$ hrs, 6 minutes; therefore sunrise is at 4:57 a.m. and sunset is at 7:03 p.m. At latitude $45°$, length of day is $223°54' = 14$ hrs, 56 minutes; therefore sunrise is at 4:32 a.m. and sunset is at 7:28 p.m.

16. Latitude is $40°53'$; position of sunrise is $32°20'$ north of east and position of sunset is $32°20'$ north of west.

18. $\lambda = 45°$: sun is $28°23'$ from zenith; $\lambda = 90°$: sun is $21°9'$ from zenith; $\lambda = 120°$: sun is $24°30'$ from zenith.

20. Midnight sun begins approximately April 30.

23. $\sqrt{3} \approx \frac{26}{15}$

Chapter 5

1. nth pentagonal number is $\frac{3n^2 - n}{2}$; nth hexagonal number is $2n^2 - n$.

2. nth (triangular) pyramidal number is $\frac{n(n + 1)(n + 2)}{6}$; nth (square) pyramidal number is $\frac{n(n + 1)(2n + 1)}{6}$.

7. 84

8. 12, 8

9. $72\frac{1}{4}$, $132\frac{1}{4}$

11. $\frac{121}{16}$

14. $x = \frac{5}{7}, y = \frac{267}{343}$

21. 336

22. $\frac{12}{25}$ of a day

23. $A : 15\frac{5}{7}, B : 18\frac{4}{7}$

Chapter 6

2. 57.5

4. 9

5. 24600

6. One solution is 1 pint of high-quality wine and 9 pints of wine dregs.

8. 36

9. 60

10. 12

11. 24

13. 23

14. 6

15. 5

18. 237

20. (a) $x = 41, y = 94$

21. $x = 2, y = 731$; $x = 20, y = 7310$

23. 59

27. $x = 9, y = 82$

29. $x = 180, y = 649$

34. $\frac{1}{14}$ of a day; $\frac{2}{14}, \frac{3}{14}, \frac{4}{14}, \frac{5}{14}$

36. 8, 28 form one answer.

37. $22\frac{1}{2}$

38. One solution is 9 peacocks, 26 pigeons, 7 swans, 30 sarasa birds.

Chapter 7

4. (a) 12; (b) 3; (c) 24

5. (a) 4

6. 6, 4

7. (a) $x = \sqrt{2\frac{1}{2} + \sqrt{1000}} - \sqrt{2\frac{1}{2}}, y = 10 + \sqrt{2\frac{1}{2}} - \sqrt{2\frac{1}{2} + \sqrt{1000}}$

(b) $x = 15 - \sqrt{125}, y = \sqrt{125} - 5$

8. (a) $x = \left(1 + \sqrt{2} + \sqrt{13 + \sqrt{8}}\,\right)^2$

(b) $x = 4\frac{1}{2} - \sqrt{8}$

(c) $x = 3, 16\frac{1}{3}$

9. $x = \sqrt[4]{\sqrt{12500} - 50}$

15. *Examples:* $c = 2$, $d = 2$ give no intersection; $c = 3$, $d = 2$ give one intersection; $c = 4$, $d = 2$ give two intersections.

26. 124°32′

28. 13,331,731 cubits, or approximately 3787 miles

Chapter 8

1. 3600, 2400, 1200

3. 375 paces

4. $6;23,21 = 6.389$

5. 13.3

6. 4.64

7. 10

9. 10

11. 10

12. $12\frac{1}{2}$

22. 1, 561, 821, 287, 609; purse has 2763

23. 119

24. 8, 2

26. $\left(\frac{25}{24}\right)^2$

27. $\left(\frac{25}{12}\right)^2$

28. $x = 5, y = 4$ or $x = 6, y = 3$

29. $x = 6, y = 4$

30. $x = 3, y = 6$

Chapter I

2. (11,20,88)

3. Let $\Delta t = t_1 - t_0 \pmod{13}$, $\Delta v = v_1 - v_0 \pmod{20}$, and $\Delta y = y_1 - y_0 \pmod{365}$. Then the minimum number of days is $365[40\Delta t - 39\Delta v - \Delta y] + \Delta y \pmod{18980}$.

4. $1,8,15,18 + 2,12,13,0 = 4,1,10,18 = 29,378$ days. Pacal lived into his 81st year.

5. A woman in section e has a mother in section m, a father in section f, a husband in section mf and children in section m^2.

Chapter 9

1. $\frac{11}{90}$ of a florin

2. $133\frac{3}{8}$

3. 1440 *denarii* = 6 *lire*

4. Third partner invested $103\frac{8}{93}$; profit of second partner was 129; profit of third partner was 153.

5. First: 10 days; second: 24 days; third: $17\frac{1}{7}$ days

6. $3\frac{1}{3}$ hours

7. $\frac{\sqrt{43}}{2} + \frac{\sqrt{11}}{2}, \frac{\sqrt{43}}{2} - \frac{\sqrt{11}}{2}$

10. $5 + \sqrt[4]{2809} - \sqrt{28}, 5 - \sqrt[4]{2809} + \sqrt{28}$

14. $\frac{2}{7}\sqrt[3]{1225}, \frac{2}{5}\sqrt[3]{1225}$

17. Daughter: $14\frac{2}{7}$; Mother: $28\frac{4}{7}$; Son: $57\frac{1}{7}$

18. $5 + \sqrt{2}$

19. 80 days

21. 15 dukes, 450 earls, 27,000 soldiers

22. 3, 5

24. $\frac{3 \pm \sqrt{5}}{2}$

25. $3, \frac{\sqrt{13} - 3}{2}; 6, \sqrt{19} - 3$

28. $4; \sqrt[3]{4} + \sqrt[3]{2}$

31. $b = 3$; Two solutions are $x = 1 \pm \sqrt{3}$.

32. Francis: -48; Dowry: 52

33. I have 8, Francis has -4.

37. $4 + \sqrt{-1}$

41. 8, 2

42. $x = 12$

43. 13⓪3①9②5③; 22⓪8①6②4③2④

Chapter 10

6 Distance from equator to 10°: 10.05 cm; distance from 10° to 20°: 10.37 cm; distance from 20° to 30°: 11.05 cm

8. $AB = 8.46$; $AG = 14.10$

9. $\angle A = 90°$; $\angle B = 63°$; $\angle C = 27°$; $a : b : c = 1.12 : 1 : 0.51$

12. Vertex angle is 35°; base angles are 72.5°.

Chapter 11

4. $x^2 + y^2 = \frac{m}{2} - a^2$ ($a = AE$, m is given area)

6. $(e - cg)y^2 + (de + fgc - bcg)xy + bcfgx^2 +$
$(dek - fg\ell c)y - bcfg\ell x = 0$

7. $2 + \sqrt{7}, 2 - \sqrt{7}, -2 + \sqrt{2}, -2 - \sqrt{2}$

8. $\frac{2\sqrt{3}}{9}, \frac{\sqrt{3}}{3}, \frac{4\sqrt{3}}{9}$

13. $x = -9 \pm \sqrt{57}$

14. $x = 1, 4$

16. $1, 16, -4\frac{1}{2} + \frac{1}{2}\sqrt{17}, -4\frac{1}{2} - \frac{1}{2}\sqrt{17}$

24. 25:11; 125:91

25. 42:22; 99:29

26. $\frac{17}{27}, \frac{5}{27}, \frac{5}{27}$

29. 1st: $\frac{37}{72}$; 2nd: $\frac{35}{72}$

30. 31:30

31. 9:6:4

32. $A : 1000$ vs. $B : 8139$

Chapter 12

4. $\frac{2b}{3}\sqrt{\frac{b}{3}}$

8. $t = \frac{pxy - 3y^3}{3x^2 - py}$

10. $t = -\frac{a^2 y^2}{b^2 x}$

25. $\frac{1024}{625}\left\{\left[\sqrt{a\left(\frac{a}{5} - \frac{q}{3}\right)}\right] + \frac{2}{15}\right\}$ where $a = 1 + \frac{25}{16}\sqrt{b}$

29. $\sqrt{1 + x} = 1 + \frac{x}{2} - \frac{x^2}{8} + \frac{x^3}{16} - \frac{5x^4}{128} + \cdots$

39. (a) $\frac{\dot{y}}{\dot{x}} = \frac{a^2 - 2x^2}{2y\sqrt{a^2 - x^2}}$

40. $y = 4x - 2x^2 + \frac{x^3}{3} - \frac{x^4}{2} - \frac{2x^5}{5} - \cdots$

Chapter 13

2. $xy = k$

6. $y = x^2 + \frac{k}{x}$

10. $y = Ae^x + Be^{2x} + Ce^{3x}$

14. $x^2 y^3 + 2x^3 y^2 + 4x^2 + 3y = k$

15. Base $= 2\sqrt{3}$, Altitude $= 3$, Side $= 2\sqrt{3}$

16. $r = \sqrt[3]{\frac{3\sqrt{2}v}{2\pi}}$, $h = \sqrt{2}r$

20. When $y = \frac{a\sqrt{2}}{2}$

21. Approximately 33 years

26. $\left(1, \frac{3}{2}\right)$ is a maximum, while $\left(\frac{1}{2}, \frac{3}{4}\right)$ is neither a maximum nor a minimum.

39. $P = \frac{1}{\sqrt{x+1} + \sqrt{x}}, p = \frac{1}{2\sqrt{x}}, Q = -\frac{1}{2\sqrt{x}(\sqrt{x+1} + \sqrt{x})^2}$,
$q = -\frac{1}{8x\sqrt{x}}$

Chapter 14

1. $B_8 = -\frac{1}{30}, B_{10} = \frac{5}{66}, B_{12} = -\frac{691}{2730}$

2. $\Sigma j^{10} = \frac{1}{11}n^{11} + \frac{1}{2}n^{10} + \frac{5}{6}n^9 - n^7 + n^5 - \frac{1}{2}n^3 + \frac{5}{66}n$

5. If $c = 100,000$, then $N = 36,966$

8. $\frac{24864}{59049} = 0.42$

9. $x = 6.6$, while the approximation is 6.3. With 7 trials, the odds are better than even, while with 6 they are less than even.

11. 150

12. 28

19. 0.91

22. 200 miles from Edinburgh

24. 7

26. 131,501

27. 5 men, 15 women

28. 70 crowns

29. 10

30. (1, 5, 8, 12), (2, 10, 3, 11), (4, 7, 6, 9)

31. Quadratic residues modulo 13 are 1, 3, 4, 9, 10, 12.

44. 19 bullocks, 1 cow, 80 sheep

46. 102.65 feet

Chapter 15

2. $2^3 \equiv 1, 3^6 \equiv 1, 4^3 \equiv 1, 5^6 \equiv 1, 6^2 \equiv 1$

3. 3, 5 are primitive roots for $p = 7$; 2, 6, 7, 11 are primitive roots for $p = 13$.

9. $3 + 5i = (1 - 4i)(-1 + i)$

13. Given two complex integers $\alpha = a + b\sqrt{2}$, $\beta = c + d\sqrt{2}$, their quotient is $\frac{\alpha}{\beta} = \frac{ac - 2bd}{c^2 - 2d^2} + \frac{bc - ad}{c^2 - 2d^2}\sqrt{2}$. Choose r, s so that $\left|\frac{ac - 2bd}{c^2 - 2d^2} - r\right| \leq \frac{1}{2}$ and $\left|\frac{bc - ad}{c^2 - 2d^2} - s\right| \leq \frac{1}{2}$. Let $\gamma = r + s\sqrt{2}$. Then $|N(\alpha - \beta\gamma)| < N(\beta)$.

15. If the group is $\{1, \alpha, \alpha^2, \ldots, \alpha^{17}\}$, then the cyclic subgroup of order 6 is $\{1, \alpha^3, \alpha^6, \alpha^9, \alpha^{12}, \alpha^{15}\}$ and the cosets are $\{\alpha, \alpha^4, \alpha^7, \alpha^{10}, \alpha^{13}, \alpha^{16}\}$ and $\{\alpha^2, \alpha^5, \alpha^8, \alpha^{11}, \alpha^{14}, \alpha^{17}\}$.

16. (a) $\frac{1}{4}(-1 + \sqrt{5}) \pm \frac{i}{2}\sqrt{\frac{5 + \sqrt{5}}{2}}, \frac{1}{4}(-1 - \sqrt{5}) \pm \frac{i}{2}\sqrt{\frac{5 + \sqrt{5}}{2}}$

   (b) Let $\rho = \sqrt[3]{\frac{7}{2}(1 + 3\sqrt{-3})}$ and $\sigma = \sqrt[3]{\frac{7}{2}(1 - 3\sqrt{-3})}$. Let $\alpha_1 = \frac{1}{3}(-1 + \rho + \sigma), \alpha_2 = \frac{1}{3}(-1 + \omega^2\rho + \omega\sigma), \alpha_3 = \frac{1}{3}(-1 + \omega\rho + \omega^2\sigma)$, where ω is a complex cube root of 1. Then the six solutions are $x = \frac{1}{2}\left(\alpha_i \pm \sqrt{\alpha_i^2 - 4}\right)$ for $i = 1, 2, 3$.

23. $x^2 + x + 1$ is irreducible modulo 5. Therefore $\{a_0 + a_1\alpha + a_2\alpha^2\}$ is a field of order 5^3, where α satisfies $\alpha^3 = -\alpha - 1$ and $0 \le a_j < 5$, for $j = 0, 1, 2$.

28. $\alpha + \beta = 5 + i + 8j, \alpha\beta = 12 - 9i + 18j + 24k$, $\frac{\alpha}{\beta} = \frac{23}{15}i + \frac{2}{3}j - \frac{4}{3}k$

31. $x\bar{y}z = 0$: There are no beasts which are clean and chew the cud, but do not divide the hoof.

40. $x = \frac{2}{\sqrt{5}}u + \frac{1}{\sqrt{5}}v, y = -\frac{1}{\sqrt{5}}u + \frac{2}{\sqrt{5}}v$

41. $\begin{pmatrix} 1 & 0 \\ 0 & 1 \end{pmatrix}, \begin{pmatrix} 0 & -1 \\ 1 & 0 \end{pmatrix}, \begin{pmatrix} i & 0 \\ 0 & -i \end{pmatrix}, \begin{pmatrix} 0 & -i \\ -i & 0 \end{pmatrix}$

Chapter 16

13. $\theta \approx 0.46$

15. $\frac{2}{3}$

19. $\phi(x) = \frac{4}{\pi}\sin x + \frac{4}{3\pi}\sin 3x + \frac{4}{5\pi}\sin 5x + \frac{4}{7\pi}\sin 7x + \ldots;$

   $1 = \frac{4}{\pi} - \frac{4}{3\pi} + \frac{4}{5\pi} - \ldots$

29. $P = \left\{\frac{1}{m} - \frac{1}{m(n+1)}\right\}$, where m, n are positive integers,

   $P' = \left\{1, \frac{1}{2}, \frac{1}{3}, \frac{1}{4}, \ldots\right\}, P'' = \{0\}$

37. Let C_1, C_2 be two paths connecting the points p_1, p_2 in 3-space. Let C be the closed path from p_1 to p_2 which

goes along C_1 from p_1 to p_2 and then goes back to p_1 along C_2. Then $\int_{C_1} \sigma \cdot dr - \int_{C_2} \sigma \cdot dr = \int_C \sigma \cdot dr = \int \int_A (\nabla \times \sigma) \cdot da = 0$. It follows that $\int_{C_1} \sigma \cdot dr = \int_{C_2} \sigma \cdot dr$.

Chapter 17

1. $\frac{1}{2}\begin{vmatrix} dx & dy \\ \delta x & \delta y \end{vmatrix} = \frac{1}{2}(dx\,\delta y - dy\,\delta x)$

3. $k = \frac{4}{(1 + 4x^2 + 4y^2)^2}$

4. If $w(x, y, z) = 0$, then $k = \frac{A}{B}$, where

   $A = w_x^2(w_{yy}w_{zz} - w_{yz})^2 + w_y^2(w_{xx}w_{zz} - w_{xz})^2 + w_z^2(w_{xx}w_{yy} - w_{xy})^2 + 2w_yw_z(w_{xz}w_{xy} - w_{xx}w_{yz}) + 2w_xw_z(w_{yz}w_{xy} - w_{yy}w_{xz}) + 2w_xw_y(w_{yz}w_{xz} - w_{zz}w_{xy})$

   and $B = (w_x^2 + w_y^2 + w_z^2)^2$.

5. $k = \frac{4^3 81(x^2 + y^2 + z^2)}{(36^2x^2 + 9^2y^2 + 4^2z^2)^2}$

12. Circumference of a circle of radius r on a sphere of radius K is $2\pi K \sin \frac{r}{K}$.

24. $(3, 4, 1), (-1, 7, 1), (1, 2, 0)$

25. $(3, 1), (2, -1), 2y + z = 0$

29. $-18[ijk]$

Chapter 18

18. Boundary: $V_1V_2V_3 - V_0V_2V_3 + V_0V_1V_3 - V_0V_1V_2$; Boundary of the boundary: $V_2V_3 - V_1V_3 + V_1V_2 - V_2V_3 + V_0V_3 - V_0V_2 + V_1V_3 - V_0V_3 + V_0V_1 - V_1V_2 + V_0V_2 - V_0V_1 = 0$.

19. Boundary is $V_1V_3 - V_0V_3 + V_0V_1$; Boundary of the boundary is $V_3 - V_1 - V_3 + V_0 + V_1 - V_0 = 0$.

22. $1 + 1 = 2$, which is not in the set.

23. If $a = r\sqrt{2}$, then $x = \frac{1}{2r}\sqrt{2}$.

General References in the History of Mathematics

Each chapter of this text provides a section of annotated references to works useful in providing further information on the material of that chapter. In general, however, if one wants to learn the history of a specific topic in mathematics, it pays to begin one's search in one of the following works.

1.   Morris Kline, *Mathematical Thought from Ancient to Modern Times* (New York: Oxford University Press, 1972). This book is the most comprehensive of the recent works in the history of mathematics and pays particular attention to the nineteenth and twentieth centuries. It provides chapter bibliographies for further help. It is, however, lacking completely in information about Chinese mathematics and is very sketchy in information about the mathematics of India and the Islamic world.

2.   Charles C. Gillispie, ed., *Dictionary of Scientific Biography* (New York: Scribners, 1970–1990). This 18 volume encyclopedia (including two recent supplementary volumes) is in essence a comprehensive history of science organized biographically. There are articles about virtually every mathematician mentioned in this text and, naturally, articles about many who are not mentioned. There are also special essays on topics in Egyptian, Babylonian, Indian, Japanese, and Mayan mathematics and astronomy. The DSB provides an extensive index which allows one to begin with a mathematical topic and find references to all the mathematicians who considered it.

3.   Kenneth O. May, *Bibliography and Research Manual of the History of Mathematics* (Toronto: University of Toronto Press, 1973). This work is an extensive bibliography of both expository and research articles on the history of mathematics written since the middle of the nineteenth century (including over 30,000 entries). It is organized not only biographically, but also by mathematical topic and historical classification. Although it is very comprehensive, the references do not include titles (to save space), so that one does not always know the exact subject of the articles referred to.

4.   Joseph W. Dauben, *The History of Mathematics from Antiquity to the Present: A Selective Bibliography* (New York: Garland, 1985). This bibliography, although with some 2000 references much more limited than May's work, is easier to use because it is carefully annotated. In addition, only the "best" work on a given subject is included (in the judgment of the editor and his staff). In any case, this is probably the best place to start on a quest for historical articles on a particular topic.

Other standard works on the history of mathematics which are useful to consult include David E. Smith, *History of Mathematics* (New York: Dover, 1958), Eric T. Bell, *The Development of Mathematics* (New York: McGraw-Hill, 1945), Eric T. Bell, *Men of Mathematics* (New York: Simon and Schuster, 1961), Edna E. Kramer, *The Nature and Growth of Modern Mathematics* (New York: Hawthorn, 1970), Dirk J. Struik, *A Concise History of Mathematics* (New York: Dover, 1967), Carl Boyer and Uta Merzbach, *A History of Mathematics* (New York: Wiley, 1989), Howard Eves, *An Introduction to the History of Mathematics* (Philadelphia: Saunders, 1990), and David Burton, *The History of Mathematics: An Introduction* (Dubuque, Ia.: William C. Brown, 1991).

For original sources, there are several collections of selections from important mathematical works, all translated into English. These include Ronald Calinger, ed., *Classics of Mathematics* (Oak Park, Il.: Moore Publishing Co., 1982), D. J. Struik, ed., *A Source Book in Mathematics, 1200–1800* (Cambridge: Harvard University Press, 1969), Garrett Birkhoff, ed., *A Source Book in Classical Analysis* (Cambridge: Harvard University Press, 1973), David Eugene Smith, ed., *A Source Book in Mathematics* (New York: Dover, 1959), and John Fuavel and Jeremy Gray, eds., *The History of Mathematics: A Reader* (London: Macmillan, 1987).

Naturally, more research continues to be done in the history of mathematics, and there are many journals which publish articles on the subject. The most important of these journals, which can be found in most major university libraries, are *Historia Mathematica* and *Archive for History of Exact Sciences.* The former publishes in each issue a list of abstracts of recent articles in the history of mathematics. To keep up fully with current literature, however, it is best to consult the *Isis Current Bibliography of the History of Science and its Cultural Influences,* published as its fifth issue every year by *Isis,* the journal of the History of Science Society. This volume contains an extensive listing, by subject, of articles published during the previous twelve months on the history of science, including, of course, the history of mathematics.

Index and Pronunciation Guide*

PRONUNCIATION KEY

a	act, bat	j	just, fudge	œ	as in German schön or in French feu
ā	cape, way	k	keep, token	R	rolled r as in French rouge or in German rot
â	dare, Mary	KH	as in Scottish loch or in German ich	sh	shoe, fish
ä	alms, calm	N	as in French bon or un	th	thin, path
ch	child, beach	o	ox, wasp	u	up, love
e	set, merry	ō	over, no	û	urge, burn
ē	equal, bee	o͝o	book, poor	y	yes, onion
ə	like a in alone or e in system	o͞o	ooze, fool	z	zeal, lazy
g	give, beg	ô	ought, raw	zh	treasure, mirage
i	if, big	oi	oil, joy		
ī	ice, bite	ou	out, cow		

Abel, Niels Henrik (ä′bəl), 586, 600–601, 604, 606, 609, 634–635, 638, 645, 651, 654, 690

Abraham bar Hiyya (KHē′yə), 266, 269–273, 275, 297, 299, 301

Abraham ibn Ezra, 277–278, 299, 301

Abstract algebra
  algebras, 746–747
  Boolean algebra, 619–621, 759–761
  categories and functors, 727, 750–751
  division algebra, 618, 747
  equivalence classes, 605, 661
  fields, 586, 601, 609–611, 718, 727, 742–746
  groups, 310–311, 556, 559–560, 586, 597, 601–609, 714–715, 718, 720, 749–751
  ideals and ideal numbers, 593, 595–596, 605, 733, 747–748
  integral domain, 748
  matrix theory, 625–628
  rings, 556, 727, 733, 746–750
  vector spaces, 716–718, 733, 751, 763

Abū Kāmil ibn Aslam (ä′bo͞o kä′məl), 234–237, 261, 265, 268, 282, 289

Abu'l-Wafā, Muhammad (ä′bo͞o l wä′fə), 254, 256–257, 265

Abū Nasr Mansūr (ä′bo͞o nä sər män′so͝or), 254, 256–257, 265

Acceleration, 150, 293–296, 384–387

Adelard of Bath, 268, 301

African mathematics, 308–311. *See also* Egyptian mathematics

Agnesi, Maria Gaetana (ag nâ′zē), 495, 511–512, 534, 536, 538

al-Battānī, Abū ʿAbdallāh (äl bä tä′nē), 254, 268

al-Bīrūnī, Abu l-Rayhān (äl bīro͞o′nē), 223, 254–260, 262, 265, 304

al-Karajī, Abū Bakr (äl kä rä′jē), 235–242, 244, 261, 265, 282, 284, 299

al-Kāshī, Ghiyāth al-Dīn (äl kä′shē), 228, 259–260, 265

al-Khayyāmī, ʿUmar (äl KHä yä′ mē), 223, 242–245, 248–250, 260–261, 265, 399, 408, 561

al-Khwārizmī, Muhammad (äl KHwär iz′mē), 225–235, 247–248, 260–261, 263, 265, 268, 282, 288
  Hisāb al-jabr wa-l-muqābala, 228–233, 247–248, 268

al-Kūhī, Abū Sahl, (äl ko͞o′ē) 252, 262, 265

al-Qabīsī, ʿAbd al-ʿAzīz (äl kä bē′zē), 239, 265

al-Samawʾal, Ibn Yahyā, (äl sä′mä wäl) 227–228, 236–238, 241–242, 244, 260–261, 265, 346, 425

al-Tūsī, Nasīr al-Dīn (äl to͞o′sē), 250, 259, 265, 370

al-Tūsī, Sharaf al-Dīn (äl to͞o′sē), 245–247, 260, 262–263, 265, 333, 399

al-Uqlīdīsī, Abu l-Hasan (äl o͝ok lə dē′zē), 226–227, 264, 265

Alberti, Leon Battista, 357–358, 391, 393

Alcuin of York (al′kwin), 267, 296–297, 301

Aleksandrov, Pavel Sergeiivich (al ig zan′drov), 749–750, 770

Alexander the Great, 50, 55

Alexander, James W., 740–742, 770

Alexandria Museum and Library, 55, 94, 157, 176–177

Algebra
  abstract. *See* Abstract algebra
  Babylonian, 14–15, 32–35, 38
  Boolean, 619–621, 630
  Chinese, 14–17, 31–32, 36, 38, 186–198
  Diophantine, 163–173
  Egyptian, 13–14, 36
  of Euler, 554–556
  French, 320–322
  geometric, 64–70, 89–90, 112, 114–116
  Indian, 204–214
  Islamic, 228–247, 314, 337–338
  Italian, 315–318, 328–337
  linear, 683
  of Maclaurin, 551–554
  medieval European, 270, 282–289
  of polynomials, 232, 235–238, 316–317
  symbolic, 611–615, 630
  and trigonometry, 341–344
  of vectors, 716–717
  of Viète, 340–345

Algebraic symbolism, 163–164, 168, 286–288, 315–317, 321–325, 327, 335–337, 340–341, 395, 398, 400–401, 406–407

Analytic geometry, 395–406, 569–570

Anasazi mathematics, 308, 311

Antenaresis, 73–74, 83

Apianus, Peter, 324–325

Apogee, 132

Apollonius (ap ə lō′nē əs), 95–96, 113, 126–127, 132–133, 338, 387, 395, 397, 401, 420, 422
  Conics, 112–119, 121, 123–124, 175–176, 243, 268, 338, 612

---

* To help the student pronounce the names of the various mathematicians discussed in the book, the phonetic pronunciation of many of them is included in parentheses after the name. Naturally, since many foreign languages have sounds that are not found in English, this pronunciation guide is only approximate. To get the exact pronunciation, the best idea is to consult a native speaker of the appropriate language.

Appel, Kenneth, 726, 766, 770
Application of areas, 67–69
Arc length, 270–271, 403, 429, 453–457
Archimedes (är kə mē′dēz), 55, 72–73, 95–97,
    100–101, 108, 126, 151–152, 175, 204, 268,
    338, 435, 456, 481, 566, 721
    On the Equilibrium of Planes, 97–100, 121–
        122, 268
    On Floating Bodies, 101, 122, 268
    On the Measurement of the Circle, 101–103,
        122, 153, 176, 268
    The Method, 104–105, 122–123
    Quadrature of the Parabola, 105–106, 123,
        268
    On the Sphere and Cylinder, 107, 123, 243,
        246, 252, 268
    On Spirals, 106–107, 123, 268
Archytas (är kī′təs), 81, 91, 94
Area, of
    circle, 18–20, 152, 247–248, 446
    polygon, 151, 154–155, 272
    region bounded by curves, 428–429, 434–450,
        456–459, 470, 648
    segment of a circle, 20, 152, 270, 273, 275
    segment of a parabola, 104–106
    segment of a spiral, 106–107
    surface, 519
    surface of a sphere, 108
    triangle, 18, 150–151, 154, 248
Argand, Jean-Robert (är′gand), 667
Aristaeus (a ri stī′əs), 109, 126, 175
Aristarchus (ar ə stär′kəs), 129, 338
Aristotle (ar′ə stot l), 43–44, 50–54, 56, 58, 94,
    127–128, 149–150, 158, 251, 289–290, 293–
    294, 368, 370, 373, 720, 722
Arithmetic computations
    Babylonian, 11–13
    Chinese, 13
    with decimals, 346–347
    Egyptian, 8–11
    Islamic, 226–228
    Mayan, 305–306
Arithmetization of analysis, 617, 658–665, 686
Artis cuiuslibet consummatio, 271–273
Āryabhata (är yä bä′tə), 202–204, 207, 216, 219–
    220, 222, 238, 248
Ashoka, 199, 215
Associative law, 76, 279, 742
Astronomy
    ancient, 21–23
    Chinese, 185–186
    Greek, 128–136
    Indian, 200, 207
    Mayan, 305
    Ptolemaic, 136–146
    Renaissance, 364–365, 368–379
Augustine, 267
Autolycus (ô tol′i kəs), 129, 149, 268, 293, 338
Axioms, 51, 608–609, 611, 614–615, 620, 664–
    665, 692, 705–706, 718–722, 727, 730–735,
    737, 739, 742–743
    completeness, 721, 727, 732, 734
    consistency, 721, 732–734
    independence, 721, 735, 742–743

Babbage, Charles, 612, 637, 728, 752–756, 761,
    768, 770
Babylonia, 2, 42, 44, 304
Babylonian mathematics, 1, 6–7, 11–15, 18–21,
    24–29, 31–35, 41, 64–67, 69–70, 89, 101,
    165–166, 171, 177, 195, 215, 229–231, 304,
    306
Bachet, Claude (ba shä′), 427
Bakhshālī manuscript, 216, 219
Banach, Stefan (bä′näкн), 733, 768, 770
Banneker, Benjamin (ban′i kər), 577, 581, 584
Barrow, Isaac, 455, 457–460, 472, 475, 488–489,
    493
Bartels, Martin, 588, 699
Bayes, Thomas, 539, 547–550, 579, 581, 584
Bayt al-Hikma (House of Wisdom) (bä ēt äl
    кнiк′mä), 224–225, 229, 233, 236
Beltrami, Eugenio, 707–709, 713, 723–725
Berkeley, George (bärk′lē), 495, 526–529, 538
Bernoulli, Daniel (bər no͞o′lē), 503–505, 522,
    524–525, 538, 636, 651
Bernoulli, Jakob (Jacques, James), 418, 481,
    495–497, 519, 538–539, 584, 754
    Ars conjectandi, 539–543, 544–548, 578
Bernoulli, Johann (Jean, John), 479, 481–486,
    493–501, 503–506, 519, 538
    differential equations, 495–499
    orthogonal trajectories, 500–501
Bernoulli, Nicolaus, 501, 538
Bessel, Friedrich Wilhelm, 667, 684
Betti, Enrico, 740, 742, 750
Bhāskara (bus kä′rə), 181, 200, 209–214, 219–
    220, 222
Billingsley, Henry, 355–356
Binomial theorem, 463–465
Boethius (bō ē′thē əs), 162, 267
Bolyai, János (bô′lyoi), 692, 699, 702–703, 709
Bolzano, Bernhard (bōlt sä′nō), 636, 641–644,
    658, 686, 690, 735–737
Bombelli, Rafael, 172, 314, 334–337, 343, 350,
    353
Boole, George, 586, 619–621, 630, 634, 759
Borel, Emile (bô rel′), 735–738, 770
Bourbaki, Nicolas (bo͞or bä kē′), 733–734
Brachistochrone problem, 497–498, 508
Bradwardine, Thomas, 289–291, 293, 301
Brahe, Tycho (brä′e), 372–374, 379, 393
Brahmagupta (brä mə go͞op′tə), 200, 204–209,
    212–213, 219, 222
Brauer, Richard, 747
Briggs, Henry, 383, 393
Brunelleschi, Filippo (bro͞o nel les′kē), 357, 393
Bürgi, Jobst (bûr′gē), 380, 393

Calculation of π, 18, 102–104, 203, 446, 476–
    477, 546
Calculus
    differential, 429–434, 459, 466–469, 474, 477–
        479, 515–517, 529–533, 566–567, 637, 646–
        647, 658
    of exponential and logarithmic functions, 381,
        450–451, 478–479, 485–486, 502–504, 506,
        512, 516
    foundations of, 526–530, 637–646
    of functions of several variables, 499–502,
        507–508, 517, 519–522, 637
    fundamental theorem of. *See* Fundamental
        theorem of calculus
    integral, 434–450, 453–458, 470–471, 474,
        479–480, 517–522, 647–650, 655–656
    texts, 482–486, 489–490, 506–512, 515–519
    of trigonometric functions, 451–453, 483, 486,
        489, 502, 504–510, 516–517
    of vectors, 675–681

Calendars
    Babylonian, 23
    Egyptian, 23
    Gregorian, 372
    Jewish, 23
    Julian, 368
    Mayan, 305–306
Cambodia, 217
Cambridge University, 355, 457, 459, 461, 511,
    607, 612–613, 615–616, 618, 623, 637, 678–
    679, 736, 757
Camorano, Rodrigo, 354
Cantor, Georg (kän′tôr), 636, 660–664, 686–687,
    690, 727, 728–732, 735, 737, 770
Cardano, Gerolamo (kär dä′nō), 313–314, 329–
    337, 343–344, 350–351, 353, 406, 410–411,
    424–425, 427, 540, 553, 558, 629
    Ars Magna, 331–334, 350
Cartan (kär taɴ′), Elie, 717
Cartan, Henri, 734
Cauchy, Augustin-Louis (kō shē′), 586, 588,
    634–636, 638, 655–658, 685–687, 690–691
    complex analysis, 667–673
    continuity, 636, 638, 640–642, 645, 654, 657–
        658
    convergence, 636, 638, 642–646, 651, 658, 660
    Cours d'Analyse, 638–646, 650
    derivatives, 640, 646–647
    determinants, 622
    differential equations, 650–651
    eigenvalues, 625–627, 630
    integrals, 636, 638, 647–650, 655, 667–672
    integral theorem, 667, 669–670
    limits, 636, 638–640
    permutations, 599
    Résumé des Leçons sur le Calcul Infinitesimal,
        638, 646–650
Cauchy-Riemann equations, 668, 673, 687
Cauchy sequence, 642–644, 649, 660–661, 738,
    745
Cavalieri, Bonaventura (kä vä lye′ʀē), 435–437,
    442–443, 493
Cayley, Arthur, 586, 606–608, 622–626, 628,
    630–631, 634, 712, 715, 765, 770
Celestial equator, 130–131, 144–146
Center of gravity, 97–100, 105
Champollion, Jean (shäɴ pô lyôɴ′), 3
Chasles, Michel (shäl), 692, 711–712, 725
Chevalier, August (shə vä lyä′), 602
China, 3–4, 181–182
Chinese mathematics, 3, 6, 13–21, 25–26, 29–32,
    41, 182–198, 215–217, 242, 247, 282, 302–
    304, 621, 682
Chinese remainder problem, 186–191, 207, 303,
    587, 589
Chrysippus (krī sip′əs), 51, 94
Chuquet, Nicolas (sho͞o kä′), 319–322, 349, 353,
    379
Church, Alonzo, 757
Chutan Hsita (jo͞o′tän dsē′tə), 185, 222
Circle, area of
    in al-Khwārizmī's *Algebra,* 247
    in Archimedes' *Measurement of the Circle,*
        101–102
    in Babylonia, 19–20
    in China, 19–20
    in Egypt, 18–19
    in Euclid's *Elements,* 86–87
    in Heron's *Metrica,* 152
    in India, 19

Circle, area of *(Cont.)*
  in *Mishnat ha-Middot,* 152
Clairaut, Alexis-Claude (klə RŌ'), 501–502, 519, 534, 535–536, 538, 540, 560, 566–568, 581, 584, 680
Clifford, William, 705–707, 725
Cohen, Paul, 734–735, 770
Collins, John, 451, 493
Commandino, Federigo, 338, 353
Compactness, 735, 738–740
Complex analysis, 648, 667–675, 693
  derivatives, 672
  integration, 668–669
  residues, 669–670
  Riemann mapping theorem, 673
  Riemann surfaces, 674–675
Complex numbers, 334–337, 407–408, 506, 554, 589–590, 593–595, 612–613, 615–617, 665–667
Computers, 727–728, 768
  Babbage's analytical engine, 753–756, 761
  Babbage's difference engine, 753–754
  early developments, 751–752
  programs for, 754–756, 762
  switching circuits for, 759–761
  Turing machine, 756–757, 768
  von Neumann and, 761–762
Cone, 85–86, 109–112, 114, 420–421
Congruences
  higher order, 587, 589
  linear, 186–191, 205–207, 303, 556–557, 586
  quadratic, 557–558, 587–589
Conic sections, 104–105, 108–121, 123–124, 243–245, 377–379, 387–388
  analytic geometry of, 397–399, 401–406, 408–409
  asymptotes, 114–115
  construction of, 359–360
  directrix, 119
  foci, 118–119
  normals, 117–118, 124
  in projective geometry, 421–422, 709–710, 712
  symptoms, 110–112, 114–115
  tangents, 116–117
Connectedness, 676, 737, 740
Continuum hypothesis, 663, 729, 734–735
Convergence, 26, 260, 465–466, 554–555, 636, 638, 642–646, 660, 740
  Cauchy criterion for, 642–644, 649, 658
  tests, 644–645
  uniform, 646, 656–657
Coordinate systems, 133–134, 144, 295, 398, 619, 710–711, 713
Copernicus, Nicolaus (kō pûr'ni kəs), 129, 136, 368–374, 376, 384, 393, 706
  De Revolutionibus, 369–372, 389
Cotes, Roger, 507
Counting board, 6, 13, 189–190, 192–195, 217, 306, 308, 315
Cramer's rule, 539, 552–553, 580–581, 623, 682
Crelle, August (kRel'ə), 600, 606, 656, 664, 726
Cube root calculations, 26, 191, 202
Cubic equations, 303, 323
  al-Khayyāmī, 243–245
  al-Tūsī, 245–247
  Archimedes, 108
  Bombelli, 336–337
  Cardano, 330–333
  Dardi, 317–318
  Descartes, 408–409

Diophantus, 168–169
  del Ferro, 328–329
  Lagrange, 558–559
  Tartaglia, 329–332
  trigonometric solution of, 343–344
  Viète, 342–345
Curvature, 468–469, 484, 568–569, 693–696, 704–707
Curves in space, 359–360, 566–568
Cycloid, 447–448, 495, 498–499

D'Alembert, Jean Le Rond (dä'ləm bâr), 522–524, 529–530, 535–536, 538, 625, 628, 630, 637, 639
da Cunha, José Anastácio (dä ko͝on'hä), 636, 643, 690
Dardi, Maestro, 317–318, 349, 353
Debeaune, Florimond (də bôn'), 404, 478
Dedekind, Richard (dā'də kint), 586, 631, 634, 690, 731–732
  algebraic number theory, 593–596, 610–611, 631, 746, 747
  cuts, 635–636, 658–660, 686
  set theory, 663–665, 687
Dee, John, 355–357, 360–361, 364–365, 383, 391, 393
Deferent, 132
Delamain, Richard, 752
Delaunay, Charles (də lô ne'), 675
del Ferro, Scipione, 314, 328–330, 353
della Francesca, Piero (del' ə frän che'skä), 348, 358, 393
della Nave, Annibale (del'ə nä'və), 329–330, 353
De Moivre, Abraham (de mwäv'), 539, 543–548, 550, 579, 584
De Morgan, Augustus, 586, 614–615, 617, 619, 631, 634, 755, 765
Desargues, Girard (dä zärg'), 395, 420–422, 427, 692
Descartes, René (dä kärt'), 121, 338–339, 394, 424, 427, 453, 472, 478, 493, 553
  analytic geometry, 395, 399–404, 422–423, 461
  normal lines, 431–433, 455, 487, 490
  theory of equations, 408–409
Determinants, 553, 586, 622–623
Dickson, Leonard Eugene, 742–743, 766, 770
Differential equations
  Daniel Bernoulli and, 524–525
  Jakob Bernoulli and, 495
  Johann Bernoulli and, 495–499, 533
  Nicolaus Bernoulli and, 501
  Cauchy and, 650–651
  D'Alembert and, 522–524, 625
  Euler and, 495, 501–502, 504–506, 518, 522, 524–525, 625
  exact, 502, 680
  Leibniz and, 480–481, 496–497, 534
  linear, 497–498, 501–502, 504–506, 518
  Newton and, 466, 468
  partial, 518, 522–525, 570–571, 651–652
Differential geometry
  Clairaut, 566–567
  of curves, 566–568
  Euler, 568–569
  Gauss, 693–696
  Monge, 569–571
  of surfaces, 568–571, 693–696
  Theorema Egregium, 695–696

Differential triangle, 448, 456, 458–459, 468, 475–476, 527–528
Diocles (dī'ə klēz), 118–119, 126
Diophantus (dī ə fan'təs), 157–158, 162–173, 176–178, 180, 235, 285, 335, 342, 420, 555
Dirichlet, Peter Lejeune (dē Rē klä'), 590, 592, 594, 634, 636, 641, 654–656, 690
Distributive law, 64, 76, 80, 742
Ditton, Humphry, 482, 485–486, 489, 493
Division
  Babylonian, 12–13
  Egyptian, 8–11, 35
Domingo Gundisalvo, 268, 301
Doubling the cube, 42, 47–48, 108–109, 120–121, 598
Duality, 709–711
Dürer, Albrecht (dyo͝or'ər), 358–360, 393
Dyck, Walther von (dīk), 586, 608, 634

Eccenter, 132–133, 146, 368
Eccentricity, 132
Ecliptic, 129–131, 144
École Polytechnique, 539–540, 569, 571, 573–574, 591, 602, 616, 625, 637–638, 648, 650–651
Egypt, 2–3, 43–44
Egyptian mathematics, 1, 5, 8–11, 13–14, 18–21, 23–24, 41, 170, 282, 304
Eigenvalues, 625–628
Eilenberg, Samuel, 727, 750–751, 770
Eisenstein, Ferdinand Gotthold (ī'zən shtīn), 622, 634
Elements of Euclid, 43, 54–92, 240, 243, 251, 267–269, 302, 347, 354–356, 461, 526, 543, 612
  Axioms, 59, 561
  Book I, 43, 56–64, 67, 85, 90–91, 560–563, 720–721
  Book II, 55–56, 64–68, 72, 91, 150, 234, 242, 270
  Book III, 56, 70–72, 85, 91, 120
  Book IV, 56, 70–72, 87, 91, 137
  Book V, 56, 72–78, 83, 91, 99, 149, 436
  Book VI, 56, 72, 74, 77–81, 84–86, 91, 104, 109, 175–176, 178, 245
  Book VII, 56, 73–74, 79–81, 83, 91, 158, 160–161, 289
  Book VIII, 56, 79, 81–82, 84, 91–92, 158, 289
  Book IX, 56, 82–83, 158, 160, 289, 418
  Book X, 56, 73–74, 83–85, 91, 107, 236, 251, 348
  Book XI, 56, 85–86, 109
  Book XII, 56, 83, 85–87, 92
  Book XIII, 56, 55–56, 83–85, 87–88, 92, 138, 174, 176, 178, 275
  Definitions, 56–57, 64, 71, 75, 77, 80, 84, 85
  Postulates, 58–59, 247, 402, 560, 721
Epicycle, 132–133, 146, 368, 372, 376–377
Equations
  construction of solutions of, 399–400, 408–409
  cubic, 108, 168–170, 243–247, 303, 317–318, 328–333, 342–345
  cyclotomic, 559, 591–593, 596–598, 600–601, 603, 629
  indeterminate, 164, 166–172, 186–191, 204–211, 284, 555–556
  linear, 13–14, 36, 164–165
  polynomial, 191–195, 197–198, 303, 317, 321, 323, 333–334, 407, 462, 553, 558–560, 586, 598, 600–601, 627

Equations *(Cont.)*
  quadratic, 31–35, 38, 65–69, 165–168, 170–172, 195–197, 212–214, 229–235, 270, 284, 287–289, 303, 324, 337
  systems of linear, 14–17, 36, 214, 284, 287–288, 303, 322, 342, 552–553
  theory of, 342–345, 407–409, 601–604
Eratosthenes (er ə tos′thə nēz), 55, 92, 97, 175
Error-correcting codes, 763–764
Ethnomathematics, 308–311
Euclid (yoo′klid), 54–55, 94, 109, 129–130, 338, 706, 720, 722
  Data, 88–90, 166, 175, 268, 286, 338
  Elements. See *Elements* of Euclid
  Optics, 148
Euclidean algorithm, 73, 80, 160, 189–190, 205–207, 303, 594
Euclidean constructions, 58–60, 66–69, 71–72, 399–400
Eudemus (yoo′də məs), 42–43
Eudoxus (yoo dok′səs), 42, 50, 74–76, 91, 94–95, 127, 131–132
Euler, Leonhard (oi′lər), 419, 486, 494–495, 501, 531, 534–536, 538–539, 560, 565, 584, 588, 637, 640–641, 647–648, 651, 668, 673, 687, 691
  differential equations, 502–506, 524–525, 625
  differential geometry, 566, 568–569
  Institutiones Calculi Differentialis, 515–517, 529
  Institutiones Calculi Integralis, 517–519
  Introductio in Analysin Infinitorum, 494, 512–515
  Introduction to Algebra, 551, 554–556, 580, 582, 590
  multiple integration, 519–522
  number theory, 556–558, 587, 590
  topology, 571–572, 581, 764
Expectation, 416–418
Exponents, Rules of, 168, 237, 292, 321–322, 327, 340, 379

False position, 14–15, 170–171, 282
Fermat, Pierre de (feR mä′), 121, 172, 396, 403–404, 427, 472, 493, 527
  analytic geometry, 394–399, 422
  areas, 437–442, 487, 490
  Last Theorem, 167, 169, 419–420, 425, 590–591, 593
  Little Theorem, 419, 424, 557
  maxima and minima, 430–431, 486
  number theory, 418–420, 424, 555, 590
  probability, 409, 412, 416
  tangents, 431, 486, 490
Ferrari, Lodovico, 330, 347, 350, 353, 540, 553, 558–559, 629
Fibonacci. *See* Leonardo of Pisa
Fibonacci sequence, 284, 298
Fields, 586, 609–611, 630, 742–746
Figurate numbers, 45–46
Finck, Thomas, 367, 393
Fluxions and fluents, 466–469, 485–486, 507, 511, 526
Forcadel, Pierre, 354
Four-color theorem, 726, 765–766, 768
Fourier, Joseph (foo Ryä′), 636, 651–654, 686, 690
Fourier series, 652–656
Fractions
  Babylonian, 11–12, 225, 345

  in Chuquet, 320–321
  decimal, 227–228, 278, 345–347
  Egyptian, 9–11, 225
  in exponents, 292, 406–407, 442–443, 445–446
  in Leonardo of Pisa, 283
  in Richard of Wallingford, 291
Fraenkel, Abraham, 732–733, 770
Franklin, Benjamin, 577
Fréchet, Maurice (fRā she′), 737–739, 767, 770
Frege, Gottlob (fRā′gə), 665, 687
French Revolution, 531, 540, 572–576, 635, 638, 642, 651, 699
Frend, William, 611, 755
Freyle, Juan Diez (fRā′lə), 578
Frobenius, Georg (fRō bā′nē əs), 586, 624, 628, 634
Functions, 147, 512–513, 523–524, 530, 653, 655
  continuous, 524, 640–642, 645, 653, 738, 740
  and graphs, 294–296
  integrable, 655–656
  representable as trigonometric series, 652–656
  spaces of, 737–739
  symmetric, 345, 407
  uniformly continuous, 657–658
Fundamental theorem of algebra, 407–408, 667
Fundamental theorem of arithmetic, 81
Fundamental theorem of calculus, 457–459, 470, 479–480, 510–511, 532–533, 649, 674, 677–678

Galilei, Galileo (gä lē le′ō gä lē le′ē), 354, 384, 390–391, 393, 400, 442, 459, 497, 519
  accelerated motion, 384–386
  projectile motion, 386–388
  Two New Sciences, 384–388, 390
Galois, Evariste (gal wä′), 586, 601–604, 609–611, 627, 634
Gauss, Carl Friedrich (gous), 558, 585, 591–592, 594, 599–603, 610, 634, 676–677, 690–691, 699, 702–705, 725
  complex analysis, 667, 669, 673
  differential geometry, 691–696, 722
  Disquisitiones Arithmeticae, 587–589, 591, 596–598, 604–605
  least squares, 636, 682–684
  number theory, 587–590, 629
  quadratic forms, 604–605, 609, 621–622
Gauss-Jordan elimination, 683
Gaussian elimination, 16, 682–683
Gaussian integers, 589–590
Geometric progression, 81–82, 380
Geometry
  in African weaving, 309
  in Archimedes, 104–108
  Babylonian, 18–21
  Chinese, 18–21, 36, 302
  of conic sections, 110–112, 114–119
  in Dee's *Mathematical Preface,* 355–356
  differential. *See* Differential geometry
  Egyptian, 18–21, 37
  Euclidean, 56–79, 83–88, 302, 700, 702–703, 705–706, 710, 714–715, 718–719, 721–722
  foundations of, 718–722
  groups in, 714–715
  Indian, 19, 37, 211
  Islamic, 247–253, 302
  medieval European, 269–275, 302
  in *n* dimensions, 704–705, 715–718

  non-Euclidean. *See* Non-Euclidean geometry
  of perspective, 357–360
  in Plato's *Republic,* 49
  projective. *See* Projective geometry
  Riemannian, 704–705
Gerard of Cremona (zhi Rärd′), 268, 286, 289, 301
Gerbert d'Aurillac (zhâr bâr′) (Pope Sylvester II), 267, 301
Germain, Sophie (zheR men′), 590–591, 634
Gibbs, Josiah Willard, 586, 619, 634, 680
Gilbert, William, 376
Girard, Albert (zhi Rärd′), 394, 406–408, 423, 425, 427
Gnomon, 23, 45
Gödel, Kurt (gœd′l), 727, 733–734, 770
Golden ratio, 67
Graphs, 309–310, 572, 764–766
Grassmann, Hermann (gRäs′män), 692, 716–718, 724–725
Great circle, 129–130
Greek Anthology, 157, 179
Green, George, 672
Greenwood, Isaac, 576, 584
Gregory, James, 451, 455–457, 460, 493
Group theory, 310–311, 556, 559–560, 586, 597, 601–609, 629–630, 714–715, 750
Guthrie, Frederick, 765

Haken, Wolfgang, 726, 766, 770
Hamilton, William Rowan, 585, 615–619, 622, 624, 630–631, 634, 665, 764–765
Hamming, Richard, 763–764, 770
Harvard University, 576
Hasse, Helmut (häs′ə), 747
Hausdorff, Felix (houz′dôrf), 739–740, 767, 770
Hayes, Charles, 482, 485–486, 489, 493
Heath, Thomas, 56, 101
Heaviside, Oliver, 586, 619, 634
Heawood, Percy, 765
Heiberg, J. L., 56
Heine, Eduard (hī′nə), 641, 657–658, 660, 690, 731, 735–738
Heliocentric system, 129, 369–372, 376–379
Helmholtz, Hermann von, 676, 692, 705–706, 725
Hensel, Kurt (hen′zəl), 727, 743–745, 767, 770
Heraclides (her ə klī′dēz), 128
Hermite, Charles (eR mēt′), 599
Heron (her′on), 128, 148–155, 157, 338
Herschel, John, 637
Heuraet, Hendrick van (fon hœ′rät), 453–455, 488, 490, 493
Heytesbury, William, 293–294, 301, 527
Hiero, 95, 97, 100–101
Hilbert, David, 692, 719–722, 724–725, 728–729, 731–732, 746, 748
Hindu-Arabic place-value system, 215–217, 225–228, 267, 278, 283, 286, 314–315, 320, 345–347, 381
Hipparchus (hi pär′kəs), 127, 133–137, 147–148, 200–201, 253
Hippocrates of Chios (hi pok′rə tēz), 48, 50, 70, 82, 86, 94, 108, 120, 126
Holzmann, Wilhelm (hōlts′män), 354
Homology, 740–742, 749–751
Hopf, Heinz (hupf), 750, 770
Horocycle, 700–701
Hudde, Johann (hœ′də), 433–434, 466, 476, 487, 493

Hundred fowls problem, 187–188, 198, 220, 283, 304

Husserl, Edmund, 729

Huygens, Christian (hoi′gens), 394, 416–418, 424, 427, 453, 472, 495–496, 539, 541, 543

Hydrostatics, 100–101

Hypatia (hī pā′shə), 157, 176–177, 180, 286

Hyperbolic functions, 514–515, 566

ibn al-Baghdādī, Abū'Abdallāh (ib ən äl bäg dä′dē), 251–252, 265

ibn al-Haytham, Abū 'Alī al-Hasan (ib ən äl hā′thəm), 239–242, 248–249, 252–253, 260–261, 263, 265, 425, 439, 453, 541

ibn Turk, 'Abd al-Hamīd (ib ən tûrk), 231–232, 265

Immanuel ben Jacob Bonfils (bōn fē′), 278, 301

Inca mathematics, 306–308, 311

Incommensurability, 47, 56, 73–74, 83–85, 94, 99, 251–252, 292–293, 348

India, 4, 199–200

Indian mathematics, 4, 7, 19, 25, 29–30, 41, 200–215, 276, 282, 302–304

Indivisibles and infinitesimals, 104–105, 252–253, 435–437, 467, 483, 527

Induction, 238–242, 279–281, 413–416, 444

Infinity, 52–54, 158, 385–386, 420, 444–446, 662–665, 728–732

Intermediate value theorem, 641, 644, 740

Irrational numbers, 46, 56, 83–85, 234–236, 289, 292–293, 347–348, 609–610, 658–659

Ishango bone, 4

Isidore of Seville, 267, 301

Islam, 223–224

Islamic mathematics, 195, 223–260, 282, 286, 289, 302–304, 333, 345–346, 434, 439, 751

Jacob Staff, 279, 304

Jacquard, Joseph (zhä kär′), 754

Jefferson, Thomas, 577

Jewish mathematics, 269–271, 276–282, 297, 303–304

Jia Xian (jyä shē′an), 191–192, 194, 222

Jiuzhang suanshu (jyōō jäng swän shōō), 3, 14–18, 20–21, 25–26, 29–32, 36–38, 182, 187, 198, 218

John of Seville, 268, 301

Jordan, Camille (zhôr dän′), 586, 603–604, 625, 627–628, 634

Jordan, Wilhelm (yôr dän′), 683

Jordanus de Nemore (jôr dā′nəs), 285–289, 298, 301, 303, 342

Jyesthadeva (jyes tə dā′ və), 451–453, 493, 541

Kempe, Alfred, 765

Kepler, Johannes, 118, 373–379, 390–391, 393, 429–430, 435–436, 481, 493, 752

laws of planetary motion, 377–379

Khayyam, Omar. See al-Khayyāmī, 'Umar

Kinematics, 289–291, 293–296, 384–388

King, Ada Byron (Lady Lovelace), 728, 754–756, 762, 768, 770

Klein, Felix, 692, 708–709, 712–715, 719, 723, 725, 750

Kossak, Ernst, 660

Kovalevskaya, Sofia (kov ə lef′skä yä), 657–658, 690

Kronecker, Leopold (krō′nek er), 586, 605–611, 631, 634, 661, 664

Kummer, Ernst (kŏŏm′ər), 585–586, 591–595, 605, 634, 656

Lacroix, Sylvestre (la kwä′), 574, 584, 612, 635, 637–639, 647–648, 690

Lagrange, Joseph Louis (la gränzh′), 495, 538, 540, 556, 574, 576, 584, 587, 591, 612, 626, 637, 676–677, 690

calculus, 519, 530–533, 536, 646–647, 635, 638, 640

solution of equations, 558–560, 580, 596, 599

Lambert, Johann (läm′bert), 515, 540, 560, 564–566, 584, 696–701, 721

Lamé, Gabriel (la mā), 590

Laplace, Pierre-Simon de (la plas′), 383, 539, 550–551, 573–576, 584, 612, 636, 638, 681, 683, 685, 690

Latitude and longitude, 361–364

Latus rectum, 110, 123

Law of cosines, 142, 150, 153, 697, 702

Law of Large Numbers, 541–542

Law of the lever, 96–100, 104–105

Law of sines, 142, 257–259, 276, 365–366, 371, 702–703

Least upper bound property, 644

Lebesgue, Henri (lə beg′), 735

Legendre, Adrien-Marie (lə zhän′drə), 557, 573, 576, 584, 587, 589–590, 636, 681–684, 690

Leibniz, Gottfried Wilhelm (līp′nīts), 428, 448, 466, 472–486, 489–490, 493–497, 499–504, 519–520, 527, 536, 538, 553, 637, 639, 648, 752

differentials, 474–480, 496–497

interchangeability theorem, 499–501

transmutation theorem, 475–477

Leonardo of Pisa, 266, 273–275, 282–285, 297–298, 301, 394

Levi ben Gerson (lā′vē ben ger′shən), 275–276, 278–282, 297, 299, 301, 303–304, 425

L'Hospital, Guillaume (lô pē tal′), 482–486, 489, 493, 495, 519, 553

Limits, 467, 471–472, 481, 528–530, 532–533, 637–640, 643, 646, 649–650, 661, 736–740, 744–745

Lindemann, Ferdinand, 599, 634

Liouville, Joseph (lyōō vēl′), 590–592, 599, 603, 634, 726

Liu Hui (lyōō hwä), 182–184, 220, 222

Li Ye (lē yōō), 182, 195–196, 222

Lobachevsky, Nikolai (lə bu chyef′skyē), 692, 698–703, 705–709, 722–725

Logarithms, 379–383, 390, 450, 464–465, 478–479, 485–486, 510, 514, 554

invention of, 380–381, 751

use of, 381–383

Logic

Aristotelian, 50–51

Boolean, 619–621

in Euclid's *Elements*, 60–62

of propositions, 51–52

Stoic, 51–52

Lunes, 48

MacLane, Saunders, 727, 750–751, 770

Maclaurin, Colin, 495, 509, 517, 532, 538–539, 584, 639

Treatise of Algebra, 551–554, 580–582

Treatise of Fluxions, 508–511, 527–530, 534, 536

Madhava (mäd hä′və), 451, 493

Maestri di'abbaco, 285, 314–319, 348–349

Mahāvīra (mə hä vēr′ə), 214–215, 217, 219–220, 222

Manifolds, 704–707, 714–715

Map-making, 361–364, 389

Maragha, 250

Maseres, Francis, 611

Mathematical games, 309–311

Mathematical model, 96–97, 127–129, 131–133, 140, 369–372, 376–379, 386–388

Matrices, 16, 586, 621–628, 630

Maxima and minima, 428–431, 433–434, 468, 478, 483–484, 486–487, 507, 511, 517, 637, 740

Maxwell, James Clerk, 618–619, 634, 675–676, 680, 690

Mayan mathematics, 304–306, 308, 310

Mayer, Walther (mī′ər), 750, 770

Mazzinghi, Antonio de', 317, 348–349, 353

Mean speed rule, 294–296, 298, 385–386

Mean value theorem, 647

Menabrea, L. F., 755

Menaechmus (mə nek′ məs), 108–109, 126

Menelaus (men ə lā′əs), 127, 143, 157, 257, 268

Meray, Charles, 660

Mercator, Gerard (mer kä′tôr), 363–364, 393

Mercator, Nicolaus, 450, 462, 476, 493

Méré, Chevalier de (mer), 394, 409, 411, 416–417, 427, 543

Mersenne, Marin (mer sen′), 399, 418–419, 427

Mesopotamia. *See* Babylonia

Method of analysis, 107–108, 119–120, 172–176, 338–340, 395

Method of exhaustion, 86–87, 101–102, 252–253, 434, 481

Method of least squares, 681–684

Metric, 704–705, 707–708, 712–713, 738–739

Metric system, 574–575

Mishnat ha-Middot, 152–153, 247–248

Modus ponens, 52, 60

Modus tollens, 50, 52, 61

Monge, Gaspard (mônzh), 540, 560, 569–571, 573–576, 581–582, 584, 651

Moon, motions of, 22–23, 128, 131–132

Moore, Eliakim H., 742

Moscow Mathematical Papyrus, 3, 13–14, 20–21, 37

Multiplication

Babylonian, 11–13

Egyptian, 8

Music, 44, 49–50, 158, 374–376

Napier, John (nā′pē ər), 380–383, 389, 393

Natural numbers, 662, 664–665, 728, 734

Negative numbers

in China, 17, 303

in Euler, 554

in Hamilton, 615–617

in India, 212–213, 303

in Islam, 236

in Italy, 316–317, 334

in Maclaurin, 552

in Peacock, 611–614

Nehemiah, Rabbi (nē ə mī′ə), 128, 152–153

Newton, Isaac, 387, 428, 451, 459–472, 480–482, 489–490, 493–495, 497, 526–527, 543, 551, 553, 636–637, 639, 665

fluxions and fluents, 466–470

Philosophiae Naturalis Principia Mathematica, 471–472, 507, 612–613

power series, 461–466

Nicomachus (ni kō mä′ kəs), 157–162, 178, 180, 267, 291
Nicomedes (ni kō mē′dēz), 103–104, 126
Nilakantha, Kerala Gargya (nē lə kän′tə), 451, 493
Nine Chapters of the Mathematical Art. See *Jiuzhang suanshu*
Noether, Emmy (nœ′tər), 726–727, 747–750, 770
Non-Euclidean geometry, 62, 251, 692, 714–715, 719, 722–723
    Bolyai, 702–703
    Lambert, 564–566, 696–697
    Lobachevsky, 699–703, 705–709, 712–714
    metrics in, 712–714
    models of, 708–709, 712–714
    Saccheri, 560–565, 696
    Taurinus, 697–699
Normal curve, 546–547, 683–685
North American Indians, 308, 311
Number Theory
    algebraic, 594–596, 610, 629
    Euclidean, 79–83
    Euler, 556–558
    Fermat, 418–420
    Gauss, 587–590
    in Leonardo of Pisa's *Liber quadratorum,* 285, 298
    Nicomachus, 158–162, 178
    in Plato's *Republic,* 49
    Pythagorean, 44–46, 49, 69, 90
Number symbols
    Babylonian, 6–7
    Chinese, 6, 35
    Egyptian, 5
    European, 215–216, 283
    Greek, 35
    Indian, 215
    Islamic, 215–216
    Mayan, 305–306
Number versus Magnitude, 47, 52–53, 79, 158, 251–252, 303, 314, 347–348, 355
Number words, 5
Nuñez, Pedro (nōōn′yəz), 319, 327–328, 353, 363, 393

Oldenburg, Henry, 428, 434, 482, 493
Oracle, 42
Oresme, Nicole (ô rem′), 292–296, 298–299, 301, 459
Osiander, Andreas, 372
Ostrogradsky, Mikhail (os trō grät′skē), 636, 677–678, 690, 715
Oughtred, William (ōō′tred), 461, 752
Oxford University, 289–291, 293–294, 298, 444

p-adic numbers, 743–745, 766
Pacific Island mathematics, 310–311
Pacioli, Luca (pä chē ō′lē), 313–314, 318–319, 327–328, 335, 349, 353, 411
Pappus (pa′pəs), 158, 173–176, 178–180, 338, 395
Parallel postulate, 58–59, 62, 248–251, 560–566, 700, 702, 709, 720
Parmenides (pär men′i dēz), 50
Pascal, Blaise (pas kal′), 394–395, 409, 411–417, 421–425, 437, 439, 448, 472, 475, 493, 692, 752
Pascal's triangle, 191–192, 194, 241–242, 304, 324–325, 411–416, 439, 446, 463–464, 473–474, 541
Peacock, George, 586, 611–615, 619, 631, 634, 637

Peano, Giuseppe (pe ä′nō), 665, 687, 692, 718, 725
Peirce, Benjamin (pûrs), 746, 770
Peirce, Charles, 765
Pell equation, 208–211, 555–556
Pentagon construction, 70–72
Perfect numbers, 82–83, 418
Perigee, 132
Permutations and combinations, 211, 214, 276–282, 293, 297–298, 558–560, 586, 599–601, 603–604
Pitiscus, Bartholomew (pi tis′kəs), 367, 389, 393
Place-value system, 7, 215–217, 225–228, 305–306
Planets, motions of, 128, 131–133
Plato (plā′tō), 42–43, 48–50, 92, 94, 127–129, 131–132
    The Republic, 49
Plato of Tivoli, 266, 268–269, 301
Plimpton 322, 27–28
Plücker, Julius (plyōō′kər), 692, 710–711, 725
Plutarch (plōō′tärk), 42, 97, 100
Poincaré, Henri (pwaN ka Rā′), 680–681, 709, 725, 727, 740–741, 770
Polygonal numbers, 45–46, 160–161, 439
Poncelet, Jean-Victor (pôns le′), 692, 709–711, 725
Postulates, 51, 97–98
Power series, 450–453, 461–466, 480–481, 509–510, 513–515, 530–533, 637–638, 745
Prime numbers, 80–82, 418–419, 592–593
Probability
    Thomas Bayes and, 547–550
    Jakob Bernoulli and, 418, 540–543
    calculations, 410–411, 415–418, 541–544, 546, 548–551, 683–684
    Cardano and, 410–411
    De Moivre and, 543–547
    of error, 683–684
    Huygens and, 416–418
    Laplace and, 685
    medieval, 409–410
    Pascal and, 411–416
    problems of de Méré and, 394, 411–412, 415–418, 541, 543–544
    and statistical inference, 547–551, 685
Proclus (prō′kləs), 42–43
Projective geometry, 692, 709, 714–715, 723
    Chasles, 711–712
    cross ratio, 711–713
    Desargues, 420–422
    Dürer, 359–360
    in map-making, 360–364, 388–389
    metric for, 712–713
    Monge, 569–570
    in painting, 357–360, 388
    Pascal, 421–422, 710
    Plücker, 710–711
    Poncelet, 709–711
Pseudosphere, 706–708, 713
Ptolemy, Claudius (tol′ə mē), 127, 130, 134, 136, 138, 157, 338, 371, 373, 376, 706, 751
    Almagest, 136–147, 153–155, 176, 200–201, 253, 268, 365, 368–370, 751
    Geography, 136, 361–363, 368
Ptolemy I, 55
Ptolemy III, 55
Pyramid, volume of, 20–21, 86, 152, 248
Pythagoras (pi thag′ər əs), 44, 94
Pythagoreans, 44–47, 67, 128, 158, 374
Pythagorean theorem, 26–31, 37–38

Chinese proof of, 30–31
and discovery of irrationality, 46–47
Euclidean proof of, 63–64
Indian proof of, 30
Pythagorean triples, 27–30, 46, 172

Qibla (kib′lə), 257–259
Qin Jiushao (chin jyōō shou), 182, 188–195, 197, 207, 218, 220, 222
Quadratic equations
    Babylonian solution of, 32–35, 38
    Chinese solution of, 31–32, 38, 195–197
    construction of solution of, 399–401
    in Diophantus' *Arithmetica,* 165–168, 170–172
    in Euclid's *Elements,* 65–69
    in India, 204, 212–214
    Islamic solution of, 229–235, 270, 288, 317
    in medieval Europe, 270, 284, 287–289
    in Renaissance Europe, 323–324, 327–328, 337
    in two variables, 398–399, 401–402, 404–406
Quadratic formula, 32–34, 204, 212–213, 230–231, 287, 323–324, 400
Quadratic reciprocity theorem, 557–558, 586–589
Quadratrix, 103–104, 403
Quaternions, 585, 618–619, 680
Quintic equation, 558, 600–601
Quipu (kē′pōō), 306–308, 311

Ratio
    compound, 79, 291–292, 386
    duplicate, 75, 78
    in Euclid's *Elements,* 72–77
    in Galileo's *Two New Sciences,* 386
    in medieval Europe, 289–293, 295
    in Nicomachus' *Introduction to Arithmetic,* 160–162, 291
    triplicate, 75
Rawlinson, Henry, 2
Real numbers, 659–661, 718
Reciprocal table, 11
Recorde, Robert (re kôrd′), 319, 325–327, 349, 353
Recursive algorithm, 26
Reductio ad absurdum, 50, 61–62, 70, 87, 98–99, 103–104, 106–107, 116, 511, 527, 529
Regiomontanus (Johannes Müller) (Rā gē ō mōn tä′ nōōs), 365–368, 371, 389, 393
Regular polyhedra, 56, 83, 87, 360
Reinhold, Erasmus, 372
Rheticus, George, 367, 369, 393
Rhind Mathematical Papyrus, 1, 3, 9, 13–14, 18–20, 35–36
Ricci, Matteo (Rēt′chē), 198, 220, 222
Richard, Jules (Rē shäR′), 766
Richard of Wallingford, 275–276, 291, 301
Riemann, Bernhard (Rē′män), 636, 686–687, 690–692, 724–725, 727
    complex analysis, 672–676
    geometry, 703–707
    integration, 655–656, 674, 740
Rings, 556, 727, 733, 746–750
Rittenhouse, David, 576, 584
Robert of Chester, 268–269, 301
Roberval, Gilles Persone de (Rō′ber väl), 433, 437–439, 447–448, 493
Robinson, Julia, 749
Roman mathematics, 159, 177
Royal Society, 444, 472, 482, 507

Rudolff, Christoff, 319, 322–325, 345, 349, 353
Ruffini, Paolo, 600, 634
Russell, Bertrand, 727, 729–732, 770

Saccheri, Girolamo (sa kâr′ē), 251, 540, 560–565, 584, 696, 698–701, 721
St. Vincent, Gregory, 449–450, 493
Sarasa, Alfonso Antonio de, 450, 493
Scheubel, Johann (shoi′bəl), 354
Schickard, Wilhelm, 752
Schooten, Frans van (shкнō′tən), 404, 416, 427, 433, 453–454, 461, 472, 496
Sea Island Mathematical Manual, 182–185, 203, 217–218, 302
Sebokht, Severus, 216
Sefer Yetsirah, 276–277
Seidel, Philip (sī del′), 656
Seki Kowa (se kī kou′wä), 553, 581, 584
Set theory, 662–665, 727, 766
  axiom of choice, 730–735
  axioms for, 731–733
  cardinality, 385, 662–663, 727–728
  paradoxes of, 729–730, 732–733
  well-ordering theorem, 728, 730–731
Shannon, Claude, 759–761, 770
Similarity, 63, 71–72, 77–79
Simpson, Thomas, 495, 506–509, 534, 538, 683
Slide rule, 752
Sluse, René François de (slōos), 433–434, 476, 487, 493
Snell's law, 497
Sphere, 21, 85–86, 108, 128–132
  coordinate systems on, 133–134
Square root calculations
  Babylonian, 24, 26
  Chinese, 25–26, 191
  Chuquet, 320–321, 349
  Egyptian, 23–24, 37
  Heron, 151
  Indian, 25
  Theon, 138, 153
  via power series, 462
Squaring the circle, 47–48, 446, 449, 463–464, 476–477, 598–599
Statistical inference, 547–551
Staudt, Christian von (shtout), 712, 725
Steinitz, Ernst (shtī′nits), 727, 745–746, 770
Stevin, Simon (stə vin′), 314, 345–348, 350–351, 353, 406
Stifel, Michael (shtē′fəl), 319, 324–325, 353, 379
Stokes, George, 656, 678–680, 690
Stonehenge, 22, 37
Strato (strā′tō), 150, 293
Sulvasutras, 4, 19, 25, 29–30, 37–38
Sums
  of arithmetic series, 203–204
  and differences, 472–474
  of geometric series, 105–106, 642
  of infinite series, 296, 298, 544–546, 642–643, 645, 653–656
  of integral powers, 106–107, 123, 204, 238–241, 253, 280, 438–439, 541
  of trigonometric series, 525
Sun, motions of, 22–23, 128, 131–132, 140–142, 144–146
Sun Zi (sōon dsē), 182, 186–189, 198, 222
Surfaces, 566, 568–571, 625–626
Surveying
  Chinese, 182–185
  Indian, 203

Islamic, 256
medieval European, 272–273, 275
Roman, 159
Sūrya-Siddhānta, 200–202, 219, 222
Syllogisms, 50–51, 60
Sylvester, James Joseph, 586, 622–624, 634

Tait, Peter Guthrie, 618–619, 634
Tangent lines, 70, 428–438, 456–458, 468, 528–529, 566–567
Tarski, Alfred, 733, 768, 770
Tartaglia, Niccoló (täR ta′glē ə), 314, 329–331, 347, 350, 353–354, 411
Taurinus, Franz (tô rē′nəs), 697–699, 701–702, 722, 724–725
Taylor, Brook, 509, 532, 538, 637
Thābit ibn Qurra (tä′bit ib ən kôR′ä), 66, 233–234, 252, 265
Thaetetus (thē ə tē′təs), 50, 73–74, 79, 83–84, 87, 94
Thales (thā′lēz), 42, 44, 90, 94
Theon of Alexandria, 55, 138, 153, 176
Thomson, William, 636, 676, 679–680
Three- and four-line locus problem, 95, 121, 124, 175–176, 395, 399, 401–402
Topology, 571–572, 733, 745
  algebraic, 742, 749–751
  axioms for, 737, 739
  Bolzano-Weierstrass theorem, 644, 735–737
  combinatorial, 727, 740–742
  differential, 671, 675–676, 681, 717
  Heine-Borel theorem, 658, 735–739, 766
  point set, 661, 727, 735–740
  well-ordering theorem, 728, 730, 746
Torricelli, Evangelista (tôR Rē chel′lē), 437, 442–443, 493
Transcendental numbers, 599, 729, 745
Translations, 267–269, 301
Tree, 308, 765
Triangles, solution of, 140–147, 257–259, 276, 365–367, 371, 382
Trigonometric
  formulas, 135–136, 140, 142–144, 153, 255–256, 275, 341–342, 382, 697–698, 701–703
  functions, 135–140, 200–202, 215, 254–256, 275, 304, 365, 367, 371, 447–448, 451–453, 465, 480–481, 488–489, 502, 504–506, 514, 524–525
  series, 525, 652–656, 660, 662
  tables, 136–140, 145, 185–186, 200–202, 255–256, 259–260, 270–271, 273–275, 304, 365, 367, 371
Trigonometry
  invention of term, 367
  non-Euclidean, 697–698, 701–703, 708
  plane, 134–142, 153–154, 255–256, 275–276, 302–303, 365–366
  proto, 185
  spherical, 134, 143–146, 154, 256–259, 276, 302–303, 366–367
Tripos exam, 612–613, 736
Trisecting the angle, 47, 119–120, 124, 598
Turing, Alan, 728, 756–757, 770

Ulūgh Beg, 259
Unique factorization, 590–596
University of Berlin, 592, 606, 654, 656–658, 661, 730
University of Bologna, 290, 328–329, 511

University of Göttingen, 588, 654, 657, 714, 719, 736, 748
University of Leiden, 346
University of Paris, 289, 290, 292, 298, 573, 741

Van der Waerden, B. L. (vän dər vâr′dən), 748, 768
Vector analysis
  divergence theorem, 677–678, 680
  Green's theorem, 671–675, 677, 679–680
  line integrals, 670–672, 674–676, 680–681
  Stokes' theorem, 679–680
  surface integrals, 676–678, 680–681
Vectors, 149, 586, 618–619, 716
Vector spaces, 692, 716–718
Velocity, 149–150, 289–291, 293–296, 303, 385–387, 458–459, 527–528
Vibrating string problem, 522–525
Viète, François (vē et′), 314, 339–345, 350–351, 353, 394–395, 398–401, 406, 461
Vietoris, Leopold, 750
Vitruvius (vi trōo′vē əs), 95, 159
Viviani, Vincenzo, 519, 538
Vlacq, Adrian (vläk), 383
Volterra, Vito, 680
Volume, of
  paraboloid, 252–253
  solids, 428–429, 434, 442–443, 519–520
  sphere, 21, 86, 520
Von Neumann, John (von noi′män), 728, 761–762, 770

Wallis, John, 443–446, 451, 453, 461, 487, 493
Wantzel, Pierre (vän′tsel), 598, 634
Weber, Heinrich (vā′bəR), 586, 608–609, 611, 630–631, 634, 742–743
Wedderburn, Joseph H. M., 747, 770
Weierstrass, Karl (vī əR shtRäs), 592, 606, 656–658, 661, 665, 690, 731, 735–738
Weil, André (vā), 734
Wessel, Caspar (ves′əl), 665–667, 690
Weyl, Hermann (vīl), 718, 725–726
Whitney, Hassler, 765, 770
Wilhelm of Moerbeke, 268, 301
Wilson's theorem, 587
Winthrop, John, 576, 584
Witt, Jan de, 404–406, 423–424, 427
Women in mathematics, 173, 176–177, 511–512, 590–591, 657–658, 736–737, 747–749, 754–756
Wren, Christopher, 453
Wright, Edward, 364, 393

Yale University, 576, 619
Yang Hui (yäng hwä), 182, 184, 191, 197–198, 218, 222
Yi Xing (yē shēng), 185, 218, 222
Young, Grace Chisholm, 736–737, 766, 770
Young, William, 736–737, 770

Zeno of Elea, 50, 53–54, 94
Zeno's paradoxes, 53–54
Zermelo, Ernst (tser me′lō), 727, 729–735, 766, 770
Zhang Qiujian (jäng chyōo jyän), 182, 187, 218, 222
Zhoubi suanjing, 3
Zhu Shijie (jōo shē jē), 181–182, 197–198, 218, 222
Zodiac, 129–130
Zorn, Max, 733